Ophiolite concept and the evolution of geological thought

Edited by

Yildirim Dilek
Department of Geology
116 Shideler Hall
Miami University
Oxford, Ohio 45056
USA

and

Sally Newcomb
13120 Two Farm Drive
Silver Spring, Maryland 20904-3418
USA

THE
GEOLOGICAL
SOCIETY
OF AMERICA

Special Paper 373

3300 Penrose Place, P.O. Box 9140 ▪ Boulder, Colorado 80301-9140 USA

2003

Published by The Geological Society of America, Inc.
3300 Penrose Place, P.O. Box 9140, Boulder, Colorado 80301 USA
www.geosociety.org

Printed in U.S.A.

GSA Books Science Editor: Abhijit Basu

Library of Congress Cataloging-in-Publication Data

Ophiolite concept and the evolution of geological thought / edited by Yildirim Dilek and Sally Newcomb
p. cm. — (Special paper ; 373)
Includes bibliographic references and index.
ISBN 0-8137-2373-6
1. 1. Ophiolites – Congresses. I. Dilek, Yildirim, 1957- II. Newcomb, Sally. III Geological Society of America. Meeting (2001 : Boston, Mass.) IV. Pardee Keynote Symposium (2001) : Boston, Mass.) V. Special papers (Geological Society of America) ; 373.

QE462.O6O58 2003
552'.1—dc21

2003049290

Cover: Sheeted dike intrusions in the upper crustal sequence of the Cretaceous Kizildag ophiolite along the Mediterranean shoreline, near Çevlik, southern Turkey. Steeply NW-dipping sheeted dikes have one- and two-sided chilled margins, are cut by dike-parallel and dike-perpendicular oceanic faults that are commonly mineralized, and occur within a structural graben on the south side of a peridotite core in the ophiolite. These kinds of wall-to-wall, 100% sheeted dike intrusions are interpreted to have formed in a narrow zone of magma injection (no more than a few hundred meters wide) beneath oceanic spreading axes. Recognition of sheeted dike complexes in ophiolites, particularly in the Troodos ophiolite (Cyprus), in the early 1970s led to the conclusion that they mark extensional, seafloor-spreading structures in ancient oceanic lithosphere represented by ophiolites. Photograph by Yildirim Dilek.

10 9 8 7 6 5 4 3 2 1

Contents

Dedication v

Acknowledgments ix

Preface xi

1. ***Ophiolite concept and its evolution***
Yildirim Dilek 1

2. ***A personal history of the ophiolite concept*** 17
Eldridge M. Moores

3. ***Identification of a mantle unit in ophiolites: A major step in the evolution of the ophiolite concept*** 31
Thierry Juteau

4. ***N.L. Bowen, H.H. Hess, and ultramafic rocks: Perspectives on ophiolites before plate tectonics*** 55
Davis A. Young

5. ***Ophiolites, ocean crust formation, and magnetic studies: A personal view*** 65
F.J. Vine

6. ***Die ophiolithischen Zonen in den mediterranen Kettengebirgen (The ophiolitic zones in the Mediterranean mountain chains)*** 77
Gustav Steinmann (Daniel Bernoulli and Gerald M. Friedman, Translators)

7. ***Where did Gustav Steinmann see the trinity? Back to the roots of an Alpine ophiolite concept*** 93
Daniel Bernoulli, Gianreto Manatschal, Laurent Desmurs, and Othmar Müntener

8. ***Ophiolites, historical contingency, and the Wilson cycle*** 111
Martin F.J. Flower

9. ***Where ophiolites come from and what they tell us*** 137
Adolphe Nicolas and Françoise Boudier

10. ***Ophiolites and lost oceans: Rifts, ridges, arcs, and/or scrapings?*** 153
John Dewey

11. ***Role of ophiolites in archipelago model of orogenesis*** 159
Kenneth J. Hsü

12. ***Ophiolites and the interpretation of marine geophysical data: How well does the ophiolite model work for the Pacific Ocean crust?*** . . . 173
James S. McClain

13. ***Development of ophiolitic perspectives on models of oceanic magma chambers beneath active spreading centers*** . . . 187
Peter Thy and Yildirim Dilek

14. ***Geology of supra-subduction zones—Implications for the origin of ophiolites*** . . . 227
James W. Hawkins

15. ***Supra-subduction zone ophiolites: The search for modern analogues*** . . . 269
Julian A. Pearce

16. ***The Troodos Massif of Cyprus: Its role in the evolution of the ophiolite concept*** . . . 295
Paul T. Robinson, John Malpas, and Costas Xenophontos

17. ***The Troodos ophiolite and the upper ocean crust; a reciprocal traffic in scientific concepts*** . . . 309
J.R. Cann

18. ***The sheeted dike complex of the Troodos ophiolite and its role in understanding mid-ocean ridge processes*** . . . 323
Robert J. Varga

19. ***The evolution of ideas for the origin and emplacement of the western Hellenic ophiolites*** . . . 337
A.G. Smith and A. Rassios

20. ***Development of the ideas on the origin of Albanian ophiolites*** . . . 351
Minella Shallo and Yildirim Dilek

21. ***High-pressure and ultrahigh-pressure metamorphic belts—Subduction, recrystallization, exhumation, and significance for ophiolite study*** . . . 365
W.G. Ernst

22. ***The repeated rediscovery of mélanges and its implications for the possibility and the role of objective evidence in the scientific enterprise*** . . . 385
A.M.C. Şengör

23. ***History of asbestos discovery and use and asbestos-related disease in context with the occurrence of asbestos within ophiolite complexes*** . . . 447
Malcolm Ross and Robert P. Nolan

Dedication

Bob Coleman pointing to outcrops of the Jurassic Coast Range ophiolite in Cuesta Ridge, near San Luis Obispo, California. The grassy valley below is in the Franciscan Complex and the foreground hills rising from it are made of serpentinized peridotite in the melange; the hills on skyline in the background are composed of Miocene strata. (July 10, 2003; photo courtesy of Clifford A. Hopson)

This volume is dedicated to Robert G. Coleman, Professor of Geology Emeritus at Stanford University, who has been at the forefront in our growing understanding of ophiolites during the past 45 years. His work on the field occurrence, petrology, geochemistry, and regional tectonics of fossil oceanic crust and upper mantle rocks exposed in the Circum-Pacific and Alpine-Himalayan orogenic belts has made significant contributions in the broad field of solid earth geophysics and has stimulated worldwide interest in the ophiolites. He actively participated in the organization of field conferences and excursions in Asia, Europe, and North America throughout the 1970s that were instrumental in melding European and North American ideas on the ophiolite concept. Bob was among the key participants of the first Penrose Field Conference on ophiolites held in September 1972 and played a major role in the preparation of the famous post-conference manifesto (Anonymous 1972, Geotimes, v. 17, p. 24–25), a consensus statement on the definition of an ophiolite. This definition has been the backbone of the ophiolite–ocean crust analogy, widely used by geologists and geophysicists alike working on the crustal architecture of modern and ancient divergent plate boundaries in ocean basins. His 1977 book, *Ophiolites*, was and still remains one of the most influential publications in ophiolite science and has been translated into Russian and Chinese. He has influenced, encouraged, guided, and inspired many generations of ophiolite researchers over the years, a colossal role that is hard to match.

An Oregonian by birth (1923) and upbringing, Bob entered Oregon State University on a football scholarship just prior to World War II. There he played middle guard (the "Monster Man" position) on a 1941 team that won the Pacific Coast Conference crown and played Duke University in the "transplanted

Rose Bowl," moved temporarily from Pasadena, California, to Durham, North Carolina, following the Pearl Harbor attack. After earning his B.S. and M.S. degrees in geology (1948, 1950) from Oregon State, Bob entered the Ph.D. program in geology at Stanford University, completing residency requirements and then combining dissertation research with employment as mineralogist with the U.S. Atomic Energy Commission (1952–1954) and as a geologist with the U.S. Geological Survey (1954–1958). He obtained his Ph.D. from Stanford in 1957.

Bob's lasting interest in oceanic rocks was spawned during his Stanford doctoral research, when he worked on the New Idria serpentinite mass within the Franciscan Complex of the California Coast Ranges. The original assignment targeted the unique mineralogy of the New Idria area, especially the beautiful gem mineral benitoite and also the recently discovered lenticular and vein jadeite. But he soon realized that understanding the geology and origin of the serpentinite mass itself was necessary to assess the mineral paragenesis in New Idria. Thus, the scope of his dissertation research grew to encompass also the geologic mapping and petrologic-tectonic evolution of the serpentinite and contiguous Franciscan rocks. His work disproved the earlier dogma that New Idria was an ultramafic igneous intrusion; it showed faulted rather than intrusive contacts with the enclosing host rocks, pointing therefore to its tectonic emplacement. His first publication on New Idria (Coleman, 1961, Journal of Petrology, v. 2, p. 209–247) dealt mainly with the jadeite deposits and petrogenesis of the serpentinite mass. This publication was also important in that it started a new look at the blueschist facies rocks and their tectonic environment. His most recent word on that subject—*New Idria Serpentinite: A land management dilemma*—discusses its geophysical signatures, three-dimensional form, tectonic emplacement, and especially the environmental problems caused by its pervasive asbestos development ("mountains of corn flakes") and widespread dispersal of asbestos fibers into surrounding areas from the early Miocene to the Present (Coleman, 1996, Environmental and Engineering Geoscience, v. 2, p. 9–22). The initial New Idria study (Bob's Ph.D. dissertation, 1957) also launched his lifelong interest in alpine-type peridotites, serpentinites, eclogites and associated high pressure-temperature metamorphism, and remnants of ancient oceanic lithosphere, *ophiolites*.

Bob remained with the USGS at Menlo Park until 1982, when he moved across town to Stanford. His USGS career, spanning 28 years, saw him advance from geologist (Geochemistry & Petrology Branch) to research geologist (1963), and included administrative duties as branch chief of Isotope Geology (1964–1968) and branch chief of Field Geochemistry & Petrology (1977–1979). Bob worked mainly out of the western regional office at Menlo Park, California, during most of his USGS tenure, interspersed with leaves for other appointments including visiting lecturer in geochemistry at Stanford (1960); visiting petrographer, New Zealand Geological Survey (1962–1963); and visiting professor of geology at Sultan Qaboos University, Oman (1988–1990). He was appointed professor of geology at Stanford University in 1982, entailing changes of title, duties, and address but hardly missing a beat in his continued collaborations and fruitful associations with USGS scientists and Stanford faculty colleagues. Bob became Professor of Geology Emeritus in 1993 but without a pause in his professional activities and commitments, which are ongoing.

Bob cut his geological teeth in the Coast Ranges of California and southwestern Oregon, and his fascination with their Mesozoic geology remains high. He also has operated on a worldwide stage, choosing to go where the rocks are well displayed and where important geological problems can best be solved. He enjoys meeting new people and appreciates their cultures. And the allure of exploration is part of it, too. As Kipling once wrote, "Go and look behind the ranges, over yonder, go you there." Bob has answered that call many times. His geological travels have taken him to the Arabian Peninsula to examine that continent's rifted southwestern margin, where early stages of the Red Sea opening are well displayed; to the Sultanate of Oman, whose mountainous spine lays bare the world's largest and best-exposed ophiolite; to the vast reaches of Central Asia, Inner Mongolia, and Xinjiang Provence of northwestern China, where Paleozoic continental plates and their bordering arcs once collided and were welded together. He is perhaps unique in bringing together geologists of China and the former Soviet Union in efforts to reconstruct tectonic features of the Paleozoic oceanic terranes that now lie across international borders.

Bob has long been a natural teacher. Many of us have our list of favorite professors from college years, those who inspired and guided us. But geologists who have worked with Bob, especially in the field and in overseas ventures, add him to their list near the top. He is remembered particularly for the "art of field geology"—how to do it, how to get it done despite obstacles, plus the many nongeological intangibles that prove crucial, especially in foreign terranes, where operating conditions are less than ideal. Bob's colleagues in Oman, for example, recall not only his obvious leadership in all aspects of that exotic operation—even in

Bob Coleman and his field assistant during fieldwork in the Cretaceous Semail ophiolite, Oman, in 1976.

cooking the meals in camp—but also for his problem-solving ability, his knack for interacting with the local populace and its leaders (winning their friendship and support) and for his remarkable personal warmth, good humor, and inspirational qualities. Many of his colleagues over the years remember and try to live by the Bob Coleman Rule: "Always Have Three Alternatives!"

Part of Bob's strength is his geological breadth. He is an expert mineralogist, petrologist, and structural geologist who delights in bringing those specialties to bear on local and regional tectonic problems. Among his strengths as a field geologist is his appreciation and grasp of geophysics, using the results of seismic, gravity, and aeromagnetic studies to view crustal features in three dimensions, and geochronological results that provide the vital fourth dimension. His many publications reveal employment of all those methods to solve local problems and to synthesize them on regional scales.

Ophiolites captured Bob Coleman's attention with the emergence of the plate tectonics concept, and the 1970s marked his own emergence as a leading authority on this rock assemblage, its origin and tectonic significance. His earlier work focused on high-pressure metamorphic rocks and their implications for tectonic processes. His 1963 paper with D.E. Lee (Journal of Petrology, v. 4, p. 260–301) on glaucophane-bearing metamorphic rock types of the Cazadero area in California is the first modern study of blueschist facies rocks "in place" detailing the mineral assemblages with mineral analyses, done with the great support of the chemists of the USGS pre-probe data. His 1965 paper, "Eclogites and eclogites—Their differences and similarities," (Coleman et al., GSA Bulletin, v. 76, p. 483–508) was aimed at showing that eclogite has many pressure-temperature settings and that eclogite facies has had little or no usefulness in petrological mapping. His 1966 paper, "New Zealand serpentinites and associated metasomatic rocks" (New Zealand Geological Survey Bulletin, v. 76, 102 p.) was among the first detailed and systematic studies of serpentinite rocks and serpentinization processes. Bob was convinced that calcium hydroxide solutions were part and parcel of serpentinization, a conviction that has been shown to be accurate as recent discoveries of the Lost City carbonate towers along the Mid-Atlantic Ridge point to similar fluid sources for their origin.

One of Bob's long-lasting imprints on ophiolite geology is his 1971 paper, "Plate tectonic emplacement of upper-mantle peridotites along continental edges" (Journal of Geophysical Research, v. 76, p. 1212–1222), in which he introduced the concept of "obduction" to explain why not all ocean crust was subducted. The obduction term (as an antonym to subduction) quickly became a key word to explain ophiolite emplacement mechanisms, and the obduction concept played a major role to convince the geologists of that time that "alpine-type" peridotites were indeed allochthonous fragments of ancient oceanic lithosphere. Although still widely used, many ophiolite researchers, including Bob Coleman, now recognize that obduction from downgoing oceanic slab (as in flake tectonics) is a least favorable mechanism in ophiolite emplacement, particularly in Tethyan-type orogenic belts. Bob's major contribution to the ophiolite science is his 1977 book, published as part of the *Minerals, Rocks, and Organic Materials* series of Springer-Verlag, following Professor Peter J. Wyllie's invitation in 1974 to write it. It was a state-of-the-art compilation of the then-available field, petrological, geochemical, and tectonic information on world ophiolites, and it has remained as an important source for many of us working on ophiolites and related rocks. Many of Bob's insights and observations on the petrogenesis and tectonic significance of ophiolites, as presented in this book, still hold true after nearly 30 years of intense research by the international ophiolite community.

Bob Coleman opened the door for the North American earth scientists to the Semail ophiolite in the Sultanate of Oman during the late 1970s by organizing and coordinating a large-scale, multi-institutional project in the Southern Oman Mountains. As he recalls, "The Oman study was a great success and was able to bring a huge amount of talent to bear on this National Science Foundation–sponsored interdisciplinary project." A special issue of the Journal of Geophysical Research (1981, v. 86, no. B4), co-edited by Bob Coleman and Clifford Hopson, was devoted to papers that reported some results of these Oman ophiolite investigations and became one of the most authoritative pieces of literature, not only on the Semail ophiolite, but also on ophiolite science in general. The Semail ophiolite itself since then has become a natural laboratory for process-oriented ophiolite investigations.

After starting in forestry as an undergraduate and then changing to geology at Oregon State, Bob found that he still had a great interest in botany. Having taught a course in geoecology with Art Kruckeberg at the Siskiyou Institute during the summers of 1999 and 2000 and inspired by the students' interest in the subject, he wrote a paper (with Kruckeberg), "Geology and plant life of the Klamath-Siskiyou mountain region" (Natural Areas Journal, v. 19, p. 318–338), for nature lovers.

Bob's publications and intellectual contributions are too numerous to list here. His honors include election to the National Academy of Science (1980), Foreign Associate of the Russian Academy of Science (since 1994), Fellow of the Geological Society of America (1961), Honorary Fellow of the American Geophysical Union (1986), and Fellow of the American Association for the Advancement of Science (since 1993). His career has been strongly supported, advanced, and made joyous by his wife, Cathryn, and enlivened by their three sons.

Bob's long and distinguished career has enriched the lives of numerous colleagues, friends, and students, who were fortunate enough to have crossed paths with him. A quintessential gentleman and an unselfish scientist, Bob Coleman is, and will remain, a role model for all of us. We are delighted to have this unique opportunity to recognize his professional achievements and significant contributions particularly in ophiolite science by dedicating this book to him in celebration of his 80 years of successful life.

Yildirim Dilek
Miami University

Clifford A. Hopson
University of California, Santa Barbara

Acknowledgments

We are indebted to the following reviewers for their timely and thorough reviews of the manuscripts; their efforts greatly improved the content of the volume and are gratefully acknowledged.

John Addison, Cottingham, East Yorkshire, U.K.; Daniel Bernoulli, Universitat Bäsel, Switzerland; Jean H. Bédard, Geological Survey of Canada, Québec; Sherman Bloomer, Oregon State University; Joe R. Cann, University of Leeds, U.K.; Mark Cloos, University of Texas at Austin; Robert G. Coleman, Stanford University; Anthony J. Crawford, University of Tasmania, Australia; John Dewey, University of California at Davis; Yildirim Dilek, Miami University; Don Elthon, University of Houston; W. Gary Ernst, Stanford University; Martin F.J. Flower, University of Illinois, Chicago; Kathryn M. Gillis, University of Victoria, Canada; Mickey E. Gunter, University of Idaho; Warren Hamilton, Colorado School of Mines; Ron Harris, Brigham Young University; Chris G.A. Harrison, University of Miami; James W. Hawkins, Scripps Institution of Oceanography; Clifford A. Hopson, University of California, Santa Barbara; Kenneth J. Hsü, ETH Zentrum, Zurich, Switzerland; Stephen Hurst, University of Illinois, Urbana; Raymond V. Ingersoll, University of California at Los Angeles; Thierry Juteau, Institut Universitaire Européen de la Mer, France; Derrill M. Kerrick, Pennsylvania State University; Yves Lagabrielle, Institut Universitaire Européen de la Mer, France; Brian T. Lewis, University of Washington; Henri Masson, Université de Lausanne, Switzerland; Alexander McBirney, University of Oregon; Eldridge Moores, University of California at Davis; Bramley J. Murton, Southampton Oceanography Centre, U.K.; Sally Newcomb, Silver Spring, Maryland; David O'Hanley, Trinity School at River Ridge, Minnesota; Michael R. Perfit, University of Florida; Giovanni B. Piccardo, Universita di Genova, Italy; John P. Platt, University College London, U.K.; James E. Quick, United States Geological Survey, National Center, Virginia; Paul T. Robinson, Dalhousie University, Canada; Matthew H. Salisbury, Geological Survey of Canada Atlantic, Nova Scotia; A.M. Celâl Şengör, Istanbul Technical University, Turkey; John W. Shervais, Utah State University; Robert J. Stern, University of Texas at Dallas; Peter Thy, University of California at Davis; Robert J. Varga, Wooster College; Fred Vine, University of East Anglia, U.K.; John Wakabayashi, Hayward, California; Fred J. Wicks, Royal Ontario Museum, Toronto, Canada; Roy H. Wilkens, University of Hawaii; An Yin, University of California at Los Angeles; Hatten S. Yoder Jr., Carnegie Institution of Washington; and Guram Zakariadze, Russian Academy of Sciences, Moscow, Russia.

Preface

For over 200 years, ophiolites—lithological assemblages of oceanic crust and mantle rocks—have played a significant role in the evolution of geological thought since their original definition by A. Brongniart (1770–1847). Ophiolite studies have progressed from observation and description of individual minerals and rocks and their spatial and temporal relations, to recognition of a recurring assemblage of mafic, ultramafic, metamorphic, and sedimentary rocks, to the connection of that assemblage to continental rifting, seafloor spreading, plate tectonics, and mountain building processes. Field observations were made with increasingly sophisticated instrumentation, and, in the circular way of science, interpretations have led to hypotheses, which in turn were tested both in the field and in the laboratory. Since their recognition as on-land remnant of fossil oceanic lithosphere, ophiolites have been a major component of the plate tectonics paradigm. Changing hypotheses about the occurrence, origin, and emplacement of ophiolites mimic the broader advancement of the many facets of geology. Virtually all subdisciplines and inquiry methods in geology have been applied to studies of ophiolites and related rocks: mineralogy and petrography, petrology and geochemistry, field studies and mapping, structural geology and tectonics, exploration geophysics, geobiology and paleontology, and marine geology and geophysics. Ophiolites thus constitute a superb natural laboratory in which to conduct integrated, interdisciplinary studies to better understand process-oriented geological questions and problems in earth history. Conversely, studies of ophiolites have both reflected and advanced the methods and theories of geology over the years.

Despite recognition of their key role in the formulation and validation of plate tectonic theory since the 1960s, ophiolites still remain controversial. In particular, their tectonic settings of origin, emplacement mechanisms, and tectonic significance of chemical fingerprinting of their magmas are the subjects of spirited debate. Continuing controversy over ophiolites raises fascinating scientific, historical, and philosophical questions in the evolution of geological thought.

In order to discuss the history and development of ideas, principles, and theories established in geological sciences as a result of investigations of ophiolites and ophiolitic rocks, we convened a Pardee Keynote Symposium at the Annual Meeting of the Geological Society of America held in Boston in November 2001. The symposium brought together a large group of international researchers with a diverse range of scientific background and interests in ophiolite geology. The presentations and discussions probed our current knowledge about ophiolites and their significance in geological studies through time, and provided a historical perspective on many fundamental problems and questions that have arisen as ophiolite studies have evolved. This volume contains contributions mainly from among the keynote, invited, and volunteered papers presented at the Pardee Symposium. These articles display how fruitful the debates on different aspects of the ophiolite concept have been since Brongniart's first definition in 1813. We believe that the historical perspective in this book brings a new and fresh dimension to our understanding of ophiolites and, through it, to related scientific and philosophical questions. Furthermore, the collection in a single volume of differing perspectives and opposing views on ophiolites and ophiolite studies through time makes this book a vivid account of the scientific process and a unique contribution to the history of science.

This historical perspective on ophiolite studies is provided in several modes in the book. Some papers present personal views of the authors on the major developments in our understanding of ophiolites and related studies. These essays make the critical connections between some of the important discoveries and the flow of events and personal communications among the scientists that led to these discoveries and major breakthroughs. It is inspiring and didactic to read how scientists arrived at their conclusions, sometimes serendipitously, at times simultaneously but independently, and how they felt about them. This aspect brings the human interests and dimensions into the science making of ophiolite research. Other contributions present a thorough chronicle of major developments in ophiolite research and provide a probing critique of how these developments aided our understanding of key scientific problems and questions.

We thank the contributors to this volume for their time and effort; without their active participation and hard work this book would not have emerged. We are grateful to the many reviewers of manuscripts for their diligent and timely work. Discussions of ideas and anecdotal information on ophiolite studies with Robert G. Coleman, Martin F.J. Flower, Clifford A. Hopson, Eldridge Moores, Julian Pearce, and Paul Robinson throughout the preparation of the book have been particularly inspiring and insightful. We extend our sincere appreciation to Abhijit Basu, Geological Society of America Books Science Editor, for his enthusiastic support that motivated us, and to the GSA Publications staff for their invaluable assistance in copyediting, formatting, and finalizing this book. We gratefully acknowledge the grants and funds from the History of Geology and the International Divisions of the Geological Society of America to support travel by Pardee Symposium participants and from Miami University to help with the preparation of this volume.

Yildirim Dilek

Sally Newcomb

Geological Society of America
Special Paper 373
2003

Ophiolite concept and its evolution

Yildirim Dilek*
Department of Geology, Miami University, Oxford, Ohio 45056, USA

ABSTRACT

The ophiolite concept, first developed in Europe in the early nineteenth century, went through several phases of evolution. Early studies of ophiolites prior to the plate tectonic revolution emphasized the development of ophiolites as in situ intrusions within geosynclines. The genetic association of mantle peridotites with volcanic and plutonic rocks in ophiolites was not considered in these studies, and the emplacement of serpentinized ultramafic rocks in orogenic belts remained a topic of debate regarding ophiolites among the North American geoscientists. Recognition of extensional sheeted dike complexes, the existence of a refractory mantle unit represented by peridotites with high-temperature deformation fabrics, fossil magma chambers in plutonic sequences, and the allochthonous nature of ophiolites by the mid 1960s was instrumental in the formulation of the ophiolite model and the ophiolite-ocean crust analogy within the framework of the new plate tectonic theory. This analogy was confirmed at the first Penrose Conference on ophiolites in 1972, whereby an ideal ophiolite sequence was defined to have a layer-cake pseudostratigraphy complete with a sheeted dike complex as a result of seafloor spreading. Ophiolites were interpreted to have developed mainly at ancient mid-ocean ridges through this model. Geochemical studies of ophiolites challenged this view as early as the beginning of the 1970s and suggested the association of magma evolution with subduction zones. This paradigm shift in the evolving ophiolite concept led to the definition of suprasubduction zone ophiolites in the early 1980s. Systematic petrological and geochemical investigations of world ophiolites throughout the 1980s and 1990s demonstrated the significance of subduction zone derived fluids and melting history in development of ophiolitic magmas; forearc, embryonic arc, and back-arc settings in suprasubduction zones became the most widely accepted tectonic environments of origin. Major differences in their internal structure and stratigraphy, extreme variations in their chemical affinities and mantle sources, and significant changes in the mode and nature of their emplacement in orogenic belts indicate that ophiolites form in a variety of tectonic environments and that they do not need to have a certain internal stratigraphy to them as defined at the 1972 Penrose Conference. A new classification scheme presented in this paper considers seven specific types of ophiolites, based on their inferred tectonic settings of igneous origin and emplacement mechanisms in different kinds of orogenic belts (i.e., collisional versus accretionary). Application of this new ophiolite classification scheme may prove helpful in recognizing the Archean oceanic crust and in better understanding the crustal and mantle processes in Earth's early history.

Keywords: ophiolite concept, Steinmann trinity, Alpine-type peridotites, ophiolite model for oceanic crust, Hess-type oceanic crust, serpentinites, ophiolite classification.

*dileky@muohio.edu

Dilek, Y., 2003, Ophiolite concept and its evolution, *in* Dilek, Y., and Newcomb, S., eds., Ophiolite concept and the evolution of geological thought: Boulder, Colorado, Geological Society of America Special Paper 373, p. 1–16. For permission to copy, contact editing@geosociety.org.

INTRODUCTION

Ophiolites, and discussions on their origin and significance in Earth's history, have been instrumental in the formulation, testing, and establishment of hypotheses and theories in earth sciences. Ophiolite studies have brought together diverse groups of international scientists on a regular basis far more than any other topic in geology, and the questions and problems raised and resolved during the course of these studies have made significant contributions to the evolution of geological thought over the years. The topic of ophiolites (their definition, tectonic origin, emplacement mechanisms, etc.) has been a dynamic and continually evolving concept since its first introduction in the geological literature by Alexandre Brongniart (1813).

This paper provides a synoptic summary of the historical, philosophical, and scientific development of the ophiolite concept during the last ~200 yr. It is written to present a chronological and conceptual backdrop for the papers on various aspects of the ophiolite concept in this book, rather than as a comprehensive overview. The papers in this volume capture, accurately and elegantly, the nature, results, and significance of many ophiolite studies, as well as the relevant scientific inquiries and their contribution to the evolution of geological thought. They also highlight the role played by scientists, institutions, and professional activities on the evolutionary course of the ophiolite concept. Thus, the book itself represents science in the making and not just a work of history.

EARLY DEFINITION OF OPHIOLITE AND STEINMANN TRINITY

The term "ophiolite" was first used in 1813 by a French mineralogist, Alexandre Brongniart (1740–1847), in reference to serpentinites in mélanges; he subsequently redefined (1821) his definition of an ophiolite to include a suite of magmatic rocks (ultramafic rocks, gabbro, diabase, and volcanic rocks) occurring in the Apennines. The coexistence of these rocks in the Alpine-Apennine mountain belts was well-recognized by the European geologists during the nineteenth century (e.g., Lotti, 1886; Suess, 1909); however, it was Gustav Steinmann (1856–1929) who elevated the rock term "ophiolite" to a new concept, one that defined ophiolites as spatially associated kindred rocks originally formed as in situ intrusions in axial parts of geosynclines (Steinmann, 1927; also see the translation in this volume, Chapter 6). Steinmann emphasized the common occurrence of peridotite (serpentinite), gabbro, and diabase-spilite (spilite = albitized, vesicular basaltic lava rock), in association with deep-sea sedimentary rocks (chert, mudstone, and limestone), in the Mediterranean mountain chains and interpreted their origin as differentiated magmatic rocks evolved on the ocean floor. He considered these rock assemblages to have developed from a consanguineous igneous process during the evolution of eugeosynclines. This interpretation subsequently led to the widely known notion of the "Steinmann trinity," consisting of serpentinite, diabase-spilite, and chert.

In his definition of the ophiolite concept, Steinmann presented some salient observations and interpretations that have contributed significantly to our understanding of ophiolites and their significance. He indicated that in the Ligurian Apennines serpentinites were predominant, an observation that later became a major topic of discussion between field geologists (i.e., Hess, 1938, 1955) and experimental petrologists (Bowen, 1927, 1928) regarding their close spatial association with volcanic and sedimentary rocks (Young, this volume, Chapter 4; Bernoulli et al., this volume, Chapter 7). While explaining the ophiolitic assemblages as a product of magmatic differentiation, Steinmann (1927) indicated that the ultrabasic rocks, peridotites, and gabbros are the early phases to solidify with more differentiated intrusions and volcanic rocks developing later and intruding into these already formed denser rocks. He offered this interpretation to refute Staub's (1922) idea of gravity-driven separation and differentiation of magma during development of ophiolites.

Although Steinmann considered peridotite, gabbro, diabase, and volcanic rocks in ophiolites as comagmatic in origin, his observation that gabbroic and diabasic rocks were intrusive bodies in the serpentinized peridotites is an extremely important one and differs from the contemporary interpretation of the layer-cake pseudostratigraphy of the "Penrose-type" ophiolite. It implies that, at least in the Apennine ophiolites, the gabbros and volcanic rocks are younger than the peridotites. This inference has been substantiated by recent petrological and geochemical studies of the Ligurian ophiolites (i.e., Rampone and Piccardo, 2000) documenting that the peridotites (Permian and older? in age) and crustal rocks (Jurassic) in these ophiolites do not have a simple melt-residue genetic relationship as expected from modern oceanic lithosphere evolved at mid-ocean ridges. Finally, Steinmann's observations of the northern Apennine ophiolites and associated deep-sea sedimentary rocks are highly important in that he correctly interpreted them as thrust sheets tectonically overlying the Tertiary sedimentary rocks in Tuscany (Steinmann, 1913) and in that this interpretation led to the delineation of allochthonous nappe sequences in the Alpine-Apennine orogenic system.

OPHIOLITE CONCEPT AND ALPINE-TYPE PERIDOTITES

The Australian geologist W.N. Benson (1926) interpreted peridotites and serpentinite occurrences in mountain belts as plutonic intrusions into folded geosynclinal sedimentary rocks of orogenic systems and called them "alpine-type" peridotites. This interpretation differed from that of Steinmann in that alpine-type peridotites were spatially and temporally (and hence genetically) unrelated to the gabbroic diabasic and volcanic (spilitic) rocks as commonly seen in an ophiolitic assemblage. Benson's term, "alpine-type" peridotites, propagated into the literature in reference to irregular to elliptical bodies of ultramafic rock that occur in mountain belts, and most researchers envisioned these bodies to be high-level, sill-like to lopolithic intrusions that differentiated in place. This idea of alpine-type peridotites caused a temporary setback early in the twentieth Century in the American geologists' understanding of the significance of the genetic link between plutonic-volcanic and ultramafic rocks in ophiolites.

It was T.P. Thayer's paper nearly 40 years later (1967) that emphasized the significance of the consanguineous relationship between ultramafic and associated mafic rocks in alpine-type peridotites, prompting American geologists to reconsider the European ophiolite concept (Coleman, 1977). Thayer had written this paper for the volume on "Ultramafic and Related Rocks," edited by Peter J. Wyllie (1967), to explain how the gabbro, diabase, and other leucocratic rocks in alpine-type peridotites could have originated from a single primary peridotitic magma. Subsequently, Jackson and Thayer (1972) distinguished harzburgite-type versus lherzolite-type alpine peridotites. According to their subgrouping, the harzburgite-type alpine peridotites represented the uppermost oceanic mantle as in ophiolites, whereas the less-depleted lherzolite-type alpine peridotites corresponded to the subcontinental mantle and/or to the deeper oceanic mantle where partial melting is much less intense. Recent studies of ophiolites have shown that both harzburgite- and lherzolite-type peridotites may occur in ophiolites and that they can be used to classify ophiolite types and their inferred spreading rates of formation in an oceanic setting (Ishiwatari, 1985; Boudier and Nicolas, 1985; Nicolas and Boudier, this volume, Chapter 9).

A different interpretation of peridotites was also introduced by the experimental petrologist Norman L. Bowen (1887–1956) almost contemporaneously with Steinmann's paper on the Alpine-Apennine ophiolites. Pointing to the lack of ultramafic lavas, Bowen (1927) argued that alpine-type peridotites and serpentinites had to form from injections of olivine-and pyroxene-rich masses in the solid state, rather than intrusions of ultrabasic magmas at crustal levels (Young, this volume, Chapter 4). Bowen did not refer to the term "ophiolite" for his discussions of peridotites in his papers, and he urged the field geologists to consider the results and findings of experimental petrological studies in addressing the origin of ultramafic rocks and serpentinites found in alpine-type peridotites.

Bowen's experimental results and interpretations on the origin of alpine-type peridotites were questioned by Harry H. Hess (1906–1969), who had studied the occurrence of ultramafic rocks and serpentinites in the Appalachians for his Ph.D. work (Moores, this volume, Chapter 2; Young, this volume, Chapter 4). Hess (1938) had observed the apparent lack of high-temperature aureoles of contact metamorphism around these Appalachian peridotites and the abundance of serpentinites associated with them. He proposed, therefore, that peridotites might have been emplaced from hydrous ultramafic magmas at lowered temperatures, and that serpentinite was magmatic in origin. This interpretation initiated a long-lasting dispute between Bowen and Hess (Young, this volume, Chapter 4). Through their experimental studies of the system MgO-SiO_2-H_2O, Bowen and Tuttle (1949) showed that serpentine mineral was only stable at low temperatures (below 1000°C) at which peridotite melt, even with water, could not exist. In their arguments on the origin of the alpine-type peridotites, neither Bowen nor Hess recognized the spatial and genetic relations between peridotites and gabbroic/volcanic rocks in Steinmann's ophiolitic assemblage, nor did they consider these ultramafic rocks as part of tectonically emplaced, exotic thrust sheets. This is interesting because the European geologists had already documented the close genetic association of peridotite, gabbro, and basaltic rocks in ophiolite assemblages and the allochthonous occurrence of ophiolites as far-traveled nappe sequences in the Alps and Apennines. Both Bowen and Hess also had difficulty with explaining the origin of serpentinites in conjunction with the evolution of ultramafic rocks.

HESS-TYPE OCEANIC CRUST

Hess elaborated on his ideas on the origin of peridotites and serpentinites in an article (Hess, 1955) he contributed to the Geological Society of America Special Paper 62, which had resulted from a symposium on "The Crust of the Earth" organized by the Department of Geology in Columbia University in 1954. He stated in this paper that Steinmann's ophiolite concept was confusing because "it obscured critical relationships of its [ophiolite] various members to the tectonic cycle" (p. 393). Linking serpentinites and alpine-type peridotites to orogeny and mountain building episodes, Hess (1955) argued that serpentinites and rocks of Steinmann's trinity are common in island arcs and that "island arcs represent an early stage in the development of an alpine-type of mountain system" (page 395). For completely different reasons, Hess was advocating an island arc origin of mafic-ultramafic rock assemblages and serpentinized peridotites found in orogenic belts. This was nearly 20 years before Miyashiro (1973) made the first formal and rather controversial call on the island arc origin of the Troodos ophiolite (Cyprus), connecting ophiolite genesis to subduction zone processes. Miyashiro's geochemical argument on the island arc origin of the Troodos ophiolite would usher in a major paradigm shift in the ophiolite concept in the wake of the plate tectonic revolution.

Although Hess did not accept the European ophiolite concept in his 1955 paper, he combined his observations from the Appalachian and Caribbean peridotites with post-war developments in marine geological and geophysical investigations to suggest that the ocean floor was extensively serpentinized by waters rising out of the mantle. He suggested that the Mid-Atlantic Ridge more likely represented "a welt of serpentine" developed as a result of volume changes accompanying serpentinization and deserpentinization of peridotite. Hess thought that the peridotite was serpentinized as it rose through the 500 °C isotherm. In his 1962 paper, which he had called an "essay in geopoetry," Hess proposed that upwelling limbs of mantle convection cells would correspond to mid-ocean ridges beneath which the isotherms were elevated. The common occurrence of serpentinized peridotite inclusions in oceanic basaltic rocks and of dredged serpentinite rocks from the Mid-Atlantic Ridge prompted Hess to suggest that the main oceanic crustal layer was made largely of serpentinite (his Fig. 2, p. 603). He further elaborated that the thickness of this crustal layer (his layer 3) would be controlled by the maximum elevation of the 500 °C isotherm beneath the Mid-Atlantic Ridge and that the seismic velocity of this layer would be highly variable, depending on the magnitude of serpentinization of the peridotite. Hess proposed that the interface between the oceanic crust (composed mainly of serpentinite) and the underlying peridotite with seismic velocities of 7.4 km/sec represented the Moho Discontinuity. Since he had interpreted serpentinites as hydrated peridotites, Hess described the Moho beneath the Mid-Atlantic Ridge as an alteration front (phase transition) rather than a sharp boundary separating the igneous crust from the underlying mantle (his Fig. 7, p. 612). Although we now know that oceanic crust is not made of 70% serpentinite, as Hess had suggested, recent marine geological and geophysical studies have documented that the slow-spreading oceanic crust along the Mid-Atlantic Ridge has a highly heterogeneous lithological make-up and thickness (Dick, 1989) and that thin-crust domains along the ridge axis (i.e. magma-poor segment ends) consist of tectonically uplifted ultramafic rocks with gabbroic intrusions and a thin basaltic cover (Cannat et al., 1995). This definition of the Mid-Atlantic crust with non-uniform thickness and a heterogeneous lithostratigraphy is remarkably similar to Steinmann's description of the Ligurian ophiolites in the Apennines and also largely corresponds to Hess' characterization of oceanic crust developed at the Mid-Atlantic Ridge. This "Hess-type crust" differs significantly from "Penrose-type" oceanic crust in terms of its internal architecture, as discussed later.

EARLY DISCUSSIONS ON THE ORIGIN OF MANTLE COMPONENT IN OPHIOLITES

While the discussions in North America focused on the origin of alpine-type peridotites and serpentinites as separate entities from the associated mafic rocks, the European geologists took Steinmann's ophiolite concept to other ophiolites in the eastern Mediterranean region to investigate the close spatial and temporal associations of peridotite, gabbro, and volcanic rocks. Dubertret (1955) in Syria and Turkey, Brunn (1956, 1960, 1961) in Greece, and Aubouin (Aubouin and Ndojaj, 1964) in Greece and Albania undertook mapping projects and considered the ophiolites there as artifacts of massive, submarine outpourings of basaltic magma on the seafloor of eugeosynclinal basins (Juteau, this volume, Chapter 3; Shallo and Dilek, this volume, Chapter 20; Smith and Rassios, this volume, Chapter 19; Thy and Dilek, this volume, Chapter 13). According to these researchers, the basaltic magma had differentiated after its emplacement on the seafloor as a gigantic extrusion such that accumulation of early crystallized minerals (olivine, pyroxene) settling out of the magma could have produced the apparent stratigraphic order from peridotites at the bottom upwards into gabbros, with basaltic lavas on top. The idea of a single, cogenetic magmatic origin of peridotite, gabbro, and volcanic rocks in an ophiolite suite, as presented in these studies, was problematic and hindered the progress of the ophiolite concept for some time.

There were some notable objections from within the European geological community, however, to the single, cogenetic magmatic origin of ophiolite suites (see Juteau, this volume, Chapter 3). The Dutch geologist de Roever presented his arguments regarding the origin of alpine-type peridotites in his 1957 paper opposing the ideas of Hess (1938, 1955) and Bowen and Tuttle (1949) and suggested that these ultramafic rocks were tectonically emplaced fragments of the peridotite layer. This solid-state, tectonic emplacement model was fundamentally different from the interpretations of in situ intrusion origin discussed in North America and from the models of differentiated seafloor outpourings of basaltic magma developed in Europe. Furthermore, de Roever (1957) reinterpreted the Steinmann trinity as a product of mantle melting, which had produced the basaltic rocks on top and the residual ultramafic rocks at the bottom.

The Swiss petrologist Vuagnat stated in his 1963 paper that the overwhelming abundance of ultramafic rocks in ophiolites compared to the small volumetric occurrence of gabbroic rocks could not simply be explained by differentiation of submarine outpourings of basaltic magma. He reasoned that this process was likely to have produced more gabbros reminiscent of continental stratiform igneous complexes. He suggested, instead, that the peridotite massifs in ophiolites were partial melting residues in the upper mantle (Vuagnat, 1963). Thus, these two papers in the literature signal the arrival of a significant shift in Steinmann's "cogenetic" ophiolite concept and of a new paradigm in oceanic crustal evolution.

PLATE TECTONIC REVOLUTION AND THE OPHIOLITE CONCEPT

Many of the ideas and observations (i.e., splitting of oceans at mid-ocean ridges, where new seafloor is produced; seafloor spreading; marine magnetic anomalies; recycling of ocean floor back into the mantle at trenches; continental mobility; seismicity of subduction zones, etc.) that formed the nucleus of the plate tectonic hypothesis, which was formulated as a synthetic, quantitative theory in 1967–68, were being discussed extensively on both sides of the Atlantic Ocean by the mid 1960s. The framework of the plate tectonic theory provided coherent and plausible explanations for different aspects of the ophiolite concept. Hess (1965) had already accepted the ophiolite concept and the interpretation of ophiolites as fragments of ocean floor found as exotic blocks or thrust sheets in mountain belts (Moores, this volume, Chapter 2; Vine, this volume, Chapter 5). The unifying plate tectonic theory helped the international geoscience community formulate the following four main conclusions about ophiolites nearly simultaneously:

1. Wall-to-wall intrusions of diabasic dikes as in recently recognized sheeted dike complexes within ophiolites signify

their extensional, seafloor spreading origin (see references in Cann, this volume, Chapter 17; Moores, this volume, Chapter 2; Robinson et al., this volume, Chapter 16; Thy and Dilek, this volume, Chapter 13; Varga, this volume, Chapter 18; Vine; this volume, Chapter 5).

2. Ophiolites are slices of fossil oceanic lithosphere composed of oceanic crust and uppermost mantle such that harzburgites and dunites represent depleted refractory mantle, whereas pyroxenites and gabbros constitute intrusive crustal bodies derived from the partial melting of this mantle. The mantle unit is not only depleted but is also tectonized, as displayed by the planar alignment and segregation of orthopyroxene and spinel grains forming a foliation and by tight to isoclinal folding of this foliation. Dunite bodies are commonly elongated in the foliation direction. This metamorphic fabric in the mantle unit is a result of high-temperature (~1000–1200 °C) deformation of the mantle rocks due to convective flow beneath the spreading axis (Nicolas, 1989). The discovery of a mantle unit in ophiolites (Juteau, this volume, Chapter 3) is a formal recognition of the fact that crustal units and mantle rocks in ophiolites are not the products of a single, comagmatic event.

3. Plutonic sequences in ophiolites represent solidified, fossil magma chambers. This conclusion has led to the formulation of various magma chamber models during the next thirty years (Thy and Dilek, this volume, Chapter 13). Ophiolitic magma chamber models were extended to the oceans, facilitating the synergy and symbiotic research efforts (but not always in a linear, two-way traffic fashion) between the ophiolite community and the marine geologists and geophysicists (Cann, this volume, Chapter 17; Thy and Dilek, this volume, Chapter 13). Funding for ophiolite research during the 1980s was justified mostly to test, to refine, and to better constrain the magma chamber models and the explanations of plate accretion processes at divergent boundaries.

4. Ophiolites are fragments of fossil oceanic lithosphere that have been thrust over or "obducted" (Coleman, 1971) into continental margins at consuming plate boundaries (Dewey, this volume, Chapter 10; Dewey and Bird, 1971; Dewey, 1976). However, mechanisms of ophiolite emplacement along continental margins immediately became a topic of strong debate among the ophiolite researchers.

OPHIOLITE MODEL FOR OCEANIC CRUST AND PENROSE DEFINITION

The inferred seafloor spreading origin of ophiolites and the magma chamber models developed based on ophiolite investigations within the framework of the plate tectonic theory considered ophiolites as oceanic crust generated at mid-ocean ridges. In a uniformitarian approach, ophiolite geologists started interpreting and reconstructing the evolution of ancient oceanic lithosphere at paleo mid-ocean ridges using the ophiolite-ocean crust analogy. On the other hand, early models of the structure and evolution of modern oceanic crust that were mainly based on geophysical modeling made little use of observations and ideas derived from ophiolites (Thy and Dilek, this volume, Chapter 13). However, as more seismic data became available from modern ocean basins, particularly from the Pacific Ocean, the ophiolite suite became an ideal analog to explain the seismic velocity structure of modern oceanic lithosphere (McClain, this volume, Chapter 12). The results of these early seismic studies suggested a profoundly uniform oceanic crustal architecture with little lateral heterogeneity and were utilized to formulate a "layer-cake" structure of oceanic crust. Combined with observations from the Troodos and Semail ophiolites in particular, this seismic velocity structure of modern oceanic crust and its inferred pseudostratigraphy with a layer-cake structure came to be known as the "ophiolite model."

This new paradigm was the driving force for the organization of an international Penrose Field Conference on ophiolites in September 1972. Participants of this conference made field observations in various ophiolite complexes in the western United States, discussed the European ophiolite concept and the ocean crust–ophiolite analogy, and produced a consensus statement on the definition of an ophiolite (Anonymous, 1972). According to those present at this GSA Penrose Conference:

> *Ophiolite* refers to a distinctive assemblage of mafic to ultramafic rocks. It should not be used as a rock name or as a lithologic unit in mapping. In a completely developed ophiolite, the rock types occur in the following sequence, starting from the bottom and working up:
>
> - *Ultramafic complex*, consisting of variable proportions of harzburgite, lherzolite and dunite, usually with a metamorphic tectonic fabric (more or less serpentinized);
> - *Gabbroic complex*, ordinarily with cumulus textures commonly containing cumulus peridotites and pyroxenites and usually less deformed than the ultramafic complex;
> - *Mafic sheeted dike complex*;
> - *Mafic volcanic complex*, commonly pillowed.
> - Associated rock types include (1) an overlying sedimentary section typically including ribbon cherts, thin shale interbeds, and minor limestones; (2) podiform bodies of chromite generally associated with dunite; and (3) sodic felsic intrusive and extrusive rocks.
>
> Faulted contacts between mappable units are common. Whole sections may be missing. An ophiolite may be incomplete, dismembered, or metamorphosed ophiolite. Although ophiolite generally is interpreted to be oceanic crust and upper mantle, the use of the term should be independent of its supposed origin. (Anonymous, 1972)

This Penrose definition of ophiolites was a sound confirmation of the inferred layered-cake pseudostratigraphy of an ideal (i.e., complete) oceanic crust. The "Penrose-type" oceanic crust with a layered internal stratigraphy is significantly different from the Hess-type oceanic crust of slow-spreading ridges that consist mainly of serpentinized peridotites capped by lavas and/or thin gabbroic rocks, and it appears to approximate the structure of modern oceanic crust developed at fast-spreading, non-rifted ridges (McClain, this volume, Chapter 12; Dilek et al., 1998). The Penrose definition of ophiolites did not include any statement about emplacement mechanism(s) of ophiolites. Characteristically, the Penrose statement did not define ophiolite based on the tectonic setting of its igneous origin, yet this aspect of the ophiolite concept became a major topic polarizing the ophiolite community for the next 30 years. The conference report ended with calling for careful field mapping and sophisticated petrologic, mineralogic, and geochemical studies of ophiolite subunits. The international ophiolite community has carried out these tasks, as charged.

SEARCH FOR GENERIC MODELS

The GSA Penrose Field Conference was a great catalyst for the initiation of international projects and workshops investigating the structure, petrology, geochemistry, and geochronology of ophiolites around the world. A two-week symposium and field excursion, held in the Soviet Union during May 31 through June 14, 1973, brought together some of the international ophiolite researchers to discuss the ophiolite concept within the framework of the global plate tectonics and to examine some of the ophiolite occurrences in the Soviet Central Asia and the Lesser Caucasus (Coleman, 1973). Participants on the field excursions observed that sheeted dikes were generally missing and the extrusive sequences were commonly thin to absent in many of the ophiolites in Central Asia and in the Lesser Caucasus. This is in fact a common feature in many Tethyan ophiolites, as later investigations have shown. Emplacement of ophiolites along continental margins and into mountain belts was the major focus of the debate during this symposium.

Following this field-based ophiolite symposium in the Soviet Union, an International Geological Correlation Program (IGCP) project on ophiolites was initiated in 1974 with funds from the United Nations Educational, Scientific and Cultural Organization (UNESCO) (IGCP-79: Ophiolites of Continents and Comparable Oceanic Rocks). The participants of this project organized and ran a series of field excursions and seminars in North America, and a collection of papers on various North American ophiolites was published in a special issue of the Bulletin of the Department of Geology and Mineral Industries of the State of Oregon, edited by Robert Coleman and Potter Irwin (1977). In the same year, Coleman published his highly acclaimed and very timely book on ophiolites (Coleman, 1977), compiling the extant information and bibliography on ophiolites thus far available in the literature. Another field conference was organized by C. Allégre and J. Aubouin in the Alps in 1977 to examine the occurrence of orogenic mafic-ultramafic rock associations. Collectively, these three field conferences after the GSA Penrose Field Conference in 1972 were instrumental in melding European and North American ideas on the ophiolite concept (R.G. Coleman, personal commun., 2003).

The Troodos massif in Cyprus played a major role in systematic ophiolite studies throughout the 1970s and 80s (Cann, this volume, Chapter 17; Robinson et al, this volume, Chapter 16; Varga, this volume, Chapter 18; Vine, this volume, Chapter 5). This ophiolite and its mineral deposits were exploited since the early days of human civilization, and the detailed geological mapping of Troodos, mainly by the British geologists in the 1950s, had accumulated a wealth of data and observations on its internal structure and stratigraphy. An international ophiolite symposium convened in Nicosia, Cyprus, in 1979 brought together a large number of researchers to discuss the existing ophiolite problems and questions and to exchange views and observations on the genesis of ophiolites around the world. The proceedings of this symposium were published in a comprehensive book (Panayiotou, 1980), which constituted a first major compilation of diverse sets of structural, petrological, and geochemical data and information available on ophiolites up to 1980.

These ophiolite projects and the results of thematic meetings emphasized the significance of ophiolites in investigating the internal structure of oceanic crust and upper mantle, as well as reconstructing the ancient plate boundaries based on the assumption that ophiolites were on-land remnants of oceanic crust. Results of the marine geological and geophysical studies of mid-ocean ridges provided new information on the structure of modern oceanic crust and strengthened the ophiolite-oceanic crust analogy. However, researchers had to look hard to find a present-day example of ophiolite emplacement, although many aspects of plate tectonics were recognized in modern plate boundaries, and actualistic models were developed based on active geodynamic processes. The incorporation of modern oceanic crust into continental margins, or ophiolite emplacement, appears to have no present-day counterparts.

Coleman (1977) urged the ophiolite community to exercise caution in correlating modern oceanic crust with ophiolites and stated that it should not be assumed that "the present-day processes that give rise to new oceanic crust are the same as those that produced ophiolites in the past" (p. 9). He asked, "Were the spreading centers that produced Jurassic oceanic crust in the Tethyan Sea the same as those now forming oceanic crust within the mid-Atlantic Ridge?" (p. 10). It is true that none of the Atlantic oceanic crust developed throughout the last 160 million years has been incorporated into any continental margins by plate tectonics and that less than 0.001% of the total oceanic crust generated at global mid-ocean ridge systems throughout the Phanerozoic history of Earth has been preserved in orogenic belts (Coleman, 1977). This observation suggests that the mid-ocean ridge generated oceanic lithosphere generally gets subducted almost entirely, and that emplacement of ancient oceanic crust in ophiolites might have resulted from unique tectonic events in Earth's history (Dewey, this volume, Chapter 10; Flower, this volume, Chapter 8).

The shift in the paradigm of mid-ocean ridge origin of ophiolites came first from a geochemist, Akiho Miyashiro, who had studied subduction-related rocks in Japan for much of his career. Miyashiro (1973) argued that "about one-third of the analyzed rocks of the lower pillow lavas and sheeted dike rocks in the Troodos ophiolite follows a calc-alkalic trend" (p. 218), suggesting that "the massif was created as a basaltic volcano in an island arc with a relatively thin ocean-type crust rather than in a mid-oceanic ridge" (p. 218). This was the first formal proposal of a subduction zone origin of the Troodos "oceanic crust" that questioned the "ruling hypothesis" of a mid-ocean ridge setting of ophiolite genesis. The ensuing scientific exchange in the form of discussions and replies to Miyashiro's 1973 paper initiated a long-lasting debate about the tectonic setting of ophiolite genesis that still continues today (Pearce, this volume, Chapter 15).

The controversy over the tectonic setting of the Troodos ophiolite was due to the difficulty of reconciling its apparent seafloor spreading structure and architecture with the arc-like chemistry of its upper crustal rocks. Pearce (1975) suggested a solution to this problem by speculating that the Troodos massif might have formed in a marginal basin during the evolution of an incipient submarine island arc. A marginal basin setting of the Troodos and other ophiolites was favored by many geochemists at this time and is still considered by some as a viable model to explain both the igneous development and tectonic accretion of ophiolites in orogenic belts (Hsü, this volume, Chapter 11).

In his 1975 paper, Miyashiro classified ophiolites into three distinct classes based on their volcanic rock series and argued that "different classes could have different origins" (p. 250). He also stated that "confining the use of the term ophiolite to masses showing a definite pseudostratigraphic sequence as observed in Troodos (Anonymous, 1972) was not accepted" (p. 250, Miyashiro, 1975). His Class-I ophiolites included volcanic rocks of both calc-alkaline and tholeiitic series and involving an island arc origin; Class-II ophiolites contained tholeiitic series volcanic rocks, having an island arc and/or mid-ocean ridge origin; Class-III ophiolites had tholeiitic and alkalic series volcanic rocks with an inferred origin in a rift along a continental edge or at or near intra-oceanic islands and seamounts. Miyashiro's classification scheme was based on limited petrological and geochemical data from a small number of ophiolites (as he pointed out), but it introduced two degrees of freedom to the prevailing ophiolite concept at that time: (1) ophiolites can form in a variety of tectonic settings; (2) ophiolites do not need to have a typical "Penrose-type" crust and pseudostratigraphy.

The major contributions to the ophiolite concept, specifically to the understanding of the geochemistry of ophiolites, came from the results of scientific cruises to the marginal basins and forearc regions of subduction zone environments in the western Pacific in the mid- to late 1970s. Ophiolitic rocks were recovered from the Lau and Mariana back-arc basins, the inner trench walls of the Yap and Mariana Trenches, and the Mariana forearc. Findings from the modern subduction zone environments in the western Pacific prompted researchers to consider more rigorously the evolution of ophiolites in spreading environments within the upper plate of subduction zones (Hawkins, 1977; Hawkins, this volume, Chapter 14; Pearce, this volume, Chapter 15). This development, which came about as a collective result of ophiolite studies on land and marine geological and geophysical investigations in modern convergent margin settings in the oceans, led to a new paradigm in the evolving ophiolite concept in the early 1980s: suprasubduction zone origin of ophiolites.

Pearce summarizes the geological and geochemical characterization of suprasubduction zone ophiolites and development of the ideas on the suprasubduction zone origin of ophiolites in an elegant paper in Chapter 15 of this volume. The geology of modern suprasubduction zones and its implications for the origin of ophiolites are discussed in great detail by Hawkins (this volume, Chapter 14). In general, two geological phenomena widely observed in the Mediterranean ophiolites became the main challenge for the concept of suprasubduction zone origin of ophiolites: (1) significant amounts of magmatic extension as recorded by sheeted dike complexes, and (2) lack of chain of arc volcanic edifices. What were the causes and mechanisms of significant amounts of extension keeping pace with robust magmatism in the upper plate above a subduction zone? If arc magmatism was responsible for the genesis of ophiolitic oceanic crust, why was it that there were no volcaniclastic rocks and volcanic edifices in ophiolites (particularly in Troodos) as typically seen in mature island arcs of the western Pacific Ocean? These questions played an important role in the search of geodynamic models for the initiation of subduction zones and for the magmatic and tectonic evolution of suprasubduction zone ophiolites.

Plausible explanations for these questions were derived, to a large extent, from the results of drilling and dredging in convergent margin settings in the Western Pacific. The Izu-Bonin-Mariana forearc region was particularly revealing in better understanding subduction initiation mechanisms and in providing possible explanations for ophiolite generation. The emerging hypothesis suggests that trench rollback during early stages of subduction of young (and hot) oceanic lithosphere induces extension and seafloor spreading in the forearc region of the upper plate. This extension and associated subduction-driven magmatism then produces oceanic crust (proto-ophiolite) with seafloor spreading structures and arc-like chemistry (including boninites). Short-lived subduction of the down-going oceanic lithosphere may explain why the nascent/incipient arc never develops into a mature island arc system with volcanic edifices (hence the lack of arc volcanic edifices in ophiolites).

Systematic geochronological studies of the igneous assemblages in ophiolites and of high-grade metamorphic rocks beneath ophiolite complexes have provided important constraints for the generic models on ophiolite formation. Thin (<500-m-thick), fault-bounded sheets of highly strained, high-grade metamorphic rocks occur structurally beneath many ophiolite complexes (Jamieson, 1986) and show inverted metamorphic field gradients from granulite/upper amphibolite to lower greenschist facies in a structurally descending order. Petrological studies and thermal modeling of these "metamorphic soles" have shown that their accretion at the base of ophiolites likely took place at the inception of an oceanic subduction such that inverted metamorphic field gradients developed at the top of the descending slab as it came into contact with progressively hotter hanging-wall rocks (Hacker, 1990 and references therein). Precise radiometric and isotopic dating of the sole rocks from various ophiolites indicates that the age difference between the ophiolites and their metamorphic soles is commonly <10 m.y. (generally ~5 m.y.; Wakabayashi and Dilek, 2000). Thus, ophiolites were apparently displaced from their original setting of igneous genesis and were incorporated into continental margins within 5–10 m.y. of their formation (Dilek et al., 1999a). This phenomenon may explain the lack of arc volcanic edifices in suprasubduction zone ophiolites because of short-lived subduction and its sudden cessation ("death" of ophiolites, as presented in Shervais, 2001).

INCORPORATION OF OPHIOLITES INTO OROGENIC BELTS

Ophiolites in orogenic systems are generally considered to mark suture zones between collided plates and/or accreted terranes. The geology of the spatially-associated lithological units and the regional tectonic history provide significant information on the mode and nature of mechanisms of the incorporation of ophiolites in orogenic systems. Moores (1982) classified ophiolites as Tethyan versus Cordilleran based upon the presence or absence of a continental substrate (i.e., passive margin of a continental plate or fragment), arc volcanic edifices, and/or accretionary mélanges. In this classification, Tethyan-type ophiolites are commonly observed to rest on passive continental margins along tectonic contacts and are considered to have formed at mid-ocean ridges; Cordilleran-type ophiolites are spatially and temporally associated with island arcs and arc volcanic edifices, volcaniclastic rocks, and accretionary mélanges and are considered to have formed in convergent margin settings (i.e., forearc, arc, or intra-arc).

Nicolas (1989) used a different approach in classifying ophiolites based on the tectonic settings of their emplacement. Those ophiolites tectonically resting on continental passive margins (i.e., Semail in Oman, Papuan ophiolite in Papua-New Guinea), ophiolites incorporated into the active continental margins of the Circum-Pacific belt (i.e., ophiolitic occurrences in the Franciscan Complex in California), and suture zone ophiolites occurring in continent-continent or arc-continent collision zones (i.e., ophiolites in the Alpine-Himalayan orogenic system, Caledonian ophiolites, Hercynian and Uralian ophiolites) constitute the three main types of ophiolites in Nicolas' classification. In this scheme, his collisional-type ophiolites correspond to the Tethyan-type ophiolites and active continental margin-related Circum-Pacific ophiolites correspond to the Cordilleran-type ophiolites of Moores (1982). However, the ophiolites emplaced onto continental passive margins (i.e., Tethyan ocean remnants in the Alpine-Himalayan orogenic system), as described in Nicolas (1989), can also be considered as Tethyan-type based on the classification criteria by Moores. Therefore, it is clear that other aspects of the regional geology, besides the tectonic basement, must be evaluated in interpreting and classifying ophiolites.

Flower (this volume, Chapter 8) suggests an actualistic model for ophiolite generation and emplacement in which proto-ophiolites are envisaged to develop in forearc settings of convergent margins involved in arc-trench (subduction) rollback. Subduction rollback may occur in response to collision-induced mantle extrusion in the region and is responsible for the suprasubduction zone origin of ophiolites (discussed above). In this model, subduction rollback and associated magmatism and extension may produce one or more episodes of arc splitting and basin opening and hence development of ophiolite belts with different lithostratigraphy and geochemical fingerprints. If the subduction rollback is arrested by a collisional event (i.e., forearc-forearc of forearc-continent collision), backarc basins in the upper plate would collapse (Hsü, this volume, Chapter 11) and the forearc oceanic crust (proto-ophiolites) would be trapped to produce ophiolites (analogy of high-tide marks or HTM of Flower [this volume, Chapter 8]). The ophiolite emplacement mechanism in this model is similar to that proposed by Moores (1982) for his Tethyan-type ophiolites. Emplacement of suprasubduction zone ophiolites involves and is facilitated by underplating and partial subduction of a continental margin (commonly a passive margin). This is why there are no modern examples of the "obduction" of oceanic crust (ophiolite emplacement) at present-day passive margins around the world, as Coleman noted in 1977; a passive margin has to arrive at a trench and underthrust a forearc region in order to be laden with an ophiolite. An actualistic example of ophiolite emplacement via passive margin subduction is at work within the eastern Sunda arc region in Indonesia, where the northwestern edge of the Australian continental shelf has underthrust, uplifted, and exposed upper crustal units of the Ocussi ophiolite of Timor.

In collision-driven orogenic belts blocks of high-pressure and ultrahigh-pressure metamorphic rocks may occur (Ernst, this volume, Chapter 21). These rocks represent exhumed sheets of deeply subducted (60–140 km) passive continental margin and/or arc crust in the downgoing plate beneath the ophiolites. Because there was no significant arc magmatism and volcanic arc edifice construction in the upper plate, these orogenic belts (i.e., the Alps) lack high-temperature metamorphic rocks, and hence, they do not contain paired metamorphic belts.

If the subduction rollback continues for a prolonged period of time eluding orogenic entrapment, accretion of proto-ophiolitic forearc assemblages may continue without any disruption, and ophiolites would become part of a growing accretionary-type orogen as in Circum-Pacific orogenic belts. In these kinds of accretionary-type orogenic belts, disrupted ophiolitic assemblages, mélanges, and deep marine sedimentary rocks become recrystallized under high-pressure conditions at depths of 15–70 km. Continued subduction of oceanic lithosphere produces a robust calc-alkaline arc inboard from the trench that is characterized by high-temperature conditions (Ernst, this volume, Chapter 21). Thus, accretionary-type orogenic belts contain paired metamorphic belts, ophiolitic mélanges, and disrupted ophiolites with disparate ages and geochemical affinities (i.e., Japan, Kamchatka, western United States).

SECOND PENROSE MEETING ON OPHIOLITES AND NEW DEVELOPMENTS IN THE OPHIOLITE CONCEPT

Interdisciplinary studies of ophiolites during the 1980s and 90s, particularly systematic structural, petrological, geochemical, and geochronological investigations, produced a wealth of new data and information on how fossil oceanic crust had evolved in various tectonic settings in terms of magmatic, structural, and hydrothermal alteration processes. The variability of internal structures and geochemical characteristics of ophiolites was well recognized and accounted for, and tectonic and geodynamic models of ophiolite generation and emplacement were refined. At the same time, large integrated and multi-institutional geophysical investigations of mid-ocean ridge systems, and petrological and geochemical studies via the Ocean Drilling Program projects of abyssal peridotites and modern oceanic crust from mid-ocean ridges and suprasubduction zone environments, greatly improved sampling, observation, and exploration of the modern ocean floor, and various theoretical and experimental modeling conducted by the international marine geology and geophysics community produced new significant data and revolutionary ideas on the evolution of oceanic lithosphere and on ocean floor processes. The scientific understanding of both ophiolites and oceanic crust had thus undergone a remarkable transformation since the 1972 Penrose Conference on ophiolites.

A new Penrose Conference on "Ophiolites and Oceanic Crust" was convened in Marshall, California, in 1998 to bring together a multidisciplinary group of geoscientists with backgrounds in ophiolites and marine geology and geophysics. The purpose of this conference was to reevaluate the existing models of oceanic crust generation and ophiolite formation and to reassess the significance of ophiolites and oceanic crust in plate-tectonic processes (Dilek et al., 1999b). The meeting generated numerous discussions and debates particularly on ophiolite-ocean crust analogy. A collection of papers presented at this meeting was subsequently published as Geological Society of America Special Paper 349 (Dilek et al., 2000).

The participants of this meeting reaffirmed the continued usefulness of the 1972 Penrose definition in ophiolite studies but concurred that the definition should be expanded to include more information about the geologic context of individual ophiolites as revealed in the overlying and underlying rock units (cover and tectonic basement, respectively) and their regional geology. The significance of the regional tectonic history of ophiolites is discussed in a paper by Moores et al. (2000), which proposes that the chemical signature of ophiolitic rocks is contingent on the prior history of plate tectonic motions and of the mantle from which the magmas were derived. This "historical contingency" model suggests that Earth's mantle is isotopically heterogeneous at all scales and chemically modified by previous subduction events so that lavas erupted at oceanic spreading centers may not necessarily reflect the characteristic chemical fingerprint of their current tectonics settings. Thus, magmatic sources with an apparent suprasubduction affinity may be tapped into at some mid-ocean ridges to produce oceanic crust displaying seafloor spreading structures and arc-like chemistry. Moores et al. (2000) suggested two modern analogs for this phenomenon in support of mantle heterogeneity and their historical contingency model: Woodlark Basin and the South Chile Ridge. This notion of historical contingency has added a new layer of debate to the current ophiolite concept (see Flower, this volume, Chapter 8).

Evidence for subduction-related contamination of the mantle beneath a mid-ocean ridge is presented in a paper by Sturm et al. (2000) that shows how the lavas erupted along the southern Chile Ridge display anomalous mid-oceanic ridge basalt (MORB) signatures with suprasubduction zone chemical characteristics. They explain this phenomenon by proposing that subduction contaminated fluids and melts may be channeled ("leaked") through an asthenospheric mantle flow into a slab window beneath the ridge segment, which is located near a ridge-trench-trench (RTT) triple junction, to generate MORB with convergent-margin geochemical signatures. This modern example and the model itself have strong implications for ophiolite genesis and emplacement at ridge-trench intersections (see Pearce, this volume, Chapter 15) but are questioned as to whether or not they represent rare examples in Earth's history (see Flower, this volume, Chapter 8).

Another unusual occurrence of a mid-ocean ridge generated crust is represented by the Macquarie Ridge Complex, an uplifted fragment of 12–9.5 m.y. old oceanic crust marking the boundary between the Australian and Pacific plates south of New Zealand (Varne et al., 2000). The Macquarie Ridge Complex was formed by slow spreading (~2 cm/yr full-rate) on a short ridge segment and was subsequently elevated above sea level by transpression as this plate boundary became obliquely convergent. Varne et al. (2000) documents that this displaced young oceanic crust exhibits a Pacific MORB signature ranging from normal (N-MORB) to enriched (E-MORB) but that it also displays more primitive, highly enriched, silica-undersaturated compositions that are extreme variants of MORB, atypical of melt evolution beneath actively spreading mid-ocean ridges. This significant deviation in chemical compositions of the Macquarie Ridge Complex magmas might have resulted from significantly reduced decompression melting of rising asthenosphere and magma mixing beneath the rift axis as the slow-spreading ridge system was being shut-off due to transpression along the plate boundary (Varne et al., 2000). Clearly, we need to better understand how magmas evolve beneath dying oceanic spreading centers because the structure and chemistry of ophiolites mostly reflect the snapshots of magmatic, structural, and hydrothermal processes operating at the time of the demise of oceanic crust and just before its "resurrection" (emplacement) on land.

In interpreting the ophiolites in California in light of new developments in the ophiolite concept, Coleman (2000) recognized five groups of ophiolites with distinct evolutionary history, stratigraphic relationships, petrological and chemical characteristics, geophysical parameters, and igneous ages of formation. These ophiolites include intra-oceanic suprasubduction zone ophiolites (e.g., Coast Range ophiolite), mafic-ultramafic slabs associated with magmatic underplating at a rifted continental margin (e.g., "Great Valley Ophiolite"), abyssal peridotites of possible fracture zone origin that were emplaced into accretionary mélanges (e.g., peridotite wedges within the

Franciscan subduction mélange), disaggregated ophiolites and ophiolitic rocks of mid-ocean ridge and/or seamount origin that were incorporated into the Franciscan accretionary wedge, and slabs of stranded oceanic crust that were tectonically underplated to the base of the continental margin in subduction zones. In classifying the Californian ophiolites into these distinctive groups, Coleman (2000) saw the need to define these ophiolites based on the tectonic affinity of their formation and the mechanical processes of their incorporation and emplacement into the North American Continental margin. His application of the term "ophiolite" deviates significantly in this case from the original Penrose Conference definition (Anonymous, 1972) because of his specific assignment of generally disassembled packages of mafic-ultramafic rocks to distinct tectonic settings of igneous origin.

NEW CLASSIFICATION OF OPHIOLITES

It is clear that there are a variety of ophiolites with different structural architecture, chemical fingerprints, and evolutionary paths, suggesting different tectonic environments of origin. Therefore, interpreting ophiolites in the strict sense of an ophiolite-ocean crust analogy as defined by the 1972 Penrose Conference description is no longer practical or fruitful. Highly dismembered and deformed ophiolitic occurrences in orogenic belts and intracontinental settings might have originated from any kind of oceanic environment in which mafic and ultramafic magmas and their derivatives may have evolved in association with spreading, extensional, and plate-accretion processes. Recognizing ophiolites with different lithological assemblages, chemical and isotopic compositions, internal structures, and regional geological characteristics can be more useful not only to identify specific tectonic settings of ophiolite generation, but also to better document the processes through which these oceanic rocks were incorporated into the continental margins. Assigning specific tectonic settings of formation to ophiolites can also be helpful to establish the nature of cogenetic relationships between the various parts of ophiolitic sequences and thus to delineate the petrological lineage of ophiolites.

The following list of ophiolite types and the inferred tectonic setting of their igneous formation is presented as a working classification scheme, which will probably be modified in the future as more structural field and petrological and/or geochemical information from world ophiolites becomes available. The geographic designation of different ophiolite types is based on some of the best documented case studies in the literature and may be subject to modification as better examples are identified in the future.

Ophiolite Types

1. Ligurian-Type Ophiolites

Typical examples of this type occur in the Liguria region of the Northern Apennines and in the Western Alps (including Corsica). These ophiolites are characterized by the widespread existence of largely serpentinized peridotites that are intruded and/or covered by small to moderate volumes of gabbros, local dikes, and pillow lavas (Bernoulli et al., this volume, Chapter 7). They do not include sheeted dike complexes, and the contacts between the mantle rocks and the crustal units may be intrusive, tectonic, and/or stratigraphic. These ophiolites have a Hess-type internal structure (as opposed to a Penrose-type, complete sequence), and they characterize the Steinmann trinity assemblages. The peridotites are mainly made of plagioclase and spinel lherzolites with clinopyroxene-rich varieties and display locally well developed high-temperature fabrics. These mantle rocks may be significantly older than the crustal components of the ophiolites. Gabbroic rocks range from cumulates to isotropic gabbros and plagiogranites, and they occur as small intrusive bodies and dikes in the peridotites. They display chemical and isotopic features and mineral assemblages characteristic of MORB affinity. Basaltic extrusive rocks occur as pillow and massive lava flows and have MORB affinities. These crustal rocks are not generally linked to the mantle units through a genetic melt and residua relationship (Rampone and Piccardo, 2000).

The Ligurian-type ophiolites represent the Class-III type of Miyashiro (1975) and Lherzolite-type (LOT) of Nicolas and Boudier (this volume, Chapter 9), and they may have formed during the early stages of opening of an ocean basin, following continental rifting and break-up. They were originally situated in a pericontinental position adjacent to rifted continental margins. An actualistic example would be the West Iberia margin (Galicia passive margin) facing the Atlantic Ocean (Boillot and Froitzheim, 2001). These ophiolites may include pieces of exhumed subcontinental lithospheric mantle and might have been bounded in their original tectonic setting by a small rift basin, an embryonic ocean basin (i.e., Red Sea type), or a mature ocean (i.e., Atlantic Ocean). The Ligurian-type ophiolites are analogous to those identified by Coleman (2000) as mafic-ultramafic slabs associated with magmatic underplating at a rifted continental margin. The emplacement of these ophiolites might have been facilitated by the inversion of seaward-dipping extensional fault systems to large landward-directed thrust faults during regional contraction and basin closure.

2. Mediterranean-Type Ophiolites

Typical examples of this type occur in the eastern Mediterranean region (extending from Albania, Greece, Cyprus, and Turkey to Oman and Tibet) and may contain Penrose-type, nearly complete pseudostratigraphy of an idealized ophiolite sequence. The sedimentary cover of these ophiolites is generally composed of pelagic rocks (limestone and/or chert) and is devoid of volcaniclastic and pyroclastic rocks typical of volcanic arcs. Best examples of this type include the Troodos (Cyprus), Kizildag (Turkey), Semail (Oman), Xigaze (Tibet), and Bay of Islands (Newfoundland, Canada) ophiolites. The ophiolites occurring in the type area (i.e., Kizildag, Semail) rest tectonically on passive margin sequences of continents or microcontinents and are characteristically underlain by metamorphic soles and mélanges, which are composed of material derived from both the ophiolites and underlying carbonate platforms. The Mediterranean-type ophiolites display seafloor-spreading generated, extensional structures within their crustal units, particularly within their sheeted dike complexes and extrusive rock assemblages. Contacts between the sheeted dike complexes and the underlying plutonic rocks may be mutually intrusive and/or faulted, depending on magma supply rates and on the mode and nature of interplay between the magmatic and amagmatic (tectonic) extension during the construction of fossil oceanic crust. Sheeted dikes are commonly the feeders to the overlying volcanic rocks, which include both pillow and massive lava flows. The nature and degree of hydrothermal alteration, seafloor metamorphism, and mineralization vary significantly among these Mediterranean-type ophiolites, depending on the history of their extensional tectonics (degree, nature, and depth of faulting and tectonically induced permeability in crustal and mantle rocks), nature and distribution of on- and off-axis magmatism, and intensity and scale of fluid flux in the environment of origin.

The Mediterranean-type ophiolites may include distinctly different mantle sequences consisting mainly of harzburgite-lherzolite and harzburgite. Compositions of harzburgite-lherzolite bearing peridotites are residual to MORB melt extraction and commonly occur within earlier (or older) mantle units (i.e., in the Mirdita ophiolites in Albania, Vourinos in Greece, Troodos in Cyprus, Semail in Oman). Harzburgite-dominant peridotite compositions reflect more rigorous melting of depleted mantle and are considered parental to boninitic and island arc (tholeiitic to calc-alkaline) magmas. Podiform chromite deposits are more common in these harzburgite-peridotites.

The Mediterranean-type ophiolites correspond to the Class-I type of Miyashiro (1975) and Harzburgite (HOT) to Harzburgite-lherzolite (LHOT)-type ophiolites of Nicolas and Boudier (this volume, Chapter 9). Most of the Tethyan ophiolites of Moores (1982) are also Mediterranean-type ophiolites as defined here. Regionally, some Ligurian ophiolites may be transitional into Mediterranean-type ophiolites within the same mountain belt, such as in the Albanides (Mirdita ophiolites in Albania; Shallo and Dilek, this volume, Chapter 20) and in the Hellenides (Pindos and Vourinos ophiolites in Greece as Ligurian- and Mediterranean-type ophiolites, respectively; Smith and Rassios, this volume, Chapter 19). The evolution of Mediterranean-type ophiolites involved seafloor spreading, on and off-axis magmatism, and tectonism in the upper plate of an intra-oceanic subduction zone at some point in time, and the ophiolites might have formed in a forearc, infant arc, and/or backarc setting. Trench rollback, basin collapse, and orogenic entrapment of "proto-ophiolites" due to passive margin-trench collision were significant tectonic processes during the evolution of Mediterranean-type ophiolites.

There is no one specific mode of generation of the Mediterranean-type ophiolites. Intra-oceanic subduction beneath mid-ocean ridges, subduction of ridge segments, and subduction-influenced mantle melting beneath oceanic spreading centers in different tectonic settings, such as the Woodlark Basin and/or the Chile Ridge, may have played a major role in the evolution of ophiolitic magmas. Actualistic examples for the Mediterranean-type ophiolites would be the modern oceanic crust ("proto-ophiolites") in the Woodlark Basin and the Chile Ridge spreading centers, Izu-Bonin-Mariana forearc region, and the Lau and East Scotia backarc basins. The emplacement of the Mediterranean-type ophiolites was mostly controlled by the relative motions of small plates and microcontinents and their interactions with the arc-trench systems within larger ocean basins (i.e., the Neo-Tethys) and might have been independent of the motions of the bounding major continental plates. However, rapid opening of some marginal basins and initiation of intra-oceanic subduction zones (and hence, igneous development of "suprasubduction zone" ophiolites) might have been driven by mantle response (i.e., mantle extrusion) to regional or global-scale tectonics and plate collisions.

3. Sierran-Type Ophiolites

These ophiolites typically occur in the Pacific Rim and have complex, polygenetic evolutionary paths. Some ophiolites in Japan, Philippines, and Cuba may belong to this group; most representative examples include the Jurassic arc ophiolite(s) exposed in the western Sierra Nevada foothills in California. Ensimatic arc ophiolites in the Sierra Nevada foothills contain volcanic, plutonic, and hypabyssal rocks and locally well developed sheeted dike swarms (i.e., the Smartville arc complex). Volcanic rocks range from basalts and basaltic andesites to dacites and rhyolites, and volcaniclastic rocks (including some subareal depositions) are widespread, indicating the construction of volcanic arc edifice(s) during the evolution of these ophiolites. The arc construction appears to have occurred on and across a pre-existing, multiply deformed and heterogeneous oceanic (ophiolitic) basement as documented by crosscutting and geochronological relations (Dilek et al., 1990). Collectively, the older ophiolitic/oceanic basement and the overlying and younger volcanic arc assemblages in the western Sierra Nevada foothills represent an ensimatic island arc terrane, which was accreted into the North American continental margin during Middle to Late Jurassic time.

The evolution of this island arc terrane was episodic (multiple magmatic pulses and extensional phases) and polygenetic. The main phase of volcanic arc construction occurred ca. 200 Ma in intra-oceanic conditions, whereas the late-stage rifting and magmatism took place ca. 160 Ma, >50 m.y. later and after accretion of the arc terrane into the North American continental margin. Oblique convergence and associated strike-slip tectonics might have played a major role in the magmatic and tectonic accretionary phases during the evolution of this island arc terrane. In this sense, this particular ophiolite type differs significantly from the suprasubduction zone–generated Mediterranean-type ophiolites, which were <20 Ma when they were emplaced following their displacement from their igneous tectonic setting (with the exception of the Troodos ophiolite in Cyprus). The Sierran-type ophiolites include Miyashiro's (1975) Class-II type island arc ophiolites and may correspond in part to the Cordilleran-type ophiolites of Moores (1982).

Similar polygenetic evolution of arc ophiolites has been reported from the Philippines (i.e., Encarnacion, 1993; Geary et al., 1989; Yumul et al., 2000) and Cuba (Cobiella-Reguera, 2002), where older ophiolitic lithologies constitute the basement of volcanic arc complexes, which underwent magmatic and tectonic extension through multiple phases of backarc basin opening. The tectonic evolution of these polygenetic arc terranes and their ophiolites in Cuba and the Philippines also experienced strike-slip deformation (and dismantling) as a result of oblique convergence during and after their accretion into continental margins.

4. Chilean-Type Ophiolites

This ophiolite type is best characterized by the Rocas Verdes ophiolites in southernmost South America that represent a relatively autochthonous fossil oceanic crust surrounded both on the east and the west by crystalline rocks of the Andes. The Rocas Verdes ophiolites, mainly the Sarmiento and Tortuga complexes in Chile, include, from top to the bottom, mafic volcanic rocks (2–3 km thick) composed of pillow lavas and volcanic breccias, sheeted dike complex (300–500 m thick), massive diabase, and coarse-grained gabbros (Stern and de Wit, 1980); mantle peridotites are not exposed. The contact between the sheeted dike complex and the underlying plutonic rocks is intrusive; individual dike swarms crosscut the gabbros, sheeted dikes, and extrusive rocks at different levels.

The Chilean ophiolites are bounded on the west by Jurassic silicic volcanic rocks (Tobifera Formation) and on the east by the Paleozoic metamorphic basement of Patagonia (pre-Andean continental crust). These continental rocks and their fault contacts with the ophiolites are intruded by basaltic dike swarms and diabasic to gabbroic stocks and sills. The dike intrusions are commonly parallel to the trend of the sheeted dike complexes in the ophiolites and have chemical compositions similar to those of the sheeted dikes. Stern and de Wit (2003) interpreted these intrusive zones flanking the ophiolites as rifted margins of a continental backarc basin, in which the Rocas Verdes ophiolites had formed (Dalziel et al., 1974). Arc volcanic and volcaniclastic rocks intercalated with mafic pillow lavas, and fine-grained deep-water turbiditic rocks overlie the ophiolites (Winn and Dott, 1978).

Mafic rocks of the Rocas Verdes ophiolites are chemically similar to MORB and display a tholeiitic differentiation trend (Saunders et al., 1979; Stern, 1980; Stern and de Wit, 2003). Their geological and geochemical characteristics, combined with the regional geology, suggest that the Chilean ophiolites formed in an extensional backarc basin, which progressively opened up as rifting of the magmatic arc propagated from south to north in the latest Jurassic and Early Cretaceous (Saunders et al., 1979; de Wit and Stern, 1981). The closure of this basin occurred in the Middle Cretaceous, possibly due to the flattening of the angle of eastward dipping subduction below the western continental margin of South America (Stern and de Wit, 2003).

The Chilean-type ophiolites differ from the Mediterranean-type ophiolites in that they are the products of backarc rifting in an "ensialic" setting within a magmatic arc, as opposed to a suprasubduction zone environment in an intra-oceanic setting, and in that they are relatively autochthonous in their current positions, as opposed to representing allochthonous thrust sheets emplaced through collisional processes. They are different from the Sierran-type ophiolites because they do not have an older ophiolitic/oceanic basement and a complex, polygenetic tectonomagmatic evolution. Thus, they appear to have a unique backarc basin origin in an active continental margin setting. An actualistic example for the Rocas Verdes basin and the Chilean ophiolites may be the Andaman Sea backarc basin behind the Burma-Sumatra magmatic arc that opened up in the latest Oligocene-Miocene through oblique, NW-SE–directed extension (Curray, 1989; Mitchell, 1993). Chilean-type ophiolites may be common in the Pontides (Turkey), Lesser Caucasus, and the Paleozoic orogenic belts in Central Asia.

5. Macquarie-Type Ophiolites

This is a unique ophiolite occurrence on the Macquarie Island in the Southern Ocean ~1500 km south-southeast of Tasmania. The 12–9.5 m.y. old oceanic crust exposed on the island formed along a nearly E-W–trending short ridge segment at the Australian-Pacific plate boundary and was subsequently displaced from this mid-ocean ridge setting and uplifted due to transpressional deformation as this plate boundary became obliquely convergent ca. 5 Ma (Sutherland, 1995; Varne et. al., 2000). Thus, the Macquarie Island ophiolite represents a relatively in situ fragment of mid-ocean ridge generated oceanic crust. The ophiolite sequence includes, from top to the bottom, basaltic extrusive rocks intermixed with volcaniclastic sedimentary rocks, sheeted dolerite dikes (~1.5 km thick), microgabbro transition zone (between sheeted dikes and underlying plutonic rocks), coarse-grained, massive, and layered gabbroic rocks with ultramafic screens, troctolite, wehrlite, dunite, and mixed dunite-wehrlite-harzburgite successions, and harzburgitic peridotites (Varne et al., 2000). Although the contacts between these lithological units are commonly faulted, transitional igneous contacts are also present, and the Macquarie Island ophiolite appears to display a typical Penrose-type layer-cake stratigraphy.

In chemical composition, basalts (including fresh glassy material) and doleritic dikes of the Macquarie Island ophiolite display a continuum ranging from N-MORB through E-MORB to more enriched (more alkalic) variants of MORB. Varne et al. (2000) suggested that this evolution toward more enriched, silica-undersaturated compositions might have resulted from reduced melting and magma mixing beneath the slow-spreading ridge system, which was converted to an oblique convergent plate boundary due to a change in the regional stress regime. In this sense, the current tectonic configuration of the Macquarie Island ophiolite is an artifact of "ridge collapse."

Although uplifted and elevated above sea-level due to transpression, the Macquarie Island ophiolite has not yet been emplaced in a continental margin. Subduction initiation is at work along segments of the collapsed Macquarie Ridge both north and south of the island (Collot et al., 1995), and it is likely that the ophiolite will be placed in the upper plate of this subduction zone in the geological future. Depending on the geodynamic evolution of this subduction zone environment, the Macquarie Island ophiolite may then become either the basement of a Sierran-type island arc ophiolite complex, or an older ophiolite complex entrapped in the forearc region of the suprasubduction zone setting west of the Campbell Plateau (with continental crust) where future Mediterranean-type ophiolites may develop. Ophiolite researchers should be on the lookout for the existence of similar mid-ocean ridge generated, Macquarie-type ophiolite occurrences in orogenic belts.

6. Caribbean-Type Ophiolites

These ophiolites represent oceanic crustal assemblages of Large Igneous Province (LIP) origin, and the best examples of LIP-generated ophiolites occur in the Caribbean region (Kerr et al., 1997; Coffin and Eldholm, 2001; Giunta et al., 2002). The internal structure and stratigraphy of tectonically emplaced fragments of oceanic plateaus are highly heterogeneous but may contain most ophiolitic subunits, including pillow and massive lava flows (ranging in composition from N-MORB through transitional-MORB [T-MORB] to E-MORB), isotropic to layered gabbros, and dunite with bands of lherzolite, olivine websterite, and olivine gabbronorites at structurally lower levels (i.e., Bolivar Complex in Western Colombia; Kerr et al., 1997). Sheeted dike complexes are generally missing, although their absence in these mafic-ultramafic complexes accreted along the periphery of the Caribbean plate does not indicate that these intrusions do not exist in the intermediate-depth crustal sections of oceanic plateaus. Seismic velocity structures of LIPs suggest that extrusive and intrusive rock units of oceanic plateaus resemble "expanded" oceanic crust, underlain by mafic cumulates (Coffin and Eldholm, 2001).

Mid-ocean ridge generated crust may progressively evolve into a plateau structure if and/or when repeated eruptions and intrusions of new basaltic and picritic magmas, generated from plume-heads, are added onto the pre-existing oceanic crust (vertical thickening; Saunders et al., 1996). When this thickened "oceanic crust" gets accreted into continental margins through complex collision and/or wrench tectonics, it becomes an ophiolite. Many of the Cretaceous ophiolites in the peri-Caribbean region, specifically those in Costa Rica, Hispaniola, Dutch Antilles, Venezuela, and Colombia, are fragments of the proto-Caribbean oceanic crust, which evolved into an oceanic plateau in the Late Cretaceous (Giunta et al., 2002).

Tectonically emplaced fragments of LIP-generated oceanic crust also occur in the Solomon Islands, the Pacific Northwest (Wrangellia), Japan (Sorachi Plateau), and Ecuador (Piñon Formation; Coffin and Eldholm, 2001). It is highly likely that some of the dismembered mafic-ultramafic rock assemblages in orogenic belts may be tectonic relics of oceanic plateau-derived ophiolite fragments. Further structural and geochemical studies of these "Caribbean-type ophiolites" in different orogenic belts should make significant contributions to the ophiolite concept.

7. Franciscan-Type Ophiolites

These ophiolite types are spatially associated with accretionary complexes of active margins and are commonly tectonically intercalated with mélanges and high-pressure metamorphic rocks characteristic of subduction zones. Different ophiolitic rock units may occur within imbricated thrust sheets that are synthetic to the paleo-subduction zone. Franciscan-type ophiolites include fragments of abyssal peridotites, gabbros, and basalts of possible fracture zone origin, disaggregated oceanic crustal slabs of mid-ocean ridge origin (pillow lavas and gabbros), and/or dismantled fragments of seamounts and island arc complexes. These oceanic rocks are locally associated with pelagic-hemipelagic sedimentary rocks (chert, limestone) and terrigenous trench-fill sediments that might have been deposited on them prior to and after their incorporation into the accretionary prism complexes. Blocks and thrust sheets of blueschist-bearing metamorphosed oceanic rocks also occur within these accretionary complexes. These high-pressure rocks might have been exhumed on the trench slope as a result of tectonic erosion of the forearc thrust front by subducting plates and/or as a consequence of syn-subduction extensional collapse of the accretionary complex (Platt, 1986).

Franciscan-type ophiolites are an integral component of the accretionary orogenic belts in the Pacific Rim. Some of the best examples occur in California (Franciscan Complex), Japanese Islands (Oeyama and Yakuno ophiolites; Shimanto and Mineoka-Setogawa accretionary complexes), Koryak Mountain belt in Kamchatka, Chugach subduction-accretion complex in Alaska, Ordovician–Middle Devonian accretionary complex in the Western Precordillera of Argentina, and Paleozoic subduction-accretionary complexes in eastern Australia and New Zealand (i.e., Kanmantoo belt, Great Serpentinite Belt, New England Belt).

Franciscan-type ophiolites and ophiolitic units in subduction-accretion complexes may have diverse lithological assemblages, metamorphic grades, and chemical affinities with no genetic links between them because they are tectonic slices of oceanic rocks scraped off from downgoing plates. These tectonically imbricated ophiolites become progressively younger in age structurally down-section within the subduction-accretionary complexes. The ones in southwestern Japan, for example, range from Paleozoic ophiolites (i.e., Oeyama and Yakuno) in structurally higher positions to Tamba (Jurassic), Shimanto (Cretaceous), and Mineoka-Setogawa (Tertiary) ophiolites and accretionary complexes in structurally lower positions, progressively downward in the orogenic belt. This inversion in crustal ages (younger at the bottom) is a result of continued subduction and accretion of oceanic plates at active continental margins. Unlike the Mediterranean ophiolites, Franciscan-type ophiolites are not underlain by passive margins of continents and continental fragments because they have not undergone trench-continent collisions in their tectonic evolutionary paths. The Franciscan-type ophiolites correspond in part to the Cordilleran-type ophiolites of Moores (1982).

CONCLUDING REMARKS

Ophiolites have played a major role in our understanding of Earth's processes ranging from seafloor spreading, melt evolution and magma transport in oceanic spreading centers, and hydrothermal alteration and mineralization of oceanic crust to collision tectonics, mountain building processes, and orogeny. They provide the essential structural, petrological, geochemical, and geochronological evidence to document the evolutionary history of ancient continental margins and ocean basins. The ophiolite concept has evolved steadily since its formulation by Steinmann in the 1920s. With more systematic structural field studies and mapping of ophiolites, isotopic delineation of their mantle sources and domains, and precise geochronology, the ophiolite concept shall further evolve, making significant contributions to our understanding of Earth's history. The synergy between ophiolite research and marine geological and geophysical investigations will always be crucial to test new ideas about the evolution of oceanic lithosphere, to make new observations toward better understanding of the nature of magmatic, metamorphic, and tectonic processes at modern and ancient plate boundaries and continental margins, and to formulate new hypotheses about Earth's behavior through time. To this end, we also need to develop a detailed ophiolite database at the international level through the use of interactive GIS (geographic information systems).

Earth's history in deep time, particularly in the Archean, is little known, and the existence of ophiolites in the Archean record has remained an enigmatic question. To a large extent, this is because the geological community has been looking for a Penrose-type complete ophiolite pseudostratigraphy in the Precambrian rock record (Condie, 1997). The Archean greenstones do not show any resemblance to this type of oceanic crust, having a 7- to 8-km-thick layer-cake stratigraphy. However, it is highly likely that other ophiolite types (i.e., Caribbean, Chilean, Franciscan, and Sierran-type ophiolites) may be well preserved in the Archean crust. Armed with new developments in the ophiolite concept, future studies of Precambrian geology should reveal new and exciting information about the nature of Archean oceanic crust and hence the mode and nature of Earth's processes and heat production during the Archean.

ACKNOWLEDGMENTS

In writing this essay, I have relied on the published work of many ophiolite researchers, marine geologists and geophysicists of the Ocean Drilling Program community, as well as colleagues investigating different orogenic belts around the world; however, I take the full responsibility for the views and interpretations expressed in this essay and for likely omissions of other pertinent references on ophiolites. I extend my sincere thanks to all members of the international ophiolite community for their inspiring work over the years, for their support of ophiolite research, and for enlightening discussions both in the field and at meetings and conferences. My work on ophiolites has been supported generously by grants from the National Science Foundation, Joint Oceanographic Institutions, NATO Science Program (CRG-970263 and EST.CLG-97617), and the National Geographic Society, which I gratefully acknowledge. I would like to thank Bob Coleman and John Alten for their insightful and constructive comments on the manuscript.

REFERENCES CITED

Anonymous, 1972, Penrose Field Conference on ophiolites: Geotimes, v. 17, p. 24–25.

Aubouin, J., and Ndojaj, I., 1964, Regard sur la geologie de l'Albanie et sa place dans la geologie des Dinarides: Bulletin de la Société Géologique de France, ser. 7, v. VI, p. 593–625.

Benson, W.N., 1926, The tectonic conditions accompanying the intrusion of basic and ultrabasic igneous rocks: Memoirs of the National Academy of Sciences, Volume XIX, First Memoir, 90 p.

Boillot, G., and Froitzheim, N., 2001, Non-volcanic rifted margins, continental break-up and the onset of sea-floor spreading: some outstanding questions, *in* Wilson, R.C.L., Whitmarsh, R.B., Taylor, B., and Froitzheim, N., eds., Non-volcanic rifting of continental margins: A comparison of evidence from land and Sea: London, Geological Society of London Special Publication 187, p. 9–30.

Bowen, N.L., 1927, The origin of ultrabasic and related rocks: American Journal of Science, v. 14, p. 89–108.

Bowen, N.L., 1928, The evolution of the igneous rocks: Princeton-New Jersey, Princeton University Press, 332 p.

Bowen, N.L., and Tuttle, O.F., 1949, The system MgO-FeO-SiO_2: Geological Society of America Bulletin, v. 60, p. 439–460.

Boudier, F., and Nicolas, A., 1985, Harzburgite and lherzolite subtypes in ophiolitic and oceanic environments: Earth and Planetary Science Letters, v. 76, p. 84–92.

Brongniart, A., 1813, Essai de classification minéralogique des roches mélanges: Journal des Mines, v. XXXIV, p. 190–199.

Brongniart, A., 1821, Sur le gisement ou position relative des ophiolites, euphotides, jaspes, etc. dans quelques parties des Apennins: Annales des Mines, Paris, v. 6, p. 177–238.

Brunn, J.H., 1956, Contribution a l'étude géologique de la Grece septentrionale: les confins de l'Epire et de la Thessalie: Annales Géologique des Pays Hellénique, v. 1, p. 1–485.

Brunn, J.H., 1960, Mise en place et différenciation pluto-volcanique du cortege ophiolitique: Revues de Géographie Physique et Géologie Dynamique, v. 3, p. 115–132.

Brunn, J.H., 1961, Les sutures ophiolitiques: contribution a l'études des relations entre phénomene magmatique et orogénique: Revues de Géographie Physique et Géologie Dynamique, v. 4, p. 89–96 and 181–202.

Cannat, M., Mevel, C., Maia, M., Deplus, C., Durand, C., Gente, P., Agrinier, P., Belarouchi, A., Dubuisson, G., Humler, E., and Reynolds, J., 1995, Thin crust, ultramafic exposures, and rugged faulting patterns at the Mid-Atlantic Ridge (22°–24°N): Geology, v. 23, p. 49–52.

Cobiella-Reguera, J., 2002, Remains of oceanic lithosphere in Cuba: Types, origins, and emplacement ages, *in* Jackson, T., ed., Caribbean geology into the third millennium: Transactions of the Fifteenth Caribbean Geological Conference, University of the West Indies Press, p. 35–46.

Coffin, M.F., and Eldholm, O., 2001, Large igneous provinces: Progenitors of some ophiolites?, *in* Ernst, R.E., and Buchan, K.L., eds., Mantle plumes: Their identification through time: Boulder, Colorado, Geological Society of America Special Paper 352, p. 59–70.

Coleman, R.G., 1971, Plate tectonic emplacement of upper mantle peridotites along continental edges: Journal of Geophysical Research, v. 76, p. 1212–1222.

Coleman, R.G., 1973, Ophiolites in the Earth's crust: A symposium, field excursions, and cultural exchange in the USSR: Geology, v. 1, p. 51–54.
Coleman, R.G., 1977, Ophiolites: New York, Springer-Verlag, 229 p.
Coleman, R.G., and Irwin, W.P., eds., 1977, North American ophiolites: Portland, Oregon, State of Oregon Department of Geology and Mineral Industries Bulletin 95, 183 p.
Coleman, R.G., 2000, Prospecting for ophiolites along the California continental margin, *in* Dilek, Y., Moores, E.M., Elthon, D., and Nicolas, A., eds., Ophiolites and oceanic crust: New insights from field studies and the Ocean Drilling Program: Boulder, Colorado, Geological Society of America Special Paper 349, p. 351–364.
Collot, J.-Y., Lamarche, G., Delteil, J., Wood, R., Sosson, M., Coffin, M., and Lebrun, J.-F., 1995, Morphostructure of an incipient subduction zone along a transform plate boundary: Puysegur Ridge and Trench: Geology, v. 23, p. 519–523.
Condie, K.C., 1997, Plate Tectonics and Crustal Evolution (Fourth Edition): Oxford, UK, Butterworth-Heinemann, 282 p.
Curray, J.R., 1989, The Sunda Arc: a model for oblique plate convergence: Journal of Sea Research, v. 24, p. 131–140.
Dalziel, I.W.D., de Wit, M.J., and Palmer, K.F., 1974, Fossil marginal basin in the southern Andes: Nature, v. 250, p. 291–294.
de Roever, W.P., 1957, Sind die alpintypen Peridotimassen vielleicht tektonisch verfrachtete Buruchstücke der Peritotschale?: Geologische Rundschau, v. 46, p. 137–146.
Dewey, J.F., 1976, Ophiolite obduction: Tectonophysics, v. 31, p. 93–120.
Dewey, J.F., and Bird, J.M., 1971, Origin and emplacement of the ophiolite suite: Appalachian ophiolites in Newfoundland: Journal of Geophysical Research, v. 76, p. 3179–3206.
de Wit, M.J., and Stern, C.R., 1981, Variations in the degree of crustal extension during the formation of a back-arc basin: Tectonophysics, v. 72, p. 229–260.
Dick, H.J.B., 1989, Abyssal peridotites, very slow-spreading ridges and ocean ridge magmatism, *in* Saunders, A.D., and Norris, M.J., eds., Magmatism in the ocean basins: London, Geological Society of London Special Publication 42, p. 71–105.
Dilek, Y., Thy, P., Moores, E.M., and Grundvig, S., 1990, Late Paleozoic-early Mesozoic oceanic basement of a Jurassic arc terrane in the northwestern Sierra Nevada, California, *in* Hardwood, D.S., and Miller, M.M., eds., Paleozoic and early Mesozoic paleogeographic relations; Sierra Nevada, Klamath Mountains, and Related Terranes: Boulder, Colorado, Geological Society of America Special Paper 255, p. 351–369.
Dilek, Y., Moores, E.M., and Furnes, H., 1998, Structure of modern oceanic crust and ophiolites and implications for faulting and magmatism at oceanic spreading centers, *in* Buck, R., Karson, J., Delaney, P., and Lagabrielle, Y., eds., Faulting and magmatism at mid-ocean ridges: Washington, D.C., American Geophysical Union Monograph 106, p. 219–266.
Dilek, Y., Thy, P., Hacker, B., and Grundvig, S., 1999a, Structure and petrology of Tauride ophiolites and mafic dike intrusions (Turkey): Implications for the Neo-Tethyan ocean: Geological Society of America Bulletin, v. 111, p. 1192–1216.
Dilek, Y., Moores, E.M., Elthon, D., and Nicolas, A., 1999b, Ophiolites and oceanic crust: New insights from field studies and Ocean Drilling Program: GSA Today, v. 9, no. 2, p. 30–32.
Dilek, Y., Moores, E.M., Elthon, D., and Nicolas, A., editors, 2000, Ophiolites and oceanic crust: New insights from field studies and the Ocean Drilling Program: Boulder, Colorado, Geological Society of America Special Paper 349, 552 p.
Dubertret, L., 1955, Geologie des roches vertes du nord-ouest de la Syrie et du Hatay (Turquie): Museum Nationale de Histoire Naturelle, Paris, Notes et Mémoires de Moyen-Orient, v. 6, 179 p.
Encarnacion, J.P., Mukasa, S.B., and Obille, E.C., 1993, Zircon U-Pb geochronology of the Zambales and Angat ophiolites, Luzon, Philippines—evidence for an Eocene arc-backarc pair: Journal of Geophysical Research, v. 98, p. 19,991–20,004.
Geary, E.E., Kay, R.W., Reynolds, J.C., and Kay, S.M., 1989, Geochemistry of mafic rocks from the Coto Block, Zambales Ophiolite, Philippines—trace element evidence for two stages of crustal growth, *in* Flower, M.F.J., and Hawkins, J.W., eds., Ophiolites and crustal genesis in the Philippines: Tectonophysics, v. 168, p. 43–63.
Giunta, G., Beccaluva, L., Coltorti, M., Mortellaro, D., Siena, F., and Cutrupia, D., 2002, The peri-Caribbean ophiolites: structure, tectono-magmatic significance and geodynamic implications: Caribbean Journal of Earth Science, v. 36, p. 1–20.
Hacker, B.R., 1990, Simulation of the metamorphic and deformational history of the metamorphic sole of the Oman ophiolite: Journal of Geophysical Research, v. 95, p. 4895–4907.
Hawkins, J.W., 1977, Petrological and geochemical characteristics of marginal basin basalts, *in* Talwani, M., and Pitman, W.C., III, eds., Island arcs, deep sea trenches and back arc basins, volume 1: Washington, D.C., American Geophysical Union, p. 355–365.
Hess, H.H., 1938, A primary peridotite magma: American Journal of Science, v. 35, p. 321–344.
Hess, H.H., 1955, Serpentines, orogeny and epeirogeny, *in* Poldevaart, A., ed., Crust of the Earth (A Symposium): New York, Geological Society of America Special Paper 62, p. 391–408.
Hess, H.H., 1962, History of ocean basins, *in* Engel, A.E.J., James, H.L., and Leonard, B.F., eds., Petrologic studies: A volume in honor of A.F. Buddington: New York, Geological Society of America, p. 599–620.
Hess, H.H., 1965, Mid-ocean ridges and tectonics of the sea floor, *in* Whittard, W.F., and Bradshaw, R., eds., Submarine geology and geophysics: Proceedings of the 17th Symposium of the Colston Research Society: London, Butterworths, p. 317–334.
Ishiwatari, A., 1985, Alpine ophiolites: Product of low-degree mantle melting in a Mesozoic transcurrent rift zone: Earth and Planetary Science Letters, v. 76, p. 93–108.
Jackson, E.D., and Thayer, T.P., 1972, Some criteria for distinguishing between stratiform, concentric and alpine peridotite-gabbro complexes: 24th International Geological Congress Section 2, p. 289–296.
Jamieson, R.A., 1986, PT paths from high temperature shear zones beneath ophiolites: Journal of Metamorphic Geology, v. 4, p. 3–22.
Kerr, A.C., Tarney, J., Marriner, G.F., Nivia, A., and Saunders, A.D., 1997, The Caribbean-Colombian Cretaceous igneous province: The internal anatomy of an oceanic plateau, *in* Mahoney, J.J., and Coffin, M.F., eds., Large igneous provinces: Continental, oceanic, and planetary flood volcanism: Washington, D.C., American Geophysical Union Monograph 100, p. 123–144.
Lotti, B., 1886, Paragone fra le rocce ofiolitiche terziarie italiene e le rocce basiche pure terziarie della Scozia e dell'Irlanda, a propisito di rue recenti pubblicazioni di J.W. Judd: Bollettino del Reale Comitato Geologico d'Italia, v. 17, no. 3-4, p. 73–86.
Mitchell, A.H.G., 1993, Cretaceous-Cenozoic tectonic events in the western Mayanmar (Burma) Assam region: Journal of the Geological Society of London, v. 150, p. 1089–1102.
Miyashiro, A., 1973, The Troodos complex was probably formed in an island arc: Earth and Planetary Science Letters, v. 19, p. 218–224.
Miyashiro, A., 1975, Classification, characteristics, and origin of ophiolites: Journal of Geology, v. 83, p. 249–281.
Moores, E.M., 1982, Origin and emplacement of ophiolites: Reviews of Geophysics and Space Physics, v. 20, p. 735–760.
Moores, E.M., Kellogg, L.H., and Dilek, Y., 2000, Tethyan ophiolites, mantle convection, and tectonic "historical contingency": A resolution of the "ophiolite conundrum," *in* Dilek, Y., Moores, E.M., Elthon, D., and Nicolas, A., eds., Ophiolites and oceanic crust: New insights from field studies and the Ocean Drilling Program: Boulder, Colorado, Geological Society of America Special Paper 349, p. 3–12.
Nicolas, A., 1989, Structures of ophiolites and dynamics of oceanic lithosphere: Dordrecht, Netherlands, Kluwer Academic Publishers, 367 p.
Panayiotou, A., ed., 1980, Ophiolites: Proceedings International Ophiolite Symposium Cyprus 1979: Nicosia, Cyprus, Geological Survey Department, 781 p.
Pearce, J.A., 1975, Basalt geochemistry used to investigate past tectonic environments on Cyprus: Tectonophysics, v. 25, p. 41–67.
Platt, J.P., 1986, Dynamics of orogenic wedges and the uplift of high-pressure metamorphic rocks: Geological Society of America Bulletin, v. 97, p. 1037–1053.
Rampone, E., and Piccardo, G.B., 2000, The ophiolite-oceanic lithosphere analogue: New insights from the Northern Apennines (Italy), *in* Dilek, Y., Moores, E.M., Elthon, D., and Nicolas, A., eds., Ophiolites and oceanic crust: New insights from field studies and the Ocean Drilling Program: Boulder, Colorado, Geological Society of America Special Paper 349, p. 21–34.
Saunders, A.D., Tayney, J., Kerr, A.C., and Kent, R.W., 1996, The formation and fate of large oceanic igneous provinces: Lithos, v. 37, p. 81–95.
Saunders, A.D., Tarney, J., Stern, C.R., and Dalziel, I.W.D., 1979, Geochemistry of Mesozoic marginal basin floor igneous rocks from southern Chile: Geological Society of America Bulletin, v. 90, p. 237–258.

Shervais, J.W., 2001, Birth, death and resurrection: The life cycle of suprasubduction zone ophiolites: Geochemistry, Geophysics, Geosystems, v. 2, paper no. 2000GC000080.

Staub, R., 1922, Über die Verteilung der Serpentine in den alpinen Ophioliten: Schweizerische Mineralogische und Petrographische Mitteilungen, v. 2, p. 78–199.

Steinmann, G., 1913, Über Tiefenabsätze des Oberjura im Apennin: Geologische Rundschau, v. 4, p. 572–575.

Steinmann, G., 1927, Der ophiolitischen Zonen in der mediterranean Kettengebirgen: 14th International Geological Congress in Madrid, v. 2, p. 638–667.

Stern, C.R., 1980, Geochemistry of Chilean ophiolites: evidence of the compositional evolution of the mantle source of back-arc basin basalts: Journal of Geophysical Research, v. 85, p. 955–966.

Stern, C.R., and de Wit, M.J., 1980, The role of spreading centre magma chambers in the formation of Phanerozoic oceanic crust: Evidence from Chilean ophiolites, *in* Panayiotou, A., ed., Ophiolites: Proceedings International Ophiolite Symposium Cyprus 1979: Nicosia, Geological Survey Department, p. 497–506.

Stern, C.R., and de Wit, M.J., 2003, Rocas Verdes ophiolites, southernmost South America: remnants of progressive stages of development of oceanic-type crust in a continental margin back-arc basin, *in* Dilek, Y., and Robinson, P.T., eds., Ophiolites in Earth History: London, Geological Society of London Special Publication (in press).

Sturm, M.E., Klein, E.M., Karsten, J.L., and Karson, J.A., 2000, Evidence for subduction-related contamination of the mantle beneath the southern Chile Ridge: Implications for ambiguous ophiolite compositions, *in* Dilek, Y., Moores, E.M., Elthon, D., and Nicolas, A., eds., Ophiolites and oceanic crust: New insights from field studies and the Ocean Drilling Program, Geological Society of America Special Paper 349, p. 13–20.

Suess, E., 1909, The Face of the Earth, v. 2, Part 3, The Sea: Oxford, Clarendon, 556 p.

Sutherland, R., 1995, The Australia-Pacific boundary and Cenozoic plate motions in the SW Pacific: Some constraints from Geosat data: Tectonics, v. 14, p. 819–831.

Thayer, T.P., 1967, Chemical and structural relations in ultramafic and feldspathic rocks in Alpine intrusive complexes, *in* Wyllie, P.J., ed., Ultramafic rocks: Wiley, New York, p. 222–238.

Varne, R., Brown, A.V., and Falloon, T., 2000, Macquarie Island: Its geology, structural history, and the timing and tectonic setting of its N-MORB to E-MORB magmatism, *in* Dilek, Y., Moores, E.M., Elthon, D., and Nicolas, A., eds., Ophiolites and oceanic crust: New insights from field studies and the Ocean Drilling Program, Geological Society of America Special Paper 349, p. 301–320.

Vuagnat, M., 1963, Remarques sur la trilogie serpentinites-gabbros-diabases dans le bassin de le Méditerranée occidentale: Geologische Rundschau, v. 53, p. 336–357.

Wakabayashi, J., and Dilek, Y., 2000, Spatial and temporal relationships between ophiolites and their metamorphic soles: A test of models of forearc ophiolite genesis *in* Dilek, Y., Moores, E.M., Elthon, D., and Nicolas, A., eds., Ophiolites and oceanic crust: New insights from field studies and the Ocean Drilling Program: Boulder, Colorado, Geological Society of America Special Paper 349, p. 53–64.

Winn, R.D., and Dott, R.H., 1978, Submarine-fan turbidites and resedimented conglomerates in a Mesozoic arc-rear marginal basin in Southern South America, *in* Standley, D.J., and Kelling, G., eds., Sedimentation in submarine canyons, fans, and trenches: Stroudsburg, Pennsylvania, Dowden, Hutchinson and Ross, p. 362–373.

Wylie, P.J., editor, 1967, Ultramafic and related rocks: New York, Wiley, 464 p.

Yumul, G.P., Dimalanta, C.B., and Jumawan, F.T., 2000, Geology of the southern Zambales Ophiolite Complex, Luzon, Philippines: Island Arc, v. 9, p. 542–555.

Manuscript Accepted by the Society May 2, 2003

Printed in the USA

Geological Society of America
Special paper 373
2003

A personal history of the ophiolite concept

Eldridge M. Moores*
Department of Geology, University of California, Davis, California, 95616, USA

ABSTRACT

Since their original recognition in the early nineteenth century, ophiolites have been an integral feature of European Alpine-Mediterranean tectonic ideas. Ophiolites were not widely accepted by English-writing geologists until the late 1960s. Prior to that time, discussions in the English-language literature centered on the nature of the peridotites and volcanic rocks as separate entities. The development of the ophiolite concept went hand-in-hand with development of ideas of plate tectonics. As with plate tectonics, ophiolites are in a post-revolutionary, "mopping up" phase of scientific development.

My personal experience in ophiolites began in the early 1960s in Greece, and has extended to the Troodos complex, Cyprus, western North America, and other localities. I recount my personal experiences with working on ophiolites during development of the plate tectonic revolution and its aftermath.

Keywords: ophiolite, history, tectonics, Mediterranean, Cordillera.

INTRODUCTION

Ophiolites, fragments of oceanic crust and mantle formed at a spreading center and preserved on land, constitute a major ingredient of the plate tectonic theory and its extension to the continents. The history of the ophiolite concept is interesting not only in its own right, but also in the context of the development and elaboration of the concepts of plate tectonics. Set in this context (see Fig. 1), early developments in ophiolites were part of the fabric of geologic theory that existed before the development of plate tectonics. The development of plate tectonics can be thought of as the outgrowth of five lines of inquiry (see Moores and Twiss, 1995, p. 251–260; Kuhn, 1970): orogenic belts and geosynclines, marine geology and geophysics, continental drift, magnetism/paleomagnetism, and seismology. At the start of the twentieth century, the first subject possessed a unifying world view—geosynclines and their role in orogenic development—which had recently eclipsed an earlier global hypothesis: contraction. The other four were then in an immature state, i.e., they possessed no unifying theory, and they developed separately from one another until the 1960s. The idea of geosynclines went through a period of crisis and overturn as the information from stratigraphy, structure, and continental tectonics were woven together with the other fields to apply the revolutionary plate tectonic ideas to geologic history.

Ophiolites are also instructive from the point of view of the difference in development of concepts with respect to nationality, principal language, and point of origin of the scientist. The ophiolite concept developed principally in Europe, from observation of exposures in western Europe. From the nineteenth century, ophiolites were an important ingredient in the discussion of any orogenic belt in European literature, but were little mentioned in English-language geologic publications until the late 1960s.

This article is a personal history of the ophiolite concept. I have been involved in ophiolite studies since 1963, starting several years before the plate tectonic revolution. I was fortunate to be at an institution (Princeton University) in the 1960s in the early days of ideas of seafloor spreading, and my ophiolite work began against a background of gathering excitement as new results and ideas cascaded in, culminating in 1968–1969 in the plate tectonic revolution.

EARLY DEVELOPMENTS

Ophiolites (from the Greek: *ophis* = snake, and *lithos* = rocks), were first defined by Brongniart (1813), as represent-

*moores@geology.ucdavis.edu

Moores, E.M., 2003, A personal history of the ophiolite concept, *in* Dilek, Y., and Newcomb, S., eds., Ophiolite concept and the evolution of geological thought: Boulder, Colorado, Geological Society of America Special Paper 373, p. 17–29. For permission to copy, contact editing@geosociety.org.

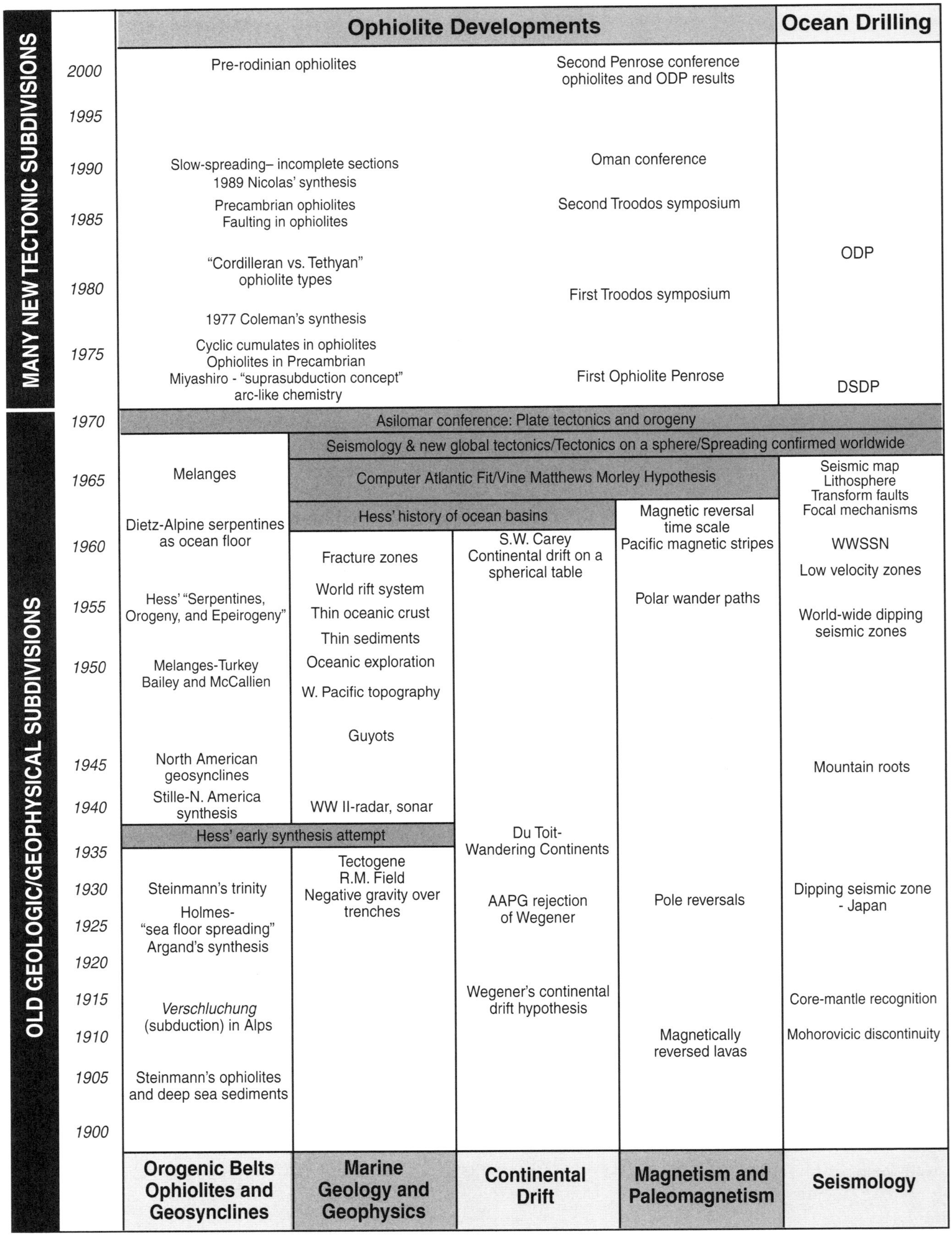

Figure 1. Historical summary of ophiolite developments and ocean drilling in the light of the plate tectonic revolution. Chronologic development of five fields from 1900 on led to the plate tectonic revolution in 1962–1970. New tectonic subdivisions arose out of the revolution.

ing the "common serpentines," the rocks that had been used as dark green-black dimension stone in many Renaissance and pre-Renaissance constructions (eight columns of serpentine breccia collected from Ephesus are present in Aghia Sophia, the church (now a museum) in Istanbul constructed in the sixth century, C.E; Ambriere, 1965, p. 214). In his first writings, Brongniart made no distinction between igneous and metamorphic rocks, but in his later classifications (e.g., Brongniart, 1827), the igneous-metamorphic rock distinction had become clear, and he grouped the ophiolites with igneous rocks. Brongniart also grouped massive altered volcanic rocks and gabbros (eufotides)[1] together with serpentines in his ophiolites, but he did not distinguish any sequence within these rocks.

By the end of the nineteenth century, the presence of "greenstones," "roches vertes," "pietri verdi," "grünsteins," was well-known throughout the world (Suess, 1909). Suess argued that these rocks commonly displayed igneous contacts, they were present as discordant or concordant sheet-like igneous masses in highly folded terranes, and they never cropped out in large batholith-sized masses, in contrast to granitic rocks.

Steinmann (1905, 1927) added considerably to the understanding of ophiolites. Beginning with his work in the late nineteenth century, Steinmann pointed to the ubiquitous association of serpentinite, "diabase" including altered volcanic rocks (spilites, keratophyres), hypabyssal and plutonic mafic rocks, and radiolarian chert. He described this association, which became known as "Steinmann's trinity" in the western Alps and Italy where the sequence abundantly crops out, but generally lacks gabbro and nowhere includes dike complexes (Bernoulli, 2001). Interestingly, Steinmann (1905) also recognized the association in the Golden Gate region, San Francisco, and on Mt. Diablo, California, based upon a field trip led by Andrew C. Lawson of the University of California, Berkeley, in 1892. Steinmann's other important contribution was drawing attention to the deep water environment of the immediately overlying radiolarites. He also observed that shallow water sediments overlay the radiolarites and argued for massive uplift in Cretaceous time and for the ophiolite emplacement during folding. After Steinmann's (1927) synthesis, nearly all workers recognized the importance of the association, but its true significance was not generally recognized until the mid-1960s (Hess, 1965). Steinmann also argued that ophiolites represented the remnants of a 500 to 700 km wide Tethyan Ocean that formerly existed between Africa and Europe (Oreskes, 1999).

Contemporaneously with Steinmann's work, two other lines of thought developed as to the origin and tectonic significance of peridotites and/or serpentines. One was the work of the American petrologist N.L. Bowen. From his experiments as early as 1914 on the crystallization of olivine and the fieldwork by R.A. Daly and others on stratiform mafic-ultramafic complexes, Bowen (1927) argued that peridotites formed chiefly by fractional crystallization of basaltic magma and stratiform accumulation of the early-formed olivine in the bottom of a magma chamber.

A contrasting view was that of the Australian geologist, W.N. Benson. Drawing upon his own fieldwork in eastern Australia and a worldwide survey of peridotite/serpentinite occurrences, Benson (1926) argued that field relations indicated that peridotites were of magmatic origin, and different from peridotites formed by fractional crystallization processes in stratiform mafic-ultramafic complexes. Benson coined the term "Alpine-type" peridotites because of the widespread presence of such peridotite/serpentinite bodies in "Alpine-type" orogenic belts. He argued that peridotites were intruded during the initiation of deformation of an orogen in its marginal parts, as opposed to the central core region where granitic intrusions predominated.

Thus, by 1927, three conflicting lines of opinion existed. The first, the continental European view, emphasized the relationship of peridotites to mafic pillow lavas and radiolarian cherts. The second, an "Anglophone" view espoused principally by workers in the USA, the UK, Australia, and New Zealand, argued that the "Alpine" peridotites were unrelated to the other rocks of Steinmann's trinity, were igneous, and were intruded into the margins of the core regions of "Alpine" mountain systems. The third view, led by Bowen, argued that the temperatures of formation of an olivine-rich magma were too high to explain the field relations of general lack of metamorphism, and that peridotites were formed by fractional crystallization of basaltic magma. This conflict raged on for four decades and was resolved only in the 1960s. With the benefit of hindsight, it is clear that all three lines of thought were partly right. Part of the controversy was cultural, and was aided and abetted by the tendency for authors to cite mainly literature in the language most familiar to them.

Vogt (1924) made an early attempt to reconcile the opposing views. He observed that naturally occurring peridotites contained minor but significant amounts of iron. As Bowen had originally dealt in his experiments with pure Mg olivine (forsterite), Vogt argued that the widespread presence of ~10% Fe might depress the melting temperature of the system below that of Mg- or Fe-olivine so that magmatic peridotites could exist at temperatures where pronounced metamorphic aureoles need not be present. In response, Bowen and Schairer (1935) investigated the system MgO-FeO-SiO_2 and found that the additions of small amounts of Fe did not reduce the melting temperature nearly as much as Vogt argued and they concluded that "dunites are formed by accumulation of crystals with perhaps some special mode of intrusion of a crystalline mass for those bodies that seem to be special intrusives" or possibly the abundant presence of volatiles (1935, p. 205).

OPHIOLITES AND THE RISE OF MODERN MARINE GEOLOGY AND GEOPHYSICS

At the turn of the twentieth century, little was known about the oceans. The expedition of HMS *Challenger* in the 1870s led

[1]Interestingly, the term "gabbro" derives from a Tuscan village of the same name, which sits on serpentine.

to recognition of something of the nature of deep-sea sediments and to the presence of relatively shallow depths in the middle of the north Atlantic. Very little was known about the oceans themselves, however, and in the second volume of Eduard Suess's monumental work *Das Andlitz der Erde* (the Face of the Earth), subtitled *Die Meere* (the Seas), Suess (1906) confined his discussion essentially exclusively the geology of the lands surrounding or exposed within the oceans. Extension of the Mid-Atlantic Ridge to the southern hemisphere was an outgrowth of the German *Meteor* expedition in the 1920s (Oreskes, 1999). Argand in 1924 (1977, p. 147) and Holmes (1928) suggested that the Mid-Atlantic Ridge was a "sialic" remnant of the separation of the continents. Dutch investigations of Indonesia in the 1920s revealed the presence of topographic depressions—deep-sea trenches. There was no unifying hypothesis, however, on the origin of the oceans. Alfred Wegener's proposal of continental drift had great implications for the nature of the oceans and their relationships with continents, but the general process was not understood (see discussion in Oreskes, 1999).

The early 1930s saw the rise of modern marine geology and geophysics. R.W. Field, a professor of geology at Princeton University, wrote a report to the American Geophysical Union arguing the scientific rationale for development of a program of exploration of the oceans. It is hard in retrospect to conceive of the ignorance at that time of the submarine realm. Little was known about the topography of the world's oceans; little or nothing was known about broad patterns of sedimentation, the nature of oceanic crust, or the thickness of the crust. Field argued in his report that as the oceans represented such a large portion of Earth's surface and were so little known, some considerable effort was appropriate. Presciently, he observed, in part responding to the debate that had raged over Alfred Wegener's concept of continental drift (Oreskes, 1999):

> The outstanding geological problem is that of the permanency of ocean basins, and it is doubtful whether this problem can be solved by the study of continents and oceanic islands alone. The pressing need is for physiographical and structural data concerning the crust under the Atlantic, Pacific, Indian, and Arctic oceans. Can geophysical methods assist us to obtain this information, by the penetration of the oceanic waters to all depths and by the determination of the composition, thickness, and stratigraphical sequence of the sediments on the sea floor? Most important of all, will these methods enable us to carry our structural studies deep beneath the ocean bottom with the same accuracy which we now attain in the exploration of the continents? (Field, 1938a, p. iv–v.).

As Field made his argument, the Dutch geophysicist F.A. Vening Meinesz was working in marine geophysics with the study of gravity at sea. Beginning first in Indonesia, where topographic trenches were discovered in the early twentieth century, Vening Meinesz extended his studies to the Caribbean in the early 1930s, borrowing a World War I-surplus U.S. submarine for the purpose—arranged for him by Field and Field's friend, Major William Bowie of the U.S. Coast and Geodetic Survey (Hess, 1962a). With him went two young graduate students who were to become giants of the field in the following decades—H.H. Hess and W.M. Ewing. Hess came to marine geology from a background in fieldwork in southern Africa and the Appalachians. He worked with Vening Meinesz on the gravity field of the Caribbean[2]. One of the features that they found during this survey was the negative gravity anomaly extending from the Puerto Rico trench to the Tobago Basin. Hess (in Field et al., 1933; cf. Oreskes, 1999) argued that these negative gravity anomalies were formed by a "tectogene" or a downbuckle in Earth's crust, possibly produced by convection currents within the mantle, a position that was accepted by Vening Meinesz and investigated experimentally by the University of California, Los Angeles geophysicist, David T. Griggs (Oreskes, 1999).

In the mid-late 1930s, Hess began to attempt to integrate his Ph.D. work in the Appalachians and with his work with Vening Meinesz. Hess delivered a paper at the 1937 International Geological Congress in Moscow, which was published in 1939 (Hess, 1939), not a good year for geologic publications from the USSR[3]. In this publication, Hess elaborated on the "tectogene" model for negative gravity anomalies, and suggested that tectogenes were the locus of geosynclines. Hess argued that peridotites were intruded in the initial stages of orogeny and, drawing on his observations from the Appalachians, that they were present in paired belts some 60 to 100 km on either side of the axis of the geosyncline/tectogene. Hess stated that peridotites are intruded in the initial stages of orogeny and are not reintruded in later stages. Ironically, Hess mentioned ophiolites only in the title, but not in the body of the paper. The meaning of Hess's observation on timing of emplacement became clear in the reinterpretation of ophiolites after the plate tectonic revolution. In a companion paper (Hess, 1938), he argued for evidence for the magmatic emplacement of peridotites, and suggested that the temperature problem documented by Bowen and Schairer (1935) could be alleviated by having a hydrous ultramafic magma of serpentine composition. The German geologist Hans Stille (1939) also argued that ophiolites were intruded in the initial stages of deformation of geosynclines, but Stille did not make any connection with the oceans, saying nothing about the oceans, gravity anomalies, or trenches. (One wonders whether Stille had attended the 1937 Moscow congress and heard Hess's presentation). Hess' (1939) work was the first to try to integrate continental geology with new marine geological and geophysical data. On Figure 1, it is illustrated as an early synthesis attempt.

[2]Hess once recounted an experience on the submarine when through a malfunction it dove to twice its designed maximum depth limit. Only drastic action by the captain saved the ship and its personnel.

[3]Hess observed in the 1960s that his 1939 article had received little attention also because he was a young person just starting out on his career, and he had not yet made his name in an accepted field. His update of same paper after he had established his reputation received wide attention in 1955.

By the eve of World War II (WWII) a great deal of progress had been made in several fields. The Mid-Atlantic Ridge was well-documented, as was the location of earthquakes along it (Heck, 1938). Earthquake seismic work suggested that the oceanic crust was thin (~7 km) relative to continents (Heck, 1938). Nevertheless, Field (1938b, p. 6) pointed out that with regard to the ocean floor that "the major portion of the earth's crust constitutes a vast 'no man's land' as yet practically unexplored..." The 1939 International Geological Congress was scheduled to convene in New York in early September with Richard Field as its Chair. Field was looking forward to a full discussion of Wegener's ideas on continental drift in the light of the new results from marine geological and geophysical work. Unfortunately, just before the Congress opened, Hitler invaded Poland, and the anticipated international colloquium on Wegener's ideas and orogeny never occurred (Oreskes, 1999).

During the Second World War, scientific progress took a back seat to the war effort. Wartime technological developments in navigation and bottom sounding were to contribute greatly to the subsequent discovery of sea-floor spreading and plate tectonics.

Having been commissioned a reserve naval officer in the 1930s to facilitate his work with Vening Meinisz, Hess reported for active duty the day following the Pearl Harbor attack. He served in the Navy throughout the war, ending up as a Captain of an attack transport in the Pacific, the USS *Cape Johnson*. The story goes that Hess kept his depth sounder going at all times, even when in close-in enemy waters. In the process, he discovered a series of flat-topped seamounts, which he named "guyots" (Hess, 1946; Arnold Guyot was the first earth scientist at Princeton, and the namesake of the Princeton geology building, Guyot Hall). Hess originally thought that these mountains were Precambrian in age, in line with the assumed permanence of oceans, but later acknowledged the Cretaceous age fossils discovered on them (Hess, 1955). Hess also acknowledged the contributions of the USS *Cape Johnson* and the battleship USS *Massachusetts* (now a museum in New Bedford, Massachusetts), as having made more soundings in WWII than any other ship. A young Princeton colleague, John C. Maxwell, spent months on New Caledonia, where he investigated and subsequently described the chromite deposits of the huge peridotite complex exposed there, and documented evidence for ultramafic intrusive relations, interpreted as magmatic dikes and sills (Maxwell, 1949).

After the war, Hess and Maxwell analyzed thousands of km of ship-track depth soundings and published large-format bathymetric maps of the western Pacific (Hess, 1948) and the SW Pacific (Hess and Maxwell, 1949). These maps represented unprecedented revelations of the topography of the western Pacific floor. Hess and Maxwell also recognized the dipping seismic zones beneath island arcs in the western Pacific, which subsequently became known as Wadati-Benioff zones.

Meanwhile, debate about the possibility of a peridotite magma continued. Challenged by Hess' 1938 paper, Bowen and Tuttle (1949) published results of an experimental study of the $MgO-SiO_2-H_20$ system. Their results indicated that pure Mg serpentine did not exist at temperatures above ~500 °C, but at higher temperatures olivine coexisted stably with H_2O vapor, and that no magma of serpentine composition existed (see Young, this volume, Chapter 4). They concluded inescapably that ultramafics were intruded only in the solid state. These results bore so heavily against Hess's (1938) hypothesis that he ultimately withdrew it (Hess, 1966). Shortly before his death in 1969, however, he was able to participate in a field trip in South Africa where he saw the copious evidence for extrusion of ultramafic magmas in the form of komatiites. In a letter to F.J. Vine, received while Vine and I were working on Cyprus, Hess described the field trip as the most interesting one that he had ever been on. He clearly felt finally vindicated on his arguments for the existence of ultramafic magma. Ironically, the Archean komatiitic magmas of South Africa did not represent the "Alpine-type" ultramafic rocks that Hess had originally considered.

CRISIS, REVOLUTION, AND OPHIOLITES

By the late 1950s it was clear that geology was in a crisis mode. Polar wander paths had been discovered, sediments in the oceans were thin, the magnetic reversal time scale was being worked out, marine seismic work confirmed that oceanic crust was relatively thin and stood isostatically 5 km or so below continents—a result that Hess called "the most momentous discovery since the war" (Hess, 1954, p. 341), magnetic stripes had been discovered off the coast of the western USA, the mid-ocean rift system had been discovered, and high heat flow was discovered on the ridge axis. None of these developments fits the geosynclinal paradigm or the dogma of permanence of ocean basins.

At the same time, the Alpine peridotite-ophiolite controversy was still unresolved. Work by Ross et al. (1954) documented the similarity of composition of fragments of Earth's mantle brought up in basaltic and many "Alpine-type" peridotites. De Roever (1957) argued that Alpine peridotites were tectonically emplaced fragments of the upper mantle. Hess (1955) essentially repeated his 1939 interpretations and argued that calling the various rocks ophiolites confused the issue, as the peridotites and volcanic rocks really were of separate origin. Hess, however, suggested that the Mid-Atlantic Ridge "represents a welt of serpentine...concentrated [by] convective circulation in the mantle..." (1955, p. 404–405). The term "ophiolite" had become almost taboo in North American circles, and remained so until the following decade. Hess acknowledged that: "So the problem remains unsolved. Some vital piece of evidence is still missing" (1955, p. 402).

One vital new idea was soon to emerge. When I enrolled as a graduate student at Princeton University in fall, 1959, one required course was called "Advanced General Geology." In it, several professors lectured on their current research. In his many comments, Hess and his colleague, A.F. Buddington, discussed polar wander paths, paleomagnetism, and rock magnetism. They emphasized the lack of understanding of the origin of rock magnetism and argued that some of the apparent polar wander data could be the result of complications of the magnetization process. Hess also mentioned the seismic results for a thin oceanic crust, the evidence for only thin sediments in the ocean basins, and the recently discovered heat flow results. Despite diligent search, no one could point to convincing modern analogues of the ultramafic-bearing "eugeosynclines" of orogenic belts, although Drake et al. (1959) had suggested the east coast of North America as a modern analogue of miogeosynclines. In retrospect, the "crisis" in the sense of Kuhn (1970) was upon the world.

Late in 1959, the Australian geologist S.W. Carey came through to deliver a lecture on continental drift and earth expansion. His ideas on expansion have been widely discounted and detract from his contributions to continental drift, however. Carey's contribution to the continental drift debate was to construct a spherical table, ~2 m in diameter, on which he plotted the 500 fathom contour, rather than the coastlines, as had Wegener. Carey gave a three-hour spell-binding lecture, ending completely spent, covered with sweat and chalk dust. At the end, we all filed numbly out of the room. Halfway through the talk, however, Hess bolted out of his seat and started pacing up and down the aisle. Thereafter in Advanced General Geology, there was no more talk of problems of paleomagnetism and polar wander paths. Within two months, Hess was circulating a manuscript, entitled "Evolution Ocean Basins" [sic], which was eventually published as "History of Ocean Basins" (Hess, 1962b). This was the key insightful paper that gave rise to the new unifying model of ocean floor spreading, just as Kuhn (1970) suggested would happen in a scientific revolution. I believe that S.W. Carey must be given the credit for "pushing Hess over the edge." In his article, Hess suggested that the oceanic crust was chiefly serpentinized mantle peridotite, that the mid-ocean ridges were the loci of upwelling and divergent motion and that the continents were passive riders on mantle material.

In the early 1960s, the advent of passenger jets made intercontinental travel relatively easy. New results from Mediterranean ophiolites caused the beginnings of a revolution in American understanding of ophiolites. In 1961–62, J.C. Maxwell traveled to Italy on a sabbatical leave, where he worked on both ophiolites and, incidentally, melanges. As detailed by many workers (e.g., Bernoulli, 2001) in Italy and elsewhere in the Mediterranean the close association of the ophiolitic lithologies in a pseudostratiform sequence is essentially unarguable. Maxwell also investigated the work of Brunn (1956, 1959, 1960, 1961) in Greece (Maxwell and Azzaroli, 1962). Brunn (1959) was the first person ever to recognize the possible association between ophiolites and sea-floor spreading centers; however, he and his associates (e.g., Aubouin, 1959), following Routhier (1946, 1953; and Dubertret, 1953; *in* Nicolas, 1989), posited that ophiolites resulted from massive outpourings of mafic magma in "geosynclines" (which they represented as sediment starved submarine troughs as seen in the Alps, but not in North America) with subsequent stratiform crystallization under a self-formed roof. Maxwell and Azzaroli (1962) essentially adopted this view.

In 1962, while finishing my Ph.D. thesis work in Nevada (see Moores et al., 1968), I persuaded Hess and Maxwell to allow me to switch to working on the Vourinos complex. I traveled to Greece in summer, 1963, on an NSF Post-Doctoral Fellowship and an NSF grant, remaining over a year. At the time I began this work, I was buffeted by three rival schools of thought: (1) the view of Hess, that Alpine-type peridotites were of very constant composition, with a Mg/Fe ratio of ~9:1, and represented either hot, diapiric intrusions (MacKenzie 1960) or solid mantle fragments, following De Roever (1957); (2) the view of Thayer (1967), who had long argued for a consanguineous relationship between the peridotite and gabbro of "Alpine-type complexes," but not for the volcanic rocks; and (3) the European (especially French) view that they represented extrusive/intrusive outpourings in geosynclinal troughs (Brunn, 1956; 1960). About this time, Dietz (1963) suggested that "Alpine serpentinites" were fragments of oceanic crust incorporated into continental margin "geosynclines" but Dietz does not mention Hess's already published article (Hess, 1962b, which was widely circulated for over a year prior to its publication) showing ocean crust as serpentinite. Dietz never mentioned the word "ophiolite." In an earlier article he stated "a preprint by H. Hess also (suggested) a highly mobile sea floor. Full credit of priority is to be accorded him for any merit which this suggestion has" (Dietz, 1962, p. 12).

Work in Vourinos led to the key discovery that the lower part of the ultramafic complex was metamorphic tectonite. This tectonite was intruded and overlain by an igneous upper ultramafic-mafic (transition) zone (Moores, 1969a). Davies (1968, 1971) made a simultaneous similar discovery in Papua-New Guinea. This discovery partly resolved the conflict between the Brunn-Aubouin view of ophiolites and that of Anglophone authors, because chemical study of the tectonite showed that they were identical with the "Alpine-type" peridotites by then well-documented by MacKenzie (1960) and Green (1964). A field trip with Harry Hess and John Maxwell to the Vourinos and the Italian ophiolites of Tuscany and Liguria led to a letter from Hess to Maxwell and myself accepting the ophiolite sequence as an inescapable fact and Hess's (1965) acceptance of ophiolites as fragments of ocean floor (Hess, 1965). I was not so sure, and in my article (Moores, 1969a; submitted in spring 1967), I proposed two hypotheses, one that it was a tectonic fragment of ocean floor, and the other that it was a solid/liquid diapiric emplacement of partially molten mantle material. I had still not made the connection to sea-floor spreading.

While writing up my Vourinos work in 1965–1966, I spent considerable time with Fred Vine and Jason Morgan, both young members of the Princeton faculty, discussing what oceanic crust at a spreading center might look like in the field. I heard of a remarkable set of rocks on Cyprus, and obtained copies of the Cyprus Geological Survey Memoirs. Opening the map of Wilson (1959), I was stunned. There in front of me lay a massif of ultramafic rocks, gabbro and related intrusive rocks, a "sheeted complex" of dike-within-dike, gradationally overlain by pillow lavas. The most arresting feature of Wilson's map is the large area of pink-colored "diabase" with dips and strikes all aligned with each other and with dikes in the overlying pillow lavas. I showed it to Fred Vine and asked, "What do you think of this as oceanic crust formed at a spreading center?" That possibility occasioned enough interest on the part of Hess and Maxwell that during the process of a circum-Mediterranean survey of peridotites in summer 1966, J.S. Dickey Jr., and I traveled to Cyprus for two-day reconnaissance of the Troodos complex.

After returning from the Mediterranean and moving immediately to Davis, California, I wrote a report to Hess and Maxwell about Troodos, arguing that it was worth a closer look. A few months later, at a plenary session of the 1966 Geological Society of America Annual Meeting in San Francisco, F.J. Vine, A. Cox, R. Doell, and G.B. Dalrymple all gave talks on summarizing the evidence for sea-floor spreading and magnetic reversals. The "Marine Geological Revolution" (Fig. 1) had arrived. During the meeting, Vine proposed to me that we do a joint study of the Troodos complex. We obtained Hess's blessing, and in early 1967 applied for and were awarded an NSF grant to study the Troodos massif as a product of sea-floor spreading. We delayed going because of political and professional conflicts in summer, 1967.

Simultaneously, Ian Gass, then of University of Leeds, clearly had similar ideas. Gass had been a staff member of the Cyprus Geological Survey and had written a memoir of part of the area (Gass, 1960). Gass (1967) had described the Troodos massif as an assemblage of pillow lavas unconformably overlying the sheeted intrusive complex and plutonic rocks, formed in an oceanic area and thrust over the African continental margin, but not as an ophiolite. In Gass (1968), he drew the analogue between the sheeted complex (Wilson, 1959) and sea-floor spreading, but he did not recognize the tectonite/cumulate contrast in the ultramafic rocks, writing that "…these plutonic rocks belong to a differentiated ultrabasic mass of batholithic dimensions" (Gass, 1968, p. 39).

Fred Vine and I traveled to Cyprus in August 1968, arriving nearly simultaneously with the Soviet invasion of Czechoslovakia and just before the anti-Vietnam War riots at the Democratic Convention in Chicago. We spent much of the first of two field seasons on the island performing field paleomagnetic vector measurements and drilling for paleomagnetic studies. Our rationale for this work was to investigate a possible magnetic reversal that was suggested by existing aeromagnetic survey maps. Though we did not find a reversal, our data were the first to document an ~90° counterclockwise rotation of the complex[4]. In summer 1969, we returned to Troodos and concentrated on the geology of the massif. Our study (Moores, 1969b; Vine and Moores, 1969; Moores and Vine, 1969; Moores and Vine, 1971; Vine and Moores, 1972) recognized the Troodos complex as an ophiolite, and was the first explicit examination of an ophiolite in the light of the concept of sea-floor spreading[5]. Our studies also reversed the sequence of ultramafic units. Following the relationships worked out in stratiform mafic-ultramafic intrusions, earlier workers, including Gass (1968) had assumed that the dunite was at the base of the ultramafic sequence. Vine and I established that the harzburgite represented tectonite mantle, and that the dunite instead was the bottom of the cumulate plutonic rocks, but above the harzburgite. We presented these results at a meeting of the Royal Society, London, in November 1969 (cf. Moores, 1969c).

In early December 1969, a GSA Penrose Conference entitled, "The meaning of the new global tectonics for magmatism, sedimentation and metamorphism in orogenic belts," was held at Asilomar, California, organized by W.R. Dickinson. At that meeting, presentations of the recently published evidence for plate tectonics were compared with evidence for orogenic history in such regions as the Appalachians and the western North America Cordillera. In the course of a few days, geology was transformed, Wegener was vindicated, active subduction zones were recognized as coinciding with dipping seismic zones, and geosynclines were reinterpreted in terms of actualistic models (Dickinson, 1970, 1971).

Ophiolites were a major topic of discussion at the Penrose conference. If they were formed by sea-floor spreading, how were they emplaced? The idea occurred to several workers that emplacement may occur by collision of a continental margin with a subduction zone. Others thought that emplacement could occur simply by "tipping up" of the oceanic edge of the overriding plate, or by incorporation of slices of the down-going plate in a subduction zone.

As the conference was ending, Dickinson rose to summarize aspects of the conference proceedings, and outlined his newly formulated ideas of geosynclines and plate tectonics. It was a tremendous excitement to hear geologic history folded into plate tectonics. A few minutes later, I formulated the general ideas of how plate tectonics, ophiolite emplacement, and the Phanerozoic evolution of western North America might fit together. It was a tremendously exciting moment. After a suitable lag time owing to the then-existing ground rules of a Penrose conference, several ophiolite-related articles appeared. I developed the model for emplacement of ophiolites by collision of a continental margin on the down-going plate with a subduction zone, emplacing a part of the upper plate as an ophiolite complex and applying the concept to a model for the

[4]At one point, we were accosted by a man who turned out to be the Cyprus Interior Minister. At the end of a long discussion about our studies and why we were drilling, he remarked, "You mean that your government pays you to come over here and drill holes in my island?" Not long afterward, in a reflection of the turbulence of the times, he was assassinated.

[5]I remember continually thinking what an extraordinary experience it was to be studying such a remarkable series of rocks and with such a remarkable person as Fred Vine.

development of the USA portion of the North American Cordillera (Moores, 1970). Coleman (1971) made a global survey of ophiolites and defined the ophiolite emplacement process as "obduction"; and Dewey and Bird (1971) discussed ophiolite emplacement, especially as exhibited in early Paleozoic rocks of Newfoundland.

"MOPPING UP": OPHIOLITES SINCE THE REVOLUTION

Historical Overview

Since the plate tectonic revolution, studies of ophiolites have been in a "mopping up" phase in the sense of Kuhn (1970) as the new paradigm is extended to global occurrences and the entire geologic record. Thousands of studies have been conducted of ophiolite complexes throughout the world, in various tectonic regions, and our understanding of them has improved enormously. Particularly important have been international conferences and field trips investigating ophiolite complexes in various regions around the world, summary monographs devoted to ophiolites, and increasingly detailed comparisons of ophiolites with oceanic crust, particularly as documented through the Ocean Drilling Project (ODP).

A first ophiolite Penrose conference in 1972 consisted of a 1600 mile (2500 km) road trip of newly-recognized ophiolite complexes in the western USA, specifically Oregon and northern California, principally the Canyon Mt. Complex, Oregon, the "type area" of Thayer's "Alpine mafic magma system," several exposures in the Klamath Mountains, including the Trinity ophiolite, exposures in the Sierra Nevada including the Feather River ultramafic belt, and the Del Puerto ophiolite of the California Coast Ranges. The 12 informal seminars during the trip culminated in the so-called "Penrose definition" of ophiolites—"a distinctive assemblage of mafic to ultramafic rocks"—consisting of an ultramafic complex, a gabbroic complex, a mafic sheeted dike complex, and a mafic volcanic complex, commonly pillowed. So-called "associated rocks types" include an overlying sedimentary section of chert, minor shale and limestone (no mention is made of volcaniclastic sediments). The report called for more careful mapping of the various members within ophiolites and more petrologic studies (Anonymous, 1972, p. 25). With the benefit of hindsight, one can note that conspicuously absent from the discussion of this field trip is any mention of the need for consideration of the ophiolite in its regional context.

A second ophiolite conference the following year (May 31–June 14, 1973) in the Soviet Union focused on Hercynian (Paleozoic) ophiolitic complexes in the Alai Range and the Kyzyl Kum Desert, as well as Mesozoic ophiolitic complexes of the Lesser Caucusus. In addition, the conference provided a detailed exchange of views between Soviet and western geologists and introduced many Soviet geologists to plate tectonic concepts (Coleman, 1973).

Several books have been devoted to ophiolites (e.g., Coleman, 1977; Nicolas, 1989). My own modest contributions have included considerations of the tectonic significance of ophiolite emplacement (Moores, 1970, 1982), the re-interpretation of all ultramafic rocks in the light of plate tectonics (Moores, 1973), an early attempt to relate the structure of oceanic crust and ophiolites to spreading rate (Moores and Jackson, 1974). Further studies with students and colleagues of the Vourinos complex led to the recognition of cyclic cumulates in the mafic-ultramafic plutonic section. These have been recognized, but not studied in detail, in other ophiolites, as well (e.g., Vourinos, Oman; Jackson et al., 1975; Harkins et al., 1980; Rassios et al., 1983). Detailed studies of dike complex of the Troodos complex, Cyprus, led to discovery and elaboration of listric normal fault systems in the Troodos (e.g., Moores et al., 1990; Varga and Moores, 1985; Verosub and Moores, 1981) and the Josephine complex, California-Oregon (Harper, 1982).

Studies of ophiolitic rocks in the western North American Cordillera have led to the use of ophiolites in a re-interpretation of the structural evolution of that margin (e.g., Dilek et al., 1990). A global review of ophiolites and their significance led to their separation into "Tethyan" and "Cordilleran" types based upon the presence or absence of a continental substrate, an island arc edifice, or other geologic criteria (Moores 1982). Exploration of the nature of Precambrian, especially Pre-Rodinian (pre 1Ga), oceanic crust has elaborated on the hypothesis of Proterozoic oceanic crustal thinning in the Neoproterozoic (Moores, 1973, 1986, 1993, 2002).

Several major conferences have contributed to our understanding, including conferences on the Troodos ophiolite in 1979 and 1987 (Panayiotou, 1980; Malpas et al., 1990), the Oman ophiolite in 1990 (Peters et al., 1991), a conference at the 29th International Geological Conference, Kyoto (Ishiwatari et al., 1994), and a second ophiolite Penrose conference comparing ophiolites and ODP results on oceanic crust in 1998 (Dilek et al., 2000). The latter conference was particularly valuable, as it brought together workers concentrating on the ODP and those more focused on land-based ophiolite studies. It is through such comparisons that new insights will develop.

A major post-revolution discussion began with Miyashiro's (1973) focus on the arc-like chemistry of the Troodos complex. Most subsequent petrological and geochemical discussions have focused on the geochemical evidence for a mantle source already depleted of its MORB components (e.g., Robinson and Malpas, 1990, Bloomer et al., 1995).

Vourinos Ophiolite

A re-interpretation of the Vourinos ophiolite, northern Greece, epitomizes changes in our understanding of ophiolites. These changes are illustrated in Figure 2. Figure 2A shows a longitudinal schematic cross-section of the Vourinos complex drawn in the late 1960s after recognition of the tectonite/cumulate contrast, but before the plate tectonic revolution. Reinterpretation of

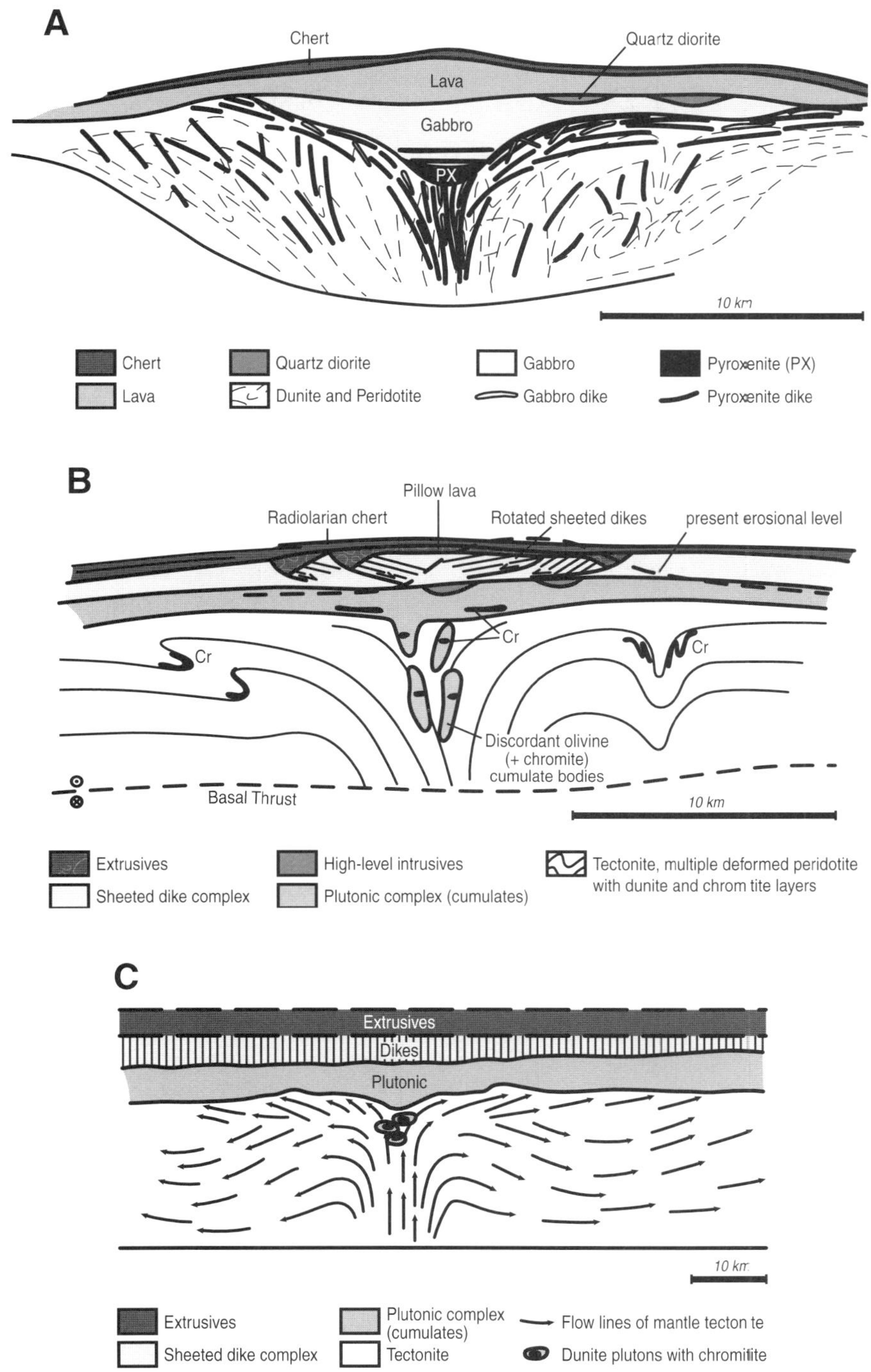

Figure 2. Evolution of ideas on the Vourinos complex, northern Greece. A: Longitudinal interpretative cross-section after Moores, 1969, showing pre-plate tectonic revolutionary interpretation of complex with tectonite dunite and peridotite and overlying magmatic plutonic/extrusive rocks. B: Re-interpretation of A based upon plate tectonic recognition of tectonic emplacement of ophiolites, a sheeted dike complex (Rassios et al., 1983), stratiform mafic-ultramafic cumulates (Jackson et al., 1975) and intrusive dunite (olivine cumulate) and chromite bodies (Harkins et al., 1980). C: Model for diapiric rise and divergence of asthenospheric mantle and development of oceanic crust at a spreading center. Cr—chromitite. (Redrawn after Nicolas and Violette, 1982, *in* Nicolas, 1989, Fig. 9.1, p. 204).

that cross-section in the light of what we know now of ophiolites as tectonic fragments of oceanic crust and mantle would lead to a revised cross-section as shown in Figure 2B. The presence of igneous dunites (Harkins et al., 1980) in the central zone and the overall structure suggests that in the Vourinos complex, one has captured an actual mantle cross-section of a spreading center, as envisioned by Nicolas and Violette (1982; see Fig. 2C).

CURRENT STATUS OF THE OPHIOLITE QUESTION

Although it is beyond the scope of this short reminiscence to discuss fully the status of the ophiolite question and the significance of ophiolites, it is worth summarizing briefly the main points concerning our present understanding. Three points stand out—the environment of formation of ophiolites and what they can tell us about oceanic crust formation; the mechanism of emplacement of ophiolites and its significance for interpretation of the tectonic development of orogenic systems; and the change in ophiolites and thus, oceanic spreading processes through time.

Environment of Formation

Ophiolites represent ocean crust and mantle formed at oceanic spreading centers. These centers occur either at mid-oceanic ridges, in pull-apart intra-arc oceanic basins, at active back-arc basins, at extensional zones in the centers of forearcs during the initial development of island arcs. The nature of the environment of formation of ophiolites cannot be obtained by geochemistry alone; rather a comprehensive set of data including geologic relations, associated deposits, internal structure, and geochemistry are necessary to evaluate the significance of ophiolite complexes. Some ophiolites from magma-rich spreading centers display a complete sequence (Penrose-type ophiolites); this sequence may imply a fast-spreading environment (Dilek et al., 1998). Other complexes, formed in a magma-starved environment, display incomplete sequences, as seen on modern slow-spreading ridges and in the Alps and Apennines where Steinmann did his work (Bernoulli, 2001, this volume, Chapter 7). These "Hess-type" complexes imply a significant portion of the oceanic crust was serpentinized peridotite, as advocated by Hess (1962) and Vine and Hess (1970; see also Moores, 2002; Dilek et al., 1998).

Discovery of faulted structures in ophiolites and oceanic crust strengthens the link between these different environments. Karson's (2002) summary of the structure within exposures of intermediate-fast spreading ridges adds a new complication to interpretation of ophiolite structure, and should be incorporated in new studies of favorably situated complexes. Many other ophiolites may display insights into mantle structure similar to that of Vourinos, as outlined above or the Semail complex, Oman (e.g.; Nicolas, 1989). Other ophiolites may contain more information about the spreading structure and oceanic crustal nature than originally assumed, following the "Penrose definition."

The environment of formation of ophiolites will continue to be controversial. Moores et al. (2000) attempted to resolve the difficulties between universal application of this model and lack of geologic evidence for any island arc in many Tethyan ophiolites by suggesting that the magma source compositions were "historically contingent," that is, a product of prior history, and not necessarily reflective, a priori, of modern environments. Metcalf and Shervais (2001) criticized the "historical contingency" concept on geochemical grounds, but they did not consider many data from the mid-oceanic ridges and from the Tethyan region. This discussion is ongoing. The issues are not whether some ophiolites are "supra-subduction zone," but all; the inadequacy of geochemistry itself to determine the tectonic environment in the absence of geologic evidence; and the increasing evidence for long-lived mantle heterogeneity. Recent discovery of silicic lavas at active mid-oceanic ridges (e.g., Stoffers et al., 2001) at the very least should encourage use of multiple working hypotheses and careful fieldwork before any dogmatic assertion of a tectonic interpretation of a given ophiolite.

Tectonic Significance

The tectonic significance of ophiolite emplacement was signaled early on, principally by Hess's (1939) and Stille's (1939) observations that ophiolites (or ultramafics) were intruded in the initial stages of orogeny; however, tectonics of ophiolite emplacement was not considered in the initial ophiolite Penrose conference (Anonymous, 1972). The tectonic significance of ophiolites have received relatively short shrift in most discussions of ophiolite complexes, which have concentrated on petrology and geochemistry.

Viewed in a plate tectonic context, the issue becomes how to get little-deformed oceanic crust and unserpentinized mantle over continental platforms (in the case of Tethyan-type ophiolites) or island arc crust (in the case of Cordilleran-type ophiolites). Many workers (starting with Temple and Zimmerman, 1969; Moores, 1970, 1973) have argued that such emplacement, given the presence of the topographic difference between continental and oceanic crust, must be by collision of a continental margin with a subduction zone dipping away from the continental margin. This is the "obduction" process of Coleman (1971) and Dewey (1976). Accordingly the basal sole thrusts of ophiolites represent the remnants of former subduction zones; their displacements are indeterminate but may be very large; they cannot be balanced. Synthetic and subsidiary sedimentary fold-thrust belts are secondary to the main ophiolite thrusts. Rare is the synthesis of any orogenic belt that has adequately taken this fact into account.

Of course, slices of down-going oceanic crust easily can have been incorporated into subduction zones. These slices, however, are almost invariably dismembered, incomplete, and can be distinguished from their better preserved counterparts.

Change of Ophiolites through Time

Many authors have suggested (e.g., Sleep and Windley, 1982; Hoffman and Ranalli, 1988) that on theoretical grounds, oceanic crust should be thicker in early earth history than at present. Moores (2002, 1993, 1986, 1973) suggested that oceanic crust thinned abruptly at ~1000 Ma. Recent reports of a ~1020 Ma ophiolite in the East Sayan belt, Siberia (Khain, et al., 2002) may alter the timing of this possible abrupt thinning. Whatever the nature of the pattern, there is increasing recognition that oceanic crustal sequences are present in Proterozoic and Archean terrains. Thus, interpretation of Proterozoic-Archean orogens can find guidance from the consideration of much better-known ophiolites in Phanerozoic orogens.

THE FUTURE OF OPHIOLITE STUDIES

Future ophiolite studies ideally will be guided by the two cardinal facts of ophiolite occurrences: (1) that they are the only guide to oceanic crust prior to 200 Ma; and (2) that ophiolite emplacement is the initial and perhaps most important tectonic event in the development of an orogenic system, as Hess first pointed out over 60 years ago. In working out oceanic spreading history, it would be encouraging to see a decline in automatic assignment of ophiolites to a back-arc origin; rather, a multiple working hypotheses approach is much preferable. In addition, orogenic models would benefit from more incorporation of models based on on-going processes in the SW Pacific and SE Asia (e.g., Hall, 2002).

Ideally, one would like to see a careful and systematic attempt to unravel the history of pre-Mesozoic ocean spreading and ophiolite emplacement processes. For example, there seem to be clusters of records of spreading and emplacement at, say, early Neoproterozoic, early Ordovician and mid-Carboniferous. These records may be biased by emplacement events, and they may apply more to Tethyan-type than Cordilleran-type occurrences. With patience and an open-minded multidisciplinary approach, we may well see a further revolution in our understanding of the orogenic process, specifically an emerging inference of oceanic spreading processes in pre-200 Ma times, and a re-evaluation of all orogenic tectonic histories using ophiolite emplacement as a primary tectonic indicator.

ACKNOWLEDGMENTS

I thank Yildirim Dilek and Sally Newcomb for encouraging me to write this paper. Comments by reviewers Yildirim Dilek, Henri Masson, Sally Newcomb, and Celâl Şengör were of great help. In addition, I thank the dozens of mentors, colleagues, and students with whom I have been fortunate to work over the past 40 years.

REFERENCES CITED

Ambriere, F., 1965, Turquie: Les guides bleus: Paris, Hachette, 792 p.

Anonymous, 1972, Penrose field conference ophiolites: Geotimes, v. 17, no. 12, p. 24–25.

Argand, E., 1977, Tectonics of Asia, Translated and edited by A.V. Carozzi: New York, Hafner, 218 p.

Aubouin, J., 1959, Contribution à l'étude géologique de la Grèce septentrionale: les confins de l'Epire et dela Thessalie: Annales Géologique des Pays Hellénique, v. 10, p. 1–483.

Benson, W.N., 1926, The tectonic conditions accompanying the intrusion of basic and ultrabasic igneous rocks: Memoirs of the National Academy of Sciences, Volume XIX, First Memoir, 90 p.

Bernoulli, D., 2001, Where did Gustav Steinmann see the trinity?: Geological Society of America Abstracts with Programs, v. 33, no. 6., p. A-172.

Bloomer, S.H., Taylor, B., MacLeod, C.J., Stern, R.J., Fryer, P., Hawkins, J.W., and Johnson, L., 1995, Early arc volcanism and the ophiolite problem: a perspective from drilling in the western Pacific: American Geophysical Union Monograph 88, p. 1–30.

Bowen, N.L., 1927, The origin of ultrabasic and related rocks: American Journal of Science, v. 14, p. 89–108.

Bowen, N.L., and Schairer, J.F., 1935, The System MgO–FeO–SiO_2: American Journal of Science, v. 29, p. 151–217.

Bowen, N.L., and Tuttle, O.F., 1949, The system MgO-FeO-SiO_2: Geological Society of America Bulletin: v. 60, p. 439–460.

Brongniart, A., 1813, Essai de classification minéralogique des roches mélanges: Journal des Mines, v. XXXIV, p. 190–199.

Brongniart, A., 1827, Classification et Caractère Minéralogique des Roches Homogène et Heterogènes: Paris, F.G. Levrault, 144 p.

Brunn, J.H., 1956, Contribution à l'etude géologique de la Grèce septentrionale: les confins de l'Epire et de la Thessalie: Annales Géologique des Pays Hellénique, v. 1, p. 1–485.

Brunn, J.H., 1959, La dorsale médio-atlantique et les épanchements ophiolitiques: Compte Rendu, Societé Géologique de France, v. 8, p. 234–236.

Brunn, J.H., 1960, Mise en place et différenciation pluto-volcanique du cortège ophiolitique: Revues de Géographie Physique et Géologie Dynamique, v. 3, p. 115–132.

Brunn, J.H., 1961, Les sutures ophiolitiques: contribution a l'études des relations entre phénomene magmatique et orogénique: Revues de Géographie Physique et Géologie Dynamique, v. 4, p. 89–96, 181–202.

Coleman, R.G., 1971, Plate tectonic emplacement of upper mantle peridotites along continental edges: Journal of Geophysical Research, v. 76, p. 1212–1222.

Coleman, R.G., 1973, Ophiolites in the earth's crust: A symposium, field excursions and cultural exchange in the USSR: Geology, v. 1, p. 51–54.

Coleman, R.G., 1977, Ophiolites: New York, Springer-Verlag, 229 p.

Davies, H.L., 1968, Papuan ultramafic belt: Prague, Czechoslovakia, XXIII International Geological Congress, v. 1, p. 209–220.

Davies, H.L., 1971, Peridotite-gabbro-basalt complex in eastern Papua: Australian Bureau of Mineral Resources Bulletin, v. 128, 48 p.

De Roever, W.P., 1957, Sind die alpintypen Peridotitmassen vielleicht tektonisch verfrachtete Buruchstücke der Peridotschale?: Geologische Rundschau, v. 46, p.137–146.

Dewey, J.F., 1976, Ophiolite obduction: Tectonophysics, v. 31, p. 93–120.

Dewey, J.F., and Bird, J.M., 1971, Origin and emplacement of the ophiolite suite: Appalachian ophiolites in Newfoundland: Journal of Geophysical Research, v. 76, p. 3179–3206.

Dickinson, W.R., 1970, The New Global Tectonics, 2nd Penrose Conference: Geotimes, v. 15. no. 4 (April), p. 18–22.

Dickinson, W.R., 1971, Plate tectonic models of geosynclines: Earth and Planetary Science Letters, v. 10, p. 165–174.

Dietz, R.S., 1963, Alpine serpentines as oceanic rind fragments: Geological Society of America Bulletin, v. 74, p. 947–952.

Dietz, R.S., 1962, Ocean-basin evolution by sea-floor spreading, *in* Macdonald, G.A., and Kuno, H., eds. The Crust of the Pacific Basin: Washington, D.C., American Geophysical Union Monograph 6, p.11–12.

Dilek, Y., Thy, P., Moores, E.M. and Grundvig, S., 1990, Late Paleozoic-early Mesozoic oceanic basement of a Jurassic arc terrane in the northwestern Sierra Nevada, California, *in* Harwood, D.S., and Miller, M.M., eds., Late Paleozoic and early Mesozoic paleogeographic relations: Klamath mountains, Sierra Nevada, and related terranes: Boulder, Colorado, Geological Society of America Special Paper 255, p. 351–369.

Dilek, Y., Moores, E.M., and Furnes, H., 1998, Structure of modern oceanic crust and ophiolites and implications for faulting and magmatism at oceanic spreading centers, *in* Buck, R., Karson, J., Delaney, P., and Lagabrielle, Y., Faulting and Magmatism at Mid-Ocean Ridges: Washington, D.C., American Geophysical Union Monograph, v. 106, p. 219–266.

Dilek, Y., Moores, E.M., Elthon, D., and Nicolas, A., eds., 2000, Ophiolites and oceanic crust: New insights from field studies and the Ocean Drilling Program: Boulder, Colorado, Geological Society of America Special Paper 349, 552 p.

Drake, C.L., Ewing, M., and Sutton, G.H., 1959, Continental margins and geosynclines: The east coast of North America north of Cape Hatteras, *in* Ahrens, L.H., Press, F., and Runcorn, S.K., eds., Physics and Chemistry of the Earth: Elmsford, New York, Pergamon, v. 5, p. 110–198.

Dubertret, L., 1959, Geologie des roches vertes du nord-ouest de la Syrie et du Hatay (Turquie): Paris, Museum Nationale de Histoire naturelle, Notes et Mémoires de Moyen-Orient, v. 6, 179 p.

Field, R.M., 1938a, Geophysical exploration of ocean basins: Proceedings of the Geological Society of London, Session 1937–38, Quarterly Journal of the Geological Society of London, v. 94, part 2, p. iv–vi.

Field, R.M., 1938b, The importance of geophysics to submarine geology: Proceedings of the American Philosophical Society, v. 79, no. 1, p. 1–7.

Field, R.M, Brown, T.T., Collins, E.B., and Hess, H.H., 1933, The Navy-Princeton gravity expedition to the West Indies in 1932: Washington, D.C., U.S. Government Printing Office, 78 p.

Gass, I.G., 1960, The geology and mineral resources of the Dhali area: Cyprus, Geological Survey Department Cyprus, Memoir No. 4. 116 p.

Gass, I.G., 1967, The ultramafic volcanic assemblage of the Troodos massif, Cyprus, *in* Wyllie, P.J., ed., Ultramafic and Related Rocks: New York, Wiley, p. 121–134.

Gass, I.G., 1968, Is the Troodos massif of Cyprus a fragment of Mesozoic ocean floor?: Nature, v. 220, p. 39–42.

Green, D.H., 1964, The petrogenesis of the high-temperature peridotite intrusion in the Lizard area, Cornwall: Journal of Petrology, v. 5., p. 134–188.

Hall, R., 2002, Cenozoic geological and plate tectonic evolution of SE Asia and the SW Pacific: computer-based reconstructions, model, and animations: Journal of Asian Earth Sciences, v. 20, p. 353–431.

Harkins, M.E., Green, H.W., and Moores, E.M., 1980, Multiple intrusive events documented from the Vourinos ophiolite complex, northern Greece: American Journal of Science, V. 280-A, p. 284–295.

Harper, G.D., 1982, Evidence for large-scale rotations at spreading centers from the Josephine ophiolite: Tectonophysics, v. 82, p. 25–44.

Heck, N.H. 1938, The role of earthquakes and the seismic method in submarine geology: Proceedings of the American Philosophical Society, v. 79, no. 1, p. 97–108.

Hess, H.H., 1938, A primary peridotite magma: American Journal of Science, v. 35, p. 321–344.

Hess, H.H., 1939, Island arcs, gravity anomalies, and serpentinite intrusions, a contribution to the ophiolite problem: International Geological Congress Moscow 1937, Report 17, v. 2, p. 263–283.

Hess, H.H., 1946, Drowned ancient islands in the Pacific: American Journal of Science, v. 244, p. 779–791.

Hess, H. H., 1948, Major structural features of the western north Pacific: An interpretation of H.O. 5485, bathymetric chart, Korea to New Guinea: Geological Society of America Bulletin, v. 59, p. 417–445.

Hess, H.H., 1954, Geological hypotheses and the Earth's crust under the Oceans: Proceedings of the Royal Society, A., v. 222, p. 341–348.

Hess, H.H., 1955, Serpentines, orogeny and epeirogeny: Boulder, Colorado, Geological Society of America Special Paper 62, p. 391–408.

Hess, H.H., 1962a, Richard Montgomery Field: Transactions, American Geophysical Union, v. 43, no. 1, p. 1–3.

Hess, H.H., 1962b, History of ocean basins, *in* Engel, A.E.J., James, H.L., and Leonard, B.F., eds., Petrologic Studies: A volume to honor A.F. Buddington: New York, Geological Society of America, p. 599–620.

Hess, H.H., 1965, Mid-oceanic ridges and tectonics of the sea floor, *in* Whittard, W.F., and Bradshaw, R., eds., Submarine Geology and Geophysics, Proceedings of the Seventeenth Symposium of the Colston Research Society: London, Butterworths, p. 317–334.

Hess, H.H., 1966, Caribbean research project, 1965, and bathymetric chart, *in* Hess, H.H., Caribbean geological investigations: Boulder, Colorado, Geological Society of America Memoir 98, p. 1–10.

Hess, H.H, and Maxwell, J.C, 1949, Major structural features of the south-west Pacific: A preliminary interpretation of H.O. 5484, bathymetric chart, New Guinea to New Zealand: Proceedings of the Pacific Science Congress, v. 2, p. 14–17, 1953.

Hoffman, P.F., and Ranalli, G., 1988, Archean oceanic flake tectonics: Geophysical Research Letters, v. 15, p. 1077–1080.

Holmes, A., 1928, Radioactivity and continental drift: Geological Magazine, v. 6, p. 236–238.

Ishiwatari, A., Malpas, J., and Ishizuka, H., 1994, Circum-Pacific Ophiolites: Proceedings of the 29th International Congress, Part D., Utrecht, the Netherlands, VSP, 286 p.

Jackson, E.D., Green, H.W., II, and Moores, E.M., 1975, The Vourinos ophiolite, Greece: Cyclic units of lineated cumulates overlying harzburgite tectonite: Geological Society of America Bulletin, v. 86, p. 390–398.

Karson, J.A., 2002, Geologic structure of the uppermost oceanic crust created at fast- to intermediate-rate spreading centers: Annual Reviews of Earth and Planetary Sciences, v. 30, p. 347–384.

Khain, E.V., Bibikova, E.V., Kröner, A., Zhuravlev, D.Z., Sklyarov, E.V., Fedotova, A.A., and Kravchenko-Berezhnoy, I.R., 2002, The most ancient ophiolite of the central Asian fold belt: U-Pb and Pb-Pb zircon ages for the Dunzhugur complex, eastern Sayan, Siberia, and geodynamic implications: Earth and Planetary Sciences Letters, v. 1399, p. 311–325.

Kuhn, T.S., 1970, The Structure of Scientific Revolutions (second edition): Chicago, University of Chicago Press, 210 p.

MacKenzie, D.B., 1960, High-temperature Alpine-type peridotite from Venezuela: Geological Society of America Bulletin, v. 71, p. 303–317.

Malpas, J., Moores, E.M., Panayiotou, A., and Xenophontos, C., eds., 1990, Ophiolites, Oceanic Crustal Analogues: Proceedings of the Troodos Symposium, Nicosia, Geological Survey Department, 733 p.

Maxwell, J.C. 1949, Some occurrences of chromite in New Caledonia: Economic Geology, v. 44, p. 525–544.

Maxwell, J.C., and Azzaroli, A., 1962, Submarine extrusion of ultramafic magma: Geological Society of America Abstracts with Programs, p. A103.

Metcalf, R.V., and Shervais, J., 2001, Supra-subduction zone (SSZ) ophiolites: Is there really an "ophiolite conundrum?": Geological Society of America Abstracts with Programs, v. 33, no. 6, p. A-173.

Miyashiro, A., 1973, The Troodos complex was probably formed in an island arc: Earth and Planetary Science Letters, v. 19, p. 218–224.

Moores, E.M., 1969a, Petrology and structure of the Vourinos ophiolite complex, northern Greece: Boulder, Colorado, Geological Society of America Special Paper 118, 74 p.

Moores, E.M, 1969b, The Troodos, Cyprus, and Vourinos, Greece, ultramafic complexes, and an evaluation of ophiolites as ocean floor slices: Program, Discussion Meeting of the Royal Society of London: London, The Royal Society, p. 12–14.

Moores, E.M., 1970, Ultramafics and orogeny, with models for the U.S. Cordillera and the Tethys: Nature, v. 228, p. 837–842.

Moores, E.M., 1973, Geotectonic significance of ultramafic rocks: Earth Science Reviews, v. 9, p. 241–258.

Moores, E.M., 1982, Origin and emplacement of ophiolites: Reviews of Geophysics and Space Physics, v. 20, p. 735–760.

Moores, E.M., 1986, The Proterozoic ophiolite problem, continental emergence, and the Venus connection: Science, v. 234, p. 65–68.

Moores, E.M., 1993, Neoproterozoic oceanic crustal thinning, emergence of continents, and origin of the Phanerozoic ecosystem: a model: Geology, v. 21, p. 5–8.

Moores, E.M., 2002, Pre-1 Ga (pre-Rodinian) ophiolites: their tectonic and environmental implications: Geological Society of America Bulletin, v. 114, p. 80–95.

Moores, E.M., and Jackson, E.D., 1974, Ophiolites and oceanic crust: Nature, v. 250, p. 136–138.

Moores, E.M., and Twiss, R.J., 1995, Tectonics: New York, W.H. Freeman, 415 p.

Moores, E.M., and Vine, F.J., 1969, Troodos Massif, Cyprus, as ocean floor: Preliminary petrologic and structural evidence: Eos (Transactions, American Geophysical Union), v. 50, p. 333

Moores, E.M., and Vine, F.J., 1971, The Troodos massif, Cyprus, and other ophiolites as oceanic crust: evaluation and implications: Philosophical Transactions of the Royal Society London, v. 268A., p. 443–466.

Moores, E.M., Scott, R.B., and Lumsden, W.W., 1968, Tertiary tectonics of the White Pine—Grant Range region, east-central Nevada, and some regional implications: Geological Society of America Bulletin, v. 79, p. 1703–1726.

Moores, E.M., Varga, R.J. , Verosub, K.L., and Ramsden, T.W. 1990, Regional structure of the Troodos dike complex, *in* Malpas, J., Moores, E.M., Panayiotou, A., and Xenophontos, C., eds., Ophiolites, oceanic crustal analogues: Proceedings of the Symposium Troodos 87, p. 27–35.

Moores, E.M., Kellogg, L.H., and Dilek, Y., 2000, Tethyan ophiolites, mantle convection, and tectonic "historical contingency:" A resolution of the "Ophiolite Conundrum," *in* Dilek, Y., Moores, E.M., Elthon, D., and Nicolas, A., eds., Ophiolites and oceanic crust: New insights from field studies and the Ocean Drilling Program: Boulder, Colorado, Geological Society of America Special Paper 349, p. 3–12.

Nicolas, A., 1989, Structure of ophiolites and dynamics of oceanic lithosphere: Boston, Kluwer, 367 p.

Nicolas, A., and Violette, J.F., 1982, Mantle flow at oceanic spreading centers: models derived from ophiolites: Tectonophysics, v. 81, p. 319–339.

Oreskes, N., 1999, The rejection of continental drift theory and method in American earth science: New York, Oxford University Press, 420 p.

Panayiotou, A., ed., 1980, Ophiolites: Proceedings International Ophiolite Symposium Cyprus 1979: Nicosia, Cyprus, Geological Survey Department, 781 p.

Peters, Tj, Nicolas, A., and Coleman, R.G., 1991, Ophiolite Genesis and Evolution of the Oceanic Lithosphere: Proceedings of the Ophiolite Conference, Muscat, Oman, 7–18 January, 1990, 903 p.

Rassios, A., Moores E.M., and Green, H.W., II. 1983, Magmatic structure and stratigraphy of the Vourinos ophiolite cumulate zone, northern Greece: Ofioliti, v. 8, p. 377–410.

Robinson, P.T., and Malpas, J., 1990, The Troodos ophiolite of Cyprus: new perspectives on its origin and emplacement, *in* Malpas, J., Moores, E.M., Panayiotou, A., and Xenophontos, C., eds., Ophiolites, Oceanic crustal analogues, Proceedings of the "Troodos 1987" Symposium: Nicosia, Cyprus, The Geological Survey Department, p. 13–26

Ross, C.S., Foster, M.D., and Myers, A.T., 1954, Origin of dunites and of olivine-rich inclusions in basaltic rocks: American Mineralogist, v. 39, p. 693–737.

Routhier, P., 1946, Volcano-plutons sous marins du cortège ophiolitique: Compte Rendu, Paris, Academie Science, v. 22, p. 192–194.

Routhier, P., 1953, Etude géologique du versant occidental de la Nouvelle-Calédonie: Paris, Memoires Societé Géologique de France, v. 67, 271 p.

Sleep, N.H, and Windley, B.F., 1982, Archean plate tectonics: constraints and inferences: Journal of Geology, v. 90, p. 363–380

Steinmann, G., 1905, Geologische Beobachtungen in den Alpen, II. Die schart'sche Überfaltungstheorie und die geologischen Bedeutung der Tiefseeabsätze und der ophiolitische massengesteine: Bericht Naturforschung Gesellschaft Freiburg 1, bd 16, p. 44–67.

Steinmann, G., 1927, Der ophiolitischen Zonen in der mediterranean Kettengebirgen, *in* Proceedings, 14th International Geological Congress, Madrid, v. 2, p. 638–667.

Stille, H., 1939, Bemerkungen betreffend die "sardische" Faltung und den Ausdruck "ophiolitisch": Zeitschrift der Deutschen Geologischen Gesellschaft, Band 91, p. 771–773.

Stoffers, P., Worthington, T., Hekinian, R., Peterson, S., Hannington, M., Türkay, M., and the SO 157 Shipboard Scientific Party, 2001, Silicic volcanism and hydrothermal activity documented at Pacific-Antarctic Ridge: Eos (Transactions, American Geophysical Union), v. 82, no. 38, p. 301, 304.

Suess, E., 1906, The face of the Earth, v.2, part 3, The Sea: Oxford, Clarendon, 556 p.

Suess, E., 1909, The face of the Earth, v. 4: Oxford, Clarendon, 678 p.

Temple, P.G., and Zimmerman, J., 1969, Tectonic significance of alpine ophiolites in Greece and Turkey: Geological Society of America Abstracts with Programs, pt. 7, p. 221–222.

Thayer, T.P., 1967, Chemical and structural relations in ultramafic and feldspathic rocks in Alpine intrusive complexes: *in* Wyllie, P.J., ed., Ultramafic and Related Rocks: New York, Wiley, p. 222–238.

Varga, R.J., and Moores, E.M., 1985, Spreading structure of the Troodos ophiolite, Cyprus: Geology, v.13, p. 846–850.

Verosub, K.L., and Moores, E.M., 1981, Tectonic rotations in extensional regimes and their paleomagnetic consequences for oceanic basalts: Journal of Geophysical Research, v. 86, p. 6335–6349.

Vine, F.J., and Hess, H.H., 1970, Sea-floor spreading, *in* Maxwell, A.E., ed., The Sea, v. 4, part 2, p. 587–622.

Vine, F.J., and Moores, E.M., 1969, Paleomagnetic results from the Troodos Complex, Cyprus: Transactions American Geophysical Union, v. 50, p.131.

Vine, F.J., and Moores, E.M., 1972, A model for the gross structure, petrology and magnetic properties of oceanic crust: Boulder, Colorado, Geological Society of America Memoir 132, p. 195–208.

Vogt, J.H.L., 1924, Role of Fe in origin of ultramafic magma: Skrifter, Norske Videnskapp Akademie Klasse i., Mathematike Naturvidenskapp Klasse, v. 14, p. 97.

Wilson, R.A.M., 1959, The geology of the Xeros-Troodos area, Nicosia, Cyprus: Geological Survey Department Cyprus, Memoir, no. 1, 184 p.

Manuscript Accepted by the Society May 2, 2003

Geological Society of America
Special Paper 373
2003

Identification of a mantle unit in ophiolites: A major step in the evolution of the ophiolite concept

Thierry Juteau*
UMR Domaines océaniques, IUEM/UBO, Place Copernic, 29280, Plouzané, France

ABSTRACT

For about 150 years prior to 1963, a number of purely magmatic models were considered to explain ophiolitic associations. The basic idea was that the puzzling association of volcanic and plutonic rocks in ophiolites was a result of magmatic differentiation in a huge and unique volume of magma. In this interpretation, ophiolite complexes were thought of as autochthonous bodies with respect to their surrounding rocks. Around 1970, several authors described a mantle sequence in ophiolite complexes that they had mapped in detail. That sequence was characterized by high-temperature ductile deformation structures, and by the residual refractory composition of its peridotites. These features were in contrast to those observed in stratigraphically higher rock types (gabbros, diabase, lavas), which were ascribed to a crustal sequence that was devoid of any evidence for high-temperature plastic deformation. This major discovery was endorsed by the 1972 Penrose Conference on Ophiolites, whose participants introduced the modern definition of a typical ophiolite.

The discovery of a mantle unit in ophiolites coincided with the emergence of plate tectonics theory, which provided a global mechanism for seafloor spreading at mid-ocean ridges. It subsequently became clear that any ophiolite complex exposed in a folded mountain range should represent a piece of fossil oceanic lithosphere on land. Many authors reexamined the contacts of ophiolites with country rocks and demonstrated that their basal contact is generally tectonic, commonly marked by the presence of metamorphic soles. This observation resulted in the interpretation that ophiolites are actually allochthonous bodies. At that time R.G. Coleman (1971) defined the term obduction as a mechanism of tectonic emplacement of oceanic lithosphere onto a continental margin. With the discovery of a mantle unit in ophiolites, the purely comagmatic model was replaced by a syngenetic model, in which the residual mantle unit is considered as a counterpart of the crustal rocks, representing the products of the upper mantle partial melting.

Keywords: ophiolites, mantle sequence, oceanic crust, oceanic ridges.

INTRODUCTION

The term, ophiolite, appeared in the scientific literature in an 1813 paper by the French mineralogist, Alexandre Brongniart (1740–1847). In 1821, he referred to a suite of magmatic rocks closely associated in the field. This puzzling association of ultramafic to mafic rocks related coarse-grained plutonic rocks to doleritic and microlitic rocks, often on the same outcrop. It appeared that this magmatic association was defying the classical distinction between plutonic and volcanic rocks, and was therefore difficult to explain.

For a century and a half, from 1813 to 1963, the ophiolitic association has been explained by a purely comagmatic assemblage, in which the rocks of the association represent

*juteau@wanadoo.fr

Juteau, T., 2003, Identification of a mantle unit in ophiolites: A major step in the evolution of the ophiolite concept, *in* Dilek, Y., and Newcomb, S., eds., Ophiolite concept and the evolution of geological thought: Boulder, Colorado, Geological Society of America Special Paper 373, p. 31–53. For permission to copy, contact editing@geosociety.org.

various parts of a single igneous body. According to this concept, ophiolites were magmatically emplaced, with the obvious consequence that they were autochthonous with respect to their country rocks, generally geosynclinal sediments. Many models were proposed, all of them satisfying the comagmatic hypothesis: intrusive laccoliths or placcoliths, subvolcanic plutons, and huge submarine outflows.

Two papers published in 1963 announced the end of the comagmatic hypothesis. In a lucid paper, Marc Vuagnat (1963) summarized the rising doubts of the scientific community concerning the comagmatic theory. The same year, Gass and Masson-Smith (1963) published a benchmark paper on the Troodos ophiolite with new geological and geophysical surveys, showing that the Troodos Massif is rooted in a slab of ultramafic upper mantle.

During the following ten years, new field studies and detailed mapping on various ophiolite complexes around the world led to the conclusion that ophiolites include a peridotitic mantle sequence, generally characterized by high-temperature plastic deformation and residual chemistry, and a comagmatic crustal sequence (gabbros, diabase dikes, and submarine basalts), weakly or not deformed. According to this new interpretation, ophiolites were allochthonous with respect to their country rocks. They were assembled during a primary accretion stage at an oceanic spreading center, and later tectonically emplaced on a continental margin or island arc.

The aim of this paper is to document the nature, chronology, and the significance of observations and findings in ophiolite studies that collectively led to the discovery and definition of the mantle sequence in ophiolites. With the recognition of residual mantle rocks in ophiolites, earlier comagmatic models suggesting an in situ origin of ophiolitic massifs were replaced by polygenetic models, which interpret the mantle rocks as residues after the extraction of basaltic magma following partial melting of the upper mantle.

THE PURELY COMAGMATIC MODELS

Brongniart's Earliest Work

Most authors agree that the earliest use of the term "ophiolite" appears in A. Brongniart's "Essai d'une classification minéralogique des roches mélangées," published in 1813 (for instance, see Juteau, 1975; Coleman, 1977; Amstutz, 1980; Nicolas, 1989). At that time, before the invention of the polarizing microscope, the main concern was to identify and give a name to all varieties of natural rocks exposed on Earth's surface. Brongniart (1813) gave the name ophiolite to a "rock species" or "mixed rock" characterized by a serpentine matrix including iron oxides or disseminated accessory minerals such as chromite, diallage, or garnet. The term, ophiolite, is composed of the Greek words "ophios" (snake) and "lithos" (rock), and refers to the snake-skin appearance of many serpentinites. Ophiolites were also given as an example for specific structures, such as the "veined structures," where "...amorphous parts are crosscut by variously coloured veins" (Brongniart, 1813, p. 24).

Brongniart was also a good field geologist. In 1821 he published a monograph on Italian ophiolites in the northern Apennines, in which he used the term ophiolite as a "collection of rocks," closely associated in outcrops (Brongniart, 1821, p. 183–184). In his field descriptions, the term was systematically used in the plural (e.g., ophiolitic terranes and ophiolitic formations), as noted by Amstutz (1980). It was a four-fold rock association, consisting of ultramafic rocks (serpentine), gabbros (locally called euphotides), diabases/vesicular spilites, and siliceous cherts or radiolarites. In his monograph, Brongniart did not give field sections or schematic drawings to illustrate the descriptions in the text. He also did not try to propose a genetic model for these ophiolites; his main concern was to give a relative age to the formations, by comparison with known outcrops in the Alps and other places. He rightly concluded that the Apennine ophiolites were of Mesozoic age.

The Italian Precursors

At the end of the 19th century, the Ligurian and Piedmont regions were remapped and re-examined by several very capable Italian field geologists, who confirmed Brongniart's ophiolitic association, and proposed the first comagmatic genetic models. In these models, the ophiolite suite was conceived of as representing a single igneous body that had differentiated and crystallized. For some of these authors, the igneous body was extruded as a huge submarine lava flow on the seafloor (Lotti, 1886a, 1910; Franchi, 1902). For others, the magma body was intrusive into marine sediments.

Figure 1 is taken from Lotti's monograph entitled "Geologia della Toscana" (1910), in which the author gave many precise descriptions of the ophiolitic association in outcrops, with abundant schematic field sections illustrating the close ophiolitic association. A geological map in color accompanied the text, showing all ophiolitic massifs in Tuscany and on the island of Elba, where Lotti made his first geological studies (reported on in Lotti, 1886b).

In his conclusions Lotti (1910) insisted on the constant association in the field of three main facies: (1) serpentinites, considered to be hydrated lherzolites; (2) euphotides, a local name given to coarse-grained to pegmatitic pyroxene gabbros; and (3) diabases, considered to be altered basalts, with various textures ranging from doleritic to microlitic and variolitic. These three facies were always superposed from bottom to top in that order. He also summarized the observed crosscutting relationships between the three facies: euphotides form intrusive dikes in the serpentinites, and diabase dikes crosscut both serpentinites and euphotides. The eruptive mass appeared interbedded within calcareous to clayey marine sediments of Eocene age, often with cherts and radiolarites at the top. No contact or thermal metamorphism was observed in the sediments, however, and Lotti concluded that these complexes "can

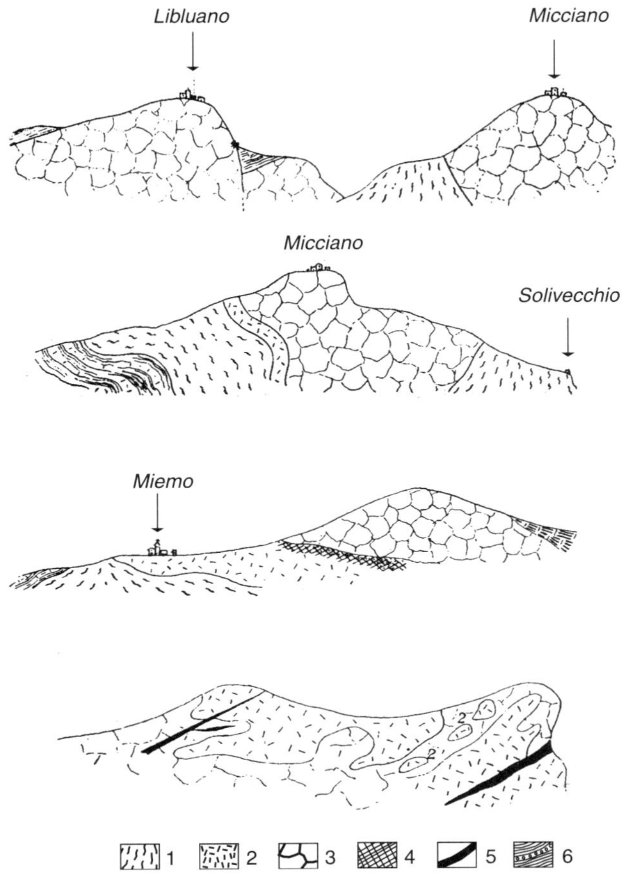

Figure 1. Some of the Lotti's field sections in the northern Apennine ophiolites (Tuscany, Italy: cross sections 1 to 3 from top to bottom), and from Elba island, near Longone village (section 4), showing the close association of serpentinites, gabbros and diabases in the field (redrawn after Lotti, 1910, p. 239, 254, and 268). (1) serpentinites. (2) euphotides. (3) diabases. (4) disseminated copper mineralizations. (5) epidote veins. (6) marine sediments.

only represent huge magmatic extrusions of lava on the Eocene seafloor" (Lotti, 1910, p. 272). He thought that the fine-grained diabases were consolidated very rapidly in the upper part of the extrusive mass, and that the coarse-grained euphotides and lherzolites crystallized slowly beneath them. After consolidation, the lherzolites were massively hydrated by seawater and transformed to serpentinites.

Steinmann's "Placcolith" Theory

The German geologist Gustav Steinmann (1856–1929) also studied the Ligurian ophiolites. As early as 1905, he published a study concluding that the Apennine ophiolites could be simatic intrusions in shallow marine sediments. But Steinmann was to become famous after his lecture at the 14th International Geological Congress (IGC) of Madrid in 1926, about the ophiolite belts of the Mediterranean mountain ranges (published in Steinmann, 1927, see translation in this volume, Chapter 6). His clear definition of the ophiolititic association, known as the "Steinmann trinity," and his coherent and well argued petrogenetic theory constitute a major stage in the development of the ophiolitic comagmatic concept.

According to Steinmann, the term, ophiolite, should be used only for the cogenetic association of mainly ultrabasic rocks such as peridotites and serpentines, with subordinate gabbros, diabases, and spilites, as well as norites and similar rocks (Steinmann, 1927). He insisted on the preponderance of ultrabasic rocks with respect to the other facies of the association. He developed this point in a chapter dedicated to volume ratios between serpentines, gabbros and diabases/spilites in which, after meticulous appraisal of the Tuscan and Ligurian ophiolites, he estimated the volumetric ratio of ultrabasic rocks to feldspathic rocks to be 5 to 2, and the ratio of gabbros to diabases/spilites to be 1 to 1.

One of the most interesting aspects of his synthesis concerns the age relationships between serpentines, gabbros and diabases/spilites. These three main groups, the famous Steinmann ophiolitic trinity, have clear crosscutting relationships, and obey the following rules: (1) serpentinites never crosscut the two other groups, but are crosscut by both of them, and are therefore the oldest rocks of the association; (2) gabbros crosscut serpentinites, never the diabases/spilites; and (3) diabases/spilites crosscut both serpentinites and gabbros, and are the youngest rocks of the association.

Steinmann insisted that there are no progressive transitions between the three groups. He also showed that their emplacement was pre-orogenic, since all these rocks have been subjected to the same tectonic events as the surrounding sediments. He argued against a popular opinion, defended by Suess (1918), saying that ophiolitic intrusions follow the stratification planes of the sediments, as well as major alpine shear zones and thrust surfaces, which implies a Tertiary age and a synorogenic emplacement for these rocks. Steinmann demonstrated clearly that in Liguria, the sedimentary host rocks of the ophiolites are of Jurassic age (and not Eocene, as proposed by Lotti), and that the whole series of "sediments plus greenstones" has passively experienced all the Tertiary tectonic phases.

To explain these observations, which later appeared to be correct, Steinmann proposed a rather complicated genetic model, based on the postulate that the three main groups of rocks of the ophiolitic association represent the fractionation products of a single ophiolitic magma. This is an example, frequent in science, of an incorrect theory developed from a set of indisputable, excellent observations. The theory has its internal logic, but does not work because it rests on an incorrect postulate which seemed evident to its author and to most of his contemporaries.

Steinmann was obliged to reconcile the crosscutting relationships observed in the field with the necessary ophiolitic magma. The result was an extremely complicated differentiation schema, needing at least six main stages (Fig. 2A). Stage

1 shows the magma emplaced as a huge, flat intrusion, bulging and convex upward, reaching more than one km in thickness, and up to 100 km in diameter forming an ophiolitic *placcolith* (a new term proposed by Steinmann, which apparently did not outlive him). Upon cooling, the magma would nucleate and precipitate. At first, the Fe-Mg-rich silicate phases at the roof of the placcolith would solidify (cold wall effect) as the main ultrabasic mass of peridotites at the top of the intrusion, over a residual magma of gabbroic composition located beneath it (R1 in Fig. 2A). This form of differentiation violated the laws of gravity, but that did not seem to have worried Steinmann. It was already known at that time, from Bowen's experiments, that olivine settles immediately in all basic magmas. But Steinmann believed in his field observations. Since there were numerous gabbro dikes crosscutting the serpentinites (see stage 3 in Fig. 2A), the residual magma (R1 in stage 2, Fig. 2A) had to be located beneath them, and could only be injected upward after fracturing its peridotite cover. After crystallization of the gabbros, a late residual magma remained, rich in volatile components, located below the peridotites and gabbros for the same reasons (R2 in stage 4, Fig. 2A): it would crosscut both of them to produce the doleritic diabases and vesicular spilites.

At the end of his discussion, Steinmann developed a number of implausible statements, such as:

(1) To explain the peridotite-gabbro-diabase succession, often observed in the field in this order from bottom to top, he was obliged to imagine that the crystallized mass, fractured by the underlying residual magma, became detached from the roof and sank progressively, finally leading to a complete spatial inversion of the residual magma and the crystallized mass. He criticized the German geologist R. Staub's model for the ophiolites of the internal Alps. Staub (1924) thought that the alpine ophiolites had been emplaced along large concordant sills or laccoliths, whose dimensions were more modest than Steinmann's placcoliths. After emplacement, these bodies differentiated by gravity into peridotites at the bottom and gabbros/diabases at the top (Fig. 2B). This model was much more reasonable than that proposed by Steinmann, but could not explain the crosscutting relationships observed in the field.

(2) Steinmann was also obliged to deny the extrusive nature of the spilitic pillow lavas, generally found overlying the coarse-grained rocks. He disagreed with the opinion of specialists such as the English geologists H. Dewey and J.S. Flett (1911) or the American J.V. Lewis (1914), who thought that pillow lavas were submarine volcanic forms. For him, the pillow lavas represented a case of magmatic fractionation into matrix, pillows, and albitic variolites, occurring at the roof of the intrusive placcolith. He discussed the consequences of his interpretation, and argued for it with lucidity. Various Italian geologists (among them Lotti), and Dewey and Flett, thought on the contrary that ophiolites were extruded on the seafloor, either in total, or at least the pillow lava component.

This mistaken interpretation of pillow lava structures is one of the weakest points of Steinmann's model; however, there are other very positive aspects of it. As early as 1905, Steinmann

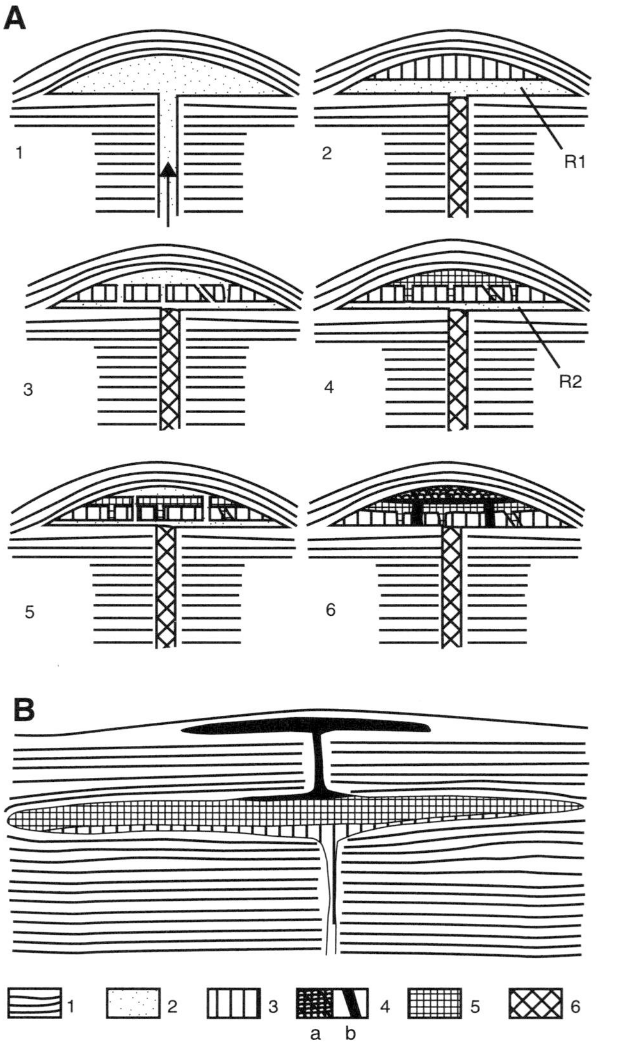

Figure 2. Two comagmatic intrusive models for ophiolites, which were very popular in the 1920s. A: Emplacement and differentiation of an ophiolitic *placcolith*, following Steinmann's conceptions (drawn after his descriptions in Steinmann, 1927). Stages 1 to 6 correspond to the crystallization and differentiation stages of the placcolith (described in the text). B: Emplacement and differentiation of an ophiolitic *laccolith*, following Staub's conceptions (drawn after his descriptions in Staub, 1924). (1) abyssal "geosynclinal" sediments. (2) basic magma. (3) ultrabasic rocks. (4) diabases and spilites, (a) pillow lavas, (b) massive dikes and sills. (5) gabbros. (6) crystallized feeder dike. R1 and R2: successive residual magmas (explanations in the text).

was the first to establish a link between ophiolites and specific "geosynclinal" (oceanic) sediments, such as radiolarites. Dewey and Flett (1911), and Davis (1918) had also shown that radiolarites are spatially linked to submarine volcanic extrusions, and proposed the hypothesis that these extrusions bring silica into the surrounding seawater, favoring the proliferation of microorganisms such as radiolarians. Later studies (Routhier, 1946b) would show that this hypothesis is valid. Steinmann, however, could not accept it, since his placcoliths were intrusive into the geosynclinal sediments, which therefore existed before the intrusion. The radiolarites found at the ophiolite roofs could not have been induced by them. He therefore gave a bathymetric significance to these sediments (of deep, abyssal water), and assigned the corresponding palaeogeographic location to the ophiolitic intrusions.

Steinmann did not use the term "initial geosynclinal magmatism" (Stille, 1938), an expression introduced later. It is clear, however, that this concept was already present in his mind. For him, the basic/ultrabasic associations found in orogenic mountain ranges were preorogenic, and contemporaneous with the geosynclinal stage.

In summary, Steinmann was above all a perspicacious observer, who was the first to state clearly: (1) a precise definition of the ophiolitic association; (2) the crosscutting relationships between the three main groups of rocks of the association; (3) the spatial link between ophiolites and radiolarites; (4) the preorogenic emplacement of these rocks and their association with the geosynclinal stage; and (5) the first classification of the main basic/ultrabasic associations in orogenic mountain ranges. His interpretations and genetic models, however, were erroneous and needlessly complicated. Steinmann missed the truth by denying the extrusive nature of pillow lavas, and by proposing differentiation processes that violated the physical rules of gravity.

The Model of Extensive Submarine Outflows (1939–1965)

When Steinmann gave his masterly lecture at the Madrid IGC in 1926, the largest and most spectacular ophiolite complexes of the Alpine range, those in the eastern Mediterranean and the Middle East, were still quite unknown, even if they were mentioned by Suess in his book *The Face of the Earth* (1918). These complexes, outcropping within unmetamorphosed sediments of Mesozoic age and exhibiting particularly well-preserved internal structures, were to be the object of new observations and ideas. These ideas, including the concept of extensive submarine outflows developed by a group of French geologists, would gradually merge between 1940 and 1960, and triumph at the World International Congress of Alger in 1952.

In November, 1939, a note from L. Dubertret about "Syrian greenstones" described the constant association of peridotites, pyroxenites, and diorites and microdiorites with volcanic tuffs and glasses, superposed in that order over more than 3000 m in thickness in the Baer Bassit region. The microlitic upper part of this complex, ~200 m thick, included glassy pillow lavas and fragments of sediments of various ages (from Triassic to Cenomanian), supposed to have been dragged along by the ophiolitic magma (Dubertret, 1939). Beyond the main body of greenstones (Baer Bassit, Kizil Dag and Karasu corridor), lenticular ophiolitic apophyses were interbedded with Maastrichtian sediments, establishing the Maastrichtian age of the Syrian greenstone masses. Dubertret concluded:

> The Syrian greenstones have their originality in the union and simultaneity of deep to shallow eruptive phenomena with a well-characterized volcanism, and also in the considerable thickness of the products emplaced, probably in a very short period of time. (translated from Dubertret, 1939, p. 764)

Two months later, J.H. Brunn (1940a) described the huge massifs of basic rocks located at both sides of the Albano-Thessalian furrow in the same way. These complexes were composed of lherzolites, diallagites, pegmatite gabbros, and amphibole gabbros with doleritic textures, of various types of olivine gabbros grading to peridotites, and of variably altered spilites and basalts. The key-section along the Haliakmon River in the Vourinos complex showed, from bottom to top (Fig. 3): Triassic limestones, then a silicified microlitic rock (interpreted as a lower chilled margin), the main harzburgitic unit, and then olivine gabbros, doleritic basalts, radiolarites, and Cretaceous limestones. These rocks rested on Triassic limestones in conditions which did not suggest thrusting, but rather a huge, massive outflow, where coarse-grained rocks would represent the rocks at depth. Four months later, Brunn explained his interpretation. Insisting on the concordance and lack of disruption of the basal contact, he proposed that the greenstones were emplaced after extensional movements preceding the Cretaceous and Tertiary compressive phases. The extensional traction probably opened at the boundary between the old Pelagonian massif and the Adriatic geosyncline, on the present emplacement of the Albano-Thessalian furrow, a wide shear across which a huge mass of simatic magma was delivered from below. This magma was poured out over the seafloor and ponded in huge masses 2 to 3 km thick, on both sides of the fracture (see Fig. 4A), over the margin of the Pelagonian continent to the east, and mainly toward the bottom of the Pindos geosyncline to the west.

The major development of coarse-grained rocks was explained by the dimensions of these extensive outpourings of magma, which had nothing to do with classical volcanic eruptions. They were the expression of the "slow solidification of the magma in the core of these masses, protected by a surficial, scoriaceous and doleritic blanket" (Brunn, 1940b, p. 735). During the following twenty years, Brunn worked hard to improve this model until the proposal of a complete theory about these "ophiolitic sutures" (Brunn, 1960, 1961).

Meanwhile, the French geologist P. Routhier (1946a) had published the geological section of the Insecca ophiolite massif in Corsica. He criticized Steinmann's model, showing the perfect concordance of radiolarites with pillow lavas and the thickness of the latter (several hundreds of meters), not allowing an interpretation as "chilled margins of intrusive laccolithes" (Routhier, 1946a, p. 194). Concerning the coarse-grained rocks serpentine and gabbro, the relations with the lavas were difficult to observe, but in some places "it was possible to observe the grading transition of gabbros to microgabbros, dolerites and fine-grained pillowed dolérites" (Routhier, 1946a, p. 194). Routhier's conclusion was the following:

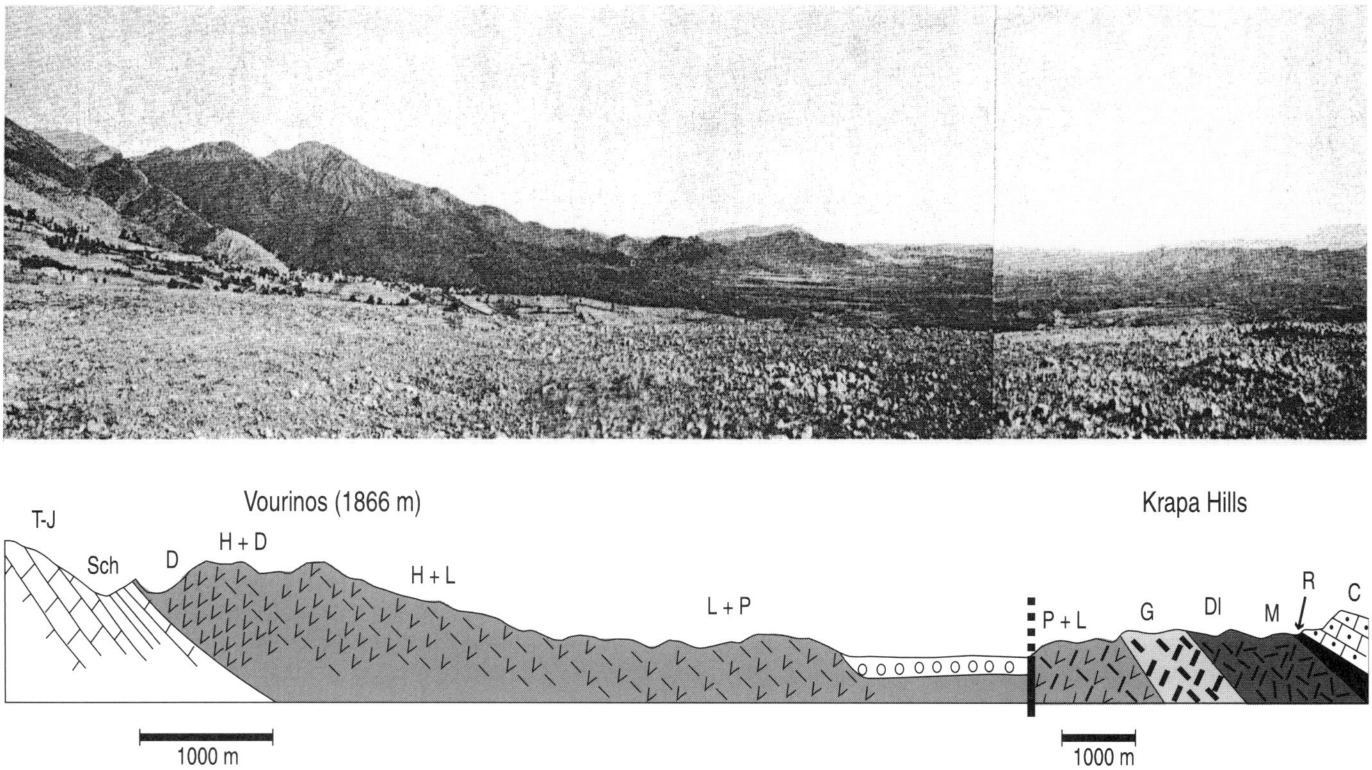

Figure 3. Top: photographic landscape of the Vourinos ophiolite massif and Krapa hills in western Macedonia, Greece (view looking southward). Taken from Brunn, 1960. Bottom: Corresponding geological section (scale deformed by perspective). From left to right: (T-J) Triassic and Jurassic limestones; Sch—interbedded schists and cherts at the top of the limestones; D—dunites; H—harzburgites; L—lherzolites; P—pyroxenites; G— gabbros; Dl—dolerites; M—microlitic rocks (spilites); R—radiolarites; C—Cretaceous limestones. Redrawn after Brunn, 1960, p. 119.

> These transitions, inside a massif whose unity is not questionable and where nothing allows a succession of eruptive stages, can only be explained by admitting that we are dealing with a magmatic complex having coarse-grained roots, and whose "hat" was extruded on the seafloor. Therefore, the Insecca complex corresponds to a *vulcano-pluton*, comparable to continental complexes of the same type. but it is a *submarine vulcano-pluton.* (translated from Routhier, 1946a, p. 194)

Routhier later demonstrated that the evidence for the deposition of radiolarian-rich jaspers associated with greenstones was geochemical, not bathymetric (abyssal), and that the formation of jasper did not imply any determined depth condition (Routhier, 1946b).

The greenstone nappes (Dubertret), the huge submarine ophiolitic outflows (Brunn), the vulcano-plutons (Routhier), these neologisms were created to offer a new model for the ophiolitic suite, which would triumph at the Algiers International Geological Congress, 1952. One of the two topics of the XVth Section dedicated to palaeovolcanism, was titled: "Ophiolitic series, greenstones and their relations to orogenesis." Brunn, Dubertret, Routhier, and also Bischopp and Vuagnat, presented their convergent interpretations, without any significant opposition. Routhier presented the huge peridotitic massifs of New Caledonia as the remains of a giant extrusion of submarine ultrabasic rocks. The new fact was that this huge "simatic outflow" closed the magmatic/tectonic history of this recent range. Therefore, ophiolitic outflows would not be restricted to the geosynclinal stage of orogens (Routhier, 1954). Glangeaud developed this point in his "reflections about the works of the XVth Section." After saying that all speakers were unanimous in thinking that ophiolite suites would represent huge simatic extrusions into deep submarine areas, he concluded that "ophiolitic suites are not more necessarily linked to orogenesis, than present-day abyssal red clays" (Glangeaud, 1954, p. 236–37).

Brunn became the theoretician of the idea of huge submarine outflows, first in his thesis, published in 1956, where he meticulously described the mechanism of the rhythmic sedimentation of the various mineral phases inside these outflows, and then in 1960 in a paper dealing with "the emplacement and differentiation of the ophiolitic plutonic-volcanic association." He proposed a complete theoretical model, thoroughly argued, summarized as follows (Figs. 4A–4C):

Following a huge tear (the *ophiolitic suture*) provoking the brutal decompression of the underlying sima, a thick ophiolitic outflow is emplaced on the seafloor. Microlitic and doleritic rocks constitute the primitive envelope of the outflow, solidifying downwards. It is a massive solidification of magma by the chilled

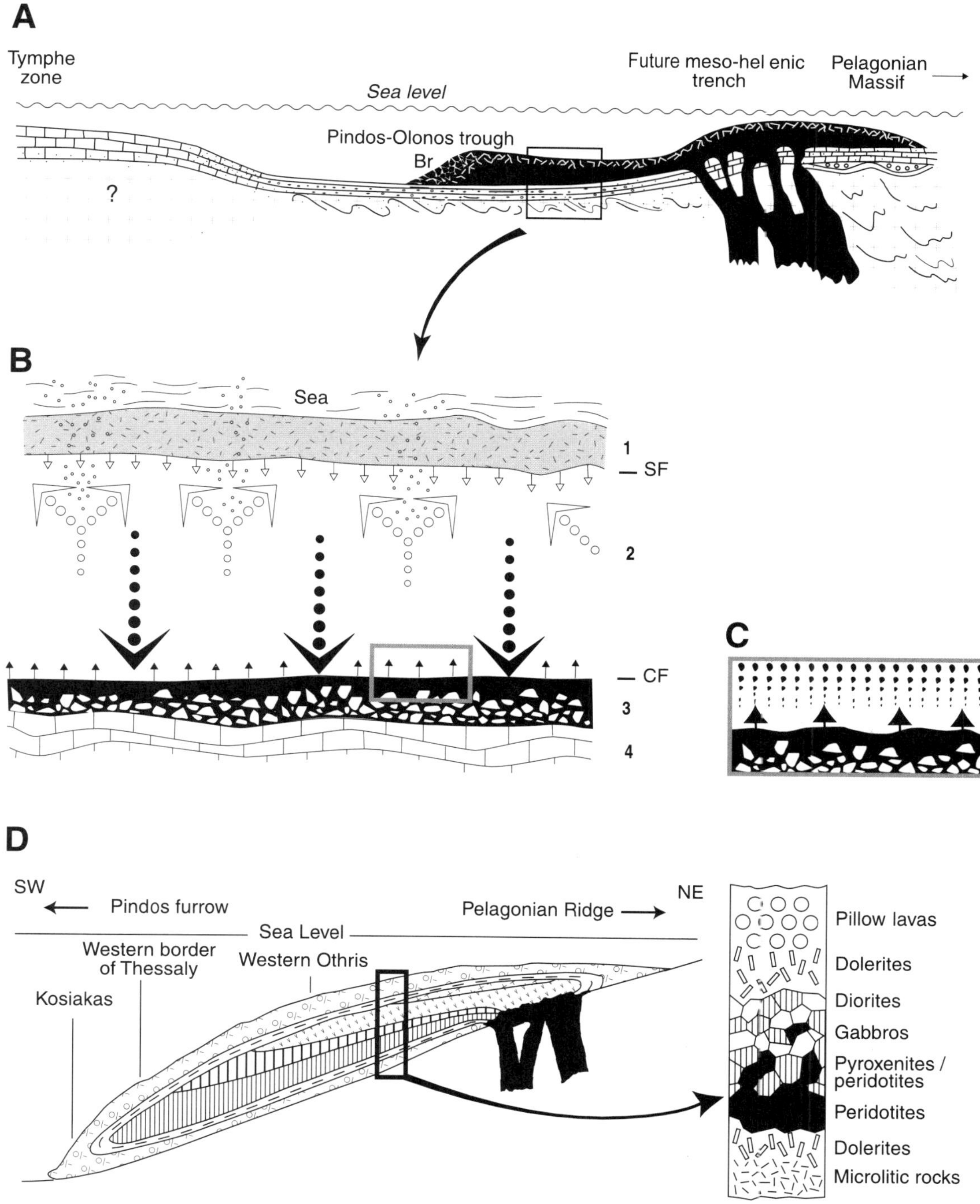

Figure 4. The model of extensive submarine outflows for ophiolites. A: Magmatic emplacement and schematic profile of an ophiolitic outflow. In black, the basic magma is extruded at the boundary between a "geosynclinal furrow" (here the Pindos-Olonos trough, western Macedonia, Greece) and an "internal zone" including a continental basement (here the Pelagonian massif). White dashed areas represent the early solidified microlitic and doleritic crust. Br—ophiolitic breccias developed at the front of the outflow. Redrawn after Brunn (1960), p. 117. B: Internal differentiation of an ophiolitic submarine outflow. (1) Doleritic and microlitic crust. SF—descending front of solidification. (2) Differentiating magma. Bubble lines with diverging arrows—rise of volatile and soluble elements (H_2O, Si, Na, S, Cl). Lines of black dots with black arrows—settling of heavy minerals. (3) Coarse-grained rocks crystallizing on the bottom. CF—rising crystallization front. (4) sedimentary substratum. Redrawn after Brunn (1960), p. 127. C: Magnification of the rectangle in B, explaining the development of magmatic layering. The predominating crystallization of one mineral phase (black) at the crystallization front reduces the potentialities of that same phase in the immediately overlying layer of the fluid phase, favoring the nucleation and growth of another mineral phase (white). Redrawn from Brunn (1960), p. 129. D: Lithological model of a submarine ophiolite flow, based on the Vourinos ophiolite. Redrawn after Aubouin (1965).

margin effect, producing rocks having an extrusive/volcanic appearance. In principle, this envelope represents the original ophiolitic magma, before differentiation. In the core of the giant outflow, a real magma chamber will differentiate slowly (Fig. 4B), following the crystal fractionation rules, crystallizing from bottom to top, producing layered and rhythmic segregation of mineral phases (Fig. 4C). These processes give birth to the coarse-grained rock units, whose magmatic layering is one of the most systematic characteristics. That part of the complex is analogous to the huge basic/ultrabasic continental stratiform complexes, such as the Bushweld, the Skaergaard, or the Stillwater complexes. Both crystallization fronts progressing in opposite directions finally meet. The convergence of crystallization conditions may blur the boundary between both domains. It seems, however, that the lower part of the dolerites is a turbulent zone due to the influx of volatile elements coming from deeper zones, hence the frequent presence of magmatic breccias at that level. Volatile elements also diffuse across the microlitic chilled margin, and spray huge quantities of silica into seawater, provoking the proliferation of radiolarians in the vicinity of the ophiolite masses. This may accentuate the overall ultrabasic composition of the stratiform part of the complex. As well, the immediate formation of a chilled crust at base and top of the extrusion, and the continuous ascent of volatile elements may explain the weakness of metamorphic impact developed at the basal contact in the sedimentary floor (Brunn, 1956; 1960).

Brunn's model postulated the comagmatism of various parts of the ophiolite suite, as Steinmann's model had done, but was more elaborate than the latter because it took into account: (1) the unity of the suite and its vertical succession (peridotites-gabbros-basalts); (2) the internal structures of the coarse-grained rocks, especially the stratiform layering and the cumulate structures; (3) the extrusive, submarine nature of the pillow lavas; (4) the spatial relationship between radiolarites and ophiolites; and (5) the absence of significant thermal metamorphism.

Marc Vuagnat's Question Marks (1963)

Although brilliant and convincing, the theory of huge submarine outflows was problematic, as discussed by the Swiss petrologist Vuagnat (1963), whose objections can be summarized in two categories: (1) Theoretically, the most serious objection concerned the relative volume ratio of ultrabasic rocks to gabbroic rocks, which is far too high, typically 2:1 to 5:1 in most ophiolites of the French and Italian Alps, and up to 9:1 in the Vourinos (Greece), the Guleman or the Hatay (Turkey) complexes. These proportions are quite different from those observed in the continental stratiform complexes, where gabbros predominate. The defenders of the theory of huge submarine outflows refer often to these complexes as a differentiation model, but as pointed out by Thayer (1967), the products of such processes should be similar to the rocks of the stratiform complexes in all their essential characteristics, since both are supposed to form by simple crystal fractionation of a basaltic magma. Also, the peridotites in all stratiform complexes usually contain plagioclase as an interstitial phase, which is extremely rare in ophiolitic peridotites. (2) In the field, a number of facts remained unexplained by this theory, particularly the penetrative high-temperature plastic deformations and flow structures observed in most ophiolitic peridotites, which were mentioned by many specialists of chromite deposits (Zachos, 1954; Thayer, 1960). The extremely restricted compositional range of the huge peridotitic masses in ophiolites was also very puzzling for a growing number of petrologists. Their origin as magmatic cumulates was simply unlikely. Alternatively, they could more likely represent partial melting residues formed within the upper mantle.

Another point was the underestimation of the network of dikes of all kinds in various parts of the ophiolitic assemblage. For instance, no dike intrusions were described in Dubertret's thesis (1955), whereas other observers mentioned many dikes of all kinds in the greenstones of Syria and Hatay (Majer, 1960). Some years later, Çogulu and Vuagnat (1967) and Parrot (1973) reported the same observation, and Thayer reproached Brunn for giving a false idea of the dense network of pyroxenite and gabbro dikes crosscutting the peridotites in the Haliakon River section of the Vourinos ophiolite (Thayer, 1967, p. 231). And, needless to say, the diabasic sheeted dike complexes were at that time totally missing from the descriptions. Last but not least, the "concordant basal contact of ophiolites with their sedimentary floor" was questioned, re-examined and found to be systematically tectonic, often marked by a sole of severely crushed mylonitic serpentinites.

All these objections were discussed by Vuagnat (1963), who remained, however, extremely prudent in his conclusion. For him, and for Aubouin as well in his famous "Geosynclines" book (Fig. 4D), there were no known alternatives to the huge submarine outflows theory. By this time it was the 1960s, and plate tectonics theory was about to emerge.

The reader may be surprised by the absence of North American examples in this first part of this review. As already pointed out by Coleman (1977) and by Nicolas (1989), this is because the European concept of an ophiolitic association was totally unsuccessful on the other side of the Atlantic. For instance, Hess (1955) summarized perfectly the common opinion of North American geologists. He recognized that Steinmann's concepts could be of some utility, but thought that gathering all "greenrocks" under the label ophiolites could lead to unfortunate confusion. Mapping plutonic and volcanic rocks together as a single ophiolitic formation could prevent recognizing time relationships of its various members with the orogenic tectonic cycle.

POLYGENETIC MODELS FOR OPHIOLITES

A First Opposition to the Comagmatic Models: The Concept of Solid Slices of Upper Mantle (de Roever, 1957)

Under a somewhat provocative title, the Dutch geologist de Roever published (in German), a paper whose title may be translated as: "Could the alpine-type peridotite massifs be tectonic fragments of the peridotite layer?" (de Roever, 1957,

p. 137). In this benchmark paper, de Roever developed his arguments in five points:

(1) He noted first that none of the common hypotheses proposed for the emplacement of alpine-type peridotites was satisfactory: *(a)* the intrusion of an ultrabasic magma had to be discarded since Bowen and Tuttle's (1949) experiments; *(b)* the intrusion of a crystal mush (the hypothesis defended by Bowen and Tuttle) was not likely, because of intact magmatic layering in the plutonic sequence; *(c)* the extrusion and differentiation of a basaltic magma was doubtful, because of the preponderance of peridotites to gabbros (as pointed out by Hess, 1938); and *(d)* metasomatic processes would require very specific conditions, obviously not realized in ophiolites. And de Roever cited Hess's sentence about the origin of alpine-type peridotites: "So the problem remains unsolved: some vital piece of evidence is still missing" (Hess, 1955).

(2) For de Roever, only one explanation remained possible: that alpine-type peridotites were simply tectonic fragments of the "peridotite layer," emplaced in the solid state by tectonic movements of huge extent and scale.

(3) This hypothesis had the merit of explaining a number of facts otherwise difficult to understand, in particular: *(a)* the metamorphic and plastic flow structures commonly observed in these peridotites, which must have developed at depth in mantle conditions; *(b)* the absence of thermal metamorphism at the contact of these rocks, since they were emplaced tectonically as cold bodies; *(c)* the consistently tectonic nature of these contacts; *(d)* the absence of ultrabasic dikes in non-ultrabasic rocks; and *(e)* the common association of ultrabasic rocks with amphibolites and other crystalline schists, supposed by de Roever to have been dragged from the metamorphic basement of geosynclines.

(4) For de Roever, the seismic structure of some oceanic basins, such as the Tonga trench, showed that peridotitic material occurred at depths of only a few km.

(5) Concerning the ophiolitic association:

> The volcanic pillow lavas of Steinmann's trinity would be a normal and concomitant product of the process explained here: this basaltic magma would be produced by melting of the peridotite layer, and extruded upward during tectonic movements of the upper part of that layer. (translated from de Roever, 1957, p. 140)

Gravity Anomalies of the Troodos Massif in Cyprus

In 1963, Gass and Masson-Smith published a long and detailed paper describing the geology and the gravity anomalies recorded in the Troodos complex of Cyprus. A strong gravity anomaly, mainly between 100 and 250 mgal, covers the whole island, and its axis lies right over the Troodos massif. This is by far the strongest gravity anomaly recorded in the eastern Mediterranean region, probably due to an extensive slab of high-density rock, ~15 km thick, underlying the island at a shallow depth.

For these authors, geological and geophysical evidence "suggest that this slab was once part of the upper mantle underlying an oceanic area between the African and Eurasian continents" (Gass and Masson-Smith, 1963, p. 419). Later, during the Alpine orogeny, this slab of mantle was underthrust by the edge of the African continent and thereby raised to its present level in the upper part of the crust (Fig. 5). In their conclusions, the authors proposed that the Troodos massif evolved in pre-Triassic times as a thick volcanic pile in an oceanic environment. An exceptional uplift and resulting deep erosion have revealed the plutonic core and the deeper part of the volcanic mass. This Troodos Plutonic Complex

> might in fact represent the sub-Mohorovicic peridotitic material which has partially fused to provide the volcanic material of the Sheeted Intrusive Complex and the Pillow Lava Series. It is also thought that while this sub-Mohorovicic material was either partially or completely fused, it was mobile enough to allow differentiation into a roughly stratiform complex ranging in composition from dunites and peridotites to overlying gabbros and granophyres. (Gass and Masson-Smith, 1963, p. 464–466)

For the first time, a direct correlation was claimed between the ultrabasic rocks of an ophiolite (the Troodos massif, curiously not yet recognized as an ophiolite in this paper) and a high-density material of "sub-Mohorovicic origin" (clearly upper mantle material), causing the main positive gravity anomaly. The Troodos ophiolite was thought to represent a pre-Triassic oceanic island eroded down to its roots, and was compared to the Hawaiian intra-oceanic islands, rather than to an oceanic ridge. In this study, high-temperature plastic deformation was not described in the lower harzburgitic unit.

Discovery, First Description, and Mapping of a Mantle Sequence in Ophiolite Complexes

A few years later, between 1968 and 1973, intensive field studies and mapping in a number of ophiolites around the world revealed the existence of a systematic and fundamental discontinuity in the plutonic sequence, and the presence of mantle rocks in an ophiolitic association. These studies were conducted mainly in the Mesozoic "Tethyan ophiolites" distributed along the alpine mountain ranges, in Italy, Greece, Cyprus, Turkey, and Oman. During the same period new stimulating data and ideas also came from remote ophiolites of the circum-Pacific orogens, such as the Papua ultramafic belt and the New Caledonia peridotite nappe. New observations from the lower Paleozoic ophiolites in the northern Appalachians, such as the Bay of Islands ophiolite in Newfoundland, Canada, also contributed significantly to our understanding of the internal structure of ophiolites.

The Mesozoic Tethyan Ophiolites

The Troodos Massif in Cyprus

Five years after the paper by Gass and Masson-Smith, Gass (1968) published a new paper in *Nature* titled: "Is the Troodos Massif of Cyprus a fragment of Mesozoic ocean floor?" In this paper, the age of the Troodos submarine lavas was corrected and attributed to the Upper Cretaceous, in agreement with the Upper

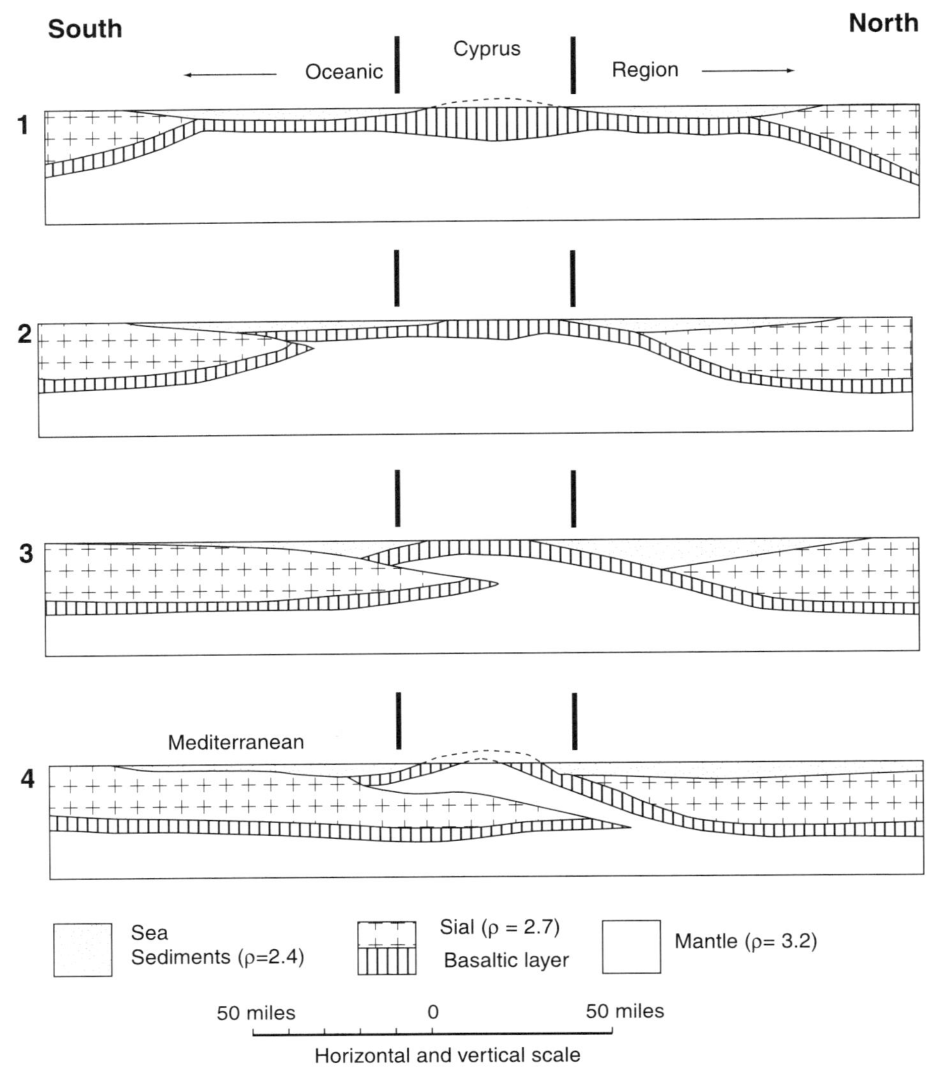

Figure 5. Gass and Masson-Smith's model (1963), showing how the Troodos ophiolite in Cyprus is rooted in a mantle slab, detached during the progressive underthrusting of Cyprus beneath the northern edge of the African continent. (1) Formation of the Troodos massif as a volcanic pile in an oceanic area between the continental masses of Africa and Eurasia (Pre-Triassic). (2) Convergence of Africa and Eurasia during the Alpine earth movements. (3) Further convergence, the African mass underthrusting the Troodos massif from the south. Possible emplacement of the Troodos serpentine at that stage. (4) Final stage–The African mass has completely underthrust Cyprus while the Eurasian mass has moved southwards compressing the rocks of northern Cyprus. At that time the Troodos ophiolite was compared to an intra-oceanic volcanic island, not to an oceanic spreading center. Redrawn from Gass and Masson-Smith (1963, p. 465).

Cretaceous microfaunas (radiolarians) found in manganese-rich cherts interbedded with the pillow lava flows (Mantis, 1970). The tholeiitic nature of the extrusive sequence, the spectacular sheeted dike complex, and the associated magnetic anomalies were the main lines of evidence used by Gass to propose that the Troodos Massif was generated by seafloor spreading at an oceanic ridge, and represented a piece of the Mesozoic Neo-Tethyan oceanic crust and mantle. A mantle sequence, however, was not identified or described as such in this paper.

The actual distinction of a mantle sequence underlying a crustal sequence in the Troodos Massif was proposed three years later by Moores and Vine (1971). For the first time, isoclinal folds, high-temperature foliations, and lineations were described and mapped in the harzburgites and dunites (Fig. 6A), and preliminary petrofabric diagrams of these rocks were presented. Moores and Vine interpreted these rocks as metamorphic tectonites "representing upper mantle material which has flowed in a solid state upward into an opening crack" (Moores and Vine, 1971, p. 461). Their interpretative cross-section of the Troodos Complex clearly showed a residual upper mantle sequence, overlain by the magmatic oceanic crust including gabbro, diabase and extrusive units. The Mohorovicic discontinuity (Moho) was indicated for the first time in an ophiolite section, separating the mantle and crustal sequences (Fig. 6B).

The Vourinos Ophiolite in Greece

In 1969, Eldridge Moores published a monograph on the "Petrology and structure of the Vourinos ophiolite complex of northern Greece." Moores improved Brunn's excellent earlier work by adding systematic mapping of all the internal structures visible in various petrological units of the complex. Banding, layering, plastic fold axes, and all kinds of dikes were care-

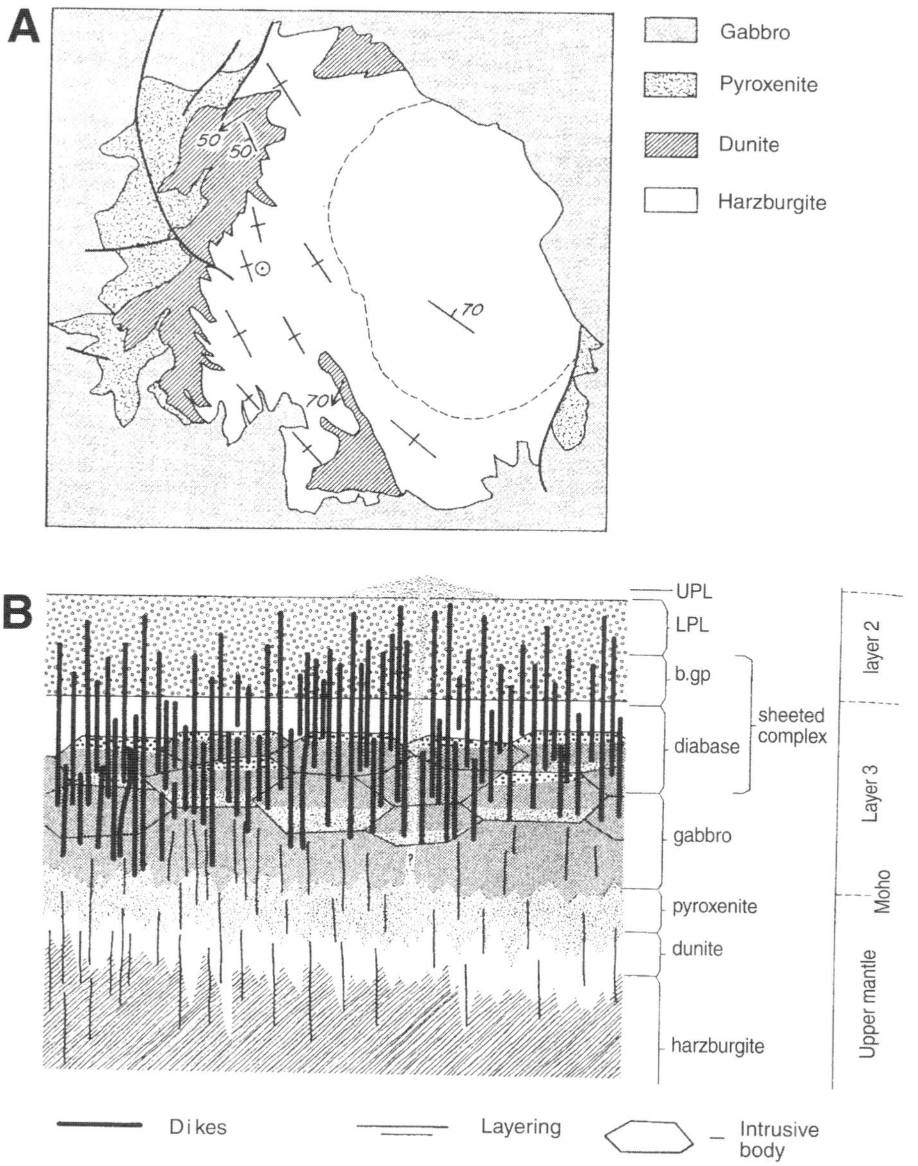

Figure 6. Mapping the mantle in the Troodos ophiolite, Cyprus. A: Moores and Vine were the first to map the internal structures of the "tectonites" in the Troodos ophiolite, showing the steep to vertical attitude of the foliations and lineations in harzburgites and dunites, interpreted as residual mantle rocks. Redrawn after Moores and Vine (1971, p. 454). B: The interpretative cross-section of the Troodos ophiolite is correlated with the seismic layers of the oceanic crust. The Moho discontinuity (seismic Moho) is shown at the boundary between ultramafic rocks and gabbros. The concept of a "petrological Moho" (boundary between residual mantle and magmatic crust), will appear later. UPL—upper lower pillows; LPL—lower pillow lavas; B.gp—basal group. Redrawn after Moores and Vine (1971, p. 462).

fully documented on a geological map. Moores interpreted the "protoclastic" textures (Fig. 7B), foliations, lineations, and isoclinal foldings generally observed in the lower harzburgites and dunites, making up two-thirds of the complex (Ultramafic Zone), as high-temperature plastic deformation fabrics. The other parts of the Vourinos complex (Transition Zone, Mafic Zone) were devoid of such high-temperature deformation fabrics.

The hypothetical reconstruction of the Vourinos Complex by Moores gave a totally new view of this ophiolite complex (Fig. 7A). Undeformed magmatic layering of the mafic rocks, once restored to horizontal position, rested discordantly over a funnel-shaped ultramafic complex, with its layering, foliation and fold axes plunging from the ends of the mass toward the center, and intruded by abundant vertical pyroxenite dikes. There was now a fundamental discontinuity between the highly deformed ultrabasic rocks (mainly harzburgites and dunites) and the overlying magmatic sequence (including pyroxenites, gabbros, diorites, diabases, and basalts).

Evaluating the volumetric ratio of each rock type, Moores showed that the calculated bulk composition of the Vourinos ophiolite was close to that of an average pyrolitic lherzolite. In the final discussion, Moores refuted Brunn's model of a large submarine overflow. For him, much evidence of high-temperature flow layering in the Ultrabasic Zone pointed to metamorphic differentiation during solid flow processes, probably in the mantle. The Vourinos complex was the result of partial melting

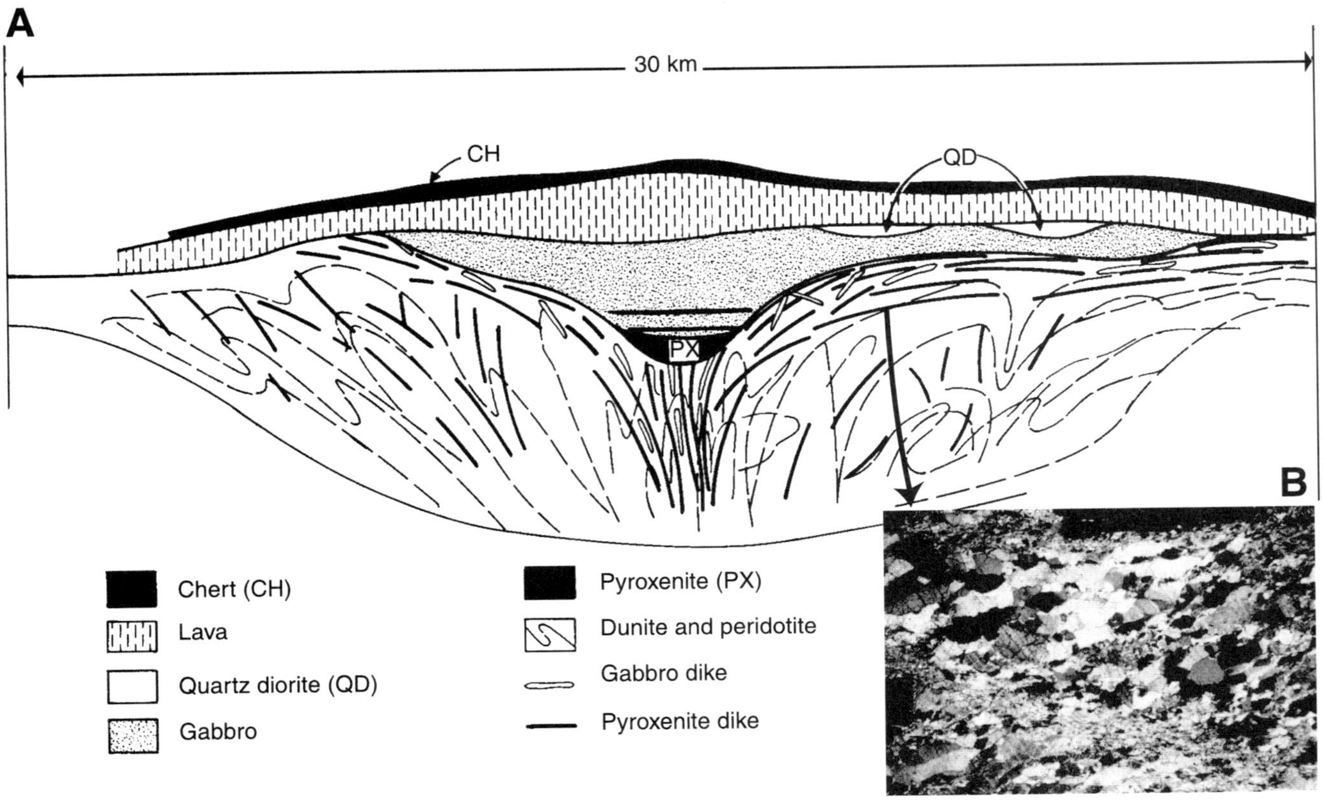

Figure 7. A: Moores' reconstruction of the Vourinos ophiolite complex in Greece showed clearly for the first time the structural discordance between foliated "tectonites" and undeformed "cumulates". When restored to an original horizontal attitude, the layered magmatic sequence rests over a funnel-shaped ultramafic complex, with its layerings, foliations and fold axes plunging from the ends of the mass toward the center, and intruded by abundant vertical pyroxenite dikes. Redrawn after Moores (1969, p. 34). B: Metamorphic texture of a typical "protoclastic" harzburgite from the Vourinos ophiolite with strained, strongly oriented olivine (with abundant kink-bands) and enstatite (with bent diopside lamellae) crystals, and disseminated chrome spinel grains. This is one of the very first published collection of microphotographs of a mantle tectonite. Taken from Moores (1969, p. 17).

and magma segregation-migration processes, starting from a lherzolitic parent rock.

The Antalya Ophiolite in Turkey

When Eldridge Moores was finishing remapping the Vourinos ophiolite in Greece, I had been mapping the Antalya ophiolites in Turkey for several years (Fig. 8A). The Cretaceous Antalya ophiolites are highly tectonized and dismembered. The whole extrusive unit and most of the sheeted dike complex are lacking, as in many ophiolites of the Taurus belt. Thick piles of submarine pillow lavas of alkali basalt, interbedded with pelagic and terrigenous sediments of an Upper Triassic age, are tectonically associated with the ophiolites. It took me a long time to realize that these extrusive rocks were a result of the early rifting stages associated with the opening of the Neo-Tethys, and therefore had nothing to do with the plutonic rocks splendidly exposed along the coastal cliffs of the Gulf of Antalya (Fig. 8B).

By the end of my fifth field season in the summer of 1969, I had many problems with Brunn's model. At that time, I was still thinking that the whole complex had the same age as the pillow lavas, that is, Upper Triassic. My main concern, however, was the fundamental discontinuity that I had found and mapped between the ultrabasic rocks (mainly harzburgites, dunites and chromitites) and the overlying layered gabbro sequence (Figs. 9A–9C). The former were strongly affected by high-temperature plastic deformation, showing metamorphic foliations, lineations, and isoclinal folds, clearly visible in the outcrops. Under the microscope, typical porphyroclastic to blastomylonitic textures were widespread (Figs. 9D and 9E), with extremely deformed and stretched olivine and orthopyroxene porphyroclasts having aspect ratios reaching 30:1. In total contrast to these rocks, the overlying layered ultramafic and gabbroic rocks showed delicate, absolutely undeformed cumulate structures and textures (Figs. 9F and 9G).

I was so impressed by these differences that I published (in French) a paper whose title translated into English is: "Petrogenesis of the ophiolites of the Antalya nappes (eastern Lycian Taurus, Turkey): their relation to an oceanic seafloor spreading

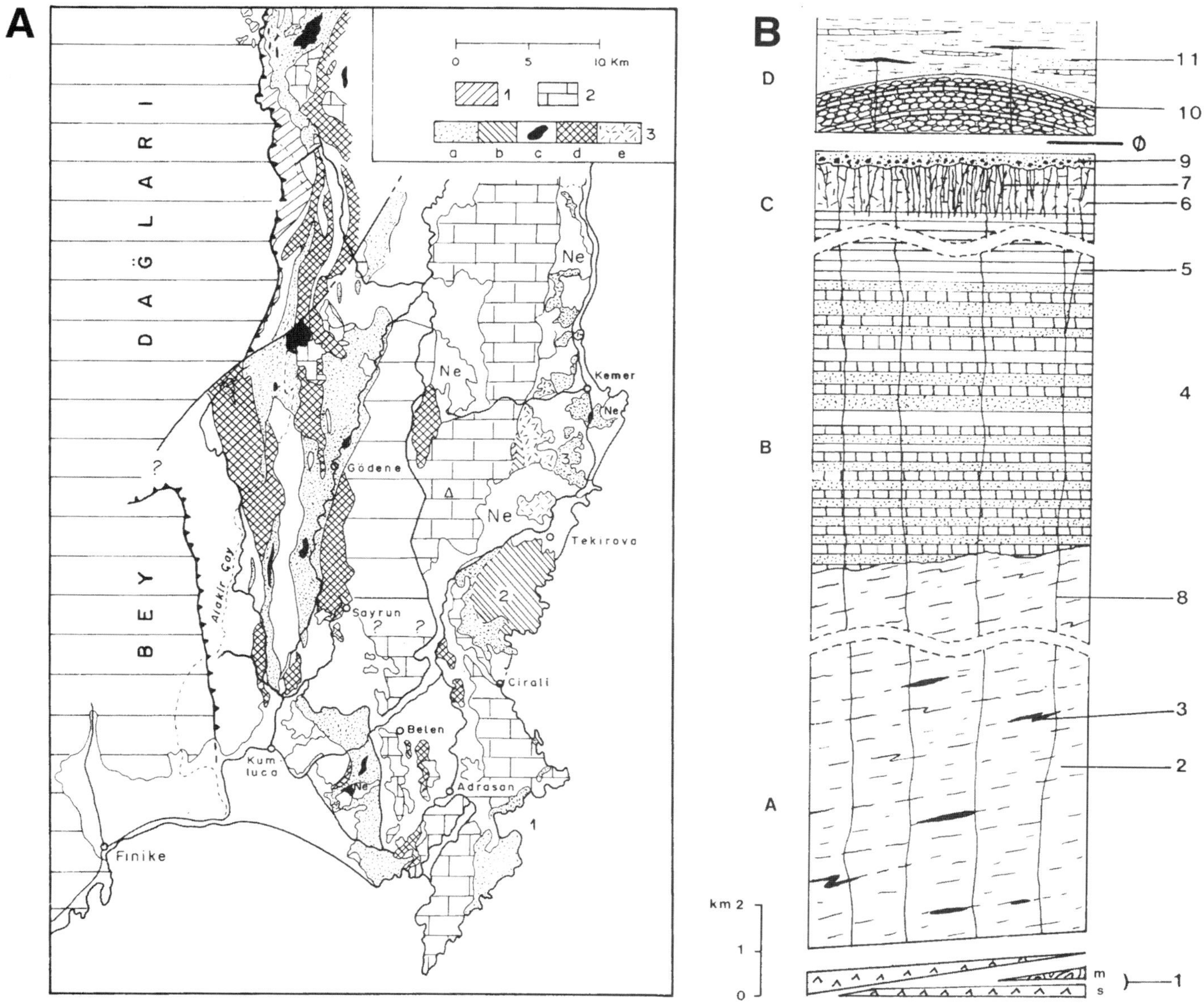

Figure 8. The highly-dismembered Antalya ophiolite complex (Gulf of Antalya, Turkey). A: Simplified geological map of the south-western segment of the Antalya nappes, showing the distribution of the ophiolitic facies in the median nappe. (1) Lower Antalya Nappe. (2) Upper Antalya Nappe. (3) Median Antalya Nappe, with (a) mantle tectonites (mainly harzburgites); (b) layered, undeformed ultramafic-gabbroic cumulates; (c) massive gabbros; (d) Upper Triassic pillow lavas (alkaline basalts); (e) diabasic sheeted dike complex of Kemer. White areas—Alakir Cay sedimentary unit. Ne—Neogene cover. B: Synthetic section of the Antalya ophiolite (Turkey). This Upper Cretaceous tholeiitic ophiolite is exposed along the Gulf of Antalya in the coastal massifs of Adrasan (A), Cirali-Tekirova (B) and Kemer (C), and also along the Alakir Cay valley, in tectonic contact with older alkaline submarine basalts and associated sediments of Upper Triassic age (D). From bottom to top: (1) metamorphic slices (m) between serpentine slices (s); (2) fresh or serpentinized harzburgites and dunites with porphyroclastic structures (tectonites); (3) podiform chromitites with dunitic matrix, deformed by plastic flow (tectonites); (4) undeformed magmatic cumulates, composed of several hectometric megacycles: layered dunites (stippled), wherlites/pyroxenites (vertical dashes) and gabbros (white); (5) upper layered gabbros; (6) massive hornblende gabbros, diorites and plagiogranites, host-rocks of the dike complex; (7) dike complex: diabases, quartz diabases and plagiogranites; (8) isolated diabase dikes cross-cutting all plutonic rocks; (9) clastic breccias and sediments, reworking ophiolitic debris and Rudist fragments (Upper Cretaceous); (10) submarine alkaline basalts (mainly pillow lavas), with interbedded *Halobia* limestones (Upper Triassic); (11) Alakir Cay unit (sediments of Upper Triassic to Upper Cretaceous age), including cherts, radiolarites, micritic limestones and terrigenous sandstones, intruded by dikes and sills of alkaline basic lavas. Redrawn after Juteau (1975).

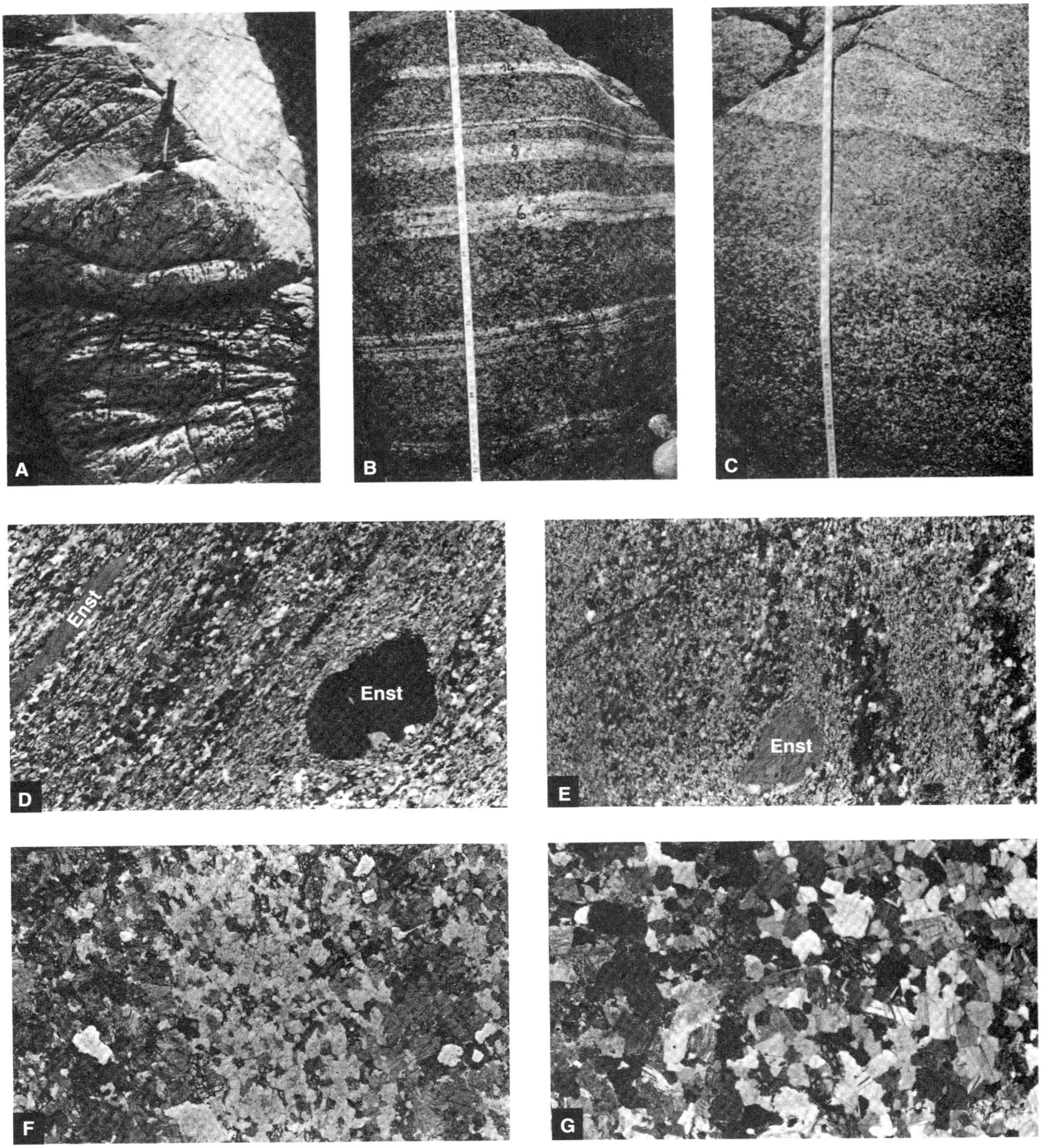

Figure 9. Contrasting rock types in the plutonic sequence of the Antalya ophiolite (Turkey). A: Harzburgite outcrop (tectonite), showing complex plastic deformations (Adrasan massif). B: Layered gabbro (thin, contrasted layers), showing delicate magmatic structures devoid of any deformation (Cirali-Tekirova massif). C: Layered gabbro showing thick, poorly contrasted undeformed layers (Cirali-Tekirova massif). D and E: Typical porphyroclastic, well foliated harzburgites, with deformed enstatite porphyroclasts (Enst) and granulated matrix. F and G: Undeformed layered gabbros, showing delicate magmatic layering. (Taken from Juteau, 1970, Plates 1 and II.).

episode of upper Triassic age" (Juteau, 1970). In this paper, I described the polygenetic nature of the Antalya ophiolites and the fundamental discontinuities existing between them. I presented the best pictures I had, both of the outcrops and under the microscope, to characterize the ultrabasic *tectonites* and the gabbroic *cumulates*.

In the discussion chapter, I wrote:

These data suggest that harzburgites and layered gabbros represent two complementary products resulting from the partial melting of a "primitive peridotite": a refractory, unmelted product, showing the imprints of intense tectonic stresses imposed on these rocks during the melting episode (or episodes), and a magmatic product, progressively extracted by filter-pressing processes along intergranular surfaces, and concentrated in "reservoir zones" where it could crystallize quietly, with rhythmic settling of the mineral phases. (translated from Juteau, 1970, p. 281)

Finally, I concluded:

After Hess (1955), de Roever (1957), Rost (1959) and more recently Reinhardt (1969), I support the idea that the "ophiolitic association" is a *polygenetic assemblage*, composed of plutonic rocks (peridotites and gabbros) directly issued from the upper mantle in a solid state, and of volcanic extrusive rocks (pillow lavas) or intrusive rocks (dolerites-diabases), issued from the same basaltic magma. (translated from Juteau, 1970, p. 282)

The last part of the paper was a comparison with the present-day oceanic ridges; this last sentence stated:

The ophiolitic assemblages of the western Taurus represent tectonic fragments of the fossil Tethysian oceanic crust, slowly transported toward the continental margin by seafloor spreading, and incorporated to the Tauric geosyncline during the main compressive movements of the alpine orogenesis. (translated from Juteau, 1970, p. 286)

Later, Adolphe Nicolas, spending a summer in Antalya with me and my students, opened my eyes to the applications of structural and petrofabric analysis as a powerful tool to analyze the effects of solid-state deformation in mantle rocks (Nicolas et al., 1971, 1972, 1973). Pioneering experimental studies on deformation of olivine and olivine aggregates had recently opened the way to the kinematic analysis of high-temperature plastic flow in peridotites (Raleigh, 1968; Carter and Avé Lallemant, 1970; Avé Lallemant and Carter, 1970; Nicolas et al., 1973). After thousands of measurements in the field and detailed petrofabric studies in the laboratory, we published the first structural model of a fossil oceanic ridge as inferred from an ophiolite complex, in the GSA Bulletin (Juteau et al., 1977) (Fig. 10).

The Semail Ophiolite in Oman

In 1969, the Swiss Society of Mineralogy and Petrography published a Memoir by B.M. Reinhardt, entitled: "On the genesis and emplacement of ophiolites in the Oman Mountains geosyncline." Reinhardt was a member of the Glennies' team of geologists, who had mapped the Oman Mountains for the Dutch Shell Oil Company. At that time, very few people had heard about the "Semail nappe," this giant piece of Upper Cretaceous ophiolite (Fig. 11A), well preserved in the desert mountains of the Sultanate of Oman.

In his memoir, Reinhardt gave a concise description of this complex and proposed an original interpretation, based on a comparison of this assemblage with modern oceanic ridges (Fig. 11B). Reinhardt concluded that the ophiolitic association of the Oman mountains was a polygenetic association of igneous and metamorphic rocks. For him, only the extrusives (pillow lavas) and subvolcanic intrusives (diabasic dikes) had been emplaced as liquid magmas, while all the other members of the assemblage (e.g., peridotites and gabbros) were uplifted from depth in a solid state. In his model, the peridotites were the oldest rocks of the association. Their mineralogy indicates that they were formed at 1400 °C and 10–20 kb, between 30 and 60 km depth. The gabbros are intrusive into the peridotites, but they were crystallized in much shallower crustal conditions. This implies an important age gap between the gabbros and the peridotites, estimated to have a pre-Permian age by the author. The gabbro-diabase-basalt suite is comagmatic; however, the considerable differences in grain sizes and textures of these rocks indicate that they crystallized under conditions which are not compatible with the huge submarine flows model.

Reinhardt's conclusion was that this polygenetic model should be similar to the internal differentiation at an oceanic ridge. The constructional stage would be followed by the "eversion of the oceanic floor" (Fig. 11B), then orogenic movements (Fig. 11B, a–e) would produce large tectonic thrust nappes. The mechanism of this "eversion" process (the term was created by Reinhardt, 1969) remains rather obscure, and is one of the weak points of Reinhardt's model. Curiously also, the word "mantle" is not mentioned in this paper.

Ligurian Ophiolites in the Northern Apennines, Italy

Ligurian ophiolites were studied again at the end of the 1960s, and re-interpreted as tectonic fragments of the ocean floor of the "northern Apennine geosyncline." Among the many papers published during this period, the most suggestive one was Bezzi and Piccardo's review of the structural features of the Ligurian ophiolites, published by the Geological Society of Italy (Bezzi and Piccardo, 1971).

In this paper, the authors described the peridotite-gabbro association in Western and Eastern Liguria. Although the Western and Eastern Ligurian associations displayed different metamorphic and tectonic features, they both included the following two main petro-structural units: (1) metamorphic lherzolites, showing penetrative deformations in outcrops (e.g., foliations and isoclinal folds) and under the microscope (porphyroclastic textures); and (2) peridotites and gabbros with relics of cumulus textures and structures, generally devoid of high-temperature penetrative deformation. In places

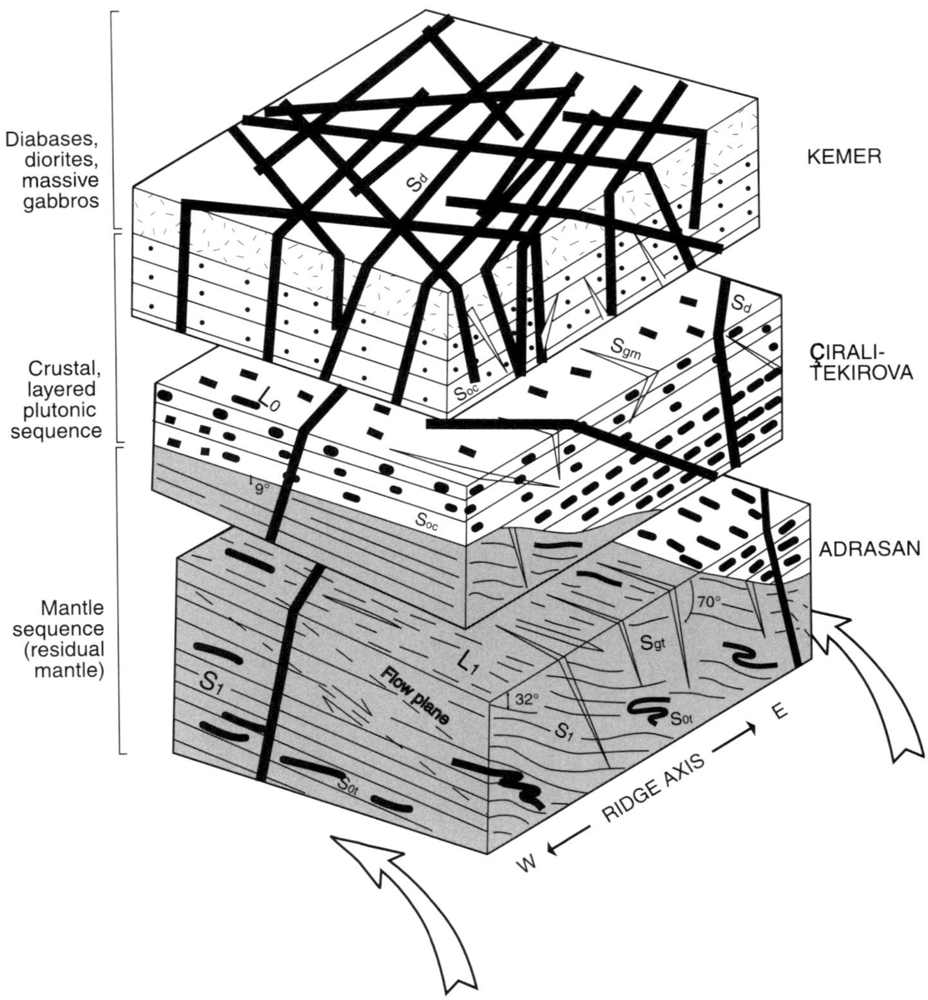

Figure 10. Structural model for an oceanic spreading center, established on the basis of the detailed field study and structural data collected along the coastal massifs of the Antalya ophiolite. The plane of magmatic accumulation (S0c) within the layered cumulates has been set back to the horizontal. The axis of the fossil ridge is considered to be perpendicular, within this horizontal plane, to the mineral lineation (L1) observed within the foliation plane (S1) of the harzburgitic tectonites. L1, supposed to mark the direction of asthenospheric plastic flow, is parallel to the fold axes formed during the pyroxenitic differentiations (S0t) accompanying the plastic flow. Small gabbro and pyroxenite dikes (Sgt), formed by partial melting during the mantle flow, are injected along planes in zone with L1. Numerous diabase dikes (Sd) crosscutting the whole complex have various directions. (Redrawn from Juteau et al., 1977).

the lherzolites contained swarms of thin plagioclase veins, similar to those described in the Lanzo lherzolitic massif by Nicolas (1969), and interpreted as the products of mobilization and segregation of gabbroic material by partial melting of a parental lherzolite. Here also, pictures taken in outcrops and under the microscope showed striking differences between lherzolitic tectonites and mafic/ultramafic cumulus rocks, generally separated by sharp contacts.

According to these authors, "the lherzolites and related rocks may represent more or less recrystallized primary material from the upper mantle, deformed and re-equilibrated during its tectonic evolution" (Bezzi and Piccardo, 1971, p. 58). They concluded that

> ...these new data favor the hypothesis that the Ligurian mafic and ultramafic rocks, characterized by the association of basic extrusive rocks, gabbro-peridotite cumulates and ultramafic tectonites, are slices of the ophiolitic material derived from the lower crust and upper mantle, and form the oceanic basaltic oceanic crust, that makes up the ancient ocean basin floor. (Bezzi and Piccardo, 1971, p. 60)

Mesozoic Papuan Ophiolites in the Circum-Pacific Orogen

At the 23rd IGC meeting held in Prague in 1968, the Australian geologist H.L. Davies presented new field data and interpretations for the Papuan Ultramafic Belt. This 400-km-long and 40-km-wide remote and exotic ophiolite belt in the wild tropical forests of Papua-New Guinea was totally unknown to most of the participants. It had been mapped starting in 1952 by the Bureau of Mineral Resources geologists, and since 1965 by a team including H.L. Davies, J.E. Tompson, J.W. Smith, D.H. Green and D.B. Dow.

Davies (1968) was extremely clear in his presentation:

> The Ultramafic Belt is thought to be a segment of oceanic upper mantle and crust which has moved westward since the Cretaceous, and has been forced to the earth's surface by collision with the sialic core of Papua-New Guinea. As the mantle segment rode up over the sialic block, it was tilted and partly eroded. The result is an exposed section (from west to east) of *(a)* sialic core, separated by a thrust fault from *(b)* ultramafic rocks (upper mantle), *(c)* gabbroic rocks (lower oceanic crust) and *(d)* Cretaceous basaltic lavas (upper oceanic crust). (Davies, 1968, p. 211)

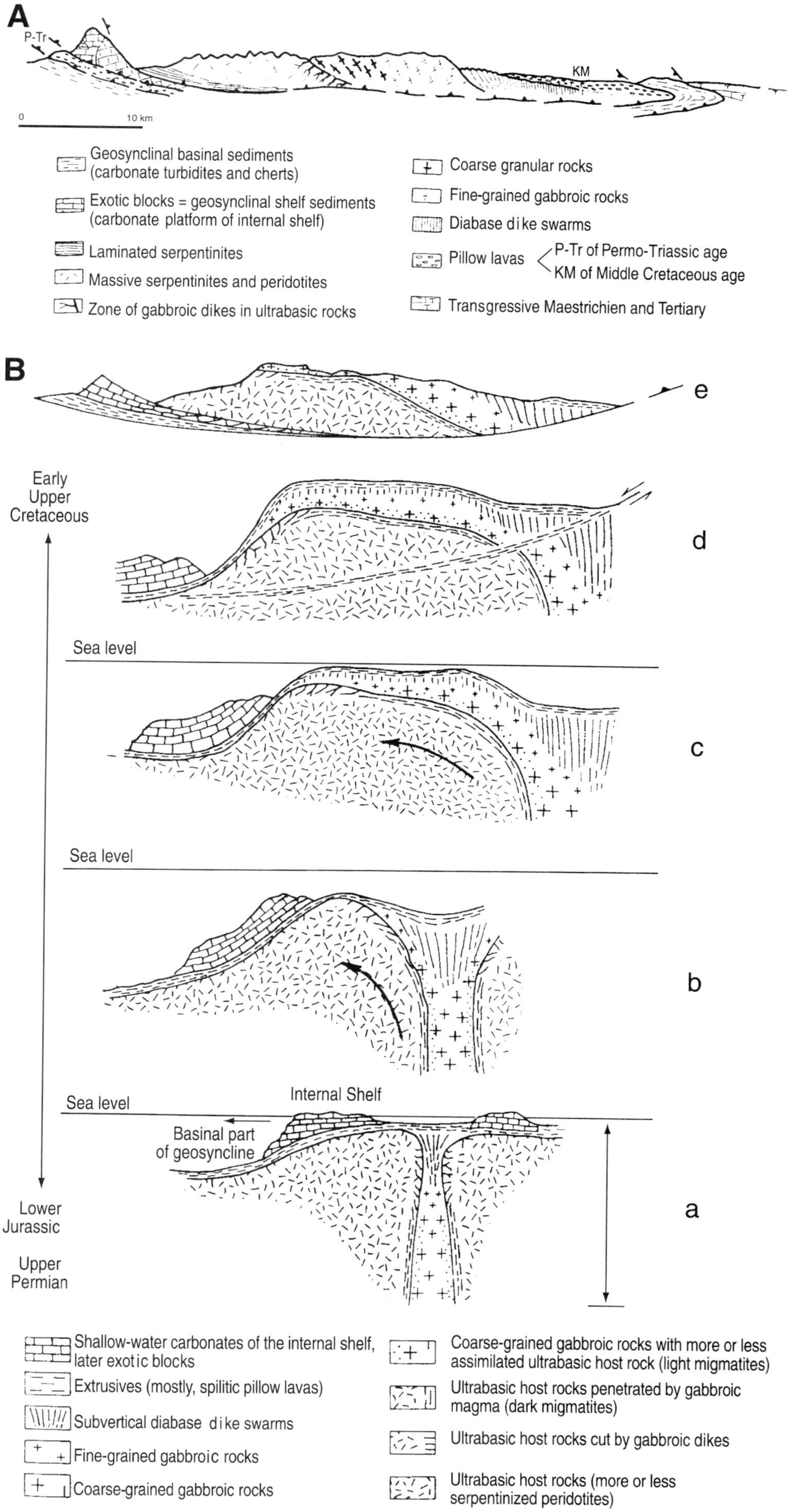

Figure 11. A: Schematic cross-section of the Semail ophiolite in Oman, redrawn after Reinhardt, 1969, p. 5. B: Model for the Oman ophiolite emplacement at a spreading ridge, followed by thrusting on a continental margin. (Redrawn after Reinhardt, 1969, p. 27).

In Davies' introduction, the ultramafic belt was interpreted as a solid fragment of oceanic lithosphere, tectonically thrust over the Papuan sialic basement (Fig. 12). The obduction concept had not yet been proposed by Coleman (1971), but H.L. Davies could have invented it!

In the belt, the ultramafic rocks make up ~50% of the association. For Davies, they represent upper mantle material:

> There is no evidence that the olivine-rich ultramafic rocks (peridotite, dunite) were ever fluid. No accumulative textures have been found and the only layering is a streaky layering probably due to shear deformation of solid rock: pyroxene grains are stretched and granulated and parallel sets of joints are developed. (Davies, 1968, p. 114)

He thought that these rocks might represent either primary mantle material, or more likely refractory peridotite material "...produced when primary pyrolitic mantle is partially melted to yield basalt magma" (Davies, 1968, p. 214). Australian geologists were, of course, aware of the experimental work performed by Green and Ringwood (1967). As in the case of the Troodos complex, preliminary gravity data conducted by P.E. St. John and J.S. Milsom showed a huge positive Bouguer anomaly over the Papuan Ultramafic Belt, extending offshore far to the NE, supporting the idea that the whole complex is a slab dipping gently (~25°) (Fig. 12).

Davies later published a complete monograph and several papers describing in detail the Papuan ophiolite belt (Davies, 1971; Davies and Smith, 1971; England and Davies, 1973), and more gravity surveys confirmed the seaward extension (St. John, 1970; Milsom, 1973a and b). In his monograph, Davies gave detailed descriptions of the high-temperature plastic deformations suffered by the "noncumulus ultramafic member" such as metamorphic foliations, protoclastic textures, abundant kink-bands in both olivine and orthopyroxene, and high-temperature annealing textures, contrasting with the delicate and undeformed structures and textures of the "cumulus member" (Davies, 1971).

Paleozoic Bay of Islands Ophiolite in Newfoundland, Canada

The Paleozoic Appalachian orogen in eastern North America includes a number of ophiolite complexes. Those outcropping in Newfoundland (Canada) are particularly well exposed (Bay of Islands, Hare Bay, Betts Cove ophiolites) and were classically interpreted as stratiform mafic/ultramafic cumulates or intrusive ultramafic bodies, emplaced during the Taconic orogeny in Late Ordovician time (Hess, 1955; Smith, 1958; Dietz, 1963; Bird and Dewey, 1970).

The mafic-ultramafic complexes of western Newfoundland are underlain by clastic sedimentary rocks, which in turn are underlain by Cambrian-Ordovician carbonates overlying the Grenville crystalline rocks of the Canadian Shield. Johnston (1941) and Kay (1945) demonstrated that the clastic series represents a transported succession. Later, Rodgers and Neale (1963) proposed that the mafic-ultramafic complexes are an integral part of the transported terrane. Stevens (1970) recognized

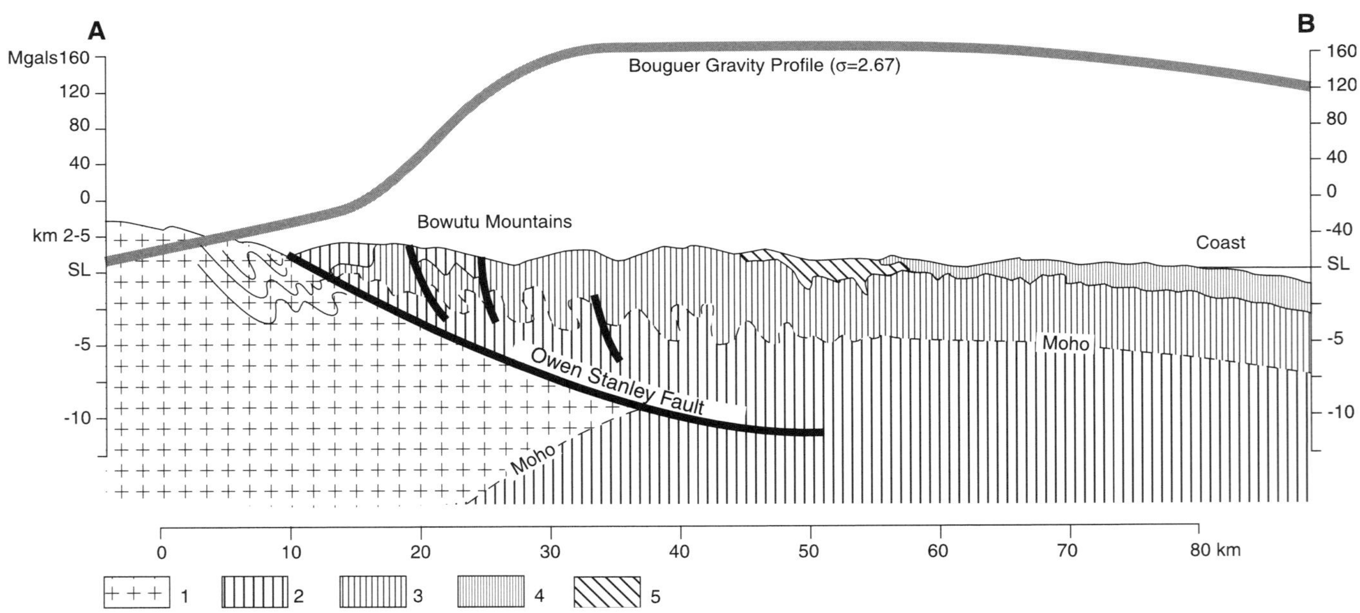

Figure 12. Schematic cross-section across the Papuan Ultramafic Belt, showing its tectonic emplacement over the Papuan shield, and its seaward extension, suggested by the Bouguer gravity anomaly profile. A—west; B—east; SL—seal level; 1—sialic core (Papuan shield); 2—ultramafic upper mantle (peridotite); 3—mafic oceanic lower crust (gabbro); 4—mafic oceanic upper crust (basalt); 5—diorite. Redrawn after Davies, 1968, p. 212.

that the mafic-ultramafic complexes, although transported, were formed in an area separated from that of the transported clastic sedimentary rocks. He demonstrated that these complexes were already tectonically emplaced by mid-Ordovician time. One year later, Church and Stevens (1971) showed very clearly that the ultramafic-mafic "igneous complexes" of the Western platform exhibit the typical ophiolitic stratigraphy, and that they were true ophiolites in the "Tethysian sense" of the term.

Further, these authors proposed that the Newfoundland ophiolite complexes represent slices of oceanic lithosphere. They stated:

> The occurrence of complex deformation structures and the presence of high-pressure mineral assemblages suggest that the peridotite is representative of the upper mantle rather than an in situ gravity cumulate. The presence of narrow high-temperature contact aureoles composed by pyroxene-garnet-amphibole-plagioclase suggests that the ultramafic was emplaced as a hot solid. (Church and Stevens, 1971, p. 1463)

The dynamothermal character of the metamorphic sole was recognized for the first time, and was compared to that of the Oman ophiolite, described by Reinhardt (1969). Church and Stevens pursued the comparison of the Bay of Islands complex with oceanic lithosphere:

> The gabbros and sheeted diabases represent layer 3, and the pillow lavas represent layer 2. At two localities at least, therefore, it is possible to directly observe the oceanic Mohorovicic discontinuity. (Church and Stevens, 1971, p. 1464)

In the same year, Williams (1971) published a long paper giving abundant stratigraphic, structural and paleontologic details confirming the allochthonous nature of the Bay of Islands complex. Also in 1971, Dewey and Bird, continuing their reflections about "mountain belts and the new global tectonics" in Dewey and Bird, 1970, proposed that the Appalachian ophiolites in Newfoundland are pieces of oceanic lithosphere, born at an oceanic spreading center, and later tectonically transported over the Canadian Shield. These new data and concepts supported the model of an opening and closing of a proto-Atlantic Ocean—an idea first proposed by Wilson (1966). Irvine and Findlay (1972), after a detailed study of the chemical and mineralogical variations in the Bay of Islands ophiolite, also concluded that the data from the lower part was consistent with a residual mantle origin.

The discovery of a mantle unit in ophiolites coincided with the emergence of the plate tectonics theory, implying seafloor spreading and giving a major role to oceanic ridges (Dietz, 1961; Hess, 1962; Vine and Mathews, 1963, Le Pichon, 1968). As early as 1959, Brunn had noted the strong lithological analogies between ophiolites and oceanic ridges (Brunn, 1959). During the 1960s, rock samples dredged along mid-ocean ridges allowed documentation of the variety of rock types represented in the oceanic crust and mantle in these zones. The whole ophiolitic suite was represented on the Mid-Atlantic Ridge or the Mid-Indian Ridge for instance: serpentinites and peridotites (dunites, harzburgites, lherzolites), gabbros of all kinds (from olivine gabbros to gabbronorites and ferrogabbros), diabases, basaltic lavas, and also a variety of metamorphic rocks such as greenschists and amphibolites (Shand, 1949; Quon and Ehlers, 1963; Muir and Tilley, 1964; Melson and Van Andel, 1966; Melson et al., 1967; Aumento, 1967; Bonatti, 1968; Engel and Fisher, 1969; Bonatti et al., 1970).

On the other hand, a growing number of seismic and gravity profiles in oceanic basins had shown the remarkable thinness of the oceanic crust (6–7 km at most) compared to that of the continental crust and its layered seismic structure. Raitt (1963) was the first to propose a subdivision of the oceanic crust into three layers, each of them characterized by its own mean seismic velocity and average thickness. He noticed that below layer 1 (made of a variable thickness of sediments), layers 2 and 3 presented almost identical thickness and seismic velocity ranges in all oceans, whatever the spreading rate. The petrographic nature of these two fundamental layers of the oceanic crust were assigned to oceanic basalts for layer 2 and to gabbros for layer 3, by compiling the numerous data obtained by dredging and by drilling, and by comparison with the ophiolitic complexes. Beyond the crust-mantle transition (oceanic Moho), the seismic velocities in the upper-mantle were compatible with a peridotitic composition.

The benchmark paper equating ophiolites with oceanic crust and upper mantle was Dewey and Bird's synthesis (1970) respecting mountain ranges and the new global tectonics. Basing their argument on geophysical data and on petrological similarities between ophiolites and oceanic hard-rock samples, they concluded that nothing distinguishes the section of the oceanic crust and upper mantle from the section of an ophiolite such as the Troodos ophiolite for instance (Gass, 1968). For Dewey and Bird, the ophiolitic association is produced presently all along the mid-oceanic ridges. The whole oceanic crust is nothing but an immense ophiolitic assemblage, and conversely, any ophiolite complex met in a folded mountain range should represent a piece of fossil oceanic lithosphere.

Many authors reexamined the contacts of ophiolites with their country rocks and demonstrated that the basal contact is always tectonic, commonly marked by the presence of metamorphic soles; this confirmed that ophiolites are basically allochthonous bodies of oceanic rocks. At that time R.G. Coleman (1971) proposed the term *obduction* for the tectonic thrusting of oceanic lithosphere onto a continental margin.

CONCLUSIONS

Ophiolites were identified and defined nearly two centuries ago (Brongniart, 1813) in the French-Italian Alps and in the Apennines. This was probably the worst place to investigate ophiolites, since they are strongly tectonized, metamorphosed, folded, and commonly upside down in those mountains!

For a century and a half, until ca. 1963, a number of purely comagmatic models were proposed to explain the ophiolitic associations. From Lotti's submarine outflows (1886a) to Staub's laccoliths (1924) or Steinmann's placcoliths (1927), from Routhier's "volcano-plutons" (1946a) to Brunn's "submarine ophiolitic outflows" (1956,

1960), the basic idea was that the puzzling association of volcanic and plutonic rocks in ophiolites was the result of magmatic differentiation in an extensive and unique volume of magma. In other words, the various parts of an ophiolite complex were supposed to be comagmatic, including the predominant peridotites (Fig. 13A). The ophiolite complexes were supposed to be autochthonous with respect to their surrounding rocks, either as igneous intrusions (Staub, Steinmann), or as giant submarine outflows (Lotti, Brunn, Aubouin), or even as both (Routhier).

Around 1970, several authors described a *mantle sequence* in the ophiolite complexes they had mapped in detail (E. Moores in the Vourinos complex in Greece; I. Gass, E. Moores and

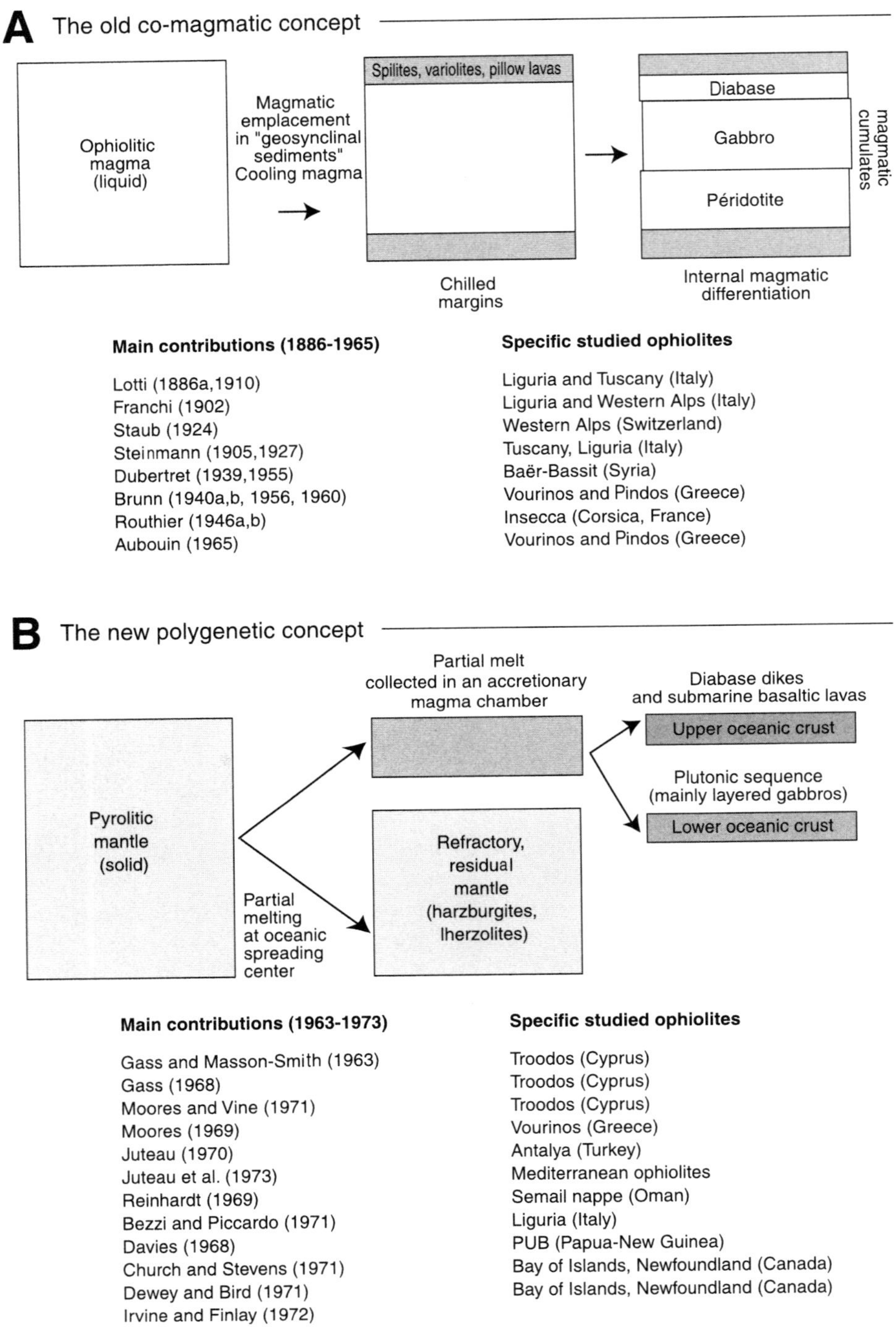

Figure 13. The old and new concepts for the genesis of the ophiolite suite, with the names of the authors and specific ophiolites that have contributed to the advancement of these concepts. A: In the old, purely comagmatic concept, the ophiolite assemblage is the result of an internal magmatic differentiation within a single, liquid mass of magma, after its intrusion or extrusion within or over "geosynclinal" sediments. B: In the new polygenetic concept, an ophiolite is a piece of oceanic lithosphere, including a refractory, residual mantle sequence, and a crustal sequence composed of a magmatic plutonic series (lower crust), and of diabases and basaltic submarine lavas (upper crust). In this concept, both sequences are the result of the partial melting of a solid mother rock, the upper mantle.

F. Vine in the Troodos complex in Cyprus; B.M. Reinhardt in the Oman ophiolite; H.L. Davies in eastern Papua, T. Juteau in the Antalya ophiolite in Turkey). This unit was characterized by high-temperature ductile deformation, and by a residual refractory composition. It contrasted markedly with the other units (gabbros, diabases, basalts), which usually did not exhibit high-temperature plastic deformation, and which were logically ascribed to a *crustal sequence*. In some cases (Cyprus, Papua), this discovery was strongly supported by gravity surveys, showing that the ophiolite is part of a mantle slab extending seaward.

With the discovery of a mantle unit in ophiolites, the purely comagmatic model was replaced by a polygenetic model, in which the residual mantle unit is the counterpart of the crustal sequence rocks, which are products of the partial melting of the mantle (Fig. 13B). This major discovery was endorsed by the 1972 Penrose Conference on Ophiolites (Anonymous, 1972), whose participants gave the modern definition of the ophiolite concept still mainly valid today, and defined the typical ophiolitic sequence. The conference participants gave strictly descriptive criteria, avoiding interpretative terms such as "oceanic crust" or "oceanic lithosphere," and even "mantle" or "crust" sequences. They emphasized the contrast between *tectonites* and *cumulates*, endorsing this fundamental intra-ophiolitic discontinuity.

We know today that things are not so simple. Many authors since then have described local or ubiquitous plastic deformation fabrics in the crustal units of several ophiolites. We know also that widespread viscous deformations affect the gabbro unit in most ophiolites, including magmatic foliations, lineations, and folds. Nobody today would describe these rocks as "undeformed." Nevertheless, these crustal deformations are generally magmatic, implying the presence of an interstitial liquid phase in a deforming crystal mush, and a strong contrast generally remains with the metamorphic mantle sequence and its plastic deformation, mainly developed in the solid state.

Finally, during this fascinating decade (1963–1973), remarkable advances in our knowledge and understanding of ophiolites, mainly due to detailed and rigorous mapping and field studies, brought considerable support to the emerging seafloor-spreading hypothesis and plate tectonics theory, which in turn provided a general frame for the comprehension of the evolution of the oceans of the past.

ACKNOWLEDGMENTS

I am extremely grateful to Jean Bédard, Alexander Mc Birney, Sally Newcomb, and Yildirim Dilek, for helping me to improve this paper on many points, and to write readable English. I am of course entirely responsible for the choice of the authors cited in this paper, selected among so many excellent publications dealing with ophiolites. I am also responsible for the translation into English of some of the quotations.

This paper is dedicated to Georges Rocci, Jan Houghton Brunn, and Adolphe Nicolas, who helped me so much when I was a young scientist.

REFERENCES CITED

Amstutz, G.C., 1980, The early history of the term ophiolites and its evolution until 1945, *in* Proceedings of the International Ophiolite Symposium: Nicosia, Cyprus, Cyprus Geological Survey Department, p. 149–152.

Anonymous, 1972, Penrose field conference on ophiolites: Geotimes, v. 17, p. 24–25.

Aubouin, J., 1965, Geosynclines: Amsterdam, Elsevier, 335 p.

Aumento, F., 1967, Magmatic evolution of the Mid-Atlantic Ridge, Earth and Planetary Science Letters, v. 2, p. 225–230.

Avé Lallemant, H.G., and Carter, N.L., 1970, Syntectonic recrystallization of olivine and modes of flow in the upper mantle: Geological Society of America Bulletin, v. 81, p. 2203–2220.

Bezzi, A., and Piccardo, G.B., 1971, Structural features of the Ligurian ophiolites: petrologic evidence for the oceanic floor of the Northern Apennine geosyncline: A contribution to the problem of the alpine-type gabbro-peridotites associations: Memorie della Società Geologica Italiana, Roma, v. 10, p. 53–63

Bird, J.M., and Dewey, J.F., 1970, Lithosphere plate: continental margin tectonics and the evolution of the Appalachian orogen: Geological Society of America Bulletin, v. 81, p. 1031–1059.

Bonatti, E., 1968, Ultramafic rocks from the Mid-Atlantic Ridge: Nature, v. 219, p. 363.

Bonatti, E., Honnorez, J., and Ferrara, G., 1970, Equatorial Mid-Atlantic Ridge: petrologic and Sr isotopic evidence for an alpine-type rock assemblage: Earth and Planetary Science Letters, v. 9, p. 247–256.

Bowen, W.N., and Tuttle, O.F., 1949, The system $MgO–SiO_2–H_2O$: Geological Society of America Bulletin, v. 60, p. 439–460.

Brongniart, A., 1813, Essai d'une classification minéralogique des roches mélangés. Journal des Mines, Paris, v. 199, p. 5–48.

Brongniart, A., 1821, Sur le gisement ou position relative des ophiolites, euphotides, jaspes, etc. dans quelques parties des Apennins: Annales des Mines, Paris, v. 6, p. 177–238.

Brunn, J.H., 1940a, Sur les roches basiques du Pinde et de la Macédoine occidentale: Comptes Rendus à l'Académie des Sciences, Paris, v. 210, p. 109–111.

Brunn, J.H., 1940b, Conditions de gisement des roches basiques en Macédoine occidentale: Comptes Rendus à l'Académie des Sciences, Paris, v. 210, p. 735.

Brunn, J.H., 1956, Contribution a l'étude géologique du Pinde septentrional et d'une partie de la Macédoine occidentale: Annales de la Géologie des Pays Helléniques, Paris, v. 7, p. 1–358.

Brunn, J.H., 1959, La dorsale médio-Atlantique et les épanchements ophiolitiques: Comptes Rendus sommaires à la Société géologique de France, v. 8, p. 234–236.

Brunn, J.H., 1960, Mise en place et différenciation de l'association pluto-volcanique du cortège ophiolitique: Revue de Géographie physique et Géologie dynamique, Paris, v. 2 (3), p. 115–132.

Brunn, J.H., 1961, Les sutures ophiolitiques. Contribution à l'étude des relations entre phénomènes magmatiques et orogéniques: Revue de Géographie physique et Géologie dynamique, Paris, v. 2 (14), p. 81–96, and 181–202.

Carter, N.L., and Avé Lallemant, H.G., 1970, High temperature flow of dunite and peridotite: Geological Society of America Bulletin, v. 81, p. 2181–2202.

Church, W.R., and Stevens, R.K., 1971, Early Paleozoic complexes of the Newfoundland Appalachians as mantle-oceanic crust sequences: Journal of Geophysical Research, v. 76, p. 1460–1466.

Çogulu, E., and Vuagnat, M., 1967, Quelques réflexions sur le masif basique-ultrabasique du Kizil Dag, Hatay, Turquie: Comptes Rendus des Séances, SPHN de Genève, NS, v. 2, p. 210–216.

Coleman, R.G., 1971, Plate tectonics emplacement of upper mantle peridotites along continental edges: Journal of Geophysical Research, v. 76, p. 1212–1222.

Coleman, R.G., 1977, Ophiolites–Ancient oceanic lithosphere?: Minerals and Rocks, Berlin, Springer-Verlag, v. 12, 229 p.

Coleman, R.G., and Irwin, W.P., 1977, North American Ophiolites, Oregon Dept. of Geology and Mineral Industries Bulletin v. 95, 183 p.

Davies, H.L., 1968, Papuan Ultramafic Belt: 23d International Geological Congress, Prague, v. 1, p. 209–220.

Davies, H.L., 1971, Peridotite-gabbro-basalt complex in eastern Papua: an overthrust plate of oceanic mantle and crust: Bureau of Mineral Resources of Australia Bulletin 128, 48 p.

Davies, H.L., and Smith, I.E., 1971, Geology of eastern Papua: Geological Society of America Bulletin, v. 82, p. 3299–3312.

Davis, E.F., 1918, The radiolarian cherts of the Franciscan Group: Bulletin of the Dept of Geology, University of California, v. 11, p. 235–432.

De Roever, W.P., 1967, Sind die alpinotypen Peridotitmassen vielleicht tektonisch verfrachtete Bruchstücke der Peridotitschale?: Geologische Rundschau, v. 46, p. 137–146.

Dewey, J.F., and Bird, J.M., 1970, Mountain belts and the New Global Tectonics: Journal of Geophysical Research, v. 75, p. 2625–2647.

Dewey, J.F., and Bird, J.M., 1971, Origin and emplacement of the ophiolite suite: Appalachian ophiolites in Newfoundland, Journal of Geophysical Research, v. 76, p. 3179–3206.

Dewey H., and Flett, J.S., 1911, On some British pillow lavas and the rocks associated with them: Geological Magazine, v. 5 (8), p. 208–209, and p. 241–248.

Dietz, R.S., 1961, Continent and ocean basin evolution by spreading of the seafloor: Nature, v. 190, p. 854–857.

Dietz, R.S., 1963, Alpine serpentines as oceanic rind fragments: Geological Society of America Bulletin, v. 74, p. 947–952.

Dubertret, L., 1939, Sur la genèse et l'âge des roches vertes syriennes: Comptes Rendus à l'Académie des Sciences, Paris, v. 209, p. 763–764.

Dubertret, L., 1955, Géologie des roches vertes du NO de la Syrie et du Hatay (Turquie): Notes et mémoires du Moyen-Orient, v. 6, p. 13–179.

Engel, C.G., and Fisher, R.L., 1969, Lherzolite, anorthosite, gabbro and basalt dredged from the Mid-Indian Ocean Ridge: Science, v. 166, p. 1136.

England, R.N., and Davies, 1973, Mineralogy of ultramafic cumulates an tectonites from eastern Papua: Earth and Planetary Science Letters, v. 17, p. 416–425.

Franchi, S., 1902, Contribuzione allo studio delle roccie a glaucophane e del metamorfismo onde ebbero origine nelle regione liguro-alpine occidentale: Bolletino della Reale Comitato Geologico d'Italia, v. 33, p. 255–318.

Gass, I.G., 1968, Is the Troodos massif of Cyprus a fragment of mesozoic ocean floor?: Nature, v. 220, p. 39–42.

Gass, I.G., and Masson-Smith, D., 1963, The geology and gravity anomalies of the Troodos massif, Cyprus: Philosophical Transactions of the Royal Society of London, A 225, p. 417–467.

Glangeaud, L., 1954, Réflexions sur les travaux de la XVe Section (palèovolcanologie et tectonique) au Congrès International d'Alger, *in*, La palèovolcanologie et ses rapports avec la tectonique, 19e Congrès Géologique International, Alger, 1952, Section XV, Fascicule XVII, p. 235–239.

Green D.H., and Ringwood, A.E., 1967, Genesis of basaltic magmas: Beiträge Mineralogie und Petrographie, v. 15, p. 103–190.

Hess, H.H., 1938, A primary peridotite magma: American Journal of Science, v. 35, 321–344.

Hess, H.H., 1955, Serpentines, orogeny and epeirogeny, *in* Poldervart, A., ed., Crust of the earth: Boulder, Colorado, Geological Society of America Special Paper 62, p. 391–408.

Hess, H.H., 1962, History of ocean basins, *in* Engel, A.E.J., et al., ed., Petrologic Studies: A volume in honor of A.F., Buddington: Geological Society of America, p. 599–620.

Irvine, T.N., and Findlay, T.C., 1972, Alpine-type peridotite with particular reference to the Bay of Islands igneous complex, *in* Irving, E., ed., The oceanic crust and identification of ancient oceanic crust on the continents: Ottawa, Ontario, Canadian Department of Energy, Mines and Resources, Special Publication, v. 42, p. 77–128.

Johnston, H., 1941, Paleozoic lowlands of western Newfoundland: Transactions of New York Academy of Science, ser. 2, v. 3, p. 141–145.

Juteau, T., 1970, Pétrogénèse des ophiolites des nappes d'Antalya (Tautus lycien oriental, Turquie). Leur liaison avec une phase d'expansion ocèanique active au Trias supérieur: Sciences de la Terre, Nancy, v. 15 (3), p. 265–288.

Juteau, T., 1975, Les ophiolites des Nappes d'Antalya, (Taurides occidentales, Turquie). Pètrologie d'un fragment de l'ancienne croûte océanique téthysienne: Sciences de la Terre, Nancy, Mèmoire no. 32, 692 p.

Juteau, T., Lapierre, H., Nicolas, A., Parrot, J.F., Ricou, L.E., Rocci, G., and Rollet, M., 1973, Idées actuelles sur la constitution, l'origine et l'évolution des assemblages ophiolitiques mésogéens: Bulletin de la Société Géologique de France, v. 7, p. 478–493.

Juteau, T., Nicolas, A., Dubessy, J., Fruchard, J.C., and Bouchez, J.L., 1977, Structural relationships in the Antalya ophiolite complex, Turkey: possible model for an oceanic ridge: Geological Society of America Bulletin, v. 88, 1740–1748.

Kay, M., 1945, Paleogeographic and palinspastic maps: Bulletin of American Association of Petroleum Geologists, v. 29, p. 426–450.

Le Pichon, X., 1968, Seafloor spreading and continental drift: Journal of Geophysical Research, v. 73, p. 3661–3697.

Lewis, J.V., 1914, Origin of pillow lava: geological Society of America Bulletin, v. 25, p. 591.

Lotti, B., 1886a, Paragone fra la roccie ofiolitiche terziare italiane e la roccie basiche pare terziare della Scozia e dell'Irlande: Bolletino della Reale Comitato Geologico d'Italia, v. 76.

Lotti, B., 1886b, Descrizione Geologica dell'Isola d'Elba, *in* Ufficio, R., ed., Memorie dexcrittive della Carta geologica d'Italia, Roma, 2.

Lotti, B., 1910, Geologia della Toscana, *in* R. Ufficio, ed., Memorie descrittive della Carta Geologica d'Italia, Roma, v. 13, p. 215–276.

Majer, V., 1960, Magmatic rocks in the region of Bassit between Latakia and Kessab in north-western Syria: Fac. Sci. Univ. Skopje, Yugoslavia, Special Edition, no. 10.

Mantis, M., 1970, Upper Cretaceous–Tertiary foraminiferal zones in Cyprus: Cyprus Res. Centre, v. 111, p. 227–241.

Melson, W.G., and Van Andel, T.H., 1966, Metamorphism in the Mid-Atlantic Ridge, 22° N latitude: Marine Geology, v. 4, p. 165.

Melson, W.G., Jarosevich, E., Bowen, V., and Thomson, G., 1967, St. Peter and Paul rocks: a high-temperature mantle derived intrusion: Science, v. 155, p. 1532.

Milsom, J., 1973a, Papuan Ultramafic Belt: Gravity anomalies and the emplacement of ophiolites. Geological Society of America Bulletin, v. 84, p. 2243–2258.

Milsom, J., 1973b, The gravity field of the Papuan peninsula. Geologie en Mijnbow, 52, p. 13–20.

Moores, E.M., 1969, Petrology and structure of the Vourinos ophiolitic complex of northern Greece: Boulder, Colorado, Geological Society of America, Special Paper 118, 74 p.

Moores, E.M., and Vine, F.J., 1971, The Troodos massif, Cyprus, and other ophiolites as oceanic crust: evaluation and implications: Philosophical Transactions of the Royal Society of London, v. 268, 443–466.

Muir, I.D., and Tilley, C.E., 1964, Basalts from the northern part of the rift zone of the Mid-Atlantic Ridge: Journal of Petrology, 5, p. 409–434.

Nicolas, A., 1969, Une vue unitaire concernant l'origine des massifs ultrabasiques des Alpes occidentales internes: Comptes Rendus à l'Académie des Sciences, Paris, v. 269, p. 1831–1834.

Nicolas, A., 1989, Structures of ophiolites and dynamics of oceanic lithosphere: Dordrecht, Kluwer Academic Publishers, 367 p.

Nicolas, A., Bouchez, J.L., Boudier, F., and Mercier, J.C., 1971, Textures, structures and fabrics due to solid state flow in some European lherzolites: Tectonophysics, v. 12, p. 55–68.

Nicolas, A., Bouchez, J.L., and Boudier, F., 1972, Interprétation cinématique des déformations plastiques dans le massif de lherzolite de Lanzo (Alpes piémontaises). Comparaison avec d'autres massifs: Tectonophysics, v. 14, p. 143–171.

Nicolas, A., Boudier, F., and Boullier, A.M., 1973, Mechanisms of flow in naturally and experimentally deformed peridotites: American Journal of Science, v. 273, p. 853–876.

Parrot, J.F., 1973, Pètrologie de la coupe du Djebel Moussa, massif basique-ultrabasique du Kizil Dag (Hatay, Turquie): Sciences de la Terre, Nancy, v. 18, p. 143–172.

Quon, S.H., and Ehlers, E.G., 1963, Rocks of the northern part of Mid-Atlantic Ridge: Geological Society of America Bulletin, v. 74, p. 1–8.

Raitt, R.W., 1963, The crustal rocks, *in* Hill, M.N., ed., The Sea, v. 3, Wiley Interscience, New York, p. 85–102.

Raleigh, C.B., 1968, Mechanism of plastic deformation of olivine: Journal of Geophysical Research, v. 73, p. 5391–5406.

Reinhardt, B.M., 1969, On the genesis and emplacement of ophiolites in the Oman mountains geosyncline: Schweizerige Mineralogishe und Petrographishe Mitteilungen, v. 49, p. 1–30.

Rodgers, J., and Neale, E.R.W., 1963, Possible "Taconic" in western Newfoundland: American Journal of Science, Memoir 1, 123 p.

Rost, F., 1959, Probleme ultrabasischer Gesteine und ihre Lagerstätten: Freiberg Forschungsh, v. 58, p. 28–64.

Routhier, P., 1946a, Vulcano-plutons sous-marins du cortége ophiolitique: Comptes Rendus à l'Académie des Sciences, v. 222, p. 192–194.

Routhier, P., 1946b, L'association Radiolarites—Roches Vertes dans les chaînes géosynclinales, Comptes Rendus sommaires à la Société géologique de France, p. 26–28, p. 66–67, and p. 108–109.

Routhier, P., 1954, Volcanicité et embryotectonique paléogènes en Nouvelle-Calèdonie, *in* La paléovolcanologie et ses rapports avec la tectonique, 19e Congrès Géologique International, Alger, 1952, Section XV, fascicule XVII, p. 43–52.

Shand, S.J., 1949, Rocks of the Mid-Atlantic Ridge: Journal of Geology, v. 57, p. 89–92.

Smith, C.H., 1958, Bay of Islands igneous complex, western Newfoundland: Ottawa, Ontario, Geological Survey of Canada Memoir 290, 132 p.

Staub, R., 1924, Über die Verteilung der Serpentine in den alpinen Ophiolithen: Schweizerige Mineralogische und Petrograpphische Mitteilungen, v. 1, p. 78–149.

Steinmann, G., 1905, Geologische Betrachtrungen in den Alpen, II: Die Schardt'sche Überfaltungstheorie und die geologische Bedeutung der Tiefseeabsätze und der ophiolitischen Massargesteine: Bericht der Naturforschenden Gesellschaft, Freiburg-in-Breisgau, v. 16, p. 44–65.

Steinmann, G., 1927, Die ophiolitische Zonen in der mediterranen Keltengebirgen: 14th International Geological Congress, Madrid, v. 2, p. 638–667.

Stevens, R.K., 1970, Cambro-Ordovician flysch sedimentation and tectonics in west Newfoundland and their possible bearing on a proto-Atlantic ocean: Ottawa, Ontario, Geological Association of Canada, Special Paper 7, p. 165–177.

St John, V.P., 1970, The gravity field and structure of Papua and New Guinea: Australian Petroleum Exploration Association Journal, v. 10 (2), p. 41–55.

Stille, H., 1938, Initialer und synorogener Magmatismus: Sitzber. Preuss. Akad. Wiss., p. 51–52.

Suess, E., 1918, La face de la Terre (*das Antlitz der Erde*), translated into French by E. Margerie, and Colin, A., ed.: Paris, 3 volumes.

Thayer, T.P., 1960, Some critical differences between Alpine type and stratiform peridotite-gabbro complexes: 21st International Geological Congress Reports, Copenhagen, v. 13, p. 247–259.

Thayer, T.P., 1967, Chemical and structural relations of ultramafics and feldspathic rocks in alpine intrusive complexes, *in* Wyllie, P.J., ed., Ultramafic and related rocks, New York, John Wiley and Sons, p. 222–238.

Vine, F.J., and Mathews, D.H., 1963, Magnetic anomalies over oceanic ridges: Nature, v. 199, p. 947–949.

Vuagnat, M., 1963, Remarques sur la trilogie serpentinites-gabbros-diabases dans le bassin de la Méditerranée occidentale: Geologische Rundschau, v. 53, 336–357.

Williams, H., 1971, Mafic-ultramafic complexes in western Newfoundland Appalachians and the evidence for their transportation: a review and interim report: Geological Association of Canada Proceedings, v. 24 (1), p. 9–25.

Wilson, J.T., 1966, Did the Atlantic close and then re-open?: Nature, v. 211, p. 676–681.

Zachos, K., 1954, Chromite deposits of Vourinon (Kozani) area, in Mineral wealth of Greece, Inst. Geol. and Subsurface Research, Athenes, v. 3, p. 1–82.

Manuscript Accepted by the Society May 2, 2003

Printed in the USA

Geological Society of America
Special Paper 373
2003

N.L. Bowen, H.H. Hess, and ultramafic rocks: Perspectives on ophiolites before plate tectonics

Davis A. Young*
Department of Geology, Geography, and Environmental Studies, Calvin College, Grand Rapids, Michigan 49546, USA

ABSTRACT

In the early twentieth century, geologists regarded ophiolites as serpentinized peridotite bodies associated with other green rocks in alpine mountain systems and island arcs. Despite the lack of contact metamorphic effects to be expected with high-temperature melts, many petrologists accepted the idea of ultrabasic magma intrusion for ophiolites. Harker and Vogt suggested that ultrabasic liquids formed by the re-melting of accumulations of olivine and pyroxene. Harker and Hess proposed that the liquidus temperatures of ultrabasic melts are significantly lowered by dissolved volatiles. Vogt maintained that the temperatures of olivine-rich melts are lowered substantially by addition of iron. Bowen, however, consistently rejected the existence of ultrabasic magma within the crust. He appealed to the lack of ultrabasic lavas, the general paucity of contact metamorphism around ultramafic plutons, and experimental data as evidence against ultrabasic magmas. In 1914, Bowen determined that forsterite melts at 1890 °C, a temperature far too high for magmas to exist in the crust. In the 1930s, Bowen and Schairer determined a significant portion of the olivine solid solution loop and, contrary to the opinion of Vogt, showed that olivine with a substantial amount of iron still requires unreasonably high temperatures to exist as melt in the crust. As a result, Bowen maintained that ultramafic plutons represent accumulations of olivine and pyroxene that were intruded as mushes of solid crystals lubricated by thin films of fluid. In his final study on ultramafic rocks with Tuttle in 1949, Bowen again showed that no liquid exists in the system MgO–SiO_2–H_2O at crustal temperatures. The mechanism of serpentinization of ophiolite also posed problems. Hess reasoned that serpentine crystallized in the waning stages of ultrabasic magma solidification. The experiments of Bowen and Tuttle, however, showed that serpentine could not crystallize directly from a magma of serpentine composition. They claimed that serpentine formed by absorption of water from wall rocks and reaction with olivine. The Bowen-Hess dispute faded with the recognition that ophiolites are fragments of oceanic lithosphere emplaced into orogens as cold slivers and that reaction of circulating sea water with new oceanic lithosphere produced the serpentinization of ophiolites.

Keywords: Bowen, Hess, serpentinite, emplacement, phase equilibria.

INTRODUCTION

The term "ophiolite" was used during the nineteenth century for serpentine-bearing rocks (e.g., Dana, 1866). Steinmann (1905) introduced the concept of an "ophiolite suite" to denote an alpine mountain belt association of serpentinized ultramafic rocks, spilitic lava flows, and other green rocks including gabbros, greenschists, and amphibolites. In addition to occurrences in the Alps, lenticular bodies of serpentinized ultramafic rocks such as dunite and peridotite were recognized in numerous alpine-type fold belts in eastern North America, New Zealand, southeastern

*youn@calvin.edu

Young, D.A., 2003, N.L. Bowen, H.H. Hess, and ultramafic rocks: Perspectives on ophiolites before plate tectonics, *in* Dilek, Y., and Newcomb, S., eds., Ophiolite concept and the evolution of geological thought: Boulder, Colorado, Geological Society of America Special Paper 373, p. 55–63. For permission to copy, contact editing@geosociety.org.

Australia, Russia, and California. Hess (1939) subsequently called attention to the presence of ophiolites in island arcs.

The serpentinized peridotites of alpine mountain systems and island arcs posed major problems. One problem concerned the mode of emplacement. Before the establishment of the theory of plate tectonics, these ultramafic rock bodies were commonly interpreted as igneous intrusions into geosynclinal sediments (e.g., Jennings and Strand, 1959). The alleged intrusions, however, generally produced little, if any, metamorphism in their host rocks. Contact alteration was commonly attributed to post-intrusion reaction between serpentine and wall rocks during the influx of secondary aqueous solutions (Phillips and Hess, 1936). Moreover, Turner (1942) and others noted the presence of pervasive tectonite fabrics in many serpentinites, a phenomenon that suggested emplacement of peridotite as a solid rather than as magma.

Another problem concerned the widespread occurrence of serpentine in alpine and island arc peridotites (Hess, 1933). In some occurrences, virtually the entire ultramafic body had been converted to serpentine. Lodochnikov (1933) believed that serpentine was a late-stage magmatic crystallization product, and Hess (1933) suggested that serpentine formed during autometamorphism of an ultrabasic intrusion that absorbed water from its surroundings. The possibility that serpentine resulted from post-emplacement hydrothermal alteration of the ultramafic rocks could not be ruled out, but the distribution of serpentine seemed to be unrelated to the contacts. Attempts to solve these two problems spawned one of the spirited petrological disputes of the first half of the twentieth century. The principals in the debate were Harry Hammond Hess (1906–1969) and Norman Levi Bowen (1887–1956).

Figure 1. Harry Hammond Hess (1906–1969). Reproduced by permission of the Archives, Department of Geosciences, Princeton University.

Hess (Fig. 1) was one of the leading students of ultramafic rocks in his generation. Hess studied an altered peridotite body in Virginia for his 1932 Ph.D. dissertation at Princeton University. His early exposure to an Appalachian peridotite initiated a lifelong interest in the problem of serpentinization and in the mode of emplacement of alpine peridotites. Early in his career, he visited numerous occurrences of such rocks, particularly in Newfoundland, New England, Quebec, Virginia, and the Caribbean. His most detailed field and mineralogical work, however, was devoted to layered ultramafic intrusions, particularly the Stillwater Complex in Montana (Hess, 1936, 1939, 1960). Although Hess generally did not carry out detailed mapping and mineralogical investigations of individual ophiolites, he gained some appreciation for the differences between the ultramafic rocks of layered intrusions and serpentinized alpine peridotites. Throughout his career, Hess continued to ponder the mode of emplacement and mechanism of serpentinization of the alpine bodies.

In a paper on serpentines, orogeny, and epeirogeny presented at a symposium on "The Crust of the Earth" on the occasion of the 1954 bicentennial of Columbia University, Hess (1955, p. 401) observed that "there has been a debate for the past 35 years between the field geologists and the laboratory investigator on how ultramafic rocks are emplaced. The field man has invariably drawn the conclusion that they appear to have been very fluid—liquid—at the time of injection. Repeatedly the laboratory investigator has ruled out each new suggestion as to how such materials could be magmas at any reasonable temperature." Hess was among those field geologists who accepted the emplacement of the ultramafic bodies by liquid injection. Bailey and McCallien (1953), too, accepted the notion of ultrabasic magma. Bowen (Fig. 2), a highly-regarded experimental petrologist at the Geophysical Laboratory, was the anti-magmatist laboratory investigator whom Harry Hess had in mind.

Whereas Hess linked alpine ultramafic rocks to the concept of an "ophiolite suite," Bowen rarely, if ever, used the term "ophiolite" in his writings. He did, however, write about dunite, peridotite, pyroxenite, serpentine, ultramafic rocks, and ultrabasic rocks in his attempts to explain monomineralic rocks in the light of his ideas about crystallization-differentiation. In an effort to understand Bowen's thoughts about ophiolites, therefore, one is limited to his views on the origin and emplacement of ultramafic rocks in general. Because Bowen died in 1956, his ideas about rocks that are now considered to be ophiolites were unaffected by either the theory of plate tectonics or the discovery of ultramafic volcanic rocks, such as komatiites.

Bowen (1909, 1910, 1911) spent several summers as a field geologist mapping diabase sills for the Ontario Bureau of Mines during his collegiate and graduate school years (Young, 1998). Having learned to appreciate the importance of field geology during these early years, Bowen took as many field trips as possible throughout his professional career in order to test his experimentally derived hypotheses. His field acquaintance with ultramafic rocks was gained in at least four localities. In the summer of 1920, Bowen (1922) collected alkalic ultramafic samples from an occurrence of alnoitic rocks at Isle Cadieux, Quebec, Canada. On

Figure 2. Norman Levi Bowen (1887–1956). Reproduced by permission of Geophysical Laboratory, Carnegie Institution of Washington.

two occasions in the 1920s, Bowen spent several days examining gabbroic and ultramafic rocks on the island of Skye with Alfred Harker and Cecil Tilley. In 1929, he participated in a ten-day field trip to the Bushveld Igneous Complex upon conclusion of the 15th International Geological Congress in Pretoria, South Africa. Unquestionably, he saw many ultramafic rocks on that occasion. So far as I am aware, Bowen's only field experience with rocks that would now most likely be considered as ophiolites took place in the summer of 1921 when he examined peridotites in western Massachusetts and Vermont with Finnish petrologist, Pentti Eskola. Given his very limited field acquaintance with ophiolites, Bowen's views about such rocks were shaped primarily by his experimental work and to a lesser extent by petrography.

When Hess called attention to the debate between the field and experimental geologists, Bowen was still embroiled in the granite controversy, and his labors were culminating in detailed studies of the granite system undertaken with his co-laborer at the Geophysical Laboratory, O. Frank Tuttle (Tuttle and Bowen, 1958). Thanks to a career spent exploring the processes that form a wide spectrum of igneous rocks and to his prominent role in the granite controversy, Bowen was widely recognized as an arch-magmatist. Ironically, when it came to ultramafic igneous rocks, he was a vigorous anti-magmatist. The question for consideration is why Bowen was so vehemently opposed to the idea of ultrabasic magmas.

BOWEN'S EARLY EXPERIMENTAL RESULTS

Bowen's interpretation of the origin of ultramafic rocks must be understood in the context of his theory of crystallization-differentiation, developed when Hess was still a young boy. Not long after joining the staff of the Geophysical Laboratory as an assistant petrologist in 1912, Bowen continued to carry out an ambitious program of research conceived and initiated by Director Arthur L. Day and aimed at determining the phase relations of common rock-forming minerals by means of carefully controlled, high-temperature experiments on chemically pure substances (Young, 1998). After completing detailed phase-equilibrium studies of the system nepheline-anorthite (Bowen, 1912) and plagioclase feldspar (Bowen, 1913), Bowen wanted to investigate magnesian olivine and pyroxene by undertaking experiments on the system $MgO–SiO_2$ (Fig. 3) with Olaf Andersen, who had just joined the Geophysical Laboratory from the University of Oslo. Bowen and Andersen (1914) discovered that the magnesian pyroxene, clinoenstatite ($MgSiO_3$), melts incongruently to the magnesian olivine, forsterite (Mg_2SiO_4), plus liquid, at 1557 °C. Although making no attempt to determine the position of the liquidus of the MgO-rich side of the diagram because of the very high temperatures involved, Bowen and Andersen did determine the approximate one-atmosphere melting point of forsterite to be 1890 °C ±20 °C by optical pyrometry. The reaction relationship between forsterite and magma at 1557 °C typified by forsterite began to play an important role in Bowen's thinking about the problem of igneous rock diversity, but he also realized that the temperatures required to form extremely olivine-rich liquids, that is, ultrabasic magmas, were much too high for crustal magmas.

Bowen investigated another common mineral, the calcium-rich monoclinic pyroxene, diopside ($CaMgSi_2O_6$), in an exploration of the ternary system, diopside–forsterite–silica (Bowen, 1914). In this paper, Bowen stressed the importance of magma cooling rate on crystallization path and described the nature of differences between equilibrium and fractional crystallization. He postulated that the rising or sinking of different crystals dur-

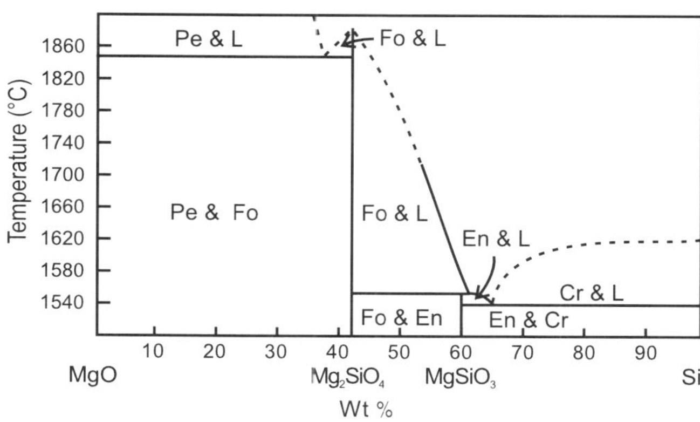

Figure 3. Phase diagram of the system $MgO–SiO_2$ at one atmosphere pressure modified from Bowen and Andersen (1914). Fo—forsterite, Pe—periclase, En—clinoenstatite, Cr—cristobalite, L—liquid. The melting temperature of forsterite (Mg_2SiO_4) was determined to be 1890 °C, a temperature far too high in Bowen's view for existence of dunitic magma in the crust.

ing very slow cooling of magma might lead to a layered igneous rock body in which the layers would contain different proportions of minerals and different assemblages of minerals. Bowen found the results of his experiments to be devastating to the widely held belief in eutectic crystallization in magmas and argued that fractional crystallization, as opposed to fractionation in the liquid state, appeared to be the prime factor in differentiation.

Bowen had proposed the possibility of differentiation by crystal settling from phase diagrams. Then, in late 1914, Bowen (1915a) conducted a few simple experiments to demonstrate the reality of differentiation by means of crystal sinking within a crucible. Upon cooling melted mixtures of diopside and enstatite, he discovered that olivine crystals began to form at ~1460 °C. After maintaining his charges at 1430 °C for varying lengths of time, Bowen quenched them in water to prevent further crystallization. He found a marked accumulation of olivine crystals toward the bottom of the crucibles. He also observed bottom accumulation of pyroxene from more viscous, silica-rich starting mixtures. Bowen further observed that the upper parts of some of his charges consisted of pyroxene and free silica, whereas the lower parts consisted of pyroxene and olivine. He had artificially produced layered differentiated masses and believed that they represented synthetic analogs of the Palisades sill in New Jersey. Lewis (1907) had interpreted the thick olivine-rich layer near the base of the Palisades sill as a mass of sunken crystals. In light of his crystal settling experiments, Bowen agreed.

After his study of the ternary system diopside–albite–anorthite, published in a paper entitled "The crystallization of haplobasaltic, haplodioritic and related magmas," Bowen (1915b) became more convinced than ever of the importance of fractional crystallization in producing igneous rock diversity. The stage was set for him to spell out his concept of differentiation in detail. Bowen (1915c) formulated his early theory of igneous rock diversity in detail in a paper dealing with the later stages of igneous rock evolution. He vigorously rejected Soret diffusion, the Gouy-Chaperon effect, in which a standing solution becomes chemically stratified by gravitation, liquid immiscibility, and assimilation as significant mechanisms of diversity. In contrast, by drawing on the wealth of data from his series of experimental studies, he argued for the dominant role of crystallization in the differentiation of sub-alkaline igneous rocks and reasoned that the crystallization-differentiation of parental basaltic liquid ultimately led to the formation of andesitic and rhyolitic liquids. Bowen alluded only briefly to the formation of monomineralic rocks such as dunite and anorthosite. He claimed that dunite did not crystallize directly from a magma of dunitic composition but formed during the earliest stages of differentiation by the more or less perfect sorting of crystals of olivine in larger bodies of magma that cooled very slowly.

PAPERS ON MONOMINERALIC ROCKS

Bowen (1917) developed his ideas about the origin of monomineralic rocks more thoroughly in a paper on anorthosite, claiming that the problem of any monomineralic rock is fundamentally the same as the problem of anorthosite. To support his contention that ultramafic rocks represent accumulations of early-formed crystals of olivine and pyroxene from basaltic liquid rather than products of crystallization from ultrabasic magmas, Bowen maintained that neither dunite nor excessively pyroxene-rich rocks generally occur as dikes. Moreover, he said, neither dunite nor pyroxenite has an effusive equivalent. Although Harker (1908) had attributed some of the banded ultramafic rocks in the Scottish Hebrides to the intrusion of a "non-uniform" magma, Bowen discounted the possibility of sustaining sharp contrasts in composition between adjacent bands of liquid and invoked the accumulation of crystals as the source of the heterogeneous banding. He also took issue with the suggestion of Harker (1916) that ultrabasic magma could be generated by re-melting masses of accumulated olivine and pyroxene crystals on the grounds that the temperatures required to re-melt such masses would be much too high.

Bowen's interest in the problem of the origin of ultramafic igneous rocks was renewed by the publication of the Mull memoir by Bailey et al. (1924) and by field trips in 1923 and 1926 to the volcanic centers of Skye and Mull with Harker and Tilley. Thus stimulated by Hebridean geology, Bowen (1927) wrote his first paper devoted entirely to "ultra-basic and related rocks." Noting that most contemporary petrologists were committed to the idea that each specimen of igneous rock corresponded to a liquid of approximately the same chemical composition, Bowen reiterated his objection to such a view in the case of rocks consisting mostly of one mineral. Given that inclusions of common rocks failed to show the expected effects that they should show if they were immersed in ultrabasic liquids, Bowen refused to accept the still too-high ~1500–1600 °C temperatures that had been postulated for ultrabasic liquids by Vogt (1926). Vogt believed that ultrabasic liquids could not maintain these high temperatures sufficiently long to be poured out at the surface from the great depths where he, like Harker, thought that they originated by the re-melting of accumulated olivine and pyroxene crystals. Bowen, however, responded that if ultrabasic liquids do exist, they should be rapidly transferred to the surface in view of their high fluidity. Harker, too, had continued his advocacy of formation of ultrabasic magmas by re-solution of accumulated crystals. He was now arguing that volatile components aided in the formation of ultrabasic liquids at lower temperatures, but both Vogt and Bowen opposed Harker's view on the grounds that olivine-rich rocks lack evidence for contained volatiles. At this point, however, Bowen still had no experimental data on the effect of volatiles on the phase relations of olivine and pyroxene.

Bowen especially drew insight from the concept of "porphyritic magma type" introduced by Bailey et al. (1924) in their memoir on the geology of Mull. The Mull authors had distinguished two magma types on chemical grounds. A "non-porphyritic central magma type" was said to exist in all gradations of texture from pure glass to crystallized gabbro. In contrast, they described a "porphyritic central magma type" as never forming glass and "always porphyritic in its finer crystallizations" with calcic plagioclase phenocrysts. From these petrographic characters, Bowen reasoned that there had existed magmatic liquids whose composition corresponded to the rocks of the non-porphyritic magma type and that there had been no magmatic liquids whose composition corresponded to the porphyritic magma type. The very calcic plagioclase phenocrysts in the rocks of the porphyritic magma type, Bowen said, represented a certain excess over any liquid that must have accumulated from a considerable mass of liquid. He asserted that rocks formed by accumulated crystals could show extreme enrichment in anorthite, whereas liquids could not.

Extending this line of reasoning to ultrabasic rocks, Bowen argued that olivine basalt, the igneous rock type most similar to ultrabasic rocks, could not contain more than a moderate amount of olivine unless the excess olivine had been present as phenocrysts. He found that basalts with large normative olivine contents always contained large amounts of olivine phenocrysts and, therefore, could not have been purely liquid at one time. Basalts rich in olivine must have become that way through the accumulation of crystals. By extension, Bowen claimed, extremely olivine-rich rocks certainly could not have formed from liquids of their own composition, but must have required accumulation of crystals from magma if that is how olivine-rich basalts formed. The paper on ultrabasic rocks was included virtually verbatim as a chapter in *The Evolution of the Igneous Rocks* (Bowen, 1928).

THE INFLUENCE OF IRON ON ULTRABASIC MAGMAS

Vogt (1921, 1926) had issued a significant challenge to Bowen by claiming that addition of substantial quantities of iron to forsterite should sufficiently lower the melting temperature of olivine that olivine-rich magmas could reasonably exist within the crust. Vogt's estimate of temperatures of ~1500–1600 °C for the formation of dunite melt was based on the assumption that the olivine contained ~20% fayalite. He assumed that the liquidus falls sharply from 1890 °C for pure Mg_2SiO_4 with even small additions of Fe_2SiO_4 (Fig. 4). To this point, Bowen's experiments pertaining to olivine-pyroxene-rich rocks were based entirely upon magnesium-bearing, iron-absent systems. Beginning in the late 1920s and extending well into the 1930s, however, he engaged in an extensive experimental program with his Geophysical Laboratory colleague, J.F. Schairer, on iron-bearing silicate systems. In a study of the system FeO–SiO_2, Bowen and Schairer (1932) established the melting point of the iron-rich olivine end member, fayalite (Fe_2SiO_4), at 1205 °C, considerably less than that of forsterite. Although this initial result might have lent support to Vogt's contention that the liquidus temperature of olivine is greatly diminished by addition of iron, further work on the system MgO–FeO–SiO_2 completely undercut Vogt's argument for the existence of ultrabasic magma in the crust.

At the 16th International Geological Congress held in Washington, D.C., in 1933, Bowen and Schairer (1936) reported new experimental results that made it glaringly clear that the melting temperatures of Fe-bearing olivines with ~20% Fe_2SiO_4 would still be above 1800 °C (Fig. 5). They claimed that olivine required 75% Fe_2SiO_4 to be completely melted at 1500 °C and that olivine with 10 to 15% Fe_2SiO_4 would melt completely at 1850 °C. If Vogt's postulated temperatures for the existence of dunite magma had been somewhat unpalatable for petrologists, the new experimentally determined temperatures would make them choke! Although recognizing that some petrologists, namely, Harker, appealed to volatiles to lower dunite magma temperatures, Bowen continued to espouse his favored notion of crystal accumulation of olivine. He also began to argue for some special mode of emplacement of a mass of solid olivine crystals as the only viable option to account for intrusive ultramafic rock bodies. The olivine diagram was discussed in greater detail in a paper on the system MgO–FeO–SiO_2 (Bowen and Schairer, 1935).

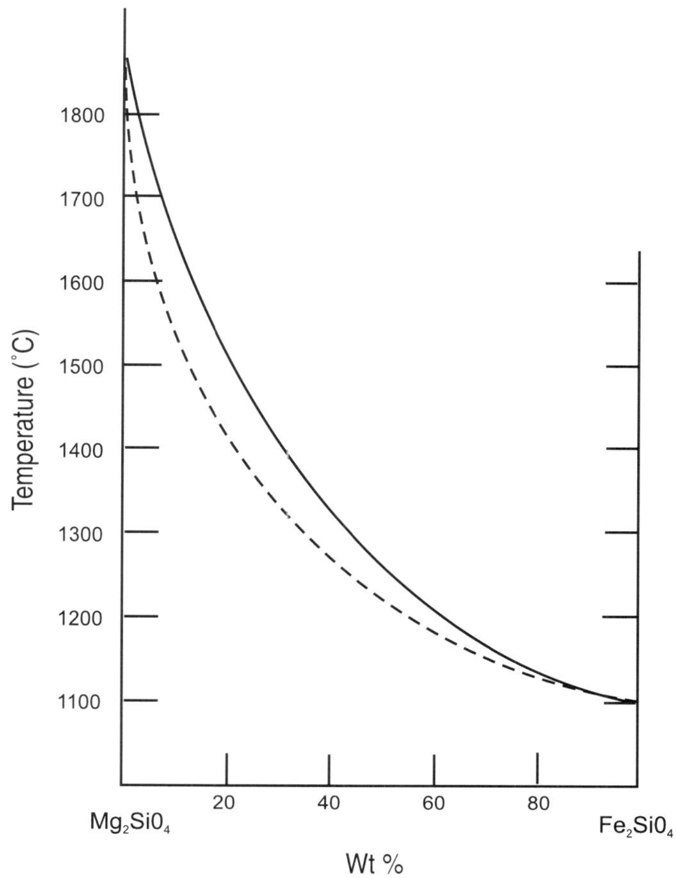

Figure 4. Schematic melting diagram of the olivine solid solution series in the system Mg_2SiO_4–Fe_2SiO_4 modified from Vogt (1921). Vogt proposed that addition of moderate amounts of iron to olivine might drastically lower the temperatures at which a dunitic melt might exist.

THE HESS-BOWEN DEBATE

In 1937, Bowen left the Geophysical Laboratory to accept a position on the geology faculty of the University of Chicago in

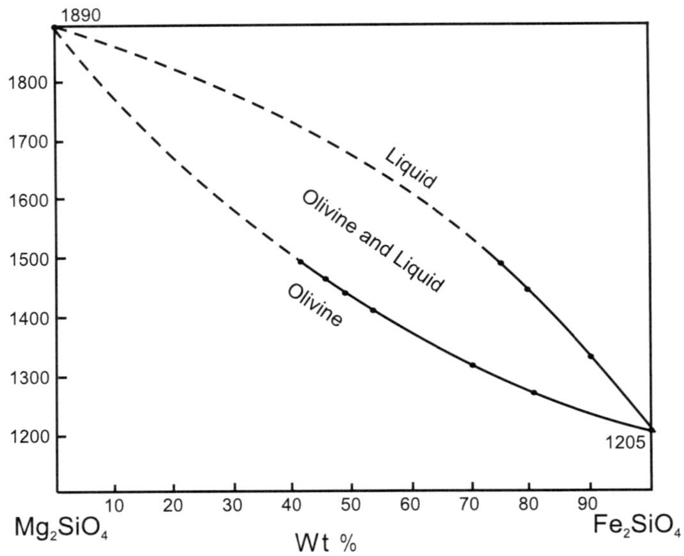

Figure 5. Phase equilibrium diagram of the olivine solid solution series in equilibrium with metallic iron at one atmosphere pressure modified from Bowen and Schairer (1935). The experimental results made it clear that olivine-rich melts would require extremely high temperatures to exist despite the addition of moderate amounts of iron.

the hopes of introducing the methods of experimental petrology into the academic world. Although Bowen's productivity as an experimentalist declined during his tenure at Chicago, he was still able, from time to time, to discuss his views on ultramafic rocks in presidential addresses. Soon after his arrival at the University of Chicago, Bowen (1938) delivered an address as the retiring president of the Mineralogical Society of America entitled "Mente et Malleo Atque Catino." Given that Bowen and his colleagues at the Geophysical Laboratory comprised a tiny minority among petrologists as practitioners of the art of experimental petrology, his address was a plea for geologists to recognize that experiment is just as important as field observation for the solution of problems in the disciplines of mineralogy and petrology. To that end, he urged cooperation between observers and experimenters. As an example of a cooperative approach in the study of igneous rocks, Bowen painted an imaginary scenario of give and take over the origin of monomineralic rocks such as dunite. Although the field observer might speak of dunitic magmas, he said that the experimenter had discovered that the melting temperature of pure magnesian olivine is extremely high and, therefore, might suggest that olivine-rich rocks represent crystal accumulations from gabbroid magma. When reminded of his own observation that there are no lavas of dunitic composition, the field observer consents to re-examine the rocks. Upon re-examination in the field, it is found that many of the examples of the olivine-rich rocks have characters that are consistent with an origin as crystal accumulations, but the re-examination also reveals variants of these rocks for which the observer is unable to visualize any method of emplacement except as a liquid. For example, a few petrologists postulated that at least some dikes of peridotite must have been liquid. Nevertheless, it was found, Bowen (1938, p. 129) asserted, that these presumed magmatic peridotite dikes had "basaltic selvages, suggesting the possibility that the flow of a crystal mush of peridotitic character was lubricated by basaltic magma." Petrologists had also argued for alleged peridotite magmas to be highly enriched in water, but such evidence was lacking. Bowen (1938, p. 129) suspected that the tenacity of life of the concept of highly magnesian magmas might be "the result of a natural reluctance on the part of the observer to abandon opinions once held."

Bowen also described a scenario in which an experimentalist proposed that serpentine is very capable of intrusion in a solid, plastic state and in which the field geologists subsequently examined intrusive serpentines with this mode of intrusion in mind. The concept of solid state intrusion would explain, in his eyes, the general lack of thermal and hydrothermal contact effects in the vicinity of serpentinized peridotites. Thus, Bowen envisioned debate over the origin of ultramafic rocks proceeding with field observers and synthesists proposing and counter-proposing hypotheses with mind, hammer, *and crucible* (atque catino).

Around the time of Bowen's address, Harry Hess emerged as the leading opponent of Bowen in the debate about the origin and emplacement of ultramafic rocks. In a paper on a primary peridotite magma, Hess (1938) said that he accepted the validity of Bowen's idea of crystal accumulation and subsequent emplacement in the solid state for many peridotite bodies. He was unable, however, to accept Bowen's hypothesis in every instance. From field evidence associated with large alpine peridotite bodies in Cuba, the East Indies, and eastern Canada, Hess argued for the existence of ultrabasic magma on several grounds. Hess reasoned that there ought to be felsic differentiates associated with ultramafic rocks, if ultramafic rocks represented early accumulations of olivine and pyroxene crystals formed from a differentiating basaltic magma as Bowen claimed, even though Bowen had envisioned such cumulates as intruded as solid crystal mushes from elsewhere. As a result, Hess regarded the absence of felsic differentiates associated with most large peridotite bodies as evidence against Bowen's view. Hess also reported that chemical analyses of chilled peridotite border facies of some gabbroic and ultramafic bodies indicated ultrabasic compositions. For Hess, such chill margins must, therefore, have represented ultrabasic liquids. To draw support for the notion of ultrabasic liquids, Hess appealed to small dikes and pipes of serpentinized dunite within an unequivocally magmatic body, the Stillwater Complex. These dikes and pipes, he believed, had formed by partial fusion of cumulate peridotites. Hess also maintained that the negligible contact metamorphic effects surrounding many peridotite bodies indicated low temperatures of magmatic intrusion, rather than the absence of magmatic intrusion altogether.

Hess rejected Bowen's concept of intrusion of a mass of cold, solid olivine crystals for shallow peridotite intrusions in

Cuba and eastern Canada because these intrusions consisted of sills with smooth, horizontal contacts. Some of the Cuban peridotites also had highly irregular contacts projecting from the main mass into sedimentary and volcanic rocks in a complicated manner. Many Cuban ultramafic bodies were intruded into nearly unconsolidated sediments, but Hess found that the slightly deformed wall rocks were not disturbed in a manner that would indicate solid flow of the country rocks around the intrusion.

Hess envisioned the nearly ubiquitous presence of serpentine in alpine peridotites as providing a solution to the problem of intrusion of low-temperature ultrabasic magma and, therefore, the lack of associated contact metamorphism. In contrast to his earlier view, he now postulated that such magmas contained between five and fifteen weight percent water and that the direct crystallization of serpentine from the magma would use up the available water, leaving none to produce hydrothermal alteration in the wall rocks. Serpentine formation might even draw water from the wall rocks. As to the lack of effusive ultrabasic equivalents, Hess supposed that volatiles would exsolve and escape from the magma with reduction in water solubility as the magma mass rose through the crust and pressure decreased. As a result, Hess envisioned that a peridotite magma would crystallize entirely before reaching the surface. Should such a magma actually reach the surface, he said, the eruption would be violent and would produce ultramafic breccias, rather than lavas.

Not until his presidential address on magmas presented before the Geological Society of America in 1946 was Bowen able to respond to the ideas that Hess had advanced. In this address, Bowen (1947) spoke about granites and ultramafic rocks, the two groups of igneous rocks that engendered the greatest amount of contentiousness concerning their magmatic nature. In regard to ultramafic rocks, Bowen replied to Hess that a magma of serpentine composition could not be retained in such a frail container as shallow, unconsolidated sediments. Moreover, he said, a serpentine magma that contained so much water would be likely to precipitate large serpentine crystals rather than the minute fibers or platy aggregates so characteristically observed in serpentinized peridotites. Bowen also thought that a water-rich silicate liquid *would* introduce water into the surrounding rocks. On the other hand, he contended that a mass of moderately hot, crystalline olivine being forced into a new position would suffer granulation and gliding, effects that should facilitate the flow of crystalline material and permit intimate access of water for lubrication. The olivine-rich mass would then absorb water from the surrounding rocks with consequent production of serpentine.

Throughout most of his career, Bowen had lacked any experimental evidence that might indicate whether or not volatiles would drastically lower the temperatures of ultrabasic melts. Upon returning to the Geophysical Laboratory from the University of Chicago in 1947, Bowen was able to resume his vigorous research activity and undertook his first hydrothermal study. He investigated the system $MgO–SiO_2–H_2O$ in collaboration with O.F. Tuttle.

Bowen and Tuttle (1949) performed experiments to temperatures as high as 1000 °C and pressures as high as 40,000 pounds per square inch (psi). They found no indication of a melt phase at any temperature (Fig. 6). Their data indicated that material of serpentine composition would be completely crystallized at high temperatures to enstatite and olivine and that H_2O would occur entirely as vapor to pressures as high as 30,000 psi. They also discovered that serpentine crystallizes only below 500 °C. The idea of Hess that an ultrabasic magma intruded at low temperature would crystallize directly to serpentine, therefore, was excluded by their experiments.

Bowen and Tuttle also rejected serpentinization of a crystallizing dunite mass by autometamorphism. Assuming the validity of crystallization-differentiation within a magma chamber, they reasoned that, with continued cooling and accumulation of crystals above early-formed ultramafic cumulates, water would be concentrated in the residual liquid at progressively higher levels within the magma chamber. When water dissolved in residual liquids eventually exsolved as a separate fluid phase, Bowen and Tuttle argued, the water would effectively be spatially removed and mechanically separated from the lower, accumulated ultramafic rocks such that residual fluid would ordinarily not have access to the ultramafic cumulates. The ultramafic rocks could not, therefore, be converted to serpentine, and we must, they reasoned, look to a totally extraneous source for the aqueous media inducing serpentinization in ultramafic rocks. To Bowen and Tuttle, a more likely hypothesis was that accumulated dunite and related ultramafic rocks were

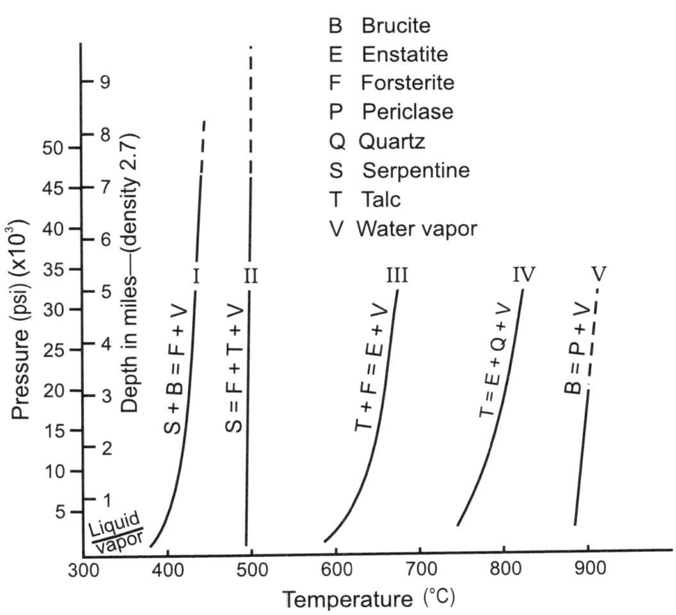

Figure 6. Pressure-temperature curves of univariant equilibrium in the system $MgO–SiO_2–H_2O$ modified from Bowen and Tuttle (1949). The experiments demonstrated that ultrabasic liquid is not stable at temperatures as high as 1000 °C and also that serpentine is stable only below 500 °C.

subsequently removed and intruded elsewhere in a crystalline state. They suggested that any vapor coexisting with a solid mass of olivine crystals might contribute to facilitating flow of the crystalline mass, and they argued that if such a mass were injected into colder wet rocks, then water might be absorbed from the surroundings and partial serpentinization might ensue if the temperature of the ultramafic mass had fallen below 500 °C, the upper limit thermal stability of serpentine.

Bowen and Tuttle further suggested that ultramafic veins, superficially resembling small intrusive dikes of magmatic dunite or pyroxenite, could be produced by the interaction of aqueous fluids streaming through fractures in ultramafic rock bodies with olivine or pyroxene. For example, they proposed that dissolved silica in an aqueous fluid passing through fractured dunite might react with olivine to produce a dike-like body of pyroxenite. And by the same token, a silica-deficient fluid flowing through fractured pyroxenite might extract silica from pyroxene, converting it into olivine and producing a dike-like mass of dunite.

AFTERMATH

After publication of their work on the system $MgO–SiO_2–H_2O$, granite completely absorbed the attention of both Bowen and Tuttle. After 1949, Bowen never again discussed ultramafic rocks in print. Throughout his career, he had consistently denied the possibility that alpine peridotites and serpentinites resulted from the intrusion of magmas of ultrabasic composition and had maintained that alpine peridotites were emplaced by the injection of olivine- and pyroxene-rich masses in the solid state. At the time of his death in 1956, Bowen remained firmly convinced of the validity of his view on the origin and emplacement of these rocks. His view would be completely reformulated only a few years after his death.

Hess went on to become one of the major architects of the new theory of plate tectonics, an outgrowth of his lifelong enthusiasm for ultramafic rocks of both the crust and mantle and their relation to mountain building. When he died in 1969, Hess had lived just long enough to see some resolution to the problem of serpentinized alpine ultramafic rocks.

The development of the theory of plate tectonics effectively rendered the debate over the views of Bowen and Hess obsolete because geologists recognized that ophiolites represent fragments of former oceanic lithosphere that were tectonically emplaced as cold slabs into continental orogens or island arcs at collisional plate margins. Moreover, deep-sea drilling confirmed the close similarities of the successions of rocks obtained from cores of oceanic crust to the successions of rocks exposed in ophiolites. The basal ultramafic sections of ophiolites are now considered to be slivers of uppermost mantle material that contain clues to the processes and history of the upper mantle.

Bowen drew some correct inferences about ophiolites from the experimental data with which he was most familiar, and Hess drew some correct inferences about ophiolites from the field and chemical data with which he was most familiar. Whatever valid inferences each may have drawn from their data, however, both Bowen and Hess applied those inferences to the wrong geological context. Because of a paucity of very detailed, well documented field, chemical, and textural data from individual ophiolites, Bowen, Hess, and most other geologists of their era failed to appreciate the spectacularly allochthonous character of ophiolites.

Ophiolites contain both igneous cumulate and metamorphic fabrics derived from accretion zone processes. Thus, detecting some field and chemical evidence for the existence of ultrabasic magmas in ophiolites, Hess emphasized the igneous parentage of such rocks. In light of plate tectonics theory, however, any ultrabasic liquids involved in ophiolite formation were more likely injected during the formation of oceanic lithosphere rather than intruded directly into continental crust as Hess had envisioned, and ophiolitic ultramafic cumulates formed in magma chambers at accretion zones rather than in alpine mountain systems. Hess's belief in ultrabasic magma was impressively confirmed by the eventual discovery of effusive ultrabasic liquids in the form of komatiites (Viljoen and Viljoen, 1969), and Hess even had opportunity to experience a sense of vindication when he saw komatiites in South Africa a month before his death (James, 1971).

In contrast, Bowen emphasized the solid-state parentage of ophiolites. He had detected some evidence for the emplacement of alpine peridotites as solid material. But rather than the lubricated mush of olivine crystals advocated by Bowen, the theory of plate tectonics accounted for the metamorphic fabrics of ophiolites in terms of hydrothermal alteration at accretion zones and subsequent shearing and tectonic emplacement at collisional margins.

The two petrologists were also led astray by incorrectly assuming that serpentinization had occurred essentially in situ. Hess maintained that serpentine had crystallized from hydrous ultrabasic melt more or less during or after emplacement of a crystallizing ultrabasic melt. Bowen's work made it clear that serpentine could not be a primary magmatic mineral, but he could not demonstrate satisfactorily that the distribution of serpentine could be accounted for by absorption of water from the wall rocks because serpentinization is generally unrelated to the water table or the contacts of ophiolite bodies. The theory of plate tectonics solved this problem by attributing serpentinization to the interaction of circulating sea water with oceanic lithosphere at newly formed spreading margins. The fact that serpentinization of the ophiolites had occurred at sites that were so far removed spatially and temporally from the loci of their emplacement in mountain belts helped explain the lack of correlation between serpentine distribution with contacts and water table at the site of emplacement.

The debate between Bowen and Hess highlighted the importance of achieving coherence between the results of field and experimental studies. The fact that Bowen's experiments were applied to rocks that were not yet very well understood in the field led to conflicting interpretations but also served as a stimulus in the search for and discovery of a new way of thinking about the long-standing problems of ophiolites.

ACKNOWLEDGMENTS

My work on Bowen and on the history of igneous petrology has been generously supported by NSF grants #SBR-9601203 and #SES-9905627, Calvin Research Fellowships, and Calvin Alumni Association Faculty Research Grants. I am much indebted to R.G. Coleman, Sally Newcomb, and Hatten S. Yoder, Jr., for their constructive reviews of the manuscript. Ellen Alderink of Calvin College's Instructional Graphics Services kindly provided the graphics for the paper.

REFERENCES CITED

Bailey, E.B., and McCallien, W.J., 1953, Serpentine lavas, the Ankara mélange and the Anatolian thrust: Transactions of the Royal Society of Edinburgh, v. 62, p. 403–442.

Bailey, E.B., Clough, C.T., Wright, W.B., Richey, J.E., and Wilson, G.V., 1924, Tertiary and post-Tertiary geology of Mull, Loch Aline, and Oban: Geological Survey of Scotland Memoir, 445 p.

Bowen, N.L., 1909, Diabase and aplite of the cobalt-silver area: Journal of the Canadian Mining Institute, p. 95–106.

Bowen, N.L., 1910, Diabase and granophyre of the Gowganda Lake District, Ontario: Journal of Geology, v. 18, p. 658–674.

Bowen, N.L., 1911, Silver in Thunder Bay District: Ottawa, Ontario, Bureau of Mines, 20th Report, p. 119–132.

Bowen, N.L., 1912, The binary system: $Na_2Al_2Si_2O_8$ (nephelite, carnegieite)—$CaAl_2Si_2O_8$ (anorthite): American Journal of Science, v. 33, 4th ser., p. 551–573.

Bowen, N.L., 1913, The melting phenomena of the plagioclase feldspars: American Journal of Science, v. 35, 4th ser., p. 577–599.

Bowen, N.L., 1914, The ternary system: diopside–forsterite–silica: American Journal of Science, v. 38, 4th ser., p. 207–264.

Bowen, N.L., 1915a, Crystallization-differentiation in silicate liquids: American Journal of Science, v. 39, 4th ser., p. 175–191.

Bowen, N.L., 1915b, The crystallization of haplobasaltic, haplodioritic and related magmas: American Journal of Science, v. 40, 4th ser., p. 161–185.

Bowen, N.L., 1915c, The later stages of the evolution of the igneous rocks: Journal of Geology, v. 23, supplement to no. 8, p. 1–91.

Bowen, N.L., 1917, The problem of the anorthosites: Journal of Geology, v. 25, p. 209–243.

Bowen, N.L., 1922, Genetic features of alnoitic rocks from Isle Cadieux, Quebec: American Journal of Science, v. 3, 5th ser., p. 1–34.

Bowen, N.L., 1927, The origin of ultra-basic and related rocks: American Journal of Science, v. 14, 5th ser., p. 89–108.

Bowen, N.L., 1928, The evolution of the igneous rocks: Princeton, New Jersey, Princeton University Press, 332 p.

Bowen, N.L., 1938, Mente et malleo atque catino: American Mineralogist, v. 23, p. 123–130.

Bowen, N.L., 1947, Magmas: Geological Society of America Bulletin, v. 58, p. 263–279.

Bowen, N.L., and Andersen, O., 1914, The system $MgO–SiO_2$: American Journal of Science, v. 37, 4th ser., p. 487–500.

Bowen, N.L., and Schairer, J.F., 1932, The system, $FeO—SiO_2$: American Journal of Science, v. 24, 5th ser., p. 177–213.

Bowen, N.L., and Schairer, J.F., 1935, The system, $MgO—FeO—SiO_2$: American Journal of Science, v. 29, 5th ser., p. 151–217.

Bowen, N.L., and Schairer, J.F., 1936, The problem of the intrusion of dunite in the light of the olivine diagram: XVI International Geological Congress, 1933 Report: Washington, DC, International Geological Congress, p. 391–396.

Bowen, N.L., and Tuttle, O.F., 1949, The system $MgO–SiO_2–H_2O$: Geological Society of America Bulletin, v. 60, p. 439–460.

Dana, J.D., 1866, Manual of Geology, 2nd ed.: Philadelphia, Theodore Bliss, 800 p.

Harker, A., 1908, The geology of the small isles of Inverness-shire: Glasgow, Geological Survey of Scotland Memoir, 210 p.

Harker, A., 1916, Differentiation in intercrustal magma basins: Journal of Geology, v. 24, p. 554–558.

Hess, H.H., 1933, The problem of serpentinization and the origin of certain chrysotile asbestos, talc, and soapstone deposits: Economic Geology, v. 28, p. 634–657.

Hess, H.H., 1936, Plagioclase, pyroxene and olivine variations in the Stillwater Complex: American Mineralogist, v. 21, p. 198–199.

Hess, H.H., 1938, A primary peridotite magma: American Journal of Science, v. 35, 5th ser., p. 321–344.

Hess, H.H., 1939, Island arcs, gravity anomalies, and serpentinite intrusions: A contribution to the ophiolite problem: *in* Report, 17th International Geological Congress, Moscow, USSR, 1937, v. 2, p. 263–283.

Hess, H.H., 1955, Serpentines, orogeny, and epeirogeny, *in* Poldervaart, A., ed., The Crust of the Earth: Boulder, Colorado, Geological Society of America Special Paper 62, p. 391–408.

Hess, H.H., 1960, Stillwater Igneous Complex, Montana: A quantitative mineralogical study: Boulder, Colorado, Geological Society of America Memoir 80, 230 p.

James, H.L., 1971, Harry Hammond Hess May 24, 1906—August 25, 1969: Biographical Memoirs of the National Academy of Sciences, v. 43, p. 109–128.

Jennings, C.W., and Strand, R.G., 1959, Geologic Map of California, Santa Cruz quadrangle: California Division of Mines and Geology, scale 1: 250 000, 1 sheet.

Lewis, J.V., 1907, The origin and relations of the Newark rocks: New Jersey Geological Survey Annual Report, p. 99–129.

Lodochnikov, W.N., 1933, Serpentine and serpentinites and the petrological problems connected with them: Problems of Soviet Geology, v. 2, p. 119–144.

Phillips, A.H., and Hess, H.H., 1936, Metamorphic differentiation at contacts between serpentinite and siliceous country rocks: American Mineralogist, v. 21, p. 333–362.

Steinmann, G., 1905, Geologische Beobachtungen in den Alpen. II: Die schardt'sche Überfaltungstheorie und die geologische Bedeutung der Tiefseeabsätze und der ophiolithische Massengesteine: Berichte der naturforschenden Gesellschaft zu Freiburg, i., v. 16, p. 44–65.

Turner, F.J., 1942, Preferred orientation of olivine crystals in peridotites with special reference to New Zealand examples: Transactions of the Royal Society of New Zealand, v. 72, p. 280–300.

Tuttle, O.F., and Bowen, N.L., 1958, Origin of granite in the light of experimental studies in the system $NaAlSi_3O_8–KAlSi_3O_8–SiO_2–H_2O$: Geological Society of America Memoir 74, 153 p.

Viljoen, M.J., and Viljoen, R.P., 1969, Evidence for the existence of a mobile extrusive peridotitic magma from the Komati Formation of the Onverwacht Group: Geological Society of South Africa Special Publication 2, Upper Mantle Project, p. 87–112.

Vogt, J.H.L., 1921, The physical chemistry of the crystallization and magmatic differentiation of igneous rocks: Journal of Geology, v. 29, p. 515–539.

Vogt, J.H.L., 1926, Magmas and igneous ore deposits: Economic Geology, v. 21, p. 207–233.

Young, D.A., 1998, N.L. Bowen and crystallization-differentiation: The evolution of a theory: Washington, D.C., Mineralogical Society of America, 276 p.

Manuscript Accepted by the Society May 2, 2003

Geological Society of America
Special Paper 373
2003

Ophiolites, ocean crust formation, and magnetic studies: A personal view

F.J. Vine*
School of Environmental Sciences, University of East Anglia, Norwich NR4 7TJ, England

ABSTRACT

This is a personal view of the seminal developments in our understanding of the nature of oceanic crust, and of ophiolites, that occurred in the 1960s and 1970s. This period followed the first measurements of the thickness and layering of the oceanic crust, using seismic refraction techniques, in the 1950s, and preceded the very detailed studies of oceanic basement, carried out within the context of the Ocean Drilling Program, and of ophiolites, notably in Cyprus and Oman, in the 1980s. During the 1960s and 1970s, I had the good fortune to work on these advances with Drummond Matthews, Harry Hess, Tuzo Wilson, Eldridge Moores, Jason Morgan, Ian Gass, and Joe Cann, among others. My own contribution related mainly to the constraints on the composition, formation and structure of oceanic crust that were provided by the interpretation of oceanic magnetic anomalies and the magnetic and paleomagnetic studies of the Troodos ophiolite of southern Cyprus. By the end of the 1970s, the essential nature of oceanic crust and ophiolites and their probable equivalence had been established, but it is only since then that the details of the variability of oceanic crust and the ways in which this might be related to the variability of ophiolites have become apparent.

Keywords: ophiolites, oceanic crust, magnetic anomalies, paleomagnetism, sea-floor spreading.

THE NATURE OF OCEANIC CRUST

My first encounter with the concept of an ophiolite suite or association was on reading the article by Harry Hess in the GSA's *Crust of the Earth* volume (Hess, 1955). I was an undergraduate at the time, and the volume had been recommended to me as a state-of-the-art overview of the subject by my geology tutor. Presumably, the choice of subject for this symposium had been prompted in part by the then very recent discovery that the Mohorovicic discontinuity beneath the ocean basins is at a much shallower depth than in continental areas, implying that oceanic crust is typically no more than 5–6 km thick (Ewing and Press, 1955). Prior to this one could only surmise, by assuming isostatic equilibrium, that oceanic crust was either relatively thick, comparable to the thickness of continental crust, in which case its average density must be greater than that of continental crust, or, if its density is similar to that of continental crust, it must be relatively thin. Either way, as Wegener had cogently argued (Wegener, 1915), oceanic crust must be systematically and significantly different from continental crust, and this is reflected in the bimodal distribution of heights and depths at Earth's surface. Wegener also argued that this fundamental difference means that continental crust cannot readily be converted to oceanic crust, or vice-versa, and that as a consequence, the concept of ephemeral "land bridges" as an alternative to continental drift, was untenable. I suspect that for most earth scientists, the demonstration that the oceanic crust is very thin came as something of a surprise; many, including Hess, had previously assumed that it was of similar thickness to continental crust but more basic in average composition.

*f.vine@uea.ac.uk

Vine, F.J., 2003, Ophiolites, ocean crust formation, and magnetic studies: A personal view, *in* Dilek, Y., and Newcomb, S., eds., Ophiolite concept and the evolution of geological thought: Boulder, Colorado, Geological Society of America Special Paper 373, p. 65–75. For permission to copy, contact editing@geosociety.org.

In his 1955 paper, Harry Hess felt that these new results, of seismic refraction experiments in the deep ocean basins, implied that the oceanic crust is mafic in composition, perhaps comparable to the lower continental crust, and underlain by peridotitic mantle. Surprisingly, however, he also maintained that the ophiolite suite of "green" mafic and ultramafic rocks does not represent oceanic crust; indeed, he seriously doubted whether the basalts, diabases, and minor gabbro are truly associated with the (typically) serpentinized ultramafics. This view presumably derived from his extensive work on what he regarded as serpentinized ultramafic intrusions in the Appalachians (Hess, 1938). Subsequently, in 1965, having seen the field relations in the Tethyan ophiolites of the Mediterranean area, he acknowledged that they are conformable. By this time, however, he had changed his view on the nature of oceanic crust, from having a composition of predominantly basalt and gabbro, to predominantly partially serpentinized peridotite, with only minor, surficial, basalts.

In January 1962, during my final year as an undergraduate, Harry Hess gave the keynote address at the annual U.K. Inter-University Geological Congress, which on that occasion was held in Cambridge. The theme of the meeting was "The Evolution of the North Atlantic," and Hess' talk was entitled *Impermanence of the Ocean Floor*. It was essentially equivalent to his subsequent paper on the *History of Ocean Basins* (Hess, 1962). For most of us in the audience, this was our first exposure to the concept of sea-floor spreading. It was clear that Hess now regarded the ocean floor as little more than the surface expression of a convecting mantle, the main crustal layer (seismic layer 3) consisting of serpentinite. He considered layers 1 and 2 to be sedimentary, a common assumption at the time, and basaltic volcanism to be restricted largely to the formation of seamounts. In 1955 he had proposed that the uplift, or subsidence, of certain oceanic ridges and swells might be explained by the hydration (i.e., serpentinization) or dehydration of the upper mantle. He considered the water required for serpentinization to be expelled from the mantle itself, and the extent of the serpentinization to be determined by the thermal regime, the upper limit of the stability field for serpentines being ~500 °C. A mechanism for subsidence, as well as uplift, was very important to Hess because of his search for an explanation of guyots. He had discovered these seamounts, with their distinctive flat tops, in the Pacific during World War II, while in command of the *USS Cape Johnston* (Hess, 1946). The flat tops, presumed to be wave-cut platforms, are now at various depths, but typically are 1–2 km below sea level.

In 1962, however, his proposal that the mid-ocean ridges are located above the upwelling limb of a mantle-wide convection cell implied that the isotherms must be elevated beneath a ridge crest, precluding the formation of serpentinite in the upper mantle. Instead, impressed by the fact that serpentinites, as well as basalts, had been dredged from the Mid-Atlantic Ridge (Shand, 1949), he proposed that the main crustal layer is serpentinite and that its thickness is determined by the maximum elevation of the 500 °C isotherm, i.e., beneath the ridge crest. This, he felt, would also provide an explanation of the remarkably uniform thickness of the oceanic crust. An integral part of his reasoning was his belief that juvenile water is constantly being expelled from the mantle and that the volume of water in the oceans is steadily increasing with time; however, these concepts have never been widely accepted. Nevertheless, if his ideas on the nature of the oceanic crust and the crust-mantle transition were controversial, the explanation that sea-floor spreading provided for the subsidence of guyots as they move down ridge flanks was undoubtedly elegant.

In October 1962, I joined the Department of Geodesy and Geophysics at Cambridge, to undertake a Ph.D. in marine geophysics. My Ph.D. supervisor was Drummond Matthews, whose own Ph.D., completed just a year or so earlier, was a study of the petrology and physical properties of basalts dredged from the floor of the northeast Atlantic (Matthews, 1961a). It seemed that in the experience of the Cambridge group, whenever they photographed and dredged hard rocks on the ocean floor, those rocks turned out to be basalts. In contrast to Hess' ideas, therefore, the received wisdom in the Department was that basalts were an important and ubiquitous constituent of oceanic crust and probably formed seismic layer 2. I never really understood why Harry Hess became so averse to the idea of a significant mafic section in the oceanic crust. In figure 7 of his 1962 paper he illustrates onlapping sediments on the flanks of a ridge, directly overlying serpentinite in seismic layer 3. Even when I first met Harry, in early 1965, he still seemed to hold this view. I think, however, that I soon managed to convince him that at least seismic layer 2, in addition to volcanic edifices on the ocean floor, consists of basalt. This was on the basis of my attempts to interpret oceanic magnetic anomalies.

Drum Matthews had set me to work on the interpretation of the Department's marine magnetic data. This had proved to be somewhat intractable because, even in areas of volcanic terrain, which were known to be strongly magnetized, the magnetic anomalies did not correlate with the topography. I developed computer-based forward and inverse modeling techniques to interpret the data, and deduced that some of the topographic features were reversely magnetized. The intensity of magnetization obtained from the inverse modeling, of both normally and reversely magnetized features, was ~5 Am^{-1}, implying that the induced component of magnetization must be very low. This agreed precisely with the magnetic property measurements Drum had made on the dredged basalt samples that he studied for his thesis. These had an average intensity of remanent magnetization of 5 Am^{-1}, and an induced component of magnetization that was more than an order of magnitude less than this (Matthews, 1961b).

It was a short step from this result to the realization that the two main features of oceanic magnetic anomalies known at the time, the pronounced anomaly associated with mid-ocean ridge crests and the linear north-south trending anomalies which had been mapped in the northeast Pacific, might be explained by a combination of sea floor spreading and reversals of Earth's magnetic field (Vine and Matthews, 1963; Vine, 2001). In illustrating our idea with some forward modeling, we acknowledged that the crust, and possibly part of the upper mantle, was potentially magnetic to the depth of the Curie point isotherm (5–600 °C), and we assumed that at a ridge crest this was at a depth of 11 km below sea level. This meant that the average intensity of magnetization we assumed was much less than that obtained for basalts. We wrote "…it is suggested that the fine-grained extrusives (basalts) of surface features are more highly magnetized than the intrusive material of the main crustal layer, which, in the absence of evidence to the contrary, we assume to be of analogous chemical composition (that is, gabbros)" (Vine and Matthews, 1963, p. 949). In contrast to Harry Hess, therefore, we were maintaining that seismic layers 2 and 3 were of mafic rather than ultramafic composition. Certainly, as others had discovered previously (e.g., Mason, 1958), I found it necessary to have essentially vertical magnetization contrasts extending up into, if not throughout, layer 2 in order to explain the observed anomaly gradients.

Just as I might have persuaded Harry Hess when I first met him in early 1965 that layer 2 consists of basalts, I think it is clear from my writings later in 1965 (Vine and Wilson, 1965;Vine, 1965) that he managed to convince me that layer 3 might well be serpentinite. This was not a problem for me in that, by this time, I had decided that, to a good first approximation, the anomalies could be explained by a strongly magnetized, basaltic layer 2. Thus, in Vine and Wilson (1965, page 488) I wrote, "…it seems increasingly probable that the observed anomalies can best be reproduced by strongly magnetized, basaltic material in layer 2 and less strongly magnetized material at depth, whether this decrease in magnetization be due to a general increase in grain size, a change in rock type, or metamorphic effects."

In August 1965, I completed my Ph.D. thesis and moved to a teaching position in Harry Hess' department at Princeton. My first year at Princeton coincided with the final year for Eldridge Moores, before he left for Davis. Eldridge and I were great admirers of Harry Hess, his work and ideas. During his time at Princeton, Eldridge had been studying the Vourinos complex of northern Greece, and had recently completed his seminal work on this ophiolite (Moores, 1969). Although my Ph.D. studies had been in the realm of marine geophysics, as an undergraduate I had specialized in petrology and mineralogy. Eldridge and I, therefore, had a lot in common, and we were keen to pursue the ophiolite analogy to ocean crust.

In finally accepting that ophiolites were fragments of ocean floor, Harry Hess had written (Hess, 1965, p. 331), "Often…they are found as large exotic blocks…or as overthrust sheets (northern Greece, Cyprus, Syria, and New Caledonia)…a re-study of this suite could give first-hand information about oceanic ridges which would be far better than the blind grappling for dredged samples." The inclusion of Cyprus in this list is rather surprising. When Eldridge and I alighted on the Troodos massif of Cyprus as the most promising ophiolite for the study of the structure and petrology of ocean floor by sea-floor spreading, Harry as good as told us we would be wasting our time. In fact, at that time, the Troodos was typically not referred to as an ophiolite (Gass, 1980). The outcrop area is predominantly mafic in composition and the plutonic section had been interpreted as a differentiated ultrabasic mass of batholithic dimensions, the parent, ultrabasic, magma being derived from the mantle by partial or complete fusion (Gass and Masson-Smith, 1963); however, the whole massif had been interpreted by Gass and Masson-Smith as an upthrust slice of ocean floor resulting from the ongoing collision of Africa and Eurasia. Harry clearly considered the mafic section to be far too thick to be consistent with his concept of oceanic crust with a main crustal layer of serpentinite.

We were attracted to the Troodos because it seemed to be relatively undisturbed tectonically, containing all the elements of the ophiolite suite—ultramafics, gabbros, diabase and pillow lavas, and overlain by deep-sea sediments. We were particularly intrigued by the presence of a unit that was, as far as we could tell, unique in the literature on ophiolites at that time. It was described as a "sheeted dike complex," a horizon consisting entirely of dikes, typically striking north-south, and implying 100% extension of the crust during formation of the massif (Wilson, 1959). Moreover, this unit had the most extensive outcrop area of any of the whole complex, and an across-strike outcrop width of 100 km.

Eldridge and I had major field seasons on Cyprus in 1968 and 1969. We were surprised by how readily one could reinterpret the Troodos as crust formed by sea-floor spreading. As anticipated, the clincher was the sheeted dike complex, but nothing was incompatible with the spreading model. We were particularly struck by the field relationships between the various units. The sheeted dikes, for example, give way, probably within no more than one hundred meters vertically, to a unit consisting of 90% pillow lavas. The full significance of this relatively abrupt transition was only realized some years later, as discussed below. At the base of the dikes there is a transition zone in which dikes intrude the gabbro and, in places, gabbro intrudes the dikes. Multiple intrusive relationships also characterize the ultramafic cumulate units, the pyroxenites, harzburgites and dunites. We deduced that the main, gabbro crustal layer must be created from a number of magma chambers rather than a single, large one (Moores and Vine, 1971, Vine and Moores, 1972). Soon, however, single magma chamber models, based on the individual units, without reference to the field relationships between them, were formulated (Greenbaum,1972; Cann, 1974). The latter gained widespread acceptance for the following fifteen years or so, until seismic and drilling data at sea (Dick et al., 1991) and further work on the Troodos (Malpas et al., 1989) suggested multiple magma chambers. For Eldridge, there were remarkable similarities between the plutonic section and that in the Vourinos complex that he had worked on previously. Particularly significant and satisfying was his reinterpretation of the ultramafic section. As on Vourinos, where he was able to demonstrate it for the first time, he recognized that the ultramafic cumulates are underlain by metamorphosed and tectonized harzburgite (and occasionally lherzolite) that represents residual mantle (i.e. mantle from which the basaltic fraction had been removed by partial melting). The boundary between the two is somewhat deformed, and characterized by dunite with chromitite pods. Appropriately, this soon became known as "the petrologic Moho."

We also undertook a paleomagnetic study during our field seasons on Cyprus, in the hope that we would find normally and reversely magnetized rocks, and, ideally, would map out the boundary between them. To my chagrin, however, the natural remanent magnetization (NRM) of everything we measured in the field, using a mini fluxgate magnetometer, was similarly directed, and probably magnetized in a field of normal polarity. Ian Gass and I subsequently obtained grants to acquire an aeromagnetic survey of the Troodos massif, which also confirmed that there are no normal/reverse magnetization boundaries within the outcrop area (Vine, Poster and Gass, 1973). Ultimately, more accurate dating of the complex (Blome and Irwin, 1985) and documentation of the Cretaceous normal superchron (Irving and Couillard, 1973) made it clear that the massif had been formed 90 m.y. ago in a field of uniform normal polarity. The magnetic properties of the different units were of considerable interest, however (Vine and Moores, 1972; Vine, Poster and Gass, 1973), as was the paleomagnetic vector, which indicated that the massif has been rotated anticlockwise through ~90° since the time of its formation (Vine and Moores, 1969).

The "ophiolite model" for oceanic crust, which was undoubtedly influenced by the Troodos section, was adopted by the Penrose Conference on ophiolites in 1972 (Anonymous, 1972), and subsequently confirmed to a depth of 2 km by deep-sea drilling at Hole 504B on the Costa Rica rift in the eastern central Pacific (Dilek, 1998).

It was during our second field season on Cyprus in 1969 that we received the devastating news of Harry Hess' sudden death from a heart attack. The news was brought to us by Augusto Gansser, the remarkable Swiss geologist with an encyclopedic knowledge of the Alpine-Himalayan ophiolites. He was visiting us to add first-hand experience of the Troodos to his collection. To the last, Harry was convinced that serpentinite played an important role in the formation of oceanic crust. In his contribution to our joint paper on *Sea-floor spreading* in *The Sea,* Volume IV (Vine and Hess, 1971), written in late 1968, he included a section on "Rocks dredged from the ocean basins," in which he documented the occurrence of serpentinite in dredge hauls to that date. Although the number and distribution of dredge hauls must have been rather limited at that time, it was striking that all the records of serpentinite were from the Atlantic and Indian oceans and none was from the Pacific. I remember wondering whether this might be related to the systematically higher spreading rates in the Pacific, compared to the Atlantic and Indian oceans. It was to be a further twenty years before the detailed differences between slow and fast spread crust were fully appreciated, notably the importance of amagmatic extension on slowly spreading ridges. This gives rise to outcrops of serpentinized ultramafics on major faults, and even areas of "Hess-type" crust (Cannat et al., 1995). Unfortunately, Harry did not live long enough to see himself vindicated, at least in part, in this way (he died at the age of 63). Fortunately, he did live to see the confirmation of his hypothesis of sea-floor spreading in terms of the magnetic anomalies and deep-sea drilling (Maxwell and Von Herzen, 1969).

For my own part, I always had problems with the concept of oceanic crust composed almost exclusively of serpentinite. I had difficulties in understanding how it was emplaced at shallow depths at mid-ocean ridge crests and also why it was always ~70% serpentinized so that it could account for layer 3 seismic velocities. Harry maintained that the serpentinites of Alpine-type peridotites were 70% serpentinized. Initially, my own view was that it only occurred at the surface as a result of diapiric intrusion along fracture zones and other major faults, or possibly within the lower part of layer 3. This latter view was largely prompted by the fact that the mafic section in most, and possibly all, ophiolites appears to be too thin to account for the total thickness of layers 2 and 3 (Vine and Moores, 1972). In 1987, in reviewing the results of the International Crustal Research Drilling Group (ICRDG) drilling program on the Troodos (Vine and Smith, 1990), I suggested that the relatively thin gabbroic section drilled (1 km), might result from the proximity of the massif as a whole, and the drill hole which penetrated the plutonic section in particular, to a transform fault. During our first field season on Cyprus, in 1968, Eldridge and I had interpreted the east-west trending Arakapas fault zone, toward the southern margin of the massif, as a transform fault. In working southward across the massif, we were struck by the sudden appearance of extensive fault and talus breccias, and even turbiditic, volcaniclastic sediments within this zone (Moores and Vine, 1971; Simonian and Gass, 1978). By the late 1980s, it was realized that the mid-ocean ridge crests are segmented, and that within a segment, as one moves away from the main magma source at its center, the crust thins toward the transform faults and other types of segment boundary. Furthermore, seismic work had shown that, not only is the mafic section thinner, but that the uppermost mantle in the vicinity of fracture zones is serpentinized, judging by the anomalously low seismic velocities (White et al., 1984). Many ophiolites are now thought to include transform fault zones or to have been emplaced because of their proximity to a transform fault. It seems probable, therefore, that the thin mafic section of many ophiolites is explicable either in this way, or by the fact that they were formed during a period of amagmatic extension, which produced "Hess-type" crust. Such extension is now known to be common on slow spreading ridges, and may be particularly characteristic of the first formed oceanic crust at slowly extending non-volcanic margins (Whitmarsh et al., 2001).

THE FORMATION OF OCEANIC CRUST

When Drum Matthews and I put forward the idea that oceanic magnetic anomalies might result from a combination of sea-floor spreading and reversals of Earth's magnetic field, we were working on magnetic profiles across the Mid-Atlantic

Ridge, in the north Atlantic, and the Carlsberg Ridge, in the northwest Indian Ocean (Vine and Matthews, 1963). None of these profiles showed an obvious symmetry of the anomalies about the ridge crests, which would be expected from a literal interpretation of the idea, and there were no published surveys from these areas suggesting that the anomalies are elongate parallel to the strike of the ridge crests. With regard to the latter point, however, we did have some new data. In 1962, Drum had carried out a detailed survey, from *H.M.S. Owen*, on the crest of the Carlsberg Ridge at 5° N (Matthews, Vine and Cann, 1965; Cann and Vine, 1966). It covered an area of just 50 by 40 nautical miles and included a fracture zone producing a small offset of the ridge crest. The remaining two-thirds of the area covered part of a spreading ridge segment. The striking feature of this latter area was that the magnetic highs and lows, and, in particular, the steep anomaly gradients between them were lineated parallel to the ridge crest and that this magnetic lineation was much more pronounced than the topographic lineation. It was this feature, coupled with the directions of magnetization deduced for the topography in this area, that convinced us that there might be some merit in the idea. Its extrapolation from this small area to the whole of the ocean basins, however, could be described as a fairly bold step.

As geologists, Drum and I were not particularly surprised or worried by the fact that there was no obvious symmetry in the magnetic anomalies. We assumed that, within the context of sea-floor spreading, new oceanic crust must be formed by a process of intermittent intrusion and extrusion of mafic material, within a fairly diffuse zone, at the ridge crest. As a consequence, with the exception of the crust formed most recently and hence during the present polarity of Earth's magnetic field, all off-axis crust would be "contaminated" with material of the opposite polarity to that in which it was initially formed. The degree of contamination would depend on the width of the zone of emplacement and the length of the polarity interval, although at the time it was thought that reversals might occur at a fixed interval of 0.5 to 1.0 m.y. (Cox, Doell and Dalrymple, 1963). The complexity of this off-axis crust would degrade the magnetic anomalies in a random way, destroying any potential symmetry. The topography on the strongly magnetized extrusives would also add "noise" to the magnetic profiles. In this way, we felt that we had an explanation of both the lack of symmetry and the high amplitude of the central anomaly at ridge crests.

As a result of further work on the magnetic anomalies across the crest of the Mid-Atlantic Ridge, at 45° N, Drum developed and refined a quantitative model of this emplacement process (Loncarevic, Mason and Matthews, 1966; Matthews and Bath, 1967; Matthews, 1969). It is clear from our paper in 1963 that we had turned to Iceland, rather than to ophiolites, for inspiration regarding ocean crust formation. Subsequently, we were both greatly influenced by the paper of Bodvarsson and Walker (1964), "Crustal Drift in Iceland." Basically, Bodvarsson and Walker postulated that Iceland is underlain by a dike swarm, individual dikes feeding lava flows at the surface via fissure eruptions. At any one time, the density of recent dikes decreases as one moves away from the mid-point of the active central rift. Initially, Drum modeled such a process of dike intrusion by rolling colored marbles (red for normal, blue for reverse magnetization) down an inclined plane into a horizontal line of pale colored marbles. The red and blue marbles were always released from the same point, but on hitting the line of marbles they bounced, and were "injected" into the line at some point, pushing the pale colored marbles to the right and left. The "average" bounce (standard deviation of the distance of the point of injection from the mid-point) could be varied by changing the inclination of the plane and/or the length of roll. A sequence of, say, ten red marbles was followed by a sequence of ten blue marbles and so on until all the pale marbles had been eliminated from the center of the line. The degrees of contamination were then compared to those required in modeling the observed magnetic anomalies over the Mid-Atlantic Ridge, in order to deduce the width of the zone of dike injection. To lend a greater air of respectability to the model, Drum and Jennifer Bath devised a comparable computer simulation which used a series of pseudo-random numbers, normally distributed about zero and having a specified standard deviation, to calculate the points of injection of the dikes. From both the analogue and numerical models, they deduced that the standard deviation that gives the best simulation of the observed magnetic profiles is 5 km, implying that almost all of the mafic material of seismic layer 2 is emplaced within the median valley at the ridge crest.

At about the same time, Jason Morgan and I (Vine and Morgan, 1967), and Chris Harrison (Harrison, 1968) developed similar computer models. I had been surprised by the remarkable symmetry of the anomalies developed over the Juan de Fuca and Pacific-Antarctic ridge crests, and their correlation with the most recent reversals of Earth's magnetic field. The latter not only implied constant spreading rates, but rates that are two or three times higher than those of the north Atlantic and northwest Indian ocean ridges (Vine and Wilson, 1965; Pitman and Heirtzler, 1966; Vine, 1966). I wondered whether the clarity and symmetry of these Pacific profiles was simply a result of the wider avenues of crust of a particular polarity, or whether it implied a narrower zone of formation of the crust at faster spreading ridge crests; thus, the simulations carried out with Jason Morgan were for a number of ridges with widely varying spreading rates. The results were presented at the Annual Meeting of the GSA in 1967 (Vine and Morgan, 1967), but were never published in detail. Figure 1 shows two of the diagrams used in the GSA presentation. We found that, in order to simulate the observed anomalies, we did need to assume a narrower zone of formation at the faster spreading ridge crests. Thus, for a normal distribution of points of injection about a ridge axis, whereas a standard deviation of 3 km or more produced good simulations for north Atlantic profiles, a standard deviation of 2 km or less was required to produce the remarkable details of the Pacific profiles—an early pointer to one of the essential differences between crust formed at fast and slow spreading ridges.

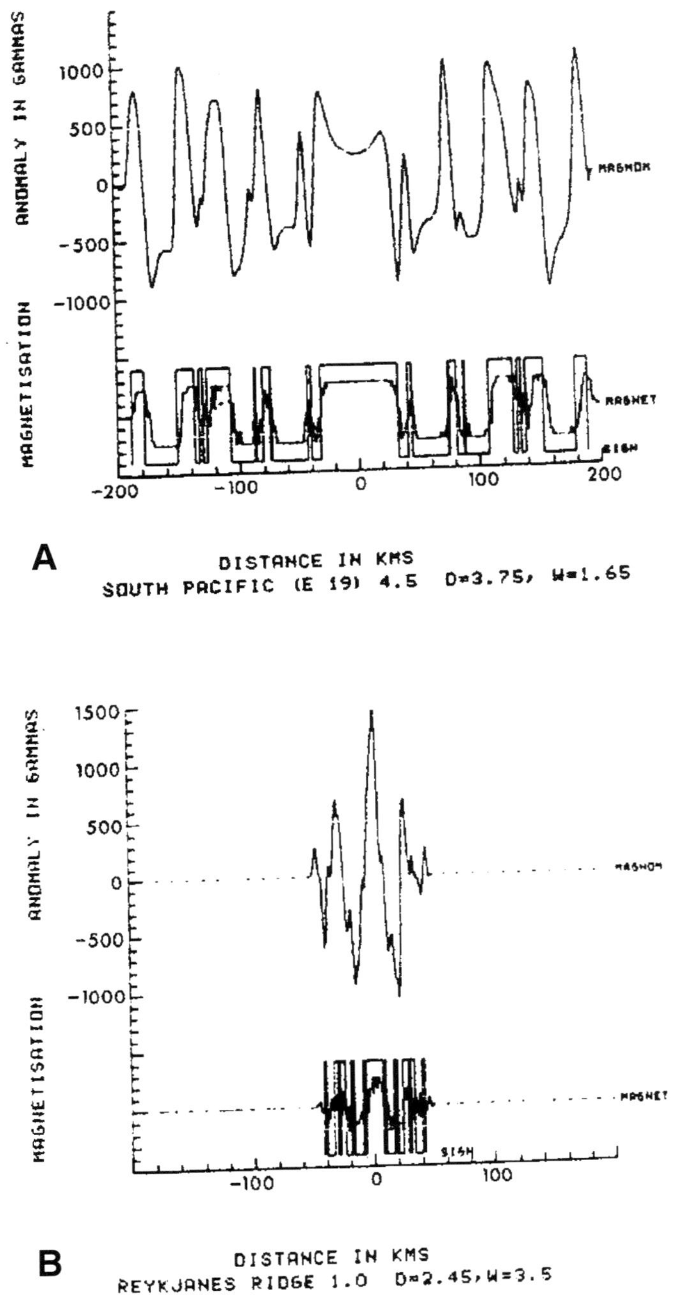

Figure 1. Examples of calculated magnetic anomalies over ridge crests, where the magnetic sources are randomly emplaced at points which are normally distributed about the ridge axis. D—depth to lines of magnetic dipoles, 100 m apart, paralleling the ridge axis, and within seismic layer 2 (in km). W—standard deviation of zone of emplacement (in km). A: for the South Pacific at 51° S, spreading rate 4.5 cm per year. B: for the Reykjanes ridge, south of Iceland, at 60° N, spreading rate 1.0 cm per year. The square wave indicates the changing polarity of Earth's magnetic field, and the "magnetization" curve a 1 km moving mean of "net" magnetization, reflecting the degree of contamination of crust of one polarity by material of the opposite polarity. 1 gamma = 1 nT (nanoTesla).

Thus, Drum Matthews and his co-workers deduced a greater "half-width" for the zone of emplacement on the North Atlantic Ridge than we did (5 km as opposed to 3 km). I think that the explanation for this can be found in the summary of my work with Jason that I included in the *History of the Earth's Crust* volume (Vine, 1968). Drum concentrated not only on degrading the anomalies so that there was no obvious symmetry, but also on obtaining a sufficient degree of contamination of off-axis crust to account for the high amplitude of the central anomaly. It is clear from my summary, however, that I felt that viscous decay of the remanent magnetization with time was also a contributing factor and that it was not necessary therefore to account for this difference in amplitude solely by contamination, as Drum and I had originally proposed. Subsequently, Ted Irving established that oxidation of oceanic basalts, much of it within half a million years of their formation, did indeed produce a decay in the intensity of remanent magnetization (Irving, 1970). It would appear that I deduced it indirectly from the gradual decay in the amplitude of the observed magnetic anomalies as one moves away from ridge crests, which was not predicted by computer modeling. An alternative explanation of this decay could be a gradual decrease in the intensity of Earth's magnetic field back in time; however, I regarded the decay of remanence with time as being more likely.

In 1970 I returned to the UK, to a position in the School of Environmental Sciences at the then very new University of East Anglia in Norwich. There, Joe Cann, whom I had known at Cambridge, had also been working on models for the formation of oceanic crust (Cann, 1968; 1970). Joe was one of relatively few people at that time who had done extensive work on dredged rocks from oceanic ridges. By 1970, he had devised a theoretical model for oceanic crust formed by sea floor spreading, constrained largely by petrological and geochemical arguments. He also pointed out the remarkable similarities between this and the model derived from the recent reinterpretation of the Troodos ophiolite by Ian Gass (Gass, 1968) and Eldridge Moores and me (Moores and Vine, 1969), and of the Oman ophiolite by Bernhard Reinhardt (Reinhardt, 1969) (Cann, 1970).

On arriving in Norwich, I discovered that Joe had been developing his ideas further, particularly relating to the geometry of the lava flows and dikes in the upper part of the crust. From some simple but elegant assumptions, he had deduced that the rate of increase in the percentage of dikes with depth in the pillow lavas is a function of the width of the zone of emplacement: a very gradual rise, followed by a very rapid increase, indicating a narrow zone of intrusion. In addition, the probability of one-way chilling in the dike complex would increase in this situation. Within the Troodos dike complex, the common occurrence of dikes with only one chilled margin, presumably a result of dikes being intruded by later dikes, had been noted by Eldridge and me. I think that Joe had been intrigued by this phenomenon when Eldridge mentioned and illustrated it in his presentation at the discussion meeting held at the Royal Society, London, in November 1969 (Moores and Vine, 1971). Clearly these ideas were a stimulus to investigate further the details of one-way chilling, and the nature of the pillow lava/dike transition on the Troodos. Joe obtained a grant to carry out further fieldwork on Cyprus, and in 1973, Joe, Rupert Kidd, then a graduate student, and I spent a month on Cyprus studying the transition, and the extent of one-way chilling in the dike complex. If the latter proved to be systematic, it would also indicate the spreading direction or directions within the Troodos. I must confess that my main, possibly sole, contribution to this project was to identify suitable localities for the work. In addition to forming the basis of Rupert's thesis (Kidd, 1976), the results of this project were published by Joe and Rupert in Kidd and Cann (1974) and Kidd (1977).

Those of us who, constrained by simulation of the magnetic anomalies, had devised random injection models to deduce the width of the zone of emplacement of the upper oceanic crust, were acutely aware that the models were very crude in terms of their representation of the structure of the crust. The points of injection were typically 100 m apart, and the magnetic profiles were calculated assuming dikes 100 m wide, or lines of dipoles 100 m apart and in the middle of seismic layer 2. The only way one could justify these assumptions, at least in part, was to note that the magnetic anomalies are measured several kilometers above the magnetic sources and, hence, the effect of an individual intrusion or extrusion is combined with that of many others to produce a net effective magnetization throughout the thickness of the magnetic layer at this point. In detail, of course, this layer consists of lava flows, dikes and, possibly, gabbros. What Joe and Rupert Kidd demonstrated, both theoretically and with reference to the Troodos, was that the key to understanding the width of the zone of intrusion was the rate at which the percentage of dikes within the pillow lavas increases with depth. If the percentage increases very gradually at first but then very rapidly—from, say, 20 to 90% within a vertical distance of ~100 m—then this implies a zone of dike injection no more than 50–100 m wide. The associated lavas, however, will flow for several kilometers from the point of extrusion; hence, it is the lavas that determine the maximum width of the zone of emplacement and the lateral extent of the intermingling of normal and reverse polarity material at a reversal boundary. This "contamination" of the vertically-sided blocks of the earlier magnetic models, however, is largely due (particularly on the faster spreading ridges) to the fact that the younger flows of one polarity overlie flows of the opposite polarity and the boundary between them dips toward the ridge axis.

Rupert and I followed up the implications of these ideas for the forward modeling of the magnetic anomalies and the extent to which such simulations differed from those obtained using vertically-sided reversal boundaries in the source layer. Rupert (Kidd, 1977) wrote about one aspect of this, the nature and shape of the magnetic sources in oceanic crust, but the actual simulations were never published. The differences were subtle and, indeed, one of the main purposes of carrying out such calculations was to convince ourselves and others that despite the fairly dramatic departure of the proposed shape of the source layer from vertically-sided blocks (Fig. 2), it has very little effect on the main features of the calculated profiles. This is essentially because the lavas are more strongly magnetized and closer to the point of measurement of the magnetic field than the dikes and gabbros. The sloping boundary in the lava layer gives rise to an offset of ~1 km between the position of the reversal boundary in the crust predicted by the dikes in this model and that deduced from vertically-sided blocks. This offset is small because of the limited horizontal extent of this slope, and is, in general, of no consequence, except for the central block at a ridge crest, i.e., the crust formed during the present polarity interval of Earth's magnetic field. According to the model, the width of this block will be slightly overestimated if one assumes vertically-sided blocks. The amplitude of the anomaly developed over the central block is slightly enhanced (Fig. 3), but in a sense, this is no different to the point made by Drum and myself in 1963, in that the sloping boundaries in the lava and gabbro layers ensure that

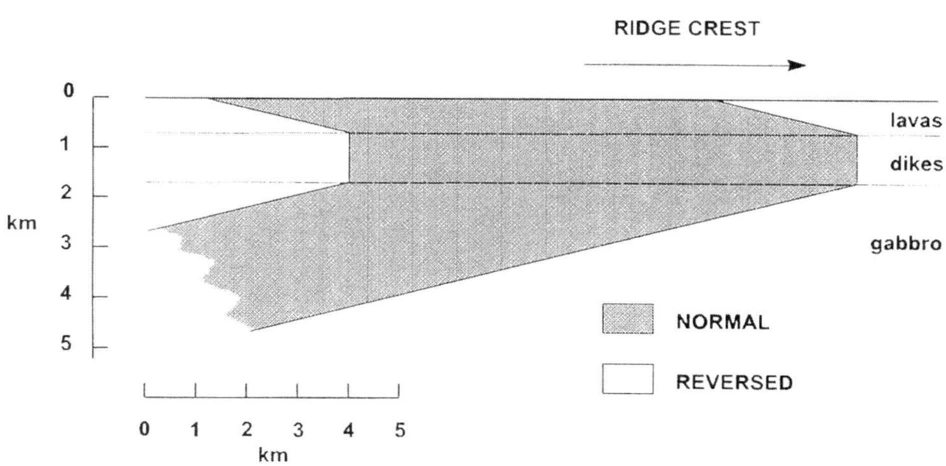

Figure 2. Shape of normal/reverse magnetization boundaries in oceanic crust proposed by Kidd (1977). Within the lava and dike layers the boundary is defined by a "transition zone," perhaps 100–200 m in width, in which some lavas and dikes have intermediate directions of magnetization, reflecting the reversing field, and others opposing polarities, as a result of random injection within a narrow zone at the ridge axis. (See Figure1 and text.) The dip on the boundary within the gabbro layer reflects the position of the Curie point isotherm in this layer. This dip is thought to be ~5/s, where "s" is the spreading rate in mm per year. The dip illustrated is for a spreading rate of 20 mm per year.

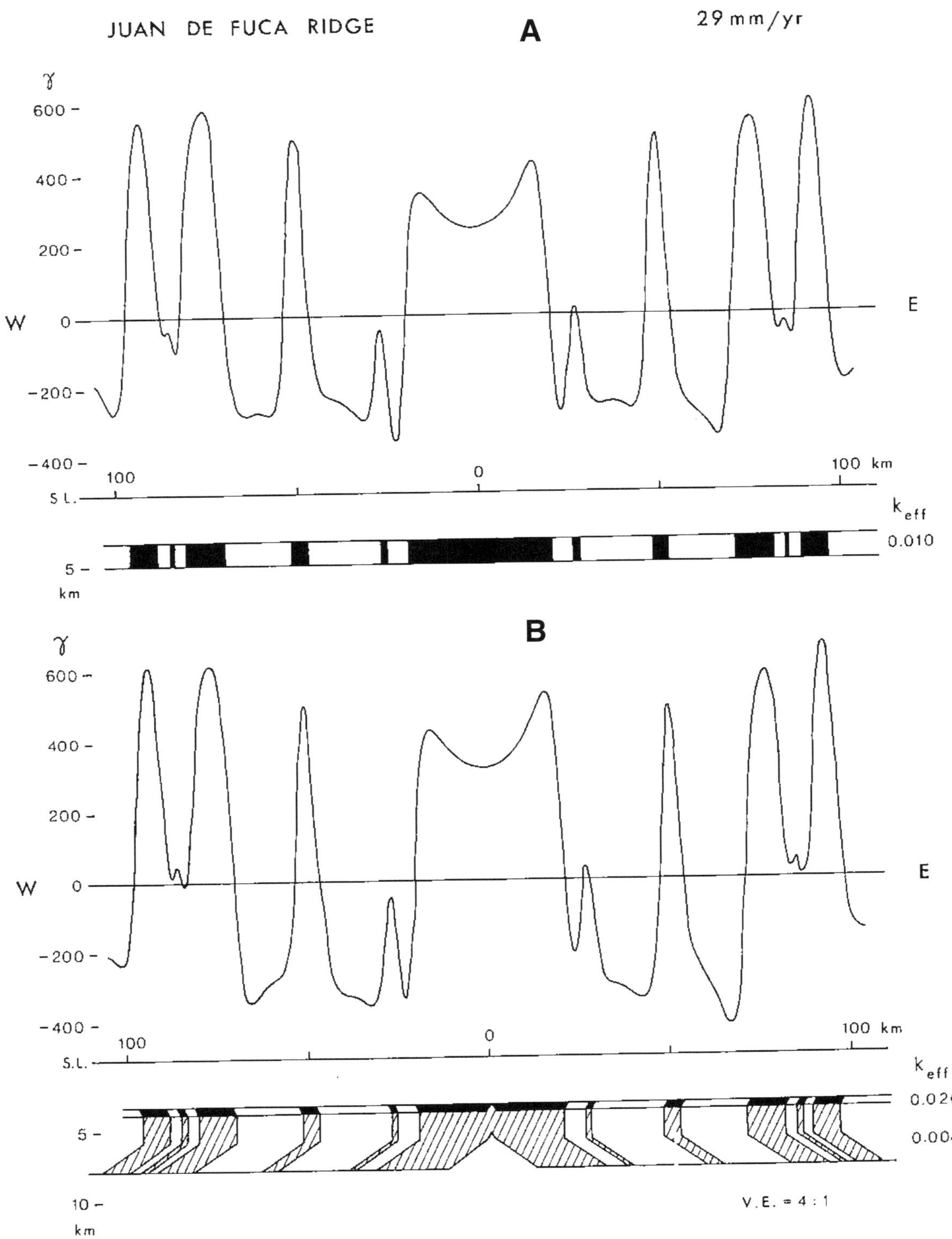

Figure 3. Computed magnetic anomaly profiles across the crest of the Juan de Fuca Ridge, southwest of Vancouver Island, assuming (A) vertically-sided blocks of normally and reversely magnetized material within seismic layer 2, and (B) magnetic sources throughout seismic layers 2 and 3, with normal/reverse boundaries analogous to those illustrated in Figure 2. 1 gamma equals 1 nT. k_{eff} × (intensity of Earth's magnetic field) equals intensity of magnetization of the layer.

all (vertically-sided) blocks other than the central block are contaminated with material of the opposite polarity.

In the gabbro layer, the normal/reverse boundary, because of its shallower slope, has a much greater horizontal extent than that in the lavas, and has the potential to produce more significant differences in the shape and amplitude of the anomalies. For narrow blocks, i.e., those generated by short polarity intervals and/or slow spreading rates, thin dipping layers of gabbro of different polarities are stacked one above the other, and their net magnetic effect is greatly reduced, thereby decreasing the amplitude of the simulated anomaly. This reduction in amplitude is only predicted for blocks less than 20 km in width, and the effect is greatest for block widths of 10 km or less (see Figures 4A and 4B). This corollary of the model might well explain the low amplitudes of the anomalies developed over crust of post-Eocene age on the flanks of the Mid-Atlantic Ridge, for the spreading rates are low and the frequency of reversals is relatively high for this period of time (Kidd, 1977).

The shape of the source layer in the dikes and gabbros also produces an asymmetry or "phase-shift" in the simulated anomaly (Fig. 4B). Tantalizingly, the sense of the phase-shift is the same as that of the "anomalous skewness" of observed anomalies (Cande, 1976); however the effect as illustrated in Figure 4, is exaggerated because the thickness of the lava layer in this model is only 500 m and, arguably, rather thin. This, however, is counteracted by the fact that the effect of the gabbro layer is reduced because of the block width (10 km), so the degree of asymmetry shown is appropriate to that produced by a wider block with a thicker lava layer. With time, as mentioned above, the remanent magnetization of the lavas decays, but the remanence in the dikes and gabbros is thought to be more stable (Bleil and Petersen, 1983). Thus, one would predict that a greater degree of skewness will be developed over older oceanic crust, particularly over wide blocks. Despite this, the magnitudes of the phase shifts produced by this model are inadequate to explain some of those observed, and undoubtedly other effects are involved in these cases (Raymond and LaBrecque, 1987).

SUBSEQUENT WORK ON THE TROODOS OPHIOLITE

In the 1980s, along with many others, I was involved in further detailed studies of the Troodos ophiolite as a participant in the Cyprus Crustal Study Project (CCSP) of the ICRDG. One of the great strengths of this project was that, in addition to the drilling, there was an extensive fieldwork program, much of it built around Ph.D. projects. For my own part, I supervised two Ph.D. students, one using paleomagnetic techniques to study structural rotations within the ophiolite (Allerton, 1988), and the other measuring a range of physical properties of the drill core (Smith, 1986).

Ken Verosub and Eldridge Moores had suggested that there are domains of rotated blocks within the sheeted dike complex, which they attributed to listric faulting associated with amagmatic extension (Verosub and Moores, 1981). As a result of further mapping, Robert Varga and Eldridge identified three grabens within the Troodos outcrop area, which they interpreted as fossil ridge crests (Varga and Moores, 1985). Perhaps the most convincing of these is the Solea graben, to the north of Mount Olympus and the central serpentinite diapir. Basically, Simon Allerton and I provided support for this by paleomagnetically confirming the implied block rotations about horizontal axes. We also documented rotations of the dikes about vertical axes, in the vicinity of the Arakapas transform fault, and suggested that these characterized areas formed at the inside corner of a

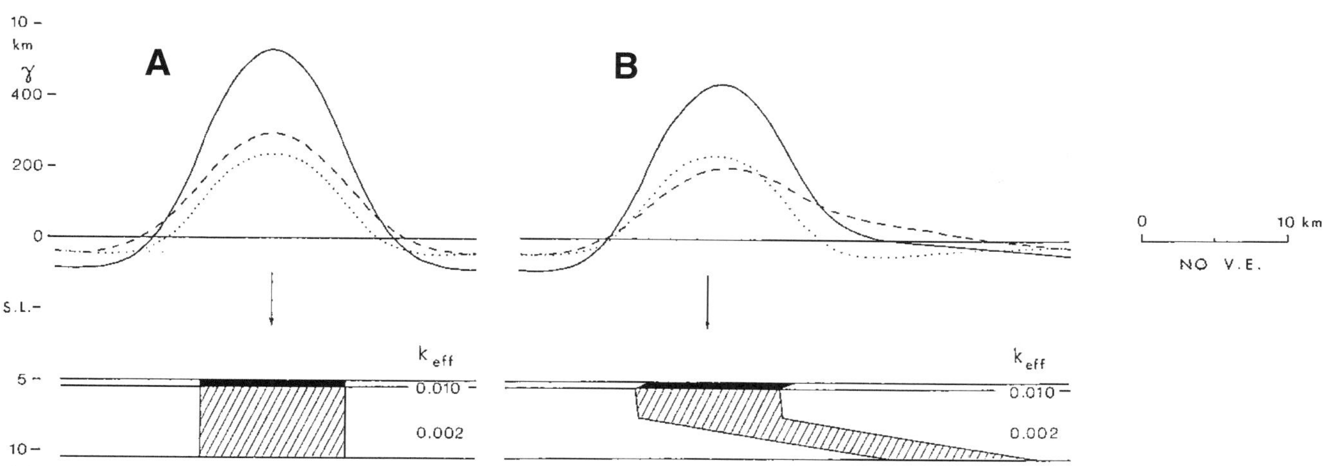

Figure 4. Computed magnetic anomalies across normally magnetized blocks at the north dip pole. A and B illustrate possible relative contributions from the lava layer (dotted curve), and the dikes plus gabbro layer (dashed curve), and (B) the asymmetry (phase shift) introduced by sloping block boundaries in the gabbro layer. For symbols, see caption to Figure 3.

ridge-transform intersection (Allerton and Vine, 1991; 1992). This implies that the South Troodos (Arakapas) fault was a dextrally-slipping transform when active; however, this and other interpretations of the structure in the vicinity of the Arakapas remain controversial (Cann et al., 2001).

Pronounced gradients in the physical properties of the drill core samples with depth, notably through the dike complex, made it possible to "stack" the three stratigraphic drill holes of the CCSP, suggesting that the crustal section was sampled to a depth of ~3.5 km (Vine and Smith, 1990). These gradients are present in the measurements of seismic velocities, density, porosity and electrical conductivity. They are more pronounced than those in in-situ ocean floor, and are a consequence of the increased volatile content of the Troodos magmas, a result of their formation in a supra-subduction zone environment (Robinson et al., 1983). Thus, the seismic velocities and densities of the pillow lavas are low, and their porosity and electrical conductivity is high, compared to oceanic crust formed at mid-ocean ridges. Differences in the geochemistry also make the magnetic properties of the pillow lavas unrepresentative of those of mid-ocean crust; the NRM, for example, is systematically higher, as previously noted by Banerjee (1980). The physical properties of the gabbros and ultramafics, however, may well be good indicators of those of in-situ oceanic crust (Vine and Smith, 1990). As with the earlier paleomagnetic study, the inclination of the NRM of the serpentinized cumulate ultramafics is consistent with formation in a Cretaceous magnetic field. That the remanent magnetization almost certainly resides in the magnetite formed during the serpentinization process again raises the possibility that the lower oceanic crust may consist of, or at least include, partially serpentinized ultramafics.

CONCLUSION

Clearly, I was extremely fortunate not only to have set out on my research career at an exciting time, in terms of new ideas and data relating to the nature of oceanic crust and of ophiolites, but also to have worked with some of the key researchers who were at the forefront of these advances. My main contributions were fairly specific—the interpretation of anomalies in Earth's magnetic field, and paleomagnetic studies. As it happened, these provided significant constraints on the nature of oceanic crust, its mode of formation, and its internal structure. As one of the few examples—indeed, the main example—of paleogeophysics, it was perhaps inevitable that paleomagnetism should play an important part in unraveling the geology of the ocean floor and of ophiolites.

ACKNOWLEDGMENTS

Undoubtedly, my greatest debt of gratitude must go to pillow lavas, whose fine-grained, single-domain titanomagnetites preserve a particularly strong and stable remanent magnetization. Without them, this would have been a very different story. In addition to my indebtedness to pillow lavas and to my many collaborators, I thank the U.S. Office of Naval Research, the U.S. National Science Foundation, and the U.K. Natural Environment Research Council for financial support, and many members of the Cyprus Geological Survey Department, in particular, George Constantinou, Andreas Panayiotou, and Costas Xenophontos, whose hospitality, friendship and assistance made visits to Cyprus so memorable and enjoyable.

REFERENCES CITED

Allerton, W.S.Q., 1988, Paleomagnetic and structural studies of the Troodos ophiolite, Cyprus [Ph.D. thesis]: Norwich, England, University of East Anglia, 243 p.

Allerton, S., and Vine, F.J., 1991, Spreading evolution of the Troodos ophiolite, Cyprus: Geology, v. 19, p. 637–640.

Allerton, S., and Vine, F.J., 1992, Deformation styles adjacent to transform faults: evidence from the Troodos ophiolite, Cyprus, *in* Parson, L.M., Murton, B.J., and Browning, P., eds., Ophiolites and their modern oceanic analogues: London, Geological Society of London Special Publication 60, p. 251–261.

Anonymous, 1972, Penrose field conference on ophiolites: Geotimes, v. 17(12), p. 24–25.

Banerjee, S.K., 1980, Magnetism of the oceanic crust: evidence from ophiolitic complexes: Journal of Geophysical Research, v. 85, p. 3557–3566.

Bleil, U., and Petersen, N., 1983, Variations in magnetization intensity and low temperature titanomagnetite oxidation of ocean floor basalts: Nature, v. 301, p. 384–388.

Blome, C.D., and Irwin, W.P., 1985, Equivalent radiolarian ages from ophiolitic terranes of Cyprus and Oman: Geology, v. 13, p. 401–404.

Bodvarsson, G., and Walker, G.P.L., 1964, Crustal drift in Iceland: Geophysical Journal of the Royal Astronomical Society, v. 8, p. 285–300.

Cande, S.C., 1976, A paleomagnetic pole from late Cretaceous marine magnetic anomalies in the Pacific: Geophysical Journal of the Royal Astronomical Society, v. 44, p. 547–566.

Cann, J.R., 1968, Geological processes at mid-ocean ridges: Geophysical Journal of the Royal Astronomical Society, v. 15, p. 331–341.

Cann, J.R., 1970, New model for the structure of the ocean crust: Nature, v. 226, p. 928–930.

Cann, J.R., 1974, A model for oceanic crustal structure developed: Geophysical Journal of the Royal Astronomical Society, v. 39, p. 169–187.

Cann, J.R., and Vine, F.J., 1966, An area on the crest of the Carlsberg ridge: petrology and magnetic survey: Philosophical Transactions of the Royal Society, London, Series A 259, p. 198–217.

Cann, J.R., Prichard, H.M., Malpas, J.G., and Xenophontos, C., 2001, Oceanic inside corner detachments of the Limassol forest area, Troodos ophiolite, Cyprus: Journal of the Geological Society of London, v. 158, p. 757–767.

Cannat, M., Mevel, C., Maia, M., Deplus, C., Durand, C., Gente, P., Agrinier, P., Belarouchi, A., Dubuisson, G., Humler, E., and Reynolds, J., 1995, Thin crust, ultramafic exposures, and rugged faulting patterns at the mid-Atlantic ridge (22°–24° N): Geology, v. 23, p. 49–52.

Cox, A., Doell, R.R., and Dalrymple, G.B., 1963, Geomagnetic polarity epochs and Pleistocene geochronology: Nature, v. 198, p. 1049–1051.

Dick, H.J.B., Meyer, P.S., Bloomer, S., Kirby, S., Stakes, D., and Mawer, C., 1991, Lithostratigraphic evolution of an in situ section of oceanic layer 3, *in* Von Herzen, R.P., Robinson, P.T., et al., eds., Proceedings of the Ocean Drilling Program, Scientific Results, v. 118, College Station, Texas, Ocean Drilling Program, p. 439–538.

Dilek, Y., 1998, Structure and tectonics of intermediate-spread oceanic crust drilled at DSDP/ODP Holes 504B and 896A, Costa Rica Rift, *in* Cramp, A., MacLeod, C.J., Lee, S.V., and Jones, E.J.W., eds., Geological evolution of ocean basins: results from the Ocean Drilling Program: London, Geological Society of London, Special Publications, v. 131, p. 179–197.

Ewing, M., and Press, F., 1955, Geophysical contrasts between continents and ocean basins, *in* Poldervaart, A., ed., Crust of the Earth: Boulder, Colorado, Geological Society of America Special Paper 62, p. 1–6.

Gass, I.G., 1968, Is the Troodos massif of Cyprus a fragment of Mesozoic ocean floor?: Nature, v. 220, p. 39–42.

Gass, I.G., 1980, The Troodos massif: Its role in the unravelling of the ophiolite problem and its significance in the understanding of constructive plate margin processes, *in* Panayiotou, A., ed., Ophiolites: Proceedings, International

Ophiolite Symposium, Cyprus, 1979: Nicosia, Cyprus, Geological Survey Department, p. 23–35.

Gass, I.G., and Masson-Smith, D., 1963, The geology and gravity anomalies of the Troodos massif, Cyprus: Philosophical Transactions of the Royal Society, London, ser. A, 255, p. 417–467.

Greenbaum, D., 1972, Magmatic processes at ocean ridges: evidence from the Troodos massif, Cyprus: Nature–Physical Science, v. 238, p. 18–21.

Harrison, C.G.A., 1968, Formation of magnetic anomaly patterns by dike injection: Journal of Geophysical Research, v. 73, p. 2137–2142.

Hess, H.H., 1938, A primary peridotite magma: American Journal of Science, v. 35, p. 321–344.

Hess, H.H., 1946, Drowned ancient islands of the Pacific Basin: American Journal of Science, v. 244, p. 772–791.

Hess, H.H., 1955, Serpentines, orogeny and epeirogeny, *in* Poldervaart, A., ed., Crust of the Earth: Boulder, Colorado, Geological Society of America Special Paper 62, p. 391–408.

Hess, H.H., 1962, History of ocean basins, *in* Engel, A.E.J., James, H.L., and Leonard, B.F., Petrologic Studies: A volume to honor A.F. Buddington: Geological Society of America, p. 599–620.

Hess, H.H., 1965, Mid-ocean ridges and tectonics of the sea-floor, *in* Whittard, W.F., and Bradshaw, R., eds., Submarine geology and geophysics: Colston Papers, v. 17, London, Butterworths Scientific Publications, p. 317–333.

Irving, E., 1970, The mid-Atlantic ridge at 45° N, 14. Oxidation and magnetic properties of basalts: review and discussion: Canadian Journal of Earth Sciences, v. 7, p. 1528–1538.

Irving, E., and Couillard, R.W., 1973, Cretaceous normal polarity interval: Nature–Physical Science, v. 244, p. 10–11.

Kidd, R.G.W., 1976, Modelling of processes at spreading plate boundaries [Ph.D. thesis]: Norwich, England, University of East Anglia, 289 p.

Kidd, R.G.W., 1977, The nature and shape of the sources of marine magnetic anomalies: Earth and Planetary Science Letters, v. 33, p. 310–320.

Kidd, R.G.W., and Cann, J.R., 1974, Chilling statistics indicate an ocean floor spreading origin for the Troodos complex, Cyprus: Earth and Planetary Science Letters, v. 24, p. 151–155.

Loncarevic, B.D., Mason, C.S., and Matthews, D.H., 1966, Mid-Atlantic ridge near 45° N, 1. The median valley: Canadian Journal of Earth Sciences, v. 3, p. 327–349.

Malpas, J., Case, G., and Moore, P., 1989, The geology of the area immediately surrounding the CY-4 borehole of the Cyprus Crustal Study Project, *in* Gibson, I.L., Malpas, J., Robinson, P.T., and Xenophontos, C., eds., Cyprus Crustal Study Project, Initial report Hole CY-4: Ottawa, Ontario, Geological Survey of Canada Paper 88-9, p. 31–37.

Mason, R.G., 1958, A magnetic survey off the west coast of the United States (Lat. 32°–36° N: Long. 121°–128° W): Geophysical Journal of the Royal Astronomical Society, v. 1, p. 320–329.

Matthews, D.H., 1961a, Rocks from the eastern North Atlantic [Ph.D. thesis]: Cambridge, England, University of Cambridge.

Matthews, D.H., 1961b, Lavas from an abyssal hill on the floor of the North Atlantic Ocean: Nature, v. 190, p. 158–159.

Matthews, D.H., 1969, Model study of the magnetic anomaly pattern across the mid-Atlantic ridge at 45° N: Tectonophysics, v. 8, p. 353–358.

Matthews, D.H., and Bath, J., 1967, Formation of magnetic anomaly pattern of mid-Atlantic ridge: Geophysical Journal of the Royal Astronomical Society, v. 13, p. 349–357.

Matthews, D.H., Vine, F.J., and Cann, J.R., 1965, Geology of an area of the Carlsberg ridge, Indian Ocean: Geological Society of America Bulletin, v. 76, p. 675–682.

Maxwell, A.E., and Von Herzen, R., 1969, The Glomar Challenger completes Atlantic track–highlights of leg III: Ocean Industry, v. 4, p. 64–65.

Moores, E.M., 1969, Petrology and structure of the Vourinos ophiolite of northern Greece: Boulder, Colorado, Geological Society of America Special Paper 118, 74 p.

Moores, E.M., and Vine, F.J., 1969, Troodos massif, Cyprus, as deep ocean floor: preliminary structural and petrological evidence (abstract): Eos, Transactions of the American Geophysical Union, v. 50, p. 333.

Moores, E.M., and Vine, F.J., 1971, The Troodos massif, Cyprus, and other ophiolites as oceanic crust: evaluation and implications: Philosophical Transactions of the Royal Society, London, Series A.268, p. 443–466.

Pitman, W.C., and Heirtzler, J.R., 1966, Magnetic anomalies over the Pacific–Antarctic ridge: Science, v. 154, p. 1164–1171.

Raymond, C.A., and LaBrecque, J.L., 1987, Magnetization of the oceanic crust: thermoremanent magnetization or chemical remanent magnetization: Journal of Geophysical Research, v. 92, p. 8077–8088.

Reinhardt, B.M., 1969, On the genesis and emplacement of ophiolites in the Oman mountains geosyncline: Schweizersche Mineralogische und Petrographische Mitteilungen, v. 49, p. 1–30.

Robinson, P.T., Melson, W.G., O'Hearn, T., and Schminke, H.-U., 1983, Volcanic glass compositions of the Troodos ophiolite, Cyprus: Geology, v. 11, p. 400–404.

Shand, S.J., 1949, Rocks of the mid-Atlantic ridge: Journal of Geology, v. 57, p. 89–92.

Simonian, K.O., and Gass, I.G., 1978, The Arakapas fault belt, Cyprus: a fossil transform fault: Geological Society of America Bulletin, v. 89, p. 1220–1230.

Smith, G.C., 1986, Physical properties of deep drillcore, Troodos ophiolite, Cyprus [Ph.D. thesis]: Norwich, England, University of East Anglia, 230 p.

Varga, R.J., and Moores, E.M., 1985, Spreading structure of the Troodos ophiolite, Cyprus: Geology, v. 13, p. 846–850.

Verosub, K.L., and Moores, E.M., 1981, Tectonic rotations in extensional regimes and their paleomagnetic consequences for oceanic basalts: Journal of Geophysical Research, v. 86, p. 6335–6349.

Vine, F.J., 1965, Interpretation of magnetic anomalies observed at sea [Ph.D. thesis]: Cambridge, England, University of Cambridge, 337 p.

Vine, F.J., 1966, Spreading of the ocean floor: New evidence: Science, v. 154, p. 1405–1415.

Vine, F.J., 1968, Magnetic anomalies associated with mid-ocean ridges, *in* Phinney, R.A., ed., History of the Earth's Crust: Princeton, New Jersey, Princeton University Press, p. 73–89.

Vine, F.J., 2001, Reversals of fortune, *in* Oreskes, N., ed., Plate tectonics: an insider's history of the modern theory of the Earth: Boulder, Colorado, Westview Press, p. 46–66.

Vine, F.J., and Hess, H.H., 1971, Sea-floor spreading, *in* Maxwell, A.E., ed., The Sea, volume IV, part 2: Wiley-Interscience, New York, p. 587–622.

Vine, F.J., and Matthews, D.H., 1963, Magnetic anomalies over oceanic ridges: Nature, v. 199, p. 947–949.

Vine, F.J., and Moores, E.M., 1969, Paleomagnetic results from the Troodos igneous massif, Cyprus (abstract): Eos, Transactions of the American Geophysical Union, v. 50, p. 131.

Vine, F.J., and Moores, E.M., 1972, A model for the gross structure, petrology, and magnetic properties of oceanic crust: Geological Society of America Memoir 132, p. 195–205.

Vine, F.J., and Morgan, W.J., 1967, Simulation of mid-ocean ridge magnetic anomalies using a random injection model (abstract): Program Geological Society of America Annual Meeting, p. 228.

Vine, F.J., and Smith, G.C., 1990, Structure and physical properties of the Troodos crustal section at ICRDG drill holes CY-1, 1a and 4, *in* Malpas, J., Moores, E.M., Panayiotou, A., and Xenophontos, C., eds., Ophiolites–oceanic crustal analogues: Nicosia, Cyprus, Geological Survey Department, p. 113–124.

Vine, F.J., and Wilson, J.T., 1965, Magnetic anomalies over a young oceanic ridge off Vancouver Island: Science, v. 150, p. 485–489.

Vine, F.J., Poster, C.K., and Gass, I.G., 1973, Aeromagnetic survey of the Troodos igneous massif, Cyprus: Nature–Physical Science, v. 244, p. 34–38.

Wegener, A., 1915, Die entstehung der kontinente und ozeane: Vieweg und Sohn, Braunschweig (1966, English translation of the fourth, revised German edition, 1929, by J. Biram: New York, Dover, 246 p.).

White, R.S., Detrick, R.S., Sinha, M.C., and Cormier, M.H., 1984, Anomalous seismic crustal structure of oceanic fracture zones: Geophysics, v. 79, p. 779–798.

Whitmarsh, R.B., Manatschal, G., and Minshull, T.A., 2001, Evolution of magma poor continental margins from rifting to sea-floor spreading: Nature, v. 413, p. 150–154.

Wilson, R.A.M., 1959, The geology of the Xeros-Troodos area, Memoir 1: Nicosia, Cyprus, Cyprus Geological Survey Department, 184 p.

Manuscript Accepted by the Society May 2, 2003

Printed in the USA

Geological Society of America
Special Paper 373
2003

Die ophiolitischen Zonen in den mediterranen Kettengebirgen (The ophiolitic zones in the Mediterranean mountain chains)

Gustav Steinmann*
*Compte-rendu, XIVe Congrès Géologique International, 1926
Madrid, Graficas Reunidas, 1927, Fascicule 2, p. 637–668*

Translated by

Daniel Bernoulli**
Department of Earth Sciences, Basel University, Bernoullistrasse 32, CH-4056 Basel, Switzerland, and Department of Earth Sciences, Swiss Institute of Technology, CH-8092 Zurich, Switzerland

Gerald M. Friedman**
Department of Geology, Brooklyn College and Graduate Center of the City University of New York, Brooklyn, New York, USA, and Northeastern Science Foundation Inc., affiliated with Brooklyn College, Rensselaer Center of Applied Geology, Troy, New York 12181, USA

Translators' Note

We followed Steinmann's text as closely as possible. Words set bold are bold in his original publication and the additions in parentheses are his. We added a few words in square brackets for clarification when needed. The numbered footnotes are by Steinmann; the lettered ones were added by us in order to explain locations, local stratigraphic terms, or petrographic terms no longer in use.

Table of contents

1. Old and young ophiolites
2. Young ophiolites in the Alps and the Apennines
3. Mass relations of serpentinite, gabbro and diabase-spilite
4. Age relations of serpentinite, gabbro and diabase-spilite
5. Granite accompanying ophiolites
6. Structural relations of ophiolites
7. Tuffs and breccias of the ophiolites
8. Age of the ophiolites
9. Geometry of the ophiolites
10. The processes of differentiation and crystallization
11. Particular significance of the ophiolites
12. Kinds of ultrabasic and basic rocks
13. Summary

*Gustav Steinmann (1856–1929)

**daniel.bernoulli@unibas.ch, gmfriedman@juno.com. Correspondence: Bernoulli, Department of Earth Sciences, Basel University, Bernoullistrasse 32, CH-4056 Basel, Switzerland.

Steinmann, G., 2003, Die ophiolitischen Zonen in den mediterranen Kettengebirgen (The ophiolitic zones in the Mediterranean mountain chains), Bernoulli, D., and Friedman, G.M., translators, *in* Dilek, Y., and Newcomb, S., eds., Ophiolite concept and the evolution of geological thought: Boulder, Colorado, Geological Society of America Special Paper 373, p. 77–91. For permission to copy, contact editing@geosociety.org.

1. OLD AND YOUNG OPHIOLITES[1]

In the Mediterranean mountain chains, there are at least two occurrences of ophiolitic rocks, whose age and geological significance differs and who therefore should be kept sharply separate.

1. An **older** one, doubtlessly pre-Permian, of early Paleozoic age or in my opinion even for the most part Precambrian in age which everywhere is intimately associated with ancient crystalline schists, mica schists, marbles, etc., and whose rocks exhibit an correspondingly high grade of metamorphism.

2. A **younger** group, of at least young Mesozoic age, that is associated with Jurassic and lower Cretaceous sediments of predominantly abyssal character. A stronger metamorphic overprint in these rocks is only evident near the root zones of the Alps, where also the associated sediments suffered from stronger metamorphism.

The **older group** is of no significance for the interpretation of the structure of the Mediterranean mountain chains, because it originated at a time when mountain building was not even in its preliminary stage, and its distribution and occurrence cannot be related to the [later tectonic] units. This becomes obvious from their [the ophiolites] distribution. [2,a] Within the Alps, components of the older group occur for the most part in the Penninic units, especially in the Piedmont zone where they are known as old "Pietri verdi."[b] In the Tuscan Apennines, they are known from eastern Elba, from several neighboring islands, and from Mte. Argentario, where they are located in the old crystalline basement of the Toscanides, i.e., in the Austroalpine realm. In Calabria, they belong to a massif that shows the characteristics of the eastern Alps and the Dinarides.[c] The Serrania de Ronda includes by far the largest occurrence, although it is impossible today to determine to which zone of the Alpine system it belongs.[d] Hammer [e] believes that likewise in the Dinaric chains younger peridotites must be distinguished from older ones which extend to Asia Minor [Anatolia]. By the way, one should add that also outside the Mediterranean chains older ophiolites are widely present which possibly, at least in part, may be of the same age.

2. YOUNG OPHIOLITES IN THE ALPS AND THE APENNINES

If one eliminates the older ophiolites as of no consequence in the zonation of the Mediterranean mountains, then the **younger** ones must be assigned more importance. Because in contrast to the older ophiolites, the younger ones with their enveloping rocks[f] define a very specific zone of these mountains, at least in the Alps, the Apennines, and the Dinaric chains. Only isolated occurrences are encountered outside this zone, and their nature is largely unexplained. Already almost 80 years ago Studer[g] suspected the importance of this zone, when he emphasized the surprising resemblance between the rocks of Lake Sils[h] and the eastern Riviera.[i] It is therefore an important task to separate sharply the old ophiolites from the younger ones everywhere, which, however, has not been done effectively, even in the Alps. Especially in the zone of the Piedmont, old and young ophiolites are in places intimately interspersed, and metamorphism which has affected both, however, in varying degrees, makes a sharp distinction difficult. A further difficulty lies in the structural complexities, because the rocks in question—including their country rocks—(as in the thrust nappes) were strongly sheared and deformed or (as in the root zones [of the nappes]) were at the same time heavily squeezed and largely altered [by metamorphism].

Distinctly more favorable are the conditions in the Dinaric chains and in the Apennines. In these areas, the rocks of the younger ophiolite zone are not like in the Alps the **footwall** of higher thrust sheets by which they were covered and crushed, but, at least in the Apennines and probably also in the Dinaric chains, they form the **uppermost thrust nappe**. Although they were transported as thrust sheets and in consequence their structural setting was highly altered, they lack the strong deformation and crushing of the Alpine rocks. One may therefore state without exaggeration that a detailed knowledge of the Apenninic and Dinaric ophiolite zones will contribute to the clarification of those questions whose solution is hampered by the particular [structural] conditions in the Alps.

While in the Dinaric chains the studies of the ophiolite zones are still under way, special maps have been drawn for

---

[1]As ophiolites one should only designate the consanguineous association of predominantly ultrabasic rocks, always composed for the most part of peridotites (serpentinites) with subordinate gabbro, diabase, spilite, or also norite and related rocks. The name ophite for this group of rocks is not appropriate because this term applies only to a diabase-like rock of specific composition and structure [texture].

[2]see Staub, R., Über die Verteilung der Serpentine in den alpinen Ophiolithen: Schweiz. miner. petrographische Mitt., I, 1924, 138–193.

[a]Steinmann's citation is not correct. The correct citation is: Staub, R., 1922, Über die Verteilung der Serpentine in den alpinen Ophiolithen: Schweizerische Mineralogische und Petrographische Mitteilungen, v. 2, p. 78–149.

[b]Correctly Pietre Verdi (= "green rocks," greenstones). The ophiolites of the Piedmont zone are now considered to be of Mesozoic age. The same holds true for the ophiolites of eastern Elba and the neighboring islands, M. Argentario and northern Calabria, which are cited later.

[c]Today, all these occurrences are allocated to the Liguro-Piemontese ophiolite zone of the Alpine-Apenninic system.

[d]The Ronda peridotite is a tectonically emplaced fragment of subcontinental mantle in the Betic Cordillera, Spain (Van der Wal and Vissers, 1993).

[e]No date cited, but with all certainty Steinmann refers to Hammer (1921, particularly p. 43).

[f]Steinmann calls the associated sediments "Hüllgesteine," "enveloping rocks." From the following it becomes clear that he thought that all the ophiolites, inluding the pillow lavas, intruded the deep-sea sediments associated with them. For this reason we shall translate "Hüllgesteine" in the following by "host rock" or "country rock."

[g]No date cited, but Steinmann obviously refers to Studer (1851, p. 327).

[h]The Platta nappe, see Bernoulli et al. (this volume, Chapter 7).

[i]The Ligurian nappes of the Apennines.

limited areas only, and uncertainties exist about the relationship between the thrust nappes and their root zones [in the Apennines], thanks to the untiring work of several Italian researchers, in particular B. Lotti, we have today special maps on the scale of 1:100,000 for extensive areas of the Tuscan and Ligurian Apennines, which are a useful foundation for additional research and will facilitate the solution of still open questions. For these reasons the following explanations are based in the first instance on my own studies in the Apennines of Tuscany and Liguria.

As is known, the **younger ophiolite zone of the Apennines** does not only form in Liguria and Tuscany, but also on Elba, a rootless thrust sheet of surprising compositional uniformity overlying [tectonically] a different stratigraphic succession that appears in smaller and larger [tectonic] windows and half-windows below. In southern Tuscany and from the Umbrian Apennines to the south, this [latter] succession occupies nearly the entire surface of the chain, and in the central Apennines only a few isolated relics of the ophiolite nappe remain recognizable. The contrast between the two tectonic units gains expression in that in the lower [unit]—which I call the **Toscanides**[j]—Permian, incomplete Triassic, Jurassic to Neocomian (in calcareous, commonly fossiliferous facies), upper Cretaceous in the facies of variegated Scaglia[k] and older Eocene as thick sandstones (Macigno)[l] may be present; however, ophiolitic igneous rocks[m] do not show **the least sign of their presence** in the Mesozoic or Tertiary succession. In contrast, in the overlying thrust sheet, which I call the **Ligurides**, Permian and Triassic seem to be entirely absent; in contrast, the presence of Jurassic and Lower Cretaceous is testified everywhere by radiolarite and koniabyssites[n] of similar or the same development as in the Tuscan succession, however, with the exception that macroscopic fossils appear to be absent entirely. The Tuscan Jurassic contains these in different stages, in part even in abundance. In the Ligurides there are in addition clay-limestone associations (including part of the so-called "argille scagliose"[o]) which may represent the lower stages of the Jurassic.

Missing [in the Ligurides] are the Scaglia, the Macigno, and the Nummulitic limestone, but ophiolites in different lithology and in considerable volume take part in building this thrust sheet and are almost never missing. **The typical, everywhere present mark of distinction of this Liguride sequence is the constant association of the ophiolites with the mentioned hypabyssites and euabyssites**[p] **of the Upper Jurassic–Lower Cretaceous**, a combination which never recurs in the Tuscan succession nor in the Abruzzides of the southern Apennines. It is truly astounding to see with which uniformity of the sedimentary as well as the igneous rocks this Ligurian thrust nappe occurs, always in the same tectonic position, over the vast areas of the Ligurian–Tuscan Apennines.

I have expressed several times my opinion about the abyssal nature of the rocks with which the ophiolites of the Ligurides are associated and into which they evidently were injected [by intrusion]. Hence I like to refer for the essential arguments to these earlier publications.[3] To date three different kinds of rock have been shown to accompany the ophiolites consistently.

1. Radiolarite and variegated deep-sea clay as typical **euabyssal** deposits which are generally **devoid of calcium carbonate**.

2. Whitish, light gray, or reddish, for the most part thick-bedded **limestones**, which overlie the radiolarite and which as Calpionella limestones are comparable to the "Felsenkalk"[q] or Maiolica[r] (= Biancone).

3. **Clays and weakly argillaceous limestones** in the form of beds or nodules embedded in the clays, which were generally called "argille scagliose," and which up-section pass into pure, lighter colored limestones with occasional calpionellids and are overlain by radiolarites.[s]

Despite the absence of megascopic fossils, one is correct if one places this 1 to 3 succession into the Jurassic and the early Lower Cretaceous.

[j]Steinmann introduced Toscanides and Ligurides as terms for the two major tectonic units (or groups of nappes) of the northern Apennines. In contrast, Liguria and Tuscany designate the geographic areas in which both units occur.

[k]A lithofacies term of Italian authors designating Upper Cretaceous to Paleogene pelagic and hemipelagic limestones and marls.

[l]The term Macigno designates the Oligocene–Lower Miocene flysch deposits of the Tuscan nappe(s).

[m]Steinmann often calls the igneous rocks of the ophiolite assemblage "eruptives." However, from all his writings, it becomes clear that, in contrast to many Italian authors, he thought that all the ophiolites, including the pillow lavas, were intrusive. Volcanic rocks he called "extrusives." We therefore translate his "eruptive" with "igneous rocks."

[n]The term abyssokonite (konia = limestone), not koniabyssite, was introduced by Steinmann (1925) for pure limestones and marly [pelagic] limestones which he compared to modern Globigerina and coccolith oozes.

[o]The term Argille scagliose is used for chaotic complexes of the Apennines which in a argillaceous matrix with a "scaly fabric" include blocks of different lithologies mostly of Liguride origin and which are interpreted as olisthostromes, tectonic mélanges, or a combination of both (Pini, 1999).

[p]Terms introduced by Steinmann (1925) for abyssal calcareous, marly, or argillaceous pelagic and hemipelagic (hypabyssal) and siliceous (radiolarites, deep-sea clays = euabyssal) deposits.

[3]Steinmann (G.), Geologische Beobachtungen in den Alpen, II. Die Schardtsche Ueberfaltungstheorie und die geologische Bedeutung der Tiefseeabsätze und der ophiolithischen Massengesteine—*Ber. Naturf. Ges. Freiburg [im Breisgau], 16*, 1905.

Ueber Tiefenabsätze des Oberjura im Apennin.—*Geolog. Rundschau, 4*, 572–575, 1913.

Giebt es fossile Tiefseeablagerungen von erdgeschichtlicher Bedeutung?:—*Geol. Rundschau,16*, 435–468, 1925. (An English translation of this article by W.H. Berger is published in Steinmann [2002]).

[q]The term "Felsenkalk" (= Calcare rupestre) was introduced by Zittel (1869) for the white pelagic limestones of latest Jurassic to early Cretaceous age of the Umbrian Apennines. Today this formation is generally called Maiolica.

[r]Lithostratigraphic term for uppermost Jurassic to Lower Cretaceous pelagic limestones with chert of Tethys-wide distribution (see Wieczorek, 1988), synonymous with Biancone.

[s]Today, these clays and limestones are included in the Argille a Palombini Formation (Lower Cretaceous). They stratigraphically overlie the Calpionella limestones, which in turn overlie the radiolarites.

Except for rare, unaltered peridotites and from likewise only rarely present pyroxenites, **three members** of the ophiolite group are distinguished already for a long time: **Serpentinite** (peridotite), **gabbro and diabase-spilite** (including variolite). That all three are consanguineous and derive from one originally homogenous magma cannot be denied, especially considering the ever recurring association [of these rocks]. This raises the following questions (a) in which mass relations do these three members occur, and (b) which are their relative age relationships?

3. MASS RELATIONS OF SERPENTINITE, GABBRO AND DIABASE-SPILITE

On the basis of limited observations in the Ligurian Apennines I previously[4] indicated the mass ratio between serpentinites on one hand and the more acidic, feldspar-bearing members (gabbro, diabase-spilite) on the other rating between 2:1 and 3:1; this ratio has been confirmed in later observations in Liguria and Tuscany. Meanwhile the geological maps of Tuscany 1:100,000 by Lotti and coworkers have been published, and these allow for further control of the ratio. I find my first estimate essentially confirmed and herewith would like to express the ratio as 5:2. A certain uncertainty remains, because we observe today only minor parts of the Ligurian thrust mass, perhaps 1/100 or even less. Everywhere, where today the Tuscan substratum occupies the [earth] surface, as in the vast Macigno areas of the Ligurian–Tuscan and Umbrian Apennines[t], the Liguride thrust nappe has been eroded , and where Tertiary sediments starting with Upper Eocene Alberese[u], or where Quaternary alluvial deposits prevail, we see nothing of it [the ophiolites] because of the presence of this cover. Further, the internal, western parts of the thrust nappes lie almost completely drowned in the Tyrrhenian [Sea] and can be seen only on Elba and several smaller islands. Despite these incomplete statistical data, the relative mass ratio of the two main groups of ophiolites is evident as soon as larger areas are considered. If one considers only a restricted area, then the **chance of preservation** becomes important, and one finds the rocks in all different proportions. This is evident from the survey which I have compiled based on Lotti's special map of Tuscany [see Table 1].

TABLE 1. MASS RELATIONS OF SERPENTINITE, GABBRO, AND DIABASE-SPILITE BASED ON LOTTI'S SPECIAL MAP OF TUSCANY

Sheet	Serpentinite	Gabbro	Diabase-spilite
Livorno	15	4	1
Volterra W	3	1	3
Volterra E	3	1	1
Massa Marittima	9	3	1
Siena	1	1	0
Elba E	10	1	40
Elba W	1	1	4
Castelnuovo di Garfagnana	0	1	50

In many other areas, however, such as in the area of the upper Tevere valley and in the Ligurian Apennines, the serpentinite prevails almost everywhere and by far over the other ophiolite members.

The **ratio of gabbro to diabase-spilite** one may generally estimate to be 1:1. There are areas in which gabbro is much more common than diabase-spilite (Livorno, Massa Marit[t]ima), and there are those where diabase-spilite prevails considerably, such as on the maps Volterra west, Elba and Castelnuovo di Garfagnana. But a specific rule—other than average ratios—cannot be determined.

Finally, I wish to emphasize a point in the distribution of the ophiolites, with which we will deal again later. One cannot speak about a regional prevalence of one or the other of the three members. In one of the cross sections of the Ligurian thrust nappe we find in the west, that is near the root zone, a tremendous excess of diabase-spilite, such as on Elba; however, at a distance from the root there is a great excess of serpentinite, such as in the upper valley of the Tevere. In another transect, in the root zone the serpentinite exceeds all other members, such as at the Riviera di Levante. Again elsewhere, the dominant masses of diabase-spilite lie approximately in the middle of the cross section like near Castelnuovo di Garfagnana. All these are only accidents of the incomplete visibility of a small part of the thrust sheet. Likewise in the long direction of the nappe no regional difference may be recognized.

Hence we may say: **The different members of the ophiolites show no regional differentiation within the Liguride thrust nappe. Generally considered, they appear to be distributed equally and without specific gaps across the nappe.**

4. AGE RELATIONS OF SERPENTINITE, GABBRO, AND DIABASE-SPILITE

[The age relationships between serpentinite, gabbro, and diabase-spilite] are, since the basic observations of the Italian researchers, especially of Lotti,[5] clear, and my own observations could only confirm the observed rules.[6] The rule is: **serpentinite** represents the **oldest product of crystallization** and exceeds in mass all the others. As a consequence, it does not cut across the

[4]Steinmann, Die Entstehung des Nephrits in Ligurien und die Schwellungsmetamorphose. —*Sitzungsb. Niederrh. Ges. f. Natur- u. Heilkunde Bonn*, 1908, 1–13.

[t]Obviously, Steinmann includes the younger, Miocene flysch sequences of the external Apennines in the Oligocene Macigno.

[u]Ancient term for fine-grained limestones, particularly the carbonate-dominated, Upper Cretaceous to Eocene Helminthoid Flysch sequences of the internal Apennines; see Trevisan in Dal Piaz and Trevisan (1956).

[5]Lotti, (B.), *Descrizione Geologica dell'Isola d'Elba*. Mem. Descrit. Carta Geol. d'Italia, 2, 1886, *Geologia della Toscana*, ibid., 13, 1910.

[6]Steinmann (G.), Die Entstehung des Nephrits in Ligurien und die Schwellungsmetamorphose. —*Sitzungsb. Niederrh. Ges. f. Natur- u. Heilkunde Bonn*, 1908.

other members, but is itself cut by the others. The **second product of crystallization** of the ophiolitic magma is the **gabbro,** which is typically cut by **diabase-spilite**. From this sequence of crystallization, one may explain the endomorphic contact phenomena in the gabbro which, in the marginal zone against the serpentinite, leads to olivine enrichment and the formation of olivine gabbro and serpentinite nodules with a saussuritic core.

In many places where the three members occur together, Lotti could in fact observe that the serpentinite forms the core of the ophiolites, the gabbro forms the cover of the serpentinite, and the diabase-spilite the cover of the gabbro. This, however, is not everywhere a recurring rule. Because the gabbro **cuts across** the serpentinite, it also occurs within and below it, and the same applies to the diabase-spilite which cuts across the two other members. Besides, one or the other member may be missing altogether, which already Lotti had noted.

If, however, Lotti and other researchers express the view that not everywhere exists a sharp separation of the three members, then I cannot agree. I rather believe that in order to explain apparent transitions or intermediate rock types the special conditions must be considered under which these rocks occur.

Apparently a transition between serpentinite and gabbro occurs when, as indicated above, gabbro becomes enriched in olivine at the contact with the serpentinite to form olivine gabbro. Near this contact, the gabbro includes large nodules of serpentinite ("**Edelserpentin**") which at times contain only minute amounts of saussuritized feldspar, or are completely devoid of it. Such nodules of the size of a head or larger remind one of the olivine nodules of basalts and like those are not simply peridotites, but marginal mineral segregations from the gabbroic melt which can be readily distinguished from the adjacent lithology of the serpentinite.

Further one encounters, not rarely, isolated nodules of gabbro within the serpentinites, which at places occur distinctly in linear rows. Or, within the gabbro, one finds similarly distributed slabs or nodules of diabase-spilite. From this one must not conclude that between the enclosing and the enclosed rocks transitions exist, nor that the enclosed rock is older than the host rock. This phenomenon is explained by the intense deformation the rocks generally underwent, on the one hand by shearing accompanying nappe transport and subsequent folding, on the other by deformation which generally occurs as a result of the volume increase of the peridotites during their transformation to serpentinite. Through these processes, originally sharply defined dikes were torn into single pieces which mimic inclusions. The same is known in nephrites which originally may have represented sharply defined dikes.[7] A "schlieren" fabric is on occasion mimicked when a younger member includes fragments of an older one.

There exists a generally recurring **principle in the sequence of crystallization** of ophiolitic magmas: first the major part [of the magma] crystallizes as an ultrabasic rock in the form of a peridotite, then follows the gabbro ("euphotide" of the Italian authors) and subsequently follows the formation of pyroxenite dikes (which are now present as nephrite[v] or carcaro[w]), finally diabase-spilite with variolitic margins ("gabbro rosso" and "verde" of the Italians), and at the end the copper-rich ore veins which cut across the gabbros and the diabase-spilite and which occur between the latter and the enclosing sediments. The heaviest, and in mineralization the poorest, rocks are the first products of crystallization; however, their formation cannot be simply explained by sinking of the heaviest particles in the melt, as Staub[8] assumed for the Alpine serpentinites. This is because in this case the still fluid acidic rest from which the feldspar-bearing rocks crystallized would have to have been located **above** the lithified peridotite, and gabbro, diabase-spilite and ore veins could not have cut across the serpentinite **from down below**. Above all, our case is an example of differentiation (not according to, but) **against the law of gravity**, the lighter masses always remained at depth. Also, the second process of differentiation occurred in this sense, as the lightest components richest in mineralizing fluids (diabase-spilite, ore veins) were emplaced **last** from down below across serpentinite and gabbro. The relative enrichment in volatile constituents of the last crystallized rock masses is distinctly emphasized by the commonly vesicular nature of the spilite pillows (whose hollow spaces were filled later with secondary minerals), by the presence of quartz in the copper ore veins and the accompanying steatitization of the country rock ("essima"), as well as through certain phenomena of contact metamorphism which are associated with the spilites.

Contact metamorphic transformations have been observed repeatedly in the enveloping sediments of the Apenninic and Alpine ophiolites, even if they cannot be easily recognized everywhere. Lotti observed on Elba the neoformation of red and green garnet, of pyroxene and epidote in limestones which are in contact with the serpentinite or are enclosed as fragments in it. More commonly the limestone is only transformed into marble or "silicified." In addition diabase-spilite appears to have caused frequent **albitization**. An especially typical occurrence of this I could observe near Aulla in Val di Magra. There, next to the village across the train station, a steep mountain slope consists essentially of diabase-spilite. This rock includes in a limited part of the old quarry numerous fragments of limestone of the Tithonian-Neocomian "Felsenkalk" which are part of a typical breccia. Its intrusive nature is quite evident, because the spilitic material envelopes (for the most part with a variolitic margin) not only the individual limestone fragments, but intrudes into very fine fractures of the limestone fragments. The limestone shows a discoloration iridescent from green to gray

[7]Steinmann (G.), Die Entstehung des Nephrits usw. 1908.

[v]Steinmann and the geologists of his generation used the term nephrite for a rock composed mainly of actinolite and chlorite (see Kalkowsky 1906).

[w]Apparently, an old term used by miners and no longer in use. Kalkowsky (1906, p. 310, 360–364, cited by Steinmann [1908]) describes it as an extremely hard, fine-grained, nephrite-like rock dominated by diopside, chlorite and garnet. Obviously the term refers to rodingites, metasomatically altered basaltic dikes which act as a sink for calcium during serpentinization.

[8]Staub (R.) Ueber die Verteilung der Serpentine in den alpinen Ophiolithen. *Schweiz. miner. u. petrogr. Mitt. I*, 78–149, 1924.

and light red, in places with spots of red and green color. Through weathering alone (but better yet through etching with acid) small crystals of an insoluble mineral, which increase in abundance in the peripheral parts of the limestone fragments, but which are less common within them, appear on the surface. In places the mineral forms a complete crust in the marginal parts [of the limestone fragments]. This mineral is albite, as Mr. Brauns confirmed. He found the rock to be on the whole identical with the contact rocks of Rovegno (upper Trebbia valley) from which Issel[9] has described such albite crystals which in places still include recognizable radiolaria. Kayser[10] attributed here mistakenly the neoformation of albite to regional and not to contact metamorphism.

According to these observations, it appears that albitization of contact metamorphic country rocks by spilite is more extensive in the Apennines than previously known. From other areas, especially the Paleozoic zones of middle and northern Europe, one has known for a long time the adinoles[x] and similar rocks which are widely occurring. The variolitic structure appears in the occurrence of Aulla along the contact with the limestone fragments; otherwise it is mostly present at the surface of the spilite pillows. Possibly the assumed "silicification" of limestones metamorphosed by ophiolites is nothing else but neoformation of albite.

5. GRANITE ACCOMPANYING OPHIOLITES

In contrast to the Alps, granite and related rocks are almost completely absent in the Apennines. The almost unaltered late Tertiary granites of Elba occur on the mainland only in very restricted areas near Gavorrano and near Campiglia Marit[t]ima; granites resembling "Augengneiss" form the core of the Toscanides near Forno in the Frigido valley (Alpi Apuane); here their pre-Permian age is evident. A third kind of granite occurs sporadically across the Ligurian-Tuscan Apennines. This kind is worth noting in several respects. For one, the granites occur only as isolated **slabs** ("schollen") or **blocks**, whose largest diameter almost everywhere barely exceeds a few hundred meters, and may decrease to very small dimensions. Secondly their occurrence is tied everywhere to the Ligurian thrust sheet, and ophiolitic rocks appear generally to be present in their immediate neighborhood, therefore one may think that they may be genetically related to them. Besides they commonly occur, especially the large blocks, in such strongly mylonitized condition that the Italian researchers did not call them granites but "Conglomerato granitico-cloritoso" [chloritic granite conglomerate]. Indeed formation of chlorite goes hand in hand with mylonitization. These granites and the occasionally associated granite- or quartz porphyries are clearly distinct from the two other granite occurrences of the Apennines and are reminiscent of the old granites that occur on Elba at the base of the Tuscan series or in Calabria.

Several researchers, including Lotti, have brought them in **genetic connection** with the ophiolites, in that they simply declared them to represent a last, acidic crystallization from the ophiolitic magma. However, this interpretation seems to me completely untenable. Perhaps one could expect such late intrusions in the form of small dikes of aplitic or pegmatitic character cutting across the other members of the ophiolites, but not of medium-grained granites which, even in the small slabs only some hundred meters long (like near Camporgiano in Garfagnana, east of the Alpi Apuane) are cut themselves by pegmatitic and aplitic dikes. Hence they prove themselves to be blocks which were torn from a larger granite massif. Nowhere can one observe that the granites cut across the ophiolites; if they were the last segregation of the ophiolitic magma, they would have to do so, like the diabase-spilite or the copper ore veins. They lie only near the ophiolites and seem to be surrounded by them on one, two, or three sides. That is why occasionally the assumption was made that they are fragments of the old crystalline basement that were brought up by the ascent of the ophiolites; however, this hypothesis was based on the wrong assumption that the ophiolites were autochthonous intrusions. After this interpretation became superseded through evidence of the nappe structure, the only conceivable interpretation remains the assumption that they are **fragments and slices of the crystalline basement of the Ligurides**, which were emplaced, like the ophiolites and their host rocks, by thrusting to their present position from root zones in the west. This interpretation is an exact analogue to the granite slices along the southern slope of the Sulzfluh in the Rhätikon which, completely floating in sediment, likewise experienced a comparable long-distance transport.[y] In places one has found in the northern Apennines fragments of gneisses with a fine parallel texture and of magnetite-bearing chlorite schists, in part as float, in part [as clasts] in breccias of unexplained stratigraphical and structural position and origin. These alone would be enough to exclude any other interpretation but that of thrust slices. Hence if I see no cause for discussing any other interpretation, the possibility still remains that the granite blocks are not simply sheared off fragments of the [original] basement of the Liguride thrust nappe, but that they were torn from their autochthonous place of origin during the ascent of the ophiolites. This would explain best their close association with the ophiolites, also in their transported state. However, more detailed research is needed to obtain clarity.[z]

[9]Issel (A.), Liguria geologica e prehistorica, Genova, 1892, *1*.

[10]Kayser (E.), Lehrbuch der Geologie, 6 Aufl., 2. Bd., 1921, 274.

[x]Fine-grained, chert-like contact metamorphic rocks consisting of quartz, chlorite, and albite.

[y]Steinmann refers to granite blocks and boulders which are part of tectonic melanges along the base of the Briançonnais Sulzfluh nappe of the central Alps.

[z]The granitoids mentioned by Steinmann were radiometrically dated as late Paleozoic (310–280 Ma), i.e., late- or post-Variscan in age (Eberhardt et al., 1962). They occur as allochthonous complexes associated with Cretaceous and Eocene flysch sequences of the external Ligurids and are interpreted to represent relics of a Jurassic ocean-continent transition (Marroni et al., 2001). On the other hand, plagiogranites of Jurassic age (153 Ma) were found to intrude the lherzolites and gabbros of the Ligurides (Borsi et al., 1996).

6. STRUCTURAL RELATIONS OF OPHIOLITES[11]

The observations of Italian researchers, in particular B. Lotti, and my own observations seem to agree that the ophiolites are, as a rule, associated with certain sediments which must be considered their **country rocks**. They are almost never absent where ophiolites are present, even when they are irregularly preserved as a result of thrusting and later erosion. Because they evidently experienced contact metamorphism by the ophiolites, they may properly be considered their "enveloping rocks" [country rocks], and for brevity may be designated as such. They must be sharply distinguished from sometimes similar rocks which are younger than the nappe transport, and which therefore were never metamorphosed by the ophiolites. Only where the original country rocks [of the ophiolites] were removed [by erosion], these sediments may unconformably overlie the ophiolites. Their younger age is evident from the fact that they contain reworked material both of the ophiolites and of the country rocks and transgress over quite different formations of the Ligurian and Tuscan successions; hence they are younger than the thrusts. Their fine-grained transgressive breccias contain the foraminiferal fauna of the Priabona Beds; their age hence begins with the late Eocene. Their calcareous rocks are known as "Alberese," the argillaceous-calcareous or argillaceous-sandy rocks as "argille scagliose" or "flysch," respectively, and the uppermost pure sandstones as "Upper Sandstone" ("Oberer Sandstein," "arenisca superiore"). The country rocks are almost everywhere different. They are radiolarites and variegated deep-sea clays ("galestro"), which can replace one another [laterally], furthermore pure limestones of gray or reddish color with radiolarians and calpionellids (Felsenkalke) overlying the radiolarites and finally argillaceous-calcareous rocks underlying the radiolarites.[ab] Since these rocks do not contain any macroscopic fossils except meaningless fucoids[ac], and since they were thrust together with the ophiolites onto certainly Tertiary sediments of the Toscanides, and are further covered by Upper Eocene marine sediments, it is clear, why they—together with the igneous intercalations—have been inferred as being of Eocene age. Only through the study of the microscopic fossils, Jurassic radiolarians and calpionellids (also typical for the Jurassic-Neocomian) their Mesozoic age could be determined. They are for the most part deep-sea deposits, partly of hypabyssal, partly of euabyssal nature, devoid of any trace of ophiolitic components, as long as they did not receive them later through contact metamorphism. In places they are cut by spilitic dikes accompanied by phenomena of contact metamorphism, or one finds fragments of them in serpentinite, gabbro, or spilite, fragments which in this case are strongly metamorphosed.

All these facts, especially in combination with the geological occurrence [of the ophiolites], certainly demonstrate that the enveloping sediments **are older than the ophiolites**. Generally the ophiolites were injected into them. This interpretation is not shared by everyone and therefore further reasons have to be given for it.

According to widespread interpretation which not only Italian geologists, such as Lotti, but also many others, like our English colleagues Dewey and Flett[12] defend, ophiolites, at least in part, and in particular the spilites, represent extrusions on the seafloor occurring at the same time the deep-sea sediments were deposited[13] According to the view of some Italian authors[ad] pneumatolytic emanations poisoned the ocean to the extent that all organisms died, which would account for the poverty in fossils of the sediments. By contrast, Dewey and Flett assumed that the components which were introduced into the sea pneumatolytically, especially silica bound to sodium, would create particularly favorable nutrient conditions for the microorganisms that need silica, such as radiolarians and diatoms. Therein would lie the cause for the prolific development of the radiolarians which build up the radiolarites. This view is also supported by E.F. Davis[14] for the interpretation of the Californian radiolarites. However, such attempts at explanation are untenable for two facts: (1) the extremely widespread **occurrence of radiolarites in ophiolite-free areas** (Toscanides, Southern and Eastern Alps),[15] and (2) the proven younger age of the ophiolites in the ophiolite-bearing areas of the Alps and Apennines.

[11]Compare: Steinmann, G., Giebt es fossile Tiefseeablagerungen von erdgeschichtlicher Bedeutung?—*Geol. Rundschau, 16*, 1925, 435–468.

[ab]Steinmann probably refers to the Argille a Palombini Formation which, however, overlies the Calpionella limestones and is of early Cretaceous age.

[ac]Trace fossils of *Chondrites*-type, at Steinmann's time thought to be the remains of algae.

[12]Dewey (H.) and Flett (J.S.): On some British Pillow-lavas and the rocks associated with them. *Geol. Magazine*, V, 8, 1911, 202–209, 241–248.

[13]It is probably not legitimate to declare the pillow-diabases only based on their structure to be submarine extrusives as this is usually done. We do not know how these may look like; however, one would expect that, in view of the resistance they would meet under the pressure of the water, they would, in spite of their low viscosity, take rather the form of a lava dome than of extended sheets. As far as I can see in the literature and as far as I can judge from my own experience, they generally have the shape of **platy sills**, as, e.g., in the Devonian of the Fichtelgebirge and of the area of the [river] Lahn. The absence of contact metamorphism is in itself no argument against a sill because this [the contact metamorphism] cannot always be proven, even in the case of intrusive stocks. R. Reuning (Diabasgesteine an der Westerwaldbahn Herborn-Driedorf. N. Jahrb. f. Mineral. etc. B.B. *24*, 1907, 390) comes with respect to the diabase occurrences investigated by him to the conclusion that dike intrusions, sills and dikes are much more frequent than hitherto admitted. Further we would expect intense deformation, folding together with the still soft marine sediments to be associated with submarine extrusions, which, however, are generally missing and appear to occur only where they can be related to the expansion of neighboring peridotites or to tectonic deformation. Likewise the glassy nature of both the chilled margins which are observed at the "Deckdiabas" [a Devonian formation of the Variscan belt in Germany] is no proof for an extrusive origin, because the **certainly** dike- or sill-like shaped pillow spilites show the same phenomenon, and it is well known that basaltic glass is produced not only in extrusive sheets, but also in the chilled margins of dike- or sill-like shaped basalts.

[ad]Pantanelli (1880).

[14]Davis (E.F.), The Radiolarian Cherts of the Franciscan Group. *Bull. Dep. Geolog. Univers. California, 11*, 235–432, 1918.

[15]Steinmann (G.), Gibt es fossile Tiefseeablagerungen von erdgeschichtlicher Bedeutung?—*Geol. Rundschau, 16*, 435–468, 1925,

For the Apenninic occurrences I have come to essentially different conclusions.

There can be hardly any doubt about the process and the place of the origin of the serpentinite and the gabbro. Both were originally plutonic rocks, crystallized below a sedimentary cover which, however, may not have been very thick; an interpretation which is totally incompatible with submarine extrusion. At most diabase and spilite may be considered as extrusive rocks. However, if these were submarine extrusions they should be under- and overlain by marine sediments, but would only in rare cases be in immediate contact with the plutonic members, i.e., there where they cut across these. The rule should be to have a **major separation in space between the plutonites and the volcanites**.

This postulate is, however, contradicted by the observations of all researchers who have worked in the ophiolites of the Apennines. Issel (l.c. 270), who has mapped the general distribution of all the Ligurian ophiolites, noted that diabase-spilite is "generalmente in contatto della serpentina" [generally in contact with serpentinite]. In contrast to Lotti he rejects that transitions occur between the individual types of ophiolites, even if they sometimes cut across each other in an intricate way. In his "Geologia della Toscana," Lotti has described all larger occurrences of the Tuscan Apennines. I will cite some examples:

Between La Botte and Mte. Ferrato near Prato Fiorentino (p. 217), shales are overlain by serpentinite, which is directly bordered by the thick mass of diabase of the Botte, for the most part with pillow structure.

The thick mass of diabase of M. Beni and Sasso di Castro near Fierenzuola in the Tuscany-Romagna Apennines is accompanied by a zone of serpentinite without any transitional material in between.

Near Conventino in the Sovara Canyon in the Monti Rognosi in the upper Tevere valley, pillow spilite more than 10 m thick is bordered by a long stretch of serpentinite and is separated from it only in places by a vein of copper ore.

Along the slopes of Bottro di Sassomagnoli and at the Sassineri near Impruneta, south of Florence, the diabase is separated from the serpenite only by gabbro.

Near Montaione, north of Volterra, diabase, in part with pillow structure, appears next to gabbro, locally, as near Gambassi, accompanied by serpentinite. At Mte. Nero diabase lies on serpentinite.

Near Paravello, in the group of Montevignoli and near Miemo in the ore district of Monte Catini, the diabase lies either on gabbro or directly on serpentinite. Likewise in the subsurface of the mines, gabbro and serpentinite are found directly underneath diabase. Similarly, near Orciatico, northwest of Monte Catini, the serpentinite is in places covered by diabase. The diabase of this area is for the most part pillow diabase (spilite).

In the vast ophiolite terrain of Castellina Marit[t]ima-Riparbella, northeast of Cecina, diabase, in part with pillow structure, is in several places in direct contact with serpentinite or gabbro; at most ore veins are present. West of Poggio Berna near Rosignano serpentinite borders diabase porphyrite and this is in contact with spilite having pillow structure.

Near the mine of Monterufoli, southeast of Cecina, the serpentinite is covered by small masses of gabbro and diabase. Near Micciano, a huge mass of pillow diabase overlies serpentinite, and in another place gabbro is intercalated between the two.

Near Roccatederighi in the Montagnola Senese, west of Siena, serpentinite, gabbro, and diabase overlie each other in this order.

The pillow diabase of Crevole and other localities in the Murlo group, south of Siena, lies on serpentinite.

Also in the smaller occurrences in southern Tuscany, like the one of Elmo, Lotti observed pillow diabase overlying serpentinite and gabbro.

Likewise the island of Elba displays several examples of a close association of these two rocks.

I have presented the indications of Lotti in such detail (without citing the numerous sites where diabase-spilite occurs as dike) because from this data-base it becomes evident that diabase and pillow-spilite are not only occasionally and accidentally but **as a rule** associated with the deep-seated rocks. In addition, the not uncommon occurrence of ore veins between diabase-spilite and gabbro and serpentinite or within the diabase-spilite itself documents clearly the subcrustal formation of the diabase-spilite. Many of the occurrences described by Lotti are known to me from personal observation, and I can simply confirm his indications. Besides I'm acquainted with many places in the Ligurian Apennines that show the same phenomena which are also evident from Issel's[16] descriptions.

7. TUFFS AND BRECCIAS OF THE OPHIOLITES

In the publications of the Italian geologists, tuffs and breccias are frequently mentioned which are said to be more or less closely associated with the ophiolites. Such rocks show indeed a rather wide occurrence, but are no proof for the extrusive nature of the ophiolites. They are without doubt of very different origin:

(a) Intrusion breccias. These are breccias of the kind I described previously from Aulla, i.e., [breccias with] sedimentary, generally limestone fragments whose interstitial spaces and joints were filled by ophiolitic material, whereby as a rule a distinct contact metamorphism has taken place with the formation of albite and garnet along the margins of the limestone fragments. Where I observed them [the breccias], they did not have the characteristics of tuff pipes or chimney fillings, but occurred in the midst of igneous rocks as at Aulla, or formed large individual slabs like near Rovegno. Their formation can only be explained assuming that the host rocks were fragmented during the ascent of the igneous material, particularly of the diabase-spilite, and injected by the melt.

(b) Reworked breccias and tuffites. These are widely distributed. In many places where the late Eocene sea (of the Priabona stage) transgressed across the ophiolites and their country rocks, the basal deposits are enriched in larger and smaller fragments of

[16]Issel, l.c.

these rocks. They are found as meter-size blocks in the Alberese, as in the Colledestro valley, northeast of Pieve Santo Stefano in the upper Tevere valley, or southeast of Pieve. In these examples their origin is obvious. However, more typically they are admixed in the sediment only as small fragments or as tiny flakes which only the microscope reveals. Then the rock becomes tuff-like, but reveals itself as a real sediment, especially as the enclosed fragments are generally serpentinite and rarely diabase. They are also accompanied by fragments of country rock.

(c) Unclear remains the nature of certain coarse breccias, which consist in part of sediment (limestone or fine-grained sandstone) in part of ophiolite fragments and which as a rule are strongly lithified. They accompany either the compact ophiolites or they are separated from them, and are embedded in younger sediments with a lenslike geometry. Whichever interpretation in detail may be relevant, in no case are their features evidence for an extrusive origin of the ophiolites, because the limestone fragments in part certainly belong to the youngest country rocks [of the ophiolites], and were mixed as solid fragments and not as a soft mud with the ophiolites. Some of these breccias may belong to a further group:

(d) Frictional breccias are not rarely seen. Especially at the boundary of the ophiolites and their host rocks such rocks are present. I saw such a breccia along the path between Monte Catini and Miemo at the contact between fine-grained diabase and radiolarite, in which fist- to head-size fragments of diabase were enveloped by the red shales of the radiolarites. That such occurrences cannot be used to prove the extrusive nature of the ophiolites need not be emphasized.

Nowhere have I seen sediments in the Tuscan–Ligurian Apennines which could be regarded as contemporaneous with the diabase or immediately postdating its crystallization. All bedded rocks that are in direct contact with ophiolites belong either to the country rocks, which are older than the ophiolites and were already lithified when these were formed, or they are much younger rocks (Alberese and Flysch), which could have formed only when the ophiolites including their country rocks were already transported [with the nappe] and largely eroded. Therefore I come to the conclusion: **The ophiolites of the Apennines, including the diabase-spilite, were injected into a system of Mesozoic deep-sea sediments and did not rise to the surface of the earth's crust. They influenced the deep-sea deposits only through contact metamorphism, but had no influence on the composition of the seawater from which the sediments were deposited.**

My studies in the Alps led to the same results[17] and I'm happy to report that Staub[18] has joined this interpretation.

Several other questions which are related to the occurrence of ophiolites need an explanation: their age, and their geometry.

[17]L.c. 1905. I emphasized there (p. 40) that "undoubted extrusive masses of ophiolitic rocks have not yet been discovered in the Alps." Nevertheless, Heim (Geol. d. Schweiz, 500) has attributed to me the opinion that they were extrusive rocks of the deep-sea floor.

[18]He likewise writes (l.c., 192) only of intrusions.

8. AGE OF THE OPHIOLITES

In the Apennines **the question of age** is hard to resolve because the younger sediments whose age can be determined belong to the late Eocene and hence a long time interval exists within which they could have intruded. The youngest host sediment, the "Felsenkalk," includes, as far as is known, parts of the Tithonian and the Neocomian, possibly up to the Barremian stage. The ophiolites cannot be older. The "Pietraforte" of the Tuscan Apennines[ae], dated by ammonites as late Cretaceous, likewise cannot resolve the question, because this formation underlies the Macigno and therefore belongs to the Toscanide succession and is nowhere associated with the Liguride series. For the determination of the age of the ophiolites, the Alps also are not helpful, because younger sediments in the ophiolite nappes are practically unknown.[19] Important, however, are the Dinaric chains.

How many details may still have to be cleared about the ophiolite zones of the Dinaric mountains, the observations of the different researchers leave no doubt that the Upper Cretaceous is found commonly in a transgressive position above the ophiolites, mostly as Gosau conglomerate. There is no lack of references according to which Lower Cretaceous stages or even Upper Jurassic deposits may already contain ophiolite clasts. Cayeux[20] has described from Nauplia [Greece] a serpentinite conglomerate which contains Upper Jurassic Kimmeridgian fossils. However, this conglomerate is almost certainly younger and the Jurassic fossils reworked. It lies discordantly on Neocomian or Barremian strata and is probably of late Eocene age, but it is important to know that in the same profile Barremian strata are cut by serpentinite **dikes**. Hence we may assume that in the Dinaric ophiolite areas the intrusion occurred either in the second half of the Lower Cretaceous or in the first half of the Upper Cretaceous. I assume that of these, the second part of the Lower Cretaceous is more probable, especially because of the occurrence of Cenomanian [strata] in the Simme nappe, and also because, as is well known, in the eastern Alps an orogenic phase predates the Cenomanian which led to considerable uplift of the Jurassic-Neocomian rocks which had been deposited at abyssal depth. We have good reasons to assume the same age for the [corresponding] events within the ophiolite zones of the Apennines, the Alps and the Dinaric chains because of the remarkable identity of the sediments as well as of the igneous rocks of all three segments of the Alpides.[21]

[ae]The Pietraforte is an Upper Cretaceous formation of fine-grained sandstones and shales (flysch); it occurs in a deeper and originally more external nappe of the Ligurian nappe edifice. Its basement is unknown.

[19]With the exception of the Simme nappe in the Prealps where Cenomanian rocks appear to unconformably overlay the ophiolites and Jurassic abyssal deposits. Compare Heim, Geologie d. Schweiz, *2*, 1922, 653, and Table opposite p. 592.

[20]Cayeux (L.): Existence du Jurassique supérieur et de l'Infracrétacé en Argolide (Grèce). —*Bull. Soc. Géol. France*, 4, sér. *4*, 1904, 87–105.

[21]Also according to Sacco (l.c.) the ophiolites of the Apennines are Cretaceous in age.

9. GEOMETRY OF THE OPHIOLITES

In the Apennines, it is not easy to develop a relevant concept for the original shape of intrusion of the igneous rocks. In general they form **lenslike** bodies which are surrounded by sediments; in part these are real country rocks, in part they are of younger age, such as Alberese and Flysch which cover them transgressively and which partially were removed by late Tertiary erosion.

Such lenses of medium size, i.e., of 3 to 10 km diameter, raise at first the impression of laccoliths, whose vertical feeder channels may of course not be easily visible in a detached and far-traveled thrust nappe. The smaller occurrences of a few kilometers diameter which are at the same time of inhomogeneous composition make one wonder how such a complex, multiphase differentiation of the magma into coarse-grained rocks like peridotites and gabbros on the one hand, and into porphyritic diabase and spilite on the other hand, and finally even into pneumatolytic ore veins could take place. Such an [extreme] differentiation is not known in smaller laccoliths. With the numerous very small occurrences of small size up to blocky masses the difficulties increase further. These considerations lead to the assumption that many of the small and middle-size occurrences visible today were part of a single larger body.

To the same result leads a view of the present distribution of the individual occurrences. Between one or two neighboring larger bodies, which are several kilometers distant from each other, we see commonly smaller bodies down to large blocks. The latter can be interpreted easily as the relics of a former connection between the larger ones, especially if they have the same lithological characteristics.

In Garfagnana in the upper Serchio valley (east of the Alpi Apuane) there is distinct complex, almost exclusively composed of diabase of 20 km in length and up to approximately 8 km wide. If one thinks of the disseminated smaller occurrences in this district, dominantly diabase, merged into one homogeneous mass, then one gets a stock of approximately 30:25 km diameter. A similar size may be obtained when the numerous dispersed occurrences of serpentinite, gabbro and diabase in the district of Rosignano–Castellina Marit[t]ima–Monte Catini are interpreted as the relics of a single ophiolite mass that was broken up by erosion or covered by younger sediments, a view which the map seems to suggest. This way one arrives at a diameter of up to 25 or 30 km of the ophiolite laccoliths. Some ophiolite masses, e.g., those grouped around Monte Penna, north of Sestri Levante, reach indeed such dimensions.

The area in the Apennines richest in ophiolites is concentrated along a stripe that extends in a north-south direction from the southern part of the Riviera di Levante between Sestri and Monterosso to the foothills of Piacenza.[22] In this zone larger and middle-size masses lie so closely together and are connected one to another by smaller occurrences, that—considering the later disintegration by thrusting and erosion—one is forced to the conclusion that they originally made up one single coherent mass 100 km long and 50 km wide. If one realizes further that continuing from the northern segment of this stripe there are numerous smaller occurrences, dispersed throughout the Apennines of [the provinces of] Parma, Modena and Bologna up to the upper reaches of the river Sieve north of Florence, which justify a similar continuity, then one imagines that these are not isolated larger or smaller laccoliths but an originally **coherent plate** of the shape of a huge, widespread **sill**.

The thickness of the ophiolites is hard to determine. Nevertheless one can conclude from the observable relationships that originally it must have been at least 1/2 km, or probably 1 km or more. I would prefer this concept of a single, of course not everywhere parallel, but in detail bumpy **plate**, for which lacking a more suitable designation the name **Plakolite** (from plax = a wide, flat body) seems appropriate, to the other similar one that the magma lithified in the form of individual, however closely neighboring laccoliths. No great difference exists, however, between these two different assumptions but the processes of differentiation of the ophiolitic magma are best explained by the first interpretation. The [observed] process of differentiation is not compatible with the concept of an ordinary sill of lesser thickness in which such a process cannot take place.

The Ligurian thrust nappe of the Apennines is almost exclusively composed of the ophiolites and their Mesozoic country sediments: the various granite imbricates and blocks do not play a significant role.

The décollement of the thrust nappe from its autochthonous basement involved hardly any older rocks which must have underlain the ophiolite plakolite and which were not detached [and transported with the thrust nappe]. From this observation one has to conclude that the décollement surface was essentially the contact surface between the Paleozoic or pre-Paleozoic basement and the lower surface of the plakolite. But this is difficult to explain if the ophiolitic intrusives and their basement were indented into each other in numerous places. From this I would like to conclude that the magma did not ascend in numerous places across the older basement, but rose only along a few or only one **linear fracture** from depth to fill a plakolitic space. This space formed as a result of orogenic processes through **detachment of the Mesozoic cover from a fundamentally different, probably old crystalline** or possibly Paleozoic, however, **typically granitic basement**.

10. THE PROCESSES OF DIFFERENTIATION AND CRYSTALLIZATION

One can visualize the processes of differentiation and crystallization based on the observed age and geometrical relationships as follows.

In the upper parts of the magma reservoir cooling was most rapid. Here the magnesium silicates were preferentially concentrated and crystallized as peridotite. The more acidic residual magma, enriched in aluminum, calcium, and alkalies, especially sodium, stayed below the crystallized shell of peridotite, and with continuing cooling, was separated during a second phase

[22]One recognizes these relationships best on the International Map of Europe where, however, the small occurrences are missing.

[of differentiation], likewise, into a higher, relatively basic partial magma, rich in alumina and calcium of gabbroic composition. This then intruded into fractures and cavities which had formed through the contraction of the [cooling] peridotite (or through ongoing orogenic movements) in the peridotite shell, but preferably filled cavities which through detachment of the peridotite shell from the overlying country rocks had formed. Thus the overwhelming mass of the gabbro crystallized **above** the peridotite.

Also within the gabbro magma a process of differentiation on a small scale occurred. Along the contact with the peridotite and perhaps also with the country rock more basic, magnesium-rich components were concentrated which in marginal parts led to the formation of olivine gabbros and olivine nodules.

The crystallization of the diabase-spilite rest magma must have been, like in the case of the crystallization of the gabbro, preceded by similar processes of mechanical nature, i.e., fracturing of the serpentinite-gabbro-shell and formation of cavities through detachment **above** the lithified shell. In this way diabase-spilite could be emplaced not only in dikes but dominantly as sills, however, for the most part in contact with the already crystallized ophiolites but **above** the latter. The diabase-spilite partial magma contained alkalies, especially sodium, and must have been rich in volatile constituents, especially water.

During this last important phase of crystallization, complications arose which cannot be overlooked in all their details. But some of them are becoming clear.

Dikes of pyroxenite, for the most part of websteritic composition, subsequently changed to **nephrite**, commonly cut—according to the observations of Kalkowsky and myself—across serpentinites and gabbros, but never across diabase or spilite what made Kalkowsky assume a large time gap between serpentinite, gabbro, and nephrite on the one hand and spilite on the other. With the formation of the websterite dikes begins the third phase which, however, according to my view is closely following the second. Hence a segregation of a magnesium-rich magma free of alkalies occurs first.

Then most probably followed the crystallization of the diabase proper with a porphyric to almost holocrystalline structure [texture] because occasionally these are cut by [still younger] spilite dikes, an observation which already Lotti made. Although there is no fundamental difference in chemical composition between the two, the spilite magma was clearly poorer in iron and much richer in sodium and volatile constituents, possibly even very rich in water. Indeed during or following these last crystallizations extensive albitization took place in the surrounding sediments and igneous rocks, not only in the Apennines, but as Dewey and Flett (l.c.) have shown, also in orogens of earlier time (Devonian). And if we wish to find a more or less satisfactory explanation for the extraordinarily widespread serpentinization, saussuritization and perhaps also for the nephritization[23] of the already present members of the ophiolites, we must look for it in the water-rich emanations of the last phase of crystallization. Every other explanation seems by comparison to fail. The enrichment in volatiles of the spilitic rest magma is proven also by the extraordinarily frequent occurrence of bubbles in the interior of the spilitic pillows.

The generally variolite-bearing **pillowed spilites** are composed of three different kinds of rocks: a so-called **interstitial matrix** which fills the spaces between the pillows and is especially common near the borders [of the spilitic bodies]. This dense and originally perhaps predominantly glassy rock contains, in addition to a water content of 6–8%, nearly 25% iron oxides and only about 34% SiO_2,[24] it is therefore a typically **basic** product of crystallization, whereas the **rock of the pillows** hardly differs from an average diabase with over 50% SiO_2, 10% iron oxides and a lower water content. The varioles, which quantitatively are only subordinate and which are confined for the most part to the outer shell of the pillows consist—as is known—for the most part of feldspar and are more acidic than the matrix and the rest of the pillows. If the varioles are not considered, two phases of crystallization remain for the spilitic magma, an **older** one which resulted apparently in the older basic matrix, and a **last** one which generated the more acidic pillows. Therefore, until the end there is a constant rule in the sequence of differentiation during each phase of magma evolution: **from basic to acidic, and against the specific weight**.

My results about the geometry of the ophiolites and their sequence of differentiation are based above all on the relatively clear occurrences in the Apennines; also in the Alps I found no contradictory observation. I wish to emphasize this, because according to R. Staub, different relationships are the rule in the Alps. Staub thinks of the process of injection as follows: the ultrabasic member, the peridotite, crystallized overwhelmingly in the fore-deeps of the embryonic thrust nappes, while the more acidic members, the gabbro and diabase-spilite—**because of their low specific weight**—moved for the most part into the higher and outer parts of the laccoliths and intruded "along bedding surfaces for 40, 50 or more kilometers into the sediments." I believe that I have convincingly, and in accordance with the observations of the Italian geologists, demonstrated that in the Apennines the differentiation was **not according**, but **against** the law of gravity. Moreover, the ophiolites fill a large synclinal area **in irregular association and without a regional pattern**. Even in the outermost parts of the syncline the heavy ultrabasic members exceed the light, basic rocks. Could the processes in the Alps have been so different, if there is such an astounding agreement in all other respects between the two areas?

I can explain the interpretations of Staub only by the later, extremely strong deformation of the entire upper Penninic nappe

[23]I wish to modify now my earlier explanation of the process of nephritization (l.c. 1908) in so far as the swelling of the peridotite during serpentinization is only the cause for the mechanical changes [observed] whereas the effect of the heated waters is the cause of the amphibolitization of the pyroxene.

[24]The analyses which were recently presented by Slavik at the International Geological Congress in Madrid of the interstitial matrix and the pillows confirm the earlier data concerning the composition of analogous rocks of the Fichtelgebirge.

system, which is the reason of the incomplete picture we have today. In addition, Cornelius[25] finds Staub's rules not confirmed in [Grau]bünden. By contrast, Staub and I agree about the interpretation of all the ophiolites as late Mesozoic **intrusives** and in the concept that they are a huge body inserted between the Mesozoic sediments. Along which pathways the intrusion occurred is not easily determined, neither in the Alps nor in the Apennines, because the [originally] autochthonous areas where these processes [the intrusions] occurred should be observable under conditions of lesser deformation which is not the case in either of the two areas (and probably not in the Dinaric chains). In addition, there is also the difficulty to distinguish between old and young ophiolites in the root zones which also in the Alps seems not to have been resolved everywhere (e.g., for the Malenco serpentinite[af], which certainly plays an important role). With regard to the question of the feeder channels, the views of Staub and myself agree, it seems, quite well. As he for the Alps, I assume for the Apennines that the magma was not derived from numerous [feeder] channels distributed over the entire geosynclinal area but from a linear strip along the **back margin** of the ophiolitic geosyncline. This intrusion fissure would be located in the Alps along the boundary between the Penninic and the Austroalpine region, in the Apennines, however, along the opposite border between the ophiolite zone and the Briançonnais zone, in both cases, however, in a **border zone of the geosyncline**.

With these results, two concepts are not in agreement which relate to the occurrence of ophiolites. One, which is only noted on occasion, seeks to fill the space by [in situ] **melting**. However, for this neither the Apenninic nor the Alpine occurrences provide the least evidence. On the contrary, the potential for assimilation and contact metamorphism appears extremely minor, especially in the case of the ophiolites. The other concept enjoys a much wider appeal and was supported also by Suess.[ag] According to this concept the intrusions follow not only bedding planes but also dislocation surfaces. If one thinks of dislocation surfaces as the thrust planes at the base of the Tertiary thrust nappes, this concept applies neither to the Alps nor to the Apennines. At that time the ophiolites were moved **entirely passively** like their country rocks and their older basement. Only the middle Cretaceous phase of movement comes into question; however, we know today still next to nothing about the nature of this process. Only one notion seems probable: a regional detachment of the Mesozoic sedimentary cover from its basement and the filling of the created flat space by the ophiolitic magma. Weak lateral compression and subsequent opening of an intrusion fissure would suffice. Strong folding or even extensive nappe movement did apparently not occur during this phase; moreover they would not explain the intrusion process. In addition, the neighboring areas should have been affected, but apparently were not. Nothing speaks against it, but much speaks for it that in the middle Cretaceous, a first preliminary stage of the great Tertiary orogenic movements is evident, as Staub assumed. However, the horizontal component could only have been minor and must have been **very different from the nappe movements and later folding**. If for this reason one assigns them [to the early orogenic movements] a special role—which, however still needs much of elucidation—amongst the mountain-building processes, then, from a geological point of view, the ophiolites likewise occupy a special place among the igneous rocks.

11. PARTICULAR SIGNIFICANCE OF THE OPHIOLITES

I have previously emphasized the widespread occurrence of ophiolites in mountains of Alpine structure and their characteristic association with Mesozoic abyssal sediments and also noted their particular setting.[26] At the same time Sacco[27] made a similar investigation and contested with good reasons a Tertiary age of the ophiolites and of the "argille scagliose" which he declared to be of Cretaceous age. He also emphasized the special geological significance of the ophiolites. Dewey and Flett (l.c.) recognized the "spilite group" likewise as equal in rank to the Atlantic and Pacific [igneous] rock groups. They declared them, as I did, to be rocks typical of areas "that have undergone a long-continued and gentle subsidence, with few or slight upward movements, and no important folding" [p. 246]. These results were, however, gained not in the young Mediterranean mountain chains but in different formations of Great Britain with widely occurring abyssal deposits which were folded during the Paleozoic. Thereby the rules of occurrence of basic and ultrabasic rocks become especially clear.

12. KINDS OF ULTRABASIC AND BASIC ROCKS

In many orogenic systems with slow and permanent subsidence, basic and ultrabasic igneous rocks have been emplaced, but their form [of occurrence] varies widely. Ophiolites represent the **abyssal type**: infracrustal rocks with the characteristics of deep-seated rocks and with prevailing magnesium silicates, associated with previously deposited thick, eu- and hypabyssal

[25]He says (*Beitr. Geol. Karte Schweiz*, N.F. *50*, 1923, 10) in his description of the Err group in [Grau]bünden: "Certainly incorrect is in our area what Staub recently said about the distribution of the various individual rock types of the ophiolite zone: the serpentinite is not absent in the deepest imbricates of the ophiolite area—already at its base east of Mühlen [Mulegns] it is present in large masses; in the highest parts it is not the only rock type because, e.g., along the western slope of Piz Nair diabase and greenschist clearly prevail. Also the presumed limitation of the gabbro to the originally northernmost parts of the nappe, i.e., the now deepest imbricates is not true. In Val [da] Natons and Val Savriez it is part of all the imbricates from top to bottom and in equal proportions."

[af]The Malenco serpentinite is a fragment of subcontinental mantle exhumed during Jurassic rifting (Müntener et al., 2000).

[ag]Steinmann did not give a reference to Suess, but most probably he meant Suess (1909, see p. 147).

[26]L.c., 1905.

[27]Sacco (F.), Les Formations ophitifères du Crétacé. *Bull. Soc. Belge de Géologie, 19,* 1905, 247–265, pl.8.

sediments[28] prevail: [these are] the ophiolites of the Mediterranean mountain chains. The **hypabyssal type** is represented by the "spilite formation," mainly rocks with a prevailing porphyric texture, (?) on occasion accompanied by extrusions[29] and occasionally by tuffs but with strongly reduced ultrabasic deep-seated rocks, however, associated with hemiabyssal, hypabyssal, and euabyssal sediments. This facies is widely occurring in the Paleozoic formations of central Europe and Great Britain. The **neritic type** is distinguished by the prevalence of extrusive rocks and tuffs and the (? complete) absence of ultrabasic rocks and is associated with neritic, more rarely hemiabyssal sediments. A classical example is the Mesozoic Porphyrite Formation of the South American Cordillera (which should more properly be called **Diabase-Melaphyre Formation**). Here extensive flows, agglomerates, and tuffs prevail which in many different ways are interbedded with marine sediments of neritic type. Spilitic lavas with pillow structure are also present on occasion; however, their main area of distribution belongs to the two other types.

The genetic differences between the three occurrences I explain as follows: where relatively strong and rapid subsidence occurred and only thin, chiefly eu- and hypabyssal sediments were deposited, neither intrusive nor extrusive emplacement of ultrabasic-basic magma occurred during the period of subsidence. An intrusion occurred only with beginning detachment and uplift of the newly formed sediments, but no separation into extrusive and intrusive masses occurred; however, repeated differentiation of the magma took place into more basic and more acidic parts, respectively. All these [rocks] crystallized during intrusion. This is the **abyssal** type.

Where subsidence was less rapid and perhaps interrupted several times by uplift and, at the same time, relatively thick sedimentary sequences were deposited, like in the Paleozoic formations of the [Variscan] mountains of [central] Europe, the magma segregated into a mainly ultrabasic part, which remained almost entirely at unobservable depth, whereas with the basic intrusive, and in part extrusive magma, small amounts of peridotitic or picritic rocks were emplaced. During this process an extensive differentiation into rock types of different composition occurred ranging from the acidic (keratophyre, sodic granite) to the ultrabasic (picrite, serpentinite) pole. On the whole, this magma was considerably richer in SiO_2 and alkalies, especially sodium, and much poorer in magnesium than the abyssal magma. This is the **hypabyssal** type.

In the case of the **neritic** type, the sinking sea floor remained at shallow depth. As in the case of the hypabyssal type, the ultrabasic part, still more sharply separated from the rest, remained at depth, and consequently exclusively **basic** material was emplaced for the most in extrusive form. Only minor differentiation of the magma occurred, because these rocks belong to a narrow [defined] group of diabase porphyrite, augite porphyrite, melaphyre, all with a porphyric texture, and of (subordinate) spilite. Ore veins, as they formed following the emplacement of the other two types [of rocks], appear to be entirely absent.[30]

Let us think that the deep-seated magma consists of layers of different composition, in the way that with depth (but without a sharp separation) always more basic parts occur, and that the process of subsidence is related to lateral flow of the shallower acidic phases. We now can understand that where also the basic component has been entirely removed, and at the same time strong subsidence had occurred, only **ultrabasic** material could intrude, as the strong pressure on the seafloor did not allow extrusion. Where a substantial part of the shallower basic layer still persisted, magmas of intermediate composition intruded and generated rocks which range between ultrabasic and basic composition. Possibly because of the lower pressure these could in part be extrusive. Where, however, only the acidic part had flowed off and the basic part was still completely present, only **basic** material crystallized in extrusive form under the lower pressure.

Such interpretations are suitable for the understanding of the differences between the three types [of occurrences], but only under certain assumptions they become plausible. One is the assumption of magma layering at depth, a further one assumes that regional subsidence is the result of lateral flow of the shallower magma layers, or, if not, at least occurs at the same time. In vain we have searched till now for an explanation why two neighboring areas with a very similar [geological] history, which underwent subsidence at the same time leading to the deposition of identical and coeval abyssites, yet in their igneous activity deviated so drastically. Staub (l.c.,1924) has raised this question for the Alps where in the upper Penninic nappes (Piedmont zone) ophiolites as a rule are associated with the abyssal rocks of the Jurassic-Lower Cretaceous, whereas in the adjacent Austroalpine nappes, which show an essentially similar sedimentary evolution, basic injections are rare and appear to be restricted regionally.

More characteristic are the circumstances in the Apennines where in the Toscanides and Ligurides, two zones bordering one another closely (corresponding to the Austroalpine and upper Penninic nappes) the same abyssites are found, however, the ophiolites are consistently lacking in the Toscanides[31] whereas they are

[28]From this general rule of association it does not follow that abyssal sediments must always be associated with abyssal igneous rocks as is commonly misunderstood. Comp. also Staub (l.c., 1924, 143).

[29]I assume that a large part of the pillow lavas, which are commonly interpreted, often without detailed study, as submarine extrusives, are in fact of infracrustal origin.

[30]The former assumption of Möricke that the Mesozoic basic igneous rocks of the South-American Cordillera are the source of the silver ores has not been found to be true. Comp. G. Steinmann, Über gebundene Erzgänge in der Kordillere Südamerikas.—*Int. Kongr. für Berg. u. Hüttenwesen, Düsseldorf*, 1910, IV, No 20.

[31]On Lotti's geological map of Toscana 1:100,000, Sheet Volterra east, an occurrence of ophiolites is shown surrounded by Triassic of the Toscanides south-southeast of Castel S. Gimignano, between Volterra and Colle Val d'Elsa. This would be the only exception to the otherwise everywhere confirmed rule of the restriction of the ophiolites to the Ligurides. Messr. Tilmann, Kuckelkorn and I have searched the area in detail, but found no trace of any ophiolite. However, as on the map of Toscana 1:500,000 this place is indicated as Triassic surrounded by Rhaetian, we apparently deal with a printing error.

invariably present in the Ligurides. To explain this difference we can either call for the effect of special, but still unclear tectonic factors, as Staub has hinted at, or there are in the early history of ophiolite-free and ophiolite-rich zones still differences which are not evident today.

Between the three types of basic-ultrabasic igneous rocks and the structure of the mountains in whose early history they were involved, there seem to exist still other relationships which may only be briefly noted here.

Within the Mediterranean mountain chains, the ophiolites occur on the one hand along narrow, highly compressed root zones, on the other hand in extremely far-traveled thrust nappes, i.e., linked with orogenic zones which represent the highest degree of horizontal nappe movement. One can designate such orogenic zones as belts of thrust nappes in the true sense of the word, as **Petasmorogenes** (Greek petasma = blanket, [nappe]). Hence the abyssal igneous type coincides with the Petasmorogene.

Where in the Paleozoic mountains the spilite formation occurs, it seems overwhelmingly tied to an orogenic structure which on occasion includes large-scale thrusts, yet on the whole includes more modest horizontal displacements and rather imbricate structures than thrust sheets. For such imbricate structures the designation **Lepidorogene** (Greek lepidos = scale) would be appropriate.

The neritic type, however, seems to coincide with orogenic zones of relatively simple folding, without major thrust nappes, and with only weak and occasional thrusting or imbricate structures. This kind would be the **ptychorogene** (Greek ptychos = fold) type with which these extrusions would be associated.

Even though sharp boundaries cannot be defined between the different structural styles, the frequent association of the [different] igneous rocks with specific types of orogens is remarkable and should be kept in mind.

13. SUMMARY

The younger ophiolites occurring in the Mediterranean mountain chains belong—not exclusively but overwhelmingly—to one zone which borders the area of Austroalpine evolution externally, i.e., toward the north, west, and east, and is itself surrounded by a belt of Pennine or Helveto-Dauphinois facies.

In the Alps and in the Apennines, where the details of the ophiolite zone are better known, only poor and incomplete sediments of the early [Triassic], and less thick, but typically abyssal deposits of the middle Mesozoic (Jurassic and Lower Cretaceous) are associated with the igneous rocks. The former show a connection to the external zones, the latter to the internal ones.

In the Alps, the ophiolite zone was called the Piedmont zone, the Rhetic nappe or the upper or south-Pennine region. Priority belongs to the name **Piedmontides** in modern terminology. In the Apennines I name these **Ligurides**, because in the Ligurian Apennines they reach their major development; in the Dinaric chains, one could call them **Bosnides**. As a comprehensive term for all three segments the term **ophiolite zone** appears to be the most suitable and understandable.

In the Alps and the Apennines, and probably also in the Dinaric chain, the ophiolites are exclusively **intrusive rocks** which were injected into the deep-sea sediments of the middle Mesozoic. Because they are unconformably covered by Upper Cretaceous sediments, they must surely be of **middle Cretaceous** age.

The ophiolitic magma seems to have spilled into a huge space **of plate-like** shape which opened by detachment between the pre-Triassic basement and the overlying Mesozoic sediments. It lithified in the form of a correspondingly shaped plate, a **plakolith**. Feeding from depth most probably occurred through one (or more?) **elongate fracture**(s) along the internal or external border of the ophiolite region.

Differentiation of the magma occurred during at least **three major phases**. During the **first**, the largest part of the magma crystallized as an **ultrabasic** rock of peridotitic composition (later mostly serpentinized). During the **second** phase, the **gabbroic** partial magma which had remained at depth of the plakolith was cutting across the lithified peridotite cover, probably simultaneously with subsidence of the latter and generated stock-, sill- and dike-like bodies of **gabbro**. During the **third** major phase (which probably consisted of several smaller ones), crystallization of the **diabase** and **spilite** with common pillow structure and the formation of variolite took place under analogous conditions, just like during the second phase. Sodium- and water-rich emanations led to serpentinization, albitization and the formation of copper ore veins.

Differentiation took place during all phases **not according to gravity** but **against** it and **in the direction of increasing acidity** and also increasing concentrations of sodium and volatile constituents. The different phases of differentiation were separated by time intervals that allowed the crystallization of the previous partial magma and the formation of fractures therein. However, the intrusions can be viewed as one single process because they are spatially and in time connected without orogenic or epeirogenic movements intervening except for the subsidence of the earlier crystallized rocks.

In all three segments of the Mediterranean mountain chains, the ophiolites, including their host rocks (subtracting the parts which had already been removed by erosion), were thrust over large distances over the neighboring orogenic units **long after their intrusion**, probably at the time of the older late Eocene, in the Alps over the deeper Pennides, in the Apennines and in the Dinaric chains over the Austroalpine areas of the Toscanides and the Albanides.

These intense [nappe] movements, as well as the later folding (Oligocene till Pliocene) have largely destroyed the original relationships between the ophiolites and their host rocks, and the rocks themselves heavily deformed and disrupted into lenticular bodies. These younger structures must, however, not be interpreted in the way that the intrusions occurred simultaneously with thrusting of the nappes or with Tertiary folding, and along the planes of dislocation.

My research in the Apennines was conducted with the support of the "Notgemeinschaft der Deutschen Wissenschaft" which is gratefully acknowledged.

REFERENCES CITED

Bernoulli, D., Manatschal, G., Desmurs, L., and Müntener, O., 2003, Where did Gustav Steinmann see the Trinity? Back to the roots of an Alpine ophiolite concept, *in* Dilek, Y. and Newcomb, S., eds., Ophiolite concept and the evolution of geological thought: Boulder, Colorado, Geological Society of America Special Paper 373, p. 93–110 (this volume).

Borsi, L., Schärer, U., Gaggero, L., and Crispini, L., 1996, Age, origin and geodynamic significance of plagiogranites in lherzolites and gabbros of the Piedmont-Ligurian ocean basin: Earth and Planetary Science Letters, v. 140, p. 227–241.

Dal Piaz, G., and Trevisan, L., 1956, Lexique Stratigraphique International, v. 1, Europe, fasc. 11, Italia: Paris, Centre Nationale de la Recherche Scientifique, 186 p.

Eberhardt, P., Ferrara, G., and Tongiorgi, E., 1962, Détermination de l'âge des granites allochthones de l'Apennin septentrionale: Bulletin de la Société Géologique de France, ser. 7, v. 4, p. 666–667.

Hammer, W., 1921, Die basischen Intrusivmassen Westserbiens, *in* Ampferer, O., and Hammer, W., Ergebnisse der geologischen Forschungsreisen in Westserbien: Denkschriften der Akademie der Wissenschaften in Wien, Mathematisch-naturwissenschaftliche Klasse, v. 98, p. 11–43.

Kalkowsky, E., 1906, Geologie des Nephrits im südlichen Ligurien: Zeitschrift der Deutschen Geologischen Gesellschaft, v. 58, p. 307–378.

Marroni, M., Molli, G., Ottria, G., and Pandolfi, L., 2001, Tectono-sedimentary evolution of the External Liguride Units (Northern Apennines, Italy): Insights in the pre-collisional history of a fossil ocean-continent transition zone: Geodinamica Acta, v. 14, p. 307–320.

Müntener, O., Hermann, J., and Trommsdorff, V., 2000, Cooling history and exhumation of lower-crustal granulite and upper mantle (Malenco, eastern Central Alps): Journal of Petrology, v. 41, p. 175–200.

Pantanelli, D., 1880, I diaspri della Toscana e i loro fossili: Atti della Reale Accademia dei Lincei, Classe di Scienze Fisiche, Matematiche e Scienze Naturali, Ser. 3, v. 8, p. 35–66.

Pini, G., 1999, Tectonosomes and olistostromes in the Argille Scagliose of the Northern Apennines, Italy: Boulder, Colorado, Geological Society of America Special Paper 335, 70 p.

Steinmann, G., 2002, Are there ancient deep-sea deposits of geological significance? English translation of: Gibt es fossile Tiefseeablagerungen von erdgeschichtlicher Bedeutung?: International Journal of Earth Sciences (Geologische Rundschau), v. 91, supplement 1, p. S18–S42.

Studer, B. 1851, Geologie der Schweiz: Bern, Stämpflische Verlagsbuchhandung and Zürich, Friedrich Schulthess, v. 1, 485 p.

Suess, E., 1909, Das Antlitz der Erde, Dritter Band, 2. Hälfte: Wien, F. Tempsky, and Leipzig, G. Freytag, 789 p.

Van der Wal, D., and Vissers, R.L.M., 1993, Uplift and emplacement of upper mantle rocks in the western Mediterranean: Geology, v. 21, p. 1119–1122.

Wieczorek, J., 1988, Maiolica—A unique facies of the western Tethys: Annales Societatis Geologorum Poloniae, v. 58, p. 255–276.

Zittel, K.A., 1869, Geologische Beobachtungen aus den Central-Apenninen: Geognostisch-Palaeontologische Beiträge: München, R. Oldenburg, v. 2, p. 91–176.

Manuscript Accepted by the Society May 2, 2003

Geological Society of America
Special Paper 373
2003

Where did Gustav Steinmann see the trinity? Back to the roots of an Alpine ophiolite concept

Daniel Bernoulli*
Department of Earth Sciences, Basel University, Bernoullistrasse 32, CH-4056 Basel, and Department of Earth Sciences, Swiss Institute of Technology, CH-8092 Zurich, Switzerland;

Gianreto Manatschal
CGS-EOST; UMR 7515 CNRS, Université Louis Pasteur, Rue Blessig 1, F-67084 Strasbourg, France

Laurent Desmurs
UMR 5570, CNRS and UBC-Lyon 1, 43 Bd 11 Novembre 1918, F-69622 Villeurbanne Cedex, France

Othmar Müntener
Institut de Géologie, Université de Neuchâtel, Rue Emile Argand 11, CH-2007 Neuchâtel, Switzerland

ABSTRACT

In 1905, Gustav Steinmann noted the close association of serpentinites, diabase and radiolarite and considered this "greenstone" or ophiolitic association as characteristic for the axial part of the "geosyncline" and the deep ocean floor. Although Steinmann considered diabase, spilite and "variolite" as intrusive rocks distinctly younger than the associated sediments, he stressed their association with deep-sea sediments, notably radiolarian cherts and pelagic limestones. In his view, the "consanguineous" association of ultramafic and mafic material was typical for suboceanic environments from where these magmas had ascended during folding of the oceanic sediments. Eventually, the importance of Steinmann's discovery was recognized, and the association of serpentinites, pillow lavas and radiolarites became known as the Steinmann trinity and, finally, a synonym for ophiolitic associations in general.

Ironically, it appears that Steinmann never saw a completely developed ophiolite like those in Oman or Troodos. In the type-area of the trinity, in the Penninic zone of the Alps where Steinmann worked, the Jurassic ophiolites are dominated by serpentinites, pillow lavas, and oceanic sediments, whereas gabbros appear to play a subordinate role and no relics of a sheeted dike complex are found. Instead, oceanic sediments, radiolarian cherts, and pelagic limestones stratigraphically overlie serpentinized mantle rocks of subcontinental origin that were exhumed along concave-downward detachment faults and exposed on the sea floor. The gabbros intruded the partially serpentinized peridotites at a shallow depth. Undeformed basaltic dikes cut across gabbros deformed at high temperatures, and pillow lavas directly overlie the exhumed peridotites and gabbros. Both types of mafic rocks are characterized by εNd values typical for an asthenospheric mid-ocean ridge–type source. They may be the products of a steady process, which combined extensional deformation with magma generation and emplacement, and appear to record the onset of sea floor spreading across an exhumed subcontinental mantle during the earliest phases of the development of a slow spreading ridge. This situation

*daniel.bernoulli@unibas.ch

Bernoulli, D., Manatschal, G., Desmurs, L., and Müntener, O., 2003, Where did Gustav Steinmann see the trinity? Back to the roots of an Alpine ophiolite concept, *in* Dilek, Y., and Newcomb, S., eds., Ophiolite concept and the evolution of geological thought: Boulder, Colorado, Geological Society of America Special Paper 373, p. 93–110. For permission to copy, contact editing@geosociety.org.

is conspicuously similar to that of the early Cretaceous ocean-continent transition along the west-Iberian passive margin where subcontinental mantle was exhumed to the sea floor prior to the onset of sea floor spreading.

Keywords: Steinmann trinity, ophiolites, ocean-continent transition, Alps, Tethys.

INTRODUCTION

In 1905, Gustav Steinmann (1856–1929; Fig. 1) noted the close association of serpentinites, diabase, and radiolarite along the western boundary of the Austroalpine nappes in eastern Switzerland. Steinmann was not the first to notice the co-occurrence of these rocks; it was already known to Italian geologists of the late nineteenth century (Pantanelli, 1880; Lotti, 1886), and also to Brongniart who in 1813 had introduced the term ophiolite for different varieties of serpentine (including gabbro and "variolite") and who had recognized the co-existence of ultramafics, gabbros, "diabase-spilite," and chert (Brongniart 1813, 1821, fide Amstutz, 1980). Steinmann, however, was probably the first to interpret this "green rock" or ophiolitic association in geodynamic terms and to consider them as characteristic for the axial part of the "geosyncline" and the deep ocean floor. Although Steinmann thought that diabase, spilite, and variolite (variolitic pillow lavas) were intrusive rocks distinctly younger than the associated sediments, he stressed their characteristic association with deep-sea sediments, notably radiolarian cherts and deep-water limestones of Maiolica facies (Steinmann 1925, 1927). In his view, the "consanguineous" association of ultramafic and mafic material was characteristic for sub-oceanic environments from where these magmas had ascended during folding of the oceanic sediments.

After 1905, Steinmann extended his observations from eastern Switzerland to the northern Apennines, where the close association of serpentinites, gabbros, basalts, and radiolarites had been recognized first. In contrast to the Italian authors, who regarded the ophiolites of the northern Apennines as intrusions into Tertiary sediments, Steinmann stressed the absence of feeder dikes in the Tertiary rocks underlying them and their association with the Jurassic radiolarites and reasoned that these Mesozoic deep-sea sediments therefore had to have been thrust together with the ophiolites onto the Tertiary sediments (of what is now the Tuscan nappe). His vision of the ophiolites as a paleotectonic element of prime importance thus led to the discovery of the nappe structure of the northern Apennines (Steinmann, 1907, 1913).

In this paper we shall briefly review the history of the concept of the Steinmann trinity introduced by Bailey and McCallien (1950). In contrast to later definitions that match the classic ophiolite stratigraphy (Anonymous, 1972), the Steinmann trinity as originally defined by these authors included serpentinites, basalts, and radiolarites. We will show that in the areas where Steinmann worked, in the nappes of the Pennine-Austroalpine boundary zone of the Alps and also in the Ligurian Apennines, these three rock types are indeed the dominant lithologies of the ophiolite nappes, whereas gabbros and dolerites are subordinate. Nevertheless, we do not plead for using the term Steinmann trinity in a puristic way, in its original sense, but rather for describing every ophiolite occurrence in its own right—indeed, "mapping the ophiolite group as a unit has often obscured critical relationship of its various members to the tectonic cycle" (Hess, 1955, p. 393). In the following, we shall summarize the characteristics of the ophiolites in Steinmann's first work-

Figure 1. Gustav Steinmann (1856–1927), from Wilckens (1930).

ing area and compare their paleotectonic evolution with that of undeformed ocean-continent transition zones. Overall, this evolution is very similar to that in the ophiolite nappes of the northern Apennines (Rampone and Piccardo, 2000; Marroni et al., 2001), the area in which Steinmann (1927) later worked.

HISTORICAL EVOLUTION OF STEINMANN'S CONCEPT

The Ophiolites

Steinmann's 1905 and 1927 papers do not include any illustrations; however, the profiles that were drawn by Argand (1916, his figure 1) and Staub (1924, his figure 61) were obviously influenced by Steinmann and illustrate his views adequately (Fig. 2). Steinmann considered the ophiolites to be passively involved in the Tertiary nappe movements and, therefore, to be older; however, he thought that their intrusion was associated with uplift and deformation of the Jurassic-early Cretaceous deep-sea floor during the "middle" Cretaceous (Steinmann, 1905, 1925, 1927). Similarly, Argand (1911, 1916) and Staub (1922, 1924) thought that the ophiolites intruded during the *early* phases of Alpine "folding" of the geosynclinal sediments and their thin sialic basement. Staub (1922) stressed their occurrences in "nappe synclines" in the back of the crystalline basement nappes of the central Alps (Adula, Suretta and Margna, compare Fig. 2), and this spatial distribution confirmed for him their intrusion into the "Bündnerschiefer" sequences "along the foredeeps of the embryonic nappes." Today, we associate these ophiolite occurrences with the scarce ophiolites of the North-Pennine Valais basin (Misox zone, ex-Adula), and with the more extensive ones of the South-Pennine Liguria-Piedmont segment of the Tethys (the ophiolites of the Avers zone (ex-Suretta) and of the Platta-Totalp nappe (ex-Margna). Similarly, Benson (1926) considered "alpine-type" peridotites and serpentinites to be plutonic and intrusive into folded geosynclinal sediments.

In the Alps, this classic interpretation did not change fundamentally before the advent of plate tectonics: The floor of the "geosyncline" was still sialic, and, although "the hypothesis of synorogenic intrusion along early thrust planes seems to be constantly loosing ground," the ophiolites were still interpreted as "pre-orogenic flows and intrusions into unconsolidated, horizontal sediments"; however, "extensional faults and flexure zones would provide the channels for the ophiolites" (Trümpy, 1960, p. 861 and 887). Since the ophiolites seemed to intrude Jurassic Bündnerschiefer or Schistes lustrés, they were generally considered to be of early to middle Cretaceous (Trümpy, 1960; Peters, 1963) or late Cretaceous age (Grunau, 1947).

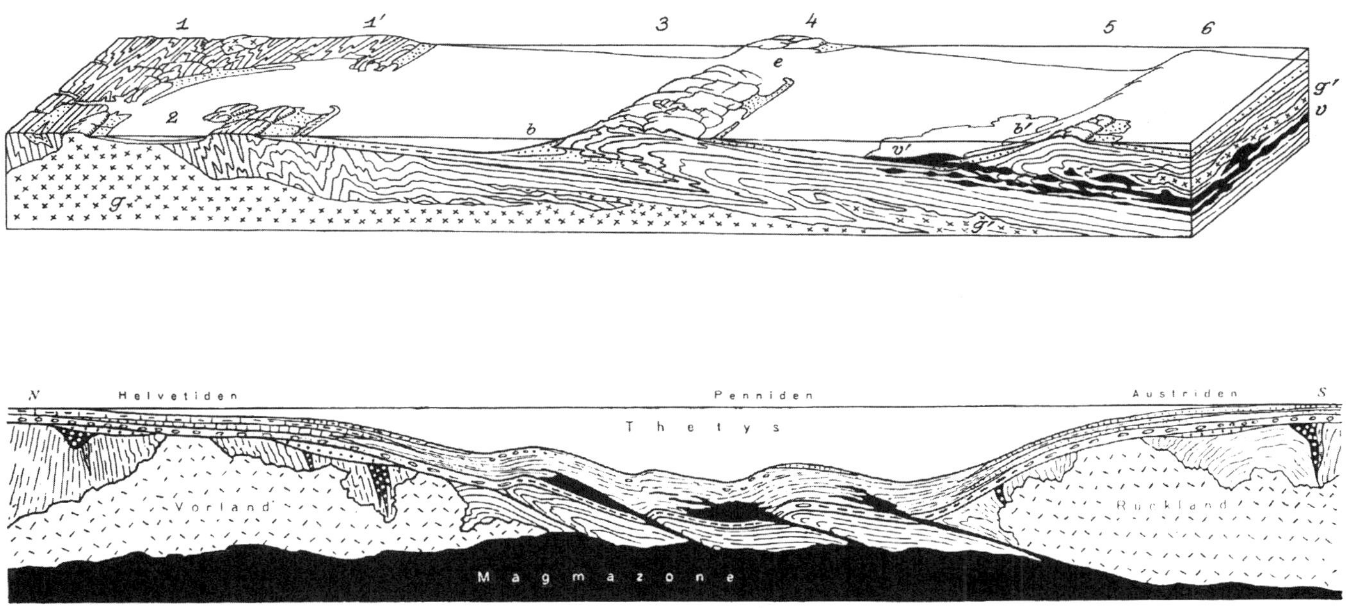

Fig. 61. **Die Grossgliederung der alten Thetys** (p. 231)

Figure 2. Cross-sections across the Alpine geosyncline showing the ophiolites as syn-tectonic intrusions along embryonic thrusts, above from Argand (1916), below from Staub (1924). 1—foreland; 1′—buckling of the foreland, due to the push of the nappes; 2—epicontinental sea; 3—foredeep; 4—frontal cordillera; 5—first [geosynclinal] through; 6—second cordillera; b and b′—slope breccias; e—marine strait at axial depression of thrust sheets; g—granitic batholiths of the foreland; g′—deformed and metamorphosed ancient batholiths, now crystalline schists; v—emplacement of basic magma along numerous listric surfaces at the base of major thrust nappes; v′—submarine extrusions, derived from the same feeders.

Eventually, the importance of Steinmann's discovery was recognized outside the Alps and the association of serpentinites, pillow lavas, and radiolarites became known as the Steinmann trinity: "Steinmann's realization of this trinity as a world-wide, age-long characteristic of geosynclinal deposition ranks among the most exciting achievements of geological research" (Bailey and McCallien, 1950). Already before this, Bailey had drawn attention to Steinmann's interpretation, which he placed into the context of continental drift.

> E. Argand [in his 1924 paper] has developed Steinmann's view, and has pictured geosynclines as determined by stretching, by continental drift apart, which attenuates the sial layer and eventually allows sima to reach the bottom of the sea at bathyal or abyssal depth. If such drift-separation continues, a new ocean bed may be developed, covered with products of submarine eruptions. (Bailey, 1936, p. 1719)

Indeed Argand (1924, p. 299) had noted that "A geosyncline will result, in general, from horizontal extension which stretches the raft of sal [continental crust] …This explains the frequent association of greenstones with bathyal or abyssal sediments. …If extension continues, …the geosyncline makes place to an ocean." Although there is no reference to Steinmann in Argand's text, his influence is likely (Henri Masson, 2002, personal commun.). Strangely, although referring to Steinmann's observations in the Alps and in the Apennines, the term Steinmann trinity became firmly established only in the Anglo-Saxon literature and made little headway in the European continental literature.

Because the original definition of the Steinmann trinity by Bailey and McCallien (1950, 1953, see also Jackson, 1997, p. 621) included igneous and sedimentary rocks, it was considered of "a somewhat hybrid nature" (den Tex, 1989) and in the following, the trinity was reinstated as the ophiolitic triad already clearly described by Steinmann (1927), viz., serpentinite or peridotite, gabbro, and dolerite (diabase) or spilite (Fig. 3, den Tex in Coleman, 1977; Vuagnat, 1963). This redefined trinity corresponds, by and large, to an ophiolite as defined in 1972 by a Penrose Conference in the light of the new paradigm of plate tectonics (Anonymous, 1972).

Before the advent of plate tectonics, "the problem [of ultramafics in mountain belts] remained unsolved. Some vital piece of evidence [was] still missing" (Hess, 1955, p. 402). Staub (1922) had considered the serpentinites as resulting from the metamorphism of intrusive peridotites, but many authors believed the serpentinites to be derived from "a hydrous ultrabasic magma" (Hess, 1938) intruded under low temperature during the early phases of orogeny (Hess, 1938, 1955), or even to represent "serpentine lavas" (Bailey and McCallien, 1950, 1953). Other authors considered the large ophiolite complexes of northern Syria and Greece as huge outpourings of mafic magma along flexures separating the continental shelf from the "eugeosyncline," whereby these basaltic magmas differentiated into peridotites and gabbros with pillow lavas forming the quenched

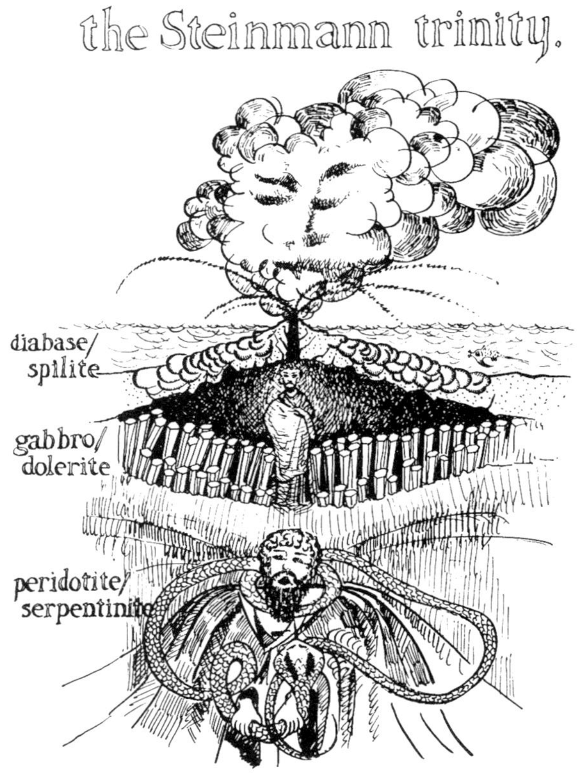

Figure 3. "Cartoon depicting a geologist's view of the Steinmann trinity." From den Tex in Coleman (1977).

surface of these enormous extrusions (Dubertret, 1953; Brunn, 1956); however, the ratio of ultramafic to mafic rocks, estimated by Steinmann (1927) to be 5:2, could not be explained by normal differentiation of a basaltic magma (Vuagnat, 1963).

Although Bowen (1927, 1928) had experimentally demonstrated "that such [peridotitic] magmas are not possible" under reasonable temperature conditions in the crust, many authors still weighted the field evidence in favor of an intrusive or even extrusive origin of the serpentinites in order to explain their close association with volcanic and sedimentary rocks (e.g., Hess, 1955). In an attempt to overcome the theoretical difficulties, other authors considered solid-state, "cold" intrusion of the peridotites (Bowen and Tuttle, 1949; Peters, 1963), or tectonic emplacement in serpentinite diapirs (Milovanovic and Karamata, 1960). With the identification of the mantle origin of peridotites in ophiolite complexes, however, the model of magmatic differentiation of a primary ultramafic or mafic magma could be abandoned in favor of a model in which the crustal rocks, gabbros, dolerites, and basalts are the products of partial melting of the mantle.

The concept of plate tectonics and the recognition that peridotites in ophiolite complexes were displaced mantle rocks, as suspected by de Roever (1957), solved the inconsistencies inherent in the interpretation of the peridotites and serpentinites as intrusive or extrusive rocks. With the new theory, the concept of a stratified oceanic lithosphere, anticipated by Hess (1954), was firmly established. It was deduced from seismic refraction studies performed in present-day oceans (Raitt, 1963) and was correlated with the stratigraphy of large ophiolitic complexes such as Troodos in Cyprus (Moores and Vine, 1971) and Semail in Oman (Glennie et al., 1974). This stratigraphy was codified during a Penrose Conference in 1972 (Anonymous, 1972) in a four-layer model matching the geophysical structure of the oceanic lithosphere. According to this model, a completely developed ophiolite should be composed of the following stratigraphic terms (from top to bottom): the sedimentary cover (layer 1) overlying a mafic volcanic complex, commonly pillowed, which together with the underlying sheeted dike complex represents the geophysical layer 2 (1–2 km). Below this basaltic sequence, a gabbroic complex (layer 3) occurs down to the Moho overlying an ultramafic complex, which corresponds to the oceanic lithospheric mantle (layer 4). The concept implies that ophiolites in Alpine-type mountain belts were tectonically emplaced onto continental margins ("obducted"), and that they are part of oceanic basement nappes or of tectonic mélanges associated with subduction complexes (e.g., Laubscher, 1969; Dewey and Bird, 1970; Coleman, 1971; Gansser, 1974). In the Alps and Apennines, it turned out that many of the tectonically emplaced ophiolite complexes and their stratigraphic cover were upside down, and the ophiolites were stratigraphically *overlain* by the Jurassic radiolarites and lower Cretaceous Maiolica limestones, and therefore they were the fossil oceanic basement on which Steinmann's deep-sea sediments were deposited (Decandia and Elter, 1972; Lagabrielle et al. 1984; Bernoulli and Weissert 1985; Weissert and Bernoulli 1985).

Ironically, it appears that Steinmann never saw a "completely developed ophiolite" such as Troodos or Semail. Over the years, detailed fieldwork in the Alps and the Apennines, where Steinmann worked, showed that the internal stratigraphy of the ophiolites in these areas does not match the classic ophiolite stratigraphy as defined by the Penrose Conference in 1972. No substantial relics of a sheeted dike complex were found, and gabbros appeared to play a subordinate role. Instead, oceanic sediments, radiolarian cherts, and pelagic limestones were observed to stratigraphically overlie (1) serpentinized mantle rocks and tectono-sedimentary breccias ("ophicalcites") derived from them; and (2) pillow lavas that were extruded onto the serpentinites previously exposed on the ocean floor (Decandia and Elter 1972; Lagabrielle et al. 1984; Bernoulli and Weissert 1985; Weissert and Bernoulli 1985). This stratigraphy is in clear contrast to the ophiolite stratigraphy as defined by the Penrose Conference. More importantly, the mantle rocks proved to have been derived from the subcontinental lithosphere (Piccardo et al., 1990; Müntener and Hermann, 1996), and no genetic link exists between the peridotites and the magmatic rocks (Rampone et al., 1998; Rampone and Piccardo, 2000; Desmurs et al., 2001).

The Association of Ophiolites and Radiolarites

Like many others, Steinmann (1905, 1927) thought that the radiolarites and pelagic limestones were older than the ophiolitic "intrusions" and, therefore, were not directly related to them; for him, these sediments merely documented the deep-sea character of the "geosyncline." He failed to appreciate the now obviously extrusive and subaqueous character of pillow lavas recognized early in the twentieth century (Anderson, 1910; Dewey and Flett, 1911). In the Apennines as early as 1880, however, Pantanelli observed radiolarites to occur as interpillow sediment and therefore suggested a genetic relationship between these rocks (Pantanelli, 1880). The "curious fact, that greenstones of a peculiar type ["diabase with a spherical or ellipsoidal structure"] are associated with [radiolarian] cherts" in Paleozoic sequences of Britain was noted also by Teall (1894, p. 563–564) who gave a very clear description of pillows "avant la lettre" and suspected them to be characteristic of "submarine and possibly deep sea lavas."

Because dissolved silica or reactive silicate material such as volcanic ash was (rightly) thought to be of prime importance for the growth of radiolarian populations, many authors (e.g., Lotti, 1886; Dewey and Flett, 1911; Davis, 1918; Wenk, 1949; Bailey and McCallien, 1953; Grunau, 1965) assumed a direct link between submarine volcanism and hydrothermal activity as sources of silica and the sedimentation and diagenesis of radiolarian ooze, the Recent analogue of radiolarite. In contrast, Steinmann (1927) questioned this link not only because he thought that the ophiolites were distinctly younger than the sediments, but also because it was, in his opinion, at odds with the regional distribution of the Jurassic radiolarites, which are not restricted to the ophiolitic units, but rather also encroach onto the deeply submerged Tethyan margins in the Apennines (Bernoulli et al., 1979; McBride in McBride and Folk, 1979), in the Alps (Winterer and Bosellini, 1981) and elsewhere, with no traces of submarine volcanism associated with them. Indeed, the facies of the deep-water pelagic sediments overlying the oceanic basement and the sunken carbonate platforms appears to be determined rather by subsidence, fluctuations of the calcite compensation depth, nutrient supply and plankton productivity, and other paleoceanographic changes (Berger and Winterer, 1974).

Are Alpine Radiolarites Deep-Sea Sediments?

The question of whether ancient deep-sea sediments could occur on land ("Gibt es fossile Tiefseeablagerungen von erdgeschichtlicher Bedeutung?," Steinmann, 1925) was part of the controversy during the early twentieth century between the advocates of permanency of the oceans and the partisans of continental drift. Steinmann (1905, 1927) compared the radiolarites and associated argillites to radiolarian oozes and red deep-sea clays of modern oceans, recognized that the Maiolica-type limestones were fossil coccolith oozes, and equated the "Argille scagliose" (our Argille a Palombini, see below) with bluish deep-sea clays (Steinmann, 1925). Such an association of "fine black shales, limestones and radiolarites" with pillow lavas was also noted for the Paleozoic of the British Isles by Dewey and Flett (1911) (see also Grunau, 1965, and den Tex, 1989).

For his part, Steinmann (1925) firmly relied on the analogy between the sediments occurring in the ophiolite units and modern oceanic sediments; however, the local association of radiolarites with "transgressive" breccias and sandstones (Saluver Formation) seemed to cast serious doubt on this interpretation (e.g., Arbenz, 1926; Grunau, 1947). Radiolarites were therefore considered by many authors to be, at least in part, shallow-water deposits (Grunau, 1947; see discussion in Cornelius, 1951). In order to save his (correct) interpretation, Steinmann (1925) incorrectly interpreted the sedimentary breccias associated with radiolarites as products of tectonic fracturing during Alpine orogeny. Interestingly, the very same breccias had been interpreted correctly one year earlier as sedimentary deep-water deposits related to submarine tectonic movements (Heim, 1924). A similar conclusion was reached by Bailey and McCallien (1953, p. 425), who suspected that the sandstones occasionally associated with the Steinmann trinity were deposited in deep water by "turbidity transport." The apparent paradox was eventually resolved when it was recognized that the syn-rift breccias underlying the radiolarites in the distal continental margin (Err nappe, e.g., Trümpy, 1975; Finger, 1978) and the breccias within the post-rift radiolarites were deposited by debris flows and turbidity currents (Manatschal and Bernoulli, 1999).

THE STEINMANN TRINITY IN ITS "TYPE-AREA"

Alpine Ophiolites and the Rhetic Nappe of Steinmann

In the Alps, remnants of an ophiolitic suture can be traced within a discontinuous belt along the entire length of the chain from Corsica and the Ligurian Apennines across the south-Pennine nappes to the eastern end of the Alps and possibly beyond (Fig. 4). Most of these ophiolites belong to the Liguria-Piedmont segment of the Mesozoic Tethys ocean that was closed by subduction and obduction during the late Cretaceous and early Tertiary. Possible remnants of a younger, smaller ocean basin, the Valais ocean, are restricted to the north-Pennine nappes of the central Alps (Fig. 4).

In eastern Switzerland and western Austria, Steinmann (1905) recognized that the ophiolitic rocks always occurred in the same tectonic unit, which he called the Rhetic nappe, a term that did not survive but that included the south-Pennine units and their outliers in the Prealpine nappes of central and western Switzerland (Iberg klippen and nappe des Gets). The relics of this south-Pennine ophiolite zone can be followed from Val Malenco (Malenco complex) across the Engadin valley to the Sursés valley (Platta nappe), and across a number of imbricates (Totalp slice) and local occurrences in tectonic mélanges (Arosa zone), to the northern margin of the Northern Calcareous Alps (Fig. 5). They occur between units derived from the Briançonnais, a possible promontory of Iberia (Stampfli, 1993), below, and units derived from the Adriatic margin (Austroalpine), above (Fig. 6). We may regard the Totalp slice of the Davos area as the northern

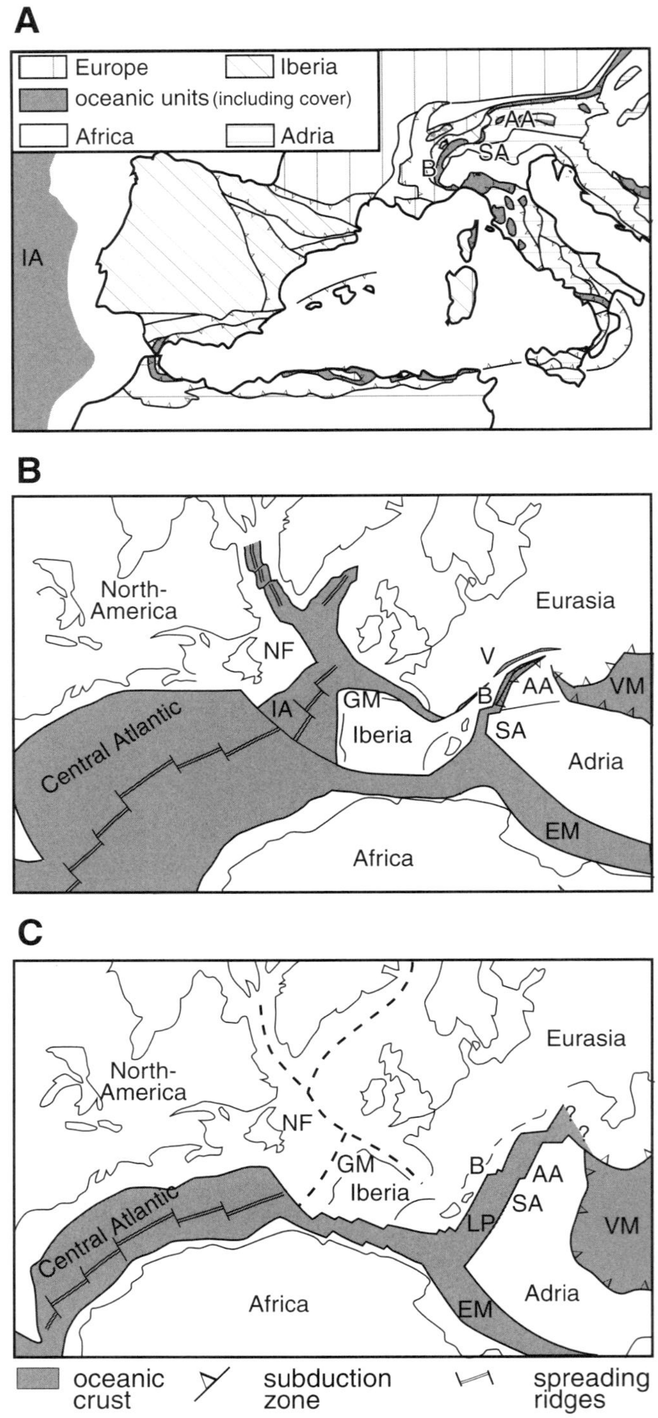

Figure 4. A: Tectonic sketch map of the present-day west-Mediterranean area, showing the occurrence of continental areas and oceanic units (ophiolites and oceanic sediments). B and C: Large-scale paleogeography reconstructed for the late Cretaceous (Fig. 4B) and for the late Jurassic (Fig. 4C). AA—Austroalpine; B—Briançonnais; EM—Eastern Mediterranean; GM—Galicia margin; IA—Iberian Atlantic; LP—Liguria-Piedmont Ocean; NF—Newfoundland; SA—Southern Alps; V—Valais Basin. VM—Vardar-Meliata Ocean. From Manatschal and Bernoulli (1999).

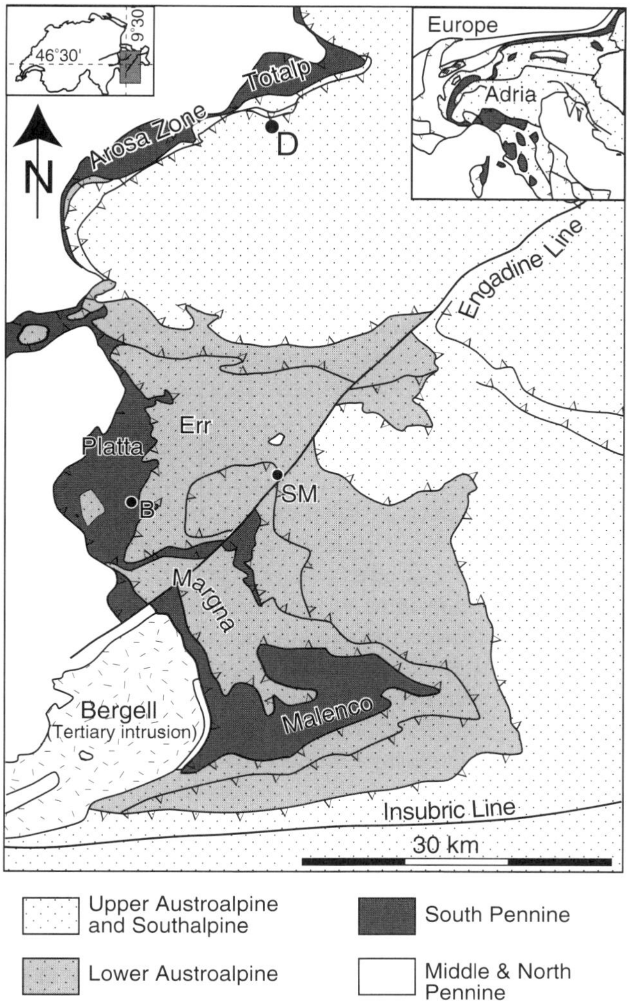

Figure 5. Simplified tectonic map of the south-Pennine-Austroalpine boundary zone in southeastern Switzerland. B—Bivio, D—Davos, SM—St Moritz. Modified after Froitzheim et al. (1994).

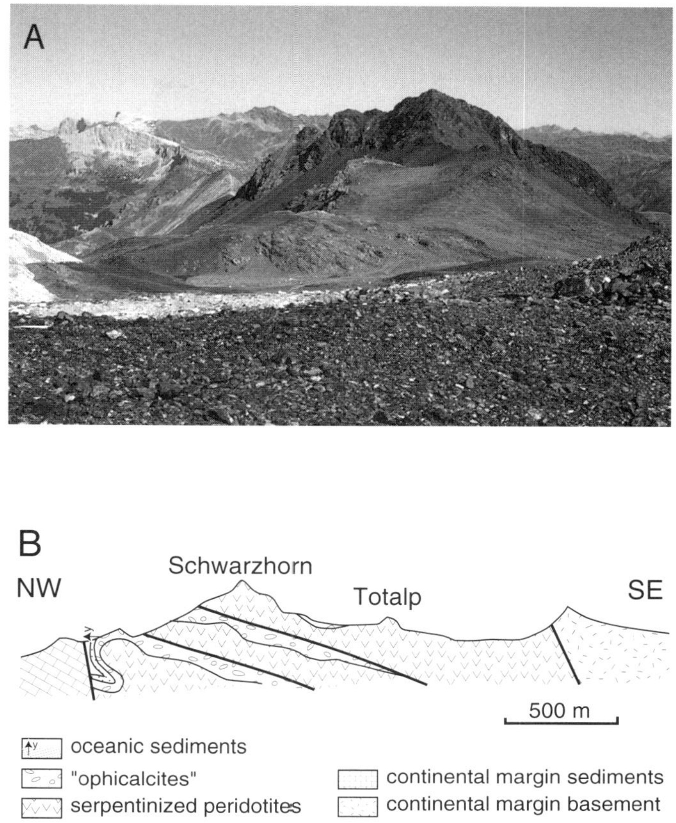

Figure 6. A: View of the Totalp nappe near Davos. Serpentinized peridotites (black) are overlain by and imbricated with ophicalcites. The Totalp nappe is sandwiched between Austroalpine (Adriatic) continental basement above (to the right, not visible) and a tectonically overturned imbricate of decolled Austroalpine sediments below (upper Triassic dolomites below, white, to the left). In the background, an ophiolite-bearing zone of imbricates (Arosa zone) is thrust onto Upper Jurassic shallow-water limestones of the Briançonnais zone (white) and overthrust by Austroalpine continental basement (Silvretta nappe). B: Tectonic profile across the Totalp nappe.

prolongation of the Platta nappe, and for simplicity we will call the combined unit the Platta-Totalp nappe.

In the western Alps, part of the ophiolites were subducted to great depths and experienced high-pressure, low-temperature metamorphism. In contrast to these largely dismembered ophiolites, the ophiolites of the Platta-Totalp nappe were always part of the upper plate of the Alpine system and were never subducted to great depths. Obviously, they belonged to the upper plate of the southeast-directed subduction zone, which had apparently first developed within the ocean (Fig. 7; Lagabrielle, 1987; Avigad et al., 1993). In the north (Totalp), the Alpine metamorphism reaches only prehnite-pumpellyite facies, whereas in the south it passes from very low-grade (anchi-) metamorphism (~250 °C, Ferreiro Mählmann, 2001) around Bivio, to the greenschist and, finally, epidote-amphibolite grade in the Malenco complex (Trommsdorff and Evans, 1977). In the Sursés valley, the Alpine movements can be kinematically inverted (Manatschal and Nievergelt, 1997), and the pre-Alpine geometrical and age relationships between the exhumed mantle rocks, the gabbroic intrusions, and the volcanic rocks can be studied in detail (Desmurs et al., 2001). In particular, the paleotectonic relationships with the distal continental margin bordering the ophiolite zone during Jurassic rifting can be established (Froitzheim and Manatschal, 1996; Manatschal and Nievergelt, 1997; Müntener and Hermann, 2001).

Structure and Stratigraphy of the Platta-Totalp Nappe

Together with the lower Austroalpine Err nappe, the Platta nappe preserves the remnants of the southeastern margin of the

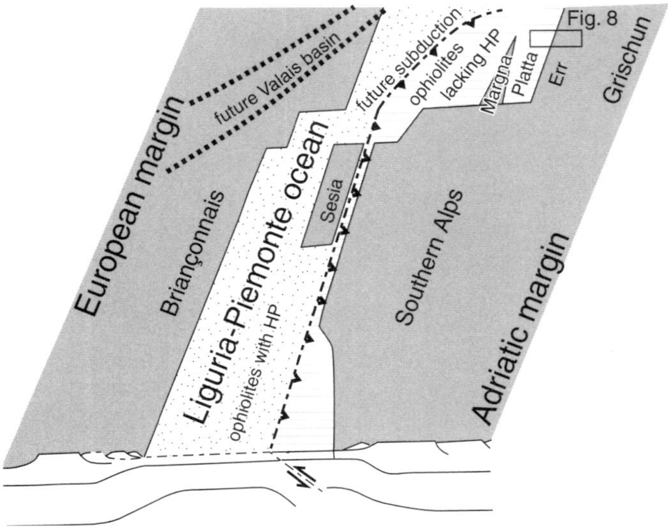

Figure 7. A possible location of the late Cretaceous–early Tertiary subduction in a late Jurassic paleogeographic scenario. Adapted from Dal Piaz (1999).

Liguria-Piedmont segment of the Alpine Tethys (Figs. 5 and 8). The Err nappe includes the remnants of the distal continental margin and was thrust onto the Platta nappe during late Cretaceous time. Both nappes are now part of a west-directed late Cretaceous thrust-wedge that during the early Tertiary continent-continent-collision was thrust northward onto the mid- and north-Pennine units (Froitzheim et al., 1994). The Err nappe consists of a thinned continental crust, overlain by a number of extensional allochthons. These stranded klippen of upper continental crust and pre-rift sediments were tectonically emplaced during the mid-Jurassic along an oceanward-dipping detachment system active during late rifting of the margin (Froitzheim and Eberli, 1990; Froitzheim and Manatschal, 1996; Manatschal and Nievergelt, 1997). The Platta nappe, originally situated oceanward from the Err distal margin, represents a fossil ocean-continent transition and exposes the Steinmann trinity (in the sense of Bailey and McCallien, 1950), i.e., serpentized peridotites, basaltic volcanic rocks, and radiolarites (Dietrich, 1969). The sequence of emplacement of these rocks on the deep-sea floor corresponds to the relative age sequence of the ophiolitic rocks as proposed by Steinmann (1927), i.e., serpentinite, gabbro, and, finally, diabase-spilite.

In the Platta nappe of the Sursés valley, two major imbricates that are floored by serpentinites and interleaved with decolled cover sequences can be distinguished: an upper serpentinite unit, originally nearer to the distal margin, and a lower one, originally situated oceanward (Fig. 9; Desmurs et al., 2001, their figure 4). The occurrence of sediments and/or basalts stratigraphically overlying the serpentinites shows that these were in an upper crustal position at the beginning of thrusting and that the Alpine décollement along the base of the nappe involved only a few hundred meter thickness of the serpentinized mantle. The apparently uniform mass of the Totalp slice is also subdivided into a number of imbricates by one or more internal thrusts (Fig. 6B). Because the mantle rocks of the Totalp slice are similar to those of the upper serpentinite unit of the Platta nappe, and lack gabbroic intrusions, we correlate the Totalp slice with the upper serpentinite unit of the Platta nappe.

The stratigraphy of the Platta-Totalp nappe is schematically illustrated in Figure 10. The basement of the imbricates of the Platta nappe and the Totalp slice is invariably made of serpentinized peridotite. The top of the serpentinites is commonly composed of ophicalcites, tectono-sedimentary breccias related to mantle exhumation along the detachment system (Lemoine et al., 1987; Desmurs et al., 2001). In the Platta nappe, extensional allochthons of continental basement rocks and their pre- and

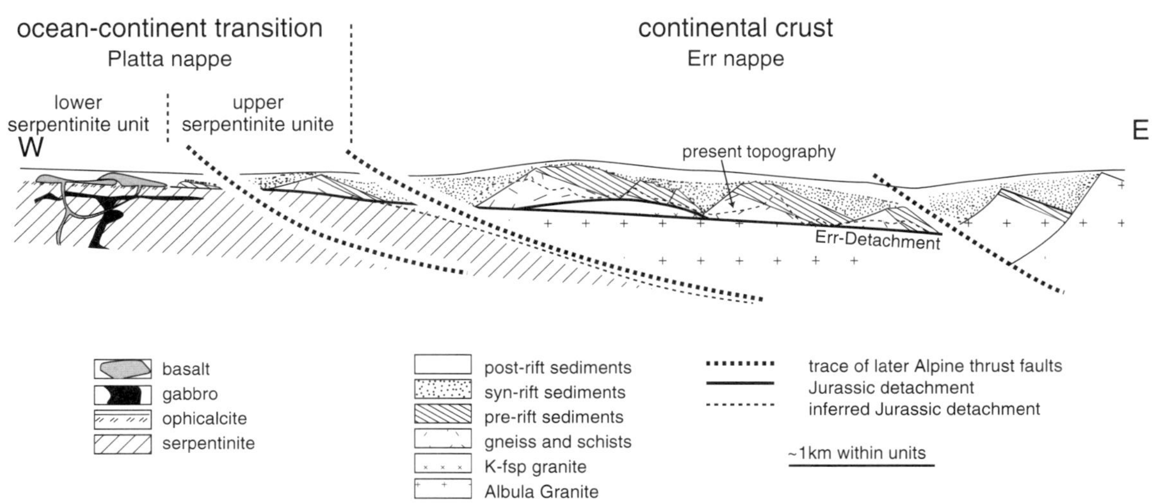

Figure 8. Palinspastic reconstruction of the Err-Platta ocean-continent transition. After Manatschal and Nievergelt (1997).

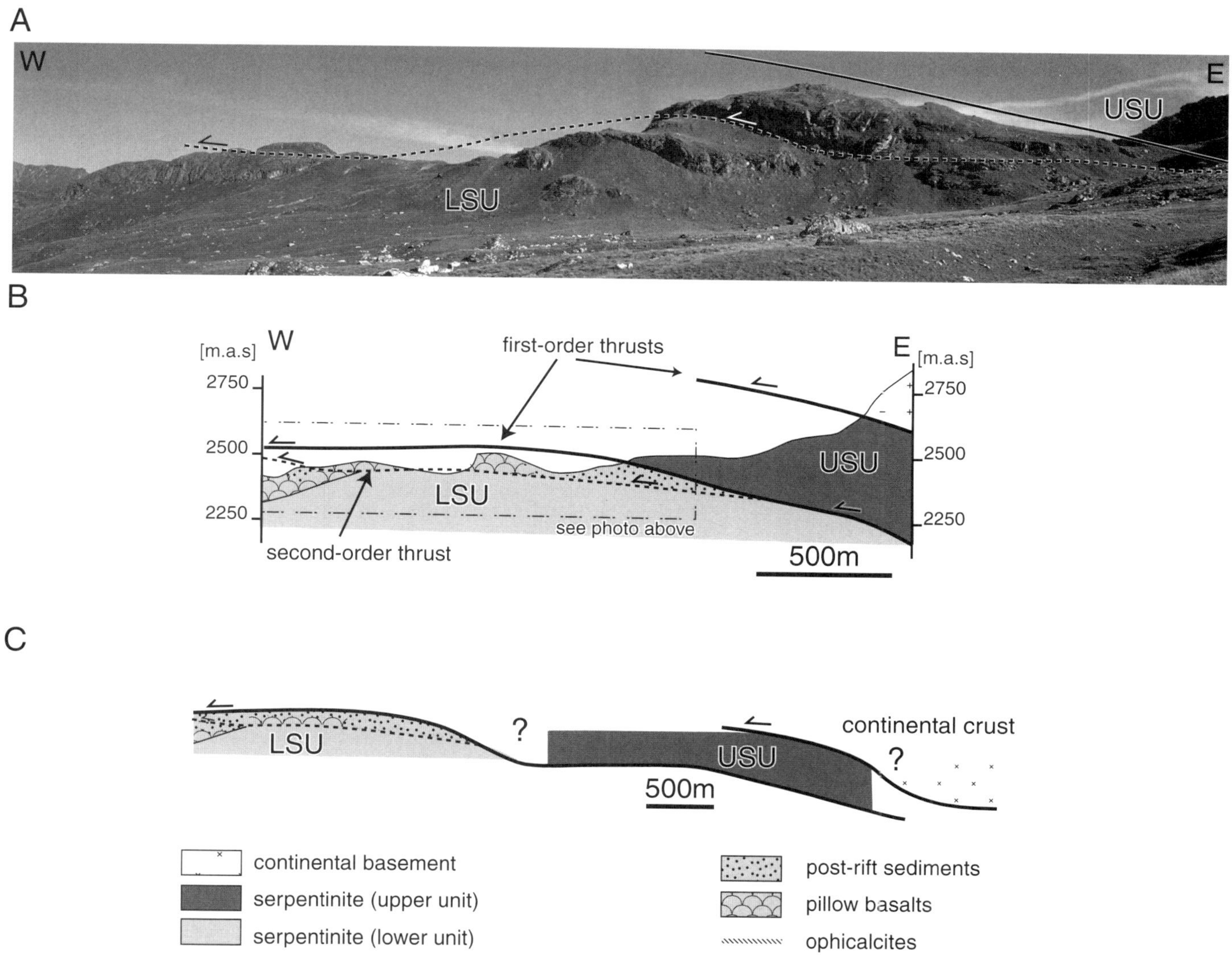

Figure 9. A: View of Falotta, north of Bivio, showing the two serpentinite units of the Platta nappe and the associated cover imbricates. USU—upper serpentinite unit; LSU—lower serpentinite unit. B and C: Qualitative palinspastic reconstruction. B—present-day situation. C—Kinematic inversion of section B. The question marks indicate the uncertainties in the estimate of the amount of thrusting. Adapted from Desmurs et al. (2001).

syn-rift sediments overlie the exhumed mantle rocks (Froitzheim and Manatschal, 1996; Manatschal and Nievergelt, 1997). In the Totalp slice, only sedimentary breccias containing clasts of continental basement rocks and gabbro mylonites (?lower continental crust) overlying the ophicalcites document the former existence of a hanging wall of continental basement rocks.

In the lower serpentinite unit of the Platta nappe, gabbros occur as planar, sill-like, or lensoidal bodies (Fig. 11) within the serpentinized peridotites. They make up <10% of the observed volume. Massive basalts, pillow lava, pillow breccias, and hyaloclastites occur in patches of varying thickness and size (Dietrich, 1969); their abundance appears to increase oceanward: they are scarce in the upper serpentinite unit and are absent in the Totalp slice. The volcanics stratigraphically overlie the serpentinites, ophicalcites, gabbros, and associated breccias, and their extrusion therefore postdates emplacement of the mantle rocks on the sea floor (Desmurs et al. 2001).

Exhumed mantle rocks, ophicalcites, pillow lavas, and extensional allochthons are all overlain by post-rift deep-water sediments (Fig. 10). These oceanic sediments include (1) red and black shales; (2) cm-bedded ribbon cherts (Radiolarite Formation, middle to upper Jurassic, Bill et al., 2001); (3) gray, thin-bedded limestones with greenish, shaly interbeds (Calpionella or Aptychus Limestone, Berriasian); (4) black shales with

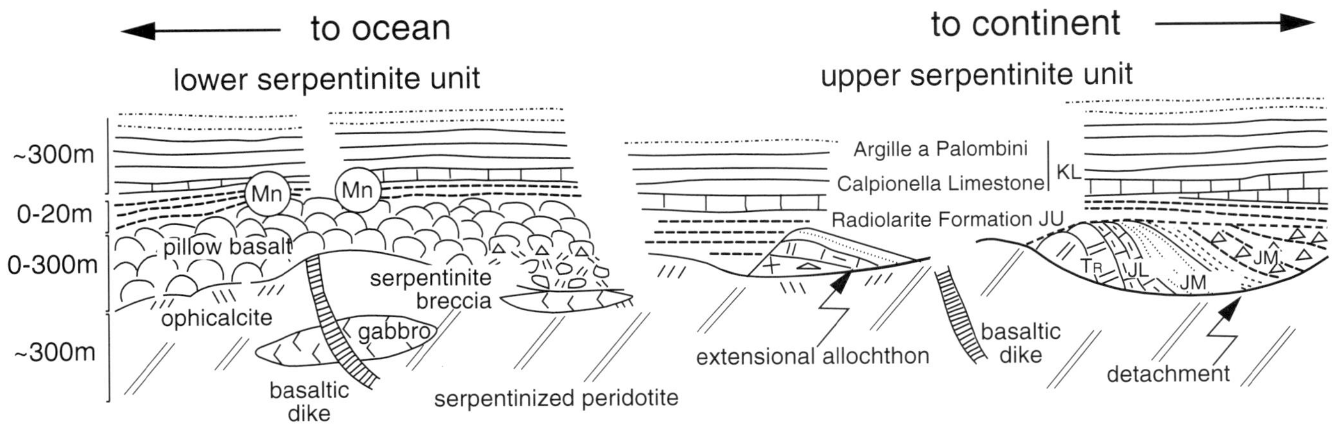

Figure 10. Generalized stratigraphy of the ocean-continent transition of the Platta-Totalp nappe. In the proximal (continent-ward) part of the ocean-continent transition (upper serpentinite unit of Platta nappe, Totalp nappe), the exhumed mantle rocks are locally overlain by extensional allochthons, stranded klippen of continental basement, pre-rift sediments and syn-rift marine breccias, emplaced along extensional detachment faults. In the distal (oceanward) part of the ocean-continent transition (lower serpentinite unit of Platta nappe), the mantle rocks are intruded by gabbros and stratigraphically overlain by pillow lavas and breccias. Deep-water sediments (Radiolarite Fm., Calpionella Limestone and Argille a Palombini) cover the entire ocean-continent transition. KL—Lower Cretaceous; JU—Upper Jurassic; JM—Middle Jurassic; JL—Lower Jurassic; Tr—Triassic; Mn—Manganese. For further explanation, see text. After Manatschal and Nievergelt (1997) and Desmurs et al. (2001).

dm-bedded fine-grained limestones (Argille a Palombini, approx. Valanginian to Barremian); and (5) hemipelagic marls with interbedded sandstone turbidites, dated by planktonic foraminifera as Aptian-Albian to ?Cenomanian (Dietrich, 1970). The red and black shales occur only locally; they are barren, associated with debris-flow deposits, and might be derived from the erosion of fault gouges associated with the detachment system or of exhumed serpentinites and gabbros. The radiolarites and the Calpionella Limestone (Maiolica) contain breccias and graded sandstones, including clasts of both serpentinite and continental basement rocks, indicating local submarine relief at the time of deposition. Hydrothermal activity was probably the source of the sulfide deposits within the ophiolites and of ferromanganese mineralizations in the oceanic sediments (Stille et al., 1989).

Peridotites and Serpentinites

The ophiolites of the Totalp-Platta nappe share with other ophiolites of the Liguria-Piedmont ocean a predominance of a largely serpentinized peridotite basement intruded by small or moderate volumes of mafic rocks, and the lack of a "complete" ophiolite stratigraphy. More importantly, the serpentinites are in places stratigraphically overlain by Jurassic radiolarites, indicating that they must have been uplifted from mantle depth to the sea floor in Mesozoic times. From field and geochemical evidence it appears that like in many areas of the Alps and Apennines (Piccardo et al., 1990; Müntener and Hermann, 1996; Rampone et al., 1995, 1998; Rampone and Piccardo, 2000), the serpentinite basement represents the former subcontinental mantle (Desmurs et al., 2001); however, the chemical composition and the conditions of equilibration are different from place to place. In the Malenco complex, the subcontinental mantle rocks are still associated with lower continental crust, and a well-preserved Permian crust/mantle boundary is observed (Hermann et al., 1997).

Serpentinized lherzolite is the major rock type all over the Platta-Totalp nappe (Peters, 1963; Dietrich 1969). Structures, petrography, and mineral chemistry show that the mantle rocks of the upper and the lower serpentinite units of the Platta record a different high-temperature tectono-metamorphic evolution (Desmurs et al., 2001). The upper serpentinites are made of heterogeneous, pyroxenite-rich, fertile lherzolites similar to those of the subcontinental mantle of the Malenco complex (Müntener and Hermann, 1996), and likewise were equilibrated in a shallow subcontinental environment (~800 °C, 7–12 kbar) near the crust/mantle boundary. In contrast, rock and mineral chemistry data show that the mantle rocks of the lower unit are fertile, but pyroxenite-poor and more homogeneous. A few samples are more refractory and represent residues of variable degrees of partial melting (Desmurs, 2002). They equilibrated at higher temperatures (≥1000 °C) and lower pressures presumably in the stability field of plagioclase (Müntener et al., 2001), which probably crystallized from impregnating melts (Müntener et al., 2002). This shows that the mantle rocks of the Platta nappe record increasing temperatures of equilibration and increasing interaction with melts oceanward. Sm-Nd model ages on clinopyroxenes from the lower serpentinites are Permian (Müntener et al., 2002); a similar Permian age was obtained by Rampone et al. (1996) from the internal Ligurides. The ages suggest that these mantle rocks became separated from the convective mantle and were accreted to the base of the continental lithosphere at a time when the mantle rocks of the upper unit were near the base of the continental crust. The mantle rocks of the upper serpentinite unit thus simply record cooling and hydration, whereas

Figure 11. Small ellipsoidal gabbro intrusion, exhumed by differential weathering, within the serpentinized peridotites of the lower serpentinite unit of the Platta nappe. B—basalt; G—gabbro; S—serpentinites. Val da Natons north of Bivio, Sursés.

the mantle rocks of the lower unit were impregnated by tholeiitic melts and equilibrated at high temperatures in the presence of plagioclase, presumably during late rifting (Müntener et al., 2002). This could mean that oceanward, deeper and deeper levels of the subcontinental mantle were exhumed during late rifting, or that the already exhumed mantle was exposed to higher temperatures toward a future sea floor spreading center. High-T (>700 °C) deformation in both units of the Platta nappe was associated with top-to-the-east, i.e., top-to-the-continent motion, as suggested by shear-sense indicators (Fig. 12).

Whereas the anhydrous evolution of the two serpentinite units of the Platta nappe is markedly different, their retrograde evolution under hydrous conditions is very similar (Desmurs et al., 2001). Hydration of the mantle rocks started under upper amphibolite facies conditions, but with decreasing temperatures, serpentinization accompanied exhumation and deformation. Ar/Ar age determination on phlogopite in a pyroxenite of the Totalp serpentinite yielded an age of 160 ± 8 Ma (Peters and Stettler, 1987), indicating that the mantle rocks were uplifted to ~10 km depth in the late middle Jurassic. Ongoing deformation under decreasing temperatures is documented by serpentinite mylonites that formed under greenschist-facies conditions that were cataclastically deformed and partially calcitized under still lower temperatures. The shear sense associated with low-temperature deformation during this late phase was top-to-the-west, i.e., top-to-the-ocean, analogous to that of the detachment system that was associated with late rifting in the distal continental margin, the emplacement of the extensional allochthons, and the final exposure of the mantle on the sea floor (Fig. 12). Because serpentinite mylonites are cut by basaltic dikes, which in turn are cut by fault gouges, it appears that movement along them was partially synchronous with MORB magmatism.

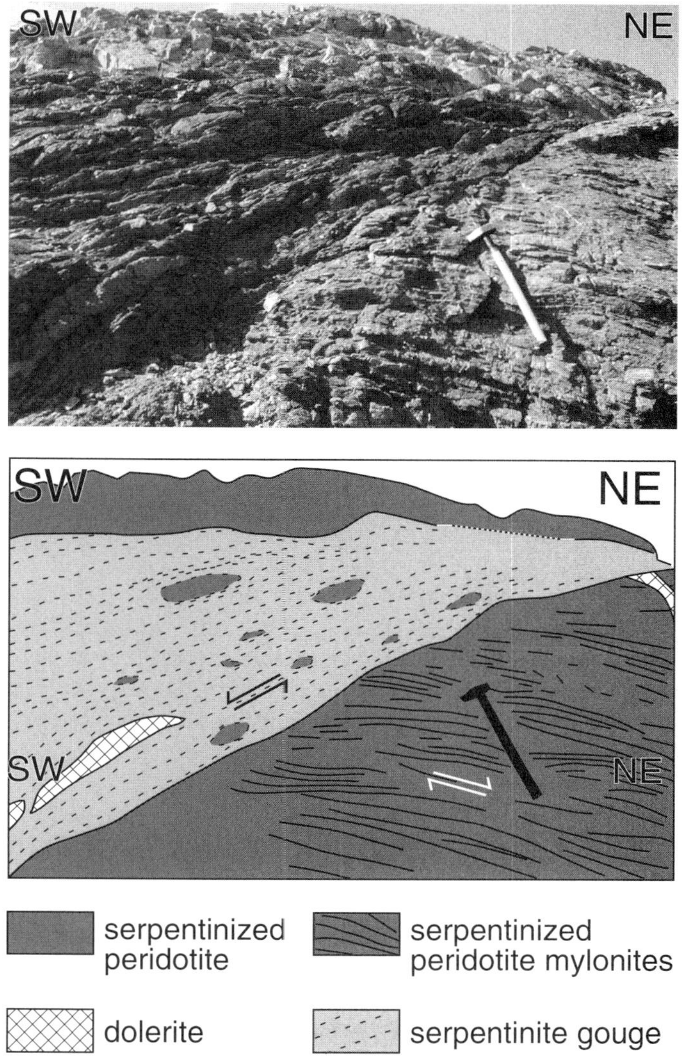

Figure 12. Mode of deformation in exhumed subcontinental mantle rocks. Serpentinized peridotites with a mylonitic high-T foliation are cut by a serpentinite gouge. Dolerite dikes cut the mylonitic foliation but are overprinted by low-T shearing. Transport direction in the peridotite mylonites is top-to-the-continent (northeast), in the serpentinite gouge it is top-to-the-ocean. Sur al Cant south of Bivio. Hammer is 60 cm long.

Ophicalcites

Steinmann (1905, 1913) did not observe contact metamorphism between the (in his opinion) intrusive ophiolites and the associated sediments, with the exception of the ophicalcites, which he wrongly interpreted as contact-metamorphic rocks. Ever since, these complex breccias, which include fragments of serpentinite in a matrix of white sparry calcite and/or of pink to red microsparite, have been interpreted in various ways, ranging from magmatic carbonatites (Bailey and McCallien, 1960) to products of subaerial weathering and caliche formation (Folk and McBride, 1976,

1978). New observations show that these breccias are the product of a combination of tectonic fracturing, replacement of serpentine minerals by calcite, calcite cementation and infill of pelagic and/or diagenetic sediment, or, in the case of many matrix-supported breccias, resedimentation by debris flows (Bernoulli and Weissert, 1985; Lemoine et al., 1987; Früh-Green et al., 1990). Indeed, in the Alps and the northern Apennines, the ophicalcites form a more or less continuous cover along the surface of the exhumed mantle rocks and mark the trace of the detachment faults along which the mantle rocks were exhumed (Lemoine et al. 1987; Florineth and Froitzheim, 1994; Manatschal and Bernoulli, 1999). Accordingly, ophicalcites are conspicuously missing in ophiolite complexes inferred to be of fast-spreading ridge origin, such as the Semail ophiolite (Glennie et al., 1974), but they occur in places where mantle rocks were brittlely deformed and exhumed to the sea floor, i.e., along slow spreading ridges such as the central Atlantic (Juteau et al., 1989), along oceanic transform faults (Bonatti et al. 1974), and along ocean-continent transition zones (Whitmarsh et al., 1998; Wilson et al., 2001).

Gabbros, Dolerites, and Basalts

In the Platta nappe, gabbros are restricted to the lower, originally oceanward serpentinite unit; in the Totalp slice, only gabbro blocks of unknown age and origin occur in mass flow breccias. The gabbros form small individual intrusions (Fig. 11); field relationships and mineral-chemistry data show that they intruded into partially serpentinized mantle rocks at a shallow depth (<8 km). The gabbros intruded during extension: their microstructures indicate syn-magmatic deformation, followed by deformation under upper amphibolite-facies conditions, whereas deformation under lower temperature conditions is only locally observed. The evidence for final exhumation and exposure on the sea floor is recorded by the occurrence of gabbro clasts in pillow and sedimentary breccias stratigraphically overlying the exhumed mantle rocks and gabbros (Desmurs et al., 2001, their figure 9). All gabbros show the same metamorphic evolution, i.e., intrusion at relatively low pressure, oceanic hydration at elevated temperature, and a subsequent static oceanic alteration.

Basaltic rocks cover large areas of the lower serpentinite unit of the Platta nappe. They are scarce in the upper serpentinite unit and absent in the Totalp slice, but they occur in other imbricates of the northern area (Arosa zone). They occur as massive flow units, pillow lavas, hyaloclastites, and breccias (Dietrich, 1969) stratigraphically overlying the exhumed mantle rocks and the associated tectono-sedimentary breccias. Undeformed dikes show chilled margins along their contacts with the serpentinites and gabbros (Manatschal and Bernoulli, 1999, their figure 6b). The extrusive rocks therefore are younger than the gabbros with which they are spatially associated *today*.

Together with the dolerites and basalts, the gabbros record the very beginning of magmatic accretion within the ocean-continent transition zone during a transitional stage between the crustal and lithospheric break-up of the continental lithosphere. All gabbros fall into a narrow time range of 161 ± 1 Ma (Schaltegger et al., 2002, concordant U/Pb ages on zircon). They range from olivine gabbros to apatite-rich ferrogabbros and diorites originating by fractional crystallization of the same parental magma. The petrology and geochemistry of the gabbros show that each continuous outcrop represents a different batch of melt that intruded the exhumed and partially serpentinized mantle. Likewise, each intrusion records different magmatic processes, ranging from predominantly fractional crystallization to solidification without fractionation (Desmurs et al., 2002). The composition of the frozen melts, as well as the composition of the liquids calculated from the chemistry of the most primitive clinopyroxene cumulate, indicates that the parental melt has the same composition as the spatially associated basalts. This finding suggests that, despite of the observation that the basalts must be locally younger, gabbros and basalts belong to the same magmatic event and were emplaced during and after exhumation of the mantle rocks on the sea floor (Fig. 13).

Most of the dolerites and basalts show a major element composition typical of tholeiitic magmas (Frisch et al., 1994); however, the chemistry of the mafic liquids (basalts and parental melts of the gabbros) appears to change from T-MORB to N-MORB composition as they crystallized away from the continental margin (Desmurs et al., 2002). Trace element and isotope chemistry suggest that these igneous rocks are the result of low degree of melting (<10%) of two distinct sources, a depleted mantle for the N-MORBs, and an enriched mantle that probably was the subcontinental mantle, for the T-MORBs. The N-MORB source of the basalts of the lower serpentinite unit is indicated by their REE concentration (Desmurs et al., 2002) and the Nd and Hf isotopic ratios determined on whole rock and zircons, respectively (Stille et al., 1989; Schaltegger et al., 2002). In contrast, the REE concentrations of the basalts of the upper serpentinite unit indicate that they originated from a source enriched in incompatible elements. Obviously, the magmatic rocks from the Platta ocean-continent transition record oceanward increasing degrees of melting and decreasing amounts of contamination of their source by an enriched component. This process is probably related to the continuous thinning of the subcontinental mantle and the associated rise of the asthenosphere during a transitional stage between the crustal and the lithospheric break-up.

A FOSSIL OCEAN-CONTINENT TRANSITION

In the Alps and northern Apennines, the particular ophiolite stratigraphy indicating exposures of mantle rocks on the sea floor was, at an early stage, postulated to be associated with oceanic transform faults in analogy with serpentinite occurrences along such faults in modern ocean basins (Gianelli, 1977; Lemoine, 1980; Bernoulli and Weissert, 1985; Weissert and Bernoulli, 1985). For the Liguria-Piedmont ocean, subsolidus exhumation of subcontinental mantle during continental rifting and its association with gabbroic intrusions and basaltic volcanics derived from partial melting of the rising asthenosphere was first

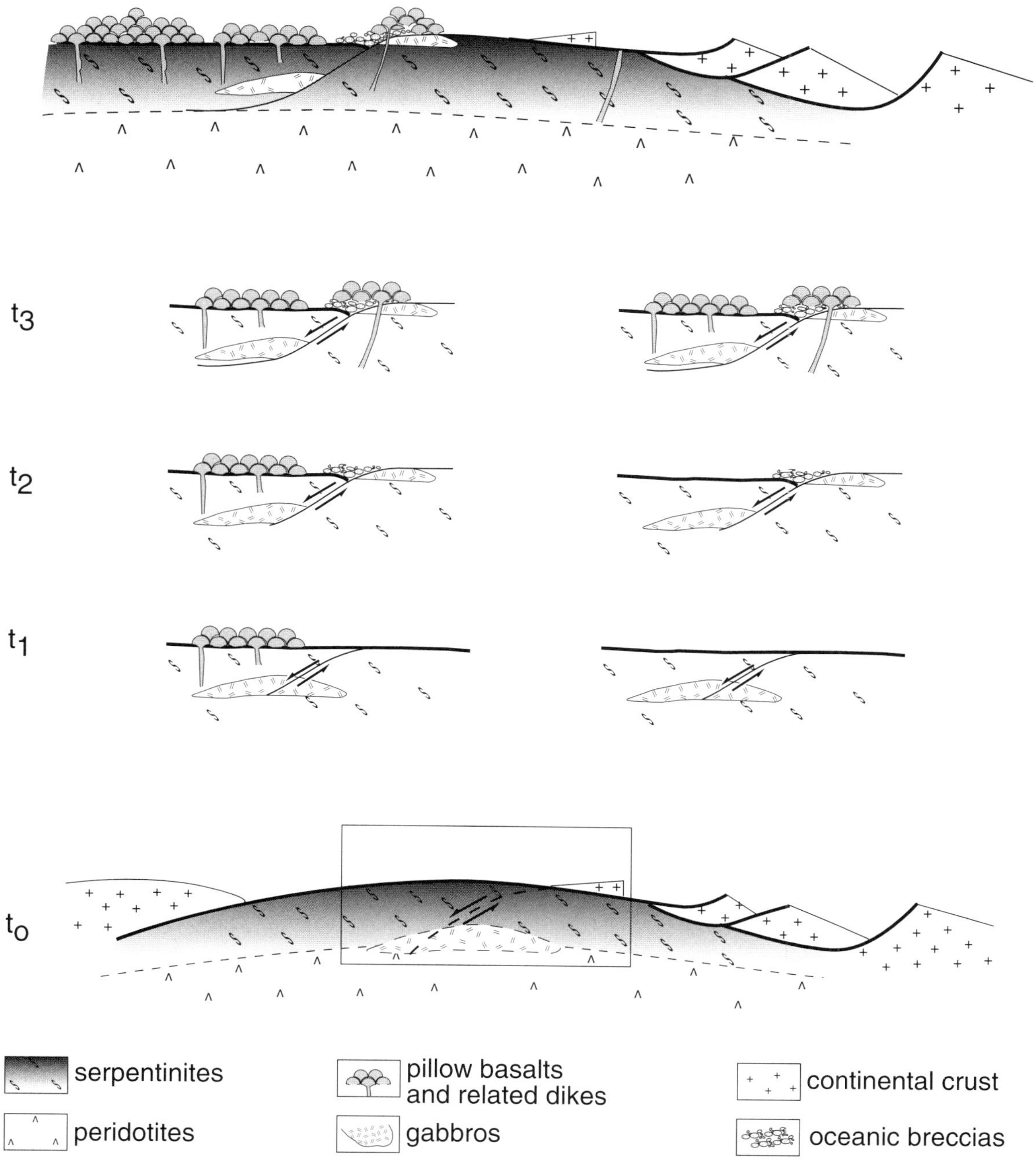

Figure 13. Sketch illustrating the temporal and spatial relationships between gabbros and basalts. t_o: detachment faulting leading to the exhumation of mantle rocks and intrusion of gabbros near the serpentinization front; t_1 to t_3: exhumation of the gabbros and emplacement of the basalts. Two hypotheses are shown: on the left, gabbros and basalts are part of the same magmatic cycle; on the right all the basalts were emplaced after the intrusion of the gabbros. Both scenarios can lead to the geometrical and stratigraphic relationships observed today (see top section). From Desmurs et al. (2001).

suggested by Piccardo (1976). Symmetrical rifting, analogous to a slow-spreading model, was proposed by Decandia and Elter (1969) and Elter (1972), whereas Lemoine et al. (1987) and Piccardo et al. (1994) envisaged mantle exhumation by asymmetric lithospheric-scale detachment faulting in analogy to the Iberian margin. The data now available from the south-Pennine-Austroalpine margin suggest that in this area subcontinental mantle (Trommsdorff et al., 1993; Hermann et al., 1997; Müntener et al., 2000) was exhumed to the sea floor along a system of extensional detachment faults (Froitzheim and Manatschal,

1996; Manatschal and Nievergelt, 1997). The resulting ophiolite stratigraphy of the Totalp-Platta nappe, with its exhumed mantle rocks, its scarce mafic rocks, and the extensional allochthons, all overlain by oceanic sediments can now be compared to the stratigraphy of ocean-continent transitions as observed along "non-volcanic" rifted continental margins (Fig. 14; Manatschal and Bernoulli, 1999). Its "serpentinite crust" is thought to grade laterally into "true" oceanic crust generated along a slow-spreading ridge. Lagabrielle and Lemoine (1997) suggest that relics of such "true" oceanic crust exist in the western Alps.

The occurrence of subcontinental mantle rocks and their exposure at the sea floor are well documented along modern magma-poor rifted margins, where exhumation occurred during the final stages of rifting and break-up of the continental lithosphere (Boillot et al.,1987, 1988; Abe, 2001; Hébert et al., 2001). Indeed, a comparison of the reconstructed southeastern margin of the Piedmont-Liguria ocean and the present-day west-Iberian margin shows that although the margins are of different ages and had different fates, their internal architectures and tectonic evolution are very similar (Fig. 14; Manatschal and Bernoulli, 1998, 1999). The similarities include the occurrence in the distal margin of detachment faults emplacing extensional allochthons of continental crust, which today occur as isolated outliers on the exhumed subcontinental mantle rocks (Manatschal et al., 2001). The tectono-sedimentary breccias accompanying the trace of the exhumed detachment faults and the base of the extensional allochthons of the Iberian margin mimic the structures and textures of the ophicalcites and related breccias (Boillot and Froitzheim, 2001, their figures 7 and 8; Wilson et al., 2001). In both margins, syn-rift melts are scarce and the volume of intrusive and extrusive rocks increases oceanward.

Along both the Pennine-Austroalpine and the west-Iberian margins, the site of rifting shifted from a broad area of the future proximal margin to a localized area in its distal part, accompanied by a change in the mode of extension (Manatschal and Bernoulli, 1998, 1999; Manatschal et al., 2001; Whitmarsh et al. 2001). On a lithospheric scale, rifting was first symmetrical and characterized by necking of the lithosphere; however, the top-to-the-continent, high-T shear zones observed in the mantle rocks may be connected to this early rifting phase (Müntener and Hermann, 2001). During a later stage of rifting, when the continental crust was already thinned to ~10 km (Manatschal et al., 2001; Whitmarsh et al., 2001), rifting shifted to the distal continental margin. During this period of advanced rifting, extension was dominated by simple shear along concave-downward detachment faults with a top-to-the-ocean sense of movement, pulling out the mantle from underneath the continental crust (Fig. 15). These shallow crustal structures formed a series of breakaways in the continental crust, from which the extensional allochthons of upper continental crust are derived, and cut into mantle rocks, exhuming them, finally, on the sea floor.

The intrusion of the gabbros and albitites of the Platta nappe is dated at 161 ± 1 Ma (Schaltegger et al., 2002). Interestingly, all gabbros from the Liguria-Piedmont ocean dated so far by reliable high-precision methods (U-Pb on zircon) show, from Corsica and the Ligurian Apennines to the eastern Alps, a surprisingly narrow window of crystallization, ca. 160 Ma (Schaltegger et al., 2002, and references therein). Most plagiogranites are distinctly younger than the gabbros (between 148 ± 2, Costa and Caby, 2001; and 152 ± 2 Ma, Lombardo et al., 2002) and may overlap in time with deposition of the radiolarites, which started in the Liguria-Piedmont ocean in the late Bathonian-early Callovian (Bill et al., 2001;

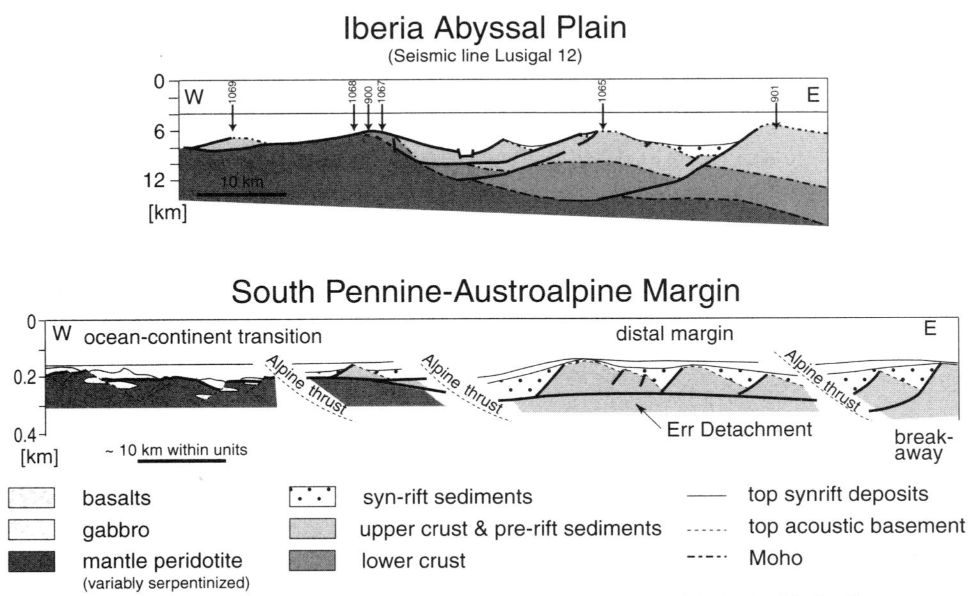

Figure 14. Comparison of the ocean-continent transition zone in the Iberia Abyssal Plain (west-Iberian margin, from Manatschal et al., 2001) and the reconstructed ocean-continent transition zone in the Err-Platta nappes.

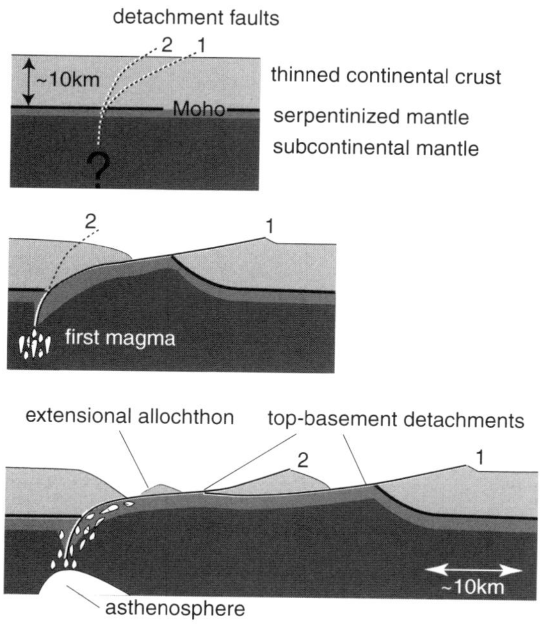

Figure 15. Kinematic concept of downward-concave faults responsible for the exhumation of subcontinental mantle and the emplacement of extensional allochthons within the ocean-continent transition zone of magma-poor margins. These downward-concave faults became active only after the continental crust was thinned to about 10 km. The model is compatible with the observed increase in magmatic activity toward the evolving ocean and the deformation of the intrusive rocks under decreasing temperature conditions.

160.4 + 1.1/–0.5 Ma in the time-scale of Pálfy et al., 2000). In any case, only a few million years or less appear to separate the intrusion of the gabbros and the onset of deposition of the radiolarites. The scarce age data thus suggest that final rifting, exhumation of the mantle along the downward-concave fault system, and coeval intrusion of late syn-rift melts occurred during a short time span.

The downward-concave geometry of the late rift faults may explain the exhumation of deeper and deeper mantle layers oceanward in the ocean-continent transition zone. Eventually, increasing melt production led to the creation of a mid-ocean ridge that accommodated further extension. Sr and Nd isotopic data of basalts and gabbros from the Platta nappe (Stille et al., 1989; Schaltegger et al., 2002) agree with values from the internal Ligurian ophiolites (Borsi et al., 1996; Rampone et al., 1998), suggesting that gabbros and basalts were derived from the same depleted mantle source. The narrow age range and the MORB-type character of the basic liquids (Desmurs et al., 2002) support the interpretation that the gabbros intruded the lithosphere close to or at the onset of active sea floor spreading.

The narrow age range of gabbros from the Liguria-Piedmont ocean (Schaltegger et al., 2002), combined with the age of the oldest radiolarites (Bill et al., 2001), further suggests that the ophiolite-bearing units of the Alps and the northern Apennines preferentially sampled the oldest parts of the Liguria-Piedmont ocean, i.e., the ocean-continent transition zone(s) and probably the oldest parts of a slow-spreading ridge (Lagabrielle and Lemoine, 1997). Because in the western Alps the ocean was closed only during the late Cretaceous to early Tertiary interval (Rubatto et al., 1998), the question arises of whether the spreading ridge of this ocean was long-lived and lasted throughout the middle Jurassic to mid-Cretaceous interval, or whether the Liguria-Piedmont ocean became a remnant ocean basin after a short time of active spreading, and remained small. Whatever the case, it appears that with a few possible exceptions, only the oldest and shallowest, more or less serpentinized parts of the Jurassic oceanic lithosphere were preserved (i.e., Steinmann's trinity of serpentinites, basalts, and radiolarites), while most of the oceanic lithosphere was consumed by subduction.

THE STEINMANN TRINITY: FROM ITS ORIGINS TO TODAY

We agree with Bailey and McCallien (1950, 1953) that Steinmann's discovery of the trinity was an important step in the evolution of a dynamic, mobilistic view of earth dynamics. The presence of igneous rocks characteristic for the deep ocean floor and of deep-sea sediments on land necessarily cast doubt upon the concept of permanency of continents and ocean basins, a concept to which, at the time of Steinmann, most Anglo-Saxon geologists adhered. Because they contravened the current dogma, and possibly because they did not write in English, the Alpine workers made little headway against the mainstream of extra-Alpine geological thought. Although Steinmann in his 1927 paper, in contrast to Argand (1924), Staub (1924), or Bailey (1936), did not mention Wegener's hypothesis of continental drift, he clearly was an advocate of large-scale orogenic movements. In retrospect, his concept of one major ophiolite zone extending from the Apennines across the Alps into the Dinarides, as a first-order paleotectonic element, was the key for the understanding of the Alpine-Mediterranean chains in terms of plate tectonics.

Steinmann's trinity of serpentinites, basalts, and oceanic sediments as defined by Bailey and McCallien (1950) includes the rock types typical of ocean-continent transitions or of slow-spreading ridges. In the area where Steinmann first noticed the trinity, in eastern Switzerland, the serpentinized peridotites show the geochemical signatures of subcontinental mantle, more and more modified oceanward by a rising asthenosphere from which Jurassic MOR-type intrusive gabbros and basalts are derived. A structural comparison between the South Pennine-Austroalpine boundary zone and the modern Iberian margin (Manatschal and Bernoulli, 1999) confirms the transitional position of Steinmann's first field area between oceanic and continental crust.

ACKNOWLEDGMENTS

We gratefully acknowledge the support of our research by the Swiss National Science Foundation (Grant 20-55284.98). We thank U. Schaltegger and F. Chalot-Prat for cooperation in the field and much information, G. Früh-Green, Y. Lagabrielle, and V. Trommsdorff for fruitful discussions, E. den Tex for information on the Steinmann trinity, Y. Dilek, Y. Lagabrielle, H. Masson, and G.B. Piccardo for constructive reviews and Y. Dilek for his patience and understanding. We are grateful to the Swiss Geological Survey, the Swiss Geological Society, Geologische Vereinigung and Springer-Verlag for permission to reproduce historical illustrations, and U. Gerber for much help in the preparation of the illustrations.

REFERENCES CITED

Abe, N., 2001, Petrochemistry of serpenitinized peridotite from the Iberia Abyssal Plain (ODP Leg 173): its character intermediate between sub-oceanic and sub-continental upper mantle, *in* Wilson, R.C.L., Whitmarsh, R.B., Taylor, B., and Froitzheim, N., eds., Non-Volcanic Rifting of Continental Margins: a Comparison of Evidence from Land and Sea: London, Geological Society of London Special Publication 187, p. 143–159.

Amstutz, G.C., 1980, The early history of the term Ophiolites and its evolution until 1945, *in* Panayiotou, A., ed., Ophiolites: Proceedings International Ophiolite Symposium Cyprus 1979, Nicosia, Cyprus Geological Survey Department, p. 149–152.

Anderson, T., 1910, The volcano of Matavanu in Savaii: Quarterly Journal of the Geological Society of London, v. 66, p. 621–639.

Anonymous, 1972, Penrose Field Conference Ophiolites: Geotimes, v. 17/2, p. 24–25.

Arbenz, P., 1926, Über das Alter der Saluverkonglomerate und die Juratransgression in den unterostalpinen Decken Graubündens: Beiblatt No. 10 zur Vierteljahresschrift der Naturforschenden Gesellschaft in Zürich, v. 76, p. 17–25.

Argand, E., 1911, Sur la répartition des roches vertes mésozoïques dans les Alpes Pennines avant la formation des grands plis couchés: Bulletin de la Société Vaudoise des Sciences Naturelles, v. 47, p. 1–2.

Argand, E., 1916, Sur l'arc des Alpes occidentales: Eclogae Geologicae Helvetiae, v. 14, p. 145–191.

Argand, E., 1924, La tectonique de l'Asie: Compte-rendu du XIIIe Congrès Géologique International 1922, Liège, Vaillant-Carmanne, p. 171–372.

Avigad, D., Chopin, C., Goffé, B., and Michard, A., 1993, Tectonic model for the evolution of the western Alps: Geology, v. 21, p. 659–662.

Bailey, E.B., 1936, Sedimentation in relation to tectonics: Bulletin of the Geological Society of America, v. 47, p. 1713–1726.

Bailey, E.B., and McCallien, W.J., 1950, The Ankara mélange and the Anatolian thrust: Nature v. 166, p. 938–943.

Bailey, E.B., and McCallien, W.J., 1953, Serpentine lavas, the Ankara mélange and the Anatolian thrust: Transactions of the Royal Society of Edinburgh, v. 62, pt. 2, p. 403–442.

Bailey, E.B., and McCallien, W.J., 1960, Some aspects of the Steinmann Trinity, mainly chemical: Quarterly Journal of the Geological Society of London, v. 116, p. 365–395.

Benson, W.N., 1926, The tectonic conditions accompanying the intrusion of basic and ultrabasic rocks: Washington, D.C., National Academy of Sciences, v. 19, 1st Memoir, 90 p.

Berger, W.H., and Winterer, E.L., 1974, Plate stratigraphy and the fluctuating carbonate line, *in* Hsü, K.J., and Jenkyns, H.C., eds., Pelagic Sediments on Land and under the Sea: Oxford, UK, International Association of Sedimentologists Special Publication 1, p. 11–48.

Bernoulli, D., and Weissert, H., 1985, Sedimentary fabrics in Alpine ophicalcites, South-Pennine Arosa zone: Geology, v. 13, p. 755–758.

Bernoulli, D., Kälin, O., and Patacca, E., 1979 A sunken continental margin of the Mesozoic Tethys: The Northern and Central Apennines, *in* Beaudoin, B., and Purser, B.H., eds., Symposium "Sédimentation Jurassique W européen": Association des Sédimentologistes Français, Publication Spéciale 1, p. 197–210.

Bill, M., O'Dogherty, L., Guex, J., Baumgartner, P.O., and Masson, H., 2001, Radiolarite ages in Alpine-Mediterranean ophiolites: Constraints on the oceanic spreading and the Tethys-Atlantic connection: Geological Society of America Bulletin, v. 113, p. 129–143.

Boillot, G., and Froitzheim, N., 2001, Non-volcanic rifted margins, continental break-up and the onset of sea-floor spreading: Some outstanding questions, *in* Wilson, R.C.L., Whitmarsh, R.B., Taylor, B., and Froitzheim, N., eds., Non-Volcanic Rifting of Continental Margins: A Comparison of Evidence from Land and Sea: Geological Society (London) Special Publication 187, p. 9–30.

Boillot, G., Recq, M., Winterer, E.L., and Shipboard Scientific Party, 1987, Tectonic denudation of the upper mantle along passive margin: a model based on drilling results (ODP Leg 103, Western Galicia Margin, Spain): Tectonophysics, v. 132, p. 334–342.

Boillot, G., Comas, M.C., Girardeau, J., Kornprobst, J., Loreau, J.-P., Malod, J., Mougenot, D., and Moullade, M., 1988, Preliminary results of the Galinaute cruise: dives of the submersible *Nautile* on the western Galicia margin, Spain, *in* Boillot G., Winterer, E.L., et al., eds., Proceedings of the Ocean Drilling Program, Scientific Results: College Station TX, Ocean Drilling Program, v.103, p. 37–51.

Bonatti, E., Ferrara G., Honnorez, J., and Rydell, H., 1974, Ultramafic carbonate breccias from the equatorial mid-Atlantic ridge: Marine Geology, v. 16, p. 83–102.

Borsi, L., Schärer, U., Gaggero, L., and Crispini, L., 1996, Age, origin and geodynamic significance of plagiogranites in lherzolites and gabbros of the Piedmont-Ligurian ocean basin: Earth and Planetary Science Letters, v. 140, p. 227–241.

Bowen, N.L., 1927, The origin of ultra-basic and related rocks: American Journal of Science, 5th, ser., v. 14, p. 89–108.

Bowen, N.L., 1928, The Evolution of Igneous Rocks: Princeton, Princeton University Press, 332 p.

Bowen, N.L., and Tuttle, O.F., 1949, The system $MgO–SiO_2–H_2O$: Geological Society of America Bulletin, v. 60, p. 439–460.

Brongniart, A., 1813, Essai d'une classification minéralogique des roches mélangées: Paris, Journal des Mines, no. 199, p. 5–48.

Brongniart, A. 1821, Sur le gisement ou position relative des ophiolites, euphotides, jaspes, etc. dans quelques parties des Apennins: Annales des Mines ou recueuil de Mémoires sur l'Exploitation des Mines, v. 6, p. 177–238.

Brunn, J.H., 1956, Contribution à l'étude géologique du Pinde septentrional et d'une partie de la Macédoine occidentale: Annales Géologiques des Pays Helléniques, 1e sér., v. 7, 358 p.

Coleman, R.G., 1971, Plate tectonic emplacement of upper mantle peridotites along continental edges: Journal of Geophysical Research, v. 76, p. 1212–1222.

Coleman, R.G., 1977, Ophiolites. Ancient Oceanic Lithosphere: Berlin, Heidelberg, New York, Springer-Verlag, 229 p.

Cornelius, H.P., 1951, Zur Frage der Absatzbedingungen der Radiolarite: Geologische Rundschau, v. 39, p. 216–221.

Costa, S., and Caby, R., 2001, Evolution of the Ligurian Tethys in the Western Alps: Sm/Nd and U/Pb geochronology and rare-earth element geochemistry of the Montgenèvre ophiolite (France): Chemical Geology, v. 175, p. 449–466.

Dal Piaz, G.V., 1999, The Austroalpine-Piedmont nappe stack and the puzzle of Alpine Tethys: Memorie di Scienze Geologiche (Padova), v. 51, p. 155–176.

Davis, E.F., 1918, The Radiolarian Cherts of the Franciscan Group: University of California Publications. Bulletin of the Department of Geology, v. 11, p. 235–432.

Decandia, F.A., and Elter, P., 1969, Riflessioni sul problema delle ofioliti nell'Appennino settentrionale, nota preliminare: Atti della Società Toscana di Scienze Naturali, v. 75, p. 1–9.

Decandia, F.A., and Elter, P., 1972, La 'zona' ofiolitifera del Bracco nel settore compreso tra Levanto e la Val Graveglia (Appennino ligure): Memorie della Società Geologica Italiana, v. 11, p. 503–530.

Den Tex, E., 1989, Steinmann Trinity, *in* Bowes, D.R., ed., The Encyclopedia of Igneous and Metamorphic Petrology: New York, Van Nostrand Reinhold, p. 549–551.

De Roever, W.P., 1957, Sind die alpino-typen Peridotitmassen vielleicht tektonisch verfrachtete Bruchstücke der Peridotitschale?: Geologische Rundschau, v. 46, p. 137–146.

Desmurs, L., 2002, Mantle evolution and magmatism in an evolving ocean-continent transition: The Platta nappe, eastern Switzerland [Ph.D. Thesis]: Zürich, Swiss Federal Institute of Technology, no. 14429, 143 p.

Desmurs, L., Manatschal, G., and Bernoulli, D. 2001, The Steinmann Trinity revisited: mantle exhumation and magmatism along an ocean-continent transition, *in* Wilson, R.C.L., Whitmarsh, R.B., Taylor, B., and Froitzheim, N., eds., Non-Volcanic Rifting of Continental Margins: A Comparison of Evidence from Land and Sea: London, Geological Society of London Special Publication 187, p. 235–266.

Desmurs, L., Müntener, O., and Manatschal, G., 2002, Onset of magmatic accretion within magma-poor passive margins: A case study from the Platta ocean-continent transition, eastern Switzerland: Contributions to Mineralogy and Petrology. v. 144, p. 365–382.

Dewey, H., and Flett, J.S., 1911, On some British pillow lavas and the rocks associated with them: Geological Magazine, v. 8., p. 202–209, 241–248.

Dewey, J.F., and Bird, J.M., 1970, Mountain belts and the new global tectonics: Journal of Geophysical Research, v. 75, p. 2625–2647.

Dietrich, V., 1969, Die Ophiolithe des Oberhalbsteins (Graubünden) und das Ophiolithmaterial der ostschweizerischen Molasseablagerungen, ein petrographischer Vergleich: Bern, Verlag Herbert Lang, 179 p.

Dietrich, V., 1970, Die Stratigraphie der Platta-Decke: Fazielle Zusammenhänge zwischen Oberpenninikum und Unterostalpin: Eclogae Geologicae Helvetiae, v. 63, p. 631–671.

Durbertret, L., 1953, Géologie des roches vertes du Nord-ouest de la Syrie et du Hatay (Turquie): Notes et Mémoires sur le Moyen-Orient, v. 4, 179 p.

Elter, P., 1972, La zona ofiolitica del Bracco nel quadro dell'Appennino settentrionale: Introduzione alla geologia delle Liguridi: 66° Congresso della Società Geologica Italiana, Guida alle Escursioni, Pisa, Pacini, p. 5–35.

Ferreiro Mählmann, R., 2001, Correlation of very low grade data to calibrate a thermal maturity model in nappe tectonic setting, a case study from the Alps: Tectonophysics, v. 334, p. 1–33.

Finger, W., 1978, Die Zone von Samaden (Unterostalpine Decken, Graubünden) und ihre jurassischen Brekzien: Mitteilungen aus dem Geologischen Institut der ETH und der Universität Zürich, Neue Folge 224, 141 p.

Florineth, D., and Froitzheim, N., 1994, Transition from continental to oceanic basement in the Tasna nappe: Evidence for early Cretaceous opening of the Valais ocean: Schweizerische Mineralogische und Petrographische Mitteilungen, v. 74, p. 437–448.

Folk, R.L., and McBride, E.F., 1976, Possible pedogenic origin of Ligurian ophicalcite: A Mesozoic calichified serpentinite: Geology, v. 4, p. 327–332.

Folk, R.L., and McBride, E.F., 1978, Radiolarites and their relation to subjacent "oceanic crust" in Liguria, Italy: Journal of Sedimentary Petrology, v. 48, p. 1069–1102.

Frisch, W., Ring, U., Dürr, S., Borchert, S., and Biehler, D., 1994, The Arosa zone and Platta nappe ophiolites (Eastern Swiss Alps): geochemical characteristics and their meaning for the evolution of the Penninic ocean: Jahrbuch der Geologischen Bundesanstalt (Wien), v. 137, p. 19–33.

Froitzheim, N., and Eberli, G., 1990, Extensional detachment faulting in the evolution of a Tethys passive continental margin (Eastern Alps, Switzerland): Geological Society of America Bulletin, v. 102, p. 1297–1308.

Froitzheim, N., and Manatschal, G., 1996, Kinematics of Jurassic rifting, mantle exhumation and passive-margin formation in the Austroalpine and Penninic nappes (eastern Switzerland): Geological Society of America Bulletin, v. 108, p. 1120–1133.

Froitzheim, N., Schmid, S.M., and Conti, P., 1994, Repeated change from crustal shortening to orogen-parallel extension in the Austroalpine units of Graubünden: Eclogae Geologicae Helvetiae, v. 87, p. 559–612.

Früh-Green, G.L., Weissert, H., and Bernoulli, D., 1990, A multiple fluid history recorded in Alpine ophiolites: Journal of the Geological Society of London, v. 147, p. 959–970.

Gansser, A., 1974, The ophiolitic mélange, a world-wide problem on Tethyan examples: Eclogae Geologicae Helvetiae, v. 67, p. 479–507.

Gianelli, G., 1977, Vara Supergroup ophiolite probably represents an ancient transcurrent fault zone: Ofioliti, v. 2, p. 115–135.

Glennie, K.W., Boeuf, M.G.A., Hughes Clarke, M.W., Moody-Stuart, M., Pilaar, W.F.H., and Reinhardt, B.M., 1974, Geology of the Oman Mountains: Verhandelingen van het Koninklijk Nederlands Geologisch Mijnbouwkundig Genootschap, v. 31, 423 p.

Grunau, H., 1947, Geologie von Arosa (Graubünden) mit besonderer Berücksichtigung des Radiolarit-Problems [Ph.D. thesis]: Bern, Switzerland, University of Bern, 108 p.

Grunau, H.R., 1965, Radiolarian cherts and associated rocks in space and time: Eclogae Geologicae Helvetiae, v. 58, 157–208.

Hébert, R., Gueddari, K., LaFlèche, M.R., Beslier, M.-O., and Gardien, V., 2001, Petrology and geochemistry of exhumed peridotites and gabbros at non-volcanic margins: ODP Leg 173, West Iberia ocean-continent transition zone, *in* Wilson, R.C.L., Whitmarsh, R.B., Taylor, B., and Froitzheim, N., eds., Non-Volcanic Rifting of Continental Margins: a Comparison of Evidence from Land and Sea: London, Geological Society of London Special Publication 187, p. 161–189.

Heim, A., 1924, Über submarine Denudation und chemische Sedimente: Geologische Rundschau, v. 15, p. 1–47.

Hermann, J., Müntener, O., Trommsdorff, V., Hansmann, W., and Piccardo, G.B., 1997, Fossil crust to mantle transition, Val Malenco (Italian Alps): Journal of Geophysical Research, B9, v. 102, p. 20123–20132.

Hess, H.H., 1938, A primary peridotite magma: American Journal of Science, 5th ser., v. 35, p. 321–344.

Hess, H.H., 1954, Geological hypotheses and the Earth's crust under the oceans: Transactions of the Royal Society, A, v. 222, p. 341–348.

Hess, H.H., 1955, Serpentines, orogeny, and epeirogeny, *in* Poldervaart, A., ed., The Crust of the Earth, A Symposium: New York, Geological Society of America Special Paper 62, p. 391–407.

Jackson, J.A., 1997, Glossary of Geology, Fourth Edition: Alexandria, Virginia, American Geological Institute, 769 p.

Juteau, T., Cannat, M., and Lagabrielle, Y., 1989, Peridotites in the upper oceanic crust away from transform zones: A comparison of the results of previous DSDP and ODP legs, *in* Detrick, R.S., Honnorez, J., Brian, W.B., Juteau, T., et al., eds., Proceedings of the Ocean Drilling Program, Scientific Results: College Station Texas, Ocean Drilling Program, v. 109, p. 303–308.

Lagabrielle, Y., 1987, Les ophiolites: marqueurs de l'histoire tectonique des domaines océaniques. Le cas des Alpes franco-italiennes (Queyras-Piémont), comparaison avec les ophiolites d'Antalya (Turquie) et du Coast Range de Californie: Thèse d'état, Université de Brest, 350 p.

Lagabrielle, Y., and Lemoine, M., 1997, Alpine, Corsican and Apennine ophiolites: the slow-spreading ridge model: Paris, Comptes Rendus de l'Académie des Sciences, Sciences de la Terre et des Planètes, v. 325, 909–920.

Lagabrielle, Y., Polino, R., Auzende, J.-M., Blanchet, R., Caby, R., Fudral, S., Lemoine, M., Mevel, C., Ohnenstetter, M., Robert, D., and Tricart, P., 1984, Les témoins d'une tectonique intra-océanique dans le domaine téthysien: analyse des rapports entre les ophiolites et leur couverture métasédimentaire dans la zone piémontaise des Alpes franco-italiennes: Ofioliti, v. 9, p. 67–88.

Laubscher, H., 1969, Mountain building: Tectonophysics, v. 7, p. 551–563.

Lemoine, M., 1980, Serpentinites, gabbros and ophicalcites in the Piemont-Ligurian domain of the Western Alps: possible indicators of oceanic fracture zones and of associated serpentinite protrusions: Archives des Sciences (Genève), v. 33, p. 103–115.

Lemoine, M., Tricart, P., and Boillot, G., 1987, Ultramafic and gabbroic ocean floor of the Ligurian Tethys (Alps, Corsica, Apennines): in search of a genetic model: Geology, v. 15, p. 622–625.

Lombardo, B., Rubatto, D., and Castelli, D., 2002, Ion microprobe U-Pb dating of zircon from a Monviso metaplagiogranite: Implications for the evolution of the Piedmont-Liguria Tethys in the Western Alps: Ofioliti, v. 27, p. 109–117.

Lotti, B., 1886, Descrizione geologica dell'Isola d'Elba: Memorie Descrittive della Carta Geologica d'Italia, v. 2, 254 p.

Manatschal, G., and Bernoulli, D., 1998, Rifting and early evolution of ancient ocean basins: the record of the Mesozoic Tethys and of the Galicia-Newfoundland margins: Marine Geophysical Researches, v. 20, p. 371–381.

Manatschal, G., and Bernoulli, D., 1999, Architecture and tectonic evolution of non-volcanic margins: present-day Galicia and ancient Adria: Tectonics, v. 18, p. 1099–1119.

Manatschal, G., and Nievergelt, P., 1997, A continent-ocean transition recorded in the Err and Platta nappes (Eastern Switzerland): Eclogae Geologicae Helvetiae, v. 90, p. 3–27.

Manatschal, G., Froitzheim, N., Rubenach, M., and Turrin, B.D., 2001, The role of detachment faulting in the formation of an ocean-continent transition: insights from the Iberia Abyssal Plain, *in* Wilson, R.C.L., Whitmarsh, R.B., Taylor, B., and Froitzheim, N., eds., Non-Volcanic Rifting of Continental Margins: A Comparison of Evidence from Land and Sea: London, Geological Society of London Special Publication 187, p. 405–428.

Marroni, M., Molli, G., Ottria, G., and Pandolfi, L., 2001, Tectono-sedimentary evolution of the External Liguride units (Northern Apennines, Italy): insights in the pre-collisional history of a fossil ocean-continent transition zone: Geodinamica Acta, v. 14, p. 307–320.

McBride, E.F., and Folk, R.L., 1979, Features and origin of Italian Jurassic radiolarites deposited on continental crust: Journal of Sedimentary Geology, v. 49, p. 837–868.

Milovanovic, B., and Karamata, S., 1960, Über den Diapirismus serpentinischer Massen: 21st International Geological Congress (Copenhagen), Report, v. 16, p. 409–416.

Moores, E.M., and Vine, F.J., 1971, The Troodos massif, Cyprus and other ophiolites as oceanic crust: evaluations and implications: Philosophical Transactions of the Royal Society London, A, v. 268, p. 443–466.

Müntener, O., and Hermann, J., 1996, The Val Malenco lower crust–upper mantle complex and its field relations (Italian Alps): Schweizerische Mineralogische und Petrogrographische Mitteilungen, v. 76, p. 475–500.

Müntener, O., and Hermann, J., 2001, The role of lower crust and continental upper mantle during formation of nonvolcanic passive margins, *in* Wilson, R.C.L., Whitmarsh, R.B., Taylor, B., & Froitzheim, N., eds., Non-Volcanic Rifting of Continental Margins: a Comparison of Evidence from Land and Sea: London, Geological Society of London Special Publication 187, p. 267–288.

Müntener, O., Hermann, J., and Trommsdorff, V., 2000, Cooling history and exhumation of lower-crustal granulite and upper mantle (Malenco, eastern Central Alps): Journal of Petrology, v. 41, p. 175–200.

Müntener, O., Desmurs, L., Pettke, T., Schaltegger, U., Manatschal, G., and Bernoulli, D., 2001, Plagioclase peridotite along continental margins: Controlled by subsolidus equilibration or melt/rock reaction?, *in* Program Volume, EUG XI, European Union of Geosciences, Strasbourg, France, 8–12 April 2001, p. 731.

Müntener, O., Desmurs, L., Pettke, T., and Schaltegger, U., 2002, Melting and melt/rock reaction in extending mantle lithosphere: Trace element and isotopic constraints from passive margin peridotites: Geochimica et Cosmochimica Acta, v. 66 (15A), supplement 1, p. A536.

Pálfy, J., Smith, P.L., and Mortensen, J.K., 2000, A U-Pb and $^{40}Ar/^{39}Ar$ time scale for the Jurassic: Canadian Journal of Earth Sciences, v. 37, p. 923–944.

Pantanelli, D., 1880, I diaspri della Toscana e i loro fossili: Atti della Reale Accademia dei Lincei, Classe di Scienze Fisiche, Matematiche e Scienze Naturali, ser. 3, v. 8, p. 35–66

Peters, T., 1963, Mineralogie und Petrographie des Totalpserpentins bei Davos: Schweizerische Mineralogische und Petrographische Mitteilungen, v. 43, p. 529–685.

Peters T., and Stettler, A., 1987, Radiometric age, thermobarometry and mode of emplacement of the Totalp peridotite in the Eastern Swiss Alps: Schweizerische Mineralogische und Petrographische Mitteilungen, v. 67, p. 285–294.

Piccardo, G.B., 1976, Petrologia del massiccio lherzolitico di Suvero, La Spezia: Ofioliti, v. 1, p. 279–317.

Piccardo, G.B., Rampone, E., and Vannucci, R., 1990, Upper mantle evolution during continental rifting and ocean formation: evidences from peridotite bodies of the western Alpine-northern Apennine system, *in* Roure F., Heitzmann, P. and Polino, R., eds., Deep Structure of the Alps: Paris, France, Mémoires de la Société Géologique de France, v. 156, p. 323–333.

Piccardo, G.B., Rampone, E., Vannucci, R., and Cimmino, F., 1994, Upper mantle evolution of ophiolitic peridotites from the northern Apennine: Petrological constraints to geodynamic processes: Roma, L'Italia, Memorie della Società Geologica Italiana, v. 48, p. 137–148.

Raitt, R.W., 1963, The crustal rocks, *in* Hill, M.N., ed., The Sea, v. 3: New York, London, Interscience Publishers, p. 85–102.

Rampone, E., and Piccardo, G.B., 2000, The ophiolite-oceanic lithosphere analogue: New insights from the Northern Apennines (Italy), *in* Dilek, Y., Moores, E.M., Elthon, D., and Nicolas, A., eds., Ophiolites and Oceanic Crust: New Insights from Field Studies and the Ocean Drilling Program: Boulder, Colorado, Geological Society of America Special Paper 349, p. 21–34.

Rampone, E., Hofmann, A.W., Piccardo, G.B., Vannucci, R., Bottazzi, P., and Ottolini, L., 1995, Petrology, mineral and isotope geochemistry of the External Liguride peridotites (Northern Apennine, Italy): Journal of Petrology, v. 36, p. 81–105.

Rampone, E., Hofmann, A.W., Piccardo, G.B., Vannucci, R., Bottazzi, P. and Ottolini, L., 1996, Trace element and isotope geochemistry of depleted peridotites from an N-MORB type ophiolite. (Internal Liguride, N. Italy): Contributions to Mineralogy and Petrology, v. 123, p. 61–76.

Rampone, E., Hofmann, A.E., and Raczek, I., 1998, Isotopic contrast within the Internal Liguride Ophiolite (N. Italy): The lack of a genetic mantle-crust link: Earth and Planetary Science Letters: v. 163, p. 175–189.

Rubatto, D., Gebauer, D., and Fanning, M., 1998, Jurassic formation and Eocene subduction of the Zermatt-Saas Fee ophiolites: implications for the geodynamic evolution of the Central and Western Alps: Contributions to Mineralogy and Petrology, v. 132, p. 269–287.

Schaltegger, U., Desmurs L., Manatschal, G., Müntener O., Meier, M., Frank, M., and Bernoulli, D., 2002, The transition from rifting to seafloor spreading within a magma-poor rifted margin: Field and isotopic constraints: Terra Nova, v. 14, p. 156–162.

Stampfli, G.M., 1993, Le Briançonnais, terrain exotique dans les Alpes: Eclogae Geologicae Helvetiae, v. 86, p. 1–45.

Staub, R., 1922, Über die Verteilung der Serpentine in den alpinen Ophiolithen: Schweizerische Mineralogische und Petrographische Mitteilungen, v. 2, p. 78–149.

Staub, R., 1924, Der Bau der Alpen: Beiträge zur Geologischen Karte der Schweiz, Neue Folge, v. 52, 272 p.

Steinmann, G., 1905, Geologische Beobachtungen in den Alpen, II. Die Schardtsche Ueberfaltungstheorie und die geologische Bedeutung der Tiefseeabsätze und der ophiolithischen Massengesteine: Berichte der Naturforschenden Gesellschaft zu Freiburg im Breisgau, v. 16, p. 18–67.

Steinmann, G., 1907, Alpen und Apennin: Monatsberichte der Deutschen Geologischen Gesellschaft, v. 59, p. 77–83.

Steinmann, G., 1913, Über Tiefenabsätze des Oberjura im Apennin: Geologische Rundschau, v. 4, p. 572–575.

Steinmann, G., 1925, Gibt es fossile Tiefseeablagerungen von erdgeschichtlicher Bedeutung?: Geologische Rundschau, v. 16, p. 435–468.

Steinmann, G., 1927, Die ophiolitischen Zonen in den mediterranen Kettengebirgen: Compte-rendu, XIVe Congrès Géologique International, 1926, Madrid, Graficas Reunidas, v. 2, p. 637–667.

Stille, P., Clauer, N., and Abrecht, J., 1989, Nd isotopic composition of Jurassic Tethys seawater and the genesis of Alpine Mn-deposits: evidence from Sr-Nd isotope data: Geochimica et Cosmochimica Acta, v. 53, p. 1095–1099.

Teall, J.J.H., 1894, On greenstones associated with radiolarian cherts: Transactions of the Royal Geological Society of Cornwall, v. 2, p. 560–565.

Trommsdorff V., and Evans B.W., 1977, Antigorite ophicarbonates: contact metamorphism in Val Malenco (Italy): Contributions to Mineralogy and Petrology, v. 62, p. 301–312.

Trommsdorff V., Piccardo, G.B., and Montrasio, A., 1993, From magmatism through metamorphism to sea floor emplacement of subcontinental Adria during pre-Alpine rifting (Malenco, Italy): Schweizerische Mineralogische und Petrographische Mitteilungen, v. 73, p. 191–203.

Trümpy, R., 1960, Paleotectonic evolution of the central and western Alps: Geological Society of America Bulletin, v. 71, p. 843–908.

Trümpy, R., 1975, Penninic-Austroalpine boundary in the Swiss Alps: a presumed former continental margin and its problems: American Journal of Science, v. 275A, p. 209–238.

Vuagnat, M., 1963, Remarques sur la trilogie serpentinites–gabbros–diabases dans le bassin de la Méditerranée occidentale: Geologische Rundschau, v. 53, p. 336–358.

Wegener, A., 1915, Die Entstehung der Kontinente und Ozeane: Sammlung Vieweg, Tagesfragen aus den Gebieten der Naturwissenschaften und der Technik, v. 23: Braunschweig, Vieweg, 94 p.

Weissert, H.J., and Bernoulli, D., 1985, A transform margin in the Mesozoic Tethys: Evidence from the Swiss Alps: Geologische Rundschau, v. 74, p. 665–679.

Wenk, E., 1949, Die Assoziation von Radiolarienhornsteinen mit ophiolitischen Errstarrungsgesteinen als petrogenetisches Problem: Experientia, v. 5, p. 226–232.

Whitmarsh, R.B., Beslier, M.-O., Wallace, P.J., et al., 1998, Proceedings of the Ocean Drilling Program, Initial Reports, Leg 173: College Station, Texas, Ocean Drilling Program, 493 p.

Whitmarsh, R.B., Manatschal, G., and Minshull, T.A., 2001, Evolution of magma-poor continental margins from rifting to seafloor spreading: Nature, v. 413, p. 150–154.

Wilckens, O., 1930, Gustav Steinmann, sein Leben und Wirken: Geologische Rundschau, v. 21, p. 389–415.

Wilson, R.C.L., Manatschal, G., and Wise, S., 2001, Rifting along non-volcanic passive margins: stratigraphic and seismic evidence from the Mesozoic successions of the Alps and western Iberia, *in* Wilson, R.C.L., Whitmarsh, R.B., Taylor, B., and Froitzheim, N. eds., Non-Volcanic Rifting of Continental Margins: A Comparison of Evidence from Land and Sea: London, Geological Society of London Special Publication 187, p. 429–452.

Winterer, E.L., and Bosellini, A., 1981, Subsidence and sedimentation on Jurassic passive continental margin, Southern Alps, Italy: American Association of Petroleum Geologists Bulletin, v. 65, p. 394–421.

Manuscript Accepted by the Society May 2, 2003

Printed in the USA

Geological Society of America
Special Paper 373
2003

Ophiolites, historical contingency, and the Wilson cycle

Martin F.J. Flower*
Department of Earth and Environmental Sciences, University of Illinois at Chicago (M/C 186), 845 W. Taylor Street, Chicago, Illinois 60607-7059, USA

Historical explanations take the form of narrative: E, the phenomenon to be explained, arose because D came before, preceded by C, B, and A. If any of these earlier stages had not occurred, or had transpired in a different way, then E would not exist (or would be present in a substantially altered form, E′, requiring a different explanation). Thus, E makes sense and can be explained rigorously as the outcome of A through D. But no law of nature enjoined E; any variant E arising from an altered set of antecedents, would have been equally explicable, though massively different in form and effect. I am not speaking of randomness (for E had to arise, as a consequence of A though D), but of the central principle of all history—contingency. A historical explanation does not rest on direct deduction from laws of nature, but on an unpredictable sequence of antecedent states, where any major change in any step of the sequence would have altered the final result. This final result is therefore dependent, or contingent, upon everything that came before—the unerasable and determining signature of history. (S.J. Gould, 1989, p. 283)

ABSTRACT

Lithologic assemblages termed "ophiolites" were at the heart of the mid-twentieth century "plate tectonics" paradigm shift when they were widely accepted as representing sections of the oceanic crust created during the early stages of "Wilson cycles." However, the past four decades have seen lengthy, if acrimonious, debates about this enigmatic phenomenon, fueled by a growing body of evidence for their association with processes of lithosphere subduction. Today, the core of the debate continues to be the coexistence of structural evidence for seafloor spreading and the chemical and petrographic signatures of subduction. More recently, the discussions have acquired a philosophical flavor, following the suggestion that a "historically contingent" model could allow subduction-related geochemical indications to be reconciled with processes at active mid-ocean ridges. Offering powerful, if not controversial insights into the nature of biological and cultural change, historical contingency is an appropriate context for resolving dilemmas in interpreting earth history, including the subject of this paper, the so-called ophiolite conundrum. Here, an actualistic model for Tethyan ophiolites is offered for testing against the geological record and contingent factors absent or underrepresented today. The model is based on processes observed in the Mediterranean and western Pacific regions and is developed on the premise that "proto-ophiolites" form as accreting forearc complexes during sub-cycles of basin opening and arc-trench rollback in closing, pre-collision stages of Wilson cycles. Rollback appears to be triggered by spontaneous subduction nucleation and may be driven by lateral mantle flow. If a rollback episode successfully evades the "jaws" of a plate collision (as observed in the western Pacific), forearc accretion may proceed indefinitely or until mantle flow dissipates. On the other hand, entrapped by a continental plate collision, forearcs are preferentially preserved in an ensuing orogeny, given their relative buoyancy relative to conjugate backarc basin

*flower@uic.edu

Flower, M.F.J., 2003, Ophiolites, historical contingency, and the Wilson cycle, *in* Dilek, Y., and Newcomb, S., eds., Ophiolite concept and the evolution of geological thought: Boulder, Colorado, Geological Society of America Special Paper 373, p. 111–135.

lithosphere. The case for studying the ophiolite conundrum from a historically contingent perspective is supported in principle, but remains accountable to petrologic constraints and evidence from the geological record.

Keywords: ophiolite, Wilson cycle, mantle, subduction zone, back-arc basin, plate collision, earth history.

INTRODUCTION

Hailed in the 1960s as one of several keys to validating the plate tectonics hypothesis, ophiolites are still considered by many as an unresolved enigma. A substantial amount of circular reasoning has dogged their study, reflecting both a skewed attention to the evidence and some basic philosophical differences. On the assumption that ophiolites represent oceanic crust, they were adopted as a model template for interpreting mid-ocean ridge dynamics and petrogenesis (Gass, 1968; Moores et al., 1968; Moores and Vine, 1971; Moores, 1974, 1982a; Nicolas and Violette, 1982). While this view persists, with substantial justification (Nicolas, 1997; Dilek et al., 1999; Benoit et al., 1999, Moores et al., 2000; Nicolas and Boudier, 2000), it had met nevertheless with serious difficulties in the mid-1970s, as the expanding database revealed key differences between ophiolites and typical sections of oceanic crust. The first of these, a series of prescient observations by Miyashiro (1973, 1975a, b, c), was that calcalkaline lithologies are a common component of ophiolites, evidence for subduction-related affinity that was initially downplayed or rejected outright by "ocean crust" protagonists (Gass et al., 1975; Moores, 1975; Hynes, 1975; cf. Miyashiro, 1977).

Further misgivings followed the revelations of early age-dating studies (e.g., Dallmeyer and Williams, 1975; Lanphere et al., 1975) and were reinforced by later research (e.g., Hacker, 1994; Parlak and Delaloye, 1999; Bill et al., 2001; Corfield et al., 2001; Costa and Caby, 2001; Dimo-Lahitte et al., 2001), that ophiolites were formed only a few million years prior to their emplacement, suggesting they may have been generated in marginal basins at later rather than early stages of ocean basin opening. By the early 1980s, a growing consensus seemed able to accept the existence of both "oceanic" and "supra-subduction" ophiolites as the rule, rather than the exception (Pearce et al., 1981, 1984), leading to what Moores et al. (2000) refer to as the "ophiolite conundrum." Debate concerning their relative significance continues with relatively little sign of consensus.

At the present time, the conundrum is best summed up as the problem posed by the coexisting structural evidence for seafloor spreading, imbricated pillow lava and plutonic sequences, sheeted dike complexes that are unambiguously associated with extensional tectonics, and show the chemical and petrographic indications of subduction-related magmatism—enrichments in H_2O and large-ion lithosphile element (LILE), depletions in high-field strength elements (HFSE), calcalkaline lithologies, and refractory mantle sequences, all of which are exclusive to compressive, subduction-related plate margin environments. However, while the majority of ophiolite specialists probably accept a generic association of ophiolites with convergent plate margins, ascribing their "oceanic" attributes to seafloor spreading in marginal basin settings (see Höck et al., 2002), Moores et al. (2000) appealed to the notion of historical contingency, a compelling view of biological evolution (Gould, 1990), to lend much-needed weight to the exclusively mid-ocean ridge ophiolite model.

CHANGING VIEWS OF EARTH HISTORY

The relationship between natural order and history has fascinated scholars since classical times. It was, however, the theories of William Smith, James Hutton, and Charles Darwin that laid the basis for a conceptual revolution that led to modern views of earth history. Smith's "Statigraphic Principle" embodied in his geological map of England and Wales and part of Scotland, published in 1815, and Hutton's "Principle of Uniformitarianism" (codified as "the present is the key to the past" with "no vestige of a beginning, no vestige of an end," published in his *Theory of the Earth* in 1795) provided the intellectual bedrock for Charles Darwin's theory, "The Evolution of Species" (1857) and, eventually, Alfred Wegener's "Theory of Continental Drift" (1912). In each case, these profoundly radical proposals were the cause of considerable pain among contemporary thinkers and, as with most intellectual paradigm shifts, the principal source of pain was resistance to the particular notion in question. While uniformitarianism is a truism for the physical sciences, it remains an important "law" in modern geoscience. At the Royal Society of Edinburgh, Hutton contended that catastrophic happenings were not responsible for Earth's landforms, which he believed had developed over unimaginable periods of time through a variety of slow processes. Above all, he rejected contemporary biblical interpretations that viewed Earth as having been created supernaturally via catastrophic events such as the biblical Flood.

Despite the lack of enthusiasm for uniformitarianism among Hutton's peers, Sir Charles Lyell published his "Principles of Geology" (1830–1833), a compendium of compelling geological evidence from the British Isles and Europe that supported Hutton's ideas. It was, moreover, implicit in Darwin's proposal that the diversity of Earth's biota results from the uniform modification of genetic traits over long periods of time. Darwin's theory evolved along two divergent paths, however, each offering potentially radical perspectives for the interpretation of earth history. The first of these, historical contingency, describes how the historical particulars of a given group of biological entities have influenced their current diversity in lineage and form, as explored through the use of phylogenetic inference. The second approach invokes "generative constraints" that affect ontogenetic process (how entities are made) and constrains the possible realizations of form, as explored through developmental biology and, more recently, studies of protein structure.

In the earth sciences, by analogy, generative constraints may be taken to be equivalent to explaining phenomena in terms of mineralogy and crystal chemistry (bonding, ligand fields, atomic site occupancies, and so on), while historical contingency can be considered as a modifier to the uniformitarian view that clearly dominated Darwin's thinking. Perhaps the most salutary example was provided by Gould (1989), who proposed that biota preserved in the Burgess Shale, British Columbia, provide a perfect illustration of the contingency challenge to Darwinian evolution, and observed that, in his view, the history of life records a "staggeringly improbable series of events, sensible enough in retrospect and subject to rigorous explanation, but utterly unpredictable and quite unrepeatable" (p. 289–290)—a far cry from Darwin's vision of inherent predictability, so dear to the ordered philosophies of the nineteenth century.

Today, the inherent unpredictability of scientific progress is self-evident and in essence conforms to Gould's perception of organic evolution. Whether such a perspective is helpful in explaining earth evolution as a whole depends on the extent to which the "fundamental laws" of our geological forefathers faithfully pre-ordain the geological record. In other words, can we view the development of our planet as a series of "staggeringly improbable" events, "utterly unpredictable and quite unrepeatable," but "explicable in retrospect," or should this view be subordinated to, or at least integrated with, the principles upon which Darwin's theory was constructed? A fundamental tenet of contingency theory is that an interpretative context is rarely that of an opposing interpretation. For example, conflict between catastrophists and uniformitarians in the early nineteenth century is redundant if the agreed context is a dynamic Earth as an interactive component in an evolving solar system. Historical contingency and classic uniformitarianism may therefore be compatible if they share the same contextual parameters. For most earth scientists today, these include the laws of thermodynamics, kinetics, and radioactive decay, along with the apparent tendency of our planet to behave as a self-regulating, steady-state system, rather than (in geological terms) a "runaway" chain reaction. If, on the other hand, earth evolution is viewed as inherently catastrophic, the effects of unpredicted (and unpredictable) events clearly outweigh those implied by conventional uniformitarian models. In either case, actualistic models developed on the basis of active geodynamic processes, in conjunction with experimental and theoretical insights, are still amenable to testing against the geological record and remain a valid approach to studying earth history. If such models are successful, a meaningful context probably exists for identifying and evaluating historically contingent factors. Alternatively, they can be reconfigured according to new contextual parameters if they fail.

In a seminal paper, Moores et al. (2000) addressed the ophiolite conundrum with a view toward reconciling ophiolite genesis with processes that characterize the global mid-ocean ridge system. Contending that debate of the conundrum hitherto has ignored historical contingency, Moores et al. (2000) observed that a correct interpretation of geochemical source "signatures" in mid-ocean ridge magmas is fundamentally contingent on prior histories of mantle convection and lithospheric plate motions, which vary *a priori* with time. Accordingly, they cite incidences of ridge subduction and the propagation of spreading centers into regions affected by subduction, and proposed that magmatic sources of apparent supra-subduction affinity may be tapped by ocean ridges along with those more familiarly characterized by mid-ocean ridge basalt (MORB) and rarely, oceanic island basalt (OIB). They further argue that ophiolites showing structural evidence for seafloor spreading should be interpreted as such, regardless of apparent indications of geochemical "fingerprints" (Pearce, 1991; Pearce et al., 1992b), the latter being explicable by a variety of contingent factors. In the present paper, the ophiolite conundrum is further scrutinized as a basis for evaluating historical contingency and the extent to which it qualifies conventional uniformitarian approaches.

AN ACTUALISTIC MODEL

Mantle Flow and the Wilson Cycle

The Wilson cycle was proposed as a way to explain opening and closing phases of ocean basins throughout earth history, in terms of: (1) continental breakup in response to subcratonic asthenosphere upwelling; (2) subsidence and sediment accumulation at thinned passive margins with the inception of seafloor spreading; (3) continued seafloor spreading on a scale comparable to that of the Atlantic, Pacific, and Indian oceans; (4) subduction inception within the basin itself or at one of its passive margins in response to distal plate collisions; (5) the consumption of remnant oceanic lithosphere by subduction, with the approach of continental plates; and (6) orogeny and ophiolite emplacement following continent-continent collision (e.g., Casey and Dewey, 1984) (Fig. 1). Wilson cycles are generally assumed to be driven by a combination of global-scale mantle convection and the effects of "slab pull" forces, as inferred from the near-contemporaneous development of the American cordilleras with opening of the Atlantic Ocean and their colinearity with the Mid-Atlantic Ridge (e.g., Russo and Silver, 1996). While studies of ophiolitic bodies played a significant part in validating Wilson cycles (Casey and Dewey, 1984; Robertson, 1994; Xiao et al., 2000), it was assumed that they represent opening, rather than closing, stages of such cycles. Given the factors that appear to preclude an exclusively ocean ridge provenance, the broader significance of ophiolites is still open to question. In fact, mantle convection paths may be quite complex toward the end of a Wilson cycle,

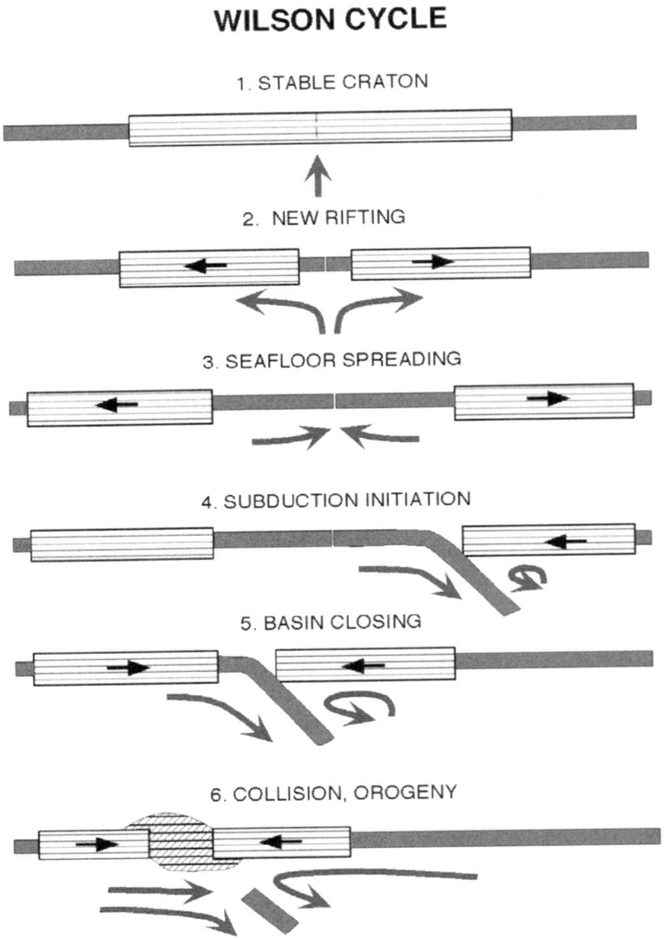

Figure 1. Schematic Wilson cycle stages, simplified from Casey and Dewey, (1984). (1) Breakup of a continent in response to subcratonic asthenosphere upwelling. (2) Rifting and development of thinned passive margins with the inception of seafloor spreading. (3) Continued seafloor spreading on a scale comparable to that of the Atlantic, Pacific, and Indian oceans. (4) The termination of basin opening in response to distal plate collision(s) triggering subduction inception at one or another passive margin. (5) Consumption by subduction of remnant oceanic lithosphere as continental plates converge. (6) Continent-continent collision, leading to orogeny and eventual exhumation and denudation of orogen roots.

when asthenosphere flow may be considered as both a driver of and a passive response to lithospheric plate motions.

This possibility may be considered in relation to collisions of the Afro-Arabian and Indian plates with Eurasia. In both cases, a process of collision-induced mantle perturbation may be inferred from the coupled kinematics of escaping continental lithosphere and newly formed marginal basins, together with spatially- and temporally-related igneous activity. (See discussions by Flower et al., 1998, 2001.) The role of mantle flow is supported by S-wave splitting data (e.g., Yang et al., 1995; Russo and Silver, 1996; Hung and Forsyth, 1999; Murdie and Russo, 1999), seismic tomography (Wortel and Spakman, 2000), and the evidence from isotopic mantle flow tracers (Flower et al., 1998, 2001). It was postulated that the approach to and eventual collision of Arabia and India with Eurasia led to the lateral extrusion of asthenosphere, in turn driving the escape of mobile continental blocks such as Eurasian Sundaland and Anatolia, with conjugate opening of marginal basins (Flower et al., 1998, 2001; cf., Briais et al., 1993; Jolivet and Faccenna, 2000). These "mirror-image" collision responses highlight the broader regional histories of these regions, and suggest that the initiation of subduction and arc-trench rollback intrinsically characterize mantle responses to the closing stages of a Wilson cycle in a constrained spatial-temporal framework.

"Proto-ophiolites" and the Forearc Analog

Because ophiolites appear to show distinctive spatial-temporal correlations with distal plate tectonic events, the question of where "proto-ophiolites" (modern or recently formed ophiolitic assemblages destined to be entrapped in orogens) are formed is fundamental. Following publication of the seafloor spreading hypothesis (Vine and Matthews, 1963), Hess (1965) proposed that ophiolites represent fragments of the oceanic crust, an observation supported by "single-chill" sheeted dikes, pervasive listric faulting, and imbricated pillow lava sequences recorded in the Troodos, Vourinos, and other ophiolites (Gass, 1968; Moores et al., 1968; Moores and Vine, 1971; Moores and Jackson, 1974). This view prevailed until the early 1970s, when the first inklings of a conundrum appeared. Miyashiro (1973) reported that in addition to MORB, calcalkaline magmatic lithologies are widespread in ophiolites, an observation which was initially downplayed and in some cases refuted (e.g., Gass et al., 1975; Hynes, 1975; Miyashiro, 1975a, b, c; Moores, 1975), cf. (Miyashiro, 1973, 1975a, b, c, 1977). In response, it was argued that calcalkaline attributes were probably an artifact of seafloor hydrothermal alteration and therefore not significant (Gass et al., 1975; cf. Pedroni et al., 1999; Gillis and Banerjee, 2000). By the 1980s, however, ocean crust protagonists had conceded ground by accepting that some ophiolites at least may have been generated in subduction-related marginal basins (e.g., Pearce et al., 1984), permitting a partial rapprochement between these positions. This view was supported by observations that MORB-like cumulates and peridotitic melt residua had been intruded or underplated by calcalkaline melts (Pearce et al., 1984; Batanova and Sobolev, 2000), and that ophiolitic lava sequences often grade upward from MORB-like to calcalkaline compositions with little or no eruptive hiatus (Alabaster et al., 1982; Pearce et al., 1984; Smewing et al., 1984).

Reports of ultramafic or strongly mafic boninite eruptives in ophiolite (Cameron et al., 1979; Upadhyay 1980; Capedri et al., 1981; Dobretsov and Kepezhinscas, 1981; Hibbard and Wright, 1982; Barringer, 1983; Walker and Cameron, 1983) were a further challenge because these rocks had hitherto been considered unique to intra-oceanic forearc settings (Cameron et al., 1979). The implication of a forearc analogy proved, however, to be critically significant. Building on the work of Casey and Dewey (1984) and Leitch (1984), Stern and Bloomer (1992) proposed that ophiolites are fundamentally linked to the inception of new subduction, whose magmatic products, typified by boninite, are commonly preserved in the lower parts of forearcs and their conjugate remnants, detached by backarc basin opening (Bloomer et al., 1995; Pearce et al., 1995; Hawkins and Castillo, 1998) (Figs. 2 and 3). On the basis of available information, Stern and Bloomer (1992) observed that ophiolites lack features that *preclude* a forearc analogy, an observation that is as valid today as it was ten years ago.

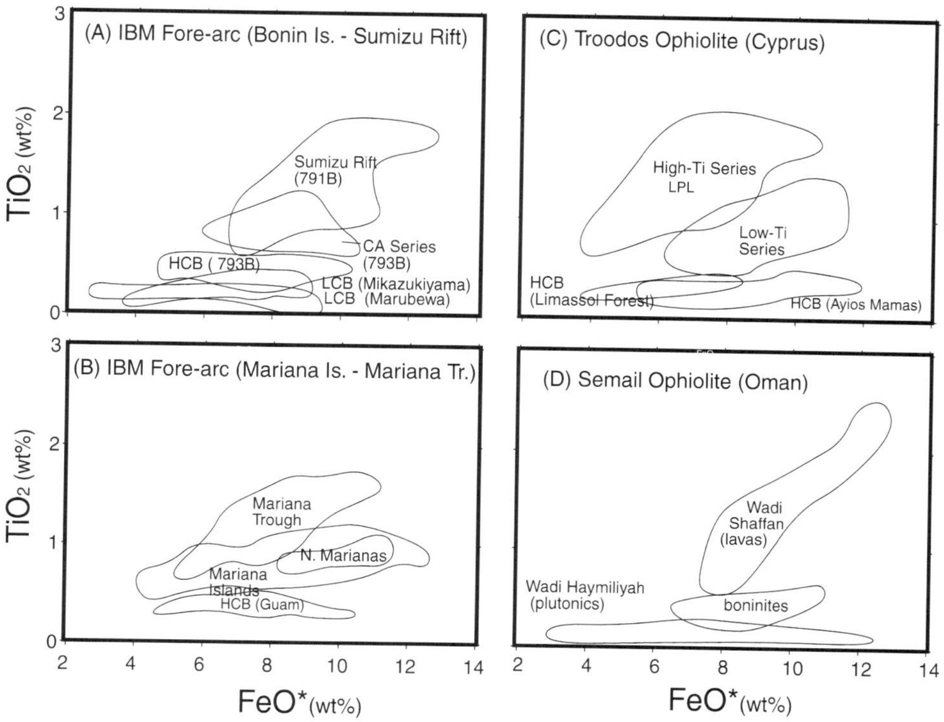

Figure 2. Variation of TiO_2 versus FeO* (in wt%) for eruptive and intrusive lithologies sampled from typical forearcs and ophiolites. (A) Bonin arc-forearc and backarc Sumisu Rift (Izu-Bonin-Mariana system [IBM]) (Pearce et al., 1992a; Taylor et al., 1994). (B) Mariana arc-forearc (Reagan and Meijer, 1984; Stern et al., 1989) and backarc Mariana Trough (Gribble et al., 1998). (C) The Troodos ophiolite, Cyprus (Flower and Levine, 1987; Gibson et al., 1987; Rogers et al., 1989; Bednarz and Schmincke, 1994; Portnyagin et al., 1997). (D) The Semail ophiolite, Oman (Lachize et al., 1996; Einaudi et al., 2000).

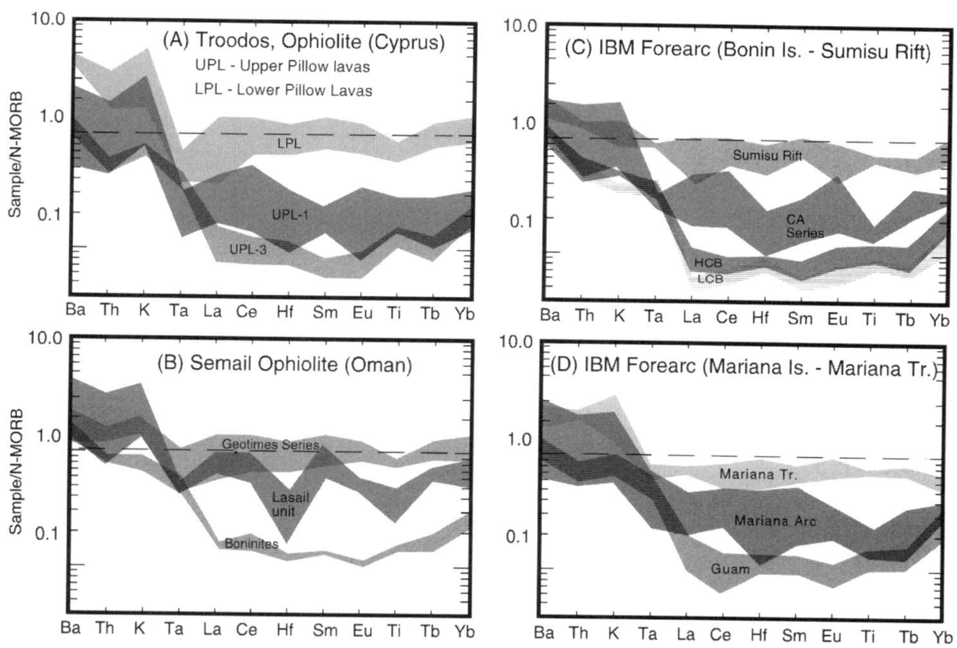

Figure 3. MORB-normalized incompatible element distributions for typical forearc and ophiolite lithologies. A: Troodos ophiolite, Cyprus—Upper Pillow Lavas (series 1 and 3), and Lower Pillow Lavas (Flower and Levine, 1987; Gibson et al, 1987; Rogers et al., 1989; Bedarz and Schminke, 1994; Portnyagin et al., 1996, 1997). B: Semail ophiolite, Oman (Umino et al., 1990; Lachize et al., 1996; Einaudi et al., 2000; Ishikawa et al., 2002). Alley Series volcanics (calcalkaline and boninitic), Geotimes Series volcanics (MORB-type volcanics). C: Izu-Bonin arc-forearc and backarc Sumisu Rift (Izu-Bonin-Mariana system) (Pearce et al., 1992a; Taylor et al., 1994). D: Mariana arc-fore-arc (Reagan and Meijer, 1984; Stern et al., 1989) and backarc Mariana Trough (Gribble et al., 1998).

More recently, Flower et al. (2001) concluded that forearc-like lithologies may range from near-exclusively oceanic types, such as in the western and southwestern Pacific (Stern and Bloomer, 1992; Danyushevsky et al., 1995; Ballantyne, 1991; Ballantyne and Parson, 1992) and the South Sandwich Islands (Pearce et al., 1995), to those in the Mediterranean and Caribbean (Cobiella-Reguera, 1996, 2000; Robertson and Grasso, 1995), which include substantial continental crustal components. Generic forearc components may therefore include MORB basement, boninitic proto-arc (including high-Mg andesite and adakite), calcalkaline arc, MORB-like backarc, and older continental relics. Even where incomplete, such assemblages are commonly juxtaposed with ultra-refractory serpentinized peridotite and other features unique to forearcs, such as serpentinite mud diapirs (Le Pichon et al., 1995a; b; Fryer et al., 2000) and anomalous, high-temperature hydrothermal deposits (Banerjee et al., 2000; Banerjee and Gillis, 2001; Gillis, 2002).

Subduction Rollback in Action

Subduction rollback is conventionally explained as a passive response to down-dip slab compression as determined by the force balance between slab pull, upper plate collision, and "ridge push" effects (Le Pichon et al., 1981). Theoretically, slab-related buoyancy forces exceed those associated with mantle convection, slab motion being largely controlled by its excess density (Royden, 1993a, b). Rollback is expected to continue until oceanic lithosphere input is terminated by the arrival of a continental plate or other obstruction. Most subduction zones behave accordingly, although the slab pull appears insufficient to explain the common coincidence of slab steepening, anomalous arc curvature, and accelerated basin opening. Moreover, the observation that some subducting slabs in Mediterranean and western Pacific arc-backarc systems exhibit tensional rather than compressional stress fields (Seno and Yamanaka, 1998) suggests that slab pull effects are subsidiary to those of trenchward mantle flow (Figs. 4A and 4B). This type of scenario presents a problem for the classic extrusion tectonics model (Tapponnier et al., 1982; 1986), which assumes that basin opening and subduction rollback are passive responses to lithosphere escape (Briais et al., 1993; Le Pichon et al., 1995a; Reilinger et al., 1997). Nevertheless, mantle-driven rollback appears to be the only mechanism capable of explaining the observation that opening of the South China and Aegean Seas began prior to and proceeded at a faster rate than the escape of their conjugate continental blocks (Indochina and Anatolia, respectively) (Seno and Yamanaka, 1998) (Figs 4 and 5).

Rollback cycles in the Mediterranean are relatively short-lived and invariably terminated by the "collapse" (or reconsumption by subduction) of backarc basins following forearc-continent collisions (Royden, 1993a, b; Jolivet and Faccenna, 2000) (Fig. 5). Following the Apulia-Eurasian collision (70–65 Ma), the Betic-Rif, Corsica-Sardinian, Calabrian, and Sicilian terrains were probably bounded by near-continuous, north-vergent subduction that subsequently evolved as discrete rollback episodes (Lonergan and White, 1997; Stampfli et al., 2002). Rollback to the west was probably concomitant with opening of the Alboran

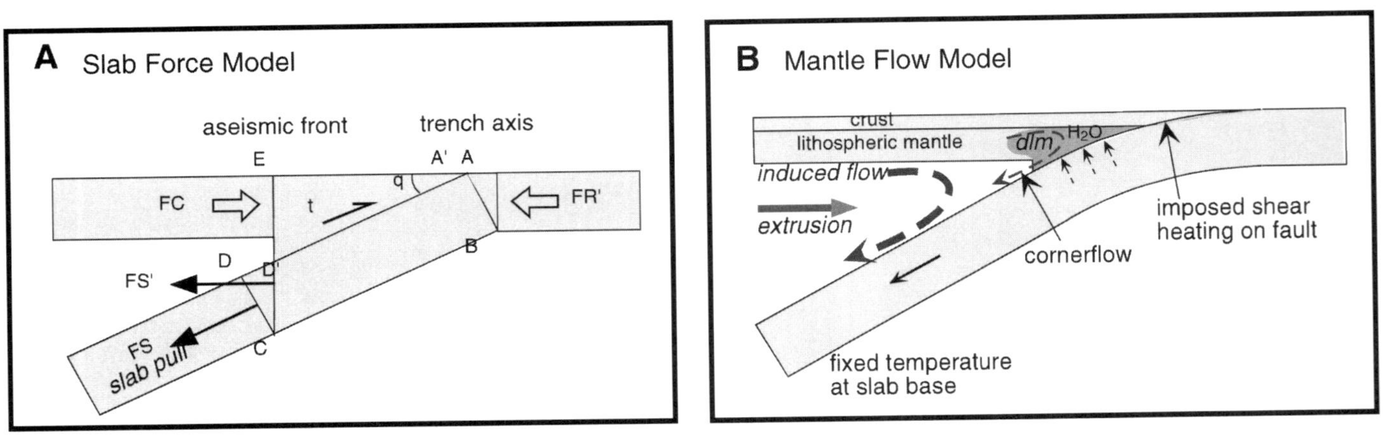

Figure 4. Hypothetical effects of "slab pull" and "mantle extrusion" in retreating arc-trench sectors (examples in Fig. 5). A: Slab force model (after Seno and Yamanaka, 1998). This shows interaction between a subducting slab and overriding plate. Where backarc basin opening represents a passive response to slab-pull effects, the slab shows down-dip compression and arc-trench rollback is relatively slow. However, in cases where basin opening is matched by down-dip extension, as recorded, for example, at the Mariana and Hellenic arcs (Seno and Yamanaka, 1998) (see Fig. 5), rollback is rapid and driven by an exogenous force such as trenchward mantle flow. The dip of the subducting plate is q, AB is the trench axis, and CE the aseismic front. PS′ is the effective ridge push, FS the slab pull, and PC the collision force. FS′ is the horizontal component of traction on CD′ and t is the shear stress at the thrust zone (see Seno and Yamanaka, 1998). B: Mantle flow model 2D (after van Keken et al., 2002). Assuming a constant rate of subduction, two flow filed components are shown: (1) "endogenous" (slab induced), and (2) exogenous (e.g., collision induced). Mantle wedge "cornerflow," combining these elements with ambient shear heating, is believed to allow for delamination of refractory (melt-depleted) lithospheric mantle, fluxed—thereby rheologically altered—by slab-derived fluids. The incorporation of hydrated, refractory peridotite is a potential mechanism, unique to this setting, for generating asthenospheric boninite magma sources (Ngyuyen and Flower, 2003).

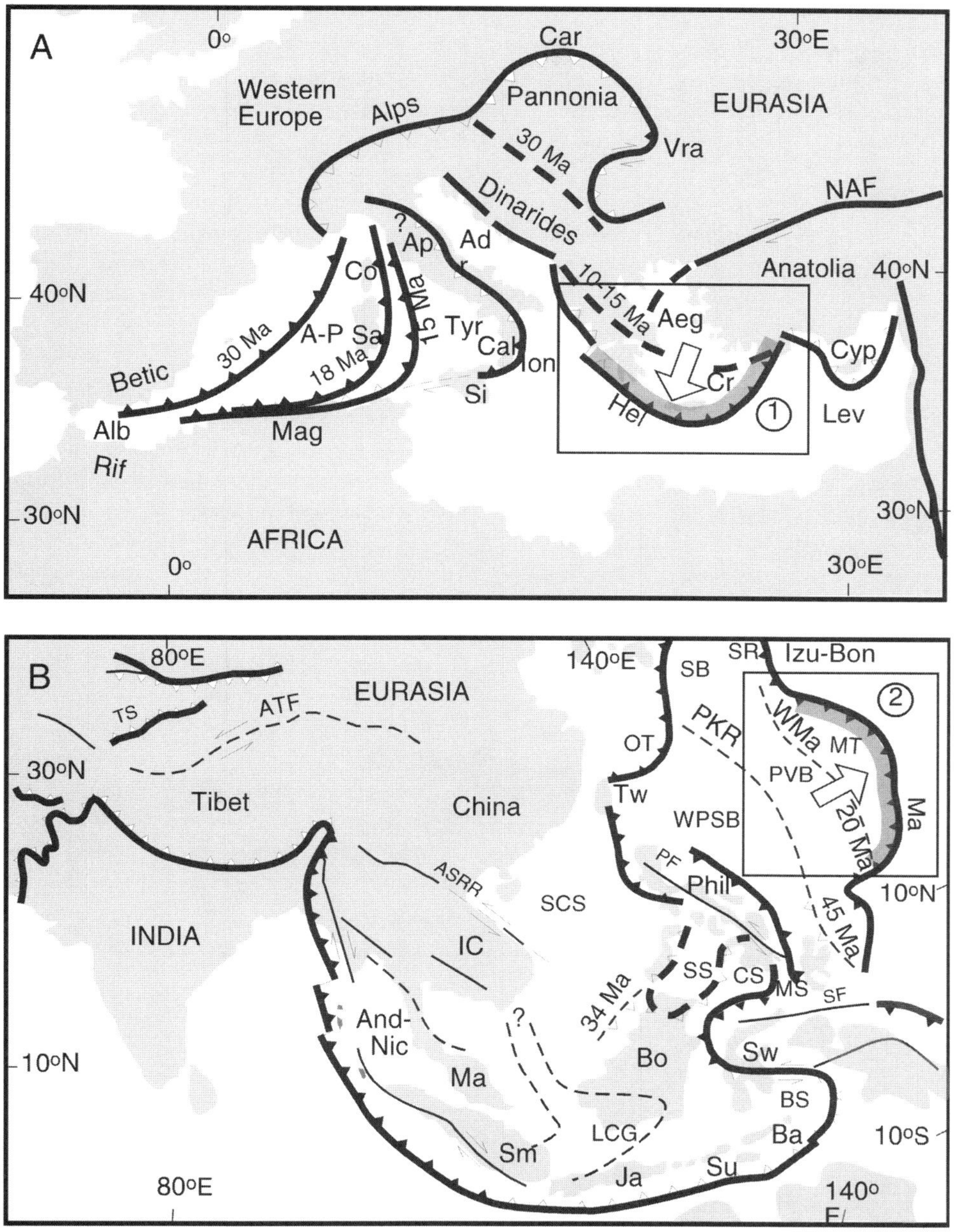

Figure 5. Plate boundary evolution and subduction rollback in active Tethyan domains. (A) The circum-Mediterranean region (after Wortel and Spakman, 2000). Arrows indicate directions of probable slab tearing beneath the Apennine-Calabrian, Hellenic, and Carpathian arcs. Adr—Adriatic Sea, Aeg—Aegean Sea, Alb—Alboran Sea, Ap—Apennines, Cr—Crete, A-P—Algero-Provencal Basin, Bet—Betics, Cal—Calabria, Car—Carpathians, Co—Corsica, Cr—Crete, Cyp—Cyprus, Din—Dinarides, Hel—Hellenic arc/trench, Ion—Ionian Sea, Lev—Levantine Basin, Mag—Maghrebides (from the Rif to Sicily), NAF—North Anatolian Fault, Pan—Pannonian Basin, Rif—Rif, Sa—Sardinia, Si—Sicily, Anat—Anatoia, Tyr—Tyrrhenian Sea. Barbs indicate subduction or thrusting vergence, black suggesting a continuous slab, and white, possible post-collision breakoff. (B) Western Pacific region (after Flower et al., 1998, 2001). Arrows indicate directions of probable slab tearing beneath the Himalayas, Sunda-Banda arcs, and Northern Luzon-Taiwan. TS—Tien Shan, ATF—Altyn Tagh Fault, IC—Indochina, RRF—Red River fault, And-Nic—Andaman-Nicobar Islands, IC—Indochina, Ma—Malay Peninsula, Sm—Sumatra, Bo—Borneo, Ja—Java, Su—Sunda, Ba—Banda, Sw—Sulawesi, Phil—Philippines, Tw—Taiwan, Izu-Bon—Izu-Bonin islands, Ma—Mariana Islands, Wma—West Mariana arc, PKR—Palau-Kyushu ridge, Ru—Ryukyu Islands, SCS—South China Sea, SS—Sulu Sea, CS—Celebes Sea, MS—Molucca Sea, WPSB—West Philippine Sea Basin, BS—Banda Sea, SB—Shikoku Basin, SR—Sumizu Rift, PVB—Parece Vela Basin. Aegean-Hellenic (1) and Izu-Bonin-Mariana (2) rollback systems are indicated (see text; Figs. 4 and 6).

Basin and terminated by collision of the retreating arc with the Betic-Rif system. (Lonergan and White, 1997). Eastward rollback of what became the Apennine-Calabrian arc system occurred in stages with the opening of the Valencia Trough and the Balearic, Algero-Provençal, and Tyrrhenean Sea Basins (Jolivet and Faccenna, 2000). In the eastern Mediterranean, the Cyprus Basin had already begun opening in the Late Jurassic, prior to the Arabia-Eurasia collision, following detachment and northward drift of the Taurus microcontinent. This basin was preserved intact following the post-collision westward escape of Anatolia. Sundering of the Hellenide and Tauride orogens (at 15–10 Ma) led to opening of the Aegean Sea and rollback of Hellenic arc-forearc complexes, with concomitant thinning of continental crust and eventual collision with the approaching African plate. Meanwhile, continued south- and east-vergent subduction of relict oceanic basins led to rapid rollback of the Carpathians and opening of the Pannonian and Transylvanian Basins from 30 Ma to the present.

In contrast, subduction rollback in the western Pacific has been much less constrained by approaching continental plates and continues propagating rapidly to the east (Fig. 5B). Despite controversies regarding kinematic reconstructions of the region, a three-fold scenario for Western Pacific Basin opening is likely: (1) a Paleocene pre-collision stage involving opening of the West Philippine Sea and Celebes Sea Basins; (2) Oligocene to Miocene opening of the South China, Sulu, and Makassar Seas and, to the north and east, the Shikoku and Parece Vela Basins; and (3) a late Miocene to Quaternary stage characterized by opening of the Banda Sea, Okinawa Trough, Mariana Trough, and Andaman Sea. A significant hiatus between the first and second episodes (50–45 Ma) either predates or is contemporaneous with the "hard" collision of India and Asia and a major reorientation of Pacific plate spreading (Lee et al., 1994). A second hiatus (15–10 Ma) appears to match microplate collisions in and around the South China Sea (North Palawan, Reed Bank, Dangerous Grounds) (Pubellier and Cobbold, 1996), while a third (Quaternary) hiatus coincides with consolidation of the Banda Sea (Milsom et al., 1996, 1999); rollback of the Japanese, Izu-Bonin, and west Mariana arcs; initiation of eastward subduction beneath the Philippines and Sulawesi; truncation of the Moluccan Sea (Rangin et al., 1995, 1996, 1999a, b, 2000); the Luzon-Eurasia collision; and the widespread appearance of "dispersed" basaltic volcanism in the western Pacific and in east and southeast Asia (Flower et al., 1998, 2001).

Thus, rollback cycles in active Tethys characteristically begin with splitting of nascent proto-arcs into active and remnant segments that become progressively detached as basin opening and arc rollback proceed, the best example in progress being that of the Izu-Bonin-Mariana (IBM) system in the western Pacific (Figs. 5 and 6). Typically signaled by boninite or high-Mg andesite magmatism (Stern and Bloomer, 1992; Monzier et al., 1993; Sigurdsson et al., 1993; Eissen et al., 1994; Crawford et al., 1997), the initial stages of rollback appear to reflect anomalous mantle thermal conditions (van der Laan et al., 1989; Falloon and Danyushevsky, 2000; Insergueix-Filippi et al., 2000). Rollback is often interrupted by additional arc splitting events, probably

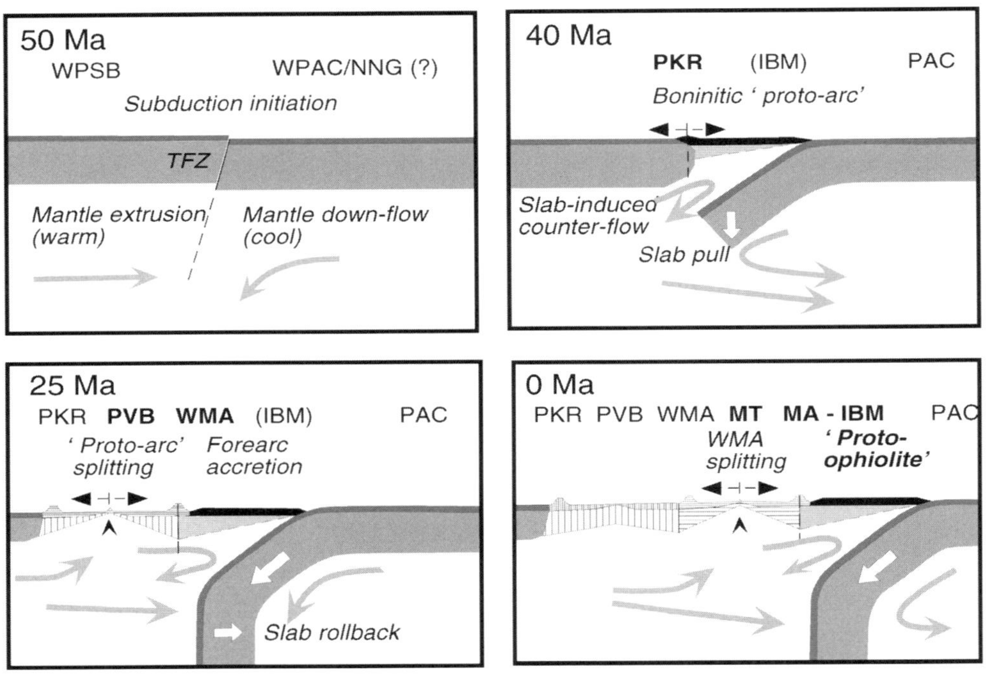

Figure 6. Subduction rollback model, from Bloomer et al. (1995), based on evolution of the IBM forearc and eastern part of the Philippine Sea Plate (following Karig, 1971; Hussong et al., 1981; and Stern and Bloomer, 1992). (A) 50–40 Ma. Subduction initiation beneath West Philippine Sea Basin (WPSB) plate either by the Pacific (PAC) plate or (hypothetical) young North New Guinea (NNG) plate along transform fracture zone in proto-WPSB spreading center. Boninite melt genesis accompanies early forearc development with inception of the calcalkaline arc forming the Palau-Kyushu ridge (PKR). (B) 40–25 Ma. Continued subduction with slab steepening of the Pacific plate, splitting of the Palau-Kyushu arc, Parece Vela Basin (PVB) opening, and rollback of the active West Mariana arc (WMA). (C) 25–0 Ma. Continued subduction, Mariana Trough (MT) opening by splitting of the West Mariana arc, and rollback of the active Mariana arc (MA). (D) 0 Ma. Subduction beneath the modern Mariana arc-trench system with continued MT opening by "unzipping" of the West Mariana-Mariana arc to the north (Iwo Jima). PKR—Palau-Kyushu Ridge, PVB—Parece Vela Basin, WMA—West Mariana Arc, MT—Mariana Trough, MA—Mariana Arc, IBM—Izu-Bonin-Mariana forearc, PAC—Pacific Plate, WPSB—West Philippine Sea Basin, WPAC—Western Pacific, NNG—"North New Guinea" Plate.

in response to plate kinematic adjustments. Evolving forearc terrains thus progressively incorporate proto-arc, arc, backarc, and (in some cases) continental lithosphere fragments (Fig. 6), and may be likened to lithologic "high tide marks" (HTMs), showing large age discrepancies within and between their crust and mantle components (Ishii et al., 1988; Ogawa, 1995; Flower et al., 1998, 2001; Kamp, 1999). If mantle flow is indeed a driver of subduction rollback, the occurrence and spatial-temporal distribution of ophiolites in the geologic record should yield new insights into mantle thermal and dynamic behavior prior to and during plate collisions.

Subduction Nucleation Versus Ridge Subduction

Forearc–Remnant Arc Pairs

The link between ophiolite subduction rollback is further supported by temporal correlations between forearc boninites and their respective proto-arc remnants. Unfortunately, active boninite-high-Mg andesite volcanism is rare, one exception being that occurring at the newly-forming Hunter Ridge proto-arc, located between southernmost Vanuatu (New Hebrides) and the Fiji islands (Falloon and Crawford, 1991; Crawford et al., 1997). Here, boninite-high-Mg andesite activity recently occurred as the southward-propagating North Fiji Basin spreading center became linked by a transform shear zone with newly active subduction (Monzier et al., 1993, 1997); however, numerous examples of coeval boninite and high-Mg andesite in forearc-remnant arc pairs confirm the effects of active rollback in the Mediterranean and western and southwestern Pacific regions (Crawford et al., 1981, 1986). The best-studied example is the Palau-Kyushu Ridge that dissects the Philippine Sea Plate, marking the locus of middle Eocene (50–42 Ma) subduction inception that occurred shortly before the India-Asia collision (45–40 Ma) and reorientation of Pacific Plate motion (ca. 43 Ma) (Stern and Bloomer, 1992; Bloomer et al., 1995; Hawkins and Castillo, 1998). Palau-Kyushu boninites match those in lower horizons of the IBM forearc that conform in all respects to in situ "proto-ophiolites" (Ogawa, 1995; Ishii et al., 1988) (see Figs. 6 and 7).

Similar effects were produced at ca. 25 Ma in the southwestern Pacific by a sharp increase in the angle and rate of Australian convergence with the northwest-moving Pacific Plate (Ballance et al., 1999). Pacific Plate subduction beneath what is now the Lau-Colville Ridge was accompanied by splitting of the Colville and Norfolk-Three Kings proto-arcs, opening of the backarc Norfolk Basin, and relatively rapid (approximately 35 mm/yr[1]) eastward retreat of the active Colville-Lau arc (Pearce et al., 1995). While basin opening temporarily ceased at ca. 15 Ma, the newly configured Colville-Coromandel-Taranaki arc was reactivated at ca. 5 Ma and split again, accommodating opening of the Lau-Havre Basin and rollback of the Tonga-Kermadec-Taupo forearc (Ballance et al., 1999).

Mid-Miocene eruptions of high-Mg andesite (ca. 13 Ma) preserved in central and southern parts of the Ryukyu forearc are matched by similar activity in the Fujian-Taiwan region and

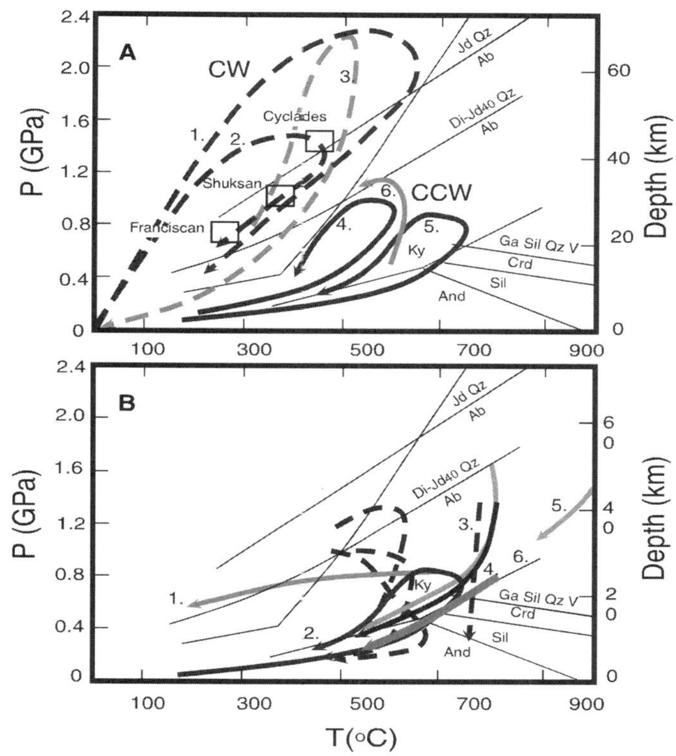

Figure 7. Pressure-temperature-time metamorphic histories in relation to a simplified petrogenetic grid. (A) *P-T*-t curves for sub-ophiolitic metamorphic soles are: 1–5—the Semail ophiolite, Oman (Hacker and Gnos, 1997); 6—the Kiziltepe ophiolite, Turkey (Dilek and Whitney, 1997). Also shown, *P-T* equilibration conditions for ophiolitic blueschist-bearing units from the Shuksan and Franciscan terrains and Cyclades (Aegean) remnant arc. (B) *P-T*-t curves for exhumed post-orogenic peridotites are: 1—the Ronda peridotite (Pearson, 1999), 2—Dabie Shan (Webb and Hacker, 1999).

marks subduction initiation, triggered by collision of the Luzon arc with Eurasia, which led to opening of the Okinawa Trough (Chung et al., 1995; Shinjo, 1999; Shinjo et al., 1999). Similarly, mid-Miocene (14–15 Ma) high-Mg andesite volcanism recorded from the islands of Evia and Skyros between the Hellenic forearc and Cycladean metamorphic core complexes (Lister et al., 1984; Pe-Piper and Piper, 1994; Forster and Lister, 1999; Migiros et al., 2000) corresponds with high-Mg andesite-bearing ophiolite fragments in the Hellenic forearc (Fortuin et al., 1997; Clift and Dixon, 1998). In addition, Oligo-Miocene high-Mg andesite volcanics (ca. 18 Ma) in Sardinia and Corsica, matched by those in Calabrian forearc ophiolite fragments (Beccaluva et al., 1982; Delaloye et al., 1984; Compagnoni et al., 1989), appear to mark proto-arc splitting associated with the inception of Tyrrhenian Sea opening (Morra et al., 1997; Padoa, 1999).

Hot Subduction Initiation

At least two lines of evidence highlight the anomalous thermal character of the mantle associated with subduction nucleation—the presence of boninite and high-Mg andesite eruptives in proto-arc formations, and the inflected "counterclockwise" *P-T*-t histories interpreted from thermobarometric studies of sub-ophiolitic metamorphic "soles" (e.g., Fig. 7). Experimental studies indicate that boninites result from the combined effects of slab-derived fluid flux, mantle decompression, and elevated mantle potential temperatures (van der Laan et al., 1989; Umino and Kushiro, 1989; Falloon and Danushevsky, 2000), consistent with seafloor spreading close to a trench and the coexistence of both upwelling and downwelling asthenosphere (Stern and Bloomer, 1992). Asthenosphere potential temperatures interpolated from H_2O-undersaturated boninite phase equilibria exceed those predicted for "normal" mid-ocean ridges by at least 150–200 °C (van der Laan et al., 1989; Umino and Kushiro, 1989; Falloon and Danyushevsky, 2000), consistent with the conditions simulated by 2D numerical models (Insergueix-Filippi et al., 1998, 2000).

Analogous thermal conditions indicated by thermobarometric data for ophiolite metamorphic soles have been interpreted by some (e.g., Crawford, 2002, personal commun.) to reflect those associated with volcanic passive margin rifting and by others (e.g., Searle and Cox, 1999) MORB-like relics of incipient subduction. Interpolated metamorphic *P-T*-t histories commonly indicate thrusting to depths >60 km with temperature increases to >900 °C, followed by rapid decompression (e.g., curves 4–6, Fig. 7) (e.g., Hacker, 1991; Wakabayashi and Unruh, 1995; Dilek and Whitney, 1997; Hacker and Gnos, 1997; Wakabayashi, 1999; Wakabayashi and Dilek, 2000; Dimo-Lahitte et al., 2001). While some ophiolites slightly post-date their high-temperature soles, most are coeval or slightly older (Parlak and Delaloyne, 1999), consistent with interpretations that sundered forearc and remnant proto-arc components were single entities prior to splitting, and that incipient subduction relics are preserved a priori in forearc proto-ophiolites (Ogawa, 1995; Ishii et al., 1988). Thermochronologic data for metamorphic soles support this model, the best documented examples being the Mirdita ophiolite in Albania (Dimo-Lahitte et al., 2001) and the Semail ophiolite, Oman (e.g., Hacker and Gnos, 1997; Searle and Cox, 1999).

Ridge Subduction

The only serious alternative to subduction initiation is the proposal that ophiolite genesis results from the consumption of recently-active spreading centers at *pre-existing* subduction zones (e.g., Crawford et al., 1986; Hacker and Mosenfelder, 1996; Hacker et al., 1996; Lytwyn et al., 1997). This rests on the presumption that newly formed (<10 my) oceanic lithosphere is too buoyant to be subducted (Cloos et al., 1994) and that high-temperature metamorphic soles represent mid-ocean ridge thermal conditions rather than those associated with "normal" subduction (e.g., Hacker and Mosenfelder, 1996; Hacker et al., 1996). Although the two interpretations are not mutually exclusive, which process dominates appears to hinge on: (1) the bulk compositional character of metamorphic soles, (2) validity of the ophiolite-forearc analog, and (3) possible spatial-temporal correlations between ophiolite formation and regional plate kinematic adjustments. In this light, ridge subduction is the less appealing model, given its absence from currently active rollback episodes. In contrast, several examples of the latter can be shown to have been triggered by subduction nucleation events characterized by boninite or high-Mg andesite volcanism (Lister et al., 1984; Falloon and Crawford, 1991; Stern and Bloomer, 1992; Monzier et al., 1993, 1997; Bloomer et al., 1995; Pe-Piper et al., 1996; Crawford et al., 1997; Hawkins and Castillo, 1998).

The MORB-like compositional character and inflected, counterclockwise *P-T*-t trajectories of many ophiolite metamorphic soles (e.g., Fig. 7, curves 4–6) (Searle and Malpas, 1980; Ghent and Stout, 1981; Malpas and Moores, 1992b; Shallo, 1992; Encarnacion et al., 1995; Hacker and Mosenfelder, 1996; Gnos, 1998; Searle and Cox, 1999; Bebien et al., 2000; Dimo-Lahitte et al., 2001) support the interpretation that they represent relict subducting slab fragments, their relatively steep thermal gradients complementing the inferred association of ultramafic boninite eruptives with fossil transforms, as represented by the Scutari-Pec lineament (Smith and Spray, 1984; Shallo, 1992; Robertson and Shallo, 2000), Arakapas fault zone (Simonian and Gass, 1977; Flower and Levine, 1987; Macleod et al., 1990; Macleod and Murton, 1995), and analogous features in the Semail ophiolite (Smewing, 1981; Boudier et al., 1988; Wakabayashi and Unruh, 1995). Together, both lines of evidence favor the notion that subduction nucleation at near-ridge transform fractures is a common precursor to subduction rollback (Stern and Bloomer, 1992; Monzier et al., 1993; Sigurdsson et al., 1993; Eissen et al., 1994; Crawford et al., 1997).

Subduction Nucleation Models

The underlying causes of subduction inception are also contentious, although critically important to evaluating the significance of ophiolites in the geological record. Is subduction initiated in response to pre-existing lithospheric heterogeneities; is it coupled to mantle flow patterns or thermal character; or does it represent a response to regional plate kinematic changes? If "spontaneous" subduction results from gravitational instabilities associated with pre-existing (e.g., transform) weak zones, a convergent stress field is not a necessary condition. On the other hand, the kinematic responses to distal plate collisions could result in the replacement of an extensional stress field by compression. For any of these cases, however, the observation that proto-arc splitting and backarc basin opening invariably follow subduction initiation events suggests that an overall extensional stress field is inherent, as suggested by Seno and Yamanaka (1998).

Smoot (1997) proposed that subduction nucleation is probably triggered by lithospheric gravitational instabilities resulting from thermal heterogeneities. On the other hand, Niu et al. (2001), citing the exceptionally low densities of refractory forearc peridotite (e.g., Parkinson et al., 1998), suggested compositional rather than thermal buoyancy contrasts may be more significant. While either or both of these factors are plausible, questions still remain as to whether pre-rollback subduction nucleation events are predictable responses to distal plate kinematic adjustments or to changes in the mantle flow field. There is evidence to support both possibilities.

Paleomagnetic data and seafloor magnetic anomaly patterns indicate that compositional boundaries between Indian Ocean and Pacific MORB mantle reservoirs—represented, for example, by the Antarctic-Australia Discordance (AAD) and retreating western Pacific arc systems—reflect a close connection between mantle flow and continental plate motions. DUPAL-like (Indian Ocean type) mantle first appeared following the late Paleozoic disaggregation of Gondwanaland and became the dominant magmatic source as the Tethyan oceans evolved, sharply demarcated from the ambient Atlantic and Pacific mantle reservoirs. As Australia (together with Africa-Arabis, India, and other Gondwana fragments) began its northward motion toward accreting Eurasia, infiltration of Pacific mantle shifted the domain boundary westward (Alvarez, 1990; Christie et al., 1998). To the north, however, as India and Australia approached, and then collided with, southern Eurasia, the domain boundary migrated to the east, as propagating marginal basins and subduction rollback accommodated displaced Indian Ocean mantle (Hickey-Vargas, 1998a, 1998b; Flower et al., 1998).

Despite the absence of active lithosphere subduction at the AAD, the evidence for preferential Pacific margin down-flow suggests that subduction nucleation associated with western and southwestern Pacific rollback episodes may have also occurred at the asthenospheric domain boundaries, implying that mantle flow traction is a significant factor. This does not preclude new subduction being triggered by plate kinematic changes resulting, for example, from collisions between retreating arc-forearc systems prior to their eventual entrapment. Subduction initiation and boninitic proto-arc formation at the Hunter Ridge, at the eastern extremity of the New Hebridies (Falloon and Crawford, 1991), appears to be linked to diachronous evolution of the New Hebrides–North Fiji Basin rollback cycle, initiated following the collision between Ontong Java and the Solomon Islands and interrupted (although not terminated) by collision of the New Hebrides forearc with the d'Entrecasteaux Ridge (Monzier et al., 1993). Following this collision, concomitant subduction and transform fracturing on the southward-propagating North Fiji Basin spreading center (Crawford et al., 1997) accommodates continued arc-forearc retreat (Monzier et al., 1997).

EMPLACEMENT MECHANISMS AND KINEMATIC IMPLICATIONS

The Obduction "Red Herring"

It is commonly remarked that the obduction of newly formed oceanic lithosphere onto passive continental margins (Coleman, 1977; Moores, 1982b; Lippard et al., 1986; Robertson, 1987; Pearce, 1991) is conspicuously absent from Earth at the present time. In contrast, the mantle-driven rollback model implies that ophiolites are emplaced shortly after their formation, following forearc-continent or forearc-forearc collisions, examples of which are briefly reviewed here.

In Kamchatka, lower Eocene oceanic crust and Late Cretaceous boninitic forearc fragments were incorporated during the Oligocene to late Miocene collision with the Aleutian forearc, accompanied by continental subduction, upper plate deformation, and reversed subduction from southeast to northwest (Geist and Scholl, 1994; Geist et al., 1994; Gaedicke et al., 2000). Similarly, the East Taiwan ophiolite was emplaced during diachronous collision between the Luzon arc (Philippine Sea Plate) and continental Eurasia (Teng, 1990; Teng et al., 2000), also interpreted by some (Hsu and Sibuet, 1995; Sibuet and Hsu, 1995) as a forearc-forearc collision with the former Ryukyu subduction system. In the Molucca Sea, a diachronous forearc-forearc collision associated with progressive backarc basin collapse represents the best actualistic example of proto-ophiolite emplacement, a process initiated in Mindanao at 4–5 Ma, between opposite-vergent subduction systems (Moore and Silver, 1983; Lallemand et al., 1998). Prior to ca. 5 Ma, eastern and western Mindanao terrains (respectively of Eurasian and Philippine Mobile Belt affinity) were also separated by collapsing backarc basins, later consumed by dual subduction (Sajona et al., 1994), whereas ophiolites in eastern Mindanao represent a further example of forearc-continent collision (Pubellier et al., 1991, 1999).

Ophiolite emplacement during the diachronous Australia-Sunda collision began at ca. 30 Ma (Audley-Charles et al., 1990; Van Bergen et al., 1993), indentation being initially absorbed by the northern edge, now represented by the Palau-Kyushu Ridge, of what later became the West Philippine Basin, Philippine Sea (Charlton, 2000). An arc-forearc complex destined to form much of the future eastern Philippines, became detached at ca. 24 Ma and removed entirely between ca. 12 and 6 Ma, by left-lateral shearing on the Sorong fault (Charlton et al., 1991a, b). At ca. 18 Ma, subduction initiation in northern New Guinea led to formation of the Maramuni proto-arc. Diverse ophiolites remain in the Sunda-Banda chain, including fragments of Indian Ocean lithosphere as large as the Semail ophiolite, welded to forearcs and commonly thrust over continental Sunda (Harris, 2001).

In western Tethys, as the Mediterranean became "locked" by the collisions of Africa and Arabia with Eurasia, first in southeastern Spain (ca. 30 Ma), then in Iran, marked by the Zagros crush (ca. 25 Ma) (Dewey et al., 1989a, b). Proto-ophiolites became entrapped by the Betic-Rif, Apennine-Calabrian, and Carpathian orogens as rollback was halted by collisions, a process also recorded in the islands of Crete and Cyprus. Metamorphic core complexes and high-Mg andesite eruptives in the Aegean Cyclades (Lister et al., 1984; Pe-Piper et al., 1996) probably represent proto-arc remnants from splitting of the Hellenide orogen (14–15 Ma), analogous to that affecting the Maures Massif that led to the detachment of Corsica-Sardinia (Jolivet et al., 1996, 1998). The Mediterranean Ridge itself evolved as a mud-dominated accretionary wedge (Robertson and Grasso, 1995) becoming trapped at 5–6 Ma, as the Hellenic Arc collided with Africa (Chaumillon and Mascle, 1997).

Analogous basin collapse and forearc collisions had closed the neo-Tethyan Pindos and Antalya Basins by the early Tertiary, the Cyprus Basin (discussed later) surviving into the Quaternary (Eaton and Robertson, 1993). The Africa-Cyprus collision is also very recent, as evidenced by subsidence record of the Eratosthenes Seamount, an inferred crustal fragment immediately to the south of Cyprus, and coupled uplift of the Limassol Forest block (Robertson et al., 1998). Paleomagnetic data indicate that Troodos rotated 90° counterclockwise (Morris et al., 1998), a possible pre-emplacement rollback effect juxtaposing the Arakapas fracture zone against arcuate, strike-slip faults.

Regional Kinematic Implications

The age distributions of Neo-Tethyan ophiolites suggest they belong to discrete temporal clusters, including an Alpine group (200–164 Ma), an "Outer-Hellenic" Pindos-Mirdita group (180–145 Ma), a "Tauride-Zagros" group (98–65 Ma), and a circum-Philippine Sea Plate group (50–43 Ma), etc. (Table 1), conceivably representing discrete subduction initiation-rollback cycles. Gnos et al. (1997), Hall (2002), and others, have observed that ophiolite ages correlate with recorded plate kinematic adjustments, which, in turn, represent regional-scale responses to distal continental splitting and collision events. Although additional paleomagnetic and geochronologic studies are needed to confirm these correlations in much of the Tethyan record, the pattern in active Tethyan regions provides a sufficiently clear basis for developing an actualistic model.

Plate kinematic adjustments in the Western Pacific are recognized at ca. 45, 34, 25, and 5 Ma (e.g., Hall, 2002). Mid-Eocene (50–42 Ma) subduction initiation at the Palau-Kyushu line shortly pre-dated the "hard" collision of India and Eurasia (45–40 Ma) and the ensuing change in Pacific Plate motion (ca. 43 Ma) (Hawkins and Castillo, 1998; Cosca et al., 1998). Construction of the IBM proto-arc began at ca. 50 Ma, giving way to "normal" calcalkaline activity as Pacific Plate motion changed at 42 Ma (Stern and Bloomer, 1992). Ophiolites rimming the Philippine Sea Plate such as the Zambales and Angat in the Philippines, and the ophiolites in Halmahera and northwestern New Guinea (Encarnacion, 1994; Ballantyne and Hall, 1990; Abbott et al., 1994) show similar ages to the IBM proto-ophiolite. Younger ophiolites may also be linked to regional kinematic changes. For example, formation of the late Eocene-early Oligocene (ca. 34 Ma) Palawan ophiolite may correspond to compressional stresses resulting from the coeval initiation of Indochina escape (32–33 Ma) (Leloup et al., 2001) and South China Sea spreading (ca. 30 Ma) (Briais et al., 1993; Encarnacion et al., 2001). Near-coeval formation of the Mindoro and East Taiwan ophiolites, emplaced during the collisions of North Palawan with Mindoro and of the Luzon arc with Eurasia (Rangin et al., 1985) and Mindanao (30–5 Ma) (Sajona et al., 1994), respectively, and those in East Sulawesi (ca. 30 Ma) (Parkinson, 1998), and Antique (Panay) (15–10 Ma) (Tamayo et al., 2001), may also correspond to common regional triggering events, although this is more speculative. A third regional plate kinematic adjustment (ca. 25 Ma) (Hall, 2002) may be

TABLE 1: TETHYAN OPHIOLITE AGES INFERRED FROM RADIOMETRIC AND PALEONTOLOGICAL DATA

Region	Ophiolite body	Time (Ma)
Betic-Calabrian	Betic-Rif	157–145
	Calabria	150–148
	Pyrenees	146–65 ?
Alpine	Balagne Corsica	200–175
	Internal Ligurides	200–165
	Montgenevre Ligurides	198–148
	Eastern Alps	ca. 187
	Ligurides, Corsica	ca. 180
	Queyras, Monte-Viso	175–145
	Gets Nappe	166–165
	Sardinia	ca. 166
Meliata	Meliata-Bodva	160–140
	Meliata-Apuseni	157–140
	Meliata-Pieniny KB	ca. 155
	Meliata-Hallstatt	136–108
Vardar "Inner-Hellenic"	Szarvasko	160–150
	Drina-Ivanjica	157–155
	Kalamaki-Mortero	155–150
	Samothraki, Guevgeli	155–154
	Evros-Lesvos	155–150
	Chalkidiki	150–140
	Makran	156–139
	Masirah	ca. 150
	Zemplen	110–66
Pindos-Mirdita "Outer-Hellenic"	Central Dinarides	190–120
	Albanides	188–168
	Mirdita	175–162
	Pindos	175–155
	Albania	174–160
	Pindos	ca. 165
	Dafnospilia-Kedros	162–153
	Eastern Pontides	160
	Vourinos, Othris	160–145
	Migdhalitsa	160–145
Tauride-Zagros	Zagros	98–75
	Semail	96–85
	Crete	95–75
	Troodos	95–75
	Kutahya	ca. 93
	Mersin	93–92
	Taurides	92–90
	Pozanti-Karsanti	90–94
	Baer-Bassit	90–65
	Central Anatolia	81–65

(*continued*)

TABLE 1: TETHYAN OPHIOLITE AGES INFERRED FROM RADIOMETRIC AND PALEONTOLOGICAL DATA (*continued*)

Region	Ophiolite body	Time (Ma)
Caucasian	Lesser Caucasus	230–225
Ladakh-Tibetan	Ras Koh	ca. 110
	NW Himalaya	ca. 80
	Kohistan	ca. 80
	Muslimbagh	76–65
	Ladakh	ca. 75
	Bela	70–65
	Yarlung-Zangbo	ca. 75
Ailaoshanian	Ailaoshan-Shuangguo Burma-Baoshan	362–328
	Jinshajiang	340–294
	Song Ma	338–290
	Bangxi-Chenxing Hainan	333
Lincangian	Baoshan-Lincang	240
	Yangtze-Lincang	210
Protosundian	Camarines Norte	ca. 140
	Meratus-Bobaris	ca. 140
	E. Sulawesi	140–115
Zambalian	Palau-Kyushu	47–45
	Zambales	50–42
	Angat	50–42
	Halmahera	50–42
Nanhaian	Palawan	ca. 34
	East Taiwan	ca. 34
	Mindoro	ca. 34
Celebian	Mindanao	30–5
	E. Sulawesi	ca. 30
	Antique-Panay	15–10

Note: Source references in Danishwar (2003).

linked to collisions of the New Guinea passive margin with the East Philippines-Halmahera-South Caroline arc system and the Australian (Bird's Head) margin, and a possible combination of factors, including further changes in Pacific Plate motion and arc-continent and arc-arc collisions in the Philippines and Indonesia.

Mesozoic and Paleocene ophiolites in the Philippines that pre-date IBM subduction initiation potentially represent rollback episodes triggered by plate collisions prior to that between India and Asia. For example, the mid to Late Cretaceous Central Cordilleran ophiolite in northern Luzon may share a common provenance with coeval ophiolites in northwestern New Guinea, signalling inception of the (present-day) West Philippine Sea or the Celebes Sea Basin (Leroy et al., 1996). Ophiolites marking distal Philippine Sea Plate boundaries highlight the potential extent and rapidity of collision-induced kinematic changes and the extreme caution required in interpreting the ophiolite record in older orogenic belts.

THE NEO-TETHYAN RECORD: TESTING THE "ROLLBACK" MODEL

The actualistic model developed (Figure 8) is predicated on the forearc analog for proto-ophiolites that is believed to characterize late, pre-collision stages of a Wilson cycle. It is contended that ophiolites represent subduction-resistant forearc assemblages entrapped in orogens following their collisions with other mobile arcs or continental plates. This model is supported by several lines of evidence: (1) the structural, lithologic, and geochemical resemblances of ophiolite to modern forearc and remnant arc terrains; (2) an association of boninitic proto-arcs with fossil transforms and their unique high-temperature character; (3) the evidence for rapid mantle-driven subduction rollback of fragmented successions of proto-arc, arc, and backarc lithosphere; (4) the common presence of relatively unmetamorphosed MORB-like basement (layers 1–4); (5) amphibolite-facies sub-ophiolitic metamorphic soles showing counterclockwise *P-T-*t histories consistent with incipient subduction; (6) age relationships suggesting that ophiolites characterize "late" rather than "early" stages of Wilson cycles (with implications for paleomagnetic plate reconstructions); and (7) the temporal coincidence of ophiolites with local and distal plate kinematic adjustments.

Several additional points may be emphasized. Tethyan ophiolites repeatedly show attributes that are rare or absent from modern spreading centers and "normal" subduction zones, but that are unique to forearcs. Features that preclude the forearc proto-ophiolite analogy have not been reported as yet. The better-preserved ophiolites invariably show developmental trends from mid-ocean ridge- to subduction-related lithologies, evidenced (e.g.) by the intrusion and underplating of lherzolitic mantle and MORB-type plutonics by boninitic and calcalkaline melts and by lava sequences showing the same stratigraphic progression with little or no eruptive break. Relatively untectonized basement sequences show classic ocean ridge attributes such as sheeted dike complexes, *en echelon* gabbroic complexes, and imbricated MORB-like pillow lavas. Below, the model is adopted as a template for interpreting neo-Tethyan ophiolite examples from Liguria, Albania (Mirdita), Cyprus (Troodos), Oman (Semail), and the Philippines (Zambales), pending publication of more detailed, comprehensive evaluations (e.g., Dilek and Flower, 2003). These are briefly reviewed in terms of their lithologic, structural, and geochemical diversity and judged, where possible, according to their internal and external chronological relations.

Liguria

Ophiolites of the Western Alps, Sardinia-Corsica, and the Apennines, that represent the Jurassic-Cretaceous Ligurian ocean, typically comprise small-volume gabbro and basalt sequences with frequent stratigraphic contact of sediments with peridotite or gabbro sequences (Lagabrielle and Lemoine, 1997). Sheeted dike complexes are mostly lacking. Despite the overall MORB-like character of Northern Apennine ophiolites, Os and Nd isotopic

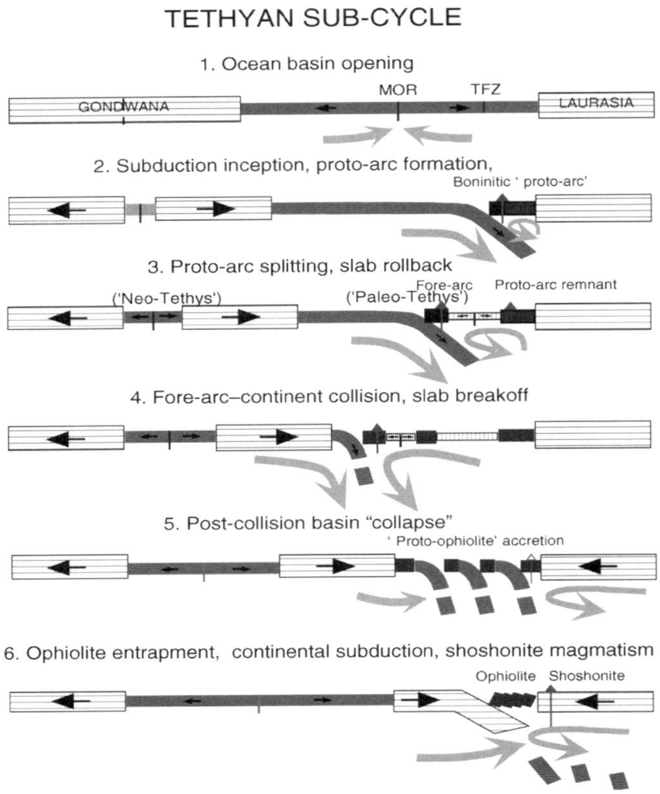

Figure 8. Mantle-driven Tethyan sub-cycle during late (pre-collision) stages of a Wilson cycle (see Fig. 1). (1) Continental plates separate as Paleo-Tethys begins opening; passive asthenosphere upwells beneath spreading axis to produce MORB-like oceanic lithosphere. (2) Laurasia blocks continued Paleotethys opening; new subduction is initiated at weak (e.g., transform) zone, with boninite magmatism forming a proto-arc on the overriding MORB plate, followed by "normal" calcalkaline arc volcanism; neo-Tethyan rifting is initiated in Gondwana. (3) "Cimmeria" microcontinent detaches from Gondwana as Palo-Tethys continues subducting; MORB-like backarc extension and arc-forearc rollback occur in response to the compression of Tethyan asthenosphere beneath Laurasia and Cimmeria. (4) Continued "Cimmerian" microcontinent migration leads to oblique collision and diachronous breakoff of the paleo-Tethyan ocean slab, accompanied by enhanced continental sediment subduction; deflected asthenosphere flow field leads to shoshonite and potassic granite magmatism derived from continent-contaminated asthenosphere. (5) Neo-Tethys continues to open until relict MORB-like backarc lithosphere is completely subducted and progressively detached from the overriding continental plate. (6) Neo-Tethys ceases opening as paleo-Tethyan basin collapse is completed, and new subduction is initiated, mélange of relict arc-forearc-backarc (ophiolite) is entrapped in the ensuing continent-continent orogeny.

compositions of the Liguride peridotites indicate significantly greater depletions than those at modern ridge systems (Rampone et al., 1996; Snow et al., 1996, 2000), while clinopyroxene and whole-rock Sr and Zr depletions relative to REE (Rampone et al., 1996) and a positive Os-Nd isotopic correlation are consistent with the addition of crustal components (Snow et al., 2000). These features led Rampone et al. (1998) and Rampone and Piccardo (2000) to conclude that the Ligurian peridotites represent relict continental lithospheric mantle, attributing its presence to the effects of continental breakup.

An alternative possibility is that, rather than representing a Red Sea–type rifted basin, the presence of refertilized, refractory peridotite may be a sign of incipient subduction that preceded subduction-related proto-arc activity. This interpretation is consistent with the unusually high equilibration temperatures (1150–1250 °C) interpreted for the Liguride peridotites and subsolidus re-equilibration histories that indicate high temperature–low pressure retrograde metamorphism prior their final emplacement (Rampone et al., 1996). It is also supported by the evidence for Jurassic intra-oceanic subduction in Corsican ophiolites and for subsequent (mid-Cretaceous) subduction of the European continental margin beneath the Adria plate (Malavieille et al., 1998). Eocene closure of the remnant Ligurian Basin that separated proto-Corsica from Adria led to a second orogenic phase and the entrapment of unmetamorphosed ophiolites over the previously exhumed (Corsican) high-pressure belt. Malavieille et al. (1998) suggested that this may be a typical effect where a continental collision is preceded by intraoceanic subduction.

Other Liguride remnants, such as the Montgenevre ophiolite in the Western Alps, were relatively unaffected by the Alpine orogeny, and preserve evidence of pre-collision intra-oceanic deformation. MORB-like gabbros yield isochron ages of 198 ± 22 Ma, older than those recorded for other Western Alpine ophiolites (ca. 165 Ma) (Costa and Caby, 2001), but close to 180 Ma ages of eclogitized gabbros from the Ligurides and Corsica. Zircon ages of 156 ± 3 and 148 ± 2 Ma for calcalkaline diorites and plagiogranites that intrude gabbros and peridotites (Costa and Caby, 2001) correlate well with stratigraphic ages (160–140 Ma) of the overlying radiolarian sediments, suggesting a life span for Ligurian ophiolites of up to ca. 30 Ma

Albania (Mirdita)

Mid to Late Jurassic ophiolites in Albania comprise subparallel eastern and western belts overthrust onto the Paleozoic Korabi (Pelagonian) block during Africa-Eurasian plate convergence (Shallo et al., 1987; Shallo, 1992; Hoxha and Boullier, 1995a, 1995b; Nicolas et al., 1999). These ophiolites probably developed within the neo-Tethyan Pindos-Mirdita marginal basin, which propagated northward as the Korabi microcontinent separated from Apulia, eventually becoming trapped by one of several diachronous arc-continent collisions (Robertson and Shallo, 2000).

The western Albanian ophiolites show fertile lherzolitic mantle sections overlain by MORB-like plutonics, sheeted dikes, pillow lavas, and minor plagiogranites (Cortesogno et al., 1998) and underlain by an amphibolite-facies metamorphic sole (Bebien et al., 2000; Dimo-Lahitte et al., 2001). These lithologies resemble those of the MORB-like Alpine-Liguride ophiolites and the Pindos ophiolite to the south, and clearly suggest a mid-ocean ridge provenance (Beccaluva et al., 1994a; Nicolas et al., 1999). In contrast, the "eastern" ophiolites are dominantly calcalkaline in character (e.g., Shebenik, Kukes) and include ultramafic boninites and refractory harzburgitic mantle sections (Hoxha et al., 1993a; Manika et al., 1997) resembling those of the Vourinos and Eastern Othris complexes to the south (Beccaluva et al., 1994b; Bebien et al., 1997); however, the commonly-observed intrusions of calcalkaline and boninite dikes into MORB-like (plagioclase-bearing) plutonic lithologies suggest that both ophiolite types in Albania are closely related (Cortesogno et al., 1998) and reflect the initiation of west-vergent subduction in oceanic lithosphere formed at a recently-active spreading system (e.g., Bebien et al., 1998).

This view is reinforced by the MORB-like character of coeval metamorphic soles (Gjata et al., 1992) which young northward from ca. 174 to 160 Ma (Dimo-Lahitte et al., 2001) and record rapid counterclockwise *P-T*-t paths (Gjata et al., 1992; Dimo-Lahitte et al., 2001; Bebien, et al., 2000). Intra-oceanic detachment and ophiolite emplacement were clearly near-contemporaneous with preceding or ongoing seafloor spreading. A fossil transform related to, or at least similar to, the northern Albanian Scutari-Pec lineament, is indicated by anisotropic mantle fabrics in the northern Kukes section, implying that the latter represents the eastern limb of a NNW-SSE fossil spreading axis (Hoxha et al., 1993a, 1993b; Hoxha and Boullier, 1995a, 1995b). Irrespective of whether Pindos-Mirdita was an intra-continental or intra-oceanic basin, there is little doubt that oceanic spreading, unaffected by subduction, characterized its initial development (Beccaluva, 1994; Nicolas et al., 1999). If seafloor spreading was interrupted by newly initiated subduction at a ridge transform, it would appear that the collision of a boninitic proto-arc (or retreating arc-forearc complex) with Pelagonia occurred shortly thereafter. Uplift, fragmentation, and partial burial of the proto-ophiolitic forearc beneath flysch (Shallo, 1992) would have been succeeded in the Eocene-Oligocene by collision between Korabi-Pelagonia and Apulia, at which point the ophiolite was thrust over the Korabi margin.

If this interpretation is correct, the MORB-like "western" ophiolites represent remnants of Pindos-Mirdita oceanic basement, whereas the inception of west-vergent subduction and ensuing development of the near-contemporaneous (162–174 Ma) is recorded by "eastern" ophiolites. The latter may be interpreted as relict proto-arc fragments formed at or close to the Pindos-Mirdita spreading axis, where subduction was nucleated at an active transform fracture zone, possibly represented by the present-day Scutari-Pec lineament (Bebien et al., 2000, Dimo-Lahitte et al., 2001). Their eventual collision with Pelagonia can be ascribed to the collapse (closure) of Pindos-Mirdita, possibly following a sequence of proto-arc splitting and arc-forearc rollback. Subsequent formation of the Periadriatic "foredeep" and Tirana Depression in the Miocene and Plio-Quaternary occurred as the continued convergence of Africa and Eurasia was accommodated by southward rollback of the Hellenic margin (Robertson et al., 1995).

Cyprus (Troodos, Mamonia)

To the south and east, Late Cretaceous ophiolites associated with the Pontide, Anatolide, and Tauride orogenies (Table 1) suggest that analogous conditions prevailed in other neo-Tethyan basins. Of these, the Troodos (Cyprus) and Semail (Oman) ophiolites are the most extensively studied. The Troodos Massif comprises a lithologic "sandwich" of variably refractory peridotite, layered, and massive plutonics, sheeted dikes, and pillow lavas with a pelagic sedimentary carapace (Moores and Vine, 1971; Robinson et al., 1983, 1984; Malpas and Robinson, 1997; Robertson, 1991; Eaton and Robertson, 1993). Age data for plagiogranites (92–90 Ma) (Mukasa and Ludden, 1987) suggest that the ophiolite post-dates the Late Jurassic initiation of Cyprus Basin opening (Garfunkel, 1998), whereas the succession of MORB-like, boninitic, and calcalkaline lithologies is strongly indicative of a forearc provenance. There is no metamorphic sole exposed and the ophiolite is underlain by continental crust (Makris et al., 1983).

Batanova and Sobolev (2000) recognize two distinct mantle sequences, consisting of spinel lherzolite (veined by dunite and clinopyroxene-bearing harzburgite) and refractory, clinopyroxene-poor harzburgite, respectively. The composition of spinel lherzolites suggest they are residual to MORB melt extraction and, while not parental to calcalkaline or boninitic magmas, were probably modified by the latter, as evidenced by veins of dunite and harzburgite (Batanova and Sobolev, 2000). Structural relationships of layered cumulates, massive gabbros, and sheeted dikes (Malpas and Robinson, 1997) suggest that seafloor spreading was largely steady-state (Allerton and Vine, 1987), with periodic shifts in the axial rift (Varga and Moores, 1985). "Lower" and "Upper" pillow lava series record a progression from dominantly MORB-like to calcalkaline and high-Ca boninitic character (Robinson et al., 1983; Schmincke et al., 1983). Hydrothermal deposits include massive sulphides and associated stockworks, often located at the overlying lava-sediment interface (Robertson and Boyle, 1983). Relict fragments of the Cyprus Basin are preserved in the Kyrenia Range in northern Cyprus, bounded on the north by a carbonate platform and on the west and southwest by tectonized MORB-like crust and pelagic sediment, part of the present-day Mamonia Complex (Robertson, 1991; Eaton and Robertson, 1993). The latter was interpreted as a deformed passive margin sequence sutured to Troodos during the Late Cretaceous (Bailey et al., 2000). Detailed mapping and microstructure studies in Mamonia suggest that an episode of pro-grade metamorphism (90–88 Ma) was followed by coeval retrograde hydration and dextral transtension, and serpentinite diapirism (83–73 Ma), prior to its exhumation at ca. 65 Ma (Malpas et al., 1992a).

Evidence for subduction nucleation in southern and southeastern Cyprus includes the presence of accretionary wedge lithologies in the Limassol Forest area (Eaton and Robertson, 1993), and high-Ca boninites associated with the Arakapas fault, a possible fossil transform (e.g., Simonian and Gass, 1977; Flower and Levine, 1987; Rogers et al., 1989; MacLeod and Murton,

1995). Similarities between extensional detachment faults in the Limassol Forest area, immediately south of the Arakapas Fault, and those in the "inside corners" of modern ocean-ridge-transform intersections appear to confirm this interpretation (Cann et al., 2001). Accordingly, the Limassol Forest block would have been formed in the western "inside corner" of the ridge-transform intersect, having spread eastward until it passed a second spreading axis. At this point magmatic crust north of the Arakapas Fault would have become welded to the Limassol Forest block as the fault evolved as a transform fracture zone. While this interpretation is supported by dike orientations by and magmatic and mantle flow fields inferred from rock fabric studies (Abelson et al., 2001, Borradaile and Lagroix, 2001), the evidence of boninite magmatic activity along the Arakapas Fault appears to confirm that a subduction nucleation event was in progress.

Formation of the Troodos ophiolite can be linked to opening of the Cyprus Basin, which began in the Mid to Late Jurassic. Despite continuing convergence of Africa and Eurasia, seafloor spreading in the Cyprus Basin continued into the Late Cretaceous, at which time (92–90 Ma) the Troodos proto-arc was formed in response, presumably, to subduction nucleation at a near-axial transform fracture zone. Proto-arc splitting and rollback of its active portion led to collision of the overriding plate with the Arabian platform at 74–70 Ma (Robertson, 2002) since when, it appears, north-vergent subduction migrated from the evolving proto-ophiolite belt to the northern edge of the basin as the Cyprus Basin began to collapse (Yilmaz, 1993). Meanwhile, the westward escape of Anatolia following the Mid-Cenozoic collision of Arabia with the Cimmerian continent (Sengör et al., 1984) and kinematic constraints imposed by opening of the Aegean Basin (Le Pichon et al., 1981, 2002; Eaton and Robertson, 1993) precluded further collapse of the basin. Such a scenario explains several hitherto-puzzling features—preservation of the Cyprus Basin, the absence of regional emplacement-related deformation in its ophiolitic forearc (Troodos-Mamonia), and the lack of post-Cretaceous magmatism at the present-day plate boundary (Moores and Vine, 1971; Malpas and Robinson, 1997; Robertson, 1991; Eaton and Robertson, 1993).

Oman (Semail)

The neo-Tethyan Semail ophiolite is the largest in the world and probably the best subaerial analog for oceanic lithosphere. There is broad agreement that the petrologic and structural attributes of the Semail ophiolite resemble those of lithosphere formed at a fast-spreading ocean ridge (Pallister and Hopson, 1981; Pallister and Knight, 1981; Pearce et al., 1981, Alabaster et al., 1982; Pallister, 1984; Umino et al., 1990; Nicolas et al., 1994, 1996; Nicolas and Boudier, 1995; Umino, 1995), as exemplified by MORB-like compositions of the Geotimes Volcanic Unit (Pearce et al., 1981; Alabaster et al., 1982; Ernewein et al., 1988; Umino et al., 1990), sheeted dikes (Nicolas and Boudier, 1991), and layered plagioclase-rich gabbros (Pallister and Knight, 1981; Browning, 1984; Benn et al., 1988; Ernewein et al., 1988; Juteau et al., 1988a, b; Nicolas et al., 1988a, b). Jousselin and Mainprice (1998) and Jousselin et al. (1998) corroborated this interpretation with structural data demonstrating the presence of a fossil subaxial mantle diaper. However, further investigation revealed an incipient second developmental stage, evidenced by the intrusion of ultramafic and calcalkaline magmas into MORB-like basement plutonics (Benn et al., 1988; Ernewein et al., 1988; Juteau et al., 1988a, 1988b), the eruption of boninites and calcalkaline tholeiites (Alabaster et al., 1982; Ernewein et al., 1988; Umino et al., 1990; Ishikawa et al., 2002), and subsequent intrusion of calcalkaline plutons, as in the Haylayn Massif and Sumail nappe regions (Pallister and Hopson, 1981; Pallister and Knight, 1981; Benn et al., 1988; Juteau et al., 1988a, b; Lachize et al., 1996; Schiano et al., 1997). The intrusives are mostly (although not always) discordant with older MORB-like layered rocks (Benn et al., 1988; Juteau et al., 1988a, 1988b; Umino, 1995) and comprise a suite of dunite, wehrlite, olivine gabbro, and diorite cumulates, and a series of lherzolite, gabbro-norite, two-pyroxene diorite, and trondhjemite (Umino et al., 1990). Previously interpreted as remobilized melt-cumulate emulsions (Pallister and Knight, 1981), these lithologies appear to reflect boninitic parent magmas (Umino et al., 1990; Schiano et al., 1997; Ishikawa et al., 2002). Evolved calcalkaline rocks also dominate in the succeeding Alley Volcanic Unit (Pearce et al., 1981; Alabaster et al., 1982; Ernewein et al., 1988) complemented by large (up to 10 km^2) intrusions of gabbro, diorite, and trondhjemite (Umino et al., 1990) genetically related to the ultramafic intrusives (Lachize et al., 1996; Schiano et al., 1997). Boninites also appear as lavas and dikes in the Alley volcanic sequence (Umino et al., 1990; Ishikawa et al., 2002), unconformably overlying, and in some cases cross-cutting, the Geotimes unit. The most recent Semail activity produced alkali basalts between ca. 85 and 90 Ma (Lachize et al., 1996).

As elsewhere, the Semail metamorphic sole is MORB-like in character (Searle and Malpas, 1980, 1982; Searle and Cox, 1999) and post-dates the ophiolite by <4 m.y. (Hacker and Gnos, 1997). Its interpolated *P-T*-t path shows peak temperatures of 775–875 °C and pressures of ca. 1.1 GPa, implying a rapid, albeit short-lived, subduction history. To explain these conditions, Boudier et al. (1998) suggested that subduction triggered within ca. 5 m.y. of ridge accretion was a counterpart to the obduction of oceanic lithosphere detached from the dying spreading axis. *P-T*-t paths for sub-ophiolitic continental rocks suggest that the Arabian margin, a potential source for two-mica, garnet-, tourmaline-, cordierite- and andalusite-bearing granites was also subducted to >70 km depth (Searle and Cox, 1999). In summary, it is abundantly clear that the bulk of the Semail ophiolite represents oceanic lithosphere, as contended by Nicolas and coworkers. It seems reasonable to conclude that subduction was initiated at 95–93.8 Ma, as indicated by the appearance of boninite volcanism (Umino et al., 1990; Umino, 1995; Kinoshita et al., 2002) and plagiogranite (adakite) intrusions (Pallister and Knight, 1981; Benn et al., 1988; Searle and Cox, 1999) following the cessation of active spreading. It is unclear whether proto-arc formation led, in turn, to an episode of subduction rollback; however, there is little doubt that calcalkaline magmatism that produced the

Lasail arc edifice ceased at ca. 87 Ma, following collision between the evolving arc-forearc complex and the Arabian continent. This was followed at ca. 85 Ma by subduction (to depths of 78–90 km) of Arabian continental crust, and then by partial melting of subducted, eclogite-facies crust, the latter as evidenced by intrusions of two-mica, cordierite-bearing granites in mantle harzburgite and cumulate sequences between 85 and 76 Ma (Searle and Cox, 1999).

Philippines (Zambales-Angat, Palawan)

The Zambales complex is one of several mid-Eocene ophiolites that rim the Philippine Sea Plate, and comprises several constituent parts (Encarnacion et al., 1993a, 1993b; Encarnacion and Fernandez, 2001). To the north, the Acoje block comprises a boninitic to calcalkaline proto-arc sequence that is bordered to the south by the "transitional MORB"-like Coto block. Farther south, the San Antonio arc terrane is separated from the Coto block by the left-lateral West Luzon shear zone (Evans and Hawkins, 1989; Hawkins 1989, 1993; Evans et al., 1991; Yumul et al., 1998a, 2000; Yumul, 2001; Encarnacion et al., 1999), apparently translated southward from the northern part of the complex (Encarnacion et al., 1993a). The Acoje and Coto blocks (44–45 Ma) appear to be near-coeval with the Angat ophiolite (ca. 48 Ma), 100 km to the east in the southern Sierra Madre, suggesting they shared a common mid-Eocene provenance (Encarnacion et al., 1993a). The recognition of older, Late Cretaceous ophiolitic lithologies southeast of the main Angat body (Encarnacion et al., 1993a) suggests that the Zambales-Angat ophiolites record opening of a backarc basin following the buildup of an Eocene arc that developed on pre-existing Cretaceous basement, and, moreover, that both ophiolites represent autochthonous Luzon basement (Geary et al., 1989; Encarnacion et al., 1993a; Yumul et al., 1998b). Interestingly, the somewhat different petrologic structure of the younger (ca. 34 Ma) ophiolite on the island of Palawan implies a similar pattern. Despite its MORB-like character, upper sequence lavas show incipient calcalkaline affinity, resembling those of peripheral parts of the Mariana and Lau Basins (Encarnacion et al., 1995). The concurrence of ophiolite and metamorphic sole ages (Encarnacion et al., 2001) supports their formation close to a spreading axis while the geochemical evidence for subduction inception is matched by the evidence for subsequent arc splitting and subduction rollback associated with development of the Cagayan-Sulu backarc system (Pubellier et al., 1991; Rangin, 1991; Serri et al., 1991; Spadea et al., 1991).

Similar petrologic-stratigraphic patterns are observed in ophiolites associated with Meso- and Paleo-Tethyan sutures in Central Eurasia, Iran, and Anatolia, and additionally, the Caribbean and Scotia Seas, and many of the older Phanerozoic cordilleras. Significantly, boninitic proto-arc sequences developed on MORB-like basement are recorded from all major Archean greenstone belts (e.g., Bateman and Costa, 2001; Polat and Kerrich, 2001; Parman and Grove, 2001; Puchtel and Brugmann, 2001), reinforcing validity of the subduction initiation-rollback model and the uniformitarian precept on which it is based.

HISTORICAL CONTINGENCY IN PERSPECTIVE

The contention of Moores et al. (2000) that ocean ridge source "signatures" are contingent on mantle convection and plate migration histories is undeniable and, with other factors, requires careful consideration in uniformitarian interpretations; however, despite the recognition of mantle isotopic heterogeneity as convincing evidence for long-term domain mixing, their conclusion that some ophiolites, at least, may be generated at mid-ocean ridges rests on two exceptional present-day cases—the Woodlark Basin, where spreading ridge magmas tapping subduction-contaminated mantle appear to have followed a change in subduction vergence (Perfit et al., 1987; Crawford, 2002, personal commun.), and the South Chile Rise, where a similar effect is produced by active ridge axis subduction (Sturm et al., 2000). It is reasonable to ask whether these and other contingent attributes are: (1) consistent with, even if not currently observed in, present-day proto-ophiolites, and (2) intrinsic to ophiolites through the known geological record. In fact, both of these examples are rare, and at least three obstacles appear to preclude their inclusion in a generic model for ophiolites.

It is hard to reconcile refractory mantle sources capable of yielding calcalkaline and boninitic magmas with mid-ocean ridge spreading systems. At the present time, elevated mantle potential temperatures needed to produce such magmas from their requisite refractory sources (as deduced from experimental studies) are confined to the loci of inferred "plume-ridge" interactions (McKenzie and Bickle, 1988) characterized largely by "fertile" OIB-like mantle source products. The apparently unique association of boninite and high-Mg andesite with initial stages of arc-trench rollback is consistent with the notion that such magmas are incompatible with mid-ocean ridge settings. It is, therefore, reasonable to suggest that while potentially significant, the idea of unpredictable contingencies in earth history need not preclude a uniformitarian approach to studying ophiolites. It is noted, finally, that the continued study of features such as the Woodlark and Mariana backarc spreading centers is likely to yield vital insights related to the development of actualistic subduction rollback models as a basis for interpreting the Tethyan record.

CONCLUSIONS

Geodynamic and petrologic studies in tectonically active sectors of the Tethyan belt suggest that unique combinations of features, rare or absent from ancient and modern mid-ocean ridges or subduction zones, have repeatedly characterized ophiolites for >2.5 b.y. These include basement sequences with typical oceanic crustal features, near-coeval MORB-like, high-temperature metamorphic soles often showing inflected, counterclockwise *P-T*-t paths; boninitic proto-arcs succeeded by calcalkaline and MORB-like eruptive sequences; refractory peridotite; and the associated effects of serpentinitic mud diapirism and high-temperature "epidosites."

Taken together, these features preclude exclusively mid-ocean ridge, "normal" arc, or backarc basin provenance for ophiolites, whereas they uniquely match those of proto-ophiolites in modern forearc settings. These appear to evolve in response to subduction nucleation—often at mid-ocean ridge transforms—signaled by boninitic and related magmatism and high-temperature hydrothermal activity, and to evolve in response to one or more episodes of arc splitting and basin opening.

Subduction nucleation may be triggered by thermal or compositional heterogeneities in the lithosphere or, more plausibly, by temperature differences across asthenospheric domain boundaries. On the other hand, subduction rollback appears to reflect sub-horizontal mantle flow, whether directly in response to collision-induced extrusion, mantle plume upwelling, or accelerated differential plate motion. In most cases rollback episodes are terminated by forearc-forearc and forearc-continent collisions ~10–30 m.y. after their commencement.

If a rollback episode successfully evades orogenic entrapment, proto-ophiolitic forearc accretion may continue indefinitely. On the other hand, if entrapped by a collision following one or more episodes of basin collapse, forearcs tend to resist subduction (in contrast to backarc basin lithosphere) and are readily preserved as ophiolites. It is, therefore, not surprising that the obduction of oceanic crust is absent from modern passive margins.

Ophiolite genesis occurs during "sub-cycles" of basin opening and arc-trench rollback at closing stages of a Wilson cycle. Modern subduction initiation precedes or is near-coeval with "hard" plate collisions and their respective plate kinematic responses. Similar relationships are discernible in Tethyan (and prior) geological record, where temporal correlations may be discerned between ophiolites and distal plate kinematic changes. If—as seems to be the case—traditional ocean ridge and supra-subduction models are only "correct" when considered together, then the ophiolite conundrum is a false dichotomy.

Historical contingency has undeniable potential as an aid to interpreting earth history; however, the Moores et al. (2001) model for ophiolite genesis is in conflict with both *a priori* petrologic constraints and the ophiolite record through geological time. Refractory mantle sources of the type yielding calcalkaline and boninitic magmas at newly-forming convergent plate margins are unlikely to have been generic to mid-ocean ridge settings. Moreover, such near-ubiquitous ophiolite components appear to be exclusive to subduction nucleation stages of rollback cycles. The model also fails because it requires us to believe that only ocean crust formed over previously modified mantle is eligible for emplacement, whether by obduction or other process, while "normal" MORB lithosphere is emplaced in accretionary complexes.

After more than 30 years, the ophiolite conundrum provides a sobering example of how ambient context (the late 1960s paradigm shift of advent of plate tectonic theory) of an observation (the enigma of ophiolites) may determine the rationale for its interpretation (as a model for seafloor spreading, "linchpin" component of plate tectonics). If, as is suggested in this paper, the conundrum is a false dichotomy—if ophiolites intrinsically are *in part* formed at spreading axes and *in part* subduction-related—then we may tentatively conclude that the debate has been fruitful.

ACKNOWLEDGMENTS

I am grateful for enlightening discussions with Deng Jinfu, Yildirim Dilek, John Encarnacion, Mohamet Ghazi, Eldridge Moores, Paul Robinson, Ray Russo, Victor Mocanu, John Wakabayashi, and Guram Zakariadze during preparation of this paper, while taking full responsibility for the views expressed. Tony Crawford and Mike Perfit are especially thanked for their thorough and constructive reviews. Support from the U.S. National Science Foundation (grants INT 0002203, IN T0129492, and EAR 9730013), United Nations Educational, Scientific and Cultural Organization (UNESCO) Earth Sciences Division, the International Union of Geological Sciences, and North Atlantic Treaty Organization (NATO) is gratefully acknowledged.

REFERENCES CITED

Abbott, L.D., Silver, E.A., and Galewsky, J., 1994, Structural evolution of a modern arc-continent collision in Papua New Guinea: Tectonics, v. 13, p. 1007–1034.

Abelson, M., Baer, G, and Agnon, A., 2001, Evidence from gabbro of the Troodos Ophiolite for lateral magma transport along a slow-spreading mid-ocean ridge: Nature, v. 409, p. 72–75.

Alabaster, T., Pearce, J.A., and Malpas, J., 1982, The volcanic stratigraphy and petrogenesis of the Oman ophiolite complex: Contributions to Mineralogy and Petrology, v. 81, p. 168–183.

Allerton, S., and Vine, F.J., 1987, Spreading structure of the Troodos Ophiolite, Cyprus—Some paleomagnetic constraints: Geology, v. 15, p. 593–597.

Alvarez, W., 1990, Geologic evidence for the plate-driving mechanism; the continental undertow hypothesis and the Australian-Antarctic Discordance: Tectonics, v. 9, p. 1213–1220.

Audley-Charles, M.G., Harris, R.A., and Dewey, J.F., 1990, Allochthonous terranes of the Southwest Pacific and Indonesia: Philosophical Transactions, of the Royal Society of London, ser. A, Mathematical and Physical Sciences, v. 331, p. 571–587.

Bailey, W.R., Holdsworth, R.E., and Swarbrick, R.E., 2000, Kinematic history of a reactivated oceanic suture; the Mamonia Complex suture zone, SW Cyprus: Journal of the Geological Society of London, v. 157, p. 1107–1126.

Ballance, P.F., Ablaev, A.G., Pushchin, I.K., Pletnev, S.P., Birylina, M.G., Itaya, T., Follas, H.A., Gibson, G.W., 1999, Morphology and history of the Kermadec trench-arc-backarc basin-remnant arc system at 30 to 32°S: geophysical profile, microfossil and K-Ar data: Marine Geology, v. 159, p. 35–62.

Ballantyne, P., 1991, Petrological constraints upon the provenance and genesis of the East Halmahera Ophiolite, Orogenesis in action; tectonics and processes at the West Equatorial Pacific margin: Journal of SE Asian Earth Science, v. 6, p. 259–269.

Ballantyne, P.D., and Hall, R., 1990, The petrology of the Halmahera Ophiolite, Indonesia; an early Tertiary forearc, *in* Malpas, J., Moores, E.M., Panayiotou, A., and Xenophontos, C., eds., Ophiolites. Oceanic Crustal Analogues. Proceedings of the Symposium "Troodos 1987:" Nicosia, Cyprus, Geological Survey Department, p. 461–475.

Ballantyne, P., and Parson, L.M., 1992, Petrology and geochemistry of the plutonic rocks of the Halmahera ophiolite, eastern Indonesia; An analogue of modern oceanic forearcs, *in* Murton, B.J., and Browning, P., eds., Ophiolites and their Modern Oceanic Analogues: London, Geological Society of London Special Publication 60, p. 179–202.

Banerjee, N.R., and Gillis, K.M., 2001, Hydrothermal alteration in a modern suprasubduction zone: The Tonga forearc crust: Journal of Geophysical Research, B-106, p. 21,737–21,750.

Banerjee, N.R., Gillis, K.M., and Muehlenbachs, K., 2000, Discovery of epidosites in a modern oceanic setting, the Tonga forearc: Geology, v. 28, p. 151–154.

Barringer, J.L.H., 1983, A boninitic series in the eastern Maryland Piedmont: Eos, Transactions of the American Geophysical Union, v. 64, p. 335.

Batanova, V.G., and Sobolev, A.V., 2000, Compositional heterogeneity in subduction-related mantle peridotites, Troodos massif, Cyprus: Geology, v. 28, p. 55–58.

Bateman, R., Costa, S., 2001, Archaean mafic magmatism in the Kalgoorlie area of the Yilgarn Craton, Western Australia; a geochemical and Nd isotopic study of the petrogenetic and tectonic evolution of a greenstone belt: Precambrian Research, v. 108, p. 75–112.

Bebien, J., Dautaj, N., Shallo, M., Turku, I., and Barbarin, B., 1997, Diversity of ophiolitic plagiogranites: the Albanian example: Comptes Rendus de l'Academie de Sciences, ser. 2, v. 324, p. 875–882.

Bebien, J., Shallo, M., Manika, K., and Gega, D., 1998, The Shebenik massif (Albania): a link between MOR- and SSZ-type ophiolites?: Ofioliti, v. 23, p. 7–15.

Bebien, J., Dimo-Lahitte, A., Vergely, P., Insergueix-Filippi, D., and Dupeyrat, L., 2000, Albanian ophiolites. I–Magmatic and metamorphic processes associated with the initiation of subduction: Ofioliti, v. 25, p. 39–45.

Beccaluva, L., 1994, Albania Ophiolites; State of the Art and Perspectives: Ofioliti, v. 19, p. 176.

Beccaluva, L., Macciotta, G., and Spadea, P., 1982, Petrology and geodynamic significance of the Calabria-Lucania ophiolites, Atti/Convegno della Societa Italiana di Mineralogia e Petrologia. Meeting of the Italian Society of Mineralogy and Petrology: Rendiconti della Societa Italiana di Mineralogia e Petrologia, v. 38, p. 973–987.

Beccaluva, L., Coltorti, M., Premti, I., Saccani, E., Siena, F., and Zeda, O., 1994a, Mid-ocean ridge and suprasubduction affinities in the ophiolitic belts of Albania: Ofioliti, v. 19, p. 77–96.

Beccaluva, L., Coltorti, M., Dostal, J., Saccani, E., Serri, G., Siena, F., and Vaccaro, C., 1994b, High-Ti and low-Ti magmatic affinities from Tethyan ophiolites, *in* Proceedings of the 76th summer meeting of the Societa Geologica Italiana; The Northern Apennines; Part 1, The Tuscan Nappe, the ophiolitic sequences, the turbiditic successions: Memorie della Societa Geologica Italiana 48, part 1, p. 149–150.

Bechennec, F., Le Metour, J., Rabu, D., Villey, M., and Beurrier, M., 1988, The Hawasina Basin; a fragment of a starved passive continental margin, thrust over the Arabian Platform during obduction of the Sumail Nappe: Tectonophysics, v. 151, p. 323–343.

Bednarz, U., and Schmincke, H.-U., 1994, Petrological and chemical evolution of the northeastern Troodos Extrusive Series, Cyprus: Journal of Petrology, v. 35, p. 489–523.

Benn, K., Nicolas, A., and Reuber, I., 1988, Mantle-crust transition zone and origin of wehrlitic magmas; evidence from the Oman Ophiolite: Tectonophysics, v. 151, p. 75–85.

Benoit, M., Ceuleneer, G., and Polvé, M., 1999, The remelting of hydrothermally altered peridotite at mid-ocean ridges by intruding mantle diapirs: Nature, v. 402, p. 514–518.

Bill, M., O'Dogherty, L., Guex, J., Baumgartner, P.O., and Masson, H., 2001, Radiolarite ages in Alpine-Mediterranean ophiolites; constraints on the oceanic spreading and the Tethys-Atlantic connection: Geological Society of America Bulletin, v. 113, p. 129–143.

Bloomer, S.H., Taylor, B., MacLeod, C.J., Stern, R.J., Fryer, P., Hawkins, J.W., and Johnson, L., 1995, Early arc volcanism and the ophiolite problem; a perspective from drilling in the western Pacific, *in* Taylor, B., and Natland, J., eds., Active Margins and Marginal Basins of the Western Pacific: Washington D.C., American Geophysical Union Geophysical Monograph, 88, p. 1–30.

Borradaile, G.J., and Lagroix, F., 2001, Magnetic fabrics reveal upper mantle flow fabrics in the Troodos ophiolite complex, Cyprus: Journal of Structural Geology, v. 23, p. 1299–1317.

Boudier, F., Ceuleneer, G., and Nicolas, A., 1988, Shear zones, thrusts and related magmatism in the Oman Ophiolite; initiation of thrusting on an oceanic ridge: Tectonophysics, v. 151, p. 275–296.

Briais, A., Patriat, P., and Tapponnier, P., 1993, Updated interpretation of magnetic anomalies and seafloor spreading stages in the South China Sea; implications for the Tertiary tectonics of Southeast Asia: Journal of Geophysical Research-B, v. 98, p. 6299–6328.

Browning, P., 1984, Cryptic variation within the Cumulate Sequence of the Oman ophiolite; magma chamber depth and petrological implications, *in* Gass, I.G., Lippard, S.J., and Shelton, A.W., eds., Ophiolites and Oceanic Lithosphere: Geological Society of London Special Publication 13, p. 71–82.

Cameron, W.E., Nisbet, E.G., and Dietrich, V.J., 1979, Boninites, komatiites and ophiolitic basalts: Nature, v. 280, p. 550–553.

Cann, J.R., Prichard, H.M., Malpas, J.G., and Xenophontos, C., 2001, Oceanic inside corner detachments of the Limassol Forest area, Troodos Ophiolite, Cyprus: Journal of the Geological Society of London, v. 158, p. 757–767.

Capedri, S., Venturelli, G., Bebien, J., and Toscani, L., 1981, Low- and high-Ti ophiolites in northern Pindos; petrological and geological constraints: Bulletin of Volcanology, v. 44, p. 439–449.

Casey, J.F., and Dewey, J.F., 1984, Initiation of subduction zones along transform and accreting plate boundaries, triple-junction evolution, and forearc spreading centres; implications for ophiolitic geology and obduction, *in* Gass, I.G., Lippard, S.J., and Shelton, A.W., eds., Ophiolites and Oceanic Lithosphere: Geological Society of London Special Publication 13, p. 269–290.

Charlton, T.R., 2000, Tertiary evolution of the Eastern Indonesia Collision Complex: Journal of Asian Earth Sciences, v. 18, p. 603–631.

Charlton, T.R., Hall, R., and Partoyo, E., 1991a, The geology and tectonic evolution of Waigeo Island, NE Indonesia, Orogenesis in action; tectonics and processes at the West Equatorial Pacific margin: Journal of SE Asian Earth Science, v. 6, p. 289–297.

Charlton, T.R., Hall, R., Partoyo, E., and Barber, A.J., 1991b, The geology and tectonic evolution of Waigeo Island, NE Indonesia, Orogenesis in action; tectonics and processes at the West Equatorial Pacific margin: orogenesis in action; tectonics and processes at the West Equatorial Pacific margin: Journal of Southeast Asian Earth Sciences, v. 6, p. 289–297.

Chaumillon, E., and Mascle, J., 1997, From foreland to forearc domains: New multichannel seismic reflection survey of the Mediterranean ridge accretionary complex (Eastern Mediterranean): Marine Geology, v. 138, p. 237–259.

Christie, D.M., West, B.P., Pyle, D.G., and Hanan, B.B., 1998, Chaotic topography, mantle flow and mantle migration in the Australian-Antarctic discordance: Nature, v. 394, p. 637–644.

Chung, S.-L., Yang, T.F., Lee, C.-y., Chen, C.-H, 1995, The igneous provinciality in Taiwan; consequence of continental rifting superimposed by Luzon and Ryukyu subduction systems: Journal of SE Asian Earth Science, v. 11, p. 73–80.

Clift, P.D., and Dixon, J.E., 1998, Jurassic ridge collapse, subduction initiation and ophiolite obduction in the southern Greek Tethys: Eclogae Geologicae Helvetiae, v. 91, p. 123–138.

Cloos, M., McMahon, T.P., Quarles van Ufford, A., Sapiie, B., Warren, P.Q., and Weiland, R.J., 1994, Collisional delamination in New Guinea: Geological Society of America Abstracts with Programs, v. 26, p. 502.

Cobiella-Reguera, J.L., 1996, El magmatismo jurasico (caloviano?-oxfordiano) de Cuba occidental; ambiente de formacion e implicaciones regionales; Jurassic (Callovian?-Oxfordian) magmatism of western Cuba; environment of formation and regional implications: Revista de la Asociasion, Geologica d' Argentina, v. 51, p. 15–28.

Cobiella-Reguera, J.L., 2000, Jurassic and Cretaceous geological history of Cuba: International Geology Review, v. 42, p. 594–616.

Coleman, R.G., 1977, Emplacement and metamorphism of ophiolites: Ofioliti, v. 2, p. 41–74.

Compagnoni, R., Morlotti, E., and Torelli, L., 1989, Crystalline and sedimentary rocks from the scarps of the Sicily-Sardinia Trough and Cornaglia Terrace (southwestern Tyrrhenian Sea, Italy); paleogeographic and geodynamic implications: Chemical Geology, v. 77, p. 375–398.

Corfield, R.I., Searle, M.P., and Pedersen, R.B., 2001, Tectonic setting, origin, and obduction history of the spontang ophiolite, Ladakh Himalaya, NW India: Journal of Geology, v. 109, p. 715–736.

Cortesogno, L., Gaggero, L., Jaho, E., Marroni, M., Pandolfi, L., and Shtjefanaku, D., 1998, The gabbroic complex of the western ophiolitic belt, northern Albania: An example of multilayered sequence in an intermediate-spreading oceanic ridge: Ofioliti, v. 23, p. 49–64.

Cosca, M.A., Arculus, R.J., Pearce, J.A., and Mitchell, J.G., 1998, $^{40}Ar/^{39}Ar$ and K-Ar geochronological age constraints for the inception and early evolution of the Izu-Bonin-Mariana arc system: The Island Arc, v. 7, p. 579–595.

Costa, S., and Caby, R., 2001, Evolution of the Ligurian Tethys in the Western Alps; Sm/Nd and U Pb geochronology and rare-earth element geochemistry of the Montgenevre Ophiolite (France): Chemical Geology, v. 175, p. 449–466.

Crawford, A.J., Beccaluva, L., and Serri, G., 1981, Tectono-magmatic evolution of the West Philippine-Mariana region and the origin of boninites: Earth and Planetary Science Letters, v. 54, p. 346–356.

Crawford, A.J., Beccaluva, L., Serri, G., and Dostal, J., 1986, Petrology, geochemistry and tectonic implications of volcanics dredged from the intersection of the Yap and Mariana trenches: Earth and Planetary Science Letters, v. 80, p. 265–280.

Crawford, A.J., Verbeeten, A., Eggins, S., Danyushevsky, L.V., Maillet, P., Sigurdsson, I.A., and Monzier, M., 1997, Hot subduction; magmatism along the Hunter Ridge, SW Pacific, State of the Arc '97: Proceedings, Island Arc Magma Genesis Workshop, v. 45, p. 9–12.

Dallmeyer, R.D., and Williams, H., 1975, $^{40}Ar/^{39}Ar$ ages from the Bay of Islands metamorphic aureole; their bearing on the timing of Ordovician ophiolite obduction: Canadian Journal of Earth Science, v. 12, p. 1685–1690.

Danishwar, S., Flower, M.F.J., Sandoval, E., Sultan, M., 2003, TETHYS: A GIS database for studying and visualizing Tethyan plate collisions: Geological Society of America abstracts with Programs, v. 35, no. 6 (in press).

Danyushevsky, L.V., Sobolev, A.V., and Falloon, T.J., 1995, North Tongan High-Ca Boninite Petrogenesis–the Role of Samoan Plume and Subduction Zone-Transform Fault Transition: Journal of Geodynamics, v. 20, p. 219–241.

Delaloye, M., Fontignie, D., and Spadea, P., 1984, Radiometric dates of Calabria-Lucania ophiolites (southern Italy) and associated continental rocks: Ofioliti, v. 9, p. 687.

Dewey, J.F., Cande, S.C., and Pitman, W.C., III, 1989a, Tectonic evolution of the India/Eurasia collision zone: Eclogae Geol. Helvetiae, v. 82, p. 717–734.

Dewey, J.F., Helman, M.L., Turco, E., Hutton, D.H.W., and Knot, S.D., 1989b, Kinematics of the western Mediterranean, *in* Coward, M.P., Dietrich, D., and Park, R.G., eds., Conference on Alpine Tectonics: Geological Society of London Special Publication, v. 45, p. 265–283.

Dilek, Y., and Whitney, D.L., 1997, Counterclockwise P-T-t trajectory from the metamorphic sole of a Neo-Tethyan ophiolite (Turkey): Tectonophysics, v. 280, p. 295–310.

Dilek, Y., Thy, P., Hacker, B., and Grundvig, S., 1999, Structure and petrology of Tauride ophiolites and mafic dike intrusions (Turkey); implications for the neo-Tethyan ocean: Geological Society of America Bulletin, v. 111, p. 1192–1216.

Dimo-Lahitte, A., Monie, P., and Vergely, P., 2001, Metamorphic soles from the Albanian ophiolites; petrology, $^{40}Ar/^{39}Ar$ geochronology, and geodynamic evolution: Tectonics, v. 20, p. 78–96.

Dobretsov, N.L., and Kepezhinscas, V.V., 1981, Three types of ultrabasic magmas and their bearing on the problem of ophiolites: Ofioliti, v. 6, p. 221–236.

Eaton, S., and Robertson, A., 1993, The Miocene Pakhna Formation, Southern Cyprus and Its Relationship to the Neogene Tectonic Evolution of the Eastern Mediterranean: Sedimentary Geology, v. 86, p. 273–296.

Einaudi, F., Pezard, P.A., Cocheme, J.J., Coulon, C., Laverne, C., and Godard, M., 2000, Petrography, geochemistry and physical properties of a continuous extrusive section from the Sarami Massif, Semail Ophiolite: Marine Geophysical Research, v. 21, p. 387–407.

Eissen, J.P., Nohara, M., Cotten, J., and Hirose, K., 1994, North Fiji Basin basalts and their magma sources: 1. Incompatible element constraints: Marine Geology, v. 116, p. 153–178.

Encarnacion, J.P., 1994, Geochronological, geochemical and geological constraints on models of ophiolite generation and arc growth; evidence from the northern Philippines [Ph.D. thesis]: Ann Arbor, University of Michigan, 144 p.

Encarnacion, J.P., Mukasa, S.B., and Obille, E.C., 1993a, Zircon U-Pb Geochronology of the Zambales and Angat ophiolites, Luzon, Philippines–evidence for an Eocene arc-backarc pair: Journal of Geophysical Research B, v. 98, p. 19,991–20,004.

Encarnacion, J.P., Mukasa, S.B., Obille, E.C., 1993b, Origin of the Zabales and Angat ophiolites, Philippines; new constraints from zircon U-Pb geochronology: Eos (Transactions, American Geophysical Union), v. 74, p. 286.

Encarnacion, J.P., Essene, E.J., Mukasa, S.B., and Hall, C.H., 1995, High-pressure and temperature subophiolitic kyanite-garnet amphibolites generated during initiation of mid-Tertiary subduction, Palawan, Philippines: Journal of Petrology, v. 36, p. 1481–1503.

Encarnacion, J., Mukasa, S.B., and Evans, C.A., 1999, Subduction components and the generation of arc-like melts in the Zambales Ophiolite, Philippines; Pb, Sr and Nd isotopic constraints: Chemical Geology, v. 156, p. 343–357.

Encarnacion, J., Fernandez, D., and Mattinson, J., 2001, Subduction Initiation by Extrusion Tectonics? Evidence From the Palawan Ophiolite, Philippines: Eos Transactions of the American Geophysical Union, v. 82, p. 47

Ernewein, M., Pflumio, C., and Whitechurch, H., 1988, The death of an accretion zone as evidenced by the magmatic history of the Sumail Ophiolite (Oman): Tectonophysics, v. 151, p. 247–274.

Evans, C., and Hawkins, J.W., 1989, Compositional Heterogeneities in Upper Mantle Peridotites from the Zambales Range Ophiolite, Luzon, Philippines: Tectonophysics, v. 168, p. 23–41.

Evans, C.A., Casteneda, G., and Franco, H., 1991, Geochemical Complexities Preserved in the Volcanic-Rocks of the Zambales Ophiolite, Philippines: Journal of Geophysical Research B, v. 96, p. 16,251–16,262.

Falloon, T.J., and Crawford, A.J., 1991, The petrogenesis of high-calcium boninite lavas dredged from the northern Tonga Ridge: Earth and Planetary Science Letters, v. 102, p. 375–394.

Falloon, T.J., and Danyushevsky, L.V., 2000, Melting of refractory mantle at 1.5, 2, 2.5 GPa under, anhydrous and H_2O-undersaturated conditions: implications for the petrogenesis of high-Ca boninites and the influence of subduction components on mantle melting: Journal of Petrology, v. 41, p. 257–283.

Flower, M.F.J., and Levine, H.M., 1987, Petrogenesis of a tholeiite-boninite sequence from Ayios Mamas, Troodos ophiolite; evidence for splitting of a volcanic arc?: Contributions to Mineralogy and Petrology, v. 97, p. 509–524.

Flower, M.F.J., Tamaki, K., and Hoang, N., 1998, Mantle extrusion: a model for dispersed volcanism and DUPAL-like asthenosphere in East Asia and the western Pacific, *in* Flower, M.F.J., Chung, S.-L., Lo, C.-H., and Lee, T.-Y., eds., Mantle Dynamics and Plate Interactions in East Asia: Washington, D.C., American Geophysical Union Geodynamics Series, v. 27, p. 67–88.

Flower, M.F.J., Russo, R.M., Tamaki, K., and Hoang, N., 2001, Mantle contamination and the Izu-Bonin-Mariana (IBM) "high-tide mark"; evidence for mantle extrusion caused by Tethyan closure: Tectonophysics, v. 333, p. 9–34.

Forster, M.A., and Lister, G.S., 1999, Detachment faults in the Aegean core complex of Ios, Cyclades, Greece, *in* Ring, U., Brandon, M.T., Lister, C.S., and Willett, S.D., eds., Exhumation Processes; Normal Faulting, Ductile Flow and Erosion: Geological Society of London Special Publication, v. 154, p. 305–323.

Fortuin, A.R., VanderWerff, W., and Wensink, H., 1997, Neogene basin history and paleomagnetism of a rifted and inverted forearc region, on- and offshore Sumba, Eastern Indonesia: Journal of Asian Earth Sciences, v. 15, p. 61–88.

Fryer, P., Lockwood, J.P., Becker, N., Phipps, S., and Todd, C.S., 2000, Significance of serpentine mud volcanism in convergent margins, *in* Dilek, Y., Moores, E., Elthon, D., Nicolas, A., eds., Ophiolites and Oceanic Crust; New Insights from Field Studies and the Ocean Drilling Program: Boulder, Colorado, Geological Society of America Special Paper, v. 349, p. 35–51.

Gaedicke, C., Baranov, B., Seliverstov, N., Alexeiev, D., Tsukanov, N., and Freitag, R., 2000, Structure of an active arc-continent collision area; the Aleutian-Kamchatka junction, in: Processes of Arc-Continent Collision: Tectonophysics, v. 325, p. 63–85.

Garfunkel, Z., 1998, Constraints on the origin and history of the Eastern Mediterranean basin: Tectonophysics, v. 298, p. 5–37

Gass, I.G., 1968, Is the Troodos massif of Cyprus a fragment of Mesozoic ocean floor: Nature, v. 220, p. 39–42.

Gass, I.G., Neary, C.R., Plant, J., Robertson, A.H.F., Simonian, K.O., Smewing, J.D., Spooner, E.T.C., and Wilson, R.A.M., 1975, Comments on "The Troodos ophiolitic complex was probably formed in an island arc," by A. Miyashiro and subsequent correspondence by A. Hynes and A Miyashiro: Earth and Planetary Science Letters, v. 25, p. 236–238.

Geary, E.E., Kay, R.W., Reynolds, J.C., and Kay, S.M., 1989, Geochemistry of mafic rocks from the Coto Block, Zambales Ophiolite, Philippines—trace element evidence for 2 stages of crustal growth: Tectonophysics, v. 168, p. 43–63.

Geist, E.L., and Scholl, D.W., 1994, Large-scale deformation related to the collision of the Aleutian Arc with Kamchatka: Tectonics, v. 13, p. 538–560.

Geist, E.L., Vallier, T.L., and Scholl, D.W., 1994, Origin, transport, and emplacement of an exotic island-arc terrane exposed in eastern Kamchatka, Russia: Geological Society of America Bulletin, v. 106, 1182–1194.

Ghent, E.D., and Stout, M.Z., 1981, Metamorphism at the base of the Samail ophiolite, southeastern Oman Mountains: Journal of Geophysical Research B, v. 86, p. 2557–2571.

Gibson, I.L., McGill, G.C., Robinson, P.T., 1987, The geochemistry and nature of unusual glassy lavas from CCSP holes CY-2 and CY-2a, Troodos Ophiolite, Cyprus, *in* Robinson, P.T., Gibson, I.L., and Panayiotou, A., eds., Cyprus Crustal Study Project; Initial Report, Holes CY-2 and 2a: Ottawa, Ontario, Geological Survey of Canada, 85–29, p. 79–85.

Gillis, K.M., 2002, The rootzone of an ancient hydrothermal system exposed in the Troodos ophiolite, Cyprus: Journal of Geology, v. 110, p. 57–74.

Gillis, K.M., and Banerjee, N.R., 2000, Hydrothermal alteration patterns in supra-subduction zone ophiolites, *in* Dilek, Y., Moores, E., Elthon, D., and Nicolas, A., eds. Ophiolites and Oceanic Crust; New Insights from Field Studies and the Ocean Drilling Program: Boulder, Colorado, Geological Society of America Special Paper 349, p. 283–297.

Gjata, K., Kornprobst, J., Kodra, A., Briot, D., and Pineau, F., 1992, Hot subduction close to a ridge—example of the garnet pyroxenite enclaves in the serpentinite breccia at Derveni, Albania: Bulletin de la Societé Géologique de France, v. 163, p. 469–476.

Gnos, E., 1998, Peak metamorphic conditions of garnet amphibolites beneath the Semail Ophiolite; implications for an inverted pressure gradient: International Geology Review, v. 40, p. 281–304.

Gnos, E., Khan, M., Mahmood, K., Khan, A.S., and Villa, I., 1997, Age and tectonic setting of the Ras Koh Ophiolite, Balochistan, Pakistan: Geological Society of America Abstracts with Programs, v. 29, p. 470.

Gould, S.J., 1989, Wonderful Life, The Burgess Shale and the Nature of History: W.W. Norton and Co., 347 p.

Gribble, R.F., Stern, R.J., Newman, S., Bloomer, S.H., and O'Hearn, T., 1998, Chemical and isotopic composition of lavas from the Northern Mariana Trough: Implications for magmagenesis in back-arc basins: Journal of Petrology, v. 39, p. 125–154.

Hacker, B.R., 1991, The role of deformation in the formation of metamorphic gradients; ridge subduction beneath the Oman ophiolite: Tectonics, v. 10, p. 455–473.

Hacker, B.R., 1994, Rapid emplacement of young oceanic lithosphere; argon geochronology of the Oman ophiolite: Science, 265, p. 1563–1565.

Hacker, B.R., and Gnos, E., 1997, The conundrum of Samail; explaining the metamorphic history, in: the Adolphe Nicolas Volume: Tectonophysics, v. 279, p. 215–226.

Hacker, B.R., and Mosenfelder, J.L., 1996, Metamorphism and deformation along the emplacement thrust of the Samail ophiolite, Oman: Earth and Planetary Science Letters, v. 144, p. 435–451.

Hacker, B.R., Mosenfelder, J.L., and Gnos, E., 1996, Rapid emplacement of the Oman ophiolite: Thermal and geochronologic constraints: Tectonics, v. 15, p. 1230–1247.

Hall, R., 2002, Cenozoic geological and plate tectonic evolution of SE Asia and the SW Pacific: computer-based reconstructions, model and animations: Journal of Asian Earth Science, v. 20, p. 353–431.

Harris, R.A., 2001, Processes of Ophiolite Genesis and Emplacement in the Nascent Collision of Asian and Northern Australia: Eos, Transactions of the American Geophysical Union, v. 82, p. 47.

Hawkins, J.W., 1989, Back-arc basins; their evolution, geological characteristics and importance in crustal evolution: International Geological Congress, 28th, Washington, D.C., Abstracts, v. 28, p. 2.41–2.42.

Hawkins, J.W., 1993, Ophiolites; misleading guides to the geology of oceanic layer 3?: Eos (Transactions, American Geophysical Union), v. 74, p. 654.

Hawkins, J.W., and Castillo, P.R., 1998, Early history of the Izu-Bonin-Mariana arc system: Evidence from Belau and the Palau Trench: The Island Arc, v. 7, p. 559–578.

Hess, H.H., 1965, Mid-oceanic ridges and tectonics of the sea-floor, submarine geology and geophysics, Proceedings 17th Symposium: Bristol, Colston Research Society.

Hickey-Vargas, R., 1998a, Geochemical characteristics of oceanic island basalts from the Philippine Sea Plate: implications for the sources of East Asian Plate margin and intraplate basalts: Journal of Geophysical Research B, v. 103, p. 20,963–20,979.

Hickey-Vargas, R., 1998, Geochemical characteristics of oceanic island basalts from the Philippine Sea Plate; implications for the sources of East Asian plate margin and intraplate basalts, *in* Flower, M.F.J., Chung, S.-L., Lo, C.-H., and Lee, T.-Y., eds. Mantle Dynamics and Plate Interactions in East Asia: Washington, D.C., American Geophysical Union Geodynamics Series, v. 27, p. 365–384.

Hibbard, J. and Wright, T.O., 1982, Obduction-related boninites, Newfoundland Appalachians, Geological Society of America, Northeastern and Southeastern combined section meetings Abstracts with Programs, v. 14, p. 25.

Höck, V., Tomek, C., Robertson, A., and Koller, F., 2002, Preface to Europrobe-Pancardi Symposium "Eastern Mediterranean Ophiolites: Magmatic Processes and Geodynamic Implications": Lithos, v. 65, p. ix–xiii.

Hoxha, M., and Boullier, A.M., 1993b, Kinematics of deformation and geometry of the ridge in the Kukes Ophiolite (Albania), Terra Abstracts, v. 5, p. 559.

Hoxha, M., and Boullier, A.M., 1995, The Peridotites of the Kukes Ophiolite (Albania)—Structure and Kinematics, Tectonophysics, 249, p. 217–231.

Hoxha, M., Boullier, A.M., and Zhupi, S., 1993a, Structure of the Peridotites from the Kukes Ophiolite (Albania): Bulletin de la Societé Géologique de France, v. 164, p. 449–458.

Hsu, S.-K., and Sibuet, J.-C., 1995, Is Taiwan the result of arc-continent or arc-arc collision?: Earth and Planetary Science Letters, v. 136, p. 315–324.

Hung, S.H., and Forsyth, D.W., 1999, Anisotropy in the oceanic lithosphere from the study of local intraplate earthquakes on the west flank of the southern East Pacific Rise: Shear wave splitting and waveform modeling: Journal of Geophysical Research B, v. 104, p. 10,695–10,717.

Hussong, D.M., Uyeda, S., and Natland, J.H., 1981, Tectonic processes and history of the Mariana arc system: Eos, Transactions of the American Geophysical Union, v. 62, p. 1090.

Hynes, A., 1975, Comment on "The Troodos ophiolitic complex was probably formed in an island arc," by A. Miyashiro: Earth and Planetary Science Letters, v. 25, p. 213–216.

Insergueix-Filippi, D., Dupeyrat, L., Dimo, A., Vergely, P., and Bebien, J., 1998, Model of subduction zone infancy at a mid-ocean ridge; magmatic and metamorphic implications, International Ophiolite Symposium and Field Excursion; Generation and Emplacement of Ophiolites through Time; abstracts, excursion guide: Espoo, Finland, Geological Survey of Finland Special Paper, v. 26, 28 p.

Insergueix-Filippi, D., Dupeyrat, L., Dimo-Lahitte, A., Vergely, P., and Bebien, J., 2000, Albanian ophiolites. II, Model of subduction zone infancy at a mid-ocean ridge: Ofioliti, v. 25, p. 47–53.

Ishii, T., Robinson, P.T., Tokuyama, H., Kobayashi, K., 1988, Relations of fore-arc proto-ophiolites, mantle and serpentine diapirs: Geological Society of America Abstracts with Programs, v. 20, p. 158.

Ishikawa, T., Nagaishi, K., and Umino, S., 2002, Boninitic volcanism in the Oman ophiolite: Implications for thermal condition during transition from spreading ridge to arc: Geology v. 30, p. 899–902.

Jolivet, L., and Faccenna, C., 2000, Mediterranean extension and the Africa-Eurasia collision: Tectonics. v. 19, p. 1095–1106.

Jolivet, L., Goffe, B., Monie, P., Truffert-Luxey, C., Patriat, M., and Bonneau, M., 1996, Miocene detachment in Crete and exhumation P-T-t paths of high-pressure metamorphic rocks: Tectonics, v. 15, p. 1129–1153.

Jolivet, L., Goffe, B., Bousquet, R., Oberhaensli, R., and Michard, A., 1998, Detachments in high-pressure mountain belts, Tethyan examples: Earth and Planetary Science Letters, v. 160, p. 31–47.

Jousselin, D., and Mainprice, D., 1998, Melt topology and seismic anisotropy in mantle peridotites of the Oman Ophiolite: Earth and Planetary Science Letters, v. 164, p. 553–568.

Jousselin, D., Nicolas, A., and Boudier, F., 1998, Detailed mapping of a mantle diapir below a paleo-spreading center in the Oman Ophiolite: Journal of Geophysical Research B, v. 103, p. 18,153–18,170.

Juteau, T., Beurrier, M., Dahl, R., and Nehlig, P., 1988a, Segmentation at a fossil spreading axis; the plutonic sequence of the Wadi Haymiliyah area (Haylayn Block, Sumail Nappe, Oman): Tectonophysics, v. 151, p. 167–197.

Juteau, T., Ernewein, M., Reuber, I., Whitechurch, H., and Dahl, R., 1988b, Duality of magmatism in the plutonic sequence of the Sumail Nappe, Oman, *in* Boudier, F., and Nicolas, A., eds., The Ophiolites of Oman: Tectonophysics, v. 151, p. 107–135.

Kamp, P.J.J., 1999, Tracking crustal processes by FT thermochronology in a forearc high (Hikurangi margin, New Zealand) involving Cretaceous subduction termination and mid-Cenozoic subduction initiation: Tectonophysics, v. 307, p. 313–343

Karig, D.E., 1971, Origin and development of marginal basins in the western Pacific: Journal of Geophysical Research B, v. 76, p. 2542–2561.

Lachize, M., Lorand, J.P., and Juteau, T., 1996, Calc-alkaline differentiation trend in the plutonic sequence of the Wadi Haymiliyah section, Haylayn massif, Semail ophiolite, Oman: Lithos, v. 38, p. 207–232.

Lagabrielle, Y., and Lemoine, M., 1997, Alpine, Corsican and Apennine ophiolites; the slow-spreading ridge model: Comptes Rendus de l'Academie des Sciences, Serie II. Sciences de la Terre et des Planetes, v. 325, p. 909–920.

Lallemand, S.E., Popoff, M., Cadet, J.-P., Bader, A.-G., Pubellier, M., Rangin, C., and Deffontaines, B., 1998, Genetic relations between the central and southern Philippine Trench and the Sangihe Trench: Journal of Geophysical Research, B, v. 103, p. 933–950.

Lanphere, M.A., Coleman, R.G., Karamata, S., and Pamic, J., 1975, Age of amphibolites associated with alpine peridotites in the Dinaride ophiolite zone, Yugoslavia: Earth and Planetary Science Letters, v. 26, p. 271–276.

Le Pichon, X., Angelier, J., Vine, F.J., and Smith, A.G., 1981, The Aegean Sea: Philosophical Transactions of the Royal Society of London, ser. A, v. 300, p. 357–372.

Le Pichon, X., Chamot-Rooke, N., Lallemant, S., Noomen, R., and Veis, G., 1995a, Geodetic determination of the kinematics of central Greece with respect to Europe; implications for eastern Mediterranean tectonics: Journal of Geophysical Research B, v. 100, p. 12,675–12,690.

Le Pichon, X., Henry, P., Goffe, B., and Chopin, C., 1995b, From crustal collision to crustal subduction; a model of evolution from the Alps to the Himalayas: Terra Abstracts, v. 7, Supplement 1, p. 114.

Lee, T.-Y., and Lawver, L.A., 1994, Cenozoic plate reconstruction of the South China Sea region: Tectonophysics, v. 235, p. 149–180.

Leitch, E.C., 1984, Island arc elements and arc-related ophiolites: Tectonophysics, v. 106, p. 177–203.

Leloup, P.H., Arnaud, N., Lacassin, R., Kienast, J.R., Harrison, T.M., Pan Trong, T.T., Replumaz, A., and Tapponnier, P., 2001, New constraints on the structure, thermochronology, and timing of the Ailao Shan-Red River shear zone, SE Asia: Journal of Geophysical Research, B, v. 106, p. 6683–6732.

Le Pichon, X., Lallemant, S.J., Chamot-Rooke, N., Lemeur, D., and Pascal, G., 2002, The Mediterranean Ridge backstop and the Hellenic nappes: Marine Geology, v. 186, p. 111–125.

Leroy, S., Mercier de Lepinay, B., Mauffret, A., and Pubellier, M., 1996, Structural and tectonic evolution of the eastern Cayman Trough (Caribbean Sea) from seismic reflection data: American Association for Petroleum Geology Bulletin, v. 80, p. 222–247.

Lippard, S.J., Shelton, A.W., and Gass, I.G., 1986, The ophiolite of northern Oman: Geological Society of London, Special Publication, v. 11, 178 p.

Lister, G.S., Banga, G., and Feenstra, A., 1984, Metamorphic core complexes of Cordilleran type in the Cyclades, Aegean Sea, Greece: Geology, v. 12, p. 221–225.

Lonergan, L., and White, N., 1997, Origin of the Betic-Rif mountain belt: Tectonics, v. 16, p. 504–522

Lytwyn, J., Casey, J., Gilbert, S., and Kusky, T., 1997, Arc-like mid-ocean ridge basalt formed seaward of a trench- forearc system just prior to ridge subduction: An example from subaccreted ophiolites in southern Alaska: Journal of Geophysical Research, B, v. 102, p. 10,225–10,243.

McKenzie, D., and Bickle, M.J., 1988, The volume and composition of melt generated by extension of the lithosphere: Journal of Petrology, v. 29, p. 625–679.

Macleod, C.J., and Murton, B.J., 1995, On the Sense of Slip of the Southern Troodos Transform-Fault Zone, Cyprus: Geology, v. 23, p. 257–260.

Macleod, C.J., Allerton, S., Gass, I.G., and Xenophontos, C., 1990, Structure of a Fossil Ridge-Transform Intersection in the Troodos Ophiolite: Nature, v. 348, p. 717–720.

Makris, J., Ben-Avraham, Z., Behle, A., Ginzburg, A., Giese, P., Steinmetz, L., Whitmarsh, R.B., and Eleftheriou, S., 1983, Seismic refraction profiles between Cyprus and Israel and their interpretation: Geophysical Journal of the Royal Astronomical Society, v. 75, p. 575–591.

Malavieille, J., Chemenda, A., and Larroque, C., 1998, Evolution model for Alpine Corsica; mechanism for ophiolite emplacement and exhumation of high-pressure rocks: Terra Nova, v. 10, p. 317–322.

Malpas, J., and Moore, P., 1992b, Ophiolites of the Dun Mountain Belt, Nelson, New Zealand: International Geological Congress, 29th, Washington, D.C., Abstracts, v. 29, p. 132.

Malpas, J., Xenophontos, C., and Williams, D., 1992a, The Ayia Varvara Formation of SW Cyprus—A Product of Complex Collisional Tectonics: Tectonophysics, v. 212, p. 193–211.

Malpas, J., and Robinson, P., 1997, Oceanic lithosphere—1. The origin and evolution of oceanic lithosphere: Geoscience Canada, v. 24, p. 100–107.

Manika, K., Shallo, M., and Gega, D., 1997, The plutonic sequence of the Shebenik ophiolite complex, Albania: evidence for dual magmatism, *in* Messiga, B., and Tribuzio, R., eds., From rifting to drifting in present-day and fossil ocean basins, International ophiolite symposium: Bologna, Italy, Pitagora Editrice, Ofioliti, v. 22 (Special Issue), p. 93–99.

Mascle, J., and Chaumillon, E., 1998, An overview of Mediterranean Ridge collisional accretionary complex as deduced from multichannel seismic data: Geo-Marine Letters, v. 18, p. 81–89.

Migiros, G., Hatzipanagiotou, K., Gartzos, E., Serelis, K., and Tsikouras, B., 2000, Petrogenetic evolution of ultramafic rocks from Lesvos Island (NE Aegean, Greece): Chemie der Erde, v. 60, p. 27–46.

Milsom, J., Hall, R., and Padmawidjaja, T., 1996, Gravity fields in eastern Halmahera and the Bonin Arc; implications for ophiolite origin and emplacement: Tectonics, v. 15, p. 84–93.

Milsom, J., Ali, J., and Sudarwono, 1999, Structure and collision history of the Buton continental fragment, eastern Indonesia: AAPS Bulletin, v. 83, p. 1666–1689.

Miyashiro, A., 1973, The Troodos ophiolitic complex was probably formed in an island arc: Earth and Planetary Science Letters, v. 19, p. 218–224.

Miyashiro, A., 1975a, Classification, characteristics, and origin of ophiolites: Journal of Geology, v. 83, p. 249–281.

Miyashiro, A., 1975b, Origin of the Troodos and other ophiolites; a reply to Hynes: Earth and Planetary Science Letters, v. 25, p. 217–222.

Miyashiro, A., 1975c, Origin of the Troodos and other ophiolites; a reply to Moores: Earth and Planetary Science Letters, v. 25, p. 227–235.

Miyashiro, A., 1977, Subduction-zone ophiolites and island-arc ophiolites, *in* Saxena, S.K., and Bhattacharji, S., eds., Energetics of Geological Processes: New York, Springer-Verlag, p. 188–213.

Monzier, M., Danyushevsky, L.V., Crawford, A.J., Bellon, H., and Cotten, J., 1993, High-Mg Andesites from the Southern Termination of the New Hebrides Island Arc (SW Pacific): Journal of Volcanology and Geothermal Research, v. 57, p. 193–217.

Monzier, M., Robin, C., Eissen, J.-P., and Cotten, J., 1997, Geochemistry vs. seismo-tectonics along the volcanic New Hebrides central chain (Southwest Pacific): Journal of Volcanology and Geothermal Research, v. 78, p. 1–29.

Moore, G.F., and Silver, E.A., 1983, Collision processes in the northern Molucca Sea, *in* Hayes, D.E., ed., The Tectonic and Geologic Evolution of Southeast Asian Seas and Islands: Part 2: American Geophysical Union Geophysical Monograph, v. 27, p. 360–372.

Moores, E.M., 1974, Re-examination of the Vourinos Ophiolite, Greece, Eos, Transactions of the American Geophysical Union, v. 55, p. 444.

Moores, E.M., 1975, Discussion of "Origin of Troodos and other ophiolites; a reply to Hynes," by Akiho Miyashiro: Earth and Planetary Science Letters, v. 25, p. 223–226.

Moores, E.M., 1982a, Origin and emplacement of ophiolites: Reviews of Geophysics and Space Physics, v. 20, p. 735–760.

Moores, E.M., 1982b, Ophiolites as indicators of ocean spreading processes, plate reorganizations, and collisional tectonics: Geological Society of America Abstracts with Programs, v. 14, p. 570.

Moores, E.M., and Jackson, E.D., 1974, Ophiolites and oceanic crust: Nature, v. 250, p. 136–139.

Moores, E.M., and Vine, F.J., 1971, The Troodos massif, Cyprus and other ophiolites as oceanic crust; evaluation and implications: A discussion on the petrology of igneous and metamorphic rocks from the ocean floor: Philosophical Transactions of the Royal Society of London, ser. A, v. 268, p. 443–466.

Moores, E.M., Hess, H.H., and Maxwell, J.C., 1968, Petrology of an ultramafic-mafic (ophiolite) sequence, *in* Abstracts for 1966: Boulder, Colorado, Geological Society of America Special Paper 101, p. 143.

Moores, E.M., Kellogg, L.H., and Dilek, Y., 2000, Tethyan ophiolites, mantle convection, and tectonic "historical contingency"; a resolution of the "ophiolite conundrum," *in* Dilek, Y., Moores, E., Elthon, D., Nicolas, A., eds. Ophiolites and Oceanic Crust; New Insights from Field Studies and the Ocean Drilling Program: Boulder, Colorado, Geological Society of America Special Paper 349, p. 3–12.

Morra, V., Secchi, F.A.G., Melluso, L., and Franciosi, L., 1997, High-Mg subduction-related Tertiary basalts in Sardinia, Italy: Lithos, v. 40, p. 69–91.

Morris, A., Anderson, M.W., and Robertson, A.H.F., 1998, Multiple tectonic rotations and transform tectonism in an intraoceanic suture zone, SW Cyprus: Tectonophysics, v. 299, p. 229–253.

Mukasa, S.B., and Ludden, J.N., Uranium-lead isotopic ages of plagiogranites from the Troodos Ophiolite, Cyprus, and their tectonic significance: Geology, v. 15, p. 825–828.

Murdie, R.E., and Russo, R.M., 1999, Seismic anisotropy in the legion of the Chile margin triple junction: Journal of South American Earth Science, v. 12, p. 261–270.

Nicolas, A., 1997, Sea floor spreading; a viewpoint from ophiolites: Comptes Rendus de l'Academie des Sciences, serie II, v. 324, p. 1–8.

Nicolas, A., and Boudier, F., 1991, Rooting of the sheeted dike complex in the Oman Ophiolite: Petrology and Structural Geology, v. 5, p. 39–54.

Nicolas, A., and Boudier, F., 1995, Mapping oceanic ridge segments in Oman Ophiolite: Journal of Geophysical Research B, v. 100, p. 6179–6197.

Nicolas, A., and Boudier, F., 2000, Large mantle upwellings and related variations in crustal thickness in the Oman ophiolite, *in* Dilek, Y., Moores, E., Elthon, D., Nicolas, A., eds., Ophiolites and Oceanic Crust; New Insights from Field Studies and the Ocean Drilling Program: Boulder, Colorado, Geological Society of America Special Paper 349, p. 67–73.

Nicolas, A., and Violette, J.F., 1982, Mantle flow at oceanic spreading centers; models derived from ophiolites, Tectonophysics, v. 81, p. 319–339.

Nicolas, A., Reuber, I., and Benn, K., 1988, A new magma chamber model based on structural studies in the Oman Ophiolite: Tectonophysics, v. 151, p. 87–105.

Nicolas, A., Boudier, F., and Ildefonse, B., 1994, Evidence from the Oman ophiolite for active mantle upwelling beneath a fast-spreading ridge: Nature, v. 370, p. 51–53.

Nicolas, A., Boudier, F., and Ildefonse, B., 1996, Variable crustal thickness in the Oman Ophiolite; implication for oceanic crust: Journal of Geophysical Research B, v. 101, p. 17,941–17,950.

Nicolas, A., Boudier, F., and Meshi, A., 1999, Slow spreading accretion and mantle denudation in the Mirdita Ophiolite (Albania): Journal of Geophysical Research B, v. 104, p. 15,155–15,167.

Niu, Y., O'Hara, M.J., and Pearce, J.A., 2001, Initiation of subduction zones: a consequence of lateral compositional buoyancy contrast within the lithosphere: Eos, Transactions of the American Geophysical Union, v. 82, p. 47.

Ogawa, Y., 1995, Some problems on emplacement of ophiolite rocks around plate subduction boundaries: Chigaku Zasshi (J. Geogr.), v. 104, p. 468–472.

Padoa, E., 1999, Les Ophiolites du Massif de l'Inzecca (Corse alpine); lithostratigraphie, structure geologique et evolution geodynamique (The Inzecca Massif ophiolites [Alpine Corsica]; lithostratigraphy, geological structure and geodynamic evolution): Geologie de la France, 1999, p. 37–48.

Pallister, J.S., 1984, Parent magmas of the Semail ophiolite, Oman, *in* Gass, I.G., Lippard, S.J., and Shelton, A.W., eds., Ophiolites and Oceanic Lithosphere: London, Geological Society of London Special Publication, v. 13, p. 63–70.

Pallister, J.S., and Hopson, C.A., 1981, Samail ophiolite plutonic suite; field relations, phase variation, cryptic variation and layering, and a model of a spreading ridge magma chamber: Journal of Geophysical Research B, v. 86, p. 2593–2644.

Pallister, J.S., and Knight, R.J., 1981, Rare-earth element geochemistry of the Samail ophiolite near Ibra, Oman: Journal of Geophysical Research B, v. 86, p. 2673–2697.

Parkinson, C., 1998, Emplacement of the East Sulawesi Ophiolite: evidence from subophiolite metamorphic rocks: Journal of Asian Earth Science, v. 16, p. 13–28.

Parkinson, I.J., C.J., Hawkesworth, A.S., Cohen, 1998, Ancient mantle in a modern arc; osmium isotopes in Izu-Bonin-Mariana Forearc peridotites: Science, v. 281, p. 2011–2013,

Parman, S.W., T.L., Grove, 2001, The production of Barberton komatiites in an Archean subduction zone: Geophysical Research Letters, v. 28, p. 2513–2516.

Parlak, O., and Delaloye, M., 1999, Precise $^{40}Ar/^{39}Ar$ ages from the metamorphic sole of the Mersin ophiolite (southern Turkey): Tectonophysics, v. 301, p. 145–158.

Parlak, O., Höck, V., and Delaloye, M., 2002, The supra-subduction zone Pozanti-Karsanti ophiolite, southern Turkey: evidence for high-pressure crystal fractionation of ultramafic cumulates: Lithos, v. 65, p. 205–224.

Pearson, D.G., 1999, The age of continental roots: Lithos, v. 48, p. 171–194.

Pearce, J.A., 1991, Ophiolites: Ocean floor comes ashore: Nature, v. 354, p. 110–111.

Pearce, J.A., and Peate, D.W., 1995, Tectonic implications of the composition of volcanic arc magmas: Annual Reviews of Earth and Planetary Sciences, v. 23, p. 251–285.

Pearce, J.A., Alabaster, T., Shelton, A.W., Searle, M.P., Vine, F.J., and Smith, A.G., 1981, The Oman ophiolite as a Cretaceous arc-basin complex; evidence and implications: Philosophical Transactions of the Royal Society of London, ser. A, v. 300, p. 299–317.

Pearce, J.A., Lippard, S.J., and Roberts, S., 1984, Characteristics and tectonic significance of supra-subduction zone ophiolites, *in* Kokelaar, B.P., and Howells, M.F., eds., Marginal basin geology; volcanic and associated sedimentary and tectonic processes in modern and ancient marginal basins: Geological Society of London Special Publication 16, v. 16, p. 74–94.

Pearce, J.A., Thirlwall, M.F., Ingram, G., Murton, B.J., Arculus, R.J., van der Laan, S.R., Fryer, P., Stokking, L.B., Ali, J.R., Arculus, R., Ballotti, D.L., Burke, M.M., Ciampo, G., Haggerty, J.A., Haston, R.B., Heling, D., Hobart, M.A., Ishii, T., Johnson, L.E., Lagabrielle, Y., McCoy, F.W., Maekawa, H., Marlow, M.S., Milner, G.J., Mottl, M.J., Phipps, S.P., Rigsby, C.A., Saboda, K.L., Stabell, B., van der Laan, S., and Xu, Y., 1992a, Isotopic evidence for the origin of boninites and related rocks drilled in the Izu-Bonin (Ogasawara) forearc, Leg 125: Proceedings of the Ocean Drilling Program, Scientific Results: College Station, Texas, Ocean Drilling Program, v. 125, p. 237–261.

Pearce, J.A., Thirlwall, M.F., Ingram, G., Murton, B.J., Arculus, R.J., van der Laan, S.R., Fryer, P., Stokking, L.B., Ali, J.R., Arculus, R., Ballotti, D.L., Burke, M.M., Ciampo, G., Haggerty, J.A., Haston, R.B., Heling, D., Hobart, M.A., Ishii, T., Johnson, L.E., Lagabrielle, Y., McCoy, F.W., Maekawa, H., Marlow, M.S., Milner, G.J., Mottl, M.J., Phipps, S.P., Rigsby, C.A., Saboda, K.L., Stabell, B., van der Laan, S., Xu, Y., 1992b, Isotopic evidence for the origin of boninites and related rocks drilled in the Izu-Bonin (Ogasawara) forearc, Leg 125, *in* Dearmont, L.H., Mazzullo, E.K., Stewart, N.J., and Winkler, W.R., eds., Proceedings of the Ocean Drilling Program, Scientific Results: College Station, Texas, Ocean Drilling Program, v. 125, p. 237–261.

Pearce, J.A., Baker, P.E., Harvey, P.K., and Luff, I.W., 1995, Geochemical evidence for subduction fluxes, mantle melting and fractional crystallization beneath the South Sandwich island arc: Journal of Petrology, v. 36, p. 1073–1109.

Pedroni, A., Hammerschmidt, K., and Friedrichsen, H., 1999, He, Ne, Ar, and C isotope systematics of geothermal emanations in the Lesser Antilles Islands Arc: Geochimica et Cosmochimica Acta, v. 63, p. 515–532.

Pe-Piper, G., and Piper, D.J.W, 1994, Miocene magnesian andesites and dacites, Evia, Greece; adakites associated with subducting slab detachment and extension: Lithos, v. 31, p. 125–140.

Pe-Piper, G., Matarangas, D., and Jacobshagen, V., 1996, The mesozoic metavolcanic rocks of Alonnisos and Kyra Panagia islands, Sporades, Greece: Neues Jahrbuch für Mineralagie, v. 37, p. 251–263.

Perfit, M.R., Langmuir, C.H., Baekisapa, M., Chappell, B.W., Johnson, R.W., Staudigel, H., Taylor, S.R., Taylor, B., and Exon, N.F., 1987, Geochemistry and petrology of volcanic rocks from the Woodlark Basin; addressing questions of ridge subduction, *in* Taylor, B., and Exon, N.F., eds., Marine Geology, Geophysics and Geochemistry of the Woodlark Basin-Solomon Islands: Houston, Texas, Circum-Pacific Council for Energy and Mineral Resources, Earth Science Series, v. 7, p. 113–154.

Polat, A., and Kerrich, R., 2001, Magnesian andesites, Nb-enriched basalt-andesites, and adakites from late-Archean 2.7 Ga Wawa greenstone belts, Superior Province, Canada: implications for late Archean subduction zone petrogenetic processes, Contributions to Mineralogy and Petrology, v. 141, p. 36–52.

Portnyagin, M.V., Danyushevsky, L.V., and Kamenetsky, V.S., 1997, Coexistence of two distinct mantle sources during formation of ophiolites: a case study of primitive pillow-lavas from the lowest part of the volcanic section of the Troodos Ophiolite, Cyprus: Contributions to Mineralogy and Petrology, v. 128, p. 287–301.

Pubellier, M., and Cobbold, P.R., 1996, Analogue models for the transpressional docking of volcanic arcs in the western Pacific: Tectonophysics, v. 253, p. 33–52.

Pubellier, M., Quebral, R., Rangin, C., Deffontaines, B., Muller, C., Butterlin, J., Manzano, J., 1991, The Mindanao collision zone; a soft collision event within a continuous Neogene strike-slip setting, *in* Barber, A.J., ed., Orogenesis in action: Tectonics and processes at the west equatorial Pacific margin: Journal of Southeast Asian Earth Science, v. 6, p. 239–248.

Pubellier, M., Bader, A.G., Rangin, C., Deffontaines, B., and Quebral, R., 1999, Upper plate deformation induced by subduction of a volcanic arc: The Snellius Plateau (Molucca Sea, Indonesia and Mindanao, Philippines): Tectonophysics, v. 304, p. 345–368.

Puchtel, I.S., and Brugmann, G.E., 2001, 187Os-enriched domain in an Archean mantle plume: evidence from 2.8 Ga komatiites of the Kostomuksha greenstone belt, NW Baltic Shield: Earth and Planetary Science Letters, v. 186, p. 513–526.

Rampone, E., and Piccardo, G.B., 2000, The ophiolite-oceanic lithosphere analogue; new insights from the Northern Apennines (Italy) *in* Dilek, Y., Moores, E., Elthon, D., Nicolas, A., eds., Ophiolites and Oceanic Crust; New Insights from Field Studies and the Ocean Drilling Program: Geological Society of America Special Paper v. 349, p. 21–34.

Rampone, E., Hofmann, A.W., Piccardo, G.B., Vannucci, R., Bottazzi, P., and Ottolini, L., 1996, Trace element and isotope geochemistry of depleted peridotites from an N-MORB type ophiolite (Internal Liguride, N Italy): Contributions to Mineralogy and Petrology, v. 123, p. 61–76.

Rampone, E., Hofmann, A.W., and Raczek, I., 1998, Isotopic contrasts within the Internal Liguride ophiolite (N. Italy): The lack of a genetic mantle-crust link: Earth and Planetary Science Letters, v. 163, p. 175–189.

Rangin, C., 1991, The Philippine mobile belt; a complex plate boundary, in: Orogenesis in action; Tectonics and processes at the west equatorial Pacific margin: Journal of SE Asian Earth Science, v. 6, p. 209–220.

Rangin, C., Stephan, J.F., and Mueller, C., 1985, Middle Oligocene oceanic crust of South China Sea jammed into Mindoro collision zone (Philippines): Geology, v. 13, p. 425–428.

Rangin, C., Silver, E.A., and Tamaki, K., 1995, Closure of western Pacific marginal basins; rupture of the oceanic crust and the emplacement of ophiolites, *in* Taylor, B., and Natland, J., Active margins and marginal basins of the western Pacific: American Geophysical Union Geophysical Monograph 88, p. 405–417.

Rangin, C., Dahrin, D., and Quebral, R., 1996, Collision and strike-slip faulting in the northern Molucca Sea (Philippines and Indonesia); preliminary results of a morphotectonic study, *in* Hall, R., and Blundell, D.J., eds., Tectonic Evolution of Southeast Asia: Geological Society of London Special Publication, v. 106, p. 29–46.

Rangin, C., Spakman, W., Pubellier, M., and Bijwaard, H., 1999a, Tomographic and geological constraints on subduction along the eastern Sundaland continental margin (South-East Asia): Bulletin de la Societé Géologique de France, v. 170, p. 775–788.

Rangin, C., Le Pichon, X., Mazzotti, S., Pubellier, M., Chamot-Rooke, N., Aurelio, M., Walpersdorf, A., and Quebral, R., 1999b, Plate convergence measured by GPS across the Sundaland/Philippine Sea Plate deformed boundary; the Philippines and eastern Indonesia: Geophysical Journal International, v. 139, p. 296–316.

Rangin, C., Pubellier, M., Chamot-Rooke, N., Bijwaard, H., and Spakman, W., 2000, Late Cenozoic reconstructions in SE Asia: New GPS and tomographic constraints, AAPG Bulletin, v. 84, p. 1480.

Reagan, M.K., and Meijer, A., 1984, Geology and geochemistry of early arc-volcanic rocks from Guam: Geological Society of America Bulletin, v. 95, p. 701–713.

Reilinger, R.E., McClusky, S.C., Oral, M.B., King, R.W., Toksoz, M.N., Barka, A.A., Kinik, I., Lenk, O., and Sanli, I., 1997, Global Positioning System measurements of present-day crustal movements in the Arabia-Africa-Eurasia plate collision zone: Journal of Geophysical Research B, v. 102, p. 9983–9999.

Robertson, A., 1987, The transition from a passive margin to an Upper Cretaceous foreland basin related to ophiolite emplacement in the Oman Mountains: Geological Society of America Bulletin, v. 99, p. 633–653.

Robertson, A.H.F., 1991, Origin and Emplacement of an Inferred Late Jurassic Subduction Accretion Complex, Euboea, Eastern Greece: Geological Magazine, v. 128, p. 27–41.

Robertson, A.H.F., 1994, Role of the tectonic facies concept in orogenic analysis and its application to Tethys in the eastern Mediterranean region: Earth Science Reviews, v. 37, p. 139–213.

Robertson, A.H.F., 1998, Tectonic significance of the Eratosthenes Seamount: A continental fragment in the process of collision with a subduction zone in the eastern Mediterranean (Ocean Drilling Program Leg 160), *in*: Tectonophysics, v. 298, p. 63–82.

Robertson, A.H.F., 2002, Overview of the genesis and emplacement of Mesozoic ophiolites in the Eastern Mediterranean Tethyan region: Lithos, v. 65, p. 1–67.

Robertson, A.H.F., and Boyle, J.F., 1983, Tectonic setting and origin of metalliferous sediments in the Mesozoic Tethys Ocean, Hydrothermal processes at seafloor spreading centers, *in* Rona, P.A., Bostroem, K., Laubier, L., and Smith, K.L., Jr, eds., Hydrothermal processes at seafloor spreading centers: New York, NY, Plenum Press, p. 595–663.

Robertson, A., and Shallo, M., 2000, Mesozoic-Tertiary tectonic evolution of Albania in its regional eastern Mediterranean context: Tectonophysics, v. 316, p. 197–254.

Robertson, A.H.F., and Grasso, M., 1995, Overview of the late Tertiary; recent tectonic and palaeo-environmental development of the Mediterranean region, Later Tertiary-Quaternary Mediterranean tectonics and paleoenvironments: Terra Nova, v. 7, p. 114–127.

Rogers, N.W., MacLeod, C.J., Murton, B.J, 1989, Petrogenesis of boninitic lavas from the Limassol Forest Complex, Cyprus, *in* Crawford, A.J., ed., Boninites: Dordrecht, Kluwer Academic Publishers, p. 288–313.

Royden, L.H., 1993a, Evolution of retreating subduction boundaries formed during continental collision: Tectonics, v. 12, p. 629–638.

Royden, L.H., 1993b, The tectonic expression of slab pull at continental convergent boundaries: Tectonics, v. 12, p. 303–325.

Russo, R.M., and Silver, P.G., 1996, Cordillera formation, mantle dynamics, and the Wilson cycle: Geology, v. 24, p. 511–514.

Sajona, F.G., Bellon, H., Maury, R.C., Pubellier, M., Cotten, J., and Rangin, C., 1994, Magmatic response to abrupt changes in geodynamic settings; Pliocene-Quaternary calc-alkaline and Nb-enriched lavas from Mindanao (Philippines): Tectonophysics, v. 237, p. 47–72.

Schiano, P., Clocchiatti, R., Lorand, J.-P., Massare, D., Deloule, E., and Chaussidon, M., 1997, Primitive basaltic melts included in podiform chromites from the Oman Ophiolite: Earth and Planetary Science Letters, v. 146, p. 489–497.

Schmincke, H.-U., Rautenschlein, M., Robinson, P.T., and Mehegan, J.M., 1983, Troodos extrusive series of Cyprus; a comparison with oceanic crust: Geology, v. 11, p. 405–409.

Searle, M., and Cox, J., 1999, Tectonic setting, origin, and obduction of the Oman Ophiolite: Geological Society of America Bulletin, v. 111, p. 104–122.

Searle, M.P., and Malpas, J., 1980, Structure and metamorphism of rocks beneath the Semail Ophiolite of Oman and their significance in ophiolite obduction: Transactions of the Royal Society of Edinburgh, v. 71, Part 4, p. 247–262.

Searle, M.P., and Malpas, J., 1982, Petrochemistry and origin of sub-ophiolitic metamorphic and related rocks in the Oman Mountains: Journal of the Geological Society of London, v. 139, p. 235–248.

Şengör, A.M.C,Yımaz Y., and Süngürlü, O., 1984, Tectonics of the Mediterranean Cimmerides: Nature and evolution of the western termination of Palaeo-Tethys, *in* Dixon, J.E., and Robertson, A.H.F., eds., The Geological Evolution of the Eastern Mediterranean: London, Geological Society of London Special Publication 17, p. 77–112.

Seno, T., and Yamanaka, Y., 1998, Arc stresses determined by slabs: Implications for mechanisms of back-arc spreading: Geophysical Research Letters, v. 25, p. 3227–3230.

Serri, G., Spadea, P., Beccaluva, L., Civetta, L., Coltorti, M., Dostal, J., Sajona, F.G., Vaccaro, C., Zeda, O., Silver, E.A., Rangin, C., von Breymann, M.T., Berner, U., Bertrand, P., Betzler, C., Brass, G.W., Hsue, V., Huang, Z., Jarrard, R.D., Lewis, S.D., Linsley, B.K., Merrill, D.L., Mueller, C.M., Nederbragt, A.J., Nichols, G.J., Pubellier, M., Scherer, R.P., Sheu, D.D., Shibuya, H., Shyu, J.-P., Smith, R.B., Smith, T., Solidum, R.U., and Tannant, D.D., 1991, Petrology of igneous rocks from the Celebes Sea basement: College Station, Texas, Ocean Drilling Program, Proceedings of the Ocean Drilling Program, Scientific Results, v. 124, p. 271–296.

Shallo, M., 1992, Geological Evolution of the Albanian Ophiolites and Their Platform Periphery: Geologische Rundschau, v. 81, p. 681–694.

Shallo, M., Kote, D., Vranai, A., 1987, Geochemistry of the volcanics from ophiolitic belts of Albanides, in: Geochemistry of Ophiolites: Ofioliti, v. 12, p. 125–136.

Shinjo, R., 1999, Geochemistry of high Mg andesites and the tectonic evolution of the Okinawa Trough-Ryukyu Arc system: Chemical Geology, v. 157, p. 69–88.

Shinjo, R., Chung, S.L., Kato, Y., and Kimura, M., 1999, Geochemical and Sr-Nd isotopic characteristics of volcanic rocks from the Okinawa Trough and Ryukyu Arc: Implications for the evolution of a young, intracontinental back arc basin: Journal of Geophysical Research B, v. 104, p. 10,591–10,608.

Sibuet, J.-C., Hsu, S.-K., 1995, Taiwan arc-arc collision model: Eos (Transactions, American Geophysical Union), v. 76, p. 614.

Sigurdsson, I.A., Kamenetsky, V.S., Crawford, A.J., Eggins, S.M., and Zlobin, S.K., 1993, Primitive island arc and oceanic lavas from the Hunter Ridge-Hunter fracture zone; evidence from glass, olivine and spinel compositions: Mineralogy and Petrology, v. 47, p. 149–169.

Simonian, K.O., and Gass, I.G., 1977, The Arakapas fault belt, Cyprus; a fossil transform fault: Eos, Transactions of the American Geophysical Union, v. 58, p. 508.

Smewing, J.D., 1981, Mixing characteristics and compositional differences in mantle-derived melts beneath spreading axes; evidence from cyclically layered rocks in the ophiolite of North Oman: Journal of Geophysical Research B, v. 86, p. 2645–2659.

Smewing, J.D., Christensen, N.I., Bartholomew, I.D., Browning, P., 1984, The structure of the oceanic upper mantle and lower crust as deduced from the northern section of the Oman ophiolite, *in* Gass, I.G., Lippard, S.J., and Shelton, A.W., eds., Ophiolites and Oceanic Lithosphere: Geological Society of London Special Publication 13, p. 41–53.

Smith, A.G., and Spray, J.G., 1984, A half-ridge transform model for the Hellenic-Dinaric ophiolites, *in* The Geological Evolution of the Eastern Mediterranean: Geological Society of London Special Publication 17, p. 629–644.

Smoot, N.C., 1997, Aligned buoyant highs, across-trench deformation, clustered volcanoes, and deep earthquakes are not aligned with plate- tectonic theory: Geomorphology, v. 18, p. 199–222.

Snow, J.E., Rampone, E., and Hofmann, A.W., 1996, The unusual Os-Nd isotopic systematics of the Ligurian ophiolite and peridotite complexes, N. Italy: Eos, Transactions of the American Geophysical Union, v. 77, p. 785.

Snow, J.E., Schmidt, G., and Rampone, E., 2000, Os isotopes and highly siderophile elements (HSE) in the Ligurian ophiolites, Italy: Earth and Planetary Science Letters, v. 175, p. 119–132.

Spadea, P., Beccaluva, L., Civetta, L., Coltorti, M., Dostal, J., Sajona, F.G., Serri, G., Zeda, O., Silver, E.A., Rangin, C., von Breymann, M.T., Berner, U., Bertrand, P., Betzler, C., Brass, G.W., Hsue, V., Huang, Z., Jarrard, R.D., Lewis, S.D., Linsley, B.K., Merrill, D.L., Mueller, C.M., Nederbragt, A.J., Nichols, G.J., Pubellier, M., Scherer, R.P., Sheu, D.D., Shibuya, H., Shyu, J.-P., Smith, R.B., Smith, T., Solidum, R.U., and Tannant, D.D., 1991, Petrology of basic igneous rocks from the floor of the Sulu Sea: College Station, Texas, Ocean Drilling Program, Proceedings of the Ocean Drilling Program, Scientific Results, v. 124, p. 251–265.

Stampfli, G.M., and Borel, G.D., 2002, A plate tectonic model for the Paleozoic and Mesozoic constrained by dynamic plate boundaries and restored synthetic oceanic isochrones: Earth and Planetary Science Letters, v. 196, p. 17–33.

Stern, R.J., and Bloomer, S.H., 1992, Subduction zone infancy; examples from the Eocene Izu-Bonin-Mariana and Jurassic California arcs: Geological Society of America Bulletin, v. 104, p. 1621–1636.

Stern, R.J., and Pearce, J., 1998, Back-arc basin TTG suites; the Mariana Trough example: Geological Society of America Abstracts with Programs, v. 30, p. 259.

Stern, R.J., Bloomer, S.H., Lin, P.-N., and Smoot, N.C., 1989, Submarine arc volcanism in the southern Mariana Arc as an ophiolite analogue: Tectonophysics, v. 168, p. 151–170.

Sturm, M.E., Klein, E.M., Karsten, J.L., and Karson, J.A., 2000, Evidence for subduction-related contamination of the mantle beneath the southern Chile Ridge; implications for ambiguous ophiolite compositions, *in* Dilek, Y., Moores, E., Elthon, D., Nicolas, A., eds., Ophiolites and Oceanic Crust; New Insights from Field Studies and the Ocean Drilling Program: Boulder, Colorado, Geological Society of America Special Paper 349, p. 13–20.

Tamayo, R.A., Yumul, G.P., Maury, R.C., Polve, M., Cotten, J., and Bohn, M., 2001, Petrochemical investigation of the Antique ophiolite (Philippines): Implications on volcanogenic massive sulfide and podiform chromitite deposits: Resource Geology, v. 51, p. 145–164.

Tapponnier, P., Peltzer, G., and Armijo, R., 1986, On the mechanics of the collision between India and Asia, *in* Coward, M.P., and Reis, A.C., eds., Collision tectonics: London, Geological Society of London Special Publication 19, p. 115–157.

Tapponnier, R., Peltzer, G., Le Dain, A.Y., Armijo, R., and Cobbold, P., 1982, Propagating extrusion tectonics in Asia; new insights from simple experiments with plasticine: Geology, v. 10, p. 611–616.

Taylor, R.N., Nesbitt, R.W., Vidal, P., Harmon, R.S., Auvray, B., and Croudace, I.W., 1994, Mineralogy, chemistry, and genesis of the boninite series volcanics, Chichijima, Bonin Islands, Japan: Journal of Petrology, v. 35, p. 577–617.

Teng, L.S., 1990, Geotectonic evolution of late Cenozoic arc-continent collision in Taiwan, *in* Geodynamic Evolution of the Eastern Eurasian Margin: Tectonophysics, v. 183, p. 57–76.

Teng, L.S., Lee, C.T., Tsai, Y.B., and Hsiao, L.-Y., 2000, Slab breakoff as a mechanism for flipping of subduction polarity in Taiwan: Geology, v. 28, p. 155–158.

Umino, S., 1995, Geology of the Semail Ophiolite, northern Oman Mountains: Chigaku Zasshi (Journal of Geography), v. 104, p. 321–349.

Umino, S. and Kushiro, I., 1989, Experimental studies on boninite petrogenesis, *in* Crawford, A.J., ed., Boninites: Dordrecht, Kluwer Academic Publishers, p. 89–111.

Upadhyay, H.D., 1980, On the boninite-komatiite connection; some observations from Betts Cove and Belingwe: Eos, Transactions of the American Geophysical Union, v. 61, p. 1140.

Umino, S., Yanai, S., Jaman, A.R., Nakamura, Y., and Iiyama, J.T., 1990, The transition from spreading to subduction, evidence from the Semail Ophiolite, northern Oman Mountains, *in* Malpas, J., Moores, E.M., Panayiotou, A., and Xenophontos, C., eds., Ophiolites, oceanic crustal analogues, Proceedings of Troodos 1987: Nicosia, Cyprus, Ministry of Agriculture and Natural Resources, p. 375–384.

Van Bergen, M.J., Vroon, P.Z., and Hoogewerff, J.A., 1993, Geochemical and tectonic relationships in the East Indonesian arc-continent collision region; implications for the subduction of the Australian passive margin: Tectonophysics, v. 223, p. 97–116.

van der Laan, S.R., Flower, M.F.J., and Koster van Groos, A.F., 1989, Experimental evidence for the origin of boninites; near-liquidus phase relations to 7.5 kbar, *in* Crawford, A.J., ed., Boninites: Dordrecht, Kluwer Academic Publishers, p. 112–147.

van Keken, P.E., Kiefer, B., and Peacock, S.M., 2002, High-resolution models of subduction zones: Implications for mineral dehydration reactions and the transport of water into the deep mantle: Geochemistry Geophysics Geosystems (G^3), no. 1056 (10 October 2002).

Varga, R.J., Moores, E.M., Spreading structure of the Troodos Ophiolite, Cyprus: Geology, v. 13, p. 846–850.

Vine, F.J., and Matthews, D.H., 1963, Magnetic anomalies over oceanic ridges: Nature, v. 199, p. 947–949.

Wakabayashi, J., 1999, Subduction and the rock record; concepts developed in the Franciscan Complex, California, *in* Moores, E., ed., Classic Cordilleran Concepts: A View from California: Boulder, Colorado, Geological Society of America Special Paper 338, p. 123–133.

Wakabayashi, J., and Dilek, Y., 2000, Spatial and temporal relationships between ophiolites and their metamorphic soles; a test of models of forearc ophiolite genesis, Ophiolites and Oceanic Crust: New Insights from Field Studies and the Ocean Drilling Program: Boulder, Colorado, Geological Society of America Special Paper 349, p. 53–64.

Wakabayashi, J., and Unruh, J.R., 1995, Tectonic wedging, blueschist metamorphism, and exposure of blueschists; are they compatible?: Geology, v. 23, p. 85–88.

Walker, D.A., and Cameron, W.E., 1983, Boninite primary magmas; evidence from the Cape Vogel Peninsula, PNG: Contributions to Mineralogy and Petrology, v. 83, p. 150–158.

Webb, L.E., and Hacker, B.R., 1999, Thermochronologic constraints on deformation and cooling history of high- and ultrahigh-pressure rocks in the Qinling – Dabie orogen, eastern China: Tectonics, v. 18, p. 621–638.

Wortel, M.J.R., and Spakman, W., 2000, Subduction and slab detachment in the Mediterranean-Carpathian region: Science, v. 290, p. 1910–1917.

Xiao, W.J., Hou, Q.L., Li, J.L., Windley, B.F., Hao, J., Fang, A.M., Zhou, H., Wang, Z.H., Chen, H.L., Zhang, G.C., and Yuan, C., 2000, Tectonic facies and the archipelago-accretion process of the West Kunlun, China: Science in China, ser. D, v. 43, p. 134–143.

Yang, X.P., Fischer, K.M., and Abers, G.A., 1995, Seismic Anisotropy beneath the Shumagin Islands Segment of the Aleutian-Alaska Subduction Zone: Journal of Geophysical Research B, v. 100, p. 18,165–18,177.

Yilmaz, Y., 1993. New evidence and model on the evolution of the Southeast Anatolian orogen: Geological Society of America Bulletin, v. 105, p. 251–271.

Yumul, G.P., 2001, The Acoje block platiniferous dunite horizon, Zambales Ophiolite Complex, Philippines: Melt type and associated geochemical controls: Resource Geology, v. 51, p. 165–174.

Yumul, G.P., Jr., Dimalanta, C.B., Faustino, D.V., and de Jesus, J.V., 1998a, Translation and docking of an arc terrane: Geological and geochemical evidence from the southern Zambales Ophiolite complex, Philippines: Tectonophysics, v. 293, p. 255–272.

Yumul, G.P., Jr., Dimalanta, C.B., Faustino, D.V., and de Jesus, J.V., 1998b, Upper mantle-lower crust dikes of the Zambales ophiolite complex (Philippines); distinct short-lived, subduction-related magmatism: Journal of Volcanology and Geothermal Research, v. 84, p. 287–309

Yumul, G.P., Dimalanta, C.B., and Jumawan, F.T., 2000, Geology of the southern Zambales Ophiolite Complex, Luzon, Philippines: The Island Arc, v. 9, p. 542–555.

Manuscript Accepted by the Society May 2, 2003

Geological Society of America
Special Paper 373
2003

Where ophiolites come from and what they tell us

Adolphe Nicolas*
Françoise Boudier
CNRS/UMII, Laboratoire de Tectonophysique Université de Montpellier II, Place Eugène Bataillon, Montpellier 34095, France

ABSTRACT

The most usual approach in ophiolite studies is based mainly on petrological and geochemical techniques and has resulted in a consensus that most ophiolites have originated from a supra-subduction zone environment. We claim that this conclusion is disputable because it is commonly based on an insufficient geological knowledge of each ophiolite and, more generally, because such a restrictive environment is incompatible with the remarkable diversity of ophiolites. Rather than from either oceanic or arc-related environments, we think that this diversity results from the physical conditions of oceanic spreading, particularly spreading rate. Dealing with fast-spreading ridges and fast-spread ophiolites, mainly the East Pacific Rise and Oman, we show how mutually enriching a few controversies between marine geophysicists and field structural geologists have been, and still are. Integrated studies that combine geophysical and geochemical investigations in oceans with structural and geochemical work in ophiolites are key for further understanding of the origin and evolution of modern and fossil oceanic lithosphere.

Keywords: ophiolites, oceanic lithosphere, mantle tectonics, spreading rate.

INTRODUCTION

Since the first written definition of ophiolites by Brongniart (1821), three distinct episodes of discovery have marked the long history of ophiolites' relation with oceanic lithosphere. An early historical period was marked by a few premonitory views developed at the turn of the nineteenth century by Italian geologists working in the western Alps and relating the "pietre verdi" to ocean seafloor (see in Nicolas, 1989, p. 3, a review on this period). This period ended in the 1960s, during the early days of plate tectonics, when the new interest in oceanic lithosphere prompted new studies in ophiolites. A milestone was the "ophiolite manifesto" (Anonymous, 1972), in which American and European geologists, starting from very distinct positions, converged to adopt a common definition of ophiolites. The idea that ophiolites represented fossil oceanic lithosphere was not explicitly mentioned, but was largely accepted. Subsequent studies were predominantly geochemical and were rapidly dominated by the question of the oceanic setting of origin. The mid-oceanic ridge (MOR) setting was favored by geologists and supported by the isotopic signature of some ophiolites (Alabaster et al., 1980; Jacobsen and Wasserburg, 1984), whereas Miyashiro (1973), on the basis of major and trace elements of basalts, claimed that the Troodos ophiolite had been generated in an island arc environment. This is conveniently referred to as a supra-subduction zone (SSZ) environment, either an island arc environment or a backarc basin. At the same time, Pearce and Cann (1973) defined the discriminant diagrams that, on the basis of minor and trace elements, were supposed to allow us to distinguish the MOR or SSZ origin of any ophiolite. These diagrams were immediately applied to many ophiolites around the world, driving to the still dominant conclusion that most of these ophiolites were generated in SSZ environments (e.g., Ishiwatari, 1994).

The last period coincides with the discovery by marine geophysicists (Morton and Sleep, 1985; Detrick et al., 1987) of a seismic reflector at 2 km below the seafloor of fast-spreading ridges that was interpreted as a melt lens at the top of a magma chamber. This intrinsically very important discovery had a side effect that is also important in our discussion. Until this point, geophysicists had been reluctant to consider ophiolites as potential fossil remnants of oceanic ridges that were very similar to

*Adolphe.Nicolas@dstu.univ-montp2.fr

Nicolas, A., and Boudier, F., 2003, Where ophiolites come from and what they tell us, *in* Dilek, Y., and Newcomb, S., eds., Ophiolite concept and the evolution of geological thought: Boulder, Colorado, Geological Society of America Special Paper 373, p. 137–152. For permission to copy, contact editing@geosociety.org.

those they were studying. Among their reasons was the difficulty of accepting the existence of a magma chamber at shallow depth below an active ridge. Their reluctance was increased by the fact that most geologists were claiming that this chamber had a very wide top, a view that admittedly is untenable from both a mechanical and a thermal point of view. Ultimately, the melt lens discovery helped to bridge the gap between these two communities. Another development helped to establish a more confident and constructive relation between the ophiolite community and the marine geology and geophysics community—the collective results of the Ocean Drilling Program (ODP) Leg 147 (to the Hess Deep) in 1993 (Gillis et al., 1993). The foliated-layered gabbros of the Hess Deep originated at the East Pacific Rise, and displayed an astonishing petrological and structural analogy with the Oman ophiolite gabbros observed at the same depth. For many geophysicists, this analogy has been a strong incentive to integrate the ophiolite data into their ridge studies.

Where do we stand now regarding the origin of ophiolites? What are the present questions about their genesis and the questions we foresee in the near future? How do we address them? We need first to clarify our motivation in studying ophiolites. Widespread at Earth's surface, ophiolites can be studied for their record of regional paleogeographic and geodynamic evolution. Here, we are more interested in their contribution to the origin and evolution of the oceanic lithosphere that they represent and in using them to better understand the dynamics of oceanic spreading centers. Slow-spreading ridges are characterized by major topographic gradients, along which the internal structure of in situ oceanic lithosphere can be directly studied and sampled by submersibles and by dredging. In fast-spreading ridges, however, (and with the exception of a few deeps), the internal structure of oceanic lithosphere is masked by a uniform blanket of basalts. The contribution of ophiolites, where the deep structure of fossil oceanic lithosphere is directly exposed, thus is mainly interesting in the latter situation. Unlike slow-spreading activity, fast-spreading activity also tends to be time-independent on the scale of at least a few million years and to generate ridges that are homogeneous over lengths of 50–100 km. To better understand the dynamics of seafloor spreading, it is thus reasonable to consider primarily fast-spreading ridges, mainly the East Pacific Rise, which is the most studied of all. In this respect, structural and petrological studies in ophiolites that originated at fast-spreading centers complement remarkably well the geophysical studies conducted on active fast-spreading ridges.

Our main goal here is to illustrate by a few examples this combined approach of structural studies in selected ophiolites and marine geophysical studies conducted mainly at the East Pacific Rise. Karson (1998) presents an important contribution on this topic that is based on a detailed review of the tectonic windows in ocean floor through which some insight on the structure of the oceanic lithosphere is obtained. Karson emphasizes the contrast between the "stratiform architecture" of the ophiolites and these tectonic windows in the oceans. On the basis of such premises, this contrast is not a surprise. Our mapping over 500 km in the Oman–United Arab Emirates ophiolite (Nicolas et al., 2000) confirms both the general stratiform architecture of this huge ophiolite and the existence of a locally intense ridge tectonic activity. For instance, steepening of the Moho to the vertical at the propagator tips is described below. The associated attenuated ophiolite crust suggests that this situation compares with that of the East Pacific Rise deeps.

Before returning to this topic, we wish to address two controversial issues, and to present viewpoints that are based on detailed structural mapping of 15 ophiolite massifs and on more limited studies in a similar number of other massifs throughout the world. Finally, we call attention to similarities and differences in ophiolites that may help to better situate them with respect to oceanic lithosphere and to enlighten their origin.

TWO CONTROVERSIAL ISSUES

We discuss here two subjects that so far have been too loosely accepted as ground truth by most geoscientists interested in ophiolites. In so doing, we update but also substantially repeat earlier discussions (Nicolas, 1989, p. 199–201 and 258).

Ophiolites Originate Dominantly in SSZ Settings

From a very general point of view, the inferred SSZ origin is difficult to sustain for all ophiolites, simply because the ~70 ophiolites for which sufficient information of chemical affinity is available exhibit too much diversity to be so narrowly constrained. For example, a few ophiolites display a felsic extrusive section, such as the Canyon Mountain in Oregon (Misseri and Boudier, 1985) or the Guana Gato in Mexico (Lapierre et al., 1992), fitting well with an island arc environment origin, but most do not. A few ophiolites, such as Karmoy in Norway (Pedersen and Hertogen, 1990), are covered by island-arc related sediments, which points to an island arc environment or backarc basin origin, but many have a deep ocean sedimentary cover that is more compatible with an MOR origin. Incidentally, a fine analysis of the associated marine sediments may turn out to be the best criterion for unraveling the environment of origin of an ophiolite (Pesagno et al., 2000).

The generalized interpretation of SSZ origin of ophiolites has been largely spread out in literature since Pearce and collaborators (Pearce and Cann, 1973; Pearce et al., 1981) pointed out that the Troodos and Oman ophiolite lavas exhibit geochemical characteristics that are transitional between mid-ocean ridge basalt (MORB) and island arc tholeiite (IAT). In Troodos, lower pillow basalts are similar to ocean floor basalts, although upper pillow lavas and subsequent Arakapas ultramafic volcanics have high-Mg, low-Ti, and other high-strength field element (HSFE) concentrations that point to affinities with IAT (Pearce, 1975), besides a large variability of these characteristics (Cameron, 1985). In Oman, the range from MORB to IAT signatures suggests a succession of volcanic episodes (Alabaster et al., 1982): the "Geotimes" ridge-axis volcanics (V1 for Ernewein et al., 1988) are very close to MORB, and subsequent seamount-type lavas (V2), "Lasail, Alley, Cpx-phyric" trace an evolution toward an "arc signature." The chronology of this evolution is well constrained to 1–2 Ma on the basis of biostratigraphy (Tippit and Pessagno, 1981), and is consistent with the time difference between V1 and the initial detachment of the ophiolite from its igneous setting (e.g., Hacker, 1994).

Actually, the arc signature covers a large spectrum of geochemical characters, variably represented in present forearc and arc volcanic compositions, and whose end member, boninite, is well defined (e.g., Juteau and Maury, 1997, p. 33–35) by (1) an extremely depleted character, i.e., high silica and magnesium content and high depletion in rare earth elements (REE) and HFSE (Zr, Ti, Nb, Ta), interpreted as resulting from a high degree of partial melting in the mantle wedge, and (2) an enrichment in highly incompatible large ion lithospheric elements (LILE), interpreted as a subsequent contamination by hydrous fluids. It is the occurrence of one or the other of these characters in volcanics overlying many ophiolites that supported the common thinking that ophiolites originated in a SSZ environment.

A major problem concerning this model is how to explain the rapid shift from an MOR to an SSZ signature, either in time, as in Oman (~1–2 Ma; Hacker, 1994), or in space, as in Albania (20 km, Shallo et al., 1995). Keeping in mind that the SSZ signature reflects fluid contamination during melting of a depleted mantle source, can we suggest possibilities to produce a SSZ signature other than the arc-related fluid flux in the mantle wedge overlying a subduction zone? Early in the study of the Oman ophiolite, we related the V2 lavas to the initial stage of oceanic thrusting of the ophiolite (Boudier et al., 1988) and explained its occurrence by a water contamination of the melt source due to the dehydration of the underthrust oceanic crust, a process akin to a subduction-related melting. Other hypotheses may be envisaged as well, such as contamination by seawater of the still hot mantle at a distance of a few tens of kilometers from the ridge, in keeping with the recent discovery of deep and hot hydrothermalism near ridge axis (Manning et al., 2000; Nicolas et al., 2003). This second hypothesis has been further developed by Godard et al. (2003) to explain the transitional trace elements and isotopic strontium signatures of V1 Geotimes and V2 Lasail in Oman volcanics. In compilations that include recent data on seamounts from the East Pacific Rise (Niu and Batiza, 1997) and North Fiji Basin (Fleutelot, 1996), the authors point to the fact that the V2 Lasail lavas, which had been assigned to SSZ origin on the basis of incompatible elements ratios, were lying in the extended field of seamounts. We wish to defend an open-minded attitude on the basis of the fact that the great variability of "arc signature" documented in volcanics from ophiolites suggests that different processes (not necessarily understood yet) may rule this diversity. When the geochemical approach does not meet with data from other origin, an SSZ setting should not be regarded as decisive on the basis only of geochemical signature of the off-axis volcanics.

Abyssal and Ophiolitic Peridotites

Many studies have been devoted to the specimens of mantle peridotites collected in the oceans, the most extensive study being that of Dick et al. (1984). Their data are still largely used, remaining the most extensive source for petrological and geochemical modeling of oceanic lithosphere. Comparing their data on abyssal peridotites to those available on peridotites from ophiolites, Dick and Bullen (1984) conclude that there are significant discrepancies. They observe that, on average, the ophiolite mantle is more depleted than the oceanic one and conclude that ophiolites should result from a higher degree of melting, and thus should be good candidates for the hydrous mantle from SSZ environments. We wish to reiterate our doubts on the relevance of the data on abyssal peridotites, recalling that they represent only slow-spreading oceanic lithosphere and associated transform faults, where mantle specimens are readily available on the seafloor. With no basalts attached to them, they may not represent constructional oceanic lithosphere. Mantle specimens from fast-spreading environments are exceptional, being collected along a few transform faults and deeps, which probably are not representative of the normal fast-spreading lithosphere. Since the Dick et al. study, more analyses have been published on the East Pacific Rise mantle. As an example, in Figure 1, we have plotted the characteristic chromium number index measured in peridotites from the fields of abyssal domains, East Pacific Rise, island arcs, and three selected ophiolites. The biased sampling of "abyssal peridotites" in favor of slow-spreading lithosphere is clearly reflected in this figure. The Bay of Islands ophiolite, thought to represent moderate spreading rate conditions (5 cm/yr; Suhr, 1992) plots in this field, and the Oman ophiolite attributed to fast-spreading conditions plots in the East Pacific Rise field. The Josephine ophiolite plots in the SSZ domain, in agreement with the geodynamic evolution of this Jurassic ophiolite in the Mesozoic California. Altogether, abyssal peridotites have a more fertile signature than most ophiolites. This is not surprising, knowing that the majority of ophiolites were derived from moderate- to fast-spreading environments (see below). From this discussion, we draw two conclusions: (1) that the high chromium number of many ophiolites does not necessarily mean that they originated in SSZ environments, and (2) that petrological and geochemical modeling of fast-spreading ridges should rather turn toward ophiolites and stay away from the abyssal peridotites reference.

MARINE GEOPHYSICAL AND OPHIOLITIC STRUCTURAL DATA: THE NECESSARY CONFRONTATION

As mentioned above and stressed by Karson (1998), the flat relief of fast-spreading ridges masks all internal structures. The geophysical, mainly seismological, tools provide only indirect information. This is where structural data from ophiolites with a clear fast-spreading center origin can provide the missing information; however, matching the results from both sources

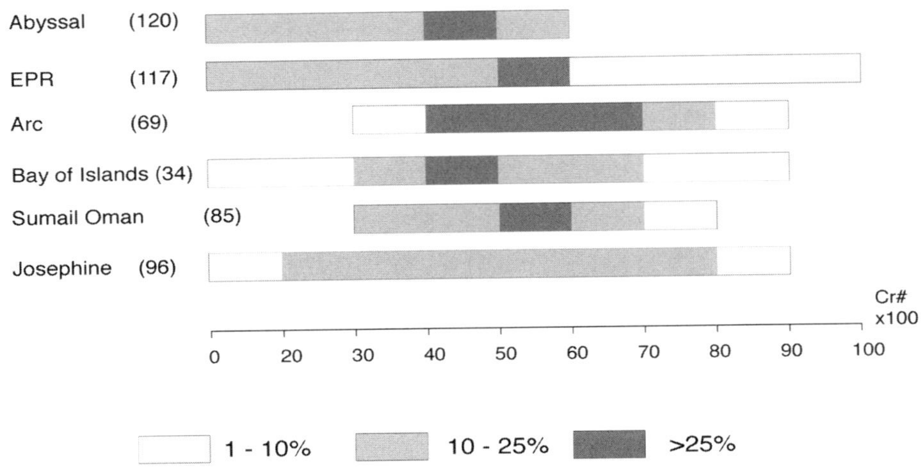

Figure 1. Fields of the chromium number in spinels from abyssal plain, East Pacific Rise (EPR), arc-related and selected ophiolite peridotites, mainly harzburgites. Data are presented in fractions of the total number of analyses bracketed in the figure. The data for abyssal peridotites are from Dick and Bullen (1984); for EPR, from Dick and Natland (1996), Nicolas et al. (1971); for arcs, from Parkinson and Pearce (1998), and for ophiolites, from Dick and Bullen (1984) regarding Bay of Islands and Josephine, and from Kelemen (1995) and Gerbert-Gaillard (2002) regarding Oman.

is difficult for two reasons. The main one deals with the difference in the scale; the resolution of seismology is poor below ~10 km, below which the scale of structural resolution usually operates. We notice here that the reason why most structural reconstructions are limited is not intrinsic, but imposed by the field conditions—the size of the massifs, and their internal dismembering. In this respect, the Oman–United Arab Emirates ophiolite constitutes an outstanding exception, with continuity in the field over many tens of kilometers. The second difficulty in matching oceanic and ophiolitic results is more apparent than real. It derives from the fact that ophiolites represent remnants of fossil lithospheres, whereas most geophysical studies deal with active spreading centers. As discussed below, many ophiolites, though fossil, have sampled spreading centers, thus keeping the structural record of their activity; this phenomenon makes the two sources complementary.

We now use a few examples drawn from our own experience to illustrate the difficulty and wealth of confronting marine geophysical and ophiolitic data.

Diapir Controversy

From the mapping of steep and concentric high temperature mantle flow structures in the Zambales (Philippines) and the Cyprus ophiolites, Nicolas and Violette (1982) have proposed that the asthenospheric mantle flow beneath oceanic ridges was diapiric (three-dimensional). The East Pacific Rise segmentation was discovered at the same time (MacDonald and Fox, 1983; Lonsdale, 1983; Whitehead et al., 1984) and its scale of 50–100 km, comparable to the spacing of transform faults along the Mid Atlantic Ridge, prompted the idea that the uprising mantle could be partitioned accordingly. This resulted in a number of three-dimensional models of mantle upwelling (Whitehead et al., 1984; Rabinovicz et al., 1984, 1987; Scott and Stevenson, 1989). The increasing confinement for lower spreading rates predicted by the last model has been confirmed by subsequent analog (Magde et al., 1996) and numerical models (Lin and Morgan, 1992; Sparks and Parmentier, 1994; Magde et al., 1997) and by the discovery of a strong mantle Bouguer negative anomaly in the middle of the Mid Atlantic Ridge segments (Detrick et al., 1995), suggesting that a hot and melting asthenosphere is rising from below. In contrast, the East Pacific Rise has no clear mantle Bouguer anomaly, suggesting a two-dimensional mantle uprising. The remarkable MELT experiment was run in the fastest and most linear part of the East Pacific Rise to test the mantle flow and melting below through a combination of seismology, gravity, and electromagnetism. With the resolution of the experiment limited to a depth of 100 km, the mantle uprising is clearly two-dimensional (Melt Seismic Team, 1998). Finer scale geophysical investigations conducted at the 9° N East Pacific Rise natural laboratory do not rule out that, on the scale of 10 km, mantle upwelling may have a three-dimensional structure (Barth and Mutter, 1996; Dunn et al., 2000; Canales et al., 2003) (Fig. 2).

In the meantime, increasingly detailed structural studies in the Oman ophiolite (Ceuleneer et al., 1988; Nicolas and Boudier, 1995; Jousselin et al., 1998) (Fig. 3) confirmed the occurrence of several frozen diapirs. Because the Oman ophiolite is regarded by all authors as being issued from a fast, possibly a super fast (Nicolas et al., 2000) spreading center, a comparison is possible with the East Pacific Rise. The most apparent explanation regarding the discrepancy between two- and three-dimensional mantle uprising is that the resolution of most East Pacific Rise geophysical data is not yet sufficient to image small diapirs, the size of those mapped in Oman. Another explanation is that the small mantle uprising diapirs in Oman, breaking just below Moho where they are mapped, may be local heads of larger three-dimensional mantle upwellings whose flow would rotate at a greater depth, below the 5–10 km depth sampled by the ophiolite. Following this idea, Nicolas and Boudier (2000) have mapped in this ophiolite the domains of intense melt circulation that should be related to mantle uprising. Such domains are characterized by the development of thick Moho transition zones

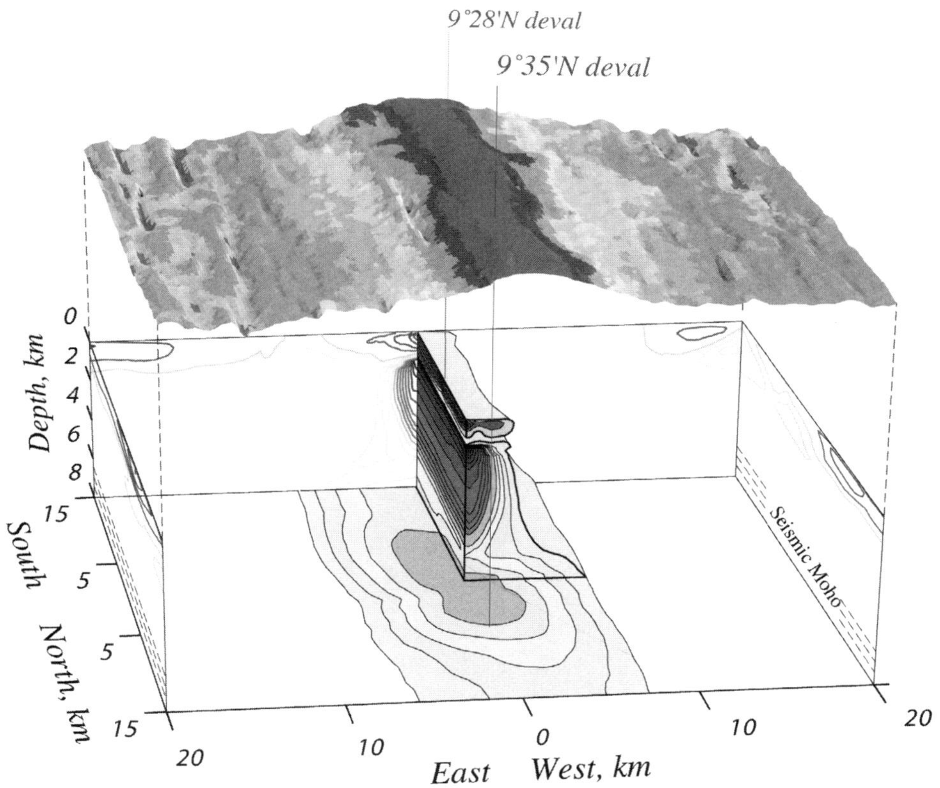

Figure 2. Seismic tomography of the crust and the uppermost mantle at 9° North along the East Pacific Rise. Contour lines are P-waves delay time; contour interval is 0.2 km s^{-1}. The darker shadings correspond to domains of increasing seismic attenuation, interpreted as being warmer. Modified from Dunn and Toomey (1997) and Dunn et al. (2000).

(MTZ) where the mantle harzburgites are replaced by dunites and by a similarly large development of olivine gabbros and wehrlitic intrusions into the lower crust. This mapping has revealed the existence of a few areas, some 50 km across, incorporating the small mapped "structural" diapirs of Figure 3. These larger domains, in which the MTZ is thicker on average and the gabbro unit is richer in olivine, could be the petrological signature of larger mantle diapirs whose size seems more compatible with the scale of segmentation expected below oceanic ridges.

Seismic Anisotropy And Mantle Tectonics

In Figure 2, P-wave anisotropy, deduced from crossed seismic lines, shows a fast-velocity axis remarkably parallel to the spreading direction at a large distance from the ridge, which is comparable to the findings of the MELT experiment. This fast seismic velocity is now interpreted as being parallel to the plastic flow direction in the mantle. At the finer scale of the ellipse of attenuated seismic velocities seen in Figure 2, which is that of the diapirs mapped in Oman, the expected mantle flow should not be so regularly oriented, but rather should be diverging. The measurements necessary to check this point seem now in view and this question may serve as an entry into a new and promising research domain.

The interest for new seismic waves, such as the SKS shear waves, which rise vertically from the core mantle boundary and whose splitting can be used to know the local seismic anisotropy in the horizontal plane (Silver, 1996), has paved the way for new studies on what has been called mantle tectonics. Because it is possible to obtain a large number of such data on a regional scale, it becomes possible to relate the mantle flow pattern deduced from them to the regional tectonics deduced from surface studies. This has been illustrated by beautiful examples incorporating the upper mantle into continental crust tectonics (Barruol and Souriau, 1995). It should not be long before such techniques are commonly applied to the oceanic lithosphere, revealing the patterns of upper mantle flow. This will make a direct comparison possible, at the same 10 km scale, with similar flow patterns deduced from structural studies in ophiolites. Spectacular progress in better understanding of mantle flow below oceanic ridges is in view.

What is a Magma Chamber?

The magma chamber in which the layered gabbros of ophiolites were generated traditionally has been regarded as a large melt pouch in which crystals would be sorted by gravity after nucleation at the colder ceiling and the heavier crystal mush settled on the floor. This view was shattered by the discovery of the perched melt lens at fast-spreading ridges <2 km wide and only a few tens of meters thick (Singh et al., 1998). Large-angle seismic reflection studies have shown that between the melt lens and the Moho there is a wide triangular domain of strong

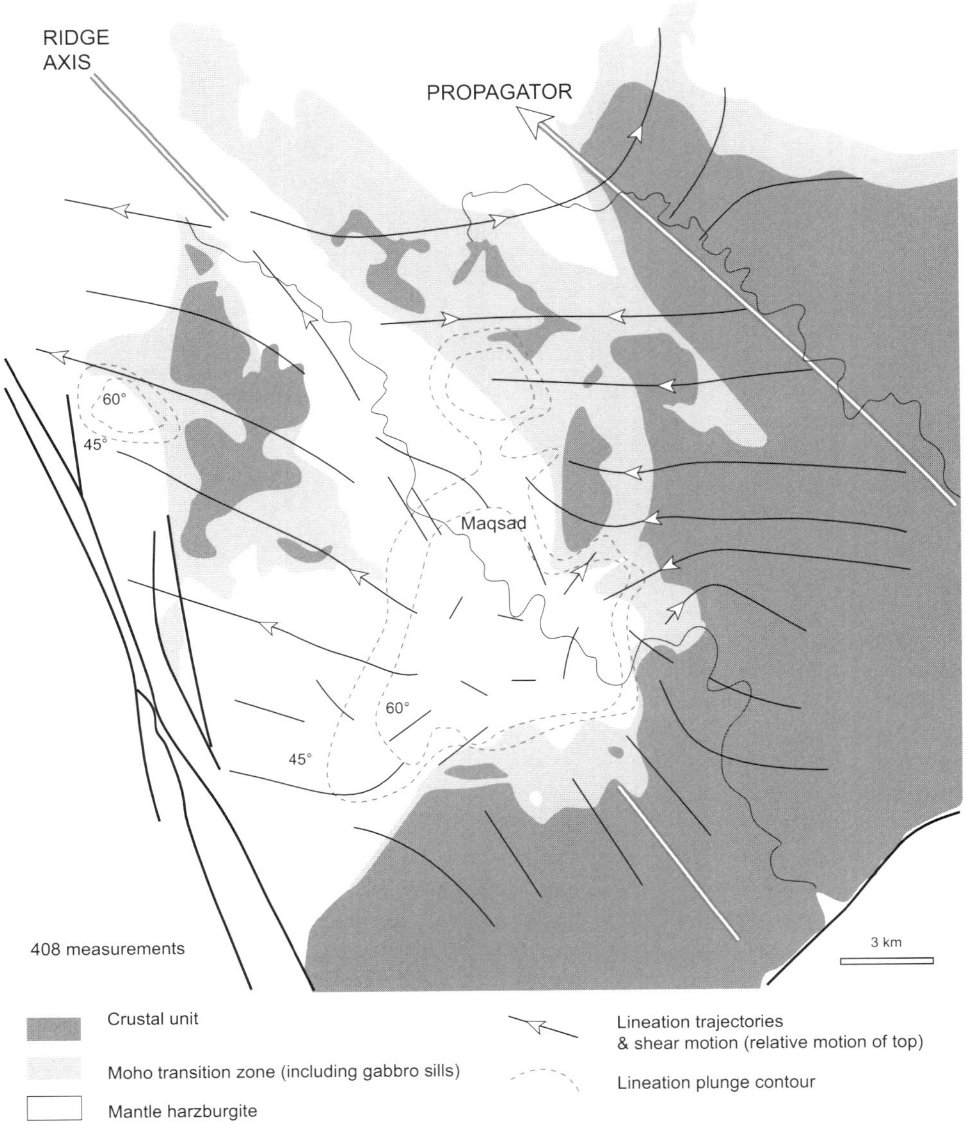

Figure 3. Trajectories of mineral lineations in the mantle equated with mantle flow trajectories in and around a mantle diapir in Oman (Maqsad in Semail massif). The Moho is nearly horizontal (10° SE dip). The dashed lines visualize the 45° and 60° plunge contours and the arrows indicate the dominant shear sense (motion of upper block/lower block). Based on 408 field stations (Jousselin et al., 1998).

seismic attenuation (Fig. 2). This attenuation has been ascribed to the presence of hot gabbros with trapped melt pockets. For marine geologists and geophysicists, the magma chamber thus is reduced to a melt lens. The vast attenuated domain underneath would have been filled with the gabbros that started crystallizing within the melt lens before subsiding as a gabbro mush and eventually evolving into a deforming solid (Quick and Dellinger, 1993; Phipps Morgan and Chen, 1993). This view is incompatible, however, with the ophiolite record showing that the gabbro pile still displays magmatic textures and not solid-state deformation textures. Structural studies conducted in the gabbro unit of the Oman ophiolite show that these gabbros were intensely deformed during their crystallization by a process akin to suspension flow, with no trace of plastic deformation, except locally at the Moho level (Nicolas, 1992). The critical melt fraction above which suspension flow is possible is ~40%, far above the fraction initially accepted for the attenuated domain below ridges, which was <10%.

By combining different approaches, a solution to this dilemma seems foreseeable. Moreover, the studies have resulted in significant advances in the fields of both petrophysical modeling of seismic data and physical properties, and rheology of gabbro mush. Taking into account the large structural anisotropy induced in the gabbros by their large deformation and by the observed flat shape of melt pockets in Oman gabbros, it is possible to calculate the seismic attenuation for any melt fraction. Melt fractions have been estimated by comparing the attitudes of the gabbros known from field studies to the seismic paths below the ridge. They are currently found to be >10% and possibly as high as 24% (Lamoureux et al., 1999) (Fig.4).

In the meantime, analog experiments have shown that the critical liquid fraction could be reduced to ~25%–30% when flat particles were submitted to a large shear flow, because of their strong alignment compared to their random initial assemblage (Nicolas et al., 1993). In these experiments, it was also observed that the flat particles would slip one on top of the other until blocked by disoriented laths. The discovery in the gabbros of microtextures of impingement of one plagioclase lath into another, indicative of pressure solution, has been interpreted as nature's response to the problem of blockage. A model has been proposed in which crystals slip by suspension flow until blockage and activation of pressure solution (Nicolas and Ildefonse, 1996). With this model, the melt fraction can be reduced drastically, the absolute limit being the few percent of melt where pressure solution would operate alone. The mush viscosity can be estimated from the model, and can be as high as 10^{15} Pa s^{-1}, a value indepen-

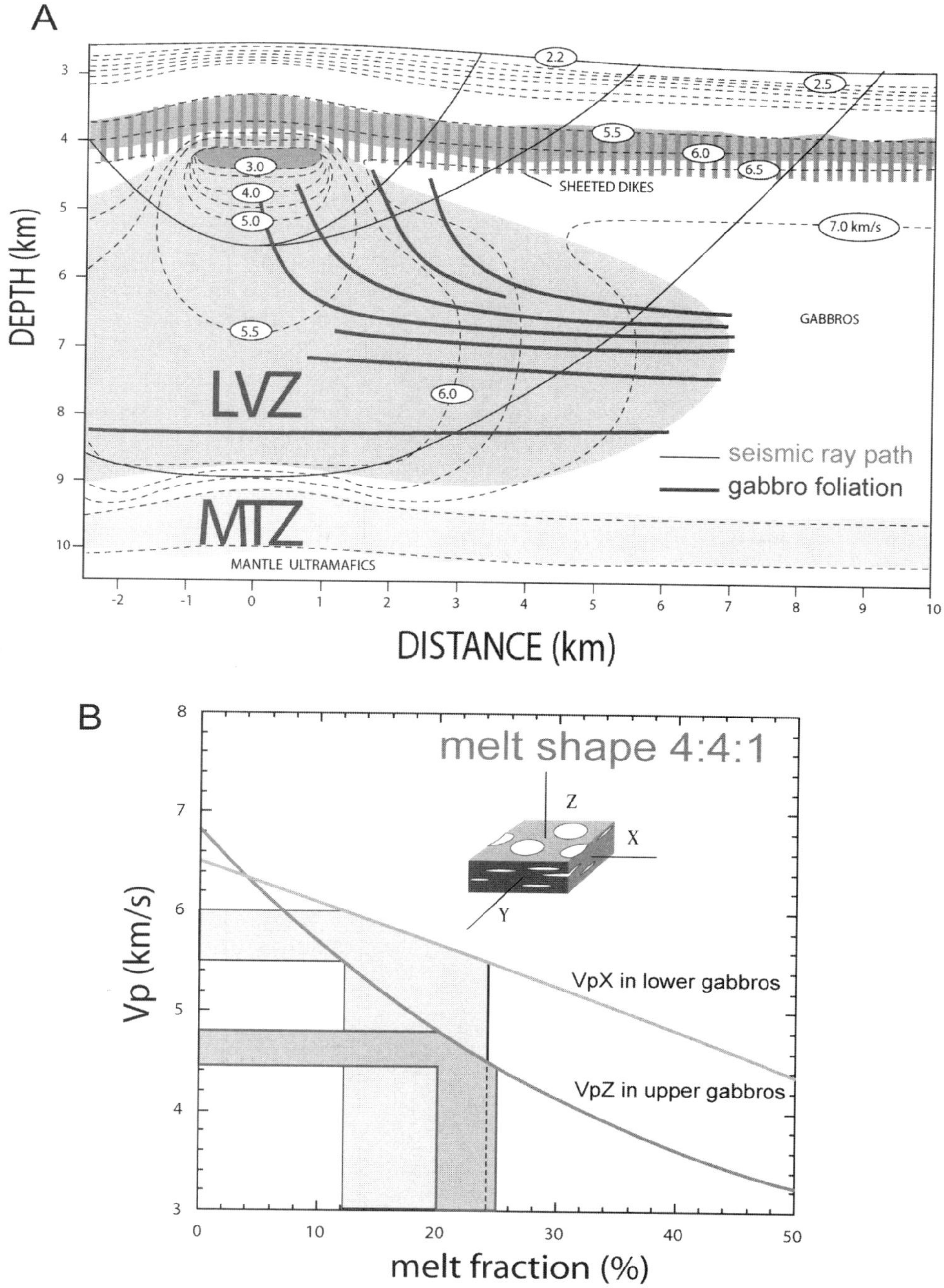

Figure 4. A: Estimations of melt fractions in the crustal low-velocity zone (LVZ) below East Pacific Rise on the basis of the East Pacific Rise seismological data and structural measurements in the Oman gabbros. B: These data are integrated with petrological modeling of the seismic anisotropy induced by both the crystallographic fabrics and the geometry of melt lenses in the gabbros. Because the lower gabbros are dominantly horizontal, the seismic velocity is close to the modeled VpX; because the upper gabbros are dominantly vertical, the seismic velocity is close to the modeled VpZ. The modeled VpX and VpZ correspond to melt fractions of 12%–24% (light gray area) and 20%–25% (dark gray area), respectively (in Mainprice, 1997).

dently derived from a physical numerical modeling of the Oman magma chamber (Chenevez et al., 1998).

We conclude that the attenuated domain below fast-spreading ridges should be considered to be the magma chamber as the continuation of the melt lens above; however, it is filled with a mush nearly as viscous as a plastically deforming body. Comparing the structures in the Oman gabbros with those of the Skaergaard Complex in Greenland, the best-documented crustal magma chamber, we observed no significant differences between them and suggested that the crystallizing gabbros deposited below the free surface in Skaergaard were rapidly forming a thick mush during their gravity-driven flow, much like the Oman gabbros (McBirney and Nicolas, 1997).

Tectonic Activity Below the Basalt Blanket of Fast-Spreading Ridges

Except for special areas such as fractures zones or microplates, the seafloor of fast-spreading ridges is remarkably flat with only few tens of meters of throw associated with normal faults parallel to the ridge, the small nodal basins between overlapping spreading centers and, of course, seamounts. The gravity signal also is very weak. These features suggest that the deep crustal and uppermost mantle structures are also simple and presumably are dominantly horizontal. Turning to Oman again, this is exactly what has been mapped there, particularly within the Aswad massif in the northern United Arab Emirates. In the western part of this massif, the Moho is absolutely horizontal with the lower gabbros defining cuestas and buttes over distances of tens of kilometers. In other parts of the belt, however, the Moho can be vertical more than 10–20 km along-strike, where it is commonly underlain by shear zones that are a few tens of meters thick, reflecting conditions of ~1000 °C in the mantle. Above, the lower crust has suffered an intense hydrous recrystallization; it is also strongly deformed and cut by amphibolite facies shear zones that are branching on the steep Moho shear zones. Such peculiar areas are also invaded by mafic dikes that are mostly noritic and that evolve to websterites in the mantle section. All these features point to an activity occurring very close to the ridge axis. In contrast to these vertical structures observed in the middle of the ophiolite section, both the basal contact and the basalt flows and sediments on top are moderately inclined and nearly parallel to each other (Fig. 5). These areas characterize the limits of segments and are most developed at the tips of propagators, as typical in the western part of the Nakhl massif, the eastern part of the Haylayn massif, and the center of the Fizh massif. The Mansah area in the Semail massif where an off-axis diapir has been mapped (Jousselin and Nicolas, 2000a) displays similar features. They are similarly interpreted to be a result of the uprising mantle punching a soft lithosphere, not older than 1 Ma. The asthenospheric pressure would tilt the Moho, fracture the crust, allow penetration of seawater to this depth, and thus favor the development of hydrous magmatism, as proposed by Boudier et al. (2000). To conclude, we suggest that the deep crust and uppermost upper mantle of fast-spreading ridges might be locally intensely tectonized, as observed in the Oman ophiolite, but invisible from the seafloor because of the very efficient ductile relaxation in this hot lithosphere.

OPHIOLITES AND SPREADING RATES

Ishiwatari (1985a) and Boudier and Nicolas (1985) independently proposed to classify ophiolites according to how the nature of their mantle constituents reflects the degree of partial melting in the mantle of origin. Ishiwatari considered petrological and chemical criteria in gabbros and basalts, whereas Boudier and Nicolas introduced more general criteria, largely structural. Going a step further, these authors related mantle depletion to spreading rate of the ridge of origin, a point that is discussed below. Boudier and Nicolas gave the names LOT (for lherzolite ophiolite type) to ophiolites whose mantle was lherzolitic and less depleted, and HOT (for harzburgite ophiolite type) to ophiolites whose mantle was harzburgitic and more depleted. Ishiwatari proposed three types: his "Liguria" type coincided with LOT and his "Papua" type with HOT, and his "Yakuno" type was in between. Several criteria are attached to the LOT and HOT types, the most prominent being the reduction or even the absence of the gabbro unit in LOT compared to HOT. The reduction affects mainly the layered gabbros that, when present, are usually poorly structured compared to those in HOT (Table 1).

Boudier and Nicolas (1985) related the degree of partial melting to the spreading rate, a position that has been and is still disputed because excess or reduced melting with respect to an average asthenospheric mantle uprising below an oceanic ridge can be explained equally well by thermal or compositional differences in the rising asthenosphere. For example, because it is under the influence of

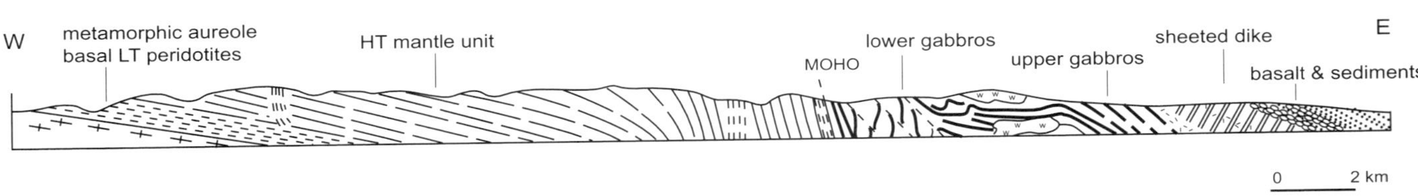

Figure 5. Cross-section through the Oman ophiolite nappe in the northern Fizh massif (modified from Boudier et al., 1988). The Moho has been sheared and tilted to the vertical, whereas the lower and upper contacts of the thrust nappe are gently east-dipping. The revealed internal tectonic activity has no visible consequence on the seafloor. LT—low temperature; HT—high temperature.

TABLE 1. OPHIOLITES TYPES

	Harzburgite ophiolite type (HOT)	Harzburgite lherzolite ophiolite type (LHOT)	Lherzolite ophiolite type (LOT)
MASSIFS CONSIDERED	Oman-UAE (1), Papua (2), New Caledonia, southern massif (3)	Bay of Islands (4), Yakuno (5), Mirdita (6), Troodos (7), Antalya (8), Kizildag (9), Aladag (10), Xigatse (11), Muslim Bagh (12), Josephine (13), Vourinos (14), Zambales (15)	Lanzo (16), Liguria (17), Trinity (18), Othris (19)
CRUSTAL SECTION	Thickness: 4–6 km	Thickness: 2–3 km	Thickness: 0–1 km
Volcanics overlying sheeted dikes	Low-alumina tholeiite	High-alumina tholeiite	Alkali basalt
Sheeted dikes	Well expressed and steeply dipping	Steeply to moderately dipping (4, 7, 9, 12, 13, 14, 15), horizontal (11, 15), absent or poorly organized (5, 6, 8)	Absent to poorly organized sills or dipping dikes
Gabbros			
exposure	Thick, continuous (1, 2)	Thin, continuous (4, 12, 13, 14, 15), nestled (7, 9), locally absent (5, 6, 9, 11)	Absent or restricted to 100-m-sized bodies
lithology	Lower: Ol-gabbro (1, 3) Upper: Ol-gabbro and gabbronorite	Dominant gabbronorite (4, 5, 8, 10, 13, 14, 15), Ol-gabbros (6, 7, 9, 11, 14) and ferrogabbro	Ol-gabbro (16, 17, 18)
structure	Lower layered gabbro Upper foliated gabbro (1,2)	Well layered (8, 9, 10, 13, 14, 15), poorly layered-foliated gabbro (4, 6, 7, 11, 12, 13), isotropic gabbro	Isotropic gabbro, foliated gabbro
penetrative deformation	Magmatic deformation	Magmatic to plastic (flasergabbro) deformation	
shear bands	Amphibolite facies, steep, horizontal motion	Granulite to greenschist facies, listric faults (6, 9, 12, 13)	Common
wehrlite intrusions	Abundant (1, 3)	Present (4, 6, 7, 12, 14)	Absent
MANTLE SECTION			
lithology	Harzburgite and dunite	Cpx-harzburgite, locally Sp-,Plag-lherzolite and dunite (5, 6, 11, 15)	Plag- (Sp-) lherzolite
temperature of deformation	Very high to high temperature	High temperature to moderate temperature	Moderate temperature to low temperature
internal structures	Flat foliation (except diapirs)	Flat (4, 6, 11, 12, 13, 14) to steep (7, 8, 14, 15) foliation	Steep foliation (14, 16, 17)
shear zones	Uncommon, vertical with flat lineation	Common, steep with steep lineation (8, 10, 11, 14)	Very common, steep
diabase occurrence	Uncommon	Common (8, 10, 11, 12, 14, 15)	Common dikes and sills in upper section
moho transition zone	Interlayering of dunite, and gabbro, websterite lenses, variable thickness <0.5km	Locally thick (up to 2 km) dunite, interlayered with wehrlite, websterite (4, 5, 6, 10, 13, 15)	Dunite lenses
chromite pods	Uncommon (1, 2), common (3)	Abundant	Uncommon, absent
serpentinization	Lizardite	Lizardite and antigorite	Antigorite

Note: Revised classification of ophiolites (Boudier and Nicolas, 1985) based on nature of their mantle section. Data from: 1—Nicolas et al. (2000); 2—Davies (1977); 3—Cassard (1980); 4—Suhr (1992); 5—Ishiwatari (1985b); 6—Hoxha and Boullier (1995); Nicolas et al. (1999); 7—George (1978); Thy and Moores (1988); 8—Juteau (1974); 9—Dilek et al. (1999); Dilek and Thy (1998); 10—Cakir (1978); Dilek et al. (1999); 11—Girardeau and Mercier (1988); 12—Mahmood (1994); 13—Dick (1977); Harper (1982); 14—Rassios et al. (1983); 15—Evans and Hawkins (1989); Abrajano et al. (1989); 16—Boudier (1978); Bodinier (1988); 17—Abbate et al. (1994); Rampone and Piccardo (2000); 18—Quick (1981); Le Sueur et al. (1984); Cannat and Boudier (1985); Boudier et al. (1989); 19—Menzies and Allen (1974); Ferrière (1985); Dijkstra et al. (2001).

the Iceland hot spot, the Reykjanes Ridge in the northern Atlantic Ocean has a fast-spreading ridge structure for a slow spreading rate. The hydrous melting of the mantle slab located above a subduction zone illustrates the role of composition, readily generating depleted mantle rocks whatever the spreading rate is. The discussion developed above on SSZ or MOR origin of ophiolites shows that it is not easy to take into account the compositional factor. The situation is more difficult with the thermal factor. We are not aware of any study of ophiolites that discusses the relation of ophiolites with hot spot or cold spot environments. Another difficulty relating ophiolite facies to spreading rate, also emphasized by Karson (1998), appears with the heterogeneous structure of slow-spreading ridges, where there are major differences between the segment centers and the vicinity of fracture zones. As documented by Karson and Rona (1990) and Tucholke et al. (1998), tectonically denudated mantle "core complexes" may occur in fractures zones that contrast with "normal" crustal structure in the middle of ridge segments. This difficulty is overcome in the Mirdita ophiolite in Albania (see below), which is large enough to encompass both a "core complex" and a "normal" lithosphere; but in a smaller ophiolite displaying either of these components, a wrong diagnosis of LOT and HOT is possible. This being acknowledged, differences between oceanic lithosphere accreted at slow- and fast-spreading ridges are so great (Karson, 1998) that Boudier and Nicolas' (1985) early classification of ophiolites in terms of their inferred spreading rates, referring mainly to their mantle signature, was largely valid. In Table 1, we compare the revisited HOT and LOT classification , and in Table 2, we summarize the new and rapidly expanding database derived from the various oceanic floors.

With more data and experience, however, we need to reappraise the meaning of LOT in terms of spreading rate. The abyssal peridotites, which are issued from slow-spreading ridges, are normally composed of harzburgites, and strictly speaking the corresponding ophiolites should be considered as HOT. A closer examination shows, however, that the abyssal peridotites are less depleted than the harzburgites that correspond to true HOT. They commonly contain a small percentage (<5%) of clinopyroxene, either as unmelted relics from the mantle of origin or as secondary melt impregnations, commonly in association with plagioclase (Cannat, 1996). The true LOTs are characterized by a mantle section of plagioclase lherzolites, still partly fertile and having preserved mantle microstructures, or issued from a secondary melt impregnation of a depleted harzburgite (Nicolas, 1989, p. 21). We are now tempted to restrict the origin of the former fertile lherzolites to very slow-spreading settings. Oceanic rifts seem to be good candidates, as illustrated by the Zabargad Island formations in the Red Sea (Nicolas et al., 1985). Another illustration could be the newly described Gakell-Lena ridge in the Arctic Ocean where similar lherzolites have been dredged extensively (Dick et al., 2002; Snow et al., 2002).

The refertilized lherzolites might be issued from fracture zones in slow-spreading environments, as illustrated by the massifs du Nord (Tiebaghi) in New Caledonia (Nicolas, 1989, p. 128) or by the Othrys ophiolite in Greece (Dijkstra et al., 2001). We propose to introduce a new category, the lherzolite-harzburgite ophiolite type (LHOT), whose mantle section is made of abyssal peridotites corresponding to slow-spreading situations. The 1985 classification of Boudier and Nicolas also needs to be revisited because a number of detailed studies in new ophiolite massifs (see Table 1) have since confirmed and enriched the criteria then proposed.

The gabbro unit, when fully developed as in the HOT Oman ophiolite, is composed of, from bottom to top, layered gabbros, foliated gabbros, and irregular patches of isotropic gabbros just below the sheeted dike complex. Layered and foliated gabbros display a magmatic fabric, with only traces of penetrative plastic deformation. In LHOT ophiolites, these features are attenuated and disappear, whereas penetrative magmatic deformation features tend to be replaced by partitioned deformation structures (flasergabbro bands). The gabbro section is plastically deformed in its lower horizons in BOI (LHOT or HOT) and devoid of plastic strain in Oman (HOT). Typically, in the LHOT Mirdita ophiolite, layered gabbros are thin and poorly layered. These gabbros are present in the eastern part of the belt, where the crust is "normal." They are represented only by highly strained lenses tectonically mixed with lherzolites in the western part, which has been interpreted as a "core complex" formed in the vicinity of a transform zone. In the Xigaze ophiolite (LHOT), the crust is composed mainly of diabase sills and basalts, and the gabbros are either absent or restricted to lenses that are a few hundred meters wide.

Dunites originated from the MTZ, and wehrlites intrusive into the crust also differ. In Oman, the MTZ varies greatly in thickness but does not exceed 400 m, whereas in the LHOT Bay of Islands ophiolite, it can attain a thickness of 2000 m; in contrast, wehrlites are much more abundant in Oman. Relating a large development of dunites and wehrlites to a high degree of melting and, thus, to a fast spreading rate, Jousselin and Nicolas (2000b) have interpreted the greater thickness of dunites in BOI compared to Oman to be a consequence of the lithosphere-asthenosphere limit being steeper in slower spreading conditions, which would result in the mantle flow and the associated, locally more constrained melt delivery (Fig. 6).

Chromite deposits may be excellent indicators. They are rare in Oman, uncommon in BOI, and remarkably abundant in the typically LHOT eastern part of Mirdita. Nicolas and Al Azri (1991) interpret this phenomenon through a combination of two causes: (1) the necessary melting-out of clinopyroxene in order to render chromium available, a condition which is achieved in both LHOT and in HOT; and (2) thermal or redox conditions that favor the crystallization of chromite in the uppermost mantle below Moho, where the chromite can also be trapped (Lago et al., 1982). These conditions would be met in LHOT uppermost mantle due to cooling related to slow-spreading and/or to seawater penetration at such depth near the ridge; in the HOT-fast-spreading setting, the conditions for chromite crystallization are normally met only above, within the magma chamber, where the chromite would be dispersed within crystallizing gabbros.

TABLE 2. OCEANIC LITHOSPHERE

	Fast-spreading	Slow-spreading	Very slow-spreading oceanic rifting
OCEANIC RIDGES	East Pacific Rise: Garrett FZ (1), Hess Deep (2), Costa Rica Rift (3) Easter microplate (4)	Mid-Atlantic Ridge: Fracture zones(5)	SW Indian Ridge (6); Gakkel-Lena Ridge (7); Zabargad (8); Galicia bank (9)
Spreading rate	10 cm/yr	3 cm/yr	<1 cm/yr
CRUSTAL SECTION			
Sheeted dikes	~1.5 km thick (3), steep to moderately dipping (2, 3)	Uncommon, thickness: 0–1 km	Diabase sills and dikes (8)
Gabbros		Thickness: 0–3 km	Thickness 0–1.5km (6)
lithology	Ol-gabbro (2, 4), ferrogabbro (4), gabbronorite (2), minor wehrlite (2)	Ol-gabbro, gabbronorite	Ol-gabbro, gabbronorite, ferrogabbro (6)
deformation	Magmatic (2, 3), plastic deformation in shear zones (2)	No deformation or high-temperature plastic (flasergabbro)	Magmatic and plastic foliation, subhorizontal or steep (6)
shear bands and faults		Cataclastic listric faults	Mylonitic bands, normal faults, granulite facies (6)
MANTLE SECTION			
lithology	Harzburgite (1, 2, 4), Plag-dunite (1, 2, 4)	Cpx-harzburgite (2–3% Cpx), dunite, Sp-, Plag-lherzolite	Plag- Sp- lherzolite (4–7%Cpx) (6,8,9), occurrence plag-rimed Sp
internal structures	High-temperature plastic foliation (1, 2, 4), gently to steeply dipping (2)	Moderate-temperature plastic foliation gently dipping	Moderate-temperature plastic foliation, high strain
shear zones and faults	Few cataclastic shear bands (2)	Low-angle detachment faults	High- to low-angle mylonitic shear bands, normal faults(6)
serpentinization	Lizardite	Lizardite and antigorite	Antigorite

Note: Criteria of Table 1 applied to oceanic ridges with different spreading rates. Data sources: 1—Hebert et al. (1983); Cannat et al. (1990); Niu and Hekinian (1997); 2—Hékinian et al. (1993); Mevel et al. (1996); 3—Alt et al. (1993); 4—Constantin et al. (1995); 5—Dick et al. (1984); Michael and Bonatti (1985); Cannat and Casey (1995); Karson (1998); 6—Dick et al. (1984); Meyer et al. (1989); Cannat (1991); Dick et al. (1999); 7—Dick et al. (2002); Snow et al. (2002); 8—Bonatti et al. (1986); Nicolas et al. (1987); 9—Evans and Girardeau (1987); Girardeau and Mercier (1988); Cornen et al. (1996).

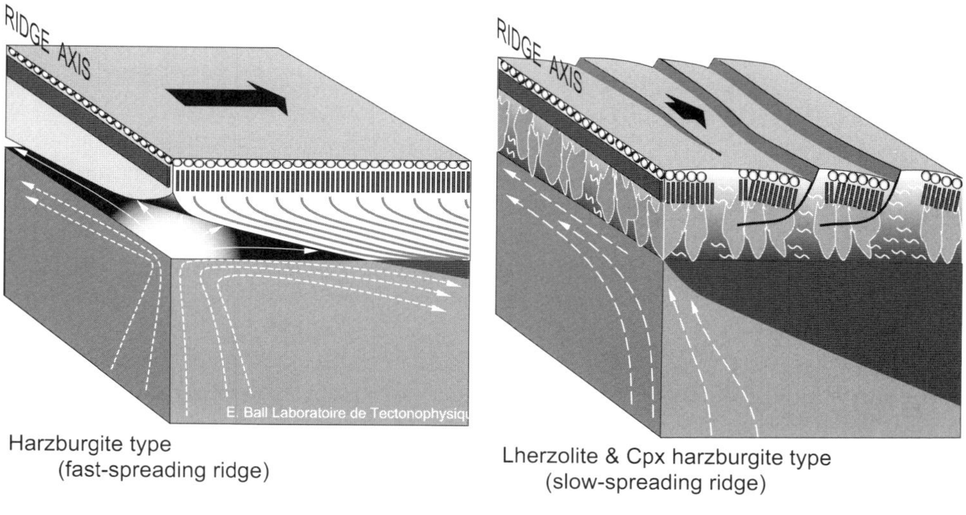

Figure 6. Sketches illustrating the main differences between fast- and slow-spreading ridges, deduced from studies in ophiolites (after Nicolas, 1999, fig. 12).

Metamorphic Aureole, Signature of Ophiolites Origin

Moores (1982) and Coleman (1984) proposed to divide the ophiolites into two families depending on whether they had been incorporated into the continental crust at a passive margin (their Tethyan ophiolites) or at an active margin (their Cordilleran ophiolites) (Fig.7). In the former case, ophiolites rest as giant thrust sheets upon a continental substrate; in the latter case, they are incorporated, often dismembered as a mélange, into accretion terranes mainly issued from trench sediments and seamount fragments (Coleman, 2000). From an earlier review (Nicolas, 1989, p. 289–309), we retain that more than 60 ophiolite massifs that comprise most, if not all, of the least dismembered ophiolites belong to the Tethyan group, with which they share a fascinating feature, the presence of a metamorphic aureole at their base. This aureole grades upsection over a few hundred meters from the underlying undeformed basalts and sediments to intensely strained amphibolites and granulites that are equilibrated above 800 °C. These amphibolites and granulites are overlain by concordant mylonitic peridotites deformed at 900–1000 °C that are grading over 1–2 km into normal mantle peridotites. Tectonic reconstructions (Boudier et al., 1988) and models (Hacker et al., 1996) show that this metamorphism was produced by the ironing effect of a ~10-km-thick lithospheric slab that represents the future ophiolite. Following thermal models of oceanic ridges (Phipps Morgan and Chen, 1993; Henstock et al., 1993; Cormier et al., 1995), we conclude that the age of detachment of the lithospheric slab cannot be older than 2 Ma. Accordingly, in Oman, the age of the metamorphic aureole is 1–2 m.y. younger than the ridge of origin. Young ages in metamorphic aureoles are the rule, with the few exceptions of 10–15 m.y. younger ages (review in Wakabayashi and Dilek, 2000). Because both the geological and the thermal modeling information are compelling, we assume that the few datations with 10–15 Ma differences are dubious (Hacker et al., 1996).

Major consequences can be drawn from the most common existence of basal aureoles below ophiolites. These ophiolites were detached at no more than 10–100 km from their ridge of origin, depending on the spreading rate and the age difference between igneous accretion and metamorphism in the basal aureole. At their time of detachment, they were thin and hot pieces of oceanic lithosphere. This extraordinary feature raises as many questions as it helps to solve. Is it not possible that the SSZ secondary signature of many ophiolites would be due to water penetration into the overriding future ophiolite (Boudier et al., 1988)? Are there such active sites in present day oceans, or do they belong to peculiar periods in earth history? Paleomagnetic studies in Oman (Perrin et al., 1993) show 30° rotations within 1 Ma, suggesting that this ophiolite might have originated as a microplate, possibly comparable to the Easter or Juan Fernandez microplates along the East Pacific Rise (Boudier et al., 1997). This model has a few appealing features: a microplate is a "precut" piece of oceanic lithosphere, confined by ridge segments and compressive boundaries. With these active boundaries, it might be more easily detached from its site of origin than a normal piece of lithosphere. This future ophiolite also contains internally active ridge segments, a feature that accounts for the presence of frozen diapirs in some ophiolites. In addition, tip of propagating segments, such as Pito or Endeavour Deeps from Easter and Juan Fernandez, respectively, create locally very slow-spreading conditions in the overall fast-spreading environment; such contrasted situation might have its equivalent in the different characteristics pointed out in the Oman ophiolite between active segments and their propagating tip (Nicolas and Boudier, 1995). In an ophiolite as small as Troodos in Cyprus, the slow-spreading situation documented on the base of evidence for a rifted valley (Varga and Moores, 1985) could be reconsidered if the ophiolite were originated from the tip of a propagator. There are, however, many pending questions. Can the compressive zones in microplates generate the large thrusts responsible for the metamorphic aureoles of ophiolites? Why and how is the microplate-ophiolite detached and overthrust?

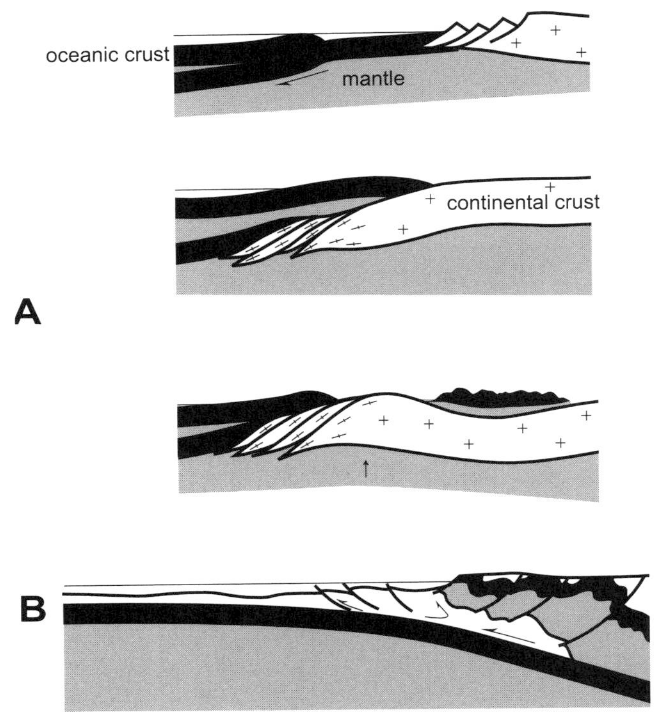

Figure 7. The two main modes of ophiolite emplacement. A: The Tethyan-type, related to obduction onto a passive margin. B: The Cordilleran-type, related to upheaval of oceanic lithosphere in an accretionary prism. After Moores (1982) and Coleman (1984).

CONCLUSION: CALL FOR AN INTEGRATED APPROACH IN OPHIOLITE STUDIES

As studies of ophiolites turn toward a better understanding of the functioning of oceanic spreading centers, the first problem is to identify the type of center where the considered ophiolite has formed. We have recalled a number of criteria that can be used to estimate the spreading rate of this center (Tables 1 and 2). Spreading rate is clearly the most important parameter to account for the diversity of the oceanic lithosphere (Karson, 1998), and it is also clearly reflected in the structure and composition of ophiolites. In the study of slow-spreading oceanic ridges, the ophiolite reference is of lesser importance because, thanks to the submarine reliefs that can be >4 km, direct observation of deep crust and mantle is possible. This reference becomes compulsory in fast-spreading situations where the subdued reliefs allow only exceptionally deep structures to be exposed. We have concluded above that many ophiolites were likely originated from spreading centers corresponding to intermediate to fast rates and, by a few examples showing the interplay between structural studies in ophiolites and geophysical studies in oceanic ridges, we have tried to illustrate the potential of this cross-fertilizing approach.

Spreading rate is not the unique parameter, however, and a closer analysis should consider the effects of different temperatures or compositions in the mantle sources. The environment of origin (MOR or SSZ) that is discussed above and that has much—too much—mobilized the ophiolite community, belongs to this category of problems. It has been approached from the standpoint of geochemistry, but with disputable results and more elementary pieces of information, e.g., the associated sedimentary record often having been overlooked.

This drives us to our main conclusion, which is a call for integrated studies. Through a few examples, we have illustrated the fruitful cooperation between structural geologists in ophiolites and marine geophysicists. A similarly rich cooperation is possible between geochemists and petrologists working in ophiolites and marine environment. Dealing with the relative roles of structural and geochemical studies in ophiolites, we insist on the common sense idea that these approaches are complementary, but that geochemistry should be applied within the framework of a thorough understanding of the structures, or at the very least, of the general geology. Indeed, many ophiolites have been dismembered extensively by collision-related tectonic deformation, making it difficult to conduct ridge-related structural studies. One might question the results of geochemical studies that are not supported by structural information in such ophiolites. In contrast, in an ophiolite whose internal structure has been preserved and decrypted, the solution to further problems depends largely on focused geochemical studies. Typically, the answer to the question evoked above regarding a mantle source outside the thermally or chemically average mantle source can be addressed by geochemistry.

As a concluding remark, we would like to encourage young scientists to undertake more structural studies in selected ophiolites, in association with and opening the way for geochemists. Evidently, they should keep an eye on corresponding marine research, but when their results are at odds with marine studies, as has happened in the past, they should remember that careful observations made on relevant outcrops in the field and the complementary laboratory studies provide the most compelling evidence.

ACKNOWLEDGMENTS

We deeply acknowledge the involvement of a large number of students and colleagues in our studies, as well as the official support received from the French research institutes (CNRS, MAE) and the friendly cooperation established in several countries, particularly Oman. The manuscript has been largely improved thanks to constructive reviews from Paul Robinson, Robert Coleman, and Yildirim Dilek. The editorial contribution of Yildirim Dilek is particularly acknowledged.

REFERENCES CITED

Abbate, E., Bortoloti, V., Passerini, P., Prijcipi, G., and Treves, B., 1994, Oceanisation processes and sedimentary evolution of the Northern Apennine ophiolite suite: A discussion: Memorie della Società Geologica Italiana, v. 48, p. 117–136.

Abrajano, T.A., Pasteris, J.D., and Bacuta, G.C., 1989, Zambales ophiolite, Philippines, I. Geology and petrology of the critical zone of the Acoje massif: Tectonophysics, v. 168, p. 65–100.

Alabaster, T., Pearce, J.A., Mallivk, D.I.J., and Elboushi, I.M., 1980, The volcanic stratigraphy and location of massive sulphide deposits in the Oman ophiolite, in International Ophiolite Symposium, Nicosia, Cyprus, p. 751–757.

Alabaster, T., Pearce, J.A., and Malpas, J., 1982, The volcanic stratigraphy and petrogenesis of the Oman ophiolite complex: Contributions to Mineralogy and Petrology, v. 81, p. 168–183.

Alt, J.C., Kinoshita, H., and Stokking, L.B., 1993, Proceedings of the Ocean Drilling Program, Initial Reports: College Station, Texas, Ocean Drilling Program, Leg 148, p. 27–121.

Anonymous, 1972, Penrose field conference on ophiolites: Geotimes, v. 17, p. 24–25.

Barruol, G., and Souriau, A., 1995, Anisotropy beneath the Pyrenees range from teleseismic shear waves splitting: Geophysical Research Letters, v. 22, p. 493–496.

Barth, G.A., and Mutter, J.C., 1996, Variability in oceanic crustal thickness and structure: Multichannel seismic reflexion results from the northern East Pacific Rise: Journal of Geophysical Research, v. 101, p. 17951–17975.

Bodinier, J.L., 1988, Geochemistry and Petrogenesis of the Lanzo peridotite body, western Alps: Tectonophysics, v. 149, p. 67–88.

Bonatti, E., Ottonello, G., and Hamlyn, P.R., 1986, Peridotites from the Island of Zabargad (St John), Red Sea: Petrology and Geochemistry: Journal of Geophysical Research, v. 91, p. 599–631.

Boudier, F., 1978, Structure and petrology of the Lanzo peridotite massif (Piedmont Alps): Geological Society of America Bulletin, v. 89, p. 1574–1591.

Boudier, F., and Nicolas, A., 1985, Harzburgite and lherzolite subtypes in ophiolitic and oceanic environments: Earth and Planetary Science Letters, v. 76, p. 84–92.

Boudier, F., Ceuleneer, G., and Nicolas, A., 1988, Shear zones, thrusts and related magmatism in the Oman ophiolite: initiation of thrusting on an oceanic ridge: Tectonophysics, v. 151, p. 275–296.

Boudier, F., Le Sueur, E., and Nicolas, A., 1989, Structure of an atypical ophiolite: The Trinity complex, eastern Klamath Mountains, California: Geological Society of America Bulletin, v. 101, p. 820–833.

Boudier, F., Nicolas, A., Ildefonse, B., and Jousselin, D., 1997, EPR microplates, a model for the Oman Ophiolite: Terra Nova, v. 9, no. 2, p. 79–82.

Boudier, F., Godard, M., and Armbruster, C., 2000, Significance of noritic gabbros in the gabbro section of the Oman ophiolite: Marine Geophysical Researches, v. 21, no. 3–4, p. 307–326.

Brongniart, A., 1821, Sur le gisement ou position relative des ophiolites, euphotides, jaspes, etc. dans quelques parties des Apennins: Annales des Mines, v. 6, p. 177–238.

Cakir, U., 1978, Pétrologie du massif ophioloitique de Pozanti-Karsanti (Taurus Cilicien, Turquie): Étude de la partie centrale: Strasbourg, France, 151 p.

Cameron, W.E., 1985, Petrology and origin of primitive lavas from the Troodos ophiolite, Cyprus: Contributions to Mineralogy and Petrology, v. 89, p. 339–255.

Canales, J.P., Detrick, R.S., Toomey, D.R., and Wilcock, W.S.D., 2003, Segment-scale variations in crustal structure of 150 to 300 k.y.-old fast spreading oceanic crust (East Pacific Rise, 8° 15' N–10° 15' N) from wide-angle seismic refraction profiles: Geophysical Journal International, v. 152, p. 766–794.
Cannat, M., 1991, Plastic deformation at an oceanic spreading ridge: a microstructural study of the site 735 gabbros (Southern Indian Ocean), *in* Von Herzen, R.P., and Robinson, P.T., eds., Proceedings of the Ocean Drilling Program, Scientific Results: College Station, Texas, Ocean Drilling Program, v. 118, p. 399–408.
Cannat, M., 1996, How thick is the magmatic crust at the slow-spreading ridges?: Journal of Geophysical Research, v. 101, p. 2847–2857.
Cannat, M., and Boudier, F., 1985, Structural study of intra-oceanic thrusting in the Klamath Mountains, Northern California: implication on accretion geometry: Tectonics, v. 4, p. 435–452.
Cannat, M., and Casey, J.F., 1995, An Ultramafic Lift at the Mid-Atlantic Ridge: Successive Stages of Magmatism in Serpentinized Peridotites from the 15° N Region, *in* Vissers, R.L.M., and Nicolas, A., eds., Mantle and Lower Crust Exposed in Oceanic Ridges and in Ophiolites: Dordrecht, The Netherlands, Kluwer, p. 5–34.
Cannat, M., Bideau, D., and Hébert, R., 1990, Plastic deformation and magmatic impregnation in serpentinized ultramafic rocks from the Garret transform fault (East Pacific Rise): Earth and Planetary Science Letters, v. 101, p. 216–232.
Cassard, D., 1980, Structure et origine des gisements de chromite du Massif du Sud (ophiolite de Nouvelle Calédonie). Guides de prospection: Nantes, France, University of Nantes, 152 p.
Ceuleneer, G., Nicolas, A., and Boudier, F., 1988, Mantle flow patterns at an oceanic spreading centre: the Oman peridotite record: Tectonophysics, v. 151, p. 1–26.
Chenevez, J., Machetel, P., and Nicolas, A., 1998, Numerical models of magma chamber in the Oman ophiolite: Journal of Geophysical Research, v. 103, p. 15,443–15,455.
Coleman, R.G., 1984, The diversity of ophiolites: Geologie en Mijnbouw, v. 63, p. 1099–1108.
Coleman, R.G., 2000, Prospecting for ophiolites along the California continental margin, *in* Dilek, Y., Moores, E.M., Elthon, D. and Nicolas, eds., Ophiolites and Oceanic Crust: New Insight from Field Studies and the Ocean Drilling Program: Boulder, Colorado, Geological Society of America Special Paper 349, p. 351–364.
Constantin, M., Hekinian, R., Ackermand, D., and Stoffers, P., 1995, Mafic and Ultramafic Intrusions into Upper Mantle Peridotites from Fast Spreading Centers of the Easter Microplate (South East Pacific), *in* Vissers, R.L.M., and Nicolas, A., eds., Mantle and Lower Crust Exposed in Ocean Ridges and in Ophiolites: Dordrecht, The Netherlands, Kluwer, p. 71–120.
Cormier, M.H., Macdonald, K.C., and Wilson, D.S., 1995, A three-dimensional gravity analysis of the East Pacific Rise from 18° to 21° 30' S: Journal of Geophysical Research, v. 100, p. 8063–8082.
Cornen, G., Beslier, M.O., and Girardeau, J., 1996, Petrologic characteristics of the ultramafic rocks from the ocean/continent transition in the Iberia abyssal plain, *in* Whitmarsh, R.B., Sawyer, D.S., Klauss, A., and Masson, D.G., eds., Proceedings of the Ocean Drilling Program, Scientific Results: College Station, Texas, Ocean Drilling Program, v. 149, p. 377–395.
Davies, H.L., 1977, Crustal structure and emplacement of ophiolite in Southern Papua, New Guinea, *in* Orogenic Mafic and Ultramafic Association: Grenoble, France, p. 17–34.
Detrick, R.S., Bulh, P., Vera, E., Mutter, J., Orcutt, J., Madsen, J., and Trocher, T., 1987, Multi-channel seismic imaging of a crustal magma chamber along the East Pacific Rise: Nature, v. 326, p. 35–41.
Detrick, R.S., Needham, H.D., and Renard, V., 1995, Gravity anomalies and crustal thickness variations along the Mid-Atlantic ridge between 33° N and 40° N: Journal of Geophysical Research, v. 100, p. 3767–3787.
Dick, H.J.B., 1977, Evidence of partial melting in the Josephine peridotite, Oregon Department of Geology and Mineral Industries, v. 96, p. 59–62.
Dick, H.J.B., and Bullen, T., 1984, Chromian spinel as a petrogenetic indicator in abyssal and alpine-type peridotites and spatially associated lavas: Contribution to Mineralogy and Petrology, v. 86, p. 54–76.
Dick, H.J.B., and Natland, J.H., 1996, Late-stage melt evolution and transport in the shallow mantle beneath the East Pacific Rise, *in* Mevel, C., Gillis, K.M., Allan, J.M., and Meyer, P.S., eds., Proceedings of the Ocean Drilling Program, Scientific Results: College Station, Texas, Ocean Drilling Program, v. 147, p. 157–172.
Dick, H.J.B., Fisher, R.L., and Bryan, W.B., 1984, Mineralogical variability of the uppermost mantle along mid-oceanic ridges: Earth and Planetary Science Letters, v. 69, p. 54–76.
Dick, H.J.B., Natland, J.H., and Miller, D.J., 1999, Proceedings of the Ocean Drilling Program, Initial Report: College Station, Texas, Ocean Drilling Program, v. 176, p. 1–314 (CD-ROM).
Dick, H.J.B., Kurras, G., Michael, P., Snow, J.E., Langmuir, C., and Jokat, W., 2002, Volcanic and Tectonic Processes along the Ultra-slow Spreading Gakkel Ridge-AMORE in the Arctic, *in* Proceedings, InterRidge South West Indian Ridge Workshop: Southampton, UK, University of Southampton, p. 22.
Dijkstra, A.H., Drury, M.R., and Vissers, R.L.M., 2001, Structural petrology of plagioclase-peridotites in the West Othrys Mountains (Greece): Melt impregnation in mantle lithosphere: Journal of Petrology, v. 42, p. 5–24.
Dilek, Y., and Thy, P., 1998, Structure, petrology, and seafloor tectonics of the Kizildag Ophiolite, Turkey, *in* Mills, R.A., ed., Modern Ocean Floor Processes and the Geological Record: London, Geological Society of London, Special Publication, p. 43–69.
Dilek, Y., Thy, P., Hacker, B., and Grundvig, S., 1999, Structure and petrology of Tauride ophiolites and mafic dike intrusions (Turkey): Implications for the Neotethyan ocean: Geological Society of America Bulletin, v. 111, p. 1192–1216.
Dunn, R.A., Toomey, D.R., and Solomon, S.C., 2000, Three-dimensional seismic structure and physical properties of shallow mantle beneath the East Pacific Rise at 9° 30' N: Journal of Geophysical Research, v. 105, p. 23,537–23,555.
Dunn, R.S., and Toomey, D.R., 1997, Seismological evidence for three-dimensional melt migration beneath the East Pacific Rise: Nature, v. 388, p. 259–262.
Ernewein, M., Pflumio, C., and Whitechurch, H., 1988, The death of an accretion zone as evidenced by the magmatic history of the Sumail ophiolite (Oman): Tectonophysics, v. 151, p. 247–274.
Evans, C., and Hawkins, J.W., 1989, Compositional heterogeneities in upper mantle peridotites from the Zambales Range Ophiolite, Luzon, Philippines: Tectonophysics, v. 168, p. 23–41.
Evans, C., and Girardeau, J., 1987, Galicia Margin Peridotites: Undepleted Abyssal Peridotites from the North Atlantic, *in* Boillot, G., Winterer, E.L., et al., eds., Proceedings of the Ocean Drilling Program, Scientific Results, College Station, Texas, Ocean Drilling Program, p. 195–208.
Ferrière, J., 1985, Nature et développement des ophiolites helléniques du secteur Othrys-Pellion: Ofioliti, v. 10, p. 255–278.
Fleutelot, C., Eissen, J.F., Tonnerre, T., Danyushevsky, L., Rory, V., Mollard, L., Juteau, T., Cotten, J., and Bohn, M., 1996, Propagation en contexte arrière-arc: premier résultats de la campagne ProFeTi (Bassin Nord-Fidjien, Pacifique SW): Comptes Rendus de l'Académie des Sciences, Paris, v. 322, p. 853–860.
George, R.P.J., 1978, Structural petrology of the Olympus Ultramafic Complex in the Troodos Ophiolite, Cyprus: Geological Society of America Bulletin, v. 89, p. 845–865.
Gerbert-Gaillard, L., 2002, Caractérisation géochimique des péridotites d'Oman: Processus magmatiques à la limite lithosphère/asthénosphère [Ph.D. thesis]: Montpellier, France, University of Montpellier II, 215 p. (wwww.ccsd.cnrs.fr).
Gillis, K., Mével, C., and Allan, J., et al., editors, 1993, *in* Proceedings of the Ocean Drilling Program, Preliminary Report: College Station, Texas, Ocean Drilling Program, v. 147, p. 45–108.
Girardeau, J., and Mercier, J.C., 1988, Petrology and texture of the ultramafic rocks of the Xigaze ophiolite (Tibet), constraints for mantle structure beneath slow spreading ridges: Tectonophysics, v. 147, p. 33–58.
Godard, M., Dautria, J.M., and Perrin, M., 2003, Geochemical variability of the Oman ophiolite lavas- Relationship with spatial distribution and paleomagnetic directions: Geochemistry Geophysics Geosystems, v. 4, no. 6, p. 1–15.
Hacker, B.R., 1994, Rapid emplacement of Young Oceanic Lithosphere: Argon Geochronology of the Oman Ophiolite: Science, v. 265, p. 1563–1565.
Hacker, B.R., Mosenfelder, J.L., and Gnos, E., 1996, Rapid emplacement of the Oman ophiolite: Tectonics, v. 15, no. 6, p. 1230–1247.
Harper, G.D., 1982, Evidence for large-scale rotations at spreading centers from the Josephine ophiolite: Tectonophysics, v. 82, p. 25–44.
Hebert, R., Bideau, D., and Hekinian, R., 1983, Ultramafic and mafic rocks from the Garret Transform Fault near 13°30'S on the East Pacific Rise: igneous petrology: Earth and Planetary Science Letters, v. 65, p. 107–125.
Hékinian, R., Bideau, D., Francheteau, J., Cheminée, J.L., Armijo, R., Londsale, P., and Blum, N., 1993, Petrology of the East Pacific Rise crust and upper mantle exposed in Hess Deep (Eastern Equatorial Pacific): Journal of Geophysical Research, v. 98, p. 8069–8094.

Henstock, T.J., Woods, A.W., and White, R.S., 1993, The accretion of oceanic crust by episodic sill intrusion: Journal of Geophysical Research, v. 98, no. B3, p. 4143–4161.

Hoxha, M., and Boullier, A.M., 1995, The peridotites of the Kukès ophiolite (Albania): structure and kinematics: Tectonophysics, v. 249, p. 217–231.

Ishiwatari, A., 1985a, Alpine ophiolites: product of low-degree mantle melting in a Mesozoic transcurrent rift zone: Earth and Planetary Science Letters, v. 76, p. 93–108.

Ishiwatari, A., 1985b, Igneous Petrogenesis of the Yakuno Ophiolite (Japan) in the context of the diversity of ophiolites: Contributions to Mineralogy and Petrology, v. 89, p. 155–167.

Ishiwatari, A., 1994, Circum-Pacific Phanerozoic Multiple Ophiolite Belts, *in* Ishiwatari, A., Malpas, J., and Ishizuka, H., eds., Proceedings of the 29th International Geological Congress, Part D: Kyoto, Japan: Utrecht, The Netherlands, VSP, p. 7–28.

Jacobsen, S.B., and Wasserburg, G.J., 1984, Nd and Sr isotopic study of the Bay of Island Ophiolite Complex and evolution of the source of mid-oceanic ridge basalts: Journal of Geophysical Research, v. 84, p. 7429–7445.

Jousselin, D., and Nicolas, A., 2000a, Oceanic ridge off-axis deep structure in the Mansah region (Sumail massif, Oman ophiolite): Marine Geophysical Researches, v. 21, no. 3–4, p. 243–257.

Jousselin, D., and Nicolas, A., 2000b, The Moho transition zone in the Oman ophiolite. Relation with wehrlites in the crust and dunites in the mantle: Marine Geophysical Researches, v. 21, p. 229-241.

Jousselin, D., Nicolas, A., and Boudier, F., 1998, Detailed mapping of a mantle diapir below a paleo-spreading center in the Oman ophiolite: Journal of Geophysical Research, v. 103, p. 18,153–18,170.

Juteau, T., 1974, Les ophiolites des nappes d'Antalya (Taurides occidentales, Turquie). Petrologie d'un fragment de l'ancienne croûte tethysienne: University of Nancy, 692 p.

Juteau, T., and Maury, R., 1997, The Oceanic Crust, from Accretion to Mantle Recycling, Series in Geophysics: Southampton, Springer-Praxis, 390 p.

Karson, J.A., 1998, Internal structure of Oceanic Lithosphere: A perspective from Tectonic Windows, *in* Buck, W.R., Delaney, P.T., Karson, J.A., Lagabrielle, Y., eds., Faulting and Magmatism at Mid-Ocean Ridges: Washington, D.C., American Geophysical Union, p. 177–218.

Karson, J.A., and Rona, P.A., 1990, Block-tilting, transfer faults, and structural control of magmatic and hydrothermal processes in the TAG area, Mid-Atlantic Ridge 26° N: Geological Society of America Bulletin, v. 102, p. 1635–1645.

Kelemen, P.B., Shimizu, N., and Salters, V.J.M., 1995, Extraction of mid-ocean-ridge basalt from mantle by focused flow of melt in dunite channels: Nature, v. 375, p. 747–753.

Lago, B.L., Rabinovicz, M., and Nicolas, A., 1982, Podiform chromite ore bodies: a genetic model: Journal of Petrology, v. 23, p. 103–125.

Lamoureux, G., Ildefonse, D., and Mainprice, D., 1999, Modelling the seismic properties of fast-spreading ridge crustal LVZ: insights from Oman gabbro textures: Tectonophysics, v. 312, p. 283–301.

Lapierre, H., Ortez, E.L., Abudrami, W., Monod, O., Coulon, C., and Zimmermann, J.L., 1992, A crustal section of intra-oceanic island arc: The Late Jurassic-Early Cretaceous Guana Guato magmatic sequence (Central Mexico): Earth and Planetary Science Letters, v. 108, p. 61–77.

Le Sueur, E., Boudier, F., Cannat, M., Ceuleneer, G., and Nicolas, A., 1984, The Trinity mafic-ultramafic complex: first result of the structural study of an atypical ophiolite: Ophioliti, v. 9, p. 487–498.

Lin, J., and Phipps Morgan, J., 1992, The spreading rate dependence of three-dimensional mid-ocean ridge gravity structure: Geophysical Research Letters, v. 19, no. 1, p. 13–16.

Lonsdale, P., 1983, Overlapping rift zones at the 5.5°S offset of the East Pacific Rise: Journal of Geophysical Research, v. 88, p. 9393–9406.

MacDonald, K.C., and Fox, P.J., 1983, Overlapping spreading centers: a new kind of accretion geometry on the East Pacific Rise: Earth and Planetary Science Letters, v. 40, p. 407–441.

Magde, L.S., Kincaid, C., Sparks, D.W., and Detrick, R.S., 1996, Combined laboratory and numerical studies of interaction between buoyant and plate-driven upwelling beneath segmented spreading centers: Journal of Geophysical Research, v. 101, p. 22,107–22,122.

Magde, L.S., Sparks, D.W., and Detrick, R.S., 1997, The relationship between buoyant mantle flow, melt migration, and gravity bull's eyes at the Mid-Atlantic Ridge 33° N: Earth and Planetary Science Letters, v. 148, p. 59–67.

Mahmood, K., 1994, Etude structurale de l'ophiolite du Muslim Bagh, Baluchistan (Pakistan) [Ph.D. thesis]: Montpellier, France, University of Montpellier, 130 p.

Mainprice, D., 1997, Modeling the anisotropic seismic properties of partially molten rocks found at mid-ocean ridges: Tectonophysics, v. 279, p. 161–180.

Manning, G.E., MacLeod, C.J., and Weston, P.E., 2000, Lower-crustal cracking front at fast-spreading ridges: Evidence from the East Pacific Rise and the Oman ophiolite, *in* Dilek, Y., et al., eds., Ophiolites and oceanic crust: New insights from field studies and the Ocean Drilling Program: Boulder, Colorado, Geological Society of America Special Paper 349, p. 261–272.

McBirney, A., and Nicolas, A., 1997, The Skaergaard Layered Series. Part II. Magmatic Flow and Dynamic Layering: Journal of Petrology, v. 38, p. 569–580.

MELT Seismic Team, 1998, Imaging the deep seismic structure beneath a mid-ocean ridge : The MELT experiment: Science, v. 280, p. 1215–1218.

Menzies, M., and Allen, C., 1974, Plagioclase lherzolite residual mantle relationships within two eastern Mediterranean ophiolites: Contributions to Mineralogy and Petrology, v. 45, p. 197–213.

Mével, C., Gillis, C., Allan, J.F., and Meyer, P.S., 1996, *in* Proceedings of the Ocean Drilling Program, Scientific Results, College Station, Texas, Ocean Drilling Program, v. 147, 543 p.

Meyer, P.S., Dick, H.J.B., and Thompson, G., 1989, Cumulate gabbros from the Southeast Indian Ridge, 54° S-7° 16' E: implications for magmatic processes at slow spreading ridge: Contributions to Mineralogy and Petrology, v. 103, p. 44–63.

Michael, P.J., and Bonatti, E., 1985, Peridotite compositions from the North Atlantic: Regional tectonic variations and implications for partial melting: Earth and Planetary Science Letters, v. 73, p. 91–104.

Misseri, M., and Boudier, F., 1985, Structures in the Canyon Mountain ophiolite indicate an island-arc intrusion: Tectonophysics, v. 120, p. 191–209.

Miyashiro, A., 1973, The Troodos Complex was probably formed in an island arc: Earth and Planetary Science Letters, v. 25, p. 217–222.

Moores, E.M., 1982, Origin and emplacement of ophiolites: Reviews of Geophysics, v. 20, p. 735–760.

Morton, J.L., and Sleep, N.H., 1985, A mid-ocean ridge thermal model: constraints on the volume of axial hydrothermal heat flux: Journal of Geophysical Research, v. 90, p. 11,345–11,353.

Nicolas, A., 1989, Structures of ophiolites and dynamics of oceanic lithosphere, Petrology and structural geology: Amsterdam, Kluwer Academic Publishers, 367 p.

Nicolas, A., 1992, Kinematics in magmatic rocks with special reference to gabbros: Journal of Petrology, v. 33, no. 4, p. 891–915.

Nicolas, A., 1999, Enjeux actuels des dorsales océaniques: Bulletin Société Géologique de France, no. spéc., p. 5–20.

Nicolas, A., and Al Azri, H., 1991, Chromite-rich and chromite poor ophiolites: the Oman case, *in* Peters, Tj., Nicolas, A., and Coleman, R.G., eds., Ophiolite genesis and Evolution of the oceanic lithosphere: Amsterdam, Elsevier, p. 261–274.

Nicolas, A., and Boudier, F., 1995, Mapping oceanic ridge segments in Oman ophiolites: Journal of Geophysical Research, v. 100, p. 6179–6197.

Nicolas, A., and Boudier, F., 2000, Large mantle upwellings and related variations in crustal thickness in the Oman ophiolite: Geological Society of America Bulletin, v. 349, p. 67–73.

Nicolas, A., and Ildefonse, B., 1996, Flow mechanism and viscosity in basaltic magma chambers: Geophysical Research Letters, v. 23, no. 16, p. 2013–2016.

Nicolas, A., and Violette, J.F., 1982, Mantle flow beneath oceanic ridges: models derived from ophiolites: Tectonophysics, v. 81, p. 319–339.

Nicolas, A., Bouchez, J.L., Boudier, F., and Mercier, J.C., 1971, Textures, structures, and fabrics due to solid state flow in some European lherzolites: Tectonophysics, v. 12, p. 55–86.

Nicolas, A., Boudier, F., Liberis, N., and Guenoc, P., 1985, L'ile de Zabargad (Saint Jean): temoin-clé de l'expansion précoce en Mer Rouge: Comptes Rendus de l'Académie des Sciences, Paris, v. 14, p. 1063–1068.

Nicolas, A., Boudier, F., and Montigny, R., 1987, Structure of the Zabargad Island; an early rifting of the Red Sea: Journal of Geophysical Research, v. 92, p. 461–474.

Nicolas, A., Freydier, C.L., Godard, M., and Vauchez, A., 1993, Magma chambers at oceanic ridges: How large?: Geology, v. 21, p. 53–56.

Nicolas, A., Boudier, F., and Meshi, A., 1999, Slow spreading accretion and mantle denudation in the Mirdita ophiolite (Albania): Journal Geophysical Research, v. 104, p. 15,155–15,167.

Nicolas, A., Boudier, F., Ildefonse, B., and Ball, E., 2000, Accretion of Oman and United Arab Emirates ophiolite—Discussion of a new structural map: Marine Geophysical Researches, v. 21, no. 3–4, p. 147–179.

Nicolas, A., Mainprice, D., and Boudier, F., 2003, High-temperature seawater circulation throughout crust of oceanic ridges: A model derived from the Oman ophiolites : Journal of Geophysical Research (in press).

Niu, Y., and Batiza, R., 1997, Trace element evidence from seamounts for recycled oceanic crust in the eastern Pacific mantle: Earth and Planetary Science Letters, v. 148, p. 471–483.

Niu, Y., and Hekinian, R., 1997, Basaltic liquids and harzburgitic residues in the Garett transform: a case study at fast-spreading ridges: Earth and Planetary Science Letters, v. 146, p. 471–484.

Parkinson, I.J., and Pearce, J.A., 1998, Peridotites from the Izu-Bonin-Mariana forearc (ODP leg 125): Evidence for mantle melting and mantle-melt interaction in a supra-subduction zone setting: Journal of Petrology, v. 39, p. 1577–1618.

Pearce, J.A., 1975, Basalt geochemistry used to investigate past tectonic environments: Tectonophysics, v. 25, p. 41–67.

Pearce, J.A., and Cann, J., 1973, Tectonic setting of basic volcanic rocks using trace elements analysis: Earth and Planetary Science Letters, v. 19, p. 290–300.

Pearce, J.A., Alabaster, T., Shelton, A.W., and Searle, M.P., 1981, The Oman ophiolite as a Cretaceous arc-basin complex: evidence and implications: Philosophical Transactions Royal Society of London, v. 300, p. 299–317.

Pedersen, R.B., and Hertogen, J., 1990, Magmatic evolution of the Karmoy Ophiolite Complex, SW Norway: relationships between MORB, IAT, boninitic-calc-alkaline and alkaline magmatism: Contributions to Mineralogy and Petrology, v. 104, p. 277–293.

Perrin, M., Prevot, M., and Bruere, F., 1993, Rotation of the Oman ophiolite and initial location of the ridge in the hotspot reference frame: Tectonophysics, v. 229, p. 31–42.

Pesagno, E.A., Hull, D.M., and Hopson, C.A., 2000, Tectonostratigraphic significance of sedimentary strata occurring within and above the Coast Range ophiolite (California Coast Ranges) and the Josephine Ophiolite (Klamath Mountains), northwest California, *in* Dilek, Y., et al., eds., Ophiolites and oceanic crust: New insights from field studies and the Ocean Drilling Program: Boulder, Colorado, Geological Society of America Special Paper 349, p. 383–394.

Phipps Morgan, J., and Chen, Y.J., 1993, The genesis of oceanic crust: magma injection, hydrothermal circulation and crustal flow: Journal of Geophysical Research, v. 98, p. 6283–6297.

Quick, J.E., 1981, Petrology and Petrogenesis of the Trinity Peridotite and upper mantle diapir in the Eastern Klamath Mountains, Northern California: Journal of Geophysical Research, v. 86, p. 11,837–11,863.

Quick, J.E., and Denlinger, R.P., 1993, Ductile deformation and the origin of layered gabbro in ophiolites: Journal of Geophysical Research, v. 98, no. B8, p. 14,015–14,027.

Rabinowicz, M., Nicolas, A., and Vigneresse, J.L., 1984, a rolling mill effect in asthenosphere beneath oceanic spreading centers: Earth and Planetary Science Letters, v. 92, p. 3474–3486.

Rabinowicz, M., Ceuleneer, M., and Nicolas, A., 1987, Melt segregation and flow in mantle diapirs below spreading centers: Evidence from the Oman ophiolites: Journal of Geophysical Research, v. 92, p. 3475–3486.

Rampone, E., and Piccardo, J.B., 2000, The ophioliote-oceanic lithosphere analogue: New insights from the Northern Apennines (Italy), *in* Dilek, Y., Moores, E.M., Elthon, D. and Nicolas, A., eds., Ophiolites and Oceanic Crust: New Insight from Field Studies and the Ocean Drilling Program: Boulder, Colorado, Geological Society of America Special Paper 349 p. 21–34.

Rassios, A., Moores, E.M., and Green, H.W., 1983, Magmatic structure and stratigraphy of the Vourinos ophiolite cumulate zone, Northern Greece: Ofioliti, v. 8, p. 377–410.

Scott, D.R., and Stevenson, D.J., 1989, A self-consistent model of melting, magma migration and buoyancy-driven circulation beneath mid-ocean ridges: Journal of Geophysical Research, v. 94, p. 2973–2988.

Shallo, M., Cina, A., and Turku, I., 1995, Outline of the metallogeny of the Albanian MOR- and SSZ-type ophiolites, *in* Ophiolites and related mineralizations: Tirana, Albania, Bureau Recherche Géologique et Minière, p. 27–47.

Silver, P.G., 1996, Seismic anisotropy beneath the continents: probing the depths of geology: Annual Reviews, Earth and Planetary Sciences, v. 24, p. 385–432.

Singh, S.C., Kent, G.M., Collier, J.S., Harding, A.J., and Orcutt, J.A., 1998, Melt to mush variations in crustal magma properties along the ridge crest at the southern East Pacific Rise: Nature, v. 394, p. 874–878.

Snow, J.E., Dick, H.J.B., Hellebrand, E., Buchl, A., Von der Handt, A., Langmuir, C., and Michael, P., 2002, Mantle peridotites of Gakkel Ridge, Arctic Ocean [abs.], *in* Orogenic lherzolites and mantle processes: Samani, Japan, p. 146.

Sparks, D.W., and Parmentier, E.M., 1994, The generation and migration of partial melt beneath oceanic spreading centers, *in* Ryan, M.P., ed., Magmatic systems: San Diego: Academic Press, p. 55–76.

Suhr, G., 1992, Upper mantle peridotites in the Bay of Islands Ophiolite, Newfoundland: formation during the final stages of a spreading center: Tectonophysics, v. 206, p. 31–53.

Thy, P., and Moores, E.M., 1988, Crustal accretion and tectonic setting of the Troodos ophiolite, Cyprus: Tectonophysics, v. 147, p. 221–245.

Tippit, P.R., and Pessagno, E.R., 1981, The biostratigraphy of sediments in the volcanic unit of the Samail ophiolite: Journal of Geophysical Research, v. 86, p. 2756–2762.

Tucholke, B.E., Lin, J., and Kleinrock, M.C., 1998, Megamullions and mullion structure defining oceanic metamorphic core complexes on the Mid-Atlantic Ridge: Journal of Geophysical Research, v. 103, p. 9857–9866.

Varga, R.J., and Moores, M.M., 1985, Spreading structure of the Troodos ophiolite, Cyprus: Geology, v. 13, p. 846–850.

Wakabayashi, J., and Dilek, Y., 2000, Spatial and temporal relationships between ophiolites and their metamorphic soles: A test of model of forearc ophiolite genesis, *in* Dilek, Y., et al., eds., Ophiolites and oceanic crust: New insights from field studies and the Ocean Drilling Program: Boulder, Colorado, Geological Society of America Special Paper 349, p. 53–64.

Whitehead, J., Dick, H., and Schouten, H., 1984, A mechanism for magmatic accretion under spreading centres: Nature, v. 312, p. 146–148.

Manuscript Accepted by the Society May 2, 2003

Printed in the USA

Geological Society of America
Special Paper 373
2003

Ophiolites and lost oceans: Rifts, ridges, arcs, and/or scrapings?

John Dewey
Department of Geology, University of California, Davis, California 95616, USA, and University College Oxford, Oxford OX1 4HH, UK

ABSTRACT

Ophiolites sensu lato originate in a variety of ways in several tectonic environments. None represents obducted sheets from the oceanic crust and mantle of large oceans; subduction destroys the oceanic lithosphere very effectively. The Jurassic Balkan and Greek ophiolites were obducted from the floors of young, very narrow oceans, and the Alpinotype ophiolites represent the strongly attenuated floors of rifts and "extreme" para-rifts in which stretched subcontinental mantle and associated mafic plutonics and volcanics are caught up in suture zones during shortening and inversion. Seamounts, plateaus, and fracture zones are clipped off subducting oceanic plates as mafic ophiolite scraps in slices and mélanges. The classic, full ophiolite complexes sensu stricto from harzburgite, through gabbro and sheeted complex, to pillow basalt, which occur in giant obducted slabs, are mainly of Ordovician or Cretaceous age and formed in supra-subduction zone environments in continent-facing arcs that were soon after obducted onto the adjacent rifted continental margin. Ophiolites sensu lato can be thought of as a simple random record of the Wilson cycle of opening, widening, narrowing, and closing oceans, but ophiolite complexes sensu stricto are events related to particular and special tectonic configurations.

Keywords: ophiolites, supra-subduction zone, island arc, rifts, Alpine-type ophiolites.

Ophiolite complexes are central to our understanding of continent/continent and arc/continent collision because they are believed to be fragments of oceanic crust and mantle generated at oceanic ridges and/or in intra-arc and back-arc supra-subduction zone environments. They may represent, therefore, a record of the processes of sea-floor spreading at oceanic ridges and/or in extensional/transtensional island arcs and the thrusting (obduction) of thin ophiolite sheets (nappes) for up to several hundred kilometers across continental margins. There is still substantial disagreement about their geology and role in tectonic processes and in their definition. I follow the 1972 Penrose definition (Anonymous, 1972) that an ophiolite complex is, ideally, an ordered sequence from harzburgite (sometimes with basal lherzolite), through banded and then isotropic gabbro with trondhjemite, to sheeted dike complex, then pillow lava, and sediment. There are many variations on this basic theme, and such sequences may be structurally dismembered. Such ophiolite sequences commonly occur in giant obducted sheets, with high-temperature, basal, garnet-two-pyroxene metamorphic soles, such as the Ordovician Bay of Islands Complex in Newfoundland and the Cretaceous Semail Complex in Oman. Such ophiolites were generated by some form of sea-floor spreading very shortly before their obduction and generally long before terminal closure of the ocean by continental collision.

Brongniart (1813, 1827) recognized that the first major deformation in mountain belts involved serpentinites, gabbros, and mafic volcanic rocks. Suess (1909) showed that greenstones (mafic/ultramafic assemblages) form narrow sheet-like bodies in orogenic belts. Steinmann (1905, 1927) argued that serpentinites, spilites/keratophyres, and radiolarian cherts (the Steinmann trinity) are characteristic of deep-water magmatism in the cores of Jurassic "geosynclines" (Tethyan Oceans) in the western Alps and Italy. In the Alpine troughs, the trinity generally lacks gabbro and never contains sheeted dike complex. Benson (1926) recognized that "Alpine-type" peridotites occur in the marginal zones of orogens, and Hess (1955) showed that serpentinites characteristically occur in paired belts marginal to orogens, exemplified by the Baie Verte and Gander Belts in Newfoundland. Hess believed

*dewey@geology.ucdavis.edu

Dewey, J., 2003, Ophiolites and lost oceans: Rifts, ridges, arcs, and/or scrapings? *in* Dilek, Y., and Newcomb, S., eds., Ophiolite concept and the evolution of geological thought: Boulder, Colorado, Geological Society of America Special Paper 373, p. 153–158. For permission to copy, contact editing@geosociety.org.

that the oceanic crust consists of serpentinite; it is interesting that recent investigations of slow-spreading ridges have shown the importance of ultramafic rocks on the ocean floor (Cann et al. 1997; Karson, 1999). De Roever (1957) argued that Alpine-type peridotites represent fragments of the upper mantle, and Dietz (1963) suggested that they are fragments of oceans incorporated into continental margin "geosynclines." An enormous breakthrough in ophiolite studies came through the work of the Cyprus Geological Survey. Wilson (1959) demonstrated that the Troodos Complex follows an ordered sequence from harzburgite through dunite, gabbro, and sheeted dike complex, to pillow basalt. Gass (1968) asseverated that the Troodos Complex is a fragment of oceanic crust and mantle generated at a spreading center and subsequently uplifted during a convergent process. Moores and Vine (1971) and Dewey and Bird (1971) argued that giant ophiolite slabs are obducted fragments of oceanic crust and mantle. Moores and Vine favored an origin at oceanic spreading centers, whereas Dewey and Bird preferred an origin in back and intra-arc oceanic basins above subduction zones, i.e., supra-subduction zone (SSZ) origin.

Stevens (1970), however, in a profoundly original paper, had already realized that Ordovician ophiolite complexes in Newfoundland are fragments of oceanic crust and mantle, for which he should be given great credit. Another great but little-known Canadian contribution to the ophiolite problem was that of Baragar (1954). In the Betts Cove Ophiolite Complex of Newfoundland (Upadhyay, Dewey and Neale, 1971), a sheeted complex is exposed, superbly, at Nippers Harbour and at Betts Cove, over a distance of several kilometers. This "intrusion" must, therefore, be composite. Baragar saw the sheeted diabase dike complex at Nippers Harbour and Betts Cove and the chilled dike margins, a profoundly brilliant discovery. Baragar (1954) argued that "over a very large area, a mafic intrusion has a diabasic texture." How can this be? Such a large intrusion should have cooled slowly with a gabbroic texture. I quote from Baragar (1954) "The volume which the dykes occupy is extremely large but individually the intrusives are small. It does not seem reasonable that numerous small intrusives would be emplaced in a large opening if it were available at the time; rather, openings available at any one time were probably small but under the compulsion of persistent, slow-acting forces, additional openings with approximately the same orientation were, perhaps, continuously being created until, eventually, the forces subsided." Of course, this was long before any knowledge of sea-floor spreading and its structural/petrological implications, and Baragar's (1954) clever and inceptive observations and ideas remain little known and rarely quoted to this day because they appeared in an open file report.

Although the term "ophiolite complex" has narrowed to the Penrose full sequence definition, the term "ophiolite" is still used, rather widely and more loosely, to include a broad range of deformed mafic and ultramafic fragments and scraps in orogenic belts, especially in suture zones. The provenance of most of these fragments is probably oceanic, including dismembered ophiolite complexes generated by some form of sea-floor-spreading, and slices of fracture zones, seamounts, and plateaus. It is likely, however, that some have other origins in tectonic environments in arcs and along and near rifted continental margins that become involved in collisional zones, including zoned ultramafic plutons (Taylor and Noble, 1960) beneath calc-alkaline volcanoes (Alaska); oceanic mafic arcs; mafic/ultramafic plutons and related flood basalt sequences (East Greenland); and exhumed subcontinental mantle beneath rifted margin detachments (Galicia margin; Boillot, Girardeau, and Kornprobst, 1988). When these assemblages are dismembered and strongly deformed in suture zones, their origin may not be obvious, although their petrology and geochemistry may give strong clues.

There has been an iterative, synergistic, and fruitful relationship between studies of ophiolite complexes of extinct oceans and studies of the structure and petrology of the oceanic crust and mantle of extant oceans, especially near and at ridge axes. For example, the Bay of Islands Complex was used by Dewey and Kidd (1977) and Casey et al. (1981) and the Troodos Complex by Greenbaum (1972) to model the plate-accretion process, especially the kinematics of magma chambers. Chilling statistics on sheeted complexes have been used to determine the dike-normal direction of the spreading axis at which the sheeted complex formed (Kidd and Cann, 1974). Fracture zones have been recognized in several ophiolite complexes (Moores and Vine, 1971; Karson and Dewey, 1978; Cann et al., 2001) and detailed structural/igneous models for fracture zone evolution have been developed from these, from modern fracture zones, and from theoretical kinematic models (DeLong et al., 1977, 1979). The ophiolite complexes and arc assemblages of western Newfoundland (Williams and Payne, 1975) and the deformed Eocene boninites of Guam (Reagan and Meijer, 1984) were used by Dewey (2002) to build a kinematic model of arc transtension; however, whether giant ophiolite complex nappes were generated at fast or slow spreading centers or in supra-subduction zone environments may matter little to developing an understanding of the spreading or plate-accretion process because sheeted dike complexes explicitly mean infinite extension at a spreading center at an oceanic ridge axis or within/behind an arc and, therefore, identical or similar processes are involved.

I suggest that there are a variety of ophiolites and ophiolite complexes, as follows:

Alpine-type ophiolites originate in the strongly attenuated floors of rifts, para-rifts, and para-oceans, such as the Early Jurassic Penninic/Austro-Alpine troughs of the Swiss Alps. Here, the continental lithosphere was extremely stretched, and subcontinental mantle (Malenco, Finero; Trommsdorff and Evans, 1974), which rose to shallow levels, was exhumed beneath extensional detachments like those of the Galicia margin. These regions were "frozen" in the early stages of continental breakup, where geometric and plate tectonic constraints prevented the troughs from widening and proceeding to organized spreading (Dewey et al. 1973). Ophiolite complexes such as the Platta Nappe (Dietrich,

Vuagnat, and Bertrand, 1974) do not contain sheeted complexes, but represent partial melting of the rising subcontinental mantle to form trough-floor gabbros and basalts. The Devonian Lizard Complex (Flett and Hill, 1946) probably represents an extreme form of this extensional process in a young narrow Rheic Ocean. It is also possible that such assemblages, formed early at the edges of much larger oceans, could be caught up in subsequent collisional sutures, after the rest of the younger oceanic lithosphere of that ocean had been subducted.

Balkan and Greek ophiolite complexes such as Vourinos (Moores, 1969) have sheeted dike complexes and were obducted from adjacent related troughs, such as the Pindos. The Pindos and similar troughs and the ophiolite complexes expelled from them with their associated "diabase-hornstein" facies may have been narrow oceans that had advanced beyond Alpine-type to the early stages of organized spreading. These ophiolites were obducted before their oceanic lithosphere had thermally thickened beyond the point where the 900 °C obduction isotherm (Dewey and Shackleton, 1984) was sufficiently shallow to allow detachment and obduction of a thin slice of ophiolite complex. Alternatively, but less likely, they may have formed in a supra-subduction zone environment during narrowing and closure of the troughs.

There are a number of ways in which a variety of oceanic fragments collide with and are clipped off in subduction zones. They include seamounts, oceanic plateaus, and fracture zone scarps, are dominated by basalts and gabbros, and rarely include ultramafic rocks. The Ordovician Annieopsquotch Complex in Newfoundland (Van Staal et al., 1998), the Eocene Olympic Peninsula in Washington, and the Ordovician South Connemara Complex in western Ireland appear to be strong candidates for this type of provenance.

Giant obducted ophiolite complexes sensu stricto (Penrose definition), such as the Semail Complex in Oman, the Troodos Complex in Cyprus, the New Guinea Ophiolite Complex (Davies, 1971), and the Bay of Islands Complex in Newfoundland are unlikely to have originated at oceanic spreading ridges. I do not believe that any have developed at ridges; rather, they are likely to have originated in supra-subduction zone island-arc settings, such as intra-arc and back-arc oceanic basins, for the following reasons, for which Miyashiro (1973) should receive most of the credit:

1. The fate of almost all, perhaps all, of the oceanic lithosphere is subduction to balance, on a constant-size Earth, the area generated by sea-floor-spreading and the modest stretching and shortening of continental lithosphere. The preservation by obduction rather than disappearance by subduction is exceptional and involves the shaving of thin (12 km) slices of oceanic crust and mantle at the 900 °C isotherm to be obducted with transported two-pyroxene granulite mafic metamorphic soles above subjacent nappes of continental margin sediments and basement (Dewey and Shackleton, 1984). The most natural plate tectonic mechanism for this is the collision of a young, continent-facing oceanic arc and its oceanic intra-arc and back-arc basins with an adjacent rifted continental margin. The notion of the obduction of newly formed oceanic crust and mantle from the flanks of spreading ridges involves arcane, peculiar, non-actualistic, rather contrived and unlikely mechanisms, such as those outlined by Dewey (1974, 1976); however, possible oceanic candidates for obduction are the closure of young, narrow strike-slip-dominated oceans such as the Gulf of California and the Cayman Trough; the oldest oceanic lithosphere adjacent to the exhumed subcontinental mantle of rifted margins; the young oceanic lithosphere at ridge/trench/trench triple junctions; and the oceanic crust and mantle at ridge subduction.

2. The interval between the generation and obduction of ophiolite complexes is short, generally <10 m.y., and obduction is long before and unrelated to terminal continent/continent collision. This suggests that arc/continent collision was involved in ophiolite obduction.

3. The petrology and geochemistry of ophiolite complexes involves boninites and calc-alkaline suites (Bédard, 1999; Bédard et al., 1998, 2000) of clearly arc origin (Elthon, 1991), both of ophiolite volcanics (Miyashiro, 1973) and immediately overlying turbidite sequences (Cousineau and Bédard, 2000; Kessler and Bédard, 2000; Metcalf et al., 2000; Wallin and Metcalf, 1998).

4. Field relationships indicate a close temporal/spatial relationship between ophiolite complexes and oceanic arcs. In western Newfoundland, the Bay of Islands Complex in the Lewis Hills cuts, magmatically, the roughly coeval island arc assemblages of the Littleport/Coastal Complex (Karson and Dewey, 1978; Suhr and Cawood, 2001; Suhr and Edwards, 2000), which confirms a supra-subduction zone setting. The arc assemblages and ophiolite complexes of western Newfoundland show some very complicated polyphase structural/magmatic relationships related to dextral transtension, transform/ridge intersections, and fracture zone evolution (Casey and Dewey, 1984; Dewey, 2002; Karson and Dewey, 1978; Suhr and Cawood, 2001; Suhr and Edwards, 2000).

The Jurassic Smartville and Devonian Trinity Ophiolite Terranes of the western U.S. Cordillera are somewhat different. The Trinity Ophiolite Complex (Metcalf et al., 2000; Wallin, and Metcalf, 1998) involves all the key Penrose ingredients and a capping thick calc-alkaline turbidite sequence, but the Smartville Ophiolite (Moores, 1982), although having sheeted complex, is more enigmatic in consisting of only gabbro, diabase, and basalt. The Smartville is in structural contact with more heavily deformed ophiolitic assemblages to the east, including ultramafic rocks, such as the Feather River Peridotite, but in extremely complicated shear zone systems without ophiolite complex sequences. These terranes are probably collided Jurassic and Devonian oceanic arcs involving both arc and intra-arc assemblages with less fully evolved ophiolite complexes (Dilek, 1989), and neither involved the obduction of an ophiolite nappe.

Sea-floor-spreading at oceanic ridges is the principal first-order plate tectonic feature of the Earth. Oceanic ridges are

~60,000 km in total length and generate the oceanic crust at ~30 km^3/yr. This fundamental process, by which the mantle partially melts to generate two-thirds of the surface of the Earth in 160 m.y., is becoming better understood by direct studies of oceanic ridges, and by drilling in the Mesozoic/Cenozoic crust of the extant oceans. It might be expected, therefore, that there should be a direct, at least partial, record of this process in suture zones where oceans have closed (Burke, Dewey, and Kidd, 1977). The five types of ophiolite outlined above might appear, in casual consideration, to be this record, and have their natural place and order related to various stages in the Wilson cycle of ocean opening and closing: type 1 is the earliest stages of continental breakup, stretching, rifting and separation; type 2 is narrow oceans; type 3 is the scrapings of subduction zones at the margins of wider oceans; and types 4 and 5 are the random collision of oceanic arcs and continental margins, with the likelihood of all types being caught up as fragments in collisional continent/continent suture zones.

The miniscule volume of ophiolite fragments—contrasted to the volume of the oceanic crust and uppermost mantle—may be considered to be a result of chance scrapings, randomly distributed in space and time, during ocean closure; however, there are temporal, physical, and spatial considerations that suggest that this view is incorrect. First, the age of origin and obduction of ophiolites appears not to be random. The giant ophiolite complex nappes are of Early Ordovician and mid-Cretaceous age, and are associated with periods of high sea level, black shales, blue-schist metamorphism, and voluminous calc-alkaline plutons (Dewey, 1988), which may have been a result of rapid sea-floor-spreading (Pitman, 1978). Second, no credible mechanisms have yet been devised for ophiolite obduction from oceanic ridges onto rifted continental margins. Third, oceanic lithosphere older than ~10 m.y. is "going downhill fast" into the realms, where detachment at the 900 °C isotherm (Dewey and Shackleton, 1984) becomes impossible because the thickness of a potentially detached slice becomes too great to obduct. Fourth, even the Jurassic ophiolites of the Tethyan realm in the Alpine, Balkan, and Greek realms were developed in rifts, para-oceans, and narrow oceans that could not develop into large oceans because they were constrained and caught in the vise of the converging African and Eurasian continents (Dewey et al., 1973). Another tectonic environment in which young oceanic lithosphere develops is during the last stages of oceanic closure between irregular continental margins. Young back-arc basins, such as the Balearic and Tyrrhenian oceans, developed behind strongly-arcuate subduction zones, which spread to fill and collide with adjacent irregular rifted margins (Dewey et al., 1989). Their lithosphere is young, thin, and hot, and a candidate for detachment and obduction during collision.

I suggest that the oceanic lithosphere generated at oceanic ridges is, effectively and wholly, subducted and that ophiolite complexes sensu stricto are a special and peculiar phenomena related to special and peculiar events in time and space in arcs and narrow oceans where the lithosphere is thin and weak. The Earth does not exhibit a random semi-continuous distribution of phenomena, events, assemblages, and facies with time. We should not expect a Lyellian uniformity of process. Processes are catastrophic at all time-scales and length scales (earthquakes, volcanoes, erosion, floods, sea level change, collisions of arcs and continents, bolides), there are "one-offs," such as the Late Cretaceous Chalk that resulted from the evolution of the Coccolithophoridae, and the late Carboniferous coal swamps. There is the secular thermal evolution of the Earth, perhaps with different tectonic processes, including types of ophiolite occurring in the Archean and Proterozoic (Burke and Dewey, 1973; Burke, Dewey, and Kidd, 1976; Moores, 1986, 1993, 2002; Sleep and Windley, 1982), and there are special tectonic situations that result from particular plate and continent configurations (Dewey et al., 1973), and perhaps from rapid short-lived changes in Earth's behavior. Ophiolites are one expression of the complexity of the Earth's behavior and history that generates a "catastrophic," seemingly non-uniformitarian pattern.

REFERENCES CITED

Anonymous, 1972, Penrose Field Conference on ophiolites: Geotimes, v. 17, no. 12, p. 24–25.

Baragar, W.R.A., 1954, Betts Pond area, Burlington Peninsula, Newfoundland: Report to Falconbridge Nickel Mines (copy lodged with Newfoundland Department of Mines).

Bédard, J.H., 1999, Petrogenesis of Boninites from the Betts Cove Ophiolite, Newfoundland, Canada: Identification of subducted source components: Journal of Petrology, v. 40, p. 1853–1889.

Bédard, J.H., Lauziere, K., Tremblay, A., and Sangster, A., 1998, Evidence for forearc seafloor spreading from the Betts Cove ophiolite, Newfoundland: Oceanic crust of boninitic affinity: Tectonophysics, v. 284, p. 233–245.

Bédard, J.H., Lauziere, K., Tremblay, A., Sangster, A., Douma, S.L., and Dec, T., 2000, Betts Cove Ophiolite and its cover rocks, Newfoundland: Geological Survey of Canada Bulletin, v. 550, 76 p.

Benson, W.N., 1926, The tectonic conditions accompanying the intrusion of basic and ultrabasic igneous rocks: Washington, D.C., National Academy of Science Memoirs, v. 19, 90 p.

Boillot, G., Girardeau, J., and Kornprobst, J., 1988, Rifting of the Galicia margin, crustal thinning and emplacement of mantle rocks on the seafloor: Proceedings of the Ocean Drilling Program, Scientific Results, College Station, Texas, Ocean Drilling Program, v. 103, p. 741–756.

Brongniart, A., 1813, Essai de classification mineralogique des roches melanges: Journal des Mines, v. 34, p. 190–199.

Brongniart, A., 1827, Classification et Charactere Mineralogique des Roches Homogene et Heterogene: Paris, F.G. Levrault.

Burke, K.C.A., and Dewey, J.F., 1973, An outline of Precambrian plate development, *in* Tarling, D.H., and Runcorn, S.K., eds., Continental Drift, Seafloor Spreading and Plate Tectonics, vol. 2 , New York, Academic Press, p. 1035–1045.

Burke, K.C.A., Dewey, J.F., and Kidd, W.S.F., 1976, Dominance of horizontal movements, arc and microcontinental collision during the later permobile regime, *in* B.F., Windley, ed., Early history of the earth (NATO Symposium, Newcastle, England, 1975), London, Wiley, p. 113–129.

Burke, K.C.A., Dewey, J.F., and Kidd, W.S.F., 1977, World distribution of sutures: the sites of former oceans: Tectonophysics, v. 40, p. 69–99.

Cann, J.R., Blackman, D.K., Smith, D.K., McAllister, E., Janssen, B., Mello, S., Avgerinos, E., Pascoe, A.R., and Escartin, J., 1997, Corrugated slip surfaces formed at ridge-transform intersections on the Mid-Atlantic Ridge: Nature, v. 385, p. 328–332.

Cann, J.R., Prichard, H.M., Malpas, J.G., and Xenophontos, C., 2001, Oceanic inside corner detachments of the Limassol Forest area, Troodos ophiolite, Cyprus: London, Geological Society of London, v. 158, p. 757–767.

Casey, J.F., and Dewey, J.F., 1984, Initiation of subduction zones along transform and accreting plate boundaries, triple junction evolution and forearc spreading centres-implications for ophiolite geology and obduction, *in* Gass, I.G., Lippard, S.J., and Shelton, A.W., Ophiolites and ocean lithosphere: London, Geological Society of London Special Publication, v. 13, p. 269–290.

Casey, J.F., Dewey, J.F., Fox, P.J., Karson, J.A., and Rosencrantz, E., 1981, Heterogeneous nature of oceanic crust and upper mantle: A perspective from the Bay of Islands Ophiolite Complex, *in* Cesare, E., ed., The sea, Volume 5: New York, Wiley Interscience, p. 305–338.

Cousineau, P.A., and Bédard, J.H., 2000, Sedimentation in a subaqueous arc/back-arc setting: the Bobby Cove Formation, Snooks Arm Group, Newfoundland: Precambrian Research, v. 101, p. 111–134.

Davies, H.L., 1971, Peridotite-gabbro-basalt complex in eastern Papua: Australian Bureau Mineral Resources Bulletin, v. 128, 48 p.

DeLong, S.E., Dewey, J.F., and Fox, P.J., 1977, Displacement history of oceanic fracture zones: Geology, v. 5, p. 199–202.

DeLong, S.E., Dewey, J.F., and Fox, P.J., 1979, Topographic and geologic evolution of fracture zones. Journal of Geological Society of London, v. 136, p. 303–310.

De Roever, W.P., 1957, Sind die alpintypen Peridotitmassen vielleicht tektonisch verfrachtete Buruchstucke der Peridotschale?: Geologische Rundschau, v. 46, p. 137–146.

Dewey, J.F., 1974, Continental margins and ophiolite obduction: Appalachian/Caledonian system, *in* Burk, C.A., and Drake, C.L., eds., Continental Margins: New York, Springer-Verlag, p. 933–952.

Dewey, J.F., 1976, Ophiolite obduction: Tectonophysics, v. 31, p. 93–120.

Dewey, J.F., 1988, Lithospheric stress, deformation and tectonic cycles: the disruption of Pangea and the closure of Tethys, *in* Audley-Charles, M.G., and Hallam, A., eds., Gondwana and Tethys: London, Geological Society of London Special Publication, v. 37, p. 23–40.

Dewey, J.F., 2002, Transtension in arcs and orogens: International Geology Review, v. 44, no. 5, p. 402–438.

Dewey, J.F., and Bird, J.M., 1971, The origin and emplacement of the ophiolite suite: Appalachian ophiolites in Newfoundland: Journal of Geophysical Research, v. 76, p. 3179–3206.

Dewey, J.F., and Kidd, W.S.F., 1977, Geometry of plate accretion: Geological Society of America Bulletin, v. 88, p. 960–968.

Dewey, J.F., and Shackleton, R.M., 1984, A model for the evolution of the Grampian Tract in the early Caledonides and Appalachians: Nature, v. 312, p. 115–121.

Dewey, J.F., Helman, M.L., Turco, E., Hutton, D.H.W., and Knott, S.D., 1989, Kinematics of the western Mediterranean, *in* Coward, M.P., Dietrich, D., and Park, R.G., Conference on Alpine tectonics: London, Geological Society of London Special Publication, v. 45, p. 265–283.

Dewey, J.F., Pitman, W.C., Ryan, W.B.F., and Bonnin, J., 1973, Plate tectonics and the evolution of the Alpine System: Geological Society of America Bulletin, v. 84, p. 3137–3180.

Dietrich, V., Vuagnat, M., and Bertrand, J., 1974, Alpine metamorphism of mafic rocks: Schweizerische Mineralogische und Petrographische Mitteilungen, v. 54, p. 291–332.

Dietz, R.A., 1963, Alpine serpentinites as oceanic rind fragments: Geological Society of America Bulletin, v. 74, p. 947–942.

Dilek, Y., 1989, Structure and tectonics of an early Mesozoic oceanic basement in the northern Sierra Nevada metamorphic belt, California: evidence for transform faulting and ensimatic arc evolution: Tectonics, v. 8, p. 999–1014.

Elthon, D., 1991, Geochemical evidence for formation of the Bay Of Islands Ophiolite above a subduction zone: Nature, v. 354, p. 140–143.

Flett, J.S., and Hill, J.B., 1946, Geology of the Lizard and Meneage: Keyworth, Nottingham, UK, Geological Survey Great Britain, 208 p.

Gass, I.G., 1968, Is the Troodos massif of Cyprus a fragment of Mesozoic ocean floor?: Nature, v. 220, p. 39–42.

Greenbaum, D., 1972, Magmatic processes at ocean ridges, evidence from the Troodos Massif, Cyprus: Nature, v. 238, p. 18–21.

Hess, H.H., 1955, Serpentinites, orogeny, and epeirogeny, in Poldevaart, A., ed., Crust of the Earth (A Symposium): New York, Geological Society of America Special Paper 62, p. 391–408.

Karson, J.A., 1999, Geological investigations of a lineated mass at the Kane transform: implications for oceanic core complexes: Philosophical Transactions, Royal Society of London, A357, p. 713–740.

Karson, J.A., and Dewey, J.F., 1978, Coastal Complex, western Newfoundland: An Early Ordovician oceanic fracture zone: Geological Society of America, Bulletin, v. 89, p. 1037–1049.

Kessler, L.G., and Bédard, J.H., 2000, Epiclastic volcanic debrites; evidence of flow transformations between avalanche and debris flow processes, Middle Ordovician, Baie Verte Peninsula, Newfoundland, Canada, *in* Mueller, W.U., Chown, E.H., and Thurston, P.C., eds., Processes in physical volcanology and volcaniclastic sedimentation; modern and ancient: Amsterdam, Elsevier, p. 135–161.

Kidd, R.G.W., and Cann, J.R., 1974, Chilling statistics indicate an ocean-floor spreading origin for the Troodos Complex, Cyprus: Earth and Planetary Science Letters, v. 24, p. 151–155.

Metcalf, R.V., Wallin, E.T., Willse, K.R., and Muller, E.R., 2000, Geology and geochemistry of the ophiolitic Trinity terrane, California: evidence of middle Paleozoic depleted supra-subduction zone magmatism in a proto-arc setting, *in* Dilek, Y., Moores, E., Elthon, D., and Nicolas, A., eds., Ophiolites and oceanic crust: New insights from field studies and the Ocean Drilling Program: Boulder, Colorado, Geological Society of America Special Paper 349, p. 403–418.

Miyashiro, A., 1973, The Troodos Complex was probably formed in an island arc: Earth and Planetary Science Letters, v. 19, p. 218–224.

Moores, E.M., 1969, Petrology and structure of the Vourinos ophiolite complex of northern Greece: Boulder, Colorado, Geological Society of America Special Paper 118, 74 p.

Moores, E.M., 1982, Origin and emplacement of ophiolites: Reviews of Geophysics, Space Physics, v. 40, p. 735–760.

Moores, E.M., 1986, The Proterozoic ophiolite problem, continental emergence, and the Venus connection: Science, v. 234, p. 65–68.

Moores, E.M., 1993, Neoproterozoic oceanic crustal thinning, emergence of continents, and origin of the Phanerozoic ecosystem: A model: Geology, v. 21, p. 5–8.

Moores, E.M., 2002, Pre-1 Ga (pre-Rodinian) ophiolites: their tectonic and environmental implications: Geological Society of America Bulletin, v. 114, p. 80–95.

Moores and Vine, 1971, The Troodos Massif, Cyprus, and other ophiolites as oceanic crust: evolution and implications: Philosophical Transactions, Royal Society of London, A286, p. 433–466.

Pitman, W.C., 1978, Relationship between eustasy and stratigraphic sequences of passive margins: Geological Society of America Bulletin, v. 89, p. 1389–1403.

Reagan, M.K., and Meijer, A., 1984, Geology and geochemistry of early arc-volcanic rocks from Guam: Geological Society of America Bulletin, v. 95, p. 701–713.

Sleep, N.H., and Windley, B.F., 1982, Archean plate tectonics: constraints and inferences: Journal of Geology, v. 90, p. 363–380.

Steinmann, G., 1905, Geologische Beobachtungen in den Alpen, II. Die schart'sche Uberfaltungstheorie und die geologischen Bedeutung der Tiefseeabsatze und der ophiolitischen massengesteine: Berichte Naturforschende Gesellschaft zu Freiburg, 1, 16, p. 44–67.

Steinmann, G., 1927, Der ophiolitischen Zonen in der mediterranean Kettengebirgen: Proceedings, International Geological Congress, 14th, Madrid, v. 2, p. 638–667.

Stevens, R.K., 1970, Cambro-Ordovician flysch sedimentation and tectonics in west Newfoundland and their possible bearing on a proto-Atlantic Ocean, *in* Flysch sedimentology in North America: Ottawa, Ontario, Geological Association of Canada Special Paper 7, p. 165–177.

Suess, E., 1909, The Face of the Earth, fourth edition: Clarendon, Oxford, 678 p.

Suhr, G., and Cawood, P.A., 2001, Southeastern Lewis Hills (Bay of Islands Ophiolite): Geology of a deeply eroded, inside corner, ridge-transform intersection: Geological Society of America Bulletin, v. 113, p. 1025–1038.

Suhr, G., and Edwards, S.J., 2000, Contrasting mantle sequences exposed in the Lewis Hills massif: evidence for the early, arc-related history of the Bay of Islands ophiolite, *in* Dilek, Y., Moores, E., Elthon, D., and Nicolas, A., eds., Ophiolites and oceanic crust: New insights from field studies and the Ocean Drilling Program: Boulder, Colorado: Geological Society of America Special Paper 349, p. 432–442.

Taylor, H.P., and Noble, J.A., 1960, Origin of the ultramafic complexes in southeastern Alaska: Proceedings of the International Geological Congress, 21st, Copenhagen, v. 13, p. 175–187.

Trommsdorff, V., and Evans, B.W., 1974, Alpine metamorphism of peridotitic rocks: Schweizerische Mineralogische und Petrographische Mitteilungen, v. 54, p. 333–352.

Upadhyay, H.D., Dewey, J.F., and Neale, E.R.W., 1971, The Betts Cove Ophiolite Complex, Newfoundland: Appalachian Oceanic crust and Mantle: Proceedings, Geological Association of Canada, v. 24, p. 27–34.

Van Staal, C.R., Dewey, J.F., MacNiocaill, C., and McKerrow, W.S., 1998, The Cambrian-Silurian tectonic evolution of the northern Appalachians and British Caledonides: history of a complex, west and southwest Pacific-type segment of Iapetus, *in* Blundell, D., and Scott, A.C., eds., Lyell: The past is the key to the present: London, Geological Society of London Special Publication 143, p. 199–242.

Wallin, E.T., and Metcalf, R.V., 1998, Supra-subduction Zone Ophiolite formed in an extensional forearc: Trinity Terrane, Klamath Mountains, California: Journal of Geology, v. 106, p. 591–608.

Williams, H., and Payne, J.G., 1975, The Twillingate Granite and nearby volcanic groups: an island arc complex in northeast Newfoundland: Canadian Journal of Earth Science, v. 12, p. 982–995.

Wilson, R.A.M., 1959, The geology of the Xeros-Troodos area: Nicosia, Cyprus, Geological Survey Department Memoir, v. 1, 184 p.

Manuscript Accepted by the Society May 2, 2003

Printed in the USA

Geological Society of America
Special paper 373
2003

Role of ophiolites in archipelago model of orogenesis

Kenneth J. Hsü*
Tarim Associates AG, Frohburgstrasse 96, Zurich, Switzerland

ABSTRACT

Ophiolites are mafic and ultramafic rock assemblages occurring in mountain belts, and they are commonly found in mélanges. Their association with deep-sea sediments suggests that they represent fragments of raised ocean floor. This conclusion is supported by marine geological and geophysical investigations.

This paper presents the evolution of my own thinking on the origin of ophiolites. I interpret ophiolites as relics of the ocean floor of backarc basins on the basis of my own investigations in the Alps and in China, as well as of literature surveys. I have concluded that only rarely do ophiolites mark the suture zones of plate-collisions.

Keywords: ophiolite, mélange, backarc basin, archipelago, orogenesis.

INTRODUCTION

The term *ophiolite* is a Greek synonym for the Latin *serpentine* and was introduced by Brongniart in the early nineteenth century; it was borrowed subsequently by Steinmann (1905), and the definition was specified to denote a suite of closely associated basic and ultramafic igneous rocks, including pillow lavas, spilite, diabase, gabbro, pyroxenite, peridotite, dunite, serpentinite, and related rocks. Steinmann recognized their association with deep-sea sediments. Suess (1909) agreed with Steinmann in that the ophiolites represent raised ocean floor. He also noted that they are not uncommonly associated with rocks of exotic origin, and that they occurred only in mountains. The ophiolite concept since has played a significant role in various models of orogenesis. Ophiolites were an indispensable component of eugeosynclinal rocks in the geosynclinal theory of orogenesis. Ophiolitic mélanges are considered as evidence of forearc accretion as a consequence of ocean-continent interaction, or of the suturing of two plates in a continent-continent collision zone.

This paper is a personal narrative of the evolution of my own thinking on the origin of mountains, progressing from the geosynclinal to the plate-tectonic theories, and culminating in the formulation of the archipelago model of orogenesis (Hsü, 1994). I finally came to the conclusion (Hsü, 1995) that ophiolite mélanges are not necessarily defining a plate-boundary. In fact, ophiolites are more commonly found in ancient subduction zone environments where backarc basins collapsed as a consequence of intraplate deformation.

*Kenjhsu@aol.com

Ophiolites in Eugeosynclines

When I first went to the Ohio State University in 1948, my mentor Ed Spieker asked me what would I like to do for my master's degree. I told him rather naively that I wanted to write a thesis on the origin of mountains. I was only 19 years old, and I knew little geology, but Spieker was very kind. He did not tell me to get lost, instead he told me to read Bucher and Holmes, and to study the German masters, Suess, Kober, Kossmat, Staub, Steinmann, among others. I checked out from the university library Staub's *Der Bewegungsmechanismus der Erde*. Little did I think that I was to be his successor in Eidgenössische Technische Hochschule–Zurich (ETH-Zurich) (Switzerland) some 30 years later.

Staub (1928, p. 209), following Argand (1922), recognized that geosynclines are formed by crustal thinning. If Staub had postulated extension as the cause as Walter Bucher did in 1933, or Rudolf Trümpy in 1960, he might have anticipated the modern theory of seafloor spreading. Instead, he followed Argand (1916)—incorrectly—and thought that geosynclines had evolved as a result of compression. Staub developed an idea that geosynclines had their origin through a mantle-density change in a region where the mantle-convection is descending (p. 255). That idea is somewhat similar to Beloussov's (1962) hypothesis of basification of crust; Staub's arrows showing the crustal and

Hsü, K.J., 2003, Role of ophiolites in archipelago model of orogenesis, *in* Dilek, Y., and Newcomb, S., eds., Ophiolite concept and the evolution of geological thought: Boulder, Colorado, Geological Society of America Special Paper 373, p. 159–171. For permission to copy, contact editing@geosociety.org.

mantle displacements are exactly opposite to those predicted by the modern plate-theory. Staub was wrong.

Staub was even more wrong to postulate that ophiolites were emplaced in an embryonic nappe as intrusives into a geosyncline under compression. In fact, the Alpine ophiolites include extrusive pillow lavas at the top, overlain by Middle Jurassic sedimentary rocks in many places. The Alpine ophiolites are thus dated as early Middle Jurassic. Trümpy (1960) voiced a consensus of the Alpine geologists that the Alpine sequences are continuous from Carboniferous to lower Tertiary. The change from Triassic shallow-marine to Jurassic deep-marine sedimentation was considered as an evidence of "leptogeosynclinal subsidence." The misidentification of the ophiolites as intrusives or as lavas in a continuously deposited sequence led Trümpy (1960) and other Alpine geologists to deny the existence of fossil oceanic crust in the Alpine region.

Silicic, intermediate, and mafic volcanic rocks are interbedded with, or intercalated in "geosynclinal" sedimentary rocks. Bubnoff (1937), Stille (1924), and Scheumann (1932), among others, invented the scheme of pre-orogenic, synorogenic, and post-orogenic stages of magmatic activities. The ophiolites were interpreted as pre-orogenic and were considered an indispensable component of eugeosynclines. The temperatures at the base of subsided sialic crust were subsequently raised high enough to produce synorogenic granite magmas. Synorogenic plutonic activities were succeeded by post-orogenic volcanism. The confusion lasted even into the earlier years of the plate-tectonics: Dewey (1969), for example, echoed the sentiments of Staub (1928) when he stated that ophiolites were emplaced through the imbrication of the top of subducting oceanic lithosphere along the inner walls of deep-sea trenches.

Ophiolites as the Crust and Upper Mantle Beneath Ocean Floor

When I first attended an annual meeting of the American Geophysical Union in 1955, I saw George Wollard's cross-section depicting the thickness of the earth crust under North America. The thickness was calculated on the basis of Airy's theory of isostasy, and the results were verified in a large part by seismic surveys. The crustal cross-section made obvious that geosynclines were simply regions underlain by thin crust. I had my first publication, *Isostasy and a theory for the origin of geosynclines* (Hsü, 1958) three years later. Miogeosynclines were continental margins where subsidence was balanced by sedimentation; eugeosynclines were simply regions underlain by oceanic crust.

I took a second step forward when I recognized Franciscan mélanges as a model for eugeosynclinal sedimentation and underthrusting tectonics (Hsü, 1971). Steinmann and Suess were right: The ophiolites are simply fragments of crust and upper mantle under ocean floors, and they are tectonically mixed with, but not intruded into, the so-called eugeosynclinal sediments. The postulate found verification by the deep-sea drilling in the eastern Atlantic. At Deep Sea Drilling Project (DSDP) Hole 120, drilling penetrated pelagic sediments on top of an ophiolitic basement. The Steinmann's trinity of radiolarian chert, pillow lava, and gabbro-serpentinite was found on modern seafloor (Hsü and Ryan, 1971).

The studies of the Troodos ophiolite have given us a "standard model" of the layered structure of oceanic crust (Moores, 1982), and this model has since been supported by observations elsewhere on land, and by drilling the in situ oceanic crust (i.e., DSDP/Ocean Drilling Program Hole 504B in Costa Rica Rift; Dilek, 1998).

The Spilite Problem

Near the turn of the last century, Harker (1896) and Becke (1903) popularized the terms "Pacific Suite" and "Atlantic Suite" for saturated and undersaturated (with respect to silica) basalts. Later, Dewey and Flatt (1911) noted the high soda content of pillow lavas associated with eugeosynclinal sediments, and they proposed a "spilitic suite" as a third major type of basalt. A controversy was born—is spilite a primary extrusive rock or an albitized basalt?

When Kennedy and Anderson reviewed the basalt problem in 1938, the geographical designations *Pacific* and *Atlantic* had proved misleading. Substitute terms *alkaline* and *calc-alkaline* were proposed. The latter is known by the now-popular name of "tholeiitic magma-type." The authors related the genesis of those two magma types to tectonisms: alkali basalts were found in stable ocean basins, whereas tholeiitic basalts and their derivatives were considered as volcanic products in orogenic belts. They found no place for spilite in their scheme; spilite was not a primary extrusive rock.

Kennedy and Anderson related the genesis of tholeiites to granitic crust, but Tilley (1950) corrected that mistake. Tholeiitic basalts are very common on the ocean floor. They have been dredged up from the ocean floor long before the start of the Joint Oceanographic Institutions for Deep Earth Sampling (JOIDES) Deep-Sea Drilling Project (Kuno, 1962). We now know that the mid-ocean ridge basalts (MORB) at active ocean-ridges are tholeiitic in composition, whereas alkali basalts are common on top of seamounts and guyots.

Spilites occur in the mountains and on ocean floor. They are basalts altered by albitization of feldspar. There are disputes as to whether spilites occurring in the mountains are albitized MORB basalts, or whether they are the alteration products of other kinds of extrusive rocks.

Miyashiro (1973) presented geochemical evidence that the Troodos ophiolite was the crust and mantle of a backarc basin. Could the chemistry of the Alpine spilites be related to their origin as a soda-rich basalt originated in a backarc basin? The studies of actualistic ocean basalts have yielded no satisfactory answer. James Hawkins (this volume, Chapter 14) believed, for example, that the ophiolite-tectonics could not be deciphered on the basis of geochemistry. We need independent lines of evidence to reconstruct the history of ophiolite emplacement.

One, Two, or More Oceans

As mentioned earlier, Steinmann and Suess voiced a consensus at the beginning of the twentieth century that alpine ophiolites were raised ocean floors. After much confusion during the next half century, the classic interpretation was accepted by all participants of the First Penrose Conference in Monterey, California (Dickinson, 1970). The conventional model of plate tectonics assumes the origin of mountains as a consequence of continental collision after the intervening ocean has been consumed. Because ophiolites are assumed to represent the remnants of fossil ocean floor, ophiolitic mélanges should mark suture zones; however, geological data are not always easily explained by such a simplistic model because more than one zone of ophiolites are present in numerous orogenic belts. Was there one, or were there two or more, oceans between the continents before their collisions?

The problem is not acute in the Western Alps. Paleogeographically, the Alpine ophiolites are mainly those of the Piedmont terrane in the South Penninic realm. Ultramafic rocks are also present in the Adula Nappe and other North Penninic units. Not much effort has been made to explain these two belts of ophiolites, except to assume a "geanticlinal," a *Zwischengebirge*, or a relic island-arc between the North and South Penninic Oceans.

The problem in the Dinarides (former Yugoslavia) and the Hellenides (Greece) is more controversial, and the debates are continuing. Steeply dipping ophiolites are present in the Vardar Zone, and a Vardar ocean is postulated between the Russian Platform and an eastern European passive margin, which was covered by shallow marine sequences of the Dinarides and Hellenides. Overthrust, apparently from east or northeast, onto the carbonate-platform deposits are ophiolite nappes. Do those ophiolites constitute rootless nappes derived from the floor of a small ocean in an archipelago of islands and basins? Or, are the ophiolites the klippen of a giant ophiolite-nappe from the Vardar Zone?

The same issue has divided the students of Anatolian geology. Are the Anatolian (northern Turkey) and the Antalya (southern Turkey) ophiolite mélanges the remnants of two different oceans, or are the Antalya nappes derived from the root zone of the Anatolian ophiolites farther to the north?

For two decades, I subscribed to the one-ocean postulation. Occam's razor favors a simpler solution, and one ocean represents a simpler mode than two or more oceans. I had to change my mind, however, when I went to northern Tibet in 1995.

The collision of Cathaysia (Mesozoic Cathaysia) and Tibet (Gondwanaland) brought two Jurassic limestones in a close juxtaposition north of Amdo on the Golmud-Lhasa Road. To the north is the Jurassic shallow limestone of the Qiangtang foreland basin of Cathaysia. To the south is the Jurassic pelagic limestone, occurring as exotic blocks in the Amdo Mélange. Aside from the Jurassic, other slabs include Lower Cretaceous pelagic limestone, radiolarian chert, as well as mafic and ultramafic rocks. The ophiolites are 75–80 m.y. old. The matrix of the mélange is made of Jurassic/Cretaceous hemipelagic and flysch sediments. The Amdo Mélange represents the accretionary complex dipping southward under a magmatic arc of Gondwanaland. The mélange crops out within the Nujiang/Bangong Zone, which marks a site of plate-collision. The contrast between the two Jurassic limestones of these two terranes is so obvious that our expedition group easily identified the spot when they drove from Cathaysia to Gondwanaland (Hsü and Chen, 1999, p.84).

There was the one ocean, but ophiolites dated at 85 to 95 Ma occur as exotic blocks in a Jurassic flysch matrix near Dongjiao, some 200 km south in Tibet. Adopting the single ocean hypothesis, the British expedition considered the Dongjiao ophiolites as the erosional remnant of a huge nappe from the root zone of the Bangong/Nujiang suture (Dewey et al., 1988), even though there is no evidence to support this audacious postulation. To the contrary, it is unlikely that the northward vergence of the Bangong/Nujiang collision could have propelled a giant ophiolite nappe nearly 200 km south (in the opposite direction).

The Dongjiao Mélange has a very limited extent. If the ophiolites are not a klippe of a large nappe, they could not be the floor of an intercontinental ocean either. The issue is no longer whether there were one or two oceans, but rather whether the Dongjiao ophiolites are fragments of a fossil oceanic crust that had evolved in a local deep-sea basin, such as a backarc basin.

Ophiolites as an Indicator of Backarc Collapsing

I first visited the Indonesian Archipelago in 1973, when I was appointed an External Examiner by the University of Malaysia. Instead of a cash-honorarium, I requested that I be taken on a field trip to see the local geology. Neville Haile knew of my interests in ophiolitic mélanges, so he arranged that I be taken to Sarawak.

The Lupa Valley Mélange of Sarawak consists of ophiolites, Mesozoic radiolarian cherts, and exotic blocks of Eocene coralline limestone. Haile and I drove from Kuching to Sibu, traversing the geological formations formed in the Oligocene/Miocene foreland basin. The ophiolite mélange extends from Lupar Valley near the western end of Sarawak, along the border region of Sarawak and Kalimantan, to Sabah, where massive ultramafic rocks are exposed. Although we could not ascertain the eastern and western continuations outside of the Island of Borneo, we did know of the ophiolite occurrences on the Islands of Anambas in South China Sea. Those might be correlative to the ophiolites of Thailand and Burma at one end and to those of the Philippines at the other end.

Assuming that ophiolite mélanges mark suture zones of inter-continent collisions, I had to postulate a paleo-Indian Ocean between South China and Sibumasu, which was assumed to have been split off from Gondwanaland. The Lupa Valley Mélange was thus considered a segment of that Southeast Asian (Donanya) suture zone. I did not think in 1973 that the mélange could represent the collapse of a local backarc basin.

My appointment in 1986 to the Tectonics Panel of JOIDES Ocean-Drilling Project broadened my horizon. I became a promoter of drilling South China Basin, and I got to know Ian Dalziel.

Marine geophysicists of the Kwangtung Regional Bureau of the Ministry of Geology in the People's Republic of China showed me their paleomagnetic and seismic results. They had submitted a proposal for drilling the continental margin and the abyssal basin of South China Sea, and I was asked to lobby for their cause.

The South China Sea is a backarc basin, floored by oceanic crust. The seafloor spreading, dated by seafloor magnetic lineation, continued during the time interval from the mid-Oligocene to mid-Miocene. Since then, the oceanic lithosphere of the eastern half of the basin has been subducted at the Philippine Trench, under the Philippine Islands. The initiation of the consumption of this basin floor via subduction was a manifestation of the collapse of the backarc basin.

Ian Dalziel was then the chairman of the JOIDES Tectonics Panel. He informed me that he had interpreted the process of backarc subduction as a "collapse of backarc basin" on the basis of his studies of the southern Andes Mountains (Dalziel, 1981). In this interpretation, the consumption of the backarc-basin ocean floor eventually leads to an arc-arc or an arc-continent collision. In the southern Andes, the eastern margin of the Pacific Ocean once had a frontal arc reminiscent of the Western Pacific. The backarc spreading started during the Jurassic, when extensive normal faulting and volcanism took place in the region behind that island arc, while the Pacific Plate was thrust under South America. Backarc basins, floored by oceanic crust, came into existence between this west-facing frontal arc and the South American continent. These basins thus had a tectonic framework similar to that of South China Sea. The Andean backarc basins were eliminated by arc-continent collisions. The suture zones of local collisions are marked by ophiolitic mélanges in the Andes. The ophiolites are present in a zone of penetrative deformation and metamorphism. East of the suture zones is the east-vergent foreland thrust belt of the Sub-Andean Foothills. Farther east are the foreland basins. In such an orogenic belt formed by backarc collapsing, the Patagonian batholith is the "motor," the ophiolite mélange and the mobilized basement are the "overridden," and the foreland thrust-belt and foreland basins are the "escape" structures in the model for Alpine Orogenesis formulated much earlier by F.E. Suess (1937).

Was There an Ocean?

I was among those who welcomed a delegation sent by China to the International Geological Congress in Sydney in 1976, and I was invited to visit China in 1977; so I made my return to the land of my birth 29 years after I had left for graduate school education in the United States.

I studied geology as an undergraduate student in China. Chinese geologists used the term North and South China Platforms, but the sedimentary formations on those platforms are not flat-lying like those on the North American platforms (or cratons). We took field trips to the vicinities of Chungking and Nanjing. We saw, and later mapped, anticlines and synclines. To contrast their tectonic style to the undeformed sedimentary rocks in North America, Chinese geologists chose the term "unstable platforms," or *diwa* (geodepressions).

I was given a geological atlas of China when I was a consultant for the Chinese government in 1979. Opening up the Hunan Sheet of the atlas volume, I was impressed by the tectonic style of the province. The structures are typical of foreland fold- and thrust-belt, analogous to those in the Appalachian Valley and Ridge province. There are also flysch formations and ophiolites. Without actually having a chance to do any fieldwork, I published at the request of Huang Jiqing, the Grand Old Man of Chinese geology, an essay speculating on a thin-skinned plate-tectonic model for collision type orogenesis (Hsü, 1981). Later I was invited by Sun Shu, Director of the Geological Institute, Chinese Academy of Sciences, to start a program to re-interpret the geology of China on the basis of the plate-tectonics theory. I spent the next 20 years doing that, with the help of Sun and his colleagues and students. We started out with an outrageous hypothesis and wrote *Huanan Alps, Not South China Platform* (Hsü, et al., 1987), and the final product was the new *Geological Atlas of China* (Hsü and Chen, 1999).

When I first started in 1983, I thought that the task would be easy. We expected to find an ophiolitic mélange marking the suture zones of plate-collisions, between the rigid-basement thrusts of the overriding continent and the foreland-deformed belts of the underthrust continent. We started in Qinling, the well-known mountain range between North and South China, and the model seemed to be applicable. Phyllites and schists constitute the matrix of an ophiolite-bearing mélange, traditionally considered as a Precambrian basement. The mélange is present in an east-west trending belt between the Archaean rigid basement to the north and the foreland folds to the south. The age of metamorphic rocks in the Qinling Mountains ranges from 500 Ma to 200 Ma. This scatter did not surprise us, because metamorphism should have taken place during subduction and collision in a time interval of a few hundred million years (Hsü et al. 1987).

The Qinling Mélange, we thought, was the remnant of an ocean separating North China from South China. Thus, crust of this ocean basin should be now represented by the Qilian and Kunlun Ophiolites in the west, and by Dabie and Jiaonan Ophiolites in the east. Our campaign thus seemed to have begun triumphantly, and we expected a quick ordering of the plates and plate-collisions by mapping and dating of ophiolites in China.

We faced a difficulty with our simplistic interpretation. Ophiolitic mélanges (Coaliangyi and Zhiyang Celtides; Fig. 1) are present in a terrane south of the foreland-deformed belts. We were thus encountering again the old dilemma of one versus two ocean basins. Choosing the ultra-nappist solution, we considered the southern ophiolites as the klippe remnants of a giant ophiolite nappe derived from a root-zone to the north (Hsü et al., 1987). It was not a very satisfactory interpretation.

Our simplistic interpretations became even less credulous when we began in 1985 to apply the collision model to interpret the geology of South China. We had to postulate the existence of two oceans in order to end up with two suture zones. (Hsü and Chen, 1999). Furthermore, the two inferred "ocean basins" separating the continental areas seemed to have a very limited geographical extent.

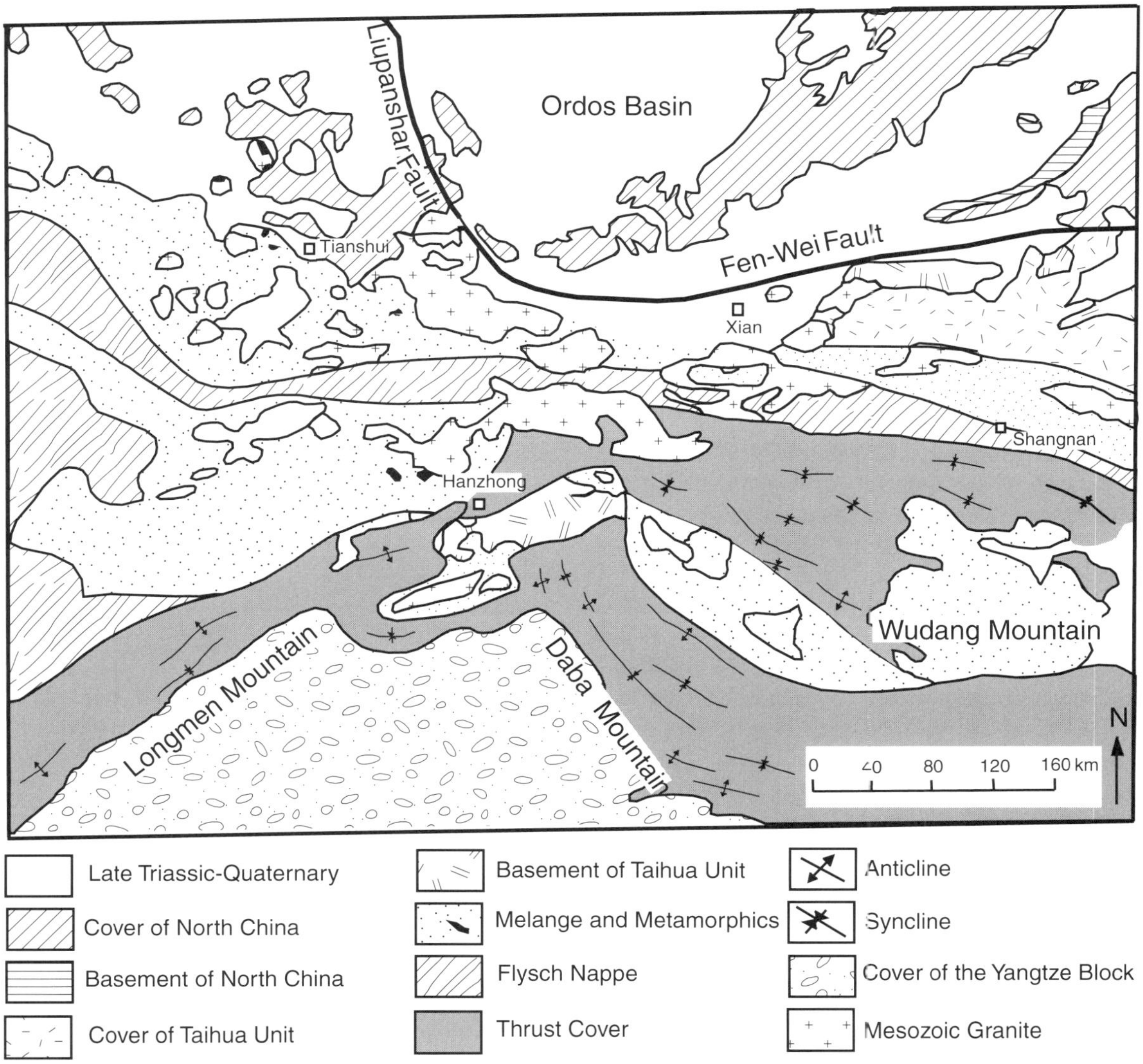

Figure 1. Tectonic facies units in Qinling Mountains, China (after Hsü and Chen, 1999).

The problem was compounded when we went to Inner Mongolia in 1988. We expected to find one suture zone between China and Siberia, as Li et al. (1982) had postulated. We found instead two belts of discontinuous outcrops of ophiolites (Ondorsom and Hegenshan Mélanges) in Inner Mongolia (Hsü et al., 1991). Şengör et al. (1993) postulated that the Hegenshan Mélange is a forearc-accretion complex on the trench wall of a subduction zone within an ocean basin separating China from Siberia (Unit C in Fig. 2). This interpretation was not consistent with our field observations: we had to suggest instead that the South Mongolia Mélange (Unit E in Fig. 2) represents the last remnant of the great Altaid Ocean (Hsü and Chen, 1999). What was the paleogeographic position of Hegenshan then?

The puzzle of multiple ophiolite belts is well known to Chinese geologists. One solution was to adhere to the simple "one ocean model," to resolve the problem through an assumption of the repeated opening (by seafloor spreading) and closing (by continental collision) of a differently positioned single ocean between the two continents. The *hypothesis of polycyclic deformations* advanced by Huang Jiqing contended, for example, that North and South Asia were pulled apart and pushed together again and again like the playing of a manual harmonium (Huang and Chen, 1987). He thus converted a spatial problem into a temporal problem: there was always only one ocean between the two continents, but the location of that ocean was different at different times. Huang and Chen (1987) applied this hypothesis to interpret the geology of Tianshan.

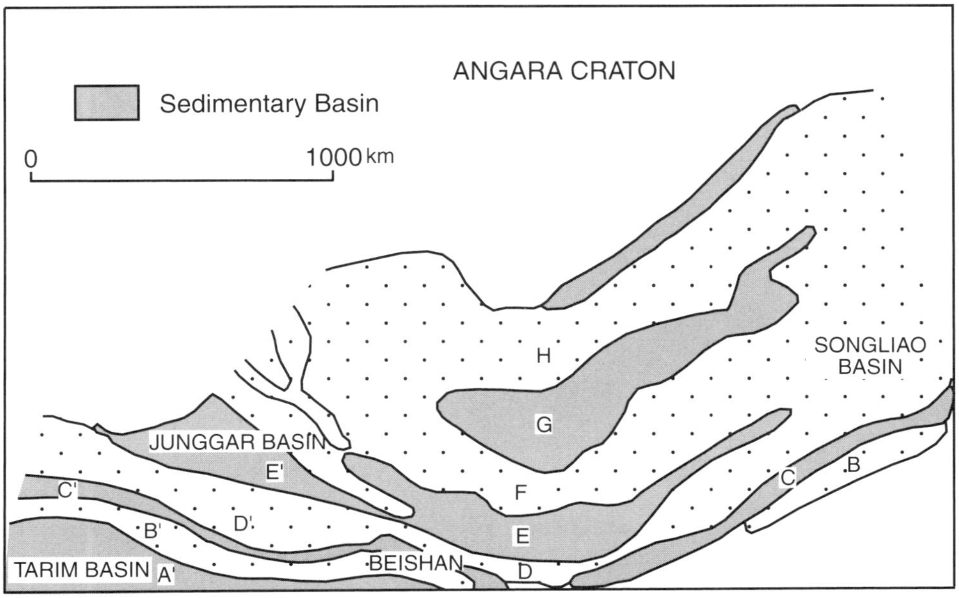

Figure 2. The tectonic facies distribution of Mongolia and the adjacent regions (after Hsü and Chen, 1999). A′—Tarim Basin; B—Sonidzuogi Rhaetide; B′—Kelpin/Kuruktag Rhaetide; C—Hegenshan Celtide; C′—South Tianshan Celtide; D—Uliasta Rhaetide; D′—Middle Tianshan Rhaetide; E—South Mongolia Celtide; E′—Junggar Basin; F—Southern Tuva Rhaeitde; G—Khanghai Celtide; H—Northern Tuva Rhaeitide.

In the company of my Chinese colleagues, I went on five expeditions across Tianshan and found that we could disprove Huang's hypothesis. Tianshan was not a "geosyncline" or an ocean between Junggar and Tarim Cratons. Tianshan was an island-arc complex between the Altaid Ocean and the Tarim Backarc-Basin. The Northern Tianshan Mélange was a forearc-accretion complex beneath a north-facing trench wall (Hsü et al., 1994). But there is still the Southern Tianshan Mélange (Fig. 3). Those ophiolites have a limited distribution and could not represent a great ocean basin. Comparing the geology of Tianshan to that of the Marianas, we came to the conclusion that there were several rows of backarc basins in Xinjiang. Between the Middle Tianshan Arc and the Kelpin Remnant Arc was a backarc basin (Fig. 3). The collapse of that backarc-basin and the subsequent arc-arc collision led to the genesis of the Southern Tianshan Mélange (Hsü et al., 1994).

"Holes in Cratons"

The sedimentary basins of the world can be grouped into three mutually exclusive and all-inclusive categories: (1) basins formed by crustal compression; (2) basins formed by crustal extension related to lithospheric stretching; and, (3) basins formed by crustal extension related to lateral movement of crustal blocks. I introduced this generic classification during an oral presentation in Xining (China) in 1985, and claimed that all basins could find their place in a pigeonhole of such a scheme. It was a rude awakening, however, when a young geologist of the Qinghai Provincial Geological Survey challenged my authority; he asked me to find a place for the Qaidam Basin of Qinghai in this classification.

Qaidam is situated near the northern edge of the Tibetan Plateau and has been considered as an "intermontane basin." The designation is literally correct, because the basin is bounded by the Kunlun Mountains to the south and by the Altun and Qilian Mountains to the north (Fig. 4). Whereas thick Neogene formations are present in pull-apart troughs along strike-slip faults, the older sediments of Qaidam were deposited in a basin underlain by thin crust.

The mountains surrounding Qaidam are not orographic features like those in the western United States, where the relief has resulted largely from normal faulting. The Qilian Mountains are underlain by deformed belts. Prior to orogenic deformations, the East Kunlun Mountains were an island arc, the Paleozoic frontal arc of the North China Plate. The Altun and Qilian Mountains were remnant arcs. Qaidam thus is not an intermontane basin as the term is commonly understood. The basin reminded me of the "holes in the ground" in the Russian Platform, such as the Caspian Sea and the Black Sea. I suspected then and my suspicion is now confirmed that very thick sediments in the so-called cratonic basins are deposited in backarc basins that became landlocked and were filled up (Hsü, 1988; Aplonov, et al., 1992).

Replying to the impertinent young man, I shot back and claimed that Qaidam was *a basin formed by crustal extension related to lithospheric stretching*. It started out as a Paleozoic backarc basin north of the East Kunlun Arc, and the stretching of the crust was caused by convection currents beneath the backarc region induced by the subduction of Prototethys under the frontal arc (Hsü and Chen, 1999, p. 62).

A similar basin between Kunlun and Tianshan is Tarim, with a size and shape comparable to the Sea of Japan. The deepest Tarim depressions are underlain by oceanic crust, the existence of which is manifested by seafloor lineations (Hsü, 1988). The Kunlun Mountains south of Tarim represent the western continuation of the East Kunlun Arc. The Southwest Tarim Depression was a

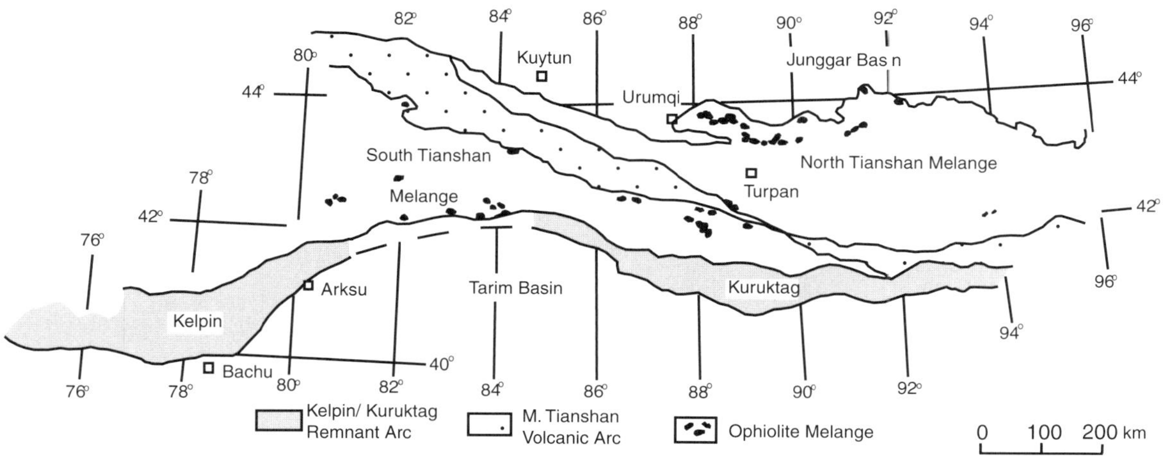

Figure 3. Tectonic facies units in Tianshan, China (after Yao and Hsü, 1994).

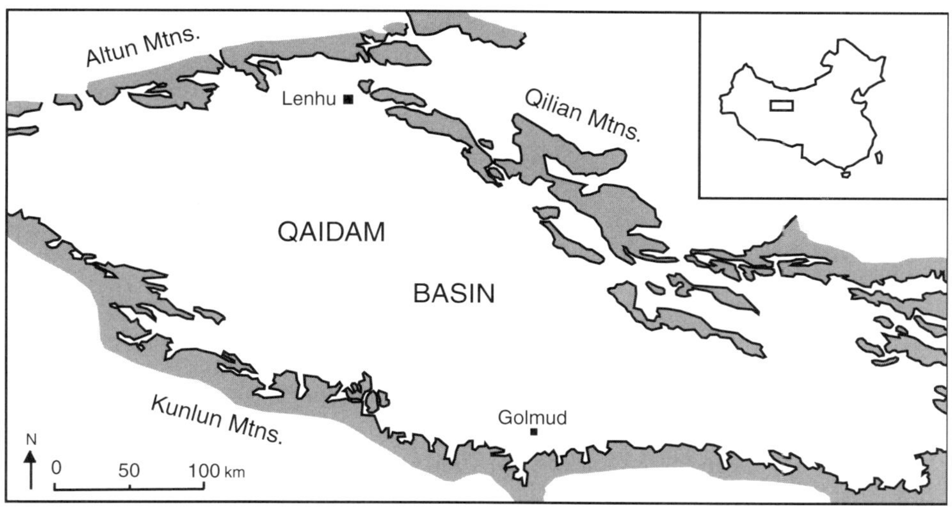

Figure 4. Qaidam Basin—a relic backarc basin surrounded by mountains, China (after Hsü, 1988).

backarc basin, and Northeast Tarim a relic backarc basin, and the two were separated by a remnant arc that is now the Central Tarim Ridge. The actualistic analogue for the Tarim backarc depressions and the Kunlun volcanic arc system can be found in the Philippine Sea region, where the Mariana Arc, Pala-Vela Basin, Kyushu-Palau Ridge, and West Philippine Basin can be considered as the modern analogues of the Kunlun volcanic arc, SW Tarim backarc basin, Central Tarim Ridge remnant arc, and the Northwest Tarim relic backarc basin, respectively (Fig. 5A, B).

The distribution of the frontal arcs, the relic arcs, and former backarc basins of Northwest China describes the paleogeography of an archipelago terrane. Not all former backarc basins have collapsed completely to form ophiolitic mélanges. The oceanic crust within Tarim and Qaidam, for example, has not been subducted completely, and the ocean floor subsided isostatically under the sedimentary load, resulting in the formation of those "holes in the craton."

Archipelago Model of Orogenesis

The first theory of the origin of mountains, the geosynclinal theory of James Hall, assumed crustal shortening of a cooling earth. The discovery of radioactive generation of heat permitted an alternative assumption of convection currents in Earth's mantle as a driving force for moving crust laterally. Wegener's theory of mountain building by collision of drifting continents was then modified into the modern plate-tectonics theory of continental collision. In classic theories, large-scale mountain building is

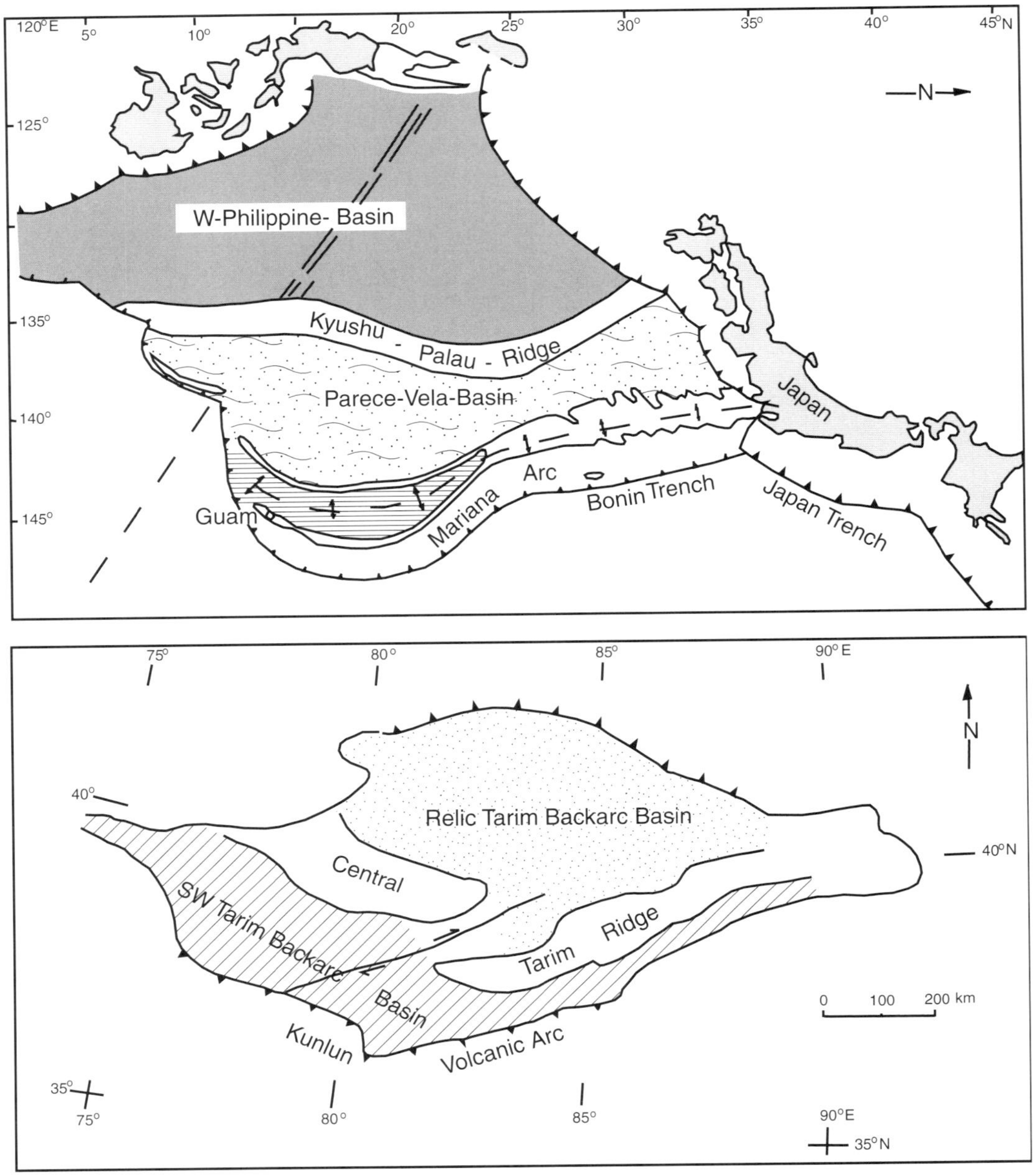

Figure 5. A: Backarc depressions of the Philippine Sea. B: Relic backarc depressions of Tarim, China. (after Hsü, 1988).

assumed to have occurred along a linear trend, although there has been controversy as to whether orogenesis is episodic and globally synchronous. As the plate tectonics theory has become the ruling paradigm, orogenic deformations are believed to have been continuous, and three types of orogenic systems have been recognized: (1) the California type, dominated by strike-slip faults; (2) the Andean type, dominated by the subduction of an oceanic lithosphere under a continental plate; and (3) the Alpine type, dominated by plate collision after the consumption of an intervening ocean basin.

When I started my project to produce a geologic atlas of China in 1983, I adopted the collision model to interpret what I considered the Alpine Type mountains in China. My experience in the Swiss Alps gave me a head start. I used Alpine terms Helvetic, Penninic, and Austroalpine to communicate with the young Chinese colleagues. I came eventually to the realization, however, that the Alpine terms are used not only to designate tectonic, but also paleogeographic units. The Helvetic and Austroalpine sedimentary sequences, mainly shallow marine, were "miogeosynclinal," laid down on European and North African margins (respectively). The Penninic were the sediments and ocean floor within the Tethyan Ocean.

The Alpine model, first proposed by Edward Suess and then refined by his son, F. Suess (1937), is a model of continental collision. Africa and Europe collided after the intervening ocean (Tethys) had been compressed together to form the Penninic Alps. The bounding continents have had different roles during this collisional orogeny. The Austroalpine nappes are thrust above the Penninic nappes, which in turn have overridden the Helvetic nappes.

Suess Junior pointed out that the three vital elements in orogenesis are: the motor, the overridden, and the escaped. Phrased in terms of the mechanics of collision between continental plates, the *motor* is represented by the overriding continental crust, and the *overridden* by the basement of underthrust continental crust and by a part of the consumed oceanic crust, whereas the *escaped* consists mainly of the sedimentary cover of the underthrust continental crust. In the Alps, the Austroalpine nappes are the motor, the Penninic the overridden, and the Helvetic the escaped.

After the adoption of the plate tectonics theory, the Alpine orogeny has been interpreted by two models of subduction: the B-type (honoring Benioff) subduction of the Tethys leading to the collision of Europe with Africa, and the A-type (honoring Amstutz) subduction of Europe beneath Africa. Substituting the Alpine paleogeographic terms by those denoting tectonic style of deformation, I proposed the names rhaetide, celtide, and alamanide to designate the three tectonic facies of the deformed rocks (Fig. 6): (1) the rhaetide facies, consisting mainly of the basement rocks of a continent crust, overthrusting onto (2) the celtide facies, consisting mainly of ophiolitic mélanges and mobilized and underthrust crust, overthrusting onto (3) the alamanide facies, consisting of thin-skinned deformation of décollement overthrusts, flysch, and molasse deposits.

The rhaetide and alamanide rocks could be the basement and the sedimentary cover of a continental margin, or of an island arc (active or remnant). The celtide rocks include those of deep-sea deposits and ocean crust.

The Helvetic and the Austroalpine sequences were the deposits of the European and African continental margins, respectively. They occupied different paleogeographic locations, but both the Helvetic and Austroalpine sedimentary sequences have been deformed to form structures of the alamanide facies. The Helvetic nappes are those of a foreland thrust belt, and the Austroalpine sedimentary rocks (e.g., the Lower Austroalpine nappes) have been deformed by décollement between underthrusting (Err-Bernina) and overthrusting (Silvretta) Austroalpine basement-nappes.

The simple single ocean model of continental collision has always been a problem, because of the presence of deformed shallow marine sequences in the Penninic realm that are represented by the Briançonnais and Schams nappes. Sandwiched between the ophiolites of the North Penninic (Valais) and South Penninic (Piedmont) nappes, those alamanide-facies rocks were supposed to have been the sedimentary cover of islands or a chain of islands in the Tethyan Ocean. Those were then detached from the underlying granitic basement and thrust northward during the Tertiary deformation. Such an oversimplified explanation has created, however, a paradox because the Schams sequences have a southerly vergence, indicating that they have been overthrust from north to south, not from south to north.

The solution of the so-called Schams paradox turns out to be very simple, after it is discovered that the Schams sequences were deformed during the Cretaceous (Hsü, 1994). The shallow marine sedimentary rocks of Schams were the sedimentary cover of a remnant arc, and the seafloor deepened to the north to form a backarc basin, alias "Valais Geosyncline." That basin was separated from a South Helvetic Flysch basin by the Habkern volcanic arc. The ocean floor of this Valais backarc basin was subducted under the Habkern arc. A Cretaceous/Paleocene

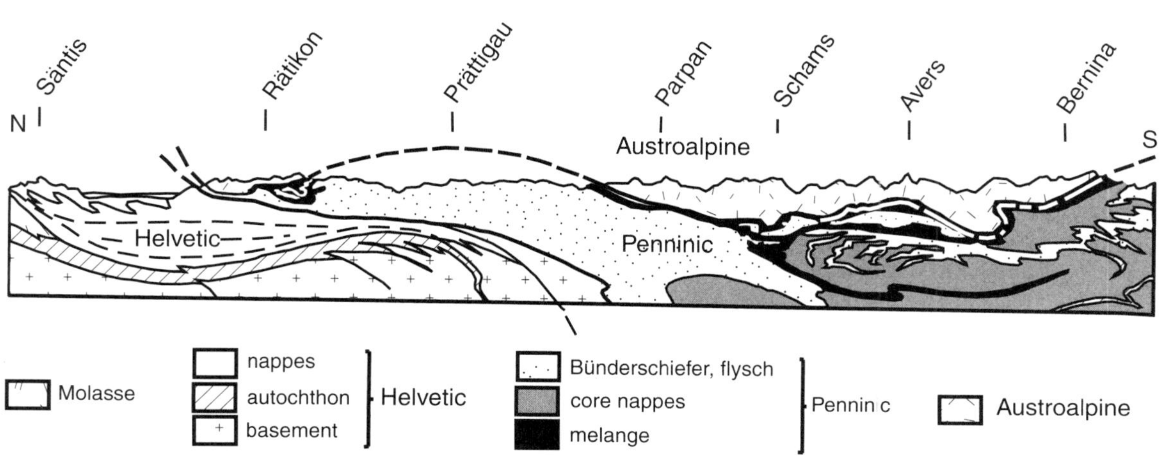

Figure 6. Tectonic facies units in the Alps, Switzerland (after Hsü, 1995). Tectonic units of the Alps, Helvetic, Penninic, and Austroalpine are designated by their paleogeographic framework. Those units have a tectonic style typical of the Alamanide, Celtide, and Rhaetide tectonic facies, respectively.

Gurnigel Flysch was deposited in a trench setting south of the arc (Fig. 7). In the back-arc basin north of the arc, the Schlieren Flysch deposits were laid down (Hsü and Schlanger, 1971). The arc volcanism of the Cretaceous subduction is evidenced by the volcanic debris in the Gurnigel and Schlieren Flysch.

The forearc basement of Habkern Arc has been metamorphosed to form the metamorphic rocks of the Adula Nappe, which is considered lower Penninic in the classic Alpine literature. Through the dating of Alpine metamorphic rocks, this Cretaceous belt of forearc metamorphism has been traced to the French Alps, to Corsica, to offshore Sardinia, and to Calabria. The north-dipping Benioff zone of the Cretaceous subduction in Switzerland has been detected by the seismic crustal studies (Frei et al., 1989).

The continental margin (Helvetic) of Europe could not have been a plate boundary, nor could the axis of the Tethyan seafloor spreading have been the southern plate margin. The northern ocean margin of Africa must have been subducted under an active margin. The European Plate margin (Cretan and Calabrian Arcs) has indeed been an island-arc margin since the middle Miocene. The topography and the magnetic anomalies indicate that the older active margin is now buried under the sediments of the Mediterranean Ridge (Hsü, 1994). Interpreted within such a paleogeographic framework, the Alpine Tethys included one or more backarc basins (Penninic, or Valais and Piedmont) beyond the European passive continental margin (Helvetic). In this configuration of an archipelago, Italy was not a passive margin on a North African promontory, but a remnant arc of carbonate sedimentation.

Adopting the archipelago model of orogenesis, I proposed that the Alpine structures were first formed as a consequence of a backarc basin collapse during the Cretaceous-early Tertiary, when the North Penninic (Valais) backarc basin was subducted down a north-dipping Benioff Zone under the Habkern Arc. Meanwhile, the South Penninic (Piedmont) floor was consumed down a south-dipping subduction zone under the Austroalpine (Italy) remnant arc. The collapse of the South Penninic basin caused a continent-arc collision between Europe and Italy. The recognition of southern Europe as a Mesozoic and Cenozoic archipelago behind an active margin of island arcs refutes the classic concept that the Alps owed their origin to the collision of two passive continental margins.

The pre-middle Miocene European plate-margin extended eastward from the Mediterranean Ridge to Florence Rise and Cyprus (Fig. 8). The ocean floor of the African Plate has been subducting beneath that arc, and the last remnant of the Tethyan Ocean between Europe and Africa is the Levantine Sea. The Permian age of the eastern Mediterranean ocean floor is indicated by a negatively magnetized quiet zone, and by the Permian and younger passive margin sequences in Levantine and Egypt. The forearc trench sediments are now covered by the sediments of Herodotus Abyssal Plain. A tectonic mélange is present under Cyprus, and ophiolitic rocks have been dredged up from the steep escarpment south of the island. A north-dipping, pre-middle Miocene subduction zone has been detected by seismic tomographic images. The European plate margin did not shift to its present position along the Cretan Arc until the late Miocene.

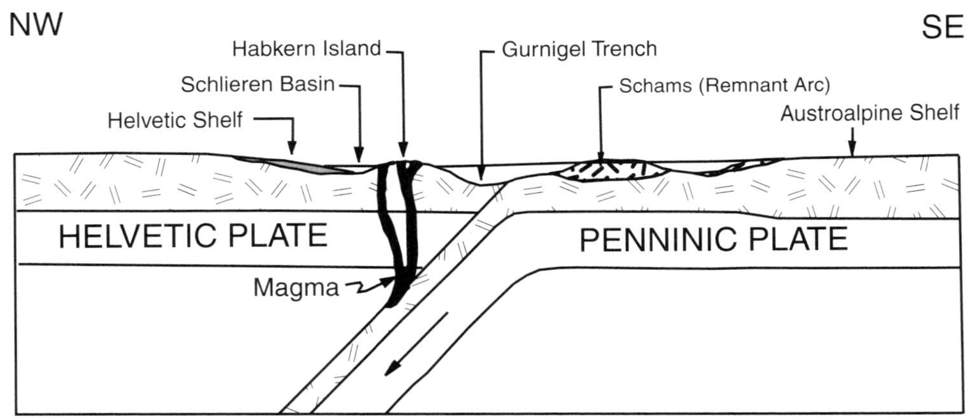

Figure 7. Plate tectonic model of Flysch deformation (after Hsü and Schlanger, 1971). The Late Cretaceous/Paleocene paleogeography of flysch sedimentation suggested to Hsü and Schlanger the presence of a magmatic arc on the southern margin of Europe. This arc was a south-facing feature bounded by a north-dipping subduction zone, which consumed the Northern Penninic backarc basin floor. The presence of volcanic debris in the flysch indicates arc volcanism during its formation. Regional tectonic considerations suggest that the Habkern arc was not the frontal arc of the European plate. The Cretaceous island arc margin of the plate had lain far to the south, and is now buried under the Mediterranean Ridge.

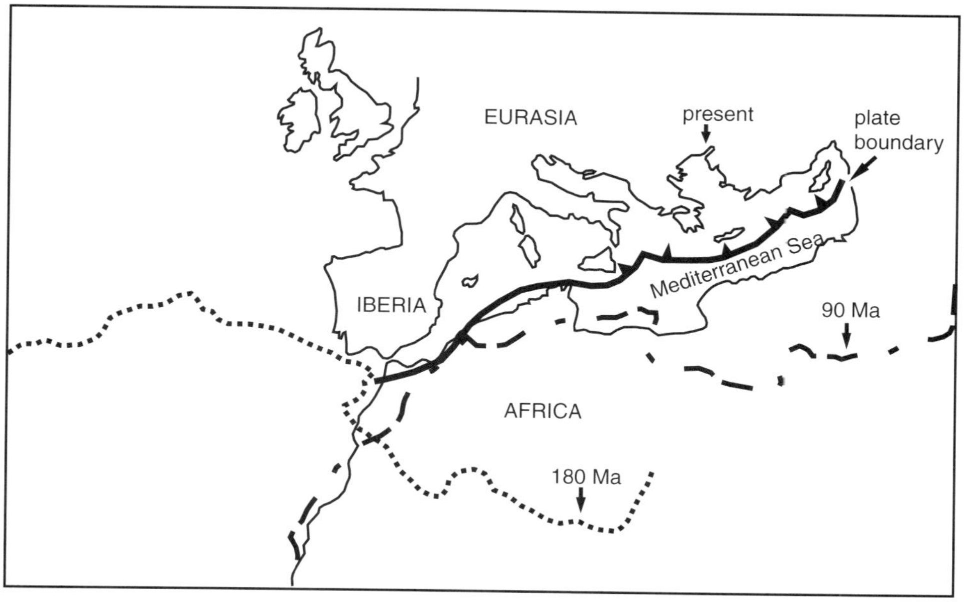

Figure 8. Plate tectonic evolution of the Alpine Mediterranean region. The southern margin of the European plate is a magmatic arc, now buried under the Mediterranean Ridge. On the basis of the data provided by magnetic lineations of the Atlantic Ocean, the African plate was found to have first moved eastward from 180 to 90 Ma, and to have rotated counterclockwise since then. Subduction of the lithospheric plate under the Paleotethys during the Mesozoic caused the change from an Andean type to an island-arc type of European margin. Subduction of the plate caused the consumption of the Paleotethys. The ocean basin south of the Mediterranean is the last remnant of the once vast Paleotethys ocean.

Franciscan Mélanges in Archipelago Model of Orogenesis

I obtained this new understanding of the Alps only after I had found that the numerous ophiolite belts of China are not suture zones of continental collisions; they are the collapsed ocean floors and sediments of former backarc basins. My tenure at ETH-Zurich was coming to an end. The model of archipelago orogenesis was formulated, but I had little chance to discuss its applications with colleagues before and after my retirement. The idea was to fade into oblivion, when the editor of this volume Yildirim Dilek asked me to contribute this article and urged me to expound my new hypothesis.

Having been encouraged to speculate, I returned to contemplate again the Franciscan-Knoxville paradox of the California Coast Ranges. I used to consider the Franciscan ophiolites as fragments of subducted crust and mantle under the Pacific Ocean. I felt uneasy with that interpretation because no remnants of the Cenozoic Pacific ocean floor have ever been found in the Franciscan. Even though all of the Mesozoic and much of the Cenozoic Pacific ocean floor east of the East Pacific Rise have been subducted under western North America, all datable Franciscan rocks are Jurassic and Cretaceous in age. There were many other problems that puzzled me when I left California for Europe: What is the structural relation of the Cretaceous broken formations to the ophiolite mélanges in the Franciscan? What is the relation of the Coastal Franciscan to the Coast Range Batholith? What is the relation of the Batholith to the Diablo Range and the Northern Franciscan? What is the relation of those mélanges to the Great Valley Sequence? What is the relation of the Franciscan/Great Valley to the Sierra Nevada Mountains?

I thought of western North America again when I wrote the chapter on Global Tectonic Facies in my introduction to tectonic facies concept (Hsü, 1995). Paleogeographic reconstructions leave little doubt that the ancient passive margin of the North American Continent could be defined by the deformed Paleozoic and Mesozoic sequences in the Foreland Thrust Belt of the Western Interior and in the Rocky Mountain region. The margin of the North American Plate was a magmatic arc active in the Sierra Nevada region during the Triassic and Jurassic, but was shifted to a position west of the Coast Range Franciscan in the late Mesozoic and Cenozoic. Considering the Continental Borderland offshore southern California as a submerged archipelago, the past magmatic margin is now buried. The topographic expression of the forearc margin should thus be the Patton Escarpment bounding the Borderland.

Adopting archipelago model of orogenesis, I propose to interpret the greenschists and ophiolites in the Mojave Desert and in the Sierra Nevada as the relics of collapsed Mesozoic backarc basins; those basins were located behind an active arc rooted in the plutonic rocks of the Sierra Nevada. The arc ceased to be active in the Cretaceous, when the active plate-margin was shifted westward. The eroded western slope of the Sierra Nevada Mountains became the site of a passive continental margin, on which the Great Valley Sequence was laid down. Farther to the west, the seafloor deepened and a basin underlain by oceanic crust evolved as a backarc basin behind an inland arc, the root of which is composed of the Coast Range Batholiths. The Cretaceous sedimentary rocks that exist in the basin are now present as exotic blocks or broken formations in the Franciscan Mélange. The island arc west of that basin was capped by the Cretaceous Calera Limestone and the associated volcanics.

In such a scheme, the Great Valley Sequence, representing the sediments of the foreland-basin, constitutes an Alamanide facies. The mélanges of the Northern Franciscan and the Diablo Range are the Celtides, which are bounded now by the San Andreas Fault of Cenozoic origin. Prior to the faulting, the Franciscan should have been separated by thrust faults from the Coast Range Batholith and its northern extension on the west side of the San Andreas Fault. The Cretaceous arc volcanism, related to this backarc subduction, was rooted in the Coast Range Batholith. The batholith should thus have been a Rhaetide in the archipelago model of orogenesis. In such an interpretation, the Franciscan/Great Valley deformation is analogous to that in the Southern Andes, where a back-arc basin floor has been subducted down a west-dipping Benioff Zone under the Andean Batholith.

The Southern Franciscan sediments of the California Coast Range were laid down in a more distant offshore backarc basin. The Jurassic Point Sal Ophiolite represents a relic of the ocean floor of that basin. Other ophiolitic rocks are present as exotic blocks in the Franciscan Mélange, which constitutes a celtide. Since no extensive shallow marine sediments were laid down in a narrow strip of the remnant arc, a foreland thrust belt does not exist. The Upper Cretaceous molasse of the Gabilan Range are the foreland-basin alamanide. Such a configuration of alamanide-celtide in the Southern Coast Ranges suggests that the offshore Channel Islands and the Transverse Ranges should have been the Rhaetide, under which the Southern Franciscan was subducted.

Are There Ophiolites at Suture Zones of Plate Collisions?

Ophiolites are not uncommon in active plate margins. Tertiary ophiolites, for example, are present along the Washington coast, where the Pacific Ocean floor is subducted under North America; they should represent a small remnant of the subducted Pacific Ocean floor. A late Mesozoic ophiolitic mélange has been sampled by dredging of the escarpment of the Mariana Trench; that ophiolite was also a remnant of the Pacific Ocean floor.

Are ophiolites common in suture zones? A few years ago, I made a survey of global tectonic facies, including those in the Alps, Appalachians, Scandinavian and British Caledonides, China, and North and South Americas, and I found that the ophiolites occurring in the mélanges within these orogenic belts were all made of the ocean floor of former backarc basins.

One notable exception could be the Zhanbo Ophiolites, commonly considered as a suture mélange developed during the collision of Tibet and India. After nine expeditions to the Tibetan Plateau, I presented my reason in *The Geologic Atlas of China* (Hsü and Chen 1999) as to why that suture zone cannot be represented by the Zhanbo ophiolites. According to the archipelago model, the former plate margin of Tibet is underlain by the metamorphic rocks of the Central Gneiss Zone on the southern slope of the Himalayas. The Zhanbo rocks constituted the ocean floor and deep-sea sediments of two backarc basins behind the Himalayan arc, which has been and is the southern boundary of the Eurasian Plate. The rocks formed by forearc accretion were metamorphosed subsequently to form the Central Gneiss Zone. After the collision of India with Eurasia, the rigid basement of the Frontal Arc Rhaetide and the Central Gneiss Celtide rose to become the Himalayan Mountains. The two collapsed backarc basins have left relics in the two ophiolite zones along the Zhanbo River Valley. The alamanides are the foreland thrust belt and the Siwalik Foreland Basin at the foot of that great mountain range (Hsü and Chen, 1999). The line of demarcation between the Eurasian and Indian plate lies buried under the Cenozoic foreland basin deposits of the Himalayas.

The Himalayan arc extends westward to the Karakorum. On an expedition to the Karakorum in 1999, we drove across a belt of mafic and ultramafic rocks (celtide) between the plutonic complex of the Karakorum (rhaetide) and the foreland fold belt of the Salt Range (alamanide). Those metamorphosed ophiolites could represent the crustal and upper mantle rocks of the Neotethyan Ocean, which separated Cathaysia from India. Farther to the west is the Bagh Celtide in the Muslim Bagh district of Pakistan, where the mélange consists of exotic blocks of ophiolites, radiolarites, and limestones in a shale matrix (Koima et al., 1994). These Neotethyan rocks crop out in the suture zone, because they have been spared from getting subducted to great depths (Hsü and Chen, 1999, p. 91).

In conclusion, I might say that I have learned a lesson after 20 years of fieldwork in China. I now think that ophiolites mostly represent the crustal and upper mantle rocks beneath the floor of former backarc basins; ophiolites are rarely present at suture zones where plates collide.

CONCLUSION

Famous Last Words

When I retired from ETH-Zurich in 1994, I planned not to do any more geology, so as to leave the field to the young people. I did not keep my promise, and the coda, like the one in Beethoven's symphonies, never ends. I thought I had sung my swan song when I completed the Geologic Atlas of China, but Yildirim Dilek urged me to make this contribution. Having given up my library and subscriptions upon my retirement, I have not been current for almost a decade. I do hope that my readers excuse me if I have not included a citation of their new discoveries or their new insights. In writing this article I summarized some personal experiences, but I could not claim to have presented any new idea. I have attempted to raise some new questions about the origin of ophiolites in association with the evolution of orogenic belts, and I do hope that new progress will be made when my hypotheses are put to test.

REFERENCES CITED

Aplonov, S., Hsü, K.J., and Ustritsky, V., 1992, Relic back-arc basins of Eurasia and their hydrocarbon potentials: The Island Arc, v. 1, p. 71–77.

Argand, E., 1916, Sur l'Arc des Alpes Occidentales: Eclogae Geologicae Helvetiae, v. 14, p. 145–191.

Argand, E., 1922, La Tectonique de l'Asie: Compte Rendu, Congrès Géologique International, 13th, Bruxelles, p. 171–372.

Becke, F., 1903, Die Eruptivgesteine der böhmischen Mittelgebirge und der amerikanischen Anden: Tschermaks Mineralogische und Petrographische Mitteilungen , v. 22, p. 209 ff.

Beloussov, V.V., 1962, Basic Problems in Geosciences: New York, McGraw-Hill, 820 p.

Bucher, W., 1933, The Deformation of the Earth's Crust: Princeton, Princeton University Press, 518 p.

Dalziel, I., 1981, Back-arc extension in the southern Andes: A review and critical appraisal: Philosophical Transactions of the Royal Society of London, ser. A, v. 300, p. 319–335.

Dewey, H., and Flett, J.S., 1911, On some British pillow-lavas and the rocks associated with them: Geological Magazine, Decade 5, v. 8, p. 202–209, 241–248.

Dewey, J., 1969, Evolution of the Appalachian/Calledonian Orogen: Nature, v. 222, p. 124–129.

Dewey, J., Shackleton, R.M., Chang, C., and Sun, Y., 1988, The tectonic evolution of the Tibetan Plateau: Philosophical Transactions of the Royal Society of London, ser. A, v. 327, p. 379–413.

Dickinson, W.R., 1970, Second Penrose Conference: The New Global Tectonics: Geotimes, v. 15, p. 18–22.

Dilek, Y., 1998, Structure and tectonics of intermediate-spread oceanic crust drilled at DSDP/ODP Sites 504B-896A, Costa Rica Rift: *in* Cramp, A., McLeod, C.J., Lee, S.V., and Jones, E.J.W., eds., Geological Evolution of Ocean Basins: Results from the Ocean Drilling Program: London, Geological Society of London Special Publication 131, p. 179–197.

Frei, W., Heitzmann, P., Lehner, P., and Vlasek, P., 1989, Die drei Alpentraversen von NFP 20: Bull. Verein Schweizersche Petroleum Geology und Inginier, v. 55, p. 13–43.

Harker, A. 1896, The Natural History of Igneous Rocks. I: Their geographical and chronological distribution: Science Progress, v. 6.

Hsü, K.J., 1958, Isostasy and a theory for the origin of geosynclines: American Journal of Science, v. 256. p. 305–327.

Hsü, K.J., 1971, Fransciscan mélanges as a model for eugeosynclinal sedimentation and underthrusting tectonics: Journal of Geophysical Research, v. 76, p. 1162–1170.

Hsü, K.J., 1981, Thin-skinned plate-tectonics model for collision type orogenesis: Scientia Sinica, v. 24, p. 100–110.

Hsü, K.J., 1988, Relic backarc basins: *in* Kleispehn, K.L., and Paola, C., eds., New Perspectives in Basin Analysis: New York, Springer Verlag, p. 246–263.

Hsü, K.J., 1994, The Geology of Switzerland: Princeton, Princeton University Press, 250 p.

Hsü, K.J., 1995, Tectonic facies in an archipelago model of intra-plate orogenesis: GSA Today, v. 4, p. 289–290.

Hsü, K.J., and Chen H., 1999, Geologic Atlas of China: Amsterdam, Elsevier, 262 p.

Hsü, K.J., and Ryan, W.B.F., 1971. Implications concerning ocean-floor genesis and destruction from deep-sea drilling in the Mediterranean Sea and eastern Atlantic Ocean: Proceedings, of the International Union of Geodesy and Geophysics, General Assembly, 15th: Moscow, sec. 1, p. 44.

Hsü, K.J., and Schlanger, S.O., 1971, Ultrahelvetic Flysch sedimentation and deformation related to plate tectonics: Geological Society of America Bulletin, v. 82, p. 1207–1218.

Hsü, K.J., Wang Qingchen, Li, J. Zhou, D., and Sun S., 1987, Tectonic evolution of the Qinling Mountains: Eclogae Geologicae Helvetiae, v. 80, p. 735–752.

Hsü, K.J., Wang Qingchen, Li, J., and Hao, J., 1991, Geologic evolution of the Neimonides: Eclogae Geologicae Helvetiae, v. 85, p. 213–225.

Hsü, K.J., Yao Y., Hao, J., Li, J., and Wang, Q., 1994, Origin of Chinese Tianshan by arc-arc collisions: Eclogae Geologicae Helvetiae, v. 87, p. 265.292.

Huang Jiqing and Chen, B., 1987, The Evolution of the Tethys in China and Adjacent regions: Beijing, Geological Publishing House, 109 p.

Kennedy, W.Q., and Anderson, E.M., 1938, Crustal layers and the origin of magmas: Bulletin Volcànologique, ser. II, v. 3, p. 24–82.

Kojima, S., Naka, T., Kimura, K., Mengel, J.M., Siddiqui, M.R.H., and Bakht, M.S., 1994, Mesozoic radiolarians from the Bagh Complex in the Muslim Bagh area, Pakistan; their significance in reconstructing the geologic history of ophiolites along the Neo-Tethys suture zone: Bulletin of the Geological Survey of Japan, v. 45, p. 63–97.

Kuno, H., 1962, Frequency distribution of rock types in oceanic, orogenic and cratogenic volcanic associations, *in* MacDonald, G.A., and Kuno, H., eds., The Crust of the Pacific Basin: Washington, D.C., American Geophysical Union Monograph Series 6, p. 135–139.

Li, C.Y., Wang, C., Liu, X.I., and Tang, Y.Q., 1982, Tectonic map of Asia: Beijing, Geological Publishing House.

Miyashiro, A., 1973, The Troodos Complex was probably formed in an island arc: Earth and Planetary Science Letters, v. 25, p. 217–222.

Moores, E.M., 1982, Origin and emplacement of ophiolites: Reviews of Geophysics, Space Physics, v. 20, p. 735–760.

Scheumann, K.H., 1932, Über die petrogenetische Abteilung des roten Erzgebirgsgneisses: Mineralogische und Petrographische Mitteilungen, v. 42, p. 423–426.

Şengör, A.M.C., Natal'in, B.A., and Burtman, V.S., 1993, Evolution of the Altaid tectonic collage and Paleozoic crustal growth in Eurasia: Nature, v. 364, p. 299–304.

Staub, R., 1928, Der Bewegungsmechanisms der Erde: Berlin, Borntraeger, 270 p.

Steinmann, G., 1905, Die geologische Bedeutung der Tiefseeabsätze und der ophiolitischen Massengesteine: Bericht naturforschenden Gesellschaft: Freiburg, v. 16, p. 44–65.

Stille, H, 1924, Grundfragen der vergleichenden Tektonik: Berlin, Borntraeger, 413 p.

Suess, E., 1909, Das Antlitz der Erde: Leipzig, Freytag, v. 3, 789 p.

Suess, F., 1937, Baustein zur einer Snhese Tektogenese: Fortschritt Geologie u. Paläontologie, v. 13, p. 1–86.

Tilley, C.E., 1950, Some aspects of magmatic evolution: Quarterly Journal of the Geological Society of London, v. 106, p. 37–61.

Trümpy, R., 1960, Paleotectonic evolution of the central and western Alps: Geological Society of America Bulletin, v. 71, p. 843–908.

von Bubnoff, S., 1937, Gebirgesgrund und Grundgebirge: Naturwissenschaften, v. 25, p. 577–588, p. 593–598.

Yao, Y., and Hsü, K.J., 1994, Origin of Kuenen Mountains by arc-arc and arc-continent collisions: The Island Arc, v. 3, p. 75–89.

Manuscript Accepted by the Society May 2, 2003

Geological Society of America
Special paper 373
2003

Ophiolites and the interpretation of marine geophysical data: How well does the ophiolite model work for the Pacific Ocean crust?

James S. McClain*
Department of Geology, University of California, Davis, California 95616, USA

ABSTRACT

The "ophiolite model" has been highly successful as an analog to explain seismic velocities in the ocean crust. The elements of the ophiolite model have all been observed on the seafloor where tectonic processes have exposed the crust. For the Pacific Ocean crust, or that which forms at non-rifted mid-ocean ridges, the development of the ophiolite model appears to resolve the contentious issues of the composition of the lower oceanic crust, and the nature of the Moho; however, studies show that the seismic structure of the crust evolves as it ages, and the depths to boundaries between different layers change. Therefore, these zones cannot represent lithological boundaries, at least in older crust. In this sense, a strict use of the ophiolite model is incorrect.

Keywords: crustal evolution, ophiolites, seismic structure, seismic interpretation.

INTRODUCTION

Oceanic crust, created at mid-ocean ridges, makes up ~60% of the earth's surface. For this paper, I define crust as that part of the earth that is bounded at the top by the water-crystalline or sediment-crystalline interface, and at the bottom by the Mohorovicic Discontinuity (Moho). Because 4–5 km of seawater overlie the crust, most of the exploration of the seafloor is carried out remotely, generally by geophysical means. But geophysical data only provide information about geophysical parameters, such as seismic velocities, densities, and magnetic susceptibility. These parameters must be interpreted to infer the lithology and physical conditions beneath the seafloor. Ophiolite studies have greatly facilitated this effort. This is particularly true for seismic data collected in the last 20 years. The "ophiolite model" has become the standard against which models of seismic velocity and density are tested. Geophysicists in recent years have tended to make the same mistake that geologists made in the 1960s and early 1970s, that is, taking the correlation between seismic velocity and lithology too literally.

There has been considerable debate over the origin of particular ophiolites, and whether any can be attributed to actual mid-ocean ridges. From the geophysical standpoint, these debates are not important. Ophiolites clearly originate in a marine extensional environment and represent mafic magmatic systems. As such, the ophiolite model stands on its own as an analogy for the oceanic crust, and it has been tremendously successful. Studies of rare ocean exposures often reveal units that can be correlated with ophiolites (e.g., Francheteau, et al., 1990). It is clear that ophiolites can be generated at mid-ocean ridges.

Figure 1 represents an idealized, widely used model that compares the ophiolite pseudostratigraphy to the velocity structure of the fast-spreading Pacific Ocean crust generated at a non-rifted mid-ocean ridge (e.g., McClain, 1992; Dilek et al., 1998). The uppermost crust, usually named Layer 2A, comprises extrusive volcanic flows, including sheet flows and pillow lavas. Recent work suggests that this layer may thicken while the crust is young, perhaps due to near-off axis volcanism (e.g., Harding et al., 1993, Toomey et al., 1994). The bulk of the upper crust (Layer 2) is generally thought of as sheeted dikes. In the ophiolite model, the lower crust (Layer 3) comprises gabbros that result from a zone of cooling magma and/or partial melt. The gabbros grade into olivine-rich cumulates and then to tectonized peridotite. This transition to ultramafic rocks results in a rapid increase in velocity that is termed the Moho and marks the base of the crust.

Velocity transition zones, where velocity gradients are high, separate the different units in the ophiolite crust. In addition to

*mcclain@geology.ucdavis.edu

McClain, J.S., 2003, Ophiolites and the interpretation of marine geophysical data: How well does the ophiolite model work for the Pacific Ocean crust?, *in* Dilek, Y., and Newcomb, S., eds., Ophiolite concept and the evolution of geological thought: Boulder, Colorado, Geological Society of America Special Paper 373, p. 173–185.

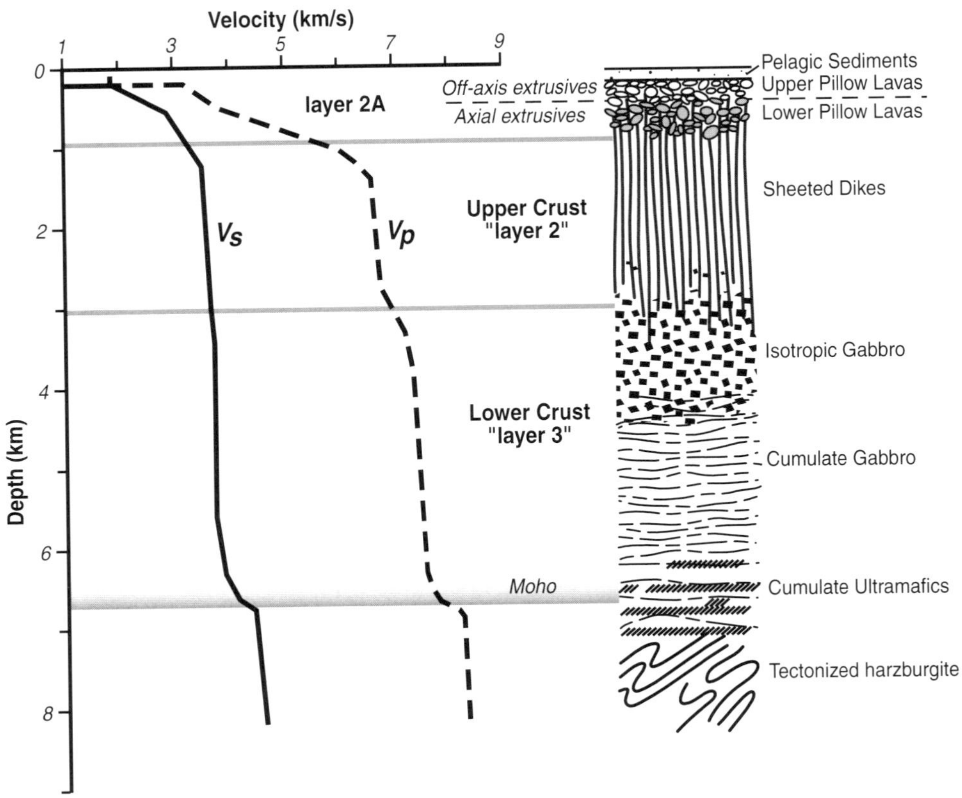

Figure 1. The seismic structure of the Pacific Ocean crust, with corresponding lithologies from an idealized "ophiolite model." Vp and Vs represent the compressional and shear velocities respectively. Individual ophiolites and individual seismic structures may vary significantly, but many exhibit these features. Figure modified from McClain (1992). The seismic structure is after McClain and Atallah (1985), and the ophiolite model is modified from Moores (1982).

the Moho, these include an upper crustal transition between the extrusives and sheeted dikes, and a mid-crustal transition that may mark the boundary between sheeted dikes and gabbros.

The analogy between ophiolite and seismic structure has been widely used by many researchers investigating the structure, composition, and evolution of modern ocean crust. The goal of this paper is to consider the implications of the ophiolite model for the ocean crust as represented in Figure 1. I will discuss its success in explaining the ocean crustal structure and, more importantly, will examine potential inconsistencies that may result from an excessive reliance on the model. I will focus on the effects of water penetration and subsequent alteration of the crust. I intend to show that the model in Figure 1 is not uniformly correct. Also, I will address the closely related topic of the role of serpentinite in crust generated on non-rifted ridges (e.g., the Pacific crust). I will discuss the potential role of serpentinite as a modification of the ophiolite model.

Rather than arrange the paper as a purely chronological sequence of observations and theory, I will first discuss the seismic studies that provide us with the greatest information about the oceanic crust. I will then discuss the lithologies and physical properties that have been used to explain the seismic observations, along with some of the debates about these models. I will then present details of the ophiolite model, with reference to marine seismic observations. Finally, I will discuss how some observations seem in conflict with a pure ophiolite model.

THE SEISMIC STRUCTURE OF THE OCEAN CRUST

Early seismic observations of the earth under the oceans were invariably tied to the concept of crust and mantle that evolved from continental studies. Thus, it is important to briefly review how these concepts came to be accepted.

What Are the Crust and Moho?

The concept of a distinct division between the crust and a mantle can at least be traced back to Boisse in the mid-nineteenth century (Hess, 1959). He believed that a model for the bulk composition of the earth could be established from meteorites, and he noted that these rocks were different from most of those observed at the surface.

The first evidence for a distinct boundary between a crust and mantle came with the work of Mohorovicic (1909). He recognized that the travel times for seismic waves required an abrupt increase in velocity at a depth of ~60 km beneath Siberia. He named the compressional and shear waves that traveled in this lower layer as P_n and S_n, respectively. The boundary between the surface layer and the deeper, high-velocity layer came to be known as the Mohorovicic Discontinuity or Moho. Mohorovicic and others accepted the correspondence of the Moho with a "crust-mantle" boundary, but this was only one of several divisions of the earth's structure that was being advocated in the early part of the twentieth century.

A number of later studies resulted in three-layer models for the earth beneath the continents (e.g., Jeffreys, 1929). The upper layer was generally thought to be a granitic zone and was often referred to as the sial. The intermediate layer, generally thought to have a mafic composition, and was often called the sima. The boundary between this intermediate layer and the "lower layer" (the mantle) was the Moho. In fact many studies used the term "lower layer," rather than "mantle," when referring to the earth beneath the Moho (e.g., Jeffreys, 1929; Leet, 1936).

By the early 1950s the concept that the Moho was the boundary between crust and mantle was accepted in the general geological literature (e.g., Gilluly et al., 1953); however, even in the 1950s this correspondence was still under discussion in the seismological literature. For example, discussing Jeffreys' division of the earth into layers "A" through "G," Bullen (1955) states, "The region A, extending from the outside surface to the Mohorovicic discontinuity, is often called the 'crust' or 'crustal layers'" (Bullen, 1955, p. 74). This implies some continued uncertainty about the concept of crust and mantle. Over the years, however, that concept has come to be fully accepted.

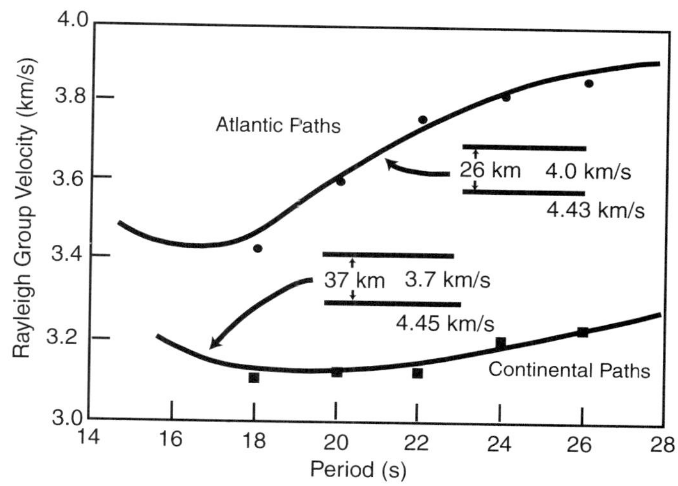

Figure 2. An example of early surface waves studies, showing the group velocity for Rayleigh as a function of wave period, i.e., the dispersion curve. Notice that, for the oceanic (Atlantic) propagation paths (circles), the velocities are substantially higher than those for continental paths (rectangles). Two theoretical curves show the dispersion curves for the ocean (upper) and continental (lower) structures given on the figure. For the oceanic "crust" the 26-km thick layer is actually a composite of the actual crust and the uppermost ocean mantle. This is why the apparent velocities are so much higher. Adapted from Wilson and Baykal (1948).

Early Marine Studies

The use of geophysical techniques in marine research began with gravity and seismic studies in the 1920s and 1930s (e.g., Angenheister, 1921; Ewing et al., 1939). Early studies usually relied on the analysis of earthquake-generated surface waves propagating across the ocean basins that were detected by seismometers on land. The resulting observations were displayed as dispersion curves, which yield group or phase velocities as a function of wave period. The propagation of surface waves is most sensitive to the shear velocities in the earth, and surface waves provide a lower resolution, or less detailed, model for earth structure than the seismic refraction studies that would later be conducted.

Early surface-wave studies revealed that waves with oceanic paths displayed higher surface-wave velocities than those found under the continents (Angenheister, 1921; Newman, 1929; Carder, 1934; Macelwane, 1951) (see Fig. 2).

Analysis and interpretation of these early studies was often incorrectly limited by an attempt to correlate the oceanic structure with that under the continents. Byerly (1930) and Carder (1934) assumed a "crustal thickness" of 35 to 40 km, which seems an astonishing error with what we know today; however, both authors correctly argued that their data were best matched by a single simatic layer, with no granitic upper crust. Wilson and Baykal (1948) made similar interpretations in a later Atlantic study (see Fig. 2). Of course, we now know that the high velocities observed on these numerous studies reflected the fact that the analysis incorrectly attributed portions of the upper mantle into the "crustal" layer. In fact, the low resolution of surface wave studies means that they are poorly suited to study the thin ocean crust.

From the 1950s to the present, studies of the oceanic crust and upper mantle have been dominated by seismic refraction experiments. In contrast to the surface waves studies, the refraction experiments are best for providing compressional wave (P-wave) velocities. This is because seismic sources in water only generate P-waves. Whereas some of the propagating compressional energy is converted to shear energy at the seafloor, identification of shear waves is often unreliable (e.g., Lewis and McClain, 1977). Thus, we usually discuss the seismic structure of the crust in terms of the P-wave velocities.

One of the first seismic refraction studies at sea was conducted in 1949, and exploited two ships (a source ship and a receiver ship) and explosives. In their analysis of the data, Ewing et al. (1950) interpreted the crust as simply a 4.5-kilometer thick section of sediments overlying the oceanic mantle or Moho, with no "basement" layer. Seismic velocities were low (7.58 km/s) compared to other (continental) measurements of P_n velocities, but high compared to the simatic crustal velocities. Because their inferred depth to Moho was so shallow, they allowed that they might not have observed P_n at all, and that they were only sampling the "intermediate" crust. They provided support for their interpretation in a companion paper, in which they analyzed surface waves (Ewing and Press, 1950). With the benefit of hindsight, it is easy to understand why Ewing et al. (1950) failed to detect basement. The shot spacing for the experiment was so large (2–4 km), and the water depths so great, that they could not hope to pick out the arrivals corresponding to waves traveling through the relatively thin crust.

In another refraction experiment, Hill and Shallow (1950) argued that they detected a thin granitic layer under the eastern

Atlantic, but their observations also suffered from a lack of resolution. We know now that their granitic layer was merely the upper basaltic crust.

It was not until the improved technology after 1950 that systematic seismic refraction studies could be made (Raitt, 1956; Shor et al., 1969). The analysis of the data collected in numerous experiments comprised picking the arrivals and computing the travel times between source and receiver. Travel times were then plotted as a function of source-receiver separation. (Fig. 3). Until the 1970s, the complex geometry of the travel time curve was approximated by a series of straight lines. Each line was assumed to represent a set of seismic paths along the top of a layer. The slope of the line was the reciprocal of the seismic velocity of the layer, and the depth of the layer could be determined from the intercept times. Thus, early models comprised homogeneous layers. These were later improved to allow for dipping layer interfaces (e.g., Shor et al., 1969).

Early compilations revealed that the earth beneath the Pacific could be approximated by a simple four-layer structure (e.g., Raitt, 1956; Shor et. al., 1969). The uppermost layer (Layer 1), with an average compressional wave velocity of ~2.2 km/s was interpreted to be sediments because pelagic sediments were known from cores and photos to cover most of the seafloor. The lowest layer (Layer 4), with velocities of ~8 km/s (corresponding to P_n), was interpreted to be the uppermost mantle.

Layers 2 and 3, lying between the sediments and the mantle, make up the actual oceanic crust. Layer 2, displayed P-wave velocities averaging 5.2 km/s, and was interpreted to represent the volcanic basement beneath the oceans. This was a natural interpretation because basalts exist virtually everywhere that sediments do not cover the seafloor. Layer 2 also displayed the widest range in velocities, and was locally subdivided into Layers 2A, 2B and 2C (e.g., Shor et al., 1969). Layer 2A was the most commonly observed subdivision, with P-wave velocity of ~2.5 to 3.2 km/s. Workers suggested that these low velocities were related to high porosities in the upper, volcanic crust (e.g., Houtz and Ewing, 1976).

The lower crust (Layer 3), with velocities of ~6.8 km/s, was found to be the major part of the crust, with a thickness averaging ~4.6 km. As we shall see, the geological interpretation of Layer 3 was controversial. Because of the lack of geological evidence, Layer 3 was ambiguously named the "oceanic layer." Shor et al. (1969) show that the oceanic layer is the most consistent and common feature for seismic profiles in the Pacific. Some studies subdivided Layer 3 into layers 3A and 3A, but these divisions were not always required by the data.

The studies revealed that the boundary between Layers 3 and 4 represented an apparently abrupt compressional velocity increase from ~7 km/s to 8 km/s. The velocities were similar to those found under the continents, and it was natural to correlate the transition zone with the Mohorovicic Discontinuity (Moho). Under the oceans, the Moho was found to be at a depth of ~10 km below sea level, instead of the highly variable depth of 20 to 30 km found under the continents. Qualitatively, this difference could be attributed to isostatic compensation mechanisms, where the continents and mountains display deep "roots."

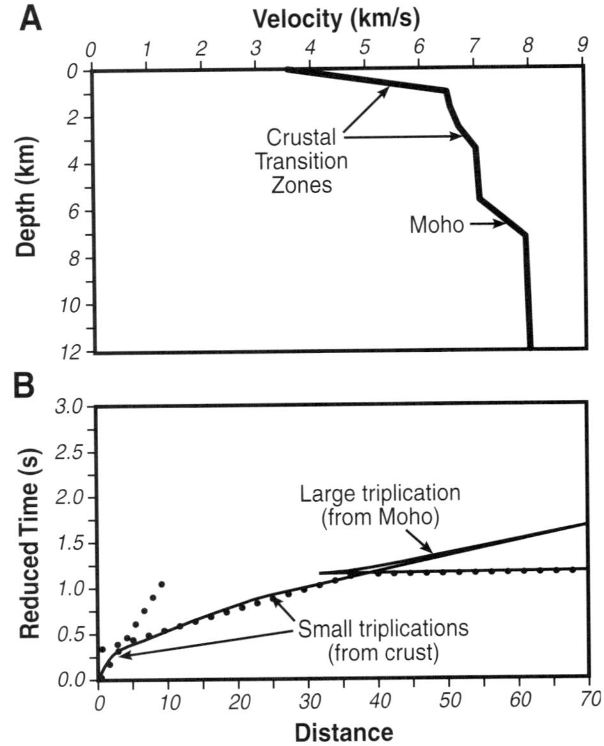

Figure 3. The relationship between structure and resulting travel times. A: A structure approximating a modern view of the ocean crust and uppermost mantle. Abrupt changes in lithology and porosity are moderated by velocity dependence on smoothly varying properties such as temperature and pressure. The relatively high velocity gradients, known as crustal transition zones, may indicate changes in lithology and/or porosity. B: Travel time curve resulting from the structure in Figure 3A. The transition zones cause multiple arrivals known as triplications. The more abrupt the transition zone, the larger the triplication. For less abrupt transition zones, however, the triplication is small and might only be detected by changes in seismic amplitudes or waveforms. Early studies did not use these properties, which led to the approximation of the travel time curve with a series of lines shown in the dotted lines. The intercept and slope of these lines were used to develop the layer-cake approximation for crustal structure.

The most important result from these early seismic studies was knowledge of the astonishing uniformity of the ocean crustal structure, particularly in the Pacific (e.g., Raitt, 1956; Shor et al., 1969). This implies that the processes that create the crust must also be uniform—an observation that foreshadowed the later discovery of seafloor spreading.

The "layer-cake" models for the ocean crust were always understood to be approximations. They required the assumption of little lateral heterogeneity (with the exception of dipping interfaces mentioned above). This is partly justified because many of the properties that affect velocity, including temperature, pressure, and porosity, are most dependent on depth; however, some of these properties (notably pressure and temperature) tend to vary smoothly with depth. Thus, even in the presence of abrupt lithological or porosity changes, seismic velocities will tend to be more gradual than those assumed for the layered approximations.

Modern Studies and Non-Layered Models

As analytic tools became more advanced, particularly with the ability to model seismic amplitudes and waveforms, a more complex seismic velocity structure was revealed (Figs. 3 and 4). Most modern studies find that the upper crust is best modeled as a zone of rapidly increasing velocities. In some cases, a thin zone with lower velocity gradients caps the uppermost crust, which is now presumed to represent Layer 2A.

In these non-layered approximations, the lower crust and mantle retained their overall characteristics of having low velocity gradients, with a Moho between them. The Moho is not a sharp boundary, but represents an increased gradient in velocity (e.g., Spudich and Orcutt, 1980). This transition, however, is still abrupt enough to produce the travel time triplication shown in Figure 3.

The boundary between the upper crust (old Layer 2) and Layer 3 is difficult to characterize; however, studies of seismic amplitudes seem to reveal a subtle transition zone at mid-crustal depths. This "mid-crustal transition zone" was first reported by Bratt and Purdy (1984) and McClain and Atallah (1985), and appears to be a common feature of the ocean crust.

The Search for the Elusive, or Not So Elusive, Magma Chamber

Early consideration of the seafloor spreading process included the expectation that mid-ocean ridges would be underlain by an axial magma chamber (AMC) (e.g., Cann, 1968, 1974; Lister, 1977). This expectation was supported by chemical analyses of seafloor basalts that indicated that the basaltic magmas had undergone low-pressure fractionation (in a magma chamber) prior to eruption. In addition, numerical thermal models indicated that a zone of partial melt or magma chamber could exist under the ridge axis. In a classic paper on the thermal structure of the mid-ocean ridges, Sleep (1975) put forth a model where the magma chamber under a fast-spreading ridge could be as much as 20 km wide (see Fig. 5a). He found that the width of the magma chamber was critically dependent on the spreading rate.

One of the key elements of the ophiolite model shown in Figure 1 is the presence of the gabbroic lower crust. With the development of the concept of seafloor spreading, the presence of gabbro in the ocean crust implied the presence of a "magma chamber" under the ridge axes, perhaps as shallow as 2 km below the seafloor. In their work on the Semail ophiolite in Oman, which is believed to have originated at a fast-spreading

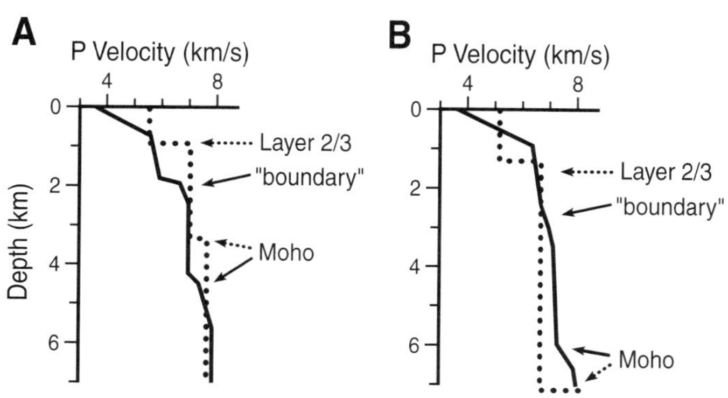

Figure 4. A comparison of layered approximations and continuously varying velocity structures for two different seismic profiles: A: Structures for 1 Ma crust. B: Structures derived for 60 Ma crust. Layered structures (dotted lines) are modified from Raitt (1956); continuously varying structures (solid lines) are after Atallah (1984) and McClain and Atallah (1986).

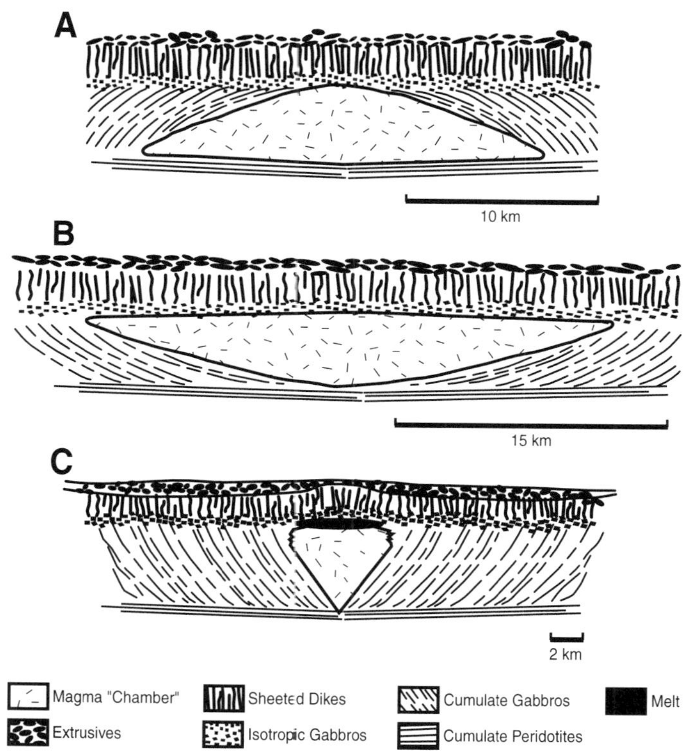

Figure 5. Three models for magma chambers on fast-spreading, non-rifted spreading centers. A: A thermal model similar to that postulated by Sleep (1975). In an early effort to detect a magma chamber, Rosendahl et al. (1976) interpreted his seismic data to indicate a magma chamber structure similar the Sleep's model. B: A large magma chamber inferred to have existed when the Samail ophiolite in Oman was generated at a fast-spreading ridge (modified from Pallister and Hopson, 1981). C: The modern view of magma chambers and structure beneath the present-day East Pacific Rise (modified from Sinton and Detrick, 1992). The chamber is a relatively narrow zone of partial melt, with a thin lens of pure melt at the top. The partial melt produces seismic velocities observed from refraction data, whereas the melt lens provides the velocity contrast that produces the "bright" reflections observed on multi-channel experiments.

ridge, Pallister and Hopson (1981) cited evidence that the AMC could be as wide as 30 km (Fig. 5B).

With these predictions, it was expected that axial magma chambers would be detectable using standard seismic techniques. Magma should have a lower compressional wave velocity than adjacent rocks. In addition, depending on the degree of melt, shear waves might be inhibited from propagating through a magma chamber. In the 1970s numerous experiments were undertaken at the axes of various spreading centers, and in particular on the northern East Pacific Rise. Orcutt et al. (1975) and Rosendahl et al. (1976) found evidence for a low-velocity zone under the axis, which they attributed to an AMC. In the latter study, the authors inferred a magma chamber that is wide and shaped like that of Sleep's (1975) numerical model, shown in Figure 5A. Reid et al. (1977) and McClain and Lewis (1980) also found evidence for magma chambers on ridges spreading at intermediate rates (50–90 mm/yr).

Acceptance of the axial magma chamber was not universal, even on fast- or immediate-spreading ridges. Lewis and Garmany (1982) point out that the absence of travel-time anomalies on a segment of the northern East Pacific Rise precludes the presence of any magma chamber wider than about two kilometers. K. McClain and Lewis (1982) argue against any magma chamber beneath the southern Juan de Fuca Ridge, which spreads at an intermediate rate. Bratt and Solomon (1984) observed shear waves on seismograms for ray paths crossing the axis of the northern East Pacific Rise. For them, this was a proof of the absence of an AMC.

Despite these doubts, experiments in the 1980s "proved" the presence of axial magma chambers. McClain et al. (1985) showed very compelling evidence for a magma chamber under the northern East Pacific Rise at 13°N. It was only ~4 km wide, narrow compared to the above-mentioned predictions. The compressional velocities were high (4.5 km/s) compared to those of a pure melt (2.5 km/s) (Murase and McBirney, 1973). This suggested that the "magma chamber" was probably a zone of partial melt. McClain and Atallah (1985) found weak evidence for a similar narrow and partial melt structure at the ultra-fast-spreading southern East Pacific Rise. Like the ridge to the north, the magma chamber under the southern EPR was narrow.

The seminal study supporting the existence of an axial magma chamber under the non-rifted East Pacific Rise was that of Detrick et al. (1987). Using a multi-channel seismic experiment, they found a large amplitude mid-crustal reflector under most of the northern East Pacific Rise. Its depth beneath the axis led them to conclude that they had imaged the AMC. It was uniformly narrow compared to the models shown in Figures 5A and B. The high amplitudes of the reflection supported a model in which the reflection was caused by a low-velocity boundary of almost pure melt (see Fig. 5C).

The results of Detrick et al. (1987) have subsequently been supported by a whole series of studies that include analyses of multichannel data (e.g., Kent et al., 1990, 1994), seismic refraction data, or seismic tomography data (e.g., Toomey et al., 1990, 1994; Vera et al., 1990; Christeson et al., 1992). Each of these studies has resulted in models with a narrow AMC. Subsequent experiments on the southern East Pacific Rise, where spreading rates are ultra-fast, also revealed that the magma chamber is narrow (Detrick et al., 1993; Kent et al., 1994).

There is little doubt now that non-rifted ridges are generally underlain by narrow magma bodies or chambers. The bulk of the body may be a zone of partial melting, thus explaining the relatively high compressional wave velocities observed from seismic refraction experiments. The large amplitude reflections can be explained by a thin lens of almost pure melt at the top of the magma chamber (see Fig. 5C and Sinton and Perfit, 1992).

Summary of the Seismic Structure of the Crust

The seismic structure of the ocean crust formed at non-rifted mid-ocean ridges is most striking because it is so uniform, with a structure characterized by that shown in Figure 4 and the left side of Figure 1. This uniformity extends to the axis itself, where a narrow axial magma chamber underlies most of the present-day East Pacific Rise.

The models discussed above are parameterized in seismic velocity (primarily P-wave velocity). The next step in understanding the crust-forming processes requires inferences about the relationship between the seismic velocity and the lithology and/or physical state (porosity and temperature) of the crust and uppermost mantle. It is here that the ophiolite model has proven its value. The development of this model did not post-date the refinement of seismic models. Instead, it was part of a parallel effort that extends back to early studies of the continental and oceanic crust.

Observations of the aging of the ocean crust are important as a test for the ophiolite model. Thus, I defer the discussion of seismic evidence for crustal evolution until the end of this paper.

THE EARLY DEBATES ABOUT THE OCEAN CRUST

While only an approximation, the layer-cake model provided substantial room for interpretation. Particularly contentious were closely linked debates about the nature of the Moho and the oceanic layer (Layer 3).

The Nature of the Moho

Models for the Mohorovicic Discontinuity have traditionally been divided into those that require a lithological change to achieve the necessary velocity transition, and those that require a phase change. For the former, most models involve a transition between a mafic or intermediate lower crust, and an ultramafic, usually peridotite, upper mantle.

In continental studies, the phase-change models centered on a transition from crustal basalt (or gabbro) to a denser eclogite phase (e.g., Loverling, 1958; Kennedy, 1959). Experimental results seemed to show that the pressures at continental Moho depths were approximately correct for this transition; however, under the oceans, the pressure and temperature conditions were not as favorable, depending on the assumed geothermal gradients.

The gabbro-eclogite model was controversial from the start, and objections were raised for a number of reasons (see Ringwood, 1969, for a review). Perhaps the most convincing argument against the eclogite hypothesis was the need to explain the chemistry of basalts that erupted at the earth's surface. A wide variety of experiments suggest that most basalts must come from the partial melting of peridotite, rather than eclogite (e.g., Ringwood, 1969).

Geophysical observations also provide important arguments against the eclogite/basalt model. Moho depths under continents are at least partly controlled by the overlying crustal load. That is, mountains have roots. If the Moho were indeed a phase transition, roots would not be preserved because mountains would push the basaltic crust into the eclogite field. Thus, topography would not be isostatically maintained, unless the above-mentioned temperature effects were very finely tuned to topography (e.g., Wyllie, 1971). In addition, a number of marine refraction experiments revealed that the mantle P-wave velocities exhibited azimuthal anisotropy (e.g., Raitt, 1969). This is unlikely to occur in an eclogite where most the minerals are isotropic, but it could occur in peridotite because olivine displays a high degree of anisotropy.

In 1939, Harry Hess suggested an alternative phase transition for the oceanic Moho. He argued that the Moho could be the boundary between serpentinized peridotite in the lower crust, and unserpentinized peridotite in the mantle. Thus, the depth of the boundary could be controlled by the depth of water penetration into the upper mantle, as well as by the pressure/temperature conditions (Hess, 1939). Hess' model received little attention at first, but he continued to argue for the model over the next two decades (Hess, 1955).

In the serpentinite model, the uniform thickness of the ocean crust must be set at the ridge axis, where geothermal gradients are high (Figs. 6A and B). As the seafloor spreads and cools, the depth to the phase change should increase; therefore, the thickness of the crust should increase with age, provided that water is available.

Unlike the case against the gabbro-eclogite model, the Hess model did preserve the peridotite mantle necessary to produce proper basalt chemistry. In addition, the model initially could not be discounted on the basis of Moho relief or mountain roots. In the Pacific, the topographic highs are usually smaller than on continents. The more topographically uniform ocean crust would not require a great deal of Moho topography.

Studies of near-ridge seamounts reveal that larger structures do indeed have roots (e.g., McCabe and McClain, 2000). In these cases, the load is clearly added after crustal formation, but while the geotherm is still high. If the Moho were the serpentinite-peridotite boundary, then the added volcanic load would result in the subsidence of the Moho. This, in turn, would cause dehydration reactions and a corresponding increase in density. Thus, the subsidence would continue until the seafloor was level again. Furthermore, the elevated temperatures caused by the off-axis volcanism

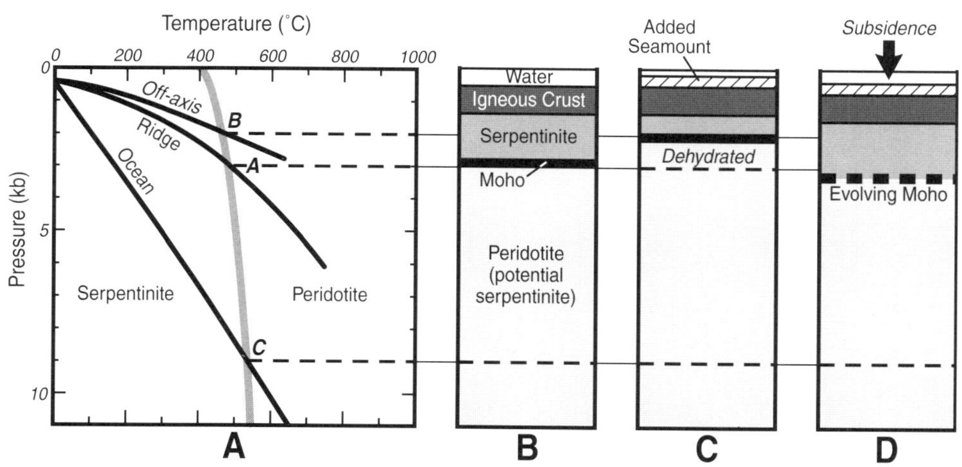

Figure 6. The serpentinite-peridotite transition as a model for the oceanic Moho (modified from Wyllie, 1971). A: Approximate phase diagram (in gray) for the peridotite-serpentinite transition, with idealized geotherms for a near-ridge site (labeled "Ridge"), an older ocean basin site (labeled "Ocean"), and an off-axis volcanic site (labeled "Off-axis"). B: As the crust cools near the ridge, it forms serpentinites, provided that water is available. As the crust ages, the geotherm decreases and the zone of potential serpentinization deepens. If water is available, the crust can thicken with age. C: If a seamount forms near a ridge, it may elevate the geotherm, causing the Moho to rise. There would be no root under a hot seamount. D: As the seamount cools, the load causes the lithosphere to subside, pushing the Moho deeper and into the peridotite field. Again, no roots would form, and isostatic subsidence would continue.

should produce dehydration of the existing serpentinite (Figs. 6C and D).

Debates about the nature of the Moho are directly linked to models for the lower ocean crust. As stated above, seismic models for the Pacific Ocean reveal a very uniform structure, with Layer 3 having P-wave velocities of ~6.8 km/s and exhibiting little internal structure. For the Hess serpentinite model, the lowest crust must be serpentinite, while for the alternative model the lower crust must be gabbro. A difficulty for the Hess model is the uniformity of Layer 3 velocities. The compressional wave velocity of serpentinite varies from 5.4 to 8.2 km/s, depending on the degree of serpentinization (Christensen, 1966). It seems unlikely that a serpentinite lower crust could be so uniform. In contrast, unfractured gabbros have a fairly narrow range of velocities, consistent with those observed for Layer 3 (Christensen, 1968, 1972; Christensen and Salisbury, 1975).

Both gabbros and serpentinites have been dredged from the seafloor; however, these dredges are from tectonically disrupted regions, where mantle rocks (peridotites) are directly exposed to seawater. These rocks will undergo serpentinization after exposure, and their presence on the seafloor cannot prove that serpentinites are present in the undissected lower crust.

The Hess serpentinite model has gained support from two lines of evidence. First, whereas the uniformity of Layer 3 velocities was difficult to reconcile with a serpentine composition, the uniform thickness of the Pacific crust appeared consistent with a phase transition model. In addition, there was evidence that the thickness of the crust may increase as the lithosphere ages. We will examine the latter issue below.

Ophiolites and the Interpretations of the Crust

Historically, the development of the ophiolite model for the ocean crust has been one of the most unifying intellectual breakthroughs related to the plate tectonic paradigm. For the first time, the results of marine geophysical studies and studies of orogenic belts had a recognizable link.

Initially, ophiolites were identified by the serpentinites in orogenic belts (Steinmann, 1927). Early studies focused on their role in these belts and in orogenesis (e.g., Dewey and Bird, 1971). It was the realm of structural geologists interested in the processes of mountain building (see Moores, 1982 for a thorough review). Early studies revealed that ophiolites typically comprised serpentinites in association with mafic volcanic rocks (Steinmann, 1927). In addition, marine sediments are usually present in ophiolite complexes, suggesting that ophiolites have originated in a marine environment.

The analogy between ophiolites and ocean crust has been a contentious issue. Brunn (1959) suggested that ophiolites provided an analogy to the mid-Atlantic ridge crust. Hess (1960, 1965) provided arguments that ophiolites formed the lower oceanic crust. The presence of sheeted dikes in ophiolites, most notably the Troodos Complex on Cyprus, led to the argument that ophiolites originated in a spreading environment, perhaps on a mid-ocean ridge (e.g., Gass, 1968; Moores and Vine, 1971; Moores and Jackson, 1974). If so, the association of serpentinites (hydrated residual mantle), igneous intrusives and extrusives (the crust), and marine sediments in ophiolites was explained, and the ophiolite model for the oceanic crust was established (see Fig. 1). Whereas serpentinized peridotites are associated with ophiolites, it is usually assumed that the serpentinization occurred because of localized processes on the seafloor, or during and after emplacement on the continents (e.g., Christensen and Salisbury, 1975). In the ophiolite model, serpentinites were not a part of the original crust.

As sensible as the ophiolite model has been, there have been difficulties. It has been noted that the mafic (crustal section) of ophiolites is thin compared to the average thickness revealed by marine seismic studies (e.g., Christensen and Salisbury, 1975; Moores, 1982). In addition, the chemistry of mafic rocks found in the Troodos ophiolite revealed a calc-alkaline trend, more compatible with an island arc origin of these rocks (e.g., Miyashiro, 1973)

Seismic Velocities and Ophiolite Rocks

A fundamental requirement of the ophiolite-ocean crust analogy is that the ophiolitic rocks have seismic velocities that are consistent with observations made at sea. A number of studies have confirmed this correspondence (e.g., Christensen, 1966, 1968, 1972; Coleman, 1971; Christensen and Salisbury, 1975).

As noted above, seismic refraction experiments reveal that the upper crust has highly variable seismic velocities, with velocities increasing rapidly with depth (e.g., Spudich and Orcutt, 1980; Fig. 4). In general, the velocities are low compared to laboratory experiments on basalts. The seminal work of N.I. Christensen revealed why (Christensen, 1968). Field experiments sample the rocks at long wavelengths. The resulting seismic velocities are averages of an upper crustal structure with high porosity. This porosity results from large fractures and voids in the volcanic rock. In contrast, small laboratory samples may not include any fractures. When they do, experiments reveal a dramatic dependence on porosity. The high-velocity gradients in the upper crust result from a pressure-induced decrease in porosity with depth.

Measurements of seismic velocities in gabbro and serpentinite are discussed above. Coleman (1971) points out that the degree of serpentinization would have to be ~40% to produce the compressional wave velocities seen in the lower crust. He thought this was unlikely, as did Christensen and Salisbury in their extensive review (1975).

One way to discriminate between a gabbro and serpentine is to measure the velocities of both shear waves and compressional waves. Unfortunately, because marine seismic sources only generate compressional waves, the detection of shear waves depends on the serendipitous conversion of compressional to shear waves at the seafloor (e.g., Lewis and McClain, 1977). As a result, reliable shear wave measurements are rare; however, when they can be made in conjunction with P wave measurements, then Poisson's ratio (σ) can be calculated by the equation,

$$\sigma = \frac{1}{2}\left(\frac{R^2-2}{R^2-1}\right), \quad (1)$$

where R is the ratio of compressional wave velocity to shear wave velocity.

Christensen (1972) made a systematic study of a number of experiments where shear waves were seen, and found that typical Poisson's ratios were between .25 and .28. He compared these results with laboratory measurements on gabbros and serpentinites, and found that the Poisson's ratio for the former was ~.25, while for serpentinite it was considerably higher (.33 or higher, depending on the degree of serpentinization). From this, he concluded that serpentinite could not be a major constituent of the lower oceanic crust.

In a very elegant reevaluation of data collected in the 1950s, Spudich and Orcutt (1981) selected the seismic profiles with the best-recorded shear waves. They conducted a very rigorous modeling effort, matching computed seismograms with those collected in the field. They found that the resulting compressional and shear wave models were most consistent with a gabbroic lower crust. They thus argue against a significant amount of serpentinite in the lower oceanic crust.

ON THE SUCCESS OF THE OPHIOLITE MODEL

The ophiolite model makes a number of predictions about the lithology beneath the seafloor. In its simplest form, the model neglects issues of variations in magma supply, major faulting on the seafloor, and significant alteration as the crust ages. Even so, its predictions about crustal lithology are widely accepted, largely because it "passes" a number of observational tests.

Ophiolites on the Seafloor

Although exposures of the deep crust are rare on the seafloor, every unit present in ophiolites—including lava flows, sheeted dikes, gabbros, and peridotites—has been identified from samples collected on the seafloor, as well as from seafloor observations. This is widely cited as evidence that the ophiolite model is correct.

Most exposures of mid- to lower crustal rocks occur on slow-spreading, rifted ridges, particularly at ridge-transform intersections (e.g., Cannat, 1993, 1996; Karson, 1998; Lagabrielle et al., 1998). The mafic sections revealed in these studies are thin compared to normal ocean crust; however, they are generally associated with "magma starved" segments of the ridge, which would not be expected to produce crust with a normal thickness.

Tectonic exposures are particularly rare on the Pacific Ocean floor, where crust formed at non-rifted ridges. The best studied exposure is the Hess deep, where the Galapagos Ridge exposes young crust generated at the nearby East Pacific Rise (e.g., Francheteau et al., 1990; Karson, 1998). A number of dives have been undertaken there by the deep-sea submersible, Alvin, but the tectonic disruption from the rifting has made it difficult to make an accurate model of the original pre-rifted crust (Hurst and Karson, 1991).

Attempted reconstructions suggest that the crust rifted at the Hess Deep (defined as the mafic rocks above the serpentinized peridotite) is thin compared to normal Pacific crust. There are three possible explanations for this discrepancy. First, the original crust may be anomalously thin, similar to that found at slower-spreading ridge-transform intersections. Alternatively, the reconstructions might be wrong because of the observational difficulties discussed above, and the crust might be normal in thickness.

The third possible explanation is that whereas gabbro comprises the upper part of Layer 3, the crustal section is augmented by serpentinites related to water penetration into the ultramafic sections. If this were true, serpentinization would have had to have occurred very rapidly while the crust was young. Lewis and Snydsman (1977) have reported seismic evidence for this rapid thickening. Thus, the correct model for the Pacific crust could be a hybrid between the ophiolite model and the Hess serpentinization model.

Hole 504B and the Penetration of Water into the Young Ocean Crust

In the end, the only true test of the ophiolite model would be to drill into undisrupted crust that was generated on a non-rifted ridge. It is unlikely that we will see this done in the immediate future, but there have been a number of efforts to drill into the upper crust. One of the most spectacular successes of the Ocean Drilling Program (ODP) and the Deep Sea Drilling Project (DSDP) has been the drilling into several hundred meters of crust at Hole 504B near the Costa Rica Rift. Over a series of eight drilling legs between 1979 and 1993, ~2111 m of crust were penetrated. The upper 570 m consisted of extrusives, below which was a transition zone into sheeted dikes that continue to the base of the hole (e.g., Dilek, 1998).

One of the critical issues related to the formation of the ocean crust is the nature and depth of hydrothermal circulation. Evidence from the ODP and DSDP studies indicates that the bulk of water circulation and subsequent alteration occurs in the extrusive layer at the top of the crust. Downhole measurements reveal a dramatic decrease in porosity and permeability at the base of the extrusive sequence (Becker et al., 1983).

There is, however, convincing evidence of alteration at greater depths in the sheeted dikes. One of the more intriguing results of the Hole 504B drilling is that depth of the mid-crustal transition zone, or the Layer 2/3 boundary, determined from seismic experiments was successfully penetrated. This did not correspond to the transition to gabbros, but existed within the sheeted dike unit (Detrick et al., 1994). This finding indicates that the mid-crustal transition may be controlled by variations in alteration and porosity (Dilek, 1998), as suggested earlier by McClain et al. (1990). If these observations are correct, they suggest that hydrothermal circulation in the sheeted dikes is accommodated by different features than those tested by downhole permeability and porosity measurements. Large-scale faulting and fracturing are likely candidates.

The presence of a narrow magma chamber or zone of partial melt is now accepted for the East Pacific Rise (e.g., Perfit and Chadwick, 1998). Some have suggested that this requires that excess cooling be provided by water penetration into the lower crust (Morton and Sleep, 1985; McClain et al., 1990). Alternatively, adequate cooling could be accommodated by water penetrating to the base of the sheeted dikes, provided that melt is concentrated at the top of the lower crustal magma chamber (Lister, 1983). The latter model seems most consistent with present seismic measurements (Fig. 5C).

The Long-Term Aging of Oceanic Crust as a Test of the Ophiolite Model

The layered approximations of the crust from seismic experiments in the Pacific Ocean were particularly well suited for examination of the long-term evolution of the ocean crust. As they age, the crust and the lithosphere cool over millions of years. In addition, the presence of water in the crust will alter the rocks of the crust, and so should change the seismic velocities.

Houtz and Ewing (1976) made an examination of Layer 2A thicknesses, whereby they observed that Layer 2A thins over time. They attributed this phenomenon to mineral precipitation in the porous extrusive rocks of the uppermost crust by way of water circulation. Wilkens et al. (1991) have made similar observations. Recent studies confirm that hydrothermal circulation continues in crust that is several m.y. old (e.g., Davis et al., 1999). Even though sediments might seal the crust as it ages, it appears that circulation will still occur in the underlying extrusive layer.

In the first systematic look at lower crustal evolution, Goslin et al. (1972) found that Layer 3 thickened by as much 1 to 2 km over 40 m.y. They argued that this process required serpentinization of the uppermost mantle, with a consequent decrease in seismic velocities down to crustal values. Presumably, this addition would form a layer beneath the lower crustal gabbro. Christensen and Salisbury (1975) also found a substantial increase in Layer 3 thickness. Although this thickening was partially offset by a thinning of Layer 2, the overall crust thickened with age. Arguing against a contribution by serpentinization, they thought this thickening could be accommodated by magmatic intrusions into the lower crust.

In a detailed reexamination of seismic data McClain (1981) and McClain and Atallah (1986) demonstrated that much of the apparent crustal thickening could be attributed to errors in the original layered models; however, some statistically significant crustal thickening was still observed. Because a number of mechanisms could result in thickening, such as localized serpentinization and volcanism, and because there are no mechanisms that thin the crust, McClain and Atallah (1986) concluded that the small amount of apparent thickening was not surprising. This apparent thickening did not indicate a widespread crustal process. They argued that if any general process exists, it must occur near the ridge axis, where it cannot be detected by these large-scale studies.

In Figure 7, we reproduce the data of Christensen and Salisbury (1975) and McClain and Atallah (1986). The latter demonstrate that total crustal thickening must be small; however, the crucial observation is the apparent transfer of mass into the lower crust, from Layer 2 to Layer 3. This can only occur because of alteration at the base of Layer 2 that raises the seismic velocities to lower-crustal values. This phenomenon requires that a substantial volume of the mid-crust be exposed to water circulation, and that the Layer 2/3 boundary probably changes depth in response to a process similar to that observed for the boundary between Layers 2A and 2.

CONCLUSIONS

The ophiolite-ocean crust analogy as represented in Figure 1 is only an approximation, and caution must be used before making too literal a correlation between seismic velocities to lithol-

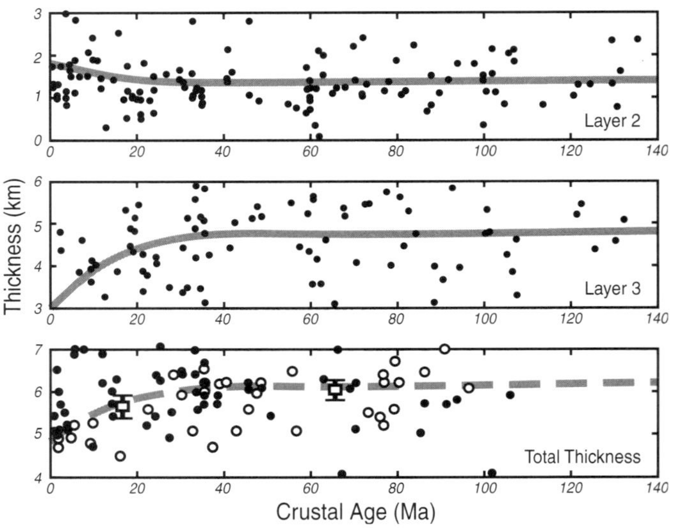

Figure 7. The evolution of the crust, using seismic refraction results. The top two panels contain the data for the thickness of Layers 2 and 3 as a function of crustal age. The gray lines represent running averages, and show a systematic thinning of Layer 2 and a thickening of Layer 3. (Modified from Christensen and Salisbury, 1975.) The bottom panel, modified from McClain and Atallah (1985), shows data for total crustal thickness as a function of age. Filled circles are the most reliable data, and open circles represent less reliable (generally older) data. Open squares show the averages for age intervals of 0–30 Ma (left square) and 30–100 Ma (right square). The error bars are for one standard deviation. The differences are statistically significant, but small. The dashed gray line combines the running averages of Christensen and Salisbury (1975), and shows a great deal more thickening. The key result is that thickening is not particularly important, but there appears to be significant transfer of material from seismic Layer 2 to seismic Layer 3. Undoubtedly, this is the result of alteration at mid-crustal levels.

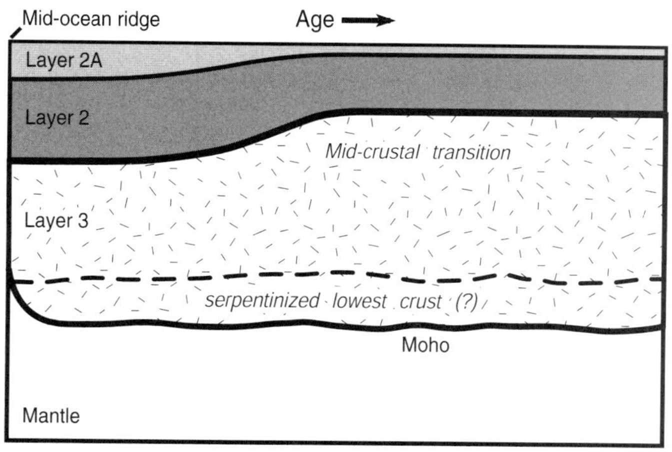

Figure 8. Cartoon showing the long-term evolution of the seismic ocean crust. The Moho changes very little, but Layer 2A thins significantly as porosity in the upper crust decreases with alteration. Similarly, Layer 3 appears to thicken because velocities increase in Layer 2, i.e., the rocks at the base of Layer 2 take on Layer 3-like velocities. These changes must be the result of ongoing alteration of crustal rocks at least down to the gabbros. Any serpentinization must take place close to the ridge axis and is unimportant to processes related to long-term aging of the crust.

ogy. The crust evolves through time. In particular, the Layer 2/3 boundary appears to shallow with age. It must represent a change in fracture density and/or alteration in the mid-crust, rather than a transition to gabbros.

There is substantial evidence for a serpentine component to the crust formed at rifted mid-ocean ridges (e.g., Cannat, 1993; Karson, 1998). The serpentine forms because of exposure of ultramafic rocks to seawater, a process accommodated by the tectonic dissection of the crust by rifting and ridge-transform faulting. In the Pacific crust, formed at non-rifted ridges, these processes are not as apparent, nor do they occur very often. Evidence provided here and previously suggests that serpentine is, at most, only a minor component of the crust. Serpentinization does not provide the dramatic long-term thickening of the crust seemingly suggested by earlier studies; however, serpentinization could explain the discrepancy between the thickness in ophiolites compared to the thickness of normal Pacific crust. The lower gabbro crust might be thickened by the addition of a thin layer of serpentinite. This process, shown in Figure 8, must have taken place very shortly after the crust was formed, similar to the model of Lewis and Snydsman (1977). The resulting modification to the ophiolite/ocean crust analogy is shown in Figure 9, which is a modified version of the accepted model shown in Figure 1.

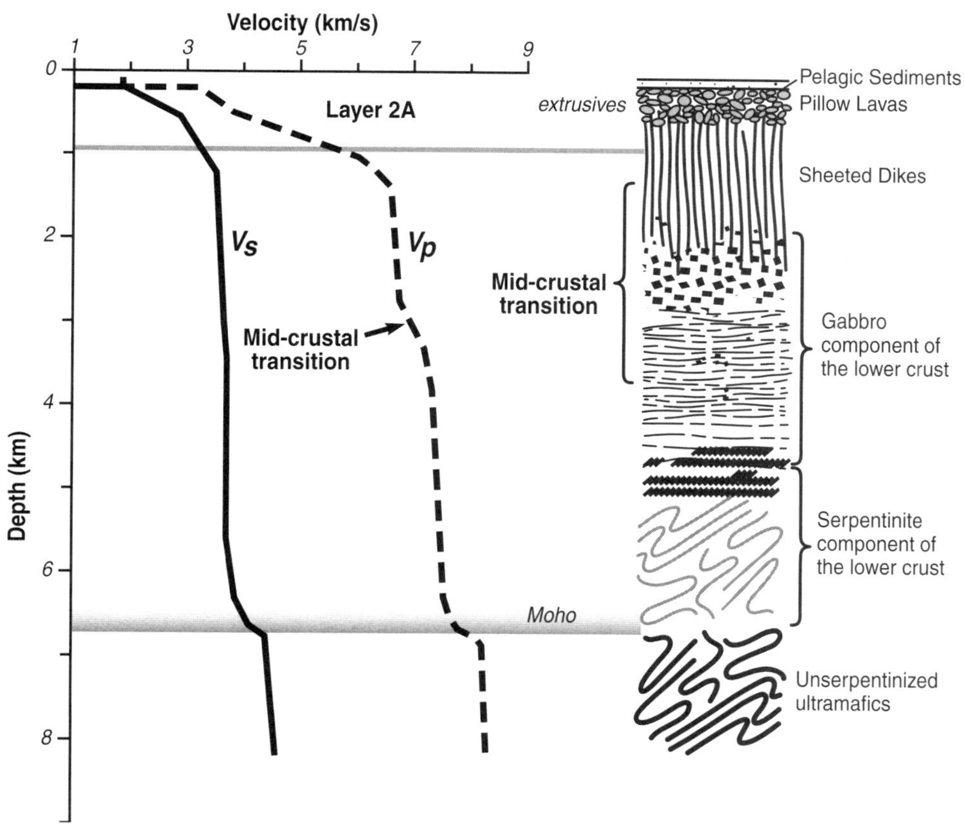

Figure 9. A sketch showing the modified correlation between ophiolite structure and seismic velocity structure. In particular, the evolution of oceanic structures revealed by seismic data suggests that the mid-crustal transition cannot always be interpreted as the boundary between sheeted dikes and gabbro. In addition, it is possible that serpentinite may form near the rise axis and constitutes a thin layer at the base of the crust.

The ophiolite model, then, is successful as a predictor of the structure of crust formed at non-rifted mid-ocean ridges, but must be used applied carefully when interpreting seismic profiles, particularly over older ocean crust.

ACKNOWLEDGMENTS

I would like to thank numerous colleagues for numerous past discussions about the issues addressed here. I would also like to thank the three reviewers for their helpful comments. The superb effort of the editors in organizing the conference and keeping all of the authors on-task was exceptional. In particular, I thank Yildirim Dilek for his enthusiastic patience in this effort. The higher quality figures in this paper were redrawn by Ms. Janice Fong of the Department of Geology at U.C. Davis.

REFERENCES CITED

Angenheister, G.A., 1921, A study of Pacific earthquakes: New Zealand Journal of Science, v. 4, p. 209–231.

Atallah, C.A., 1984, Compressional and shear wave structure of the Pacific ocean crust and upper mantle: implications for crustal evolution [M.S. thesis]: Davis, University of California, 327 p.

Becker, K., Langseth, M., Von Herzen, R.P., and Anderson, R., 1983, Deep crustal geothermal measurements, Hole 504B, Costa Rica Rift: Journal of Geophysical Research, v. 88, p. 3447–3457.

Bratt, S.R., and Purdy, G.M., 1984, Structure and variability of oceanic crust on the flanks of the East Pacific Rise between 11° and 13° N: Journal of Geophysical Research v. 89, p. 6111–6125.

Bratt, S.R., and Solomon, S.C., 1984, Compressional and shear wave structure of the East Pacific Rise at 11°20'N: Constraints from three-component ocean-bottom seismometer data: Journal of Geophysical Research, v. 89, p. 6095–6110.

Brunn, J.H., 1959, La dorsale medi-atlantique et les epanchement ophiolites: Compte Rendu; Société Géologique de France, v. 8, p. 234–236.

Bullen, K.E., 1955, Seismology and the broad structure of the Earth's interior: Physics and Chemistry of the Earth, v. 1, p. 68–93.

Byerly, P., 1930, The dispersion of seismic waves of the Love type and the thickness of the surface layer of the earth under the Pacific: Gerlands Beitrage zur Geophysik, v. 26, p. 27–33.

Cann, J.R., 1968, Geological processes at mid-ocean ridge crests: Geophysical Journal of the Royal Astronomical Society, v. 15, p. 331–341.

Cann, J.R., 1974, A model for oceanic crustal structure developed: Geophysical Journal of the Royal Astronomical Society, v. 29, p. 169–187.

Cannat, M., 1993, Emplacement of mantle rocks in the seafloor at mid-ocean ridges: Journal of Geophysical Research, v. 98, p. 4163–4172.

Cannat, M., 1996, How thick is the magmatic crust at slow-spreading ridges?: Journal of Geophysical Research, v. 101, p. 2847–2857.

Carder, D.S., 1934, Seismic surface waves and the crustal structure of the Pacific region: Bulletin of the Seismological Society of America, v. 24, p. 231–303.

Christensen, N.I., 1966, Elasticity of ultrabasic rocks: Journal of Geophysical Research, v. 71, p. 5921–5931.

Christensen, N.I., 1968, Compressional wave velocities in basic rocks: Pacific Science, v. 22, p. 41–44.

Christensen, N.I., 1972, The abundance of serpentinites in the ocean crust: Journal of Geology, v. 80, p. 709–719.

Christensen N.I., and Salisbury, M.H., 1975, Structure and constitution of the lower oceanic crust: Reviews of Geophysics and Space Physics, v. 13, p. 57–86.

Christeson, G.L., Purdy, G.M., and Fryer, G.J., 1992, Structure of young oceanic crust at the East Pacific Rise near 9°30' N: Geophysical Research Letters, v. 19, p. 1045–1048.

Coleman, R.G., 1971, Petrologic and geophysical nature of serpentinites: Geological Society of America Bulletin, v. 82, p. 897–917.

Davis, E.E., Chapman, D.S., Wang, K., Villinger, H., Fisher, A.T., Robinson, S.W., Grigel, J., Pribnow, D., Stein, J., and Becker, K., 1999, Regional heat flow variations across the sedimented Juan de Fuca Ridge eastern flank constraints on lithospheric cooling and lateral hydrothermal heat transport: Journal of Geophysical Research, v. 104, p. 17,675–17,688.

Detrick, R.S., Buhl, P., Vera, E., J. Mutter, J., Orcutt, J., Madsen, J., and Brocher, T., 1987, Multi-channel imaging of a crustal magma chamber: Nature, v. 326, p. 35–41.

Detrick, R.S., Harding, A.J., Kent, G.M., Orcutt, J.A., Mutter, J.C., and Buhl, P.,1993, Seismic structure of the southern East Pacific Rise: Science, v. 259, p. 499–503.

Detrick, R.S., Collins, J., Stephen, R., and Swift, S., 1994, In situ evidence for the nature of the seismic layer 2/3 boundary in oceanic crust: Nature, v. 370, p. 288–290.

Dewey, J.F., and Bird, J.M., 1971, Origin and emplacement of the ophiolite suite: Appalachian ophiolites in Newfoundland: Journal of Geophysical Research, v. 76, p. 3179–3206.

Dilek, Y., 1998, Structure and tectonics of intermediate spread oceanic crust drilled at DSDP/ODP Holes 504B and 896A, Costa Rica Rift, *in* Cramp, A., MacLeod, C.J., Lee, S.V., and Jones, E.J.W., eds., Geological Evolution of Ocean Basins: Results from the Ocean Drilling Program: London, Geological Society of London Special Publications 131, p. 179–197.

Dilek, Y., Moores, E.M., and Furness, H., 1998, Structure of modern oceanic crust and ophiolites and implications for faulting and magmatism at oceanic spreading centers, *in* Buck, W.R., Delaney, P.T., Karson, J.A., and Lagrielle, Y., eds., Faulting and Magmatism at Mid-Ocean Ridges: Washington, D.C., American Geophysical Union, p. 219–265.

Ewing, M., and Press, F., 1950, Crustal structure and surface wave dispersion: Bulletin of the Seismological Society of America, v. 40, p. 271–280.

Ewing, M., A. Woollard, G.P ., and Vine, A.C., 1939, Geophysical investigations in the emerged and submerged Atlantic coastal plain, 3, Barnegat Bay, New Jersey section: Geological Society of America Bulletin, v. 50, p. 257–296.

Ewing, M., Worzel, J.L., Hersey, J.L., Press, F., and Hamilton, G.R., 1950, Seismic refraction measurements in the Atlantic Ocean basin: Bulletin of the Seismological Society of America, v. 40, p. 233–242.

Francheteau, J., Armijo, R., Cheminee, J.L., Hekiniian, R., Lonsdale, P., and Blume, N., 1990, 1 Ma East Pacific Rise oceanic crust and upper mantle expose by rifting in Hess Deep (equatorial Pacific Ocean): Earth Planetary Science Letters, v. 101, p. 281–295.

Gass, I.G., 1968, Is the Troodos massif of Cyprus a fragment of Mesozoic ocean floor?: Nature, v. 220, p. 39–42.

Gilluly, J., Waters, A.C., and Woodford A.O., 1953. Principles of Geology: San Francisco, W.H. Freeman and Company, 631 p.

Goslin, J., Beauzart, P. Franchetau, J. and Le Pichon, X., 1972, Thickening of the oceanic layer in the Pacific Ocean: Marine Geophysical Researches, v. 1, p. 418–427.

Harding, A.J., Kent, G.M., and Orcutt, J.A., 1993, A multichannel seismic investigation of upper crustal structure at 9°N on the East Pacific Rise: Implications for crustal accretion: Journal of Geophysical Research, v. 98, p. 13,925–13,944.

Hess, H.H., 1939, Island arcs, gravity anomalies and serpentinite intrusions: A contribution to the ophiolite problem: International Geological Congress, 17th, Moscow, Report, v 17, v. 2, p. 263–283.

Hess, H.H., 1955, The oceanic crust: Journal of Marine Research, v. 14, p. 423–439.

Hess, H.H., 1959, The AMSOC hole to the Earth's Mantle: Eos, Transactions, American Geophysical Union, v. 40, p. 340–345.

Hess, H.H., 1960, Caribbean Research Project: Progress report: Geological Society of America Bulletin, v. 71, p. 235–240.

Hess, H.H., 1965, Mid-oceanic ridges and tectonics of the sea floor, *in* Whittard, A., and Bradshaw, R., eds., Submarine Geology and Geophysics: Proceedings of the 17th Symposium of the Colston Research Society: London, Butterworths, p. 317–334.

Hill, M.N., and Shallow, J.C., 1950, Seismic experiments in the Atlantic: Nature, v. 165, p. 193–194.

Houtz, R.E., and Ewing, J., 1976, Upper crustal structure as a function of plate age: Journal of Geophysical Research, v. 81, p. 2490–2498.

Hurst, S.D., and Karson, J.A., 1991, Cross sections of the Hess Deep Rift: A critical analysis: Eos, Transactions, American Geophysical Union, v. 72, p. 488.

Jeffreys, H., 1929, The Earth: Its Origin, History, and Physical Constitution, second edition: London, Cambridge University Press, 574 p.

Karson, J.A., 1998, Internal structure of oceanic lithosphere: a perspective from tectonic windows, *in* Buck, W.R., Delaney, P.T., Karson, J.A., and Lagrielle, Y., eds., Faulting and Magmatism at Mid-Ocean Ridges: Washington D.C., American Geophysical Union, p. 177–218.

Kennedy, G.C., 1959, The origin of continents, mountain ranges, and ocean basins: American Scientist, v. 47, p. 491–504.

Kent, G.A., Harding, A.J., and Orcutt, J.A., 1990, Evidence for a smaller magma chamber beneath the East Pacific Rise at 9°30' N: Nature, v. 344, p. 650–653.

Kent, G.A., Harding, A.J., Orcutt, J.A., Detrick, R.S., Mutter, J.C., and Buhl, P., 1994, Uniform accretion of oceanic crust south of the Garrett transform at 14°15' S on the East Pacific Rise: Journal of Geophysical Research, v. 99, p. 9097–9116.

Lagabrielle, Y., Bideau, D., Cannat, M., Karson, J.A., and Mevel, C., 1998, Ultramafic-mafic plutonic rock suites exposed along the Mid-Atlantic Ridge (10°N-30°N), Symmetrical-asymmetrical distribution and implications for seafloor spreading processes, *in* Buck, W.R., Delaney, P.T., Karson, J.A., and Lagrielle, Y., eds. Faulting and Magmatism at Mid-Ocean Ridges: Washington D.C., American Geophysical Union, p. 153–176.

Leet, L.D., 1936, Seismological data on surface layers in New England: Bulletin of the Seismological Society of America, v. 26, p. 129–146.

Lewis, B.T.R., and Garmany, J.D.,1982, Constraints on the structure of the East Pacific Rise from seismic refraction data: Journal of Geophysical Research, v. 87, p. 8417–8425.

Lewis, B.T.R., and McClain, J., 1977, Converted shear waves as seen by ocean bottom seismometers and surface buoys: Bulletin of the Seismological Society of America, v. 67, p. 1291–1302.

Lewis, B.T.R., and Snydsman, W.E., 1977, Evidence for a low velocity layer at the base of the oceanic crust: Nature, v. 226, p. 340–344.

Lister, C.R.B., 1977, Qualitative models of spreading-center processes, including hydrothermal penetration: Tectonophysics, v. 37, p. 203–218.

Lister, C.R.B., 1983, On the intermittency and crystallization mechanisms of sub-seafloor magma chambers: Geophysical Journal of the Royal Astronomical Society, v. 73, p. 351–365.

Loverling, J.F., 1958, The nature of the Mohorovicic Discontinuity: Eos, Transactions, American Geophysical Union, v. 39, p. 947–955.

Macelwane, J.B., 1951, Evidence on the interior of the Earth derived from seismic sources, in Gutenberg, B., ed., Internal Constitution of the Earth, second edition: London, Dover Publications, p. 227–304.

McCabe, J.R., and McClain, J.S., 2000, A comparison of large and small seamounts formed near the East Pacific Rise : Eos, Transactions, American Geophysical Union, v. 72, p. F1075.

McClain, J.S., 1981, On long term thickening of the oceanic crust: Geophysical Research Letters, v. 8, p. 1191–1194.

McClain, J.S., 1992, Oceanic crust: Encyclopedia of Earth System Science: San Diego, Academic Press, v. 3, p. 415–425.

McClain, J.S., and Atallah, C.A., 1985, The structure of young oceanic crust near a very fast spreading ridge: Geophysical Research Letters, v. 12, p. 680–692.

McClain, J.S., and Atallah, C.A., 1986, Thickening of the ocean crust with age: Geology, v. 14, p. 574–576.

McClain, J.S., and Lewis, B.T.R., 1980, A seismic experiment at the axis of the East Pacific Rise: Marine Geology, v. 35, p. 147–169.

McClain, J.S., Orcutt, J.A., and Burnett, M., 1985, The East Pacific Rise in cross section: a seismic model: Journal of Geophysical Research, v. 90, p. 8627–8639.

McClain J.S., Chin, C., and Moores, E.M., 1990, Seismic velocities in young ocean crust: constraints on models for the development of permeability and water penetration, *in* Malpas, J., Moores, E.M., Panayiotou, A., and Xenophontos, C., Ophiolites Ocean Crustal Analogues: Nicosia, Geological Survey of Cyprus, p. 577–581.

McClain, K.J., and Lewis, B.T.R., 1982, Geophysical evidence for the absence of a crustal magma chamber under the northern Juan de Fuca Ridge: A contrast with ROSE results: Journal of Geophysical Research, v. 87, p. 8477–8489.

Miyashiro, A., 1973, The Troodos complex was probably formed in an island arc: Earth and Planetary Science Letters, v. 19, p. 218–224.

Mohorovicic, A., 1909, Das Beben vom 8.X.1909, Jahrbruch met. Obs. Zagreb (Yearbook of the Meteorological Observatory of Zagreb), v. 9, p. 1–63.

Moores, E.M., 1982, Origin and emplacement of ophiolites: Reviews of Geophysics and Space Physics, v. 20, p. 735–760.

Moores, E.M., and Jackson, E.D., 1974, Ophiolites and oceanic crust: Nature, v. 250, p. 136–138.

Moores, E.M., and Vine, F.J., 1971, Troodos massif, Cyprus, and other ophiolites as oceanic crust: Evaluation and implications: Transactions of the Royal Society of London, ser. A, v. 269, p. 443–466.

Morton, J.L., and Sleep, N.H., 1985, A mid-ocean ridge thermal model: constraints on the volume of axial hydrothermal heat flux: Journal of Geophysical Research, v. 90, p. 11,345–11,353.

Murase, T., and McBirney, A.R., 1973, Properties of some common rocks and their melts at high temperatures: Geological Society of America Bulletin, v. 84, p. 3563–3692.

Newman, F., 1929, The velocity of seismic surface waves over Pacific Paths: Bulletin of the Seismological Society of America, v. 19, p. 63–76.

Orcutt, J.A., Kennett, B.L.N., Dorman, L.M., and Prothero, W.A., 1975, Evidence for a low-velocity zone underlying a fast-spreading rise crest: Nature, v. 256, p. 475–476.

Pallister, J.S., and Hopson, C.A., 1981, Samial ophiolite suite: Field relations, phase variation, cryptic variation and layering, and a model of a spreading ridge magma chamber: Journal of Geophysical Research, v. 86, p. 2593–2644.

Perfit, M.R., and Chadwick, Jr., W.W., 1998, Magmatism at mid-ocean ridges: constraints from volcanological and geochemical investigations, *in* Buck, W.R., Delaney, P.T., Karson, J.A., and Lagrielle, Y., eds., Faulting and Magmatism at Mid-Ocean Ridges: Washington D.C., American Geophysical Union, p. 177–218.

Raitt, R.W., 1956, Seismic-refraction studies of the Pacific Ocean basin, Part I: Crustal thickness of the central equatorial Pacific: Geological Society of America Bulletin, v. 67, p. 1623–1640.

Raitt, R.W., 1969, Anisotropy of the upper mantle, *in* Hart, P.J., ed., The Earth's Crust and Upper Mantle: Washington D.C., American Geophysical Union, Geophysical Monograph 13, p. 250–256.

Ringwood, A.E., 1969, Composition and evolution of the upper mantle, *in* Hart, P.J., ed., The Earth's Crust and Upper Mantle: Washington D.C., American Geophysical Union, Geophysical Monograph 13, p. 1–17.

Reid, I.D., Orcutt, J.A., and Prothero, W.A., 1977, Seismic evidence for a narrow zone of partial melting underlying the East Pacific Rise at 21° N: Geological Society of America Bulletin, v. 88, p. 678–682.

Rosendahl, B.R., Raitt, R.W., Dorman, L.M., Bibee, L.D., Hussong, D.M., and Sutton, G.H., 1976, Evolution of the oceanic crust, 1: A physical model of the East Pacific Rise crest derived from seismic refraction data: Journal of Geophysical Research, v. 81, p. 5294–5304.

Shor, Jr., G.G., Menard, H.W., and Raitt, R.W., 1969, Structure of the Pacific Basin, *in* Maxwell, A.E., ed. The Sea, Volume 4, New York, John Wiley & Sons, p. 3–27.

Sinton, J.M., and Detrick, R.S., 1992, Mid-ocean ridge magma chambers: Journal of Geophysical Research, v. 97, p. 197–216.

Sleep, N.H., 1975, Formation of the oceanic crust: Some thermal constraints: Journal of Geophysical Research, v. 80, p. 4037–4042.

Spudich, P., and Orcutt, J., 1980, Petrology and porosity of an oceanic crustal site: results from wave form modeling of seismic refraction data: Journal of Geophysical Research, v. 85, p. 1409–1433.

Steinmann, G., 1927, Die ophiolitischen Zoneen in der mediterranean Kettingebirgen: International Geological Congress, 14th, Madrid, Proceedings, v. 2, p. 638–667.

Toomey, D.R., Purdy, G.M., Solomon, S.C., and Wilcock, W.S.D., 1990, The three-dimensional seismic velocity structure of the East Pacific Rise, near latitude 9°30' N: Nature, v. 347, p. 639–645.

Toomey, D.R., Solomon, S.C., and Purdy, G.M., 1994, Tomographic imaging of the shallow crustal structure of the East Pacific Rise at 9° 30' N: Journal of Geophysical Research, v. 99, p. 24,135–24,157.

Vera, E.E., Mutter J.C., Buhl, P., Orcutt, J.A., Harding. A.J., Kappus, M.E., Detrick, R.S., and Brocher, T.M., 1990, the structure of the 0-to 0.2-m.y.-old oceanic crust at 9°N on the East Pacific Rise from expanded spread profiles: Journal of Geophysical Research, v. 95, p. 15,529–15,556.

Wilkens, R.H., Fryer, G.J., and J. Karsten, J., 1991, Evolution of porosity and seismic structure of upper oceanic crust: Importance of aspect ratios: Journal of Geophysical Research, v. 96, p. 17,981–17,995.

Wilson, J.T., and Baykal, O., 1948, Crustal structure of the North Atlantic basin as determined from Rayleigh-wave dispersion: Bulletin of the Seismological Society of America, v. 38, p. 41–54.

Wyllie, P.J., 1971, The Dynamic Earth: Textbook in Geosciences: New York, John Wiley and Sons, 416 p.

Manuscript Accepted by the Society May 2, 2003

Printed in the USA

Geological Society of America
Special Paper 373
2003

Development of ophiolitic perspectives on models of oceanic magma chambers beneath active spreading centers

Peter Thy*
Department of Geology, One Shields Avenue, University of California, Davis, California 95616, USA

Yildirim Dilek
Geology Department, Miami University, Oxford, Ohio 45056, USA

ABSTRACT

Ophiolites played a minor role in the formulation of the ideas of the history and evolution of the ocean floor because initially they were not interpreted as oceanic in origin. Early researchers working on ophiolites were late in claiming their oceanic origin. Ophiolites were universally considered to be large sill intrusions or thick lava flows emplaced from an abyssal substratum into subsiding, marginal sedimentary basins. Such large pools of basaltic magma were proposed to have differentiated internally as a result of crystallization within an outer veneer of a giant lava pillow. It was first with the emergence of the theory of plate tectonics that it became possible to distinguish between mantle rocks and crustal intrusive plutons. This development led to the recognition of fossil magma chambers in ophiolites. Plate tectonic models implied that basaltic magma either erupted directly on the seafloor or accumulated in crustal chambers beneath spreading centers. The first models developed in the late 1960s for the structure of modern oceanic crust did not include lower crustal magma chambers; however, subsequent models formulated in the early 1970s incorporated the existence of magma chambers, but with little reference to ophiolite studies. Oceanic magma chambers were seen as large, continuously growing entities keeping pace with seafloor spreading and steady-state magma supply. This magma chamber interpretation was very influential and contained the key elements soon adapted by many of the subsequent studies of ophiolites. Such ophiolitic magma chamber models were extended to the oceans in the 1980s, inferring the size, shape, and evolution of magma chambers beneath active spreading centers.

The challenge to the steady-state magma chamber model came from geophysical and theoretical modeling of modern oceanic crust. These studies revealed that large magma chambers could not exist beneath slow-spreading centers, and that they might be present beneath fast-spreading centers only as a thin magma lens on top of a large crystal mush. Other predictions suggested the presence of small magma chambers in broad zones under the axial valley and emphasized the intermittent nature of crustal chambers as a result of the interplay between tectonic and magmatic processes. This modeling of oceanic crust made little use of observations from ophiolites. The general image that emerged in the 1990s was that magma chambers beneath fast-spreading centers were largely filled with crystal mush containing only small percentages of melt. The large molten chambers envisioned during the proceeding three decades have thus vanished and been replaced by crystal mush chambers. The most important implication

* thy@geology.ucdavis.edu

Thy, P., and Dilek, Y., 2003, Development of ophiolitic perspectives on models of oceanic magma chambers beneath active spreading centers, *in* Dilek, Y., and Newcomb, S., eds., Ophiolite concept and the evolution of geological thought: Boulder, Colorado, Geological Society of America Special Paper 373, p. 187–226.

of this interpretation for the origin of ophiolites is that layering commonly observed in gabbros is likely to have formed as a result of crustal strain, not from direct deposition or crystallization on chamber walls.

Deep drilling during the last ten years into in situ modern oceanic crust near active spreading centers has supported the existence of crystal mush zones. A surprising discovery has been the preservation of evolved gabbros within primitive host gabbro sequences. The evolved gabbros were interpreted to represent the interstitial melt that migrated upward and laterally as a result of compaction and/or tectonic deformation. The interaction between new magma seeping up from beneath the chamber and the lower strata of the mush makes the mush zone grow. Detailed field studies of ophiolites in recent years have demonstrated that new magma may form sill complexes in crystal mushes. The mush and sill complexes act as a reactive filter, modifying the migrating residual melt and controlling the composition of the residual melt, which eventually reaches the magma lens or gets erupted as lavas on the seafloor.

Our understanding of oceanic magma chambers has been much influenced by direct observations and careful fieldwork in plutonic complexes of ophiolites. The revival of systematic field and petrographic studies in ophiolites is a necessary approach to further advance our understanding of oceanic magma chambers and the magmatic and tectonic processes associated with the evolution of oceanic crust.

Keywords: ideas of magma chambers, conceptual models, ophiolite studies, history of ocean floor, plate tectonics, crustal accretion, ophiolites, oceanic crust, active spreading centers, , abyssal substratum, magma differentiation, steady-state magma chambers, crystal-mush magma chambers, melt lens, slow-spreading magma chambers, fast-spreading magma chambers, sill complexes, magmatic processes, revival of field and petrographic studies

INTRODUCTION

Development of the plate tectonic theory and the early models for the accretion of oceanic crust predicted the presence of magma chambers beneath the rift axes of mid-ocean ridge systems (Cann, 1970; Moore et al., 1974; Christensen and Salisbury, 1975; Dewey and Kidd, 1977; Bryan and Moore, 1977; Macdonald, 1989). Because magma chambers cannot directly be observed (Macdonald, 1982; Orcutt, 1987), ophiolite studies have become one of our principal sources of information about magma chambers and magma chamber processes. Fossil magma chambers have been identified in many ophiolite complexes (Gass, 1968) and ophiolitic plutons have been, and still are, central to our understanding of crustal accretion and magmatic processes in active oceanic spreading settings, although the ideas and conceptual models for magma chamber evolution and associated magmatic processes have varied considerably over the years since the structural views of ophiolitic and oceanic crust formation first emerged (Fig. 1). Studies of continental layered intrusions (Wager and Brown, 1967) have been another important source for understanding both ophiolitic and oceanic magma chambers and their solidification processes.

In this paper, we explore how ideas of ophiolitic magma chambers have developed through time and have been influenced by our geochemical, petrological, tectonic, and geophysical understanding of modern oceanic crust. We show that early models of ophiolitic magma chambers were influenced by studies of layered intrusions and other stratiform continental plutons. Development of the theory of plate tectonics resulted in a widely accepted idea of continually replenished magma chambers, an idea that, for years, strongly influenced ophiolite investigations. Studies of core samples drilled from gabbros near active oceanic spreading centers and their results have provided the critical information that stimulated a re-examination of oceanic magma chamber models. Thus, current models of oceanic magma chambers are fundamentally different than the early models, reflecting major changes and shifts in our geological thought.

After a brief introduction to oceanic magma chambers, we review the development of ideas on ophiolitic magma chambers, using observations from the Troodos (Cyprus), Bay of Islands (Newfoundland, Canada), and Oman (Sultanate of Oman) ophiolites. Next, we discuss the ophiolitic chamber models in the context of ideas developed concurrently for magma chambers beneath modern oceanic spreading centers. We then assess the mutual influence of ophiolitic and oceanic studies on the development of ideas for magma chamber evolution at active spreading centers. In our discussion, we use some of the most important original illustrations of ophiolitic and oceanic magma chambers (only slightly edited or redrawn for the purpose of our presentation).

MAGMA CHAMBERS

Magma forms by partial melting in adiabatically upwelling mantle (e.g., McKenzie and Bickle, 1988); it then migrates upward and concentrates beneath ridge crests (Morgan, 1987).

Figure 1. Modern artist's drawing of continental rifting and magmatism as inspired by eighteenth century illustrations. This illustration was used in connection with a conference, convened by M.A. Menzies at Royal Holloway, University of London in 1987, on *Oceanic and Continental Lithosphere: Similarities and Differences*. The conference resulted in a special volume of Journal of Petrology (1998). Reproduced with the permission of Martin Menzies.

The differential buoyancy between melt and the residual mantle peridotite, or the host crustal rocks, drives the upward movement to a level of neutral buoyancy (Turcotte and Morgan, 1992; Bryan, 1993). The melt is thought to migrate by channeled porous flow along grain boundaries and along fractures ranging in width from meters to kilometers (see review by Turcotte and Morgan, 1992). Magma migration may also be controlled by hydraulic overpressure and dike formation above a plume head (e.g., Maaløe, 2002). The melt differentiates in transient dikes and sills or in chambers of various shapes and sizes as a result of gradual cooling, and leaves behind solid crystallization and reaction products that may significantly modify host crustal and mantle properties, as well as the composition of the melt. The storage and the flux of magma through oceanic crust depend on the spreading rate (Dilek et al., 1998; Karson, 1998, 2002). The magma that migrates through the crustal "plumbing system" can change composition to an extent that the derived magmas may have little resemblance to the parental magma that originally separated from the host peridotite (Grove et al., 1992). Magma trapped in high-level plutons or erupted as lava flows on the seafloor therefore may not easily be "fingerprinted" to infer the source of melting and the mode of differentiation processes (Langmuir et al., 1992; Grove et al., 1992; Bédard, 1993).

It was mainly petrographic studies of gabbroic and ultramafic intercontinental-layered plutons (McBirney, 1993; Marsh, 2000) that shaped our early views and resulted in widely publicized models for spreading-related magma chambers (e.g., Bryan and Moore, 1977; Brown and Musset, 1981; Best, 1982; Bott, 1982; Fig. 2). The early ophiolitic magma chamber models were strongly influenced by studies of layered intrusions and other continental plutons, which were believed to represent trapped magma chambers solidified at crustal levels (Hopson and Frano, 1977; Pallister and Hopson, 1981; Nicolas, 1989, p. 261; McBirney, 1993; Marsh, 2000). The internal constitution of plutons records evidence for crystallization and consolidation processes (Marsh, 1996) that can only be unraveled by careful field mapping and detailed petrographic and geochemical studies (Wager and Brown, 1967; Turner and Campbell, 1986; Parsons, 1987; Cawthorn, 1996; Irvine et al., 1998). Such studies of stratigraphically well constrained sequences may not always be feasible at active spreading centers, where complex thermal and tectonic processes may have shaped or modified the internal constitution of plutons (Sleep, 1975; Lister, 1977) and where perhaps only a few dredges and drill core are available.

Development of the plate tectonic theory and the direct sampling of lava flows near active ridges have revealed systematic variations in lava compositions that are interpreted as related to the bathymetry of the ridge crest (Klein and Langmuir, 1987). Temporal and lateral ridge segmentations are likely to be controlled by the lack and/or the presence of the underlying axial magma chambers, magma supply rates, and the depth to the crust-mantle transition zone (Macdonald, 1982; Klein and Langmuir, 1987; Solomon and Toomey, 1992; Batiza, 1996). Such views, together with theoretical and geophysical studies, have markedly changed the classic view of oceanic magma chambers and the processes that control magma transport and differentiation. Most recently, major advances have resulted from direct sampling by drilling into in situ gabbroic complexes exposed near active oceanic spreading centers. The most complete section of oceanic gabbros has been sampled in Ocean Drilling Program Holes 735B and 1105A, near the very slow-spreading Southwest Indian Ridge (Dick et al., 2000; Natland and Dick, 2001; Thy, 2002). These results have highlighted the important role of spreading rate on magma chamber development and the interplay between the tectonic extension and magma chamber processes on the evolution of oceanic crust.

Despite the advances made in our understanding of the evolution of the plutonic foundation of modern oceanic crust, it is still important to realize that direct information on the nature and sizes of magma chambers and on the processes that control magma crystallization and differentiation in them is limited. Because magma chambers and magma solidification processes cannot be

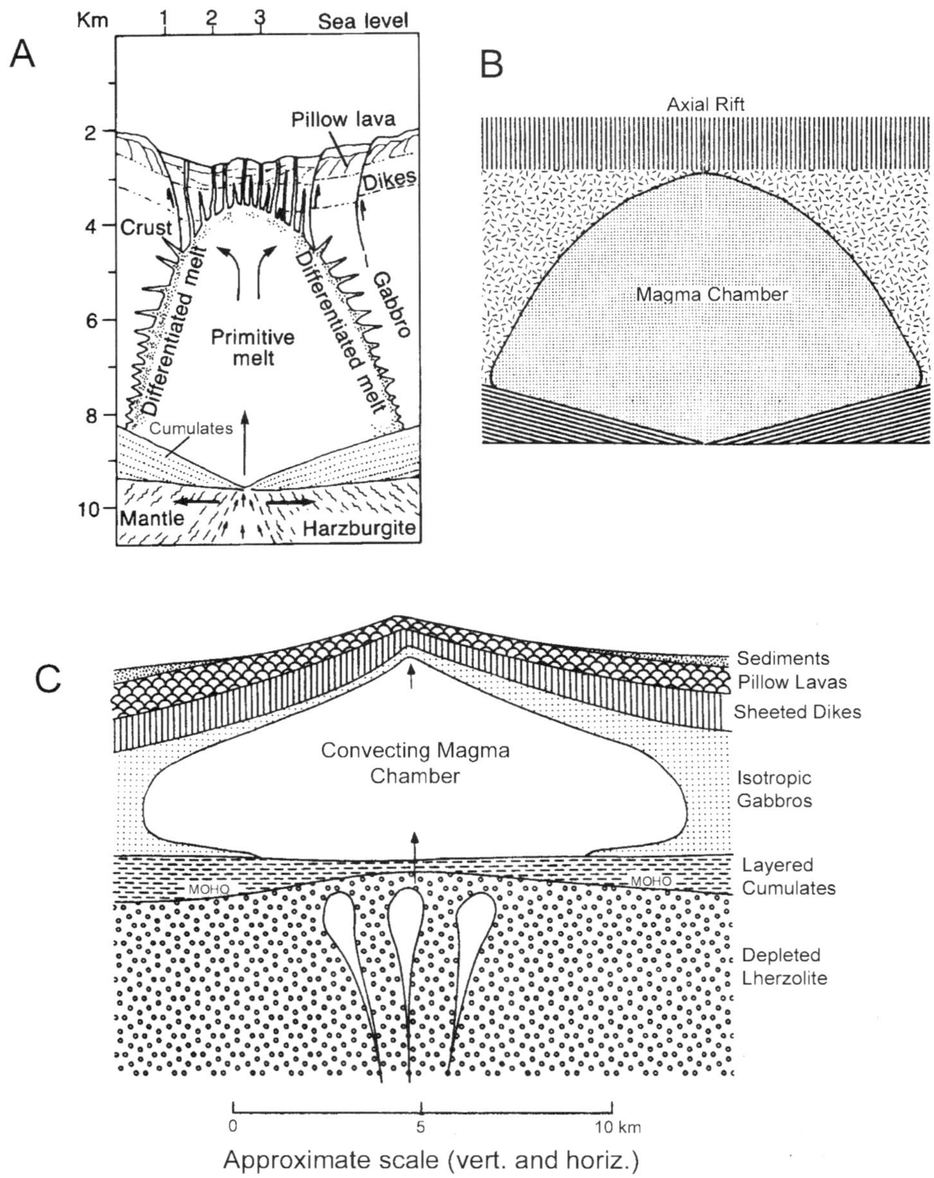

Figure 2. Schematic illustrations of oceanic spreading chambers. A: Spreading magma chamber after Bryan and Moore (1977, Fig. 16). The model is based on the petrology and geochemistry of dredge samples and was drawn to scale. B: Schematic spreading center magma chamber redrawn by Pallister and Hopson (1981, Fig. 17) after Cann (1974, Fig. 6). Note the similarity with Bryan and Moore's (1977) chamber model in Figure 2A. Reproduced by permission of Royal Astronomical Society. C: Textbook cross-section of a mid-ocean spreading center magma chamber (Brown and Mussett, 1981, Fig. 7.8).

observed directly, the formulated models often combine limited observations with current theoretical views of crustal accretion and mid-ocean ridge processes. As a result, magma chamber models naturally change as our conceptual ideas of ocean crust formation and magma crystallization processes evolve.

OPHIOLITE STUDIES

The term "ophiolite" was originally used to describe serpentinites in orogenic belts (cf., Coleman, 1977). After Steinmann (1927), the term ophiolite was used to identify a special association of rocks that included serpentinite, gabbro, diabase, spilite, and chert (see Bernoulli et al., this volume, Chapter 7). Ophiolites were believed to have formed as massive mafic and ultramafic submarine outpourings of magma onto the floor of eugeosynclinal basins during early stages of their development (Kay, 1951; Brunn, 1960; Aubouin, 1965). Differentiation in such thick outpourings was thought to have resulted in a variation of rocks ranging from peridotite to gabbro within an outer selvage of diabase and pillow lavas (like a gigantic basalt pillow; cf., Aubouin, 1965, p. 148–159). Ophiolitic peridotites were considered to have formed by accumulation of early crystallizing minerals, such as olivine and pyroxene, in the lower part of these in situ crystallizing massifs.

Although it was well known that the Earth's mantle was of peridotitic composition (MacGregor, 1967) and that melting of this peridotite would produce typical oceanic basalts (Bowen, 1928; Daly, 1933; Hess, 1955; Yoder and Tilley, 1962; Wyllie, 1967), the connection between these observations and the basaltic oceanic layer was not always made (Hess, 1955). Hess (1962) hypothesized that the oceanic layer formed by serpentinization of peridotite beneath ridge crests located above the rising limbs of mantle convection cells. This view was so influential that it actually might have held back the development of modern views of ocean ridges and the generation of basaltic crust (cf., Heezen, 1962). For this reason, ophiolite studies initially struggled with the origin of basaltic lavas and their relations with tectonized peridotites and serpentinites (Gass and Masson-Smith, 1963; Gass, 1967). It was first with the emergence of plate tectonics in the mid- to late 1960s that it became possible to distinguish between mantle tectonites and intrusive plutonic rocks as separate components of oceanic lithosphere. This development further made it clear that ophiolites were mostly basaltic in composition and that the internal variation in their pseudostratigraphy was caused by fractional crystallization forming solid cumulates and differentiated melts (Brunn, 1960; Dewey and Bird, 1971; Moores and Vine, 1971; Church, 1972). Thus, it became possible to identify fossil magma chambers in ophiolitic sequences. This was achieved convincingly, for the first time, in the Troodos ophiolite (e.g., Gass, 1968).

Troodos Ophiolite (Cyprus)

The Troodos ophiolite is part of a mountain range extending east-west across the island of Cyprus, and is composed of serpentinites, peridotites, mafic and ultramafic plutons, sheeted dikes, and lavas (Fig. 3). The general geology and the historical background of studies in the Troodos ophiolite can be found in several review papers (Gass, 1980, 1989, 1990; Gass and Smewing, 1981; Robertson and Xenophontos, 1993). Although the first geological maps of the Troodos Mountains were published in 1906, these early maps and accompanying reports contained little information on the plutonic complex. It was first with the systematic mapping initiated in the early 1950s by the Cyprus Geological Survey (Bishopp, 1952, 1954) that a detailed stratigraphy of the massif was achieved (e.g., Wilson, 1957; Bear, 1960; Gass, 1960). The result of this mapping project guided the development of the early models for the origin of the ophiolite (Gass and Masson-Smith, 1963; Gass, 1967) and laid the foundation for our current understanding of the Troodos massif.

The early models for the origin of the Troodos ophiolite were developed in the context of the geosynclinal theory, which dominated the geodynamic thinking of that time (i.e., Aubouin, 1965). The contact between the volcanic pile and the peridotitic rocks and the apparent lack of sialic crust beneath the ophiolite led to the

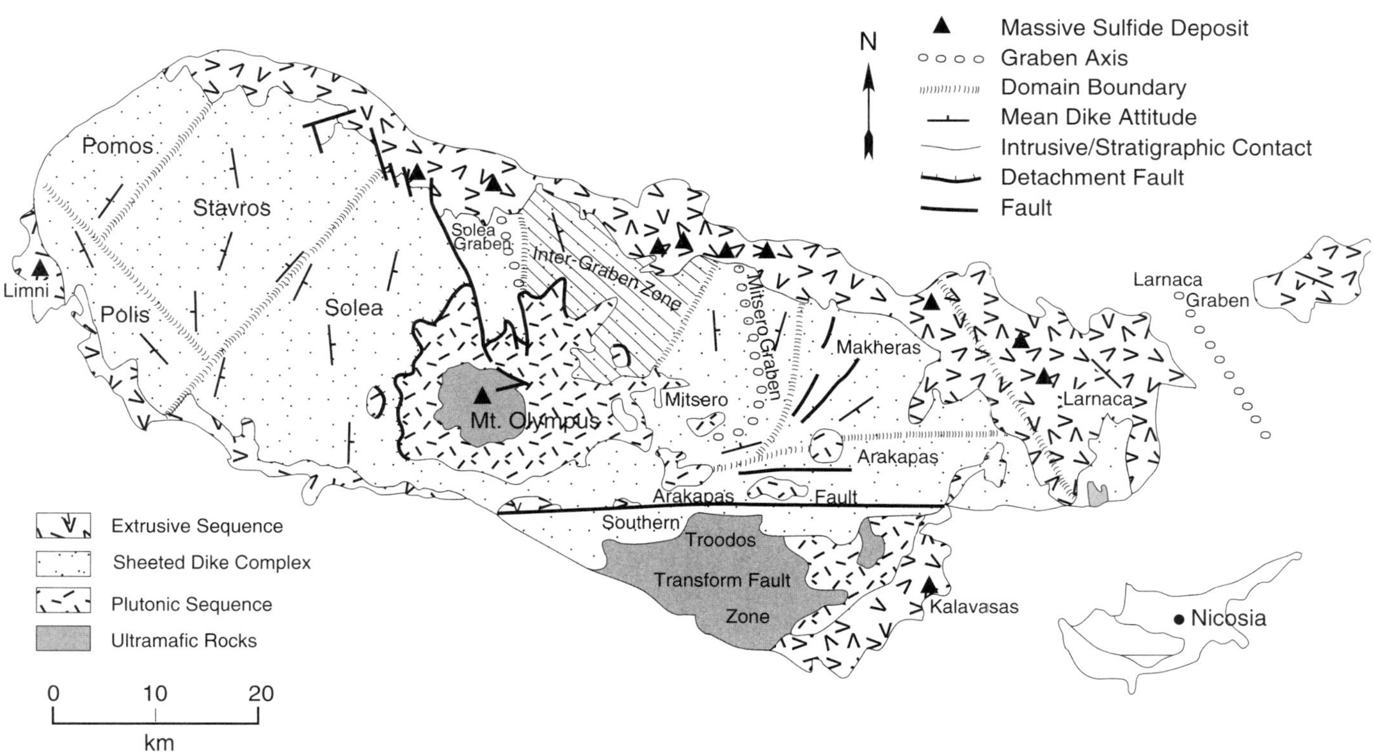

Figure 3. Simplified geological and structural map of the Troodos massif from Eddy et al. (1998, Fig. 1). Shown are the main lithological and tectonic features and orientations of structural domains within the sheeted dike complex. Locations of the structural grabens and graben axes are after Moores et al. (1990). Reproduced by permission of Elsevier.

suggestion that the volcanic outpouring occurred in a subaqueous oceanic setting (Gass and Masson-Smith, 1963) between the continents of Africa and Eurasia (Fig. 4A). The plutonic complex was considered to have formed by internal differentiation in a large, laccolith-shaped outpour (Fig. 4B) of submarine lava in a subsiding marginal sedimentary basin. The model suggested that this volcanic pile and a slice of the upper mantle were uplifted and thrust over the Eurasian continent during the Alpine orogeny.

Gass and Masson-Smith (1963) deviated from the pure eugeosynclinal view by suggesting that the Troodos complex represented a partly or completely fused upper mantle fragment, and that this resultant melt pocket differentiated into a stratiform complex composed of dunites, peridotites, gabbros, and granophyres. This view might have been influenced by a contemporaneous paper of Hess (1962) that dealt with the relationship between mantle convection cells and the mid-ocean ridge systems. It might also have been influenced by Heezen's (1962) paper on the evolution of the deep-sea floor. Both authors had assumed an oceanic layer composed of serpentinites. Gass (1967) elaborated on the idea of peridotite melting, suggesting that the Troodos plutonic rocks belonged to a "differentiated ultrabasic mass of batholithic dimensions" (p. 126). Gass (1967), thus, subscribed to the widely held view that ultramafic melts were primary melts (Hess, 1938). Gass went further by pointing out that the "structure, particularly of the sheeted intrusive complex, is of the type that could be formed on the crest of a mid-ocean ridge"

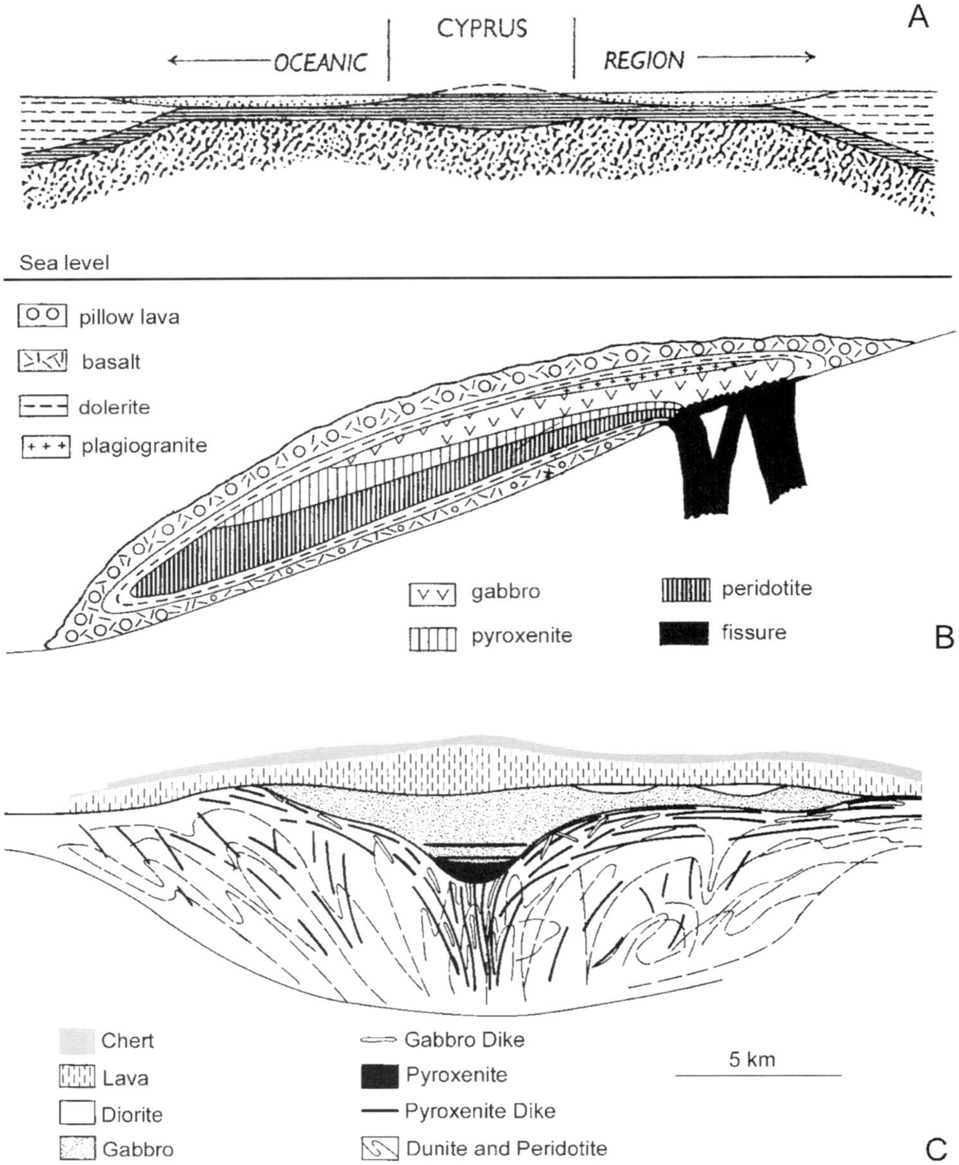

Figure 4. Early Troodos ophiolite models. A: Formation of the Troodos ophiolite as a volcanic pile on ocean floor (Gass and Masson-Smith, 1963, Fig. 14). Modified by permission of Royal Society of London. B: Troodos ophiolite formed from a massive volcanic outpouring in a geosynclinal basin (Aubouin, 1965, Fig. 37). C: Reconstructed cross-section of the Vourinos ophiolite in Greece (Moores, 1969, Fig 13).

(p. 133). Finally, he noted that "it is possible that the Troodos massif represents a volcanic edifice formed on the median ridge of Tethys" (p. 133). Thus, Gass (1967) was the first to see the similarities between the global mid-ocean ridge system and the Troodos massif, despite expressing at the same time his reservations for an ophiolite-ocean crust analogy because of the lack of direct evidence. Only a year later, Gass (1968) again expressed his reservation in a paper titled: "Is the Troodos Massif of Cyprus a Fragment of Mesozoic Ocean Floor?" Despite this skepticism, he proceeded with a comparison of the Troodos ophiolite with new oceanic crust generated at mid-ocean ridges. Gass (1968) proposed that the Troodos was indeed an uplifted fragment of fossil oceanic crust, and that its plutonic complex might represent oceanic layer 3 formed beneath a former spreading axis.

This idea was subsequently more fully developed by Moores and Vine (1971) and Vine and Moores (1972) in now-classic Troodos papers. Moores and Vine (1971) concurred with Gass (1968) in that the Troodos "represents a slice of oceanic crust and uppermost mantle" (p. 446). They inferred that the harzburgites and dunites represent depleted mantle, whereas the pyroxenites and gabbros are intrusive bodies that locally contained cumulates. This was the first time that fossil magma chambers were recognized within the plutonic sequence of the Troodos massif. Moores and Vine (1971) suggested that the lower part of the sheeted dike complex, isotropic gabbros, and the gabbroic and pyroxenitic cumulates were equivalent to ocean layer 3 (Fig. 5). They further advocated the high-level emplacement of multiple magma chambers in a slow-spreading oceanic setting characterized by varied rates of magma supply and extension (Moores and Vine, 1971, p. 461). Kidd (1977) modeled the zonation, subsidence, and rotation of the sheeted dike complex and extrusive lavas, and noted a good correspondence between the field observations and the theoretical predictions. The magma chamber modeled by Kidd (1977) was narrow, with steeply dipping walls and a general shape of a dike, broadening downward toward the base of the crust.

The significance of the Troodos plutons subsequently came into focus again with the pioneering work of Greenbaum (1972). Greenbaum (1972) accepted the proposals of Gass (1968) and Moores and Vine (1971) by suggesting that the Troodos was a slice of fossil ocean floor. Both the gabbros and pyroxenites were interpreted to represent cumulates similar to rocks found in layered intrusions (Hess, 1960; Wager et al., 1960) and also dredged from the ocean floor (Muir and Tilley, 1964; Engel and Fisher, 1969). The magma chamber shape proposed by Greenbaum (1972) was that of a large, infinite, steady-state magma chamber continuously replenished from below by undifferentiated magma (Fig. 6A). The chamber was envisioned to have a trough-like shape, 20-km wide and 1-km deep, and to be continuous along the entire length of the spreading axis. The internal constitution of this chamber was seen to contain a series of concentric shells of cumulates as referred to in "infinite onion" or "infinite leek" models (Cann, 1991). Greenbaum (1972) argued that new magma was injected at the bottom center of the chamber and that crystallization occurred along the margins in response to a lateral temperature gradient. Crystals accumulating on the outward moving chamber floor were in time being buried beneath lower temperature minerals. The Troodos observations argued for similar, laterally continuous chambers beneath active spreading centers of the ocean floor. The magma chamber model of Greenbaum (1972) was widely advocated for the Troodos, as well as for the Oman ophiolites, in the years to follow (Fig. 6B; Gass, 1980; Pallister and Hopson, 1981).

Several problems immediately became apparent with the large, steady-state chamber model of Greenbaum (1972). The principal concern was that the large single-chamber model proposed by Greenbaum (1972) required a non-convective magma, although thermal considerations predicted convection in such chambers (Bartlett, 1969; Irvine, 1970). Sleep (1975) brought up another problematic issue by pointing out that the roof over large magma chambers of the dimensions envisioned by Greenbaum (1972) would be mechanically unstable and would thus be prone to collapse. For this reason, strengthening of the roof structure by underplating and chilling of gabbro was considered necessary to prevent collapse (Dewey and Kidd, 1977). The diverse composition of the extruded lavas also appeared more consistent with multiple-chamber models, instead of large, long-lived, and zoned single-chamber models (Allen, 1975; Smewing et al., 1975; Greenbaum, 1977); however, it was still widely believed that the thick plutonic rock sequence of the Troodos ophiolite and other ophiolites was the product of cumulus processes that occurred on the floor and along the walls in a melt column that moved away from the hot axial rift zone into progressively cooler flanks (Gass and Smewing, 1981).

In the mid 1970s, Cameron Allen at Cambridge University (U.K.) proposed an episodic chamber model for the Troodos crust in an unpublished, but widely circulated Ph.D. thesis (Allen, 1975). He observed repeated cyclic variations in the cumulate sequence and inferred this phenomenon to reflect multiple episodes of replenishments of the chamber. He further suggested the existence of axial plutons, 4-km long, 2-km wide, and with an estimated depth of 4 km (Allen, 1975, p. 109). The plutons contained layering that could be traced for distances of >1.5 km. Allen (1975, p. 110) suggested that at any time, a number of discrete, possibly interconnected magma cells of variable stages of differentiation existed along the spreading axis (Fig. 7A). The emplacement and formation of new axial cells were episodic and led to splitting of the preexisting and perhaps partially solidified cells and to their migration away from the active rift axis (Fig. 7B). Allen's (1975) magma chamber model is significant because it assumes episodic and discontinuous chambers along the spreading axis. The model further allows chamber solidification by convection and formation of bottom and marginal cumulates. A somewhat different axial structure was proposed by Smewing et al. (1975), principally based on the compositional variability in the extruded lavas of the Troodos ophiolite. Smewing et al. (1975) inferred magma eruptions from discrete chambers and mantle diapirs over a broad zone, laterally extending ~30 km on both sides of the spreading axis

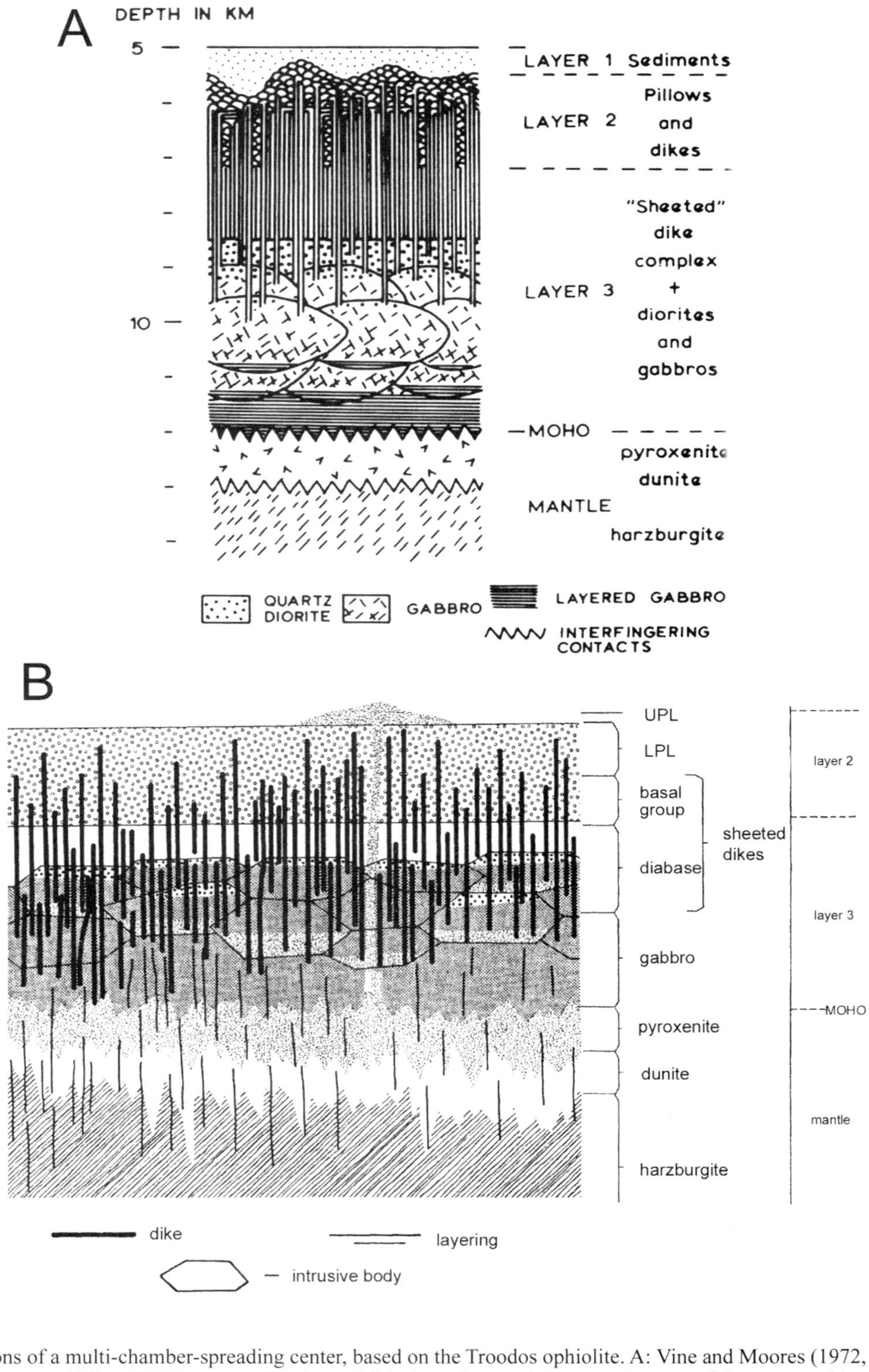

Figure 5. Cross-sections of a multi-chamber-spreading center, based on the Troodos ophiolite. A: Vine and Moores (1972, Fig. 1). Similar to the section drawn by Moores and Vine (1971), except that the multiple intrusions and the nested nature of the upper gabbro cells are marked and the lower gabbros are shown as layered. B: Moores and Vine (1971, Fig. 20). The upper gabbros are drawn as individual magma cells and the lower gabbros as stratiform gabbros. The dotted horizons on top of the magma cells are diorite/granophyre segregations; the fine dotted base of some cells is composed of ultramafic cumulates. Heavy vertical lines represent dikes, and thin vertical lines depict banding of lower gabbros and mantle tectonites. LPL—lower pillow lavas; UPL—upper pillow lavas. Reproduced by permission of Royal Society of London.

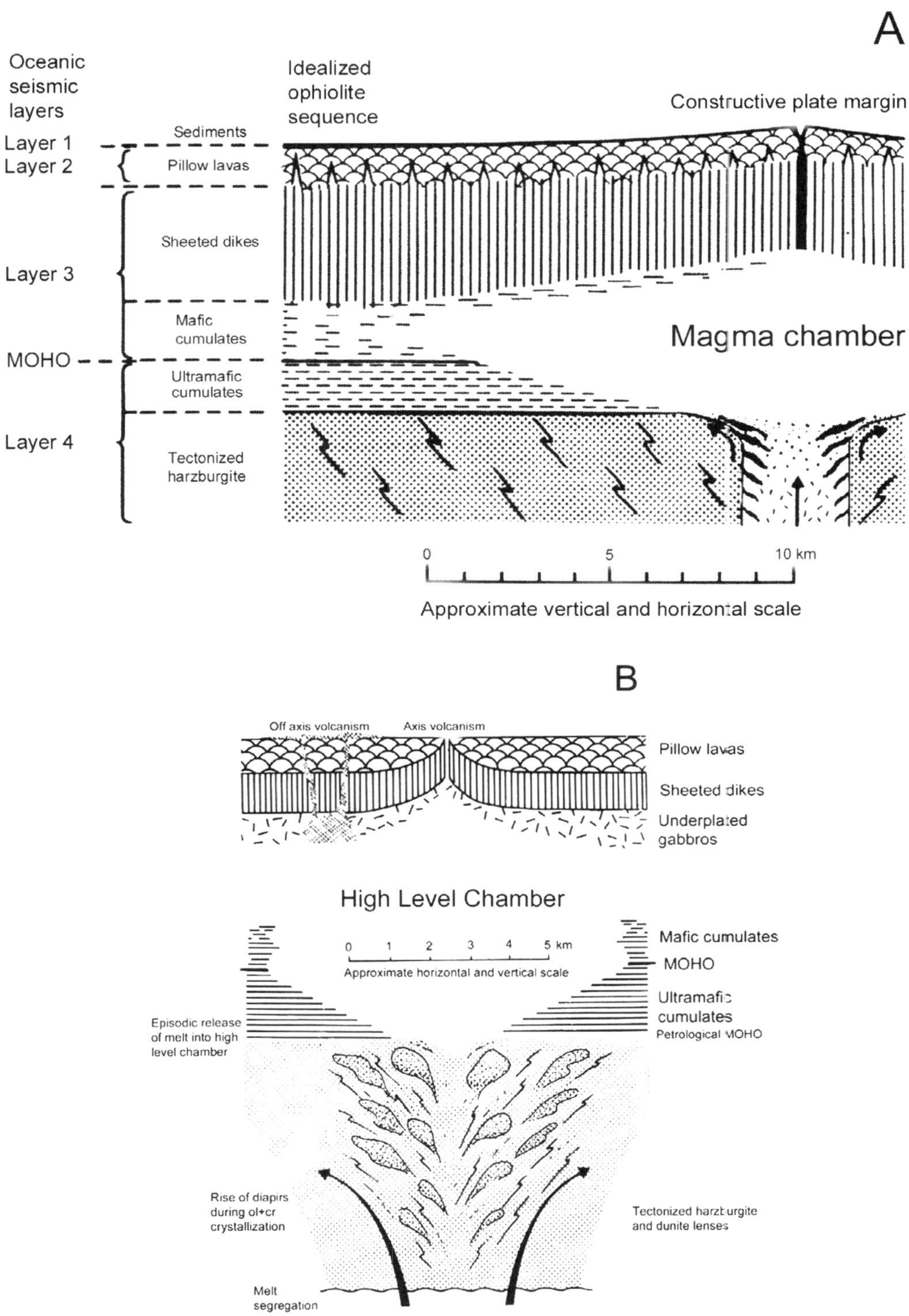

Figure 6. Cross-sections of a single-chamber-spreading center, based on the Troodos ophiolite. Reproduced by permission of the Geological Survey Department of Cyprus. A: After Gass (1980, Fig. 3), drawn from an illustration in Greenbaum (1972, Fig. 3). B: Chamber cross-section, based on a composite of ophiolites (simplified from Gass, 1980, Fig. 4).

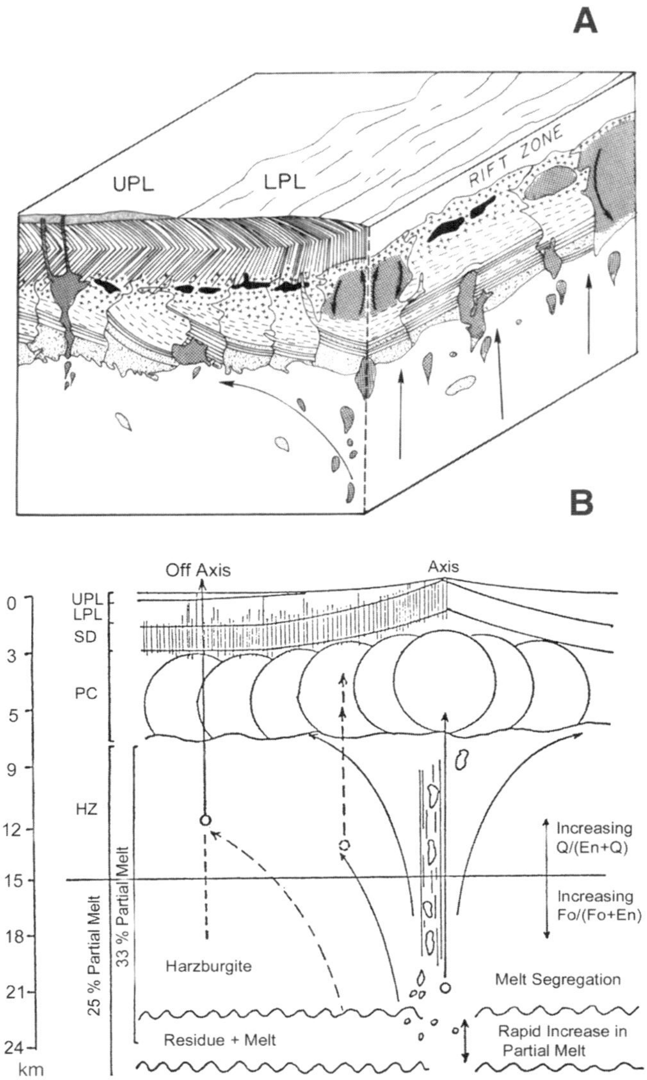

Figure 7. Cross-sections of a multi-chamber-spreading center, based on the Troodos ophiolite. A: Allen (1975, Fig. 49). Individual magma cell boundaries are illustrated to emphasize the periodic, along-axis construction of the crust. Also shown are off-axis intrusions and volcanics. B: Schematic model of spreading center. Simplified and redrawn from Allen (1975, Fig. 51).

(Fig. 8). The axial rift zone model of Smewing et al. (1975) is in essence similar to the subsequent models by Eldridge Moores and coworkers (Varga and Moores, 1985; Moores et al., 1990), suggesting several successive rift jump events during the formation of the Troodos ophiolite. This structure added an additional "degree of freedom" to the axial structure of magma chambers depicted by Allen (1975).

The plutonic section of the Troodos ophiolite was sampled in 1986 by drill hole CY4 of the Cyprus Crustal Study Project. The hole penetrated the transition between the sheeted dike complex and the gabbros at 630 m below the surface, continued through gabbroic cumulates to a depth of 1750 m, and was terminated in ultramafic cumulates at a depth of 2260 m, without reaching tectonized peridotites (Fig. 9A; Thy et al., 1989; Malpas et al., 1989). This 1630-m continuous core of the gabbroic and ultramafic cumulates provides a direct look into the nature of the fossil upper Troodos magma chambers and their crystallization processes. The lithologic units defined in the CY4 drill core are mainly based on the observed petrography and the compositional variation of pyroxene (Fig. 9B). The upper gabbros are isotropic and are composed of plagioclase, augite, and orthopyroxene cumulus phases. Fe-Ti oxides are not early crystallization phases, except in a short interval between the 1196 and 1213 m (Thy et al., 1989). The contact at 1300 m below the surface (between Units 3 and 4, lower and upper gabbros, respectively) was described as intrusive with the lower gabbro showing grain size fining upward against the upper gabbro (Thy et al., 1989). The lower gabbros, down to the 1750-m depth, commonly show grain-size and modal layering of gabbronorite, gabbro, olivine gabbro, and websterite on a scale ranging from 1 to 10 m. Olivine only locally joins the cumulus assemblage. The lowermost part of the core is composed dominantly of websterite with minor amounts of clinopyroxenite, wehrlite, lherzolite, and gabbronorite. Olivine is consistently a minor cumulus phase in the websterite. The lower cumulates display well developed modal and grain-size layering on scales ranging from a few centimeters to several meters. Medium-grained to pegmatitic gabbroic rocks are locally present as layers and patches in the websteritic cumulates (Thy et al., 1989; Browning et al., 1989).

The cryptic variation of the cumulus minerals is illustrated in Figure 9B. The lower gabbroic to websteritic cumulates (Units 4 and 5) show a systematic upward increase in the iron content for all three mafic phases. In the upper part of the lower gabbros (Unit 4), a reversal to more primitive compositions (and olivine saturation) is followed by a decrease in the iron content of pyroxene. Plagioclase shows a weak compositional variation in the albite content that mimics the variation of the mafic phases in the gabbroic cumulates. The unsystematic variation of plagioclase composition in the websteritic cumulates is related to the intercumulus origin of plagioclase in these rocks. These first order observations from the lower cumulate series led Thy et al. (1989) to suggest that the variations were in general consistent with fractional crystallization in a mostly closed magma chamber.

The intrusive contact at the 1300-m depth is a major compositional boundary in the core. The gabbros below this boundary are more primitive in terms of their major and minor element ratios than the gabbros above. The pyroxenes, for example, have markedly lower Al and Ti contents in the lower gabbros compared to the upper gabbros (Thy et al., 1989; Malpas et al., 1989). The upper gabbros show systematic variations that allow three units to be defined (Units 1–3), each having a systematic upward regression in magnesium contents of the pyroxenes, followed by a trend toward decreasing iron contents. Fe-Ti-oxide minerals occur

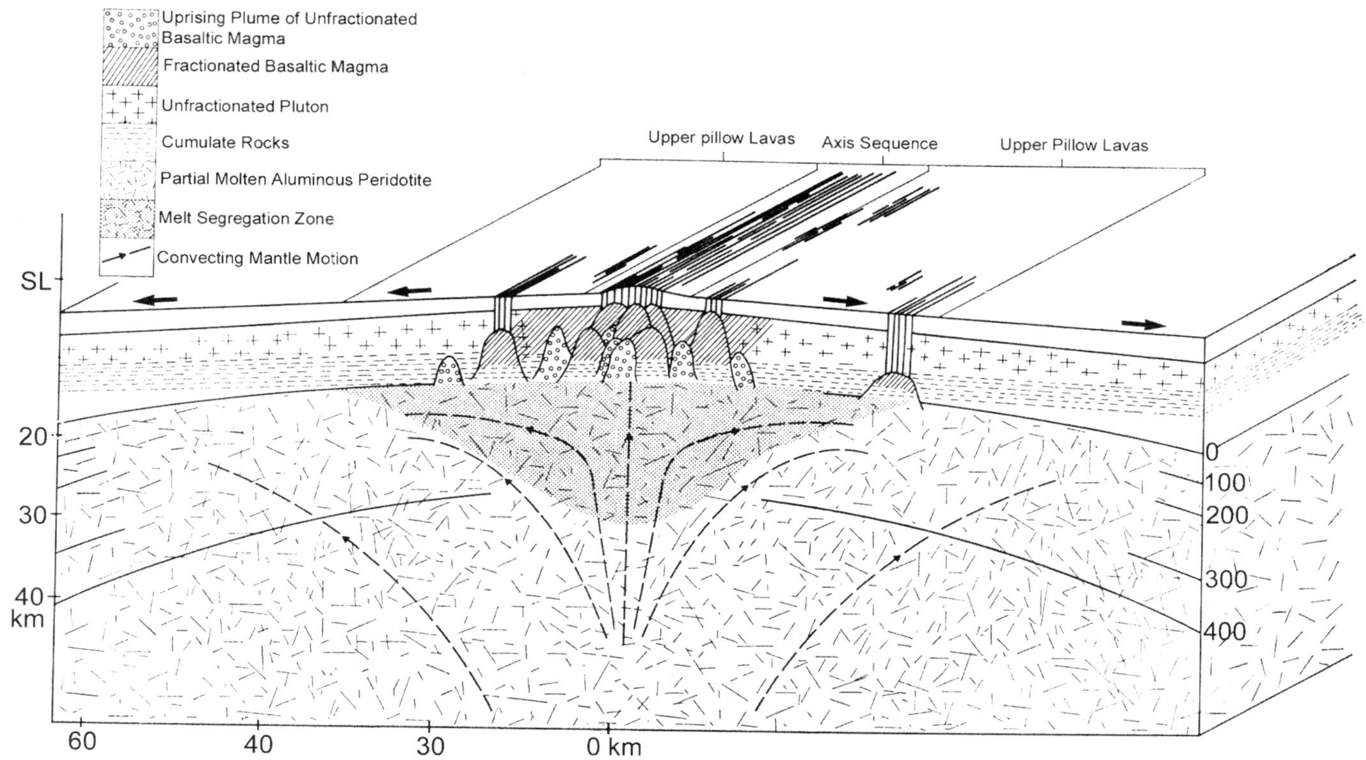

Figure 8. Spreading processes forming the Troodos ophiolite (after Smewing et al., 1975, Fig. 4). Shown are mantle flow lines and zone of partial melting (shaded). The vertical scale on the right-hand side of the diagram shows the excess temperatures relative to the fusion of olivine tholeiite. Reproduced by permission of Springer-Verlag.

intermittently in the most evolved gabbros at the top of Unit 3, just prior to initiation of the regression, marking the beginning of Unit 2. These units were interpreted (Thy et al., 1989) to reflect an intermittently open magma chamber system with three cycles, each showing gradational filling and mixing between resident magma and a new influx of a relatively more primitive magma before closed system fractionation was resumed.

Field studies have elaborated on the spatial and temporal relationships between intrusive episodes and the high temperature deformation in the Troodos plutonic complex (Malpas et al., 1989; Malpas, 1990; Dilek and Eddy, 1992). Malpas et al. (1989) mapped structural details in the Mount Olympus area of the Troodos complex and observed two major plutonic suites on the basis of relative ages of deformation and magmatism. An early suite consists mainly of harzburgite, dunite, layered olivine pyroxenite, and gabbro and displays a penetrative high-temperature deformation fabric. This tectonic fabric is generally parallel to the compositional layering in the plutonic suite. Malpas (1990) suggested that these structures of the early suite reflect the intrusion of plutons in an upwelling mantle column at, or near, a spreading center, and that they might have resulted from upward and lateral solid flow. A late suite of plutons includes gabbroic and websteritic cumulates and displays locally well preserved primary cumulus textures containing xenolithic inclusions of the early suite (Benn and Laurent, 1987; Malpas, 1990). These field observations thus illustrate the complex relations between the gabbroic and ultramafic plutons of the Troodos ophiolite and strongly support the multiple magma chamber evolution model (e.g., Allen, 1975) at different crustal levels associated with intermittent episodes of magmatic and tectonic spreading (Malpas, 1990). The upper parts of the CY4 drill core sampled the early plutonic suite, which records a relatively open magmatic system (Thy et al., 1989; Malpas, 1990). The lower part of the drill core sampled a late-stage pluton, which may represent one of several episodes of plutonism that constituted the late suite (Malpas, 1990). Geochemical studies of cumulus clinopyroxene and melt inclusions in plagioclase from the southern flank of the Troodos massif have suggested that the latest plutons originated from the most depleted melts (Batanova et al., 1995). This multiple and intermittently replenished type of magma chambers, seen in the Troodos ophiolite, are not found in the plutonic complexes of the Bay of Islands and the Oman ophiolites.

Bay of Islands Ophiolite (Newfoundland)

The Bay of Islands ophiolite complex is a 100-km long discontinuous belt of mafic and ultramafic rocks emplaced onto

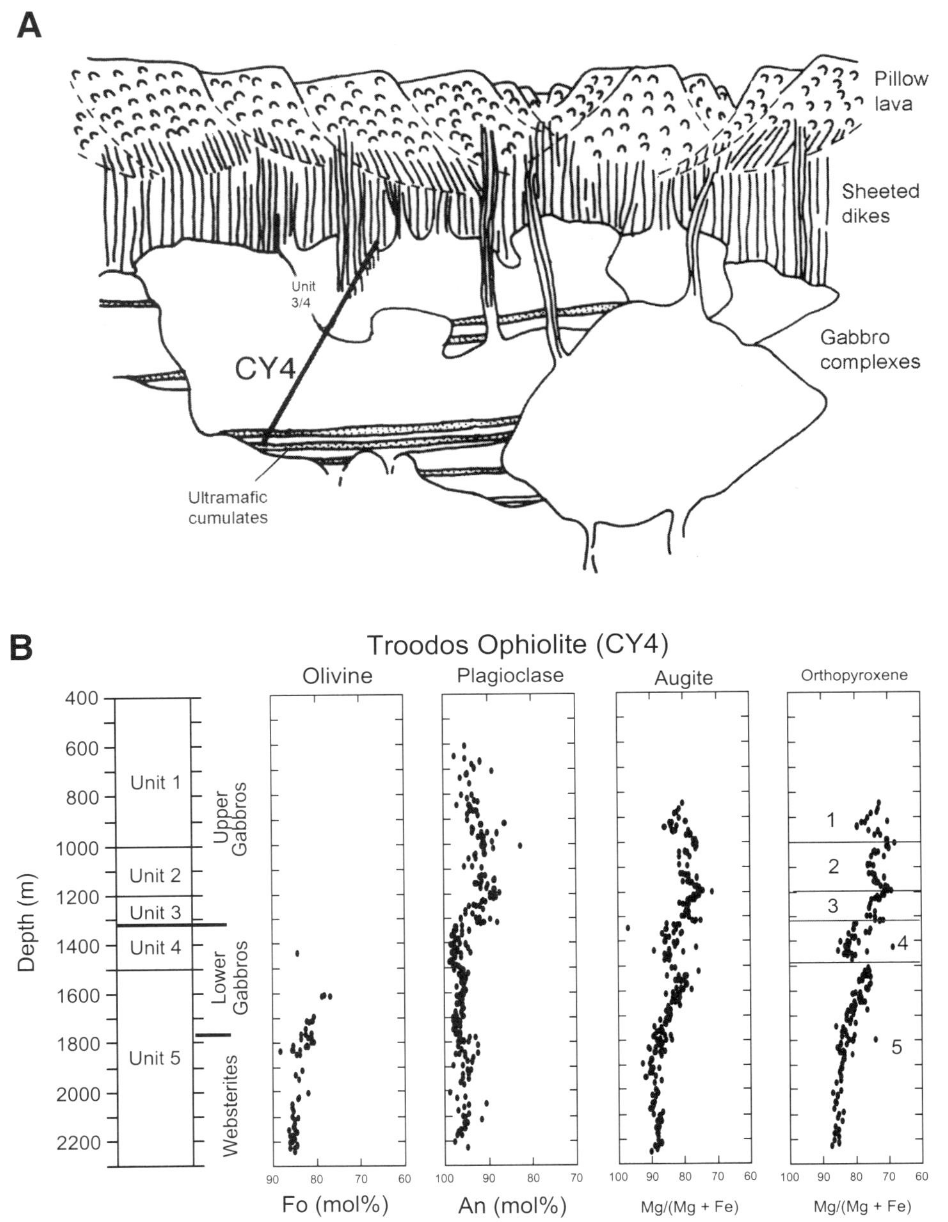

Figure 9. Cross-section and compositional variation in the multiple intrusive gabbros and ultramafic cumulates of drill core CY4 through the Troodos ophiolite. A: Schematic cross-section illustrates the inferred location and position of drill Hole CY4 (Malpas, 1990, Fig. 12). Reproduced by permission of the Geological Survey Department of Cyprus. B: Cryptic variation in the plutonic complex penetrated by CY4. The units are based on the systematic variations and discontinuities in the cryptic variation. The contact between Units 3 and 4 is interpreted to be intrusive. See Thy et al. (1989) and Thy and Dilek (2000) for details.

the Paleozoic continental margin of southwestern Newfoundland during the early stages of the tectonic evolution of the Appalachian orogen (Fig. 10). Structural and petrological studies (Cooper, 1937; Smith, 1958; Church and Stevens, 1971; Dewey and Bird, 1971; Church and Riccio, 1977) revealed the existence of a typical ophiolite succession composed of tectonized peridotite, ultramafic and mafic plutons, sheeted dikes, extrusive rocks, and overlying sedimentary rocks. The early studies of the Bay of Islands complex interpreted this ophiolite as part of the western Newfoundland eugeosynclinal belt (Fig. 10) and suggested that the internal ophiolite differentiation mostly resulted from accumulation in a syn-orogenic pluton (Smith, 1958). Our present understanding of this ophiolite is based mainly on the results of the systematic mapping programs in the late 1970s (Dewey and Kidd, 1977; Casey and Karson, 1981; Casey et al., 1981), well after the formulation of the general plate tectonic models (e.g., Casey and Karson, 1981; Casey et al., 1981).

In 1977, Dewey and Kidd proposed a general steady-state magma chamber model for accreting plate margins (Fig. 11) and based this model mainly on field evidence from the Bay of Islands ophiolite. Their model depicted five components that included, from top to base: (1) a lid composed of extruded lavas and an

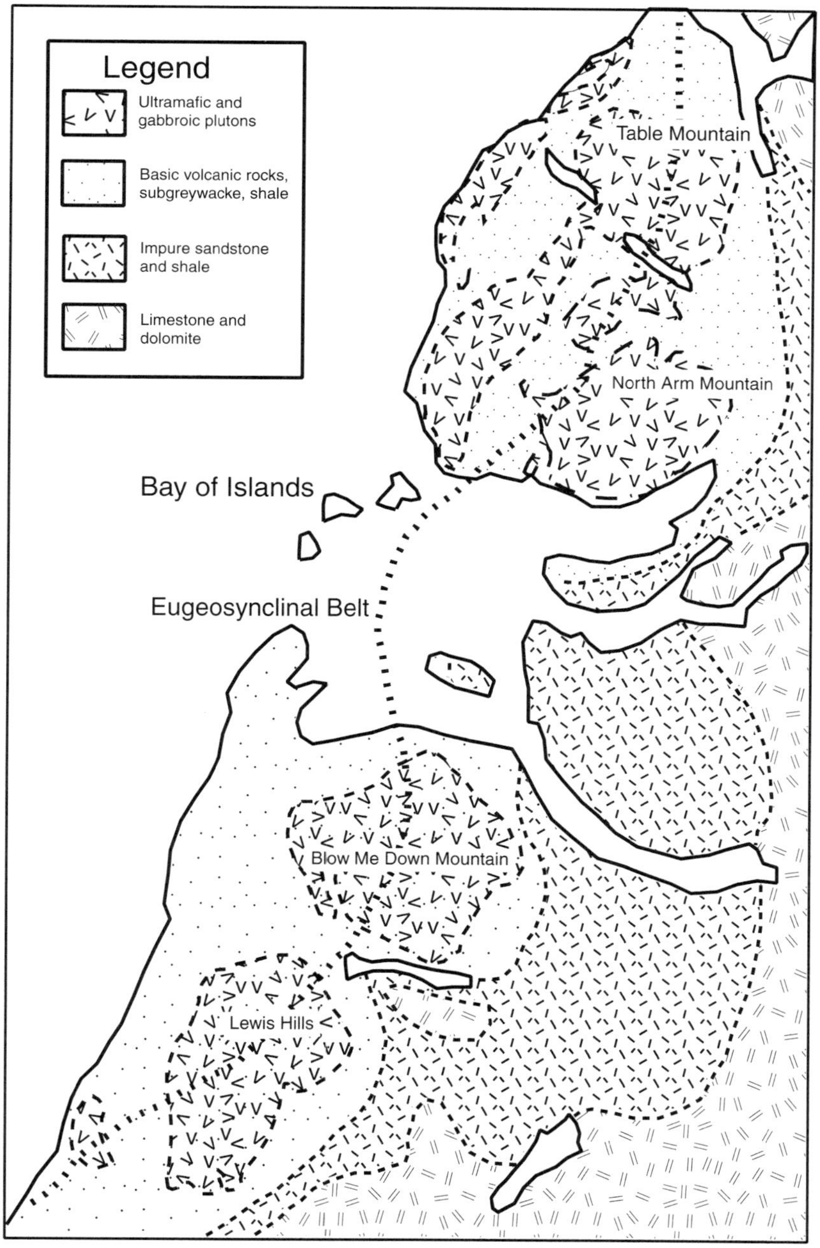

Figure 10. Simplified geological map of the Bay of Islands ophiolite complex, based on Smith (1958). Dotted line marked as the "eugeosynclinal belt" is the inferred axis of basaltic magma outpour into the Appalachian eugeosynclinal marginal belt of western Newfoundland.

underlying sheeted dike complex; (2) isotropic gabbros accreted to the lid and progressively thickened away from the axial rift; (3) a tent-shaped magma chamber with a flat floor; (4) differentially subsided cumulates beneath the chamber floor; and (5) a narrow axial dome of upwelling lherzolite, from which magma was segregated into the magma chamber within the crust, and from which a harzburgitic mantle was accreted beneath the lower crust. Girardeau and Nicolas (1981) expressed very similar views on the Bay of Islands ophiolite. In this model the depth of the magma chamber controls the thickness of the underplated upper gabbros; the width of the chamber (<10 km) is determined by the cooling and spreading rates that similarly control the thickness of the cumulates and the orientation of banding (Dewey and Kidd, 1977). Hydrothermal circulation is responsible for the thickness of the sheeted dike complex and for the accretion of the gabbros along the roof of the axial chamber. These gabbros contribute to the stabilization of the axial lid over the chamber (Rosencrantz and Nelson, 1982; Rosencrantz, 1983). The model of Dewey and Kidd (1977) assumes a near crystal-free magma chamber and infers that crystal mushes are confined to the underplated gabbros and to the subsiding mafic and ultramafic cumulates. Perhaps one of the most interesting aspects of Dewey and Kidd's axial cham-

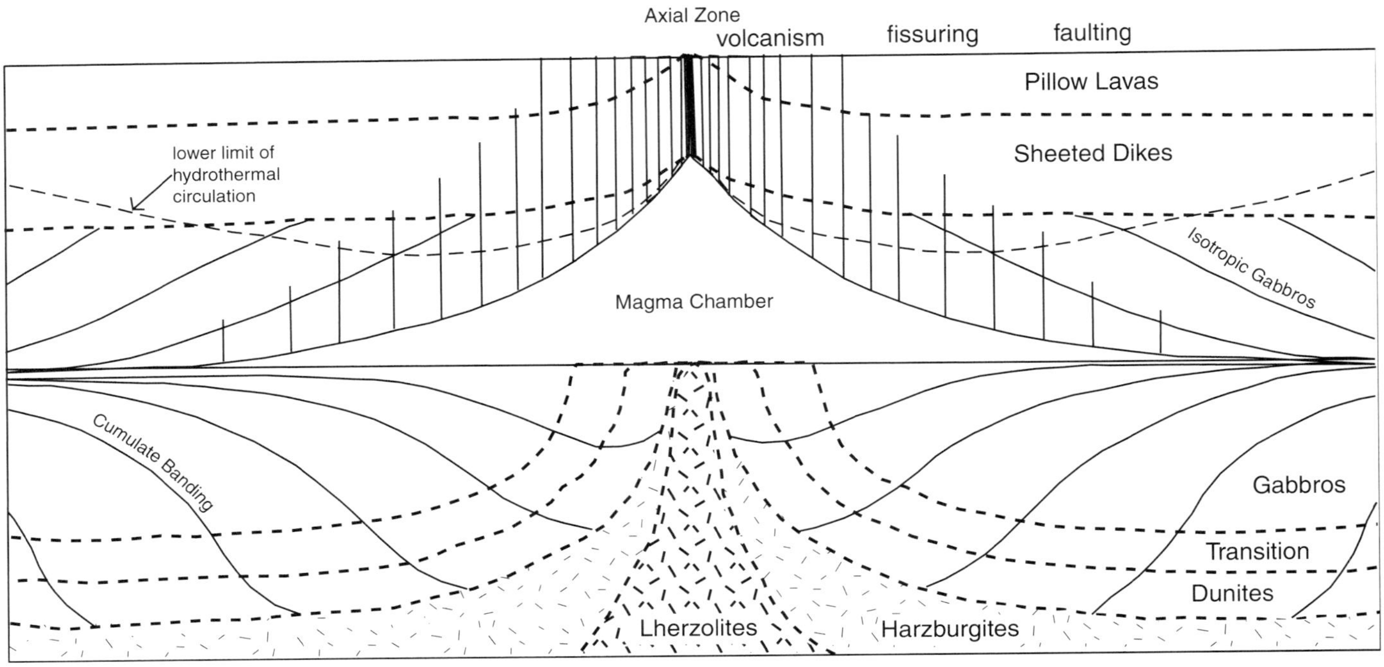

Figure 11. Constructive plate accretion model for axial spreading center, formulated from observations in the Bay of Islands ophiolite. Simplified from Dewey and Kidd (1977, Fig. 1).

ber model is that it differentiates between the upper, underplated gabbros and the lower, gabbroic and ultramafic cumulates.

Casey and Karson (1981) and Casey et al. (1981) subsequently reported the results of their field mapping in the Bay of Islands ophiolite. The axial magma chamber they envisioned differed significantly from that of Dewey and Kidd (1977). They proposed a large, steady-state chamber extending along the ridge axis and only terminating against the transform fault intersections. They inferred a chamber depth of 6–10 km and a width of 14–16 km. Banding in the gabbros was inferred to be parallel to the inclined margins of the chamber and was attributed to primary igneous layering that outlined the original shape of the axial chamber (Fig. 12). Steepening of the cumulate banding away from the axial center, inferred by Dewey and Kidd (1977) to have resulted from differential subsidence, was interpreted as the primary layering formed along the chamber margins. This interpretation required that the cumulates formed principally by in situ crystallization with little interstitial trapped melt and that they were unaffected by postcumulus deformation. The magma chamber suggested by Casey and Karson was very similar to the large continuous, steady-state chamber envisioned by Greenbaum (1972) for the Troodos ophiolite; however, Casey and Karson (1981) found no indication of a flat chamber floor, and suggested instead a large chamber keel extending into the tectonized mantle sequence (Fig. 12). Elthon et al. (1982, 1984) and Casey et al. (1983) expanded this keel to a depth of 30–40 km (Fig. 13) to account for the early appearance of clinopyroxene (Church and Riccio, 1977), which they interpreted to record high-pressure fractionation. With the exception of a suggestion in a short note by Strong and Malpas (1975), the occurrence of multiple magma chambers was not proposed for the Bay of Islands ophiolite.

The "Blow Me Down" and "North Arm" cumulate sections in the Bay of Islands ophiolite (Fig. 10) are composed of interlayered dunite, wehrlite, harzburgite, pyroxenite, and gabbro. A thick gabbro sequence overlies the ultramafics and contains layers of troctolite, anorthosite, and gabbronorite. Various domains of porphyritic, pegmatitic and poorly layered gabbros, hornblendite, and evolved rocks also exist. These plutonic rocks collectively occur beneath the sheeted dike complex and pillow lavas (Church and Riccio, 1977; Casey and Karson, 1981; Bédard 1993). The cumulate section displays cumulus phase assemblages and cumulus compositions that apparently evolved upward due to fractionation processes (Fig. 14). The compositions recorded by the cumulates correspond to the total variation in the sheeted dike complex and the extrusive rocks (Church and Riccio, 1977). This observation was used to suggest an open nature of the Bay of Islands magma chamber. Despite the systematic upward cumulus evolution toward more evolved compositions, the ultramafic and gabbroic cumulates show reversals to more primitive compositions on various scales (Church and Riccio, 1977; Elthon et al., 1982, 1984; Komor and Elthon, 1990; Komor et al., 1985, 1987). These small-scale variations may encompass the near total compositional variation seen in the complete stratigraphic section (Fig. 14). Elthon et al. (1982) attributed this phenomenon to in situ crystallization processes along vertical chamber margins or conduits (Fig. 13).

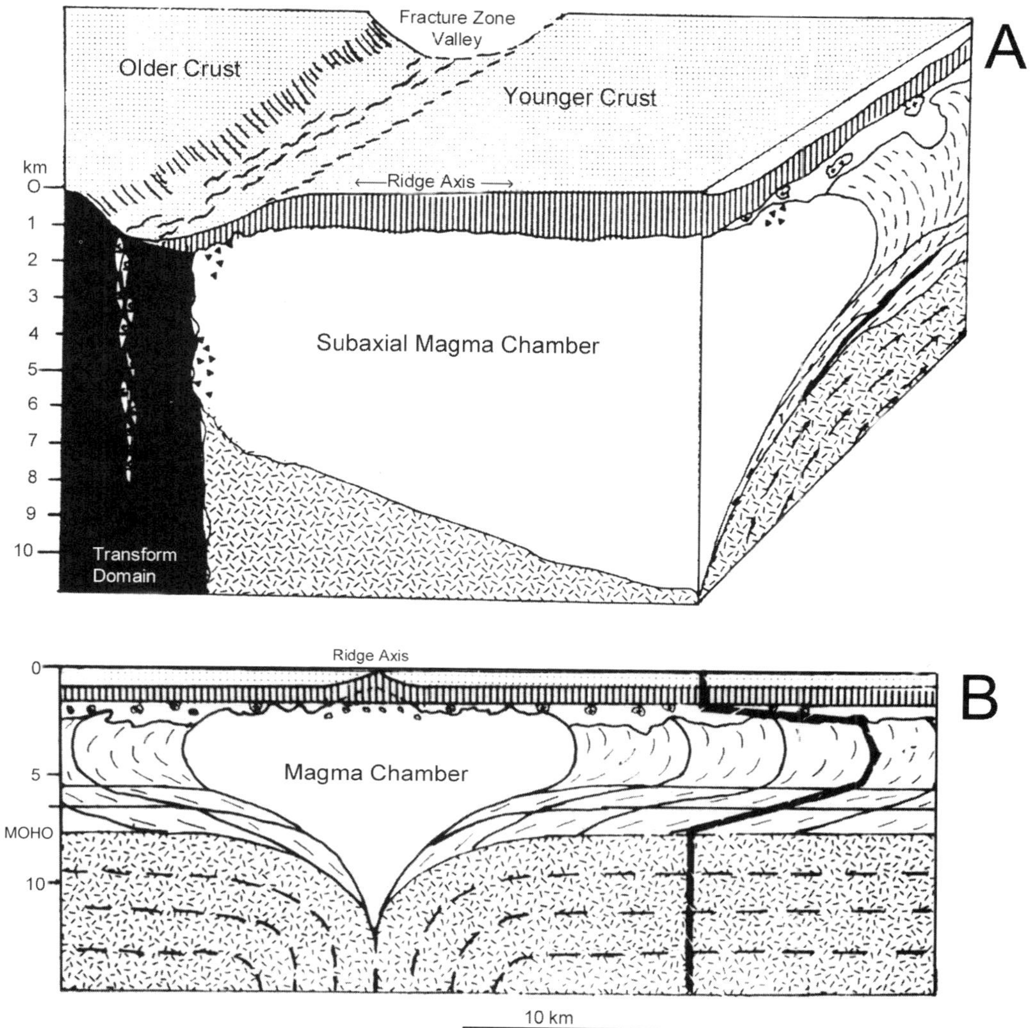

Figure 12. Magma chamber model based on the Bay of Islands ophiolite (from Casey and Karson, 1981, Figs. 4 and 5). Reproduced by permission of *Nature*. A: Block diagram illustrating the axial chamber and its intersection with a transform zone (Casey and Karson, 1981, Fig. 4). B: Cross-section through an idealized axial magma chamber for the Bay of Islands ophiolite. The thick line is an isochron at 1.25 m.y. for a 10 cm/yr half-spreading rate. Note the deep keel of the chamber.

A detailed field and petrographic study of the ultramafic and mafic cumulates of the North Arm massif by Jean Bédard and coworkers did not find strong evidence for large magma chambers, as were previously reported. Bédard (1991, 1993) and Bédard and Hébert (1996) recognized near-pervasive syn-magmatic deformation and proposed that the entire Bay of Islands crust be formed as a synkinematic sill complex. Residual magma from the sills and from the host mush/gabbro was expelled laterally or obliquely along shear zones. The ultramafic cumulate section was proposed to be an underplating boninitic sill complex genetically unrelated to the overlying gabbros (Bédard and Hébert, 1996). Upward regressions in cryptic mineral variations in the gabbroic section were interpreted not as replenishments into an open chamber, but either as tectonic inversion produced by deformation, or by injection of primitive sills into the cumulate mush (Bédard, 1993). The petrographic and compositional variations were thus attributed to assimilation and reaction processes between primitive intrusive sills and host cumulates, and not to an open, frequently replenished magma chamber, as previously argued. These observations have significant implications for how primitive magma migrates through cumulate mushes and how this migration may affect the residual solid product and the liquid line of descent of primitive melts beneath oceanic spreading centers. This theme was subsequently taken up by other researchers for the Oman ophiolite.

Oman Ophiolite (Sultanate of Oman)

The Oman ophiolite constitutes a significant component of a nearly 800-km long, arcuate mountain belt that borders the

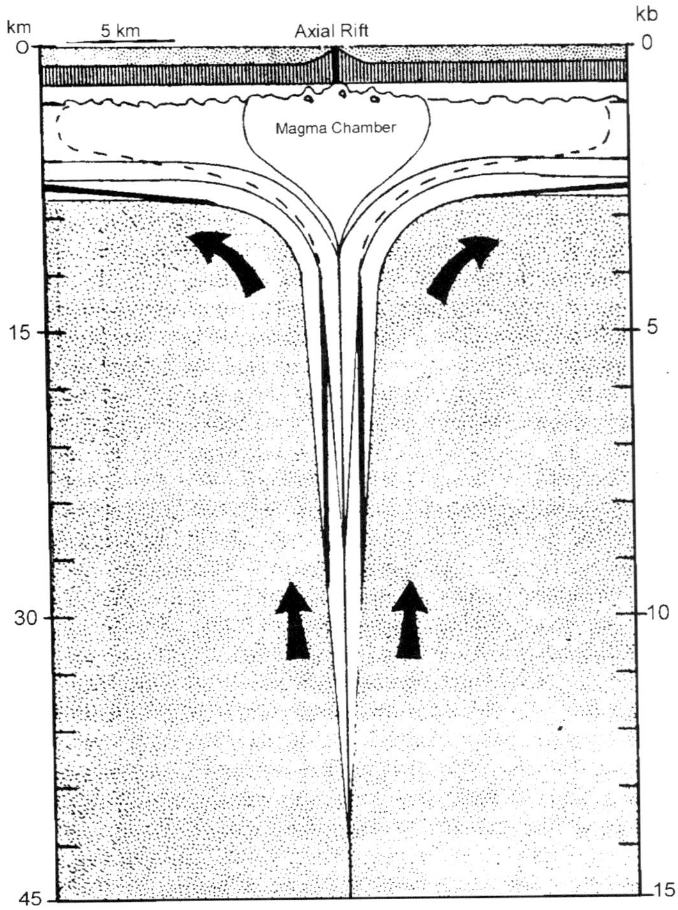

Figure 13. Bay of Islands magma chamber modified from Elthon et al. (1982, Fig. 8). Elthon et al. (1982) observed the effects of high-pressure fractionation in the North Arms massif and suggested that crystallization occurred at deep levels in the mantle. A chamber keel extending deep down into the mantle was proposed. Reproduced by permission of American Geophysical Union.

southern coast of the Gulf of Oman (Fig. 15). The ophiolite represents a fragment of the fossil Tethyan oceanic crust emplaced as a nappe system onto the passive continental margin of the Arabian Peninsula (Lees, 1928; Glennie et al., 1974; Coleman, 1981). The Oman ophiolite contains tectonized peridotites and subordinate gabbroic cumulates with lesser amounts of dunite and wehrlite (Reinhardt, 1969; Allemann and Peters, 1972; Glennie et al., 1974). Overlying the plutonic sequence are the basaltic sheeted dike complex and pillow lavas. Early studies culminated in typical geosynclinal views for the origin of the ophiolite (Wilson, 1969; Reinhardt, 1969). Wilson (1969) saw the igneous complex as being autochthonous and suggested that the peridotites and gabbros formed from a slowly cooling, large, abyssal ultrabasic magma flood erupted in a eugeosynclinal marginal basin. Other studies recognized an age hiatus between the host tectonized peridotites and basaltic intrusive and extrusive rocks. Of particular note is the model of Reinhardt (1969), illustrated in Figure 16. This model depicted the ophiolite as a large basaltic intrusion or a lava lake emplaced into the host tectonized peridotite and extruded on the sea bottom, forming sheeted dikes and extrusive spilitic pillow lavas. The internal constitution of the massif was believed to have resulted from differentiation of the magma into fine- and coarse-grained gabbros with a marginal gabbroic-peridotitic zone.

A Dutch group mapping in the Oman Mountains in the late 1960s contributed significantly to our understanding of the internal structure and stratigraphy of this ophiolite. The first comprehensive geological map as a result of this project was published by Glennie et al. (1974). This study demonstrated the similarities between the Oman ophiolite and the general structure of modern oceanic crust. The gabbroic sequence (or layer 3) was interpreted to have formed from a large dike-like intrusion that had extended beneath an oceanic ridge to a depth of 50 km above an upwelling mantle. The layers of oceanic lower crust formed on the walls of this dike and were transported upward and laterally as if on large conveyer belts driven by mantle convection (Fig. 17).

More detailed mapping in the 1970s by American and British research groups followed this early reconnaissance mapping. These international efforts resulted in detailed maps and produced significant petrological and geochemical information that permitted detailed magma chamber models to be formulated (Coleman and Hopson, 1981; Lippard et al., 1996). The most influential chamber models were published by Pallister and Hopson (1981) and Smewing (1981) as shown in Figure 18. Pallister and Hopson (1981) made several important observations that they used to infer magma chamber processes and the shape of the magma chamber of the Oman ophiolite. In this model, the typical cumulate sequence includes a basal section composed of dunite that grades downward into a transitional zone made of harzburgitic tectonites. The main part of the cumulates consists of interlayered wehrlite, melanogabbro, and gabbro rocks, followed upward by layered gabbro with recurring zones of wehrlite and melanogabbro. The uppermost plutonic rocks are made of laminated to non-laminated, cumulus gabbros and isotropic hornblende-bearing gabbros. The cumulates, composed dominantly of adcumulates (Wager et al., 1960), were accumulated on the floor of the magma chamber (Pallister and Hopson, 1981). Downward crystallization from the roof of the chamber was minor, as indicated by a rather thin unit of isotropic gabbros below the chamber roof and the overlying sheeted dike complex. Magmatic cyclicity in the chamber was seen as petrographic repetitions in the occurrence of olivine-rich intervals, commonly related to reversals in the cryptic variation. Despite such reversals, the gabbro sequence shows very limited cryptic variation (Fig. 19). This was taken to indicate frequent replenishments restricting differentiation of magma in the chamber. Pallister and Hopson (1981), thus, inferred that the Oman gabbros were the product of crystallization in a large, long-lived, frequently replenished magma chamber (Fig. 18A), as opposed to the small transient chambers inferred for the Troodos ophiolite (Smewing et al., 1975). Smewing (1981) suggested

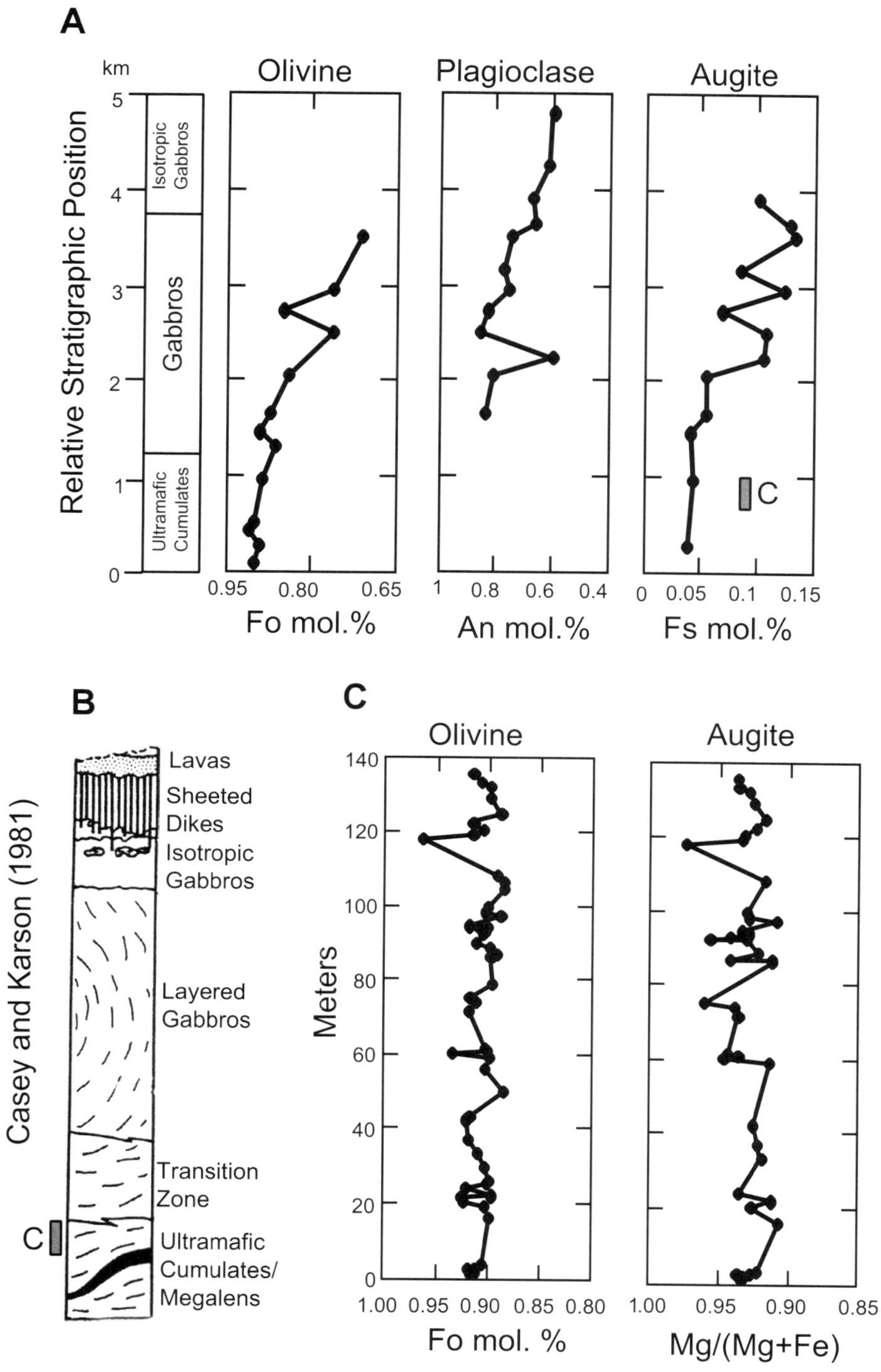

Figure 14. Cryptic variation of the main minerals of the Bay of Islands ultramafic and gabbroic cumulates. A: Composite of the "Blow Me Down" and "North Arm" sections, from Church and Riccio (1977). "Fs" is ferrosilite component of augite (Fe/(Fe+Mg+Ca)). B: Generalized stratigraphic section through the Bay of Islands ophiolite from Casey and Karson (1981, Fig. 2). The vertical bar marked as "C" shows the position of the section illustrated in greater details. C: Small-scale cryptic variation in a 140-meter layered section of ultramafic cumulates (Elthon et al., 1984, Table 2).

that the cyclicity in the Oman cumulates resulted from influx of relatively undifferentiated picritic melt entering the chamber at the base and being retained along the floor because of a density contrast with the residual, relatively evolved melt in the magma chamber (Fig. 18B). In Smewing's model, olivine-rich cumulates formed along the floor and away from the axial center until the bottom layer approached the composition and density of the overlying magma. Eventually, mixing between the bottom layer and the main chamber reservoir homogenized the remaining melt in the chamber.

The chamber shape was interpreted to have been controlled mainly by upward growth from the floor and to a lesser extent by

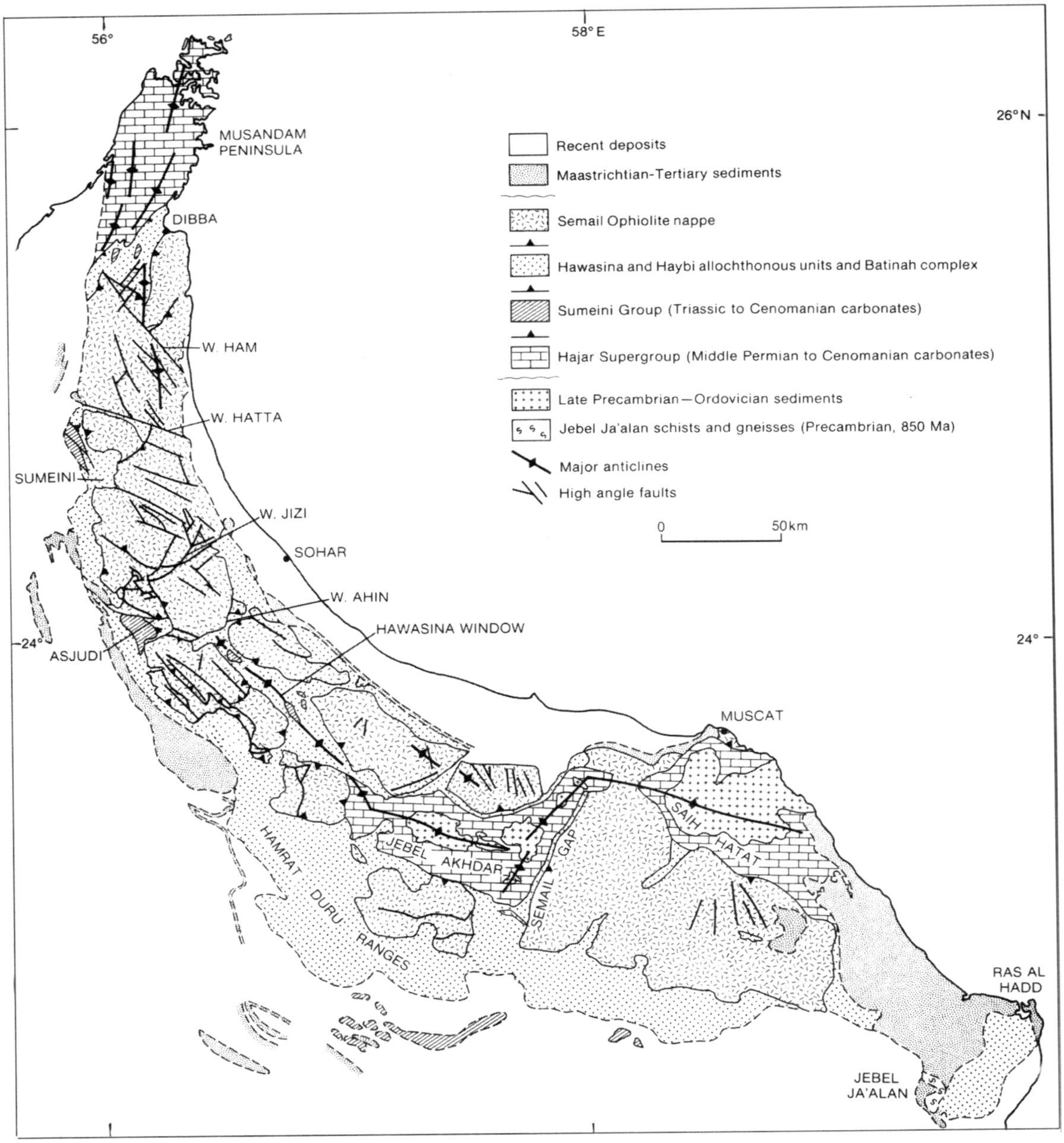

Figure 15. Geological and structural map of the Oman ophiolite (Lippard et al., 1986). Reproduced by permission of Geological Society of London.

downward growth from the ceiling (Fig. 18A). The layering in the gabbros was observed to approach banding in the peridotitic tectonites at a 20° angle. Pallister and Hopson (1981) envisioned a broad, funnel-shaped chamber with floors sloping gently inward toward the axial center and laterally terminating in a sandwiched position against and within the upper gabbros (Fig. 18A). They estimated a minimum width of 30 km for this funnel-shaped chamber by using a 15° slope of the layering in the ~5-km-thick cumulate package. These dimensions are relatively similar to those inferred by Smewing (1981; Figure 18B) and Lippard et al. (1986; Figure 20) from the northern part of the Oman ophiolite (see also Browning, 1984; Ernewein et al., 1988).

In a highly influential paper published in 1984, Browning described the occurrence of cyclic variations in the cumulus phase compositions of the cumulate sequence in the Oman ophiolite. His results were used to estimate the height of the melt column for each cyclic unit as ~100 m. Browning (1984) discussed two ways of achieving this cyclic variation (Fig. 21). He briefly

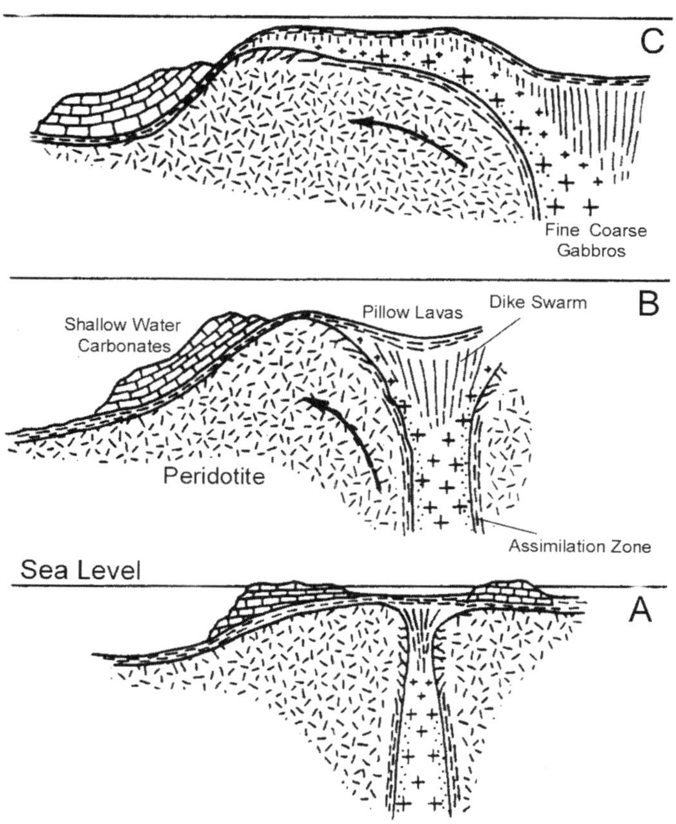

Figure 16. Eugeosynclinal emplacement and structural development (A to C) of the Oman ophiolite, as suggested by Reinhardt (1969, Fig. 17).

mentioned the possibility of a subsiding cumulate pile and stated that the true depth of the subaxial chamber was no more than the 100 m estimated for each cyclic cumulate unit (e.g., Dewey and Kidd, 1977); however, Browning (1984) indicated his preference for the idea of a zoned magma chamber and the breakdown of the melt column into double-diffusive convection cells, each with approximately a 100-m depth. This latter idea had been formulated from scaled tank experiments (e.g., Turner and Chen, 1974; Turner and Campbell, 1986) and applied to explain some features of layered intrusions (McBirney and Noyes, 1979; Irvine et al., 1983; Wilson and Larsen, 1985). The important outcome of Browning's (1984) observations is that magma cells of very limited depths can be mapped in the layered gabbroic cumulates in Oman, and that they may similarly exist in magma chambers beneath oceanic spreading centers.

Structural mapping by French groups in the Oman Mountains during the early 1980s led to a detailed understanding of the mantle-gabbro transition that markedly changed magma chamber models (Nicolas, 1989). Nicolas et al. (1988) showed that the crustal section was affected by magmatic flow and that this flow was coupled with a solid-state mantle flow. They also showed that the orientation of gabbro layering resulted in part from tectonic deformation, and that it, therefore, did not necessarily reflect primary magmatic layering. The chamber model proposed by Nicolas et al. (1988) is shown in Figure 22. The principal component of this model is a small central, tent-shaped chamber constrained by the upwelling mantle at the mantle-crust transition. The chamber is also drawn to concur in size and shape with the upper crustal seismic results from the East Pacific Rise (Detrick et al., 1987). The magma chamber and the surrounding mush zone are controlled in this model by cooling, which results in a mush below a basalt fraction of 0.70 (~1185 °C), and by marked changes in the rheological properties near the basalt solidus (~1100 °C). An additional observation, also made by Nicolas et al. (1988), is that the chamber is fed from two main sources: (1) the dominant axial basaltic melt injected from greater depths, and (2) subordinate wehrlitic mush released by compaction of material within the transitional zone at the mantle-crust boundary.

In the model of Nicolas et al. (1988), the layering of the gabbro section dips away from the axial center in a concave-up geometry as a result of the outward drag induced by deeper mantle flow. This latter observation is in sharp contrast with the previous chamber models for the Oman and other ophiolites (Greenbaum, 1972; Pallister and Hopson, 1981), as well as with the subsiding cumulate pile explanation of Dewey and Kidd (1977). This Oman chamber model has changed over the years since first proposed by Nicolas et al. (1988). Its essential features are preserved, however, in latter models (Nicolas et al., 1993; Nicolas, 1992, 1994; Boudier et al., 1996; Nicolas and Ildefonse, 1996; Chenevez et al., 1998; Nicolas and Poliakov, 2001).

Nicolas (1994) points out that the differences between the Oman magma chamber model and the oceanic model proposed by Phipps Morgan et al. (1994) can be explained by whether the mantle is upwelling in an active or passive manner (Fig. 23C). The observations and modeling for the Oman chamber suggest active upwelling (Nicolas, 1994), which results in diversion of the mantle below the axial zone. This mantle flow causes a drag in the overlying lower crust that, in turn, deforms the gabbroic rocks along planes parallel to the mantle foliation. Another important point is that the chamber is filled by a mush with perhaps 10%–20% melt, or less, and hence is able to support magmatic suspension flow (Nicolas, 1992; Nicolas and Poliakov, 2001). The magma chamber, thus, is defined as a combination of melt lens and thick mush (Nicolas et al., 1993; Nicolas and Ildefonse, 1996).

A further development in the magma chamber models for the Oman ophiolite was presented in a paper by Boudier et al. (1996). Previous models were based on the assumption that the lower crustal section was built from a subsiding floor of a perched magma lens beneath the axial rift (Fig. 23A). With an actively upwelling mantle (Nicolas, 1994), the gabbro mush pile subsides while being dragged out of the low-velocity zone, and then freezes parallel to the chamber margin. Previous observations had noted the existence of sills (Benn et al., 1988) and intrusive wehrlite bodies (Nicolas et al., 1988) at or near the mantle-

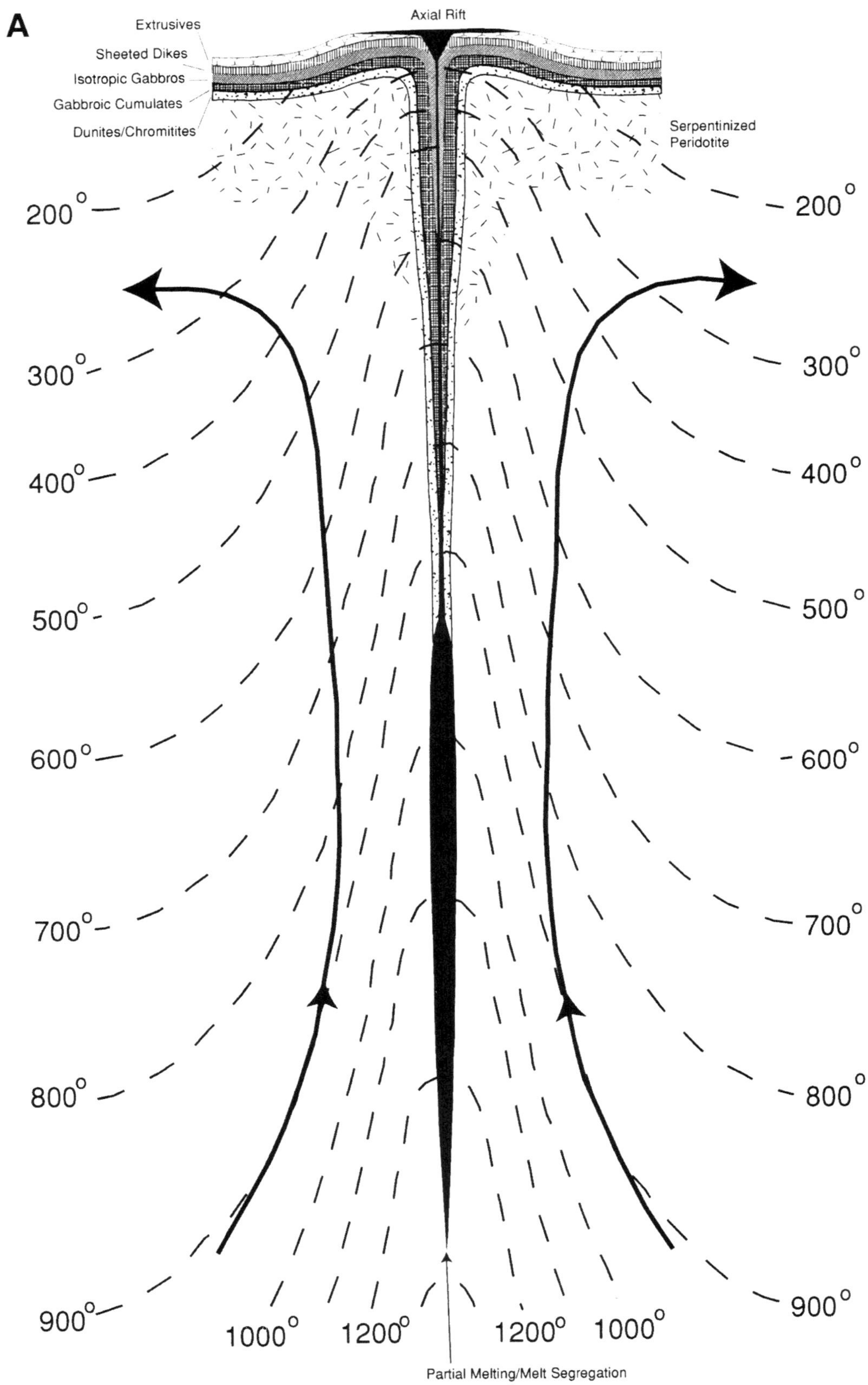

Figure 17. Crustal construction of the Oman ophiolite, as interpreted by Glennie et al. (1974, Fig. 6.6.3).

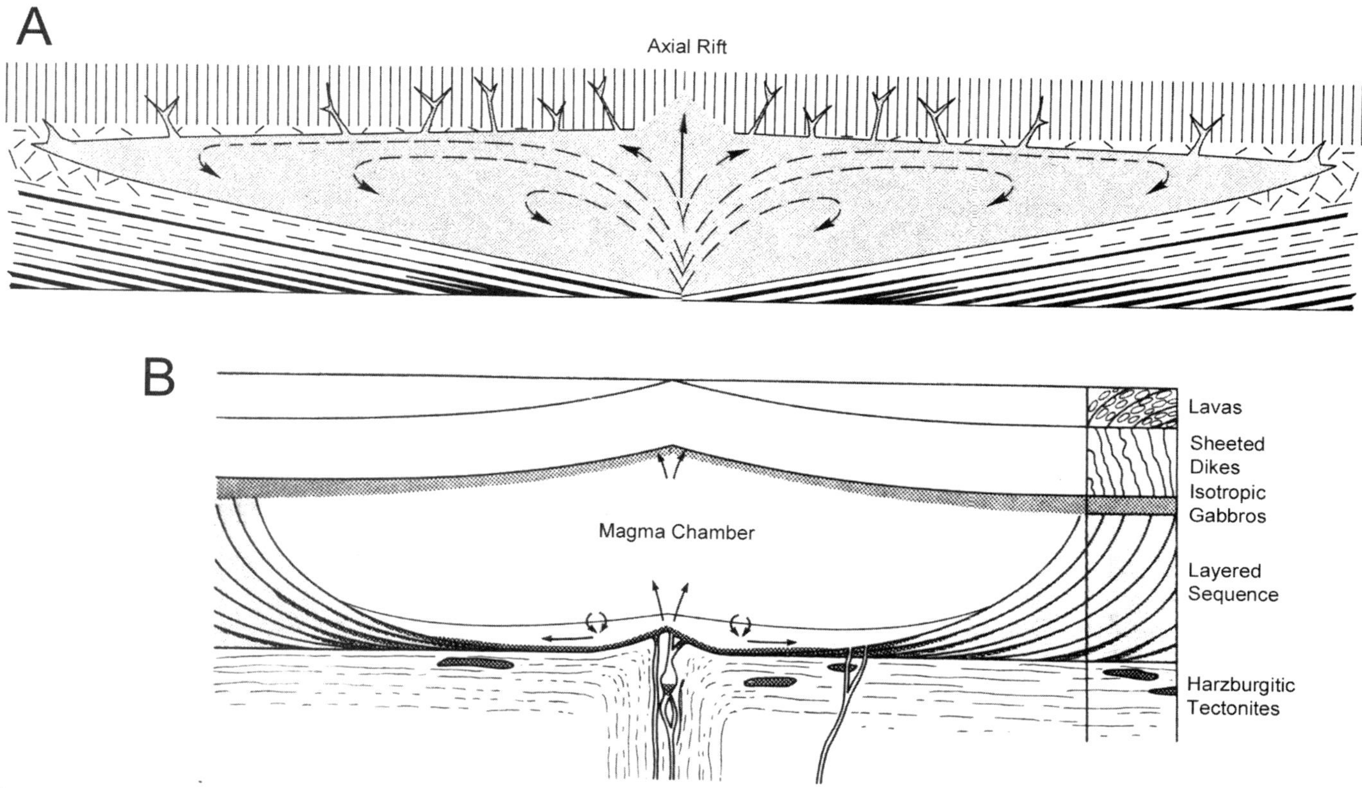

Figure 18. Steady-state magma chambers of the Oman ophiolite. A: Pallister and Hopson (1981, Fig. 19). See Hopson and Franco (1977) for an earlier version of this model. Modified by permission of American Geophysical Union. B: Smewing (1981, Fig. 9). Reproduced by permission of American Geophysical Union.

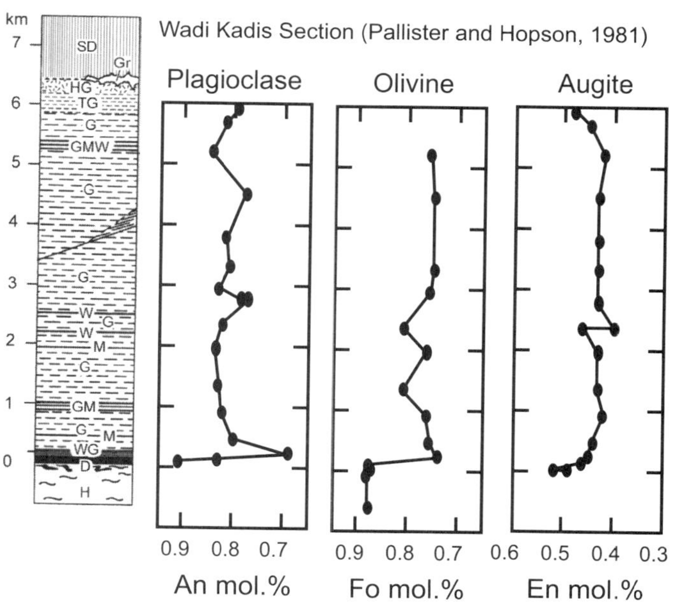

Figure 19. Cryptic variation in the Wadi Kadis section of the Oman ophiolite. Stratigraphic section and mineral compositions from Pallister and Hopson (1981, Fig. 5; Tables 2 and 3). Figure 19. Stratigraphic column reproduced by permission of American Geophysical Union.

crust transition. Boudier et al. (1996) provided further evidence for sill intrusions in the deeper parts of the crust and presented the implications of this phenomenon for magma chamber models. They suggested that the crustal section was built up both from the top via crystallization on the floor and the walls of a perched magma lens, and from the bottom through sill intrusions in the lower crustal section and within the transitional zone (Fig. 23B). In this model, the upper gabbros crystallized on the floor of the subsiding lens as a magmatic mush and obtained their foliated nature from subsidence and outward flow. The lower banded gabbros were interpreted as a composite of foliated gabbros and gabbroic or wehrlitic sill intrusions. The suggestion was based in part on the observation that graded layers were present in both sills intruded into the mantle, as well as in the transitional zone and within the lower gabbros. The strong deformation was suggested to have transformed gabbro layers into bands and lenses and to have obliterated the evidence for intrusive contacts and igneous banding (including graded layers). Numerical modeling (Chenevez et al., 1998) has supported field evidence for actively upwelling mantle and has shown that whether the magma was fed from the top or the bottom is of little consequence for the shape of the Oman magma chamber. The shape of the chamber is mainly controlled by the thermal boundary conditions.

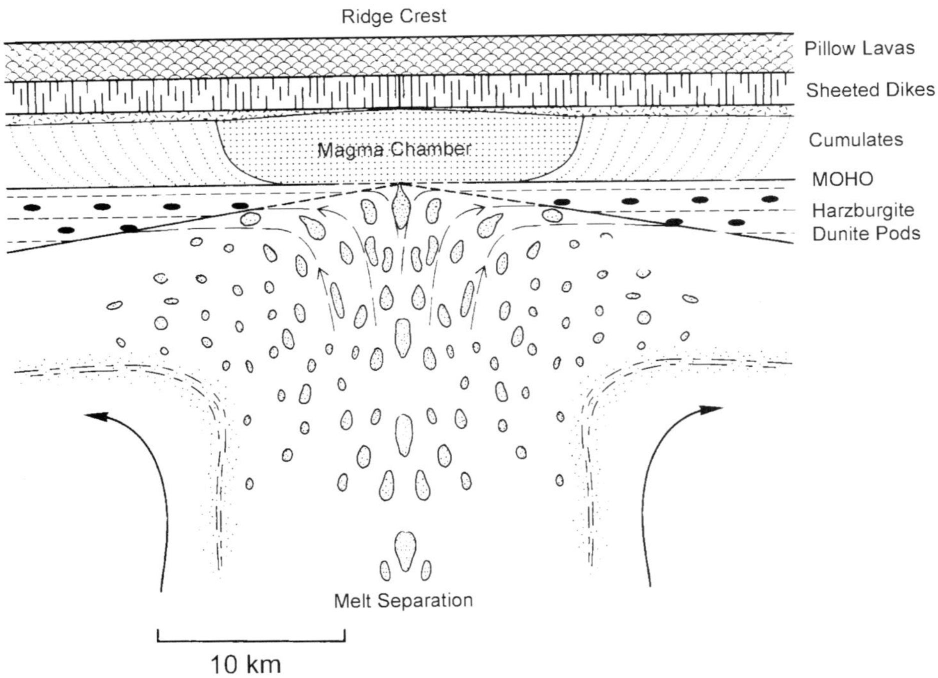

Figure 20. Schematic cross-section of the northern Oman spreading center (from Lippard et al., 1986). Reproduced by permission of the Geological Society of London.

The most recent development in magma chamber models signals the disappearance of the classic spreading center magma chambers from our conceptual thinking (Fig. 24). Kelemen et al. (1997) studied the petrography and geochemistry of gabbroic sills at the crust-mantle transition zone in the Oman ophiolite and extended the observation of Boudier et al. (1996) to the main crustal section. They proposed that primitive and residual melts migrated upward through a combination of porous flow and hydrofracturing to form the upper gabbros, the sheeted dike complex, and the lava flows (Fig. 24). The upper gabbros appear to have crystallized largely in situ (cf., MacLeod and Yaouancq, 2000), without contributing to the construction of the sheeted dike complex. The melt migration is seen in these recent models as controlled by porous flow, permeability barriers, and hydrofracturing (Korenaga and Kelemen, 1997, 1998; Kelemen and Aharonov, 1998).

DISCUSSION

Pre–Plate Tectonic Models on Oceanic Magma Chambers

Our earliest understanding of the nature of modern oceanic crust came mostly from studies of ocean islands. The rocks of these islands were found to be compositionally similar to flood basalts and other continental basaltic eruptions (Daly, 1933). The submerged ocean ridges were sampled for the first time in 1947 on an expedition to the Mid-Atlantic Ridge led by Maurice Ewing of the Woods Hole Oceanographic Institution. A petrographic examination of the dredged lavas, gabbros, and serpentinite rocks confirmed the relatively global uniformity of basalts (Shand, 1949). During the 1960s, the collection of dredge samples significantly increased our knowledge of the petrology of the ocean floor (e.g., Muir and Tilley, 1964; Christensen and Salisbury, 1975; Fox and Stroup, 1981). This advance made it possible to show that the oceanic crust of gabbroic and peridotitic compositions was associated with escarpments on the seafloor, where deeper parts of the lithosphere were interpreted to have been uplifted to the surface by transform faulting.

Marine gravimetric and seismic surveys had established very early that the earth's mantle was composed of peridotite (Hess, 1955). The oceanic crust was assumed to be composed of gabbros and lavas with a thin veneer of marine sediments (Daly, 1942). It was also well understood that melting of peridotite would produce basalts and gabbros, and that pressure release could result in decompressional melting of peridotite (Bowen, 1928; Daly, 1933; Barth, 1962). Daly (1933) believed that the lower part of the gabbroic crust and the mantle peridotites was vitreous or glassy. The mid-ocean ridges were seen as compressional zones, where the oceanic crust was bent downward into the vitreous layer, causing melting. "If…the depressed crust was broken into blocks, forced away from one another, the rising melt would occupy large chambers within the broken crust and then solidify there as crystalline peridotite or its derivatives" (Daly, 1942, p. 96).

The idea that the newly formed oceanic crust moved apart at mid-ocean ridges can be attributed to Hess (1962) and Dietz (1961, 1962) both of whom proposed that upwelling mantle beneath the ridges would laterally spread away. In contrast to his

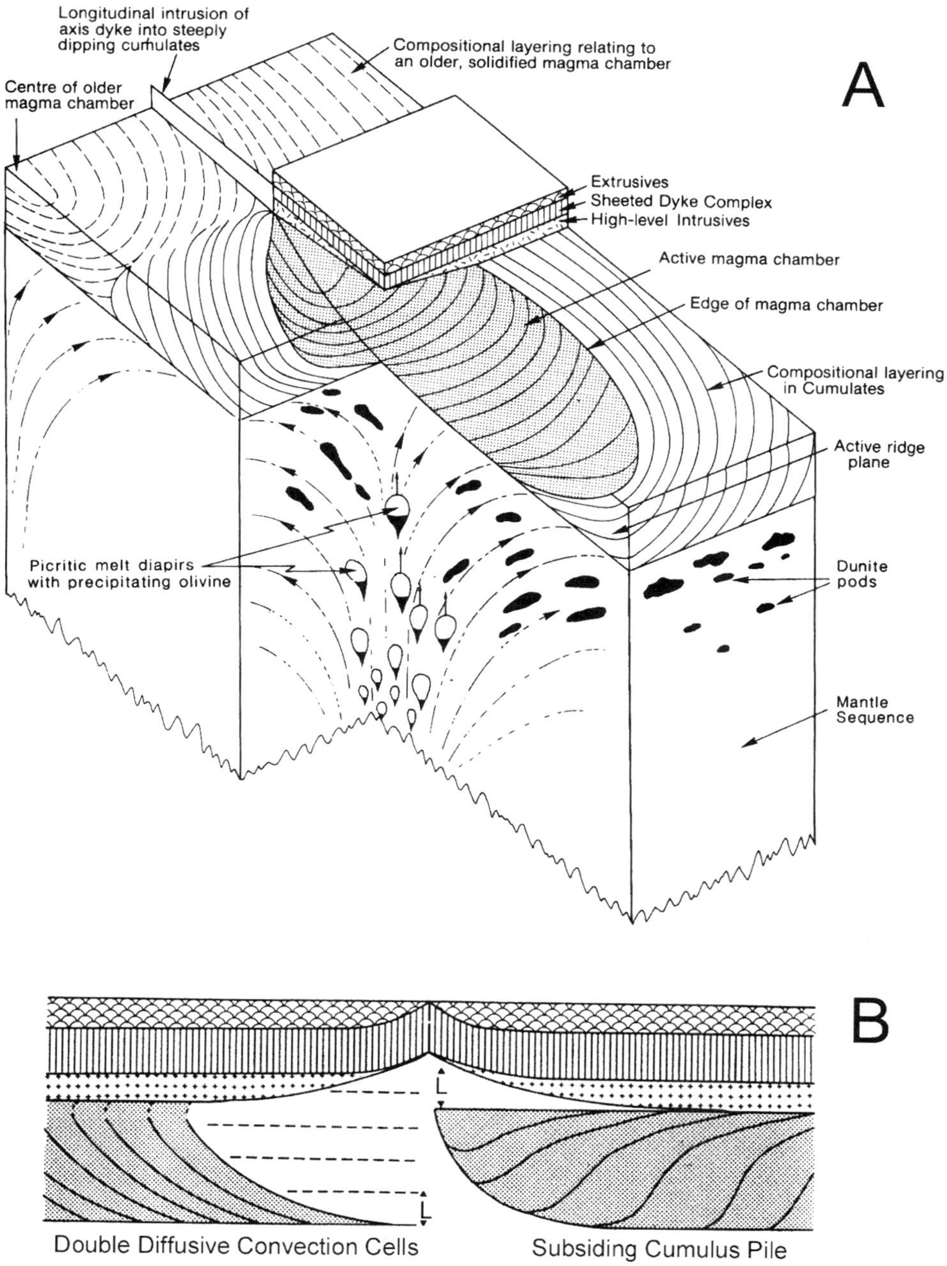

Figure 21. The Oman spreading center and axial magma chamber. After Browning (1984). Reproduced by permission of the Geological Society of London. A: Block diagram illustrating the spreading center and the overlying magma chamber (from Gass, 1989, Fig. 5). B: Cross-section through an axial chamber, illustrating alternative models to explain the calculated magma column heights: (1) double diffusive convection and (2) subsiding cumulate pile (Browning, 1984, Fig. 5).

earlier views, Hess (1962) assumed that the crust was composed of serpentinites produced by high-level hydration of peridotite. Although it was known for a long time that the oceanic crust was basaltic in composition and that melting of the mantle peridotite would produce basalt (Daly, 1914; Bowen, 1928; Hess, 1955; Erwing and Erwing, 1959; Barth, 1962; Holmes, 1965; MacGregor, 1967; Green, 1968), basalts or gabbros were not mentioned in the ocean-spreading model of Hess (1962). The view of Harry Hess that oceanic crust was composed of serpentinites was so influential at the time that when in 1962 Bruce Heezen summarized the results of seismic studies of the Mid-Atlantic Ridge, he based his interpretation on Hess' views and attributed the anomalous crest seismicity to serpentinization above the upwelling limbs of mantle convection cells (Fig. 25). Although it was accepted that melting of peridotite would produce basalts, the connection between the seismically distinct oceanic layer (layer 3) and peridotite melting was not made until much later. Alpine peridotites, serpentinites, and gabbros were thought to have first been intruded into their hosts as crystal mushes (the Alpine Mafic Magma Stem of Thayer, 1967), and then to have

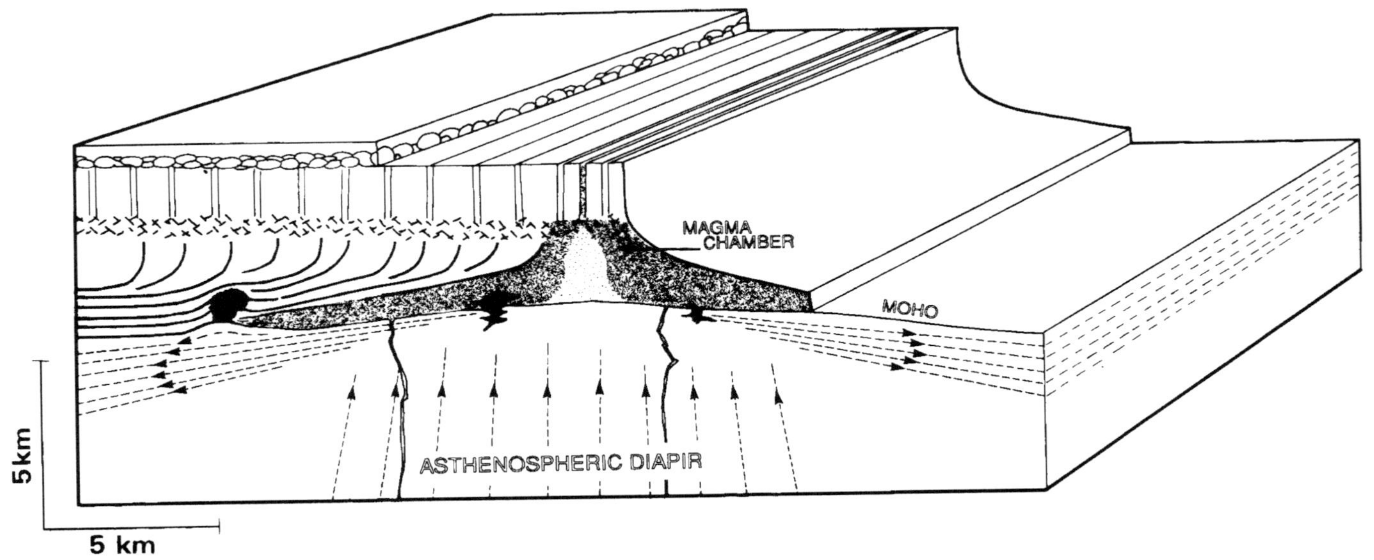

Figure 22. Magma chamber model based on structural studies of the Oman ophiolite (Nicolas et al., 1988, Fig. 10). Reproduced by permission of Elsevier.

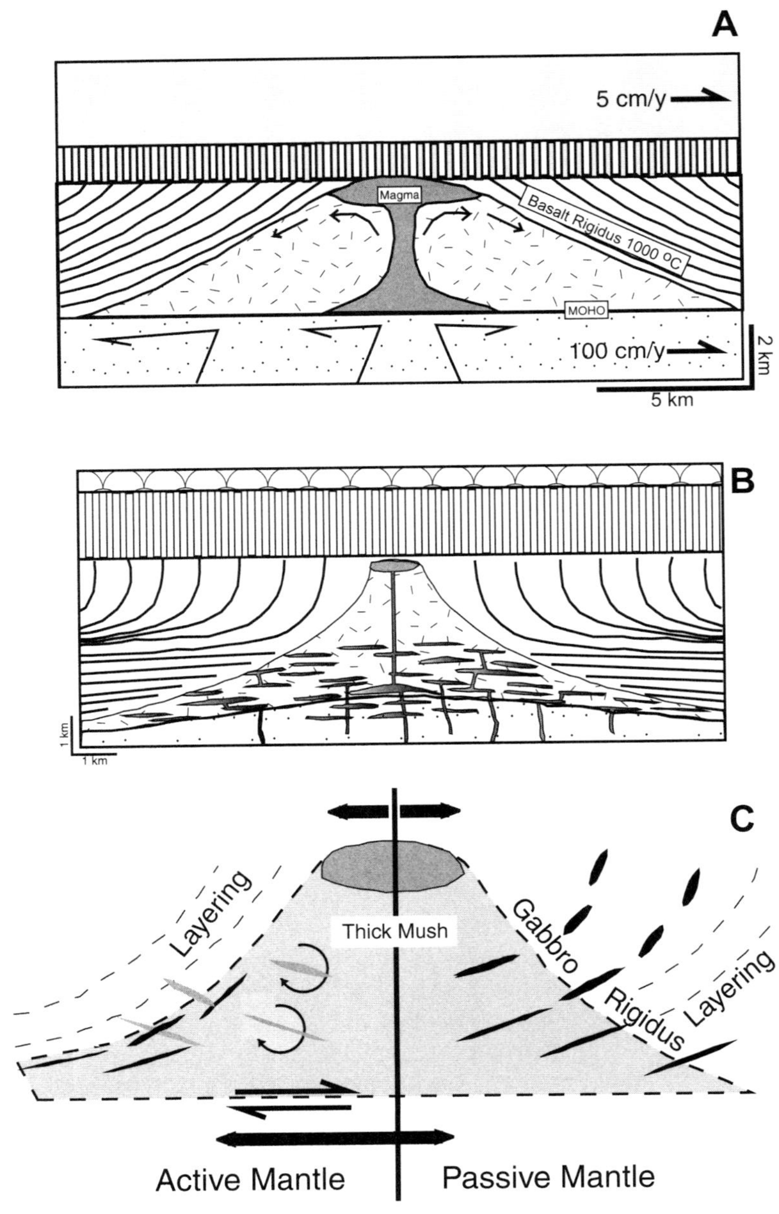

Figure 23. Models of Oman magma chambers. A: After Nicolas (1992, Fig. 15). Modified by permission of Oxford University Press. B: After Boudier et al. (1996, Fig. 7). Modified by permission of Elsevier. C: Illustration of the effects of active versus passive magma upwelling on subaxial magma chamber and gabbro banding (Nicolas, 1994, Fig. 1). Modified by permission of American Geophysical Union.

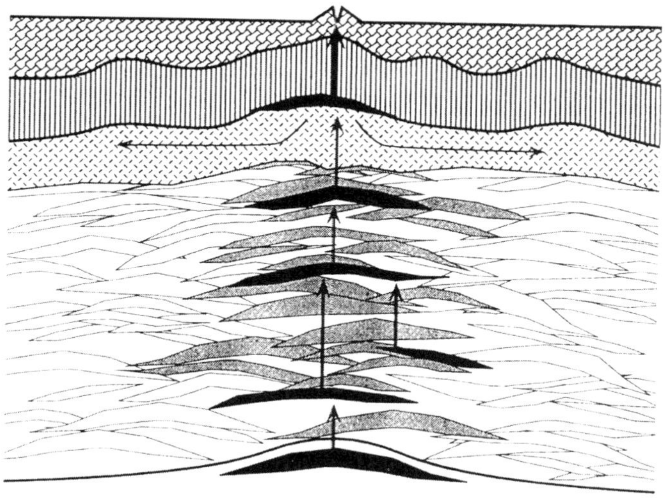

Figure 24. Schematic illustration of sill complex formation in the lower crustal section of the Oman crust. (After Kelemen et al., 1997, Fig. 6). Reproduced by permission of Elsevier.

been tectonized. Hence, tectonized peridotitic complexes in the continental crust were not always interpreted as being slices of oceanic mantle rocks.

Ophiolites had minor role, if any, in the early formulation of the history of the ocean floor during the 1950s and 1960s. They were universally considered to be thick lava flows erupted into subsiding marginal basins (geosynclines; see also Moores, this volume, Chapter 2; Shallo and Dilek, this volume, Chapter 20; Smith and Rassios, this volume, Chapter 19). This is perhaps best illustrated by Kay (1951) in a paleogeographical section of the Caledonian geosyncline of eastern North America (Fig. 26). The basaltic magma was viewed as having been injected from an abyssal substratum as a result of geosynclinal down warping (Daly, 1914, 1942). The presence of sedimentary rocks associated with many ophiolitic complexes was taken to confirm the suggestion that ophiolitic complexes were intruded into developing sedimentary basins. Thus, these views saw ophiolitic complexes as having formed from large intrusive laccoliths or sill intrusions emplaced into subsiding sedimentary basins (Steinmann, 1927, and this volume, Chapter 6). The ideas that subsequently dominated the evolving ophiolite concept considered ophiolites as products of fissure eruption-generated flood basalts (e.g., Aubouin, 1965). The internal differentiation in ophiolites was seen as having occurred in situ by crystal accumulation in large basaltic outpours (Gass and Masson-Smith, 1963; Aubouin, 1965).

Tectonized harzburgites in many ophiolite complexes were observed to have been intruded by ultramafic and gabbroic plutons that were part of these ophiolite sequences. This observation clearly was a problem for the traditional view of a eugeosynclinal eruptive environment suggested by earlier field studies in ophiolites. Ophiolite researchers dealt with this problem in various ways. In a paper published relatively late in 1969, Reinhardt suggested that the Oman ophiolite was emplaced through tectonized peridotite and onto the seafloor at a mid-ocean ridge. Gass and Masson-Smith (1963), perhaps having completed their manuscript just before the publication of Harry Hess' influential 1962 paper, considered the Troodos ophiolite as an oceanic pond of ultramafic magma. Moores (1969) interpreted the Vourinos ophiolite, northern Greece, as being a result of partial fusion of lherzolite. A reconstructed Vourinos cross-section showed ultramafic to mafic cumulates as a stratiform complex nested within a structural trough in the lherzolite tectonites (Fig. 4C; Moores, 1969, and this volume, Chapter 2). In a 1967 review paper, Ian Gass suggested that low-pressure partial fusion of the peridotite produced a liquid fraction of tholeiitic composition. Without quoting any of the papers supporting ocean-spreading models, Gass (1967) pointed out that it was possible that the Troodos formed as a "volcanic edifice" at a Tethyan median ridge, similar to the situation in modern ocean basins. Gass cited several of the early ocean-spreading and continental drift papers, published since 1958, for the first time in his 1968 paper, suggesting that the Troodos complex likely represented a slice of a Mesozoic ocean floor.

It is clear from these early developments that ophiolitic studies had little direct influence on the formulation of ocean floor

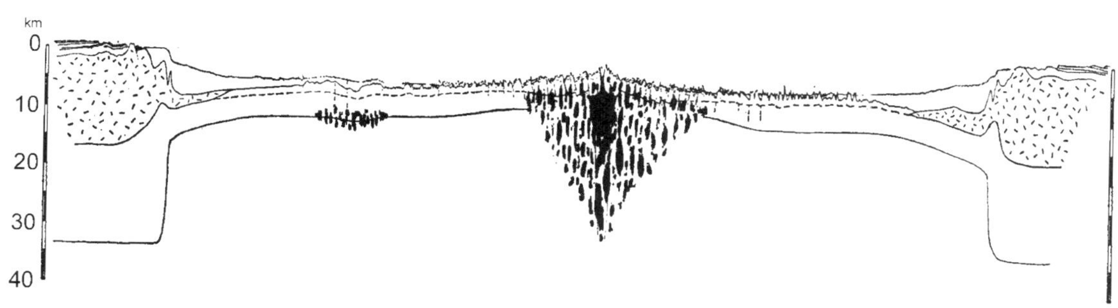

Figure 25. Cross-section through the Atlantic Ocean basin (after Heezen, 1962, Fig. 23). The dark shaded areas and dots are the 7.2 to 7.4 km/s material beneath the Mid-Atlantic Ridge. This material is interpreted as serpentinized upper mantle. The figure is constructed to conform to Hess' (1962) view of mantle convection and serpentinite composition of the lower crust (layer 3). Reproduced by permission of Elsevier.

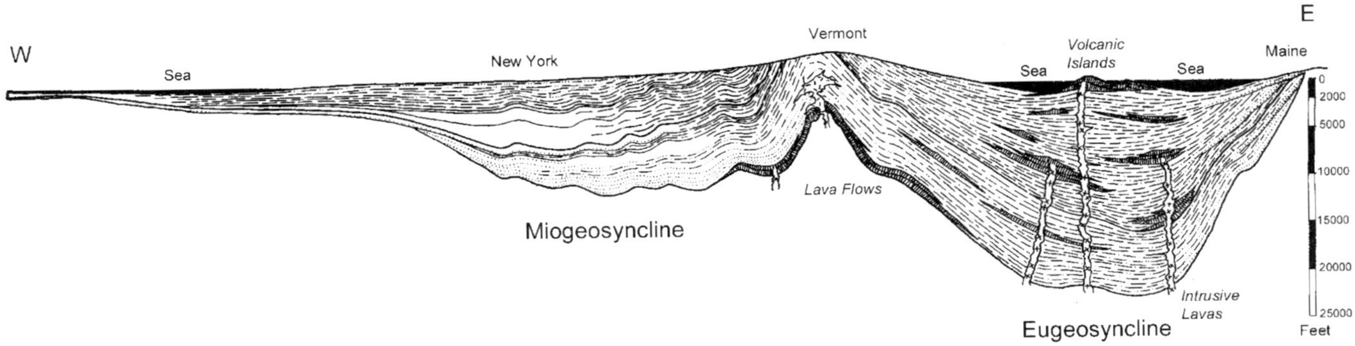

Figure 26. Paleogeographic reconstruction of the Maine (east) to New York (west) geosynclinal basins (after Kay, 1951, plate 9).

spreading models. Neither Dietz (1961, 1962) nor Hess (1962) based their theories on ophiolite studies for the simple reason that ophiolites were not seen as oceanic in origin. Similarly, ophiolite researchers were late in claiming the oceanic origin of ophiolite complexes. This concept emerged first in the late 1960s.

Development of Steady-State Magma Chambers

The plate tectonic theory was formulated during the late 1960s, principally on the basis of geophysical evidence (cf., Glen, 1982). Ophiolite studies had little direct impact on the development of plate tectonic ideas. Oxburgh and Turcotte (1968) modeled the thermal structure at mid-ocean ridges to be result of a convecting mantle due to thermal conduction. They proposed that basaltic melt was separated from upwelling zones of partially molten peridotite beneath the ridge crest, and that it was then either erupted on the seafloor or accumulated in magma chambers situated at neutral buoyancy levels within the crust and/or in the upper mantle. A consequence of this phenomenon was that a basaltic oceanic layer 3 was being continuously generated at a mid-ocean ridge and then laterally transported away.

In 1968, Cann proposed a model for the structure of oceanic crust that did not include lower crustal magma chambers. Only two years later, however, Cann (1970) proposed a "new model for the structure of ocean crust," now involving the presence of lower crustal magma chambers and gabbros. Cann illustrated this concept of ocean floor spreading by a remarkably simple and elegant drawing that in a few strokes gave the fundamental aspects of mid-ocean ridge spreading and subaxial magma chambers (Fig. 27A). It is quite illuminating to observe the impact of ophiolite studies on oceanic crust models during this time, as reflected in the three papers published by Joe Cann between 1968 and 1974. His 1968 paper contains no references to ophiolites. In the 1970 paper, mentioned above, the discussion of ophiolites is restricted to a short paragraph stating the similarities between the ocean crust model based on geophysical and kinematic reasoning, and observations from ophiolites. "The fact that (ophiolitic) complexes very much like the (ocean crust) model exist reinforces both the model and the interpretation of ophiolite complexes as pieces of ocean crust" (Cann, 1970, p. 930). In a following paper in 1974, Cann proposed an elaborate model for the genesis of oceanic crust involving an axial magma chamber. This chamber was seen with cooling along the roof to form upper isotropic gabbros and lower cumulates with layers dipping toward the chamber center (Fig. 27B). A three-dimensional version of this chamber model was subsequently developed by Robson and Cann (1982) (Fig. 27C). This chamber model was very influential and contained the key elements used in many later studies of ophiolites and modern oceanic crust (Kusznir and Bott, 1976; Pallister and Hopson, 1981; Casey and Karson, 1981; Smewing, 1981; Browning, 1984). The principal difference between the oceanic model of Cann (1974) and that developed for the Troodos ophiolite by Greenbaum (1972) is, respectively, the inward dipping versus the horizontal layering in the lower cumulates. Greenbaum (1972) assumed a zoned chamber, whereas Cann (1974) assumed magma convection.

Cann's (1974) oceanic crust model, principally derived from marine geological and geophysical information, was used to infer the origin of ophiolites, but ophiolite-derived models were not readily accepted by marine geologists and geophysicists to explain the origin of modern oceanic crust. Cann (1974) pointed out that although ophiolites contained all the fundamental elements inferred for modern ocean crust, it would not be possible to determine the origin of ophiolites without documenting the dip of slumping directions in layered gabbros, dike chilling statistics, dip of lava flows, paleo-horizontal, and orientation and kinematics of conjugate faults, among other key elements. Christensen and Salisbury (1975) expressed a similar view in a survey of the modern lower oceanic crust, concluding that ophiolites could not be fossil remnants of normal oceanic crust. They rejected the ophiolitic view that ocean crust would form in a narrow zone under the median valley of a mid-ocean ridge, as expressed by Greenbaum (1972). Instead, they proposed that the lower levels of layer 3 are thickened under the ridge flanks by

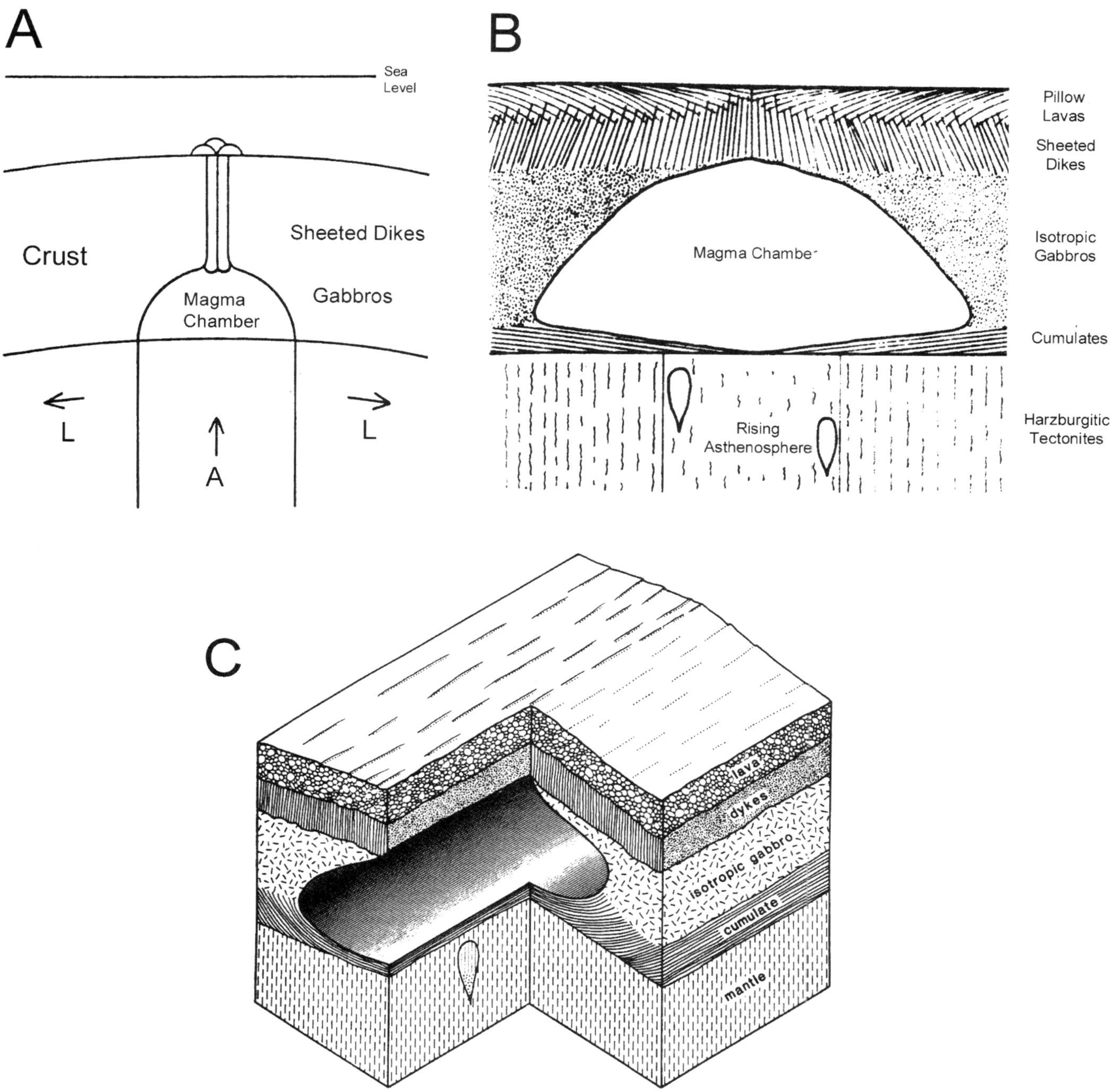

Figure 27. The ocean crust according to Joe Cann. A: Schematic sketch published in 1970 (Fig. 2) Reproduced by permission of *Nature*. B: Schematic cross-section of a spreading center chamber (Cann, 1974, Fig. 6). The illustration was used by Pallister and Hopson (1981) to develop their chamber model for the Oman ophiolite. Reproduced by permission of Royal Astronomical Society. C: Spreading chamber model based on ophiolite studies (Robson and Cann, 1982, Fig. 1). Reproduced by permission of Elsevier.

gabbro intrusions added to the crust from the thick anomalous mantle, inferred to exist beneath the ridge areas. The ophiolitic layer 3 was attributed, in this model, to development of immature oceanic crust lacking the effect of off-axis thickening.

The observations from ophiolites might thus be interpreted in the light of observations from the ocean floors, but they would not necessarily serve to formulate models for normal oceanic crust (Christensen and Salisbury, 1975). The reason for this thought was that ophiolites might have formed either in an off-axis setting or in an unusual tectonic setting, such as back-arc basins. Geochemical studies of extrusive rocks in various ophiolite complexes have strongly substantiated this view and led to the formulation of the supra-subduction zone concept for the origin of ophiolites (e.g., Pearce, this volume, Chapter 15; Taylor and Nesbitt, 1988; Elthon, 1991). Although this concept certainly has a strong validity, it is not clear how magma chamber processes in supra-subduction zone environments would differ from those in mid-ocean ridge environments. It is expected that it principally would be active spreading, with its specific spreading rates and its interplay of tectonic and magmatic activities, that will control magma chamber formation and evolution, and not the presence or absence of an underlying subduction zone.

Kusznir and Bott (1976) used thermal and heat balance calculations to model magma chamber size and shape as a function of spreading rate (Fig. 28A). They assumed that layer 2 was cooled by water circulation and that layer 3 was formed from magma solidifying to gabbro as a result of thermal conduction. Cumulates forming along the floor would narrow the magma chamber and would result in a relatively flat roof. They further showed that the chamber width was a function of spreading rate (at 3 cm/yr, a total width of 40 km was predicted). For spreading rates below 0.5 cm/yr, the heat flux was too fast to develop cham-

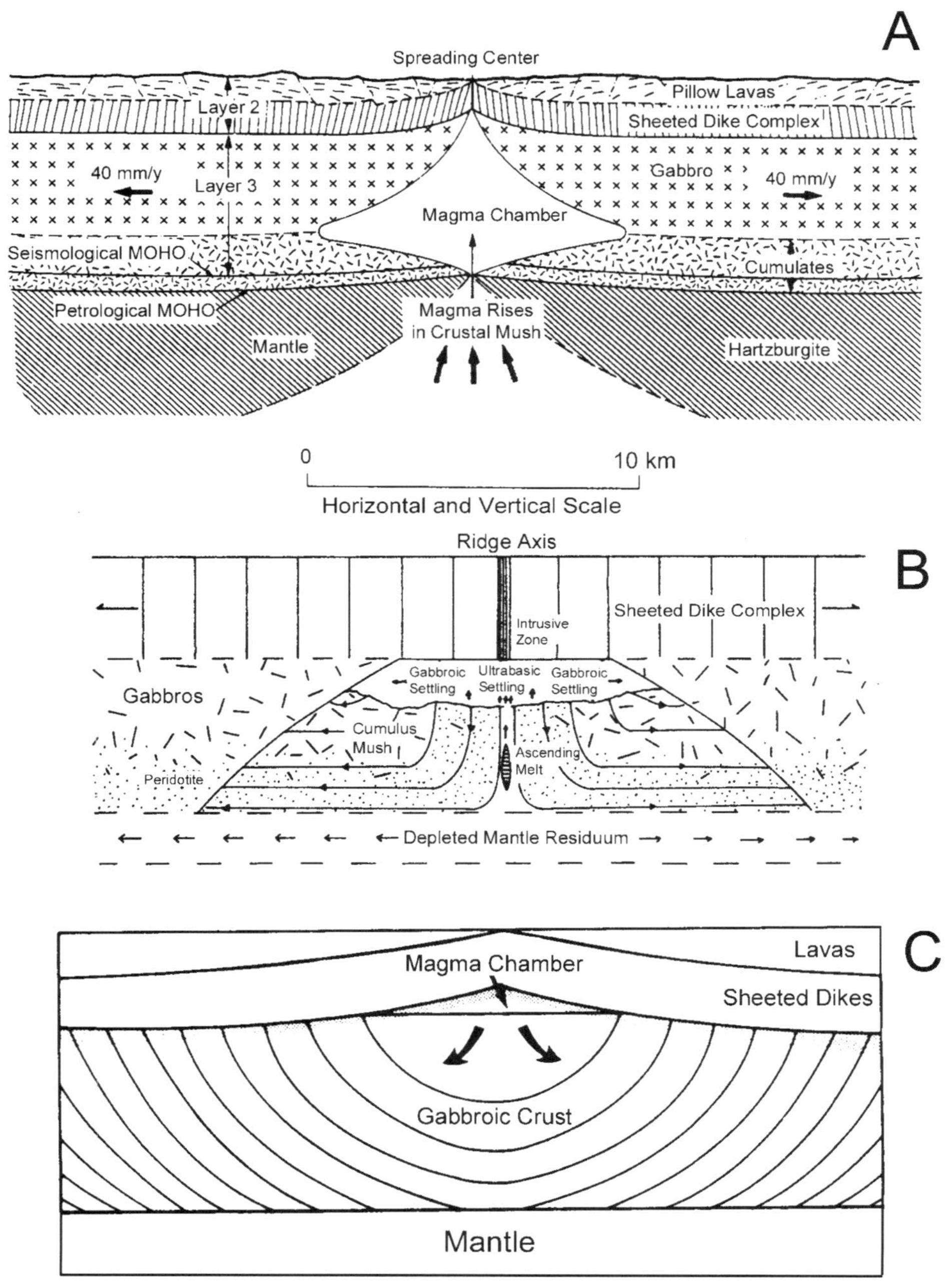

Figure 28. Thermal models for the constitution of ocean crust. A: Kusznir and Bott (1976) as illustrated by Bott (1982, Fig. 3.22). B: Sleep (1975, Fig. 4). Reproduced by permission of American Geophysical Union. C: Henstock et al. (1993, Fig. 6b). Reproduced by permission of American Geophysical Union.

bers and, instead, magma would solidify as dikes. The model of Kusznir and Bott (1976) shows similarities to the one proposed by Cann (1974), by reflecting a decreasing cooling rate toward the accretionary boundary. Other chamber models have been proposed more recently that differ mainly in their relative sizes and shapes (e.g., Lewis, 1983; Wilson et al., 1988).

In a similar fashion, but using heat budget constraints, Sleep (1975, 1978) deduced the size and shape of mid-ocean ridge magma chambers for relatively fast-spreading rates. Sleep concluded that magma chambers exist beneath the axial zone as a thin melt lens just below the sheeted dike complex, and that they extend downward into the mantle as a crystal-liquid mush (Fig. 28B). Fresh magma would have to penetrate this mush in order to enter the chamber, which is laterally zoned with ultramafic cumulates forming in the center and gabbroic cumulates forming near its margins, as a consequence of cooling and crystal settling. Cumulates formed on the floor of the melt lens would subside and flow laterally such that the centrally formed ultramafic cumulates eventually would constitute the base of the crust (Fig. 28B). The horizontal layering of the cumulate section would be dipping toward the axial rift as a result of differential flow outward from the axis, as had been proposed previously for oceanic crust (Cann, 1974; Kusznir and Bott, 1976; Dewey and Kidd, 1977). This explanation contrasted with the dominating view from ophiolite studies at the time, which suggested the occurrence of large magma chambers with little mush and with layering dipping toward the axial rift without strong indications for flow deformation (Smewing, 1981; Pallister and Hopson, 1981). Despite this major difference, it was the general geometry of magma chambers beneath the oceanic spreading centers, as proposed by Sleep (1975), that was adopted in the late 1980s by Adolphe Nicolas for the Oman ophiolite, with some important modifications necessitated by detailed petrographic and structural observations (Nicolas et al., 1988; Nicolas, 1992; Boudier et al., 1996). On the basis of detailed structural mapping, Nicolas et al. (1988) observed that the layering in the gabbros was dipping away from the inferred spreading center in the Oman ophiolite. This geometry with cumulate layering dipping away from the spreading axis had not been reported from the Oman ophiolite before, or from any other ophiolite, for that matter.

Only to a certain extent have ophiolite studies served as a test for the theoretical and geophysical modeling of oceanic crust. On the other hand, it is fair to say that many, if not all, ophiolite studies that present magma chamber models extended these ideas also to oceanic spreading centers (Greenbaum, 1972; Allen, 1975; Casey and Karson, 1981; Glennie et al., 1974; Pallister and Hopson, 1981; Smewing, 1981; Browning, 1984; Nicolas et al., 1988; Nicolas, 1992; Boudier et al., 1996). There are several outstanding studies that developed spreading center models based mainly on the observations derived from ophiolite studies. Most notably, Kidd (1977) used field data from the Troodos ophiolite to propose a crustal construction model involving a narrow zone of dike injection at a ridge axis, subsidence and faulting of accumulating lavas, and rotation of dikes and lava flows. The magma chamber envisioned by Kidd (1977) was narrow and had very steep contacts (<80°), such that its shape approached that of a downward broadening dike. While this model integrated the existing observations from the Troodos ophiolite effectively, it was also used successfully to discuss some oceanic features, such as hydrothermal convection cells, seafloor metamorphism, and magnetic reversals. Dewey and Kidd (1977) developed a comprehensive kinematic model for the Bay of Islands ophiolite. This model involved a subsiding floor of a centrally located chamber. Despite the simplicity and beauty of the model, it was rarely applied in subsequent studies, not even to the Bay of Islands ophiolite complex (Casey and Karson, 1981; Elthon et al., 1982).

The large, steady-state chamber idea was the dominating conceptual model for magma chambers beneath oceanic spreading centers until the early 1990s. Typically, it was based on the model of Cann (1970), and was in general applied to Troodos (Greenbaum, 1972), Bay of Islands (Casey and Karson, 1981), and Oman (Pallister and Hopson, 1981; Smewing, 1981) ophiolites. Suggestions involving the subsidence of magma chamber floor and flow in crystal mushes (Dewey and Kidd, 1977; Sleep, 1975) were not seriously considered (e.g., Browning, 1984). Solidification of the chamber was principally seen as having proceeded by crystal accumulation on the floors and along the walls of a commonly zoned magma chamber. Despite the fact that crystal mushes were known to have existed during formation of continental layered intrusions, this possibility was not considered in the formulation of oceanic magma chambers. The reason for this omission was simply that ophiolitic cumulates were believed to have solidified in equilibrium with the main melt volume, without retainment of any trapped melt (adcumulates; Pallister and Hopson, 1981).

Magma Lenses and Crystal Mushes

The conundrum between the two opposite views—the presence of large, steady-state magma chambers beneath the axial centers of mid-ocean ridges, as predicted by ophiolite studies, and the absence of such magma chambers, as suggested by the geophysical studies—was not easily resolved. One important reason was the strong influence among ophiolite researchers of early magma chamber solidification models developed from studies of layered continental intrusions. Seismic studies of the fast-spreading East Pacific Rise had shown the presence of small, shallow magma pockets and low-velocity zones along ridge segments (Fig. 29A) (Harding et al., 1989; Toomey et al., 1990; Vera et al., 1990; Kent et al., 1990, 1993). The existing evidence for axial magma chambers at fast-spreading centers has been reviewed by Sinton and Detrick (1992), Henstock et al. (1993), Quick and Denlinger (1993), Phipps Morgan and Chen (1993), and Phipps Morgan et al. (1994), among many others. The image that has emerged from these studies is illustrated in Figure 29B (Sinton and Detrick, 1992; Kent et al., 1990, 1993). The seismic results have suggested magma chambers beneath fast-spreading centers to be mainly filled with crystal mushes containing only

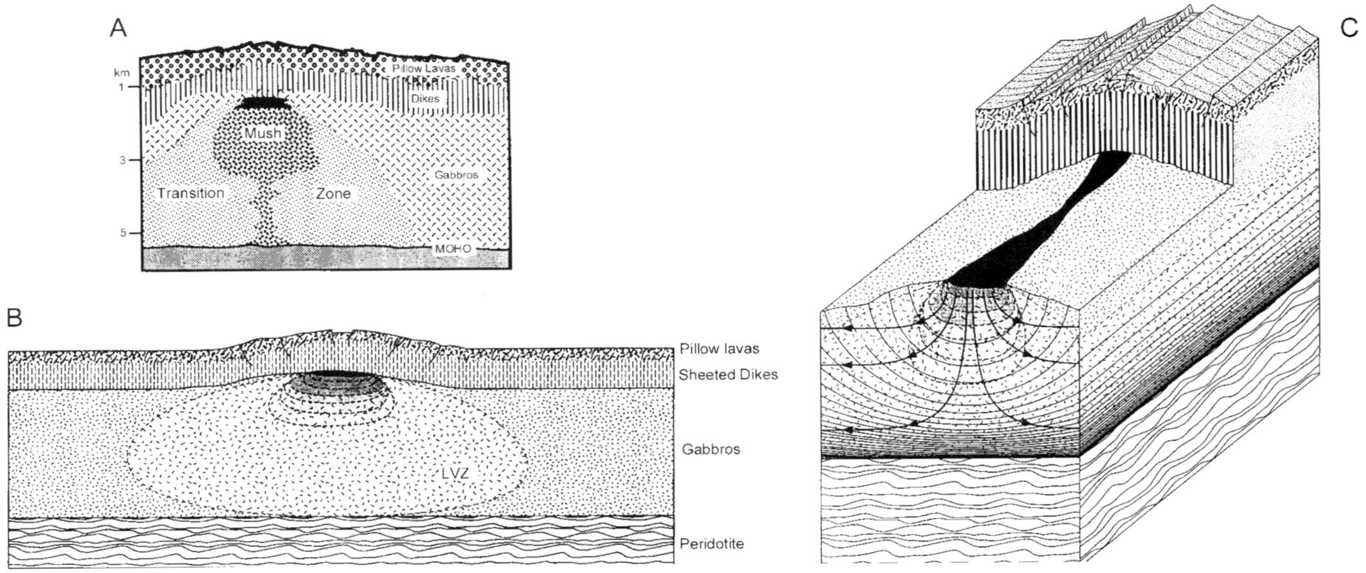

Figure 29. Geophysical models of ocean spreading center magma chambers. A: Sinton and Detrick (1992, Fig. 9). Reproduced by permission of American Geophysical Union. B: Phipps Morgan et al. (1994, Fig. 1b). Reproduced by permission of Elsevier. C: Kent et al. (1990) and Kent et al. (1993, Fig. 21). Reproduced by permission of American Geophysical Union.

small percentages of melt extending to the base of the crust. The large molten chambers, inferred from ophiolite studies, have thus vanished and have been replaced by chambers filled with crystal mushes with a thin melt lens (or sill) on top of the mush zone (Fig. 29C). Despite the fact that this view was anticipated by Sleep (1975) and Dewey and Kidd (1977), it was the work of Nicolas et al. (1988) that introduced and incorporated these ideas effectively into ophiolite studies (Nicolas, 1992, 1994; Boudier et al. 1996). MacLeod and Yaouancq (2000) have demonstrated the existence of a ferrobasaltic fossil magma chamber as a thin melt lens overlying the cumulates and perched just beneath the sheeted dike complex in the Oman ophiolite. They inferred that this lens was fed by melt seeping up from below through the underlying mush zone. The most important implication of the mush interpretation for the ophiolite concept has been that layering could have formed as a result of crustal strain, and not from direct deposition or crystallization on chamber walls.

Lately, these ophiolite studies-based views on magma chambers have been extended to continental layered intrusions. It has been documented convincingly that some of the classic layered intrusions, such as Skaergaard, consolidated from a mush by compaction and migration of interstitial melt toward an upper zone of largely molten chamber (Nicolas, 1992; McBirney, 1995; McBirney and Nicolas, 1997). It is worth noting at this point that crystal mushes should be viewed as part of a magma chamber, and that processes during their solidification are fundamental for the mode and nature of magma crystallization. Thus, there is no foundation for the view that studies of continental layered intrusions have outlasted their relevance for understanding the magma chamber processes at spreading centers (e.g., Sinton and Detrick, 1992). The existence of mushes has long been recognized in layered intrusions (Wager and Brown, 1967). The models for large, completely molten magma chambers beneath spreading centers were the outcome of marine studies of oceanic spreading centers and were readily adopted by the students of ophiolites.

Several important questions remain to be answered. How do fresh pulses of magma migrate through the mantle-crust transition and through the crystal mush, and to what extent does this migration affect the magma composition? More than anything, it has been the deep drilling into in situ oceanic crust near active spreading centers undertaken by various projects of the Ocean Drilling Program that has contributed some partial answers to these questions.

Slow-Spreading Magma Chambers

In contrast to the studies of the fast-spreading East Pacific Rise, similar studies of the slow-spreading Mid-Atlantic Ridge has provided little evidence for the presence of low-velocity zones and magma chambers (Purdy and Detrick, 1986). This difference in the amount of detectable melt fractions as a function of spreading rate is surprising, considering the similarities in the overall seismic structures (Henstock et al., 1993). The geophysical observations were supported by theoretical modeling that showed that steady-state magma chambers did not, and could not, exist beneath spreading centers with slow-spreading rates (Sleep, 1975). This observation was often in conflict with petrographic and geochemical studies of the rocks recovered from the slow-spreading centers, such as the Mid-Atlantic Ridge system (~1 cm/yr), which was assumed to have a central, steady-state magma chamber under the inner floor of its median valley (e.g., Bryan and Moore, 1977).

The reason for the difference between fast- and slow-spreading centers is most likely due to the dynamics of slow-spreading ridges, which make them more directly controlled by the availability and arrival of intermittent batches of magma from the mantle (Sinton and Detrick, 1992). A wide range of processes that are locally determined by thermal, magmatic, and tectonic regimes thus might control the crystallization and the evolution of magma chambers at slow-spreading centers. Nisbet and Fowler (1978) suggested, on the basis of the Mid-Atlantic Ridge seismic and petrographic data, that small melt pockets migrate upward by crack propagation and form magma chambers in a broad zone under the axial valley and at a horizon beneath the sheeted dike complex (Fig. 30A). They described the shape and the geometry of these inferred small chambers as "infinite leeks." Other studies of the formation of oceanic crust have pointed out that magma chambers are intermittent features (Lister, 1983), and that magmatic and tectonic activities likely shift laterally and along-axis of the spreading centers (Pálmason, 1973; Ramberg and van Andel, 1977; Nisbet and Fowler, 1978). Smith and Cann (1992) used the distribution of seamounts on the Mid-Atlantic Ridge to propose that crustal accretion resulted from the "pile-up" of seamounts and lava flows. They related this eruption style to the buoyant rise of magma chambers with limited size and frequency and inferred that this phenomenon would form a gabbroic crust composed of multiple frozen chambers over a broad interval (Fig. 30B).

Such small intermittent and laterally limited chambers have certainly had their followers among the researchers of the Troodos ophiolite (Allen, 1975; Smewing et al., 1975; Varga and Moores, 1985; Thy et al., 1989; Malpas, 1990; Moores et al., 1990), but have rarely been considered for other ophiolites (i.e., Strong and Malpas, 1975).

In Situ Oceanic Crust

Our view of oceanic magma chambers has been strongly influenced by the results of deep drilling by the ODP into in situ lower oceanic crust. Of particular significance is Hole 735B, drilled ~1500 mbsf into a gabbroic massif that formed beneath the median valley of the slow-spreading Southwest Indian Ridge near the Atlantis II Fracture Zone (Dick et al., 2000; Natland and Dick, 2001). An association of evolved Fe-Ti-oxide gabbros with relatively primitive olivine gabbros characterizes the upper ~500 m of gabbros in Hole 735B. The Fe-Ti-oxide gabbros occur as the major constituent of some plutonic units or as a minor component of layers and vein networks in dominantly olivine gabbro units. The contacts between the gabbros and the Fe-Ti-oxide gabbros are, when preserved, mostly gradational and only rarely intrusive. The apparent bimodal variation in the core, indicated by all cumulus phases, suggests limited mixing and compositional re-equilibrium between the two magmatic components. This phenomenon may suggest that one cumulate component was solidified, or partially solidified, before the other component was formed or was introduced into a preexisting mush pile.

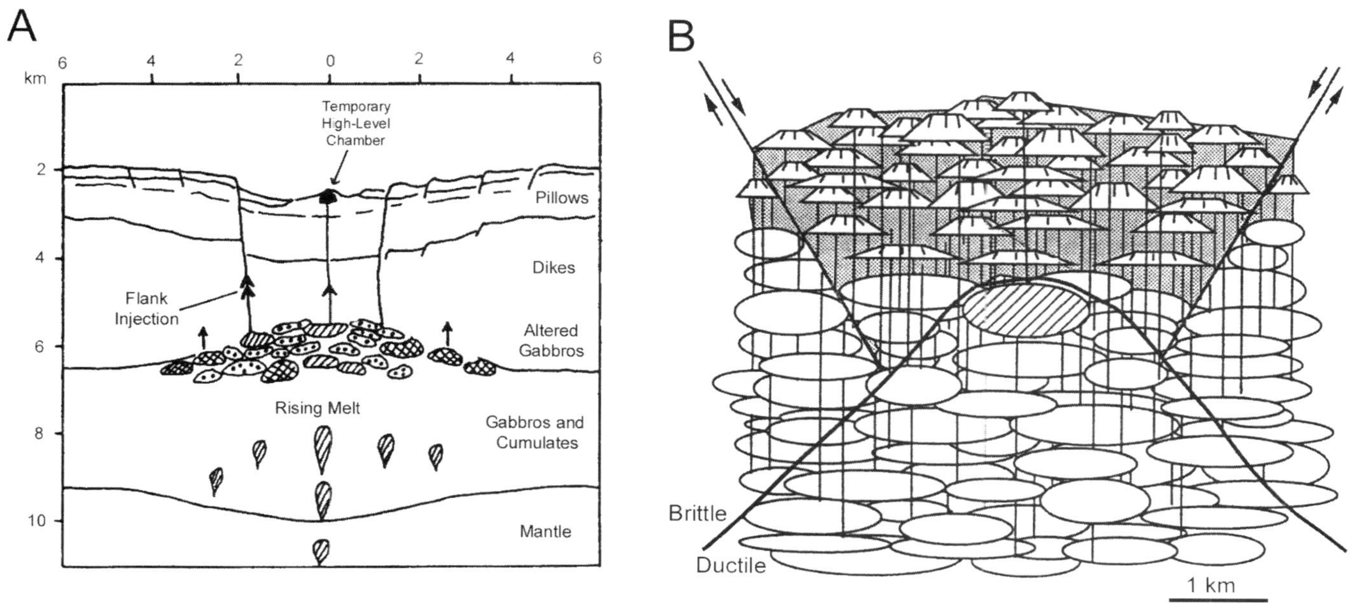

Figure 30. Magma chamber models depicted for the Mid-Atlantic Ridge. A: Multiple chamber model of Nisbet and Fowler (1978, Fig. 6b), which they referred to as the "infinite leek" model. Reproduced by permissions of the Royal Astronomical Society. B: Multiple chamber model of Smith and Cann (1992, Fig. 10). Reproduced by permission of American Geophysical Union.

Below 500 m and toward the bottom, coarse-grained olivine gabbro dominates, with subordinate amounts of Fe-Ti-oxide gabbro (Natland and Dick, 2001; Niu et al., 2002). ODP Hole 1105A was drilled in a location offset by 1 to 2 km NNW of Hole 735B, and recovered gabbro sequences quite similar to those from the upper part of Hole 735B (Shipboard Scientific Party, 1999; Thy, 2003). The important observation was the preservation in the olivine gabbro sequence of an evolved component of Fe-Ti-oxide gabbros believed to represent migrating interstitial melt. The distribution of this interstitial melt and the processes responsible for its preservation and migration have naturally been a main theme for development of our ideas for spreading-related magma chambers (e.g., Sinton and Detrick, 1992).

The compositional variation of the main cumulus phases in Hole 735B (Bloomer et al., 1991; Ozawa et al., 1991; Natland et al. 1991; Hébert et al., 1991; Niu et al., 2002) is illustrated in Figure 31. The main features shown by the cryptic variation are that the olivine gabbros are composed of at least three segments, each showing upward trends toward more primitive compositions (Niu et al., 2002). A more detailed stratigraphy has been constructed on the basis of the whole-rock composition (Natland and Dick, 2001), which has not yet been confirmed by the still limited, extant mineral composition data (Niu et al., 2002). The two upper segments of host gabbros recovered during Leg 118 are characterized by associated Fe-Ti-oxide gabbros that show steep downward evolution trends toward more evolved compositions, in the opposite direction of the upward evolution of the host gabbros (Fig. 31). It is also significant to note that the upper segments of host gabbros are relatively more primitive than the lower segment recovered during Leg 176. The segmentation of the host gabbros is in many respects similar to that seen from ophiolitic gabbros (Troodos; Thy et al., 1989) and layered intrusions (Wilson and Larsen, 1985). Such large-scale, vertical gabbro segmentation has been related to replenishment and influx of fresh magma into an existing and partially crystallized chamber. It is also possible that the segmentation results from the intrusion of multiple chambers at approximately the same crustal level.

The systematic upward increase in the amount of Fe-Ti-oxide gabbros suggests that the gabbro cells were connected, acted, and developed as one crystal mush, perhaps formed from a

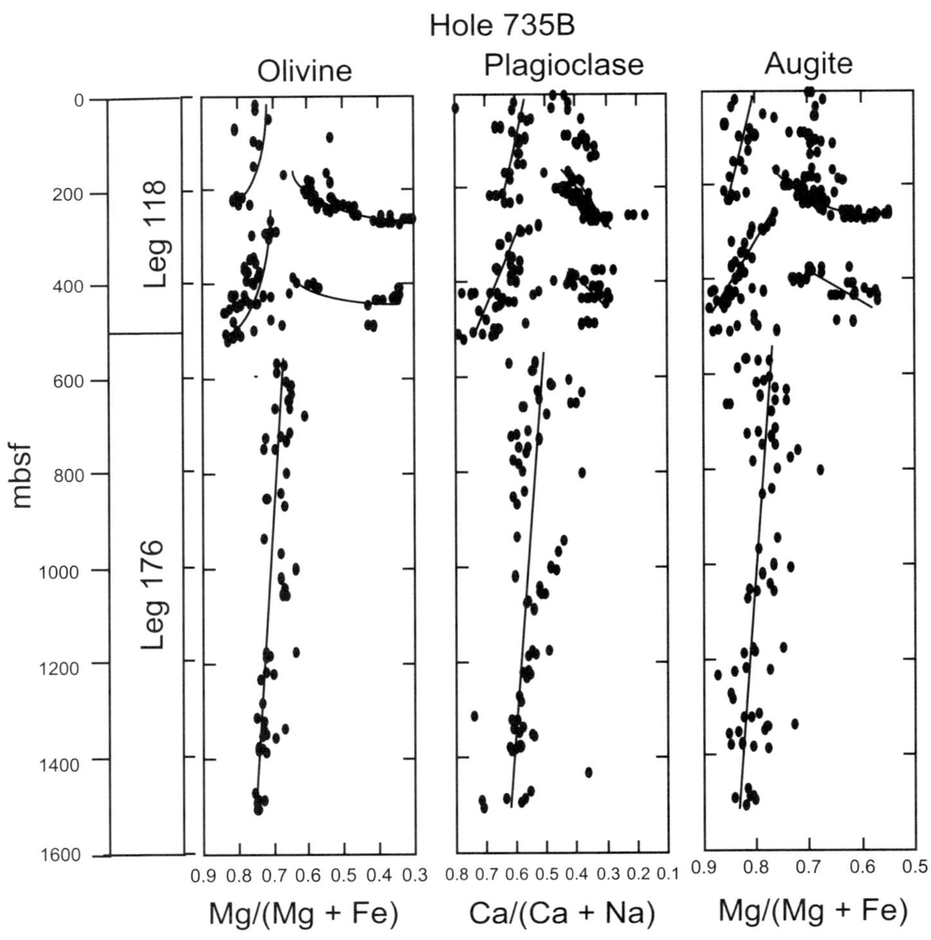

Figure 31. The cryptic variation in the composition of the main silicate minerals in the gabbros of Hole 735B (Bloomer et al., 1991; Ozawa et al., 1991; Natland et al. 1991; Hébert et al., 1991; Niu et al., 2002).

subsiding magma lens beneath the active ridge. The Fe-Ti-oxide gabbros have been inferred to represent interstitial melt that has migrated upward due to compaction, or laterally due to tectonic deformation and shear in a crystal mush. The observation that the evolved gabbro component appears to be concentrated upward in the gabbro section of Hole 735B has been interpreted to indicate that compaction was the dominating process on which the effects of ductile deformation and the lateral accumulation and flow of the interstitial melt were locally superimposed. These observations are not so different from the observations from continental layered intrusions (Thy and Dilek, 2000) and are predictable consequences of the solidification of any crystal mush pile at the bottom of a magma chamber (McBirney and Nicolas, 1997).

How the crystal mush grows and interacts with the new and fresh mantle-derived magma seeping up from beneath the chamber and entering the lower strata of the mush is an obvious and challenging question that has been addressed only most recently (e.g., Sinton and Detrick, 1992). It has been suggested from detailed field studies in the mantle-cumulate transitional zones of the Oman (Benn et al., 1988; Nicolas et al., 1988; Boudier et al., 1996) and the Bay of Islands (Bédard, 1991, 1993; Bédard and Hébert, 1996) ophiolites that new magma can form sill complexes in the crystal mush, perhaps in a similar fashion to sill intrusions in magmatically active, evolving sedimentary basins (Einsele, 1982). This concept has been developed by Bédard (1993) and more recently has been used by Kelemen et al. (1997) to suggest that the whole of the lower crust in Oman is built by a sill complex. It has also been proposed that the lower crust acts as a reactive filter modifying the residual melt migrating through the mush and sill complex, and, thus, buffering the composition of the melt that reaches the magma lens above the mush (Bédard, 1993; Bédard et al., 2000; MacLeod and Yaouancq, 2000).

These latest results have clearly demonstrated that the further development of our understanding of oceanic magma chambers has been strongly modified and improved by careful fieldwork of plutonic complexes in ophiolites accompanied by detailed petrographic and geochemical studies. This development substantiates the notion that the revival of systematic field studies in ophiolites is a forward approach to further advance our understanding of the workings of oceanic spreading centers and the related magmatic and tectonic processes during the evolution of oceanic crust.

EPILOGUE

It is tempting to incorporate near the conclusion of this review of the historical development of ophiolitic and oceanic magma chambers our own preferred idea of the "nature" of ophiolitic and oceanic magma chambers. We do this with the help of several simple diagrams depicted in Figure 32. From the outset, we would like to caution against a complete rejection of closed magma chambers and the traditional cumulus theories, and to point out the potential pitfalls for the adoption of a singular view of oceanic magma chambers. We think that this caution is well founded in studies of ophiolites and continental layered plutons. The problems with the poor resolution of seismic imaging and the difficulties of direct sampling of in situ oceanic crust contribute to our limited understanding of magma chamber processes.

The evolution of slow-spreading crust is characterized by small, ephemeral chambers that were emplaced at various crustal levels, and is controlled by localized and varied magma supply rates, sporadic episodes of tectonic extension, and possible rift axis relocations. Chambers emplaced by a single or a few pulses of magma may, therefore, be common features of slow-spreading centers. The relative volume of crystal mushes in such ephemeral magma chambers and the nature of crystallization processes in them may vary as a function of cooling rate. In many aspects, plutonic rocks that formed in such chambers may show many similarities to plutonic rocks in continental layered intrusions. Some continental layered intrusions seem to have been dominated by thick cumulate (or crystal mush) sections with significant amounts of trapped melt that were expelled slowly by compaction and migrated toward the main reservoir (Fig. 32A). Other intrusions might have been dominated by the inward migration of the crystallization front along a boundary layer, leaving behind a solid product in the bottom and along the margins of magma chambers (Fig. 32A). Similar processes can be envisioned for slow- to intermediate-spreading oceanic crust (Fig. 32B). This would account for the lack of deformation banding in some ophiolite complexes, such as in the Troodos, and for the evidence of multiple intrusive events that alternated with strong differentiation processes in the absence of steady-state equilibria. Because the actual mode of solidification of continental layered intrusions is still controversial, there is no objective reason to assume that this task should be easier for ophiolitic plutons or for oceanic magma chambers.

The high rates of magma flux at fast-spreading centers are likely to support development of large chambers with steady-state mushes (Fig. 32C) and crustal thermal gradients that are well above those observed from slow-spreading centers. It is important to emphasize at this point that because the seismically imaged melt lens is fed and maintained by the underlying mush zone, these two entities, the melt lens and the mush, should not be viewed as spatially isolated systems. The chambers underlying the intermediate- to fast-spreading centers thus may have substantial volumes of magma, far exceeding those beneath slow-spreading centers. Hence, the definition and discussion of a magma chamber should include the melt lens, as well as the surrounding and underlying mush zone.

Magma chambers beneath fast-spreading centers are likely controlled by the processes of compaction, subsidence, and near-solidus deformation in thick mush zones, which results in flow banding rather than the modally graded magmatic layering commonly observed in continental layered intrusions. At fast-spreading centers, fresh magma is continually added to the mush zone from the underlying mantle. The addition of this new magma to the mush and eventually to the melt lens above it helps the chamber attain its steady-state volume and makes the crust grow. The migration and the interaction of the fresh magma with the existing

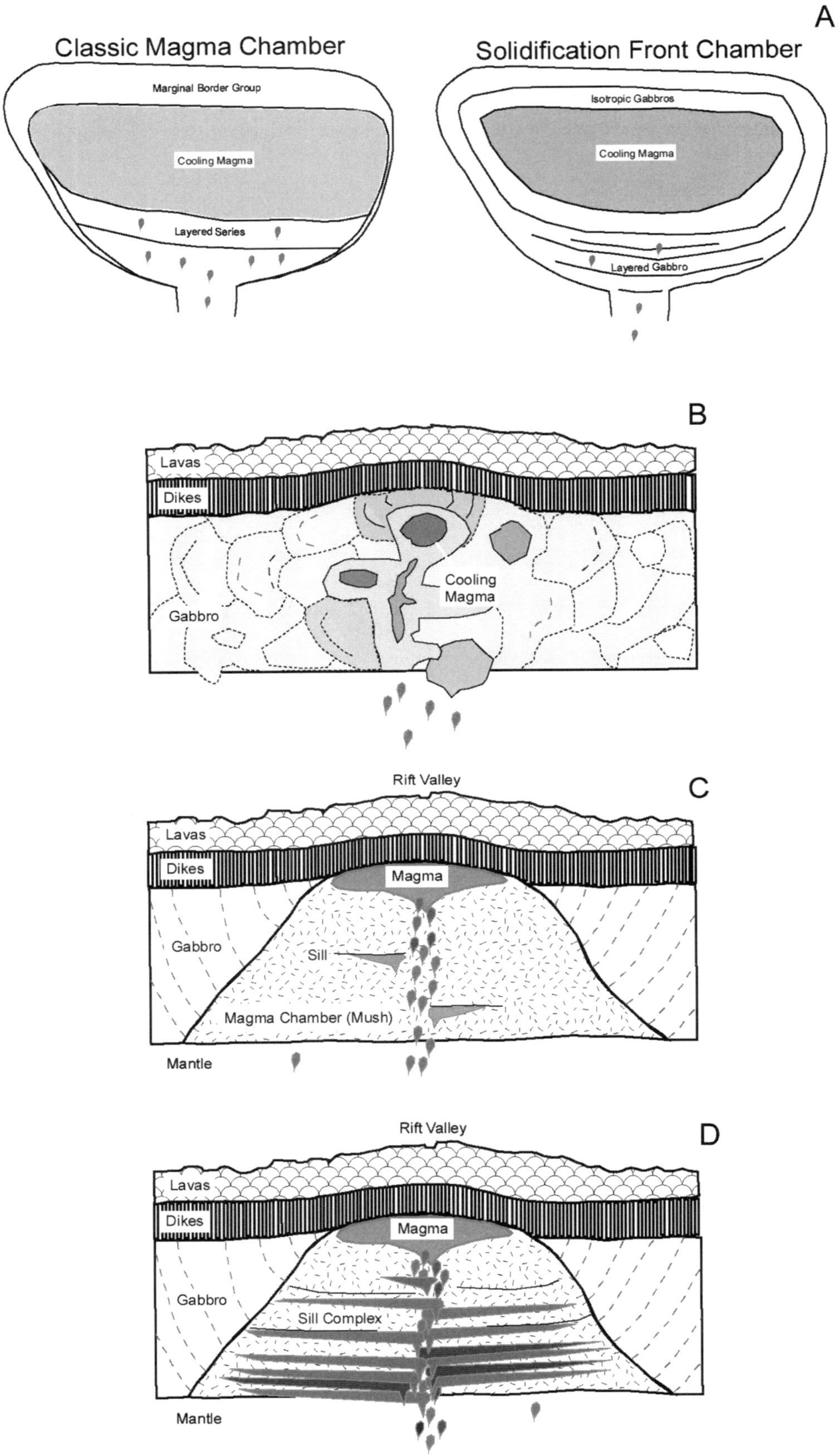

Figure 32. Tentative end-member models for magma chambers. A: Closed chamber crystallizing by either gravitational settling and formation of a crystal mush along the floor with little influx of new magma, or by crystallizing along the margins and inward without forming a mush. The latter model incorporates initial phenocrysts settling along the bottom of the chamber (after Marsh, 2000). B: Slow-spreading crust with intermittent magma supply and tectonic extension. Magma chamber formation and evolution are viewed as being very similar to what we envision for continental layered intrusions. C: Open, spreading-controlled magma chamber with thick mush on the floor and a thin magma lens on top. Compaction and migration of residual melt, as well as of fresh magma seeping up from the mantle into the mush, fill up the chamber. Local development of sills in the mush is determined by local porosity variation. D: Mush-filled, spreading-controlled magma chamber replenished with incoming fresh magma, which builds a sill complex in the mush concurrently with compaction and upward migration of the residual melt.

mush are critical processes during the evolution of fast-spreading chambers. In some cases, the new magma percolates rather freely through the existing mush while crystallizing and reacting on its way to the melt lens (Fig. 32C). In some other cases, the new influx of magma may form sill complexes in the existing mush (Fig. 32D), as a function of magma flux rate, permeability variation, and/or contrasting physical properties between the mush and the migrating magma. We see evidence for the artifacts of complex assimilation and percolation processes in some ophiolite complexes, such as in the Bay of Islands and Oman, whereas other ophiolites, e.g., Troodos, show no evidence for them.

Our three "end-member" types of magma chambers, as illustrated in Figure 32, are not always mutually exclusive. For example, whereas strong evidence for flow deformation exists in the Oman gabbros, modally graded layers more reminiscent of classic cumulus processes are also observed in these gabbros. Similarly, upward migrating ferrobasaltic melt appears to have existed in gabbros near the Atlantis Fracture Zone (Southwest Indian Ridge), that also contain evidence for magma replenishment and differentiation, as inferred for the Troodos gabbros and for some continental layered intrusions. Thus, we suggest that different aspects of these three end-members may occur in individual ophiolite complexes, depending on the mode and nature of solidification processes that dominated during their formation.

We shall always be limited in our interpretations of oceanic magma chambers because we cannot directly observe the responsible processes. We can, however, make significant advances in understanding oceanic magma chambers and processes by conducting detailed petrographic and geochemical studies of ophiolites, as well as of continental layered intrusions, combined with systematic structural field investigations. In the meantime, future deep drilling into in situ oceanic crust at and near modern spreading centers, and associated geophysical measurements and modeling, should still provide critical information, observations, and stimulating questions about oceanic crust formation and related processes.

SUMMARY

Conceptual models for oceanic magma chambers have evolved continually over the years since the first structural observations and interpretations from ophiolites and modern oceanic crust. We can examine this evolution of ideas in three overlapping time frames.

Prior to the advent of plate tectonics, mafic-ultramafic and sedimentary rock associations, which we today know as ophiolites, were considered generally as large submarine eruptions into subsiding marginal basins. The genetic connection between mantle peridotites and the commonly associated basaltic lavas, as in the Steinmann Trinity, was not yet made. This was the time when marginal basins (or geosynclines) were considered as the precursors to mountain belts.

Recognition of a sheeted dike complex in the Troodos ophiolite in the 1960s prompted the early researchers to think about seafloor spreading processes and to suggest a possible ophiolite-ocean crust analogy. The identification of harzburgites and dunites as mantle rocks, and gabbros and pyroxenites as crustal rocks in the Troodos, Bay of Islands, and Oman ophiolites (see Juteau, this volume, Chapter 3) led to the recognition of fossil magma chambers in the plutonic sections of ophiolite complexes. This major step in the formulation of the ophiolitic magma chamber concept was prompted by the newly established plate tectonics theory, which predicted that basaltic melt, separated from upwelling zones of partially molten peridotites beneath the crest of mid-ocean ridges, would accumulate in crustal magma chambers.

Early models of the structure and evolution of modern oceanic crust, based mainly on geophysical modeling, did not incorporate magma chambers and made little use of observations and ideas derived from ophiolites. The ruling geophysical interpretation of magma chambers was the classic, steady-state large chamber model that dominated the models of ocean crust up to the early 1990s. Ophiolite researchers adopted this chamber model and readily extended their ophiolite-derived interpretations of magma chambers to the oceans, suggesting various models for magma chamber evolution beneath active spreading centers throughout the 1970s and 1980s. Ophiolite-derived magma chamber models evolved from single-chamber to multiple-chamber configurations with various sizes and geometries during the 1980s to explain the thickness of gabbroic rocks and their layering, the observed fractionation patterns, and the distribution of cumulus phase assemblages.

Following the early ideas, geophysical modeling and interpretations of the late 1970s envisioned oceanic magma chambers as a thin melt lens located below the sheeted dike complex and extending downward to the mantle as a crystal-liquid mush zone. These models eventually predicted that steady-state magma chambers could not exist beneath slow-spreading centers, as thermal calculations had suggested, and that small crustal chambers beneath the axial valleys of these spreading centers were likely to be intermittent (ephemeral) in nature.

These geophysical interpretations of oceanic magma chambers were substantiated by seismic studies of the fast-spreading East Pacific Rise and the slow-spreading Mid-Atlantic Ridge in the 1990s. Whereas through these studies small, shallow magma pockets and low-velocity zones were detected along ridge segments of the East Pacific Rise, there was no indication of the presence of low-velocity zones and magma chambers beneath the Mid-Atlantic Ridge. The significance of the seafloor spreading and magma supply rates for magma chamber evolution and, hence, for oceanic crust generation was thus underscored by the mid-1990s.

Observations and interpretations from ophiolites and modern spreading centers started converging in the early 1990s, and this "two-way traffic" in the development of scientific concepts modified the magma chamber models significantly. This era in the evolution of geological thought regarding the magma chamber concept coincides with the productive attempts of the ODP to drill into in situ lower oceanic crust at or near active spreading centers.

Various projects of the ODP and the related workshops brought together the members of the ophiolite and marine geology and geophysics communities, facilitating the exchange and cross-pollination of ideas and observations on oceanic crust formation.

The new magma chamber model that emerged in the mid-1990s envisioned the magma chambers beneath fast-spreading centers to be filled mainly with crystal mush containing only small percentages of melt and to be capped by a thin melt lens perched on top of a mush zone. Thus, a magma chamber of a fast-spreading system would consist of a melt lens on top and the mush zone below that is made of a plastically deforming body of high viscosity material. The operation of mush zones during the crystallization of gabbros had long been predicted for the igneous development of continental layered intrusions; thus, the idea of magma chambers containing mush zones was not necessarily new. Core samples of gabbroic rocks recovered from deep drilling into in situ lower oceanic crust include evolved Fe-Ti-oxide gabbros representing the interstitial melt trapped in the host gabbro. Upward and lateral migration of this melt within the host gabbro network is facilitated by compaction and/or tectonic deformation occurring in the spreading system.

Detailed and systematic studies in recent years of ophiolitic plutons and plutonic complexes have highlighted the significance of syntexis processes, by which new magmas react with their host rocks, during the evolution of oceanic crust. New batches of magma seeping up from beneath the chamber form networks of sill-like bodies within the crystal mush that act as a reactive filter to the residual melt migrating upward through the mush. Thus, the composition of this melt changes as it passes through the mush zone on its way to the melt lens on top. This current thinking in melt and magma chamber evolution has been significantly improved by the revival of systematic petrographic and geochemical studies in ophiolites, and the synergy and mutual exchange between the ophiolite and marine geology and geophysics communities have helped further advance our understanding of oceanic magma chambers and related processes.

ACKNOWLEDGMENTS

Critical and thorough reviews by Jean Bédard, Sherman Bloomer, Don Elthon, and Sally Newcomb greatly improved the text and our understanding of ophiolitic and oceanic magma chambers and are gratefully acknowledged. In writing this historical review paper, we have made an effort to be objective in acknowledging the contributions of all major papers and the literature involving oceanic and ophiolitic magma chamber studies; we bear the responsibility of any remaining omissions. We would like to extend our sincere thanks to numerous colleagues and scientists in the international ophiolite and marine geology and geophysics communities with whom we have had collaborated and exchanged ideas and information on different aspects of the evolution of ophiolites and modern oceanic crust. We appreciate the permission granted from publishers and authors to use many of the illustrations shown in this review. Our studies of ophiolites, modern oceanic crust, and layered continental intrusions have been supported by research grants from the National Science Foundation (EAR-9219064, EAR-9796011, OCE-9813451 to Y. Dilek), Joint Oceanographic Institutions, the U.S. Science Advisory Committee, the Ocean Drilling Program, and the NATO Science Programme (Environmental and Earth Science & Technology Collaborative Linkage grant, EST.CLG-97617).

REFERENCES CITED

Allemann, F., and Peters, T., 1972, The ophiolite-radiolarite belt of North-Oman Mountains. Eclogae Geologicae Helvetiae, v. 65, p. 657–697.

Allen, C.R., 1975, The Petrology of a portion of the Troodos plutonic complex, Cyprus [Ph.D. Thesis]: Cambridge, UK, Cambridge University, 161 p.

Aubouin, J., 1965, Geosynclines: Amsterdam, Elsevier, 335 p.

Barth, T.F.W., 1962, Theoretical Petrology, second edition: New York, John Wiley, 416 p.

Bartlett, R.W., 1969, Magma convention, temperature distribution, and differentiation: American Journal of Science, v. 267, p. 1067–1082.

Batanova, V.G., Sobolev, A.V., and Schmincke, H.-U., 1995, Parental melts of the intrusive cumulates of the Troodos massif, Cyprus: A study of clinopyroxenes and melt inclusions in plagioclase: Petrology, v. 4, p. 255–264.

Batiza, R., 1996, Magmatic segmentation of mid-ocean ridges: A review, *in* MacLeod, C.J., Tyler, P.A., and Walker, C.L., eds., Tectonic, Magmatic, Hydrothermal and Biological Segmentation of Mid-Ocean Ridges: London, Geological Society of London Special Publication 118, p. 103–130.

Bear, L.M., 1960, The Geology and Mineral Resources of the Akaki-Lythrodhouda Area: Nicosia, Cyprus, Geological Survey of Cyprus, Memoir 3, 208 p.

Bédard, J.H., 1991, Cumulate recycling and crustal evolution in the Bay of Islands ophiolite: Journal of Geology, v. 99, p. 225–249.

Bédard, J.H., 1993, Oceanic crust as a reactive filter: synkinematic intrusion, hybridization, and assimilation in an ophiolitic magma chamber, western Newfoundland: Geology, v. 21, p. 77–80.

Bédard, J.H., and Hébert, R., 1996, The lower crust of the Bay of Islands ophiolite, Canada: petrology, mineralogy, and the importance of syntexis in magmatic differentiation in ophiolites and at ocean ridges: Journal of Geophysical Research, v. 101, p. 25,105–25,124.

Bédard, J.H., Hébert, R., Berclaz, A., and Varfalvy, V., 2000, Synthexis and the genesis of lower oceanic crust, *in* Dilek, Y., Moores, E.M., Elthon, D., and Nicolas, A., eds., Ophiolites and oceanic crust: New insights from field studies and the Ocean Drilling Program: Boulder, Colorado, Geological Society of America Special Paper 349, p. 105–119.

Benn, K., and Laurent, R., 1987, Intrusive suite documented in the Troodos ophiolite plutonic complex, Cyprus: Geology, v. 15, p. 821–824.

Benn, K., Nicolas, A., and Reuber, I., 1988, Mantle-crust transition zone and origin of wehrlitic magmas: Evidence from the Oman ophiolite: Tectonophysics, v. 151, p. 75–85.

Best, M.G., 1982, Igneous and Metamorphic Petrology: New York, W.H. Freeman, 630 p.

Bishopp, D.W., 1952, The Troödos massif, Cyprus: Nature, v. 169, p. 489–490.

Bishopp, D.W., 1954, Some new features of the geology of Cyprus: Congrès Géologique International, 19th, Alger, Proceedings, v. 17, p. 13–18.

Bloomer, S.H., Meyer, P.S., Dick, H.J.B., Ozawa, K., and Natland, J.H., 1991, Textural and mineralogical variations in gabbroic rocks from Hole 735B, *in* Von Herzen, R.P., Robinson, P.T., eds., Proceedings of the Ocean Drilling Program, Scientific Results: College Station, Texas, Ocean Drilling Program, v. 118, p. 21–39.

Bott, M.H.P., 1982. The Interior of the Earth: Its Structure, Constitution and Evolution, second edition: Amsterdam, Elsevier, 402 p.

Boudier, F., Nicolas, A., and Ildefonse, B., 1996, Magma chambers in the Oman ophiolite: Fed from the top and the bottom: Earth and Planetary Science Letters, v. 144, p. 239–250.

Bowen, N.L., 1928, The Evolution of the Igneous Rocks: Princeton, Princeton University Press, 332.

Brown, G.C., and Mussett, A.E., 1981, The Inaccessible Earth: London, George Allen & Unwin, 235 p.

Browning, P., 1984, Cryptic variation within the cumulate sequence of the Oman ophiolite: Magma chamber depth and petrological implications, *in* Gass, I.G.,

Lippard, S.J., and Shelton, A.W., eds., Ophiolites and Oceanic Lithosphere: Geological Society London, Special Publication 13, p. 71–82.

Browning, P., Roberts, S., and Alabaster, T. 1989, Fine-scale modal layering and cyclic units in ultramafic cumulates from the CY-4 borehole, Troodos ophiolite: evidence for open system magma chamber, *in* Gibson, I.L., Malpas, J., Robinson, P.T., and Xenophontos, C., eds., Cyprus Crustal Study Project, Hole CY-4, Geological Survey Canada Paper, v. 88–89, p. 193–220.

Brunn, J.H., 1960, Mise en place et differenciation pluto-volcanique de cortège ophiolitique: Revue de Geologie Dynamique et de Geographie Physique, v. 3, p. 115–132.

Bryan, M.P., 1993, Neutral buoyancy and the structure of mid-ocean ridge magma reservoirs: Journal of Geophysical Research, v. 98, p. 22,321–22,338.

Bryan, W.B., and Moore, J.G., 1977, Compositional variations of young basalts in the mid-Atlantic ridge rift valley near lat 36° 49' N: Geological Society of America Bulletin, v. 88, p. 556–570.

Cann, J.R., 1968, Geological processes at mid-ocean ridge crests: Geophysical Journal of the Royal Astronomical Society, v. 15, p. 331–341.

Cann, J.R., 1970, New model for the structure of the ocean crust: Nature, v. 226, p. 928–930.

Cann, J.R., 1974, A model for oceanic crustal structure developed: Geophysical Journal of the Royal Astronomical Society, v. 39, p. 169–187.

Cann, J.R., 1991, Onions and leaks: magma at mid-ocean ridges. A very personal view: Oceanus, v. 34 (4), p. 6–41.

Casey, J.F., and Karson, J.A., 1981, Magma chamber profiles from the Bay of Islands ophiolite complex: Nature, v. 292, p. 295–301.

Casey, J.F., Dewey, J.F., Fox, P.J., Karson, J.A., and Rosencrantz, E., 1981, Heterogeneous nature of oceanic crust and upper mantle: a perspective from the Bay of Islands ophiolitic complex, *in* Emiliani, C., ed., The Oceanic Lithosphere, The Sea, Volume 7: New York, John Wiley, p. 305–338.

Casey, J.F., Karson, J.A., Elthon, D., Rosencrantz, E., and Titus, M., 1983, Reconstruction of the geometry of accretion during formation of the Bay of Islands ophiolite complex: Tectonics, v. 2, p. 509–528.

Cawthorn, R.G., ed., 1996, Layered Intrusions: Amsterdam, Elsevier, 531 p.

Chenevez, J., Machetel, P., and Nicolas, A., 1998, Numerical models of magma chambers in the Oman ophiolite: Journal of Geophysical Research, v. 103, p. 15,443–15,455.

Christensen, N.I., and Salisbury. M.H., 1975, Structure and constitution of the lower oceanic crust: Reviews of Geophysics and Space Physics, v. 13, p. 57–86.

Church, W.R., 1972, Ophiolite: its definition, origin as oceanic crust, and mode of emplacement in orogenic belts, with special reference to the Appalachians: Ottawa, Ontario, Canada, Department of Energy, Mines and Resources Publications 42, p. 71–85.

Church, W.R., and Riccio, L., 1977, Fractionation trends in the Bay of Islands ophiolite of Newfoundland: polycyclic cumulate sequences in ophiolites and their classification: Canadian Journal of Earth Science, v. 14, p. 1156–1165.

Church, W.R., and Stevens, R.K., 1971, Early Paleozoic ophiolite complexes of the Newfoundland Appalachians as mantle-oceanic crust sequences: Journal of Geophysical Research, v. 76, p. 1460–1466.

Coleman, R.G., 1977, Ophiolites. Ancient Oceanic Lithosphere?: Berlin, Springer-Verlag, 229 p.

Coleman, R.G., 1981,Tectonic setting for ophiolite obduction in Oman: Journal of Geophysical Research, v. 86, p. 2497–2508.

Coleman, R.G., and Hopson, C.A., eds., 1981, Oman Ophiolite: Journal Geophysical Research, v. 86, p. 2497–2782.

Cooper, J.R., 1937, Geology of the Southern Half of the Bay of Islands Igneous Complex: St. Johns, Newfoundland, Canada, Newfoundland Department of Natural Resources Bulletin, v. 4, 35 p.

Daly, R.A., 1914, Igneous Rocks and Their Origin: New York, McGraw-Hill, 563 p.

Daly, R.A., 1933, Igneous Rocks and the Depth of the Earth: New York, McGraw-Hill, 598 p.

Daly, R.A., 1942, The Floor of the Ocean. New Light on Old Mysteries: Chapel Hill, University of North Carolina Press, 177 p.

Detrick, R.S., Buhl, P., Vera, E., Mutter, J, Orcutt, J.A., Madsen, J., and Brocher, T., 1987, Multichannel seismic imaging of an axial magma chamber along the East Pacific Rise between 9°N and 13°N: Nature, v. 326, p. 35–47.

Dewey, J.F., and Bird, J.M., 1971, Origin and emplacement of the ophiolitic suite: Appalachian ophiolites in Newfoundland: Journal of Geophysical Research, v. 76, p. 3179–3206.

Dewey, J.F., and Kidd, W.S.F., 1977, Geometry of plate accretion: Geological Society of America Bulletin, v. 88, p. 960–968.

Dick, H.J.B., Meyer, P.S., Bloomer, S., Kirby, S., Stakes, D., and Mawer, C., 1991b, Lithostratigraphic evolution of an in-situ section of oceanic layer 3, *in* Von Herzen, R.P., and Robinson, P.T., eds., Proceedings Ocean Drilling Program, Scientific Results, 118, College Station, Texas, p. 439–538.

Dick, H.J.B., Natland, J.H., Alt, J C., Bach, W., Bideau, D., Gee, J.S., Haggas, S., Hertogen, J.G.H., Hirth, G., Holm, P.M., Ildefonse, B., Iturrino, G.J., John, B.E., Kelley, D.S., Kikawa, E., Kingdon, A., LeRoux, P.J., Maeda, J., Meyer, P.S., Miller, D.J., Naslund, H.R., Niu, Y.-L., Robinson, P.T., Snow, J., Stephen, R.A., Trimby, P.W., Worm, H.-U., and Yoshinoby, A., 2000, A long in situ section of the lower ocean crust: Results of ODP Leg 176 drilling at the Southwest Indian Ridge: Earth Planetary Science Letters, v. 179, p. 31–51.

Dietz, R.S., 1961, Continent and ocean basin evolution by spreading of the sea floor: Nature, v. 190, p. 854–857.

Dietz, R.S., 1962, Ocean-basin evolution by sea-floor spreading, *in* Runcorn, S.K., ed., Continental Drift, New York, Academic Press. p. 289–298.

Dilek, Y., and Eddy, C.A., 1992, The Troodos (Cyprus) and Kizildag (S. Turkey) Ophiolites as structural models for slow-spreading ridge segments: Journal of Geology, v. 100, p. 305–322.

Dilek, Y., Moores, E.M., and Furnes, H., 1998, Structure of modern oceanic crust and ophiolites and implications for faulting and magmatism at oceanic spreading centers: *in* Buck, W.R., Delaney, P.T., Karson, J.A., and Lagabrielle, eds., Faulting and Magmatism at mid-ocean ridges: Washington, D.C., American Geophysical Union, Monograph 106, p. 219–265.

Eddy, C.A., Dilek, Y., Hurst, S., and Moores, E.M., 1998, Seamount formation and associated caldera complex and hydrothermal mineralization in ancient oceanic crust, Troodos ophiolite (Cyprus): Tectonophysics, v. 292, p. 189–210.

Einsele, G., 1982, Mechanism of sill intrusion into soft sediment and expulsion of pore water, *in* Curray, J.R., Moore, D.G., et al., Initial Reports of the Deep Sea Drilling Project, v. 64: Washington D.C., U.S. Government Printing Office, p. 1169–1176.

Elthon, D., 1991, Geochemical evidence for formation if the Bay of Islands ophiolite above a subduction zone: Nature, v. 354, p. 140–143.

Elthon, D., Casey, J.F., and Komor, S., 1982, Mineral chemistry of ultramafic cumulates from the North Arm Mountain massif of the Bay of Islands ophiolite: evidence for high-pressure crystal fractionation of oceanic basalts: Journal of Geophysical Research, v. 87, p. 8717–8734.

Elthon, D., Casey, J.F., and Komor, S., 1984, Cryptic mineral-chemistry variations in a detailed traverse through the cumulate ultramafic rocks of the North Arm Mountain massif of the Bay of Islands ophiolite, Newfoundland, *in* Gass, I.G., Lippard, S.J., and Shelton, A.W., eds., Ophiolites and Oceanic Lithosphere: Geological Society London Special Publication 13, p. 83–97.

Engel, C.G., and Fisher, R.L., 1969, Lherzolite, anorthosite, gabbro and basalt dredges from the mid-Indian ocean ridge: Science, v. 166, p. 1136–1141.

Ernewein, M., Pflumio, C., and Whitechurch, H., 1988, The death of an accretion zone as evidenced by the magmatic history of the Sumail ophiolite (Oman): Tectonophysics, v. 151, p. 247–274.

Erwing, J., and Ewing, M., 1959, Seismic refraction measurements in the Atlantic Ocean basin, in the Mediterranean Sea, on the Mid-Atlantic Ridge, and in the Norwegian Sea: Geological Society of America Bulletin, v. 70, p. 291–318.

Fox, P.J., and Stroup, J.B., 1981, The plutonic foundation of the oceanic crust, *in* Emiliani, C., ed., The Oceanic Lithosphere, The Sea, Volume 7: New York, John Wiley p. 119–217.

Gass, I.G., 1960, The Geology and Mineral Resources of the Dhali Area. Geological Survey of Cyprus, Memoir 4, 116 p.

Gass, I.G., 1967, The ultrabasic volcanic assemblage of the Troodos massif, Cyprus, *in* Wyllie, P.J., ed., Ultramafic and Related Rocks: New York, John Wiley and Sons, p. 121–134.

Gass, I.G., 1968. Is the Troodos massif of Cyprus a fragment of Mesozoic ocean floor?: Nature v. 220, p. 39–42.

Gass, I.G., 1980, The Troodos massif: its role in the unraveling of the ophiolite problem and its significance in the understanding of constructive plate margin processes, *in* Panayiotou, A., ed., Ophiolites, Proceedings, 1979 International Ophiolite Symposium, Cyprus, Nicosia, Cyprus Geological Survey Department, p. 23–35.

Gass, I.G., 1989, Magmatic processes at and near constructive plate margins as deduced from Troodos (Cyprus) and Semail Nappe (N. Oman) ophiolites, *in* Saunders, A.D., and Norry, M.J., eds., Magmatism in the Ocean Basins: Geological Society London Special Publication 42, p. 1–15.

Gass, I.G., 1990, Ophiolites and oceanic lithosphere, *in* Malpas, J., Moores, E.M., Panayiotou, A., and Xenophontos, C., eds., Ophiolites: Oceanic Crustal Analogies, Proceedings, 1987 Troodos, Nicosia, Cyprus, Geological Survey Department, p. 1–10.

Gass, I.G., and Masson-Smith, D., 1963, The geology and gravity anomalies of the Troodos massif, Cyprus: Royal Society of London, Philosophical Transactions, ser. A, v. 255, p. 417–467.

Gass, I.G., and Smewing, J.D., 1981, Ophiolites: obducted oceanic lithosphere, *in* Emiliani, C., ed., The Oceanic Lithosphere. The Sea, Volume 7:New York, John Wiley and Sons, p. 339–362.

Girardeau, J., and Nicolas, A., 1981, The structures of two ophiolite massifs, Bay-of-Islands, Newfoundland: A model for the oceanic crust and upper mantle: Tectonophysics, v. 77, p. 1–34.

Glen, W., 1982, The Road to Jaramillo. Critical Years of the Revolution in Earth Science. Stanford, Stanford University Press, 459 p.

Glennie, K.W., Boeuf, M.G.A., Hughes Clarke, M.W., Moody-Stuart, M., Pilaar, W.F.H., and Reinhardt, B.M., 1974, Geology of the Oman Mountains. Verhandelingen van het Koninklijk Nederlands: Geologisch Mijnbouwkunding Genootschap, v. 31, p. 423 p.

Green, D.H., 1968, Origin of Basaltic Magmas, *in* Hess, H.H., and Poldervaart, A., eds., Basalts: The Poldervaart Treatise on Rocks of Basaltic Composition, Volume 2: New York, Interscience Publishers, p. 835–862.

Greenbaum, D., 1972, Magmatic processes at ocean ridges: evidence from the Troodos massif, Cyprus: Nature, v. 238, p. 18–21.

Greenbaum, D., 1977, The chromitiferous rocks of the Troodos ophiolite, Cyprus: Economic Geology, v. 72, p. 1175–1194.

Grove, T.L., Kinzler, R.J., and Bryan, W.B., 1992, Fractionation of mid-ocean ridge basalt, *in* Morgan, J.P., Blackman, D.K., and Sinton, J.M., eds., Mantle Flow and Melt Generation at Mid-Ocean Ridges: Washington, D.C., American Geophysical Union Monograph 71, p. 281–310.

Harding, A.J., Orcutt, J.A., Kappus, M.E., Vera, E.E., Mutter, J.C., Buhl, P., Detrick, R.S., and Brocher, T.M., 1989, Structure of young oceanic crust at 13° N on the East Pacific Rice from expanding spread profiles: Journal of Geophysical Research, v. 94, p. 12,163–12,196.

Hébert, R., Constantin, M., and Robinson, P.T., 1991, Primary mineralogy of Leg 118 gabbroic rocks and their place in the spectrum of oceanic mafic igneous rocks, *in* Von Herzen, R.P., Robinson, P.T., et. al., eds., Proceedings, Ocean Drilling Program, Scientific Results, 118, College Station, Texas, p. 3–20.

Heezen, B.C., 1962, The deep-sea floor, *in* Runcorn, S.K., ed., Continental Drift, New York, Academic Press, p. 235–288.

Henstock, T.J., Woods, A.W., and White, R.S., 1993, The accretion of oceanic crust by episodic sill intrusion: Journal of Geophysical Research, v. 98, p. 4143–4161.

Hess, H.H., 1938, A primary peridotite magma: American Journal of Science, v. 35, p. 321–344.

Hess, H.H., 1955, The oceanic crust: Journal of Marine Research, v. 14, p. 423–439.

Hess, H.H., 1960, Stillwater Igneous Complex, Montana: a Quantitative Mineralogical Study: Geological Society of America, Memoir 80, 230 p.

Hess, H.H., 1962, History of ocean basins, *in* Engel, A.E.J., James, H.L., and Leonard, B.F., eds., Petrologic Studies: A Volume in Honor of A.F., Buddington: Boulder, Colorado, Geological Society of America Monograph, p. 599–620.

Holmes, A., 1965, Principles of Physical Geology, second edition: New York, Ronald Press, 1288 p.

Hopson, C.A., and Frano, C.J., 1977, Igneous history of the Point Sal ophiolite, southern California, *in* Coleman, R.G., and Irwin, W.P., eds., Northern American Ophiolites: Corvalis, Oregon Department of Geology and Mineral Industries Bulletin, v. 95, p. 161–183.

Irvine, T.N., 1970, Heat transfer during solidification of layered intrusions. 1. Sheets and sills: Canadian Journal of Earth Science, v. 7, p. 1031–1061.

Irvine, T.N., Keith, D.W., and Todd, S.G., 1983, The J-M platinum-palladium reef of the Stillwater complex, Montana: II. Origin by double-diffusive convective magma mixing and implications for the Bushveld complex: Economic Geology, v. 78, p. 1287–1334.

Irvine, T.N., Andersen, J.C.O. and Brooks, C.K., 1998, Included blocks (and blocks within blocks) in the Skaergaard intrusion: Geologic relations and the origins of rhythmic modally graded layers: Geological Society of America Bulletin, v. 110, p. 1398–1447.

Karson, J.A., 1998, Internal structure of oceanic lithosphere: a perspective from tectonic windows, *in* Buck, W.R., Delaney, P.T., Karson, J.A., and Lagabrielle, Y., eds., Faulting and Magmatism at mid-ocean ridges: Washington, D.C., American Geophysical Union Monograph 106, p. 177–218.

Karson, J.A., 2002, Geological structure of the uppermost oceanic crust created at fast- to intermediate-rate spreading centers: Annual Reviews of Earth Sciences v. 30, p. 347–348.

Kay, M., 1951, North American Geosynclines: Boulder, Colorado, Geological Society of America Memoir 48, 143 p.

Kelemen, P.B., and Aharonov, E., 1998, Periodic formation of magma fractures and generation of layered gabbros in the lower crust beneath oceanic spreading ridges, *in* Buck, W.R., Delaney, P.T., Karson, J.A., and Lagabrielle, Y., eds., Faulting and Magmatism at mid-ocean ridges: Washington, D.C., American Geophysical Union Monograph 106, p. 267–289.

Kelemen, P.B., Koga, K., and Shimizu, N., 1997, Geochemistry of gabbro sills in the crust-mantle transition zone of the Oman ophiolite: implications for the origin of the oceanic lower crust: Earth and Planetary Science Letters, v. 146, p. 475–488.

Kent, G.M., Harding, A.J., and Orcutt, F., 1990, Evidence for a smaller magma chamber beneath the East Pacific Rise at 9° 20′ N: Nature, v. 344, p. 650–653.

Kent, G.M., Harding, A.J., and Orcutt, F., 1993, Distribution of magma beneath the East Pacific Rise near the 9° 03′ N overlapping spreading center from forward modeling of CDP data: Journal of Geophysical Research, v. 98, p. 13,971–13,998.

Kidd, R.G.W., 1977, A model for the process of formation of the upper oceanic crust: Geophysical Journal of the Royal Astronomical Society, v. 50, p. 149–183.

Klein, E.M., and Langmuir, C.H., 1987, Global correlation of ocean ridge basalt chemistry with axial depth and crystal thickness: Journal of Geophysical Research, v. 92, p. 8089–8115.

Komor, S.C., and Elthon, D., 1990, Formation of anorthosite-gabbro rhythmic phase layering: an example at North Arm Mountain, Bay of Islands ophiolite: Journal of Petrology, v. 31, p. 1–50.

Komor, S.C., Elthon, D., and Casey, J.F., 1985, Mineralogic variation in a layered ultramafic cumulate sequence at the North Arm Mountain massif, Bay of Islands ophiolite, Newfoundland: Journal of Geophysical Research, v. 90, p. 7705–7736.

Komor, S.C., Elthon, D., and Casey, J.F., 1987, Petrology of a leucogabbroic interval within basal layered gabbros at North Arm Mountain massif, Bay of Islands ophiolite: Contributions to Mineralogy and Petrology, v. 95, p. 278–300.

Korenaga, J., and Kelemen, P.B., 1997, Origin of gabbro sills in the Moho transition zone of the Oman ophiolite: implications for magma transport in the oceanic lower crust: Journal of Geophysical Research, v. 102, p. 27,729–27,749.

Korenaga, J., and Kelemen, P.B., 1998, Melt migration through the oceanic lower crust: a constraint from melt percolation modeling with finite solid diffusion: Earth and Planetary Science Letters, v. 156, p. 1–11.

Kusznir, N.J., and Bott, M.H.P., 1976, A thermal study of the formation of oceanic crust: Geophysical Journal of the Royal Astronomical Society, v. 47, p. 83–95.

Langmuir, C.H., Klein, E.Y., and Plank, T., 1992, Petrological systematics of mid-ocean ridge basalts: constraints on melt generation beneath ocean ridges, *in* Morgan, J.P., Blackman, D.K., and Sinton, J.M., eds., Mantle Flow and Melt Generation at Mid-Ocean Ridges: Washington, D.C., American Geophysical Union Monograph 71, p. 183–280.

Lees, G.M., 1928, The geology and tectonics of Oman and of parts of south-eastern Arabia: Quaternary Journal of the Geological Society, v. 84, p. 585–670.

Lewis, B.T.R., 1983, The process of formation of ocean crust: Science, v. 220, p. 151–157.

Lippard, S.J., Shelton, A.W., and Gass, I.G., 1986, The Ophiolites of Northern Oman: Geological Society of London, Memoir 11, 178 p.

Lister, C.R.B., 1977, Qualitative models of spreading center processes, including hydrothermal penetration: Tectonophysics v. 37, p. 203–218.

Lister, C.R.B., 1983, On the intermittency and crystallization mechanisms of sub-seafloor magma chambers: Geophysical Journal of the Royal Astronomical Society, v. 73, p. 351–365.

Maaløe, S., 2002, Physical behavior of the plume source during intermittent eruptions of Hawaiian basalts: Contributions to Mineralogy and Petrology v. 142, p. 653–665.

Macdonald, K.C., 1982, Mid-ocean ridges: fine scale tectonics, volcanic and hydrothermal processes within the plate boundary zone: Annual Review of Earth Sciences, v. 10, p. 155–190.

Macdonald, K.C., 1989, Anatomy of the magma reservoir: Nature, v. 339, p. 178–190.

MacGregor, I., 1967, Mineralogy and model mantle compositions, *in* Wyllie, P.J., ed., Ultramafic and Related Rocks: New York, John Wiley, p. 381–392.

MacLeod, C.J., and Yaouancq, G., 2000, A fossil melt lens in the Oman ophiolite: implications for magma chamber processes at fast spreading ridges: Earth and Planetary Science Letters, v. 176, p. 357–373.

Malpas, J., 1990, Crustal accretionary processes in the Troodos ophiolite, Cyprus: evidence from field mapping and deep crustal drilling, *in* Malpas, J., Moores, E.M., Panayiotou, A., and Xenophontos, C., eds., Ophiolites: Oceanic Crustal Analogues, Proceedings, 1987 Troodos, Nicosia, Cyprus, Geological Survey Department, p. 65–74.

Malpas, J., Brace, T., and Dunsworth, S.M., 1989, Structural and petrological relationships of the CY-4 drill hole of the Cyprus Crustal Study Project, *in* Gibson, I.L., Malpas, J., Robinson, P.T., and Xenophontos, C., eds., Cyprus Crustal Study Project, Hole CY-4: Geological Survey Canada Paper, v. 88–9, p. 39–67.

Marsh, B.D., 1996, Solidification fronts and magmatic evolution: Mineralogical Magazine, v. 60, p. 5–10.

Marsh, B.D., 2000, Magma chambers, *in* Sigurdsson, H., ed., Encyclopedia of Volcanoes: San Diego, Academic Press, p. 191–206.

McBirney, A.R., 1993, Igneous Petrology, second edition: Boston, Jones and Bartlett, 520 p.

McBirney, A.R., 1995, Mechanisms of differentiation in the Skaergaard intrusion: Journal of Geological Society of London, v. 152, p. 421–435.

McBirney, A., and Noyes, R.M., 1979, Crystallization and layering of the Skaergaard intrusion: Journal of Petrology, v. 20, p. 487–554.

McBirney, A.R., and Nicolas, A., 1997, The Skaergaard layered series. Part II. Magmatic flow and dynamic layering: Journal of Petrology, v. 38, p. 569–580.

McKenzie, D. and Bickle, M.J., 1988, The volume and composition of melt generated by extension of the lithosphere: Journal of Petrology, v. 29, p. 625–679.

Moore, J.G., Fleming, H.S., Phillips, J.D., 1974, Preliminary model for extrusion and rifting at the axis of the mid-Atlantic ridge, 36° 48' N: Geology, v. 2, p. 437–440.

Moores, E.M., 1969, Petrology and Structure of the Vourinos Ophiolite Complex of Northern Greece: Boulder, Colorado, Geological Society of America Special Paper 118, 74 p.

Moores, E.M., and Vine, F.J., 1971, The Troodos massif, Cyprus and other ophiolites as oceanic crust: evaluation and implications: Royal Society of London, Philosophical Transactions, ser. A, v. 268, p. 443–466.

Moores, E.M., Varga, R.J., Verosub, K.L., and Ramsden, T.W., 1990, Regional structure of the Troodos dyke complex, *in* Malpas, J., Moores, E.M., Panayiotou, A., and Xenophontos, C., eds., Ophiolites: Oceanic crustal analogues: Proceedings, 1987 Troodos, Nicosia, Cyprus, Geological Survey Department, p. 27–35.

Morgan, J.P., 1987, Melt migration beneath mid-ocean spreading centers: Journal of Geophysical Research Letters, v. 14, p. 1238–1241.

Muir, I.D., and Tilley, C.E., 1964, Basalts from the northern part of the rift zone of the mid-Atlantic ridge: Journal of Petrology, v. 5, p. 409–434.

Natland, J.H., and Dick, H.J.B., 2001, Formation of the lower ocean crust and the crystallization of gabbroic cumulates at a very slowly spreading ridge: Journal of Volcanology and Geothermal Research, v. 110, p. 191–233.

Natland, J.H., Meyer, P.S., Dick, H.J.B., and Bloomer, S.H., 1991, Magmatic oxides and sulfides in gabbroic rocks from Hole 735B and the later development of the liquid line of descent, *in* Von Herzen, R.P., Robinson, P.T., et. al., eds., Proceedings, Ocean Drilling Program, Scientific Results, 118, College Station, Texas, p. 75–111.

Nicolas, A., 1989, Structures of Ophiolites and Dynamics of Oceanic Lithosphere: Dordrecht, Kluwer Academic Publishers, 367 p.

Nicolas, A., 1992, Kinematics in magmatic rocks with special reference to gabbro: Journal of Petrology, v. 33, p. 891–915.

Nicolas, A., 1994, Comment on "The genesis of oceanic crust: magma injection, hydrothermal circulation, and crustal flow" by Morgan, J.P., and Chen, Y.J.: Journal of Geophysical Research, v. 99, p. 12,029–12,030.

Nicolas, A. and Ildefonse, B., 1996, Flow mechanism and viscosity in basaltic magma chambers: Geophysical Research Letters, v. 23, p. 2013–2016.

Nicolas, A. and Poliakov, A., 2001, Melt migration and mechanical state in the lower crust of oceanic ridges: Terra Nova, v. 13, p. 64–69.

Nicolas, A., Freydier, C., Godard, M., and Vauchez, A., 1993, Magma chambers at oceanic ridges: How large?: Geology, v. 21, p. 53–56.

Nicolas, A., Reuber, I., and Benn, K., 1988, A new magma chamber model based on structural studies in the Oman ophiolite: Tectonophysics, v. 151, p. 87–105.

Nisbet, E.G., and Fowler, C.M., 1978, The Mid-Atlantic Ridge at 37° and 45° N: Some geophysical and petrological constraints: Geophysical Journal of the Royal Astronomical Society, v. 54, p. 631–660.

Niu, Y., Gilmore, T., Mackie, S., Greig, A., and Bach, W., 2002, Mineral chemistry, whole-rock compositions, and petrogenesis of Leg 176 gabbros: Data and discussion: *in* Natland, J.H., Dick, H.J.B., Miller, D.J., and Von Herzen, R.P., eds., Proceedings, Ocean Drilling Program, Scientific Results 176, p. 1–60. Available from World Wide Web: http://www-odp.tamu.edu/publications/176_SR/VOLUME/CHAPTERS/SR176_08.PDF.

Orcutt, J.A., 1987, Structure of the Earth: oceanic crust and uppermost mantle: Reviews of Geophysics, v. 25, p. 1177–1196.

Oxburgh, E.R., and Turcotte, D.L., 1968, Mid-ocean ridges and geotherm distribution during mantle convection: Journal of Geophysical Research, v. 73, p. 2643–2661.

Ozawa, K., Meyer, P., and Bloomer, S.H., 1991, Mineralogy and textures of iron-titanium oxide gabbros and associated olivine gabbros from Hole 735B: *in* Von Herzen, R.P., Robinson, P.T., et al., Proceedings, Ocean Drilling Program, Scientific Results, 118, College Station, Texas, Ocean Drilling Program, p. 41–73.

Pallister, J.S., and Hopson, C.A., 1981, Semail ophiolite plutonic suite: field relations, phase variation and layering and a model of a spreading ridge magma chamber: Journal of Geophysical Research, v. 86, p. 2593–2644.

Pálmason, G., 1973, Kinematics and heat flow in a volcanic rift zone with application to Iceland: Geophysical Journal of the Royal Astronomical Society, v. 15, p. 331–341.

Parsons, I., 1987, Origins of Igneous Layering, NATO ASI ser. C: Mathematical and Physical Sciences v. 196: Dordrecht, D. Reidel, 666 p.

Phipps Morgan, J., and Chen, Y.J., 1993, The genesis of oceanic crust: magma injection, hydrothermal circulation, and crustal flow: Journal Geophysical Research, v. 98, p. 6283–6298.

Phipps Morgan, J., Harding, A., Orcutt, J., Kent, G., and Chen, Y.J., 1994, An observational and theoretical synthesis of magma chamber geometry and crustal genesis along a mid-ocean ridge spreading center, *in* Ryan, M.P., ed., Magmatic Systems: San Diego, Academic Press, p. 139–178.

Purdy, G.M., and Detrick, R.S., 1986, Crustal structure of the Mid-Atlantic Ridge at 23° N from seismic refraction studies: Journal of Geophysical Research, v. 91, p. 3739–3762.

Quick, J.E., and Denlinger, R.P., 1993, Ductile deformation and the origin of layered gabbro in ophiolites: Journal of Geophysical Research, v. 98, p. 14,015–14,027.

Ramberg, I.B., and van Andel, T.H., 1977, Morphology and tectonic evolution of the rift valley at Lat. 36° 30' N, Mid-Atlantic Ridge: Geological Society America Bulletin, v. 88. p. 577–586.

Reinhardt, B.M., 1969, On the genesis and emplacement of ophiolites in the Oman Mountains geosyncline: Schweiz Mineral. Petrogr. Mitt. v. 49, p. 1–30.

Robertson, A., and Xenophontos, C., 1993, Development of concepts concerning the Troodos ophiolite and adjacent units in Cyprus, *in* Pricard, H.M., Alabaster, T, Harris, N.B.W., and Neary, C.R., eds., Magmatic Processes and Plate Tectonics, Geological Society London Special Publication 75, p. 85–119.

Robson, D., and Cann, J.R., 1982, A geochemical model of mid-ocean ridge magma chambers: Earth and Planetary Science Letters, v. 60, p. 93–104.

Rosencrantz, E., 1983, The structure of sheeted dykes and associated rocks in North Arm massif, Bay of Islands ophiolite complex, and the intrusive process at oceanic spreading centers: Canadian Journal of Earth Science, v. 20, p. 787–801.

Rosencrantz, E., and Nelson, K.D., 1982, Floating lids on large mid-ocean spreading center magma chambers: Journal of Geology, v. 90, p. 455–458.

Shand, S.J., 1949, Rocks of the mid-Atlantic ridge: Journal of Geology, v. 57, p. 89–92.

Shipboard Scientific Party, 1999, Hammer drill site (1104 and 1106) and Site 1105, *in* Pettigrew, T.L., Casey, J.F., Miller, D.J., et al., Proceedings, Ocean Drilling Program, Initial Reports, 179: College Station, Texas, Ocean Drilling Program, p. 1–183 [CD-ROM].

Sinton, J.M., and Detrick, R.S., 1992, Mid-ocean ridge magma chambers: Journal of Geophysical Research, v. 97, v. 197–216.

Sleep, N.H., 1975, Formation of oceanic crust: some thermal constraints: Journal of Geophysical Research, v. 80, p. 4037–4042.

Sleep, N.H., 1978, Thermal structure and kinematics of mid-oceanic ridge axis, some implications to basaltic volcanism: Geophysical Research Letters, v. 5, p. 426–428.

Smewing, J.D., 1981, Mixing characteristics and compositional differences in mantle-derived melts beneath spreading axes: evidence from cyclically layered rocks in the ophiolite of North Oman: Journal of Geophysical Research, v. 86, p. 2645–2659.

Smewing, J.D., Simonian, K.O., and Gass, I.G., 1975, Metabasalts from the Troodos massif, Cyprus: genetic implication deduced from petrography and trace element geochemistry: Contributions to Mineralogy and Petrology, v. 51, p. 49–64.

Smith, D.K., and Cann, J.R., 1992, The role of seamount volcanism in crustal construction at the Mid-Atlantic Ridge (24°-20° N): Journal of Geophysical Research, v. 97, p. 1645–1658.

Smith, G.H., 1958, Bay of Islands Igneous Complex, Western Newfoundland: Ottawa, Ontario, Geological Survey of Canada Memoir 290, 132 p.

Solomon, S.C., and Toomey, D.R., 1992, The structure of mid-ocean ridges: Annual Reviews of Earth Planetary Science, v. 20, p. 329–364.

Steinmann, G., 1927, Die ophiolithischen Zonen in dem mediterreanen Kettengebige. International Geological Congress, 14th, Madrid, Proceedings, v. 2, p. 638–667.

Strong, D.F., and Malpas, J., 1975, The sheeted dike layer of the Betts Cove ophiolite complex does not represent spreading: Further discussion: Canadian Journal of Earth Science, v. 12, p. 894–896.

Taylor, R.N., and Nesbitt, R.W., 1988, Light rare-earth enrichment of supra subduction-zone mantle: evidence from the Troodos ophiolite, Cyprus: Geology, v. 16, p. 448–451.

Thayer, T.P., 1967, Chemical and structural relations of ultramafic and feldspatic rocks in alpine intrusive complexes, *in* Wyllie, P.J., ed., Ultramafic and Related Rocks: New York, John Wiley, p. 222–239.

Thy, P., 2003. Igneous petrology of gabbros from Hole 1105A: oceanic magma chamber processes, *in* Casey, J.F., and Miller, D.J., eds., Proceedings, Ocean Drilling Program, Scientific Results, 179: College Station, Texas, Ocean Drilling Program, p. 1–76 (online; http//www-odp.tamu.edu/publications/179_SR/volume/chapters/SR179_02.pdf).

Thy, P., and Dilek, Y., 2000, Magmatic and tectonic controls on the evolution of oceanic magma chambers at slow-spreading ridges: perspectives from ophiolitic and continental layered intrusions, *in* Dilek, Y., Moores, E.M., Elthon, D., and Nicolas, A., eds., Ophiolite and Oceanic Crust: New insights from field studies and the Ocean Drilling Program: Boulder, Colorado, Geological Society of America Special Paper 349, p. 87–104.

Thy, P., Schiffman, P., and Moores, E.M., 1989, Igneous mineral stratigraphy and chemistry of the Cyprus Crustal Study Project drill core in the plutonic sequences of the Troodos ophiolite, *in* Gibson, I.L., Malpas, J., Robinson, P.T., and Xenophontos, C., eds., Cyprus Crustal Study Project, Hole CY-4: Geol. Surv. Can. Paper, v. 88–89, p. 147–185.

Toomey, D.R., Purdy, G.M., Solomon, S.C., and Wilcock, W.S., 1990, The three-dimensional seismic velocity structure of the East Pacific Rise near latitude 9° 30′N: Nature, v. 347, p. 639–645.

Turcotte, D.L., and Morgan, J.P., 1992, The physics of magma migration and mantle flow beneath a mid-ocean ridge: *in* Morgan, J.P., Blackman, D.K., and Sinton, J.M., eds., Mantle Flow and Melt Generation at Mid-Ocean Ridges: Washington, D.C., American Geophysical Union Monograph 71, p. 155–182.

Turner, J.S., and Campbell, I.S., 1986, Convection and mixing in magma chambers: Earth Science Reviews, v. 23, p. 255–352.

Turner, J.S., and Chen, C.F., 1974, Two-dimensional effects in double-diffusive convection: Journal of Fluid Mechanics, v. 63, p. 577–592.

Varga, R.J., and Moores, E.M., 1985, Spreading structure of the Troodos ophiolite, Cyprus: Geology, v. 13, p. 846–850.

Vera, E.E., Mutter, J.C., Buhl., P., Orcutt, J.A., Harding, A.J., Kappus, M.E., Detrick, R.S., and Brocher, T.M., 1990, The structure of 0 to 0.2-m.y.-old oceanic crust at 9° N on the East Pacific Rise from expanded spread profiles: Journal of Geophysical Research, v. 95, p. 15,529–15,556.

Vine, F.J., and Moores, E.M., 1972, A model for the gross structure, petrology, and magnetic properties of oceanic crust, *in* Shagam, R., Hargraves, R.B., Morgan, W.J., Van Houten, F.B., Burk, C.A., Holland. H.D., and Hollister, L.C., eds., Studies in Earth and Space Sciences: A Memoir in Honor of Harry Hammond Hess: Boulder, Colorado, Geological Society of America Memoir 132, p. 195–205.

Wager, L.R., and Brown, G.M., 1967, Layered Igneous Rocks. San Francisco, Freeman, 588 p.

Wager, L.R., Brown, G.M., and Wadsworth, W.J., 1960, Types of igneous cumulates: Journal of Petrology, v. 1, p. 73–85.

Wilson, D.S., Clague, D.A., Sleep, N.H., and Morton, J., 1988, A model for narrow, steady state magma chambers on fast spreading ridges: Journal of Geophysical Research, v. 93, p. 11,974–11,984.

Wilson, H.H., 1969, Late Cretaceous eugeosynclinal sedimentation, gravity tectonics, and ophiolite emplacement in Oman Mountains, southeast Arabia: American Association of Petroleum Geologists Bulletin, v. 53, p. 626–671.

Wilson, J.R., and Larsen, S.B., 1985, Two-dimensional study of a layered intrusion—the Hyllingen series, Norway: Geological Magazine, v. 122, p. 97–124.

Wilson, R.A.M., 1957, The Geology of the Xeros-Troodos Area: Nicosia, Cyprus, Geological Survey of Cyprus, Memoir 1, 135 p.

Wyllie, P.J., 1967, Ultramafic and Related Rocks: New York, John Wiley and Sons, 416 p.

Yoder, H.S., and Tilley, C.F., 1962, Origin of basaltic magmas: An experimental study of natural and synthetic rock systems: Journal of Petrology, v. 3, p. 342–532.

Young, D.A., 1998, N.L., Bowen and Crystallization-Differentiation: The evolution of a theory: Washington, D.C., Mineralogical Society of America, 276 p.

Manuscript Accepted by the Society May 2, 2003

Geological Society of America
Special paper 373
2003

Geology of supra-subduction zones—Implications for the origin of ophiolites

James W. Hawkins
Geosciences Research Division, Scripps Institution of Oceanography, La Jolla, Califorinia 92093-0220, USA

ABSTRACT

Ophiolites are rock assemblages recognized as relict fragments of oceanic crust and upper mantle. They provide insights into petrologic and seismic layering of in situ oceanic lithosphere at mid-ocean ridges and into deep oceanic crustal layers not easily sampled. Ocean crust also forms at other settings. At convergent oceanic plate margins, aged oceanic lithosphere is subducted into depleted mantle. Above the subduction zone, new oceanic crust is formed in the forearc, volcanic arc, and backarc basins. The magmas, products of supra-subduction zone (SSZ) processes, carry chemical and isotopic signatures reflecting SSZ sources and can be distinguished from mid-ocean ridge magmas. Ophiolite assemblages within belts of accreted terranes (e.g., the North American Cordillera) indicate an oceanic origin for some terranes, but their close geologic association with arc volcanic and volcaniclastic material, granitoids, and silicic extrusives makes a mid-ocean ridge origin suspect. There is much similarity between Cordilleran ophiolites, many other ophiolites that are associated both with tectonically juxtaposed volcanic arc complexes, and the Cenozoic SSZ systems of the Western Pacific Basin. The latter includes immature SSZ systems, such as the Tonga-Kermadec, Vanuatu, Mariana-Izu-Bonin, and also more evolved systems, such as Palau-Kyushu Ridge, Fiji, Luzon, and Mindanao. All geologic units that characterize ophiolites are found in SSZ rock assemblages, as well as distinctive rock types, such as boninite and island arc series rocks. Backarc basin crust is predominantly formed of rocks similar to mid-ocean ridge basalt, but commonly includes variants gradational to island arc tholeiites. Forearc and backarc basins may have thick ponds of arc-derived tuffs and siliciclastics. These may lie on nearly coeval oceanic crust. Parts of backarc basins, starved of clastics, may have pelagic sediments, metalliferous sediments, "umbers" and protoliths for cherty argillites. Mature arcs may be underlain by plagiogranites and tonalites that are plutonic complements of silicic arc volcanism.

Key Words: ophiolite, boninite, backarc basin, trenches, subduction.

INTRODUCTION

It is widely accepted that ophiolites are fragments of oceanic crust and depleted peridotitic mantle, the latter being the depleted residue from which magmas that formed the crustal units were derived (cf. Moores and Vine, 1971; Moores and Jackson, 1974; Coleman, 1977). Recognition of ophiolite assemblages in orogenic belts and collages of accreted terranes (Coney et al., 1980) establishes the incorporation of former oceanic crust into continental margins. Many geologists accept the notion that this ocean crust formed at deep-water "mid-ocean ridge" spreading centers (e.g., Bailey et al., 1970; Moores and Vine, 1971; Moores and Jackson, 1974; Coleman, 1977; Hopson et al., 1981; Moores et al., 2000). Under this notion, oceanic crust, destined for recycling into the earth's mantle at subduction zones, is inferred to have been emplaced onto sialic crust by a process termed "obduction."

*jhawkins @ucsd.edu

Hawkins, J.W., 2003, Geology of supra-subduction zones—Implications for the origin of ophiolites, *in* Dilek, Y., and Newcomb, S., eds., Ophiolite concept and the evolution of geological thought: Boulder, Colorado, Geological Society of America Special Paper 373, p. 227–268. For permission to copy, contact editing@geosociety.org.

Miyashiro (1973) proposed the highly controversial idea that the Troodos ophiolite formed in an island arc. Since then, the number of papers in which a subduction zone origin for ophiolites is presented has increased steadily. This discussion summarizes some of the evidence supporting supra-subduction zones (SSZ) as likely sources for many, if not most, ophiolites.

ON SUPRA-SUBDUCTION ZONES

Our understanding of SSZ systems has greatly benefitted from results of the Deep Sea Drilling Project (DSDP) and Ocean Drilling Program (ODP) studies focused on the Tonga, Mariana, and Izu-Bonin arc systems. Many of the facts and interpretations summarized here come from observations and syntheses of drilling results presented by Kroenke et al. (1980); Hussong and Uyeda (1981); B. Taylor et al. (1992); Fryer et al. (1992); and Hawkins et al. (1994), and from regional syntheses of data by Hayes (1980, 1983); B. Taylor (1995); B. Taylor and Natland (1995); Fryer (1992); Stern and Arima (1998); and Stern (2002). There are many modern SSZ systems, as illustrated in Figure 1. All are potential sites for generating ophiolite-type rock series.

I will use the Mariana and Tonga arc-trench systems of the Western Pacific Basin as type examples of regions where new oceanic crust is formed by magmatism in the forearc and backarc-basin, as well as in the nascent stage of arc volcanism, in the region above the Wadati-Benioff zone. This is the SSZ tectonic setting, a term first proposed by Pearce et al., 1984. Figure 2 is a schematic cross-section of a generic intra-oceanic subduction system. It is based largely on the distribution of rock series, morphologic features, and geophysical data for the region between the Lau Ridge remnant arc and the Tonga Trench, but also applies generally to most other regions having island arc-trench systems. The entire region between the trench and backarc basin lies above the array of deep seismic events, which define the Wadati-Benioff zone. In the Lau-Tonga system, the seismic zone extends under the remnant arc. This arrangement, however, is not always observed in other SSZs. The seismic zone marks the presumed location of the subducted plate of oceanic lithosphere; hence, this is the supra-subduction zone environment. Some may choose to call this a "convergent plate margin," but in fact, the geologic processes distinctive for this setting extend across the entire region and are not limited to the site of convergence at the trench that marks the convergent margin. An important observation about SSZ environment is that although it is spatially associated with the *convergent* margin, the entire region between remnant arc and the outer trench swell is undergoing *extension*, and the hinge line outboard of

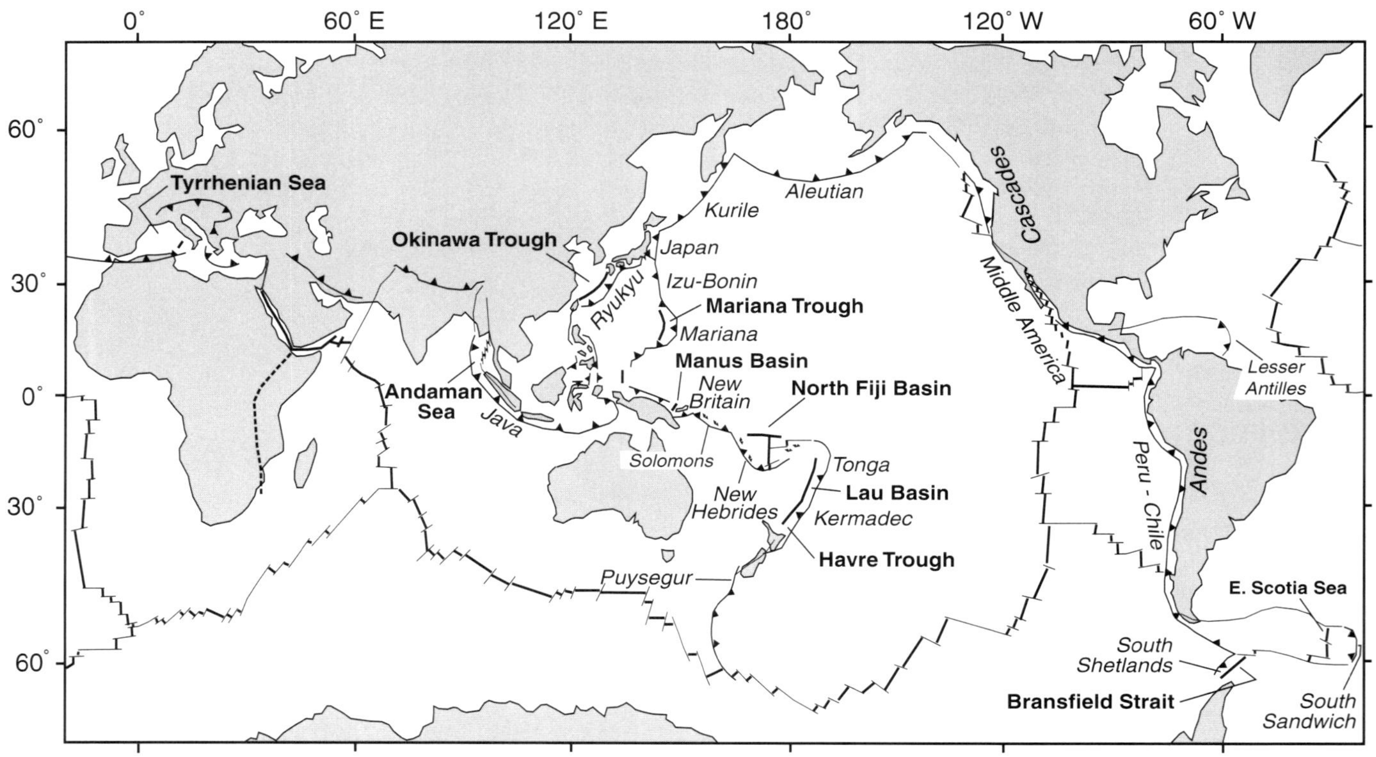

Figure 1. Active backarc basins (bold type) and their related ocean trenches/subduction zones (italics). Courtesy of Robert Stern (cf. Stern, 2002).

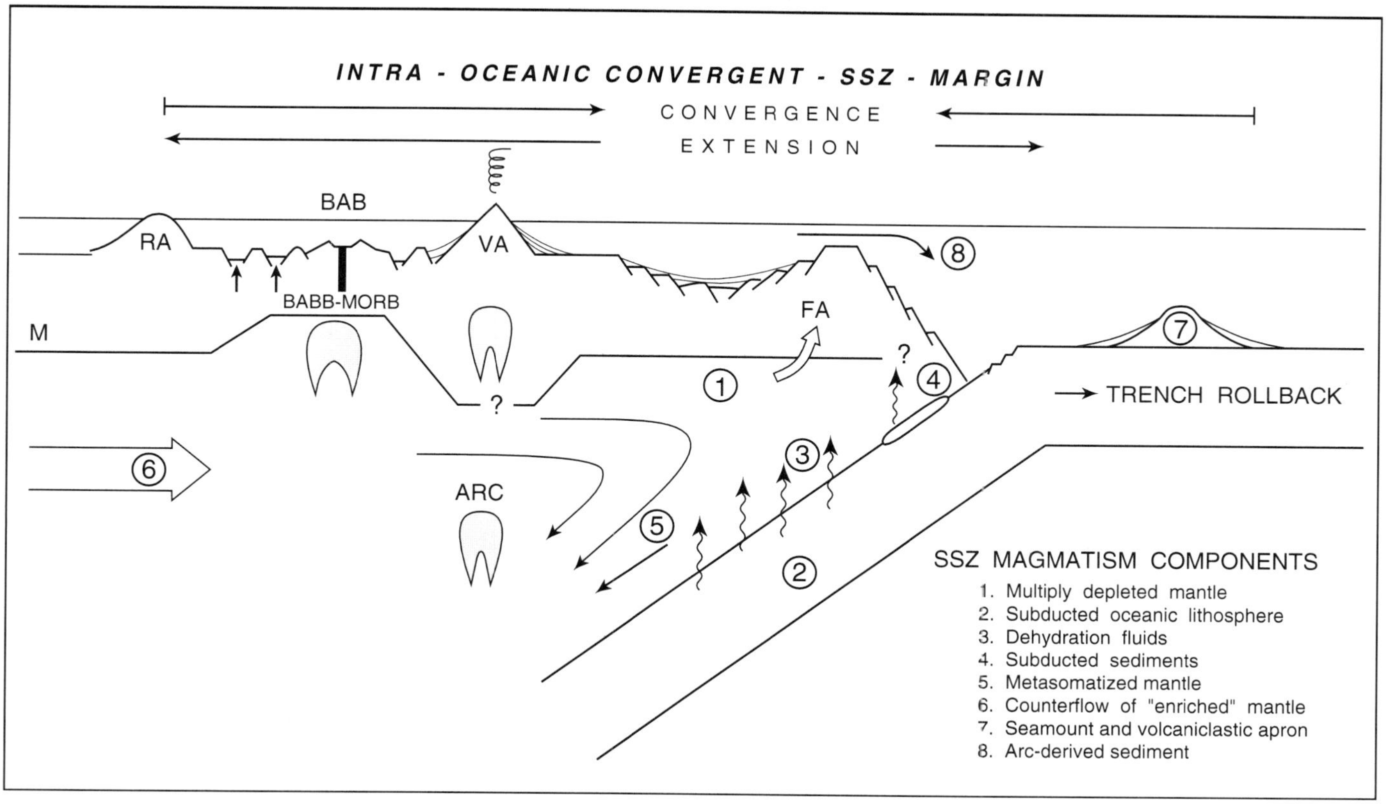

Figure 2. Schematic cross-section of an intra-oceanic SSZ system based on the Lau Basin-Tonga Trench region. Vertical dimension is at varied scale in order to conveniently show all parts. Crustal features are: RA—remnant arc, e.g., Lau Ridge; BAB—backarc basin, e.g., Lau Basin; VA—active arc, e.g., Tofua Arc; and FA—forearc. The arcs mainly erupt low-K island arc tholeiitic basalt (IAT) and their differentiates, whereas the BAB is dominated by tholeiite ranging from MOR-type basalt to basalts having chemistry transitional to arc tholeiite, i.e., backarc basin basalt (BABB). Initial stage of development, in which boninites form in the forearc, is not represented. Forearc magmatism shown here represents limited later stage dikes of arc magma intruded into forearc sediments. Potential source materials for melts are identified on the sketch. Mantle counterflow provides a continual supply of fertile mantle. Not shown are backarc basin seamounts, which may include OIB, MORB, and IAT.

the trench retreats by rollback (cf. Oxburgh and Turcotte, 1970; Elsasser, 1971; Uyeda and Kanamori, 1979; Hamilton, 1995, 2002). Furthermore, geophysical data indicate that there is a broad region of strong seismic wave attenuation (low-Q), probably from high temperature (*T*) and some partial melting, underlying the backarc basin (Barazangi and Isacks, 1971; Crawford et al., 2003). Sleep and Toksoz (1971) presented a model for opening of backarc basins showing convective overturn in SSZ mantle. They stated that "The down going slab generates a flow pattern by 'dragging' the low velocity layer material in the aesthenosphere. The convective cell thus driven exerts a stress at the bottom of the lithosphere behind the island arc forcing it in the direction of the trench." They also observed that "Heat transport by conduction alone cannot explain the high heat flow in the marginal basin." Influx and upwelling of relatively enriched mantle, driven by convective drag of the subducting plate (cf. Ida, 1983; Ewart and Hawkesworth, 1987), provides a continuous source for basaltic melts. These features, as well as the various components potentially available to form SSZ magmas, are discussed in more detail below.

The main purpose of this essay is to summarize the nature of the rocks formed in the SSZ setting and the processes important in their petrogenesis. Inasmuch as there is an extensive literature on details of petrogenesis, I will emphasize here the petrology of certain rock types and rock associations that are distinctive for the SSZ setting and that distinguishes these rocks from those formed at mid-ocean ridge spreading centers. The goal is to show how we can distinguish between ophiolites and associated rocks of the SSZ environment from ocean crust formed at mid-ocean ridges. Most backarc basin rocks could successfully masquerade as "true" mid-ocean ridge basalt (MORB), especially if they have experienced low-grade metamorphism. Rocks formed in arcs and forearcs have some fundamental differences from most in backarc basins, yet there is some gradation and overlap. Collectively, the arc and forearc assemblages are distinctive and are unlike crust formed at mid-ocean ridge spreading centers. A key point I wish to make is that in order to distinguish between ocean crust (ophiolite) formed at a mid-ocean ridge and ocean crust formed in an SSZ setting, one must evaluate all aspects of the geology of the ophiolite assemblage.

A Brief History—The Nature of the SSZ Setting

Introduction

The concept of the SSZ as a location of generation of new crust and recycling of aged oceanic lithosphere back to the mantle forms an important part of our present understanding of plate tectonic processes and crustal evolution. There was considerable knowledge of the geology of what we now call the SSZ setting, long before the advent of the modern plate tectonics model in 1961, but the data had not been integrated to form an understanding for, let alone to frame a model for, the geologic processes occurring in these regions. As with any formulation of a model for earth processes, there were many studies that laid the groundwork. What follows here is a brief summary of the key observations and interpretations pertinent to the SSZ setting, dating back to 1911, but mostly from between 1940 and 1980. My examples are mainly from the Western Pacific Basin, although related research was conducted in all major oceanic trenches. These years can be considered the "era of pioneer exploration" in arc-trench-backarc systems. The examples cited here were important antecedents to the extensive work of the last 20 years, which has produced a voluminous database and body of literature. Elsewhere in this volume, Julian Pearce (Chapter 15) presents a complementary discussion of the historical development of thought on the SSZ origin of ophiolites.

The first presentation of data and a description for the SSZ setting was by Marshall (1911), in which he summarizes: "the Tonga-Kermadec-New Zealand line is an important structural feature in the southwestern Pacific. The soundings throughout the ridge are shallow, the summits that project above the ocean along its crest are nearly all volcanic on the west side of the axis and raised limestone on the east. The whole line is bounded by a profound deep on its eastern side…in all of these islands the volcanic rocks of Cainozoic age are mainly andesite as in Tonga, Kermadec, and northern New Zealand. Everywhere to the east of this curved line the whole unbroken depths of the Pacific extend to the American coast…" Marshall also notes that "the island arcs….are largely formed of andesitic rocks, though they are usually associated with basaltic rocks" (1911). He describes the known great deeps of Kermadec-Tonga, New Hebrides, and Mariana, and suggests that they are "foldings" or a "synclinorium."

In 1961, Dietz (1961) and Hess (1962) presented their landmark papers proposing that new seafloor crust was being generated at ocean ridges, that seafloor spread from these ridges, and that older crust was "disappearing" (not being subducted!) into the mantle at the ocean trenches. According to Dietz (1961) oceanic "median rises mark the up-welling sites of divergence and the trenches are associated with the convergences or down-welling sites." Prior to 1961, the location and spatial association of volcanic island arcs, zones of deep seismicity, and deep ocean trenches had been recognized. The common development of arcuate arrays of island arc volcanoes is one of the more striking features of the SSZ setting. As mentioned above, Marshall (1911) recognized an "andesite line." This line separated "more acid lavas, especially andesites" from "the truly basaltic central area of the deep Pacific seafloor" (Keunen, 1950). The trace of the andesite line in the Western Pacific followed the eastern limit of the great trench systems. Hess (1948) proposed that drawing the andesite line on the outer, convex, side of the trenches would make it a useful structural, as well as petrographic, boundary.

Alfred Wegener (1915, see also 1966) gets credit for the first formal presentation of a model for continental drift. In his book, he also presented a model pertinent to opening of back arc basins and showed a cross-section sketch not unlike many that have been drawn for SSZ systems. Wegener proposed that "the island arcs, and particularly the eastern Asiatic ones, are marginal chains which were detached from continental masses, when the latter drifted westwards and remained fast in the old seafloor, which was solidified to great depths. Between the arcs and the continental margins later, still-liquid areas of seafloor were exposed as windows" (Wegener, 1966). These "windows" of seafloor are the backarc basins.

Our understanding of trenches, especially those of the Western Pacific, profited by the thousands of kilometers of bathymetric data collected in naval operations in World War II. Harry Hess (1948) presented a synthesis of seafloor bathymetry and fabric for the Northwestern Pacific Basin in which the curvilinear array of trenches, known to be "deeps," were identified as "axes of tectogenes" (down-buckling of the crust), whereas their adjacent island arc systems were labeled as "geanticlines."

Trenches and Benioff Zones

The exploration of the Western Pacific arc-trench systems in the 1950s led to a better understanding of the bathymetry of trenches and of the geophysical data for crust and upper mantle (see summary in Fisher and Hess [1963]). Hess (1948) presented profiles across the Mariana-Bonin, Kamchatka-Kurile, and Tonga-Kermadec trench-arc systems. He noted the previously unrecognized continuity of the deep trenches and their parallelism to island arcs; the close spatial association of seismic activity, gravity anomalies, and volcanism along the arc-trench trends; and the common occurrence of peridotite and blueschist metamorphism. Benioff (1949) showed the relation between ocean trenches and deep seismic zones that were delineated by an array of seismic events that he interpreted as representing the traces of great faults. "The oceanic deeps associated with these faults are surface expressions of the downwarping of their oceanic blocks" (Benioff, 1949). He noted that in some regions, such as Tonga, the array dipped uniformly at ~45° to nearly 700 km depth, whereas in Kamchatka–Kuriles, the array defined a zone convex upward that varied from nearly horizontal near the surface to ~60°. In a discussion of Dietz (1961), Bernal (1961) presented an interpretative sketch showing an ocean-continent interface with a thrust plane separating the plates, and proposed that the Pacific trenches formed by overthrusting along an inclined planar zone represented by deep earthquakes. He also suggested that "vulcanicity associated with mountain building is assumed to arise from activity of the fault plane itself" (Bernal, 1961).

Early models for plate tectonics (e.g., Morgan, 1968; Isacks et al., 1968) recognized three types of plate boundaries, one of which is "trench or young fold mountain." Morgan noted that "the compressive boundary seems to be the most difficult to delineate." For the Tonga–New Zealand–Macquarie system, he proposed that "fast compression leads to the trench-type structure" whereas "the [Macquarie] ridge is the result of slow compression." Neither backarc nor forearc basins were recognized as significant components of early plate tectonic models.

Extensive exploration of the crustal structure and rock types of trenches began in the 1950s. For example, Raitt et al. (1955) made seismic profiles parallel to the axis of the Tonga Trench and estimated the depth to the Moho (P-wave velocity >8.1 km sec^{-1}) along the trench axis at 20 km below sea level, whereas outboard of the trench, the same velocity was measured at 12 km below sea level. The highest velocities recorded beneath the Tofua Arc were 7.6 km sec^{-1}. They then extrapolated to a depth of 20 km under the arc before reaching mantle velocity. Samples recovered from the Tonga Trench wall at depths of 9 km included both unaltered and partly serpentinized peridotite (Fisher and Engel, 1969). Bowin et al. (1966) previously had collected serpentinite and altered basalt from the Puerto Rico Trench.

Backarc Basins

As noted above, Wegener's prescient suggestion for the origin of backarc basins was published before much was known about the trenches themselves. Holmes (1931, 1965) presented a trench arc system cross section that showed but did not discuss convective overturn in the mantle under arc systems, and a rift basin, equivalent to a backarc basin. For an interesting discussion of early evolution of ideas about mantle convection see Allwart (1988).

The first specific suggestion for extension and formation of new seafloor inboard of a trench-arc system appears to have been Rodolfo's (1969) proposal that the Andaman Sea had formed as a rhombochasm resulting from south to southeastward rifting of the southeast Asia margin. Karig (1970) observed that the Tonga-Kermadec, New Hebrides, and Mariana island arc systems displayed similar geologic characteristics. He concluded that inter-arc basins (e.g., Lau Basin and Havre Trough) are "underlain by crust of oceanic character and…[are] the locus of high heat flow." Karig probably was first to postulate "a process of island arc tectonism…during which the frontal arc and trench position migrate, generally away from the continents, concomitantly creating new oceanic crust, without mid-ocean ridges, in the ridge bounded basins behind the arcs." Packham and Falvey (1971) presented a similar observation; their main new contribution was data for heat flow, magnetics, and sediment thickness. They concluded that "genesis of oceanic crust takes place in a way that is analogous to seafloor spreading, but not at a mid-ocean ridge"; however, they suggested that "upwelling behind the Benioff zone, by a mechanism similar to…mid-ocean ridge crustal generation" was involved.

Moberly expanded on the backarc basin model and showed how various possible tectonic configurations for relative plate motions could develop different crustal fabric in the backarc basin. He, too, recognized the importance of trench and arc migration, proposing that "trench and arc migrate seaward against the retreating line of flexure of the suboceanic lithosphere…warm aesthenosphere…migrates up…forms new lithosphere in the extensional region behind the advancing arc" (Moberly, 1972). He also discussed causes and consequences of arc migration, and made some reconstructions of past trench configurations for the Western Pacific.

Barazangi and Isacks (1971) determined that a zone of strong seismic attenuation, possibly the result of high-temperature mantle, lay beneath the Lau Basin. The relation between subduction and mantle counterflow above the subduction zone was addressed by Sleep and Toksoz (1971), who showed how "marginal basin" extension and elevated heat flow may be related to convection in low-velocity aesthenosphere.

All backarc basin models postulated that some type of ocean crust underlies backarc basins. The first verification that the backarc basin crust was indeed basaltic and had MORB-like chemistry was in 1970, when we sampled the axis of what is now called the East Lau Spreading Center, while on the Scripps Institution of Oceanography 7-TOW expedition (Hawkins et al., 1970). Soon afterward, Hart et al. (1972) established that Mariana Trough basalts also were like MORBs.

Island Arc Magmas

The notion that the arc systems include andesitic rocks goes back at least to Marshall (1911) as cited above. In their synthesis of data for trench systems, Fisher and Hess (1963) suggested that "andesitic magmas might be produced where deformation is enough to force the basalt, or lower density crust layer, downward sufficiently to cause partial fusion." The most significant advances to understanding petrogenesis in arc systems came in the 1970s when Ringwood and colleagues made fundamental contributions based on laboratory melting experiments. Ringwood (1974) succinctly observes that "The most important petrologic problem relating to the development of island arc systems is the origin of the basalt-andesite-rhyolite volcanic suite."

In my opinion, the work of Coats (1962) was one of the most significant early advances in relating arc magma chemistry to the addition of water into the region of arc magma generation via altered seafloor rocks. Coats' work predated publication of the plate tectonic hypothesis. Owing to time lags in publication, his work followed the concept of geosynclines and was not integrated with plate tectonics concepts; however, in hindsight, we can recognize that his model anticipated many modern ideas about relations between subduction and arc volcanism. He observed that Aleutian Arc volcanoes ranged in composition from basalt to rhyolite, but were largely pyroxene andesite. He proposed that "andesitic rocks (formed) through addition of water and hyperfusible materials from eugeosynclinal deposits to eruptible basaltic material in the mantle" (Coats, 1962). The *eruptible basaltic material* was in a deep mantle layer containing "vitreous interstitial basalt." The water-bearing phases were dragged down along inclined faults systems, which he interpreted as reverse faults dipping under the arc. At that time, first motion studies of P waves were "inherently ambiguous…two possible orthogonal planes either of which might be the actual fault plane…two solutions for each earthquake" (Coats, 1962). Coats based his model on the interpretation that allowed N-S compression. His ideas were still couched in the now defunct geosynclinal dogma, and he proposed that the entrained ("subducted") material included eugeosynclinal assemblages of serpentine blocks and breccia, basalt, eclogite, greenstone and eugeosynclinal sediments.

Dickinson and Hatherton (1967) compared magma chemistry with the depth to the seismic zone beneath arcs. Data for different arc systems showed a positive correlation between K_2O and SiO_2. Using comparable values for SiO_2 (e.g., 55 and 60%), they plotted K_2O against depth to the seismic zone for which they found a positive correlation. They proposed that volcanic arcs were "located above the intersection of the Benioff zone with sub-horizontal zones favorable for melting within the Gutenberg low velocity zone" (Dickinson and Hatherton, 1967). Baker (1968), in an early synthesis of arc petrology, proposed that the relative maturity of island arcs was reflected in their chemistry; relatively young arcs such as Tonga, Mariana, and South Sandwich had a spectrum of magma types, from basaltic to rhyolitic but mainly basalt and basaltic andesite, whereas more mature arcs such as the Indonesian, Aleutians, and Kamchatka displayed the same range in composition, but were mainly andesitic.

Ringwood and co-workers made fundamental contributions to subduction system magmagenesis; they addressed the distribution of arc volcanoes relative to Benioff zones and the apparent temporal evolution from arc tholeiite to calc-alkaline melts, a rock series he called "orogenic volcanic series." The mantle wedge and basalt, sediment, or both, of the subducted plate were considered as sources. Green and Ringwood (1968) proposed that the (parental) basalt series of island arcs results from melting of pyrolite in the mantle wedge above the subduction zone. Considering the role of subducted crust, Ringwood (1969) also proposed that "partial melting, at high *P* [pressure] and shear stress, of subducted ocean crust leads to generation of calc-alkaline magmas." The role of water was recognized in Ringwood (1974), who noted that it is "most plausible that water is introduced along the Benioff zone via the subsiding lithosphere plate" and that it is "unlikely that the source of this water is to be found in subducted sediments" but rather it is "introduced as hydrated minerals occurring in the subducted oceanic crust" (Ringwood, 1974).

Gill (1970) used his data for Viti Levu, Fiji to show how during Cenozoic time there had been three successive periods of volcanism and plutonism, all of which could be related to subduction. Another landmark paper on arc volcanism was Miyashiro's (1974) presentation, which discussed variations in magma chemistry of island and continental arcs. He showed that they could be separated into calc-alkaline, tholeiitic, and alkaline series. These compositional differences could be related to crustal thickness, and crustal thickness could be correlated with the relative abundance of calc-alkaline series to tholeiitic series in different ocean arcs.

Perfit et al. (1980) presented a compendium of island arc magma chemistry and discussed implications for mantle sources. Jim Gill (1981) summed up the state-of-the-art thought on arc magmatism in his book, *Orogenic Andesites and Plate Tectonics.* The literature on SSZ petrology has expanded greatly in the last 20 years. Key papers are cited in the section on island arc petrology.

OCEAN CRUST

The fundamental magma type that has erupted to form earth's crust since mid-Proterozoic time has been Mg-rich melts ranging in composition from komatiite to picrite and basalt. We find the evidence for this in ancient rocks preserved within and on the margins of continents. Direct sampling of present day ocean crust shows that the ocean basins have been floored with basalt since at least 200 Ma, and most would agree that this process has prevailed throughout most of earth's history.

The primary site of origin of ocean crust has been at mid-ocean ridge spreading centers. It is estimated that ~80% or more of the basaltic crust of present ocean basins was generated at mid-ocean ridge systems. Perhaps another 10% of this crust was generated at SSZ in the backarc or forearc of intra-oceanic convergent plate margin systems. The remainder formed as oceanic plateaus, linear volcanic chains, and isolated seamounts. Basalts that form mid-ocean ridge crust are referred to as mid-ocean ridge basalts (MORBs); a subset of this broad group is termed normal mid-ocean ridge basalt (N-MORB) (cf. Viereck et al., 1989). N-MORB represents relatively unfractionated tholeiitic basalts having trace element and isotopic chemistry indicative of small amounts of melting of a Ca- and Al-enriched peridotite source that, relative to the composition of estimates of the bulk silicate earth, is depleted in Th, U, alkali metals, and light rare earth elements. (cf. Ringwood, 1975; Hart and Zindler, 1986). N-MORBs, as defined, have distinctive trace element and isotopic characteristics and generally low concentrations of magmaphilic elements. They are distinctive in their marked depletion in alkali metals, alkaline earths, light rare earth elements, and water, relative to other basalt types. Mid-ocean ridge crust is not formed exclusively of N-MORB; commonly, there are regional compositional variations that may be attributed to variation in source chemistry, extent of melting, and fractionation, but on most ridge segments the least fractionated rocks are similar to the model N-MORB chemistry. The petrogenesis of MORB is not the focus of this discussion, but inasmuch as their chemistry gives insight to their mantle source and melting processes, MORB serves as a useful point of comparison for the magmas of the SSZ setting and for understanding their mantle sources and melting processes. The chemistry of SSZ magmas indicates that their sources may include mantle even more depleted than the mid-ocean ridge source. There are rare SSZ occurrences of rocks resembling intra-plate, or off-axis, magmas that are indicative of mantle sources enriched, compared to the MORB source.

Present annual mid-ocean ridge basalt magma production is estimated to be ~24 km^3 yr^{-1}. It is estimated that, since solid earth accreted, nearly three-fourths of the upper mantle has been partly melted, and that crust formed from those Mg-rich melts has been recycled back to the mantle (O'Neill and Palme, 1998). Since Paleozoic time, the maximum residence time of ocean crust has been ~200 Ma. Lithosphere formed at mid-ocean ridges ages in transit to the subduction zones; it thickens, subsides, gets altered by hydrothermal processes, is injected by off-axis magmas, and receives a thin veneer of pelagic, terrigenous, and meteoritic sediments. In the grand geologic scheme of crustal evolution, ocean lithosphere is destined to return to the mantle via subduction in ocean trench systems. Yet, we find remnants of old ocean crust and its subjacent melt-depleted upper mantle preserved on margins of continents, as uplifted masses of ocean lithosphere in some arc systems, and on micro-continents in ocean basins (e.g., New Caledonia, Fiji, Luzon, Mindanao). These are the ophiolites. Are these truly remnants of crust formed at mid-ocean ridges? I will present evidence to support the hypothesis that these allochthonous fragments of fossil oceanic lithosphere are predominantly from the SSZ environment.

THE NATURE OF OPHIOLITES

Ophiolites, first defined as a rock assemblage by Steinmann (1927), have been the topic of many extensive discussions (e.g., Coleman, 1977; Panayiotou,1980; Peters et al., 1991; Dilek et al., 2000), all of which support the contention that they are fragments of oceanic lithosphere. The strongest evidence for their oceanic origin is their siliceous and carbonate sedimentary rocks bearing oceanic fossil assemblages, and the biochemical precipitates interbedded with and overlying pillow basalts. Umbers and metalliferous sediments (Robertson and Hudson, 1973) similar to those that form at ocean ridge hydrothermal systems (Bostrom and Peterson, 1966) also are common. Pillow basalts of ophiolites, morphologically identical to those dredged from and photographed on the seafloor, overlie a sheeted-dike series. This same petrologic layering is found on the seafloor, as evidenced by ocean drilling (e.g., ODP Site 504). Rocks from deeper levels of ophiolite assemblages resemble dredged samples of gabbros, cumulate rocks, and fresh or partly serpentinized peridotites. The peridotites show varying degrees of depletion in the components of the basalts (e.g., they are depleted lherzolite, harzburgite, and dunite) that comprise the probable lower oceanic crustal and upper mantle rocks.

Interpretations of oceanic seismic velocity layering, based on seismic velocities determined for ophiolite samples (e.g., Salisbury and Christensen, 1978; Brocher et al., 1985), have supported a lithologic and velocity stratigraphy as shown in Figure 3. This is a somewhat circular argument that ophiolite equals ocean lithosphere, which in turn is ophiolite, but most geologists will agree that it is qualitatively correct; however, "ophiolites are defined largely in terms of petrologic and structural field relations, whereas the majority of the oceanic crust, owing to its inaccessibility, is best defined in geophysical terms, specifically, its velocity structure as revealed by seismic refraction" (Salisbury and Christensen, 1978). A major problem is that many ophiolites have rock types, stratigraphic and temporal relations, and details of rock chemistry that are contrary to what is known about ocean lithosphere formed at mid-ocean ridge

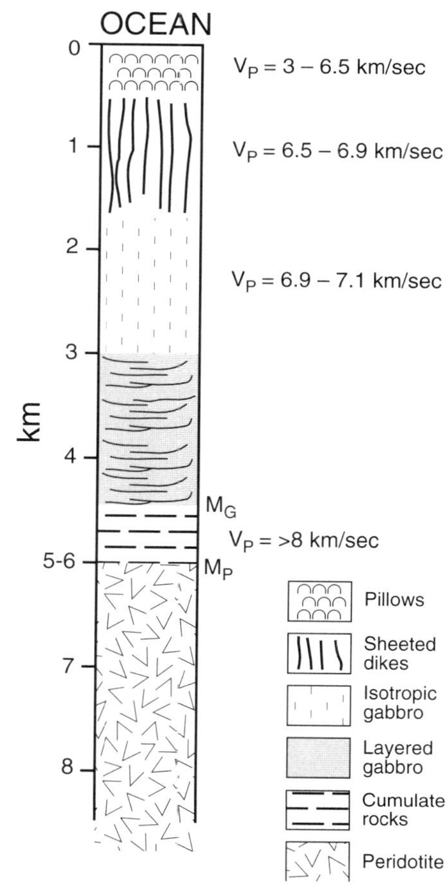

Figure 3. Ideal rock column for ocean crust and upper mantle. P–wave velocities (V_p) from Crawford et al. (2003).

spreading centers (e.g., Saunders et al., 1980; Hawkins, 1980; Pearce et al., 1981; Harper, 1984). Thus, we have the "ophiolite conundrum" (Moores et al., 2000). The problem that gives rise to this conundrum is the a priori assumption by many geoscientists that ophiolites are formed at mid-ocean ridges, when in fact we well know that ocean crust may be formed at other sites, as described above. With a proper understanding of SSZ geology, the supposed ophiolite conundrum can be eliminated.

GEOLOGY OF SUPRA-SUBDUCTION ZONE SETTINGS

Introduction

Convergent plate margins in intra-oceanic settings have a broad region that is several hundred kilometers wide across strike and that extends from the trench axis to the backarc basin. The major features of intra-oceanic convergent plate margins, the tectonic environment for SSZ systems, are shown in Figure 2. In addition to the trench, the major features are the backarc basin, active volcanic arc, and forearc. The forearc may include fragments of older convergent margin systems, old deep-sea floor ocean crust, or both. This region is situated above the trace of the inclined array of seismicity that defines the Wadati-Benioff Zone (WBZ), or subduction zone. The modern SSZ systems of the Western Pacific Basin form a nearly continuous array of trenches, arcs, and backarc basins extending across more than 100 degrees of latitude from Kamchatka to the nascent Puysegur and Hjort trenches south of New Zealand. In some settings (e.g., Tonga), the WBZ may extend beneath the remnant arc. In others (e.g., Mariana), the WBZ steepens to near vertical under the backarc basin or (e.g., New Hebrides) the WBZ may extend only to beneath the volcanic arc (Isacks and Barazangi, 1977). The term supra-subduction zone aptly describes the region in which there is concurrent extension, magmatism, and sediment accumulation in the backarc, arc, and forearc, as well as serpentinite diapirism in the forearc, and extension on the inner wall of the trench.

It is important to note that in Figure 2, the entire region from the remnant arc to the outboard side of the trench is under extension, even though on a regional scale it is the locus of plate convergence. The subducted plate itself is extended and broken up into a series of en-echelon rift basins as the incoming plate bends into the subduction zone (Hawkins et al., 2000). Note also the lack of an accretionary sedimentary prism in this generic cross-section. Accretionary prisms do, indeed, exist at some convergent margins (e.g., Sumatra, Antilles, eastern Aleutians), but the largest are in those regions where large drainage systems (e.g., Brahmaputra and Orinoco Rivers) bring in massive amounts of terrigenous sediments that funnel into the trench. Active volcanic arcs are close to newly formed crust of the backarc basin; siliciclastic arc-derived sediments are deposited directly onto crust having essentially the same age (e.g., <3 Ma) as the sediments. This would be most unusual for "normal" mid-ocean ridge crust. Extensional basins of the forearc serve both as depocenters for arc-derived clastics and as potential sites for the emplacement of mantle-derived melts. An important factor is that basaltic oceanic crust, arc rocks, granitoid roots of the arcs, and arc-derived clastics all form in close proximity and are essentially coeval.

The crust and mantle of SSZ settings constitute dynamic systems that evolve through the complex interplay of subduction of cold "rigid" lithosphere, upwelling hot mantle, transfer of heat and mass by magmatism, and crustal extension and rifting of SSZ lithosphere, coupled with, or driven by, rollback of the hinge line on the incoming subducted plate (e.g., Elsasser, 1971; Uyeda and Kanamori, 1979; Hawkins et al., 1984; Hawkins, 1994, 1995a, 1995b; Hamilton, 1988, 1995, 2002). Heat and mantle mass transfer may be accomplished by "ascending diapir (that) induces a descending return flow and results in a circulation" (Ida, 1983). Hamilton (2002) proposed that the incoming lithosphere plate "falls" into the mantle as the hinge line rolls back, that "the slab sinks more steeply than it dips…the magmatic arc follows the retreating hinge, extension affects the arc and backarc, and new hot mantle rises into the wedge from the backarc." Unloading of the crust by thinning and rifting has a feedback relationship to the rise and decompression melting of fertile mantle diapirs (McKenzie and Bickle, 1988; Hawkins, 1995a, 1995b). The tectonic setting is different, but the dynamic processes suggest similarities to the rise of mantle, ductile deformation of crust, tectonic denudation, and the close association of rifting and volcanism seen in detachment structures of the Basin and Range Province of North America (cf. Rehrig, 1986; Gans et al., 1989).

The first use of the term supra-subduction zone for this type of setting was by Pearce et al. (1984). Alabaster et al. (1982) used "supra-Benioff zone" for the same type of setting. In both papers, the term was used in reference to the site of origin of ophiolites and specifically for ophiolites derived from immature arc systems. Since 1984, there have been numerous uses of the term supra-subduction zone, both to describe present arc-trench system geometry and for interpretations of sites of origin for ophiolites. Hawkins and Evans (1983) and Hawkins et al. (1984) expanded the implications of the term to include an array of crust/mantle material that included forearc and backarc crust, as well as crust formed in a nascent arc. Part of the support for this extended use of the term was owed to recognition of the role and site of origin of boninites in the early stages of SSZ evolution and to the range in chemistry of backarc basin crust that ranged from MORB-like to island arc tholeiitic compositions.

In the northern hemisphere, the Western Pacific Basin (Fig. 4) has three concentric SSZ systems dating from mid- to late-Eocene time to the present, indicating that for this region, the present is the key to the past, at least through Cenozoic time (cf. Fryer, 1992). There has been sequential development of volcanic arcs and backarc basins on the eastern edge of the Philippine Sea since the inception of the ancestral "Mariana Trench" that formed the Palau-Kyushu Ridge. From oldest to youngest,

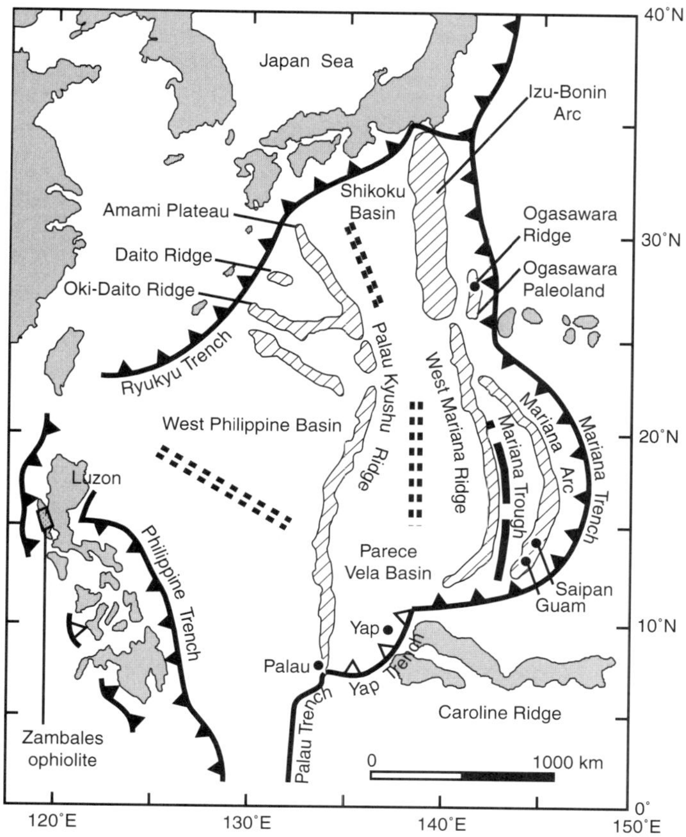

Figure 4. Schematic map of part of the northwestern Pacific, showing features of subduction systems discussed in text. Solid barbs denote active subduction systems. Open barbs are dormant or fossil systems. The actively spreading axis of the Mariana Trough is shown as solid lines. Inactive "fossil" spreading centers in Parece Vela-Shikoku Basin and West Philippine Basin are dashed.

from west to east, these features are: (1) Palau-Kyushu Ridge, active from mid-Eocene to late Oligocene (~48–30 Ma), and now a remnant arc; (2) Parece Vela and Shikoku backarc basins, which opened from 32 to 17 Ma and 25 to 15 Ma, respectively, and which both are now inactive; (3) West Mariana Ridge, active from late Oligocene to early Miocene, and now a remnant arc; (4) Mariana Trough backarc basin, active since ~6–7 Ma (?) to present; Fryer (1995) proposes 3.67 Ma for the beginning of seafloor spreading, but this was preceded by a rifting phase; (5) Mariana Arc, active since ~3–6 (?) Ma to present. Note the overlap in times of volcanic activity between "remnant arcs," backarc basins, and "frontal volcanic arcs," and that backarc spreading precedes appearance of the "frontal volcanic arcs." Outboard of these arcs is the Mariana forearc, which includes composite fragments of older arcs (e.g., Eocene and Oligocene volcanic units of Guam and Saipan), equivalent to volcanic rocks on the Palau-Kyushu and West Mariana Ridges. The volcanic rocks are partially capped by Miocene carbonate bank deposits and Pleistocene reef, bank, and lagoonal deposits (Cloud et al., 1958). The southwestern Pacific basin has at least three systems with similar configuration, petrology, and age ranges, but the geometry is more complex. A record of these is preserved on Fiji (cf. Gill, 1970). At present, there is an east-dipping subduction zone at the Vanuatu-New Hebrides Trenches and a west-dipping subduction zone at the Tonga-Kermadec Trenches. We find mid- and late-Eocene arc material and ophiolites on Fiji (Gill, 1970), and on the outer part of the Tonga Ridge, e.g., 'Eua (Ewart and Bryan, 1972), and at ODP Site 841 (Bloomer et al., 1994). This great assemblage of SSZ systems spans >45 Ma of history and lies waiting to be accreted to a continent, where it will constitute an assemblage of terranes like those of the western North America Cordillera (cf. Coney et al., 1980). Some of these Cordilleran terranes constitute ophiolites (e.g., Harper, 1984). It is reasonable to infer that pre-Cenozoic SSZ systems developed similar concentric arc-backarc systems. The Cenozoic Western Pacific Basin may be a mirror-image analogue for the Eastern Pacific in Paleozoic-early Mesozoic time (cf. Hawkins et al., 1984).

SSZ Magma Systems

The geologic and geometric features listed above help characterize active SSZ systems; however, there are additional factors that are critical for identifying remnants of SSZ systems preserved in the geologic record. Magmas erupted in modern SSZ settings have distinctive chemical, isotopic, and mineralogic characteristics that help distinguish them from major magma types such as N-MORB or ocean island basalt (OIB). These differences reflect the more complex history of the SSZ source relative to the MORB sources, and offer constraints on the *nature* of the SSZ magma sources. These data give useful insights into the extent of melting, chemistry, and mineralogy of the mantle source, and into eruption processes; however, the mineralogy and chemistry, if used alone, are probably insufficient for inverting the problem and understanding the tectonic setting in which magmas developed. For example, rocks from the Chile Rise carry an SSZ trace element signature, but they formed at an oceanic spreading center (Sturm et al., 2000). When petrology and geochemistry are coupled with other data, such as field associations, rock series, age, stratigraphy, and regional structures—in short, the geology—we should be able to make well-reasoned inferences about the tectonic setting in which an ophiolite series was formed.

Sources of SSZ Magmas

There are three main source components important for SSZ petrogenesis (Pearce et al., 1984; Hawkins et al., 1984; Hawkins and Melchior, 1985; Ellam and Hawkesworth, 1988; Stern et al., 1991). The primary source is the SSZ mantle wedge situated above the Wadati-Benioff Zone (Fig. 2); in nearly all SSZ settings, this should be depleted mantle (cf. Woodhead et al., 1993).

The subduction component is the oceanic lithosphere, of which the most important contributions are from altered ocean crust, altered upper mantle, and a thin sediment carapace, all of which are enriched in volatiles (mainly H_2O, CO_2, and CH_4) and elements mobile in aqueous fluids (cf. Tatsumi and Eggins, 1995). The fluid-mobile elements are the large ionic radius lithophile elements (LILE). These are also known as low-ionic potential elements (i.e., they have a small value for charge/ionic radius). A third potential component is mantle material convectively entrained above the subduction zone by viscous drag exerted by the subducted plate, and brought in under the backarc basin and arc (cf. Ewart and Hawkesworth, 1987). This "new" mantle may be relatively fertile and capable of generating N-MORB, enriched MORB (E-MORB), or OIB (Hawkins, 1976, 1995a, 1995b; Hawkins et al., 1990; Volpe et al., 1987, 1988, 1990; Ikeda and Yuasa, 1989).

The melting process preferentially extracts magmaphilic elements from the mantle source, as controlled by their bulk distribution coefficients (D_O) for mantle source materials. Both high field strength elements (HFSE) and LILE have low values for D_O and thus will be concentrated in melts, with the further constraints being element abundance in the source and the extent of melting. There are abundant data from active SSZ systems that show that HFSE are depleted in all magma types relative to N-MORB, whereas the LILEs show variable enrichments. This variable enrichment in LILE is attributed to the subduction component (fluids, melts, or both, derived from subducted sediments and lithosphere), whereas the HFSE depletion stems from previous melt extraction.

Mantle wedge. The petrology of the SSZ mantle wedge (SSZ-MW) is arguably the fundamental control on the chemistry of magmas erupted and on the mineral composition of the rocks formed on cooling. Initially, the SSZ-MW probably is dominated by extensively depleted peridotitic mantle from which basalt previously was extracted (e.g., at a mid-ocean ridge or in a backarc basin). Relative to the MORB-source mantle, which itself is depleted relative to bulk earth mantle, it is even more depleted in all magmaphilic or low partition coefficient elements, including both LILE and HFSE. We find examples of this depleted SSZ-MW in rocks dredged from trench walls, such as the abundant dunite and harzburgite, both serpentinized to various degrees (e.g., Fisher and Engel, 1969; Dietrich et al., 1978; Bloomer and Hawkins, 1983). Peridotites of ophiolite terranes exhibit similar depleted mantle material (e.g., Hawkins and Evans, 1983; Harper, 1984; Wallin and Metcalf, 1998). Because of this previous depletion, the depleted peridotites of the initial SSZ-MW are unlikely sources for extensive melt generation because they have low abundances of Al and Ca (which reside mainly in clinopyroxene) and alkali metals. This greatly limits the amount of melt they can form. The most likely melts would be the class of high-Mg SSZ rocks known as boninites, discussed below, but boninites require high thermal gradients and addition of water to the source to allow further melting of already depleted mantle (cf. Crawford, 1989).

Partial melting is promoted by convective transport of hot mantle, adiabatic decompression from upwelling, and lowering of solidus temperature *(T)* of hydrated (e.g., serpentinized) harzburgite mantle.

Subduction component. Subduction of altered oceanic crust, serpentinized upper mantle peridotite, and sediment all serve to convey water, as well as mobile, low ionic potential elements into the SSZ-MW (Peacock, 1990; Tatsumi and Eggins, 1995). The low ionic potential trace elements include Li, K, Rb, Cs, B, Ba, Sr, La, Ce, U, and Pb. Th is not considered to be mobile in fluids, and Th enrichments observed in some SSZ magmas are likely from mobilization in melt. The chemistry of the LILE-enriched fluids is controlled by dehydration of structural water from hydrous minerals and by fluid/mineral partitioning of mobile elements. Extent of re-enrichment may be time-dependent or may be controlled by abundance of fluid-bearing and LILE-bearing minerals in the subducted crust. In addition to metasomatizing the depleted mantle with fluid-mobile elements, the added hydrous phase affects the melting relationships of minerals and has important influence on Al, Fe, Ti, and Na abundances in SSZ magmas (cf. Fryer et al., 1981). "Melting of the mantle wedge is induced by the influx of H_2O released from pressure-sensitive dehydration reactions in hydrous peridotite as it is dragged downwards by subduction along the base of the mantle wedge—preferential transport of elements soluble in H_2O-rich fluids gives rise to the distinctive geochemical characteristics of subduction zone magmas and their mantle wedge source regions" (Tatsumi and Eggins, 1995). Much of the original fluid is lost by compaction and dewatering of sediments at the onset of subduction; progressive dehydration occurs as the subducted material is heated and compressed. Much of the subducted fluid escapes and is returned to the ocean; some is taken up by newly formed metamorphic minerals (cf. Peacock, 1990). Hydrous minerals (e.g., zeolites, chlorites, micas, serpentines, epidote, chloritoid, amphiboles, lawsonite, and clinohumite) persist to different pressures and temperatures, but some are stable, and fluids may be retained to depths of 80–100 km, depending on thermal gradients and rate of heat transfer. Calcite may be an effective medium for deep subduction of CO_2. The progressive dehydration of hydrous phases transfers the low ionic potential elements to the depleted mantle peridotite.

Sediment subduction may be important in some settings, especially those having large accretionary prisms (e.g., Sunda Arc), but most intra-oceanic subduction systems lack large accretionary prisms of sediment. This is likely the result of minimum input of sediment to the trench and effective subduction of the small amount of sediment that is present. Subduction of small amounts of sediment need not result in delivery of sediment to depths and temperatures where the (nearly anhydrous) sediment would melt; however, direct or indirect (via fluids) contributions may explain some SSZ Sr and Pb isotope data. For example, Meijer (1976) proposed that Pb isotope data for Mariana Arc lavas may be explained by <1% sediment contribution. Pb data for the Tonga Arc may permit <2% sediment contribution (Oversby and Ewart, 1972; Ewart and Hawkesworth, 1987). Sr isotope data for both arcs are consistent with similar small contributions, although fluid phase transfer rather than direct melting of sediments is highly probable. For the Tonga Arc, influx of more radiogenic mantle sources may be involved (Hergt and Hawkesworth, 1994). Cosmogenic ^{10}Be (half-life 1.6 Ma) has proved useful to show sediment contributions in many arcs (Tera et al., 1986; Ben Othman et al., 1989; Morris et al., 1990). The short half-life requires rapid transfer from ocean sediments, where it accumulated from atmospheric precipitation, to deeper levels having low ^{10}Be, where it is incorporated into the magma source. The transfer mechanism is debated, but many geoscientists consider that Be is relatively insoluble, and thus immobile, at near surface T and P (Leeman, 1996). Hydrothermal experiments (You et al., 1996) show that Be is mobile in fluids at 300 °C, thus obviating the need to invoke sediment melting to explain Be data. Sediments clearly play a role in SSZ magmatism, but most data support a role for sediment-derived fluid, rather than direct melting of sediment (e.g., Stern et al., 1991); however, the aqueous fluids cannot transport HFSE. Melting is required for that.

Partial melting of the subducted ocean lithosphere must be considered as another source and, in view of its "limitless" supply, a potentially significant one; however, many geochemical data indicate that the subducted lithosphere rarely is a source of melts in most subduction systems (cf. Gill, 1981). Melting of the subducted basalt would mobilize HFSE, but SSZ magmas are low in HFSE. In some places, a case can be made for melting of young, and hot, subducted ocean crust (cf. Defant and Drummond, 1990; Maury et al., 1996), but the resulting magmas (adakites) are rare in the intra-oceanic SSZ setting (e.g., they should have exceptionally high Sr/Y and La/Yb, and show HFSE enrichments). Thermal configurations of subduction systems apparently limit the situations in which low-P melting of the subducted slab can occur to settings where very young, hot, crust is subducted. Evidence for high-P melting (eclogite field of stability) in oceanic arc settings is not supported by rare-earth element (REE) data, which should show strong negative slopes on chondrite-normalized plots. It is unlikely that melting of subducted oceanic crust plays a major role as a magma source for the intra-oceanic SSZ, although it may be important in magmatic arcs of Andean-type margins.

Mantle counterflow. Counterflow of hot mantle above the subduction zone (cf. Fig. 2) was proposed by Holmes (1931, 1965) "to illustrate a speculative arrangement of convection currents which might account for the origin of andesitic lavas….and for the concentration of volcanoes over the region of intermediate earthquakes." See also a discussion by Allwart (1988). Most modern ideas about the dynamics of the SSZ mantle utilize some form of convecting mantle (cf. Ewart and Hawkesworth, 1987). Sleep and Toksoz (1971), Toksoz and Bird (1977), and Davies and Stevenson (1992) proposed that counterflow responsible for extension and arc magmatism is a consequence of viscous drag induced by the subducted ocean lithosphere. Alternatively, the driving force may be upward circulation of hot mantle driven by descent of cooled mantle (Ida, 1983) or a passive response to seaward retreat of the subducted lithosphere accompanying rollback of the hinge (Hamilton, 1995, 2002).

The entrained "new" mantle may be N-MORB source or more enriched material, such as a source for E-MORB or OIB magmas. Evidence for mantle counterflow into the SSZ comes from isotope studies of Western Pacific Basin systems (e.g., Hergt and Hawkesworth, 1994; Hickey-Vargas et al., 1995), that show that young SSZ systems having Pb and Sr isotope systematics more typical of an "Indian" MORB source have replaced older sources with "Pacific" MORB source isotope signatures. For SSZ systems with active backarc basins, this convective overturn of mantle above the WBZ is significant because it brings in mantle that is fertile enough to generate the MORB-like melts characteristic of backarc basin crust (Hawkins, 1995a, 1995b). Most models envision counterflow in a direction opposite to the incoming subducted plate; however, complex flow under the Lau Basin is indicated by seismic anisotropy (Smith et al., 2001). Their data suggest that in addition to trench-normal flow directions, there is north-south flow. Independent evidence for the latter is found in ^{3}He/^{4}He data showing high ratios at some northern Lau Basin seamounts, which may represent a "plume" component from Samoa (Poreda and Craig, 1993).

Extraction of MORB magmas from the convectively overturned mantle beneath backarc basins leaves a partially depleted mantle residue that may provide continuous influx of mantle that is still capable of making melts to form island arc tholeiite basalts and basaltic andesite (cf. Ewart and Hawkesworth, 1987).

BACKARC BASINS

Introduction

It is my thesis that few, if any, ophiolites represent fragments of true mid-ocean ridges. The Macquarie Ridge ophiolite (Varne et al., 2000) is a rare occurrence that clearly is derived from crust formed at a mid-ocean ridge, but there are few other unequivocal examples. In the case of Macquarie Ridge, rock chemistry ranges from N-MORB to E-MORB and all rocks have relatively high alkali metal content. Few ophiolites have this enriched chemistry, but in this location it has been emplaced by crustal contraction on a former fracture zone and clearly represents an ophiolite assemblage.

If most ophiolites are not from a mid-ocean ridge, then from where do they come? The geology of backarc basins may be the key to understanding the source for the MORB-like crust of many ophiolites. In this section, I summarize the petrologic characteristics of backarc basin crust and show its close geochemical and mineralogic similarity to crust formed at mid-ocean ridges. I use the Lau Basin and Mariana Trough as type examples, although any of the backarc basins shown on Figure 1 would suffice. The North Fiji Basin (Price et al., 1990) and Bransfield Strait (Weaver et al., 1979) are nearly identical in terms of petrology and crustal fabric. Backarc basin crust forms at spreading centers that, in many backarc basins, erupt tholeiitic melts that have mineral and chemical signatures ranging from N-MORB to island arc tholeiite (Weaver et al., 1979; Hawkins and Melchior, 1985; Hawkins, 1995 a, b; Gribble et al., 1996, 1998). Price et al. (1990) showed that backarc lavas from the North Fiji Basin range in composition from N-MORB to rocks more enriched in LILE and having "transitional alkalic chemistry (up to 0.5% Ne in the Norm)." This range in rock chemistry may be related to the maturity of the basins, which in turn may be owing to the relative importance of subduction-derived fluids in the mantle source. In their early stage, basalts retaining an SSZ signature may form, but as the backarc system evolves, the main magma becomes basalt similar to MORB in nearly all respects. The early SSZ or "alkalic" signature (cf. Price et al., 1990) may be a consequence of source heterogeneity on various scales that retain memory of the SSZ setting (e.g., Volpe et al., 1987, 1988, 1990). An alternate explanation, well developed by Gribble et al. (1996) for the southern part of the Mariana Trough, is that the source is fluids derived from the subduction component. The southern Mariana Trough lies 40–60 km west of the trace of the steeply inclined seismic zone; this suggest that the the subduction-derived fluids must be entrained in convecting mantle material.

It is important to note that relatively "mature" backarc basins such as the Lau Basin presently erupt MORB on the axial ridges, whereas in the Mariana Trough, we find a range from arc-like (IAT) basalt to MORB on the present axial ridge. Both basins have about the same chronologic age, but clearly have different mantle sources (Hawkins, 1995 a, b). Can this be related to differences in the geometry of the WBZ, as discussed above? There is no *a priori* reason why magma evolution must follow a common pattern; mantle source composition, including the nature of mantle convectively entrained above the WBZ, must be a controlling factor.

Seamounts in the backarc basins range in composition from island arc chemistry to ocean island basalt. Backarc basins may be blanketed with arc-derived clastic sediments on their flanks. This variety of geologic features is part of the key to recognizing the backarc basin component in SSZ ophiolites. I postulate that if crust from the Lau Basin or Mariana Trough were incorporated among rocks of an orogenic belt, they would be identified as ophiolite and, if considered out of context with related rocks, might well be interpreted as having formed at a mid-ocean ridge spreading center. Important to the study of ophiolites is the observation that backarc basins may have heterogeneous compositions in which MORB may dominate, but in which both arc-like and OIB-like rocks also are present.

Lau Basin—Form and Geological Setting

The Lau Basin separates the Lau Ridge remnant arc from the active Tofua (Tonga) Arc (Fig. 5). It has a trapezoidal shape, with the Tofua Arc and Tonga Trench on its eastern side. The Tonga Trench curves sharply westward along the northern edge, where it becomes part of the proposed transform system linking the Tonga and Vanuatu-New Hebrides trenches. The backarc basin narrows southward to near 26° S, where it becomes a narrow basin, the Havre Trough, which is the backarc basin of the Kermadec trench-arc system. Farther south, the Havre Trough merges with the Taupo graben of New Zealand. The shape of the Lau Basin may in part be a consequence of backarc crustal extension being hindered by the thickened lithosphere of the Louisville Seamount Chain, which enters the Tonga-Kermadec Trench near 26° S. Rapid extension (165 mm/yr^{-1}, Bevis et al., 1995) at the unconstrained north end of the Lau Basin is a consequence of rapid trench rollback manifested as crustal extension on both the North Lau spreading center (NLSC) and the Mangatolu Triple Junction (MTJ). The central part of the Lau Basin, near 19° S, is situated ~250 km above the seismic zone (Isacks and Barazangi, 1977) and above a zone of extremely low-Q (seismic attenuation) that presumably is because of elevated mantle temperature (Barazangi and Isacks, 1971; Zhao et al., 1997). Seismic refraction studies (Crawford et al., 2003) show that the crust of the basin varies from 5.5 to 6.5 km thick on the east side, with a relatively thick "sheeted dike" section (2–3 km) overlying a thin section (2 km) of lower crust. The crust thickens to the west, being 7–8 km thick near the Central Lau spreading center (CLSC) and 8.9–9.5 km thick at ~50 km west of the CLSC.

The Lau Basin began to open at ~6.5 Ma, as determined from ODP Site 834 data (Hawkins et al., 1994). At that time the Lau Ridge volcanic arc (>14 Ma to 2.5 Ma) was still active. The forearc was extended and small rift basins began to open. At ~4.5 Ma, a southward propagating ridge—the present East Lau spreading center (ELSC)—began to replace former crust with new seafloor having N-MORB composition and developing symmetric magnetic anomalies. A second propagator started at 1.5–1 Ma and formed the CLSC, which also has symmetric magnetic anomalies. At about this time, the North Lau spreading center (NLSC) and the complex of three separate ridges that radiate from the MTJ also formed. As discussed below, all of these spreading centers are dominated by N-MORB-type chemistry, although the MTJ lavas show a subduction component (Table 1). All but the CLSC have a distinctive depletion in Nb and Ta, which points to their SSZ heritage. Lau Basin seamounts and some of the seafloor include both arc-like lavas, as evidenced by a strong SSZ signature in rocks and detritus (e.g., ODP Site 839, Table 1), as well as some have an enriched or E-MORB signature (e.g., Rochambeau Bank and Niua Fo'ou Island, Table 1).

The western side of the Lau Basin has many small rift basins, up to 25 km long and 8–14 km wide, partly filled with volcaniclastic gravity flow deposits (turbidites) derived from local high-standing fragments of rifted arc crust or nearby submarine arc-composition volcanoes (Parson et al., 1994). The basins have on the order of 100 m of fill overlying basaltic basement. Lower in the section, clastic sediments are interbedded with and intruded by tholeiitic flows and sills having the general

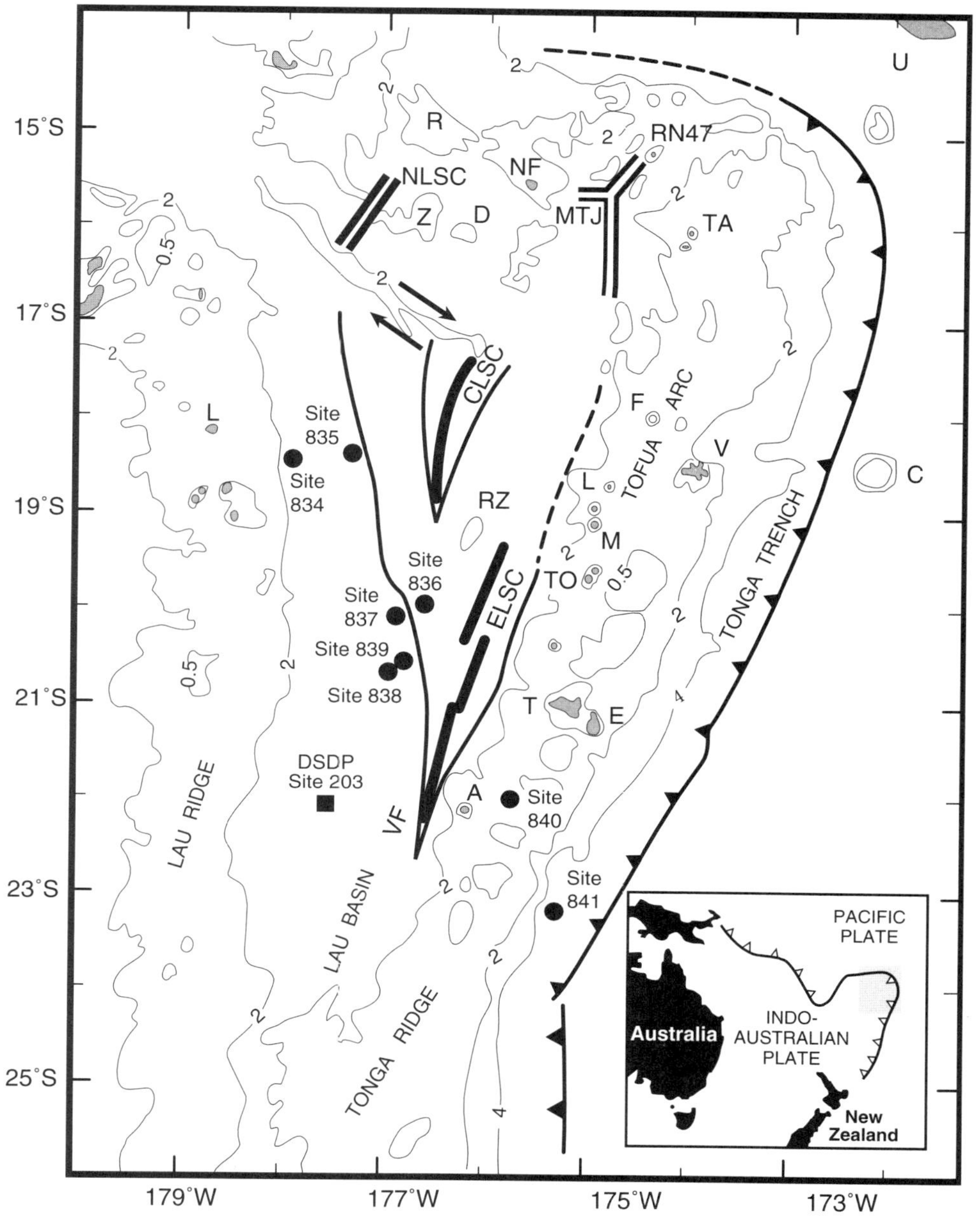

Figure 5. Map of Lau-Tonga arc-trench system, showing features discussed in text. Active spreading centers are: ELSC—East Lau Spreading Center; CLSC—Central Lau Spreading Center; RZ—relay zone between them; NLSC—North Lau Spreading Center; and MTJ—Mangatolu Triple Junction. Sites 834–841 are ODP Leg 135 drill sites; DSDP 203 is from DSDP Leg 21. Seamounts are: D—Donna Seamount, R—Rochambeau Bank, Z—Zephyr Shoal, and NF—island of Niua Fo'ou. Features on Tonga Ridge are: volcanic islands, TA—Tafahi; RN 47—an unnamed seamount; L—Late; F—Fonualei; M—Metis Shoal; TO—Tofua; A—Ata; and raised coralline-capped platforms, V—Vava'u; and T—Tongatapu. Other features are E—'Eua Island; U—Upolu, Samoa; and C—Capricorn Seamount.

mineralogy and chemical composition of MORB. The volcaniclastics are overlain by hemipelagic clay-nannofossil oozes stained by ferro-manganese oxy-hydroxide (Parson et al., 1994; Rothwell et al., 1994). In the central part of the Lau Basin, the igneous basement has only a discontinuous cover of pelagic and hemi-pelagic sediments, many of which in their basal section are heavily stained with ferro-manganese oxyhydroxides from hydrothermal circulation at the axial ridge (Hodkinson and Cronan, 1994). Arc-derived volcaniclastic rocks from the Lau Ridge and the Tofua Arc partially blanket both sides of the basin and partly fill the rift basins (Clift et al., 1998).

Mariana Trough—Form and Geologic Setting

The Mariana Trough (Fig. 4) separates the West Mariana Ridge remnant arc from the active Mariana volcanic arc. The age of initial opening is uncertain, but the best estimate for the initiation of rifting is ca. 10 Ma (Fryer, 1992). The crescent shaped Mariana Trough narrows northward to ~25° N, where its extension may be pinned by thickened crust of the Magellan seamounts that lie outboard of the Mariana Trench. The ~200 km wide south end of the Mariana Trough is unconstrained, and basin depths descend to the ~10 km-deep, west-trending, part of the Mariana Trench.

TABLE 1. LAU BASIN

Location:	ELSC	ELSC	ELSC	ELSC	VALU FA	CLSC	CLSC	CLSC	CLSC	CLSC
Site:	ODP 834	R-5	R-11	R-7	VF-1	ST-64	R-26	R-25	P-33	R-13
Sample:	3H7 54-55	RNDB 5-3 GL	RNDB 11-1 GL	RNDB 7-2 GL	Lee 1-5	STO 64-3 GL	RNDB 26-2 GL	RNDB 25-1 GL	PPTU 33-1 GL	RNDB 13-2 GL
Mg#:	69.3	62.9	61.8	35.6	39.9	66.3	63.3	59.5	45.5	25.2
					Wt%					
SiO_2	47.13	51.44	50.62	52.58	56.69	48.58	50.25	50.96	50.45	60.41
TiO_2	0.69	0.90	1.02	2.11	1.38	0.70	0.94	1.05	1.93	1.62
Al_2O_3	19.18	15.44	15.18	13.43	14.57	16.85	15.86	14.65	14.11	12.93
FeO*	8.08	9.39	9.54	14.76	11.51	9.54	9.79	10.6	13.29	12.16
MnO	0.16	0.19	0.18	0.21	0.22	0.19	0.19	0.20	0.25	0.22
MgO	8.88	8.02	7.51	3.99	3.73	8.93	8.25	7.60	5.41	1.99
CaO	13.03	12.87	12.83	8.55	7.65	12.77	13.06	12.76	10.17	6.15
Na_2O	1.90	1.91	1.90	3.24	3.36	2.42	2.05	2.26	2.85	3.91
K_2O	0.01	0.03	0.13	0.14	0.52	0.01	0.07	0.03	0.08	0.42
P_2O_5	0.14	0.06	0.12	0.20	0.16	0.03	0.08	0.07	0.13	0.32
Total	99.20	100.25	99.03	99.21	99.79	100.02	100.54	100.18	98.67	100.13
					Trace elements (ppm)					
Cr		337	342		4	430	325	303	20	
Ni	111	110	103	8	25	189	149	116	37	7
Co		43	43			70	43		44	
V	216	307	338	502	350	202	323	332	455	147
Zr	47	53	57	127	81	33	51	45	128	441
Y	19	27	26	55	36	20	2929	30	59	138
Nb	2	1	1	1		2	2	1	1	
Hf		1.16	1.55		2.3			1.89	3.57	
Ta		<.05	0.08					0.04	0.16	
Rb	0.1	1	6	4	8.6	0.5	0.3	0.1	1	10
Ba	17	4	8	27	96	6.2	4	2	2	75
Sr	140	69	111	95	168	62	63	49	114	86

Locations: ELSC—East Lau Spreaading Center; CLSC—Central Lau Spreading Center

(*continued*)

The axial ridge of the Mariana Trough is asymmetric, offset to the east of the center of the basin; however, the axis lies slightly to the west of the deepest part of the inclined seismic zone, which here is nearly vertical (Isacks and Barazangi, 1977). Fryer (1992) recognized an early stage of rifting of small basins, followed by a later stage of seafloor spreading, similar to the early stages of evolution of the Lau Basin. Fryer (1995) interpreted the fabric of the northern Mariana Trough as representing an early stage of rifting, whereas the central region represents a more mature stage typical of slow spreading. The crustal fabric is dominated by north-south-trending abyssal hills. The western side of the basin, next to the remnant arc, is complexly faulted with numerous rift basins that record the earliest stage of opening (Fryer, 1992; Martinez et al., 1995). Cross-chains of arc volcanic rocks cut though the modern volcanic arc and transect the axial ridge.

A significant difference between the Lau Basin and the Mariana Trough is that the axial ridge of the slowly spreading Mariana Trough is a relatively simple feature, apparently propagating northward and having only minor offsets of the axis, whereas the fast spreading Lau Basin has six separate spreading ridges, three of which meet at a triple junction, and contains evidence for sequential development of propagating rifts since inception of rifting at ~6.5 Ma. Both sides of the Mariana Trough are blanketed with arc-derived clastic rocks and, as in the Lau Basin, the sediments lie on nearly coeval seafloor basalts. The occurrence of arc-derived clastic sediments, deposited on back-

TABLE 1. LAU BASIN (*continued*)

Location:	RZ	WB	WB	WB	CB	NLSC	MTJ	MTJ	Donna	Rochambeau
Site:	R-31	834B	834B	835	839A	P-19	R-17	R-19	Seamount	Bank
Sample:	RNDB	29R1	56R1	6R1	24x1	PPTU	RNDB	RNDB	123	PPTU
	31-4GL	/07-12	/01-15	/32-37	/00-06	19-1 GL	17-2 GL	19-1 GL	89-4 GL	23-2
Mg#:	68.3	64.6	5-Feb	65	67.1	69.1	27.2	64.4	53.5	54.5
					Wt%					
SiO_2	49.25	50.17	52.20	50.35	53.03	49.17	59.98	50.89	50.53	48.88
TiO_2	0.79	1.42	2.11	1.09	0.65	0.82	1.37	0.94	1.26	1.89
Al_2O_3	17.36	16.33	15.71	16.47	15.06	17.15	15.16	15.93	14.30	14.94
FeO*	8.65	8.67	12.41	8.46	9.04	8.72	10.50	8.53	11.97	11.35
MnO	0.17	0.20	0.20	0.16	0.17	0.18	0.23	0.17	0.25	0.19
MgO	9.10	7.73	3.40	7.67	9.00	9.51	1.91	7.52	6.73	6.63
CaO	12.82	11.87	8.58	12.90	11.66	12.75	5.62	13.10	11.72	11.79
Na_2O	2.20	3.08	3.87	1.82	1.35	2.11	3.89	2.14	2.88	3.10
K_2O	0.03	0.09	0.73	0.28	0.23	0.03	0.97	0.21	0.08	0.26
P_2O_5	0.03	0.11	0.21	0.11	0.07	0.04	0.42	0.11	0.12	0.17
Total	100.40	99.67	99.42	99.31	100.26	100.48	100.05	99.54	99.84	99.20
					Trace elements (ppm)					
Cr	334		0	179	618	741	39	309	150	238
Ni	163	102	12	78	123	312	6	90	45	72
Co	44					56	15	41	50	47
V	212	227				212	51	226	334	355
Zr	68	102	146	44	32	44	238	64	77	127
Y	24	29	40	22	16	20	78	24	32	31
Nb	3	3	1	1	1	1	9	1	3	7
Hf	1.14					1.15	6.09	1.59		3.27
Ta	<.05					0.045	0.65	0.133		0.375
Rb	1	3	13	5	5	0.3	27	7	16	5
Ba	3	22	60	42	78	1	147	27	11.1	40
Sr	93	176	195	130	148	79	135	133	90	215

Locations: WB—Western Basin; CB—Central Basin; NLSC—North Lau; MTJ—Mangatolu Triple Junction.

arc crust of essentially the same age, is an important feature of arc-backarc systems and important in recognizing an SSZ origin for some ophiolites. In the case of the Lau Basin, the Lau Ridge "remnant arc" remained active for ~4 my after initial rifting that formed the Lau Basin, and overlapped in time the volcanism on the Tofua (Tonga) Arc.

Petrology and Chemistry of Backarc Basins

Introduction

The primary rock type forming the crust of both the Lau and Mariana Trough backarc basins is tholeiitic basalt that is similar in most respects to N-MORB (Table 1). For detailed discussions of data supporting this claim, see Hart et al. (1972); Hawkins (1974, 1976, 1977, 1994, 1995a, b); Hawkins et al. (1984, 1990); Fryer et al. (1981); Sinton and Fryer (1987); Volpe et al. (1987, 1988, 1990); Stern et al. (1990); Falloon et al. (1992); Pearce et al. (1995); and Gribble et al. (1996, 1998). As discussed below, both basins have a range in rock composition that, in addition to N-MORB, includes IAT, backarc basin basalt (BABB, cf. Fryer et al., 1981), and enriched basalts that include OIB, as well as fractionated rocks of each of these series.

In both basins, the freshest samples occur on the axial ridge on presumed "zero age" crust and on seamounts close to the ridge axes. Axial ridge samples collected with ALVIN in the Mariana Trough were picked from fresh accumulations of pillows having essentially no sediment cover (Hawkins et al., 1990). Deeper crustal levels sampled at DSDP and ODP sites include older and more altered material, and some off-axis dredge sites at depths >4000 m also have altered rocks. Nearly all dredged rocks are wedge-shaped fragments of pillow basalts, many of which have rinds of fresh vitrophyre; minor amounts of microdiabase have been dredged, but these are more common in drill cores, and represent dikes, sills, or interiors of large pillows.

Gabbro is rare from back-arc basins, but one Lau Basin site (ANT 233, for details see Hawkins, 1976) yielded plagioclase rich, clinopyroxene gabbro with primary brown hornblende. DSDP Site 453, on the western edge of the Mariana Trough, recovered plagioclase-rich, two-pyroxene gabbro and mylonitized gabbro; Natland (1981) interpreted these rocks as likely to be from the remnant arc. Stern et al. (1996) and Ohara et al. (2002) recovered probable mid- or lower crust and upper mantle rocks at an amagmatic rift basin in the northern Mariana Trough. Ultramafic rocks are even less common; Ohara et al. (2002) recovered serpentinized harzburgite from this same Mariana Trough rift basin. It is mainly olivine (Fo 90.5), with orthopyroxene (En 90). Chrome spinel has Cr# 25–40, where Cr# = 100 × (Cr/Cr+Al) and the rocks have up to 2% modal clinopyroxene. These rocks are depleted mantle residue, but are less depleted than peridotites from forearcs, discussed below, which commonly have spinel with higher Cr#. The deep crustal rocks from this rift basin include anorthosite, diorite, tonalite, and granitoids. The latter have up to 79.7% SiO_2, and may be plutonic equivalents of rhyolitic glasses found in the Lau Basin and discussed below. Arai (1991) proposed that upper mantle rocks of the Circum-Izu Massif peridotite of central Japan represent backarc basin mantle of the Shikoku Basin. It, too, is mainly harzburgite with lesser amounts of dunite and clinopyroxene-poor lherzolite. Olivine in harzburgite is Fo 90–92 and in dunite is as high as Fo 94; the Cr# for chrome spinel ranges from 20 to 70, but is mostly 40 to 60. Some rocks have minor plagioclase ranging in composition from bytownite to anorthite. The spinel is unusual in having hydrous inclusions of pargasite or phlogopite + orthopyroxene. The peridotites have low-Ti, like many depleted peridotites but are unusual in having primary amphibole and mica. These minerals are never found in "abyssal" peridotites and have not been reported in depleted peridotites of trenches. Arai concludes that the peridotite composition is consistent with an SSZ origin.

The crystallization sequence commonly found in backarc basin rocks, as determined from vitrophyres and crystalline rocks, is: olivine + chrome spinel-plagioclase-clinopyroxene-iron-titanium oxides, just as we find for MORB. Orthopyroxene is not found in basalts. We can infer that backarc basin basal cumulates are likely to be clinopyroxene-bearing gabbros, or troctolite (i.e., plagioclase and olivine), rather than norites (olivine and orthopyroxene) or wehrlites (olivine and clinopyroxene), which are typical of arc series.

Backarc basin magma trace element chemistry (Hawkins et al., 1990) shows a small range from arc-like basalt (IAT) to N-MORB, with variable depletion in HFSE relative to N-MORB (e.g., Ti, Zr, Y, Hf, heavy rare earth elements, and Nb, and Ta in particular, Woodhead et al., 1993). These depletions in elements considered to be incompatible indicate magma derivation from a highly depleted mantle source. The LILE (e.g., K, Rb, Cs, Ba, Sr, and, to a lesser extent, La, Ce) show varied enrichments relative to N-MORB stemming from addition of low ionic potential elements mobile in aqueous fluids (cf. Tatsumi and Eggins, 1995). The presumed source of the LILE and fluid phase is altered ocean crust and subducted sediments.

Thus, in backarc basins we have evidence for samples petrographically and chemically equivalent to presumed oceanic seismic layers 2a and 2b, and a few samples representative of layer 3, and depleted upper mantle. These rocks are petrographically comparable to their putative "equivalents" in ophiolites. In addition, metalliferous sediments and hydrothermal vents with polymetallic sulfide deposits have been found in both basins (Hawkins et al., 1990; Hawkins and Helu, 1986; von Stackelberg et al., 1988).

Lau Basin. The early eruptions of backarc magmas in the Lau Basin at ~6.5 Ma were in rift basins seaward of the (still active) Lau Ridge. They were erupted in what then was the forearc of the Lau Ridge. At ODP Site 834, we drilled through 112.5 m of sediments with clayey nannofossil ooze overlying pyroclastic airfall ashes and epiclastic turbidites, and then 323 m into basalt having 13 texturally different units having aphyric to diabasic textures. These comprise a series of basalt flows or sills with rare interbedded layers of silts and clays presumed to have come from the Lau Ridge volcanic arc. Table 1 has representative data for Site 834 and other features discussed below. At Site 835, on crust ~3.4 m.y. old and in a setting like that of Site 834, we recovered similar sediments and basalt. The distribution of rock types in the Lau Basin suggests that the early eruptions of basalt at ODP Site 834 and 835 are MORB in most respects, but have a distinctive arc-like depletion in HFSE, especially Nb and Ta, as seen in the abundances of trace elements normalized to values for N-MORB (Fig. 6A). Site 835 is even more depleted in HFSE than are the older rocks of Site 834. Somewhat younger samples at Sites 837 and 838 (Fig. 6B) are even more "arc-like" than the oldest samples. In contrast, the propagating rift that formed the ELSC (Fig. 6C) has trace element values more like true N-MORB, suggesting that MORB-source mantle had mixed with older SSZ mantle sources. Sample VF-1 (Table 1) is a differentiated rock (andesite) from the propagating rift tip of ELSC (Valu Fa). The rocks from CLSC and NLSC (Fig. 6D) have a strong MORB pattern. Basalts from Sites 834 and 835 have Pb and Sr isotope ratios that are indicative of a "Pacific" mantle MORB source, in contrast to other Lau Basin rocks, which have ratios typical of "Indian" MORB-source mantle (Hergt and Hawkesworth, 1994). The data suggest that convective influx of relatively fertile (Indian mantle-type) MORB-source has mixed with and swamped the original strongly depleted SSZ "Pacific" mantle (cf. Fig. 2). This retention of an arc signature in early melts is not necessarily a general evolutionary path for backarc basins,

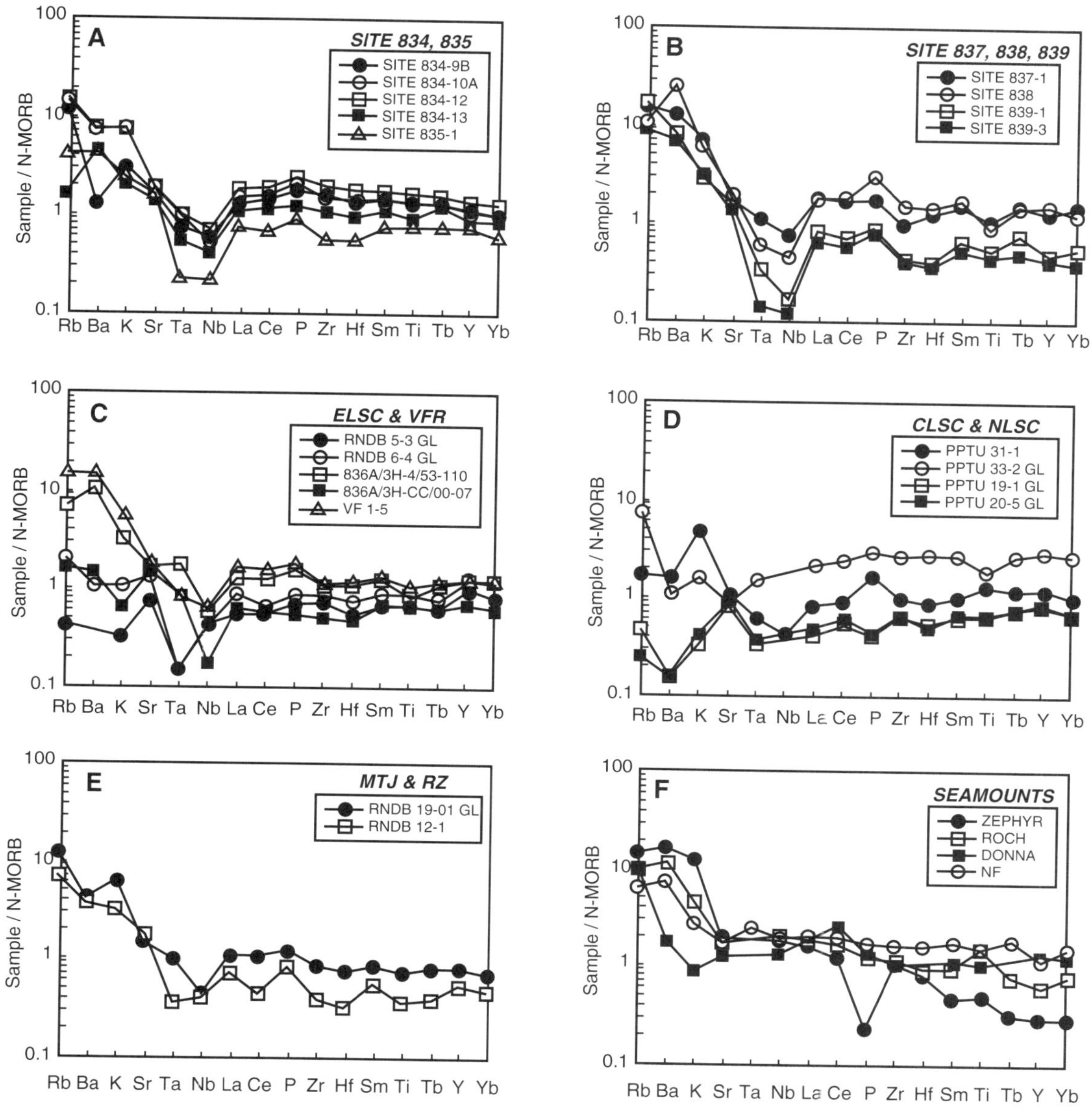

Figure 6. Trace elements for Lau Basin rocks normalized to N-MORB values (Sun and McDonough, 1989). Locations of ODP sites and spreading ridges are in Figure 5. SSZ magmas are characterized by variable LILE enrichments and HFSE (especially Nb and Ta) depletions relative to N-MORB. A: Data from ODP Sites 834 and 835 are from crust formed in rift basins near Lau Ridge remnant arc early in opening of the basin (~6.5–5 Ma). B: Sites 837, 838, 839 are from younger crust on west side of ELSC. C: Data from ELSC and its propagating rift tip (VF-1). Most ELSC data are close to N-MORB abundances, but Nb and Ta depletions are retained. D: CLSC, samples PPTU 31–1 and 32–2, and NLSC, samples PPTU 19–1 and 20–5 have MORB-like abundances with only minor variations in Rb, K, and Ba. E: RNDB 12–1 is from the relay zone between CLSC and ELSC and shows arc-like depletions in HFSE and enriched LILE. Mangatolu Triple Junction, sample RNDB 19–1, is from within 80 km of Tofua arc and shows pattern similar to relay zone. F: Data for enriched MORB of Niuafo'ou Island (NF) and Rochambeau Bank (ROCH); Donna Seamount sample is altered MORB crust, whereas Zephyr Shoal sample is a hybrid of silicic glass and mafic phenocrysts.

as shown by the presence of N-MORB-like basalt in the nascent Sumisu Rift of the Izu system (Fryer et al., 1992).

The Mangatolu Triple Junction (RNDB 19–01) and Relay Zone (RNDB 12–1) basalts (Fig. 6E) show the SSZ pattern. That both are from within ~50 km of the Tofua arc must reflect an arc source contribution. Seamount data (Fig. 6F) for Rochambeau Bank (ROCH) and the island of Niua Fo'ou (NF) are relatively enriched in LILE, but have MORB-like HFSE; these probably reflect an E-MORB contribution. Donna Seamount is a large uplifted block of moderately altered MORB-like seafloor, whereas Zephyr Shoal is a hybrid of rhyolitic glass and mafic phenocrysts.

The important point to be made from these data is that in backarc basins, some seafloor mimics mid-ocean ridge crust in those elements (HFSE) least likely to be modified by low-grade metamorphism or alteration, whereas in other parts of the same backarc basin, the HFSE may have a more arc-like signature. These should be important criteria for determining the provenance of ophiolites.

Mariana Trough. Mariana Trough basalts also are broadly similar to N-MORB in terms of mineralogy, major elements, HFSE, and isotope ratios for Sr, Nd, Pb, He, and O (Fryer et al., 1981; Sinton and Fryer, 1987; Hawkins et al., 1990; Volpe et al., 1987, 1990; Gribble et al., 1996, 1998). Element ratios for both HFSE and LILE resemble MORB, but the LILE and volatiles, especially H_2O, are variable, with many being higher than for MORB. Typical values for H_2O in N-MORB are close to 0.2%, whereas for the Mariana Trough, Stolper and Newman (1994) determined a range for H_2O from 0.49% to 2.1%; the CO_2 range for the same samples is from <10 ppm to171 ppm. The data display a negative correlation between CO_2 and H_2O and a negative correlation between depth of samples and H_2O. Garcia et al. (1979) determined an average H_2O value of 1.02% and 0.21% for CO_2 for a different set of samples.

Some chondrite-normalized REE patterns differ from MORB in having flat to slightly enriched light rare earth elements (LREE) patterns (e.g., La/Sm_{ef} ~1; Hawkins et al., 1990). Samples were divided into N-MORB type and Mriana Trough Basalts (MTB) type (Hawkins et al., 1990) on the basis of volatile content, LILE, and ratios of oxides of Al, Ca, Ti, and Fe relative to MgO (cf. BABB of Fryer et al., 1981; Sinton and Fryer, 1987). For 177 vitrophyric samples having Mg# of 60 to 70, the two types are approximately equally represented. Some representative data are in Table 2, where glass analyses were averaged for different

TABLE 2. MARIANA TROUGH

Sample:	MARA	MARA	MARA	MARA	MARA	MARA	MARA	ALV
Type:	N	N	N	M	M	M	"ARC"	1846-9
Mean:	n=50	n=42	n=26	n=43	n=33	n=10	n=10	
	Mg# >65	Mg# 60–65	Mg <60	Mg# >65	Mg# 60–65	Mg <60	Mg# 63–68	Mg# 67
				Wt%				
SiO_2	50.63	51.38	52.67	50.79	51.52	54.51	50.15	49.67
TiO_2	1.13	1.41	1.23	1.01	1.21	1.75	0.58	0.57
Al_2O_3	17.21	16.34	16.51	17.41	16.71	16.05	17.37	17.45
FeO*	7.64	8.58	8.57	7.28	7.98	9.53	7.48	7.03
MnO	0.15	0.18	0.18	0.15	0.16	0.19	0.16	0.16
MgO	7.72	6.94	5.51	7.37	6.59	3.92	6.83	6.95
CaO	11.65	11.11	10.26	11.86	11.19	8.2	13.06	13.46
Na_2O	2.79	3.11	3.03	2.67	2.92	3.21	2.13	2.07
K_2O	0.21	0.25	0.49	0.32	0.36	0.73	0.76	0.71
P_2O_5	0.13	0.17	0.19	0.14	0.17	0.31	0.13	0.13
Total	99.26	99.47	98.64	99	98.81	98.4	98.65	98.2
				Trace elements (ppm)				
Cr	310	230	57	250	260	247		198
Ni	135	110	30	115	95	56		115
Co	35	35	65	30	30	57		69
V	220	250	271	220	250	292		195
Zr	90	100	81	75	95	139		44
Nb	1	2	1	1	2	2		1
Y	26	33	28	24	32	39		15
Ta	0.18	0.19		0.16	0.22			0.054
Th	0.31	0.24		0.4	0.36			0.53
Sr	195	170	222	205	220	193		314
Rb	2.6	1.5	9.3	4.7	4.7	5.4		25
Ba	25	25	69	45	50	38		49
La	4.8	4.6	7.14	5.1	5.1	6.79		8.2
Nd	9.45	8.1	11.33	9.2	10.9	13.42		10.48
Sm	3.1	2.7	3.22	2.8	3.5	4.02		2.26

Note: Blank—no data; Type N—N-MORB; Type M—MTB (BABB).

ranges of Mg# and are separated into N-MORB types (N) and MTB (M). Table 2 also has examples of arc-like lavas from the axial ridge. Element ratios plotted on Harker diagrams (cf. Hawkins, 1995b) show that Mariana Trough axial ridge data from near 18° N define fields different from N-MORB. Mean value for Al_2O_3 is 17% versus an upper limit of 16% for N-MORB, CaO is higher (mean value is 12%), thereby retaining the N-MORB ratio of 0.5–0.65 for CaO/Al_2O_3. The FeO* (mean value of 7%) is higher than for N-MORB at comparable MgO content.

Our ALVIN samples collected along the axial ridge have Mg# ranging from >65 to 50 over distances of 1 to 2 km. Data for these samples are in Table 2. Axial ridge seamount samples labeled "ARC" and ALV 1846–9 on Table 2 have island arc tholeiite chemistry; they are surrounded by crust with MORB chemistry having Mg# = 62. The most fractionated axial ridge sample we found is a basaltic andesite having 56% SiO_2, 4.3% MgO, and Mg# = 36, suggesting that shallow-level fractionation was important in the Mariana Trough, as in other ocean ridges.

Mariana Trough mantle source history may differ from that of the Lau Basin. In the Mariana Trough, we found that both N-MORB type lavas and MTB, a variety of transitional to arc-like lavas, erupted synchronously at closely spaced sites along the axial ridge (Table 2). Volpe et al. (1987, 1990) proposed mixing of sources. At its northernmost part, 22° N, the axial ridge lavas display arc-like chemical and isotopic signatures (Stern et al., 1990).

On backarc magmas. The HFSE abundances and element ratios are the most useful chemical signatures likely to be preserved, even if rocks are altered or have experienced low-grade metamorphism. The cause of the HFSE depletion is a matter of debate, e.g., theoretically, a titanate residual in the mantle helps explain this, but the implicit Ti concentrations for the mantle make this unlikely. The simplest explanation is that the SSZ mantle has been depleted in HFSE relative to MORB. These depletions are imparted to SSZ melts and appear to be diagnostic for the setting. Several classification schemes and discriminant diagrams are useful. Pearce and Cann (1973), Pearce and Norry (1979), and Pearce et al. (1984) used multi-element ratios such as Ti, Zr, Y, and Cr (and Sr for unaltered rocks) to define fields for mafic magma from various sources, including "island arc" and MORB. Shervais (1982) used Ti vs. V to define fields for arc (Ti/V <20), ocean ridge (Ti/V 20–50) and ocean island (Ti/V >50). Cr-Y variations (Fig. 7) prove useful for separating IAT magmas from N-MORB and from boninites. Pearce et al. (1981, 1984) proposed that abundances of a compatible element, such as Cr or Ni, relative to high field strength (high ionic potential) incompatible elements, such as Y or Zr, would indicate effects of fractionation or extent of melting. Cr and Y are "not significantly affected by the processes that cause heterogeneity in the convecting upper mantle." Cr would decrease with melt fractionation because it is accommodated in olivine and pyroxene, whereas Y is excluded and would increase in residual melt; the converse would apply for fractional melting. Data for MORB and island arc lavas show that, for a given Cr content, Y and other HFSE are relatively depleted in island arc lavas. MORB-source mantle is assumed to have >2500 ppm Cr and ~4 ppm Y; primitive N-MORB has ~340 ppm Cr and 25 ppm Y, with a negative correlation between them for fractionated magmas (e.g., 60 ppm Y at 100 ppm Cr). In contrast, melts from IAT settings commonly have markedly lower Y than N-MORB at comparable Cr content (e.g., 10–20 ppm Y). Boninites have high Cr and very low Y relative to Cr. Inasmuch as Cr and Y are relatively unaffected by weathering and low-*T* metamorphism they are useful in understanding the origin of ophiolites, and in particular, help recognize SSZ ophiolites. Discriminant diagrams show that most backarc basin basalts plot in the field for N-MORB, but some have an SSZ signature. The IAT and boninite data require derivation from a source that has lost a basalt component and thus is depleted in Y relative to bulk earth mantle. Figure 7 shows data for the Lau Basin, Mariana Trough, and their nearby contemporary arcs, compared with fields for boninite, island arcs, and MORB. Also shown are data for the Palau-Kyushu Ridge. Note that for the Lau Basin, the ODP drill data from the older, pre-spreading crust partly overlaps the IAT field, whereas most of the axial ridge dredged samples lie in the MORB field. The Mariana Trough data are all from near-zero age crust and lie in the MORB field. The Palau-Kyushu Ridge data include IAT and MORB, as well as some samples that overlap onto the boninite field. Pearce et al. (1981) and Wallin and Metcalf (1998) used the Cr-Y plot to demonstrate an SSZ origin for the Oman and Trinity ophiolites, respectively.

The ternary plot for Hf/3-Ta-Th (Wood et al., 1981) is another useful example (Fig. 8). In Figure 8A, we see that Site 834 data lie in the N-MORB field along with much younger Site 838, whereas Site 835 data lie in the SSZ field and are displaced toward Th, an effect of sediment admixture. The field for the Tofua Arc and RN 47, a primitive Tofua arc seamount, are shown for comparison. Figure 8B shows that Site 839 data, from the central part of the basin, lies on the Tofua field, whereas Site 836, adjacent to the ELSC, plots in the N-MORB field. It is interesting that ODP Site 834 data, some of the oldest Lau Basin crust, plot in the N-MORB field, as do Site 836 data, young rocks drilled close to the ELSC. Dredged samples from the axial ridges (Fig. 8C) plot mainly in the N-MORB field, whereas the MTJ samples trend into the SSZ field. The MTJ is situated within 50 km of the Tofua arc and likely is influenced by the subduction component added to the mantle wedge. Two seamounts show an E-MORB component, but the other ODP drill sites lie in the SSZ field, together with samples from the Tofua Arc and RN 47.

Mariana Trough ALVIN samples and dredged basalts lie partly in the N-MORB field and partly in the SSZ field on Figure 8D. ODP drill data for Site 454 from the west flank of the axial ridge has an SSZ signature, whereas Site 456 on the east flank plots in the N-MORB field. The same diagram show two Mariana forearc sites, DSDP Sites 458 and 459, which plot in the SSZ field.

In summary, backarc basin crust, especially crust of "mature" basins such as the Lau Basin, is dominated by MORB-like

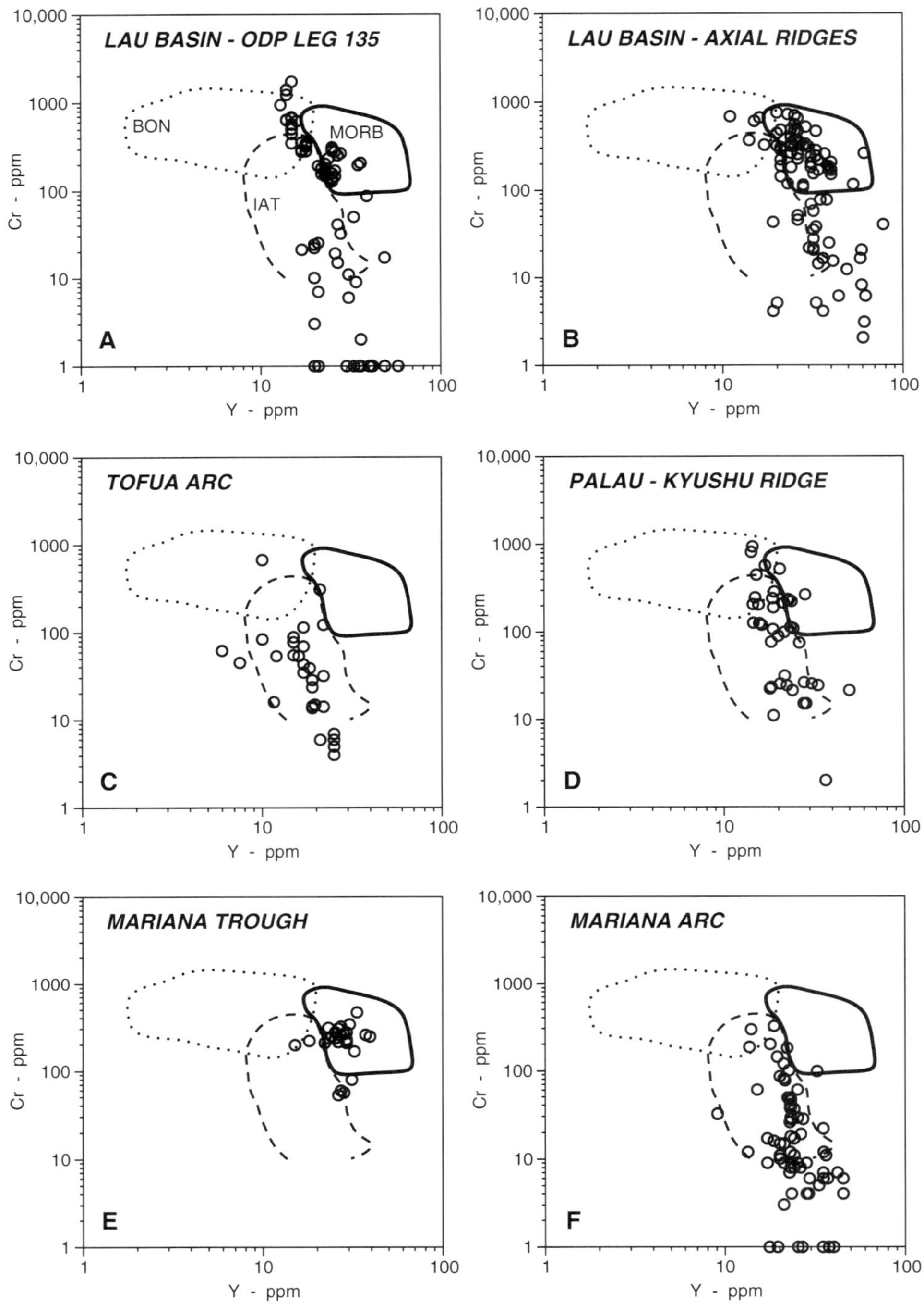

Figure 7. Cr-Y plot after Pearce et al. (1981), showing fields for BON—boninite, IAT—island arc tholeiite, and MORB—mid-oceanic ridge. A: ODP Leg 135, Lau Basin–off axis samples (data from Hawkins and Allan, 1994); B: Lau Basin axial ridge (data from Hawkins and Melchior, 1985; Sunkel, 1990; Falloon et al., 1992; Hawkins, 1995 a, b); C: Tofua Arc (data from Ewart et al., 1973; Ewart et al., 1977; Ewart and Hawkesworth, 1987; Vallier et al., 1985); D: Palau-Kyushu Ridge (data from Hawkins and Castillo, 1998); E: Mariana Trough axial ridge (data from Hawkins and Melchior, 1985; Hawkins et al., 1990); F: Mariana Arc (data from Woodhead, 1989; Peate and Pearce, 1998).

basalt. Ophiolite remnants of these could easily be misinterpreted as having formed at a mid-ocean ridge spreading center; however, many backarc basin basalts also carry the chemical imprint of the SSZ mantle, as seen, for example, in Figure 8 and in comparing relative abundances of HFSE and LILE. As will be shown in a section to follow, many "classic" ophiolites plot in the SSZ or IAT fields on the Hf/3-Ta-Th diagram.

GEOLOGY OF FOREARCS

Introduction

Clearly, the term "forearc" has meaning only if there is a nearby volcanic arc to serve as a reference point. In most modern arc-trench systems, the forearc includes ocean crust

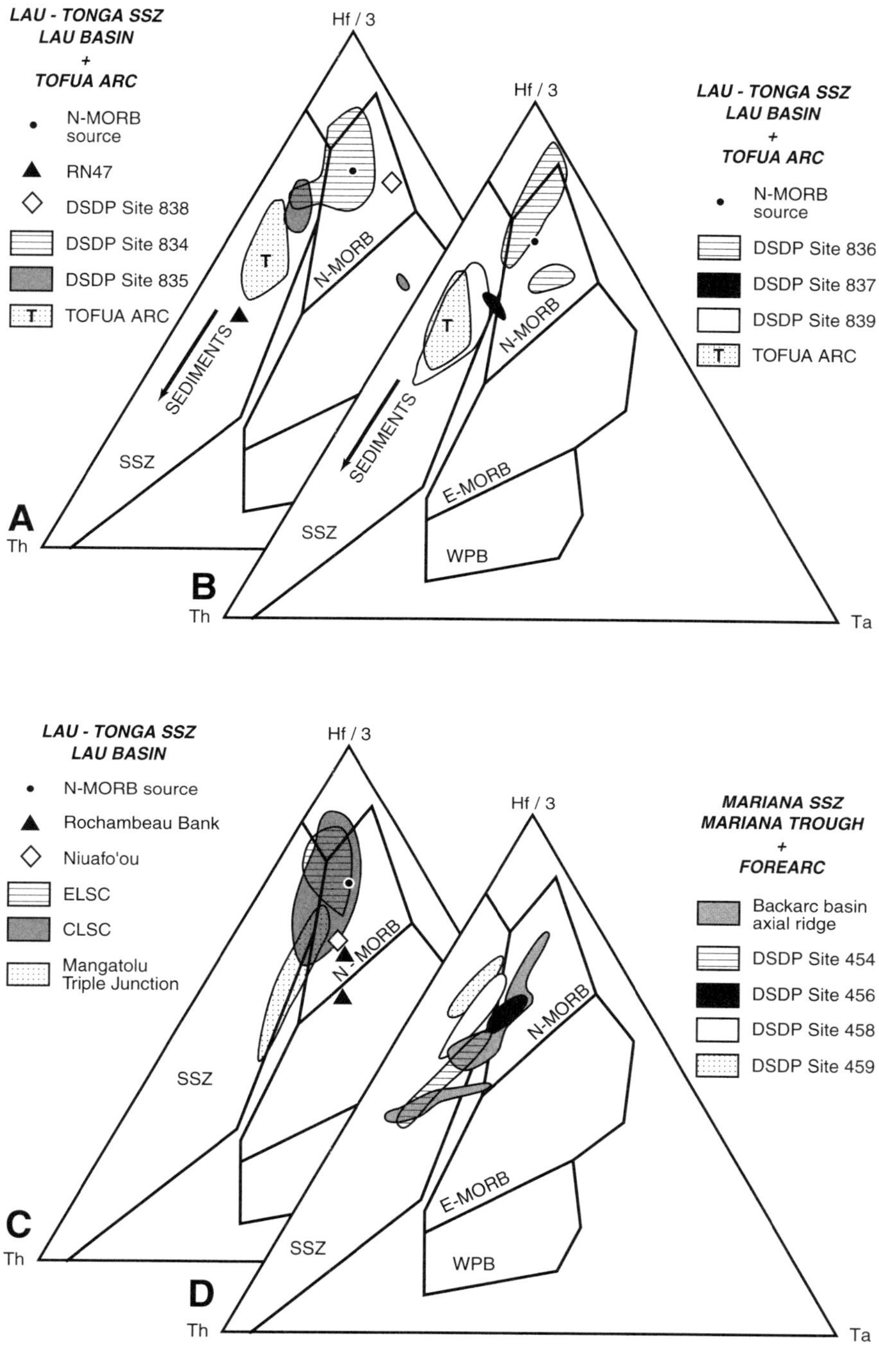

Figure 8. Plot of Hf/3-Ta-Th after Wood et al. (1981), showing fields for N-MORB, E-MORB, WPB (within plate basalts), and SSZ magmas. Arrow labeled "sediment" shows effect of adding Th from sediment. Data for arc and backarc samples are discussed in text.

extending outboard from the volcanic arc as much as 100 km or more to the break in seafloor slope down into the trench. For purposes of this discussion, I will also use it for the seafloor that will become part of the upper oceanic plate upon the initiation of a subduction system. In a simplified view of SSZ geometry, forearcs may be interpreted as trapped fragments of aged, cold, amagmatic, oceanic lithosphere on which island arcs are built.

Our studies of Western Pacific Basin SSZ systems show that they are far more complex and interesting. The amagmatic inference is supported by their (present) low heat flow and their position in the thermal setting envisioned for subduction systems (i.e., the cooling effect of penetration of "cold" oceanic lithosphere into the ambient mantle thermal regime; cf. Stein and Stein, 1996). "Isotherms in the slab extend downward such that the maximum depth reached by an isotherm is proportional to the product of the vertical descent rate…and the square of plate thickness." However, this strong downward deflection of isotherms may not always be the case throughout SSZ history. At the onset of subduction, oceanic crust above the nascent subduction zone may be extended and magmas erupted into the extension zone. Hawkins et al. (1984) proposed that subduction might begin on a former, serpentinized, fracture zone trace because of a change in relative plate motion. We suggested that "differences in crustal age, density, and thickness, plus changes in relative plate motion cause[s] the failure along the fracture zone," and that "[mantle] counterflow heats the serpentinized peridotite, causes an uparch of the lithosphere, and promotes dilation of…the forearc," "boninite[s] …are the first melts to leak out." We called this the boninite stage of an "immature arc." Stern and Bloomer (1992), in their more refined model, showed how this helps explain generation of both boninite and high-Mg arc tholeiite magmas that form in the earliest stages of SSZ magmatism, referring to this as an "infant arc." They also proposed that an initial phase of extension, or transtension, would result in foundering of a more dense plate that eventually becomes the subducted plate. R.N. Taylor and Nesbit (1995) presented evidence for the "broad region of volcanism generated in an extensional regime" in the Eocene-Oligocene Izu-Bonin arc-forearc. They visualized the nascent forearc as consisting of a "series of small rift basins, punctuated and bordered by coherent volcanic edifices."

As discussed below, boninites require extensive melting of depleted mantle, which implies high thermal gradients. Recognition that boninite occurs interlayered with arc tholeiite supports the idea that forearcs are potential sites for the origin of island arcs and are a likely source for at least some ophiolites (Hawkins and Evans, 1983; Hawkins et al., 1984; Stern and Bloomer, 1992; Bloomer et al., 1995).

The geography of forearcs differs; for example, the Tonga forearc is only ~80 km wide, whereas the Mariana forearc is upward of 200 km wide. The early magmatic, tectonic, and sedimentary history of an SSZ system is recorded in forearc sediments and igneous rocks, as well as in successive overprints of magmatic activity. The Tonga and Mariana SSZ systems have evolved by trench rollback and sequential development of arc and backarc systems through Cenozoic time. This history is represented by fragments of pre-existing lithosphere that was rifted in the extensional and magmatic episodes that generated arc and backarc systems. Some of this older forearc crust could be ancient seafloor that formed at a mid-ocean ridge, but in the Tonga and Mariana SSZ systems, the composite crust comprises remnants of older arc and forearc terranes. In both arc systems, parts of the forearc include uplifted platforms covered with Miocene and younger reef limestones.

The Tonga forearc preserves mid- to late-Eocene age fragments of arc crust that are similar to rocks of Viti Levu, Fiji. These may have formed adjacent to either the "ancestral" Tonga Trench (prior to opening of the Lau Basin) or the now inactive Vitiaz Trench. The Mariana forearc includes a frontal arc, comprising islands (Guam, Saipan) with arc volcanic rocks as old as Eocene and similar rock series that are exposed on the outer arc high, near the trench. These probably are fragments of the Palau-Kyushu Ridge that formed in response to subduction in the "ancestral" Mariana Trench, when it lay several hundred kilometers west of its present location (cf. Hawkins and Castillo, 1998).

Guam and Saipan—Insight to Forearc Geology

The broad forearc of the Mariana arc-trench system has been well explored; I will use it as a "type" example. The frontal arc (Guam, Saipan, Rota, Tinian) is separated from an outer arc high that has rocks of similar age and composition. The forearc has a complicated structure that shows evidence of extension in the development of half-grabens. Seismic data show that sedimentary layers are cut by dipping reflectors that are interpreted as traces of normal faults (Mrzowski and Hayes, 1980). Fryer (1992) suggested three possible explanations for the origin of the Mariana forearc. It could have been a continuous broad zone of Eocene volcanism later overprinted by forearc volcanism; the frontal arc and outer high might be separate features; or the forearc terrane might have originated by extension and rifting that dispersed parts of a formerly contiguous Eocene arc. The latter seems to be the most likely explanation.

Mariana forearc crust is well exposed on Guam and Saipan. The oldest rocks on Guam are the late-middle Eocene Facpi Formation comprising "largely interbedded boninite series pillow lavas, pillow breccias and dikes"; these are capped with arc-tholeiitic lavas. The Facpi Formation is overlain by "interbedded volcanic breccia, tuffaceous sandstone, lava flows and sills" of the late Eocene to Oligocene Alutom Formation (Reagan and Meijer, 1984). These range in composition from boninitic to arc tholeiitic and calc-alkaline. Miocene limestones and minor flows and pyroclastics overlie the Alutom Formation. Saipan also has a volcanic core overlain by platform carbonates (Cloud et al., 1958). The oldest rocks are probable late Eocene dacitic flows and pyroclastic rocks. These are overlain by andesitic flows and pyroclastic rock, which in turn are overlain by late Eocene limestone. Probable Oligocene andesitic flows are interbedded with fossiliferous carbonates. Miocene and Pliocene(?), Pleistocene, and Holocene bioclastic, foraminiferal, and coral-algal limestones, with occasional interbeds of tuffaceous limestone and calcareous tuffs complete the section.

For the student of ophiolites, the important geologic factors are the generation of boninite, arc volcanic rocks, and platform carbonates all in a geologic setting in close proximity to MORB-like crust of the backarc basin, all having formed in a geologically short time span (e.g., late Eocene to early Miocene or ca. 15–20 Ma).

Forearc Sediments

DSDP and ODP drill cores from the forearc complement exposures on Guam and Saipan. The drill hole data show that the sediments are largely calcareous ooze with varying amounts of vitric ash and volcanic debris ranging in age from middle Eocene to Recent. The seismic data that show half grabens, together with evidence from DSDP Leg 60, Sites 459 B, and 460 in the forearc (Latouche et al., 1981), support an interpretation of episodic periods of extension that were accompanied by arc volcanism that shed sediment having varying amounts of vitric ash into developing rift basins. From youngest to oldest these are: middle to upper Eocene reworked volcanic detritus, representing early volcanism on the Palau-Kyushu Ridge; upper Oligocene through lower Miocene, representing volcanism on the West Mariana Ridge; and middle Pliocene to present vitric ash from the modern Mariana Arc. Between early Oligocene and early Miocene time, there is a marked diminution of volcanic debris, suggesting a hiatus or minimum in volcanism. The forearc has moved continually trenchward, carrying the rift basins that serve as depocenters for arc detritus, tracking seaward rollback of the trench axis (cf. Elsasser, 1971; Hamilton, 2002). Note that the feature we call a backarc basin, because of its relative position, is initiated in the forearc of an active arc that eventually becomes extinct. There is a lag time of a few million years before the new arc gets established.

Forearc Seamounts

The Mariana forearc has more than 50 seamounts, which rise as much as 2.5 km above the seaflooor and range from 10 to 30 km in diameter; some are irregular in form, but many are conical. The conical features are serpentinite mud volcanoes (diapirs); others are uplifted blocks of serpentinized peridotite. Together they form an array aligned parallel to the trench-slope break and to the Mariana Trench axis and set back from the slope break ~50 to 120 km (Fryer, 1996). They have been sampled by dredging (Bloomer and Hawkins, 1983), and some were drilled on ODP Leg 125 (Fryer et al., 1992). The mud volcanoes formed from unconsolidated slurries of serpentinite mud and rock fragments. The latter gives insight to forearc basement, the SSZ-MW, and low-grade metamorphism under the forearc. The entrained rocks are mostly serpentinized harzburgite, lesser amounts of serpentinized dunite, and minor amounts of meta-gabbro, meta-basalt, and metasediments. The harzburgite is highly depleted, probably representing the residue from (multi-stage?) >20% melting. The harzburgite has only traces of TiO_2, and both CaO and Al_2O_3 are <0.6%. Olivine and orthopyroxene have Mg# = 92, and there are only traces of diopside. Chrome spinels have Cr# up to 83 (Ishii et al., 1992); these authors concluded that the peridotites are more depleted than abyssal peridiotites from mid-ocean ridges. Metamorphic assemblages are typical of prehnite-pumpellyite or greenschist facies. Particularly notable is the occurrence of blueschist facies minerals in pebbles of aphyric basalt entrained in the mud. The diagnostic minerals are lawsonite, aragonite, blue amphibole (winchite), and Na-pyroxene (chloromelanite with 30%–35% jadeite) and aegerine-augite (with <2% jadeite) (Maekawa et al., 1992). On the basis of thermal modeling and presence of serpentinite debris in Eocene sediments, Fryer (1996) proposed that extensive forearc metamorphism probably occurred within the first several hundred thousand years of initiation of subduction, and continues today.

Forearc Magmas

The main development of forearc magmatism may occur at the onset of subduction to form nascent arc systems (Hawkins et al., 1984; Stern and Bloomer, 1992; Fryer, 1996). Forearc eruptions of "middle and late Eocene supra-subduction zone magmatism formed a vast terrane of boninites and arc tholeiites that is unlike active arc systems but is similar to many ophiolites" (B. Taylor, 1992). In addition to this major eruption of magmas at the onset of subduction, later basaltic flows and intrusions are found in forearc sedimentary units.

Arc-basalt intrusives. There is evidence from ODP drill sites on the Bonin, Mariana, and Tonga forearcs that arc composition magmas have intruded volcaniclastic sediments; "these intrusives appear to be a common feature of Western Pacific intra-oceanic forearcs" (R. Taylor et al., 1994). There is strong petrologic evidence indicating that these are derived from the volcanic arc rather than forearc mantle (e.g., at ODP Site 841 in the Tonga forearc we recovered basaltic dikes and sills intrusive into lower and middle Miocene sediments, Table 3). Compositions ranged from basalt to andesitic (50–54% SiO_2); they are enriched in K, Rb, Ba, and Sr, but depleted in HFSE, relative to MORB. They resemble Lau Ridge volcanic rocks and show no geochemical evidence to suggest that they were derived by melting of the highly depleted mantle inferred to underlie the forearc. The dikes were likely propagated from the Lau Ridge (Bloomer et al., 1994), which required transport under forearc crust for ~100 km. Marlowe et al. (1992) described similar sills in the Mariana forearc for which R. Taylor et al. (1995) concluded that they are "sills… most likely to have been fed by dikes which propagated from the arc though basement rocks at deeper crustal levels. They do not resemble backarc basalts…." ODP drilling in the Bonin forearc encountered similar intrusives (cf. B. Taylor, 1992). If these intrusive bodies are derived from the volcanic arc, the long transit of relatively unfractionated melts under the "cold" forearc poses problems in view of the thermal setting.

Boninites. Boninites are one of the most distinctive rock types found in the SSZ environment. They never have been found on or near mid-ocean ridges or intra-plate settings, but they are commonly found in close geological association with many ophiolites and arc series rocks. The close geological association of boninites with ophiolites offers unequivocal evidence that the ophiolite had an SSZ origin. Boninites are found in some "classic" ophiolite examples, including Troodos, Pindos, California Coast Range, and Zambales (cf. Pearce et al., 1984; Thy and Xenophontos, 1991; Jones and Robertson, 1991; Shervais and Kimbrough, 1985; Hawkins and Evans, 1983).

Boninites are olivine - chrome spinel - pyroxene rocks in which plagioclase is conspicuously absent. Clinoenstatite is highly distinctive; both clinopyroxene and Mg-rich orthopyroxene are common. Commonly, boninites are vitrophyric with the

TABLE 3. MARIANA FOREARC BONINITES, TONGA FOREARC INTRUSIVES

Sample:	MARA 50-14	MARA 28-18	MARA 28-1	MARA 28-9	MARA 28-43	MARA 51-8	MARA 27-3	ODP 841 Unit 1 Ave
				Wt%				
SiO_2	53.60	57.12	59.46	61.90	67.45	55.22	55.33	51.52
TiO_2	0.18	0.18	0.16	0.17	0.19	0.40	0.74	1.05
Al_2O_3	10.77	10.18	11.16	11.66	13.20	14.45	16.31	16.33
FeO*	8.07	7.70	7.13	5.97	6.27	6.11	9.68	11.40
MnO	0.15	0.16	0.13	0.11	0.09	0.11	0.16	0.24
MgO	15.53	15.25	11.26	8.92	3.55	7.76	4.26	4.67
CaO	6.98	4.36	4.97	4.13	3.97	8.44	8.56	9.59
Na_2O	1.00	1.96	2.19	3.48	3.78	2.60	2.25	3.09
K_2O	0.66	0.98	0.64	0.70	1.37	0.76	0.35	0.28
P_2O_5	0.05	0.02	0.04	0.05	0.06	0.09	0.12	0.16
H_2O	2.60		2.92	1.90	1.40	3.26	2.49	
Total	99.59	97.91	100.06	98.99	101.33	99.20	100.25	98.33
Type:	OL LowCa 3	OL LowCa 1	B LowCa 1	A	Bd	T	At	At
				Trace elements (ppm)				
Cr	1152	1310	722	477	77	3313	29	20
Ni	384	337	2345	150	24	134	14	20
V	161	119	126	100	85	160	321	376
Zr	32	42	50	63	76	60	64	67
Y	7	6	5	3	7	15	30	25
Nb	4	4	6			5	4	1
Rb	10	18	10	10	7	5	7	2.7
Ba	17		36	44	42	17	45	157
Sr	71	93	154	220	283	196	140	234
				Rare earth elements (ppm)				
Ce	2.9		4.01	6.3	7.02	6.24	7.46	
Nd	1.59		2.18	3.22	3.97	4.39	6.6	
Sm	0.46		0.58	0.79	1	1.31	2.28	
Eu	0.15		0.2	0.25	0.3	0.47	0.85	
Gd	0.58		0.71		0.98	1.68	3.26	
Dy	0.68		0.69	0.73	0.86	1.75	3.33	
Er	0.49		0.46	0.48	0.48	1.21	2.6	
Yb				0.45		1.24	2.63	

Note: Blank—no data; A—Boninitic andesite; At—Arc tholeiite; B—Boninite; Bd—Boninitic dacite; OL—Olivine boninite; T—Transitional; LowCa 3—Low Ca Type 3; LowCa 1—Low Ca Type 1.

glass having a broadly andesitic to dacitic composition (Meijer, 1980; Bloomer and Hawkins, 1987). Holocrystalline varieties are also found. The chemistry of boninites is distinctive—rocks considered boninitic "have >53% SiO_2 and Mg# >60, or are demonstrably derived from parental magmas meeting these compositional requirements" (Crawford et al., 1989). R. Taylor et al. (1994) expanded on this by adding that TiO_2 is <0.6%, 7% < MgO < 25%, and orthopyroxene forms phenocrysts and microphenocrysts. These authors propose that "presence of clinoenstatite, or absence of plagioclase, is diagnostic of boninites sensu stricto" (R. Taylor et al., 1994). A complete discussion of boninite petrology is beyond the scope of this paper. I will summarize some of their distinctive features and tabulate data for representative samples in Table 3. See *Boninites and related rocks,* Crawford (1989); Hickey-Vargas (1989); Pearce et al. (1992); and R. Taylor et al. (1994) for more thorough presentations.

Boninites were first described by Kikuchi (1890), from their type locality on Chichijma in the Bonin arc. The type boninites are distinctive in having phenocrysts of olivine, chrome spinel, clinoenstatite, and orthopyroxene enclosed in silica-rich glass. Olivine boninite from the Mariana Trench has glass with 59% SiO_2, and dacite boninite glass has 75% SiO_2 (Bloomer and Hawkins, 1987). Other boninite varieties have both low- and high-Ca pyroxene, but plagioclase in all varieties is restricted to the groundmass of more highly evolved melts. The combination of olivine, chrome spinel, and orthopyroxene, as phenocrysts, the general absence of plagioclase, and the high-silica glass matrix are distinctive features that characterize boninites.

Olivine typically ranges from Fo 86 to 92, chrome spinel has Cr# ranging from 68 to 75, and the clinoenstatite is En 90–96 with <5% Wo. Clinopyroxene, when present, usually is diopside. The distinctive mineralogy is matched with distinctive chemistry. The range in SiO_2 for boninites is 53% to >58%, whereas MgO ranges from 14% to 18%. Fe, Al, and Ca are low, compared to N-MORB. The ratio CaO/Al_2O_3 is variable (e.g., 0.6–0.9) because of the low Al_2O_3 content (commonly <12%) and varied CaO. This probably serves to inhibit crystallization of plagioclase. The HFSEs are depleted relative to MORB, the transition metals are high, as would be expected with the high Mg, and bulk rock Cr# ranges from 70 to 98. The LILE are variable but enriched relative to MORB. The REE are greatly depleted relative to MORB, and chondrite normalized patterns have either negative slope or a U-shaped pattern. Chondrite normalized REE patterns for Bonin and Mariana Trench samples, and for a Zambales Range boninite discussed in a separate section, are shown in Figure 9. Boninites have variable enrichment in volatiles, alkali metals, and alkaline earth trace elements, as well as in water. Using combined deuterium and oxygen isotope ratios to discriminate between primary and secondary water, Dobson and O'Neill (1987) determined that primary boninite and bronzite andesite melts from Chichijima had from 1.6 to 2.4 wt % water. This water is attributed to dehydration fluids derived from subducted oceanic crust.

Boninites may be divided into low-Ca and high-Ca classes; low-Ca boninites are further subdivided into three types on the basis of their SiO_2, CaO, FeO, alkalies, and the CaO/Al_2O_3 ratio. High-Ca boninites do not develop clinoenstatite. Table 3 gives data for Type 1 and Type 3 low-Ca boninites from the Mariana Trench and for a rock transitional between boninite and arc tholeiite. Low-*P* crystal fractionation forms a boninite series (Meijer, 1980; Bloomer and Hawkins, 1987) comprising olivine boninite, boninite, boninitic bronzite andesite, and boninitic hypersthene dacite. Some important criteria for boninites are high Cr and Ni, which is consistent with high Mg, very low HFSE, especially in relation to their high SiO_2 and low Al_2O_3 contents. These relationships are retained through the boninite series, and distinguish it from calc-alkaline series rocks. The boninites of the Hunter fracture zone region (Sigurdsson et al., 1993) are layered with low-K rhyolites having high Na and Si. Similar low-K rhyolites of the late Eocene Sankakuyama Formation on Saipan comprise the oldest arc-related volcanic rocks in the Marianas. Meijer (1983) attributed these to extreme crystal fractionation of high-Mg andesites of the boninite series.

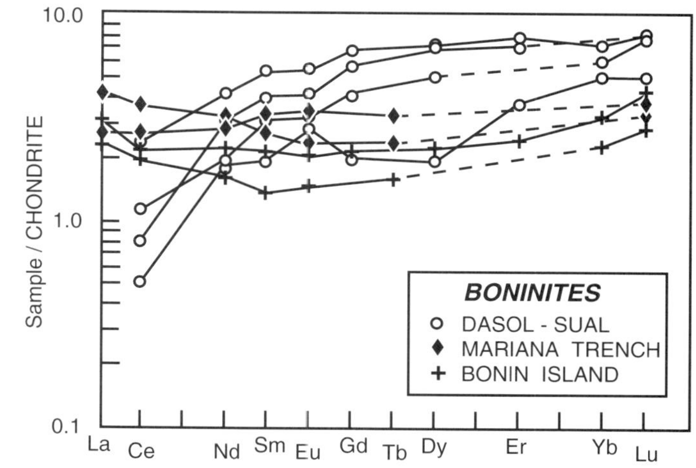

Figure 9. Chondrite normalized REE for boninites of Dasol–Sual series of Zambales Range Luzon (Hawkins and Florendo, 1997, personal commun); Mariana Trench (Bloomer and Hawkins, 1983); and type boninite (Crawford, 1989).

Meijer pointed out their chemical resemblance to some trondhjemites, plagiogranites, and quartz keratophyres that are common constituents of ophiolite assemblages.

Meta-boninites might not be readily recognized as such but would have abundant serpentine, talc, or tremolite, with relict high Cr# spinel; anorthite substitute minerals would likely be rare owing to low Al_2O_3 content of boninites.

The distinctive mineralogy and chemistry (Table 3 and Fig. 9) point to an origin in highly depleted mantle harzburgite, e.g., mantle from which MORB or IAT basalt has been extracted. Note the distinctive field for boninites in the Cr–Y diagram (Fig. 7), where high Cr indicates extensive melting and low Y indicates a depleted source. This melt loss could have been at a mid-ocean ridge or in a backarc basin. Sr and Nd isotope data point to a source with high $^{143}Nd/^{144}Nd$ and low $^{87}Sr/^{86}Sr$, i.e., like MORB-source mantle. This depleted source has been reenriched in LILE, LREE, Zr, Hf, water, Si, and Na, presumably by subduction fluids. The last three constituents probably help to promote melting of this refractory source material.

The origin of boninites is still not completely resolved, although there seems to be a consensus on several points. A depleted peridotite source such as harzburgite or dunite is required in order to explain the low HFSE and REE and the low Ca and Al. Moderately extensive melting (e.g., up to 25%, Pearce et al., 1992) is required to derive the high Mg and transition metals, and the source must have been enriched in aqueous fluids carrying LILE, LREE, Si, and alkalies in order to facilitate melting and to give the LILE signatures. Addition of Zr may be required to explain high Zr/Ti and Zr/Sm. Melting must have been at moderately low-*P* but high-*T* (e.g., 1250° C; Pearce et al., 1992). The unusual conditions required are most likely to have developed at convergent margins that are not in thermal equilibrium, such as during the initial stages of subduction (e.g., Hickey and Frey, 1982; Bloomer and Hawkins, 1983; Kobayashi, 1983; Crawford et al., 1989; Hawkins et al., 1984; Stern et al., 1991). Influx of high-*T* mantle is essential to provide the heat; thus, conventional models involving cold/cold subduction as modeled for most subduction zones (Minear and Toksoz, 1970; Oxburgh and Turcotte, 1970; Hasebe et al., 1970) are inadequate. Tomographic imagery for the mantle wedge indicates that "slow" velocities are under the arcs, whereas velocity structures are too "fast" near the subduction zone to indicate high temperatures (van der Hilst, 1995). Thus, subduction of high-temperature lithosphere or mantle upwelling, rather than "cold" subduction, is required. Pearce et al. (1992) summarizes several of the more plausible geologic settings for boninite genesis. They include: "partial melting of serpentinized harzburgite during initiation of subduction on a former transform fault," as suggested by Hawkins et al. (1984); "partial melting owing to subduction of young hot crust" (e.g., subduction of a ridge lying parallel to the subduction zone), as suggested by Tatsumi (1982) and Crawford et al. (1989); and "contact melting of hydrous mantle wedge by injection of a hot mantle diapir during arc rifting," as suggested by Crawford et al. (1989). Stern and Bloomer (1992) proposed upwelling of hot mantle into a rift formed by transtension at the initiation of a subduction zone. Yet another possibility is propagation of a ridge into the forearc from the *backarc* regions, as suggested by Hawkins and Castillo (1998). An example of the latter is found on the Hunter fracture zone where a propagating ridge from the North Fiji Basin encounters depleted mantle above an incipient oblique subduction zone (Sigurdsson et al., 1993). Boninite is layered with primitive high-Mg arc tholeiite and low-K rhyolite. The Hunter fracture zone setting may be a unique place, where the earliest stage of subduction zone initiation is found.

The forearc of the north end of the Lau Basin has a similar, but less well understood, occurrence of boninites (Falloon et al., 1989). The authors concluded that these lavas were formed as primary melts of upper-mantle peridotite. The boninite was collected from the upper slopes of the westward trending part of the Tonga Trench. One limb of the active Mangatolu Triple Junction in the Lau (backarc) Basin penetrates into the forearc near these boninites; this may have been important as a heat source. Depleted peridotite of the mantle wedge was the melt source.

The various models for boninite genesis involve extensive low-*P* melting (input of heat) into depleted mantle of the forearc region, this requires upwelling of hot mantle into rifted oceanic crust. Boninites are found in forearcs. Apparently they are in situ; it is likely that this is where they formed. The common occurrence of arc tholeiite interlayered with boninite argues that they represent the nascent stages of arc genesis.

GEOLOGY OF TRENCHES

The 10- to 10.9-km deep Tonga and Mariana Trenches are far removed from major continental drainage systems or heavily glaciated regions and, thus, receive an insignificant amount of sediment, except for the thin layer of siliceous and carbonate pelagic material and the volcaniclastic material from within-plate volcanic edifices, that forms oceanic seismic layer 1. A small amount of airborne and suspended material from the continents and nearby volcanic arcs, as well as a very small amount of meteoritic material, is also added. As a consequence, there is no accreted wedge of sediments, such as is found in the Sunda, eastern Aleutian, and Middle America trenches. The Tonga and Mariana trenches resemble other Western Pacific Basin trenches, such as those of Yap, Palau, Philippine, New Hebrides and northern Kermadec. Rock dredges collected over short depth increments have given us a qualitative understanding of rock types and their probable layering on trench walls. Intuitively, we might expect to find the classic ophiolite sequence neatly displayed in cross-section. Such is not the case; although all parts of the ophiolite sequence are found, they have been shuffled by deformation (Fisher and Engel, 1969; Dietrich et al., 1978; Sharaskin et al., 1980; Bloomer and Hawkins, 1983; Bloomer, 1983; Bloomer and Fisher, 1987; Hawkins and Castillo, 1998). In both the Tonga and Mariana Trenches, and indeed in all Western Pacific Basin trenches, samples from the inner trench wall include large amounts of highly depleted, variably serpentinized peridotite, presumably representing sub-arc mantle. We have the best idea of the range in rock types and their relative "stratigraphy" from the Mariana Trench (Bloomer and Hawkins, 1983; Bloomer, 1983), and we use it as our model for an inner trench wall, as in Figure 10, which is based on dredge sampling at ~18° N. Note that Figure 10 is a composite sketch based on several dredge collections over short depth increment. The inferred "petrologic layering" is based on the first appearance with depth of a rock type (rocks do not roll upslope) or the unique composition of a single dredge increment. The same rock types and similar relative "layering" are found in the Tonga Trench (e.g., Fisher and Engel, 1969; Bloomer and Fisher, 1987; Hawkins et al., 2000). It should be emphasized that we have no data for the nature of contacts between rock types, and nothing to indicate that an "ophiolite" layering or any other kind of primary layering is preserved. In fact, the abundance of serpentinized peridotite found at all depths on trench walls points to the highly disrupted and interlayered distribution of rock types. The presumed layers are shown dipping inward, but this is an inference based on the likelihood that rocks have been imbricated because of relative overthrusting of the hanging wall of the subduction zone. The diagram has great vertical exaggeration in order to show the layers; true slopes angles are shown on the sketch.

It is highly significant that crustal rock assemblages exposed on the inner trench wall carry the chemical signature of SSZ magmatism and thus are likely to be from the *forearc*. DSDP and ODP drilling of the Mariana forearc support this inference. Minor scraps of MORB-composition seafloor or seamounts have been dredged, but for the most part trench wall rocks *are not* fragments of mid-ocean ridge-derived lithosphere. They do, however, represent all parts of classic ophiolite assemblages.

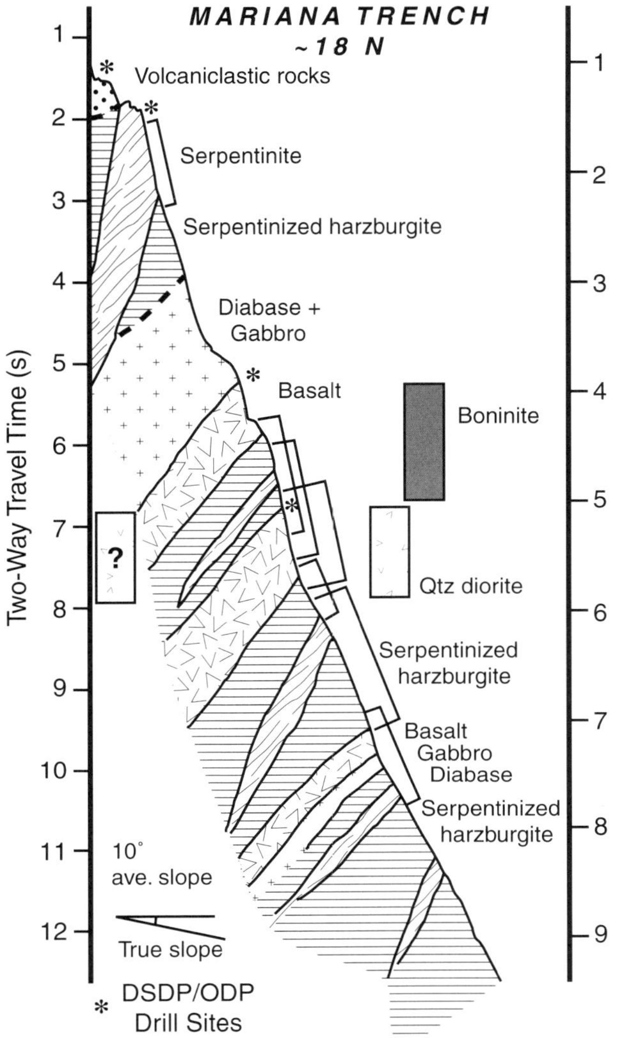

Figure 10. Schematic section based on the author's fieldwork, showing distribution of rock types and their probable depth range of exposure on the inner wall of the Mariana Trench at 18° N (cf. Bloomer and Hawkins, 1983). Boninite and tonalite were collected from depth ranges shown, but their relation to other types is not known. Inward dip of layers is inferred and their lateral extent is not known. Note that the slope of the trench wall has been greatly exaggerated to show the rocks. True slope angle is ~10°.

Our data for the Mariana Trench wall indicates that it mainly exposes ultramafic rocks between depths of 2 and 8.8 km; these are largely serpentinized, with lizardite as the main serpentine mineral and with lesser amounts of antigorite. The protoliths, as indicated by relict grains are (in decreasing order of abundance) harzburgite, dunite, and lherzolite. Most are tectonites, but some are cumulates. Many show effects of deformation, such as flattened olivine and kink-bands in olivine, indicating high-T deformation. At the southern end of the trench, southwest of Guam, we collected many samples having cataclastic textures; some are recrystallized to talc-tremolite schists.

The following summary of rock types is based on a more extensive discussion given in Bloomer and Hawkins (1983). Harzburgite is the main non-cumulate rock type found, and consists of olivine, Fo 91–92; orthopyroxene with En 90–92 and Wo 0.2–2; and chrome spinel with Cr# 62–84. Lherzolite is rare, and has Fo 87–91; orthopyroxene with En 85–89 and Wo 1–3; clinopyroxene with En 50–52 and Wo 42–45; and chrome spinel with Cr# 43–58. The clinopyroxene is diopside or diopsidic augite. Lherzolite commonly has small patches of plagioclase and clinopyroxene, suggesting either that they have been impregnated with melt or that they retain trapped melt. Cumulate textured ultramafic rocks, orthopyroxenites, and websterites have Fo 76–85; orthopyroxene En 67–85 and Wo 1–2; and clinopyroxene En 43–48 and Wo 45–46. One cumulate textured harzburgite has olivine poikilitically enclosed in orthopyroxene; calcic plagioclase; and primary, intercumulus, brown pargasitic amphibole (Mg# 74).

Massive and cumulate textured gabbros constitute an important part of the Mariana collection and were collected at eight dredge sites. They are most abundant between 3 and 4 km depth and between 7 and 7.5 km depth. The massive gabbros have hypidiomorphic textures, consisting mainly of plagioclase and clinopyroxene; noritic gabbro is less common and comprises plagioclase-clinopyroxene-orthopyroxene-Fe-Ti oxide. Both types grade texturally into micro-gabbro and diabase. Cumulate textured anorthositic norite, clinopyroxene norite, and orthopyroxene gabbro have heteradcumulate texture comprising plagioclase-clinopyroxene-orthopyroxene-Fe-Ti oxide. All of the gabbros have minor amounts of secondary minerals (e.g., uralite, chlorite, talc, serpentine, calcite). Many gabbro samples display metamorphic textures reflecting moderately high-T shearing and recrystallization. Some are blasto-mylonitic hornblende gabbro gneisses.

The crystallization sequence of most gabbros was olivine-chrome spinel-orthopyroxene-clinopyroxene-plagioclase; less common was olivine-plagioclase-clinopyroxene. The dominance of orthopyroxene in most samples indicates derivation from silica-saturated melts. Bulk rock chemistry of the cumulate gabbros shows very low concentrations of K_2O, P_2O_5, Rb, Ba, Zr, and generally low TiO_2 (Bloomer and Hawkins, 1983). We interpreted the cumulate gabbros as probable derivatives of boninite series magmas. This is consistent with the interpretation that most of the Mariana Trench collection of mafic and ultramafic rocks formed in the forearc and may have formed contemporary with the boninites.

Tonalite and quartz diorite are rare, but significant rock types collected from the Mariana Trench in the depth range of 5 to 6 km (Bloomer and Hawkins, 1983), as shown schematically on Figure 10. These rock types have been collected from the Tonga Trench wall from depths between 5.9 and 5.2 km (Hawkins et al., 2000). Suyehiro et al. (1996) proposed, on the basis of geophysical data, that tonalite, like the tonalite of the Tanzawa (Japan) arc, may underlie the Izu-Ogasawara ridge. The occurrence of tonalite-quartz diorite on intra-oceanic trench walls is significant in terms of arc evolution, but we have no idea of its abundance or distribution and attach no significance to the similar depth of occurrence in the Mariana and Tonga Trenches.

Rocks representing higher crustal level include variously altered basalt, diabase, andesite, dacite, and *boninite*. The only evidence for accreted material is scraps of alkalic basalt, chert, hyaloclastite, shallow water limestone, and upper Cretaceous radiolarian silicic sediment. It is likely that these were swept from a seamount riding in on the incoming plate. Highest samples on the trench slope are volcaniclastic siltstones containing Miocene discoasters and glass shards. The glass shards have refractive indices that are consistent with 54%–56% SiO_2 (Bloomer and Hawkins, 1983). Lee and Stern (1995) presented a comprehensive study of the 40 m.y. record of tephra glass from DSDP and ODP drill sites for the region from the West Philippine Sea to the Mariana forearc. Their data are mainly for low and medium-K basaltic to rhyolitic compositions, but they noted a pronounced minimum in the 65% to 66% SiO_2 range. Johnson et al. (1991) reported samples of MORB and OIB, together with arc-like material, dredged from the forearc. They interpreted the MORB and OIB as oceanic plate material accreted to the forearc.

INTRA-OCEANIC VOLCANIC ARCS

Introduction

The Plio-Pleistocene to Recent island arc volcanoes erupted on the Tonga-Kermadec and Mariana Ridges can serve as type examples of subduction related intra-oceanic volcanism. Inasmuch as there is a voluminous literature on island arc magmas (cf. Barker, 1979; Perfit et al., 1980; Gill, 1981), I will summarize some of their distinctive features and associations that may be useful in recognizing them as components of SSZ arc-ophiolite terranes. One of the keys is the close geologic association of island arc volcanics with MORB-like crust of backarc basins and depleted rocks of forearcs. A mature arc system, with 20–30 km thickness of plutonic and volcanic rocks, which had evolved to a calc-alkaline series, would not be interpreted as an ophiolite. An objective of this discussion is to show how deep crust remnants of such a complex, or remnants of a nascent arc that was terminated in its earliest stages, could mimic the petrology, geochemistry, and layer thickness much of what we term an ophiolite.

In early stages of arc volcanism (the "infant arc" stage of Stern and Bloomer [1992]), eruption of high-Mg melts—boninite, picrite, and high-Mg basalts—forms pillows, dikes, and plutons. Their chemistry requires high thermal gradients and extensive melting of the mantle sources (Hawkins et al., 1984; Stern and Bloomer, 1992; Eggins, 1993). Depth of melt separation may be <50 km. Boninites form in the forearc concurrently with Mg-rich, low-K arc tholeiite, as shown by their interbedded occurrence in ophiolites and their in situ presence in forearcs. As arcs evolve, they become more enriched in K and other LILE, and depleted in Mg by fractionation. Thus, arcs may evolve from an early boninite plus arc tholeiite stage, to a basalt-dominated arc tholeiitic stage (Hawkins et al., 1984; Stern and Bloomer, 1992), to the "classic" calc-alkaline volcanic series (cf. Gill, 1981). The earliest stages of "arc volcanism" obviously involve eruption of magmas onto the seafloor, either in rift systems (cf. Hawkins and Evans, 1983; Hawkins et al., 1984; Stern and Bloomer, 1992) or as point sources. Several tectonic factors (e.g., a change in relative plate motions from transtension to transpression) may be involved in the early termination of a nascent arc system, such that it never develops a chain of robust emergent edifices that evolve to calc-alkaline series lavas or shed large aprons of volcaniclastic material. Remnants of nascent arc crust would resemble an ophiolite. A strong argument can be made that nascent arc crust is the likely protolith for many ophiolites (cf. Stern and Bloomer, 1992).

Active volcanic island arcs are separated from extinct remnant arcs by backarc basins. It is debatable as to whether or not the *active volcanic arc* was rifted from the remnant arc at the initiation of backarc opening, or whether a wave of ephemeral arc volcanoes marched seaward across the forearc until it was established in a new fixed position. In my opinion, the latter seems to be most likely. The evidence from the Lau Basin demonstrates that the Lau Ridge (now a remnant arc) remained active well into the opening history of the basin. Local intra-basinal arc volcanoes were active as the basin opened; these were less voluminous than the basalt eruptions that formed *new* Lau Basin seafloor and filled rift basins.

Tofua arc volcanic debris first appears on the Tonga Ridge (forearc) at ca. 3 Ma (Clift et al., 1994, 1998), long after rifting of the basin began and at about the same time as last volcanism on the Lau Ridge. If the ancestral Lau Ridge arc had been split, there should be an ~6–7 m.y. record of Tofua Arc volcanism on the forearc. The arc is constructed along a trend parallel to the Tonga Ridge, ~50 km west of its inboard edge and on the outboard edge of the Lau Basin. In most places, it is separated from the Tonga Ridge by small rift basins. The arc may overlie part of the Tonga Ridge crust, but there is no evidence for this; my prejudice is that arc magmas penetrated along deep zones of weakness that correspond to the extensional faults that bound the backarc basin on each side. This observation may be of use interpreting arc-ophiolite assemblages. In the case of the Tofua Arc, the arc edifice and its underlying plutonic section may lie directly on depleted mantle rather than on older crust. Hawkins and Evans (1983) reached a similar conclusion for the island arc part (Acoje block) of the Zambales ophiolite.

The Tonga Ridge comprises basement formed of older arc volcanic, plutonic, and arc-derived clastic rocks that are overlain by six seismic sequences inferred to range in age from late Eocene to late Miocene (Austin et al., 1989). These comprise volcaniclastic turbidites and hemipelagic carbonates. They are overlain by post–Late Miocene marls, reefal deposits, and shelf limestones. Between Vava'u and 'Eua are uplifted and westward-tilted blocks of forearc volcaniclastics, marls, and carbonates capped by shoals, reefs, and coralline islands. The oldest known rocks of the Tonga Ridge rocks are mid-to late-Eocene (46–40 Ma) arc volcanic and plutonic material exposed on 'Eua Island (Ewart and Bryan, 1972). Gabbro dredged at the north end of the Tonga Ridge gave an Ar/Ar radiometric age of 50±9 Ma, as cited in Clift et al. (1998). ODP Site 841, drilled on the forearc near 23° S, recovered low-K rhyolite, rhyolite tuff, breccias, welded tuffs, and lapilli tuffs of a rhyolite caldera. The pyroclastic rocks are interbedded with mid- to late-Eocene reef material that includes the foraminifer *Discocyclina.* A K-Ar age determination for rhyolite glass gave an age of 44±2 Ma (McDougall, 1994). The forearc igneous basement probably was rifted from the late Eocene arc series of Viti Levu, Fiji Islands (cf. Whelan et al., 1985; Gill, 1970, 1987). This igneous basement also includes Oligocene lavas (33–31 Ma) and early Miocene dikes (19–17 Ma) that probably represent Lau Ridge volcanism (Hawkins and Falvey, 1985; Duncan et al., 1985). The Oligocene and Miocene arc magmas intrude, or are emplaced on, late-Eocene arc material of the Tonga Ridge.

The present Mariana Arc probably formed after the ca. 10 Ma initiation of rifting and the subsequent episode of spreading that formed Mariana Trough crust (e.g., by mid-Pliocene; Latouche et al., 1981; Fryer, 1996). The Mariana Arc is constructed in part on thinned forearc crust (e.g., like crust exposed on Guam and Saipan), and in part on backarc basin crust (Stern et al., 1989). The Mariana Trough has cross-chains of arc volcanism that persisted during basin rifting (Fryer, 1996), and at least one arc composition volcano sits on "zero age" crust of the axial ridge (Table 2, this paper; Hawkins et al., 1990). The data for both the Mariana and Tonga systems suggest that as trenches and their hinge lines retreat seaward because of rollback, a continual igneous carpet of arc, backarc, and forearc magmas is laid down. This is seen in the fabric of seafloor features of the northwestern Pacific basin (Fig. 4).

Island Arc Petrology

The petrologic peculiarities of boninites have been discussed and an argument made that their most likely site of origin is in the forearc. In the earliest stages of arc evolution, as seen on Guam and on the Mariana forearc, we find boninite interlayered with arc tholeiite. An important implication of this association is that eruption of high-Mg island arc lavas may begin anew in the forearc and, as discussed above, that in the course of SSZ evolution, "new" arcs need not have split off from an ancestral "remnant"arc.

*Origin of High-*Al *Basalts*

Occurrences of low-Ti, high-Mg basalts and picrites (e.g., MgO >8%, Mg# >70, Ni >200 ppm) are rare in island arc systems, but important for understanding the petrogenesis of arc magmas. An explanation for this rarity is that high-Mg magmas are relatively dense (e.g., >2.7 gm cm^{-3}); this traps them under less dense crust (e.g., 2.6–2.65 gm cm^{-3}), which effectively acts as a filter (Stolper and Walker, 1980; Gust and Perfit, 1987). Stolper and Walker (1980) showed that there is a liquid density minimum in fractionation of picritic melts. Melt density is controlled by molar ratio (Fe/(Fe + Mg)); minimum melt density is achieved at a molar ratio of ~0.3 to 0.4. Fractional crystallization of olivine obviously decreases Mg, whereas it increases Si, Al, Ca, and Fe. Plagioclase and clinopyroxene form next as liquidus phase, but elevated water content (as envisioned for arc magma systems) would suppress plagioclase crystallization, causing further increase in Al and Ca. Thus, high-Mg basalts could be parental magmas for low-K, low-Ti, high-Al basalts common in many immature island arcs. The probable parental role for picrite can be demonstrated by fractionation models using high-Mg olivine and clinopyroxene phenocrysts, and putative derivative melts. For example, Kay and Kay (1985) modeled the evolution of high-Al arc tholeiite of Adak Island, Aleutian Arc from primitive tholeiitic melts, using the mineralogy of olivine clinopyroxenite xenoliths as the presumed cumulate assemblage. One can visualize a sequence of parental picritic magmas being stripped of high-Mg olivine to form high-Al basalt, followed by crystallization of both olivine and clinopyroxene; this reduces Mg and the Ca/Al ratio, and leads to basaltic andesite compositions. The ultramafic cumulates would comprise dunite, harzburgite, wehrlite, websterite, and clinopyroxenite. As discussed below, these are common in cumulate horizons of what are interpreted as "roots" of island arcs.

Arc picrites are known from the Solomon Islands (Ramsay et al., 1984); Aleutians (Gust and Perfit, 1987); and Vanuatu (Eggins, 1993). Seamount RN-47 from the northern Tofua Arc (Fig. 5 and Table 4) also has high-Mg chemistry. Petrographic evidence and melting experiments demonstrate that the liquidus phases for these high-Mg lavas were olivine and spinel, closely followed by clinopyroxene. Crystal cumulates would form wehrlite and olivine clinopyroxenite, or websterite in melts more saturated in Si. Complementary fractionated melts would be high-Al basalt. These ultramafic rock types form layered cumulates of some ophiolite terranes. Some examples are the Acoje block of the Zambales ophiolite, Luzon (Hawkins and Evans, 1983), and the Tonsina-Nelchina segment of the Talkeetna arc, Alaska (DeBari and Coleman, 1989; DeBari and Sleep, 1991). The Talkeetna cumulates form thin layers of wehrlite, websterite, and olivine clinopyroxenite beneath a 5- to 6-km-thick series of layered two-pyroxene gabbro and gabbro-norite. DeBari and her colleagues interpreted the complex as the remains of a deep (30–35 km) magma chamber that was the roots of a mature Jurassic island arc. Upper levels of the complex include plutonic rocks ranging from gabbro to tonalite; these are overlain by basaltic andesite and silicic ignimbrite.

Early in their histories the Tofua and Mariana arcs erupted low-K island arc tholeiite series lavas, primarily basaltic and basaltic andesite in composition (e.g., Ewart and Bryan, 1972; Woodhead, 1988; Table 4). Andesite and silica-rich varieties become more abundant as arcs mature and evolve into the K-enriched calc-alkaline series (cf. Gill, 1981). Many Mariana arc

TABLE 4. TOFUA AND MARIANA ARC

Sample:	RNDB 47-19	TAFAH1 116	LATE L-20	LATE L -5	LATE L-13	TOFUA N32	ATA 8-12	ASCUNCION AS-3	AGRIGAN 9-1A	AGRIGAN 7-1	SARIGAN SA-52	SARIGAN SA-53	S. HYOSHI ave	N. HYOSHI ave
								Wt%						
SiO_2	49.12	52.52	53.45	53.86	57.57	56.91	51.90	53.09	48.70	60.10	52.59	57.06	50.68	45.79
TiO_2	0.85	0.36	0.51	0.58	0.80	0.77	0.66	0.84	0.49	0.81	0.74	0.89	0.98	1.01
Al_2O_3	16.39	16.85	15.86	16.95	14.14	14.77	16.30	17.88	20.00	16.00	16.50	17.74	18.74	18.52
FeO*	8.48	9.68	9.83	9.62	11.86	10.94	9.72	9.42	8.31	8.32	9.22	7.48	8.97	10.82
MnO	0.19	0.18	0.19	0.16	0.21	0.19	0.19	0.17	0.16	0.22	0.18	0.20	0.18	0.18
MgO	7.42	6.04	6.06	4.85	3.29	3.81	6.72	3.63	6.19	1.87	6.76	3.87	3.33	6.43
CaO	13.12	11.95	11.38	11.23	8.42	8.85	12.10	10.01	13.32	5.81	10.75	8.68	8.79	11.56
Na_2O	2.17	0.95	1.65	1.74	2.42	2.22	1.83	2.75	1.57	4.25	3.13	3.11	3.48	2.32
K_2O	0.53	0.15	0.38	0.49	0.70	0.58	0.41	0.54	0.37	2.06	0.53	0.89	2.07	1.40
P_2O_5	0.21	0.04	0.08	0.07	0.12	0.11	0.12	0.15	0.12	0.35		0.15	0.25	0.29
Total	98.48	98.72	99.39	99.55	99.53	99.15	99.95	98.48	99.23	99.79	100.4	100.07	97.47	98.32
								Trace elements (ppm)						
Cr	544	62	78	55	6	15	89	13	28	22			11	49
Ni	154	23	26	25	9	10	28		26	13	33		17	30
Co	50	41	33	29	31	36	39	30	28					
V		315	295	320	410	392	550						260	398
Sc	36	45	40	40	41	41	48						23	35
Zr	41	8	25	24	38	43	40		45		64		88	31
Y		6	15	15	25	20	15				23		25	21
Nb	<1	0.47				0.4								
Hf	1.33	0.38				1.65	0.859	1.7						
Ta	<0.1	0.03				0.04		1.7						
Rb	14	2	6	6	9	8	8	8.58	8	39	10		48	26
Ba	34	47.9	96	125	165	173	133	178	86	333	271		672	714
Sr	120	108	215	230	225	207	214	293	305	314	350		674	1134
B			10	10	18									
								Rare earth elements (ppm)						
La	3.95	1.05	2.7	1.4	3	3.1	2.91	4.65						
Ce	8.46	2.28	6.6	3.4	7.6	8.48	7.74	10.94						
Nd	6	1.98				8.15	5.42							
Sm	2.35	0.75				2.73	1.78	2.77						
Eu	0.825	0.31				0.95	0.692	0.09						
Gd		1.12				3.47	2.33							
Tb	0.387	0.22				0.62	0.43	0.65						
Yb	1.85	1.03				2.45	1.68	2.66						
Lu	0.242	0.17				0.4	0.263	0.39						
Source:	JWH	EH	EB	EB	EB	EB	V	DB	S	S	MR	MR	B	B

Note: B—Bloomer et al., 1989; DB—Dixon and Batiza, 1979; EB—Ewart et al. 1973; EH—Ewart and Hawkesworth, 1987; JWH—J. Hawkins, this paper; MR—Meijer and Reagan,1981; S—Stern, 1979; V—Vallier et al., 1985; blank—no data.

volcanoes are arc tholeiitic, but some have evolved to medium-K tholeiitic and low-K calc-alkaline lavas that include andesite (e.g., Asuncion, Agrigan, and Sarigan, Stern, 1979; Meijer and Reagan, 1981; Table 4). The northernmost volcanoes of the Mariana Arc, Northern Seamount Province, are unusual in that they are LILE-enriched shoshonites (Bloomer et al. 1989; Lin et al., 1989). Quaternary air-fall tephra collected from the backarc basin record melts that are more silicic than are found as flows in the arc volcanoes. Straub (1995) described a range in tephra chemistry from basaltic to dacitic (48%–71% SiO_2; 0.8%–3.2% K_2O). These silica-rich sediments could serve as protoliths for "cherty argilites" common in many arc-ophiolite assemblages.

Tofua and Mariana Arcs

Many Tofua arc volcanoes are basalt or basaltic-andesite (e.g., Tafahi, Late, Tofua, Ata; Table 4), yet there is evidence for bimodalism represented by low-K dacite and rhyodacite (Btryan et al., 1972; Ewart et al., 1973; Hawkins, 1985). Tofua arc volcanoes (e.g., Fonualei Island and Metis Shoal, Fig. 5 and Table 5) have erupted low-K rhyolite pumice and low-K, high-Ca dacites, respectively. Bryan (1979) noted that Tongan dacites are among the most calcic in the world. Metis Shoal has phenocrysts of An 84 and hypersthene, and xenocrysts of bronzite and olivine, enclosed in high-silica glass. Fonualei Island dacite has labradorite, hypersthene, and augite phenocrysts. Both lack modal alkali feldspar or quartz, even though they have up to 20% normative quartz. These low-K, high-Ca dacites are not unique to Tonga; they are found in the Mariana and Bonin arcs, as well in as several others.

Zephyr Shoal, a Lau Basin seamount (Fig. 5), has andesitic bulk rock chemistry (Table 5). It has vitrophyric texture consisting of a web of low-K rhyolite glass (73% SiO_2) that encloses abundant phenocrysts of An 65, hypersthene, and augite. Clearly high-silica melts have been important, even in the juvenile stages of arc genesis, and may erupt in backarc basins, as well, as shown by Lau Basin sediments that have glass shards ranging in composition from basaltic to rhyolitic, with up to 76% SiO_2 (Clift and Dixon, 1994).

The plutonic equivalents of low-K dacites and rhyolites are the trondhjemite-tonalite plutons known from several arc systems (Table 5). The mid- to late-Miocene gabbro to tonalite plutons of the Coto plutonic suite of Viti Levu, Fiji are an example (Gill and Stork, 1979). The Miocene Tanzawa gabbro-tonalite plutonic complex may underlie much of the Izu arc (Kawate and Arima, 1998). Suyehiro et al. (1996), using seismic velocity data, proposed that as much as 30% of arc crust might be composed of rocks similar to the Tanzawa tonalite. The depth to the Moho under the Tonga Ridge has not been determined, but Crawford et al. (2003) suggest that it may be 20–25 km. They recognized a layer having relatively low P-wave velocities (6.0–6.5 km sec^{-1}) that begins at 4–5 km below seafloor and is as much as 5 km thick. They proposed that these velocities are consistent with the layer being tonalite, as was interpreted for similar observations on the ~22 km thick Izu-Ogasawara ridge (Suyehiro et al., 1996). Granitoids are rarely exposed in young arcs (e.g., post-Pliocene), but are found in some Aleutian volcanoes, and are more common in those older than late Miocene (cf. Byers, 1961; Vallier et al., 1994). Minor amounts of tonalite and quartz diorite have been dredged from the Tonga Trench inner wall near 15° S (Table 5; Hawkins et al., 2000), as well as from the Mariana Trench near 18° N (Bloomer and Hawkins, 1983). A possible explanation for the origin of the tonalitic-dacitic rocks is fractionation of silicic melts formed by dehydration partial melting of hydrous (amphibolitic) basalt, as suggested by Kawate and Arima (1998).

EXAMPLES OF SSZ OPHIOLITES

The objective of this presentation has been to summarize the geology of SSZ with special emphasis on the petrology and geochemistry of the igneous rocks. A primary goal was to show the broad petrologic similarity of backarc basin crust to ocean crust formed at mid-ocean ridges. There are many similarities, but there also are subtle differences in trace element chemistry that serve to distinguish between them as possible sources of ophiolites. Of these, the HFSE are the most important in that they will survive effects of alteration and low-T metamorphism that may be imposed during terrane accretion and deformation. Another objective was to show that backarc basin crust forms nearby, and concurrently with, other rock units such as boninites, arc volcanic and volcaniclastic rocks, granitoids, as well as pelagic sediments. Thus, there is a rock association—*geology*—that helps to recognize SSZ ophiolites. Using similar lines of reasoning, others have concluded that some ophiolites, including many "classic" examples first thought to be from mid-ocean ridges, probably originated in an SSZ setting. Dewey and Bird (1971) based on their study of Newfoundland ophiolites, were among the first to recognize that "The ophiolite suite is probably generated at oceanic ridges, and by diffuse spreading in marginal basins behind and within island arc complexes."

Table 6 is a partial list of ophiolites that have abundant evidence for their assignment to an SSZ origin. Data for several of these ophiolites are plotted on the Hf/3-Ta-Th diagram (Fig. 11). They form a virtually complete overlay with data for SSZ systems shown in Figure 8. Trace element patterns for different segments of the Zambales Range, Luzon ophiolite closely resemble patterns for forearc, arc, and backarc rocks from the Mariana and Tonga SSZ systems as shown in Figures 9, 12, 13. Estimates of crustal layer thickness for ophiolite sections are compared to oceanic crust in Figures 14 and 15. There are large differences in layer thickness especially for those interpreted as being from an arc ophiolite. In several instances (e.g., Zambales, Luzon and Bay of Islands, Newfoundland), the ophiolite probably is a composite of arc, and backarc crust. For the Zambales, we recognized a probable forearc (boninite) unit as well. Part of the reasoning behind these interpretations was the ophiolite geochemistry, but the primary factors were the *geology* of the ophiolites and their associated rocks.

TABLE 5. SUPRA-SUBDUCTION ZONE SILICA-RICH ROCKS

Sample	FON F-30	FON F-4	METIS ROCK	METIS GL	S.P. LEE DR-4	ZEPHYR SHOAL-rk	ZEPHYR SHOAL-gl	TONGA TR. PPT 2-15	TONGA TR. PPT 2-5	TONGA TR. PPT 2-14	FIJI COLO	FIJI COLO	TANZAWA KU-1	TANZAWA FJ-1
								Wt%						
SiO_2	60.33	65.09	63.66	73.6	73.15	63.26	73.16	59.30	67.23	58.57	64.40	72.40	60.72	71.13
TiO_2	0.65	0.56	0.39	0.52	0.44	0.53	0.71	0.83	0.25	0.76	0.51	0.43	0.61	0.30
Al_2O_3	14.6	14.39	12.42	12.29	14.85	14.56	13.54	15.59	16.08	15.64	16.90	14.10	17.45	15.55
FeO*	9.83	7.76	6.44	3.89	3.25	5.33	3.29	7.85	4.59	6.78	5.10	3.10	6.58	2.83
MnO	0.21	0.17	0.12	0.06	0.16	0.09	0.07	0.07	0.06	0.09	0.14	0.07	0.16	0.12
MgO	2.59	1.32	5.10	1.07	0.72	3.42	0.71	4.81	2.01	7.40	2.20	0.90	3.09	0.93
CaO	7.48	5.90	6.97	3.61	3.30	5.16	2.31	7.67	6.55	6.86	6.20	3.20	7.18	3.58
Na_2O	2.69	3.00	2.58	3.17	2.85	4.03	4.40	3.62	2.85	3.80	3.20	4.60	3.54	4.46
K_2O	0.85	1.13	0.90	1.47	0.50	0.98	1.63	0.17	0.37	0.16	0.67	1.00	0.57	0.99
P_2O_5	0.17	0.17	0.07	0.07	0.06	0.05	0.17	0.09	0.02	0.03	0.10	0.10	0.09	0.10
Total	99.40	99.49	98.65	99.75	99.28	97.41	99.99	100.00	100.01	100.09	99.42	99.90	99.99	99.99
								Trace elements (ppm)						
Rb	10	15	14	21				5	7	5	8	9	7	16.1
Ba	195	315	360	610				9	124	21	171	160	181	317
Sr	295	305	140	130				104	163	119	349	210	254	247
Zr	40	46	46	68				67	40	48	82	125	87	99
Y	20	25	21	25				27	11	27	22	35	24	17
Hf			1.22					1.65		1.21				
Ta			0.04					0.132		0.159				
Nb			0.44					4	2	3			1.8	.0.9
Cr		5	230	7				12	27	4	7	4	20	9
Ni		8	53	4				22	16	33	<5	<5	7	2
Co	21	13	25	10				26		22	12	5	19	
Sc	34	25	29	12				29	24	31	12	9		
V	175	105	175	130				233	165	312	66	32	163	29
B	12	15	53	105										
								Rare earth elements (ppm)						
La	5.50	3.50	3.24					2.04		1.59			4.6	2
Ce	11.00	10.50	7.46					5.63		4.59			13.7	11.6
Nd			5.58					3.7		3.1			8.5	4.1
Sm			2.25					1.95		1.99			2.94	1.43
Eu			0.84					0.692		0.619			0.9	0.45
Tb			0.39					0.444		0.434				
Yb			1.64					2.34		2.26			2.25	0.85
Lu			0.24					0.291		0.332			0.39	0.2
Source:	EBG	EBG	EBG	EBG	JH	JWH	JWH	JWH	JWH	JWH	GS	GS	KA	KA

Note: EBG—Ewart, Bryan, and Gill,1973; JH—J. Hawkins, 1985; JWH—J. Hawkins, this paper; GS—Gill and Stork, 1979; KA—Kawate and Arima, 1998; blank = no data.

TABLE 6. SUPRA-SUBDUCTION ZONE OPHIOLITES

Location	References
Troodos[#] *	Miyashiro, 1973; Pearce, 1975; Pearce, 1980; Schmincke et al., 1983; Robinson et al, 1983; Pearce et al., 1987; Flower and Levine, 1987
Oman*	Pearce, 1980; Pearce et al., 1981; Alabaster et al., 1982; Pearce et al., 1984
Vourinos*	Noiret et al., 1981; Pearce et al., 1984
Pindos[#] *	Pearce et al., 1987; Jones and Robertson, 1991; Capedri et al., 1982
Bay of Islands	Coish and Church, 1979; Elthon, 1991; Jenner et al., 1991
Josephine[#]	Harper,1984; Alexander and Harper, 1992
Trinity	Wallin and Metcalf, 1998; Metcalf et al., 2000
Coast Range[#]	Menzies et al., 1977; Shervais and Kimbrough, 1985
Zambales[#] *	Hawkins, 1980; Hawkins and Evans, 1983; Hawkins and Florendo, 1994
Marum	Davies and Smith, 1971

[#]Ophiolites having boninite as part of the assemblage.
*Ophiolites having island arc tholeiites as part of the assemblage.

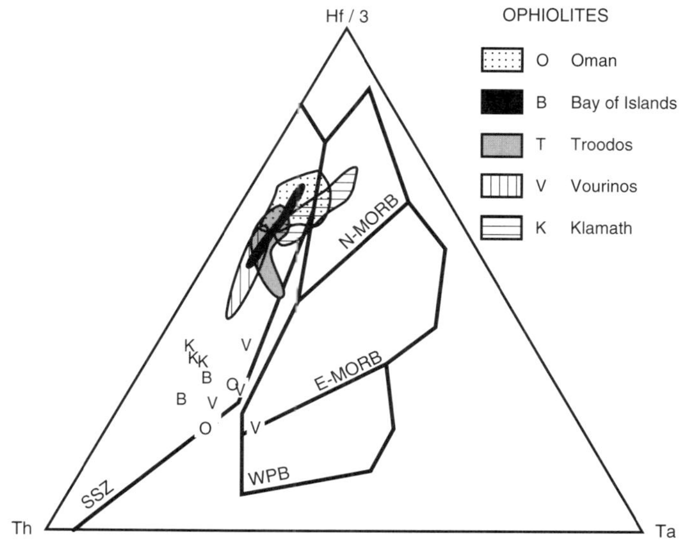

Figure 11. Plot of Hf/3-Ta-Th (after Wood et al., 1981), showing fields for N-MORB, E-MORB, WPB (within plate basalts) and SSZ magmas. Addition of sediment as fluid or melt will shift magma compositions toward Th apex of diagram. Data for ophiolites are from references cited in text. Letters are for isolated datum lying outside fields shown for main groups of data.

OPHIOLITE PROTOLITH—BACKARC OR FOREARC?

In the preceding discussion, I have made the case that forearc, backarc, and nascent arc crust all could be progenitors for the *igneous* part of ophiolites. Is there a way to distinguish between these potential sources? Mature arcs, especially those that have matured beyond the low-K arc tholeiitic stage, are unlikely to be called an ophiolite, although basal cumulates might be confused if not tied to upper level petrology. It is important to recognize that arc assemblages of varied stage in their evolution are to be expected in close association with SSZ ophiolites.

In their nascent stages, at least some arcs are initiated in rift basins (cf. R. Taylor and Nesbitt, 1995). If this nascent stage occurs on crust formed at a mid-ocean ridge (i.e., at the initiation of subduction as described elsewhere in this chapter), boninite, high-Mg basalts, island arc tholeiites, and low-K rhyolites would constitute a distinctive assemblage that is highly unlikely to form in a backarc basin, i.e., these have not been found in backarc basins, to date. The occurrence of this assemblage on the Hunter fracture zone (as described above) is in the forearc of what probably is an incipient subduction zone. This may be an "actualistic" example of a forearc ophiolite in genesis. Guam and the Mariana forearc display examples of parts of a forearc ophiolite, but we do not see them as an integral unit. They are dispersed across the forearc and exposed on walls of the Mariana and Palau trenches. Figure 10 shows all of these constituents.

Forearc ophiolites would differ from those derived from backarc basins in that the latter would resemble mid-ocean ridge crust in terms of petrology and chemistry, as well as in having basalts transitional to arc tholeiite. This compositional range distinguishes them from MORB formed at a true mid-ocean ridge. The lack of boninite and general rarity of high-Mg basalts would separate them from forearc assemblages. The close spatial and temporal association of backarc basin rocks with arc rocks and volcanic detritus would be another important factor. This may not be unique, but a true nascent forearc ophiolite might not have evolved into an emergent volcanic arc. Even if it had, there should be upward continuity without intervening layers of pelagic sediments as found in backarc basins (cf. Hawkins and Evans, 1983).

It is obvious that crustal contraction would tectonically juxtapose arc, backarc, and forearc crust. Rather than hoping to identify a particular ophiolite as having a forearc or backarc origin, it may be more important to recognize it as having an SSZ origin and then attempt to separate out units having petrologic features distinctive for the original setting. I would be surprised to find an SSZ ophiolite associated with an arc terrane that came only from a forearc or a backarc setting, but I would let the geology lead me to the interpretation.

ON THE EMPLACEMENT OF OPHIOLITIES

Classic ideas of ophiolite emplacement by obduction have faced the problem of how oceanic lithosphere, which was

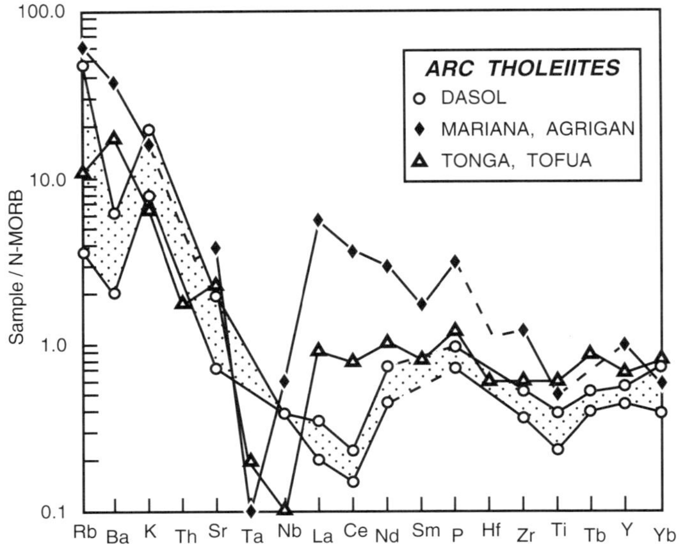

Figure 12. Trace elements for arc tholeiitic rocks from Agrigan Island, Mariana Arc; Tofua Island, Tofua Arc; and Dasol series of Zambales Range, Luzon (Hawkins and Evans, 1983). Data are normalized to N-MORB values (Sun and McDonough, 1989).

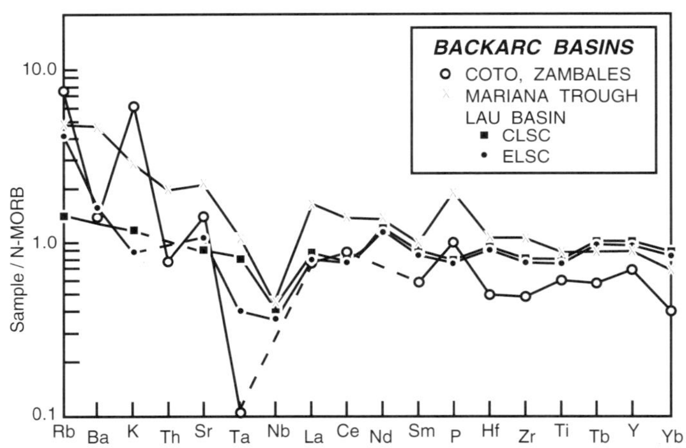

Figure 13. Trace elements for Mariana Trough, Lau backarc basin tholeiitic rocks, and probable backarc basin crust in Zambales Range, Luzon (Hawkins and Evans, 1983). Data are normalized to N-MORB values (Sun and McDonough, 1989).

destined to return to the mantle in a subduction zone, could be diverted and make its way onto continental margins. Coleman (2000) discussed five different mechanisms pertinent to California ophiolites, one of which concerns SSZ ophiolites. He observes that "obduction of ophiolites probably is a rare event when compared to the stranding of oceanic-crust slabs in subduction zones along convergent margins." I agree; this probably could explain the origin of dismembered fragments of ophiolite material, but it seems hard to visualize this process for explaining large, intact ophiolites.

Clearly, some unquestionably oceanic crust is presently being thrust onto other ocean crust (on Macquarie Ridge) in the case of the incipient Puysegur trench; however, many of the characteristic constituents of arc-ophiolite assemblages, such as those in the North American Cordillera, are missing (e.g., boninites and arc plutonic-volcanic series).

In the case of young intra-continental ocean basins such as the Red Sea and Gulf of California, it seems likely that a change in stresses from extension to crustal contraction would lead to emplacement of true ocean crust onto continental crust. Some of the Tethyan ophiolites may have formed this way. These ophiolites would be lacking in the associated arc and forearc rocks common to many ophiolites, but they too would represent ocean crust.

The purpose of this essay is to discuss geologic tracers pointing to an SSZ origin for many ophiolites. The emphasis has been on the *geologic association* of rocks that are formed in this setting and that are common to many ophiolites. If we consider the spatial relations shown in Figures 1 and 2, it should be easy to see how a change from crustal extension to contraction—say, by the blocking of subduction by an oceanic plateau or a microcontinent such as Fiji—could lead to the imbrication of the different rock series. The weakest part of the system would be the axial ridges of the backarc basin. Low-angle thrusts would emplace young, hot, ocean crust over colder ocean crust and set up conditions for inverted metamorphic gradients on the sole of the thrust zone. If the backarc crust were thrust onto a nascent arc, normal, prograde downward metamorphism would prevail. The backarc crust could be thrust onto either older remnant arc crust or younger active arc crust.

There are strong arguments for ophiolites being derived from nascent arcs formed in what might be the forearc to an older arc, or at the inception of subduction (cf. Stern and Bloomer, 1992). Several models for the inception of subduction propose an initial phase of transtension on a fracture zone to set the stage for foundering of older lithosphere. If this were to be followed by transpression before subduction got under way, it could imbricate the young crust and transport it onto thicker crust, e.g., a remnant arc. There are many possible permutations. The key point is that ophiolites have been emplaced onto older crust, many ophiolites carry the SSZ petrologic signature, and their geology indicates a temporal and genetic association with island arc systems. Geometrically, it seems not to be a problem to bring these geologic components together if they are initially arranged as shown in Figures 1 and 2.

SUMMARY

The geology of SSZ systems, as exemplified by Western Pacific Basin backarc-arc-trench systems such as the Lau-Tonga and Mariana zones of plate convergence, exhibits a spectrum of rock types, including boninites (formed in forearcs as initial stages

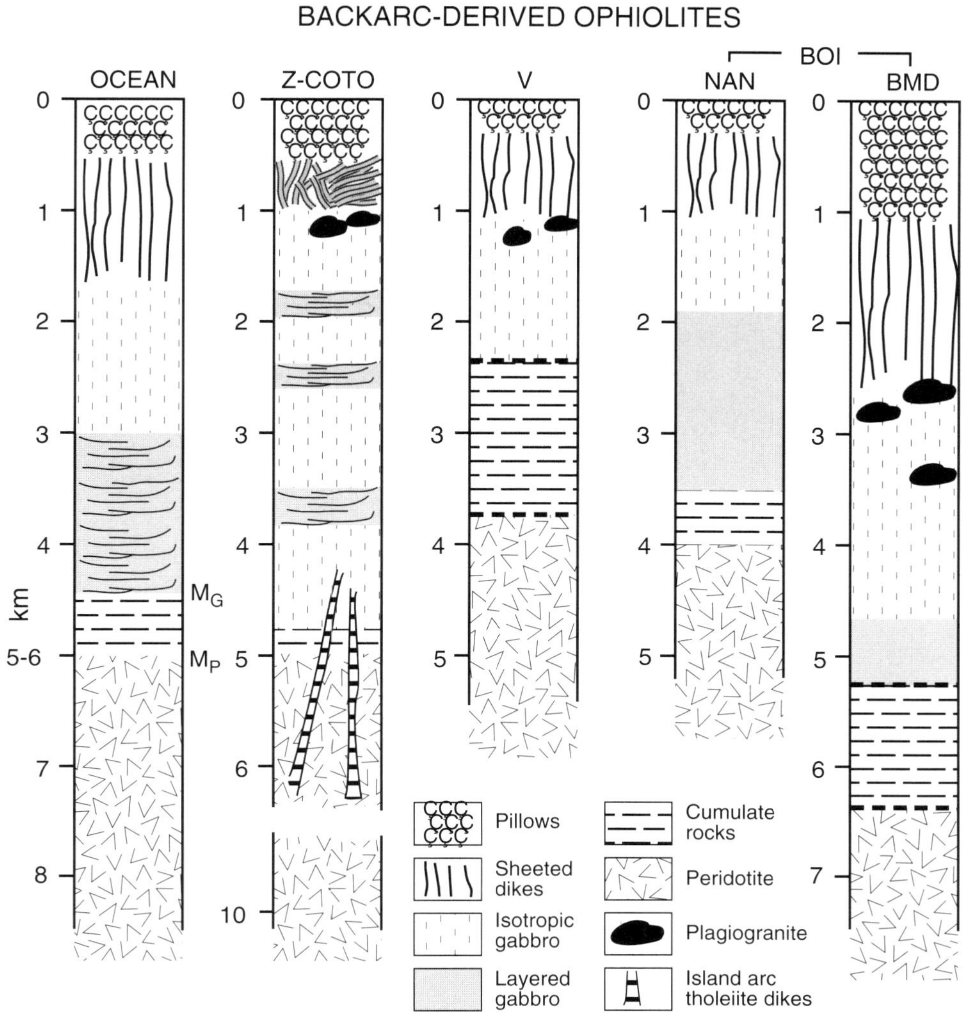

Figure 14. Comparison of oceanic crust with probable backarc basin ophiolite crust for: Z-COTO—Coto, Zambales Range, Luzon; V—Vourinos, Greece; BMD—Blow-me-down Mountain; BOI—Bay of Islands, Newfoundland. References are in Table 6.

of arc volcanism); mafic ocean crust (formed in early stages of arc volcanism and in backarc basins); arc volcanic series ranging from picritic to rhyolitic, and including arc tholeiitic, high-K calc-alkaline; and shoshonitic magma series. Deeper-level arc crust comprises various types of layered and isotropic gabbro, as well as granitoid plutons (e.g., tonalite and quartz diorite). There is a wide range of sediment types, from arc-derived tephra and volcaniclastic turbidites to silicic and carbonate oozes. Crustal units are underlain by mafic cumulates and by mantle rocks comprising depleted harzburgite, dunite, and their serpentinized equivalents. Ultramafic cumulates from beneath island arcs may be dominated by pyroxene and may lack plagioclase. All of the crustal units may be coeval, although some may overlie fragments of an older substrate that comprises similar SSZ material. Arc-derived clastics are deposited on new seafloor of essentially the same age. In the early rift stage of backarc basins, as well as in the forearc, basalt flows, sills, and dikes are interspersed with arc-derived clastic and hemipelagic sediments. Carbonate nanno-fossil ooze, as well as metalliferous sediments and hydrothermal vent material, are deposited on and among basalt pillows.

The geology of SSZ plate margins illustrates a dynamic system comprising converging plates, mantle upwelling, generation of new oceanic crust and sialic proto-continental crust. On the 100–200 km scale across strike, SSZ systems are regions of lithosphere extension. On a broad regional scale (100s of kms), they are regions of lithosphere contraction; this leads to their eventual shortening and incorporation into tectonic belts at a continental, micro-continental, or oceanic plateau backstop. Details of rock chemistry and isotope systematics, coupled with the *geology* known for SSZ, are criteria that help distinguish the ophiolite assemblage from N-MORB crust formed at "bluewater" true mid-ocean ridges. The ophiolite conundrum, the puzzle, is solved once you understand the nature of SSZ systems. Yes, some ophiolites may be derived from "bluewater" spreading systems and emplaced on continents by obduction, although unusual tectonic settings must be involved. Consider-

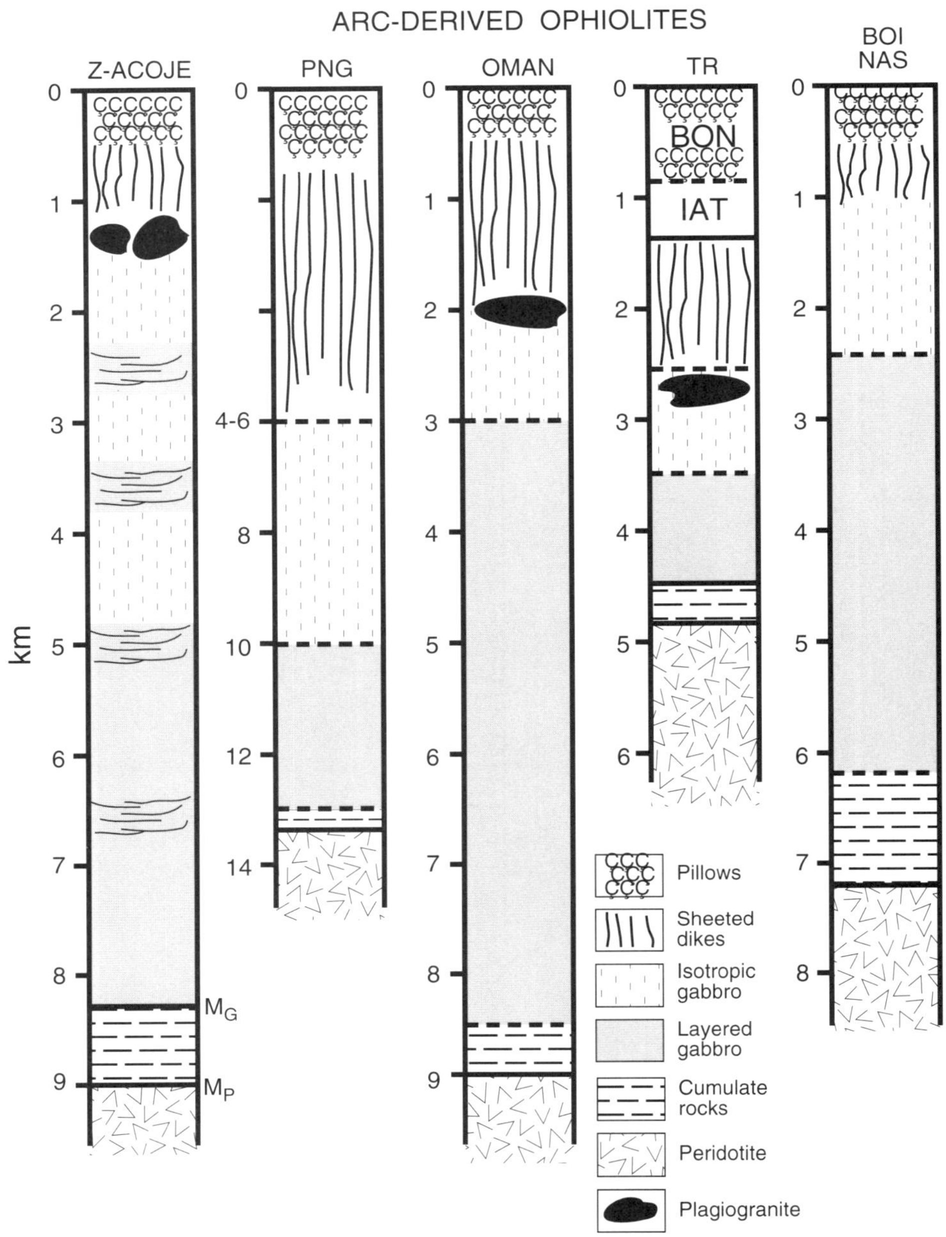

Figure 15. Comparison of oceanic crust with probable nascent arc ophiolite crust for: Z-ACOJE—Acoje, Zambales Range, Luzon; PNG—Papua, New Guinea; Oman; TR—Troodos; and NAS—Southern North Arm; BOI—Bay of Islands, Newfoundland. References are in Table 6.

ation of the geology and chemistry of ophiolites, their common rock associations, and the geology of SSZs leads me to reiterate an old claim of mine that many, perhaps most, ophiolites come from intra-oceanic plate convergence systems.

ACKNOWLEDGMENTS

This synthesis has benefitted from discussions, over many years, with co-workers and students. In particular I thank Bob Stern, Sherman Bloomer, Patty Fryer, Jim Gill, Brian Taylor, Jim Natland, and Bob Fisher for sharing their ideas and insights. The original manuscript has been improved by helpful comments and suggestions provided by Robert Stern, Warren Hamilton, and Yildirim Dilek. Special thanks to Jo Griffith for her help with illustrations. The author's research has been supported by research awards from the National Science Foundation, OCE, and the University of California—San Diego, USA.

REFERENCES CITED

Alabaster, T., Pearce, J.A., and Malpas, J., 1982, The volcanic stratigraphy and petrogenesis of the Oman ophiolite complex: Contributions to Mineralogy and Petrology, v. 81, p. 168–183.

Allwart, A.O., 1988, Working at cross-purposes; Holmes and Vening Meinesz on convection: Eos (Transactions, American Geophysical Union), v. 69, p. 899–906.

Arai, S., 1991, The circum-Izu massif peridotite, central Japan, as back -arc mantle fragments of the Izu-Bonin arc system, *in* Peters, Tj., Nicolas, A., and Coleman, R.G., eds., Ophiolite Genesis and Evolution of the Oceanic Lithosphere: Dordrecht, Kluwer Academic Publishers, p. 801–816.

Alexander, R.J., and Harper, G.D., 1992, The Josephine ophiolite: an ancient analogue for slow- to intermediate-spreading oceanic ridges, *in* Parson, L.M., Murton, B.J., and Browning, P., eds., Ophiolites and their Modern Oceanic Analogues: London, Geological Society of London Special Publication 60, p. 3–38.

Austin, J., A., Taylor, F.W., and Cagle, C.D., 1989, Seismic stratigraphy of the central Tonga Ridge: Marine and Petroleum Geology, v. 6, p. 71–92.

Bailey, E.H., Blake, M.C., and Jones, D.L., 1970, On-land Mesozoic oceanic crust in California coast ranges: Geological Survey Research, 1970, U.S., Geological Survey, Prof. Paper 700-C, p. C-70–C-81.

Baker, P.E., 1968, Comparative volcanology and petrology of the Atlantic island arcs: Bulletin of Volcanology, v. 32, p. 189–206.

Barazangi, M., and Isacks, B., 1971, Lateral variations of seismic-wave attenuation in upper mantle above the inclined earthquake zone of the Tonga Island Arc: Journal of Geophysical Research, v. 76, p. 8493–8516.

Barker, F., ed., 1979, Trondhjemites, Dacites, and Related Rocks: Elsevier, New York, 659 p.

Benioff, H., 1949, Seismic evidence for the fault origin of oceanic deeps: Geological Society of America Bulletin, v. 60, p. 1337–1356.

Ben Othman, D., White, W.M., and Patchett, J., 1989, The geochemistry of marine sediments, island arc magma genesis, and crust-mantle recycling: Earth and Planetary Science Letters, v. 94, p. 1–21.

Bernal, J.D., 1961, Continental and oceanic differentiation: Nature, v. 192, p. 123–124.

Bevis, M., Taylor, F.W., Schutz, B.E., Recy, J., Isacks, B.L., Hely, S., Singh, R., Kendrick, E., Stowell, J., Taylor, B., and Calmant, S., 1995, Geodetic observations of very rapid convergence and back-arc extension at the Tonga arc: Nature, v. 374, p. 249–251.

Bloomer, S.H., 1983, Distribution and origin of igneous rocks from the landward slope of the Mariana Trench: implications for its structure and evolution: Journal of Geophysical Research, v. 88, p. 7411–7428.

Bloomer, S.H., and Fisher, R.L., 1987, Petrology and geochemistry of igneous rocks from the Tonga Trench-A non-accreting plate boundary: Journal of Geology, v. 95, p. 469–495.

Bloomer, S.H., and Hawkins, J.W., 1983, Gabbroic and ultramafic rocks from the Mariana Trench, *in* Hayes, D.E., eds., The Tectonic Evolution of Southeast Asian Seas and Islands, pt. 2: Washington, D.C., American Geophysical Union Memoir 27, p. 294–317.

Bloomer, S.H., and Hawkins, J.W., 1987, Petrology and geochemistry of boninite series volcanic rocks from the Mariana Trench: Contributions to Mineralogy and Petrology, v. 97, p. 361–377.

Bloomer, S.H., Stern, R.J., Fisk, E., and Geshwind, C.H., 1989, Shoshonitic volcanism in the northern Mariana Arc 1. Mineralogic and major and trace element characteristics: Journal of Geophysical Research, v. 94, p. 4469–4496.

Bloomer, S.H., Ewart, A., Hergt, J.M., and Bryan, W.B., 1994, Geochemistry and origin of igneous rocks from the outer Tonga forearc (site 841), *in* Hawkins, J., Parson, L., Allan, J., et al., eds., Proceedings, Ocean Drilling Program, Scientific Results, 135: College Station, Texas, Ocean Drilling Program, p. 625–646.

Bloomer, S.H., Taylor, B., MacLeod, C.J., Stern, R.J., Fryer, P., Hawkins, J.W., and Johnson, L., 1995, Early arc volcanism and the ophiolite problem: A perspective from drilling in the Western Pacific, *in* Taylor, B., and Natland, J., eds., Active Margins and Marginal Basins of the Western Pacific: Washington, D.C., American Geophysical Union Monograph 88, p. 1–30.

Bostrom, K., and Peterson, M.N.A., 1966, Precipitates from hydrothermal exhalation on the East Pacific Rise: Economic Geology, v. 61, p. 1258–1265.

Bowin, C.O., Nalwalk, A.J., and Hersey, J.B., 1966, Serpentinized peridotite dredged from the north walls of the Puerto Rico Trench: Geological Society of America Bulletin, v. 77, p. 257–270.

Brocher, T.M., Karson, J.A., and Collins, J. A, 1985, Seismic stratigraphy of the oceanic Moho based on ophiolite models: Geology, v. 13, p. 62–65.

Bryan, W.B., 1979, Low K_2O dacite from the Tonga-Kermadec arc: Petrography, chemistry and petrogenesis, *in* Barker, F., ed., Trondhjemites, Dacites, and Related Rocks: Amsterdam, Elsevier, p. 581 –600.

Bryan, W.B., Stice, G.D., and Ewart, A.E., 1972, Geology, petrography, and geochemistry of the volcanic islands of Tonga: Journal of Geophysical Research, v. 77, p. 1566–1585.

Byers, F.M., Jr., 1961, Petrology of three volcanic suites, Umnak and Bogoslof Islands, Alaska: Geological Society of America Bulletin, v. 72, p. 93–128.

Capedri, S., Venturelli, G., and Toscani, L., 1982, Petrology of an ophiolitic cumulate sequence from Pindos, Greece: Geological Journal, v. 17, p. 223–242.

Clift, P.D., and Dixon, J.E., 1994, Variations in arc volcanism and sedimentation related to rifting of the Lau Basin, *in* Hawkins, J.W., Parson, L.M., Allan, J.F., et al., eds., Proceedings, Ocean Drilling Program, Scientific Results, 135: College Station, Texas, Ocean Drilling Program, p. 23–50.

Clift, P.D., MacLeod, C.J.,Tappin D.R., Wright, D.J., and Bloomer, S.H., 1998, Tectonic controls on sedimentation and diagenesis in the Tonga Trench and forearc, southwest Pacific: Geological Society of America Bulletin, v. 110, p. 483–496.

Cloud, P.E., Schmidt, R.G., and Burke, H.W., 1958, Geology of Saipan Island: U.S. Geological Survey Professional Paper 280-A, p. 1–111.

Coats, R.R., 1962, Magma type and crustal structure in the Aleutian Arc, *in* Macdonald, G.A., and Kuno, H., eds., The Crust of the Pacific Basin: American Geophysical Union, Geophysical Monograph 6, p. 92–109.

Coish, R.A., and Church, W.R., 1979, Igneous geochemistry of mafic rocks in the Betts Cove ophiolite, Newfoundland: Contributions to Mineralogy and Petrology, v. 70, p. 29–39.

Coney, P.J., Jones, D.A., and Monger, J.W.H., 1980, Cordilleran suspect terranes: Nature, v. 288, p. 329–333.

Coleman, R.G., 1977, Ophiolites, Ancient Oceanic Lithosphere?: New York, Springer-Verlag, 229 p.

Coleman, R.G., 2000, Prospecting for ophiolites along the California continental margin, *in* Dilek, Y, Moores, E.M., Elthon, D., and Nicolas, A., eds., Ophiolite and Oceanic Crust: New insights from field studies and the Ocean Drilling Program: Boulder, Colorado, Geological Society of America Special Paper 349, p. 351–364.

Crawford, A, J., ed., 1989, Boninites: London, Unwin Hyman, 465 p.

Crawford, A.J., Falloon, T.J., and Green, D.H., 1989, Classification, petrogenesis and tectonic setting of boninites, *in* Crawford, A.J., ed., Boninite: London, Unwin Hyman, p. 2–49.

Crawford, W.C., Hildebrand, J.A., Dorman, L.M., Webb, S.C., and Wiems, D.A., 2003, Tonga Ridge and Lau Basin crustal structure from seismic refraction data: Journal of Geophysical Research, v. 108, EPM 6, DOI 10.10291/2001JB00 1435.

Davies, H.L., and Smith, E.I., 1971, Geology of Eastern Papua: Geological Society of America Bulletin, v. 82, p. 3299–3312.

Davies, J.H., and Stevenson, D.J., 1992, Physical model of source region of subduction zone volcanics: Journal of Geophysical Research, v. 97, p. 2037–2070.

DeBari, S.M., and Coleman, R.G., 1989, Examination of the deep levels of an island arc: Evidence from the Tonsina ultramafic-mafic assemblage, Tonsina, Alaska: Journal of Geophysical Research, v. 94, p. 4373–4391.

DeBari, S.M., and Sleep, N.H., 1991, High-Mg, low-Al bulk composition of the Talkeetna island arc, Alaska: Implications for primary magmas and the nature of arc crust: Geological Society of America Bulletin, v. 103, p. 37–47.

Defant, M.J., and Drummond, M.S., 1990, Derivation of some modern arc magmas by melting of young subducted lithosphere: Nature, v. 347, p. 662–665.

Dewey, J.F., and Bird, J.M., 1971, Origin and emplacement of the ophiolite suite: Appalachian ophiolites of Newfoundland: Journal of Geophysical Research, v. 76, p. 3179–3206.

Dickinson, W.R., and Hatherton, T.R., 1967, Andesitic volcanism and seismicity around the Pacific: Science, v. 157, p. 801–803.

Dietrich, V.J., Emmerman, R., Oberhansli, R., and Puchelt, H., 1978, Geochemistry of basaltic and gabbroic rocks from the west Mariana Basin and the Mariana Trench: Earth and Planetary Science Letters, v. 38, p. 127–144.

Dietz, R.S., 1961, Continent and ocean basin evolution by spreading of the sea floor: Nature, v. 190, p. 854–857.

Dilek, Y., Moores, E.M., Elthon, D., and Nicolas, A., eds., 2000, Ophiolites and Oceanic Crust: New insights from field studies and the Ocean Drilling Program: Boulder, Colorado, Geological Society of America Special Paper 349, 552 p.

Dixon, T.H., and Batiza, R., 1979, Petrology and chemistry of Recent lavas in the northern Marianas: Implications for the origin of island arc basalts: Contributions to Mineralogy and Petrology, v. 70, p. 167–181.

Dobson, P.F., and O'Neill, J.R., 1987, Stable isotope compositions and water content of boninite series volcanic rocks from Chichi-jima, Bonin islands, Japan: Earth and Planetary Science Letters, v. 82, p. 75–86.

Duncan, R.A., Vallier, T.L., and Falvey, D.A., 1985, Volcanic episodes at 'Eua, Tonga Islands, *in* Scholl, D.W., and Vallier, T.L., 1985, Geology and Offshore Resources of Pacific Island Arcs-Tonga Region: Houston, Texas,

Circum-Pacific Council for Energy and Mineral Resources Earth Science Series, v. 2, p. 281–290.

Eggins, S.M., 1993, Origin and differentiation of picritic arc magmas, Ambae (Aoba) Vanuatu: Contributions to Mineralogy and Petrology, v. 114, p. 79–100.

Ellam, R.M., and Hawkesworth, C.J., 1988, Elemental and isotopic variations in subduction related basalts: Evidence for a three component model: Contributions to Mineralogy and Petrology, v. 98, p. 72–80.

Elsasser, W.M., 1971, Sea-floor spreading as thermal convection: Journal of Geophysical Research, v. 76, p.1101–1112.

Elthon, D., 1991, Geochemical evidence for formation of the Bay of Islands ophiolite above a subduction zone: Nature, v. 354, p. 140–143.

Ewart, A., and Bryan, W.B., 1972, Petrography and geochemistry of the igneous rocks from "Eua, Tongan Islands: Geological Society of America Bulletin, v. 83, p. 3281–3298.

Ewart, A., and Hawkesworth, C.J., 1987, The Pleistocene-Recent Tonga-Kermadec arc lavas: Interpretation of new isotopic and rare earth data in terms of a depleted mantle source model: Journal of Petrology, v. 28, p. 495–530.

Ewart, A., Bryan, W.B., and Gill, J.B., 1973, Mineralogy and geochemistry of the younger Tongan Islands, S.W., Pacific: Journal of Petrology, v. 14, p. 429–466.

Ewart, A., Brothers, R.N., and Mateen, A., 1977, An outline of the geology and geochemistry, and the possible petrogenetic evolution of the volcanic rocks of the Tonga-Kermadec-New Zealand island arc: Journal of Volcanology and Geothermal Research, v. 2, p. 205–250.

Falloon, T.J., Green, D.H., and McCullough, M.T., 1989, Petrogenesis of high-Mg and associated lavas from the northern Tonga Trench, *in* Crawford, A.J., ed., Boninite: London, Unwin Hyman, p. 357–395.

Falloon, T.J., Malahoff, A., Zonenschain, L.P., and Bogdanov, Y., 1992, Petrology and geochemistry of back-arc basin basalts from Lau Basin spreading ridges at 15, 18, 19° S: Mineralogy and Petrology, v. 47, p. 1–36.

Fisher, R.L., and Engel, C.G., 1969, Ultramafic and basaltic rocks dredged from the nearshore flank of the Tonga Trench: Geological Society of America Bulletin, v. 80, p. 1373 –1378.

Fisher, R.L., and Hess, H.H., 1963, Trenches *in* Hill, M.N., ed., The Sea, v. 3: New York, Wiley Interscience, p. 411–436.

Flower, M.F.J., and Levine, H.M., 1987, Petrogenesis of a tholeiite-boninite sequence from Ayios Mamas, Troodos ophiolite: evidence for splitting of a volcanic arc: Contributions to Mineralogy and Petrology, v. 97, p. 509–524.

Fryer, P., 1992, A synthesis of Leg 125 drilling of serpentinite seamounts on the Mariana and Izu-Bonin forearcs, *in* Fryer, P., Pearce, J.A., Stokking, L.B., et al., eds., Proceedings, Ocean Drilling Program, Scientific Results, 125: College Station, Texas, Ocean Drilling Program, p. 593–614.

Fryer, P., 1995, Geology of the Mariana Trough, *in* Taylor B., ed., Backarc Basins: Tectonics and Magmatism: New York, Plenum Press, p. 237–279.

Fryer, P., 1996, Evolution of the Mariana convergent plate margin system: Reviews of Geophysics, v. 34, p. 89–125.

Fryer, P.J., Sinton, J.M., and Philpotts, J.A., 1981, Basaltic glasses from the Mariana Trough, *in* Hussong, D.M., Uyeda, S., et al., Initial Reports of the Deep Sea Drilling Project, v. 60: Washington, D.C., U.S. Government Printing Office, p. 601–610.

Fryer, P., Pearce, J.A., Stokking, L.B., et al., eds., 1992, Proceedings, Ocean Drilling Program, Scientific Results, 125: College Station, Texas, Ocean Drilling Program, 716 p.

Gans, P. B, Mahood, G.A., and Schermer, E., 1989, Synextensional magmatism in the Basin and Range province: A case study in the eastern Great Basin: Boulder, Colorado, Geological Society of America Special Paper 233, p. 1–53.

Garcia, M.G., Liu, N.W.K., and Muenow, D.W., 1979, Volatiles in submarine volcanic rocks from the Mariana Island arc and trough: Geochimica et Comochimica Acta, v. 43, p. 305–312.

Gill, J.B., 1970, Geochemistry of Viti Levu, Fiji, and its evolution as an island arc: Contributions to Mineralogy and Petrology, v. 27, p. 179–203.

Gill, J.B., 1981, Orogenic andesites and plate tectonics: New York, Springer-Verlag, 390 p.

Gill, J.B., 1987, Early geochemical evolution of an oceanic island arc and back-arc: Fiji and the South Fiji Basin: Journal of Geology, v. 95, p. 589–615.

Gill, J.B., and Stork, A.L., 1979, Miocene low-K dacites and and trondhjemites of Fiji, *in* Barker, F., ed., Trondhjemites, Dacites, and Related Rocks: New York, Elsevier, p. 629–649.

Green, T.H., and Ringwood, A.E., 1968, Genesis of the calc-alkaline igneous rock suite: Contributions to Mineralogy and Petrology, v. 8, p.105–162.

Gribble, R.F., Stern, R.J., Bloomer, S.H., Stuben, D.H., O'Hearn, T., and Newman, S., 1996, MORB mantle and subduction components interact to generate basalts in the southern Mariana Trough back-arc basin: Geochimica et Cosmochimica Acta, v. 60, p. 2153–2166.

Gribble, R.F., Stern, R.J., Bloomer, S.H., Newman, S, and O'Hearn, T., 1998, Chemical and isotopic composition of lavas from the northern Mariana Trough: Implications for magmagenesis in back-arc basins: Journal of Petrology, v. 39, p. 122–154.

Gust, D.A., and Perfit, M.R., 1987, Phase relations of a high-Mg basalt from the Aleutian Island Arc: Implications for primary island arc basalts and high-Al basalts: Contributions to Mineralogy and Petrology, v. 97, p. 7–18.

Hamilton, W.B., 1988, Plate tectonics and island arcs: Geological Society of America Bulletin, v. 100, p. 1503–1527.

Hamilton, W.B., 1995, Subduction systems and magmatism, *in* Smellie, J.L., ed., Volcanism Associated with Extension at Consuming Plate Margins: Geological Society Special Publication 81, p. 3–28.

Hamilton, W.B., 2002, The closed upper-mantle circulation of plate tectonics, *in* Stein S., and Freymuller, J.T., eds., Plate Boundary Zones: American Geophysical Union, Geodynamics Series, v. 30, p. 359–410.

Harper, G.D., 1984, The Josephine ophiolite, northwestern California: Geological Society of America Bulletin, v. 95, p.1009–1026.

Hart, S.R., and Zindler, A., 1986, In search of a bulk-earth composition: Chemical Geology, v. 57, p. 247–267.

Hart, S.R., Glassley, W.E., and Karig, D.E., 1972, Basalts and seafloor spreading behind the Mariana Island Arc: Earth and Planetary Science Letters, v. 15, p. 12 –18.

Hasebe, K., Fujii, N., and Uyeda, S., 1970, Thermal processes under island arcs: Tectonophysics, v. 10, p. 335–355.

Hawkins, J.W., 1974, Geology of the Lau Basin a marginal sea behind the Tonga Arc, *in* Burk, C., and Drake, C., eds., Geology of Continental Margins: New York, Springer, p. 505–520.

Hawkins, J.W., 1976, Petrology and geochemistry of basaltic rocks of the Lau Basin: Earth and Planetary Science Letters, v. 28, p. 283–298.

Hawkins, J.W., 1977, Petrological and geochemical characteristics of marginal basin basalts, *in* Talwani, M., and Pitman, W., eds., Island Arcs, Deep Sea Trenches and Back-Arc Basins: Washington, D.C., American Geophysical Union, Maurice Ewing Series 1, p. 355–365.

Hawkins, J.W., 1980, Petrology of back-arc basins and island arcs: Their possible role in the origin of ophiolites, *in* Panayiotou, A., ed., Ophiolites: Proceedings, International Ophiolite Symposium, 1979: Nicosia, Cyprus, Geological Survey Department, p. 244–255.

Hawkins, J.W., 1985, Low-K rhyolitic pumice from the Tonga Ridge, *in* Scholl, D.W., and Vallier, T.L., 1985, Geology and Offshore Resources of Pacific Island Arcs-Tonga Region: Houston, Texas, Circum-Pacific Council for Energy and Mineral Resources Earth Science Series, v. 2, p. 171–178.

Hawkins, J.W., 1994, Petrologic synthesis, ODP Leg 135-Lau Basin transect, *in* Hawkins, J.W., Parson, L.M., Allan, J.F., et al., eds., Proceedings, Ocean Drilling Program, Scientific Results, 135: College Station, Texas, Ocean Drilling Program, p. 879–908.

Hawkins, J.W., 1995a, The geology of the Lau Basin *in* Taylor B., ed., Backarc Basins: Tectonics and Magmatism: New York, Plenum Press, p. 63–138.

Hawkins, J.W., 1995b, Evolution of the Lau Basin- Insights from ODP Leg 135, *in* Taylor, B., and Natland, J., eds., Active Margins and Marginal Basins of the Western Pacific: American Geophysical Union Monograph 88, p. 126–174.

Hawkins, J.W., and Allan, J.F., 1994, Petrologic evolution of Lau Basin Sites 834 through 839, *in* Hawkins, J.W., Parson, L.M., Allan, J.F., et al., eds., Proceedings, Ocean Drilling Program, Scientific Results, 125: College Station, Texas, Ocean Drilling Program, p. 427–470

Hawkins, J.W., and Castillo, P.R., 1998, Early history of the Izu-Bonin-Mariana arc system: Evidence from Belau and the Palau Trench: The Island Arc, v. 7, p. 559–578.

Hawkins, J.W., and Evans, C.A., 1983, Geology of the Zambales Range, Luzon, Philippines: ophiolite derived from an island arc-backarc basin pair, *in* Hayes, D.E., ed., The Tectonic Evolution of Southeast Asian Seas and Islands (pt 2): Washington, D.C., American Geophysical Union Memoir 27, p. 124–138.

Hawkins, J.W., and Falvey, D.A., 1985, Petrology of andesite dikes and flows from 'Eua, Tonga, *in* Scholl, D.W., and Vallier, T.L., 1985, Geology and Offshore Resources of Pacific Island Arcs-Tonga Region: Houston, Texas, Circum-Pacific Council for Energy and Mineral Resources Earth Science Series, v. 2, p. 269–280.

Hawkins, J.W., and Florendo, F., 1994, Supra-subduction zone magmatism: Implications for the origin of Philippine ophiolites: Acta Geologica Taiwanica, v. 30, p. 63–171.

Hawkins, J.W., and Helu, S., 1986, Polymetallic sulfide deposit from"black smoker" chimney, Lau Basin: Eos, Transactions of American Geophysical Union, v. 67, p. 378.

Hawkins, J.W., and Melchior, J.T., 1985, Petrology of Mariana Trough and Lau Basin basalts: Journal of Geophysical Research, v. 90, p. 11,431–11,468.

Hawkins, J.W., Sclater, J., and Hohnhaus, G., 1970, Petrologic and geophysical characteristics of the Lau Basin Ridge: A spreading center behind the Tonga Arc: Geological Society of America Abstracts with Programs, v. 2, no. 7, p. 571.

Hawkins, J.W., Bloomer, S.H., Evans, C.A., and Melchior, J.T., 1984, Evolution of intra-oceanic arc-trench systems: Tectonophysics, v. 102, p. 175–205.

Hawkins, J.W., Lonsdale, P.F., Macdougall, J.D., and Volpe, A.M., 1990, Petrology of the axial ridge of the Mariana Trough backarc spreading center: Earth and Planetary Science Letters, v. 100, p. 226–250.

Hawkins, J.W., Parson, L.M., Allan, J.F., et al., eds., 1994, Proceedings, Ocean Drilling Program, Scientific Results, 135: College Station, Texas, Ocean Drilling Program, 984 p.

Hawkins, J.W., Castillo, P., and Lonsdale, P., 2000, Geology of the Tonga Arc-Forearc-Trench system: Eos (Transactions, American Geophysical Union), v. 81, p. F1119.

Hayes, D.E., ed., 1980, The Tectonic Evolution of Southeast Asian Seas and Islands: Washington, D.C., American Geophysical Union Memoir 23, 326 p.

Hayes, D.E., ed., 1983, The Tectonic Evolution of Southeast Asian Seas and Islands (pt 2): Washington, D.C., American Geophysical Union Memoir 27, 396 p.

Hergt, J.M., and Hawkesworth, C.J., 1994, Pb-, Sr-, and Nd-isotopic evolution of the Lau Basin: implications for mantle dynamics during backarc opening, *in* Hawkins, J., Parson, L., Allan, J., et al., eds., Proceedings, Ocean Drilling Program, Scientific Results, 135: College Station, Texas, Ocean Drilling Program, p. 505–518.

Hess, H.H., 1948, Major structural features of the western North Pacific: An interpretation of H.O., 5485, bathymetric chart, Korea to New Guinea: Geological Society of America Bulletin, v. 59, p. 417–466.

Hess, H.H., 1962, History of ocean basins, *in* Engel, A.E.J., James, H.L., and Leonard, B.F., eds., Petrologic Studies: A Volume in honor of A.F., Buddington: New York, Geological Society of America, p. 599–620.

Hickey, R.L., and Frey, F.A., 1982, Rare earth element geochemistry of Mariana forearc volcanics, *in* Hussong, D.M., Uyeda, S., et al., Initial reports of the Deep Sea Drilling Project, v. 60: Washington, D.C., U.S. Government Printing Office, p. 735–742.

Hickey-Vargas, R., 1989, Boninites and tholeiites from DSDP Site 458, Mariana forearc, *in* Crawford, A.J., ed., Boninite: London, Unwin Hyman, p. 340–357.

Hodkinson, R.A., and Cronan, D.S., 1994, Variability in the hydrothermal component of the sedimentary sequence in the Lau backarc basin ODP Leg 135 Sites 834A, 835A, 835A, 837A, 838A and 839A, *in* Hawkins, J., Parson, L., Allan, J., et al., eds., Proceedings, Ocean Drilling Program, Scientific Results, 135: College Station, Texas, Ocean Drilling Program, p. 75–86.

Holmes, A., 1931, Radioactivity and earth movements: Transactions, Geological Society of Glasgow, v. 18, p. 559–606.

Holmes, A., 1965, Principles of Geology, second edition: New York, Ronald Press, 1288 p.

Hopson, C.A., Mattinson, J.M., and Pessagno, E.A., 1981, Coast Range ophiolite, western California, *in* Ernst, W.G., ed., The Geotectonic Development of California, Rubey Volume I: Englewood Cliffs, New Jersey, Prentice Hall, p. 418 –510.

Hussong, D.M., and Uyeda, S., editors, 1981, Initial reports of the Deep Sea Drilling Project Leg 60: Washington, U. S Government Printing Office, 929 p.

Ida, Y., 1983, Convection in the mantle wedge above the slab and the tectonic processes in subduction zones: Journal of Geophysical Research, v. 88, p. 7449–7456.

Ikeda, Y., and Yuasa, M., 1989, Volcanism in nascent back-arc basins behind the Shichito Ridge and adjacent areas in the Izu-Ogasawara arc, northwest Pacific: evidence for mixing between E-type MORB and island arc magmas at the initiation of back-arc rifting: Contributions to Mineralogy and Petrology, v. 101, p. 377–393.

Isacks, B.L., and Barazangi, M., 1977, Geometry of Benioff zones: Lateral segmentation and downwards bending of the subducted lithosphere, *in* Talwani, M., and Pitman, W., eds., Island Arcs, Deep Sea Trenches and Back-Arc Basins: Washington, D.C., American Geophysical Union, Maurice Ewing Series 1, p. 99–114.

Isacks, B.L., Oliver, J., and Sykes, L.R., 1968, Seismology and the new global tectonics: Journal of Geophysical Research, v. 73, p. 5855–5899.

Ishii, T., Robinson, P.T., Maekawa, H., and Fiske, R., 1992, Petrological studies of peridotites from diapiric serpentinite seamounts in the Izu-Ogasawara-Mariana forearcs, *in* Fryer, P., Pearce, J.A., Stokking, L.B., et al., eds., Proceedings, Ocean Drilling Program, Scientific Results, 125: College Station, Texas, Ocean Drilling Program, p. 445–486.

Jenner, G.A., Dunning, G.R., Malpas, J., Brown, M., and Brace, T., 1991, Bay of Islands and Little Port complexes, revisited: age, geochemical and isotopic evidence confirms suprasubduction-zone origin: Canadian Journal of Earth Science, v. 28, p. 1635–1652.

Johnson, L.E., Fryer, P., Taylor, B., Silk, M., Jones, D.L., Sliter, W.V., Itaya, T., and Ishii, T., 1991, New evidence for crustal accretion in the outer Mariana fore arc: Cretaceous radiolarian cherts and mid-ocean ridge basalt-like lavas: Geology, v. 19, p. 811–814.

Jones, G., and Robertson, A.H.F., 1991, Tectono-stratigraphy and evolution of the Mesozoic Pindos ophiolite and related units, northwestern Greece: Journal of the Geological Society of London, v. 148, p. 267–288.

Karig, D.E., 1970, Ridges and basins of the Tonga-Kermadec island arc system: Journal of Geophysical Research, v. 75, p. 239–254.

Kawate, S., and Arima, M., 1998, Petrogenesis of the Tanzawa plutonic complex, central Japan: Exposed felsic middle crust of the Izu-Bonin-Mariana arc: The Island Arc, v. 7, p. 342–358.

Kay, S.M., and Kay, R.W., 1985, Role of crystal cumulates and the oceanic crust in the formation of the lower crust of the Aleutian arc: Geology, v. 13, p. 461–464.

Keunen, Ph. H., 1950, Marine Geology: New York, Wiley, 568 p.

Kikuchi, Y., 1890, On pyroxene compositions in certain volcanic rocks from Bonin Island: Journal of College of Science, Imperial University of Japan, v. 3, p. 67–89.

Kobayashi, K., 1983, Fore-arc volcanism and cycles of subduction: *in* Shimozuru, D., and Yokoyama, I., eds., Arc Volcanism: Physics and Tectonics: Tokyo, Terra Scientific Publishing Co., p. 153–163.

Kroenke, L., Scott, R., et al., eds., 1980, Initial reports of the Deep Sea Drilling Project Leg 59: Washington, D.C., U. S Government Printing Office, 820 p.

Latouche, C., Maillet, N., and Blanchet, R., 1981, X-ray mineralogy studies, Deep Sea Drilling Project Leg 60, *in* Hussong, D.M., Uyeda, S., et al., eds., Initial reports of the Deep Sea Drilling Project Leg 60, p. 437–454.

Lee, J., and Stern, R.J., 1995, Forty million years of magmatic evolution in the Mariana arc: The glass tephra record: Journal of Geophysical Research, v. 100, p. 17,671–17,687.

Leeman, W.R., 1996, Boron and other fluid-mobile elements in volcanic arc lavas: Implications for subduction processes, *in* Bebout, G.E., Scholl, D.W., Kirby, S.H., and Platt, J.P., eds., Subduction Top to Bottom: Washington, D.C., American Geophysical Union Monograph 96, p. 269 –276.

Lin, P-N., Stern, R.J., and Bloomer, S.H., 1989, Shoshonitic volcanism in the northern Mariana Arc 2. Large-ion lithophile and rare earth element abundances: Evidence for the source of incompatible element enrichments in intra-oceanic arcs: Journal of Geophysical Research, v. 94, p. 4497–4514.

Maekawa, H., Shozui, M., Ishii, T., Saboda, K., and Ogawa, Y., 1992, Metamorphic rocks from the serpentinite seamounts in the Mariana and Izu-Ogasawara forearcs, *in* Fryer, P., Pearce, J.A., Stokking, L.B., et al., eds., Proceedings, Ocean Drilling Program, Scientific Results, 125: College Station, Texas, Ocean Drilling Program, p. 415–430.

Marlow, M.S., Johnson, L.E., Pearce, J.A., Fryer, P.B., Pickthorn, L.G., and Murton, B.J., 1992, Pleistocene volcanic rocks in the Mariana forearc arc revealed by drilling at Site 781, *in* Fryer, P., Pearce, J.A., Stokking, L.B., et al., eds., Proceedings, Ocean Drilling Program, Scientific Results, 125: College Station, Texas, Ocean Drilling Program, p. 293–310.

Marshall, P., 1911, Oceania: *in* Handbuch der regionalen Geologie, v. 7, 36 p.

Martinez, F., Fryer, P., Baker, N., and Yamakazi, T., 1995, Evolution of backarc rifting: Mariana Trough, 20°–24° N: Journal of Geophysical Research, v. 100, p. 3807–3827.

Maury, R.C., Sajona, F.G., Pubellier, M., Bellon, H., et Defant, M.J., 1996, Fusion de la croute ocearique dans les zones de subduction/collision recentes: l'exemple de Mindanao (Philippines): Bulletin de la Societe Geologique de France, t. 167, p. 579–595.

McDougall, I., 1994, Dating of rhyolite glass in the Tonga forearc (Hole 841 B), *in* Hawkins, J., Parson, L., Allan, J., et al., eds., Proceedings, Ocean Drilling Program, Scientific Results, 125: College Station, Texas, Ocean Drilling Program, p. 923–924.

McKenzie, D., and Bickle, M.J.,1988, The volume and composition of melt generated by extension of lithosphere: Journal of Petrology, v. 29, p. 625–679.

Meijer, A., 1976, Pb and Sr isotopic data bearing on the origin of volcanic rocks from the Mariana island arc system: Geological Society of America Bulletin v. 87, p. 1358–1369.

Meijer, A., 1980, Primitive arc volcanism and a boninite series: Examples from western Pacific island arcs, *in* Hayes, D.E., ed., The Tectonic Evolution of Southeast Asian Seas and Islands: Washington, D.C., American Geophysical Union Memoir 23, p. 272–282.

Meijer, A., 1983, The origin of low-K rhyolites from the Mariana frontal arc: Contributions to Mineralogy and Petrology, v. 83, p. 45–51.

Meijer, A., and Reagan, M., 1981, Petrology and geochemistry of the island of Sarigan in the Mariana Arc: Calc-alkaline volcanism in an oceanic setting: Contributions to Mineralogy and Petrology, v. 77, p. 37–354.

Metcalf, R.V., Wallin, E.T., Willse, K.R., and Muller, E.R., 2000, Geology and geochemistry of the ophiolitic Trinity terrane, California: Evidence of middle Paleozoic depleted supra-subduction zone magmatism in a proto-arc setting, *in* Dilek, Y., Moores, E.M., Elthon, D., and Nicolas, A., eds., Ophiolite and Oceanic Crust: New insights from field studies and the Ocean Drilling Program: Boulder, Colorado, Geological Society of America Special Paper 349, p. 403–418.

Menzies, M., Blanchard, D., and Jacobs, J., 1977, Rare earth and trace element chemistry of metabasalts from the Point Sal ophiolite: Earth and Planetary Science Letters, v. 37, p. 203–215.

Minear, J.W., and Toksoz, M.N., 1970, Thermal regime of a downgoing slab and new global tectonics: Journal of Geophysical Research, v. 75, p. 1397–1419.

Miyashiro, A., 1973, The Troodos ophiolite was probably formed in an island arc: Earth and Planetary Science Letters, v. 19, p. 218–224.

Miyashiro, A., 1974, Volcanic rock series in island arcs and active continental margins: American Journal of Science, v. 274, p. 321–355.

Moberly, R., 1972, Origin of lithosphere behind island arcs with reference to the western Pacific: Boulder, Colorado, Geological Society of America Memoir 132, p. 35–55.

Moores, E.M., and Jackson, E.D., 1974, Ophiolites and oceanic crust: Nature, v. 250, p. 136 –138.

Moores, E.M., and Vine, F.J., 1971, The Troodos massif Cyprus as oceanic crust: evaluation and implications: Philosophical Transactions Royal Society of London, ser. A, v. 268, p. 443–466.

Moores, E.M., Kellog, L.H., and Dilek, Y., 2000, Tethyan ophiolites, mantle convection, and tectonic "historical contingency": A resolution of the "ophiolite conundrum," *in* Dilek, Y., Moores, E.M., Elthon, D., and Nicolas, A., eds., Ophiolite and Oceanic Crust: New insights from field studies and the Ocean Drilling Program: Boulder, Colorado, Geological Society of America Special Paper 349, p. 3–12.

Morgan, W.J., 1968, Rises, trenches, great faults, and crustal blocks: Journal of Geophysical Research, v. 75, p. 1397–1419.

Morris, J.D., Leeman, W.P., and Tera, F., 1990, The subducted component in island arc lavas: constraints from Be isotopes and B–Be systematics: Nature, v. 344, p. 31–36.

Mrzowski, C.L., and Hayes, D.E., 1980, A seismic reflection study of faulting in the Mariana fore arc, *in* Hayes, D.E., eds., The Tectonic Evolution of Southeast Asian Seas and Islands: Washington, D.C., American Geophysical Union Memoir 23, p. 223–234.

Natland, J.H., 1981, Petrography and mineral composition of gabbros recovered in DSDP Hole 453 on the western side of the Mariana Trough, *in* Hussong, D.M., Uyeda, S., et al., Initial Reports of the Deep Sea Drilling Project, v. 60, p. 579–599.

Noiret, G., Montigny, R., and Allegre, C., 1981, Is the Vourinos Complex an island arc ophiolite ?: Earth and Planetary Science Letters, v.56, p. 357–386.

Ohara, Y., Stern, R.J., Ishii, T., Yurimoto, H., Yamakazi, T., 2002, Peridotites from the Mariana Trough: first look at the mantle beneath an active back-arc basin: Contributions to Mineralogy and Petrology, v. 143, p. 1–18.

O'Neill, H.C., and Palme, H., 1998, Composition of the silicate earth: Implications for accretion and core formation, *in* Jackson, I., ed., The Earth's Mantle: New York, Cambridge University Press, 566 p.

Oversby, V.M., and Ewart, A., 1972, Lead isotopic compositions of Tonga-Kermadec volcanics and their petrogenetic significance: Contributions to Mineralogy and Petrology, v. 37, p. 181–210.

Oxburgh, E.R., and Turcotte, D.L., 1970, Thermal structure of island arcs: Geological Society of America Bulletin, v. 81, p. 1665–1688.

Packham, G.H., and Falvey, D.A., 1971, An hypothesis for the formation of marginal seas in the western Pacific: Tectonophysics, v. 11, p. 79–109.

Panayiotou, A., ed., 1980, Ophiolites: Proceedings, International Ophiolite Symposium, 1979: Nicosia, Cyprus, Geological Survey Department, 781 p.

Parson, L.M., Rothwell, R.G., and MacLeod, C.J., 1994, Tectonics and sedimentation in the Lau Basin (southwest Pacific), *in* Hawkins, J.W., Parson, L.M., Allan, J.F., et al., eds., Proceedings, Ocean Drilling Program, Scientific Results, 135: College Station, Texas, Ocean Drilling Program, p. 9–22.

Peacock, S.M., 1990, Fluid processes in subduction zones: Science, v. 248, p. 247–336.

Pearce, J.A., 1975, Basalt geochemistry used to investigate past tectonic environments on Cyprus: Tectonophysics, v. 25, p. 41–67.

Pearce, J.A., 1980, Geochemical evidence for the genesis and eruptive setting of lavas from Tethyan ophiolites, *in* Panayiotou, A., ed., Ophiolites: Proceedings, International Ophiolite Symposium, 1979: Nicosia, Cyprus, Geological Survey Department, p. 261–272.

Pearce, J.A., and Cann, J.R., 1973, Tectonic setting of basic volcanic rocks determined using trace element analyses: Earth and Planetary Science Letters, v.19, p. 290–300.

Pearce, J.A., and Norry, M.J., 1979, Petrogenetic implications of Ti, Zr, Y and Nb variations in volcanic rocks: Contributions to Mineralogy and Petrology, v. 37, p. 33–47.

Pearce, J.A., Alabaster, T., Shelton, A.W., and Searle, M.P., 1981, The Oman ophiolite as a Cretaceous arc-basin complex: evidence and implications: Philosophical Transactions Royal Society of London, ser. A, v. 300, p. 299–317.

Pearce, J.A., Lippard, S.J., and Roberts, S., 1984, Characteristics and tectonic significance of supra-subduction zone ophiolites, *in* Kokelaar, B.P., and Howells, M.F., eds., Marginal Basin Geology: Geological Society Special Publication 16, London, Blackwell, p. 77–94.

Pearce, J.A., van der Laan, S.R., Arculus, R.J., Murton, B.J., Ishii, T., Peate, D.W., and Parkinson, I., 1992, Boninite and harzburgite from Leg 125 (Bonin-Mariana forearc): A case study of magmagenesis during the initial stages of subduction, *in* Fryer, P., Pearce, J.A., Stokking, L.B., et al., eds., Proceedings, Ocean Drilling Program, Scientific Results, 125: College Station, Texas, Ocean Drilling Program, p. 623–659.

Pearce, J.A., Ernewein, M., Bloomer, S.H., Parson, L.M., Murton, B.J., and Johnson, L.E., 1995, Geochemistry of Lau Basin volcanic rocks: influence of ridge segmentation and arc proximity, *in* Smellie, J, ed., Volcanism Associated with Extension at Consuming Plate Margins: London, Geological Society of London Special Publication 81, p. 53–75.

Peate, D.W., and Pearce, J.A., 1998, Causes of spatial compositional variations in Mariana arc lavas: Trace element variations: The Island Arc, v. 7, p. 479–495.

Perfit, M.R., Gust, D.A., Bence, A.E., Arculus, R.J., and Taylor, S.R., 1980, Chemical characteristics of island-arc basalts: Implications for mantle sources: Chemical Geology, v. 30, p. 227–256.

Peters, Tj., Nicolas, A., and Coleman, R.G., eds., 1991, Ophiolite Genesis and Evolution of the Oceanic Lithosphere: Dordrecht, Kluwer Academic Publishers, 903 p.

Poreda, R., and Craig, H., 1993, He and Sr isotopes in the Lau basin mantle: Depleted and primitive mantle components: Earth and Planetary Science Letters, v. 119, p. 319–329.

Price, R.C., Johnson, L.E., and Crawford, A.J., 1990, Basalts of the North Fiji Basin: the generation of back arc basin magmas by mixing of depleted and enriched mantle sources: Contributions to Mineralogy and Petrology, v. 105, p. 106–121.

Raitt, R.W., Fisher, R.L., and Mason, R.G., 1955, Tonga Trench *in* Poldervaart, A. ed. Crust of the Earth: Boulder, Colorado, Geological Society of America Special Paper 62, p. 217 –254.

Ramsay, W.R.H., Crawford, A.J., and Foden, J.F., 1984, Field setting, mineralogy, and genesis of arc picrites, New Georgia, Solomon Islands: Contributions to Mineralogy and Petrology, v. 88, p. 386–402.

Reagan, M.K., and Meijer, A., 1984, Geology and geochemistry of early arc-volcanic rocks from Guam: Geological Society of America Bulletin, v. 95, p. 701 –713.

Rehrig, W.A., 1986, Processes of regional Tertiary extension in the western Cordillera: Insights from the metamorphic core complexes, *in* Mayer, L. ed., Extensional Tectonics of the Southwestern United States: A Perspective on Processes and Kinematics: Boulder, Colorado, Geological Society of America Special Paper 208, p. 97–122.

Ringwood, A.E., 1969, Composition and evolution of the upper mantle, *in* Hart, P.J., ed., The Earth's Crust and Upper Mantle: American Geophysical Union, Geophysical Monograph 13, p. 1–17.

Ringwood, A.E., 1974, The petrological evolution of island arc systems: Journal of Geological Society of London, v. 130, p. 183–204.

Ringwood, A.E., 1975, Composition and Petrology of the Earth's Mantle: New York, McGraw Hill, 604 p.

Robertson, A.H.F., and Hudson, J.D., 1973, Cyprus umbers: Chemical precipitates on a Tethyan ocean ridge: Earth and Planetary Science Letters, v.18, p. 93–101.

Robinson, P.T., Melson, W.G., O'Hearn, T., Schmincke, H.U., 1983, Volcanic glass compositions from the Troodos ophiolite, Cyprus: Geology, v.11, p. 400–404.

Rodolfo, K.S., 1969, Bathymetry and marine geology of the Andaman Basin, and tectonic implications for southeast Asia: Geological Society of America Bulletin, v. 80, p. 1203–1230.

Rothwell, R.G., Bednarz, U., Boe, R., Clift, P.D., Hodkinson, R.A., Ledbetter, J.K., Pratt, C.E., and Soakai, S., 1994, Sedimentation and sedimentary processes in the Lau backarc basin: Results from Leg 135, *in* Hawkins, J.W., Parson, L.M., Allan, J.F., et al., eds., Proceedings, Ocean Drilling Program, Scientific Results, 135: College Station, Texas, Ocean Drilling Program, p. 829–842.

Salisbury, M.H., and Christensen, N.K., 1978, The seismic velocity structure of a traverse through the Bay of Islands ophiolite complex, Newfoundland, an exposure of oceanic crust and upper mantle: Journal of Geophysical Research, v. 83, p. 805–817.

Saunders, A.D., Tarney, J., Marsh, N.G., and Wood, D.A., 1980, Ophiolites as ocean crust or marginal basin crust: A geochemical approach, *in* Panayiotou, A., ed., Ophiolites: Proceedings, International Ophiolite Symposium, 1979: Nicosia, Cyprus, Geological Survey Department, p. 193–204.

Sharaskin A. Ya., Dobretsov, N.L., and Sobolev, N.V., 1980, Marianites: the clinoenstatite bearing pillow-lavas with the ophiolite assemblage of Mariana Trench, *in* Panayiotou, A., ed., Ophiolites: Proceedings, International Ophiolite Symposium, 1979: Nicosia, Cyprus, Geological Survey Department, p. 473–479.

Schmincke, H.U., Rautenschlein M., Robinson, P., and Mehegan, J.M., 1983, Troodos extrusive series of Cyprus: A comparison with oceanic crust: Geology, v. 11, p. 405–409.

Shervais, J.W., 1982, Ti -V plots and the petrogenesis of modern and ophiolitic lavas: Earth and Planetary Science Letters, v. 59, p. 101–118.

Shervais, J.W., and Kimbrough, D.L., 1985, Geochemical evidence for the tectonic setting of the Coast Range ophiolite: A composite island arc-oceanic crust terrane in western California: Geology, v. 13, p. 35–38.

Sigurdsson, I.A., Kamenetsky, V.S., Crawford, A.J., Eggins, S.M., and Zlobin, S.K., 1993, Primitive island arc and oceanic lavas from the Hunter ridge-Hunter fracture zone. Evidence from glass, olivine, and spinel compositions: Mineralogy and Petrology, v. 47, p. 149–169.

Sinton, J., and Fryer, P., 1987, Mariana Trough lavas from 18 N; Implications for the origin of backarc basin basalts: Journal of Geophysical Research, v. 92, p. 12,782–12,802.

Sleep, N., and Toksoz, M.N., 1971, Evolution of marginal basins, Nature, v. 233, p. 548–550.

Smith, G., Wiens, D.A., Fischer, K.M., Dorman, L.M., Webb, S.C., and Hildebrand, J.A., 2001, A complex pattern of mantle flow in the Lau backarc: Science, v. 292, p. 713–716.

Stein, S., and Stein, C.A., 1996, Thermo-mechanical evolution of oceanic lithosphere: Implications for the subduction process and deep earthquakes, *in* Bebout, G., E., Scholl, D.W., Kirby, S.H., and Platt, J.P., eds., Subduction Top to Bottom: American Geophysical Union Monograph 96, p. 1 –18.

Steinmann, G., 1927, Die ophiolitischen Zonen in der mediterranean Kettengebirgen: International Geological Congress, 14th, Madrid, Proceedings, v. 2, p. 638–667.

Stern, R.J., 1979, On the origin of andesite in the northern Mariana Island Arc: Implications from Agrigan: Contributions to Mineralogy and Petrology, v. 68, p. 207–219.

Stern, R.J., 2002, Subduction zones: Reviews of Geophysics, v. 40, 10.1029/2001 RG 000108.

Stern, R.J., and Arima, M., editors,1998, Special Issue; Geophysical and geochemical studies of the Izu-Bonin-Mariana Arc: The Island Arc, v. 7, p. 295–607.

Stern, R.J., and Bloomer, S. H, 1992, Subduction zone infancy: Examples from the Eocene Izu-Bonin-Mariana and Jurassic California arcs: Geological Society of America Bulletin, v.104, p. 1621–1636.

Stern, R.J., Bloomer, S.H., Lin, P-N., and Smoot, N.C., 1989, Submarine arc volcanism in the southern Mariana Arc as an ophiolite analogue: Tectonophysics, v. 168, p. 151–170.

Stern, R.J., Lin P. N, Morris, J.D., Jackson, M.C., Fryer, P., Bloomer, S.H., and Ito. E., 1990, Enriched back-arc basin basalts from the northern Mariana Trough: implications for the magmatic evolution of back-arc basins: Earth and Planetary Science Letters, v. 100, p. 210–225.

Stern, R.J., Morris, J., Bloomer, S.H., and Hawkins, J.W., 1991, The source of the subduction component in convergent margin magmas: Trace element and radiogenic isotope evidence from Eocene boninites, Mariana forearc: Geochimica et Cosmochimica Acta, v. 55, p. 1467–1481.

Stern, R.J., Bloomer, S.H., Martinez, F., Yamazaki, Y., and Harrison, T.M., 1996, The composition of back-arc basin lower crust and upper mantle in the Mariana Trough: A first report: The Island Arc, v. 5., p. 354–372.

Stolper, E., and Newman, S., 1994, The role of water in the petrogenesis of Mariana trough magmas: Earth and Planetary Science Letters, v. 121, p. 293–325.

Stolper, E., and Walker, D., 1980, Melt density and the average composition of basalt: Contributions to Mineralogy and Petrology, v. 74, p. 7–12.

Straub, S.M., 1995, Contrasting compositions of Mariana Trough fallout tephra and Mariana Island arc volcanics: a fractional crystallization link: Bulletin of Volcanology, v. 57, p. 403–421.

Sturm, M.E., Klein E.M., Karsten, J.L., and Karson, J.A., 2000, Evidence for subduction- related contamination of the mantle beneath the southern Chile Ridge: Implications for ambiguous ophiolite compositions, *in* Dilek, Y., Moores, E.M., Elthon, D., and Nicolas, A., eds., Geological Society of America Special Paper 349, p. 13 –20.

Sun, S.S., and McDonough, W.F., 1989, Chemical and isotopic systematics of oceanic basalts: Implications for mantle composition and processes, *in* Magmatism in the Ocean Basins, Saunders, A.D., and Norry, M.J., eds., Geological Society Special Publication, London, v. 42, p. 323–346.

Sunkel, G., 1990, Origin of petrological and geochemical variations of Lau Basin lavas (SW Pacific): Marine Mining, v. 9, p. 205–234.

Suyehiro, K., et al., 1996, Continental crust, crustal underplating, and low-Q upper mantle beneath an oceanic island arc: Science, v. 272, p. 390–392.

Tatsumi, Y., 1982, Origin of high-magnesian andesites in the Setouchi volcanic belt, southwest Japan, II, Melting phase relations at high pressure: Earth and Planetary Science Letters, v. 60, p. 305–317.

Tatsumi, Y., and Eggins, S.E., 1995, Subduction Zone Magmatism: Blackwell Science, Oxford, England, 211 p.

Taylor, B., 1992, Rifting and the volcanic-tectonic evolution of the Izu-Bonin-Mariana Arc, *in* Taylor, B., Fujioka, K., et al., eds., Proceedings ODP Science Results, 126, Ocean Drilling Program: College Station Texas, p. 627–651.

Taylor, B., ed., 1995, Backarc Basins: New York, Plenum Press, 524 p.

Taylor, B., and Natland, J., eds., 1995, Active Margins and Marginal Basins of the Western Pacific: Washington, D.C., American Geophysical Union Monograph 88, 417 p.

Taylor, B, Fujioka, K., et al., editors, 1992, Proceedings, Ocean Drilling Program, Scientific Results, 126: College Station, Texas, Ocean Drilling Program, 709 p.

Taylor, R.N., and Nesbitt, R.W., 1995, Arc volcanism in an extensional regime at the initiation of subduction: a geochemical study of Hahajima, Bonin Islands, Japan, *in* Smellie, J, ed., Volcanism Associated with Extension at Consuming Plate Margins: Geological Society of London Special Publication 81, p. 115–134.

Taylor, R.N., Murton, B.J., and Nesbitt, R.W., 1992, Chemical transects across intra-oceanic arcs: implications for the tectonic setting of ophiolites, *in* Parson, L.M., Murton, B.J., and Browning, P., eds., Ophiolites and their Modern Oceanic Analogues Geological Society of London Special Publication 60, p. 117–132.

Taylor, R.N., Nesbitt, R.W., Vidal, P., Harmon, R.S., Auvray, B., and Croudace, I., 1994, Mineralogy, chemistry, and genesis of the boninite series volcanics, Chichijima, Bonin Islalnds, Japan: Journal of Petrology, v. 35, p. 577–617.

Tera, F., Brown, L., Morris, J., Sacks, S., Klein J., and Middleton, R., 1986, Sediment incorporation in island arc magmas: Inferences from ^{10}Be: Geochimica et Cosmochimica Acta, v. 50, p. 535–550.

Thy, P., and Xenophontos, C., 1991, Crystallization orders and phase chemistry of glassy lavas from the pillow sequences, Troodos ophiolite, Cyprus: Journal of Petrology, 32, p. 403–424.

Toksoz, M.N., and Bird, P., 1977, Formation and evolution of marginal basins and continental plateaus, *in* Talwani, M., and Pitman, W., eds., Island Arcs, Deep Sea Trenches and Back-Arc Basins: Washington, D.C., American Geophysical Union, Maurice Ewing Series 1, p. 379–394.

Uyeda, S., and Kanamori, H., 1979, Backarc opening and the mode of subduction: Journal of Geophysical Research, v. 84, p.1049–1061.

Vallier, T.L., Stevenson, A.J., and Scholl, D.W., 1985, Petrology of igneous rocks from Ata Island, Kingdom of Tonga, *in* Scholl, D.W., and Vallier, T.L., 1985, Geology and Offshore Resources of Pacific Island Arcs-Tonga Region: Houston, Texas, Circum-Pacific Council for Energy and Mineral Resources Earth Science Series, v. 2, p. 301–306.

Vallier, T.L., Scholl, D.W., Fisher, M.A., Bruns, T.R., Wilson, F.H., von Heune, R., and Stevenson, A.J., 1994, Geologic framework of the Aleutian arc, Alaska, *in* Plafker, G., and Berg, H.C., eds., The Geology of Alaska: Boulder, Colorado, Geological Society of America, The Geology of North America, v. G-1, p. 367–388.

van der Hilst, R., 1995, Complex morphology of subducted lithosphere in the mantle beneath the Tonga trench: Science, v. 374, p. 154–157.

Varne, R., Brown, A., and Falloon, T., 2000, Macquarie Island: Its geology, structural history and the timing and tectonic setting of its N-MORB to E-MORB magmatism, *in* Dilek, Y., Moores, E.M., Elthon, D., and Nicolas, A., eds., Ophiolites and Oceanic Crust: Geological Society of America Special Paper 349, p. 301–320.

Viereck, L.G., Flower, M.F.J., Hertogen, J., Schmincke, H.-U., and Jenner, G.A., 1989, The genesis and significance of N-MORB sub-types: Contributions to Mineralogy and Petrology, v. 102, p. 112–126.

Volpe, A.M., Macdougall, J.D., and Hawkins, J.W., 1987, Mariana Trough basalts (MTB): trace element and Sr-Nd isotopic evidence for mixing between MORB-like and Arc-like melts: Earth and Planetary Science Letters, v. 82, p. 241–254.

Volpe, A.M., Macdougall, J.D., and Hawkins, J.W., 1988, Lau Basin basalt (LBB): trace element and Sr-Nd isotopic evidence for heterogeneity in backarc basin mantle: Earth and Planetary Science Letters, v. 90, p.174–186.

Volpe, A.M., Macdougall, J.D., Lugmair, G., Hawkins, J.W., and Lonsdale, P.F., 1990, Fine-scale isotopic variation in Mariana Trough basalts: evidence for heterogeneity and a recycled component in backarc basin mantle: Earth and Planetary Science Letters, v. 100, p. 251–264.

von Stackleberg, U., and Shipboard Party, 1988, Active hydrothermalism in the Lau backarc basin (SW Pacific): First results from the *Sonne 48* cruise (1987): Marine Minerals, v. 7, p. 431–442.

Wallin, E.T., and Metcalf, R.V., 1998, Supra-subduction zone ophiolite formed in an extensional forearc: Trinity Terrane, Klamath Mountains, California: Journal of Geology, v. 106, p. 591–608.

Weaver, S.D., Saunders, A.D., Pankhurst, R.J., and Tarney, J., 1979, A geochemical study of magmatism associated with the initial stages of back-arc spreading: Contributions to Mineralogy and Petrology, v. 68, p. 151–169.

Wegener, A. 1915, Die Entstehung der Kontinente und Ozeane, Braunschweig, F. Vieweg & Sohn, 242 p.

Wegener, A., 1966, The Origin of Continents and Oceans (translated by J. Biram): New York, Dover Publications, 246 p.

Whelan, P.M., Gill, J.B., Kollman, E., Duncan, R., and Drake, R.E., 1985, Radiometric dating of magmatic stages in Fiji, *in* Scholl, D.W., and Vallier, T.L., 1985, Geology and Offshore Resources of Pacific Island Arcs-Tonga Region: Houston, Texas, Circum-Pacific Council for Energy and Mineral Resources Earth Science Series, v. 2, p. 415–440.

Wood, D.A., Marsh, N.G., Tarney, J., Joron, L.J., Fryer, P., and Treuil, M., 1981, Geochemistry of igneous rocks recovered from a transect across the Mariana trough, arc, fore-arc, and trench, sites 453 through 461, Deep Sea Drilling Project Leg 60, *in* Hussong, D.M., Uyeda, S., et al., eds., Initial Reports of the Deep Sea Drilling Project, v. 60, p. 611–645.

Woodhead, J.D., 1988, The origin of geochemical variations in Mariana lavas: A general model for petrogenesis in intra-oceanic island arcs: Journal of Petrology, v. 29, p. 805–830.

Woodhead, J.D., 1989, Geochemistry of the Mariana arc (western Pacific): Source composition and processes: Chemical Geology, v. 76, p. 1–24.

Woodhead, J.D., Eggins, S., and Gamble, J., 1993, High field strength and transition element systematics in island arc and back-arc basin basalts: evidence for multi-phase melt extraction and a depleted mantle wedge: Earth and Planetary Science Letters, v. 114, p. 491–504.

You, C.F., Castillo, P.R., Gieskes, J.M., and Spivack, A.J., 1996, Trace element behavior in hydrothermal experiments: Implications for fluid processes at shallow depths in subduction zones: Earth and Planetary Science Letters, v. 140, p. 41–52.

Zhao, D., Xu,Y., Wiens, D. A, Dorman, L., Hildebrand, J., and Webb, S., 1997, Depth extent of the Lau back-arc spreading center and its relation to subduction processes: Science, v. 278, p. 254–257.

Manuscript Accepted by the Society May 2, 2003

Printed in the USA

Geological Society of America
Special paper 373
2003

Supra-subduction zone ophiolites: The search for modern analogues

Julian A. Pearce
School of Earth, Ocean, and Planetary Sciences, Cardiff University, Cardiff CF10 3YE, UK

ABSTRACT

The development of ideas on modern analogues for ophiolite complexes can be divided into three parts. In the first (1963 to 1972), ophiolites were mainly viewed as on-land analogues of the crust and upper mantle formed at mid-ocean ridges. The second (from 1972 to 1984) is a period of paradigm shift to the view that the majority of ophiolites forms above subduction zones, and can be grouped as the class known as supra-subduction zone ophiolites. The third (from 1984 onward) is a period that has led to a much better understanding of what these supra-subduction zone settings represent. The realization that many ophiolites were linked to subduction resulted from a combination of advances in geochemical discrimination methods, exploration through dredging and drilling of arc-basin systems of the Western Pacific and South Atlantic, and further field studies in ophiolite terranes. The transition in ideas from a mid-ocean ridge to a subduction origin was prolonged and much-debated, primarily because many of the best-developed ophiolites had the geochemical characteristics of subduction environments, yet none of the geological characteristics (arc volcanics and overlying volcanogenic sediments) thought to characterize those environments. This in turn led to a need for a non-generic term, supra-subduction zone, to describe this ophiolite type. Since the supra-subduction zone concept became established, precise analogues continued to be debated. Subduction initiation and ridge-trench intersections are probably the most important settings, with back-arc rifting and spreading and oblique subduction also significant.

Keywords: ophiolites, geochemistry, mid-ocean ridges, subduction zones, marginal basins.

INTRODUCTION

Supra-subduction zone (SSZ) ophiolites have the structures or inferred structures of oceanic crust, yet a geochemical composition which indicates that they formed not at mid-ocean ridges but at spreading centers associated with subduction zones. Such ophiolites are common in the geological record; in fact, they are much more common than ophiolites from mid-oceanic ridges themselves.

The recognition of SSZ ophiolites in the geological record has not been a sudden (Kuhnian) jump in knowledge and understanding. From the time that ophiolites were first recognized as land-based analogues of oceanic lithosphere, some 10 years elapsed before the presentation of evidence that ophiolites may not all form at mid-ocean ridges. Another 10 years were occupied by controversy over whether they had a mid-ocean ridge or subduction-related origin. By the end of this time, accumulation of new data from land and the oceans provided sufficient evidence that both options were possible and led to the publication of a formal definition of supra-subduction zone ophiolite. The remaining period up to the time of this writing has seen the emergence of the significance and diversity of the SSZ ophiolite type.

My account begins in 1971, at which time ophiolites had been highlighted as on-land slices of oceanic lithosphere with the capacity to ground-truth marine geophysical models of oceanic crust and document the seafloor spreading process. The "jewel in the crown" was the Troodos Massif of Cyprus, mapped in detail by the Cyprus Geological Survey, and exhibiting the full ophiolite stratigraphy, including an extensive sheeted dike swarm, the key piece of evidence for a seafloor spreading origin (Gass, 1963; Moores and Vine, 1971). Ophiolites were about to be defined formally by the Penrose Conference Participants (Anonymous, 1972).

Pearce, J.A., 2003, Supra-subduction zone ophiolites: The search for modern analogues, *in* Dilek, Y., and Newcomb, S., eds., Ophiolite concept and the evolution of geological thought: Boulder, Colorado, Geological Society of America Special Paper 373, p. 269–293. For permission to copy, contact editing@geosociety.org.

In this paper, I document the increasing awareness of the diversity of ophiolites and the subduction signal that many exhibit. I report how exploration of marginal basins transformed our knowledge of ophiolite analogues, and describe the increasing part played by geochemical discriminant diagrams in fingerprinting the subduction signal. I explain how the increase in knowledge of ophiolites, modern analogues, and geochemistry led, by 1984, to the supra-subduction zone concept. I then trace the subsequent efforts to find more exact analogues for this class of ophiolites. Finally, I attempt an analysis of the evolution of ideas from Kuhnian and Bayesian perspectives.

I have also included two figures that present the conclusions reached in the text in graphical form. Figures 1A and B illustrate some of the key advances in our concept of modern analogues. Figure 2 illustrates some of the key advances in the fingerprinting of SSZ settings through geochemical analysis of the various rocks that make up the ophiolite sequence.

RECOGNITION OF OPHIOLITE DIVERSITY AND SUBDUCTION INFLUENCE

The Troodos Debate

The ophiolite definition published by the Penrose Conference participants in 1972 was the culmination of a decade of study of ophiolite complexes, reflecting the consensus that ophiolites are fragments of oceanic lithosphere formed at mid-ocean ridges by seafloor spreading (Anonymous, 1972). So strong was this consensus that, in our initial attempts at geochemical discrimination of tectonic setting, Joe Cann and I (Pearce and Cann, 1971, 1973) used lavas from known tectonic settings (arcs, ridges, and intra-plate) to define discriminant fields, and used lavas from ophiolites to test the method. We assumed that, if the method of discrimination were correct, the ophiolites would plot in the "ocean floor basalt" field. In the event, our first discriminant diagrams were based solely on the elements Ti, Zr, and Y. On these diagrams, the Troodos and Oman ophiolites, as well as greenstones from the "Steinmann trinity" in the Austrian Alps, did plot predominantly in the "ocean floor basalt" field. Only a subset of Greek ophiolite samples plotted in the island arc field, and we interpreted these as not part of a true ophiolite. Thus, we assumed that the method had worked successfully, the results having shown that "ophiolites [including the Troodos Massif] had ocean floor affinities."

Whereas we had worked primarily on dredged rocks from ocean ridges, Akiho Miyashiro had worked for much of his career on subduction-related rocks from Japan. He saw strong similarities in major element geochemistry between the Troodos lavas and the volcanic rocks of the Izu-Bonin arc south of Japan and concluded that the Troodos Massif formed in an island arc (Miyashiro, 1973). He developed this theme further in *Classification, characteristics, and origin of ophiolites* (Miyashiro, 1975a). Essentially, he argued that the Troodos lavas have a greater range of silica contents than do dredged rocks from ocean ridges, and a lesser degree of iron-enrichment that resembles the calc-alkaline rather than tholeiitic series. He was the first to propose a subduction-related origin for the Troodos Massif, and thus raise the question of whether all ophiolites really formed at mid-ocean ridges.

Miyashiro had never visited Cyprus, however, nor viewed the rocks in the field or in thin section. He abstracted published major element data and focused on the elements (Si, Mg, and Fe) that were successful in characterizing island arc lavas in the 1960s. Many of the principal ophiolite scientists of the time, most of whom had worked on Cyprus, queried his approach in a series of discussion papers in *Earth and Planetary Science Letters* by Hynes (1975); Moores (1975); Gass et al. (1975a), and Church and Coish (1976). Miyashiro attempted to deflect these criticisms in his replies to Hynes (Miyashiro, 1975b) and Moores (Miyashiro, 1975c). Some of the key elements of this scientific exchange are reported below.

> ...about one-third of the analyzed rocks of the lower pillow lavas and sheeted complex in it follows a calc-alkalic trend. This strongly suggests that the massif was created as a basaltic volcano in an island arc with a relatively thin oceanic-type crust rather than a mid-ocean ridge. Miyashiro (1973, p. 218)

> In view of the alteration described it is not surprising that the data from Troodos produce a broad scatter on Miyashiro's diagrams. Clearly there is a strong possibility of chemical mobility. Any direct comparison of these chemical data with an island-arc igneous suite and with unaltered rocks from the ocean floors can have little validity. Hynes (1975, p. 213)

> In spite of his denial of its possibility, much of the bulk chemical variations cited as evidence for calc-alkaline origin by Miyashiro appears to be due in fact to metasomatic alteration. Moores (1975, p. 223)

> The Sheeted Intrusive Complex, measured across strike, is 120 km of nothing but near-vertical dykes. Miyashiro dismisses this as "overlapping vent volcanoes." Having, between us seen several hundred central vent volcanoes in various states of erosion and tectonic setting, we venture to suggest that if Professor Miyashiro were to visit Troodos he might change his opinion. Gass et al. (1975a, p. 237)

> It is not clear whether the sheeted complex in Troodos may be regarded as a result of creation of new crust by spreading or some spreading-like process. Even if we assume it to be, it is not consistent with the island arc setting of the Troodos Massif. Spreading or some spreading-like process may well occur along island arcs, as was suggested by many authors in order to explain the origin of marginal seas and inter-arc basins. Miyashiro (1975a, p. 221)

> [Miyashiro's] conclusion is, however, contradicted by the observation that the cumulate units of all such ophiolites directly overlie mantle peridotite rather than basaltic oceanic crust (gabbro, dikes, pillow lavas). Church and Coish (1976, p. 13)

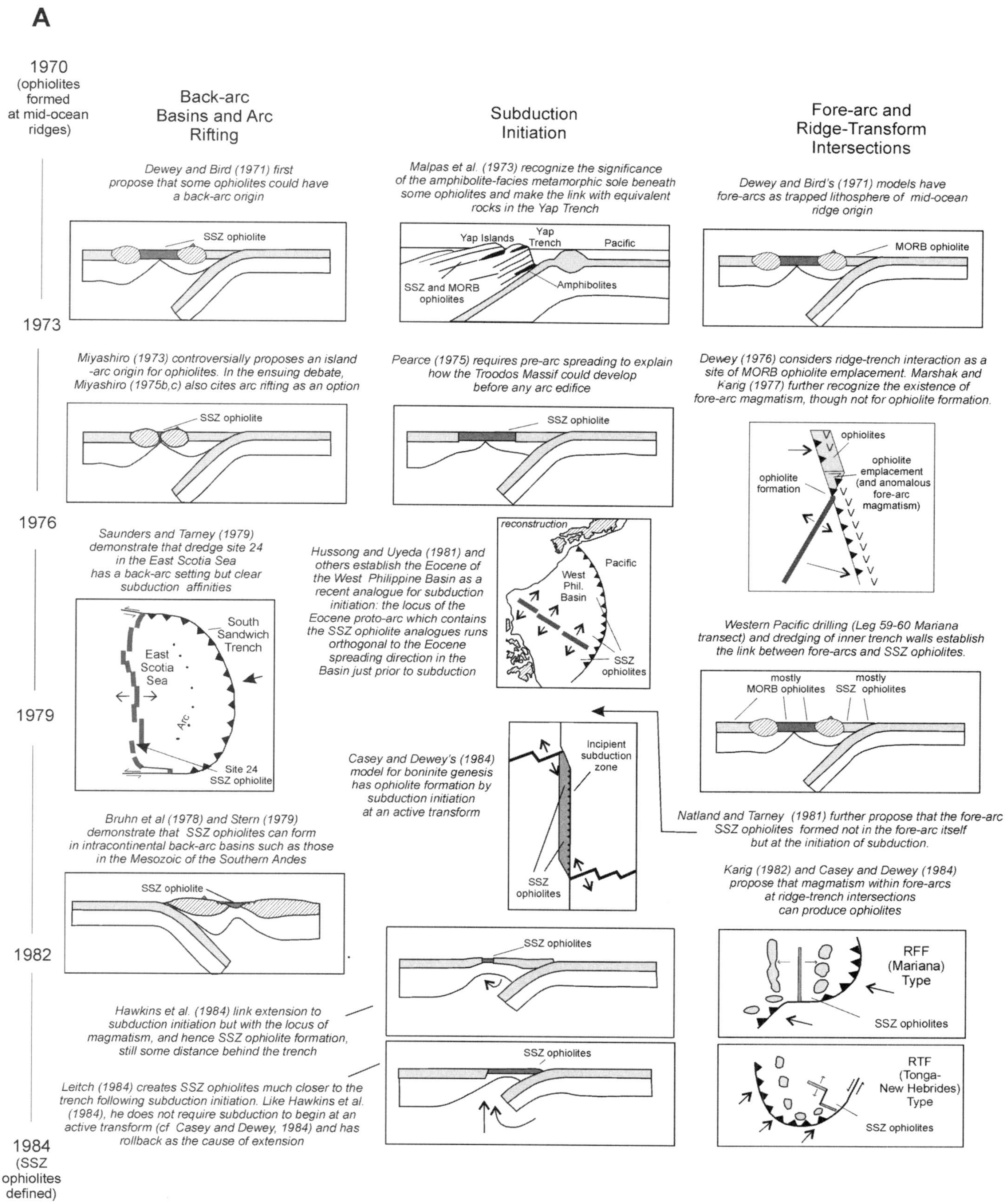

Figure 1 (*on this and following page*). Evolution of ideas on modern analogues for supra-subduction zone (SSZ) ophiolite formation, 1970 to 1984 (A), and 1984 to 2000 (B). Illustration of some of the key advances in ideas on the tectonic setting of SSZ ophiolites. In the diagrams showing triple junctions, R is ridge, T is trench, and F is fracture zone. Note that the figures have been simplified. Refer to the original papers for full details.

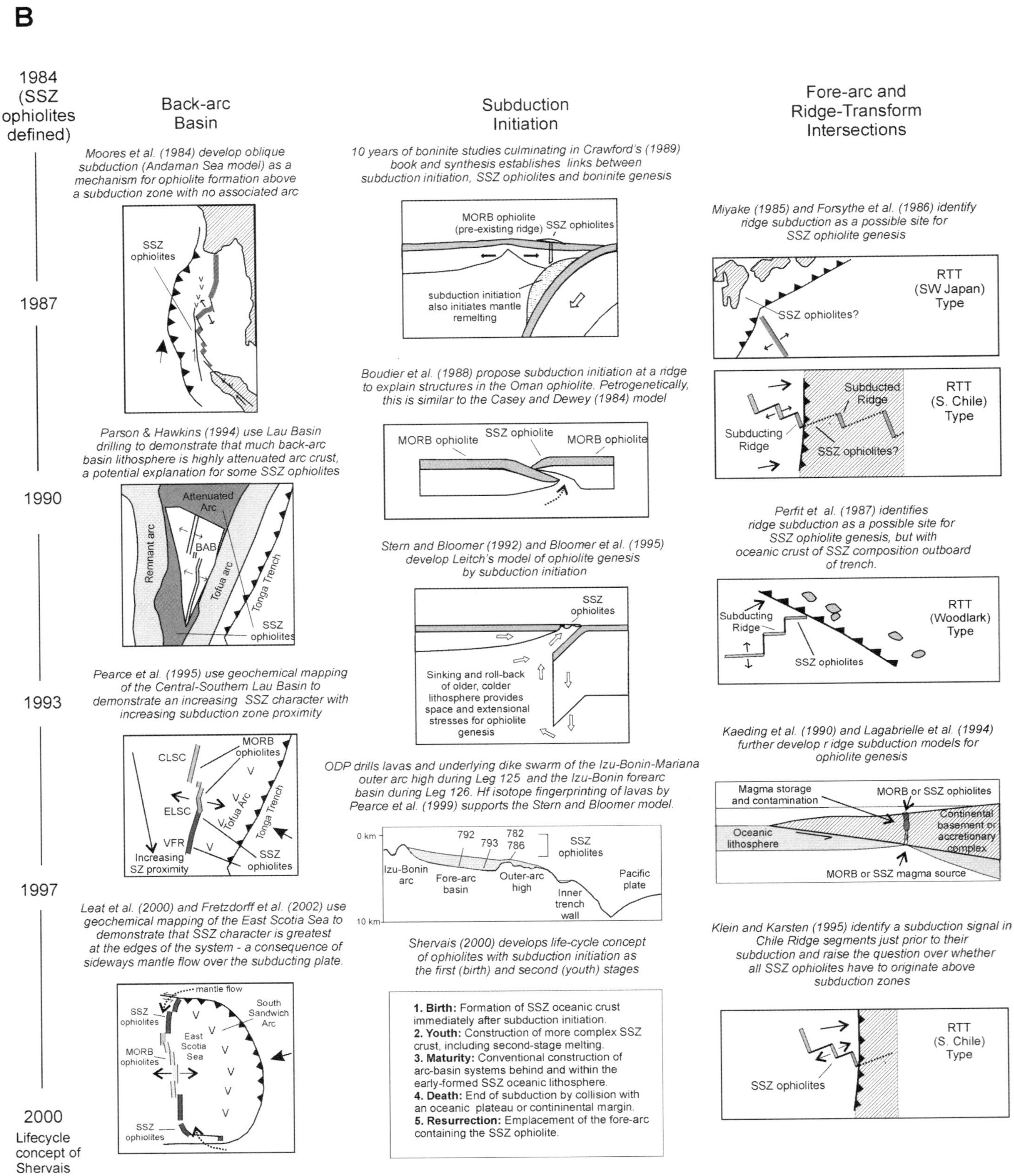

Figure 1 (*continued*).

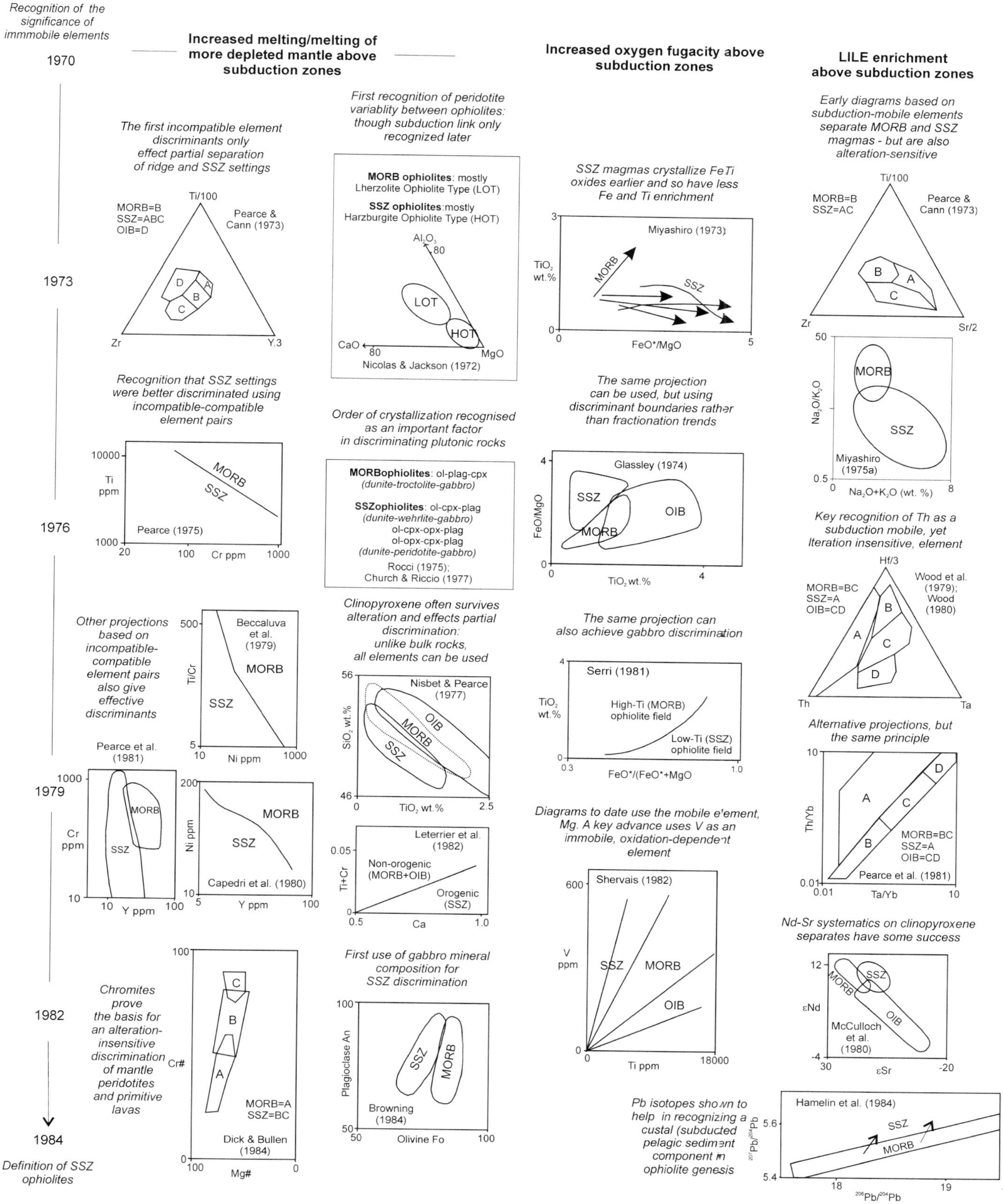

Figure 2. Fingerprinting of supra-subduction zone (SSZ) character in ophiolites—evolution of methodologies. Illustration of some of the key advances (up to 1984) in the geochemical identification of SSZ ophiolites and their discrimination from ophiolites of mid-ocean ridge basalt (MORB) and ocean island basalt (OIB) affinities. LILE—large ion lithophile elements.

Miyashiro's (1975b, 1975c) replies did address the effects of seafloor weathering, metamorphism, and vesicle infilling, and their impact on his methods of discrimination. In particular, he noted that weathered and metamorphosed basalts dredged from the ocean ridges were still not isochemical with the Troodos lavas. He also addressed the problem that the Troodos ophiolite sequence (pelagic sediments, thin crust, kilometers of sheeted dikes) was not that expected for discrete volcanoes, even overlapping volcanoes. He noted that radial dikes within volcanic centers may give rise to extension-parallel dikes between centers, and that extension parallel-dikes may develop during the transition from arc proper to back-arc basins. He pointed out that the Troodos dike orientations showed some variation, and argued that they could represent something more complex than a mid-ocean ridge environment.

It is an interesting exercise to review the Miyashiro interpretation in the light of present knowledge and databases. They demonstrate that the histograms published by Miyashiro were quite unrepresentative. Essentially, the published data he used were few and were designed to illustrate the range of rock types, not their distribution. Unlike the broad silica distribution shown by Miyashiro, the true plot (uncorrected for alteration) has a much lower standard deviation, more like ocean ridges, in fact. Nonetheless, the mean silica content is ~3% higher than for ocean ridges (basaltic andesite, rather than basalt), although the rocks are mostly very altered and silica certainly mobile. However, the remarkable discovery of fresh glass on the Troodos Massif (Robinson et al., 1983; Rautenschlein et al., 1985) does indicate that silica really is higher than in lavas from mid-ocean ridges. The "calc-alkaline" affinities identified by Miyashiro are somewhat misleading. By present arc terminology, the samples would be termed "island arc tholeiites" rather than "calc-alkaline" because of their low potassium content. Iron enrichment is essentially intermediate between ridges and arcs. As far as the structure is concerned, hindsight in the form of seismic evidence for arc crustal thickness (e.g., Suyehiro, 1996) shows that even the most primitive arcs are underlain by crust on the order of 15–30 km thick, many times the ~6 km crustal thickness of the Troodos Massif.

To be fair to Miyashiro, however, only three tectonic settings of oceanic magmatic activity were recognized at the time he wrote his paper: divergent plate boundaries (ridges), convergent plate boundaries (arcs), and intra-plate settings (continental rifts and oceanic islands). It was not unreasonable in that context to infer that, if the Troodos Massif had subduction characteristics, it must have formed in an arc.

Miyashiro himself "could not understand" the fuss his paper had created (Miyashiro, 1975, personal commun.). Despite the fact that few ever believed that the Troodos Massif formed in an island arc sensu stricto, he gets due credit for initiating the paradigm shift away from the view that all ophiolites formed at mid-ocean ridges. This was much easier to do from outside the ophiolite community, and emphasizes the need for any scientific consensus to be open to scrutiny. On the other hand, the ophiolite community might have received the paper better had Miyashiro based his work on systematic sampling and consideration of petrographic and field constraints! The scientific community also made its own judgment on the question of alteration: very few papers thereafter relied on the major elements to characterize ophiolites.

Miyashiro (1975b) had, however, correctly pointed out that Ti, Zr, and Y, although usefully immobile, were by themselves insufficient to assign the Troodos Massif uniquely to a mid-ocean ridge setting, as Pearce and Cann (1971, 1973) had inferred. The same argument applies to the rare-earth elements (REEs), as Smewing et al. (1975) and Kay and Senechal (1976) discovered. By this time, though, we had ourselves improved this aspect of the discrimination by extending the range of immobile elements to those with compatible behavior (see later). The resulting plot of Ti-Cr (Pearce, 1975) demonstrated to an extremely high level of probability that the Troodos Lavas had an arc, rather than an ocean ridge, affinity. Like the respondents to the Miyashiro paper, however, I could not envisage the Troodos Massif as an arc sensu stricto because its structure was quite different from that of overlapping shield volcanoes, as Gass et al. (1975a) had so graphically pointed out.

The obvious solution was that the Troodos Massif formed in a marginal basin. Dewey and Bird (1971) had stressed the potential of marginal basins for ophiolite genesis well before the Troodos debate. They had concluded (p. 3179):

> The ophiolite suite is probably generated by axial plate accretion at oceanic ridges and by diffuse slow spreading in marginal basins within and behind island arc complexes.

Bird et al. (1971) explained most of the Caledonian/Appalachian ophiolites in terms of marginal basins. For example (p. 31):

> In the western Pacific the marginal basins behind, and between, island arc complexes are fairly certainly the result of a sea-floor spreading process by which island arcs move slowly away from continental margins. We suggest that the Early Ordovician Appalachian/Caledonian ophiolites resulted from a similar process during the establishment of marginal basins within, and north-west of, the Early Ordovician metamorphic terrain."

Dewey and Bird (1971) had proposed that marginal basin ophiolites could be distinguished from mid-ocean ridge ophiolites by a number of features that included their apron of volcanogenic sediments (see later). The Troodos Massif differed from the Caledonian/Appalachian ophiolites described by Bird et al. (1971) because it had no contemporaneous, overlying volcanogenic sediments and no nearby arc. As a result, the Troodos Massif did not fit the Karig (1971) model of back-arc or intra-arc basin formation that formed the basis for Dewey and Bird's reconstructions. The marginal basin interpretations of Dewey and Bird (1971) and Bird et al. (1971) therefore avoided the subsequent controversy surrounding the Troodos Massif.

Essentially, then, the Troodos lithosphere had arc chemistry, yet ridge geology, something for which there was no analogue at that time. I speculated that the Troodos Massif formed in a marginal basin before, and at the inception of, arc volcanism (Pearce, 1975, p. 64):

> The Troodos Massif apparently contained rocks which have not been well sampled at the present time, the upper pillow lavas probably representing submarine island arc eruptions while the lower pillow lavas may have been produced by spreading in a marginal basin directly overlying a subduction zone.

Thus, according to my interpretation, the main Troodos Massif was a marginal basin formed by extension above the subducting plate, while the upper lavas represented the first constructional stages of an island arc edifice. My view was that collision with a continental margin had prevented the arc-basin system from evolving into a "proper" volcanic arc. As will be seen, the part of this hypothesis that required spreading to predate arc volcanism after the initiation of subduction successfully anticipated subsequent studies of marginal basins. However, the idea that the upper lavas are the first stages of arc activity was not supported by later studies. These are better viewed as off-axis eruptions (e.g., Smewing and Potts, 1976). Thus, it is likely that the Troodos Massif really did form by seafloor spreading, with no arc in evidence either before or after its formation.

The Troodos debate divided the ophiolite community into two groups. One group (mainly geochemists) was convinced of a marginal basin setting on the basis of the immobile element arguments. The other (mainly geologists) was convinced of a mid-ocean ridge setting on the basis of the crustal structure and the absence of any arc volcano or arc-derived sediments.

Ophiolite Diversity

The predictions of Dewey and Bird in 1971, coupled with the debate over the origin of the Troodos Massif that was initiated by Miyashiro in 1973, made it clear that a mid-ocean ridge origin for ophiolites could not be taken for granted. In parallel with the Troodos debate, therefore, geochemical fingerprinting of ophiolite complexes increased hugely as more X-ray fluorescence (XRF) instruments became installed worldwide, and as these machines became more automated. Major analytical and field programs were carried out on ophiolites, greenstones, and other lavas from orogenic belts, and there is no space to cite them all here. Notable among them were the Caledonides in Norway, the Newfoundland Appalachians, the Western Alps, Greece, and the Coast Ranges and Klamath Mountains in Western USA. By the mid-to-late 1970s, subduction chemistries outnumbered mid-ocean ridge basalt (MORB) chemistries by some margin, and it was becoming apparent that marginal basins may represent a better analogue for most ophiolites than mid-ocean ridges.

Meanwhile, even without immobile element geochemistry, scientists drew attention to the fact that ophiolites appeared to subdivide into distinct groups according to their geological and petrological characteristics. This was particularly evident in the Mediterranean region, where ophiolites of contrasting characteristics were closely associated in space and time. Mesorian (1973) divided Alpine ophiolites into three groups (a "Troodos-type," a "Pindos-type," and an "Antalya-type"), according to the distribution of dikes and consideration of whether the lavas had tholeiitic, calc-alkaline, or alkaline characteristics. Nicolas and Jackson (1972) and Menzies and Allen (1974) recognized that there was a bimodal distribution of peridotites in the Mediterranean region, one mode being a lherzolite-harzburgite subtype and the other a harzburgite subtype. Rocci et al. (1975) also recognized two groups, his "dualité" of ophiolite types:

- Type I: tectonized lherzolite, plagioclase peridotites, troctolites, gabbros, Fe-gabbros.
- Type II: tectonized harzburgite, dunite, chromitite, pyroxenites, norites, olivine gabbros.

Without immobile element geochemistry, however, it was not possible to resolve whether subduction or normal ridge crest variables such as spreading rate is responsible for the observed diversity. For example, Nicolas and Jackson (1972) and Menzies and Allen (1974) proposed that the lherzolites formed in small (Red Sea type) ocean basins and the harzburgites formed in major oceans—without discussing the subduction option. Rocci et al. (1975) additionally mentioned the possible role of marginal basins, but declared that "it was still impossible to distinguish between the various options." Thus, it was geochemistry, rather than petrology or geology that led to the subduction link.

By the middle to late 1970s, some scientists had moved to the opposite extreme from that represented by those still holding out for a mid-ocean ridge origin for all ophiolites.

> Most ophiolites so far studied in detail seem to represent allochthonous fragments of the crust of small oceans or marginal seas. No convincing representative from a major ocean has been identified. Gass et al. (1975b, p. 64)

> ...although we do not rule out the possibility that ophiolites could originate in more than one tectonic setting, the structural, stratigraphic and trace element data we have summarised show that the most logical site for the origin of many ophiolites around the world was in the Western Pacific marginal seas bordered by island arcs on one side and continental margins on the other. Upadhyay and Neale (1979, p. 100).

Perhaps the most balanced view of the state-of-the-art at the time was presented by Coleman (1977) in his book, *Ophiolites*. On page 10, he states:

Formation of oceanic lithosphere in other tectonic situations besides mid-ocean ridges should also be carefully considered when comparisons are being made between ophiolites and oceanic lithosphere. For instance, it is quite reasonable to expect some oceanic crust to form in marginal basins behind oceanic island arcs and presumably above subduction zones... Unfortunately, at this time, there are not enough detailed studies on marginal basins on which to formulate a model which can adequately explain the formation of oceanic crust in this situation.

So it was clear that many ophiolites had subduction zone affinities, but many of these still did not have a known analogue that had both ridge geology and subduction geochemistry. A geologist suspecting a tectonic setting but having no modern analogue was like a detective suspecting a murder but having no body. In both cases, direct evidence is all-important. Thus, the principal requirement was now a better coverage of rocks from the ocean basins, and the marginal basins in particular.

EMERGENCE OF POTENTIAL SSZ OPHIOLITE ANALOGUES

At the time of the controversy over the origin of the Troodos Massif, there were very few geochemical data from marginal basins. Those that were published were scarcely distinguishable from mid-ocean ridge chemistry (in fact, the Pearce and Cann (1973) MORB database included four samples from the Mariana Trough). Thus, it was not clear that mid-ocean ridges differed chemically from marginal basins. By 1975, the MORB database was fairly representative (though biased to slow-spreading ridges), as was the oceanic arc database (though biased to the SW Pacific). By contrast, construction of the marginal basin database had scarcely begun.

The key to understanding ophiolite geochemistry came from a series of cruises to the marginal basins of the western Pacific and South Atlantic. The first modern geochemical studies of back-arc basins were published by Hart et al. (1972), Gill (1976), and Hawkins (1976, 1977), and focused on the Lau and Mariana Basins. However, the vast majority of their published samples remained typical MORB, while the remainder was MORB with a small subduction component that scarcely influenced the discrimination. This subsequently led to the term back-arc basin basalt (BABB) to describe MORB with a small subduction zone modification. Gill (1976, p. 1394) describes such samples from the Lau Basin as follows:

Basin extension is accompanied by intrusion of basalt generally similar to mid-ocean ridge tholeiite but having higher Sr^{87}, Ba, light REE and possibly alkali and H_2O contents, and lower Y, heavy REE, Zr, Hf and possibly Ti contents.

These still did not, however, resemble those of the Troodos Massif or the vast majority of the other ophiolites with distinctive subduction chemistry that were beginning to be identified. The first back-arc lavas to do so were the lavas from Site 24 from the East Scotia Sea back-arc basin, published by Saunders and Tarney (1979). Although not typical of the East Scotia Sea as a whole, they did finally demonstrate that lavas erupted at a marginal basin spreading axis could have a significant subduction component. Here, then, was the first back-arc site where the crustal structure could be "ophiolitic" while the composition of the lavas was arc-like.

At about the same time, Bruhn et al. (1978) and Stern (1979) demonstrated that mafic complexes with ophiolitic structures could also form in intracontinental back-arc basins. Their studies of the Tortuga and Sarmiento back-arc complexes of Early Cretaceous age in Southern Chile revealed selective enrichments in fluid-mobile elements, but lacked sufficient evidence to distinguish between a subduction-modified mantle source and a MORB mantle source in which the resulting MORB magma had undergone crustal contamination.

Back-arc basins proved, however, to be less important than fore-arcs in the search for analogues to ophiolites with subduction characteristics. There were good geological arguments for a link between ophiolites and fore-arcs because fore-arcs are among the first terranes to be emplaced during continent-arc collision (e.g., Dewey, 1976). Moreover, there are many structural and metamorphic similarities between fore-arc complexes and ophiolites (e.g., Malpas et al., 1973). However, their significance was not so apparent geochemically, as the first ophiolite-like sequences sampled and analyzed from fore-arcs, from the Yap islands south of the Marianas (e.g., Shiraki, 1971; Hawkins and Batiza, 1977), had Pacific Ocean provenance and MORB-like characteristics.

The major breakthrough in identifying fore-arc ophiolites with subduction characteristics came from two projects. The first was the Soviet-sponsored RV *Dimitry Mendeleev* cruise to the Western Pacific in 1976, organized scientifically by members of the International Geological Correlation Programme (IGCP) "Ophiolites" project led by Nikita Bogdanov (International Working Group of the IGCP Project "Ophiolites," 1977). The cruise carried out a series of dredges in the Philippine Sea, including the inner trench walls of the Yap and Mariana Trenches. Scientists obtained a range of rocks (peridotites, gabbros and basalts) that they recognized as part of a typical ophiolite suite. Importantly, they demonstrated that ocean basins in subduction settings could both have subduction characteristics and yield rocks similar to those recovered from ophiolites such as the Troodos Massif.

The results of this cruise first reached the international literature in the paper by Dietrich et al. (1978), and also resulted in a series of less "accessible" books and publications in Russian and conference proceedings (e.g., Sharaskin et al., 1980). Dietrich et al. (1978, p. 142) discussed the implications as follows:

An interesting parallel may be drawn with the controversy about the origin of the "Troodos Complex." This ophiolite complex has been variously interpreted as representing oceanic crust or an island arc and contains many rocks similar to those dredged from the Mariana arc slope.

However, Dietrich et al. (1978) did not interpret the rocks as the product of spreading above a subduction zone. They assumed, as Dewey and Bird (1971) and Coleman (1971) had done, that the rocks represented slices of Pacific Ocean lithosphere that had become incorporated into the fore-arc and thus formed the basement to the subsequent island arc.

The *Dimitry Mendeleev* scientists also had the distinction of being the first to find and correctly identify boninites from the ocean floor, likening them to the (then) two reported localities from Bonin Island and Papua New Guinea. However, they did not immediately recognize their huge significance for identifying and interpreting subduction-related ophiolites. Dietrich et al. (1978), for example, assigned an ocean ridge origin to the boninites. In fact, at the time of the cruise, boninites were known in ophiolites but viewed as part of the komatiite lineage (e.g., Simonian and Gass, 1978). Sun and Nesbitt (1978) demonstrated that there are major geochemical differences between komatiites and "low-Ti ophiolitic basalts," and made the key prediction that the latter were linked to subduction. They also made the first link between high-Mg lavas from ophiolites and from the Mariana Trench, though without mentioning "boninite" as a rock type.

> We suggest that these low-Ti basalts are not formed at mid-ocean ridges, but are products of a spreading center close to a subduction zone, for example, in an interarc basin or incipient island arc site. In this model, remelting of the refractory source is induced by the introduction of water from subducted oceanic crust. (Sun and Nesbitt, p. 689)

This led Cameron et al. (1979) to reclassify as boninites the "basaltic komatiites" from the "fossil transform fault" region of the Troodos Massif. A number of papers rapidly followed to describe boninites from ophiolites in other parts of the world, and so the link between boninites, ophiolites and fore-arcs was clear by the end of the 1970s. Importantly, the concept of boninite origin by remelting of a refractory source was clearly demonstrated. Boninites, therefore, appeared to require petrogenetic processes that were quite distinct from those of normal arc volcanism or of extensional processes well behind the trench.

The second breakthrough came from the Western Pacific program of the Deep Sea Drilling Project (DSDP). DSDP had drilled the Mariana Trough (the back-arc basin) during DSDP Leg 6 and the West Philippine Basin during DSDP Leg 31 but, like early dredging programs, had sampled only MORB and oceanic island basalts (OIB). The drilling program during Legs 59 and 60 drilled a transect across the arcs and marginal basins of the Izu-Bonin-Mariana system. The transect began in the West Philippine Basin in the west, crossing the Palau Kyushu Ridge remnant arc, the Shikoku and Parece Vela fossil back-arc basins, the West Mariana Ridge remnant arc, the active Mariana Trough back-arc basin and the Mariana arc. It finished in the Mariana fore-arc and inner trench wall. The drill sites yielded a bimodal distribution of samples: MORB in the active and fossil back-arc basins and arc-like in the active and remnant arcs (Wood et al., 1981). The only lavas that might resemble those of the Troodos Massif and many other ophiolites were the boninites and island arc tholeiites recovered from the fore-arc basin.

This fore-arc basin link even led Natland and Tarney (1981) to refer to the fore-arc terrane as the "Mariana Fore-arc Ophiolite" and to discuss the viability of fore-arcs as ophiolite analogues. They conclude (p. 895):

> We suggest that an appropriate equivalent of Troodos in an island arc setting is the Mariana fore-arc region and that at the earliest stages of arc volcanism, a type of sea-floor spreading may have occurred before volcanism focused along the line of the arc. ... Whether or not true sea-floor spreading produced the Mariana fore-arc region remains to be seen in the light of more detailed geophysical experiments, careful dredging in the trench, and perhaps drilling. Regardless, the occurrence of boninites in certain ophiolites is good evidence that spreading has occurred in several ancient arc systems, probably early in their history.

Bloomer and Hawkins (1983) and Sharaskin et al. (1983) reached similar conclusions through renewed dredging of the Mariana and Tonga Trenches, respectively. The former concluded (p. 313):

> The trench slope assemblage in general is similar to that of an ophiolite. In particular it is similar to ophiolites containing low Ti, Si-saturated volcanic rocks and [orthopyroxene-rich] cumulate sections. It is suggested that these ophiolites represent oceanic crust formed in the early stages of island-arc volcanism.

Thus, for the first time, we had a convergence of interpretations from ophiolites and marginal basins that spreading could take place above a subduction zone before any arc had formed. This was not a full consensus, but it paved the way for the development of the supra-subduction zone ophiolite concept. The link between many ophiolites and subduction zones was made more rigorous through the development of discriminant diagrams that enabled the two groups of rock to be compared even when strongly altered.

EARLY DEVELOPMENT OF SUBDUCTION ZONE DISCRIMINANTS

Geological Discrimination

The first attempt to discriminate between a mid-ocean and marginal basin origin for ophiolite complexes was geological. Dewey and Bird (1971) identified a series of features that might distinguish a marginal basin origin. These included: overlying volcanogenic sediments; complex intrusive and extrusive relationships; irregular thicknesses of layers 2 and 3; and a large proportion of andesites mixed with the basalts.

These remain effective discriminants for oceanic lithosphere closely associated with active arc volcanism. As the Troodos controversy demonstrated, however, oceanic crust could have an arc signature even when there was no associated arc volcanism. Thus, geological criteria were not always sufficient and geochemical discriminants also had to be developed.

Immobile Element Fingerprinting of Lavas

Much of the conventional igneous geochemistry of the 1960s utilized major elements, particularly the highly mobile elements such as Na, K, and Mg. As the Troodos debate was to illustrate, however, this approach was ineffective (or, at best, difficult to apply rigorously) in interpreting and classifying the often highly altered rocks from ophiolite complexes. This led to the sudden surge in interest in elements that were not transported during deep-sea weathering and metamorphism. The key technological development was the XRF method of geochemical analysis, which opened a range of such elements to routine analysis. Cann (1970) provided the basis for this immobile element approach to ophiolite geochemistry. Making use of one of the first mass-produced XRF machines, he analyzed a large number of dredged basalts from mid-ocean ridges. He discovered that when he plotted pairs of immobile incompatible elements, such as Ti-Zr, the basalts formed highly significant linear correlations. If, on the other hand, he plotted a mobile element against an immobile element, such as K-Zr, the correlation became insignificant—more precisely, the fresh rocks still formed a good correlation but the weathered and metamorphosed rocks gave a large scatter. Thus, Cann (1970) was able to demonstrate that only immobile elements were likely to be effective in studying lavas and dikes from ophiolite complexes. This led to the immobile element (Ti-Zr-Y) fingerprinting of Pearce and Cann (1971, 1973)

As already pointed out, neither the Ti-Zr-Y combination nor REE patterns were very effective in discriminating between MORB and subduction-related basalts: the field of overlap was too large. The reason was that the fractional crystallization and partial melting variations were sub-parallel on these projections, so that crystallization masked the key variations in mantle melting that provided the basis for discriminating the ridge setting from the wetter subduction zone setting. The key was to take one of these three immobile incompatible elements and plot it against a compatible element. This meant that partial melting and fractional crystallization vectors were close to orthogonal and a much more effective discriminant resulted.

Glassley (1974) produced the first plot of this type for interpreting rocks from the Olympic Peninsula in Washington State. He used Ti as the immobile, incompatible element and FeO/MgO as the (inversely) compatible index. This approach differs from the equivalent plot used by Miyashiro in that the latter is based on fractionation trends, whereas the Glassley plot is based on compositional fields. Its use of the mobile element, Mg, presented potential problems in very altered rocks, however. As noted earlier, the Ti-Cr plot of Pearce (1975) used to interpret Troodos lavas, adopted the more immobile element, Cr, as the compatible index. Variations on a theme appeared, including Ti/Cr-Ni (Beccaluva et al., 1979), Ni-Y (Capedri et al., 1980), and my own Y-Cr projection (Pearce et al., 1981, Pearce, 1982). The point of the Y-based diagrams is that Y is a potential improvement on Ti because it is not affected by the crystallization of Ti-bearing oxides that is common in subduction-related magmas. In all these diagrams, displacement to lower incompatible element content (Ti, Y) at a given compatible element content (Cr, Ni, Mg) was a predictable consequence of subduction (fluid) influence, i.e. due to more melting or melting of more depleted mantle.

The second discriminant of this type to be used extensively by the ophiolite community was the Shervais (1982) plot of Ti-V. Subduction-related lavas have lower Ti than MORB at a given V content, whereas intra-plate lavas have higher Ti. As noted by Shervais, this diagram also had a strong petrogenetic base. Ti is more incompatible than V, so it is more depleted in magmas produced by increasing degrees of melting or more depleted sources. In this way, it resembles Ti-Cr, except that the discrimination due to degree of melting is less. The "bonus," however, is that V is highly sensitive to oxygen fugacity. For melting under hydrous (subduction) conditions, more vanadium is in the higher oxidation states 4+ and 5+ relative to the common 3+ state. This means that V is more incompatible during subduction zone melting, and so the lavas have higher V than under anhydrous (MORB) conditions. Thus, subduction-related lavas have both higher V and lower Ti, which improves discrimination.

Ti, Cr, Y, and V all had the advantage that they could be analyzed by the analytical technology of the moment, namely XRF, a simple and accessible technique. The disadvantage of XRF elements was that there was no analyzable element which was both mobile in subduction zones yet immobile during seafloor weathering and metamorphism. As studies of K, Ba, Rb, and Sr showed, a high mobility in subduction zone fluids provided an excellent basis for discrimination, but also meant a high mobility during the weathering and metamorphic events experienced by ophiolitic lavas and dikes. This mobility commonly masked the discrimination.

The breakthrough came with the development of routine (instrumental) neutron activation analysis (INAA). Although it required access to a nuclear reactor for irradiating samples, and so was less universally available, it opened up a range of "new" elements to routine analysis. The first applications of INAA to trace element discrimination began in the mid-1970s (e.g., Smewing et al., 1975; Kay and Senechal, 1976). As with most INAA-based geochemical studies, these focused on the rare-earth elements, which did not discriminate well between ocean ridges and island arcs. INAA did, however, permit analysis of the element Th at the low abundance levels found in basic, oceanic rocks. The significance of Th only became apparent several years later, when the Paris-based group of Treuil, Joron, and co-workers set up a semi-automated system that permitted accurate analysis of Th in large numbers of samples.

Th is a key element in subduction zone discrimination. As we now know, it is enriched in all arc lavas, and so is mobilized in subduction zones, but is immobile until temperatures close to melting temperatures are reached (Johnson and Plank, 1999). INAA is also effective at analyzing Hf (which behaves like Zr) and Ta (which behaves like Nb). From this emerged the Th-Ta-Hf discriminant devised by Wood and the Paris group (Wood et al., 1979, Wood, 1980). Subduction-related lavas have enriched Th relative to Ta and Hf. In addition, many intra-plate lavas have higher Th and Nb relative to Hf. Pearce et al. (1981) and Pearce (1983) used a variant of this theme for a projection of Th/Yb versus Ta/Yb, which is possibly easier to interpret petrogenetically, but has the same theoretical basis.

Geochemical Fingerprinting of Clinopyroxenes

Clinopyroxene discriminants provided an alternative approach. Glass and most minerals are easily altered in volcanic rocks, but clinopyroxene is commonly resistant to alteration because of the greater energy required to break the strong Si-O bonds in its Si_2O_6 chains. Nisbet and Pearce (1977) attempted the first set of clinopyroxene discriminants using discriminant analysis based on major and minor elements, a Ti-Si bivariate plot and a Ti-Mn-Na triangular plot. Leterrier et al. (1982) produced several projections, the most used being Ti+Cr versus Ca, which is 80% successful in discriminating "non-orogenic" from "orogenic" basalts. In all these methods, the key element is Ti, which (as bulk rock discriminants have shown) has lower abundances in subduction-related magmas at a given concentration of a compatible element such as Cr. In general, though, the clinopyroxene discriminants did not provide much new discriminant information. The advantage of using clinopyroxene was that major elements could be used for discrimination. This was counterbalanced by two disadvantages. First, ocean ridge basalts rarely contain clinopyroxene phenocrysts, so the chemically more variable groundmass clinopyroxene had to be used instead. Second, the constraints of stoichiometry, charge balance, crystallization rate, and temperature dependence tend to mask some of the most useful variations. In volcanic rocks, therefore, one might as well discriminate using bulk rock compositions. The most important use of clinopyroxene is thus likely to be in the interpretation of plutonic rocks, where crystal cumulation makes bulk compositions difficult to interpret, although this has yet to be exploited fully.

Isotopic Discrimination

Pb, Sr, and Nd isotope ratios had begun to be used to discriminate between tectonic settings by the late 1970s. Their major advantage is that, unlike trace elements, they can be used to fingerprint both volcanic and plutonic rocks. Sr and Pb, however, are mobile. Nd is much less mobile, but it is less useful without another isotope ratio for co-variation.

Richard and Allègre (1980) carried out the first systematic Nd isotope study of ophiolites, though with a maximum of two analyses from any one complex. They found that most of the ophiolites they studied had εNd values within the range of MORB, and that a few had lower values. They concluded that Nd isotopes were "not in contradiction" with a mid-ocean ridge origin. They plotted the data in Nd-Sr isotope space, and found that many exhibited an isotopic shift from within the diagonal MORB array toward higher Sr isotope ratios. However, it was not clear whether this shift was a consequence of subduction or of sub-seafloor alteration. McCulloch et al. (1980), Noiret et al. (1981) and McCulloch and Cameron (1983) applied Nd-Sr isotope systematics to more detailed studies of the Semail, Vourinos and Troodos ophiolites. Again, the method established that some samples (the Troodos boninites) had εNd values that were too low for MORB, but did not uniquely fingerprint the setting. McCulloch et al. (1980) and Pearce et al. (1981) demonstrated the importance of using clinopyroxene separates rather than bulk rocks to overcome the alteration problem and allow Nd-Sr isotope covariations to be used. However, the discriminating power of this projection is still limited by the fact that many lavas from subduction-related settings also plot along the MORB array.

Because seawater contains so little Pb, Pb isotopes are, in fact, surprisingly insensitive to alteration. The principal problem is that age-corrections are difficult to carry out because of the low abundances and high mobilities of Pb and U. Essentially, discrimination is based on observing whether the ophiolite complex plots within the MORB mantle array in Pb isotope space, or whether it plots toward pelagic sediments with their high $^{207}Pb/^{206}Pb$ ratios. Pb isotope ratios are not especially good at discriminating tectonic settings, however. This is partly because recycled subduction components (as found in "Indian" MORB, for example) can be difficult to separate isotopically from the effects of recent subduction. Moreover, subducted lead can derive from a number of sources—from subducted MORB crust and volcanogenic sediments derived from oceanic islands, as well as from the pelagic sediments upon which the discrimination is based.

The earliest Pb isotope data focused on the easy-to-analyze metalliferous sediments. Thus, Gale et al. (1981) analyzed samples from the Troodos, Syrian, and Oman ophiolites. They found that the first two of these formed a trend between MORB and pelagic sediments. However, they interpreted their results in terms of the mixing of basalt lead and seawater Pb during sub-seafloor fluid circulation, rather than in terms of subduction processes. Hamelin (1984) carried out a systematic study of the igneous rocks from a number of the world's best-known ophiolites, and demonstrated a large variability, some of it certainly linked to tectonic environment. Significantly, the lavas of the Troodos Massif and some other ophiolites plotted above the MORB array toward pelagic sediment compositions, whereas other ophiolites plotted within the MORB array, pointing to a potential contribution of subducted pelagic sediments to their origin.

Despite some successes, however, no clear isotope-based discriminants of tectonic setting have yet been devised, although isotopes (including, most recently, those of Hf and Os) can clearly improve an overall understanding of the petrogenesis of an ophiolite. Trace elements, therefore, remain the discriminants of choice for lavas.

Gabbro Discrimination

Plutonic rocks in ophiolites tend to be much less altered than lavas, which means that some "mobile elements" can be used in their discrimination. Unfortunately, however, the variable and complex processes of crystal cumulation and melt trapping mean that conventional discriminants cannot easily be applied. Nonetheless, fractionation trends can be reliable if they are chosen so that the melt and the crystals plot approximately on the same trends. Serri (1981) used the same elements as Miyashiro (1973) and Glassley (1974), though a slightly different projection. Serri plotted gabbros on a graph of TiO_2 against a Mafic Index [MI = $FeO_{tot}/(FeO_{tot}+MgO)$] to identify two ophiolite types, which he termed high-Ti ophiolites and low-Ti ophiolites. He interpreted the high-Ti ophiolites as those formed at mid-ocean ridges, and the low Ti ophiolites as "oceanic crust generated by spreading processes occurring above a dehydrating subducted oceanic plate in the early stage of opening of intra-oceanic back-arc basins."

Serri's further observation was that these two geochemically-defined ophiolite types were matched by differences in the order of crystallization and hence the type of plutonic rock. These differences, originally recognized by Rocci et al. (1975), were extended by Riccio (1977) and Church and Riccio (1977) into a four-fold classification of ophiolites based on crystallization sequence:

(A) (ol+chr)-opx-cpx-plag
(B) (ol+chr)-cpx-plag-opx
(C) (ol+chr)-cpx-opx-plag
(D) (ol+chr)-plag-opx-cpx

Serri (1981) pointed out that these differences in order of crystallization must be a function of the chemistry of the lavas, and whether the gabbros were of high-Ti (ocean ridge-type) and low-Ti (subduction-related) type. Beccaluva et al. (1983) extended Serri's definition to include a very-low-Ti type, which was subduction-related and associated with boninite lavas. They also made the important link between orders of crystallization and ophiolite type. They noted that the very-low-Ti ophiolite type followed Church and Riccio's orders A and C, giving them an associated cumulate sequence made up of a thick ultrabasic unit of olivine and two pyroxenes in varying proportions, with an overlying sequence of gabbro norite. The low-Ti ophiolite type (which included the Troodos Massif) followed Church and Riccio's order B, giving an ultrabasic unit mainly of olivine and clinopyroxene overlain by a gabbro norite sequence. The high-Ti type followed Church and Riccio's order D, comprising a thin olivine-rich (dunite) ultrabasic sequence overlain by olivine-plagioclase (troctolite) and then gabbro (sensu stricto).

Serri (1981) and Beccaluva et al. (1983) had to use lava geochemistry to develop their plutonic rock discriminants because so few cumulate rocks from known settings had been collected. Detailed ground-truthing of this method began when Bloomer and Hawkins (1983) compared the orders of crystallization of their dredged rocks from the Mariana fore-arc with orders of crystallization in ophiolites. They confirmed the link between subduction zone magmas and early crystallization of pyroxenes.

The reason for the different order of crystallization in low- and high-Ti ophiolites was only discussed in general terms, although the greater degree of melting and higher water content in subduction environments were generally considered responsible for earlier crystallization of pyroxenes with respect to plagioclase feldspar. The point, therefore, is that there is a clear link between subduction processes (increase in degree of melting, or melting of depleted sources) and both lava chemistry (low Ti and Y at a given degree of crystallization) and order of crystallization in the plutonic rocks.

Few attempts were made to use mineral chemistry as a discriminant. However, Browning (1984) did instigate the use of a plot of anorthite content of plagioclase (An) versus forsterite content of olivine (Fo), a projection developed further by Thy (1987). For a given value of An, SSZ lavas have lower Fo.

The granitic rocks from ophiolites, though minor in proportion, had also received attention, and their diversity was the subject of several papers. Pearce et al. (1984b) included them in their granite discrimination. Plagiogranites from ophiolites with subduction affinities typically had lower Y and Nb concentrations than those with ridge affinities, and plotted as volcanic arc granites (VAG), rather than ocean ridge granites (ORG).

Peridotites

Discrimination of ophiolitic peridotites in a modern framework started with Nicolas and Jackson (1972) and Menzies and Allen (1974). They both studied the so-called Alpine peridotites from the Mediterranean region and recognized that they could be separated into two subtypes, a lherzolite-harzburgite subtype and a harzburgite subtype. [Boudier and Nicolas (1985) subsequently gave these the acronyms LOS (lherzolite ophiolite subtype) and HOS (harzburgite ophiolite subtype). Nicolas (1989) renamed them LOT (lherzolite ophiolite type) and HOT (harzburgite ophiolite type).] They discriminated between these on a triangular Al_2O_3-CaO-MgO (ACM) plot, the more residual harzburgite subtype plotting closer to the MgO apex. They explained the differences in terms of geodynamic processes, but not to the presence or absence of subduction. Instead, they considered the more fertile lherzolite-harzburgite subtype to be the residue from melting in small ocean basins, and the harzburgite subtype to be the residue from melting at an established mid-ocean ridge. There was, of course, a very good reason why no link had been made to subduction: a significant number of peridotites had been recovered through dredging at ocean ridges, but there simply were no samples from arcs or marginal basins.

Developments of discriminants initially focused on the harzburgite and lherzolite peridotite types. Because ACM plots were alteration-sensitive, attention switched to the chemically resistant chrome spinels (e.g., Malpas and Strong, 1975; Coleman, 1977). It did, however, take some time for effective discriminant diagrams to be developed. Studies instead focused on the origin of the peridotites; whether they were melt residues, diapirs, or cumulates; and whether they had similar origins to mantle xenoliths.

The realization that harzburgite was a common rock type in subduction-related settings became apparent through inner trench wall dredging. However, the connection between peridotite type and geochemical affinities of the associated lavas provided the best clue. Thus, Beccaluva et al. (1983, p. 307) reported:

> In many ophiolites, parental basalts and mantle ultramafics may be interpreted in terms of generated melts and refractory residua. These are of lherzolitic type in high-Ti ophiolites, whereas mantle harzburgite/dunite commonly occur in low-Ti and very-low Ti ophiolites.

The culmination of the peridotite studies of the 1970s and early 1980s is the paper by Dick and Bullen (1984). They addressed the discrimination of ophiolitic peridotites through a series of projections, the most used being Cr# versus Mg# in chrome spinels. Chrome spinels are remarkably resistant to weathering and alteration, so this discriminant was both easy to obtain and easy to apply.

Dick and Bullen found that ocean ridge (abyssal) peridotites all occupied a well defined field with Cr# ranging from 0.6 to 0.1, as did their associated lavas. In contrast, lavas and plutonic rocks from arcs and layered intrusions typically have Cr# >0.6. Dick and Bullen reported that spinels from "Alpine-type peridotites" and their associated lavas could be divided into three types: Type I with Cr# <0.6, Type III with Cr# >0.6, and Type II with Cr# that spanned the 0.6 discriminant boundary. In terms of setting, they related Type I to mid-ocean ridges (including most back-arc basins), Type III to the earliest stages of arc formation, and Type II to composite settings such as an island arc overlying oceanic crust. The Troodos Massif was Type III on this classification, and the Semail ophiolite was Type II. Dick and Bullen concluded (p. 72):

> Based on analogy to spinels in lavas and plutonic rocks of known provenance, we feel in agreement with Miyashiro (1973) that the provenance of the Type III peridotites is related to the earliest stages of arc-formation on oceanic crust environments which may be preserved in the tectonized material found in the fore-arc regions of many modern island arcs.

Dick and Bullen (1984) also provided the petrogenetic explanation for their observations, namely that the Cr# of the spinels is a function of the degree of melting that the mantle has undergone. Above subduction zones, the added water means that the degree of melting is higher, so the Cr# in the peridotite residue from melting is greater. Moreover, it fits in with the mineralogy of the peridotites. Type I peridotites typically (though not always) leave residues that still contain some clinopyroxene (lherzolites and clinopyroxene-rich harzburgites). Type III peridotites, by contrast, contain very little clinopyroxene as a result of the greater melting (clinopyroxene-poor to clinopyroxene-free harzburgites).

Dick and Bullen's work was the prelude to several other peridotite discriminants. Some extended the range of tectonic settings under consideration (e.g., Bonatti and Michael, 1989), and some replaced Mg# with a more effective axis (e.g., the Arai (1994) plot Cr# against forsterite content of olivine). These all have the same explanation, namely that subduction induces more melting (or melting of more depleted mantle).

DEVELOPMENT OF THE SUPRA-SUBDUCTION ZONE DEFINITION

Thus, by the early 1980s, the Y-Cr, Ti-V and Th-Ta-Hf discriminant diagrams (and their various morphs) were all effective in recognizing subduction signatures in lavas and dikes. The Ti-MI diagram could fingerprint gabbros and the Cr–Mg# diagram was effective at fingerprinting peridotites. Other methods, including mineral compositions and isotope systematics, could add further information. With the exception of a few small improvements, the main methods of discrimination were in place by this time. It was also apparent that there were links between tectonic setting and rock type. Moreover, by this time, a substantial database had been built up, both on ophiolite complexes and on modern analogues. It was clearly the right time to synthesize the data and provide a model for characterizing ophiolites, and it is no coincidence that a series of such papers were published almost simultaneously.

The syntheses were not, however, entirely straightforward to write. The principal impediment was that some ophiolites remained controversial, and these included the world's two biggest and best-exposed ophiolites, the Troodos Massif of Cyprus and the Semail ophiolite of Oman. Both of these were important for analogue studies of mid-ocean ridges, and the role of subduction in their origin became a major issue.

The problem with the Troodos Massif remained that, if one accepted (as most did by this time) that it formed at a spreading axis above and behind a subduction zone, it was difficult to explain the absence of an associated island arc. All modern analogues, Scotia Sea Site 24, the more arc-like lavas of the Western Pacific marginal basins, and even the fore-arc terranes, had associated island arcs. The concept of spreading at the initial stages of subduction was popular, but there was still no modern analogue where subduction had recently initiated.

The Semail (Oman) ophiolite debate was more complex. Two groups began working in Oman in the mid-1970s. The UK led by Ian Gass (of which I was part), worked in the north, where the lavas were best exposed. The USA group, led by Bob Coleman and Cliff Hopson, worked in the south, where the lower crust and upper mantle were best exposed. In the north, a late basalt-andesite-dacite sequence with distinct subduction chemistry on all discriminants formed small volcanic edifices above the ophiolite, and we suggested that they might represent an incipient arc (Pearce et al., 1981; Alabaster et al., 1982). The underlying ophiolite was close to MORB in its geochemistry, but we did detect a small subduction signature based on Y-Cr covariations. In the south, there were no upper lavas, and the gabbros were equivalent to those sampled in the main ocean basins, and so there was no indicator of subduction. To the USA group, the ophiolite was typical of lithosphere formed at a fast-spreading mid-ocean ridge unrelated to subduction (e.g., Coleman, 1981). The USA group rejected a subduction link for the same reason that the setting of the Troodos Massif continued to be debated: there was no chain of arc volcanic edifices to prove a subduction-related setting geologically.

Finding a term to describe ophiolites with a subduction-related geochemistry but no associated chain of arc volcanoes proved a surprisingly difficult problem. We did not view as accurate terms such as "fore-arc ophiolite," "back-arc ophiolite," "pre-arc ophiolite," or "island arc ophiolite" because there was no arc to pre-date or post-date. Moreover, even in ophiolites that were associated with island arc activity, it was usually impossible to tell geochemically or geologically where the ophiolite originally lay with respect to the arc. "Marginal basin ophiolite" was equally inappropriate, because many marginal basins were floored by crust of MORB composition formed before subduction started. Thus, a term was needed that described the subduction influence in generic terms without defining exactly where it lay within the subduction system.

Resolution of this semantic problem began when John Malpas (1981, personal commun.) introduced the term "supra-Benioff zone" to explain our inferred setting of formation of the Semail ophiolite, and this was first used in a paper on Oman geochemistry by Alabaster et al. (1982). The term "supra-subduction zone (SSZ) ophiolites" evolved from this and was formally named and defined by Pearce et al. (1984a, p. 77) as follows:

> Supra-subduction zone (SSZ) ophiolites have the geochemical characteristics of island arcs but the structure of oceanic crust and are thought to have formed by sea-floor spreading directly above subducted oceanic lithosphere. They differ from "MORB" ophiolites in their mantle sequences, the more common presence of podiform chromite deposits, and the crystallization of clinopyroxene before plagioclase which is reflected in the high abundance of wehrlite relative to troctolite in the cumulate sequence. Most of the best-preserved ophiolites in orogenic belts are of this type.

Drawing on the work so far described, we demonstrated that geochemical discriminant diagrams could be used together with information on the order of crystallization and nature of mantle peridotite to distinguish two ophiolite types: a MORB type and an SSZ type. According to our paper, SSZ ophiolites could occupy one of several settings:

- The initial stages of spreading (pre-arc ophiolites)
- The spreading-arc transition
- The arc-back-arc spreading transition

To unravel the precise setting of any given ophiolite, if possible at all, would (we argued) require a combination of geochemistry, lava stratigraphy, and field geology. We proposed that the geochemical stratigraphy of the lavas (e.g., subduction-zone affinity increasing upward for the spreading arc transition, but downward for the arc-back-arc spreading transition) might be one important characteristic.

At about the same time, Leitch (1984) divided what he termed "arc related ophiolites" into four classes:

- Type A: ophiolites derived from oceanic lithosphere in existence prior to island arc inception and trapped within the arc system at the onset of overthrusting
- Type B: ophiolites derived from oceanic lithosphere formed by rifting of the overriding plate after the start of slab-related magmatic activity
- Type C: ophiolites produced at the inception of island arc formation
- Type D: ophiolites produced by slab-related magmatism during arc evolution

As explained above, I view SSZ as a better term than "arc-related" because some ophiolites have no arc to which to relate. Nonetheless, Leitch's concept and our own are very similar. Leitch's Type A is actually a MORB ophiolite according to our definition, having been incorporated in the subduction system without having any genetic relationship to a subduction zone. Type B is an SSZ ophiolite, which is equivalent to pre-arc ophiolites as we defined them. Type C is an SSZ ophiolite created at the start of subduction by near-trench processes. Type D is an SSZ ophiolite, which encompasses our spreading-arc and arc-spreading transitions.

These papers represent syntheses of the work of many scientists since the Penrose Conference definition was published in 1972 (Anonymous, 1972). By 1984, it was clear to almost all geoscientists that ophiolites had diverse affinities and that many, and probably most, had subduction affinities. Enough data had accumulated that several groups of scientists were able more-or-less independently to make the subduction link, define the geochemical and petrological characteristics of subduction-related ophiolites, and define the types of setting at which they formed. The question remained, however: to precisely what setting above a subduction zone did a given ophiolite belong, and how could one tell? And what can SSZ ophiolites then tell us about global dynamics and subduction processes? These now became important questions for ophiolite scientists.

DEVELOPMENT OF THE SUBDUCTION INITIATION MODEL

One of the most significant advances after 1984 was the development of the subduction initiation model for SSZ ophiolites. This model does not apply to all SSZ ophiolites, but may apply to many, including some of the largest and best preserved. Ideas essentially evolved along three tracks, all requiring rifting and spreading as the first manifestation of crustal accretion following initiation of subduction and all assuming that, given time and sufficient lithosphere to subduct, the new lithosphere would form the substrate to subsequent arc volcanism.

- Rifting and extension of lithosphere of the overriding plate at the locus of normal arc magmatism following subduction initiation
- Rifting of young, near-trench lithosphere of the overriding plate following subduction initiation at an old-young lithosphere boundary
- Rifting of young, near-trench lithosphere of the overriding plate following subduction initiation at an active ridge-transform boundary

The hypothesis that spreading might predate arc volcanism was originally developed by Pearce (1975), Pearce et al. (1981), and Alabaster et al. (1982). Its aim was to explain the aforementioned observation that ophiolites such as the Troodos Massif and Semail ophiolite of Oman had the geochemical characteristics of island arcs, but the structure of oceanic crust and no associated arc edifice or overlying volcanogenic sediment. Our view was that they did not evolve into an arc proper because there was not enough oceanic lithosphere to subduct. We argued that many ophiolites (including Troodos and Semail) formed by closure of small oceans, and so froze the arc-basin system in its early, extensional stages of formation.

Drilling and dredging of the Western Pacific trenches supported the concept of extension associated with subduction initiation, but changed the perception that it must take place some distance behind the trench. The important observation was that the ophiolite sequences in the inner trench wall formed very soon after subduction initiation. Thus, either spreading must have taken place close to the trench in the first place, or 100 km or more of old, MORB-like lithosphere in the fore-arc had to be scraped off the inner trench wall and subducted. Karig (1982, p. 571) was one of the first to raise doubts:

> Although it is almost certain that some oceanic arc systems were initiated along transforms or fracture zones, it is less obvious that this process traps a 60–100 km wide belt of typical oceanic crust between the trench and arc.

Of course, this crust need not be trapped; it could be lost by trench erosion or development of a second subduction zone. While not impossible, this was unlikely and so ophiolite formation by extension at the locus of normal arc volcanism did not survive much longer as a model. Whatever model did apply had to explain extension before arc volcanism, the common presence of boninites, and the creation of new oceanic crust near the trench.

Hawkins et al. (1984), started from a different perspective, that of Western Pacific marginal basins. They also emphasized the importance of rollback-driven extension during the initiation and evolution of arc-basin systems, pointing out that this enabled ophiolites to originate in fore-arc, intra-arc, and back-arc settings. They also demonstrated that spreading before active arc volcanism was a common feature of arc-basin systems. Like us, they formed the earliest, extensional lithosphere behind the trench in response to upwelling of hot mantle. They then made the key, additional deduction that the locus of this crustal accretion must be nearer the trench than the active arc. Their argument was that boninites require a depleted mantle source and shallow depths of melt segregation, which in turn implies an origin at shallow subduction depths. In their words:

> [Mantle counterflow following subduction initiation] heats up the serpentinised peridotite, causes an uparch of the lithosphere and promotes dilation of a zone in the fore-arc. Boninite magmas, derived from melting of the peridotite, are the first melts to leak out along the zone of fore-arc dilation. Although the actual depth of boninite generation cannot be determined at this time, a depth of 50 km or less for melt-residue separation has been proposed. Thus, if a relation to the subduction process is inferred, boninite eruption may occur in (or be restricted to) the fore-arc region. (Hawkins et al., 1984, p. 194)

The Hawkins et al. (1984) mechanism still placed the first-formed oceanic lithosphere some distance from the trench, however. Leitch (1984) went a step further and proposed a mechanism for generating oceanic lithosphere still closer to the new trench:

> Where, after initiation of subduction, convergence keeps apace with rollback of the subduction hinge the rocks on the leading edge of the overriding plate will form the basement. However, if roll back rate exceeds convergence rate extensional conditions exist that may lead to ophiolite development in the forearc region. It is envisaged that in this situation asthenosphere displaced by the sinking slab flows around the leading edge of the slab into the region above it. Partial fusion produces basaltic melts that rise to form a magma body similar to that beneath one side of a mid-ocean rift, and new lithosphere fills the space between the retreating subduction hinge and the edge of the upper plate (p. 192–194.)

The Izu-Bonin-Mariana fore-arc provided the natural laboratory for investigating subduction initiation models. Although subduction initiation was of Eocene and not Recent age, the system was the only one to have retained the geometric elements of subduction initiation and so it was the nearest to an analogue for this ophiolite type. If we could understand the Mariana system, we should make significant advances in our understanding of ophiolites.

Uyeda and Ben-Avraham (1972), Hilde at al. (1977), and Hussong and Uyeda (1981) had already noticed that the Western Pacific arc terranes were orthogonal to the relict spreading axis in the West Philippine Basin. This led them to propose that subduction initiated at a transform fault associated with a change in Pacific plate motion. In this model, old Pacific lithosphere was juxtaposed against the young West Philippine Basin lithosphere of the Philippine plate at the time of subduction initiation. Although initiation at a transform fault was more or less consensus, the plate reconstruction was not. For example, Seno and Maruyama (1984) required a plate between the Pacific

Plate and Philippine Plate (their New Guinea plate) at the time of subduction initiation. Casey and Dewey (1984) raised the additional possibility that the transform was not pure strike slip, but contained active ridge-transform-ridge segments at which subduction could more readily initiate.

Casey and Dewey (1984) also made the link to boninite genesis by pointing out that subduction at a ridge or transform between two ridge segments would initiate melting beneath hot, shallow, and depleted lithosphere—all the criteria for boninite formation. Usefully, the model also explained how new ocean crust could be generated almost at the trench itself, as this region was weak and underlain by asthenosphere at shallow depth. The Casey and Dewey model and variants of it initially proved attractive. For example, Boudier et al. (1988) explained the Oman ophiolite by subduction beginning at an active ridge. In this way, they argued, it could begin life as a MORB ophiolite and evolve into a SSZ ophiolite, thus reconciling the ongoing debate about its tectonic setting of formation. This currently remains a matter of debate.

The second phase of drilling in the Western Pacific also contributed to our understanding of subduction initiation and its link to ophiolites. ODP Leg 125 drilled holes in fore-arc serpentinite seamounts and the outer arc high of the Izu-Bonin-Mariana fore-arc (Fryer et al., 1989) and Leg 126 drilled holes in the Izu-Bonin fore-arc basin (Taylor et al., 1990). Hole 786B, drilled during Leg 125 in the outer-arc high of the Izu-Bonin fore-arc, was particularly informative. It penetrated an arc edifice into the underlying oceanic crust, reaching sheeted dikes (the first drilled in SSZ crust). Both the arc edifice and the underlying, possibly oceanic, crust were boninitic in composition.

Our interpretation of the ODP Leg 125 drill core (Pearce et al., 1992) was that the fore-arc crust formed by subduction beneath the ridge or young crust of the West Philippine Basin. The boninites carried an unusual geochemical signal in which Zr was strongly enriched relative to many other elements, and Sr/Y ratios were high. We interpreted this as an "adakitic" slab-melting fingerprint, which would then support the Casey and Dewey model because this required subduction of very young lithosphere.

At around the same time, Stern and Bloomer (1992) published their model for the link between subduction initiation and ophiolite formation. They favored the original models of Uyeda and co-workers, in which older, colder, and thicker Pacific oceanic lithosphere is juxtaposed against younger, hotter, and thinner West Philippine Basin lithosphere before subduction starts. In terms of mechanism, their starting point was the fore-arc extensional model of Hawkins et al. (1984) and, particularly, the Leitch (1984) Type C ophiolite model. The Stern and Bloomer (1992) model provided the link between Leitch's model and the Uyeda picture of subduction initiation in the Western Pacific. They extended the Leitch ideas as follows:

> Melting in such an extensional environment would lead to infant-arc crust formation by sea-floor spreading. This spreading would in many ways be similar to that of mid-ocean ridge and back-arc basin spreading regimes with two significant differences. First spreading is likely to be poorly organised, with many discrete ridge segments and asymmetric spreading. The infant arc crust nearest the precursor transform fault might be first to form, although this may be replaced by younger infant-arc crust during later rifting events. (Stern and Bloomer, 1992, p. 1626)

The important component in their model was the buoyancy contrast between the young, hot lithosphere of the West Philippine Basin in the west and the old, cold Pacific lithosphere in the east. Their idea is that, on initiation of subduction, the older, colder slab sinks and the trench rolls back, thus creating extension close to the trench and allowing asthenosphere to rise into the space created. Stern and Bloomer used the term "arc infancy" to describe this setting. Bloomer et al. (1995) developed this concept further, linking it to the fore-arc drilling from ODP Legs 125 and 135. They concluded (p. 1):

> This early or "infant" arc volcanism was characterised by the eruption of very depleted boninitic and arc tholeiitic lava compositions and occurred in extensional environments. This infant arc volcanism was built on, or displaced, the pre-existing oceanic crust and had igneous production rates much higher than those of mature arcs—eruption rates on the order of those from in slow-spreading ridges. This initial arc volcanism is unlike that developed during the "normal or mature phases of arc activity and is a plausible mechanism for developing supra-subduction zone ophiolites.

Geologically, it is impossible to distinguish between the Uyeda model and the Casey and Dewey model because the evidence (the crust immediately outboard of the incipient subduction zone) has all been subducted. Pearce et al. (1999) tested the two models using isotopic fingerprinting of the earliest supra-subduction zone magmas. The starting point was the Hickey-Vargas (1991) discovery that the marginal basins of the Western Pacific exhibit an "Indian" Pb isotope signature, whereas the subducting Pacific Plate exhibits a "Pacific" signature. In the Uyeda model, therefore, the subduction signature should be Pacific, and the mantle signature Indian. In the Casey and Dewey model, the subduction and mantle signatures should be the same unless subduction initiation fortuitously began at a mantle domain boundary. In the event, the Pb signature was Pacific and the Hf signature was Indian, clearly supporting the Uyeda, and Stern and Bloomer, model of Pacific plate subduction beneath the West Philippine Basin.

Overall, therefore, we can use the Mariana analogue to build up quite a detailed picture of how ophiolites might form at the start of subduction. But is this the only analogue?

RIDGE-TRENCH INTERSECTION (RTI) MODELS FOR SSZ OPHIOLITE FORMATION

Subduction initiation models for SSZ ophiolite formation require that trench rollback take place in the early stages of subduction and that this will lead to seafloor spreading near the trench above the nascent subduction zone. Near-trench seafloor spreading is therefore a direct consequence of rollback that is linked to subduction initiation. The RTI models have a very different mechanism because the subduction zone imparts a subduction characteristic to ridges that are near-trench for reasons unrelated to subduction initiation. In some cases, as I shall explain, subduction initiation may take place at an RTI. RTI intersections capable of producing SSZ ophiolites can be divided into three types:

- Subduction beneath a ridge
- Subduction of a ridge (SSZ ophiolite formation in the fore-arc)
- Subduction of a ridge (SSZ ophiolite formation in the subducting plate)

In the first two of these, the active spreading takes place in the fore-arc. In the third, the spreading takes place in the subducting plate.

Subduction Beneath a Ridge

Ridge-trench intersections have many possible geometries. The first to be invoked as settings for ophiolite genesis is the class that involves subduction beneath a ridge. This setting was first invoked for ophiolite emplacement by Dewey, 1976, and then for anomalous, near-trench magmatism by Marshak and Karig, 1977. Subsequently, Karig (1982) and Casey and Dewey (1984) focused on its potential importance for ophiolite generation. Typically, the spreading axes are orthogonal to the trench, in contrast to subduction initiation models, where the spreading is trench-subparallel. There are many such examples at the present time, mostly at the edges of arc-trench systems, such as the northern termination of the Lau Trench, the southern termination of the Mariana Trench, the southern termination of the Vanuatu Trench, and both north and south terminations of the South Sandwich Trench. With these in mind, Casey and Dewey concluded:

> Therefore, whilst some ophiolites may have "island arc" petrological affinities, they should not necessarily be discounted as the end products of the sea-floor spreading process. Ophiolite formation by sea-floor spreading, whether at a normal mid-ocean ridge or at a back-arc spreading centre that traverses a usually non-volcanic forearc region, would seem to best explain the geologic character of many obducted ophiolite complexes. (Casey and Dewey, 1984, p. 286)

Compositionally, these settings provide good analogues for some ophiolites. The principal rock types are island arc tholeiites, back-arc basin basalts, and boninites, especially those of high-Ca type. Those studying dredged rocks from modern analogues have commented on their similarity with many supra-subduction zone ophiolites, including the Troodos Massif. For example, Falloon et al. (1987, p. 487) interpreted the dredged boninitic rocks from the northern termination of the Tonga Trench as follows:

> [These rocks] are similar to low-Ti ophiolitic basalts such as those of the Upper Pillow Lavas of the Troodos ophiolite, Cyprus. The dredged Tongan high-Mg lavas are therefore an in situ occurrence of a low-Ti ophiolitic lava suite, and support an intra-oceanic island arc origin for the Troodos and other ophiolites with these lavas.

Falloon et al. (1987) interpreted the setting of boninite formation as an island arc, but it is probably more complicated than that. Both the ridge (Northern Lau Spreading Center) and arc (Tofua arc) converge at that point, and it is not clear whether the lithosphere is oceanic or arc-like. The southern termination of the Vanuatu arc similarly converges on the southernmost spreading center of the North Fiji Basin. The submarine islands have the composition of high-magnesium andesites (Monzier et al., 1993). The southern termination of the Mariana system (Crawford et al., 1986; Stern et al., 1989) and the southern and northern terminations of the Scotia Sea (Leat et al., 2000; Fretzdorff et al., 2002) both contain depleted island arc tholeiites resembling those found in many SSZ ophiolites.

In each of these settings, there is the potential to modify the mantle by addition of a subduction component before it rises beneath the ridge. In addition, the residual lithosphere at that ridge would be hot, and therefore able to further melt by fluid addition, to give boninites. Finally, because the setting is at the fracture-zone edge of a subduction system, it would be the first to be emplaced by continent collision.

It is also important to note that the reconstruction of the Western Pacific at the time of subduction initiation had an RTI where the West Philippine Basin spreading center intersects the embryonic subduction zone. Therefore, over some of its length at least, subduction initiation and RTI mechanisms can overlap.

Subduction of a Ridge: SSZ Ophiolite Formation in the Fore-arc

The second type of RTI invoked for ophiolite genesis is that of ridge subduction. The concept of ridge subduction beneath accretionary complexes as a means of explaining near-trench magmatism was put forward by Marshak and Karig (1977). They also demonstrated how the migration of triple junctions could produce the trench-parallel belts of near-trench magmatism observed in Sumatra, Alaska and Japan. Miyake (1985) published one of the first papers to investigate in detail the potential link between ridge subduction and ophiolite genesis. He studied the Shionomisake igneous complex, which formed autochthonously in the fore-arc basin on Honshu in the Miocene. Miyake identified two of the characteristic geochemical features of this type of setting: a MORB character when plotted on diagrams such as Ti-Cr (which are based on degree of melting) and TiO_2-FeO/MgO (which also depend on oxygen fugacity). He also identified enrichment in Th and LREE, but, having no Ta or Nb data, was not able to attribute this to a SSZ character. He concluded:

> The Shionomisake igneous complex is analogous petrologically and petrochemically to abyssal igneous association, though the basic rocks are more enriched in LREE and Th than N-MORB and are associated with relatively more voluminous acidic rocks than …the oceanic crust. The very existence of the Shionomisake igneous complex and other igneous masses in the modern forearc basins suggest another template for the formation of ancient ophiolite complexes. (Miyake, 1985, p. 32)

The Taitao ophiolite in Southern Chile presents an even better example, being younger (Plio-Pleistocene) and having a more complete ophiolite structure. It is located in the fore-arc above the site of Chile Ridge subduction. Forsyth et al. (1986) demonstrated that the Taitao ophiolite has a composition intermediate between MORB and SSZ character on a Ti-Cr plot and they also noted a high proportion of silicic rocks compared with normal oceanic crust. Kaeding et al. (1990) and Lagabrielle et al. (1994) carried out more complete geochemical studies. They reached similar conclusions, namely that ridge subduction leads to decompression melting of MORB mantle, but with the top of the melting column located beneath the continental edge so that the degree of melting is less than normally experienced at mid-ocean ridges. Moreover, the resulting enriched MORB then undergoes extensive assimilation and fractional crystallization in the crust to produce the high proportion of silicic rocks and the enrichments in Th and LREE.

A series of subsequent papers found near-trench plutonic and ophiolite complexes of various ages around much of the Pacific Rim. In Japan and Alaska, in particular, ridge subduction took place beneath accretionary complexes rather than continental crust. The current view is that a wide range of ophiolites and related rock types can be emplaced in fore-arc terranes. These may be derived both from crust formed outboard of the trench and emplaced in the fore-arc (as first envisaged by Dewey, 1976) and from in situ injection of basic magma into the accretionary complex (e.g., Lytwyn et al., 1997, 2000).

Both subduction and crustal assimilation are capable of giving an apparent subduction signature on plots such as Th-Hf-Ta, particularly for the most fractionated rocks. Whether or not this type of ophiolite should be termed SSZ is a semantic problem. The key question should be whether or not the mantle source has a subduction signature. If the mantle source lay beneath the subducting plate and has flowed into the zone of melting without residence in the mantle wedge, then the ophiolite is strictly of MORB type (albeit contaminated by crust). If the mantle source has flowed laterally from the adjacent mantle wedge into the zone of melting, then it is strictly of SSZ type. Evidence to date is ambiguous, and it is quite possible that both MORB and SSZ mantle sources are potentially available, i.e., that both ophiolite types are present.

Subduction of a Ridge: SSZ Ophiolite Formation Before Subduction

Another scenario for SSZ ophiolite formation is when subduction-influenced mantle can melt to form oceanic crust of apparent SSZ affinity before it is subducted. This effect was first recorded in the Woodlark Basin, where the spreading center approaches the Solomon Trench. The crust formed in this setting has a distinct SSZ character (Perfit et al., 1987; Johnson et al., 1987). The Solomon Island region is complicated by the fact that the subduction zone originally faced south, and then changed polarity to its present northerly direction. In consequence, it is possible that the SSZ composition is not related to the subduction of the ridge, but to the presence of mantle that once overlay the earlier subduction zone. Perfit et al. (1987, p. 114) conclude:

> The Woodlark Basin is apparently chemically zoned orthogonal to the current locus of subduction. Crust farther than 150 km from the subduction zone is normal MORB. Nearer the subduction zone, the submarine volcanic rocks acquire an increasing arc-like chemistry…The chemical zonation is difficult to explain in terms of the current plate configuration but may be explicable if the mantle beneath the Woodlark Basin was previously affected by subduction of reversed polarity.

The Woodlark Basin has proved an attractive analogue for SSZ ophiolite genesis. In fact Saleeby (1982) invoked the analogue for the Sierra foothills ophiolites well before Perfit et al. and Johnson et al. published the analyses of their dredged samples. Following the publication of the analyses, Dilek and Moores (1990) carried out a more detailed assessment of the merits of the eastern Woodlark Basin as an SSZ ophiolite analogue, concluding that it was a viable option for some Eastern Mediterranean ophiolites.

Subsequently, Klein and Karsten (1995) identified a similar signature where the Chile Ridge enters the South American Trench. Here, reversal of subduction polarity was clearly not a factor. They recognized the potential significance to ophiolite studies, as follows:

> …the Chile Ridge lavas are the first mid-ocean ridge basalts with convergent-margin characteristics suggests that caution is warranted in using trace-element systematics to identify the provenance (mid-ocean ridge or supra-subduction zone) of ophiolites. (Klein and Karsten, 1995, p. 52)

The implication of their observation is that asthenosphere can flow from the mantle wedge and beneath and along the subducting mid-ocean ridge to impart a SSZ chemistry. There are, though, differences with most SSZ ophiolites: although elements such as Th and LREE are enriched relative to Nb and other HFSEs, the rocks still plot as MORB on diagrams such as Cr-Y, that are based on degree of melting or oxygen fugacity. This setting does, however, provide a small piece of ammunition for those still holding out for a mid-ocean ridge origin for all ophiolites.

BACK-ARC BASIN ANALOGUES

Following the discovery of Site 24 of the East Scotia Sea, the detailed geochemical mapping of the various marginal basins that have been carried out over the past 20 years has revealed that all back-arc basins had highly variable compositions ranging from MORB to SSZ type, with all gradations in between. A key site is the Valu Fa Ridge at the south of the Lau Basin. It has an arc-like chemistry (Jenner et al., 1987), but an oceanic crustal structure and a sub-axial magma chamber. Andesite-dacite lavas are exposed at the ocean floor. More regional geochemical mapping showed a progressive increase in SSZ character from MORB in the center of the basin to the Valu Fa Ridge in the south as the arc is approached (Pearce et al., 1995).

Gradual accumulation of data essentially showed that back-arc basins could have MORB or SSZ character according to the precise setting of the fragment of oceanic lithosphere with respect to the subducting plate and the direction of mantle flow. For example, in the Central and Southern Lau Basin, the subduction affinities increase toward the south, where the spreading center converges on the arc. In the East Scotia Sea, as already noted, subduction signatures are highest in the north and south, where mantle flows onto the system over the subducting plate (Leat et al., 2000; Fretzdorff et al., 2002). In the Mariana Trough, all parts of the spectrum between MORB and SSZ compositions have been identified, with a general increase to the north as the ridge changes to a rift zone and then merges with the arc itself (e.g., Stolper and Newman, 1994; Gribble et al., 1998).

Within this phase of data accumulation, a major and unexpected advance was made during ODP Leg 135. Scientists discovered that much of the Lau Basin is floored not by new oceanic lithosphere formed by back-arc spreading, but by strongly attenuated arc lithosphere (Parson and Hawkins, 1994). In other words, the sequence of rifting and drifting that applies to continental separation also applies to arc separation. From the ophiolite perspective, it raises the question of whether some fragments of SSZ ophiolite represent the products of true back-arc spreading or whether they result from rifting of arc terrane. Given that many MORB ophiolites represent incipient spreading of "normal" oceans, it would not be surprising to find their equivalents in marginal basins. A further point of note is that the rifting process also gives rise to magmatism by decompression. If the rifting is advanced, and the subduction zone distal, the overlying magmas can be of MORB composition while the underlying oceanic lithosphere is of SSZ type.

All the scenarios listed above assume that, if subduction is sufficiently long-lived, an arc and back-arc basin will result. This need not be the case, however. Moores et al. (1984) pointed out that oblique subduction could lead to the formation of an active ridge-transform system above a subduction zone, without associated arc volcanoes. They proposed that this process could explain the absence of arc volcanoes in many Eastern Mediterranean SSZ ophiolites and cited the Andaman Sea as a modern analogue. This is an attractive hypothesis, but still has to be tested by sampling and analysis of Andaman Sea lavas.

"LIFE CYCLES" OF SSZ OPHIOLITES

Although space has restricted this article to the identification and origin of SSZ ophiolites, Shervais (2000) has presented a more holistic synthesis, incorporating information on their evolution and emplacement. He discusses what he terms the "life cycle" of SSZ ophiolites. There are five stages to his cycle:

- Birth (formation in a nascent or reconfigured subduction zone)
- Youth (continued mantle wedge melting and crustal accretion)
- Maturity (onset of arc volcanism)
- Death (sudden cessation of subduction, usually because of continent collision)
- Resurrection (emplacement of the ophiolite onto a passive continental margin)

Shervais (2000) explains this sequence in more detail in paragraph [13]:

> These events generally progress in an orderly fashion from birth through death and resurrection, but not all ophiolites display all stages of this life cycle. In particular some SSZ ophiolites never reach maturity but skip directly to death and resurrection. In others, death and resurrection are coincident, with no evidence for a prolonged interval between these events. The "birth" and "youth" stages seem to be common to all SSZ ophiolites and reflect their original formation in the upper plate of a subduction zone.

In many respects, this life-cycle concept began with the tectonic diagrams of Dewey and Bird (e.g., 1971), who were (as recounted earlier) the first to examine the possibility that ophiolites might form in marginal basins. Shervais' synthesis incorporates the 30 years of work that have subsequently been carried out to define and explain the origin of SSZ ophiolites. It is interesting to compare these papers, which are currently among the first and last articles written about subduction-related ophiolites. The comparison shows how far ideas on subduction-related ophiolites have changed over the period of time discussed in this paper. For example, in Dewey and Bird (1971), subduction-related ophiolites are restricted to back-arc and intra-arc basins, whereas fore-arcs contain trapped ocean crust once formed at mid-ocean ridges. In Shervais (2000), back-arc and intra-arc basins are much less important: it is the fore-arcs that provide the source of most subduction-related ophiolites, and subduction initiation that is the principal mechanism for their formation.

A KUHNIAN ANALYSIS OF THE EVOLUTION OF IDEAS ON SSZ OPHIOLITES

Kuhn (1962) postulated that scientists, at any one time, adhere to a paradigm—a set of fundamental beliefs. Under conditions of "normal science," this paradigm shifts very slowly. Under conditions of "extraordinary science," however, the paradigm shifts rapidly and we have a "scientific revolution." The revolution need not be instantaneous but may be preceded by years of debate. Kuhn's ideas on scientific revolutions have already been applied with apparent success to the development of the plate tectonic hypothesis (e.g., Hallam, 1973). In this context, global tectonic studies have been, since the mid-1960s, in a period of "normal science" following the plate tectonic revolution.

Development of ideas on ophiolites can be viewed on one level as part of the post-plate tectonic "mopping-up" operation whereby geological features are reinterpreted in terms of the new paradigm. As a subject on their own, however, they reveal their own paradigm shifts. The first of these in recent times (but predating the period discussed in this paper) was the shift from ophiolites as components of geosynclines to fragments of oceanic lithosphere. This shift was extremely rapid and occupied the immediate post-plate tectonic period, culminating in the Penrose Conference definition published in 1972 (Anonymous, 1972).

The second shift, discussed in much detail here, was from ophiolites as fragments of oceanic crust formed at mid-ocean ridges to ophiolites as fragments of oceanic crust formed at supra-subduction zone ridges. This shift was slower and, as discussed in this paper, developed through geological studies on ophiolites, development of analytical and techniques and geochemical methods, and drilling and dredging of potential analogues at sea. It was the product of some 12 years (1972–1984) of debate and data acquisition, and culminated in the supra-subduction zone concept. In fact, 1984 saw a series of key papers that established this concept, not just my own that defined the term (Pearce et al., 1984a), but others that placed this group of ophiolites clearly in a subduction context (e.g., Leitch, 1984; Hawkins et al., 1984; Casey and Dewey, 1984; Dick and Bullen, 1984).

The third (post-1984) stage of evolution of ideas is the post-SSZ ophiolite stage. The SSZ definition makes no statement about the precise setting of the ophiolite other than its link with subduction. A period of "normal science" is now taking place to pinpoint the precise settings in which these could happen. This could be viewed as a "mopping up operation," but that description would hide a series of small revolutions that have led to the establishment of subduction initiation and ridge-transform intersections as prime candidates for SSZ ophiolite formation, and have raised the possibility that strongly attenuated arc terranes and oblique subduction present other candidates for SSZ settings.

A BAYESIAN ANALYSIS OF THE EVOLUTION OF IDEAS ON SSZ OPHIOLITES

It is also possible to analyze the chain of events using Bayesian methods, specifically the Bayes Decision Rule. This is because the assignment of a given ophiolite to its most probable tectonic setting of formation involves decision-making processes. Decision making requires two components: a knowledge base comprising the evidence, and an inferencing engine comprising the decision-making method, in this case the Bayes Decision Rule (see Pearce, 1987 for a more detailed explanation). The Bayes Decision Rule can be written as:

$$p(T_j|E_i) = \frac{p(E_i|T_j)p(T_i)}{p(E_i)} \quad (1)$$

where $p(T_j)$ is the a priori probability of the unknown sample x belonging to tectonic setting T_j; $p(E_i/T_j)$ is the likelihood that the unknown sample x belongs to setting T_j, given the evidence E_i; $p(E_i)$ is the total probability of the piece of evidence, E_i; and $p(T_j/E_i)$ is the posterior probability of the unknown sample x belonging to setting T_j, given the evidence E_i.

To apply this rule for ophiolites, we start with the a priori probabilities (the probabilities before any discrimination criteria are applied). Then, for the first piece of geochemical, geological, or petrological evidence, we use equation 1 to determine the posterior probability of each setting, i.e., the probabilities after that piece of evidence has been considered. This posterior probability is then treated as the a priori probability for consideration of the next piece of evidence, and so on. The eventual outcome is a set of probabilities that the ophiolite in question belongs to each of the designated magma types.

Two other points are important in making the assignment. First, it is possible to incorporate the measure of belief into the equation. Measure of belief of a given piece of evidence (E_i) can be written (MB/E_i), and follows a scale of 1 (complete belief) to 0 (no belief, equivalent to nonconsideration of the evidence). The posterior probabilities then become:

$$p(T_j|E_i) = \frac{p(E_i|T_j)p(T_i)(MB|E_i)}{p(E_i)} - (1 - MB|E_i)p(M_j) \quad (2)$$

The second point has to do with whether or not the assignment can be made with confidence. Essentially, the most likely setting is the one with the highest probability. For this assignment to be made with confidence, the value of this probability has to lie above a statistically determined threshold.

Now consider the situation at the time of the Penrose ophiolite definition. Essentially, the database at the time comprised only three potential settings for creating oceanic igneous crust: mid-ocean ridge, island arc, and oceanic island.

The knowledge base in the 1960s comprised crustal structure, the sheeted dike evidence for extension, and the rock associations. If all existed (as in completely developed ophiolites, such as the Troodos Massif), a mid-ocean origin was reasonable and inevitable. If only some components were present (as in some fragmentary ophiolites), then the overall probability may not exceed the threshold, so that the conclusion may not be made with confidence. If pieces of evidence were present but in a state that they had to be inferred (as in other fragmentary ophiolites), then a mid-ocean ridge origin was possible only if the MB of the various pieces of evidence were high.

When geochemistry became available (in the 1970s), ophiolites such as the Troodos Massif gave mixed signals. Geological evidence gave highest probabilities for the mid-ocean ridge setting, whereas geochemical evidence gave the highest probability to the island arc setting. In detail, this should mean that no overall probability would exceed the threshold, so that no definite conclusion could be made. However, measure of belief, subjectively applied, could become highly significant. If a geologist could set the MB of the geochemical evidence to 0, but the MB of geological evidence to 1, then a mid-ocean ridge origin possible. Similarly, a geochemist could set the MB of the geological evidence to 0, but the MB of geochemical evidence to 1, so making an island arc origin possible. This, of course, is the essence of the Troodos controversy, with two groups questioning the validity of the evidence used by the other.

The problem was that the tripartite division of tectonic settings did not fully describe the possibilities. By 1975, the knowledge base had increased and the possible settings had become four: mid-ocean ridge, island arc, oceanic island, and marginal basin ridge.

Essentially, the marginal basin setting would have similar probabilities to mid-ocean ridges for geological pieces of evidence, but probabilities overlapping those of island arcs for geochemical lines of evidence. Thus, an ophiolite with ridge geology and arc geochemistry could exceed the probability threshold as a marginal basin ophiolite. Even so, the controversy could (and, to a certain extent still does) continue for anybody wishing to set a low MB for some of the lines of evidence. It could also reasonably continue for ophiolites for which the geochemical signals were insufficiently strong (such as the Semail ophiolite) and those that were too fragmentary for confident application of geological criteria.

In detail, however, this was not efficient. A marginal basin ridge could chemically span the spectrum of compositions from arcs (if subduction-zone influenced) to ridges (if not subduction-zone influenced). A much more effective discrimination thus is achieved by making a further subdivision of magma types into MORB ridge, SSZ ridge, island arc, and oceanic island.

This was the effective recommendation of the Pearce et al. (1984a) paper. By this time, too, the knowledge base had increased enough that the petrology and geochemistry of all components of the ophiolite sequence could be used in the discrimination.

The post-1984 work then focused on whether the MORB ridge and SSZ ridge could be subdivided in a way that would improve decision-making. This work has made it possible to define tectonic settings as:

MORB ridge	**SSZ ridge**
Major ocean	Subduction initiation
Ocean margin	Subduction beneath a ridge
Plume-influenced ridge	Attenuated arc

Back-arc basin
Ridge subduction
Oblique subduction

Clearly, however, neither geochemistry alone nor the combination of geology and geochemistry can always separate these settings at above-threshold probability. The aim must still be to recognize the SSZ characteristics and then examine the petrogenesis and geological setting in some detail to make the final assignment.

SUMMARY

1. The concept of ophiolite formation in marginal basin settings was first proposed by Dewey and Bird in 1971. They presented a set of geological criteria for distinguishing ophiolites formed in marginal basin ridges from those formed at mid-ocean ridges.

2. The use of immobile elements to fingerprint ophiolites began with Cann (1970) and continued with Pearce and Cann (1971, 1973). The projections used at the time had a large overlap between island arc and mid-ocean ridge basalts, and so did not contradict a mid-ocean ridge origin for ophiolites.

3. The first use of geochemical criteria to demonstrate a subduction-related setting was that of Miyashiro (1973), who proposed an island arc origin for the Troodos Massif, on the basis of major element fractionation trends. This attracted much criticism because of the absence of geological evidence and fact that his criteria were sensitive to weathering and metamorphism.

4. Improvements in immobile element discrimination supported and extended many geochemical aspects of Miyashiro assertion. However, they did not explain how a fully developed ophiolite such as the Troodos Massif could have subduction-zone geochemistry but ocean-crust structure and no associated arc or arc-derived sediments. It led to the inference that the Troodos Massif and comparable ophiolites formed in a marginal basin setting. Even so, debate continued because of the absence of a modern analogue.

5. Exploration of marginal basins in the late 1970s helped locate these potential analogues. Back-arc basins continued to yield rocks with compositions close to those of mid-ocean ridges, with the notable exception of East Scotia Sea Site 24. More significantly, the RV *Dimitry Mendeleev* cruise and drilling on DSDP Legs 59 and 60 recovered all components of the ophiolite section from the Tonga and Mariana fore-arcs—and these had the necessary subduction signature. They also recovered the first boninites from the ocean floor, which led to the recognition of the essential link between fore-arcs, boninites, and oceanic crust with subduction characteristics.

6. By the early 1980s, accumulation of petrologic and geochemical data had demonstrated that there were two principal types of ophiolites, one with subduction geochemistry, early crystallization of clinopyroxene, and harzburgitic residual mantle, and another with MORB geochemistry, early crystallization of plagioclase, and lherzolitic-harzburgitic residual mantle. Most differences could be explained by the subduction influence on magma genesis: addition of subduction-mobile elements, increased fluid-induced melting, and increased oxygen fugacity.

7. The term "supra-subduction zone (SSZ) ophiolites" was devised to describe the larger group because, although the geochemical and petrological link to subduction was evident, the

precise setting was rarely clear. Moreover, terms such as "arc-related" did not take into account the observation that oceanic crust with a subduction signature could exist without any associated arc. This definition, published in 1984 at the same time as a number of equivalent syntheses, is an important landmark in the evolution of ideas on ophiolites. Before 1984, the emphasis was on debating and establishing the subduction link to ophiolites. After 1984, the emphasis was on establishing the precise settings in which SSZ ophiolites form, and the mechanisms by which they do so.

8. Subduction initiation soon became the setting of choice for deriving many of the SSZ ophiolites. The principal question was the precise mechanism. The absence of relict crust between the ophiolite and the trench favors a model in which spreading takes place near the trench itself. Two models were proposed and investigated: subduction initiation at an active ridge-transform system, and subduction initiation at a transform separating young and old oceanic lithosphere. In the type example of the Izu-Bonin-Mariana system, data favors the second of the two models, hence the mechanism for ophiolite formation summarized by Stern and Bloomer in 1992.

9. Subduction initiation did not emerge as the only mechanism for producing oceanic lithosphere in a fore-arc-type setting. Ridge-trench intersections can achieve this, either by subduction beneath a ridge or subduction of a ridge. Other settings hitherto unconsidered also were proposed, including oblique subduction and attenuation of arc lithosphere at the margins of back-arc basins.

10. A recent synthesis by Shervais (2000) demonstrates how those SSZ ophiolites formed by subduction initiation in fore-arcs can be considered to follow a "life cycle" from birth, through youth, maturity, and death, to resurrection.

11. Evolution of ideas can be considered in both a Kuhnian and Bayesian context. From a Kuhnian perspective, the definition of SSZ ophiolites was the final event in a paradigm shift that depended on a series of advances in geochemistry, marginal basin exploration, and land-based geology. This paradigm shift—from ophiolites as fragments of lithosphere created at mid-ocean ridges to lithosphere created mainly at a variety of subduction-related settings (some with no Recent analogues)—took ~10 years to achieve. From a Bayesian perspective, the assignment of an ophiolite to its correct tectonic setting is a question of probabilities, in which the necessary and correct knowledge base and decision-making apparatus has to evolve before the probability of assignment can exceed the decision-making threshold and give a robust interpretation.

ACKNOWLEDGMENTS

I thank the many ophiolite workers, geochemists, and marine geologists with whom I have discussed ophiolite problems over the past 30 years. In writing this review, I have attempted to be objective in acknowledging the achievements of all people and groups involved, but there are only so many references I can cite and I cannot fully submerge my personal perspective. I therefore take this opportunity to acknowledge the work of numerous scientists not mentioned here who have contributed to the development of ideas on ophiolites and their relationship to subduction zones. I am also grateful to Yildirim Dilek, John Shervais, Paul Robinson, and Bob Stern for their constructive comments. This is contribution 5, School of Earth, Ocean, and Planetary Sciences, Cardiff University.

REFERENCES CITED

Alabaster, T., Pearce, J.A., and Malpas, J., 1982, The volcanic stratigraphy and petrogenesis of the Oman ophiolite complex: Contributions to Mineralogy and Petrology, v. 82, p. 168–183.

Anonymous, 1972, Penrose field conference on ophiolites: Geotimes, v. 17, p. 24–25.

Arai, S., 1994, Characterization of spinel peridotites by olivine-spinel compositional relationships: Review and interpretation: Chemical Geology, v. 113, p. 191–204.

Beccaluva, L., Ohnenstetter, D., and Ohnenstetter, M., 1979, Geochemical discrimination between ocean-floor and island arc tholeiites—application to some ophiolites: Canadian Journal of Earth Sciences, v. 16, p. 1874–1882.

Beccaluva, L., Di Girolamo, P., Macciotta, G., and Morra, V., 1983, Magma affinities and fractionation trends in ophiolites: Ofioliti, v. 8, p. 307–324.

Bird, J.M., Dewey, J.F., and Kidd, W.S.F., 1971, Proto-Atlantic oceanic crust and mantle: Appalachian/Caledonian ophiolites: Nature Physical Sciences, v. 231, p. 28–31.

Bloomer, S.H., and Hawkins, J.W., 1983, Gabbroic and ultramafic rocks from the Mariana Trench: An island arc ophiolite, *in* Hayes, D.E., ed., The Tectonic and Geologic Evolution of Southeast Asian Seas and Islands (Part II): Washington, D.C., American Geophysical Union Geophysical Monograph Series, v. 27, p. 294–317.

Bloomer, S.H., Taylor, B., MacLeod, C.J., Stern, R.J., Fryer, P., Hawkins, J.W., and Johnson, L., 1995, Early arc volcanism and the ophiolite problem: a perspective from drilling in the Western Pacific, *in* Taylor, B., and Natland, J., eds., Active Margins and Marginal Basins of the Western Pacific: Washington, D.C., American Geophysical Union Geophysical Monograph Series, v. 88, p. 1–30.

Bonatti, E., and Michael, P.J., 1989, Mantle peridotites from continental rifts to ocean basins to subduction zones: Earth and Planetary Science Letters, v. 91, p. 297–311.

Boudier, F., and Nicolas, A., 1985, Harzburgite and lherzolite subtypes in ophiolitic and oceanic environments: Earth and Planetary Science Letters, v. 76, p. 84–92.

Boudier, F., Ceuleneer, G., and Nicolas, A., 1988, Shear zones, thrusts and related magmatism in the Oman ophiolite: initiation of thrusting on an oceanic ridge: Tectonophysics, v. 151, p. 275–296.

Browning, P., 1984, Cryptic variation within the cumulate sequence of the Oman ophiolite: magma chamber depth and petrological implications: London, Geological Society of London Special Publication, v. 13, p. 71–82.

Bruhn, R.L., Stern, C.R., and De Wit, M.J., 1978, Field and geochemical data bearing on the development of a Mesozoic volcano-tectonic rift zone and back-arc basin in southernmost South America: Earth and Planetary Science Letters, v. 41, p. 32–46.

Cameron, W.E., Nisbet, E.G., and Dietrich, V.J., 1979, Boninites, komatiites and ophiolitic basalts: Nature, v. 280, p. 550–553.

Cann, J.R., 1970, Rb, Sr, Y, Zr, and Nb in some ocean-floor basaltic rocks: Earth and Planetary Science Letters, v. 19, p. 7–11.

Capedri, S., Venturelli, G., Bocchi, G., Dostal, J., Garuti, G., and Rossi, A., 1980, The geochemistry and petrogenesis of an ophiolitic sequence from Pindos, Greece: Contributions to Mineralogy and Petrology, v. 74, p. 189–200.

Casey, J.F., and Dewey, J.F., 1984, Initiation of subduction zones along transform and accreting plate boundaries, triple-junction evolution and forearc spreading centres - implications for ophiolitic geology and obduction: London, Geological Society of London Special Publication, v. 13, p. 269–290.

Church, W.R., and Coish, 1976, Oceanic versus island arc origin of ophiolites: Earth and Planetary Science Letters, v. 31, p. 8–14.

Church, W.R., and Riccio, L., 1977, Fractionation trends in the Bay of Islands ophiolite of Newfoundland: polycyclic cumulate sequences in ophiolites and their classification, Canadian Journal of Earth Sciences, v. 14, p. 1156–1165.

Coleman, R.G., 1971, Plate tectonic emplacement of upper mantle peridotite along continental edges: Journal of Geophysical Research, v. 76, p. 1212–1222.

Coleman, R.G., 1977, Ophiolites: New York, Springer-Verlag, 288 p.

Coleman, R.G., 1981, Tectonic setting for ophiolite obduction in Oman: Journal of Geophysical Research, v. 86, p. 2497–2508.

Crawford, A.J., 1989, Boninites: London, Unwin Hyman, 465 p.

Crawford, A.J., Beccaluva, L., Serri, G., and Dostal, J., 1986, Petrology, geochemistry and tectonic implications of volcanics dredged from the intersection of the Yap and Mariana Trenches: Earth and Planetary Science Letters, v. 80, p. 265–280.

Dewey, J.F., 1976, Ophiolite obduction, Tectonophysics, v. 31, p. 93–120.

Dewey, J.F., and Bird, J.M., 1971, Origin and emplacement of the ophiliolite suite: Appalachian ophiolites in Newfoundland: Journal of Geophysical Research, v. 76, p. 3179–3206.

Dick, H.J.B., and Bullen, T., 1984, Chromian spinel as a petrogenetic indicator in abyssal and Alpine-type peridotites and spatially associated lavas: Contributions to Mineralogy and Petrology, v. 86, p. 54–76.

Dietrich, V.R., Emmerman, R, Oberhansli, R., and Puchelt, H., 1978, Geochemistry of basaltic and gabbroic rocks from the West Mariana Basin and the Mariana: Earth and Planetary Science Letters, v. 39, p. 127–144.

Dilek, Y., and Moores, E.M., 1990, Regional tectonics of the eastern Mediterranean ophiolites, *in* Malpas, J., Moores, E.M., Panayiotou, A., and Xenophontos, C., eds., Ophiolites, Oceanic Crustal Analogues: Nicosia, Cyprus, Geological Survey Department, p. 295–309.

Falloon, T.J., Green, D.H., and Crawford, A.J., 1987, Dredged igneous rocks from the northern termination of the Tofua magmatic arc, Tonga and adjacent Lau Basin: Australian Journal of Earth Sciences, v. 34, p. 487–506.

Forsythe, R.D., Nelson, E.P., Kaeding, M.E., Carr, M.J., Herve, M., Mpodozis, C., Soffia, J.M., and Harambour, S., 1986, Pliocene near-trench magmatism in southern Chile: a possible manifestation of ridge collision: Geology, v. 14, p. 23–27.

Fretzdorff, S., Livermore, R.A., Devey, C.W., Leat, P.T., and Stoffers, P., 2002, Petrogenesis of the Back-arc East Scotia Ridge, South Atlantic Ocean: Journal of Petrology, v. 43, p. 1435–1467.

Fryer, P., Pearce, J.A., and the ODP Leg 125 Shipboard Party, 1989, ODP Leg 125 drills forearc crust, mantle: Geotimes, v. 34, p. 18–20.

Gale, N.H., Spooner, E.T.C., and Potts, P.J., 1981, The lead and strontium geochemistry of metalliferous sediments associated with Upper Cretaceous ophiolitic rocks in Cyprus, Syria and the Sultanate of Oman: Canadian Journal of Earth Sciences, v. 18, p. 1290–1302.

Gass, I.G., 1963, Is the Troodos Massif of Cyprus a fragment of Mesozoic ocean floor?: Nature, v. 220, p. 39–42.

Gass, I.G., Neary, C.R., Plant, J., Robertson, A.H.F., Simonian, K.O., Smewing, J.D., Spooner, E.T.C., and Wilson, R.A.M., 1975a, Comments on "The Troodos ophiolitic complex was probably formed in an island arc," by A. Miyashiro and subsequent correspondence by A. Hynes and A. Miyashiro: Earth and Planetary Science Letters, v. 25, p. 236–238.

Gass, I.G., Smith, A.G., and Vine, F.J., 1975b, Origin and emplacement of ophiolites, *in* Geodynamics today: A review of the Earth's dynamic processes: London, Royal Society of London, p. 54–65.

Gill, J.B., 1976, Comparison and age of Lau Basin and Ridge volcanic rocks: implications for evolution of an interarc basin and remnant arc: Geological Society of America Bulletin, v. 87, p. 1384–1395.

Glassley, W., 1974, Geochemistry and tectonics of the Crescent Volcanic Rocks, Olympic Peninsula, Washington: Geological Society of America Bulletin, v. 85, p. 785–794.

Gribble, R.F., Stern, R.J., Newman, S., Bloomer, S.H., and O'Hearn, T., 1998, Chemical and isotopic composition of lavas from the Northern Mariana Trough: Implications for magmagenesis in back-arc basins: Journal of Petrology, v. 39, p. 125–154.

Hallam, A., 1973, A revolution in the Earth Sciences: From continental drift to plate tectonics: Oxford, Oxford University Press, 127 p.

Hamelin, B., Dupre, B., and Allegre, C.J., 1984, The lead isotope systematics of ophiolite complexes: Earth and Planetary Science Letters, v. 67, p. 351–366.

Hart, S.R., Glassley, W.A., and Karig, D.E., 1972, Basalts and seafloor spreading behind the Mariana island arc: Earth and Planetary Science Letters, v. 15, p. 12–18.

Hawkins, J.W., 1976, Petrology and geochemistry of basaltic rocks of the Lau Basin: Earth and Planetary Science Letters, v. 28, p. 283–297.

Hawkins, J.W., 1977, Petrological and geochemical characteristics of marginal basin basalts, *in* Talwani, M., and Pitman W.C., III, eds., Island Arcs, Deep Sea Trenches and Back Arc Basins, v. 1: Washington, D.C., American Geophysical Union, p. 355–365.

Hawkins, J.W., and Batiza, R., 1977, Metamorphic rocks of the Yap arc-trench system: Earth and Planetary Science Letters, v. 37, p. 216–229.

Hawkins, J.W., Bloomer, S.H., Evans, C.A., and Melchior, J.T., 1984, Evolution of intra-oceanic arc-trench systems: Tectonophysics, v. 102, p. 175–205.

Hickey-Vargas, R., 1991, Isotope characteristics of submarine lavas from the Philippine Sea: implications for the origin of arc and basin magmas of the Philippine plate: Earth and Planetary Science Letters, v. 107, p. 290–304.

Hilde, T., Uyeda, S., and Kroenke, L., 1977, Evolution of the western Pacific and its margin: Tectonophysics, v. 38, p. 145–165.

Hussong, D.M., and Uyeda, S., 1981, Tectonic processes and the history of the Mariana arc: a synthesis of the results of Deep Sea Drilling Project Leg 60, *in* Hussong, D.M., Uyeda, S. et al., eds., Initial Reports of the Deep Sea Drilling Project v. 60: Washington, U.S. Government Printing Office, p. 909–929.

Hynes, A., 1975, Comment on "The Troodos complex was probably formed in an island arc:" Earth and Planetary Science Letters, v. 25, p. 213–216.

International Working Group on the IGCP Project "Ophiolites," 1977, Initial report of the geological study of oceanic crust of the Philippine Sea floor: Ofioliti, v. 2, p. 137–168.

Jenner, G.A., Cawood, M., Rautenschlein, M., and White, W.M., 1987, Composition of back-arc basin volcanics, Valu Fa Ridge, Lau Basin: evidence for a slab-derived component in their mantle source: Journal of Volcanology and Geothermal Research, v. 32, p. 209–222.

Johnson, M.C., and Plank, T., 1999, Dehydration and melting experiments constrain the fate of subducted sediments: Geochemistry, Geophysics, Geosystems, v. 1, Paper number 1999GC000014.

Johnson, R.W., Jaques, A.L., Langmuir, C.H., Perfit, M.R., Staudigel, H., Dunkley, P.N., Chappell, B.W., Taylor, S.R., and Baekisapa, M., 1987, Ridge subduction and forearc volcanism: petrology and geochemistry of rocks dredged from the Western Solomon arc and Woodlark Basin, *in* Taylor, B., and Exon, N.F., eds., Marine Geology, Geophysics and Geochemistry of the Woodlark Basin-Solomon Islands: Houston, Texas, Circum-Pacific Council for Energy and Mineral Resources, p. 155–226.

Kaeding, M., Forsyth, R.D., and Nelson, E.P., 1990, Geochemistry of the Taitao ophiolite and near-trench intrusions from the Chile margin triple junction: Journal of South American Earth Sciences, v. 3, p. 161–177.

Karig, D.E., 1971, Origin and development of marginal basins in the western Pacific: Journal of Geophysical Research, v. 76, p. 2542–2559.

Karig, D.E., 1982, Initiation of subduction zones: implications for arc evolution and ophiolite development: London, Geological Society of London Special Publication, v. 10, p. 563–576.

Klein, E.M., and Karsten, J.L., 1995, Ocean-ridge basalts with convergent-margin geochemical affinities from the Chile Ridge: Nature, v. 374, p. 52–57.

Kay, R.W., and Senechal, R.G., 1976, The rare earth geochemistry of the Troodos ophiolite complex: Journal of Geophysical Research, p. 81, v. 964–970.

Kuhn, T.S., 1962, The structure of scientific revolutions: Chicago, University of Chicago Press, 212 p.

Lagabrielle, Y., Le Moigne, J., Maury, R.C., Cotton, J., and Bourgois, J., 1994, Volcanic record of the subduction of an active spreading ridge, Taitao Peninsula (southern Chile): Geology, v. 22, p. 515–518.

Leat, P.T., Livermore, R.A., Millar, I.L., and Pearce, J.A., 2000, Magma supply in back-arc spreading segment E2, East Scotia Ridge: Journal of Petrology, v. 41, p. 845–866.

Leitch, E.C., 1984, Island arc elements and arc-related ophiolites: Tectonophysics, v. 106, p. 177–203.

Leterrier, J., Maury, R.C., Thonon, P., Girard, P., and Marchal, M., 1982, Clinopyroxene composition as a method of identification of the magmatic affinities of paleo-volcanic series: Earth and Planetary Science Letters, v. 59, p. 139–154.

Lytwyn, J., Casey, J., and Gilbert, S., 1997, Arc-like mid-ocean ridge basalt formed seaward of a trench-forearc system just prior to ridge subduction: an example from subaccreted ophiolites in southern Alaska: Journal of Geophysical Research, v. 102, p. 10,225–10,243.

Lytwyn, J., Lockhart, S., Casey, J., and Kusky, T., 2000, Geochemistry of near-trench intrusives associated with ridge subduction, Seldovia Quadrangle, southern Alaska: Journal of Geophysical Research, v. 105, p. 27,957–27,978.

Marshak, R.S., and Karig, D.E., 1977, Triple junctions as a cause for anomalously near-trench igneous activity between the trench and volcanic arc: Geology, v. 5, p. 233–236.

Malpas, J., and Strong, D.F., 1975, A comparison of chrome-spinels in ophiolites and mantle diapirs of Newfoundland: Geochimica et Cosmochimica Acta, v. 39, p. 1045–1060.

Malpas, J., Stevens, R.K., and Strong, D.F., 1973, Amphibolite associated with Newfoundland ophiolite: its classification and tectonic significance: Geology, v. 1, p. 45–47.

McCulloch, M.T., and Cameron, W.E., 1983, Nd-Sr isotopic study of primitive lavas from the Troodos ophiolite, Cyprus: Evidence for a subduction-related setting: Geology, v. 11, p. 727–731.

McCulloch, M.T., Gregory, R.T., Wasserburg, G.J., and Taylor, H.P., 1980, A neodymium, strontium and oxygen isotopic study of the Cretaceous Semail ophiolite and implications for the petrogenesis and seawater-hydrothermal alteration of oceanic crust: Earth and Planetary Science Letters, v. 46, p. 201–211.

Menzies, M., and Allen, C., 1974, Plagioclase lherzolite-residual mantle relationships within two eastern Mediterranean ophiolites: Contributions to Mineralogy and Petrology, v. 45, p. 197–213.

Mesorian, H., 1973, Idées actuelles sur la construction, l'origine et l'evolution des assemblages ophiolitiques mésogenées, Bulletin of the Geological Society of France, v. 15, v. 478–493.

Miyake, Y., 1985, MORB-like tholeiites formed within the Miocene forearc basin, Southwest Japan: Lithos, v. 18, p. 23–34.

Miyashiro, A., 1973, The Troodos ophiolitic complex was probably formed in an island arc: Earth and Planetary Science Letters, v. 19, p. 218–224.

Miyashiro, A., 1975a, Classification, characteristics, and origin of ophiolites: Journal of Geology, v. 83, p. 249–281.

Miyashiro, A., 1975b, Origin of the Troodos and other ophiolites: A reply to Hynes: Earth and Planetary Science Letters, v. 25, p. 217–222.

Miyashiro, A., 1975c, Origin of the Troodos and other ophiolites: a reply to Moores: Earth and Planetary Science Letters, v. 25,p. 227–235.

Monzier, M., Danyushevsky, L.V., Crawford, A.J., Bellon, H., and Cotten, J., 1993, High-Mg andesites from the southern termination of the New Hebrides island arc (SW Pacific): Journal of Volcanology and Geothermal Research, v. 57, p. 193–217.

Moores, E.M., 1975, Discussion of "Origin of the Troodos and other ophiolites: a reply to Hynes" by Akiho Miyashiro: Earth and Planetary Science Letters, v. 25, p. 223–226.

Moores, E.M., and Vine, F.J., 1971, Troodos Massif, Cyprus and other ophiolites as oceanic crust: evidence and implications: Philosophical Transactions of the Royal Society of London, v. A268, p. 443–466.

Moores, E.M., Robinson. P.T., Malpas, J., and Xenophontos, C., 1984, Model for the origin of the Troodos Massif, Cyprus and other mideast ophiolites: Geology, v. 12, p. 500–503.

Natland, J.H., and Tarney, J., 1981, Petrologic evolution of the Mariana arc and back-arc system: A synthesis of drilling results in the southern Philippine Sea, *in* Hussong, D.M., Uyeda, S. et al., eds., Initial Reports of the Deep Sea Drilling Project v. 60: Washington, U.S. Government Printing Office, p. 681–708.

Nicolas, A., 1989, Structures of ophiolites and dynamics of oceanic lithosphere: Dordrecht, Kluwer Academic Publishers, 367 p.

Nicolas, A., and Jackson, M., 1972, Répartition en deux provinces des peridotites des chaînes alpines longeant la méditerranée: Implications géotectoniques: Schweizische Mineralogische Petrographische Mitteilungen, v. 52, p. 479–495.

Nisbet, E.G., and Pearce, J.A., 1977, Clinopyroxene composition in mafic lavas from different tectonic settings: Contributions to Mineralogy and Petrology, v. 63, p. 149–160.

Noiret, G., Montigny, R., and Allegre, C.J., 1981, Is the Vourinos ophiolite complex an island arc ophiolite?: Earth and Planetary Science Letters, v. 56, p. 375–386.

Parson, L.M., and Hawkins, J.W., 1994, Two-stage ridge propagation and the geological history of the Lau backarc basin, *in* Hawkins, J.W., Parson, L.M., Allan, J.F., et al., eds., Proceedings of the Ocean Drilling Program Scientific Results, v. 135: Washington, U.S. Government Printing Office, p. 819–828.

Pearce, J.A., 1975, Basalt geochemistry used to investigate past tectonic environments on Cyprus: Tectonophysics v. 25, p. 41–67.

Pearce, J.A., 1982, Trace element characteristics of lavas from destructive plate boundaries, *in* Thorpe, R.S., ed., Orogenic Andesites: Chichester, Wiley, p. 528–548.

Pearce, J.A., 1983, Role of the sub-continental lithosphere in magma genesis at active continental margins, *in* Hawkesworth, C.J., and Norry, M.J., eds., Continental basalts and mantle xenoliths: Nantwich, Shiva, p. 230–249.

Pearce, 1987, An expert system for the tectonic characterization of ancient volcanic rocks: Journal of Volcanology and Geothermal Research, v. 32, p. 51–65.

Pearce, J.A., and Cann, J.R., 1971, Ophiolite origin investigated by discriminant analysis using Ti, Zr and Y: Earth and Planetary Science Letters, v. 12: p. 339–349.

Pearce, J.A., and Cann, J.R., 1973, Tectonic setting of basic volcanic rocks determined using trace element analyses: Earth and Planetary Science Letters, v. 19, p. 290–300.

Pearce, J.A., Alabaster, T., Shelton, A.W., and Searle, M.P., 1981, The Oman ophiolite as a Cretaceous arc-basin complex: evidence and implications: Philosophical Transactions of the Royal Society of London, v. A300, p. 299–317.

Pearce, J.A., Lippard, S.J., and Roberts, S., 1984a, Characteristics and tectonic significance of supra-subduction zone ophiolites: London, Geological Society of London Special Publication, v. 16, p. 77–94.

Pearce, J.A., Harris, N.B.W., and Tindle, A.G., 1984b, Trace element discrimination diagrams for the tectonic interpretation of granitic rocks: Journal of Petrology, v. 25, p. 956–983.

Pearce, J.A., Van der Laan, S.R., Arculus, R.J., Murton, B.J., Ishii, T., Peate, D.W., and Parkinson, I.J., 1992, Boninite and harzburgite from ODP Leg 125 (Bonin-Mariana forearc): A case study of magma genesis during the initial stages of subduction, *in* Fryer, P., Pearce, J.A., and Stokking, L.B., et al., eds., Proceedings of the Ocean Drilling Program Scientific Results, v. 125: College Station Texas, Ocean Drilling Program, p. 623–659.

Pearce, J.A., Ernewein, Bloomer, S.H., Parson, L.M., Murton, B.J., and Johnson, L.E., 1995, Geochemistry of Lau Basin volcanic rocks: London, Geological Society of London Special Publication, v. 81, p. 53–75.

Pearce, J.A., Kempton, P.D., Nowell, G.M., and Noble, S.R., 1999, Hf-Nd element and isotope perspective on the nature and provenance of mantle and subduction components in Western Pacific arc-basin systems: Journal of Petrology, v. 40, p. 1579–1611.

Perfit, M.R., Langmuir, C.H., Baekisapa, M., Chappell, B., Johnson, R.W., Staudigel, H., and Taylor, S.R., 1987, Geochemistry and petrology of volcanic rocks from the Woodlark Basin: addressing questions of ridge subduction, *in* Taylor, B., and Exon, N.F., eds., Marine Geology, Geophysics and Geochemistry of the Woodlark basin-Solomon Islands: Houston, Circum-Pacific Council for Energy and Mineral Resources, p. 113–154.

Rautenschlein, M., Jenner, G.A., Hertogen, J., Hofmann, A.W., Kerrich, R., Schmincke, H.-U., and White, W.M., 1985, Isotopic and trace element composition of volcanic glasses from the Akaki Canyon, Cyprus: implications for the origin of the Troodos ophiolite: Earth and Planetary Science Letters, v. 75, p. 369–383.

Richard, P., and Allègre, C.J., 1980, Neodymium and strontium isotope study of ophiolite and orogenic lherzolite petrogenesis: Earth and Planetary Science Letters, v. 47, p. 65–74.

Riccio, L., 1977, Ophiolites and ophiolites and the nature of the oceanic crust: Ofioliti, v. 2, p. 169–180.

Robinson, P.T., Melson, W.G., O'Hearn, T., and Schmincke, H.-U., 1983, Volcanic glass compositions of the Troodos ophiolite: Geology, v. 11, p. 400–404.

Rocci, G., Ohnenstetter, D., and Ohnenstetter, M., 1975, La dualité des ophiolites téthysiennes: Pétrologie, v. 1, p. 172–174.

Saleeby, J.B., 1982, Polygenetic ophiolite belt of the California Sierra Nevada: Geochronological and tectonostratigraphic development: Journal of Geophysical Research Series, v. 87, p. 1803–1824.

Saunders, A.D., and Tarney, J., 1979, The geochemistry of basalts from a back-arc spreading center in the East Scotia Sea. Geochimica et Cosmochimica Acta, v. 43, p. 555–572.

Seno, T., and Maruyama, S., 1984, Paleogeographic reconstruction and origin of the Philippine Sea: Tectonophysics, v. 102, p. 53–84.

Serri, G., 1981, The petrochemistry of ophiolitic gabbroic complexes: a key for the classification of ophiolites into low-Ti and high-Ti types: Earth and Planetary Science Letters, v. 52, p. 203–212.

Sharaskin, A. Ya., Dobretsov, N.L., and Sobolev, N.J., 1980, Marianites: The clinoenstatite bearing pillow lavas associated with the ophiolitic assemblage of the Mariana Trench, *in* Panayiotou, A., ed., Ophiolites: Proceedings of the International Ophiolite Symposium, Cyprus 1979: Nicosia, Cyprus, Geological Survey Department, p. 473–479.

Sharaskin, A. Ya, Pustchin, I.K., Ziobin, S.K., and Kolesov, G.M., 1983, Two ophiolitic sequences from the basement of the northern Tonga arc: Ofioiliti, v. 8, p. 411–430.

Shervais, J.W., 1982, Ti-V plots and the petrogenesis of modern and ophiolitic lavas: Earth and Planetary Science Letters, v. 59, p. 101–118.

Shervais, J.W., 2000. Birth, death and resurrection: The life cycle of suprasubduction zone ophiolites: Geochemistry, Geophysics, Geosystems, v. 2, Paper number 2000GC000080.

Shiraki, K., 1971, Metamorphic basement rocks of Yap Islands, western Pacific: possible ocean crust beneath an island arc: Earth and Planetary Science Letters, v. 13, p. 167–174.

Simonian, K.O., and Gass, I.G., 1978, Arakapas fault belt, Cyprus: A fossil transform fault: Geological Society of America Bulletin, v. 89, p. 1220–1230.

Smewing, J.D., and Potts, P.J., 1976, Rare earth abundances in basalts and metabasalts from the Troodos massif, Cyprus: Contributions to Mineralogy and Petrology, v. 57, p. 245–258.

Smewing, J.D., Simonian, K.I., and Gass, I.G., 1975, Metabasalts from the Troodos Massif, Cyprus: genetic implication deduced from petrology and trace element composition: Contributions to Mineralogy and Petrology, v. 51, p. 49–64.

Stern, C., 1979, Open and closed system igneous fractionation within two Chilean ophiolites and the tectonic implication: Contributions to Mineralogy and Petrology, v. 68, p. 243–258.

Stern, R.J., and Bloomer, S.H., 1992, Subduction zone infancy: examples from the Eocene Izu-Bonin-Mariana and Jurassic California Arcs: Geological Society of America Bulletin, v. 104, p. 1621–1636.

Stern, R.J., Bloomer, S.H., Lin, P.-N. and Smoot, N.C., 1989, Submarine arc volcanism in the southern Mariana Arc as an ophiolite analogue: Tectonophysics, v. 168, p. 151–170.

Stolper, E., and Newman, S., 1994, The role of water in the petrogenesis of Mariana Trough magmas: Earth and Planetary Science Letters, v. 121, p. 293–325.

Sun, S.-S., and Nesbitt, R.W., 1978, Geochemical regularities and genetic significance of ophiolitic basalts: Geology, v. 6, p. 689–693.

Suyehiro, K., Takahashi, N., Ariie, Y., Yokoi, Y., Hino. R., Shinohara, M., Kanazawa, T., Hirata, N., Tokuyama, H., and Taira, A., 1996, Continental crust, crustal underplating, and low-Q upper mantle beneath an oceanic island arc: Science, v. 272, p. 390–392.

Taylor, B., Fujioka, K., et al., 1990, Proceedings of the Ocean Drilling Program Scientific Results, v. 126: College Station Texas, Ocean Drilling Program, 709 p.

Thy, P., Petrogenetic implication of mineral crystallization trends of Troodos cumulates, Cyprus: Geological Magazine, v. 124, p. 1–11.

Upadhyay, H.P., and Neale, E.R.W., 1979, On the tectonic regimes of ophiolite genesis: Earth and Planetary Science Letters, v. 43, p. 93–102.

Uyeda, S., and Ben-Avraham, Z., 1972, Origin and development of the Philippine Sea: Nature, v. 240, p. 176–178.

Wood, D.A., 1980, The application of a Th-Hf-Ta diagram to problems of tectonomagmatic classification and to establishing the nature of crustal contamination of basaltic lavas of the British Tertiary volcanic province: Earth and Planetary Science Letters, v. 50, p. 11–30.

Wood, D.A., Joron, J.-L., and Treuil, M., 1979, A re-appraisal of the use of trace elements to classify and discriminate between magma series erupted in different tectonic settings: Earth and Planetary Science Letters, v. 45, p. 326–336.

Wood, D.A., Marsh, N.G., Tarney, J., Joron, J.-L., Fryer, P., and Treuil, M., 1981, Geochemistry of igneous rocks recovered from a transect across the Mariana Trough, arc, fore-arc and trench, Sites 453 through 461, Deep Sea Drilling Project Leg 60, *in* Hussong, D.M., Uyeda, S. et al., eds., Initial Reports of the Deep Sea Drilling Project, v. 60: Washington, D.C., U.S. Government Printing Office, p. 611–630.

Manuscript Accepted by the Society May 2, 2003

Printed in the USA

Geological Society of America
Special Paper 373
2003

The Troodos Massif of Cyprus: Its role in the evolution of the ophiolite concept

Paul T. Robinson*
Royal Flemish Academy of Belgium for Science and the Arts, Hertogsstraat 1, 1000 Brussels, Belgium

John Malpas
Costas Xenophontos
Department of Earth Sciences, The University of Hong Kong, Pokfulam Road, Hong Kong

ABSTRACT

The Troodos Massif has played a unique role in the development of the ophiolite concept. The importance of Troodos derives from its coherent nature, its excellent exposures and easy access, and the fact that it was one of the first such bodies to be studied in detail. Early studies demonstrated the genetic relationship between ophiolitic units and elucidated the nature and significance of sheeted dikes, leading to the recognition that ophiolites are fragments of ocean lithosphere formed in spreading environments. Information derived from Troodos played a key role in formulating a new definition of ophiolites in 1972, one that has remained essentially unchanged for 30 years. The suprasubduction zone nature of ophiolitic lavas was first recognized in Troodos, providing a new understanding of where ophiolites originate in the ocean basins and raising questions as to the validity of applying the ophiolite model to the greater part of the oceanic lithosphere. Detailed mapping and drilling of Troodos has shown that the ophiolite was constructed from multiple subrift magma chambers, and has elucidated the relationships between the plutonic complex and the sheeted dikes. The base metal orebodies of Troodos were shown to be hydrothermal deposits before black smokers were discovered on the seafloor, and they now provide opportunities for detailed, three-dimensional studies of fossil hydrothermal systems. Finally, study of Troodos provided convincing evidence for uplift of oceanic lithosphere by partial underthrusting of continental lithosphere, a model now widely favored for many cases of ophiolite emplacement. Although our understanding of ophiolites has been greatly enhanced by studies of Troodos, many questions remain and we expect that the massif will continue to play a significant role in ophiolite research.

Keywords: ophiolite, Troodos, suprasubduction, obduction, hydrothermal deposits.

INTRODUCTION

Our understanding of ophiolites has evolved considerably over the last 60 years, during which time the comagmatic nature of the igneous components of the ophiolite assemblage has been recognized, ophiolite complexes have been identified as uplifted fragments of ocean lithosphere, and new definitions of the term ophiolite have been formulated. Our present interpretations of ophiolites have allowed us to further understand the construction of oceanic lithosphere, particularly the nature of the seismic layering of the crust, and the magmatic, structural, and hydrothermal processes at oceanic ridges. Equally important, however, has been their key role in identifying ancient sutures between tectonic plates, and in delineating the mechanisms of plate collision. Among the many ophiolite suites now recognized

*p.robinson@ns.sympatico.ca. Permanent address: Department of Earth Sciences, Dalhousie University, Halifax, Nova Scotia, Canada B3H3J5

Robinson, P.T., Malpas, J., and Xenophontos, C., 2003, The Troodos Massif of Cyprus: Its role in the evolution of the ophiolite concept, *in* Dilek, Y., and Newcomb, S., eds., Ophiolite concept and the evolution of geological thought: Boulder, Colorado, Geological Society of America Special Paper 373, p. 295–308. For permission to copy, contact editing@geosociety.org.

worldwide, the Troodos Massif of Cyprus has played a key role in developing the modern definition of ophiolites and in elucidating their petrologic and tectonic significance. The importance of the Troodos Massif stems from its complete and coherent nature, its excellent exposure of all the constituent rock types, its ease of accessibility, and from the fact that it was the first such body to be studied in detail.

The term ophiolite was first used nearly 200 years ago (Brongniart, 1813) to describe occurrences of serpentinite found in orogenic belts, but soon afterward was expanded to include an assemblage of volcanic rocks, gabbros and cherts associated with the serpentinites (Brongniart, 1821; 1827). During the nineteenth century, work in the Apennines and Western Alps confirmed a widespread association of radiolarian chert, basic igneous lavas, intrusive rocks, and serpentinized peridotite that became known as the Steinmann trinity (Steinmann, 1927). Although this assemblage was well established, there was no consensus as to the origin of the rocks, particularly the serpentinites, which subsequently became known as Alpine peridotites. Influenced by his experimental work on the differentiation of basaltic magmas and by field evidence from stratiform complexes, Bowen (1927) proposed an igneous origin by crystal fractionation for the ultramafic rocks, whereas Benson (1926) argued for some type of tectonic intrusion of ductile crystalline serpentine to explain the Alpine peridotites. Ophiolite suites were seen to represent the intrusive and volcanic products of a eugeosynclinal environment that later became involved in the mountain building episode of the orogenic cycle. A magmatic model for both the mafic and ultramafic rocks was favored for many years (Hess, 1938; Routhier, 1946, 1953; Dubertret, 1953; Brunn, 1956, 1960), although it was recognized that intrusion of high-temperature magmas should have produced significant zones of contact metamorphism that were conspicuously absent. This lack of contact metamorphism around the serpentinite bodies led to the resurrection of the Benson concept of cold intrusion, i.e., tectonic emplacement of ultramafic slices from the mantle (De Roever, 1957; Hess, 1960). In this model, the associated basaltic rocks were attributed to partial melting of the mantle rocks prior to their emplacement. In hindsight, this is a model not too dissimilar to the currently accepted view.

The Role of the Troodos Massif in the Definition of "Ophiolite"

Systematic geological study in Cyprus began shortly after World War II, and the major lithologic units on the island had already been identified by the late 1940s (Henson et al., 1949). Over the next two decades detailed mapping of the Troodos Massif by the Cyprus Geological Survey produced a number of 1: 25000 scale maps and accompanying memoirs (Wilson, 1959; Bagnall, 1960; Bear, 1960; Bear and Morel, 1960; Carr and Bear, 1960; Gass, 1960; Moore, 1960; Pantazis, 1967). These provided the geologic foundation on which many later interpretations were based, including the fact that the Troodos Massif was an uplifted elongate dome with the structurally deepest rocks exposed at Mount Olympus, the highest topographic point (Fig. 1). The sequence was divided into a plutonic complex, including both ultramafic and mafic rocks, a "sheeted" complex, and an extrusive sequence. This stratigraphy was later refined by Constantinou (1980) who identified, from the base upward, partially serpentinized peridotite, layered ultramafic rocks, layered gabbro, massive gabbro and plagiogranite, sheeted diabase, a "basal group" of mixed hypabyssal volcanic rocks, and mafic to siliceous pillowed and massive lavas. A thin layer of iron-manganese sediment (umbers) is locally present at the top of the lavas, intercalated with marls (Fig. 2).

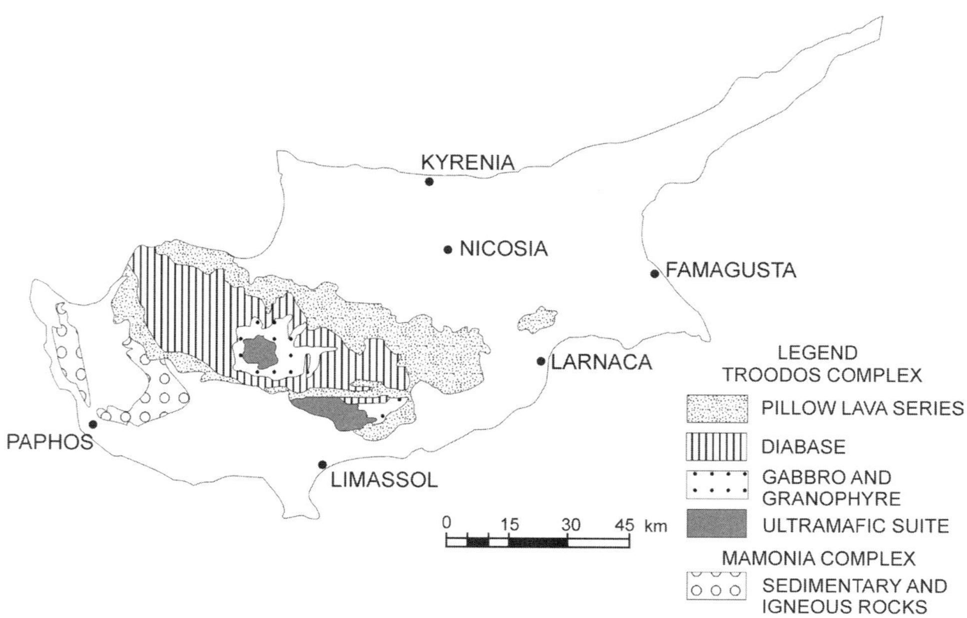

Figure 1. Generalized geologic map of the Troodos ophiolite illustrating its concentric outcrop pattern with a core of ultramafic rocks centered on Mount Olympus (after Robinson and Malpas, 1990).

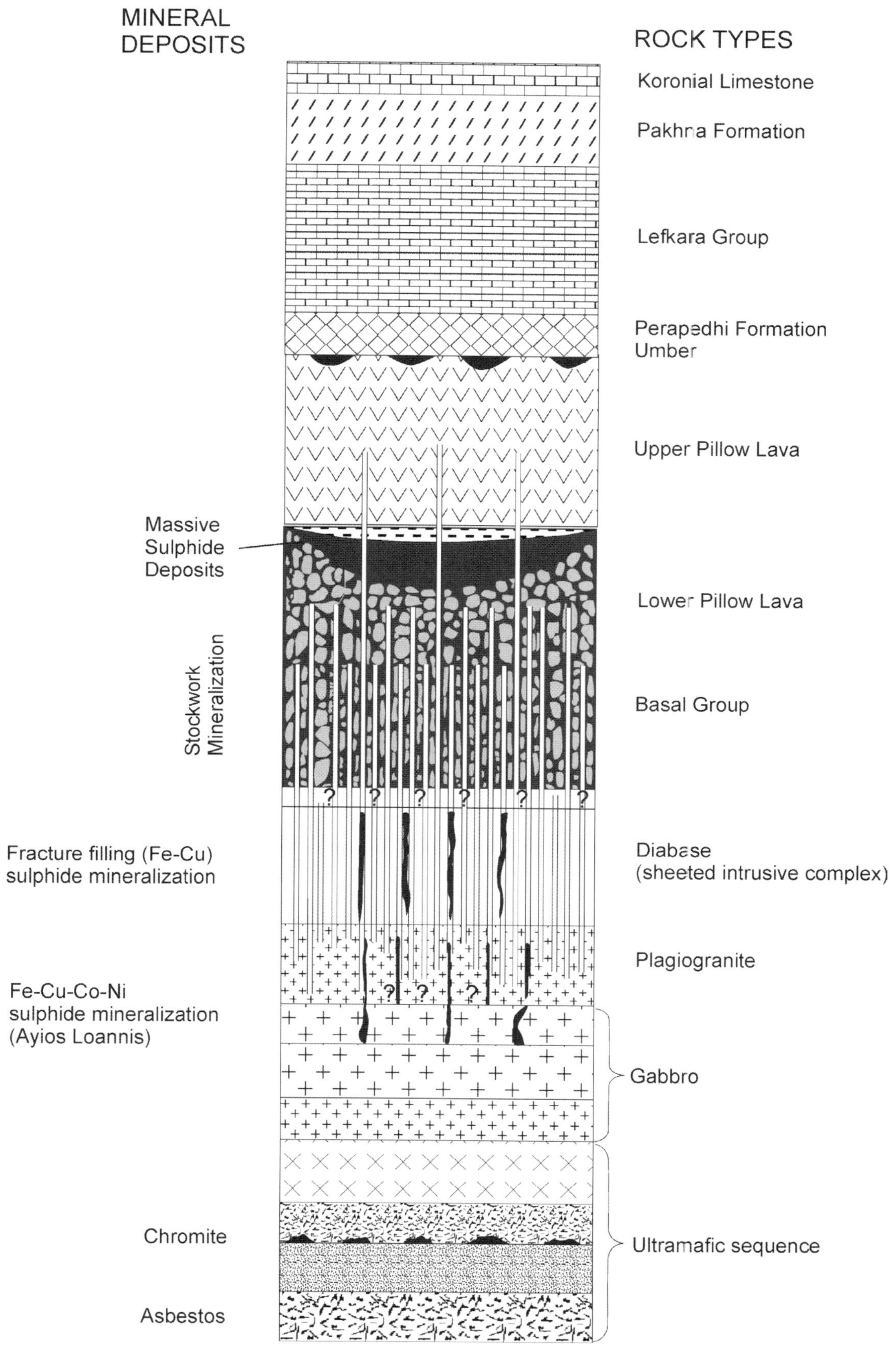

Figure 2. Diagrammatic composite geologic column of the Troodos ophiolite, showing the major lithologic units and the distribution of orebodies in the complex (after Constantinou, 1980). Not to scale.

However, more recent investigations (e.g., Malpas, 1990) have shown that even this sequence is somewhat simplified and that layered and isotropic gabbros and plagiogranites occur as variably sized intrusive bodies throughout the plutonic sequence, including the peridotites.

During early stages of investigation by the Cyprus Geological Survey, the relationships between the various lithologic units and their origins were not clear. Although the pillow lavas were properly identified as being submarine and volcanic in origin, the sheeted diabase was considered to be constructed from isoclinally folded lava flows (Bishopp, 1952). They were only subsequently interpreted as "a series of intrusive sheets which over a long period of time were successively injected and acted as feeders to the overlying flows" by Wilson (1959, p. 60), mainly on the basis of their massive nature and the clearly visible chilled margins. Soon afterward, Gass and Masson-Smith (1963) identified the cogenetic nature of the igneous units in the massif and suggested that the entire complex had formed as a volcanic pile in an extensional environment between Africa and Eurasia.

Thus, much was already known about the Troodos ophiolite complex when the concept of sea-floor spreading was first proposed (Vine and Matthews, 1963). This hypothesis provided a conceptual model in which the Troodos Massif could be understood and explained. Gass (1968) was the first to propose a possible link between Troodos and seafloor spreading when he suggested that the complex might represent a fragment of Mesozoic ocean floor formed at an extensional axis between Africa and Eurasia. By the early 1970s, the Troodos Massif was widely accepted as a coherent assemblage of rocks formed on the sea floor. Similar interpretations were being established for ophiolite assemblages elsewhere in the world (Dietz, 1963; Hess, 1964; Reinhardt, 1969; Church and Stevens, 1971; Williams and Malpas, 1972) and together with the conclusions reached by Moores and Vine (1971) for the Troodos Massif, these led to the formal redefinition of the term ophiolite during a Penrose Conference. This new definition stated that:

> Ophiolite refers to a distinctive assemblage of mafic to ultramafic rocks. It should not be used as a rock name or a petrologic unit in mapping. In a completely developed ophiolite, the rock types occur in the following sequence, starting from the bottom and working up:
>
> Ultramafic complex, consisting of variable proportions of harzburgite, lherzolite and dunite, usually with a metamorphic tectonic fabric (more or less serpentinized);
>
> Gabbroic complex, ordinarily with cumulus textures commonly containing cumulus peridotites and pyroxenites and usually less deformed than the ultramafic complex;
>
> Mafic sheeted dike complex;
>
> Mafic volcanic complex, commonly pillowed;
>
> Associated rock types include (1) an overlying sedimentary section typically including ribbon cherts, thin shale interbeds, and minor limestones; (2) podiform bodies of chromite generally associated with dunite; and (3) sodic felsic intrusive and extrusive rocks
>
> Faulted contacts between mappable units are common. Whole sections may be missing. An ophiolite may be incomplete, dismembered, or metamorphosed. Although ophiolite generally is interpreted to be oceanic crust and upper mantle, the use of the term should be independent of its supposed origin. (Anonymous, 1972)

It is clear that this definition was based largely on a few well-known ophiolite complexes, among which the most significant was the Troodos Massif. Despite this, it is basically a good definition that has been applied widely and which has stood the test of time. Very few ophiolites are as well preserved and complete as Troodos, however, and many are tectonic mélanges composed primarily of blocks of peridotite, minor gabbro, and mafic to intermediate volcanic rocks. Cherts and siltstones may overlie or be interbedded with the lavas, but they more commonly form the mélange matrix together with serpentinite. In some well preserved complexes, metamorphic soles occur at the base of the peridotite sections, but these are absent beneath mélanges. Sheeted dike complexes are rare and, where they do occur, are typically small and discontinuous. The paucity of sheeted dikes in many ophiolites might be explained in one of two ways: either they did not form originally or they were selectively removed during ophiolite emplacement. Selective tectonic removal seems unlikely as a general explanation, suggesting that they were not formed originally. If this interpretation is correct, ophiolites may record relatively small degrees of extension and/or form in regimes characterized by low magma supply.

Even though the Penrose definition of an ophiolite is fundamentally sound, it is commonly difficult to determine whether any given mafic-ultramafic body is, in fact, an ophiolite. This is particularly true for highly dismembered bodies where the relationships between lithostratigraphic units are not clear. In particular, better criteria are needed for properly identifying and classifying Precambrian bodies that have been proposed as ophiolites (e.g., Kusky et al., 2001, Zhai et al., 2002).

The Environment of Formation of Ophiolites

The Troodos Massif is part of a belt of ophiolitic fragments that lies within the Bitlis-Zagros suture zone between the African and Arabian plates (Dilek et al., 1990). When the Troodos Massif was first interpreted as a piece of ocean lithosphere, it was assumed to have formed at a mid-ocean ridge during spreading in the Tethyan Ocean. There were several reasons for this. First, the well-developed sheeted complex feeding a pillow lava sequence indicated formation in a submarine extensional environment (Gass, and Masson-Smith, 1963; Gass, 1968). The only pragmatic (and topical!) solution to this in the early 1970s was that of formation at a major oceanic spreading center. Second, the pillow lavas were thought to be basaltic in composition, and the presence of more silicic lavas was attributed to submarine alteration and silica metasomatism (Gass and Smewing, 1973; Smewing et al., 1975). Third, the ophiolitic rocks of the Troodos Massif are overlain by fine-grained pelagic limestones, suggesting that the complex had formed in an open ocean environment (Robertson, 1976).

Miyashiro (1973) was the first to challenge this view. He pointed out that the major element geochemistry of the lavas in Troodos displayed a calc-alkaline affinity, indicating formation in an island arc. His arguments were vigorously attacked by Gass et al. (1975), Hynes (1975), and Moores (1975) on the grounds that the lava geochemistry had been modified by silica metasomatism, the sheeted dikes indicated an extensional environment, the complex lacked any pyroclastic rocks, and the ophiolite was overlain by deep-water limestones.

In the early 1970s, however, trace element geochemistry of basaltic lavas was recognized as a useful tool in determining past tectono-magmatic affinities (Pearce and Cann, 1971; 1973). Accepting that certain trace elements are relatively immobile during alteration and that element ratios could be used to discriminate tectonic environments of volcanic eruption (e.g., Beccaluva et al., 1979), it was suggested that the Troodos ophiolite had formed in a marginal basin environment (Pearce, 1975; 1980). This view gradually won wider acceptance (e.g., Gass, 1980), and was subsequently supported by detailed geochemical investigations (e.g., Robinson et al., 1983; McCulloch and Cameron, 1983; Malpas and Langdon, 1984; Cameron, 1985; Rautenschlein et al., 1985).

During the early phases of the Cyprus Crystal Study Project in 1980, it was discovered that, contrary to earlier reports (Gass and Smewing, 1973; Smewing et al., 1975), the Troodos lavas are not pervasively altered and that fresh volcanic glass is common throughout the extrusive section (Robinson et al., 1983). Fresh glass samples are particularly useful in determining the compositions and origins of volcanic rocks because they represent snapshots of liquid compositions unaffected by alteration or crystal accumulation. Early studies in Troodos had identified two lava suites, the Upper and Lower Pillow Lavas, generally identified as olivine-rich and olivine-poor, respectively, although the boundary between them was not clearly defined. Relatively siliceous rocks were recognized in the Lower Pillow Lavas, but these were attributed to alteration and silica metasomatism (Gass and Smewing, 1973).

Detailed investigations of the volcanic glasses in Troodos lavas indicated the presence of three suites, an arc tholeiite sequence of intermediate to silicic rocks, a depleted tholeiite sequence of basalt and picrite and a highly depleted boninitic suite (Fig. 3) (Robinson et al., 1983; Schmincke et al., 1983; Malpas and Langdon, 1984; Rautenschlein et al., 1985; Cameron, 1985; Mehegan, 1988; Rogers et al., 1989; Taylor, 1990; Gass et al., 1994). These lavas are compositionally indistinguishable from those of the Marianas arc and forearc, and demonstrate convincingly that the ophiolite is unlikely to have formed at a mid-ocean ridge. Rather, the geochemical data clearly show that the ophiolitic rocks of Troodos formed in a "suprasubduction zone" (SSZ) environment, possibly

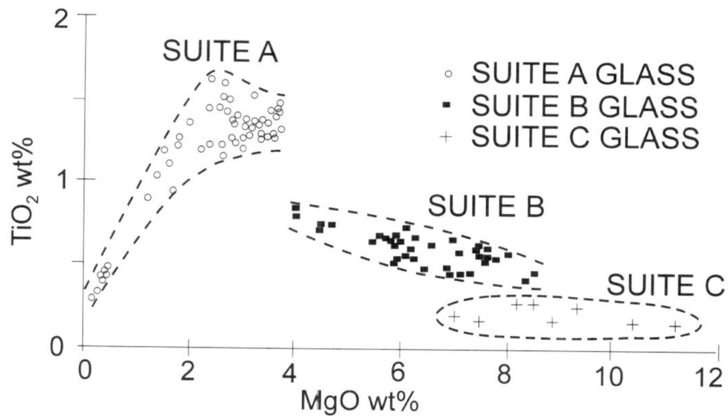

Figure 3. Plot of volcanic glass composition of extrusive rocks of the Troodos ophiolite, illustrating the presence of highly evolved, silicic lavas (Suite A), depleted arc tholeiites (Suite B), and boninitic lavas (Suite C) (after Robinson et al., 1983; Mehegan, 1988).

beneath a nascent arc (Moores et al., 1984). The boninitic lavas are particularly important because modern examples of such rocks are found only in forearc regions. Such rocks have now been found in many ophiolites around the world, suggesting that the ophiolites formed in forearc regions during subduction rollback (Stern and Bloomer, 1992). The term suprasubduction zone was introduced by Alabaster et al. (1982) to describe lavas formed above a subducting oceanic lithosphere slab, without specifying their exact tectonic position with respect to the overlying island arc. Indeed, in the case of the Troodos Massif, there appears to have been no concomitant arc edifice, and the ophiolite simply represents suprasubduction spreading of the oceanic lithosphere.

These findings from Troodos prompted reinvestigation and reinterpretation of many other ophiolites. Most were found to contain lavas with similar suprasubduction zone signatures (e.g., Alabaster et al., 1982; Shervais and Kimbrough, 1985; Jenner et al., 1991; Shervais, 1990). Thus, as a result of studies originally carried out on Troodos and Oman, it is now generally accepted that most ophiolites were formed in suprasubduction zone environments, although relict fragments of lithosphere from mid-ocean ridges may be incorporated in some complexes (e.g., Batanova and Sobolev, 2000, Malpas and Zhou, 2001).

Oceanic Crustal Structure

By the early 1960s, the seismic structure of the ocean crust had been reasonably well established for the Atlantic Ocean (Ewing, 1965). A three-layer structure was recognized above the Moho (Fig. 4). Layer 1, which is clearly composed of sediment and has an average thickness of ~0.5 km, is characterized by P-wave velocities ~2 km/s. Layer 2 has P-wave velocities between 4.0 and 6.0 km/s and averages ~1.75 km in thickness. This layer is exposed on mid-ocean ridges where dredging indicated a basaltic composition. The range of seismic velocities, however, suggests

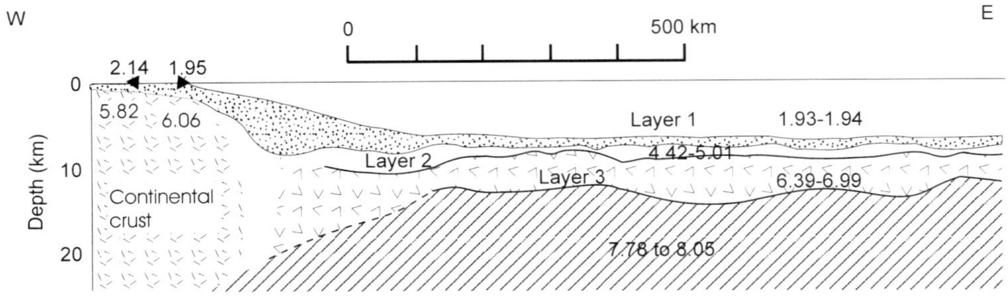

Figure 4. Model of the oceanic crustal structure of the southwest Atlantic Ocean showing, the seismic layers and their thicknesses and P-wave velocities (after Ewing, 1965).

that Layer 2 either consists of a wide variety of rock types or that there are large differences in the degree of crack and fracture filling. Making up the bulk of the crust is Layer 3, with an average thickness of nearly 5 km and a P-wave velocity between 6 and 7 km/s. Little was known about the composition of this layer, although dredging in fracture zones suggested that it might consist of gabbro and peridotite.

Moores and Vine (1971) were the first to correlate lithostratigraphic units in ophiolites with the oceanic crustal layers. Using Troodos as a guide, they suggested that Layer 2 consists of pillow lavas and minor dikes, whereas Layer 3 was correlated with the sheeted dike complex and the gabbro section (Fig. 5). Within Troodos, they placed the Moho at the base of the gabbros and included the pyroxenites, dunites, and harzburgites in the mantle; however, working in the Bay of Islands ophiolite of Canada, Malpas (1973) recognized that the dunites overlying the harzburgite tectonite are actually cumulate in origin and hence are part of the crustal sequence. Thus, he made a distinction between the "seismic Moho," which lies at the top of the cumulate ultramafic rocks, and the "petrologic Moho," which lies at the dunite-harzburgite contact and represents the base of the crustal sequence.

Christensen and Salisbury (1975) carried out a detailed study of the seismic properties of various mafic and ultramafic rocks for comparison with in situ ocean crustal velocities and concluded that Layer 2 consists of pillow lavas and dikes whereas Layer 3 consists largely of gabbro. Deep drilling by the Ocean Drilling Program (ODP) has generally supported the original correlations of ophiolite lithology with ocean crustal structure. Numerous holes penetrating Layer 2 have recovered chiefly pillowed or massive lavas and sheeted dikes. Gabbros recovered from ODP Site 735 have seismic velocities and physical properties identical to those of Layer 3 (Robinson, von Herzen, et al., 1989; Dick, Natland, Miller, et al., 1999). However, drilling in ODP Hole 504B in the Eastern Pacific Ocean suggests that the Layer 2–3 boundary is a metamorphic front, rather than a lithologic transition (Detrick et al., 1994). The Moho has yet to be sampled by drilling, although the crust-mantle boundary is exposed on the walls of some fracture zones.

Fossil Transform Faults

In the Troodos Massif, the Arakapas Fault is considered by many to represent a fossil oceanic transform fault exposed on land (Moores and Vine, 1971; Simonian and Gass, 1978). The well defined and orderly lithostratigraphic sequence of the ophiolite complex to the north of the fault is not mirrored to the south in the Limassol Forest, where mantle sequence rocks are juxtaposed against pillow lavas erupted in extensional graben, and subsequently covered by volcanic and sedimentary breccias. Although considered in part a leaky transform fault domain and in part a portion of the "Anti-Troodos Plate" by MacLeod and Murton (1993) and Gass et al. (1994), the ophiolitic rocks of the Limassol Forest remain an enigma. Malpas et al. (1992, 1993) recognized that both the lithologies and relationships displayed by the rocks in the Limassol Forest extend farther to the west, to the area north and northeast of Paphos, but that here, they are dominated by normal faults with little evidence of strike-slip movements. Malpas et al. (1992; 1993) therefore interpreted these rocks, and by analogy those in the main Limassol Forest, to have formed in a major extensional environment in a nascent forearc. More recently, Cann et al. (2001) proposed an alternative model in which the Limassol Forest complex formed at an ancient inside corner of a ridge-transform intersection. In this model, the intricate lithological and structural relationships are considered to be the result of extreme extension with the development of flat-lying or gently dipping detachment faults similar to those imaged in the oceans (Cann et al., 1997). Indeed, if this model proves to be correct, it provides an opportunity to study in great detail the processes of shallow detachment at an oceanic transform.

The acceptance of any one model has important implications for the tectonic environment in which the ophiolite formed. Thus, MacLeod and Murton (1993) and Gass et al. (1994) placed the Troodos and Limassol Forest complexes on opposite sides of a transform fault with dextral slip, and suggested that they formed on two different plates (the Troodos and Anti-Troodos plates). In the model of Cann et al. (2001), the Troodos and Limassol Forest complexes are considered to have formed on the same plate, juxtaposed across a fracture zone associated with a sinistrally slipping transform that lay to the west. Dilek et al. (1990) pointed to the similarities between the inferred ridge-transform intersection (RTI) of the Mitsero graben and Arakapas fault with the Vema Ridge transform–Mid-Atlantic Ridge RTI, and argued for left-lateral slip on the Arakapas fault. The model envisaged by Malpas et al. (1992; 1993), calls into question the existence of significant strike-slip motion of any kind on the Arakapas fault.

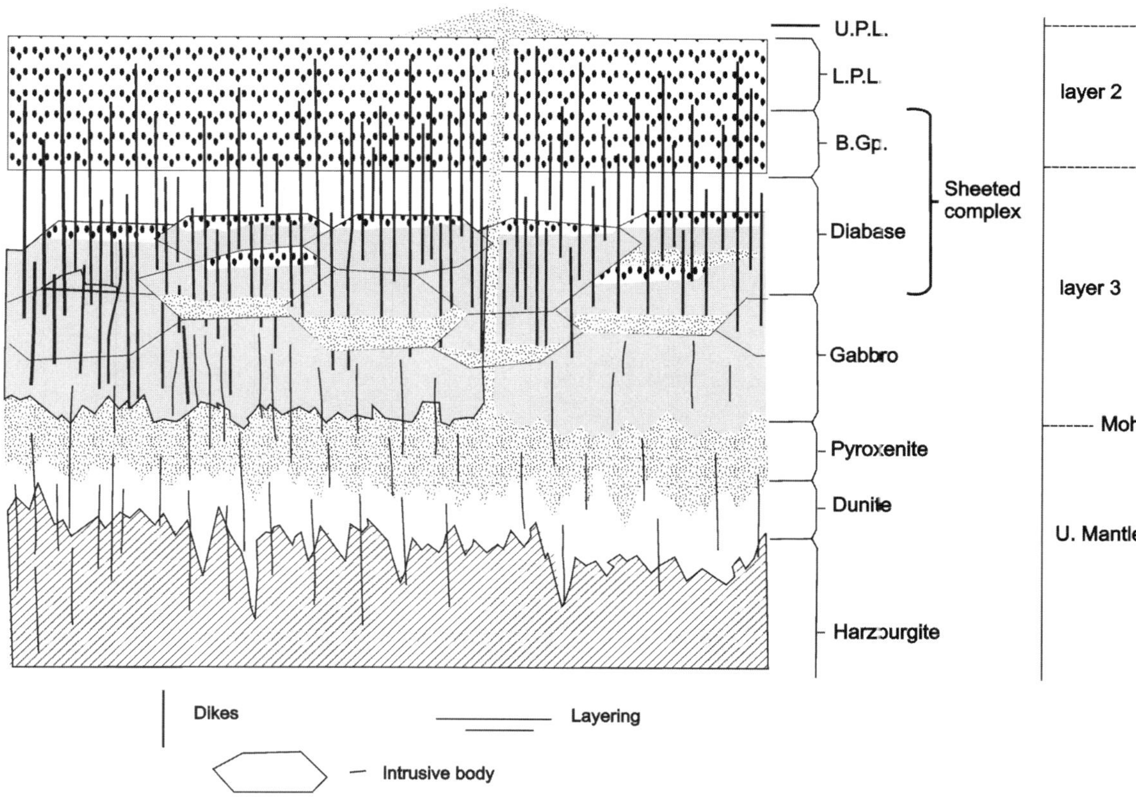

Figure 5. Interpretative diagrammatic cross-section of the Troodos Complex, showing numerous individual plutons, suggesting multiple magma chambers and crosscutting relationships between dikes and gabbros (after Moores and Vine, 1971). Not to scale. U.P.L.—Upper Pillow Lavas; L.P.L—Lower Pillow Lavas; B.Gp.—Basal Group.

Oceanic Magma Chamber Models

As one of the few ophiolites with a complete crustal section that can be studied in detail, the Troodos Massif has been important in elucidating the nature of oceanic sub-rift magma chambers. Figures 5, 6, and 7 represent some of the evolving ideas on the size and shape of magma chambers beneath Troodos. In their landmark study of Troodos, Moores and Vine (1971) observed abundant chilled margins in the plutonic complex and contradictory relations between the dikes and gabbros. From these observations, they suggested that the upper part of the plutonic section is composed of numerous small plutons, each representing an isolated magma chamber (Fig. 5). Subsequent workers, such as Greenbaum (1972), impressed by the abundant cumulate mafic and ultramafic rocks and influenced by models for mid-ocean ridges, proposed a single, large, continuously evolving magma chamber from which the entire crustal section was constructed (Fig. 6). Such a model also found favor for the Oman ophiolite, where, it was suggested, individual cumulate layers could be traced over several kilometers (Pallister and Hopson, 1981). Structural studies of the Oman ophiolite also led Nicolas et al. (1988) to propose a single, tent-shaped chamber beneath a spreading axis.

Later studies of Troodos, however, found abundant evidence of multiple independent and ephemeral chambers similar to those originally proposed by Moores and Vine (1971) (cf. Allen, 1975; Malpas and Brace, 1987; Malpas, 1990; Robinson and Malpas, 1990). The complex lava chemistry in Troodos alone argues strongly against a single, continuously evolving chamber, and such an interpretation is clearly not supported by intrusive relations between dikes and gabbros (Fig. 7). Several generations of dikes have been recognized, with dikes cutting ultramafic rocks, gabbros, and other dikes (Varga and Moores, 1985; Malpas and Brace, 1987; Baragar et al., 1987; Malpas, 1990; Robinson and Malpas, 1990). Gabbroic plutons locally cut one another and intrude the base of the sheeted dike complex (Fig. 7). Such features argue strongly for multiple, discrete magma chambers. The common occurrence of several magma types in other ophiolites, such as Oman (Alabaster et al., 1982), supports similar models elsewhere.

Hydrothermal Circulation and Ore Deposition

The numerous massive sulfide orebodies of Cyprus are world famous and have been mined for copper since pre-Roman times. By the early 1970s, the orebodies had been studied extensively but their origin was still being debated. Some (e.g., Hutchison and

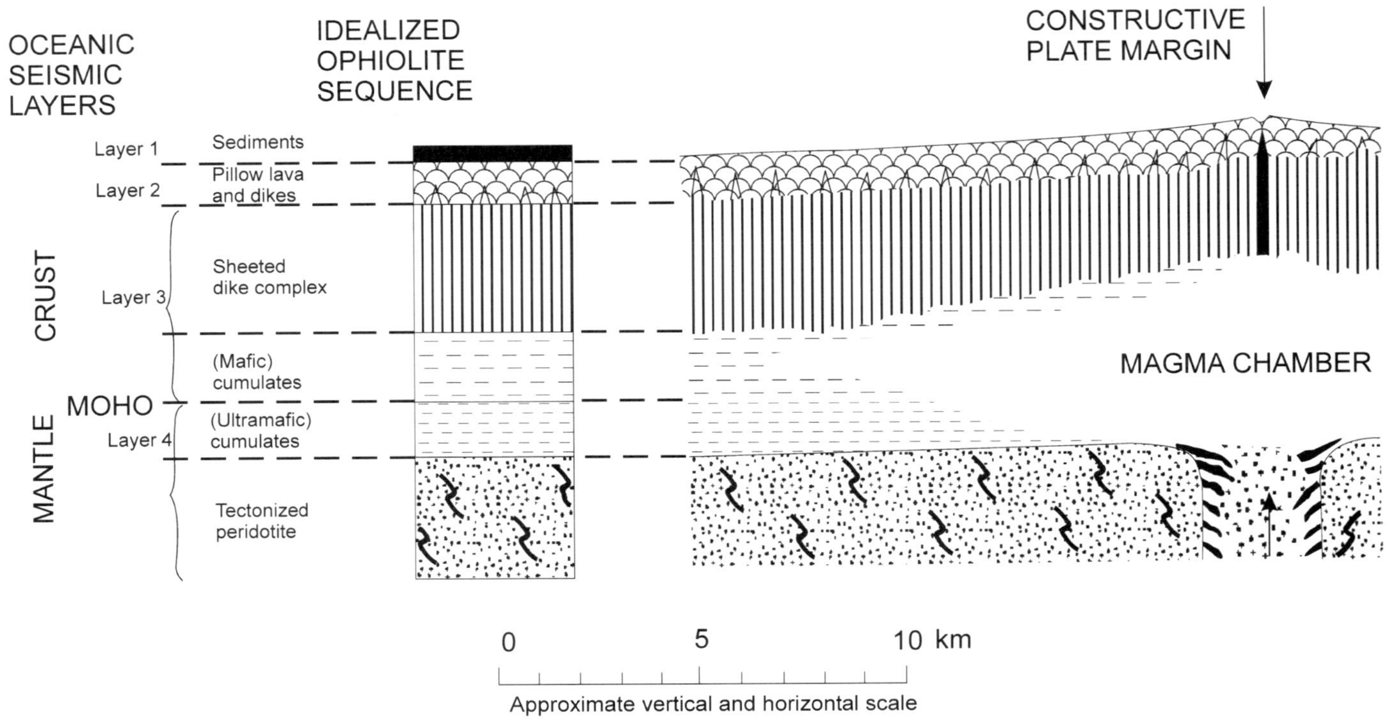

Figure 6. Model of a single, continuously evolving magma chamber beneath the Troodos spreading axis (after Greenbaum, 1972).

Searle, 1971; Johnson, 1972) favored a magmatic origin, whereas others (e.g., Bear, 1963, Pantazis, 1967; Constantinou and Govett, 1972; 1973) argued for a hydrothermal origin. The discovery of massive sulfide mounds on the sea floor (CYAMEX Scientific Team. 1979) and active black smokers (Spiess et al., 1980) provided direct evidence of the link between hydrothermal circulation and the formation of massive sulfide deposits. The Troodos orebodies provided an excellent opportunity for comparison with those formed by sea-floor hydrothermal systems. As a result, a number of detailed studies of the orebodies and associated rocks were undertaken (Constantinou, 1980; Oudin and Constantinou, 1984; Panayiotou, 1986), including a combined field and drilling investigation of the Agrokipia deposit (Robinson et al., 1987).

The Troodos deposits generally lie within the lava pile, although the largest body, Skouriotissa, is located near the top of the extrusive sequence and some small sulfide deposits also occur in the gabbros (Panayiotou, 1980; 1986). The massive ores have typical conglomeratic textures in which blocks of solid pyrite are embedded in a matrix of friable, highly porous, colloform pyrite. The ores consist almost entirely of sulfur and iron with minor amounts of copper, zinc, cobalt, nickel and gold (Constantinou, 1980). On the basis of similarities in petrography, mineralogy and composition, Oudin and Constantinou (1984) concluded that the Troodos orebodies are exhalative deposits similar to those being formed on modern ridge axes. Spooner (1977, 1980) demonstrated that the ore-forming fluid of the Troodos deposits was originally seawater, and proposed a hydrothermal circulation system reaching down to the top of the gabbros (Fig. 8). Structural studies demonstrated that the orebodies formed along normal faults in graben (Adamides, 1980), some of which were associated with former spreading axes (Varga and Moores, 1985). In many cases, the deposits formed during periods of renewed

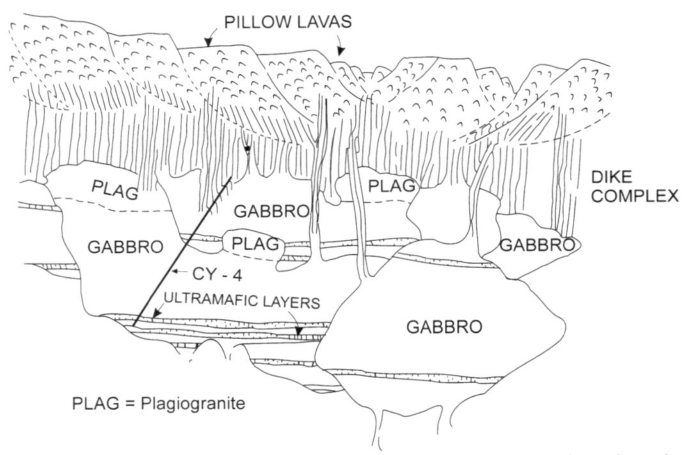

Figure 7. Conceptual model of the structure of the Troodos plutonic complex and its relationship to the sheeted dikes, based on field and drilling data. Hole CY–4, drilled to a depth of 2300 m by the Cyprus Crustal Study Project in 1984 penetrated the dike-gabbro transition, the upper homogeneous gabbros, the lower layered gabbros, and the uppermost layered ultramafic rocks (after Malpas, 1990). Not to scale.

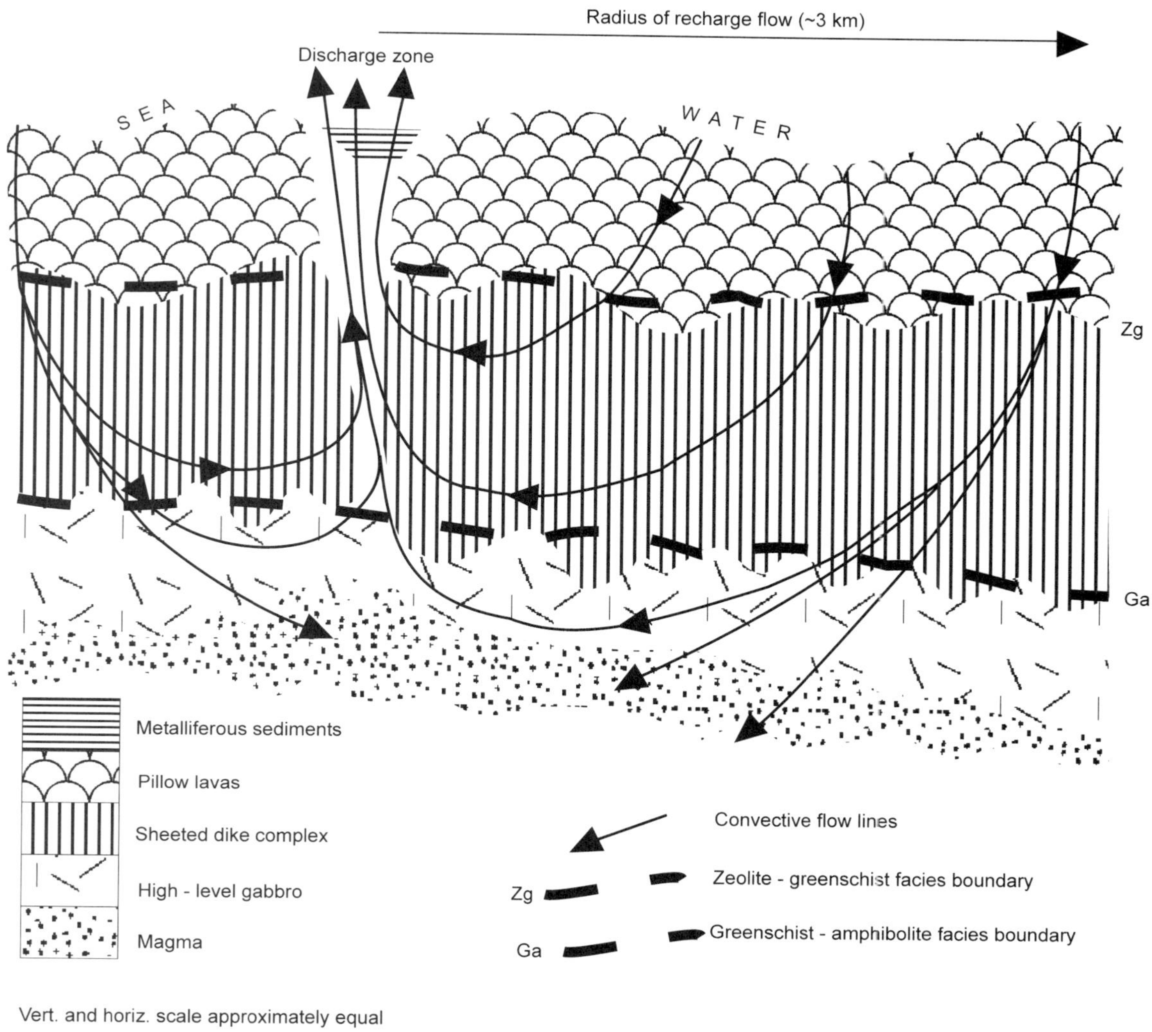

Figure 8. Conceptual model of a hydrothermal system in Troodos leading to the formation of a copper sulfide orebody. This model proposes a broad downwelling zone and a narrow upwelling zone (stockwork) below the orebody (after Spooner, 1977; Gass, 1980). In this diagram, the boundary between amphibolite and greenschist facies metamorphism (Ga) lies approximately at the dike-gabbro boundary, and the boundary between greenschist and zeolite facies metamorphism (Zg) lies at the base of the pillow lavas. Drilling in Hole CY-4, however, showed that these boundaries are not sub-horizontal, but rather vary greatly in depth, depending on local intrusive relationships.

volcanism, which followed crustal attenuation by normal faulting (Varga and Moores, 1985). This renewed volcanism appears to have been related to intrusion of small, high-level magma chambers into the upper gabbros and lowermost sheeted dikes.

Drilling of the Agrokipia deposit by the Cyprus Crustal Study Project revealed that the orebody is directly underlain by a highly silicified and argillized zone below which there is a propylitic zone that, in turn, grades downward into a sub-greenschist zone (Robinson et al., 1987). Most of the massive sulphide orebodies in Troodos appear to have formed as exhalative deposits on the seafloor (Oudin and Constantinou, 1984); however, the Agrokipia deposit is a replacement deposit formed by mixing of high-temperature (~350 °C) hydrothermal fluids with cool seawater in the upper part of the lava pile. The stockwork zone beneath the Agrokipia deposit tapers downward, and the boundary between fresh and altered rocks is sharp, indicating that alteration is restricted to narrow hydrothermal pathways.

A number of relatively large epidosite zones at or near the base of the sheeted dikes are believed to represent hydrothermal reaction zones directly above the high-level plutons that fueled the hydrothermal systems. Fluid inclusions in these rocks suggest reaction temperatures between 350 °C and 400 °C (Cann, et al., 1987; Cowan et al., 1987; Schiffman et al., 1987; 1990; Kelley and Robinson, 1990; Kelley et al., 1992). The epidosites are strongly depleted in Ca, Mn, Na, K, Rb, Cu, and Zn, such that a zone ~2 km long could provide the necessary metals for a large sulfide orebody (Richardson et al., 1988). Calculated fluid compositions are remarkably similar to black smoker fluids, but somewhat enriched in Fe, Cu, and Zn, which may explain why most of the Troodos orebodies are larger than their modern analogues. There appears to have been significant input to the hydrothermal circulation of metal-charged magmatic fluids, perhaps released during formation of plagiogranites (Malpas, 1990).

Drilling of active hydrothermal systems on modern ridge crests has proven to be extremely difficult, and little is known about their three-dimensional geometry, their temperature gradients, and their interaction with the host lavas. Mapping and drilling of such systems in Troodos have provided a wealth of information on the relationship between hydrothermal circulation and ore depositions, which aids in the interpretation of modern systems and can help in the exploration of ophiolitic massifs for massive sulfide deposits.

Ophiolite Emplacement

As well as providing insights into the mechanisms of formation of oceanic crust, studies of the Troodos Massif have contributed to a better understanding of the complex processes that take place when lithospheric plates and microplates collide and ophiolitic rocks are emplaced on continental margins. In the early 1970s, when ophiolites were considered fragments of normal ocean lithosphere, it was recognized that emplacement took place as a result of plate collision and it was assumed that the ocean lithosphere segments, originally formed at a mid-ocean spreading center, were brought to continental margins by subduction of the oceanic plate. The problem, however, was how to uplift fragments of dense ocean lithosphere from water depths of 5–6 km, transport them across the subduction zone, and emplace them onto the continental margin. The term "obduction" was coined for this process (Coleman, 1971), and a number of models were subsequently put forth to explain the emplacement of such bodies (for summaries, see Dewey, 1976; Malpas and Stevens, 1978).

It was later suggested that emplacement processes could vary on the basis of whether obduction was thought to have occurred at a passive (Tethyan-type) continental margin or at an active (Cordilleran-type) margin (Moores, 1982; Coleman, 1984). Classic examples of the former include the Oman and Bay of Islands ophiolites, where the massifs were emplaced by gravity sliding over continental margin sedimentary rocks along low-angle thrust faults (Searle and Stevens, 1984). The Coast Range ophiolite of California is the type example of the Cordilleran-type, in which lithosphere fragments are dismembered and incorporated into an accretionary terrane (Shervais and Kimbrough, 1985).

Again, studies of the Troodos Massif and the associated Mamonia Complex in southwest Cyprus have helped to resolve these questions. Puzzled by the elevation of the Troodos Massif, Gass and Masson-Smith (1963) conducted a gravity survey of the ophiolite and concluded that it must be underlain by lower-density crustal material. They postulated that a fragment of the African continental margin had been underthrust beneath Troodos, a model that is now widely accepted (Fig. 9).

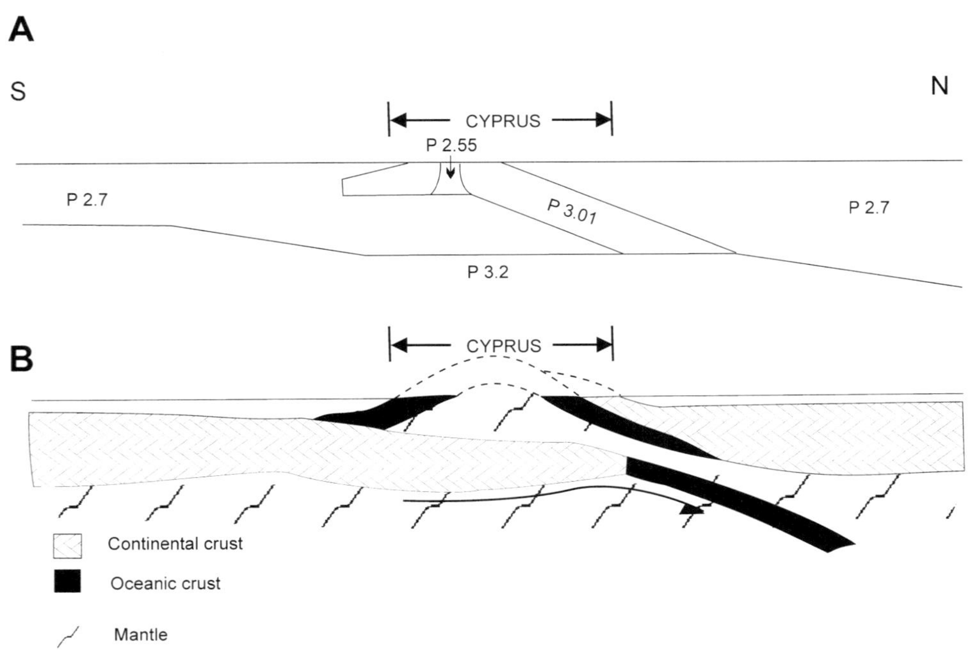

Figure 9. A: Calculated densities of Troodos rocks and surrounding lithosphere, based on a gravity survey by Gass and Masson-Smith (1963). B: Geologic interpretation of the gravity data, showing underthrusting of Troodos by the leading edge of the African continent (after Gass, 1980).

The Mamonia Complex in southwestern Cyprus consists of pillow lavas, sedimentary rocks, serpentinite, and minor amphibolite arranged in a complex sequence of thrust slices juxtaposed along the southwestern margin of the Troodos Massif (LaPierre, 1975; Swarbrick, 1980; Clube and Robertson, 1986; Malpas et al., 1987; 1992; 1993). The sedimentary rocks were shown to be of passive margin origin, probably deposited on the edge of a microcontinent within Tethys (Robertson and Woodcock, 1979). Pillow lavas within the Mamonia Complex are different both in age and chemistry from those of the Troodos Massif, and include both MORB and alkalic varieties. The latter suggests formation of volcanic seamounts in Neo-Tethys, as do the reef-type limestones that occur as olistoliths within the complex (Malpas et al., 1993). The complex presumably was scraped off the down-going slab during subduction and accreted to the leading edge of the overriding Troodos microplate. This model is supported by back-thrusting of rocks of the Troodos Massif over those of the Mamonia Complex, a process that was concomitant with the initiation of a 90° counterclockwise rotation of the Troodos microplate (Moores and Vine, 1971; Robertson and Xenophontos, 1993; Malpas et al., 1993).

The evidence of ophiolite emplacement provided by the Troodos Massif supports the notion of Malpas and Stevens (1978) that uplift and emplacement of typical ophiolites requires underplating by less dense continental crust, a process that can only take place in a subduction zone. They argued that partial subduction of low-density continental lithosphere and its subsequent isostatic rebound can provide the uplift needed to emplace ophiolites on continental margins. Whereas the Troodos Massif has remained essentially in place, other ophiolites, such as the Oman and Bay of Islands complexes, appear to have been uplifted enough to allow them to gravity-slide onto continental margins along low-angle fault planes (Searle and Stevens, 1984).

Cordilleran ophiolites typically consist of lithospheric slabs emplaced in forearc complexes and are probably formed and emplaced in a similar manner. In long-lived ocean basins, a series of arc and forearc complexes may be accreted to the continental margin during subduction rollback (Stern and Bloomer, 1992). During closure of the ocean basin, these complexes may be uplifted to form broad suture zones between continental masses (cf. Şengör et al., 1993).

Thus, regardless of where an ophiolite complex is originally formed, it is most easily emplaced in a suprasubduction zone environment. If, as seems to be the case, most ophiolite complexes are formed at a convergent margin, formation is dependent upon significant extension taking place in an overall convergent regime. As a result of this extension, particularly above an intra-ocean subduction zone, the crust and mantle might include fragments of normal ocean lithosphere, as well as the material generated within the suprasubduction zone environment. These fragments of old mid-ocean ridge lithosphere were trapped during lithosphere extension and incorporated in the ophiolite complex with the new lithosphere. Again, the Troodos Massif provides an excellent example of such a mixture of lithologies of different origins, as demonstrated by Batanov and Sobelev (2000). They showed that a block of lherzolite, presumed to be old, normal ocean lithosphere, is preserved within the harzburgitic mantle sequence of the Troodos Massif on the east side of Mount Olympus. This mantle material was not involved in the production of the ophiolitic lavas, but was affected by suprasubduction zone melts passing through it.

The formation of Troodos by extension above a subducting slab, a process believed to apply to many other ophiolites, suggests that only certain features in ophiolites may be comparable to oceanic crust produced at mid-ocean ridges (Malpas and Robinson, 1997; 2000). In a suprasubduction zone environment, lithospheric extension or spreading rate is not controlled by the same factors as at a mid-ocean ridge, although the processes of magmatism and amagmatic lithosphere thinning are similar. Extension in suprasubduction zone environments is mainly facilitated by subduction rollback, which creates space independent of the availability of magma. Thus, the fastest spreading rates and greatest lithospheric extension would occur where subduction rollback is rapid, thus favoring amagmatic spreading. This would result in thin crust, a paucity of sheeted dikes, and a dominance of extensional faults and large-scale crustal block rotation. These are features that many ophiolite complexes have in common, but which would be more likely associated with slow-spreading ridges in a mid-ocean setting. Rapid subduction rollback would be facilitated if the subducting slab were composed of old and cold lithosphere that was "dropping" through the asthenosphere rather than being forced at a shallow angle beneath the overriding plate. Suprasubduction zone ophiolites might therefore be expected to form above the margins of older, longer-lived oceanic plates. These are scenarios that have yet to be fully investigated, but the Troodos Massif is likely to play a significant part in their development.

SUMMARY

For the past 50 years the Troodos Massif has played a preeminent role in development of the ophiolite concept and in understanding the processes of crustal accretion within the ocean basins. Recognition of the true nature of the sheeted dike complex led to the understanding that ophiolites form in extensional environments, and it was in Cyprus that the connection between ophiolites and oceanic lithosphere was first made. The modern definition of ophiolites is based heavily on the geology of the Troodos Massif. The suprasubduction zone origin of ophiolitic lavas was first recognized there, leading to the understanding that most ophiolites are formed in nascent arcs or forearcs. The study of massive sulfide deposits of the Troodos Massif and the hydrothermal systems that produced them has provided a better understanding of active systems on modern spreading ridges, and helped to formulate better exploration models for ancient ocean lithosphere.

Recognition of the suprasubduction origin of the Troodos Massif greatly simplified the problems of ophiolite uplift and emplacement. Because the ophiolite is currently being uplifted and emplaced, it provides an actualistic model for these processes.

The Troodos Massif owes its pre-eminent role in ophiolite study to its coherent, well-exposed character and to the wealth of detailed information that has accumulated over the years. These factors ensure that it will continue to play a leading role in ophiolite studies for years to come.

REFERENCES CITED

Adamides, N., 1980, The form and environment of formation of the Kalavasos ore deposits, Cyprus, *in* Panayiotou. A., ed., Ophiolites: Proceedings of the International Ophiolite Symposium, Cyprus, 1979. Nicosia, Cyprus, Geological Survey Department, p. 117–128

Alabaster, T., Pearce, J., and Malpas, J.G., 1982, Volcanic stratigraphy and petrogenesis of the Oman ophiolite complex: Contributions to Mineralogy and Petrology, v. 81, p. 168–183.

Allen, C.R., 1975, The petrology of a portion of the Troodos Plutonic Complex, Cyprus [Ph.D. thesis]: Cambridge, UK, University of Cambridge, 315 p.

Anonymous, 1972, Penrose field conference on ophiolites: Geotimes, v. 17, p 24–25.

Bagnall, P.S., 1960, The geology and mineral resources of the Pano Lefkara-Larnaca area: Nicosia, Cyprus, Geological Survey Department Memoir no. 5, 116 p.

Baragar, W.R.A., Lambert, M.B., Baglow, N., and Gibson, I.L., 1987, Sheeted dikes of the Troodos ophiolite, Cyprus, *in* Halls, H.C., and Fahrig, W.C., eds., Mafic Dike Swarms: Ottawa, Ontario, Geological Association of Canada Special Paper 34, p. 257–272.

Batanova, V.G., and Sobolev, A.V., 2000, Compositional heterogeneity in subduction-related mantle peridotites, Troodos massif, Cyprus: Geology, v. 28, p. 55–58.

Bear, L.M., 1960, The geology and mineral resources of the Akaki-Lythrodondha area: Nicosia, Cyprus, Geological Survey Department Memoir no. 3, 122 p.

Bear, L.M., 1963, The mineral resources and mining industry of Cyprus: Nicosia, Cyprus, Geological Survey Department Bulletin no. 1, 208 p.

Bear, L.M., and Morel, S.W., 1960, The geology and mineral resources of the Agros-Akrotiri area: Nicosia, Cyprus, Geological Survey Department Memoir no. 7, 88 p.

Beccaluva, L., Ohnenstetter, D., and Ohnenstetter, M., 1979, Geochemical discrimination between ocean floor and island arc tholeiites: Application to some ophiolites: Ofioliti, v. 4, p. 67–72

Benson, W.N., 1926, The tectonic conditions accompanying the intrusion of basic and ultrabasic igneous rocks: Washington, D.C., National Academy of Science, v. 19, p.1–90.

Bishopp, D.W., 1952, Some new features of the geology of Cyprus: International Geological Congress, 19th, Algiers, Algeria, v. 17, p. 13–18.

Bowen, N.L., 1927, The origin of ultrabasic and related rocks: American Journal of Science, v. 14, p. 89–108.

Brongniart, A., 1813, Essai d'une classification mineralogique des roches mèlangèes: Journal des Mines, Paris, no. 199 (Juillet 1813) p. 5–48.

Brongniart, A., 1821, Sur le gisement ou position relative des ophiolites euphotides, jaspes, etc. dans quelques partie des Apennins: Annual del Mines ou Recueil de Memoires sur l'Exploitation des Mines, v. 6, p. 177–238.

Brongniart, A., 1827, Classification et caracteres mineralogiques des roches homogenes et heterogenes: Paris, Levrault, 144 p.

Brunn, J.H., 1956, Contribution a l'etude geologique du Pinde septentrional et d'une partie de la Macedoine occidentale: Annales geologique Pays helleniques, v. 7, 358 p.

Brunn, J.H., 1960, Mise en place et differentation pluto-volcanique du cortege ophiolitique: Revue de Geographie Physique et de Geologie Dynamique, v. 3, p. 115–132.

Cameron, W.E., 1985, Petrology and origin of primitive lavas from the Troodos ophiolite, Cyprus: Contributions to Mineralogy and Petrology, v. 89, p. 239–255.

Cann, J.R., Blackman, D.K., Smith, D.K., McAllister, E., Janssen, B., Mello, S., Avgerinos, E., Pascoe, A.R., and Escartin, J., 1997, Corrugated slip surfaces formed at ridge-transform intersections on the Mid-Atlantic Ridge: Nature, v. 385, p. 329–332.

Cann, J.R., Oakely, R.J., Richards, H.G., and Richardson, C.J., 1987, Geochemistry of hydrothermally altered rocks from Cyprus Drill Holes Cy-2 and Cy-2a compared with other Cyprus stockworks, *in* Robinson, P.T., Gibson, I.L., and Panayiotou, A., eds., Cyprus Crustal Study Project: Initial Report, Holes Cy-2 and 2a. Ottawa, Ontario, Geological Survey of Canada Special Paper 82–29, p. 87–102.

Cann, J.R., Prichard, H.M., Malpas, J.G., and Xenophontos, C., 2001, Oceanic inside corner detachments of the Limassol Forest area, Troodos ophiolite, Cyprus: Journal of the Geological Society of London, v. 158, p. 757–767.

Carr, J.H., and Bear, L.M., 1960, The geology and mineral resources of the Peristerona-Lagoudhera area: Nicosia, Cyprus, Geological Survey Department Memoir no. 2, 79 p.

Christensen, N.I., and Salisbury, M.H., 1975, Structure and constitution of the lower oceanic crust: Reviews of Geophysics and Space Physics, v. 13, p. 57–86

Church, W.R., and Stevens, R.K., 1971, Early Paleozoic ophiolite complexes of the Newfoundland Appalachians as Mantle-Oceanic Crust sequences: Journal of Geophysical Research, v. 76, p 1460–1466.

Clube, T.M.M., and Robertson, A.H.F., 1986, The paleorotation of the Troodos microplate, Cyprus, in the Late Mesozoic–Early Cretaceous plate tectonic framework of the Eastern Mediterranean: Surveys in Geophysics, v. 8, p. 375–434.

Coleman, R.G., 1971, Plate tectonic emplacement of upper mantle peridotite along continental edges: Journal of Geophysical Research, v. 76, p. 1212–1222

Coleman, R.G., 1984, The diversity of ophiolites: Geologie en Mijnbouw, v. 63, 141–150

Constantinou, G., 1980, Metallogenesis associated with Troodos ophiolite, *in* Panayiotou, A., ed., Ophiolites: Proceedings of the International Ophiolite Symposium, Nicosia, Cyprus, 1979, Geological Survey Department, p. 663–674.

Constantinou, G., and Govett, G.J.S., 1972, Genesis of sulphide deposits, ochre and umber of Cyprus: Transactions of the Institution of Mining and Metallurgy, v. 81, p. B34–B46.

Constantinou, G., and Govett, G.J.S., 1973, Geology, geochemistry and genesis of Cyprus sulphide deposits: Economic Geology, v. 68, p. 843–858.

Cowan, J.G., Richardson, C.J., and Cann, J.R., 1987, Trace and major element geochemistry of epidosite zones and the formation of black smoker fluids in the Troodos ophiolite, Cyprus, *in* Malpas, J. and Xenophontos, C., eds., Troodos 87, Ophiolites and Ocean Lithosphere: Nicosia, Cyprus, Geological Survey Department, p. 127.

CYAMEX Scientific Team, 1979, Massive deep-sea-sulfide ore deposits discovered on the East Pacific Rise: Nature, v. 277, p. 523–528.

De Roever, W.P., 1957, Sind die alpinotypen peridotitmassen vielleicht tektonisch verfrachtete bruchstucke der perdotitschale?: Geologische Rundschau, v. 46, p. 137–146.

Detrick, R.S., Collins, J., Stephen, R., and Swift, S., 1994, In situ evidence for the nature of the seismic layer 2/3 boundary in oceanic crust: Nature, v. 370, p. 288–290.

Dewey, J.F., 1976, Ophiolite obduction: Tectonophysics, v. 31, p. 93–120.

Dick, H.J.B., Natland, J.H., and Miller, J., eds., 1999, Proceedings of the Ocean Drilling Program, Initial Reports 176: College Station, Texas, Ocean Drilling Program, 823 p.

Dietz, R.S., 1963, Alpine serpentinites as oceanic rind fragments: Geological Society of America Bulletin, v. 74, p. 947–952.

Dilek, Y., Thy, P., Moores, E.M., and Ramsden, T.W., 1990, Tectonic evolution of the Troodos ophiolite within the Tethyan framework. Tectonics, v. 9, p. 811–823.

Dubertret, L., 1953, Geologie des roches vertes du Nord-Ouest de las Syrie et due Hatay: Notes et Memoires Moyen-Orient 6, 179 p.

Ewing, M., 1965, The sediments of the Argentine Basin: Quarterly Journal of the Royal Astronomical Society, v. 6, p. 10–27.

Gass, I.G., 1960, The geology and mineral resources of the Dhali area: Nicosia, Cyprus, Geological Survey Department Memoir no. 4, 116 p.

Gass, I.G., 1968, Is the Troodos Massif of Cyprus a fragment of Mesozoic ocean floor?: Nature, v. 220, p. 39–42.

Gass, I.G., 1980, The Troodos massif: Its role in the unraveling of the ophiolitic problem and its significance in the understanding of constructive plate margin processes, *in* Panayiotou, A., ed., Ophiolites, Proceedings of the International Ophiolite Symposium, Cyprus, 1979, Nicosia, Cyprus Geological Survey Department, p. 23–35.

Gass, I.G., MacLeod, C.J., Murton, B.J., Panayiotou, A., Simonian, K.O., and Xenophontos, C., 1994, The geology of the Southern Troodos Transform Zone: Nicosia, Cyprus, Geological Survey Department, Memoir 9, 218 p.

Gass, I.G., and Masson-Smith, D., 1963, The geology and gravity anomalies of the Troodos massif, Cyprus: Royal Society of London Philosophical Transactions, v. A255, p. 417–467.

Gass, I.G., Neary, C.R., Plant, J., Robertson, A.H.F., Simonian, K.O., Smewing, J.D., Spooner, E.T.C., and Wilson, R.A.M., 1975, Comments on "The Troodos ophiolite complex was probably formed in an island arc" by A. Miyashiro, and subsequent correspondence by H. Hynes and A. Miyashiro: Earth and Planetary Science Letters, v. 25, p. 236–238.

Gass, I.G., and Smewing, J.D., 1973, Intrusion, extrusion, and metamorphism at constructive margins; Evidence from the Troodos massif, Cyprus: Nature, v. 242, p. 26–29.

Greenbaum, D., 1972, Magmatic processes at ocean ridges: evidence from the Troodos massif, Cyprus: Nature, v. 238, p. 18–21,

Henson, F.R.S., Browne, R.V., and McGinty, J., 1949, A synopsis of the stratigraphy and geological history of Cyprus: Quarterly Journal of the Geological Society, v. 105, p. 1–41.

Hess, H.H., 1938, A primary peridotite magma: American Journal of Science, v. 35, p. 321–344.

Hess, H.H., 1960, Caribbean Research Project: Progress report: Geological Society of America Bulletin, v. 71, p. 235–240.

Hess, H.H., 1964, The oceanic crust, the upper mantle and the Mayagues serpentinised peridotite. I: A study of serpentine: the AMSOC core hole near Mayagues, Puerto Rico: Washington. D.C., National Academy of Sciences, Natural Resources Council Publication 1188, p. 169–175.

Hutchinson, R.W., and Searle, D.L., 1971, Stratabound pyrite deposits in Cyprus and relations to other sulphide ores, *in* Proceedings of IMA-IAGOD meetings, 1970, Tokyo, 198–205.

Hynes, A., 1975, Comment on the "The Troodos ophiolite complex was probably formed in an island arc" by A. Miyashiro: Earth and Planetary Science Letters, v. 25, p. 213–216.

Jenner, G.A., Dunning, G.R., Malpas, J., Brown, M, and Brace, T., 1991, Bay of Islands and Little Port complexes, revisited: Age, geochemical and isotopic evidence confirms suprasubduction-zone origin: Canadian Journal of Earth Sciences, v. 28, p. 1635–1652.

Johnson, A.E., 1972, Origin of Cyprus pyrite deposits: International Geological Congress, 24th, Montreal, sec. 4, p. 291–298.

Kelley, D.S., and Robinson, P.T., 1990, Development of a brine-dominated hydrothermal system at temperatures of 400–500 °C in the upper level plutonic sequence, Troodos ophiolite, Cyprus: Geochemica et Cosmochimica Acta, v. 54, p. 653–661.

Kelley, D.S., Robinson, P.T., and Malpas, J., 1992, Processes of brine generation and circulation in the oceanic crust: fluid inclusion evidence from the Troodos ophiolite, Cyprus: Journal of Geophysical Research, v. 97, p. 9307–9322.

Kusky, T.M., Li, J.H., and Tucker, R.D., 2001, The Archean Dongwanzi Ophiolite Complex, North China Craton: 2.505-Billion-Year-Old Oceanic Crust and Mantle: Science, v. 292, p. 1142–1145.

LaPierre, H., 1975, Les formations sedimentaires et eruptive des nappes de Mamonia et leurs relations avec le massif de Troodos (Chypre occidentale): Memoire Societe geologique France 123, 420 p.

MacLeod, C.J., and Murton, B.J., 1993, Structure and tectonic evolution of the Southern Troodos Transform fault Zone, Cyprus, *in* Prichard, H.M., Alabaster, T., Harris, N.B.W., and Neary, C.R., eds., Magmatic Processes and Plate Tectonics: London, Geological Society of London Special Publication no. 76, p. 141–176.

Malpas, J., 1973, A restored section of oceanic crust and mantle in western Newfoundland: Geological Society of America Abstracts with Programs, v. 5, p. 191–192.

Malpas, J., 1990, Crustal accretionary processes in the Troodos Ophiolite, Cyprus: Evidence from field mapping and deep crustal drilling, *in* Malpas, J., Moores, E.M., Panayiotou, A., and Xenophontos, C., eds., Ophiolites: Oceanic Crustal Analogues: Proceedings of the Symposium "Troodos 1987": Nicosia, Cyprus, Geological Survey Department, p. 65–74.

Malpas, J., and Brace, T., 1987, Geological map of the Amiandos-Palekori area, Cyprus: 1:12,500 map with marginal notes, Geological Survey Department, Cyprus.

Malpas, J., Calon, T., and Squires, G., 1993, The development of a Late Cretaceous microplate suture zone in S.W., Cyprus, *in* Prichard, H., Alabaster, T., Harris, N.B.W., and Neary, C.R., eds., Magmatic processes and plate tectonics: London, Geological Society of London Special Publication 76, p. 177–195.

Malpas, J., and Langdon, G., 1984, Petrology of the upper pillow lava suite, Troodos ophiolite, Cyprus, *in* Gass, I.G., Lippared, S.J., and Shelton, A.W., eds., Ophiolites and Oceanic Lithosphere: London, Geological Society of London Special Publication 13, p. 155–167.

Malpas, J., and Robinson, P.T., 1997, Oceanic lithosphere 1: the origin and evolution of oceanic lithosphere, processes at spreading centres: Geoscience Canada, v. 24, p. 100–107.

Malpas, J., and Robinson, P.T., 2000, Oceanic lithosphere 4: The origin and evolution of oceanic lithosphere: magmatic processes at oceanic spreading centres: Geoscience Canada, v. 27, p. 131–146.

Malpas, J., and Stevens, R.K., 1978, Origin and emplacement of the ophiolite suite: Geotectonics, v. 11, p. 453–466.

Malpas, J., Xenophontos, C., and Robertson, A.H.F., 1987, Mamonia complex and its relationship to the Troodos Complex, *in* Xenophontos, C., and Malpas, J., eds., Field Excursion Guidebook, Troodos 1987, Ophiolites and Oceanic Lithosphere: Nicosia, Cyprus, Geological Survey Department, p. 234–259.

Malpas, J., Xenophontos, C., and Williams, D., 1992, The Ayia Varvara Formation of S.W. Cyprus: A product of complex collisional tectonics: Tectonophysics, v. 212, p. 193–241.

Malpas, J., and Zhou, M.F., 2001, The significance of paired ophiolite belts, examples from the western Pacific and Caribbean: Eos, Transactions of the American Geophysical Union, v. 82, p. 47.

McCulloch, M.T., and Cameron, W.E., 1983, Nd–Sr isotopic study of primitive lavas from the Troodos ophiolite, Cyprus: evidence for a subduction related setting: Geology, v. 11, p. 727–731.

Mehegan, J., 1988, Temporal, spatial and chemical evolution of the Troodos ophiolitic lavas in Cyprus [Ph.D. thesis]: Halifax, Nova Scotia, Dalhousie University, 863 p.

Miyashiro, A., 1973, The Troodos ophiolite complex was probably formed in an island arc: Earth and Planetary Science Letters, v. 19, p. 218–224.

Moore, P., 1960, The geology and mineral resources of the Astromeritis-Kormakiti area: Nicosia, Cyprus, Geological Survey Department Memoir no. 6, 96 p.

Moores, E.M., 1975, Discussion of 'Origin of Troodos and their ophiolites: A reply to Hynes' by A. Miyashiro: Earth and Planetary Science Letters, v. 25, p. 223–226.

Moores, E.M., 1982, Origin and emplacement of ophiolites: Reviews of Geophysics and Space Physics, v. 20, p. 735–760.

Moores, E.M., Robinson, P.T., Malpas, J., and Xenophontos, C., 1984, Model for the origin of the Troodos Massif, Cyprus, and other Mideast ophiolites: Geology, v. 12, p. 500–503.

Moores, E.M., and Vine, F.J., 1971, The Troodos Massif, Cyprus and other ophiolites as oceanic crust: evaluation and implications: Philosophical Transactions Royal Society of London, v. A268, p. 433–466.

Nicolas, A., Reuber, I., and Benn, K., 1988, A new magma chamber model based on structural studies in the Oman ophiolite: Tectonophysics, v. 151, p. 97–105.

Oudin, E. and Constantinou, G., 1984, Black smoker chimney fragments in Cyprus sulphide deposits: Nature, v. 308, p. 349–353.

Pallister, J.S., and Hopson, C.A., 1981, Semail ophiolite plutonic suite: Field relations, phase variation, cryptic variation and layering, and a model of a spreading ridge magma chamber: Journal of Geophysical Research, v. 86, p 2593–2644.

Pantazis, Th. M., 1967, The geology and mineral resources of the Pharmakas–Kalavassos area: Nicosia, Cyprus, Geological Survey Department Memoir no. 8, 190 p.

Panayiotou, A., 1980, Cu–Ni–Co–Fe sulphide mineralization, Limassol Forest, Cyprus, *in* Panayiotou, A., ed., Ophiolites: Proceedings of the International Ophiolite Symposium, Cyprus, 1979: Nicosia, Cyprus, Geological Survey Department, p. 102–116.

Panayiotou, A., 1986, Sulphide and arsenide mineralization associated with the basic and ultrabasic plutonic rocks of the Troodos ophiolite, Cyprus, *in* Carter, B., et al., eds., Metallogeny of the basic and ultrabasic rocks: Athens, Theophrastus Publications S.A., p. 417–447

Pearce, J.A., 1975, Basalt geochemistry used to investigate past tectonic environments on Cyprus: Tectonophysics, v. 25, p. 41–67.

Pearce, J.A., 1980, Geochemical evidence for the genesis and eruptive setting of lavas from Tethyan ophiolites, in Panayiotou, A., ed., Proceedings of the International Symposium, Cyprus, 1979: Nicosia, Cyprus, Geological Survey Department, p. 261–272

Pearce, J.A., and Cann, J.R., 1971, Ophiolite origin investigated by discriminant analysis using Ti, Zr and Y: Earth and Planetary Science Letters, v. 12, p. 339–349.

Pearce, J.A., and Cann, J.R., 1973, Tectonic setting of basic volcanic rocks determined using trace element analyses: Earth and Planetary Science Letters, v. 19, p. 290–300.

Rautenschlein, M., Jenner, G.A., Hertogen, J., Hofmann, A., Kerrich, R., Schmincke, H.-U., and White, W.M., 1985, Isotopic and trace element compositions of volcanic glasses from the Akaki Canyon, Cyprus: implication for the origin of the Troodos ophiolite, Cyprus: Earth and Planetary Science Letters, v. 75, p. 369–383.

Reinhardt, B.M., 1969, On the genesis and emplacement of ophiolites in the Oman Mountains geosynclines: Schweizerische Mineralogische und Petrographische Mitteilungen, v. 49, p. 1–30

Richardson, C.J., Cann, J.R., Richards, H.G., and Cowan, J.G., 1988, Metal-depleted root zones of the Troodos ore-forming hydrothermal systems, Cyprus: Earth and Planetary Science Letters, v. 94, p. 243–253.

Robertson, A.H.F., 1976, Pelagic chalks and calciturbidites from the Lower Tertiary of the Troodos Massif: Journal of Sedimentary Petrology, v. 46, p. 1007–1016.

Robertson, A.H.F., and Woodcock, N.H., 1979, The Mamonia Complex, southwest Cyprus: the evolution and emplacement of a Mesozoic continental margin: Geological Society of America Bulletin, v. 90, p. 651–665

Robertson A.H.F., and Xenophontos, C., 1993, Development of concepts concerning the Troodos ophiolite and adjacent units in Cyprus, *in* Prichard, H.M., Alabaster, T., Harris, N.B.W., and Neary C.R., eds., Magmatic Processes and Plate Tectonics: Boulder, Colorado, Geological Society Special Publication no. 76., p. 85–119.

Robinson, P.T., Gibson, I.L., and Panayiotou, A., editors, 1987, Cyprus Crustal Study Project Initial Report, Holes CY-2 and CY-2a: Ottawa, Ontario, Geological Survey of Canada Special Paper 85–29, 381 p.

Robinson, P.T., and Malpas, J., 1990, The Troodos ophiolite of Cyprus: New perspectives on it origin and emplacement, *in* Malpas, J., Moores, E.M., Panayiotou, A., and Xenophontos, C., eds., Ophiolites: Oceanic Crustal Analogues: Nicosia, Cyprus, Geological Survey Department, p. 13–36.

Robinson, P.T., Melson, W.G., O'Hearn, T., and Schmincke, H.-U., 1983, Volcanic glass compositions of the Troodos ophiolite, Cyprus: Geology, v. 11, p. 400–404.

Robinson, P.T., Von Herzen, R. et al., 1989, Proceedings of the Ocean Drilling Program, Initial Reports 118: College Station, Texas, Ocean Drilling Program, 823 p.

Rogers, N.W., MacLeod, C.J., and Murton, B.J., 1989, Petrogenesis of boninitic lavas from the Limassol Forest Complex, Cyprus, *in* Crawford, A.J., ed., Boninites: London, Unwin Hyman, p. 288–313.

Routhier, P., 1946, Volcano-plutons sous marines du cortege ophiolitique: Comptes Rendu Academie Science Paris, v. 22, p. 192–194.

Routhier, P., 1953, Etude geologique du versant occidental de la Nouvelle-Caledonie: Paris, Memoir Societe geologique France, v. 67, 271 p.

Schiffman, P., Bettinson, L.A., and Smith, B.M., 1990, Mineralogy and geochemistry of epidosites from the Solea graben, Troodos ophiolite, Cyprus, *in* Malpas, J., Moores, E.M., Panayiotou, A., and Xenophontos, C., eds., Ophiolites: Oceanic Crustal Analogues: Nicosia, Cyprus, Geological Survey Department, p. 673–684.

Schiffman, P., Smith, B.M., Varga, R.J., and Moores, E.M., 1987, Geometry, conditions and timing of the off-axis hydrothermal metamorphism and ore deposition in the Solea graben, N. Troodos ophiolite, Cyprus: Nature, v. 325, p. 423–425.

Schmincke, H.-U., Rautenschlein, M., Robinson, P.T., and Mehegan, J., 1983, Troodos extrusive series of Cyprus: A comparison with oceanic crust: Geology, v. 11, p. 405–409.

Searle, M.P., and Stevens, R.K., 1984, Obduction processes in ancient, modern and future ophiolites, *in* Gass, I.G., Lippard, S.J., and Shelton, A.W., eds., Ophiolites and oceanic lithosphere. London, Geological Society of London Special Publication, 13, p. 303–319.

Şengör, A.M.C., Natal'In, B.A., and Burtman, V.S., 1993, Evolution of the Altaid tectonic collage and Palaeozoic crustal growth in Eurasia: Nature v. 364, p. 229–307.

Shervais, J.W., 1990, Island arc and ocean crust ophiolites: Contrasts in the petrology, geochemistry and tectonic style of ophiolite assemblages in the California Coast Range, *in* Malpas, J., Moores, E.M., Panayiotou, A., and Xenophontos, C., eds., Ophiolites: Oceanic Crustal Analogues: Nicosia, Cyprus, Geological Survey Department, p. 507–520.

Shervais, J.W., and Kimbrough, D.L., 1985, Geochemical evidence for the tectonic setting of the Coast Range ophiolite: A composite island arc-oceanic crust terrane in western California: Geology, v. 13, p. 35–38.

Simonian, K.O., and Gass, I.G., 1978, Arakapas fault belt, Cyprus: a fossil transform fault: Geological Society of America Bulletin, v. 89, p. 1220–1230.

Smewing, J.D., Simonian, K.O., and Gass, I.G., 1975, Metabasalts from the Troodos Massif, Cyprus: Genetic implications deduced from petrography and trace element geochemistry: Contributions to Mineralogy and Petrology, v. 51, no. 1, p. 49–64.

Spiess, F.N., Macdonald, K.C., Atwater, T., Ballard, R., Carranza, A., Cordoba, D. Cox, C., Diaz Garcia, V.M., Francheteau, J., Guerrero, J., Hawkins, J., Haymon, R., Hessler, R., Juteau, T., Kastner, M., Larson, R., Luyendyk, B., Macdougall, J.D., Miller, S., Normark, W., Orcutt, J., and Rangin, C., 1980, East Pacific Rise: Hot springs and geophysical experiments: Science, v. 207, p. 1421–1433.

Spooner, E.T.C., 1977, Hydrothermal model for the origin of the ophiolitic cuperiferous pyrite ore deposits of Cyprus, in Volcanic Processes in Ore Genesis: London, Geological Society London Special Publication 7, p. 58–71.

Spooner, E.T.C., 1980, Cu-pyrite mineralization and seawater convection in oceanic crust; the ophiolitic ore deposits of Cyprus, *in* The Continental Crust and its Mineral Deposits: St. John's, Newfoundland, Geological Association of Canada Special Paper 20, p. 685–704.

Steinmann, G., 1927, Die ophiolithischen Zonen in dem mediterranen Kettengebirge: International Geological Congress, 14th, Madrid, v. 2, p. 638–667.

Stern, R.J., and Bloomer, S.H., 1992, Subduction zone infancy: Examples from the Eocene Izu-Bonin-Mariana and Jurassic California arcs: Geological Society of America Bulletin, v. 104, p. 1621–1636.

Swarbrick, R.E., 1980, The Mamonia Complex of SW Cyprus and its relationship with the Troodos Complex, *in* Panayiotou, A., ed., Ophiolites, Proceeding of the International Ophiolite Symposium, Cyprus, 1979: Nicosia, Cyprus, Geological Survey Department, p. 86–92.

Taylor, R.N., 1990, Geochemical stratigraphy of the Troodos extrusive sequence: temporal development of a spreading centre magma chamber, *in* Malpas, J., Moores, E.M., Panayiotou, A., and Xenophontos, C., eds., Ophiolites: Nicosia, Cyprus, Oceanic Crustal Analogues, Geological Survey Department, p. 173–184.

Varga, R.J., and Moores, E.M., 1985, Spreading structure of the Troodos ophiolite, Cyprus: Geology, v. 13, p. 846–850.

Vine, F.J., and Matthews, D.H., 1963, Magnetic anomalies over oceanic ridges: Nature, v. 199, p. 947–949.

Williams, H., and Malpas, J., 1972, Sheeted dykes and brecciated dyke rocks within transported igneous complexes, Bay of Islands, western Newfoundland: Canadian Journal of Earth Sciences, v. 9, p. 1216–1229.

Wilson, R.A.M., 1959, The geology and mineral resources of the Xeros-Troodos area: Nicosia, Cyprus, Geological Survey Department Memoir no. 1, 136 p.

Zhai, M., Zhao, B., and Zhang, Q., 2002, Is the Dongwanzi complex an Archean ophiolite?: Science, v. 295, p. 923–925.

Manuscript Accepted by the Society May 2, 2003

Printed in the USA

Geological Society of America
Special Paper 373
2003

The Troodos ophiolite and the upper ocean crust; a reciprocal traffic in scientific concepts

J.R. Cann*
School of Earth Sciences, University of Leeds, Leeds LS2 9JT, UK

ABSTRACT

Current understanding of the processes that construct and modify the oceanic upper crust is the result of 40 years interplay between concepts derived from ophiolites and those derived from active mid-ocean ridges. In the oceans, rates of processes can be measured, and the spatial distribution of current activity determined. In ophiolites the resulting structure is exposed in three dimensions. The Troodos ophiolite in Cyprus has played a crucial role in this exchange, since there the upper ophiolitic crust is preserved for over 100 km across strike, and has scarcely been deformed or heated more than a few tens of degrees since it formed in a supra-subduction zone environment ~90 million years ago. The presence of the sheeted dike complex in Troodos, demonstrating 70 km of 100% extension, was crucial in the early stages of the plate tectonic debate. Later, the recognition of the exhalative nature of the ore deposits within the volcanics prepared the way for the discovery of black smokers in the oceans. The oceanic observations in turn brought new insights into the hydrothermal systems that had fed the ophiolitic sulfide deposits. Research on these ophiolitic systems then illuminated the oceanic systems again. Demonstration of the narrowness of the zone of crustal construction in the oceans and determination of the structure of upper crust at oceanic spreading axes led to a reassessment of the structure of the upper crust in Troodos. This in turn generated, and is continuing to generate, new approaches to the construction of the upper crust in the oceans. Because of chemical differences between ophiolitic crust and that of major ocean basins, parallels must be drawn with care. Creative interplay between the ophiolitic and oceanic environments, so crucial early in the plate tectonic revolution, shows no sign of abating. This topic is not solely history; it is also science in progress.

Keywords: ophiolite, mid-ocean ridge, ocean crust, hydrothermal systems.

INTRODUCTION

When the Cyprus Geological Survey began its program of detailed mapping of the Troodos ophiolite in the early 1950s, plate tectonics and ocean floor spreading still lay in the future. Though Du Toit (1937) had given a firm geological basis to the speculations of Wegener (1915), continental drift was regarded as unsound by most professional geologists of the northern hemisphere. By the 1950s, paleomagnetic work (e.g., Runcorn, 1959) was beginning to show evidence for major changes in continental latitude, and geological evidence existed for relative crustal displacements of tens of kilometers (e.g., Kennedy, 1946). But there were real problems about finding evidence for major extensional movement, and this was critical for accepting continental drift. Even 100 km of compression could be explained by thermal contraction of the Earth, and strike slip motion could be linked to contraction. It was not clear, for example, if the East African Rift Valley is extensional or compressional. Ophiolites were

*joe@earth.leeds.ac.uk

Cann, J.R., 2003, The Troodos ophiolite and the upper ocean crust; a reciprocal traffic in scientific concepts, *in* Dilek, Y., and Newcomb, S., eds., Ophiolite concept and the evolution of geological thought: Boulder, Colorado, Geological Society of America Special Paper 373, p. 309–321. For permission to copy, contact editing@geosociety.org.

considered to be the products of major eruptive events within geosynclinal basins (Aubouin, 1965). Plate-tectonic geology was still in its infancy.

Previous research in Cyprus had shown the Troodos mountains to be composed of igneous rocks, with submarine volcanics around the fringes and plutonics in the center, all overlapped by marine sediments. The first product of the detailed mapping of the Cyprus Geological Survey, Memoir 1 (Wilson, 1959), covering the Xeros-Troodos area, gave substance to these outlines. It laid the foundation of the lithostratigraphy of the ophiolite, and also recognized for the first time the nature of the sheeted dike complex, a unit that was to play a crucial role in the evolution of plate tectonic ideas in the following years. The mapping of the island has progressed greatly since those early times and is still continuing. Figure 1 shows a simplified map of the geology of the ophiolite, based on the accumulated experience of the Geological Survey Department in Cyprus.

The first formal identification of ophiolites as slices of crust and upper mantle formed by ocean floor spreading was made by Gass (1968) on the basis of the Geological Survey Department mapping of Troodos. This was reinforced by the work of Moores and Vine (1971) also on Troodos, which tested that relationship by making critical field observations. Since that time research on both ophiolites and ocean crust has proceeded as two strongly-linked pathways.

Starting from this point, I want to explore the development since the 1960s of the understanding of the processes that shape the oceanic lithosphere, and that shaped the Troodos ophiolite. I shall concentrate on areas in which there has been significant interplay between the two sources of evidence.

This is not a comprehensive review of past research on Troodos (for that, see Robertson and Xenophontos (1993)), nor does it deal with the lower crust and upper mantle, since these are much better exposed in Oman. Instead I concentrate on discussing the strongest links between research on the upper crust of the Troodos ophiolite and that on the upper oceanic crust.

IGNEOUS AND TECTONIC PROCESSES

The Sheeted Dike Complex

The sheeted dike complex is a unit of the ophiolite, more than 90 km across strike, made of tens of thousands of dikes, each one intruding earlier dikes, which themselves are intruded into even earlier dikes. There is no sign of any non-ophiolitic country rocks into which the earliest dikes might have been intruded. The dikes near the top of the complex are intruded into lavas, and those near the base either intrude ophiolitic plutonic rocks, or are intruded by them. This complex makes up a major part of the outcrop of the ophiolite.

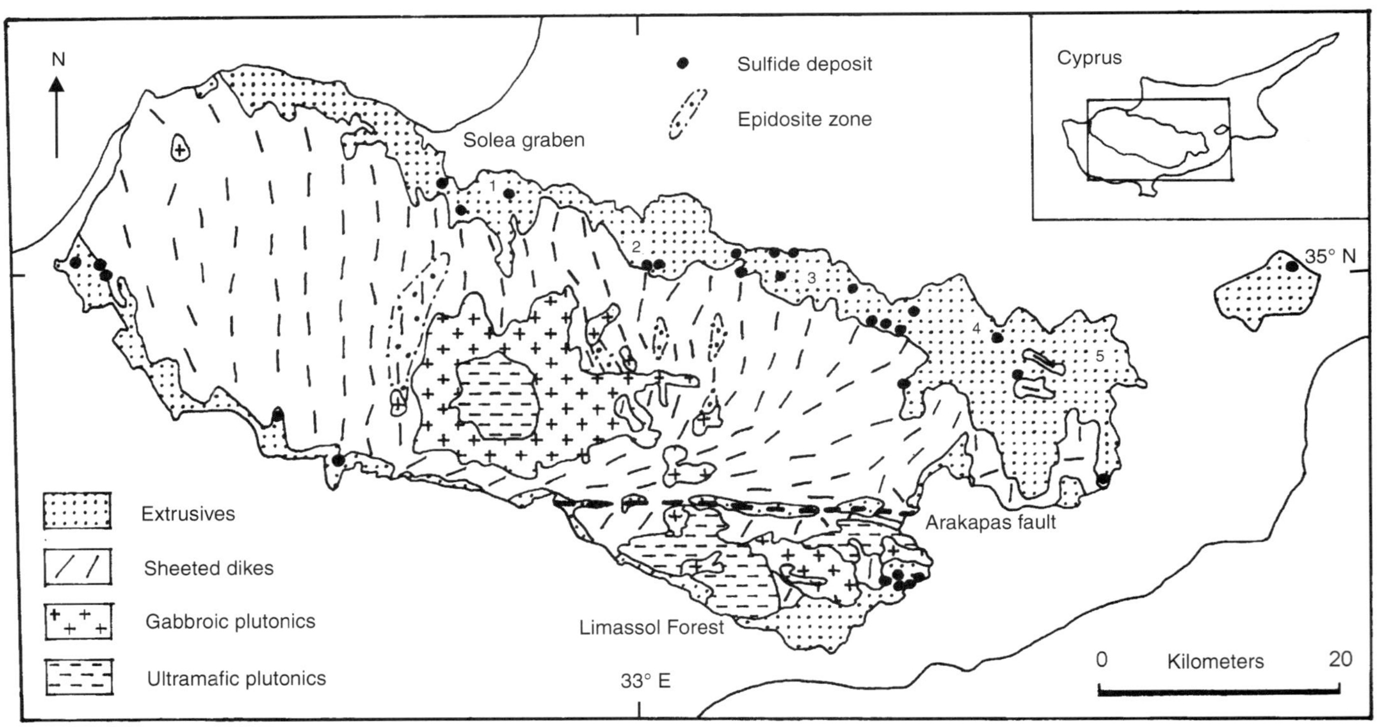

Figure 1. Simplified outline of the geology of the Troodos ophiolite, based on the mapping of the Cyprus Geological Survey, and showing the lithological units and other features mentioned in the text. The boundary between the extrusives and sheeted dikes shown is that mapped between the lavas and the basal group. Localities referred to in the text are: 1. Skouriotissa, 2. Alestos, 3. Akaki Canyon, 4. Mathiati, 5. Mosphiloti.

The discussion in Memoir 1 (Wilson, 1959) shows that this unit caused serious conceptual problems; indeed, Bishopp (1952) had proposed instead that this is a unit of isoclinally-folded lava flows. Wilson wrote (p. 60–61): "When it is considered, however, that the diabase [sheeted dike] outcrop extends for ~60 miles, to suppose that it is entirely composed of dikes would appear to be an improbable conclusion. …[discussion of some of the evidence] … For these reasons, the writer concludes, on consideration of the evidence at present available, that the intrusive hypothesis is to be preferred…." Hesitantly, but in the end firmly, Wilson came down in favor of sheeted dikes. These gave a physical reality to ocean floor spreading several years before that hypothesis had been framed by Hess (1962).

In the Troodos ophiolite, for the first time, crustal extension of more than a few kilometers could be demonstrated directly. Equally significant was the recognition, forced by field observation, that extension could reach 100% over wide areas and that in the center of the sheeted dike complex there is no country rock to the dikes except earlier dikes. This is the most dramatic demonstration of the actuality of ocean floor spreading, depending on no models, but on simple field observation, and it still provides one of the most inspiring geological experiences.

It is unclear whether Hess knew about the Troodos sheeted dikes as he developed his ideas of ocean floor spreading. He was a frequent visitor to the UK, and news of the sheeted dike complex was being actively discussed in the UK in the late 1950s. I do not recall that the two ideas, of ocean floor spreading and sheeted dikes, were linked at that time, though they may have been. But whether Hess was influenced or not, Vine and Matthews (1963) may have been influenced by Troodos in their use of vertical-sided blocks of crust to simulate magnetic anomalies. True, models involving vertical-sided blocks were about as complex as could be handled by the computers of the time, but the concept of a series of dikes seems clearly to underlie their paper.

The Ophiolite Model of the Ocean Crust

The early seismic refraction experiments in the 1950s and 1960s (Raitt, 1963) rapidly established two important results: the oceanic crust is far thinner than continental crust (6 km as opposed to 35 km), and the structure of the oceanic crust is similar for all of the major ocean basins. Layer 2, the upper layer, ~2 km thick, of relatively low seismic velocity, overlies Layer 3, ~4 km thick, of high and constant seismic velocity, and this in turn overlies the mantle (seismic Layer 4). Modern experiments have added refinements to the structure, but have essentially confirmed the early results. Nowadays Layer 2A is seen as the uppermost part of the crust, a few hundred meters thick close to the spreading axis, with a very steep velocity gradient. It is separated by a sharp boundary, which often gives rise to seismic reflections, from Layer 2B, with a less steep velocity gradient, and ~1.5 km thick. This overlies Layer 3, which has a very constant P-wave velocity of close to 7.2 km/s, and which is separated from the mantle by a sharp Moho, which again gives rise to seismic reflections (e.g., Vera et al., 1990).

But what lithologies do these seismic layers represent? Hess (1962) argued that the lower crust must be partly serpentinized peridotite. That argument is periodically revived, especially in discussing the Mid-Atlantic Ridge, where outcrops of serpentinite are abundant, mostly at the ends of ridge segments, but also in the centers of segments. Cann (1968) made the case that the lower crust might be amphibolite facies metabasalts, but in the same year Gass (1968) proposed that the Troodos ophiolite might be a section of the ocean crust, and that the lithological sequence seen there might be that of Layers 2, 3 and 4, with Layer 2 corresponding to lavas and sheeted dikes, Layer 3 to gabbro, and Layer 4 to dunites and harzburgites. When very similar sequences were reported shortly afterwards by Reinhardt (1969) in the Oman ophiolite, and by Davies (1968; 1971) in the Papua New Guinea ophiolite, the momentum behind this proposal grew, and the ophiolite model of the oceanic lithosphere became firmly consolidated

Currently Layer 2A is generally seen as correlated with extrusives, Layer 2B with sheeted dikes and Layer 3 with gabbros that typically form a large fraction of the lower parts of ophiolite sections. Deep ocean drilling has shown, in Hole 504B on the Costa Rica Rift in the east Pacific (Adamson, 1985) and in Hole 395A on the Mid-Atlantic Ridge (Aumento, Melson et al., 1977) that the extrusives are 500–600 m thick. Sheeted dikes and gabbros have been abundantly observed and recovered from submarine scarps that yield deeper sections into the crust (for a review see Karson, 1998), so the appropriate lithologies are present, though this lithological identity still has to be confirmed in detail. The ophiolite model of the ocean crust is thus firmly entrenched in the geological literature, periodically challenged, but hard to dislodge. It is especially strong for ocean crust formed at fast-spreading ridges for two reasons: 1. Rock sampling from slow-spreading ridges indicates a much more complex geological structure than the simply layered ophiolite model (e.g., Cannat et al., 1995), and 2. High-resolution seismic observations on fast-spreading ridges, by contrast, do show a simple, laterally extensive layered structure (e.g., Hooft et al., 1996).

Tectonic Environment of Formation of the Troodos Spreading Center

For a few years after Gass (1968) recognized that the Troodos ophiolite had formed by ocean floor spreading, there was no question about using its structure as a model for the ocean crust. But Miyashiro (1973) called that simple equivalence into question by showing that the chemistry of lavas from Troodos is different in ways that he considered crucial from the chemistry of the lavas from mid-ocean ridges, with basaltic andesites as the dominant lava type. Instead he suggested that Troodos had originated in an island arc. Gass et al. (1975) made a robust reply to this, attributing the differences in composition to weathering and hydrothermal alteration. However further geochemical evidence, including analyses of trace elements less susceptible to alteration (Pearce, 1975; 1980), and of fresh volcanic glasses (Robinson et al., 1983; Rautenschlein et al., 1985) indicated that there is a strong island-arc character to the chemistry of many of the Troodos lavas. Pearce et al. (1984) incorporated this evidence into the concept of supra-subduction zone (SSZ) ophiolites; ophiolites formed by seafloor spreading, but fed by magmas with an island arc chemistry. The supra-subduction zone character of an ophiolite is seen not only in the trace element geochemistry, but also in the petrology of the cumulates and of the mantle section, all resulting from the nature of the primary magmas. All of these characteristics are strongly developed in the Troodos ophiolite. They are shared, less strongly, by the Oman ophiolite, and in varying degrees by many other ophiolites (Pearce, et al., 1984).

The demonstration of tens of kilometers of spreading in the sheeted dike complex appeared to show that Troodos could not have formed as an island arc volcano. But the arc-related geochemistry seemed to show a tectonic environment closely related to a subduction zone. This apparent paradox led to a controversy, poorly-documented in the literature, but very prominent in discussion at the time. At one end of the spectrum were those who concluded that the geochemistry showed that the Troodos ophiolite had formed part of an island arc, and therefore was irrelevant to understanding the ocean crust. At the other end of the spectrum were those who knew that ophiolites had clearly formed by spreading, felt that geochemistry was irrelevant, and were happy when they could show that occasionally an andesite could be found at a mid-ocean ridge. Most of us occupied a position somewhere between those extremes.

The controversy was partly resolved when the Valu Fa spreading center behind the Tonga Arc was investigated. The Valu Fa spreading center is the southernmost section of the Lau Basin back-arc spreading center complex, lying in the back-arc basin behind the Tonga Arc. From north to south, the successive sections of the Lau Basin spreading center complex approach closer and closer to the Tonga island arc. In the north, the Central Lau Spreading Center is over 200 km from the arc, while at the southern end the Valu Fa Spreading Center is less than 100 km from the island arc chain of volcanoes. Ewart et al. (1994) review the regional lava geochemistry. While the Central Lau Spreading Center erupts basalts with an essentially MORB geochemistry, the lavas of the Valu Fa spreading center have a strong arc geochemical signature (Jenner et al., 1987), and show great similarities with the lavas of Troodos. It thus seems likely that many ophiolites formed at spreading centers in back-arc basins, and that those with more arc-like geochemistry formed from spreading centers close to the arc, while those with more MORB-like geochemistry formed farther from the arc.

Has this resolved the controversy? In part it has, though problems remain with the sedimentary cover (Dilek et al., 1990). In the Valu Fa area, the close proximity of the spreading center to the island arc means that arc-derived volcaniclastic materials, both airfall deposits and submarine deposits, accumulate rapidly on top of the crust formed by spreading (Bednarz and Schmincke, 1994). In Troodos, such contemporaneous arc-derived volcaniclastics are absent. The Kannaviou Formation (Robertson, 1977) overlies the ophiolite and is composed of calc-alkaline volcaniclastics, but dates from ~20 million years after the ophiolite was formed. Was the subduction zone above which the Troodos spreading center formed very short lived, so that arc volcanoes did not have time to grow? It is possible.

How does this discussion affect the use of observations from Troodos to constrain processes at mid-ocean ridges? Clearly the overall chemistry of the ophiolite is different from that of a typical section of crust in a major ocean basin, so that inferences based on geochemistry and petrology of Troodos (and most other ophiolites) must be treated with care. But the Troodos crust has clearly formed by more than 70 km of spreading in an oceanic (though near-arc) environment, and many of the physical processes involved are unaffected by magma chemistry, so that it should be possible to transfer physical insights directly from one environment to the other.

Constructing the Upper Crust

Assembling a Pile of Lavas and Dikes

Soon after the recognition of ocean floor spreading, questions arose over how the upper ocean crust is built at mid-ocean ridges. How are lavas and dikes organized into a single structure, and how does that structure evolve through time? Central to this is estimation of the width of the zone of crustal construction at a spreading center. At first geophysical evidence from the oceans was most important; the shape of the marine magnetic anomalies was used to derive estimates of the transition width from crust of one polarity to that of another, leading to values of 2–3 km, and hence to a rather broad zone of crustal construction at spreading axes.

From a geological point of view, the first insights into the construction of the upper crust at spreading centers came from Iceland (Bodvarsson and Walker, 1964; Palmasson 1973, 1980), and were introduced to the oceans by Cann (1974). Observations on Troodos and other ophiolites allowed these ideas to be developed further by Kidd (1977).

The early mapping of the Troodos ophiolite (Cyprus Geological Survey Memoirs 1 (Wilson, 1959), 2 (Carr and Bear, 1960), 3 (Bear, 1960) and 4; (Gass, 1960)) established some important observations about the lithostratigraphy of the lavas and dikes. The lavas in general have dips of less than a few tens of degrees, and contain very few dikes, even toward their base. At the base of the lavas there is a rapid and mappable transition to the basal group, a unit composed predominantly of dikes with subordinate screens of lavas. With increasing depth in the complex, the screens of lavas become rarer, until none can be detected, at which point the unit was mapped as sheeted dikes. The transition from basal group to sheeted dikes is much less easily mapped than that from lavas to basal group, and depends on the recognition of the rare screens of lava. In terms of the broad distinction between lava-dominated and dike-dominated units, the significant boundary is that between the lavas and the basal group. A recent survey by Hall and Yang (1995) has mapped the percentage of dykes across the northern outcrop of the lavas and basal group. The width of the transition varies across the outcrop, perhaps reflecting variation in the vertical thickness of the transition zone, or perhaps variation in the dip of the units.

Kidd (1977) showed by numerical modeling coupled with field studies that two observations from Troodos are critical for understanding the width of the zone of crustal construction: 1. the lava section of the ophiolite contains very few dikes, and 2. the vertical transition from lava with few dikes to basal group (dikes with very few screens of lava) is very rapid. The transition from lavas with less than 10% dikes to dikes with less than 10% lavas is complete typically in ~200 m of vertical section in the sections examined by Kidd. These observations can only be explained if the upper crust is constructed by eruption from a zone of fissures that is much narrower than the extent of the lava flows. These conclusions were confirmed by a combination of observation from submersibles and sidescan sonar mapping of the East Pacific Rise (Crane, 1987), which showed a zone of fissuring at the axis only a few hundred meters wide, corresponding to Kidd's zone of dike injection, and lava flows extending commonly for more than a kilometer from the axis.

An apparently contradictory piece of evidence comes from the observation of cross-cutting dikes in Troodos, which may be taken to indicate a broader zone of crustal construction. However, there is good evidence that rotation of blocks of sheeted dikes occurred very early in the evolution of the crust (see later), and also strong evidence for along-axis intrusion of dikes (Staudigel et al., 1992; Tauxe et al., 1998), both of which explain the presence of cross-cutting dikes.

Reduced to a simple statement, formation of a unit of 100% sheeted dikes requires that the dikes be intruded in a zone no more than a few hundred meters wide at the spreading axis. If the zone of dike injection is much wider than a few hundred meters, then a unit of 100% sheeted dikes will not form. Abundant screens of lava will be found deep within the dikes and screens of plutonics will be found at high levels in the dikes (Cann, 1974). These ideas have recently been re-visited using new data from the oceans by Hooft et al. (1996). If the newly-recognized seismic reflector from the Layer 2A/2B boundary represents a sharp transition from lavas to sheeted dikes such as is seen in Troodos, then the geometry of this reflector at the spreading axis can constrain the process of crustal construction. The presence of the reflector suggests that the vertical transition from lavas to dikes is rapid, and hence that the upper crust is constructed from a very narrow zone of dike intrusion. The shape of the reflector close to the spreading axis can be interpreted to give information about the spectrum of lengths of lava flows. The observed shape indicates flows of a range of lengths, from short flows that end close to the axis to longer flows that extend several kilometers from the axis. A further extension of these ideas to Troodos, using paleomagnetic inclinations of lava flows has been made by Schouten and Denham (2000).

One important consequence of this work is that, in upper crust constructed at a spreading center with a narrow zone of dike intrusion, a vertical section of mature crust will reflect the lateral stratigraphy of crustal construction (Fig. 2). The top of the crust will be made of the distal tips of the lava flows, not covered by any subsequent flow, while the base of the lava section will have formed very close to the center of spreading, within or just next to the zone of dike intrusion. Depth in the lava pile correlates with distance from the spreading axis at the time of formation. Similarly, dikes that penetrate high in the lava pile

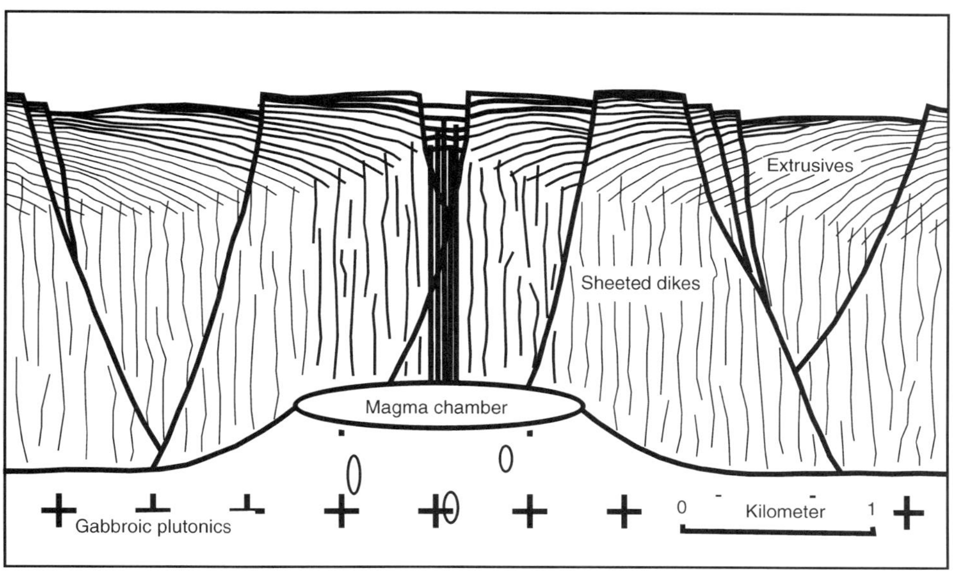

Figure 2. Cross-section of the upper crust at an unrifted spreading center, based on a combination of evidence from Troodos and from the East Pacific Rise. The younger dikes and lavas are drawn in heavier lines, and the older dikes and lavas in thinner lines. The structure shown is the result of intrusion of dikes in a zone much narrower than the width of the zone covered by lavas (Cann, 1974). Note that in mature crust, the deeper lavas are those that cooled closest to the spreading axis, while the seafloor is made of the distal ends of flows.

must have been intruded away from the axis, while dikes that do not reach high levels must have been intruded very close to the center of spreading.

Upper and Lower Pillow Lavas in Troodos

Though the concepts of the construction of the upper ocean crust originated in Troodos, there are still inconsistencies in their application there. This is best shown by the view that the extrusive series must be divided into two units, the Upper and Lower Pillow Lavas, which goes back to Wilson (1959), in pre-plate tectonic days. The division of the extrusive series has been made over the years on a number of grounds. The upper part of the extrusive section has a distinctively different color in the field (orange/pale gray) as opposed to the greenish gray of the lower part. Lavas with olivine phenocrysts are more abundant in the upper part, and these lavas commonly have a chemistry that is depleted in incompatible elements, while the lower part of the extrusives is dominated by flows of more andesitic chemistry (Robinson et al., 1983). The massive sulfide deposits occur within the lava section, and their deposition has been related to a hiatus in volcanism. Occasionally there are unconformities, or talus accumulations, or interlava hydrothermal sediments within the extrusives. Not only do these criteria give different positions for the Upper Pillow Lava/Lower Pillow Lava boundary in the same part of the extrusive section, such as when sulfide deposits occur well within the greenish gray, olivine-poor lower part of the section, but none of them has any force individually or collectively.

The difference in color between the upper and lower parts of the extrusive section has been shown to be the result of long-term (tens of millions of years) of sea floor weathering (Gillis and Robinson, 1985; Staudigel et al., 1986; Gallahan and Duncan, 1994), not of primary eruptive processes. The differences in chemistry and petrography between the upper and lower lavas are not quantitative, as ought to be the case if two separate volcanic units were erupted, but statistical; incompatible-element-depleted lavas occur close to the base of the section as well as close to the top, and incompatible-element-depleted dikes are found both early and late in the sequence of injection at any one site (Baragar et al., 1990). Interlava sediments and unconformities do not occur at a consistent horizon in the extrusives, and, instead of indicating a prolonged hiatus in volcanism, are the result of syn-volcanic faulting, as can be demonstrated on a small scale in the Akaki Canyon (Schmincke and Bednarz, 1990), and/or local increases in hydrothermal sedimentation rate because of proximity to active black smokers.

Instead the extrusive unit must be interpreted as a coherent volcanic unit erupted from a narrow zone of dike intrusion in a spreading environment, from which a variety of magma types were available at any one time. The upper part of this unit represents the distal ends of the lava flows, not reached by later eruptions, while the lower parts represent the parts of the flow units that solidified closest to the dike intrusion zone, and were buried deepest in the lava pile. The statistical differences in chemistry and petrography through the pile must then represent differences in the length of flows of different compositions, with the more andesitic units having flowed less far from the intrusion zone, on average, than the more incompatible-element-depleted units (Schouten and Kelemen, 2002). The varying depth within the unit of exhalative sulfide deposits must then represent differences in the distance of hydrothermal venting from the zone of dike intrusion.

Within this framework, the variation in magma chemistry within a single section of lavas, such as is seen within the CY-1 drillhole (Robinson et al., 1983) cannot be the result of magma compositions varying through time at one locality. Instead, the geochemical variation seen must result from along-axis transport of magma within dikes from isolated bodies of magma with very different chemistries existing at the same time at different points along the ancient spreading axis.. Staudigel et al., (1992) and Tauxe et al., (1998) showed, using both imbrication of vesicles in dike margins and anisotropy of magnetic susceptibility, that many dikes have a substantial component of along-axis magma flow recorded within them, reinforcing this possibility.

There is some evidence for off-axis igneous activity within the ophiolite. Eddy et al. (1998) consider that the area centered on the Alestos hill, near Ayia Marina in the center of the northern outcrop of the extrusives represents a seamount constructed on crust formed by spreading. Another candidate, in urgent need of re-examination, is an area of rhyolitic intrusives at the eastern end of the ophiolite, close to the village of Mosphiloti (Gass, 1960). But over most of the ophiolite the extrusives seem to form a coherent unit created as an integral part of the spreading process.

Spreading Rate of the Troodos Spreading Center

In the oceans, spreading direction and spreading rate can be determined directly by using magnetic anomalies. There is a strong correlation of properties of the mid-ocean ridges and the crust created from them with spreading rate. Slow-spreading mid-ocean ridges are rifted, characterized by a major axial valley, 1–1.5 km deep, bordered by high, consistently inward-facing fault scarps up to several hundred meters high, and rugged topography on the flanks, with an amplitude of hundreds of meters. The full range of rock types characteristic of the oceanic lithosphere outcrops at the seafloor, including gabbros and serpentinized peridotites, in many places toward the ends of segments, but also at segment centers. Exposure of deep-seated lithologies is frequently linked to the outcrop at the ocean floor of low-angle detachment faults, with apparent displacements of several kilometers, but plutonic outcrops occur that are apparently not linked to such faults. Fast-spreading ridges are non-rifted, topographically smooth at the axis and on the flanks, and are cut by low fault scarps, typically tens of meters high that face both inwards and outwards (Small, 1998; Carbotte and Macdonald, 1990). Almost all of the seafloor is made up of extrusive outcrops, though plutonic rocks may outcrop in major transform fault valleys. The axial magma chamber and underlying mush zone is continuous at fast-spreading ridges, and the crustal thickness changes only slowly along axis, while at slow-spreading ridges, magmatic systems appear to be confined in space and in time, and crustal thickness varies markedly over a few tens of kilometers (see Small, 1998 for a review). The most marked change in these properties occurs on either side of a full spreading rate of ~70 mm/yr, though this is not a sharply-defined value. If ophiolites are to be used as a guide to processes at spreading centers, then there is clearly a great advantage in being able to determine spreading rate and spreading direction. This turns out to be surprisingly hard. Both the Troodos and Oman ophiolites were created during the long Cretaceous normal polarity interval, so there are no magnetic anomalies in either ophiolite to provide a reference frame.

This uncertainty has led to a prolonged debate on the spreading rate of Troodos. One group of researchers (Varga and Moores, 1985; Dilek et al., 1990; Dilek and Eddy, 1992) has proposed a slow to intermediate spreading rate. Their conclusion is primarily based on structures within the ophiolitic crust. The sheeted dike complex is cut by abundant brittle faults, and blocks of sheeted dykes have been rotated by tens of degrees from their original orientation. These features are most evident close to a structural graben in the ophiolite, the Solea Graben, considered to be an extinct spreading axis (see later). The Solea Graben is associated with inward-facing normal faults and with low-angle detachment faults at the sheeted-dike gabbro boundary. Similar detachments have been identified away from the Solea Graben as well (Agar and Klitgord, 1995). Such intense deformation, these authors argue, must demonstrate a slow spreading rate. But in linking these observations to spreading rate, they compared the internal structure of the ophiolite with an inferred internal structure of the ocean crust.

A different set of indicators of spreading rate are features that can be observed at the ancient sea floor, and can be compared directly with observations at the modern ocean floor. The most critical of these features are the surface relief of the ophiolite, the lithological units outcropping at the ophiolitic sea floor, and the scarp height and facing direction of the faults that cut the surface. The surface relief of the northern margin of the ophiolite (north of the Arakapas Transform Fault) is very small. Fault scarp heights are usually 30 m or less. Some areas, in which umbers are more common, and early cherts are preserved, must represent seafloor lows, while other areas, not covered by sediments until well after the onset of chalk sedimentation, must have been seafloor highs (see Gass, 1960, for example). These highs and lows are separated by a few kilometers across strike, and may have had an amplitude of as much as 100–200 m, similar to the wavelength and amplitude of abyssal hills in the Pacific today The low fault scarps face both to the east and to the west. All over the ophiolite north of the Arakapas Transform Fault, the unit that outcropped at the ancient sea floor is the extrusive unit. There is no evidence for outcropping of sheeted dikes or any deeper lithostratigraphic units. All of these indicators suggest a relatively fast spreading rate (Allerton and Vine, 1987; Hurst et al., 1994).

Apparently contradictory indicators come from the plutonic part of the section. Extensive exposures of the upper part of the plutonics are seen, and these were shown by Moores and Vine (1971) to be very complex. Their conclusions were confirmed by detailed mapping by Malpas and colleagues (Malpas, 1990). These observations seem inconsistent with the idea that at fast spreading rates there should be a nearly permanent magma chamber present at the spreading axis, and hence that the structure of the plutonics should be simple. Yet the upper part of the plutonic section may not be simple even at fast spreading rates. The structure of the plutonics constructed at a spreading center has a similar logic to it to the structure of the extrusives. While many of the structural complexities depend on the way in which the lower crust is constructed, and may be an indicator of the processes of crustal construction, there are some interpretations which are independent of models. For example, in a single exposure of the upper plutonics, the latest intrusive phase must have been emplaced furthest from the spreading axis, and the complexity of the section may reflect the evolution of small pockets of melt trapped at the fringes of the magma chamber, rather than the main trends of evolution of the magma chamber itself. Additional evidence from the plutonics comes from the near-absence of crystal-plastic deformation of the plutonics in Troodos. Gabbros from slow and ultraslow spreading ridges are commonly deformed to augen gneisses, as was seen, for example, in the rocks of ODP Leg 118 in the southwest Indian Ocean and Leg 153 in the Atlantic Ocean, but such deformation is rare in gabbros from fast spreading ridges, such as those from Hess Deep in the Equatorial Pacific drilled on ODP Leg 147.

South of the Arakapas Transform Fault, the ophiolite within the Limassol Forest shows a very different set of relationships. There the layer-cake lithostratigraphy of the northern outcrops is absent; instead the same lithologies are juxtaposed apparently at random (MacLeod and Murton, 1993; Gass et al., 1994). This relationship is linked to major extensional faulting which may have taken place after the construction of the crust (Gass et al., 1994), but may alternatively be the result of detachment faulting in the inside corner area of the transform fault.(Cann et al., 2001). This latter model would set further limits on spreading rate, since inside corner detachments have so far been seen only on crust formed at slow or intermediate spreading rates.

Taking all of this evidence together, it seems likely that the spreading rate was close to 70 mm/yr. Perhaps a parallel can be drawn with the contrast between the rifted Costa Rica Rift and the unrifted Ecuador and Galapagos Rifts across a transform in the equatorial Pacific close to 2°N, 86°W. The spreading rate there is ~70 mm/yr on both sides of the transform, and at that spreading rate it seems that other factors may determine whether an individual segment of a spreading center is rifted or non-rifted. In Troodos the non-rifted segment would be the section of the ophiolite north of the Arakapas Fault, while the rifted segment would be that south of the transform.

The Solea Graben

The Solea Graben has been identified as an extinct spreading center (Varga and Moores, 1985; Varga, 1991; MacLeod et al., 1990; Hurst et al., 1992) on the grounds of the structural thinning of sheeted dike and gabbro units within it, the presence of inward-facing fault scarps that displace the lithological units by hundreds of meters, and the low-angle detachments that separate sheeted dikes from gabbro on its western margin. There are major rotations of blocks of sheeted dikes on the western side of the graben away from the axis of the graben, consistent with rotation in the hanging wall of listric faults throwing down toward the graben axis. The graben is asymmetric, with much greater evidence for tectonic extension on its western side than on the east. The structural evidence for tectonic rotation has been confirmed paleomagnetically (Allerton and Vine, 1987; Hurst et al., 1992).

Despite the thinning of the dike and gabbro units, there is no obvious topographic valley where the extrusive unit meets the sea floor. Instead the thinning of the deeper units seems to be compensated by thickening of the extrusive unit above them.

Extinct spreading centers in the oceans are marked by an increase in the amplitude of topography consistent with a decrease in spreading rate during extinction, ending in a major topographic valley at the site of extinction. This signature is most easily seen on fast-spreading crust, cutting through the regional low topographic relief. Good examples can be seen on the satellite gravity map of Sandwell and Smith (1997), and the corresponding predicted topography map of Smith and Sandwell (1997) at 45°S, 15°E (SW of S Africa), 45°S, 155°E (Tasman Rise), 55°S, 95°W (Galapagos Rise), 15°N, 110°W (Mathematicians Ridge) and 57°S, 60°W (in the Drake Passage). The valley that marks the site of extinction is typically a kilometer or more deep and 15–30 km wide, regardless of the original spreading rate of the ridge (see for example Vanko and Batiza, 1982; Mammerickx et al., 1988; Fujioka et al., 1999; Livermore et al., 2000). Plutonic rocks have been recovered from the extinction valley of the Mathematicians Ridge (Vanko and Batiza, 1982).

Comparison of the Solea graben with oceanic examples of extinct spreading centers indicates some parallels, and especially in the evidence for an increase in the degree of tectonic extension during its formation. But the topographic depression that typically marks an oceanic example is not noticeably present at Solea. A valley 800 m deep would result in a deflection in the outcrop of the seafloor at the top of the Solea graben of ~5 km, given the overall regional dip of the ophiolite to the north of ~10°. Such a deflection is not seen. In addition there is no evidence that plutonic rocks outcropped at the seafloor within the graben. The comparison with oceanic examples suggests that the Solea Graben may not be an extinct spreading center, but may represent a period of lower magma supply and increased tectonic extension preserved on one flank of a spreading ridge, as suggested by Allerton and Vine (1987).

Structural Evolution Close to the Spreading Axis

The coexistence in Troodos of very small topographic relief at the ancient seafloor together with intense deformation within the sheeted dikes has important consequences for understanding processes at non-rifted mid-ocean ridges. Often the contribution of faulting to seafloor spreading in the oceans has been estimated by summing the scarp heights at the ocean floor away from the spreading axis and calculation of extension from this sum by assuming some appropriate fault dip. This is clearly not appropriate in Troodos, and is probably not appropriate either in the modern oceans.

The evidence in Cyprus is very clear. Faults cut the ancient sea floor frequently, but the scarp heights are a few tens of meters at the most, and the lavas close to the ancient seafloor dip in general gently to the north, following the regional dip of the ophiolite. Deeper in the section similar faults may displace lithostratigraphic boundaries by hundreds of meters, and may be responsible for tens of degrees of rotation of blocks of sheeted dikes (Varga and Moores, 1985; Allerton and Vine, 1987; Moores et al., 1990). The low-angle detachment faults may have seen even more movement.

This apparent paradox can be resolved if the upper crust was deformed as it was being constructed. In such a model, growth of the faults would be nearly complete within the zone resurfaced by lava flows, so that the faults would slip synvolcanically, and the throw on the faults would increase with depth in the crustal section. Within the sheeted dikes, both structural evidence (Schiffman et al., 1987) and paleomagnetic evidence (Varga et al., 1999) show that much of the movement on the faults happened after the high-temperature circulation associated with black smoker activity. The presence of cross-cutting dikes within the sheeted dikes (Moores and Vine, 1971; Varga and Moores, 1985) shows that dike rotation had already happened close enough to the axis for new dikes to intrude already rotated blocks of dikes.

The implications of these observations for the oceans are fundamental. The absence of large fault scarps cutting the sea floor cannot be taken as evidence that all of the spreading has taken place by dike intrusion. A substantial component of tectonic extension may also be present, which has taken place within a kilometer or so of the spreading axis, and which is concealed beneath undeformed lavas. Models of unrifted spreading centers ought to be adjusted to take this possibility into account.

HYDROTHERMAL SYSTEMS

The Massive Sulfide Deposits

It was one of the triumphs of ophiolite studies that the existence of black smoker hot springs in the oceans, the temperature of the hot water and the nature of the sulfide deposits surrounding them were predicted from studies of the sulfide deposits of Troodos.

The concept that ores might be formed by exhalation of hot hydrothermal fluids onto the ocean floor has a long pedigree.

Back in the 19th century, ore deposits such as the massive sulfides of the Trondheim district in Norway, the Røros type of deposit, the great Rio Tinto deposits of Spain, and the Rammelsberg deposit of the Harz mountains in Germany had been considered by some to be exhalative in origin (for example: Vogt, 1894, 1899; Carstens 1919), though this was strongly contested. For example, the influential ore geologist Waldemar Lindgren (1933) dismissed these ideas briefly and firmly. Similar exhalative concepts were developed from the 19th century onwards to account for the Kuroko and Besshi types of deposits of Japan (for an account in English, see Tatsumi, 1970). These ideas were revived in the 1950s (for example: Oftedahl, 1958) and became influential without commanding general support.

It was the work of the United Nations team in Cyprus in the late 1960s that was able to demonstrate conclusively that ores could be formed by exhalation (Constantinou and Govett, 1972, 1973). In Cyprus, deposits of massive pyrite ("massive" meaning made up almost entirely of pyrite, including no country rock), up to a few million tonnes in size, and grading a few percent of copper and zinc were being mined within the volcanic rocks of the Troodos ophiolite (Bear, 1963). These lenticular massive sulfide deposits were enclosed above and below by seafloor volcanics, of which those above had not been altered by hydrothermal solutions, while those beneath had been altered. Particularly critical in this was the recognition that the yellow ochres associated with several deposits, and especially the Skouriotissa deposit, had been formed by submarine weathering of the sulfide ore, and the resedimentation of the yellow iron oxides formed in this way on the surrounding ocean floor. This demonstrated that the ores had been exposed, and partly weathered , at the ocean floor before the succeeding lavas had been deposited. A network of sulfide veins cutting the altered lavas (stockwork), could be found beneath the deposits, representing the channelway up which the hydrothermal fluids had risen. The temperature of the fluids could be estimated from the mineral assemblages in the stockworks, and also from fluid inclusions (Spooner and Bray, 1977).

The result of this was that in the oceans through the 1970s, exploration of the mid-ocean ridges included as one of its main targets the discovery of these metal-bearing hot springs, which must be emerging at the sea floor at temperatures of more than 300 °C and precipitating large masses of iron, copper and zinc sulfides. Such hot springs and associated deposits were eventually discovered in 1978 at the crest of the East Pacific Rise (Spiess et al., 1980). The black smoker fluids turned out to emerge through chimneys a few centimeters in diameter, lined with coarsely-crystalline chalcopyrite encased in anhydrite. Areas of warm diffuse flow emerging close to the black smoker chimneys were oases of abundant life, including hitherto unknown organisms. Both of these features had not been anticipated from the observations in Cyprus, and this led some to challenge the exhalative nature of the ores there. However, further investigation in Troodos discovered within sulfides in the massive deposits both chalcopyrite-lined chimneys and fossils of tube worms and gastropods (Oudin and Constantinou, 1984; Little et al., 1999), thus confirming the earlier conclusions, and demonstrating close parallels between the black smokers of Cyprus and the ocean floor.

The Hydrothermal Plumbing System

Chemical studies of the fluids emerging from the vents rapidly established in outline the conditions under which black smoker fluids are generated (for example, Von Damm, et al. 1985). About 10 km^3 of black smoker fluid would be needed to produce each of the large sulfide deposits of Troodos. The fluids could be shown to originate as cold sea water, which penetrates into the ocean crust close to the spreading axis, becomes heated, and reacts chemically with the enclosing rock. In the course of this reaction the fluid loses almost all of its magnesium to hydrous silicates, and sulfate to precipitation of anhydrite. The fluid gains iron, copper and zinc, as well as hydrogen sulfide, while becoming highly acid. These chemical changes must involve reaction between heated sea water and basaltic lava and dikes, but observation at the seafloor alone cannot determine the processes that transform cold ocean water to hot, metalliferous hydrothermal fluid.

Research in Troodos (Constantinou and Govett, 1972, 1973; Constantinou, 1980; Adamides, 1980, 1990; Lydon and Galley, 1986; Richards et al., 1989) established that the stockworks beneath the massive sulfides are underlain by pipe-like feeder zones of highly altered lava, with a greenschist facies mineralogy. These alteration pipes are mineralogically zoned, from assemblages typical of low water/rock ratios at their periphery to assemblages formed at high water/rock ratios at their cores. The mineralogy of the high water/rock alteration assemblages varies from pipe to pipe, apparently depending on the chemistry of the hydrothermal solutions. The volumes of fluid passing through each alteration pipe would be far greater than the volume of the pipe itself, so that the alteration would be very much fluid-dominated. Stockworks of this kind have been discovered exposed on fault scarps in the oceans (Embley et al., 1988), and were drilled beneath the TAG hydrothermal field in the Mid-Atlantic Ridge (Humphris et al., 1995).

Farther down, in the sheeted dikes close to the plutonic unit, and hence close to the axial magma chamber, extensive areas of sheeted dikes are transformed to a variety of greenschist facies assemblages including epidosites (epidote-quartz rocks) (Richardson et al., 1987; Schiffman et al., 1987, 1990; Schiffman and Smith, 1988; Bettison-Varga et al., 1995). The epidosite forms as stripes within the sheeted dikes, replacing the basaltic fabric, and showing that hot water had penetrated into the body of the dike rock. In these epidosite zones, the greenschist facies mineral assemblages show flow of water at 300–400 °C. The rocks contain fluid inclusions of approximately seawater salinity, which give trapping temperatures of 350–400 °C when pressure corrected. The epiodosites are strongly depleted in copper, zinc and manganese, elements that are enriched in the ore deposits and their surroundings. The transformation of the basaltic dikes to epidosite demonstrates a very large water/rock ratio because of the profound chemical changes that have taken place. Bettison-Varga et al. (1995) showed that this transformation had to happen in two stages, one from basalt to a rock with a 5 or 6 phase greenschist facies assemblage to epidosites with only two phases. All of these features point toward the epidosite zones being the reaction zones in which cold seawater is transformed to hot, acid, metal-bearing hydrothermal fluids. Alt (1994) would include the upper gabbros as well within any reaction zone, based on balance of sulfur and sulfur isotopes, though these still possess multi-phase assemblages, and do not seem to been the site of major fluid flow. Such reaction zones are inaccessible in the oceans, yet in Troodos they can be studied readily in the field, and their geometry and relations with the surrounding sheeted dikes can be determined (see Figure 2).

Troodos thus allows us to investigate the plumbing system beneath black smokers in a way quite impossible in the oceans, and to discover the complex processes that transform cold seawater to hot hydrothermal fluid. But it is a problem that epidosites are very rare in the oceans. One was reported from the Mid-Atlantic Ridge in one of the very first papers about ocean floor rocks (Quon and Ehlers, 1963), but the only other discoveries to date have been from the Tonga forearc (Banerjee et al., 2000). Is it because epidosites form only a very small proportion of the sheeted dikes of the oceans, and that further sampling of deep sheeted dike outcrops will eventually yield more? Or is it because the hydrothermal reaction zones of the oceans are not made of epidosites, but of another assemblage such as quartz + chlorite?

Black Smoker Fallout

Certain types of iron ore associated with submarine volcanics have been thought by some, since the 19th century, to have formed by exhalation, though this conclusion was always controversial at the time. Similar Fe and Mn-rich metalliferous sediments were discovered in the 1960s on the floor of the Pacific Ocean (Boström and Peterson, 1966; Bonatti and Joensuu, 1966), and were considered to be hydrothermal in origin.

In Troodos the equivalent rocks are the umbers, dark brown, Fe and Mn-rich sediments, quarried for pigments. These both overlie the volcanics, accumulating in lenses up to 20 m thick in local depressions, and occur as continuous layers within the volcanics between individual submarine flows (Robertson, 1975; Boyle and Robertson, 1984; Boyle, 1990). In some cases, such as close to the Mathiati sulfide deposit in eastern Troodos, it is possible to trace an inter-lava umber horizon right up to a black smoker deposit. By combining observations of black smoker plumes in the oceans (see Mills and Elderfield, 1995 for a review) and the stratigraphy seen in Cyprus, it is possible to interpret umbers as fallout from black smoker plumes. As the plumes of black smoke rise above the chimneys on the ocean floor, the minute flakes of sulfide that make up the smoke are rapidly oxidized, and the newly-formed iron oxides adsorb manganese and many trace elements from the hydrothermal fluid and from seawater. The cloud of oxide particles gradually sinks toward the ocean floor and accumulates within a few kilometers of the vent to form the umbers. Here the interpretation of oceanic and Cypriot examples has proceeded hand in hand, resulting in an integrated model for umber genesis.

OUTSTANDING PROBLEMS OF THE UPPER OCEAN CRUST

Unraveling the processes that create the upper ocean crust continues to require interaction between evidence from ophiolites and from the oceans. Recently, for example, the energy and chemical balances in the formation of a large massive sulfide deposit were clarified by co-interpretation of evidence from the Mid-Atlantic Ridge and Troodos (Humphris and Cann, 2000). And Gillis (2002) has been able to determine the nature of the interface between hydrothermal circulation and magma, by supplementing remote-sensing geophysical data from the oceans with detailed field and petrological evidence from Troodos.

But there are still many problems, for the solution of which evidence must be drawn either from ophiolites, or from the oceans, or both. Thus, to use the structural evidence from Troodos to understand oceanic crustal construction requires a reliable indicator of both spreading direction and spreading rate in ophiolites without magnetic reversals (as is the case with both Troodos and Oman). At that point a number of questions become accessible. Was the Troodos ophiolite formed at several spreading centers (Varga and Moores, 1985) that each produced 10–30 km of crust? Or on one flank of a single spreading center, as might seem likely from a simple view of modern mid-ocean ridges? How much of the early rotation of the lavas and dikes was caused by rotation on listric faults (Varga and Moores, 1985), and how much by lava loading (Cann, 1974; Schouten and Denham, 2000)? Answers to these questions would have great applicability to mid-ocean ridge dynamics.

In hydrothermal systems, it is important to test further the similarities and differences between hydrothermal activity in Troodos and at mid-ocean ridges. The main problem at present is that of the epidosites. Is the absence of epidosites from the oceans a sampling problem only; are they there, but not sampled? Or are the high-temperature hydrothermal reaction zones of mid-ocean ridges very different in character from those of Cyprus? And how does the geochemistry of black smoker solutions relate to the possible presence of epidosites at depth?

Signatures of off-axis hydrothermal circulation have been studied both in Troodos and on the ocean floor. The problem at present is to disentangle the products of off-axis circulation from those of the axial black smoker systems, especially within the sheeted dikes, where it seems that the two types of system overlap successively within the same rocks. Is the alteration pattern of the sheeted dikes generated in the recharge zones of black smokers? Or is it formed after the main black smoker circulation has switched off, and if so where, very close to the spreading axis, or at some distance away?

Answers to these questions will not only improve our understanding of the Troodos ophiolite, but will also impact our models of the dynamics of the generation of the ocean crust that covers three-fifths of the surface of the Earth.

ACKNOWLEDGMENTS

This paper rests on the work not only of those cited here, but all of the other scientists who have worked on Troodos. All of those would agree, too, that the greatest debt is owing to Dr. Costas Xenophontos, who has supported research of all kinds on the Troodos ophiolite over very many years. I am grateful to Bob Varga, Kathy Gillis, Yildirim Dilek and an anonymous referee, whose comments on the first draft of this paper led to substantial improvements in the text.

REFERENCES CITED

Adamides, N.G., 1980, The form and environment of formation of the Kalavasos ore deposits—Cyprus, *in* Panayiotou, A., ed., Ophiolites: Proceedings International Ophiolite Symposium, Cyprus 1979: Nicosia, Cyprus, Ministry of Agriculture and Natural Resources, Geological Survey Department, p. 117–127.

Adamides, N.G., 1990, Hydrothermal circulation and ore deposition in the Troodos ophiolite, Cyprus, *in* Malpas, J., Moores, E.M., Panayiotou, A. and Xenophontos, C., eds., Ophiolites: oceanic crustal analogues: Nicosia, Cyprus, Ministry of Agriculture and Natural Resources, Geological Survey Department, p. 685–704.

Agar, S.M., and Klitgord, K., 1995, A mechanism for decoupling within the oceanic lithosphere revealed in the Troodos ophiolite: Nature, v. 374, p. 232–238.

Adamson, A.C., 1985. Basement stratigraphy, Deep Sea Drilling Project Hole 504B, Initial Reports, Deep Sea Drilling Project 83: Washington, D.C., U.S. Government Printing Office, p. 121–127.

Allerton, S., and Vine, F.J., 1987, Spreading structure of the Troodos ophiolite, Cyprus: Some paleomagnetic constraints: Geology, v. 15, p. 593–597.

Alt, J.C., A sulfur isotopic profile through the Troodos ophiolite, Cyprus, 1994, Primary composition and the effects of seawater hydrothermal alteration: Geochimica et Cosmochimica Acta, v. 58, p. 1825–1840.

Aumento, F., Melson, W.G., et al., 1977, Initial Reports of the Deep Sea Drilling Project, Volume 37: U.S. Government Printing Office, Washington, D.C., 423 p.

Aubouin, J., 1965, Geosynclines: Amsterdam, Elsevier, 335 p.

Banerjee, N.R., Gillis, K.M., and Muelenbachs, K., 2000, Discovery of epidosites in a modern oceanic setting, the Tonga forearc: Geology, v. 28, p. 151–154.

Baragar, W.R.A., Lambert, M.B., Baglow, N. and Gibson, I.L., 1990, The sheeted dike zone in the Troodos ophiolite, *in* Malpas, J., Moores, E.M., Panayiotou, A., and Xenophontos, C., eds., Ophiolites: Oceanic crustal analogues: Nicosia, Cyprus, Ministry of Agriculture and Natural Resources, Geological Survey Department, p. 37–52.

Bear, L.M., 1960, The geology and mineral resources of the Akaki-Lythrodondha area: Nicosia, Cyprus, Geological Survey Department, Memoir 3, 122 p.

Bear, L.M., 1963, The mineral resources and mining industry of Cyprus: Nicosia, Cyprus, Geological Survey Department, 208 p.

Bednarz, U., and Schmincke, H.-U., 1994, Composition and origin of volcaniclastic sediments in the Lau Basin (southwest Pacific), Leg 135 (Sites 834–839), *in* Hawkins, J., Parson, L.M., Allan, J. et al., eds., Proceedings of the Ocean Drilling Program, Scientific Results: College Station, Texas, Ocean Drilling Program, v. 135, p. 51–74.

Bettison-Varga, L., Schiffman, P., and Janecky, D., 1995, Fluid-rock interaction in the hydrothermal upflow zone of the Solea Graben, Troodos ophiolite, Cyprus, *in* Schiffman, P., and Day H., eds., Low grade metamorphism of mafic rocks: Boulder, Colorado, Geological Society of America Special Paper 296, p. 81–100.

Bishopp, D.W., 1952, The Troodos massif, Cyprus: Nature, v. 169, p. 489.

Bodvarsson, G., and Walker, G.P.L., 1964, Crustal drift in Iceland: Geophysical Journal of the Royal Astronomical Society, v. 9, p. 285–300.

Bonatti, E., and Joensuu, O., 1966, Deep sea iron deposits from the South Pacific: Science, v. 154, p. 643–645.

Boström, K., and Peterson, M.N.A., 1966, Precipitates from hydrothermal exhalations on the East Pacific Rise: Economic Geology, v. 61, p. 1258–1265.

Boyle, J.F., and Robertson, A.H.F., 1984, Evolving metallogenesis at the Troodos spreading axis, *in* Gass, I.G., Lippard, S.J., and Shelton, A.W., eds., Ophiolites and oceanic lithosphere: Geological Society of London Special Publication 13, p. 169–181.

Boyle, J.F., 1990 The composition and origin of oxide metalliferous sediments from the Troodos ophiolite, Cyprus, *in* Malpas, J., Moores, E.M., Panayiotou, A., and Xenophontos, C., eds., Ophiolites: Oceanic crustal analogues: Nicosia, Cyprus, Ministry of Agriculture and Natural Resources, Geological Survey Department, p. 705–717.

Cann, J.R., 1968, Geological processes at mid-ocean ridge crests: Geophysical Journal of the Royal Astronomical Society, v. 15, p. 331–341.

Cann, J.R., 1974, A model for oceanic crustal structure developed: Geophysical Journal of the Royal Astronomical Society, v. 39, p. 169–187.

Cann, J.R., Prichard, H.M., Malpas, J.G., and Xenophontos, C., 2001, Oceanic inside corner detachments of the Limassol Forest area, Troodos ophiolite, Cyprus: Journal of the Geological Society of London, v. 158, p. 757–767.

Cannat, M., Mevel, C., Maia, M., Deplus, C., Durand, C., Gente, P., Agrinier, P., Belarouchi, A., Dubuisson, G., Humler, E. & Reynolds, J., 1995. Thin crust, ultramafic exposures, and rugged faulting patterns at the Mid-Atlantic Ridge (22°–24°N): Geology, v. 23, p. 49–52.

Carbotte, S.M., and Macdonald, K.C., 1990, Causes of variation in fault-facing direction on the ocean floor: Geology, v. 18, p. 749–752.

Carr, J.M., and Bear, L.M., 1960, The geology and mineral resources of the Peristerona–Laghoudhera area: Nicosia, Cyprus, Geological Survey Department, Memoir 2.

Carstens, C.W., 1919, Oversight over Trondhjemsfeltets bergbygning (Review of mountain building in the Trondheim district): Det Kongelig Norske Videnskabelig Selskabets Skrifter 1919-1, 152 p.

Constantinou, G., and Govett, G.J.S., 1972, Genesis of sulfide deposits, ochre and umber of Cyprus: Transactions of the Institution of Mining and Metallurgy, Section B: Applied Earth Science, v. 81, p. B34–B46.

Constantinou, G., and Govett, G J.S., 1973, Geology, geochemistry, and genesis of Cyprus sulfide deposits: Economic Geology, v. 68, p. 843–858.

Constantinou, G., 1980, Metallogenesis associated with the Troodos ophiolite, *in* Panayiotou, A., ed., Ophiolites: Proceedings, International Ophiolite Symposium, Cyprus 1979: Nicosia, Cyprus, Geological Survey Department, p. 663–674.

Crane, K., 1987, Structural evolution of the East Pacific Rise axis from 13°10′N to 10°35′N: interpretations from SeaMARC 1 data: Tectonophysics, v. 136, p. 65–124.

Davies, H.L., 1968, Papuan Ultramafic Belt: Proceedings of the 23rd International Geological Congress, v. 1, p. 209–220.

Davies, H.L., 1971, Peridotite-gabbro-basalt complex in Eastern Papua: an overthrust plate of oceanic mantle and crust: Bureau of Mineral Resources, Geology and Geophysics, Australia, Bulletin 128, 48 p.

Dilek, Y., Thy, P., Moores, E., and Ramsden, T., 1990, Tectonic evolution of the Troodos ophiolite within the Tethyan framework: Tectonics, v. 9, p. 811–823.

Dilek, Y., and Eddy, C.A., 1992, The Troodos (Cyprus) and Kizildag (S. Turkey) ophiolites as structural models for slow-spreading ridge segments: Journal of Geology, v. 100, p. 305–322.

Du Toit, A.L., 1937, Our wandering continents: Edinburgh, Oliver and Boyd, 366 p.

Eddy, C.A., Dilek, Y., Hurst, S., and Moores, E., Seamount formation and associated caldera complex and hydrothermal mineralization in ancient oceanic crust, Troodos ophiolite (Cyprus): Tectonophysics, v. 292, p. 189–210

Embley, R.W., Jonasson, I.R., Perfit, M.R., Franklin, J.M., Tivey, M.A., Malahoff, A., Smith, M.F., and Francis, T.G.J., 1988, Submersible investigation of an extinct hydrothermal system on the Galapagos Ridge; sulfide mounds, stockwork zone and differentiated lavas: Canadian Mineralogist, v. 26, p. 517–539.

Ewart, A., Bryan, W.B., Chappell, B.W., and Rudnick, R.L., 1994, Regional geochemistry of the Lau-Tonga arc and back-arc systems, *in* Hawkins,

J., Parson, L.M., Allan, J., et al., eds., Proceedings of the Ocean Drilling Program, Scientific Results: College Station, Texas, Ocean Drilling Program, v. 135, p. 385–425.

Fujioka, K., Okino, K., Kanamatsu, T., Ohara, Y., Ishizuka, O., Haraguchi, S., and Ishii, T., 1999, Enigmatic extinct spreading center in the West Philippine backarc basin unveiled: Geology, v. 27, p. 1135–1138.

Gallahan, W.E., and Duncan, R.A., 1994, Spatial and temporal variability in crystallization of celadonites within the Troodos ophiolite, Cyprus: implications for low-temperature alteration of the oceanic crust: Journal of Geophysical Research, v. 99, p. 3147–3161.

Gass, I.G., 1960, The geology and mineral resources of the Dhali area: Nicosia, Cyprus, Cyprus Geological Survey Department Memoir 4.

Gass, I.G., 1968, Is the Troodos massif of Cyprus a fragment of Mesozoic ocean floor?: Nature, v. 220, p. 39–42

Gass, I.G., Neary, C.R., Plant, J., Robertson, A.H.F., Simonian, K.O., Smewing, J.D., Spooner, E.T.C., and Wilson, R.A.M., 1975, Comments on "The Troodos ophiolite complex was probably formed in a volcanic arc" by A. Miyashiro: Earth and Planetary Science Letters, v. 25, p. 236–238.

Gass, I.G., MacLeod, C.J., Murton, B.J., Panayiotou, A., Simonian, K., and Xenophontos, C., 1994, The geology of the Southern Troodos Transform Fault Zone: Nicosia, Cyprus, Geological Survey Department Memoir 9, 218 p.

Gillis, K.M., and Robinson, P.T., 1985, Low-temperature alteration of the extrusive sequence, Troodos ophiolite, Cyprus: Canadian Mineralogist, v. 23, p. 431–441.

Gillis, K.M., 2002. The rootzone of an ancient hydrothermal system exposed in the Troodos ophiolite, Cyprus: Journal of Geology, v. 110, p. 57–74.

Hall, J.M., and Yang, J.S., 1995, Constructional features of Troodos type oceanic crust; relationships between dike density, alteration, magnetization, and ore body distribution and their implications for in situ oceanic crust: Journal of Geophysical Research, B, Solid Earth and Planets, v. 100, p. 19,973–19,989.

Hess, H.H., 1962, History of ocean basins, *in* Engel, A.E.J., James, H.L., and Leonard, B.F., eds., Petrologic studies: A volume in Honor of A.F. Buddington: Boulder, Colorado, Geological Society of America, p. 599–620.

Hooft, E.E.E., Schouten, H., and Detrick, R.S., 1996, Constraining crustal emplacement processes from the variation in seismic layer 2A thickness at the East Pacific Rise: Earth and Planetary Science Letters, v. 142, p. 289–309.

Humphris, S.E., Herzig, P.M., Miller, D.J., Alt, J.C., Becker, K., Brown., D., Bruegmann, G., Chiba, H., Fouquet, Y., Gemmell, J.B., Guerin, G., Hannington, M.D., Holm, N.G., Honnorez, J.J., Iturrino, G.J., Knott, R., Ludwig, R., Nakamura, K., Petersen, S., Reysenbach, A.L., Rona, P.A., Smith, S., Sturz, A.A., Tivey, M.K., and Zhao, X., 1995, The internal structure of an active seafloor massive sulfide deposit: Nature, v. 377, p. 713–716.

Humphris, S.E., and Cann, J.R., 2000, Constraints on the energy and chemical balances of the modern TAG and ancient Cyprus seafloor sulfide deposits: Journal of Geophysical Research, v. 105, p. 28,477–28,488.

Hurst, S.D., Moores, E.M., and Varga, R.J., 1994, Structural and geophysical expression of the Solea Graben, Troodos ophiolite, Cyprus: Tectonics, v. 13, p. 139–156.

Jenner, G.A., Cawood, P.A., Rautenschlein, M., and White, W.M., 1987, Composition of back-arc basin volcanics, Valu Fa Ridge, Lau Basin—evidence for a slab-derived component in their mantle source: Journal of Volcanology and Geothermal Research, v. 32, p. 209–222

Karson, J.A., 1998, Internal structure of oceanic lithosphere: a perspective from tectonic windows, *in* Buck, W.R., Delaney, P.T., Karson, J.A., and Lagabrielle, Y., eds., Faulting and magmatism at mid-ocean ridges: Washington, D.C., American Geophysical Union Geophysical Monograph 106, p. 177–218.

Kennedy, W.Q., 1946, The Great Glen Fault: Quarterly Journal of the Geological Society of London, v. 102, p. 41–76.

Kidd, R.G.W., 1977, A model for the process of formation of the upper oceanic crust: Geophysical Journal of the Royal Astronomical Society, v. 50, p. 149–183.

Lindgren, W., 1933, Mineral deposits (4th Ed): New York, McGraw Hill, 930 p.

Little, C.T.S., Cann, J.R., and Morriseau, M., 1999, Late Cretaceous hydrothermal vent communities from the Troodos ophiolite, Cyprus: Geology, v. 27, p. 1027–1030.

Livermore, R., Balanya, J.C., Maldonado, A., Martinez, J.M., Rodriguez-Fernandez, J., de Galdeano, C.S., Zalzivar, J.G., Jabaloy, A., Barnolas, A., Somoza, L., Hernandez-Molina, J., Surinach, E., and Viseras, C., 2000, Autopsy on a dead spreading center: The Phoenix Ridge, Drake Passage, Antarctica: Geology, v. 28, p. 607–610.

Lydon, J.W., and Galley, A., 1986, The chemical and mineralogical zonation of Mathiati alteration pipe, Cyprus, and its genetic significance, *in* Gallagher, M.J., Ixer, R.A., Neary, C.R., and Prichard, H.M., eds., Metallogeny of basic and ultrabasic rocks: London, Institute of Mining and Metallurgy, p. 49–71.

MacLeod, C.J., and Murton, B.J., 1993, Structure and tectonic evolution of the Southern Troodos Transform Fault Zone, Cyprus, *in* Prichard, H.M., Alabaster, T., Harris N.B.W., and Neary, C.R., eds., Magmatic processes and plate tectonics: London, Geological Society of London Special Publication 76, p. 141–176.

Malpas, J.G., 1990, Crustal accretionary processes in the Troodos ophiolite, Cyprus: evidence from field mapping and deep crustal drilling, *in* Malpas, J., Moores, E.M., Panayiotou, A., and Xenophontos, C., eds., Ophiolites: oceanic crustal analogues: Nicosia, Cyprus. Ministry of Agriculture and Natural Resources, Geological Survey Department, p. 65–74.

Mammerickx, J, Naar, D.F., and Tyce, R.L., 1988, The Mathematician paleoplate: Journal of Geophysical Research, v. 93, p. 3025–3040.

Mills, R.A., and Elderfield, H., 1995, Hydrothermal activity and the geochemistry of metalliferous sediments, *in* Humphris, S.E., Zierenberg, R.A., Mullineaux, L.S., and Thomson, R.E., eds., Seafloor hydrothermal systems: Washington, D.C., American Geophysical Union Geophysical Monograph 91, p. 392–407.

Miyashiro, A., 1973, The Troodos ophiolite complex was probably formed in a volcanic arc: Earth and Planetary Science Letters, v. 19, p. 218–224.

Moores, E.M., and Vine, F.J., 1971, The Troodos Massif, Cyprus and other ophiolites as oceanic crust: evaluation and implications: Philosophical Transactions of the Royal Society of London, v. 268A, p. 443–466.

Moores, E.M., Varga, R.J., Verosub, K.L., and Ramsden, T., 1990, Regional structure of the Troodos dyke complex, *in* Malpas, J., Moores, E.M., Panayiotou, A., and Xenophontos, C., eds., Ophiolites: oceanic crustal analogues: Nicosia, Cyprus, Ministry of Agriculture and Natural Resources, Geological Survey Department, p. 27–35.

Oftedahl, Chr., 1958, A theory of exhalative-sedimentary ores. Geologiska Föreningen i Stockholms Förhandlingar, v. 80, p. 1–19.

Oudin, E. and Constantinou, G., 1984. Black smoker chimney fragments in Cyprus sulfide deposits: Nature, v. 308, p. 349–353.

Palmason, G., 1973, Kinematics and heat flow in a volcanic rift zone, with application to Iceland: Geophysical Journal of the Royal Astronomical Society., v. 33, p. 451–481.

Palmason, G., 1980, A continuum model of crustal generation in Iceland: Kinematic aspects: Journal of Geophysics, v. 47, p. 7–18.

Pearce, J.A., 1975, Basalt geochemistry used to investigate past tectonic environments on Cyprus: Tectonophysics, v. 25, p. 41–67.

Pearce, J.A., 1980, Geochemical evidence for the genesis and eruptive setting of lavas from the Tethyan ophiolites, *in* Panayiotou, A., ed., Ophiolites; Proceedings Ophiolite Symposium, Cyprus 1979, Ministry of Agriculture and Natural Resources: Cyprus, Nicosia, Geological Survey Department, p. 261–272.

Pearce, J.A., Lippard, S.J., and Roberts, S., 1984, Characteristics and tectonic significance of supra-subduction zone ophiolites, *in* Kokelaar, B.P., and Howells, M.F., eds., Marginal basin geology: London, Geological Society of London Special Publication 16, p. 77–93.

Quon, S.H., and Ehlers, E.G., 1963, Rocks of northern part of Mid-Atlantic Ridge: Geological Society of America Bulletin, v. 74, p. 1–8.

Raitt, R., 1963, The crustal rocks, *in* Hill, M.N., ed., The Sea, Volume 3: New York, Wiley-Interscience, p. 85–102.

Rautenschlein, M., Jenner, G.A., Hertogen, J., Hoffman, A.W., Kerrich, R., Schmincke, H.-U., and White, W.M., 1985, Isotopic and trace element composition of volcanic glasses from the Akaki Canyon, Cyprus: implications for the origin of the Troodos ophiolite: Earth and Planetary Science Letters, v. 75, p. 369–383.

Reinhardt, B.M., 1969, On the genesis and emplacement of ophiolites in the Oman mountains geosyncline: Schweizer Mineralogische und Petrologische Mitteilungen , v. 49, p. 1–30.

Richards, H.G., Cann, J.R., and Jensenius, J., 1989, Mineralogical zonation and metasomatism of the alteration pipes of Cyprus sulfide deposits: Economic Geology, v. 84, p. 91–115.

Richardson, C.J., Cann, J.R., Richards, H.G., and Cowan, J.G., 1987. Metal-depleted root zones of the Troodos ore-forming hydrothermal systems, Cyprus: Earth and Planetary Science Letters, v. 84, p. 243–253.

Robertson, A.H.F., 1975, Cyprus umbers: basalt-sediment relationships on a Mesozoic ocean ridge: Journal of the Geological Society of London, v. 131, p. 511–531.

Robertson, A.H.F., 1977, The Kannaviou Formation, Cyprus: volcanic sedimentation of a probable late-Cretaceous volcanic arc: Journal of the Geological Society of London, v. 134, p. 269–292.

Robertson, A.H.F., and Xenophontos, C., 1993, Development of concepts concerning the Troodos ophiolite and adjacent units in Cyprus, *in* Prichard, H.M., Alabaster, T., Harris N.B.W., and Neary, C.R., eds. Magmatic processes and plate tectonics: London, Geological Society of London Special Publication 76, p. 85–119.

Robinson, P.T., Melson, W.G., O'Hearn, T., and Schmincke, H.-U., 1983, Volcanic glass composition of the Troodos ophiolite, Cyprus: Geology, v. 11, p. 400–404.

Runcorn, S.K., 1959, Rock magnetism: Science, v. 129, p. 1002–1011.

Sandwell, D.T., and Smith, W.H.F., 1997, Marine gravity anomaly from Geosat and ERS 1 satellite altimetry: Journal of Geophysical Research, v. 102, p. 10,039–10,054.

Schiffman, P., and Smith, B.M., 1988, Petrology and oxygen isotope geochemistry of a fossil seawater hydrothermal system within the Solea Graben, northern Troodos Ophiolite, Cyprus: Journal of Geophysical Research, v. 93, p. 4612–4624.

Schiffman, P., Smith, B.M., Varga, R.J., and Moores, E.M., 1987, Geometry, conditions and timing of off-axis hydrothermal metamorphism and ore-deposition in the Solea graben: Nature, v. 325, p. 423–425.

Schiffman, P., Bettison, L.A., and Smith, B.M., 1990, Mineralogy and geochemistry of epidosites from the Solea graben, Troodos ophiolite, Cyprus, *in* Malpas, J., Moores, E.M., Panayiotou, A., and Xenophontos, C., eds., Ophiolites: oceanic crustal analogues: Nicosia, Cyprus, Ministry of Agriculture and Natural Resources, Geological Survey Department, p. 673–684.

Schmincke, H.-U., and Bednarz, U., 1990, Pillow, sheet flows and breccia flow volcanoes and volcano-tectonic hydrothermal cycles in the extrusive series of the northeastern Troodos ophiolite, Cyprus, *in* Malpas, J., Moores, E.M., Panayiotou, A., and Xenophontos, C., eds., Ophiolites: oceanic crustal analogues: Nicosia, Cyprus, Ministry of Agriculture and Natural Resources, Geological Survey Department, p. 207–216.

Schouten, H., and Denham, C.R., 2000, Comparison of volcanic construction in the Troodos ophiolite and oceanic crust using paleomagnetic inclinations from Cyprus Crustal Study Project (CCSP) CY-1 and CY-1A and Ocean Drilling Program (ODP) 504B drill cores, *in* Dilek, Y., Moores, E., Elthon, D. and Nicolas, A., eds., Ophiolites and oceanic crust: New insights from field studies and the Ocean Drilling Program: Boulder, Colorado, Geological Society of America Special Paper 349, p. 181–194.

Schouten, H., and Kelemen, P., 2002, Melt viscosity, temperature and transport processes, Troodos ophiolite, Cyprus: Earth and Planetary Science Letters, v. 201, p. 337–352.

Small, C., 1998, Global systematics of mid-ocean ridge morphology, *in* Buck, W.R., Delaney, P.T., Karson, J.A., and Lagabrielle, Y., eds., Faulting and magmatism at mid-ocean ridges: Washington, D.C., American Geophysical Union Geophysical Monograph 106, p. 1–26.

Smith, W.H.F., and Sandwell, D.T., 1997, Global seafloor topography from satellite altimetry and ship depth soundings: Science, v. 277, p. 1956–1962.

Spiess, F.N., Macdonald, K.C., Atwater, T., Ballard, R., Carranza, A., Cordoba, D., Cox, C., Diaz Garcia, V.M., Francheteau, J., Guerrero, J., Hawkins, J., Haymon, R., Hessler, R., Juteau, T., Kastner, M., Larson, R., Luyendyk, B., Macdougall, J.D., Miller, S., Normark, R., Orcutt, J., and Rangin, C., 1980, East Pacific Rise: hot springs and geophysical experiments: Science, v. 207, p. 1421–1433.

Spooner, E.T.C., and Bray, C.J., 1977, Hydrothermal fluids of seawater salinity in ophiolitic sulfide deposits in Cyprus: Nature, v. 266, p. 808–812.

Staudigel, H., Gillis, K., and Duncan, R., 1986, K/Ar and Rb/Sr ages of celadonites from the Troodos ophiolite, Cyprus: Geology, v. 14, p. 72–75.

Staudigel, H., Gee, J.S., Tauxe, L., and Varga, R.J., 1992, Shallow intrusive directions of sheeted dikes in the Troodos Ophiolite: Anisotropy of magnetic susceptibility and structural data: Geology, v. 20, p. 841–844.

Tauxe, L., Gee, J.S., and Staudigel, H., 1998, Flow directions in dikes from anisotropy of magnetic susceptibility data; the bootstrap way: Journal of Geophysical Research, v. 103, p. 17,775–17,790.

Tatsumi, T.,1970, Volcanism and ore genesis: Tokyo, University of Tokyo Press, 450 p.

Vanko, D.A., and Batiza, R., 1982, Gabbroic rocks from the Mathematician Ridge failed rift: Nature, v. 300, p. 742–744.

Varga, R.J., and Moores, E.M., 1985, Spreading structure of the Troodos ophiolite, Cyprus: Geology, v. 13. p. 846–850.

Varga, R.J., Gee, J.S., Bettison-Varga, L., Anderson, R.S., and Johnson, C.L., 1999, Early establishment of seafloor hydrothermal systems during structural extension: paleomagnetic evidence from the Troodos Ophiolite: Earth and Planetary Science Letters, v. 171, p. 221–235.

Vera, E.E., Mutter, J.C., Buhl, P., Orcutt, J.A., Harding, A.J., Kappus, M.E., Detrick, R.S., and Brocher, T.M., 1990, The structure of 0 to 0.2 m.y.-old oceanic crust at 9°N on the East Pacific Rise from expanded spread profiles: Journal of Geophysical Research, v. 95, p. 15,529–15,556.

Vine, F.J., and Matthews, D.H., 1963, Magnetic anomalies over oceanic ridges: Nature, v. 199, p. 947–949.

Vogt, J.H.L., 1894, Über die Genesis der Kieslagerstätten vom typus Røros–Rammellsberg: Zeitzschrift für Praktalisches Geologie, v. 2, p. 173.

Vogt, J.H.L., 1899, Das Huelva kiesfeld: Zeitzschrift für Praktalisches Geologie, v. 7, p. 241–254.

Von Damm, K.L., Edmond, J.M., Grant, B., Measures, C.I., Walden, B. and Weiss, R.F., 1985, Chemistry of submarine hydrothermal solutions at 21°N, East Pacific Rise: Geochimica et Cosmochimica Acta, v. 49, p. 2197–2220.

Wegener, A., 1915, Die Entstehung der Continente und Ozeane, Braunschweig, 94 p.

Wilson, R.A.M., 1959, The geology of the Xeros-Troodos area: Nicosia, Cyprus, Geological Survey Department Memoir 1, 184 p.

Manuscript Accepted by the Society May 2, 2003

Printed in the USA

Geological Society of America
Special Paper 373
2003

The sheeted dike complex of the Troodos ophiolite and its role in understanding mid-ocean ridge processes

Robert J. Varga*
Department of Geology, The College of Wooster, Wooster, Ohio 44691, USA

ABSTRACT

Geologic research on the Troodos ophiolite flourished during two distinct periods following general recognition of the importance of ophiolites as fragments of ocean crust in the late 1960s. During the first period, the "petrologic revolution," research focused on the petrological and geochemical aspects of the ophiolite pseudostratigraphic sequence. In part, this focus is attributed to the early recognition of the enriched, non-MORB character of Troodos lavas, which set off a controversy that continues to the present. The period that followed was a "structure revolution," and was a direct consequence of the Cyprus Drilling Project in the early 1980s. Much research during this time period focused on the extensively exposed sheeted dike complex and was directed toward gaining an understanding into the structure of modern mid-ocean ridges. Field, paleomagnetic, and petrographic work on the dike complex have yielded insights into the interplay between magmatism and structural extension during crustal accretion, the nature of normal faulting at ridges, the depth-dependent changes in extensional style, the importance of structural extension for focused hydrothermal circulation at black smoker vents, and the importance of along-axis migration of melts above axial magma chambers. Future insights into ridge crest dynamics will be enhanced through iterative exchange between field studies of ophiolites and observations of well-exposed cross sectional windows into modern ocean crust.

Keywords: Troodos, ophiolite, dike, spreading centers, ocean crust.

INTRODUCTION

Since its recognition by Wilson (1959), the sheeted dike complex of the Troodos ophiolite (Fig. 1A) has played a key role in the characterization of the petrologic composition of ocean crust and of the structure of oceanic spreading centers. The generally intact nature of the ophiolite and, in particular, the extensive exposure of its sheeted dike section inspired the initial proposal that ophiolites represent fossil ocean crust (Gass, 1968; Moores and Vine, 1971) and that study of these terranes might provide insights into ocean crustal processes.

The purpose of this paper is to outline the historical development of ideas concerning the recognition and significance of the Troodos ophiolite as an ocean crustal analog, with emphasis on the importance of the sheeted dike complex in this process. Readers interested in a recent account of the development of historical ideas concerning the greater Troodos ophiolite and its tectonic context are directed toward the excellent review by Robertson and Xenophontos (1993). Studies of the sheeted dike section have revealed details of the nature of extension within ocean crust and of the interplay between magmatism and structural extension. Excellent exposure of all levels of the sheeted dike complex has also allowed study of the depth-dependent nature of structural extension, the role of faulting and hydrothermal alteration, and the nature of magma migration through the dike complex. As with any account of this nature, this retrospective is admittedly personal and reflects the experience, background, and biases of the author. The conclusion drawn from this account is that, although the exact environment of formation of the Troodos ophiolite is controversial (e.g., Moores et al., 1984, 2000), the sheeted dike section of the ophiolite has, and will, substantially

*E-mail: rvarga@wooster.edu

Varga, R.J., 2003, The sheeted dike complex of the Troodos ophiolite and its role in understanding mid-ocean ridge processes, *in* Dilek, Y., and Newcomb, S., eds., Ophiolite concept and the evolution of geological thought: Boulder, Colorado, Geological Society of America Special Paper 373, p. 323–336. For permission to copy, contact editing@geosociety.org.

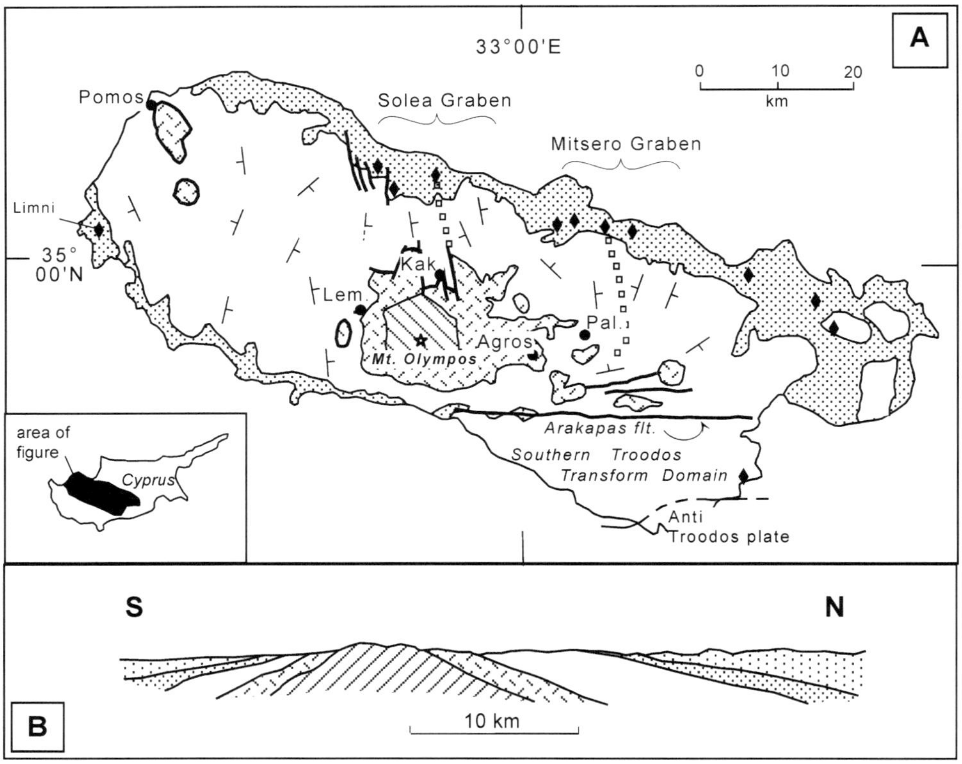

Figure 1. A: Generalized geologic map of the Troodos ophiolite (after Varga et al., 1999a). B: Generalized N–S cross section. Ruled area near Mount Olympos—ultramafic rock; area shown with short line segments—gabbro; white area with strike and dip symbols north of the Arakapas fault—sheeted dike complex; dark shading—volcanics; light shading—primarily Mesozoic to Recent sediment. Generalized north/south cross section after Gass and Masson-Smith (1963); symbols same as those for geologic map. Dots—villages (Lem.—Lemithou, Kak.—Kakopetria, Pal.—Palekhori), diamonds—massive sulfide mines.

contribute to an understanding of processes of crustal accretion at modern oceanic spreading centers.

General Geologic Setting of the Sheeted Dike Section

The ca. 92 Ma Troodos ophiolite (Mukasa and Ludden, 1987) is one of several fragments of ocean crust emplaced throughout the Mediterranean-Himalayan region during closing of the Paleotethyan and Neotethyan oceans (e.g., Moores, 1982; Dilek et al., 1990a). Upon emplacement, the ophiolite was subjected to north/south compression and gently warped into a broad, east/west-trending anticline. Subsequent erosion of this anticline has resulted in exposure of the deepest parts of the ophiolite pseudostratigraphy in the central, and highest, parts of the Troodos Mountains. Successively higher crustal layers are exposed toward the topographically lower peripheries of the complex (Fig. 1A, 1B). Between the Campanian and early Eocene and prior to emplacement, the Troodos ophiolite was rotated about ~90° counterclockwise, as documented with paleomagnetic data (Moores and Vine, 1971; Clube et al., 1985). Considering this rotation, the Troodos apparently formed along one segment of a series of east/west-trending ridges within the Neotethys (Dilek et al., 1990a). Today, the spreading direction of the ophiolite is considered to be oriented east/west (Wilson, 1959; Moores and Vine, 1971), perpendicular to the dominant north/south strike of dikes and parallel to the long axis of the anticline and trend of the Troodos Mountains. This geometry, coupled with the broad geometry of the anticline and later erosion of the complex, has resulted in today's extensive exposure of the sheeted dike complex. Emplacement-related uplift of the Troodos complex continues to the present (Robertson, 1977; Poole and Robertson, 1991).

The 1–2 km thick sheeted dike complex in the Troodos ophiolite consists of 100% dikes in its lower parts and 90%–100% dikes cutting pillow and sheet flow screens in its upper part; this upper section has historically been referred to as the Basal Group, though its upper and lower boundaries are transitional (Wilson, 1959; Baragar et al., 1987, 1990). Typically, dikes within the sheeted zone are 1–2 m thick and have steep orientations (Fig. 2A), except where rotated by later deformation (Figs. 2C, 2D). The strike of dikes is generally northward about an average of ~340°. Much of the local strike variation in the dikes can be accounted for by extensional faulting (Varga, 1991). More regionally, however, dike strikes swing into parallelism with the east/west-trending Arakapas fault zone (Fig. 1A). This deviation in strike directions is generally attributed to drag related to dextral shear along the Arakapas (Bonhommet et al., 1988; MacLeod and Murton, 1993; Gass et al., 1994), though sinistral slip has also been suggested (Moores and Vine, 1971; Varga and Moores, 1985; Murton and Gass, 1986; Dilek et al., 1990a, 1990b). Geochemistry of the diabase dikes reveal depleted and nondepleted types that match those found in overlying lavas, although no age distinction between these geochemical types is apparent in the dike section (Baragar et al., 1990).

Figure 2. A: Vertical sheeted dikes near Palekhori. B: Highly tilted (to right) pillow lavas near Pomos. Arrow shows pillow with ideal morphology. C: Moderately tilted dikes in Mitsero area. D: Highly tilted (to left) dikes east of Kakopetria. Outlined area is fault zone breccia preferentially altered to epidosite assemblage (ep). Relative motion arrows indicate normal offset along fault.

HISTORICAL PERSPECTIVE

A time-line for scientific recognition of the sheeted dike complex in Cyprus as related to seafloor spreading processes generally follows the development and acceptance of ideas that now constitute what we refer to as plate tectonics. Prior to general acceptance of plate tectonic theory as the marriage of continental drift and seafloor spreading—the 1966 Goddard Conference provides a convenient datum for this transition (Steward, 1990)—an oceanic origin of the Troodos and other ophiolites had been suggested (e.g., Cullis and Edge, 1922), but was only one of several competing theories.

Pre–Plate Tectonic Recognition of the Sheeted Dike Complex

The early work on the Troodos complex is well summarized in the excellent review by Wilson (1959), writing at the dawn of the plate tectonic revolution. The reader is referred to this work for a general account of the development of ideas for Cyprus geology. Gaudry (1862) published some of the first accounts of Troodos geology, describing the igneous core of the mountain range, which he inferred was intruded *en masse* into the surrounding sediments. Gaudry (1862) also noted the characteristic steep joints within what he termed "affanites," an obvious reference to the sheeted dike complex. Subsequent mapping of the island and publications by Bellamy (1905) and Bellamy and Jukes-Browne (1905) showed the widespread extent of the "diabase" sequence (sheeted dikes) within the Troodos complex. Cullis and Edge (1922) first recognized the pillow lava section to represent magma erupted onto the seafloor, but apparently felt that the underlying diabase section was much older and unrelated to pillow formation.

Some of the most important early references to the origin of the Troodos dike section were made by D.W. Bishopp (1952;

1954). Bishopp was a geologist with the fledgling Geological Survey of Cyprus, formed in late 1950 under Britain's Directorate of Colonial Geological Surveys. Their charge was largely to map the Troodos complex, with the particular goal of defining its well-known resources of copper, chromite, and asbestos. In his 1952 paper, Bishopp (1952) summarized some important aspects of the Troodos complex as recognized at the time. Among these were his belief that the diabase section of the Troodos complex represented a series of folded lava flows and intrusive sheets (sills?); the presence of isoclinal folds within the section was also offered as a possibility. One of the more interesting aspects of Bishopp's account is his description of the folds within the diabase. He wrote:

> The outstanding feature of this formation is its structure: it contains several major anticlines and synclines with an axial trend a few degrees either side of north and south—frequently north-north-west—and with dips of the order of 60°–70° to east or west, rarely less than 45°. (Bishopp, 1952, p. 489)

Though we now recognize Bishopp's (1952) folded lava sections (called by him "The Folded Diabase Series") to represent sheeted dikes, his salient observations presaged the much later recognition of structural tilting within the dikes and presence of major graben (his synclines), such as the Solea Graben (Fig. 1), believed to be of oceanic origin and perhaps representing a fossil axial valley (Varga and Moores, 1985). Bishopp (1952) also recognized the low-grade metamorphism within the diabase complex (presence of "epidiorites"), but continued the belief of Cullis and Edge (1922) that the pillow section is younger and unconformably overlies the diabase section.

In this era prior to plate tectonics, much was made of the regional nature of structural trends and of the correlation of structural trends over large distances. Bishopp (1952) suggested that the north/south trend of structures within The Folded Diabase Series of the Troodos is an extension of the north/south-trending African rift system. In his view, this was the first of two orogenic episodes to affect the Troodos, the second being the shattering of the diabase section due to north/south-directed compression that produced the Kyrenia range. Interestingly, Bishopp (1952) was the first to recognize what we now call the Kykko thrust, west of the Solea Graben area, and offered this structure as evidence for his second orogenic phase. Other late features (thrust and strike-slip faults) representing late north/south compression have since been recognized and related to post-accretion emplacement of the Troodos ophiolite onto Cyprus (e.g., Biju-Duvall and Lapierre, 1976; Lapierre et al., 1988; Varga, 1991; Grand et al., 1993). In a later paper, Hiessleitner (1957) also related both structural trends in the diabase and mineral banding within the ultramafic section of Mount Olympos to pre-Mesozoic north/south trends in Africa.

Without a doubt, the defining pre–plate tectonic paper on the Troodos dike section is the well-known Cyprus Geological Survey Memoir 1 by R.A.M. Wilson (Wilson, 1959). Working on the Survey's mapping project of the Troodos Mountains initiated in 1950, Wilson mapped the Xeros-Troodos region in the center of the range. In selecting this particular region to map, Wilson could not have chosen a more representative section of an ideal ophiolite as defined years later (Anonymous, 1972). His 270-mi^2 map region includes the summit region of Mount Olympos, covering much of the ultramafic section, and a complete crustal package that comprised the plutonic, sheeted dike, volcanic, and overlying sediment sections (Fig. 1). The region also includes several of the largest massive sulfide mines (Skouriotissa, Apliki, Mavrovouni) on Cyprus (Fig. 1). Wilson's (1959) two-sheet map of the Xeros-Troodos region is a stunning example of keen observation and insight. These maps and his discussion of the geology of the area directly influenced later workers in their recognition of ophiolites as ocean crust.

In addition to his maps, Wilson (1959) made many key observations that would guide subsequent study in many areas of ophiolite research. With respect to the origin of the diabase complex, several of Wilson's observations are particularly noteworthy. Taking issue with Bishopp's (1952) earlier work, Wilson concluded that the diabase complex comprises a series of concordant dikes, rather than a series of folded lava flows. In a further departure with previous studies (Cullis and Edge, 1922; Bishopp, 1952), Wilson (1959) described detailed relationships between the diabase and volcanic sequences, noted the transitional and complex nature of the boundary between these units, and suggested that they might be coeval. In a statement that would set the stage for recognition of ophiolite layers as a "pseudostratigraphy," he wrote:

> It is therefore possible that the diabase is a series of intrusive sheets which, over a long period of time, were successively injected and acted as feeders to the overlying lava flows. (Wilson, 1959, p. 60)

Wilson clearly recognized that the sheeted dike complex required large extension and even speculated that "the lavas probably formed a submarine mountain of which, in the tensional areas of the Pacific Ocean, there are numerous examples" (Wilson, 1959, p. 134). Lastly, Wilson noted the presence of banded epidosite in the diabase where dikes were severely altered to largely quartz and epidote (with minor chlorite). The main mass of epidosite northwest of the village of Kakopetria (Fig. 1A), would later become the focus of research leading to characterization of hydrothermal upflow zones beneath black smoker-type orebodies (Richardson et al., 1987; Schiffman et al., 1987; Bettison-Varga et al., 1995).

Recognition of the Sheeted Dike Complex as a Product of Mid-Ocean Ridge Processes

Despite the many salient observations made by these early geologists studying Troodos geology, their work was generally unappreciated in their time. The rapid formulation of plate tectonic theory during the early and mid-1960s, however, gave context to ophiolite studies, and the importance of the Troodos complex as a possible ocean crustal analog was quickly realized. A sense of the explosion in interest in Troodos geology during the plate tectonic revolution is seen in Figure 3, which is a graph showing numbers of papers mentioning either "Troodos" or "Cyprus and dikes" over the past 45 years. Although admittedly flawed in detail, this graph generally reflects the rise and fall in interest in the Troodos ophiolite in recent time.

One of the most striking aspects of Figure 3 is the exponential increase in publications about Troodos geology once plate tectonics became reasonably established as a theory (again, the 1966 Goddard Conference provides a useful date for comparison). Two papers in particular were instrumental in establishing the importance of the Troodos ophiolite in studies of ocean crust formation: Gass (1968) and Moores and Vine (1971). These papers together were directly responsible for interest in Troodos and other ophiolites as possible ocean crustal analogs (Fig. 3), and set the stage for the next 30 years of research and controversy surrounding these complexes.

In suggesting that the Troodos complex might represent an on-land piece of ocean crust and mantle, Ian Gass drew upon his past experiences as a member of the Geological Survey Department of Cyprus during their formative years in the 1950s. In contrast, Eldridge Moores and Fred Vine recognized a potential link between the Troodos complex and ocean crust while looking at a copy of Wilson's (1959) map in 1966, while both were at Princeton. What impressed them at the time was the vast expanse of subparallel sheeted dikes depicted on Wilson's map. Recognizing the link between these dikes and spreading that was indicated by magnetic reversal lineations in the Pacific (Vine and Matthews, 1963), Moores and Vine applied for and received NSF funding to visit the Troodos in 1968 (E. Moores and F. Vine, personal communication, 2002). Though they searched for but failed to document a magnetic reversal in the Troodos ophiolite, their paper contains a detailed account of how the various units within the complex might fit into a seafloor spreading context. They also speculated for the first time that what we now refer to as the Arakapas fault zone (Fig. 1) might represent a fossil transform fault (Moores and Vine, 1971).

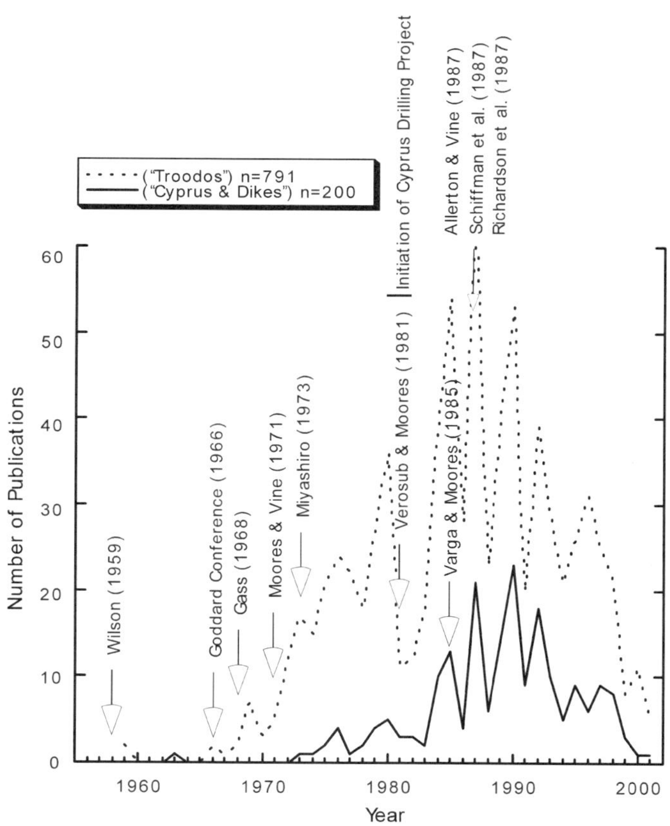

Figure 3. Graph showing number of publications versus time, for publications that mention either "Troodos" (dotted line) or both "Cyprus" and "dikes" (solid line). Search was done in Georef. Selected papers mentioned in text are shown at their date of publication.

Although the sheeted dike complex in the Troodos Mountains played a pivotal role in the early recognition of, and interest in, ophiolites as ocean crust, the focus of research subsequent to publication of Gass' (1968) and Moores and Vine's (1971) papers was largely on petrology. This petrologic focus was, of course, further fueled by Miyashiro's (1973) paper questioning the geochemical link between Troodos lavas and what was known at the time of the composition of oceanic basalts. This and subsequent petrologically based papers are responsible for the first wave of interest in the Troodos during the decade ~1970–1980, as seen on Figure 3. We could refer to this decade as the "petrologic revolution" for Troodos ophiolite research.

THE CYPRUS DRILLING PROJECT AND THE "STRUCTURE REVOLUTION"

Interest in Troodos, as evidenced crudely by numbers of publications (Fig. 3), began to wane in ca. 1980, after a decade of heated debate about the petrologic origin of the Troodos ophiolite—debate that continues, of course, to this day. This short-lived lull rather abruptly ended in ~1982, and was followed by a spectacular rise in publications that would continue well into the mid-1990s. The impetus for this second round of intense Troodos research was, undoubtedly, the Cyprus Drilling Project. Initiated in 1980 by the International Crustal Research Drilling Group (ICRDG) and in cooperation with the Cyprus Geological Survey, the Cyprus Drilling Project involved the drilling of several core holes into offset sections of the Troodos ophiolite, with the goal of comparing these to results from the Deep Sea Drilling Project and the Ocean Drilling Program. Many participating countries provided support to the drilling program and supported ancillary research efforts, including field studies of the ophiolite (see Malpas et al., 1990). Perhaps more than the drill holes themselves, these derivative studies resulted in new insights into Troodos geology and directly impacted the way marine geologists now view mid-ocean ridge processes.

Surely, many of the papers published on Troodos geology during the years following initiation of the Cyprus Drilling Project focus on the petrology of the complex, including notably those concerning the geochemical signature of Troodos magmas (e.g., Rautenschlein et al., 1985) and the geometry of subaxial magma chambers (e.g., Malpas, 1990); however, this second wave of Troodos research was also characterized by a distinct rise in interest in the sheeted dike complex, in the structure of the Troodos ophiolite as largely deduced from the dike complex, and in the hydrothermal alteration of the ophiolite. One might refer to this second period of intense Troodos research as the "structure revolution."

A key paper in initiating the structure revolution in Troodos research was published by Eldridge Moores and paleomagnetist Ken Verosub, both at University of California at Davis. Verosub and Moores (1981) attempted to explain anomalous magnetic inclination data from oceanic drill cores by using a model of rotational normal faulting, much like what (they noted) was being used to explain widespread regional tilting in the U.S. Basin and Range. Their paper contained several prescient observations and predictions that would drive later research, including: (1) tilting of ocean crust is accomplished through rotational normal faulting, perhaps involving low-angle detachment faults; (2) non-vertical dikes in ophiolites, such as the Troodos, may result from structural tilting; and (3) the field of paleomagnetism can provide an important constraint on structural tilting in oceanic igneous rocks. In addition to suggesting paleomagnetism as a valuable tool for use in ophiolite terranes that lack a reliable paleohorizontal reference, their suggestions provided concrete hypotheses that could be easily tested in reasonably intact ophiolites such as Troodos.

Inspired initially by Verosub and Moore's (1981) paper, subsequent work on the sheeted complex has concentrated in several areas of research that, in rough order, has proceeded as follows:

1. Structural mapping using dike orientation as an approximate proxy for tilt magnitude/direction and based on the recognition of compatible normal fault systems
2. Recognition of "oceanic structure" as reflected in dike patterns (e.g., grabens, the Arakapas transform)
3. Use of paleomagnetism as a general tool in demonstrating that, in fact, non-vertical dikes are indeed structurally tilted as well. Paleomagnetism also used to assess other structural rotations, including rotations due to emplacement and vertical-axis rotations due to transform movements
4. Use of structure and paleomagnetism in assessing hydrothermal alteration history of the sheeted dike complex
5. Structural and paleomagnetic studies to assess movement of axial magmas

Sheeted Dike Orientations as a Proxy for Ridge Crest Structure

To test several of the predictions made by Verosub and Moores (1981), Moores and I initiated a study of sheeted dike structure in the summer of 1984. At the time, the prevailing view of the structure of the Troodos ophiolite was that it comprised a series of thrust nappes and that the dominant structural features within the terrane reflected north/south compression during emplacement (Biju-Duvall and Lapierre, 1976). A significant exception to this view was that by Simonian and Gass (1978), who interpreted the deviations in dike strikes near the Arakapas fault zone to represent the effects of transform shear. The most significant observation made by Varga and Moores (1985) and within subsequent papers (Moores et al., 1990; Varga and Moores, 1990; Varga, 1991) was that, within the large regions of the sheeted dike section containing non-vertical dikes, dike inclinations were closely associated with well-exposed normal faults (Fig. 2). The normal faults were interpreted as kinematically coordinated with the dike tilts. Dikes dip in the same direction as adjacent normal faults while, in shallow sections, pillow (Fig. 2B) and sheet lava flows dip in opposing directions. While a limited number of listric-shaped normal faults occur, the preponderance of faults are subparallel to dike margins (Fig. 2D), suggesting a rotational-planar ("bookshelf") extension model (Varga, 1991; Dietrich and Spencer, 1993; Varga et al., 1999a). Varga and Moores (1985) also noted the spatial pattern of inward-dipping dikes and related normal faults in three areas and suggested that these might be oceanic grabens (Fig. 1). The further association of large massive sulfide bodies within several of these, most notably the Solea Graben, suggested that these might represent abandoned axial valleys.

A significant problem with the model of Varga and Moores (1985) has been that, until recently, there were no good constraints on the *timing* of the dike tilting and related normal faulting. East/west extension during emplacement-related, north/south compression is a plausible alternative mechanism for these structures; the Rhine graben (Şengör et al., 1978) and Ancestral Rocky Mountains (Kluth and Coney, 1981) are examples of this geometry. The principal evidence cited by Varga and Moores (1985) to support a ridge crest origin for the sheeted dike structure were hydrothermal alteration along fault zones (Fig. 2D) with a seawater signature (Schiffman et al., 1987), the presence of tilted dikes crosscut by less-tilted dikes of similar composition (Fig. 4D), and the generally much less deformed nature of overlying Cretaceous marine sediments.

A variety of studies that followed Varga and Moores (1985) have questioned the significance of the graben structures outlined by dike patterns within the Troodos ophiolite. Allerton and Vine (1987) and van Everndingen and Cawood (1995) concur that these graben probably reflect oceanic structures, but suggest that they formed in off-axis positions relative to an active, Cretaceous spreading center. Of the three grabens noted by Varga and Moores (1985), the Solea Graben (Fig. 1) remains the most viable of probable fossil axial valleys. This graben rests above mantle rocks of the Mount Olympos region that possess steep, high-temperature foliations that are thought by some workers to reflect mantle upwelling (Wilson, 1959; George, 1978; Bartholomew, 1993; Martinez, 1996). Modeling of asthenospheric flow patterns suggests that such steep foliations would be quickly overprinted as material moves away from a spreading center

Figure 4. A: Generally low-angle fault near village of Lemithou (dashed line separating highly tilted dikes) (D) from sheared gabbro (G). B: Small, lozenge-shaped blocks (phacoids) plucked from gabbro beneath the fault zone at Lemithou. Gabbro is deformed at a larger scale into similar-shaped blocks. Note slickenlines on the surfaces of each block (knife for scale). C: Steeply dipping dike margin (arrow) with steeply raking, *hot slickenlines* (Varga et al., 1998), indicating subvertical magma flow during dike injection. D: Crosscutting dikes in Mitsero area. Old, relatively fractured dikes dip moderately to right and are cut by late, relatively unreformed, vertical dike (pen for scale).

(Nicolas and Rabinowicz, 1984). The pattern of foliations within the Mount Olympos region suggests that it formed as a mantle diapir at a spreading center, much like diapirs mapped in the Semail ophiolite (Nicolas et al., 1988).

Mapping of dike orientations was also important in framing discussions about the origin of the Arakapas transform fault. Moores and Vine (1971) first noted the clockwise swing in dike orientations near the Arakapas fault zone (Fig. 1). Two end-member models have been offered to describe this pattern. The first model (Moores and Vine, 1971; Varga and Moores, 1985; Murton and Gass, 1986; Dilek et al., 1990a, 1990b) suggests that dikes were originally injected in this pattern as a reflection of rotated stress axes known to exist near ridge-transform intersections (Fox and Gallo, 1984). A second model ascribes the dike pattern to post-injection rotation during dextral simple shear along the Arakapas zone (Simonian and Gass, 1978). Subsequent paleomagnetic study of dikes near the Arakapas (e.g., Bonhommet et al., 1988) and detailed structural studies (e.g., MacLeod and Murton, 1993; Gass et al., 1994) within the Arakapas appear to support the latter model.

Paleomagnetism as a Tool in Defining the Structure of the Sheeted Dike Complex

In addition to suggesting that dike orientations are important in discerning the oceanic structure of ophiolites, Verosub and

Moores (1981) suggested that paleomagnetism might be a valuable tool in assessing dike rotations. The obvious problem working within the igneous components of an ophiolite is the lack of a reliable paleohorizontal reference frame against which to assess structural rotations. On the basis of the observation, at shallow levels in the ophiolite, of a high angle between dikes and tilted lava flows, Varga and Moores (1985) and Varga (1991) assumed a generally steep original dike orientation. Potentially significant original inclinations of lavas and non-vertical dike injections are serious potential problems with this simplifying assumption. Paleomagnetic remanence provides a means for assessing whether or not non-vertical dikes have been structurally rotated.

For paleomagnetism to be useful in assessing structural rotations in igneous rocks, it is critical that the rotation axis lay at a high angle to the remanence direction. In the context of oceanic spreading centers, paleomagnetic deviations from expected directions due to extensional faulting should be greater, and more easily discerned, at east/west-trending ridges than at north/south-trending ridges. Moores and Vine (1971) and subsequent studies (e.g., Clube et al., 1985) have demonstrated that the Troodos ophiolite possessed the ideal geometry of an originally nearly east/west-trending ridge during the Cretaceous. These studies have defined a paleomagnetic "north" remanence direction (the Troodos mean direction (TMD) = 36°/273°), indicating an ~90° counterclockwise rotation of the complex since its formation (Clube et al., 1985; Abrahamsen and Schoenharting, 1987). Additionally, the current large deviation between the TMD and current magnetic north allows clear assessment of post-cooling, magnetic overprinting.

Allerton and Vine (1987) and subsequent studies (e.g., Hurst et al., 1994) used paleomagnetism to test the rotational faulting hypothesis proposed earlier for Troodos dikes (Verosub and Moores, 1981; Varga and Moores, 1985). Their results show clearly that dikes within the sheeted complex were, indeed, intruded at steep angles (Fig. 2A) and that most moderately and shallowly dipping dikes (Figs. 2C, 2D) have been structurally rotated, and not injected at moderate to shallow angles.

In working with the paleomagnetic data from the sheeted dike or any other igneous complex, however, exactly how to accomplish rotational corrections of data is not readily apparent. Although the target of the restoration is well known, the TMD in the case of Troodos, the *path* or *sequence* of rotation(s) to restore remanence directions is not constrained by independent and subjective measurements. The first to tackle this important problem was Allerton and Vine (e.g., 1990), who offered a single rotation, trigonometric solution to remanence rotation, using the strict constraint of requiring an originally vertical dike attitude. Their routine appears to provide reasonable solutions to rotational axes in many areas of the sheeted dike complex with simple histories. As pointed out by Varga (1991) and Varga et al. (1999a), however, much of the sheeted dike complex has suffered multiple rotations, each about different rotational axes. Much of the tilting of the dikes is coordinated about associated normal faults with highly raking slickenlines, indicating rotational axes more or less parallel to dike strike directions (Fig. 2D). However, north/south cross sections (Wilson, 1959; Gass and Masson-Smith, 1963) across the complex and the dip of overlying Cretaceous marine sediments indicate that much of the exposed sheeted complex north of the Arakapas zone was tilted northward at ~10° during emplacement (Fig. 1). Additionally, at fairly large distances (up to ~10 km) from the Arakapas, dikes have suffered variable amounts of clockwise vertical-axis rotation due to transform motion (Fig. 1). To account for these additional rotations, Varga et al. (1999a) suggested a scheme for correction of paleomagnetic remanence data that uses these geologically observed rotations and removes them in reverse order of formation. Corrected in this manner, sheeted dikes generally restore to steep angles with northward strike directions. To summarize, the correction method of Allerton and Vine (1987) assumes an initial dike orientation and solves for a single rotation angle that, in many cases, represents a composite rotation axis for several structural rotations. The correction scheme proposed by Varga et al. (1999a) assumes geologic knowledge of several successive rotation axes and solves for initial dike orientations.

Nature of the Sheeted Dike/Plutonic Contact

Demonstration of the usefulness of dike orientations and paleomagnetism to gain insights into possible ridge-axis structure of the Troodos ophiolite has been used to study the depth-dependent nature of extension and the possibility for mid-crustal detachment between zones undergoing varying amounts of extension or varying structural styles. These studies quickly focused on the nature of the contact between the plutonic and sheeted dike sections of the ophiolite, and have demonstrated important structural differences between these parts of the pseudostratigraphy.

Varga and Moores (1985) documented a low-angle zone of structural discontinuity between dikes and high-level gabbros of the plutonic section in the Xeros-Troodos region. This structural contact is best exposed near the hill village of Lemithou, west of Mount Olympos (Fig. 4A). There, dikes of the sheeted section dip shallowly to the northeast and contact underlying gabbro along a low-angle fault. Named the "Kakopetria detachment" by Varga and Moores (1985), this fault zone was envisioned to have operated as a zone of simple-shear decoupling between differentially extending crustal zones, much like detachments in continental regions (e.g., Davis and Lister, 1988). A subsequent study by Hurst et al. (1994) showed paleomagnetic rotations of the hanging-wall dikes compatible with this model. In a study of this contact west of Mount Olympos near Agros (Fig. 1), Agar and Klitgord (1995) also suggested that the dike/plutonic contact is indeed a zone of structural decoupling, but suggested that, instead of comprising a zone of simple shear, the zone represents decoupling between regions undergoing differing styles of deformation. In their model, uniform tilting of sheeted dikes above this zone is separated from gabbros across a "crush zone" characterized by a more complex fault array that approximates pure shear deformation.

Subsequent paleomagnetic studies across the dike/plutonic contact attempted to resolve these differing models for mid-crustal decoupling. Varga and Gee (1995) compared pairs of paleomagnetic data sets from dikes and gabbros at various locations in the central Troodos Mountains. Their data indicate that the nature of decoupling changes from place to place. Observed relationships include: (1) areas where no differential rotation has taken place (e.g., gabbros rotated in same manner and magnitude as overlying dikes, or not at all); (2) areas where dikes are rotated more than the gabbros, but in the same sense; and (3) areas where dikes and gabbros are rotated in the opposite sense. One result of this study was the observation that the footwall of the "classic" exposure of the Kakopetria detachment near Lemitou (Fig. 4A) is a good example of relationship (1) above. There, because the footwall gabbro and hanging-wall dikes are similarly rotated, the fault separating them is probably a highly rotated rotational-planar normal fault. Finally, Gerbi (1996) and Gerbi et al. (1996) studied the paleomagnetism of gabbros throughout the Kakopetria region and showed that, while remanence in dikes indicates single-sense rotation indicative of rotational-planar normal faulting, remanence in gabbros is variable and rotated about inclined axes that differ substantially from those in overlying dikes. Coupled with observations of complex fault arrays within the gabbro, Gerbi (1996) suggest that gabbro in that area deformed brittlely along a series of faults defining lozenge-shaped blocks (Fig. 4B); together this pattern of deformation approximated a pure shear extension (Fig. 5). Thus, the weight of evidence to date supports the general model of Agar and Klitgord (1995), in which much of the dike/plutonic contact separates regions undergoing differing structural styles. This model has recently been used to explain disparate structural observations between dikes and underlying gabbros exposed at the Hess Deep in the equatorial Pacific (Karson and al., 1999; Karson et al., 2002).

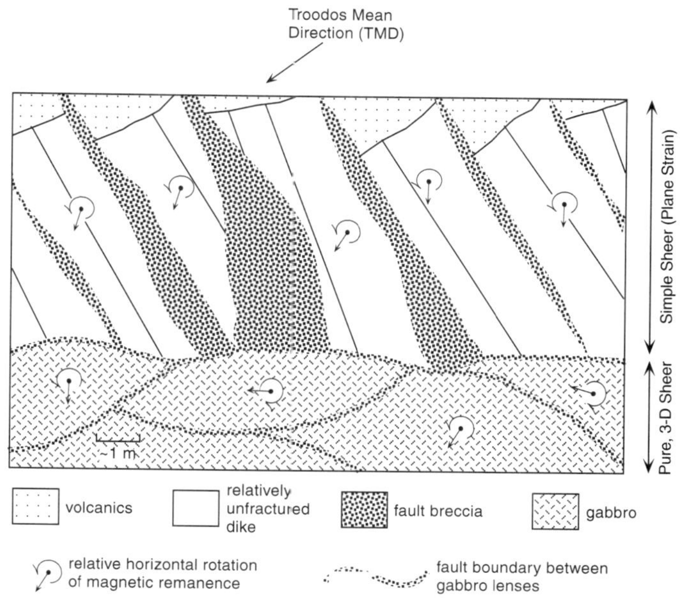

Figure 5. Generalized model for deformation of gabbro and overlying dike/volcanic section in Troodos ophiolite, based on structural and paleomagnetic studies. Drawing represents a vertical section oriented ~276° (parallel to the Troodos average magnetic north direction in current coordinates) and approximately parallel to the ridge-axis extension direction; view is to the north. In this model, the dike/gabbro contact represents a fundamental boundary between areas undergoing extension by different mechanisms. Volcanics (lightly stippled areas) and dikes (ruled areas) experience plane strain, simple shear approximating a rotational-planar fault model. Faults are subparallel to dike margins and are characterized by breccia zones (coarse stippling) of varying width. Wide breccia zones allow differential tilting between dike panels. Tilting of dikes is shown by deviation of magnetic remanence directions (arrows) from TMD. In contrast, gabbro section undergoes three-dimensional (3-D) pure shear by differential movement between lozenge-shaped, fault-bounded blocks. Each block can have significantly different tilts (as shown by magnetic remanence arrows), compared to neighboring blocks. Note that block rotation can also be in or out of the plane of drawing.

The Sheeted Dike Complex and Hydrothermal Alteration

Studies of the sheeted dike complex of the Troodos ophiolite have led to insights into the plumbing systems of black-smoker hydrothermal systems at mid-ocean ridges. Wilson (1959) first noted the occurrence of regions within the dike section where diabase has been completely recrystallized to assemblages of quartz, epidote, and minor chlorite. These epidosites overprint an earlier, widespread low-temperature greenschist alteration that generally preserves original diabase textures (Wilson, 1959; Gass and Smewing, 1973; Bettison-Varga et al., 1995). Later studies of the epidosite alteration in a number of areas showed that they represent high-temperature (270–370 °C) interaction of dikes with seawater, and that dikes in these regions have been stripped of elements now concentrated in overlying massive sulfide ore deposits (Richardson et al., 1987; Schiffman et al., 1987; Schiffman and Smith, 1988; Schiffman et al., 1990; Bettison-Varga et al., 1995). The orebodies themselves share many characteristics with exhalative massive sulfides at modern ridges (Constantinou and Grovett, 1973; Constantinou, 1980).

A major debate surrounding the origin of the epidosites has been the relationship between the hydrothermal circulation that created them and fault-related permeability. Is faulting essential in providing the necessary permeability structure for concentration of fluid flow in these relatively narrow regions? One model considers faulting relatively unimportant in this process. Richardson et al. (1987) suggest that epidosites form as an integral part of dike injection at the ridge crest, and that little deformation is required. Bettison-Varga et al. (1992) found that epidosite assemblages are concentrated within fault zones (Fig. 2D) in some areas and suggested that faulting along dike margins provides the essential permeability pathways for hydrothermal fluids. To test these end-member models, Varga et al. (1999a) compared magnetic remanence directions in wholly recrystallized (and remagnetized) epidosites to those in surrounding, relatively unaltered dikes. Their results show that little tilting occurred prior to epidosite alteration. Taking these results together with the preferential occurrence in some areas of epidosite within fault zones, Varga et al. (1999a) suggest a hybrid model in which epidosites formed early in the deformational history of the sheeted dikes during periods of amagmatic extension. During the early stages of extension, tensional fractures open parallel to steep dike margins, providing subvertical fluid paths. During continued structural extension, the early-formed fractures act as faults as dikes rotate in a rotational-planar model, shutting off hydrothermal circulation as the fault zones become less permeable during frictional grain size reduction.

Melt Migration Within the Sheeted Dike Complex

The dynamics of migration of magmas along ridge crests has become increasingly important with the recognition of along-axis variations in magma budget at most ridges (Macdonald et al., 1988) and with the apparent scarcity of discernable magma chambers at slow-spread ridges such as the Mid-Atlantic Ridge (Macdonald, 1982; Karson et al., 1987; Detrick et al., 1990). An attractive model to explain these observations involves lateral migration of melts from isolated magma chambers, similar to that which occurs beneath Hawaii (Rubin, 1990), Iceland (Sigurdsson and Sparks, 1978; Sigurdsson, 1987), and the Juan de Fuca Ridge (Axial Seamount) (e.g., Dziak et al., 1995).

Ophiolites with well-exposed sheeted dike sections, such as Troodos and the Semail ophiolites, have been useful in assessing the applicability of the lateral melt migration model for formation of ocean crust. Magma flow direction in dikes can be determined by interpretation of a variety of field (Fig. 4C) and petrographic criteria (Delaney and Pollard, 1981; Smith, 1987; Varga et al., 1998), as well as through determination of anisotropy of magnetic susceptibility (AMS) near dike margins. Rochette (1991) first demonstrated that AMS can give reliable results for flow direction in dikes of ophiolites and is significantly more time-effective than petrographic methods. Staudigel et al. (1992) compared both field indicators of flow and AMS for a number of dike exposures in the Troodos ophiolite and showed that subhorizontal flow of magma is as important as subvertical flow. Lateral flow of magma along the Troodos ridge crest may help explain the mutually crosscutting relationships between petrologically unrelated dikes seen in some dike sections (Baragar et al., 1987, 1990). Subsequent studies (Varga et al., 1998; Staudigel et al., 1999) showed that AMS reflects petrographic alignment in elongate silicate fabric and that magma flow was preferentially toward the Arakapas fault zone. These results support the notion of movement of magma away from centralized chambers located between transform segments (Macdonald et al., 1988).

CONTINUING CONTROVERSIES AND RELATIONSHIP TO MARINE STUDIES

While much has been learned about the nature of oceanic structure of the Troodos ophiolite through studies of its sheeted dike complex, discussions continue to focus on the relationship and relevance of this structure and activity to modern ridge crests. One important area of disagreement between workers is the relationship of tilting and faulting as observed in the sheeted complex and ridge crest deformation. Tilted dikes and lavas in the Troodos ophiolite have commonly been interpreted to imply that normal faults dip inward at spreading centers while lavas dip away (Varga and Moores, 1985; Varga, 1991). Certainly, this pattern is commonly observed at many structurally extended ridges (e.g., Macdonald, 1982; Karson and Rona, 1990). Shallow lava flows on Iceland (Bodvarsson and Walker, 1964) and within Hole 504B south of the Costa Rica rift (Faruta and Levi, 1983; Allerton et al., 1995; Dilek, 1998), however, dip *toward* their respective spreading centers. Similarly, direct observation of faults and tilted sheeted dikes at Hess Deep show that these features dip away from the East Pacific Rise where they have been presumed to have formed (Karson and al., 1999; Karson et al., 2002).

These observations have led many workers to suggest a model that combines outward-dipping normal faulting and depositional loading of lavas at oceanic ridge crests (e.g., Cann, 1974; Carbotte et al., 1997; Karson et al., 2002). Schouten and Denham (2000) have recently applied this model to interpret data from several of the Cyprus Crustal Drilling Project core holes. In that study, they suggest that surface loading due to lava extrusion caused tilting within the extrusive section, which was compensated at depth by shear along vertical fractures within the sheeted dike section. The main evidence for this is the observation that dikes in the area of the drill holes remain vertical, whereas overlying lavas are moderately tilted. While this model cannot be generally applied to areas of the complex where dikes at depth are highly tilted (e.g., in the Solea Graben region), the possibility of a component of tilting due to loading within the lava section is a potential complication to previous models.

The question of the timing of tilting of dikes in the sheeted dike section has long posed a difficult challenge in interpreting the significance of the internal structure of the Troodos ophiolite. As mentioned above, a near-axis position for much of the tilting has been suggested because in many areas, the upper lavas and immediately overlying marine sediments are far less tilted and deformed than underlying parts of the ophiolite. Additionally, fault zones within the sheeted dike section appear to have acted as fluid conduits for hydrothermal fluids early in their formation. More convincing, however, is the observation of crosscutting dikes, where highly fractured, shallowly dipping dikes are generally cut by less fractured, steep dikes of the same composition (Fig. 4D). Until recently, however, there was no independent evidence to demonstrate *differential* tilting between crosscutting dikes that would suggest that magmatism was concurrent with faulting and tilting. New paleomagnetic data from the sheeted complex in the Mitsero graben area (Fig. 1) have provided this crucial link. There, remanence directions within younger, crosscutting dikes are demonstrably less tilted than are remanence directions in the older dikes (Varga and Gee, 2000), thus supporting a near-axis origin for normal faulting and related tilting in that area. Additional study of crosscutting dike relationships need to be done in other areas of the complex to further document the timing of dike tilting.

The marine community has been slow to embrace structural models developed from the Troodos ophiolite over the past ~15 years, in part, because of the Troodos' non-MORB geochemistry. Although many of the structural features documented within the ophiolite have now been observed within well-studied oceanic sections (Karson, 1990; Karson and al., 1999; Karson et al., 2002), the geochemistry of the Troodos lavas (Miyashiro, 1973; Schmincke et al., 1983) continues to inspire suprasubduction zone models for generation of the ophiolite (see Moores et al., 1984). Although this debate continues, recent studies have suggested scenarios in which geochemical signatures similar to those in some ophiolites may be generated at conventional ridge crests. Such magmas may be produced where ridges intersect continents or near ridge-subduction zone intersections (Trull et al., 1990; Sturm et al., 2000). Alternatively, the notion of "historical contingency" has been suggested, wherein oceanic spreading can occur in mantle enriched within a previous tectonic setting (Moores et al., 2000).

Despite the geochemical debate about the tectonic setting of the Troodos ophiolite, observations made there have guided research and interpretation of modern spreading centers. The concept of amagmatic spreading at spreading centers in which the crustal section structurally is highly distended is well-documented within the Troodos ophiolite (Varga and Moores, 1985, 1990; Dilek and Eddy, 1992) and is applicable to slow-spreading ridges such as the Mid-Atlantic Ridge, where magmatism appears ephemeral (Karson et al., 1987; Karson, 1990). The style of faulting at some spreading centers has also been found to be similar to that documented in the Troodos. Recent *Argo II* and *Alvin* investigations of exposed crust within the Hess Deep has clearly shown that tilting there is accomplished primarily along rotational-planar, dike parallel faults (Fig. 6) (Karson and al., 1999; Varga et al., 1999b; Karson et al., 2002). These same studies at the Hess Deep also document primarily brittle deformation within the gabbro section and hint at a profound change in structural style at the dike/plutonic contact. Fracture patterns within Hess Deep gabbro, as in the Troodos ophiolite, are much more complex than in the overlying sheeted dike section and may indicate pure-shear extension along a network of anastomosing faults that outline less-deformed, lozenge-shaped blocks (Fig. 5).

Figure 6. *Argo II* digital collage images of tilted sheeted dikes exposed along the walls of the Hess Deep, equatorial Pacific. A: Moderately right-dipping (to east) dikes. Note well developed cross (cooling) joints that transect dikes and dip moderately to left (west). B: West (to right) tilted dike (D) cut and partially transected by fault breccia (dashed line shows approximate contact). Note that fault breccia is subparallel to dike margin (compare to Fig. 5). In both collages, the white-colored bar at bottom is ~5 m in length.

DIRECTION OF FUTURE STUDIES OF THE TROODOS SHEETED COMPLEX

The well exposed sheeted dike complex of the Troodos ophiolite affords an excellent opportunity to study the interplay between magmatism and extension at divergent plate margins, and will continue to provide field-based analogs for interpretation of observations made in the modern oceans. What has become clear is that neither study of ophiolites nor of modern ocean crust can efficiently proceed without detailed knowledge of the other.

Outstanding problems in the study of both ophiolites and ocean crust include documentation of a structural mechanism for accumulation of thick volcanic sequences at ridge crests and the nature of accommodation of extensional strain at depth. How do several kilometers of lava accumulate in relatively narrow magmatic zones at spreading centers, and how is the implied substantial subsidence at ridges accommodated within deeper crustal levels? Is the critical space created by extensional faulting and tilting or through vertical shear along closely space fractures within the sheeted dike section (Schouten and Denham, 2000)? Also, is uniform-sense shear along rotational-planar normal faults within the upper crust generally accommodated through detachment faulting, as suggested for the MARK area of the Mid-Atlantic Ridge (Karson, 1990), or through transition to pure shear, as may be the case at Hess Deep (Karson et al., 2002). These and other structural problems will be solved through iteration between studies of well exposed crustal sections in the oceans and studies of intact ophiolites such as the Troodos.

ACKNOWLEDGMENTS

The author thanks the many collaborators who have shared their ideas and expertise throughout the years and who have contributed to the ideas reviewed in this paper. In particular, I acknowledge the insights of Eldridge Moores, Jeff Gee, Lori Bettison-Varga, Steve Hurst, Peter Schiffman, Brian Smith, Fred Vine, Simon Allerton, and Jeff Karson. Financial support from National Science Foundation (grants EAR-830695, EAR-8610503, EAR-9017670, EAR-942393), the Keck Geology Consortium, The College of Wooster, and Unocal Corporation made the author's research on Cyprus possible. I also thank my many friends at the Cyprus Geological Survey Department, especially George Constantinou and Costas Xenophontos, for their substantial logistical support. Thoughtful reviews by Yildirim Dilek, Bramley Murton, and Fred Vine are greatly appreciated.

REFERENCES CITED

Abrahamsen, N., and Schoenharting, G., 1987, Palaeomagnetic timing of the rotation and translation of Cyprus: Earth and Planetary Science Letters, v. 81, p. 409–418.

Agar, S.M., and Klitgord, K.D., 1995, A mechanism for decoupling within the oceanic lithosphere revealed in the Troodos ophiolite: Nature, v. 374, p. 232–238.

Allerton, S., and Vine, F.J., 1987, Spreading structure of the Troodos ophiolite, Cyprus: Some paleomagnetic constraints: Geology, v. 15, p. 593–597.

Allerton, S., and Vine, F.J., 1990, Palaeomagnetic and structural studies of the southeastern part of the Troodos complex, *in* Malpas, J.G., et al., ed., Ophiolites and Oceanic Crustal Analogues, Proceedings of the Symposium "Troodos 1987": Nicosia, Cyprus, Geological Survey Department, p. 99–109.

Allerton, S., McNeill, A., Stoking, L., Pariso, J., Tartarotti, P., Marton, F., and Pertsev, N., 1995, Structures and magnetic fabrics from the lower sheeted dike complex of Hole 504B reoriented using stable magnetic remanence, *in* Erzinger, J., Becker, K., Dick, H., and Stokking, L., eds., Proceedings of the Ocean Drilling Programs, Scientific Results: College Station, Texas, Ocean Drilling Program, v. 137/140, p. 245–252.

Anonymous, 1972, Penrose field conference on ophiolites: Geotimes, v. 17, p. 24–25.

Baragar, W.R.A., Lambert, M.B., Baglow, N., and Gibson, I.L., 1987, Sheeted dikes of the Troodos ophiolite, Cyprus, *in* Halls, H.C., and Fahrig, W.F., eds., Mafic Dyke Swarms: Vancouver, British Columbia, Geological Association of Canada, p. 257–272.

Baragar, W.R.A., Lambert, M.B., Baglow, N., and Gibson, I.L., 1990, The sheeted dike zone in the Troodos ophiolite, *in* Malpas, J.G., et al., ed., Ophiolites and Oceanic Crustal Analogues, Proceedings of the Symposium "Troodos 1987": Nicosia, Cyprus, Geological Survey Department, p. 37–51.

Bartholomew, I.D., 1993, The interaction and geometries of diapiric uprise centres along mid-ocean ridges; evidence from mantle fabric studies of ophiolite complexes, *in* Prichard, A., Alabaster, T., Harris, N., and Neary, C., eds., Magmatic processes and plate tectonics: London, Geological Society of London Special Publication 76, p. 245–256.

Bellamy, C.V., 1905, Geological Map of Cyprus: scale 1:348 480, 1 sheet.

Bellamy, C.V., and Jukes-Browne, A.J., 1905, The Geology of Cyprus: London.

Bettison-Varga, L., Schiffman, P., and Janecky, D.R., 1995, Fluid-rock interaction in the hydrothermal upflow zone of the Solea graben, Troodos ophiolite, Cyprus, *in* Schiffman, P., and Day, H.W., eds., Low-grade metamorphism of mafic rocks: Boulder, Colorado, Geological Society of America Special Paper 296, p. 81–100.

Bettison-Varga, L., Varga, R.J., and Schiffman, P., 1992, Relationship between ore-forming hydrothermal systems and extensional deformation in the Solea graben spreading center, Troodos ophiolite, Cyprus: Geology, v. 20, p. 987–990.

Biju-Duvall, B., and Lapierre, H., 1976, Is the Troodos massif (Cyprus) allochthonous?: Bulletin de la Société Géologique de France, v. 7, no. 1, p. 1347–1356.

Bishopp, D.W., 1952, The Troödos massif, Cyprus: Nature, v. 169, no. 4299, p. p. 489–490.

Bishopp, D.W., 1954, Some new features of the geology of Cyprus, *in* Proceedings: International Geological Congress, 19th, Algiers, Algeria, p. 13–18.

Bodvarsson, G., and Walker, G., 1964, Crustal drift in Iceland: Geophysical Journal of the Royal Astronomical Society, v. 9, p. 285–300.

Bonhommet, N., Roperch, P., and Calza, F., 1988, Palaeomagnetic arguments for block rotations along the Arakapas fault (Cyprus): Geology, v. 16, p. 422–425.

Cann, J.R., 1974, A model for oceanic crustal structure developed: Geophysical Journal of the Royal Astronomical Society, v. 39, p. 169–187.

Carbotte, S., Mutter, J., and Xu, L., 1997, Contribution of volcanism and tectonism to axial and flank morphology of the southern EPR 17° 10'-17° 40' S, from a study of layer 2A geometry: Journal Geophysical Research, v. 102, p. 10,165–10,184.

Clube, T.M.M., Creer, K.M., and Robertson, A.H.F., 1985, Paleorotation of the Troodos microplate, Cyprus: Nature, v. 317, p. 522–525.

Constantinou, G., 1980, Metallogenesis associated with the Troodos ophiolite, *in* Panayiotou, A., ed., Ophiolites: Proceedings of the International Ophiolite Symposium, Cyprus, 1979: Nicosia, Cyprus, Geological Survey Department, p. 663–674.

Constantinou, G., and Grovett, G.J.S., 1973, Geology, geochemistry and genesis of Cyprus sulphide deposits: Economic Geology, v. 68, p. 843–858.

Cullis, C.G., and Edge, A.B., 1922, Report on the cupriferous deposits of Cyprus: London, Crown Agents for Overseas Governments and Administrations.

Davis, G.A., and Lister, G.S., 1988, Detachment faulting in continental extension: Perspectives from the southwestern U.S. Cordillera, *in* Clark, S.P., Burchfiel, B.C., and Suppe, J., eds., Processes in continental lithospheric deformation: Boulder, Colorado, Geological Society of America Special Paper 218, p. 133–159.

Delaney, P., and Pollard, D., 1981, Deformation of host rock and flow of magma during growth of minette dikes and breccia-bearing intrusions near Ship Rock, New Mexico: Washington, D.C., U.S. Geological Survey Professional Paper 1202, 61 p.

Detrick, R.S., Mutter, J.C., Buhl, P., and Kim, I.I., 1990, No evidence from multichannel reflection data for a crustal magma chamber in the MARK area on the Mid-Atlantic Ridge: Nature, v. 347, p. 61–64.

Dietrich, D., and Spencer, S., 1993, Spreading-induced faulting and fracturing of oceanic crust: examples from the sheeted dyke complex of the Troodos ophiolite, Cyprus, *in* Prichard, A., Alabaster, T., Harris, N., and Neary, C., eds., Magmatic Processes and Plate Tectonics: Boulder, Colorado, Geological Society of America Special Publication 76, p. 121–139.

Dilek, Y., 1998, Structure and Tectonics of Intermediate-Spread Oceanic Crust Drilled at DSDP/ODP Holes 504B and 896A, Costa Rica Rift, *in* Jones, E., Lee, S., and MacLeod, C.J., eds., Geological evolution of ocean basins; results from the Ocean Drilling Program: London, Geological Society of London Special Publication 131, p. 179–197.

Dilek, Y., and Eddy, C., 1992, The Troodos (Cyprus) and Kizilday (S. Turkey) ophiolites as structural models for slow-spreading ridge segments: Journal of Geology, v. 100, p. 305–322.

Dilek, Y., Thy, P., Moores, E.M., and Ramsden, T., 1990a, Tectonic evolution of the Troodos ophiolite within the Tethyan framework: Tectonics, v. 9, no. 4, p. 811–823.

Dilek, Y., Thy, P., Moores, E.M., and Ramsden, T., 1990b, Tectonic evolution of the Troodos ophiolite within the Tethyan framework: Reply: Tectonics, v. 11, no. 4, p. 916–923.

Dziak, R.P., Fox, C.G., and Schreiner, A.E., 1995, The June-July 1993 seismo-acoustic event at CoAxial segment, Juan de Fuca Ridge: evidence for a lateral dike injection: Geophysical Research Letters, v. 22, p. 135–138.

Faruta, T., and Levi, S., 1983, Basement paleomagnetism of Hole 504B and 896A, Costa Rica rift, *in* Cann, J.R., Langseth, M., Honnorez, J., Von Herzen, R., and White, S., eds., Initial Reports of the Deep Sea Drilling Project: Washington, D.C., U.S. Government Printing Office, p. 697–703.

Fox, P., and Gallo, D., 1984, A tectonic model for ridge-transform-ridge plate boundaries: Implications for the structure of oceanic lithosphere: Tectonophysics, v. 104, p. 205–242.

Gass, I.G., 1968, Is the Troodos Complex, Cyprus formed by sea floor spreading?: Nature, v. 220, p. 39–42.

Gass, I.G., and Masson-Smith, D., 1963, The geology and gravity anomalies of the Troodos massif, Cyprus: Philosophical Transactions of the Royal Society of London, ser. A, v. 141, p. 417–467.

Gass, I.G., and Smewing, J.D., 1973, Intrusion, extrusion and metamorphism at constructive margins: evidence from the Troodos Massif, Cyprus: Nature, v. 220, p. 39–42.

Gass, I.G., MacLeod, C.J., Murton, B.J., Panayiotou, A., Simonian, K.O., and Xenophontos, C., 1994, The geology of the southern Troodos transform fault zone: Nicosia, Cyprus, Geological Survey Department, Memoir 9, 218 p.

Gaudry, A., 1862, Géologie de l'ile de Chypre: Paris, Memoir Société de Geologie France, v. ser. 2, vii, p. 149.

George, R.P., 1978, Structural petrology of the Olympus ultramafic complex in the Troodos ophiolite, Cyprus: Geological Society of America Bulletin, v. 89, p. 845–865.

Gerbi, C., 1996, Structural interpretation of the gabbro in the Solea graben, Troodos ophiolite, Cyprus, *in* Mendelson, C.V., and Mankiewicz, C., eds., The Ninth Keck Research Symposium in Geology: Beloit, The Keck Geology Consortium, p. 217–220.

Gerbi, C., Crowley, P., and Varga, R.J., 1996, Structural interpretation of the gabbro of the Solea graben, Troodos ophiolite, Cyprus: Geological Society of America Abstracts with Programs, v. 28, no. 3, p. 57.

Grand, T., Lapierre, H., and Mascle, G.H., 1993, Superimposed tectonics of the Cyprus ophiolitic massifs: Tectonics, v. 12, no. 1, p. 93–101.

Hiessleitner, G., 1957, Beitrag zur Geologie von Cypern: Jahrbuch der Geologischen Bundesanstalt, v. 100, no. 2.

Hurst, S., Karson, J.A., and Verosub, K.L., 1994, Paleomagnetism of tilted dikes in fast spread oceanic crust exposed in the Hess Deep Rift: Implications for spreading and rift propagation: Tectonics, v. 13, no. 4, p. 789–802.

Karson, J.A., 1990, Seafloor spreading on the Mid-Atlantic Ridge: Implications for the structure of ophiolites and oceanic lithosphere produced in slow-spreading environments, *in* Malpas, J.G., et al., ed., Ophiolites and Oceanic Crustal Analogues, Proceedings of the Symposium "Troodos 1987": Nicosia, Cyprus, Geological Survey Department, p. 547–555.

Karson, J.A., and Rona, P.A., 1990, Block-tilting, transfer faults, and structural control of magmatic and hydrothermal processes in the TAG area, Mid-Atlantic Ridge 26° N: Geological Society of America Bulletin, v. 102, p. 1635–1645.

Karson, J.A., Thompson, G., Humphris, S., Edmond, J., Bryan, W., Brown, J., Winters, A., Pockalny, R., Casey, J., Campbell, A., Klinkhammer, G., Palmer, M., Kinzler, R., and Sulanowska, M., 1987, Along axis variations in seafloor spreading in the MARK area: Nature, v. 328, p. 681–685.

Karson, J.A., Hurst, S.D., Klein, E.M., Gillis, K., Miller, J., Varga, R.J., et al., 1999, Large-scale ARGO II digital images of upper crustal structures at Hess Deep: Ridge Events, v. 10, p. 29–32.

Karson, J.A., Klein, E., Hurst, S., Lee, C., Rivizzigno, P., Curewitz, D., Morris, A., Miller, D.J., Varga, R.G., Christeson, G.L., Cushman, B., O'Neill, J.M., Brophy, J.G., Gillis, K.M., Stewart, M.A.; and Sutton, A.L., 2002, Structure of uppermost fast-spread oceanic crust exposed at the Hess Deep Rift: Implications for subaxial processes at the East Pacific Rise: Geochemistry, Geophysics, Geosystems, v. 3, paper no. 2001GC000155.

Kluth, C.F., and Coney, P.J., 1981, Plate tectonics of the Ancestral Rocky Mountains: Geology, v. 9, p. 10–15.

Lapierre, H., Angelier, J., Cogne, X., Grand, T., and Mascle, G.H., 1988, Tectonique superposée de la zone de Faille d'Arakapas (Massif de Troodos, Chypre): Geodinamica Acta, v. 2, no. 4, p. 197–206.

Macdonald, K., 1982, Mid-ocean ridges: fine scale tectonic, volcanic and hydrothermal processes within the plate boundary zone: Annual Reviews of Earth and Planetary Science, v. 10, p. 155–190.

Macdonald, K.C., Fox, P.J., Perram, L.J., Eisen, M.F., Haymon, R.M., Miller, S.P., Carbotte, S.M., Cormier, M.-H., and Shor, A.N., 1988, A new view of the mid-ocean ridge from the behavior of ridge-axis discontinuities: Nature, v. 335, p. 217–225.

MacLeod, C.J., and Murton, B.J., 1993, Structure and tectonic evolution of the southern Troodos transform fault zone, Cyprus, *in* Prichard, A., Alabaster, T., Harris, N., and Neary, C., eds., Magmatic processes and plate tectonics: London, Geological Society of London Special Publication 76, p. 141–176.

Malpas, J.G., 1990, Crustal accretionary processes in the Troodos ophiolite, Cyprus: evidence from field mapping and deep crustal drilling, *in* Malpas, J.G., et al., ed., Ophiolites and Oceanic Crustal Analogues, Proceedings of the Symposium "Troodos 1987": Nicosia, Cyprus, Geological Survey Department, p. 65–74.

Malpas, J.G., Moores, E.M., Panayiotou, A., and Xenophontos, C., 1990, Ophiolites and Oceanic Crustal Analogues, Proceedings of the Symposium "Troodos 1987": Nicosia, Cyprus, Geological Survey Department, 733 p.

Martinez, C., 1996, Flow patterns in a mantle diapir: A look at peridotites of the Troodos ophiolite, Cyprus: Geological Society of America Abstracts with Programs, v. 28, no. 3, p. 78.

Miyashiro, A., 1973, The Troodos ophiolite was probably formed in an island arc: Earth and Planetary Science Letters, v. 19, p. 218–224.

Moores, E.M., 1982, Origin and emplacement of ophiolites: Reviews of Geophysics and Space Sciences, v. 20, p. 735–760.

Moores, E.M., and Vine, F.J., 1971, Troodos massif, Cyprus and other ophiolites as ocean crust: evaluations and implications: Philosophical Transactions of the Royal Society of London, ser. A., v. 268, p. 443–466.

Moores, E.M., Robinson, P., Malpas, J.G., and Xenophontos, C., 1984, Model for the origin of the Troodos massif, Cyprus and other Mideast Ophiolites: Geology, v. 12, p. 500–503.

Moores, E.M., Varga, R.J., Verosub, K.L., and Ramsden, T., 1990, Regional structure of the Troodos dike complex, *in* Malpas, J.G., et al., ed., Ophiolites and Oceanic Crustal Analogues, Proceedings of the Symposium "Troodos 1987": Nicosia, Cyprus, Geological Survey Department, p. 27–35.

Moores, E.M., Kellog, L., and Dilek, Y., 2000, Tethyan ophiolites, mantle convection, and tectonic "historical contingency": A resolution of the "ophiolite conundrum", *in* Dilek, Y., Moores, E.M., Elthon, D., and Nicolas, A., eds., Ophiolites and Oceanic Crust: New Insights from Field Studies and the Ocean Drilling Program: Boulder, Colorado, Geological Society of America Special Paper 349, p. 3–12.

Mukasa, S., and Ludden, J.N., 1987, Uranium-lead isotopic ages of plagiogranites from the Troodos Ophiolite, Cyprus, and their tectonic significance: Geology, v. 15, p. 825–828.

Murton, B.J., and Gass, I., 1986, Western Limassol Forest Complex, Cyprus: Part of an Upper Cretaceous leaky transform fault: Geology, v. 14, p. 255–258.

Nicolas, A., and Rabinowicz, M., 1984, Mantle flow patterns at oceanic spreading centres: Relation with ophiolitic and oceanic structures, *in* Gass, I., Lippard, S.J., and Shelton, A., eds., Ophiolites and oceanic lithosphere: London, Geological Society of London Special Publication 13, p. 147–151.

Nicolas, A., Ceuleneer, G., Boudier, F., and Misseri, M., 1988, Structural mapping in the Oman ophiolites: mantle diapirism along an oceanic ridge: Tectonophysics, v. 151, p. 27–56.

Poole, A.J., and Robertson, A., 1991, Quaternary uplift and sea-level change at an active plate boundary, Cyprus: Journal of the Geological Society of London, v. 148, p. 909–921.

Rautenschlein, M., Jenner, G., Hertogen, J., Hofmann, A.W., Kerrich, J., Schmincke, H.-U., and White, W.E.M., 1985, Isotopic and trace element composition of volcanic glass from the Akaki Canyon, Cyprus: Earth and Planetary Science Letters, v. 75, p. 369–383.

Richardson, C.J., Cann, J.R., Richards, H.G., and Cowan, J.G., 1987, Metal-depleted root zones of the Troodos ore-forming hydrothermal systems, Cyprus: Earth and Planetary Science Letters, v. 84, p. 243–253.

Robertson, A., 1977, Tertiary uplift history of the Troodos Massif, Cyprus: Geological Society of America Bulletin, v. 88, p. 1763–1772.

Robertson, A., and Xenophontos, C., 1993, Development of concepts concerning the Troodos ophiolite and adjacent units in Cyprus, *in* Prichard, A., Alabaster, T., Harris, N., and Neary, C., eds., Magmatic Processes and Plate Tectonics: London, Geological Society of London Special Publication 76, p. 85–119.

Rochette, P., Jenatton, L., Dupuy, C., Boudier, F., and Rueber, I., 1991, Diabase dike emplacement in the Oman ophiolite: A magnetic fabric study with reference to geochemistry, *in* Peters, T., and Coleman, R.G., eds., Ophiolite Genesis and Evolution of the Oceanic Lithosphere: Norwell, Massachusetts, Kluwer, p. 55–82.

Rubin, A.M., 1990, A comparison of rift-zone tectonics in Iceland and Hawaii: Bulletin of Volcanology, v. 52, p. 302–319.

Schiffman, P., and Smith, B.M., 1988, Petrology and O-isotope geochemistry of a fossil seawater hydrothermal system within the Solea graben, northern Troodos ophiolite, Cyprus: Journal of Geophysical Research, v. 93, p. 4612–4624.

Schiffman, P., Smith, B.M., Varga, R.J., and Moores, E.M., 1987, Geometry, conditions and timing of off-axis hydrothermal metamorphism and ore deposition in the Solea graben, Troodos ophiolite, Cyprus: Nature, v. 325, p. 423–425.

Schiffman, P., Bettison, L.A., and Smith, B.M., 1990, Mineralogy and geochemistry of epidosites from the Solea graben, Troodos ophiolite, Cyprus, *in* Malpas, J.G., et al., ed., Ophiolites and Oceanic Crustal Analogues, Proceedings of the Symposium "Troodos 1987": Nicosia, Cyprus, Geological Survey Department, p. 673–683.

Schmincke, H.-U., Rautenschlein, M., Robinson, P., and J., M., 1983, Troodos extrusives series of Cyprus: A comparison with oceanic crust: Geology, v. 11, p. 405–409.

Schouten, H., and Denham, C., 2000, Comparison of volcanic construction in the Troodos ophiolite and oceanic crust using paleomagnetic inclinations from Cyprus Crustal Study Project (CCSP) CY-1 and CY-1A and Ocean Drilling Program (ODP) 504B drill cores, *in* Dilek, Y., Moores, E.M., Elthon, D., and Nicolas, A., eds., Ophiolites and Oceanic Crust: New Insights from Field Studies and the Ocean Drilling Program: Boulder, Colorado, Geological Society of America Special Paper 349, p. 181–194.

Şengör, A.M.C., Burke, K., and Dewey, J.F., 1978, Rifts at high angles to orogenic belts: Tests for their origin and the Upper Rhine graben as an example: American Journal of Science, v. 278, p. 24–40.

Sigurdsson, H., 1987, Dyke injection in Iceland: A review, *in* Halls, H.C., and Fahrig, W.F., eds., Mafic Dyke Swarms: Vancouver, British Columbia, Geological Association of Canada, p. 55–64.

Sigurdsson, H., and Sparks, S.R.J., 1978, Lateral magma flow within rifted Icelandic crust: Nature, v. 274, p. 126–130.

Simonian, D.O., and Gass, I.G., 1978, Arakapas fault belt, Cyprus: A fossil transform fault: Geological Society of America Bulletin, v. 89, p. 1220–1230.

Smith, R.P., 1987, Dyke emplacement at Spanish Peaks, Colorado, *in* Halls, H.C., and Fahrig, W.F., eds., Mafic dyke swarms: Ottawa, Ontario, Geological Association of Canada, p. 47–54.

Staudigel, H., Gee, J., Tauxe, L., and Varga, R.J., 1992, Shallow intrusive directions of sheeted dikes in the Troodos ophiolite: Anisotropy of magnetic susceptibility and structural data: Geology, v. 20, p. 841–844.

Staudigel, H., Tauxe, L., Gee, J., Bogaard, P., Haspels, J., Kale, G., Leenders, A., Meijer, P., Swaak, B., Tuin, M., Van Soest, M., Verdurmen, E., and Zevenhuizen, A., 1999, Geochemistry and intrusive directions in sheeted dikes in the Troodos ophiolite: Implications for mid-ocean ridge spreading centers: Geochemistry, Geophysics, Geosystems, v. 1, paper no. 1999GC000001.

Steward, J.A., 1990, Drifting continents & colliding paradigms: Perspectives on the geoscience revolution: Bloomington, Indiana University Press, 302 p.

Sturm, M., Klein, E., Karsten, J., and Karson, J.A., 2000, Evidence for subduction-related contamination of the mantle beneath the Chile Ridge: Implications for ambiguous ophiolite compositions, *in* Dilek, Y., Moores, E.M., Elthon, D., and Nicolas, A., eds., Ophiolites and Oceanic Crust: New Insights from Field Studies and the Ocean Drilling Program: Boulder, Colorado, Geological Society of America Special Paper 349, p. 13–20.

Trull, T., M., P., and Kurtz, M., 1990, He and Sr isotopic constraints on subduction zone contributions to Woodlark Basin volcanism, *in* V.M. Goldschmidt conference, Baltimore, Maryland, May 2–4, 1990, Program and Abstracts: Baltimore, Maryland, Geochemical Society, p. 87.

van Everdingen, D., and Cawood, P., 1995, Dyke domains in the Mitsero graben, Troodos ophiolite, Cyprus: an off-axis model for graben formation at a spreading centre: Geological Society of London Journal, v. 152, p. 923–932.

Varga, R.J., 1991, Modes of extension at oceanic spreading centers: evidence from the Solea graben, Troodos ophiolite, Cyprus: Journal of Structural Geology, v. 13, p. 517–537.

Varga, R.J., and Gee, J., 1995, Paleomagnetic evaluation of the detachment model for crustal extension within the Troodos ophiolite, Cyprus: Eos (Transactions, American Geophysical Union), v. 76, no. 46, p. 616.

Varga, R.J., and Gee, J., 2000, Use of paleomagnetism to solve fundamental geological problems in the evolution of the Troodos ophiolite, Cyprus: Geological Society of America Abstracts with Programs, v. 32, no. 7, p. 247.

Varga, R.J., and Moores, E.M., 1985, Spreading structure of the Troodos ophiolite, Cyprus: Geology, v. 13, no. 12, p. 846–850.

Varga, R.J., and Moores, E.M., 1990, Intermittent magmatic spreading and tectonic extension in the Troodos ophiolite: implications for exploration for black smoker-type ore deposits, *in* Malpas, J.G., et al., ed., Ophiolites and Oceanic Crustal Analogues, Proceedings of the Symposium "Troodos 1987": Nicosia, Cyprus, Geological Survey Department, p. 53–64.

Varga, R.J., Gee, J.S., Staudigel, H., and Tauxe, L., 1998, Dike surface lineations as magma flow indicators within the sheeted dike complex of the Troodos ophiolite, Cyprus: Journal of Geophysical Research, v. 103, p. 5241–5256.

Varga, R.J., Gee, J., Bettison-Varga, L., Anderson, R.A., and Johnson, C.L., 1999a, Early establishment of seafloor hydrothermal systems during structural extension: paleomagnetic evidence from the Troodos ophiolite, Cyprus: Earth and Planetary Science Letters, v. 171, p. 221–235.

Varga, R.J., Karson, J.A., and Lee, C., 1999b, Styles of ridge-crest extensional faulting: Comparison of the Troodos ophiolite with *Argo II* imagery and *Alvin* investigations of the Hess Deep North Wall, Equatorial Pacific: Eos (Transactions, American Geophysical Union), v. 80, no. 46, p. 984.

Verosub, K.L., and Moores, E.M., 1981, Tectonic rotations in extensional regimes and their paleomagnetic consequences for oceanic basalts: Journal of Geophysical Research, v. 86, p. 6335–6349.

Vine, F.J., and Matthews, D.H., 1963, Magnetic anomalies over oceanic ridges: Nature, v. 199, p. 947.

Wilson, R.A.M., 1959, The geology of the Xeros-Troodos area: Nicosia, Cyprus, Geological Survey Department, Memoir 1, 135 p.

Manuscript Accepted by the Society May 2, 2003

Printed in the USA

Geological Society of America
Special paper 373
2003

The evolution of ideas for the origin and emplacement of the western Hellenic ophiolites

A.G. Smith*
Department of Earth Sciences, Downing Street, Cambridge CB2 3EQ, England

A. Rassios*
Institute of Geological and Mineral Exploration, Kozani Branch, Lefkovrisi, Kozani, Greece

ABSTRACT

Reference to some of the western Hellenic ophiolites (Vourinos, Pindos and Othris) as serpentine and dolerite eruptions by Boué (1794–1881) dates from at least 1840. The observation of tuffs and pillow lavas (?), possibly not ophiolitic, in Othris in 1880 led Neumayr (1845–1890) to suggest their origin as large submarine eruptions. A submarine setting was reinforced by Brunn's mapping of the Pindos and Vourinos ophiolites in the 1930s and 1950s. He concluded that they were in situ, formed an entity entirely magmatic in origin, had been extruded onto the Pelagonian continent in the Jurassic along feeders concealed under the Meso-Hellenic trough and that the ophiolitic "magmas" had flowed outward to the NE (Vourinos) and SW (Pindos). Brunn interpreted the low-grade metamorphism at the base of the Vourinos ophiolite as contact metamorphism that provided additional evidence for their origin as huge submarine flows. In 1959, Brunn made the first analogy between these ophiolites and the rocks of the Mid-Atlantic Ridge—the oceanic rocks and ophiolitic rocks were formed by similar processes of submarine extrusion, but were not necessarily equivalent. Moores (1969) recognized the fundamental difference between magmatic and tectonized ultramafics in the Vourinos and suggested that many of the ultramafic rocks had been emplaced by solid flow to produce structures that were symmetric about a NE-trending axis. All of these interpretations were made in an essentially "fixist" framework.

Subsequent interpretations assumed a plate tectonic setting. As with many other ophiolites, the main problems have been the "root zone" of the ophiolites; the location and dip direction of associated subduction zones; the tectonic setting(s) indicated by their marked variations in composition; the spatial and temporal relationships between extension and convergence as recorded in ophiolitic rocks; how the ophiolites were emplaced onto a continent; the significance of the metamorphic sole; and the spatial and temporal relations of the Hellenic ophiolites to contiguous ophiolites in Albania and former Yugoslavia.

A modified model for the emplacement of the western Hellenic ophiolites is set out in this paper. In outline, the model starts with the creation of the Pindos Ocean, a branch of the Tethyan Ocean in Triassic time, separating western Greece from the Pelagonian continent to the east. A west-dipping subduction zone was established in mid-Jurassic time in western Greece, located in the south against the eastern margin of the Parnassos platform and in the north located within the Pindos Ocean. Initially, the hinge line

E-mails: Smith—ags1@esc.com.ac.uk; Rassios—blather@grevenanet.gr

Smith, A.G., and Rassios, A., 2003, The evolution of ideas for the origin and emplacement of the western Hellenic ophiolites, *in* Dilek, Y., and Newcomb, S., eds., Ophiolite concept and the evolution of geological thought: Boulder, Colorado, Geological Society of America Special Paper 373, p. 337–350. For permission to copy, contact editing@geosociety.org.

of the subduction zone retreated east at a faster rate than the overall convergence. The difference in velocities caused extension between the hinge line and the continental margin that generated a back-arc basin contemporaneously with subduction. Convergence was stopped by the attempted subduction of the Pelagonian continent under the eastern part of the new ocean basin that was then emplaced as the western Hellenic ophiolites, leaving the remainder of the supra-subduction ocean basin and the Pindos Ocean to be subducted in a later phase of convergent orogeny. The model requires only one subduction zone that starts at a continental margin. It may be of general application, as sketched in a model for the emplacement of the Semail ophiolite in Oman.

Keywords: ophiolites, emplacement, Greece, supra-subduction zone, models.

INTRODUCTION

This paper focuses on the ideas on ophiolite emplacement that have been derived from, or applied to, the western Hellenic ophiolites (Vourinos, Pindos, and Othris) of Greece (Figs. 1 and 2), mostly by the authors and their collaborators. More than 30 years have elapsed since the first tentative steps were made to place the western Hellenic ophiolites in a plate tectonic setting (Hynes et al., 1972). It has been interesting to look back at the old emplacement models to see what new data or new concepts caused them to be modified. Despite the passage of time, there is still no entirely acceptable plate tectonic model for their emplacement, although a new model is presented at the end of this paper. The lack of agreement is partly due to the controversies that still exist about the relationships between the western and eastern Hellenic ophiolites, which partly stems from the lack of detailed knowledge, particularly about the eastern ophiolites. Any model must also be applicable to the Dinaric ophiolites in Albania and former Yugoslavia (Fig. 1). Advances in understanding the Hellenic ophiolites will probably come from the following: (1) the study of their along-

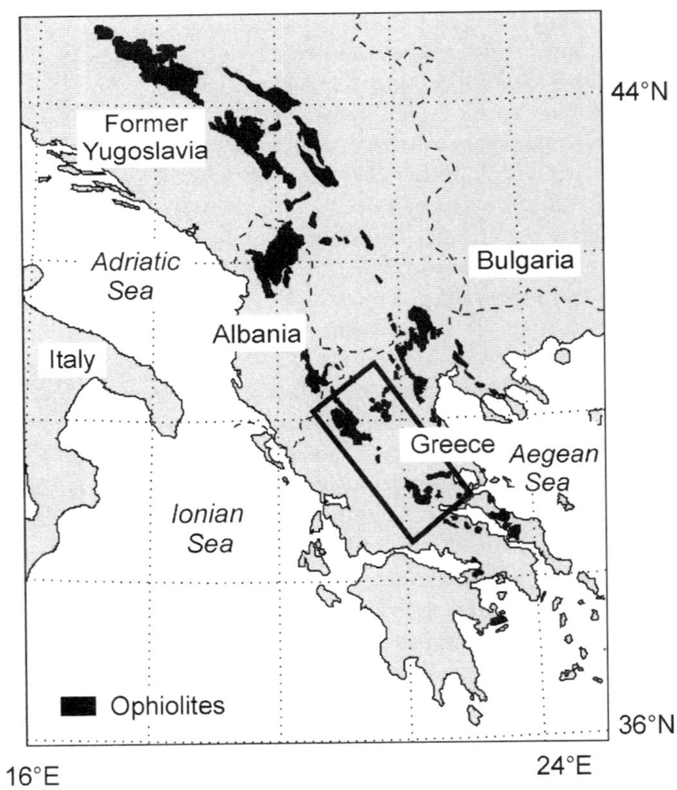

Figure 1. Map of Hellenic-Dinaric ophiolite outcrops in the Balkan Peninsula. The box outlines the area discussed in the paper.

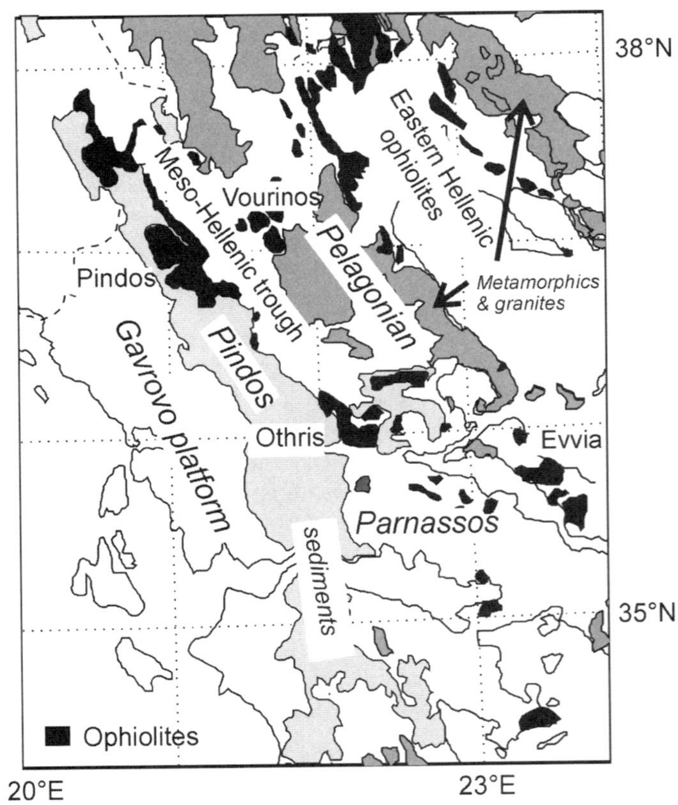

Figure 2. Map showing distribution of ophiolites (black), metamorphic rocks and granites (gray), and Pindos sediments (light gray), together with some locations.

strike equivalents in Albania (Shallo, 1992), where there are well-exposed, extensive ophiolite outcrops; (2) from a dating program (there are no reliable isotopic dates for the ophiolites); and (3) from modern geochemical studies of the ophiolites, which are integrated with the wealth of structural data that are available.

Historically, one of the first allusions to the Hellenic ophiolites was by Boué (1834), who referred to "serpentine eruptions" in Greece. He traveled widely in Greece (which was then part of "Turkey in Europe"), through areas where the Pindos, Vourinos, and Othris ophiolites are exposed (Boué, 1840). The observation that tuffs and pillow lavas (?), possibly not ophiolitic, were present in Othris led Neumayr (1880) to suggest these were large submarine eruptions.

It is impossible to discuss all of the important observations and ideas about ophiolites here, but among the more important is Steinmann's (1905) classic observation of the common association in the Alpine-Himalayan belt of serpentine, spilites, and cherts, and Hess's (1955) paper on serpentines, orogeny, and epeirogeny, in which he discusses the apparent paradox between field observations that suggested the existence of an ultramafic magma, and lab experiments that indicated that such a magma did not exist. The consensus that most ophiolites are probably slices of oceanic lithosphere with a characteristic tectonic-magmatic-sedimentary stratigraphy was the outcome of a Penrose conference (Anonymous, 1972), though several researchers had already suggested that ophiolites were slices of oceanic lithosphere (see review in Moores and Vine, 1971).

A submarine setting was reinforced by Brunn's mapping of the Pindos and Vourinos ophiolites in the late 1930s and early 1950s. He concluded that they were formed in situ as an entity, and had been extruded in late Jurassic time along feeders concealed under the Meso-Hellenic trough, and that the ophiolite magma had flowed outward to the NE (Vourinos) and SW (Pindos). Brunn also inferred that the ophiolitic magma was differentiated into ultramafic and mafic portions, much like layered intrusions. Sheared serpentinite at the base of the ophiolites was attributed to post-consolidation effects related to later tectonics, with the required water possibly being derived from the underlying sediments (Brunn, 1956). The amphibolites and garnet-mica schists at the base of the ophiolites were interpreted originally as slivers of the underlying Paleozoic metamorphic basement. Brunn made the first analogy between these ophiolites and the rocks of the Mid-Atlantic Ridge (Brunn, 1959).

Moores (1969) distinguished between the magmatic ultramafic rocks and the metamorphic tectonized ultramafic rocks in the Vourinos Complex, but was uncertain of the significance of the differences. He recognized structures within the tectonites that were symmetric about a NE-trending axis and attributed these to solid-state flow. Two of his models for the origin of the Vourinos were made—as were all other previous interpretations, including Brunn's analogy with the Mid-Atlantic Ridge—in an essentially *fixist* framework; however, Moores also speculated that the symmetry axis in the Vourinos structures might represent a spreading axis at mantle depth.

The authors' own experience with the evolving ophiolite concept spans the period 1965 to the present day. It comes mostly from the western Hellenic ophiolites, principally the Vourinos (AR), Pindos (AR), and Othris (AR and AGS) ophiolites (Fig. 2). At the beginning of this period, ocean-floor spreading had already been proposed (Vine and Matthews, 1963), but the first papers on plate tectonics were a few years in the future (Morgan, 1968; McKenzie and Parker, 1967). Ideas have evolved since then such that at the present time, the western Hellenic ophiolites are viewed as having formed in the upper plate above a subduction zone, where spreading related to extension was essentially contemporaneous with the emplacement of the ophiolites (Rassios and Smith, 2000). The contributions that the study of the western Hellenic ophiolites have made to the ophiolite concept include the following: (1) the initiation of ophiolite studies elsewhere as a result of work on the Vourinos Complex (Brunn 1956, 1959; Moores, 1969); (2) the recognition that the ultramafic rocks in many ophiolites can be separated into magmatic rocks and tectonites (Moores, 1969); and (3) the realization that all the tectonite sections of the western Hellenic ophiolites have been overprinted by a much stronger, emplacement fabric which obscures any earlier pre-emplacement fabrics (Grivas et al., 1993).

All the emplacement models discussed below assume the following about Western Hellenic ophiolites:

1. That they are slices of fossil oceanic lithosphere (here including back-arc basins)
2. That they originated as magmatic rocks (partially melted peridotite) and as residual peridotite at a spreading center
3. That they were emplaced at a convergent plate margin

Changes in geological time-scales and in biostratigraphic dating schemes, particularly for radiolaria, have also been instrumental in changing emplacement models. In this paper, the time-scale of Gradstein and Ogg (1996) is used throughout, unless otherwise stated.

THE OPHIOLITE EMPLACEMENT PROBLEM

Although the emplacement of the Hellenic ophiolites is the main topic of this paper, their evolution is similar to that of many other Tethyan ophiolites, thus similar emplacement models may apply to all of them. Important features common to most Tethyan ophiolites are:

- Oceanic lithosphere gets emplaced ("obducted") tectonically on top of a continental margin, rather than being subducted under a continental margin.
- Ophiolites are commonly emplaced early in the development of an orogenic belt.
- Initial emplacement is commonly <10 m.y. after ophiolite formation.
- A metamorphic sole is developed under the ophiolite.
- Sole rocks are metamorphic rocks commonly of oceanic origin but may include continental margin material, forming metamorphic slices tectonically sandwiched between

the base of the ophiolite and the underlying continental platform or continental margin.

- Emplacement is not accompanied by regional metamorphism and large-scale plutonism.
- After ophiolite emplacement, there is commonly a hiatus of some tens of millions of years before a second phase of deformation takes place.
- The composition of many ophiolites shows that they formed above subduction zones (supra-subduction zone [SSZ]).

Several of these features were discovered long after the start of detailed geological mapping in Greece in 1963. Thus, the earliest models do not deal with all the features mentioned above. The Hellenides are contiguous with the Dinarides in Albania and former Yugoslavia, forming one branch of the Alpine orogenic belt, sensu lato (Fig. 1). Apart from an abrupt 90°-bend in the outcrop pattern at the Scutari-Peč line in northern Albania, the western ophiolites form a relatively narrow (~50 km across), discontinuous strip that is >700 km long. It is assumed that, to a first approximation, cross-strike (NW-SE) sections vary only slowly along the length of the belt.

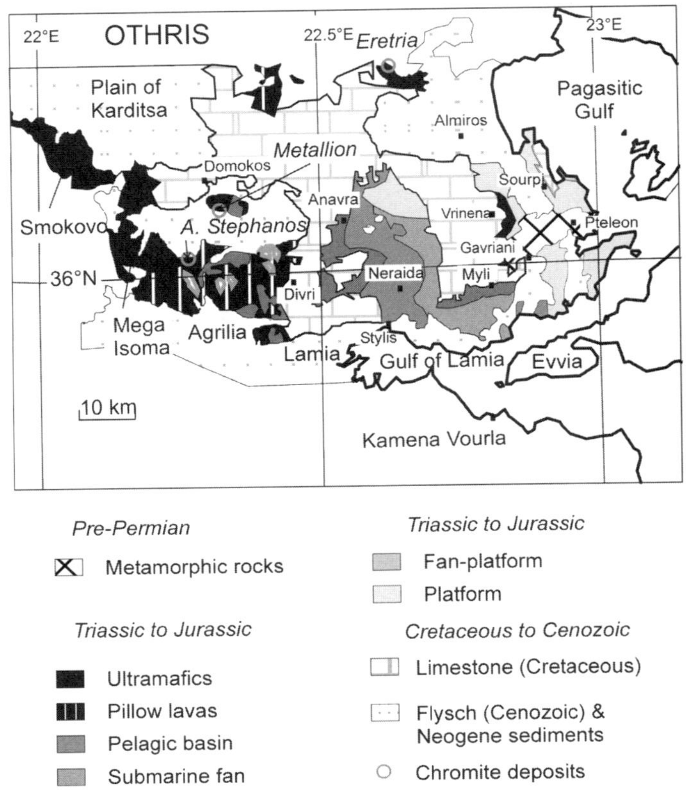

Figure 3. Summary map of Othris. The Triassic-Jurassic thrusts dip west and are exposed beneath an unconformable cover, consisting mostly of Cretaceous platform limestone and Cenozoic sediments. The westernmost and structurally highest thrusts are ophiolitic. The underlying thrusts include non-ophiolitic igneous rocks whose Triassic-Jurassic cover becomes shallower to the east. The structurally lowest thrust lies tectonically on a thick Triassic-Jurassic carbonate platform. Three chromite deposits are present (Aghios Stephanos, Metallion, and Eretria) and two small outcrops in central Othris.

A TECTONIC FLAKE FROM AN OCEAN SOUTHWEST OF THE PELAGONIAN PASSIVE MARGIN?

One of us (Smith) initiated a detailed (1:10,000 to 1:20,000) geological mapping program in Greece in 1965, principally as a result of the enthusiasm and encouragement of Eldridge Moores, who at the time was mapping the Vourinos Complex (Figs. 1 and 2). The Othris Mountains in eastern central Greece were selected for detailed study because of the presence in a relatively small area of Permian shallow-water, fusulinid-bearing limestones, shown to Smith by John Cutbill during a field expedition, and because of Mesozoic ophiolites, an association that was not well understood at the time.

The Othris Mountains of eastern central Greece extend from the Pagasitic Gulf east of Almiros to the Pindos Mountains west of Lamia (Fig. 3). Geologically, the eastern part lies in the Pelagonian zone, a metamorphic complex that includes Triassic and Jurassic marbles resting unconformably on a Hercynian (late Carboniferous) basement. The Sub-Pelagonian (Othris) zone, in which the western Hellenic ophiolites are exposed, lies west of the Pelagonian zone. After emplacement in the Jurassic the ophiolites were eroded and covered by a Cretaceous platform cover. West of the Sub-Pelagonian zone is the Pindos zone, which includes a thrust complex of pelagic Triassic to Cretaceous sediments and Cenozoic flysch. This, in turn, passes west into the carbonate platform of the Gavrovo zone, which was not affected by deformation until the later Cenozoic.

Mapping in the Othris mountains in the Sub-Pelagonian zone revealed a sequence of thrust sheets of presumed Jurassic age that had been emplaced onto the Triassic-Jurassic carbonate platform overlying the late Carboniferous Hercynian basement (Hynes et al., 1972; Ferrière, 1974, 1982; Smith et al., 1975). The carbonate platform is laterally equivalent to the Triassic and Jurassic marbles of the Pelagonian zone. The sections in the Othris thrust sheets formed an ordered progression from a submarine fan into structurally higher and penecontemporanous pelagic basin sections that, in turn, were overridden by slices of a dismembered ophiolite (Fig. 4).

The Triassic igneous sequence in the pelagic basin sections is up to 250 m thick and thins eastward. It is only a few meters thick at the base of the fan, where it lies on Permian sedimentary rocks, and absent at the base of the platform. The thickest sections are in the west and are terminated at their base by thrust faults. The igneous rocks are undersaturated with respect to silica, mildly alkaline, and include picritic basalt, diabase, and keratophyric tuff that are compositionally distinct from the structurally overlying ophiolitic suite, and that were interpreted as the products of early rifting at new passive margins (Hynes, 1974; Menzies, 1976). Detailed work on plagioclase lherzolite and harzburgites

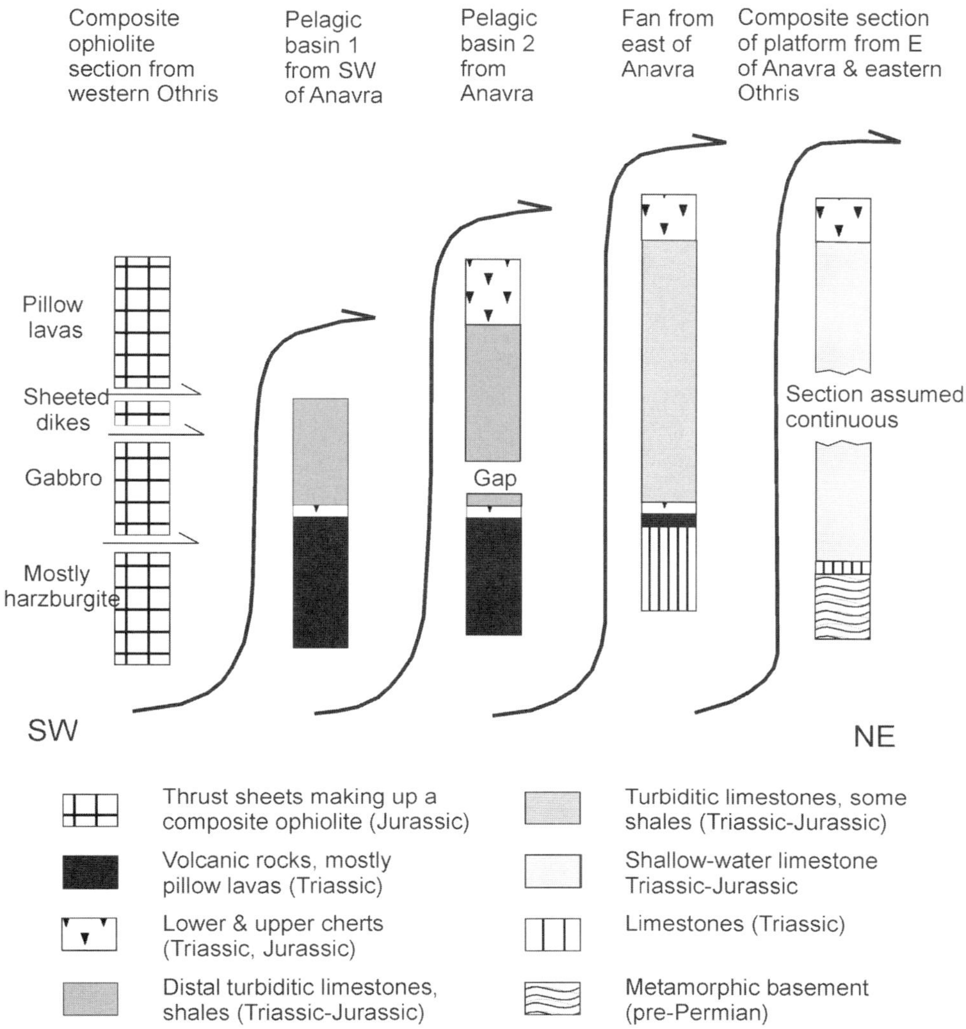

Figure 4. Othris thrust sheets. Five sections are shown, illustrating the lateral facies changes from the ophiolite to the platform carbonate.

in the ophiolites of western Othris suggested that these rocks also formed in the early stages of continental separation (Menzies and Allen, 1974).

Whole rock K-Ar dates were determined from the ophiolitic lavas and from a volcanic unit that is older than the ophiolite (Agrilia Formation). Most of the lavas are altered, show dates ranging from 187 to 70 Ma, and have experimental errors of ±7 to ±20 m.y. (Hynes et al., 1972; Ferrière, 1982, p. 338). K-Ar whole rock dates on altered rocks are commonly unreliable and are virtually useless for precise work. In the absence of any other isotopic data and with no paleontological control, a cluster of four old dates in the range 187–178 Ma, with none older, was tentatively taken to indicate that the Othris ophiolite crystallized in Early to Middle Jurassic time. There was certainly no indication of a Triassic age. It is unfortunate that there are still no comprehensive modern isotopic dating studies of any of the western Hellenic ophiolites to confirm this conclusion.

The entire sequence of thrust sheets, together with the underlying platform, was interpreted as a passive margin sequence that had been destroyed by convergence in late Jurassic time (Hynes et al., 1972). The existence of thrust sheets clearly showed that the Othris ophiolites could not be interpreted as Brunn (1956) had interpreted the Vourinos and Pindos ophiolites, i.e., as in situ masses with a little disturbed, planar contact. The detailed structural evidence in Othris showed that the inferred root zone of the Othris ocean lay to the SW (Smith et al., 1979). Although this SW origin is *fundamental* to all plate tectonic models for the emplacement of the western Hellenic ophiolites, it has not been accepted by every geoscientist in the Hellenides. A summary of the arguments for and against a SW origin is given in Smith (1993), with additional evidence for a SW origin from the structures in the ultramafics in Rassios and Smith (2000). The data suggested that the ophiolites were younger than the probable Carnian (Late Triassic) age for some of the pelagic basin volcanics (Ferrière, 1982, p 192), and older than the beginning of ophiolite emplacement, then assigned to Late Jurassic time (Smith et al., 1979), and possibly of Early to Middle Jurassic age, as suggested by the K-Ar dates.

Hynes et al. (1972) assumed that the emplacement of the Othris ophiolite and the telescoping of the continental margin were related to convergent margin tectonics at a subduction zone. At the time, no evidence was found for any west-dipping mid-Jurassic subduction zone. Thus, it was concluded that the inferred paleo-subduction zone had to dip to the NE. This conclusion formed the cornerstone of most western Hellenic emplacement models for many years. Because the thrust stack seemed to form an ordered tectonic progression from a continental shelf to an ocean basin, it was also concluded that the ophiolite represented a strip of oceanic lithosphere that had lain adjacent to the continental margin. This conclusion could not be tested by the available age data because one could argue that a hiatus between the Triassic volcanics and the mid-Jurassic ophiolites could have formed just prior to break-up, but, as shown above, the dating evidence was inconclusive. Thus, the initial fracture at the base of the ophiolite was regarded as a sole thrust that continued beneath the base of the continental rise and slope sequences until it reached the surface near the continentward limit of the submarine fan (Fig. 5).

The model developed further when the mechanism of ophiolite emplacement onto the continental margin was considered (Smith and Woodcock 1976). As subduction progressed, the shear at the base of the ophiolite wedge was envisaged as the driving force that propelled that wedge onto the continental margin. Geometrically, it was, in fact, an example of "flake tectonics," with the ophiolite resembling a wood shaving in front of an advancing chisel (Oxburgh, 1972). Analogy with the present-day setting of the Troodos massif in Cyprus suggested that after the creation of the initial emplacement fracture, a zone of serpentinization was formed in the mantle above the subduction zone, which would reduce the density of the ocean slab at depth and aid emplacement (Smith and Woodcock, 1976). Thus, the evolving model represented a partial attempt to solve one of the problems of how a slice of dense oceanic lithosphere ends up on top of a low-density continental margin (Fig. 6). The plausibility of the proposed detachment and emplacement mechanism was assumed, rather than scrutinized further.

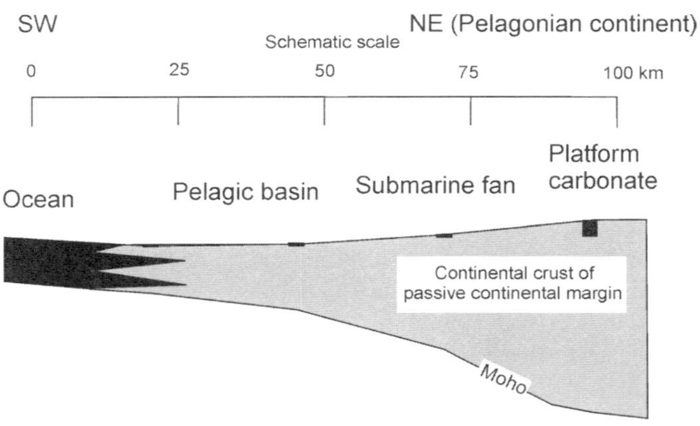

Figure 5. The sections from the thrust sheets in Othris are restored to their possible original positions prior to thrusting. The adopted scale could be incorrect by a factor of 0.5–2. In this model, the oceanic lithosphere (future ophiolite) passed laterally and continuously into the thinned continental crust assumed to underlie the pelagic basin and the submarine fan. Thus, the ophiolite next to the passive continental margin would be the same age as the margin. The black bars and black rectangle give the thickness of the sections on which the model is based.

While this simple flake model accounted for many of the observations, it failed to explain the geometry of detachment of the slab itself via creation of a major fracture that would eventually constitute the base of the ophiolite. The detachment fault dipped to the SW, in the opposite direction to the NE-dipping subduction zone, though apparently merging with it some 150 km away from the continent (Fig. 6). The model did not give a reason why, once started, subduction did not result in a late Jurassic or early Cretaceous continent-continent collision between western Greece and the Pelagonian zone. Among the informal suggestions made was that the main sub-ophiolitic fracture had formerly been a transform fault that had been exploited during emplacement. This was considered unlikely: it would imply that transform faults existed along the strike length of the Dinaric-Hellenic ophiolites, with spreading at right angles to their trend, which was considered improbable and was in disagreement with the limited information on dike trends within the western Hellenic ophiolites, most of which are parallel to the orogenic belt.

Flakes are bounded below by a low-angle fault. If the flake does not undergo much internal deformation, its boundary is a low-angle reverse fault (thrust) at its front and a low-angle normal fault at its rear. This geometry is characteristic of gravity slides, an origin that has been suggested for some Tethyan ophiolites (Stoneley, 1975). Gravity sliding has not been considered as a viable mechanism for the emplacement of the western Hellenic ophiolites, principally because the inferred geometry (Fig. 5) suggests that work has been done against gravity to emplace the ophiolite (i.e., with compressional ramp and imbrication structures), though this conclusion has also not been subjected to a detailed analysis.

METAMORPHIC SOLE

Slices of amphibolites and garnet mica schist at the base of the Vourinos ophiolite had been assigned to the Paleozoic Pelagonian basement (Brunn, 1956). Brunn interpreted the lower grade metamorphism at the base of the Vourinos ophiolite as contact metamorphism that provided additional evidence for the origin of the ophiolite as a whole as a huge submarine flow. In some cases a chilled margin was present; in others, it was concluded that the peridotites themselves had provided the heat for the metamorphism. Scrappy slices of foliated amphibolite are

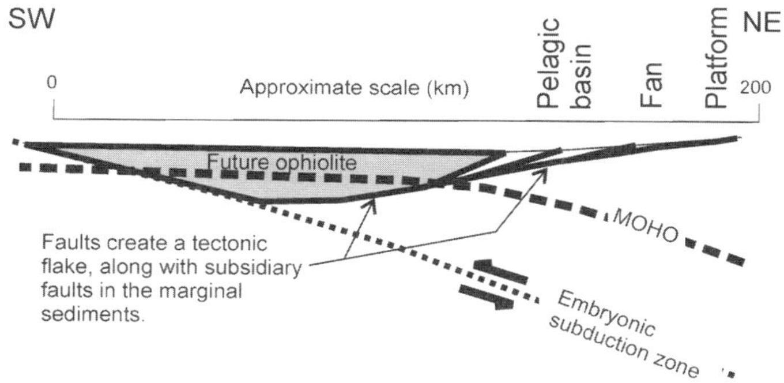

Figure 6. Othris reconstruction showing different components of the flake model. The ophiolite is the oldest oceanic lithosphere next to the passive continental margin. Detachment faults create a flake that is pushed onto the continent by shear from the subduction zone, together with thrust sheets of continental margin sediments.

known from Othris (Hynes, 1972), where they had yielded Jurassic K-Ar dates that were not published because it was unclear at the time how they fit into the scheme of things; however, it was becoming known that the base of several ophiolites located elsewhere displayed a zone of intensely sheared rocks with high-temperature minerals that passed downward into undeformed and unmetamorphosed rocks ("inverted metamorphism"). The tectonic importance of these rocks was being recognized, but was not well understood (e.g., Williams and Smyth, 1973).

With the recognition that these amphibolitic rocks were metamorphic soles created during emplacement by heating beneath an ophiolite sheet, the soles under the Othris, Pindos, Vourinos, and Evvia ophiolites were studied in detail (Spray and Roddick, 1980). Argon-argon (^{40}Ar-^{39}Ar) dates on amphiboles were found initially in the range of 180 to 170 Ma. By contrast, the top of the Anavra Chert in the Othris region, the youngest unit under the ophiolite, was dated as early Cretaceous using radiolaria whose age ranges at the time were still being worked out by paleontologists (Smith et al., 1975). This implied that the time interval between sole formation and the arrival of the ophiolite on the continental margin was perhaps as much as 40 m.y. Fortunately, no models were based on this conclusion, which is now known to be incorrect.

The original ^{40}Ar-^{39}Ar sole dates were shown to need a correction that placed them in the 171 to 165 Ma age range (Spray et al., 1984), giving Middle Jurassic (latest Bajocian to Bathonian) stratigraphic ages (Gradstein and Ogg, 1996). The radiolarian biostratigraphy has also been significantly revised, giving an age range from late Bajocian to middle Callovian for both the Anavra Chert on top of the fan and pelagic basin sequences (Danelian, 1994), a biostratigraphic age range which is now estimated as ca. 172–162 Ma. The same unit on top of the platform has not yet been re-examined (Danelian, 1994). Given that the field evidence shows the top of the Anavra Chert to just precede the arrival of the ophiolite (Smith et al., 1979), then the time interval between the formation of the initial fracture (sole formation) and the arrival of the ophiolite on the continental margin is at most a few million years. The late Bajocian to middle Callovian is also the period during which most of the Pelagonian platform was covered by pelagic sediments, followed by ophiolite emplacement onto the platform during the Oxfordian to Kimmeridgian (Danelian and Robertson, 2001). Analogy with Oman also suggested that the spreading ages would be at most a few million years older than the sole rocks, ca. 176–170 Ma.

Trace element compositions of the bulk of the amphibolites suggested that they derived from oceanic rather than from continental basalts (Spray and Roddick, 1980). Spray and Roddick attributed the heat source to residual heat conducted downwards from the base of hot, young oceanic lithosphere that was tectonically overriding adjacent oceanic lithosphere on its way to the continent, possibly augmented by shear heating. Their results led to a refinement of the emplacement model because the fracture along which the ophiolite had moved must have emerged within the ocean, otherwise there would have been no oceanic amphibolites. Thus, unless the thrust was overriding debris eroded from its leading edge, the ophiolite was not from the marginal strip immediately next to the continent, but was more distant to allow the base of the sheet to override oceanic lithosphere.

Most of the sole rocks are in epidote to low amphibolite facies, but one sample from Othris is in the high amphibolite to granulite facies, recording temperatures as high as 860°C and a pressure of ~7 kb. (Spray and Roddick, 1980). Garnets have been found in amphibolite in the Pindos (personal observations of Rassios and Karamata), though field observations of mineralogic variation suggest internal variation in the temperature of the sole. The origin of such a high temperature is unclear. If a hot slab is placed on top of a cold slab, the contact temperature will be the mean temperature of the two. Given that the highest temperature at a ridge is never higher than the near-surface asthenospheric temperature of ~1250 °C, it is unclear what process(es) can raise the contact temperature by more than 200°C above the mean contact temperature. Among the suggestions made are frictional heating and, possibly, a need to re-calibrate the garnet-clinopyroxene geothermometer. A third suggestion draws an analogy with an iron on an ironing board in which the rear of the iron (ophiolite) moves over an area already heated by the front of the iron (leading edge of the ophiolite sheet), raising the contact temperature.

A TECTONIC FLAKE THAT IS THE WESTERN HALF OF THE OTHRIS OCEANIC RIDGE SYSTEM?

An opportunity to review ideas on the emplacement of the Hellenic ophiolites arose at the Edinburgh Meeting of the Geological Society of London in 1982 (Smith and Spray, 1984). The new findings that were incorporated into the model were the age of the metamorphic sole and the possible relationship of this age to the regional plate boundary network of Jurassic time. The relationship to the plate network will be mentioned only briefly because it is not central to the main theme of this paper, but it is interesting because it suggests that the opening of the central Atlantic was contemporaneous with the western Dinaric-Hellenic ophiolite emplacement, and was a driving force for emplacement. It was speculated that the opening of the central Atlantic between North America and Africa started at the same time as the creation of the metamorphic soles beneath the western Hellenic ophiolites because both were part of the same plate boundary, connected via segments that ran between Iberia and Africa, continued west of Adria (Italy), around the Alps and then turned south into the Dinaric-Hellenic belt.

At about the same time, new age dates suggested that the metamorphic sole under the Semail ophiolite in Oman formed no more than 5 m.y. after its crystallization (Coleman, 1981). The Semail ophiolite was interpreted to have been emplaced as the result of deformation in an ocean bordered by passive margins, as in Greece. The weakest part of the lithosphere that would be affected by deformation less than 5 m.y. after spreading had ceased was thought to be the mid-ocean ridge (Coleman, 1981, fig. 6, p. 2504). This interpretation led to the suggestion that the Semail nappe was one-half of a mid-ocean ridge system.

The continued absence of what was considered convincing evidence for a west-dipping subduction zone in the western Hellenides meant that the tectonic flake model was retained for these ophiolites, but that the detachment fracture was placed at the ridge crest (Smith and Spray, 1984), a location supported by Spray and Roddick's (1980) recognition that ocean-floor amphibolites had formed as part of the sole. The revised flake model showed the metamorphic soles to have been formed by the overriding of the eastern half of the Othris oceanic lithosphere onto the western half. The eastern half then broke to become a tectonic flake, which was then emplaced eastward as a flake onto the Pelagonian continent (Fig. 7). This revised model seems rather clumsy compared with Coleman's elegant model, but is elegance always a property of a correct model?

The principal advance on the previous model was that it predicted that the western Hellenic and Dinaric ophiolites should be asymmetric, representing the eastern half of a former ridge system with the oldest oceanic lithosphere in the east, but no data were available to test this hypothesis. The same model also attempted to account for the apparently identical nature of the eastern Hellenic ophiolites: it was speculated that the eastern and western had originally been part of a much more extensive ophiolite belt that had subsequently separated by a left-lateral transform of late Jurassic age (Smith and Spray, 1984), but this proposal is not considered directly relevant to the emplacement problem discussed here.

At about this time Rassios (1981) completed her Ph.D. on the Vourinos ophiolite and began what would be nearly two decades of detailed structural and petrologic study of the western Hellenic ophiolites, much of it in unpublished EEC reports. In the mid-1980s, structural investigations for chrome ore exploration within the mantle section of Vourinos took a practical turn when unsuccessful attempts to relate ridge-crest extensional deformation to the positions of ore bodies were replaced by a model in which ridge-crest extension was followed almost immediately by compression and initial obduction. Unlike the ridge-crest model, this new model successfully predicted the sub-surface location of chromite bodies (Grivas et al., 1993; Rassios et al., 1994).

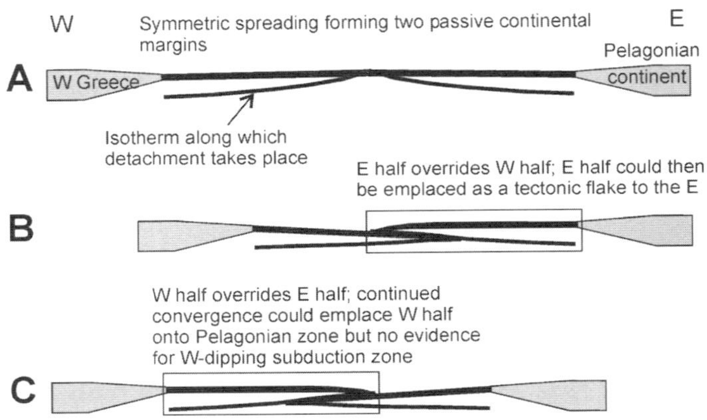

Figure 7. Half-ridge model, with and without a tectonic flake. A: Symmetric spreading generates MORB and forms two passive continental margins. Detachment of the ophiolite takes place at the ridge crest along an isotherm. B: A subduction zone starts at the spreading ridge, dipping east. The eastern half of the ridge overrides the western half. As subduction progresses, the shear exerted on the base of the eastern half of the ridge could break it at the base, forming a flake that is then emplaced onto the Pelagonian zone (as in Fig. 6). There is no obvious mechanism to prevent a continent-continent collision between western and eastern Greece. C: The western half overrides the eastern half and is eventually emplaced onto the Pelagonian zone. The subduction zone would come to a halt when the sinking plate attempts to subduct the Pelagonian continent. This model was rejected because of the lack of evidence for any subduction effects west of the Pelagonian continent. In 7B and 7C, subduction zone magmas could be generated by partial melting of the sinking slab, but these would cut the previously formed MORB or the adjacent continent, an effect that is not observed.

SUPRA-SUBDUCTION ZONE OPHIOLITES?

At the same time that Othris was being mapped in the late 1960s and early 1970s, geochemical work by Pearce and Cann (1973) suggested that trace elements provided a unique fingerprint for the present tectonic setting of basaltic rocks. The method was extended by Beccaluva et al. (1979) and shown to be capable of separating most present-day island arc tholeiites (IAT) from ocean-floor tholeiites (OFT). Their application of this method to ophiolites placed basaltic rocks from lavas, sills, and dikes of the Vourinos and Pindos complexes in the island arc field. Capredi et al. (1980) confirmed this geochemistry for the Pindos ophiolite.

Subsequently, more detailed work confirmed this conclusion and suggested that two magmatic series with island arc affinities could be recognized within the Vourinos ophiolite (Beccaluva et al., 1984). These results implied that the Pindos and Vourinos ophiolites had been formed above a subduction zone as SSZ ophiolites (Pearce et al., 1984); however, Dupuy et al. (1984) suggested that partial melting of a rising mantle diapir could account for the geochemical variations of the Pindos ophiolite without the need for a subduction zone. Partly for these reasons, and partly because the Othris ophiolite appeared to be a strike continuation of the Pindos and Vourinos ophiolites (Smith, 1979), which at the time seemed to be mostly MORB in origin, the possibility of a subduction zone origin was not incorporated into the 1984 emplacement model. Nicolas (1989) has continued to doubt the SSZ origin of many ophiolites, suggesting that the many of their supposed SSZ characteristics result from fast spreading.

Additional evidence for an SSZ origin, however, comes from chromite deposits in the ophiolites. Economically important chromite deposits are rare in MORB ophiolites, but relatively abundant in ophiolites with SSZ characteristics (Roberts, 1988, 1992). Although chromite is present in all the western Hellenic ophiolites (Rassios et al., 1990), in the Pindos ophiolite it has been redistributed by shearing and does not give rise to economically viable deposits (Economou et al., 1986). The major chromite deposits are in the Vourinos Complex; Othris itself has three small exploited deposits, which are associated with the more depleted portions of the Othris mantle rocks (harzburgitic nappe, Fig. 4).

THE WESTERN HELLENIC OPHIOLITES AS THE WESTERN HALF OF AN SSZ OCEAN RIDGE SYSTEM

The acceptance of a SSZ origin for the western Hellenic ophiolites led to radical changes in emplacement models (Fig. 8). Jones and Robertson (1991, fig. 13D, p. 284) illustrated an earlier model of theirs showing an east-dipping intra-oceanic subduction zone within the Pindos Ocean basin, but concluded that this would not give "a feasible mechanism for the eastward emplacement of the ophiolites when viewed on an orogen-wide scale" (Jones and Robertson, 1991, p. 285), though it is not clear to what evidence they refer. They favored a west-dipping intra-oceanic subduction zone that started in the middle of the Pindos Ocean (Jones and Robertson, 1991, fig. 13E, p. 284). Subduction of its eastern half, coupled with "rollback" (discussed below), created an SSZ environment in which a back-arc basin was created. The eastern part of the back-arc basin was then emplaced onto the Pelagonian continent from the west as the Pindos-Vourinos ophiolite, though the details are unclear.

Robertson et al. give a second, more detailed model (1991, figs. 6 and 7, p. 301 and p. 305) in which a west-dipping subduction zone starts in the middle of the late Triassic Pindos Ocean. The Vourinos ophiolite forms to the east of the ridge in the supra-subduction ocean basin and the Pindos ophiolite to the west. Attempted subduction of the Pelagonian continent stopped the process, leaving the western half of the Pindos Ocean available for subduction during Cenozoic orogenesis. The model is also illustrated in Jones et al. (1991, fig. 12, p. 792); however, in the Jones et al. illustration, the ductile deformation in both ophiolites has a similar orientation, with mantle folds and brittle structures verging to the NE, an observation that is difficult to fit into their suggestion that an oceanic ridge separated the Pindos and Vourinos complexes. The two ophiolites are not mirror images: the entire mantle and crustal sequence in Pindos has much less of a SSZ composition than Vourinos, including mantle sections that grade to lherzolite and even plagioclase lherzolite.

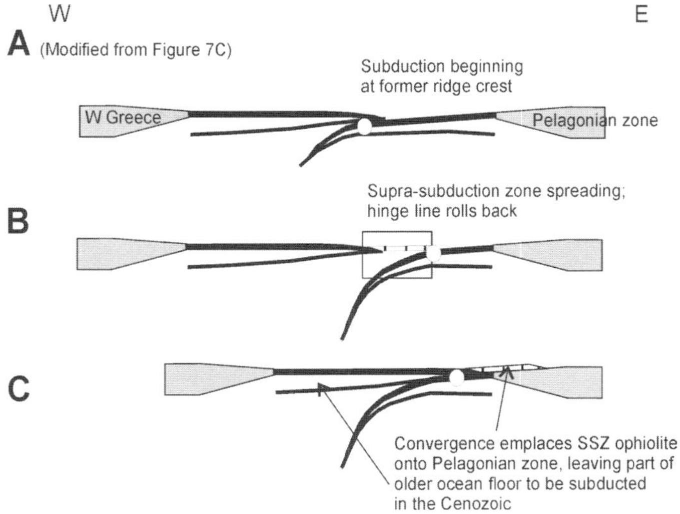

Figure 8. West-dipping subduction zone with rollback, modified from Robertson et al. (1991). A: As with Figure 7C, after some convergence, the overridden slab bends as it sinks into the mantle. The hinge line of the subduction zone is shown as a circle. B: The hinge line of the sinking slab retreats to the east; convergence is less than the motion of the hinge line and the space opened up is occupied by supra-subduction zone magma (vertical stripes). C: Convergence continues until the sinking plate attempts to subduct the Pelagonian continent. The SSZ oceanic lithosphere is emplaced as an ophiolite onto the western edge of the Pelagonian continent. The remainder of the older oceanic lithosphere in the upper plate is available for subduction in the Cenozoic.

Smith (1993, figs. 13A and 13B, p. 234) also accepted the SSZ evidence and somewhat reluctantly postulated a west-dipping subduction zone, but positioned it east of the Pelagonian zone, simply because there seemed to be no evidence for a mid-Jurassic subduction west of the Pelagonian zone. The ophiolites were simply part of a back-arc basin that lay above this subduction zone and had then been subjected to convergence, with a fracture at the spreading center and emplacement of the western half of this spreading system. It must be admitted, however, that, apart from the SSZ composition of the ophiolites, it is still difficult to point to any evidence for a west-dipping mid-Jurassic subduction zone in Greece! The thrust sheets between the ophiolite and the continental margin do mark a former subduction zone in the sense that pre-existing oceanic lithosphere had been destroyed along that zone. The absence of other evidence for a subduction zone east of (or west of) the Pelagonian zone can be rationalized, somewhat arbitrarily, by postulating overprinting or destruction by later (Cenozoic) orogenesis, or by postulating that rollback detaches the evidence of subduction from the overriding slab and carries it with the hinge line of the subducting slab.

The "half-ridge" model seemed an even more likely explanation for the origin of the Hellenic ophiolites if one's viewpoint were expanded to include the Dinaric ophiolites in former Yugoslavia (Fig. 1). There are significant lateral changes in the nature of the ultramafics associated with the ophiolites: those in the west tend to be lherzolitic and those in the east tend to be harzburgitic (Pamic, 1983). These lateral compositional changes indicate a lateral asymmetry to the ophiolites of the kind implied by the half-ridge model. The lherzolitic ultramafics were speculated as being parental to and therefore older than the depleted harzburgite via melting events associated with a subduction zone (Smith, 1993). This inference implied that the original ridge lay to the east. It gave rise to a speculative half-ridge model for the bulk of the Hellenic-Dinaric ophiolites, in which the eastern half of the presumed symmetric oceanic lithosphere had been overridden by the western half, which in turn formed the current ophiolite belt: the eastern half had been subducted. The field evidence for this view was weak, principally because all the mantle sections have been overprinted by a similar, much stronger, emplacement fabric that obscures earlier pre-emplacement fabrics (Grivas et al., 1993); however, a west-dipping subduction zone no longer requires the western Hellenic ophiolites to be tectonic flakes. (One might conclude that the flake model has lived up to its name well.)

The half ridge model incorporates another important element that is a general property of all emplacement models that postulate an intra-oceanic detachment fracture evolving into a subduction zone: oceanic lithosphere forms the hanging wall of the subduction zone. When an Atlantic-type ocean is destroyed as a result of a subduction zone starting somewhere within it, this inevitably leads to the arrival of a continental margin in the footwall of the subduction zone. Because continental lithosphere cannot be subducted to any great extent, its arrival at the subduction zone acts as a brake on the system, causing subduction to cease, as well as emplacing the pre-existing oceanic lithosphere onto the continental margin. Thus, the model appears to account for the metamorphic soles, for the cessation of subduction, and for ophiolite emplacement. Emplacement of dense oceanic lithosphere onto a continental margin is simply a byproduct of a much broader-scale process—the attempted subduction of continental lithosphere at an ocean-continent collision zone. Zimmerman (1972) recognized most of these consequences in one of the earliest plate tectonic models for Vourinos emplacement.

ARE HALF-RIDGE MODELS REALLY APPLICABLE TO GREECE AND OMAN?

There are strong similarities between the events leading to the emplacement of the Jurassic ophiolites of western Greece and the mid-Cretaceous Semail ophiolite of Oman. Because of its excellent exposure and the detailed studies made on it, the Semail ophiolite might be expected to yield an emplacement model that would be applicable to the Hellenic ophiolites. One example is the adoption of the half-ridge model on the basis of Coleman's (1981) interpretation of the location of the metamorphic sole, discussed above.

Evidence for an SSZ setting for the Semail ophiolite became available at about the same time (Pearce et al., 1981), but was not incorporated into Coleman's model. One of the most detailed Oman emplacement models that incorporated an SSZ marginal basin interpretation for the emplacement of the northern Semail ophiolite was that of the Open University group (Lippard et al., 1986, fig. 4-9, p. 153 and fig. 4-11, p. 155). In their model, convergence in the Tethyan ocean between Oman and Makran creates a NE-dipping subduction zone at the Tethyan ridge crest, whose hinge line retreats to the SW. The retreat is faster than convergence, and the resulting space is occupied by an SSZ back-arc basin. Attempted subduction of the Oman continental margin leads to emplacement of the Semail ophiolite onto the margin and the cessation of subduction; however, although Tethyan subduction is initiated at the ridge crest, the fracture that leads to ophiolite emplacement is located at the boundary between the old Tethyan oceanic lithosphere and the new back-arc basin, rather than at the spreading ridge of the back-arc basin, which seems inconsistent. Nevertheless, the fracture that emplaced the ophiolite does not appear to have been at the SSZ spreading center: the available evidence supports a center that lies to the present NE of the Semail outcrop, well away from the SW boundary of the ophiolite. The evidence includes a preponderance of west-sided chilling in the sheeted dikes, though this could be a local effect (Lippard et al., 1986, p. 99); the sense of shear in some mantle tectonites (Lippard et al., 1986, p. 99); and the dip of the layering in cumulus gabbros, which is thought to be toward-ridge (Lippard et al., 1986, fig. 3-3, p. 42). That part of the back-arc basin on the NE side of the ridge, which becomes older to the NE, is assumed to be located offshore beneath the Gulf of Oman (Lippard et al., 1986, fig. 4-11, p. 155). Dating in Oman is not sufficiently precise to confirm this model. In Greece, similar evidence is not available: the cumulate sections in the Hellenic ophiolites are always affected by later rotations related to obduction.

Thus, Coleman's elegant model does not appear to be applicable to the area for which it was proposed. It cannot be rejected as inapplicable to Greece because the relevant data to determine where the ocean ridge lay for the Greek ophiolites are not available; however, if Oman is a good analogue for Greece, the half-ridge model is unlikely to be relevant, and those parts of emplacement models that use it are questionable.

EXTENSION IN A CONVERGENT SETTING: A SIMPLE EMPLACEMENT MODEL

As noted above, the initial estimate for the time difference between ocean lithosphere crystallization and metamorphic sole formation for Oman was ~5 m.y. Recent dating shows that this time difference is actually <1 m.y. (Hacker et al., 1996). The same may be true of Greece, but the existing data are unconvincing. It is quite unclear how such a rapid change in plate motions can come about, unless it is an inherent part of the ophiolite emplacement system.

In effect, any SSZ system is an example of rollback, which is simply the retreat of the hinge line of a subduction zone initially at a continental (or island-arc) margin. Elsasser (1971) discussed the kinematics involved, referring to the hinge line retreat as "retrograde" motion of the lithospheric slab. As the hinge line retreats, the potential space that is created is occupied by a supra-subduction zone marginal basin. If the subducting slab is attached to a converging continent (A) that is approaching a second continent (B), then extension in the SSZ basin is contemporaneous with convergence. Extension will cease once the hinge line no longer migrates away from B. It will pass imperceptibly into convergence if the hinge line starts to move back to B, starting the emplacement of the edge of the SSZ margin basin onto the continental margin. Convergence will cease once the attempted subduction of the edge of A at the subduction zone ceases. This model says nothing about the reasons for changes in the motion of the hinge line. It is merely a kinematic description.

In Greece, the Pelagonian continent is assumed to have broken away from western Greece in Triassic time, with the creation of two passive margins separated by the Pindos Ocean (Fig. 9A). The slope and rise of the eastern margin are represented by the Othris fan and pelagic basin sediments, and its shelf is represented by the Triassic-Jurassic platform carbonates (Fig. 5). The conjugate western margin is formed by the carbonate platforms of western Greece. The Triassic and younger pelagic sediments in the Pindos Ocean are represented by the Pindos zone (Green, 1982). Triassic MORB basalt clasts in the Abdella mélange support this general interpretation.

However, a large continental fragment, the Parnassos platform, lies within the Pindos Ocean (Robertson et al., 1991, figs. 6A, B, p. 301; fig. 2) and seems to have played a pivotal role in determining the location of the convergent plate margin. There are no ophiolites to the west of the Parnassos platform, but there are extensive outcrops of Pindos sediments (Fig. 2), which at the time were lying on Triassic ocean floor. It is assumed that the mid-Jurassic subduction zone was initiated at the eastern edge of the Parnassos platform (Fig. 9B) and that its northern continuation lay within the Pindos Ocean as shown by Robertson et al. (1991, figs. 6A, B, p. 301), but with a location that was not necessarily at the Triassic spreading center. This subduction zone began to roll back and, as convergence continued, metamorphic soles were created at the subduction zone (Fig. 9C). Eventually, the rate of convergence exceeded the rate of retreat of the hinge line of the subduction zone and ophiolite emplacement began (Fig. 9D). In essence, the thrust sheets in the Abdella mélange are small accretionary prisms formed at a subduction zone (Robertson et al., 1991). The more easterly thrust stack in central Othris shows that the Othris ophiolite overrode the subducted continental margin sequence of the Pelagonian continent. Further east still, in eastern Othris and in Vourinos, the ophiolite has moved over the presumed continental margin sediments and now lies tectonically on the Pelagonian platform.

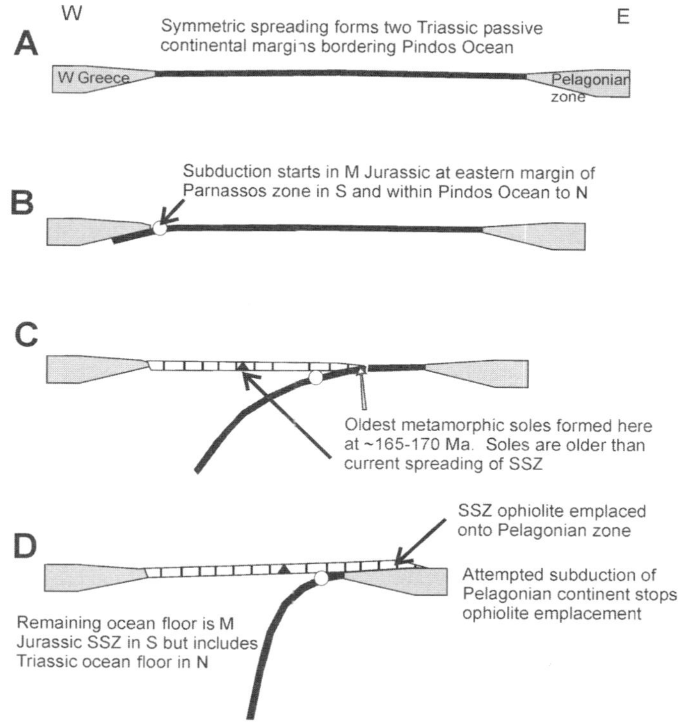

Figure 9. The simplest emplacement model that appears compatible with the data. A: Symmetric spreading in Triassic time to create a pair of conjugate passive margins (as in Fig. 7A). B: Subduction starts at the western continental margin. C: Convergence of continent A and simultaneous rollback of the subduction zone creates supra-subduction zone oceanic lithosphere in the upper plate. D: Continued convergence emplaces the SSZ oceanic lithosphere as ophiolites onto the Pelagonian continent, whose attempted subduction brings the subduction process to a halt.

The hanging wall of the subduction zone was the mid-Jurassic oceanic lithosphere created above the subduction zone; the footwall was the presumed Triassic Pindos oceanic lithosphere and its sedimentary cover. An intra-oceanic fracture is not fundamental to the model, but arises as an accidental consequence of the limited geographic extent of the Parnassos platform. Mid-Jurassic subduction left behind a wedge-shaped remnant of the Pindos Ocean, which narrowed to the north. Much of the Cenozoic tectonics of western Greece results from the subduction of the remaining Pindos Ocean with a pronounced narrowing of the Pindos outcrop to the north (Fig. 2). This new model is a modification of one recently applied to the western Hellenic ophiolites (Rassios and Smith, 2000).

WHY WAS SUCH A SIMPLE MODEL THOUGHT TO BE INAPPLICABLE TO HELLENIC OPHIOLITES?

The major stumbling block to adopting such a simple model has always been the lack of convincing evidence for a west-dipping subduction zone west of the Pelagonian zone. In western

Greece, we are driven to conclude from geochemical evidence that parts of the Othris, Pindos, and Vourinos ophiolites formed above a subduction zone, and to conclude from structural evidence that all have been emplaced from the west. There certainly is an abundance of mélange sediments to the west, but "the model" demands that there be some further subduction zone evidence, such as arc magmatism, somewhere west of (or possibly east of) the Pelagonian zone. In reality, we can find no trace of arc formation in either direction. We can *assume* that this evidence has been conveniently destroyed or that there was insufficient subduction to generate the volcanics, but perhaps it was never there—and the SSZ ophiolites themselves actually are that missing evidence.

At first, not all of the relationships in the new model were known, but those that were, were explained by flake tectonics, then by a half-ridge model. Neither of these models provides a satisfying account for either the SSZ nature of the ophiolites or for the rapid transition from extension to convergence or their possible contemporaneity; however, their geochemistry and geometry are almost inevitable consequences of extension in the convergent setting discussed above.

Petrologically, the basaltic and ultramafic components of all ophiolites that were first investigated showed what appeared to be normal MORB characteristics. Only subsequently was the SSZ nature of a significant part of them recognized. The Pindos and Vourinos ophiolites show much more evidence of subduction activity. In them and in the harzburgites of western Othris, the subduction components that express themselves as andesitic magmas in a normal island arc appear to have been incorporated into the mafic and ultramafic rocks of these ophiolites, rather than intruded as later magmas into pre-existing oceanic lithosphere. It is speculated that the SSZ nature of the ophiolites may reflect the fact that the two magma sources that are present in a "normal" island arc and back-arc basin—an andesitic source and a spreading ridge—have not yet clearly separated from one another in the Hellenic ophiolites. The western Hellenic ophiolites may therefore represent early stages in the growth of an island arc and a back-arc basin in which two magma sources partly shared the same plumbing system.

General Properties of This Simple Model (Fig. 9)

- The initial deformation is due to the initiation of subduction at a passive margin (Fig. 9B).
- The initial record of deformation may be detached from the margin and may migrate with the retreating hinge line of the subduction zone (Fig. 9C).
- The space created by rollback will be occupied by newly formed oceanic lithosphere, which is likely to have SSZ characteristics.
- The initial phase of subduction will automatically emplace an ophiolite onto the conjugate margin (Fig. 9D).
- The model provides an automatic brake on the first phase of an orogeny—due to the attempted subduction of the conjugate continental margin beneath the ophiolite.
- In the first phase of orogeny, there is no continent-continent collision and no large-scale regional metamorphism, although an island arc may be present if the ocean is sufficiently wide.
- Because convergence and extension are contemporaneous, it is possible to envisage that some soles that were created in the initial phase of convergence are older than the youngest spreading in the SSZ basin.
- Given that the net effect of orogeny is to annihilate oceanic lithosphere, a second orogenic phase is required where the initial orogeny fails to emplace the entire back-arc basin. The entire sequence of subduction, rollback, and ophiolite emplacement could be repeated. Alternatively, the remaining oceanic lithosphere could be eliminated by convergence without rollback, a process that is likely to lead to continent-continent collision and regional metamorphism, and may also involve strike-slip faulting.

The model may be a critical link in the reconstruction of Tethyan convergence: modifications should be applicable along the entire western Hellenic-Dinaric ophiolite belt and may be applicable to other Mesozoic ophiolites in the Tethys. For example, we outline here a new model for the emplacement of the Oman ophiolite, in which subduction of the Tethys starts at the Makran continental margin rather than within the Tethys, and which assigns the Permian and Triassic "exotics" and associated Triassic volcanics to pieces detached from the Makran margin, rather than to oceanic islands within the Tethys (Lippard et al., 1986). The ophiolite is emplaced by the continued action of this subduction zone, rather than by a required second fracture at the boundary between the Tethyan oceanic lithosphere and the new supra-subduction lithosphere, as in the model of the Open University group. The rather awkward mechanism for hiding that part of the old Tethyan lithosphere lying between the Oman continental margin and the new oceanic lithosphere of the ophiolite is then unnecessary. There is no half-ridge and only one subduction zone—elegance returns, but field testing is required to see if it is true.

ACKNOWLEDGMENTS

A. Rassios gives special thanks to E. Grivas, I. Vacondios, G. Konstantopoulou, Y. Dilek, and E.M. Moores. A.G. Smith thanks the organizers of the Pardee Symposium for the invitation to attend, and St. John's College, Cambridge, UK, for financial support. Both authors thank J.F. Dewey, Y. Dilek, M.F.J. Flower, and S. Newcomb for their encouragement and helpful reviews.

REFERENCES CITED

Anonymous, 1972, Penrose Field Conference on ophiolites: Geotimes, v. 17, no. 12, p. 24–25.

Beccaluva, L., Ohnenstetter, D., and Ohnenstetter, M., 1979, Geochemical discrimination between ocean floor and island arc tholeiites: application to some ophiolites: Canadian Journal of Earth Sciences, v. 16, p. 1874–1882.

Beccaluva, L., Ohnenstetter, D., Ohnenstetter, M., and Paupy, A., 1984, Two magmatic series with island arc affinities within the Vourinos ophiolite: Contributions to Mineralogy and Petrology, v. 85, p. 253–271.

Boué, A., 1834, Résumé des progrès des sciences géologiques pendant l'année 1832: Bulletin de la Société géologique de France, v. Première série, p. 346–383.

Boué, A., 1840, La Turquie d'Europe: Paris.

Brunn, J.H., 1956, Contribution a l'étude géologique du Pinde septentrional et de la Macédoine occidentale: Annales Géologiques des Pays Helléniques, v. 7, p. 1–358.

Brunn, J.H., 1959, La dorsale médio-atlantique et les épanchements ophiolitiques: Compte rendues sommaire des sciences de la Société géologique de France, p. 234–237.

Capedri, S., Venturelli, G., Bocchi, G., Dostal, J., Garuti, G., and Rossi, A., 1980, The geochemistry and petrogenesis of an ophiolite sequence from Pindos, Greece: Contributions to Mineralogy and Petrology, v. 74, p. 189–200.

Coleman, R.G., 1981, Tectonic setting for ophiolite obduction in Oman: Journal of Geophysical Research, v. 86, p. 2497–2508.

Danelian, T., 1994, Palaeo-oceanographic significance of mid-Jurassic radiolarites from the Maliac (Sub-Pelagonian) margin of Othris (Greece): Bulletin of the Geological Society of Greece, v. XXX, p. 165–175.

Danelian, T., and Robertson, A.H.F., 2001, Neotethyan evolution of eastern Greece (Pagondas Mélange, Evia island) inferred from radiolarian biostratigraphy and the geochemistry of associated extrusive rocks: Geological Magazine, v. 138, p. 345–363.

Dupuy, C., Dostal, J., Capedri, S., and Venturelli, G., 1984, Geochemistry and petrogenesis of ophiolites from northern Pindos (Greece): Bulletin Volcanologique, v. 47, no. 1, p. 39–46.

Economou, M., Dimou, E., Economou, G., Migiros, G., Vacondios, I., Grivas, E., Rassios, A., and Dabitzias, S., 1986, Chromite deposits of Greece, UNESCO IGCP 197: Metallogeny of Ophiolites: Athens, Greece, Theophrastus Publications, S.A., 129–159 p.

Elsasser, W.M., 1971, Sea floor spreading as thermal convection: Journal of Geophysical Research, v. 76, p. 1101–1112.

Ferrière, J., 1974, Etude géologique d'un secteur des zone helléniques internes subpelagonienne et pelagonienne (massif de l'Othrys-Grece continentale). Importance et signification de la période orogenique ante-Cretace superieur: Bulletin de la Societe Geologique de France, v. XVI, p. 543–562.

Ferrière, J., 1982, Paléogéographie et tectoniques superposées dans les Hellénides internes: les massif de l'Othrys et du Pelion (Grèce septentrional): Société Géologique du Nord, v. 8, p. 1–970.

Gradstein, F.M., and Ogg, J.M., 1996, A Phanerozoic time scale: Episodes, v. 19.

Green, T.J., 1982, Structural and sedimentological studies of the Pindos zone, central Greece [Ph.D. thesis]: Cambridge, University of Cambridge, 238 p.

Grivas, E., Rassios, A., Konstantpoulou, G., Vacondis, I., and Vrahatis, G., 1993, Drilling for "blind" podiform chrome orebodies at Voidolakkos in the Vourinos ophiolite complex: Economic Geology, v. 88, p. 461–468.

Hacker, B.R., Mosenfelder, J.L., and Gnos, E., 1996, Rapid emplacement of the Oman ophiolite: thermal and geochronologic constraints: Tectonics, v. 15, p. 1230–1247.

Hess, H.H., 1955, Serpentines, orogeny and epeirogeny: Boulder, Colorado, Geological Society of America Special Paper, v. 62, p. 391–408.

Hynes, A.J., 1972, The geology of part of the western Othris Mountains [Ph.D. thesis]: Cambridge, University of Cambridge, 196 p.

Hynes, A.J., 1974, Igneous activity at the birth of an ocean basin in eastern Greece: Canadian Journal of Earth Sciences, v. 11, p. 842–853.

Hynes, A.J., Nisbet, E.G., Smith, A.G., Welland, M.J.P., and Rex, D.C., 1972, Spreading and emplacement ages of some ophiolites in the Othris region (eastern central Greece): Zeitschrift Deutsche Geologische Gesellschaft, v. 123, p. 455–468.

Jones, G., and Robertson, A., H.F., 1991, Tectono-stratigraphy and evolution of the Pindos ophiolite and associated units: Journal of the Geological Society of London, v. 148, p. 267–268.

Jones, G., Robertson, A.H.F., and Cann, J.R., 1991, Genesis and emplacement of the supra-subduction zone Pindos ophiolite, northwestern Greece, *in* Peters, T., Nicolas, A., and Coleman, G., eds., Ophiolite genesis and the evolution of the oceanic lithosphere: London, Kluwer Academic Publishers, Petrology and Structural Geology, vol. 5, p. 771–800.

Lippard, S.J., Shelton, A.W., and Gass, I.G., 1986, The ophiolite of northern Oman: London, Geological Society of London Memoir v. 11, p. 178.

McKenzie, D.P., and Parker, R.L., 1967, The north Pacific: an example of tectonics on a sphere: Nature, v. 216, p. 1276–1280.

Menzies, M., 1976, Rifting of a Tethyan continent—rare earth evidence of an accreting plate margin: Earth and Planetary Science Letters, v. 28, p. 427–438.

Menzies, M., and Allen, C., 1974. Plagioclase lherzolite-residual mantle: relationships within two eastern Mediterranean ophiolites: Contributions to Mineralogy and Petrology, v. 46, p. 454–487.

Morgan, W.J., 1968, Rises, trenches, great faults and crustal blocks: Journal of Geophysical Research, v. 73, p. 1959–1982.

Moores, E.M., 1969, Petrology and structure of the Vourinos ophiolitic complex of northern Greece: Boulder, Colorado, Geological Society of America Special Paper 118, p. 74.

Moores, E.M., and Vine, F.J., 1971, Troodos massif, Cyprus and other ophiolites as oceanic crust: evaluation and implications: Philosophical Transactions of the Royal Society of London, v. 268, p. 443–466.

Neumayr, M., 1880, Der geologische Bau des westlichen Mittelgriechenland: Denschriften der Osterreichischen Akademie der Wissenschaften Mathematisch-Naturwissenschaftliche Klasse, v. 40, p. 91–128.

Nicolas, A., 1989, Structures of ophiolites and dynamics of oceanic lithosphere: London, Kluwer Academic, 367 p.

Oxburgh, E.R., 1972, Flake tectonics and continental collisions: Nature, v. 239, p. 202–204.

Pamic, J., 1983, Considerations on the boundary between lherzolite and harzburgite subprovinces in the Dinarides and northern Hellenides: Ofioliti, v. 8, p. 153–164.

Pearce, J.A., and Cann, J.R., 1973, Tectonic setting of basic volcanic rocks using trace element analysis: Earth and Planetary Science Letters, v. 19, p. 290–300.

Pearce, J.A., Alabaster, T., Shelton, A.W., and Searle, M.P., 1981, The Oman ophiolite as a Cretaceous arc-basin complex: Philosophical Transactions of the Royal Society of London, v. A3, p. 299–317.

Pearce, J.A., Lippard, S.J., and Roberts, S., 1984, Characteristics and tectonic significance of supra-subduction zone ophiolites, *in* Kokelaar, B.P., and Howells, M.F., eds., Marginal basin geology: Oxford, Blackwell Scientific Publications, Special Publication, Geological Society of London, p. 77–94.

Rassios, A., 1981, Geology and evolution of the Vourinos Complex, northern Greece [Ph.D. thesis]: University of California (Davis), 499 p.

Rassios, A., and Smith, A.G., 2000, Constraints on the formation and emplacement age of western Greek ophiolites (Vourinos, Pindos and Othris) inferred from deformation structures in peridotites, *in* Dilek, Y., Moores, E.M., Elthon, D., and Nicolas, A., eds., Ophiolites and oceanic crust: new insights from field studies and ocean drilling program: Boulder, Colorado, Geological Society of America Special Paper 349, p. 473–483.

Rassios, A., Konstantopoulou, G., and Vakondis, I., 1990, Tectonic controls on chrome ore localization in ophiolites: Athens, Greece, Institute of Mining and Mineral Exploration, p. 228.

Rassios, A., Grivas, E., Konstantopoulou, G., and Vacondios, I., 1994, The geometry of structures forming around the ductile-brittle transition in the Vourinos-Pindos-Othris oceanic slab: Bulletin of the Geological Society of Greece, v. XXX, no. 2, p. 109–121.

Roberts, S., 1988, Ophiolite chromite formation: a marginal basin phenomenon: Economic Geology, v. 83, p. 1034–1036.

Roberts, S., 1992, Influence of the partial melting regime on the formation of ophiolitic chromitite, *in* Parson, L.M., Murton, B.J., and Browning, P., eds., Ophiolites and their modern oceanic analogues: London, Volume Special Publication 60, Geological Society of London Special Publication, v. 60, p. 203–217.

Robertson, A.H.F., Clift, P.D., Degnan, P.J., and Jones, G., 1991, Palaeogeographic and palaeotectonic evolution of the Eastern Mediterranean Neotethys: Palaeogeography, Palaeoclimatology, Palaeoecology, v. 87, p. 289–343.

Shallo, M., 1992, Geological evolution of the Albanian ophiolites and their platform periphery: Geologische Rundschau, v. 81, p. 681–694.

Smith, A. G., 1979, Othris, Pindos and Vourinos ophiolites and the Pelagonian zone, in Kallergis, G., ed., Proceedings of the VI Colloquium on the geology of the Aegean region, Athens 1977: Athens, Greece, Institute of Geological and Mining Research, v. 3, p. 1369–1374.

Smith, A.G., 1993, The tectonic significance of the Hellenic-Dinaric ophiolites: London, Special Geological Society of London Special Publication, v. 76, p. 213–243.

Smith, A.G., and Spray, J.G., 1984, A half-ridge transform model for the Hellenic-Dinaric ophiolites: London, Geological Society of London Special Publication, v. 17, p. 629–644.

Smith, A.G., and Woodcock, N.H., 1976, Emplacement model for some 'Tethyan' ophiolites: Geology, v. 4, p. 653–656.

Smith, A.G., Hynes, A.J., Menzies, M., Nisbet, E.G., Price, I., Welland, M.J.P., and Ferrière, J., 1975, The stratigraphy of the Othris mountains, eastern central Greece: a deformed Mesozoic continental margin sequence: Eclogae Geologicae Helvetiae, v. 68, p. 463–481.

Smith, A.G., Woodcock, N.H., and Naylor, M.A., 1979, The structural evolution of a Mesozoic continental margin, Othris mountains, Greece: Journal of the Geological Society of London, v. 136, p. 589–603.

Spray, J.G., and Roddick, J.C., 1980, Petrology and 40Ar/39Ar geochronology of some Hellenic sub-ophiolitic metamorphic rocks: Contributions to Mineralogy and Petrology, v. 72, p. 43–55.

Spray, J.G., Bebien, J., Rex, D.C., and Roddick, J.C., 1984, Age constraints on the igneous and metamorphic evolution of the Hellenic-Dinaric ophiolites, *in* Dixon, J.E., and Robertson, A.H.F., eds., The geological evolution of the Eastern Mediterranean: London, Geological Society of London Special Publication, v. 17, p. 619–627.

Steinmann, G., 1905, Geologische Beobachtungen in den Alpen. II. Die Schardt'sche Uberfaltungstheorie und die Bedeutung der Tiefseeabsatze und der ophiolithischen Maasengesteine: Berichte der Naturforschenden Gesellschaft zu Freiburg, v. i, Bd. 16, p. 44–65.

Stoneley, R., 1975, On the origin of ophiolite complexes in the southern Tethys region: Tectonophysics, v. 25, p. 303–322.

Vine, F.J., and Matthews, D.H., 1963, Magnetic anomalies over oceanic ridges: Nature, v. 199, p. 947–949.

Williams, H., and Smyth, W.R., 1973, Metamorphic aureoles beneath ophiolite suites and Alpine peridotites: tectonic implications with west Newfoundland examples: American Journal of Science, v. 273, p. 594–621.

Zimmerman, J., Jr., 1972, Emplacement of the Vourinos ophiolitic complex, northern Greece, in Shagam, R., Hargraves, R.B., Morgan, W.J., Van Houten, F.B., Burk, C.A., Holland., H.D., and Holister, L.C., eds., Studies in earth and space sciences: Boulder, Colorado, Geological Society of America Memoir 132, p. 225–239.

Manuscript Accepted by the Society May 2, 2003

Printed in the USA

Geological Society of America
Special Paper 373
2003

Development of the ideas on the origin of Albanian ophiolites

Minella Shallo
Fakulteti i Gjeologjise dhe Minierave, Universiteti Politeknik, Tirana, Albania

Yildirim Dilek
Department of Geology, Miami University, Oxford, Ohio 45056, USA

ABSTRACT

The Jurassic Albanian ophiolites represent the remnants of the Mesozoic Neo-Tethyan ocean and occur within the Dinaride-Hellenide segment of the Alpine orogenic system. Studies of Albanian ophiolites have produced significant results to better understand the magmatic, metamorphic, and tectonic processes involved in the formation of Tethyan ocean floor and its emplacement on continental margins. Albanian ophiolites form two subparallel, adjacent belts that display different pseudostratigraphy, internal architecture, and chemical affinities. The western belt consists of several massifs that are ~3 km in thickness and composed of lherzolitic peridotites, gabbros, and extrusive rocks with a MORB affinity. The eastern belt ophiolites are ~10 km in thickness and include a Penrose-type pseudostratigraphy complete with a ~1.5-km-thick sheeted dike complex in the Mirdita region. The upper mantle rocks of the eastern-type consist mainly of harzburgite and dunite, and the extrusive sequence includes basalt, basaltic andesite, dacite, and rhyodacite showing an island arc tholeiite affinity. The differences between the upper mantle sequences of the paired ophiolite belts were first recognized in the late 1960s, mainly through careful field mapping and petrographic studies, which led in the 1980s to the discovery of the chemically distinct western- and eastern-type ophiolites. Fixist models prior to the late 1960s considered these ophiolites as submarine eruptions and intrusions on and across the Pelagonian carbonate platform. With the late arrival of the plate tectonic revolution in Albania in the 1980s came the recognition of a sheeted dike complex indicating a seafloor spreading origin, and the root zones of the Albanian ophiolites were established. Additional systematic petrological studies in the mid-1980s confirmed a subduction zone setting for the eastern-type ophiolites. A number of models have been advanced since the 1990s to explain the occurrence of these contemporaneous paired ophiolite belts with MORB and suprasubduction zone affinities. The location, polarity, and timing of the initiation of the inferred subduction zone are critical questions for the origin of the Albanian ophiolites, which provide an excellent opportunity to investigate spatial and temporal relationships in melt evolution of fossil oceanic crust within Tethyan-type small ocean basins.

Keywords: Albanian geology, Dinarides-Hellenides, Jurassic oceanic crust, paired ophiolite belts, MORB vs. suprasubduction zone ophiolites, Neo-Tethys, Pindos-Mirdita ocean basin

INTRODUCTION

This paper is an essay documenting the nature, chronology, and development of ideas on the origin of the Albanian ophiolites and how these ideas have kept pace with and complemented the evolving ophiolite concept during the last century. This historical account of ophiolite studies in Albania is important mainly for two reasons. First, the Albanian ophiolites form a critical transition from the Jurassic Alpine ophiolites in the northwest to the Jurassic-Cretaceous Neo-Tethyan ophiolites in the eastern

Shallo, M., and Dilek, Y., 2003, Development of the ideas on the origin of Albanian ophiolites, *in* Dilek, Y., and Newcomb, S., eds., Ophiolite concept and the evolution of geological thought: Boulder, Colorado, Geological Society of America Special Paper 373, p. 351–363. For permission to copy, contact editing@geosociety.org.

Mediterranean region to the southeast. The Alpine ophiolites represent the remnants of intracontinental rift basins (Caby, 1995; Froitzheim and Manatschal, 1996; Manatschal and Nievergelt, 1997), whereas the eastern Mediterranean ophiolites are fragments of fossil oceanic lithosphere developed by seafloor spreading within restricted ocean basins (Dilek and Moores, 1990; Dilek et al., 1999). Therefore, the Albanian ophiolites form a significant geological and geodynamic link between these two different domains within the Neo-Tethyan realm. Thus, a better understanding of the evolution of these ophiolites in Albania can help us elucidate the magmatic, tectonic, and chemical processes involved in oceanic crust generation in Tethyan-type small ocean basins. The second reason for the importance of this historical account is that Albania as a country has lived through major political changes and autocratic regimes, which at times have resulted in extreme political and economic isolation of its people and scientists. This restriction, in turn, strongly influenced the mode of scientific inquiry and the nature of scientific investigations in the country. A strong push for self-reliance in search of natural resources and ore bodies in the country helped the ophiolite investigations advance at a fast pace during the 1960s through 1980s. Field mapping and petrological (mainly petrographic) studies in ophiolites by Albanian geologists led to some crucial observations and objective evidence that to this day have significant implications for the modern ophiolite concept. Although the plate tectonic revolution was nearly 15 years late arriving in Albania, independent observations by Albanian field geologists prior to this time were on a par with the then current thoughts on ophiolites and related processes. This phenomenon underscores the significance of careful field observations in ophiolite investigations and the "role of objective evidence in the scientific enterprise" (see Şengör, this volume, Chapter 22). Opening of the Albanian political system to the western world in the early 1990s ushered in an era of mutually beneficial scientific collaboration with western scientists on the Albanian ophiolites and some potentially major advances in our understanding of the evolution of ophiolitic magmas in Tethyan rift basins and subduction-accretion cycles in Tethyan-type orogenic belts.

In the first part of the paper we present a geological outline of the Albanian ophiolites. In the second part we document the development of ideas on the origin of Albanian ophiolites in two main time frames, prior to plate tectonics and after the arrival of the plate tectonic revolution in Albania. The final part of the paper presents a discussion on recent findings and models of the Albanian ophiolites, and their contribution to the evolving ophiolite concept.

GEOLOGICAL OUTLINE OF THE ALBANIAN OPHIOLITES

The Jurassic Albanian ophiolites are part of the Dinaride-Hellenide segment of the Alpine orogenic system and occur in a 250-km-long, northerly zone extending through the Mirdita region between the Pelagonian and Apulian microcontinents (Fig. 1). This belt continues southward into the Subpelagonian-Pindos zone ophiolites in the Hellenides (Greece; also see Smith and Rassios, this volume, Chapter 19) and northward into the Dinaride ophiolites in Serbia and Macedonia. The age of the Albanian ophiolites is determined as Middle-Late Jurassic on the basis of paleontological and radiometric age data (Shallo, 1992). The fault contacts between the Albanian ophiolites and the platform carbonates in the east and the west are commonly marked by thin slivers of metamorphic rocks ("metamorphic soles"; Carosi et al., 1996), with ages ranging from 160 to 174 Ma (Middle Jurassic) based on $^{40}Ar/^{39}Ar$ dates (Vergély et al., 1998; Dimo-Lahitte et al., 2001), and a mélange composed of Upper Triassic volcano-sedimentary rocks and blocks of the platform carbonates (Figs. 2 and 3).

The Albanian ophiolites display great lithological, structural, and chemical differences both along- and across-strike of this zone, and they are grouped into two main types (Shallo et al., 1985; Shallo, 1990a; Figs. 2 and 3). The western-type ophiolites are up to 3 km in thickness and consist of a thick upper mantle unit composed of lherzolite-harzburgite, plagioclase lherzolite, plagioclase dunite, and rare amphibole peridotite, overlain by a plutonic sequence that includes troctolite, olivine gabbro, ferrogabbro, gabbro, and scarce plagiogranite (Fig. 4). The common crystallization order in these plutonic rocks is ol-pl-px (Beccaluva et al., 1994). Volcanic rocks making up the upper crustal unit in the western-type ophiolites are represented by massive and pillow lavas with basaltic to basaltic-andesitic compositions having a MORB affinity. These extrusive rocks are overlain by radiolarites of Bajocian-Bathonian age; locally the lavas rest directly on partially serpentinized upper mantle rocks.

The eastern-type ophiolites have an intact ophiolite succession, following the 1972 Penrose definition (Anonymous, 1972), with a total thickness of 10 km. They consist, from bottom to top, of harzburgite, harzburgite-dunite interlayers, and dunite overlain by pyroxenite, layered gabbronorite, massive gabbronorite, gabbro and amphibole gabbro, quartz diorite-plagiogranite, sheeted dikes, and volcanic rocks (Fig. 4). The common crystallization order in these plutonic rocks is ol-cpx-pl (Beccaluva et al., 1994). The extrusive rocks include basalt, basaltic-andesite, andesite, dacite, and rhyodacite displaying an island arc tholeiite (IAT) chemistry. Locally in the northern Mirdita zone, boninitic rocks occur within both the sheeted dike complex and the extrusive sequence (Shallo, 1990b). The volcanic rocks in the eastern-type ophiolites are, in turn, overlain by chert and radiolarites of the late Bathonian-Oxfordian age (Marcucci et al., 1994).

Initial displacement and partial emplacement of the Albanian ophiolites onto the peripheral carbonate platforms occurred during the Tithonian and were accompanied by the formation of flyschoidal sedimentary rocks (Tithonian-Lower Cretaceous) onlapping both ophiolite types and the platform carbonates (Shallo, 1990c; Fig. 4). The Pindos-Mirdita basin remained partly open until the Early Tertiary, when continental convergence (Alpine orogeny) caused its final closure and resulted in the westward thrusting of the Albanian ophiolites over the Apulian terranes (Robertson and Shallo, 2000).

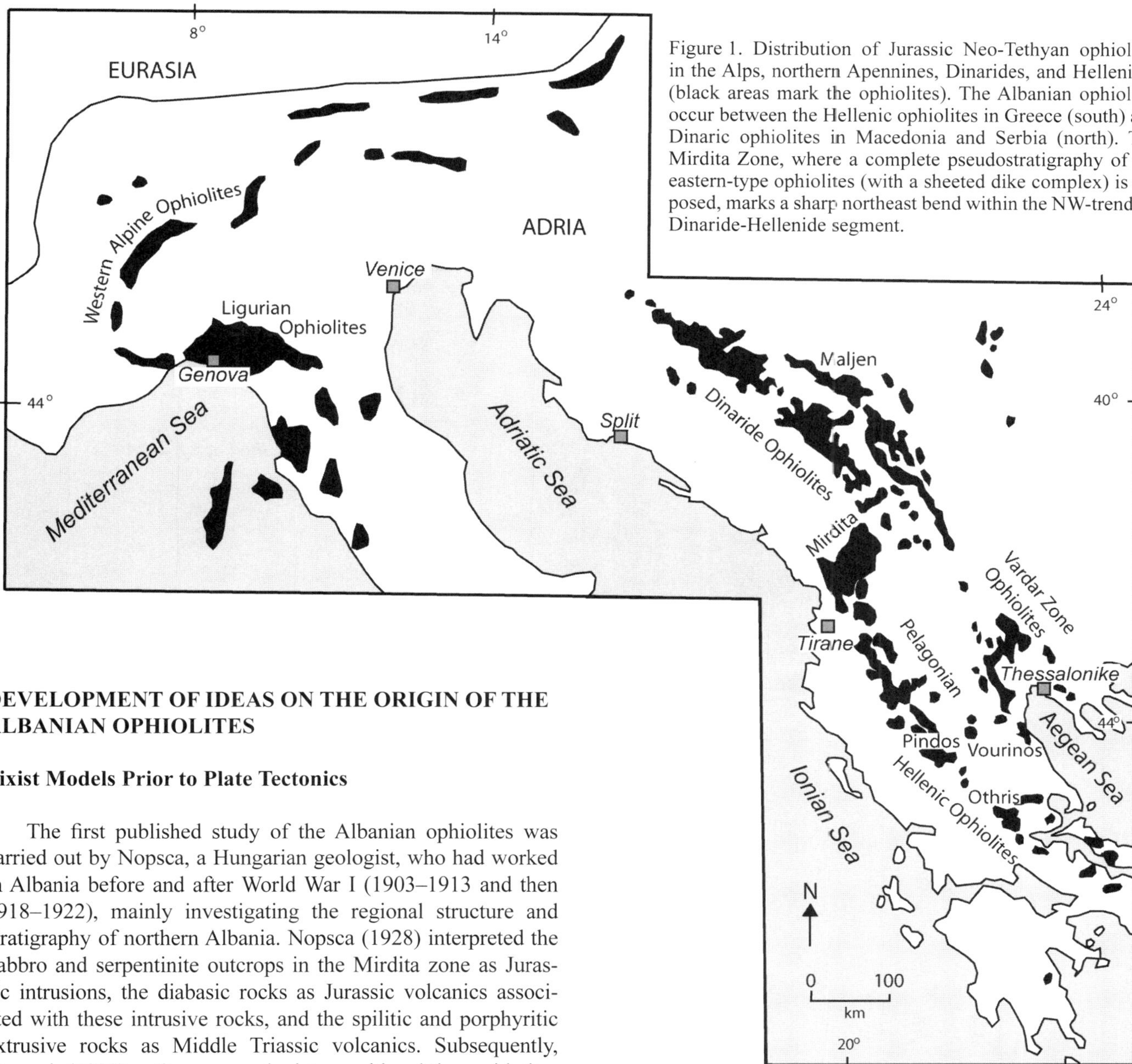

Figure 1. Distribution of Jurassic Neo-Tethyan ophiolites in the Alps, northern Apennines, Dinarides, and Hellenides (black areas mark the ophiolites). The Albanian ophiolites occur between the Hellenic ophiolites in Greece (south) and Dinaric ophiolites in Macedonia and Serbia (north). The Mirdita Zone, where a complete pseudostratigraphy of the eastern-type ophiolites (with a sheeted dike complex) is exposed, marks a sharp northeast bend within the NW-trending Dinaride-Hellenide segment.

DEVELOPMENT OF IDEAS ON THE ORIGIN OF THE ALBANIAN OPHIOLITES

Fixist Models Prior to Plate Tectonics

The first published study of the Albanian ophiolites was carried out by Nopsca, a Hungarian geologist, who had worked in Albania before and after World War I (1903–1913 and then 1918–1922), mainly investigating the regional structure and stratigraphy of northern Albania. Nopsca (1928) interpreted the gabbro and serpentinite outcrops in the Mirdita zone as Jurassic intrusions, the diabasic rocks as Jurassic volcanics associated with these intrusive rocks, and the spilitic and porphyritic extrusive rocks as Middle Triassic volcanics. Subsequently, Nowack (1929), a German geologist, considered the peridotites and serpentinites in the Mirdita zone as the product of a primary magma, and the other associated ophiolitic rocks as the derivatives of these ultramafic rocks. Based on limited geological mapping and field observations, Nowack expressed the idea that peridotites might represent the exhumed "sima" now exposed at the surface in contact with the sedimentary rocks, and that the Jurassic gabbros and diabasic rocks were the intrusive and extrusive end products of this sima, respectively. Nowack further suggested that limited eruptions of the ophiolitic lavas might have occurred until the Paleogene, although he recognized that the main eruptions of the ophiolitic series were of Jurassic age. At the end of World War II (1944), Albania became liberated from the German-Italian occupation and fell under Communist Party rule; Enver Hoxha became the first leader of this party and also the Prime Minister of the country at this time.

After World War II, strong political and military relationships developed between Albania and the former Soviet Union, and geologists from the USSR started working in Albania. They were basically doing systematic geological mapping and exploration of ore deposits, mainly chromite and massive sulfide deposits in the ophiolite massifs. The Soviet geologists (i.e., Nejellov, 1951; Zhadan, 1959; Kollmak, 1959) carried out geological mapping at a scale of 1:50,000 in the main part of the Mirdita zone during a

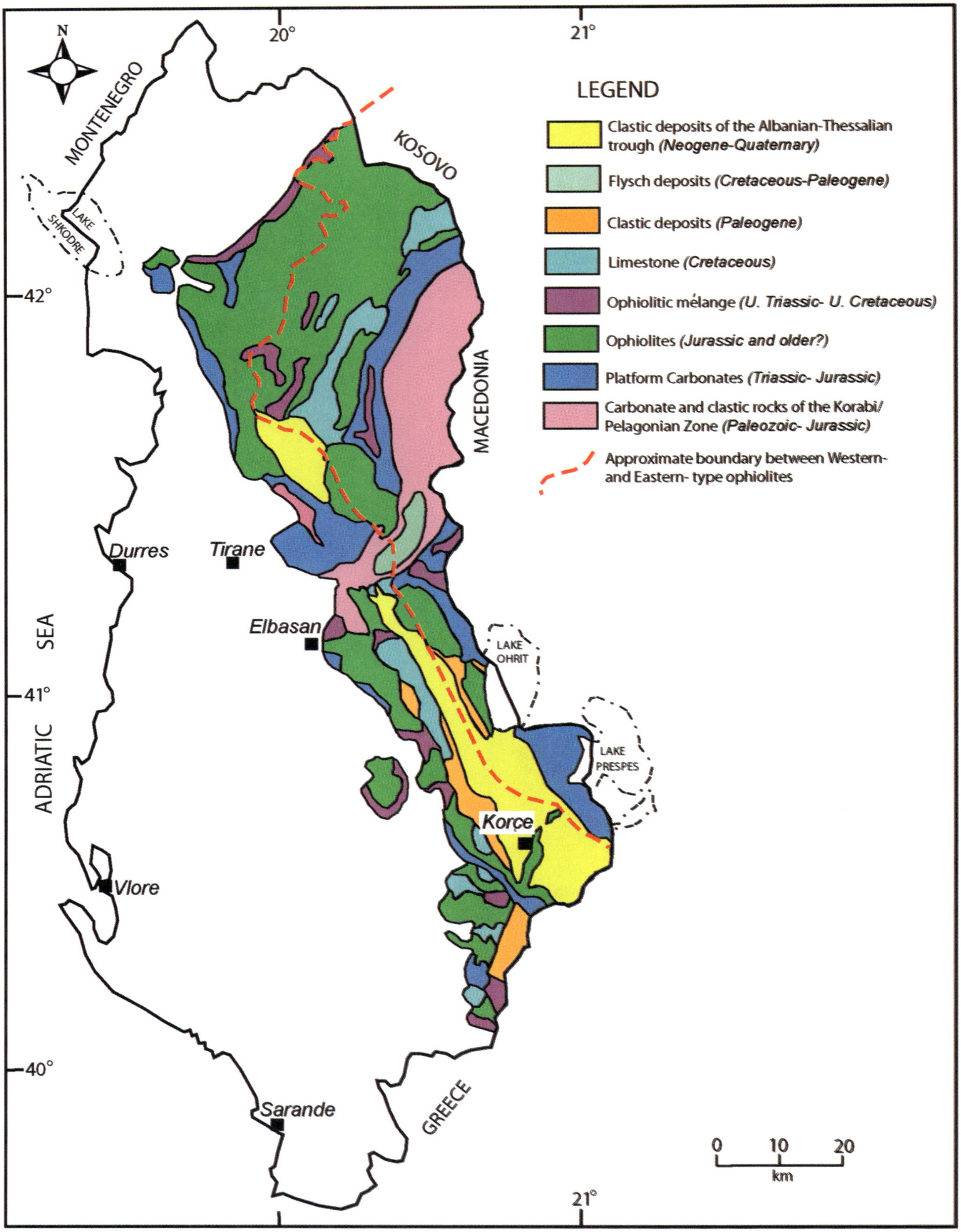

Figure 2: Geological map of the Albanian ophiolites and their immediate vicinity. The approximate boundary between the western- and eastern-type ophiolites, based mainly on the changes in the chemical affinities of the crustal and upper mantle rocks from MORB to SSZ (respectively), is shown with a red dashed line.

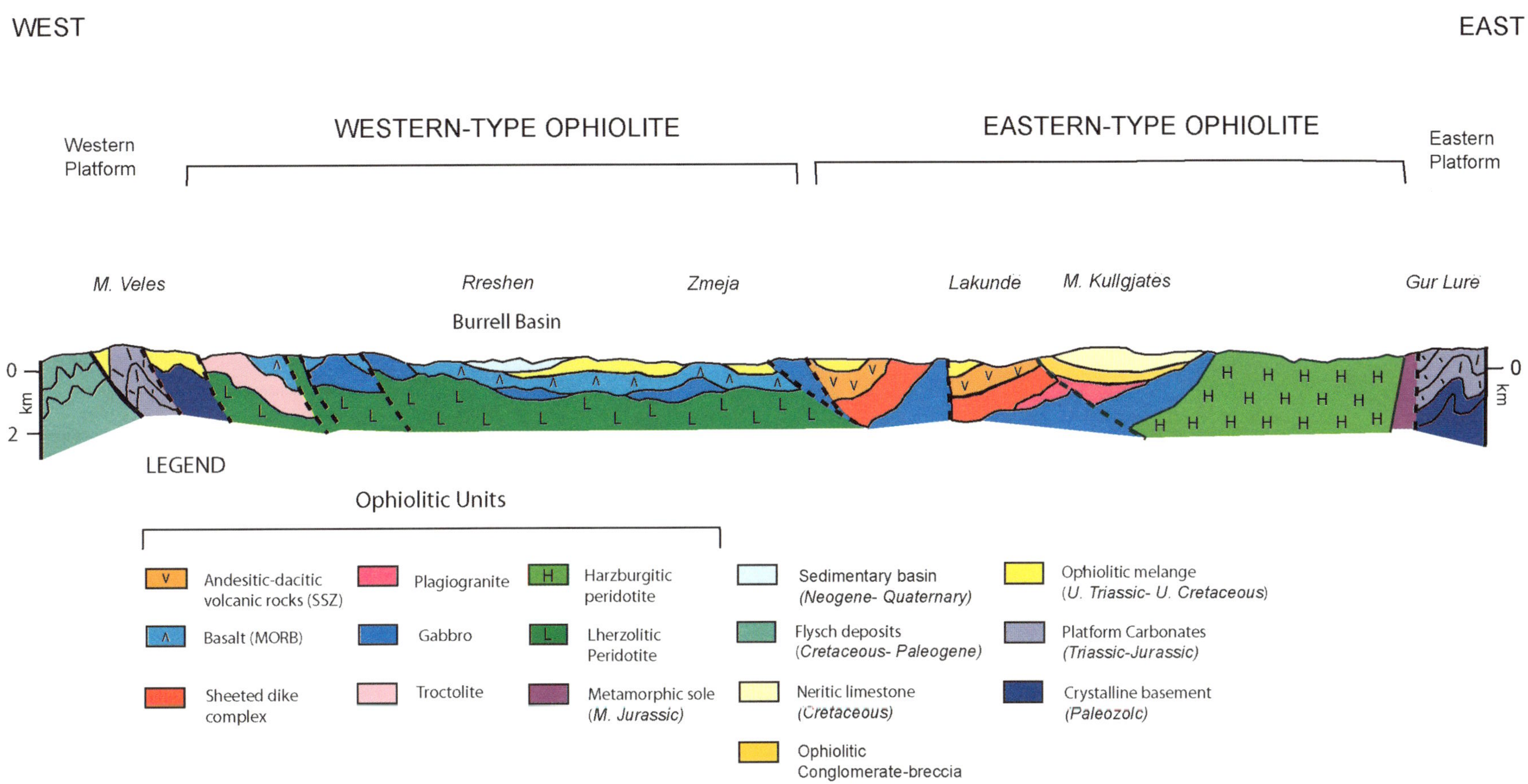

Figure 3: Interpretive geological cross-section across the Albanian ophiolites in the Mirdita zone at the latitude of ~41°40’N. The Western and Eastern platforms represent the Cukali-Apulian and Korabi-Pelagonian microcontinents, respectively.

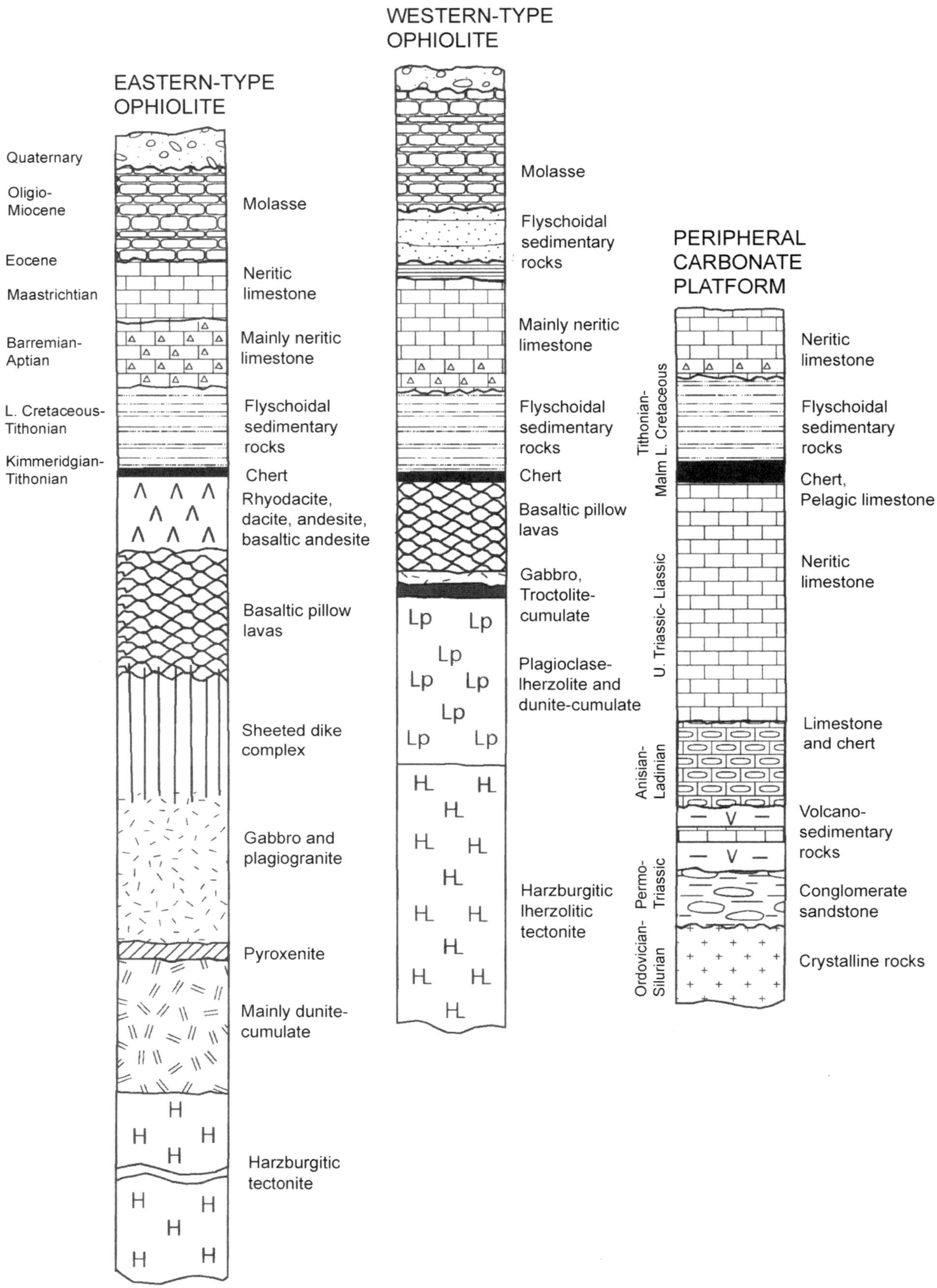

Figure 4: Schematic generalized columnar sections of the eastern- and western-type ophiolites and their peripheral platform carbonates (Korabi-Pelagonian in the east and Cukali-Apulian in the west). Flyschoidal sedimentary rocks of Tithonian-Lower Cretaceous age cover both the ophiolites and the carbonate platforms.

10-year period between 1951 and 1961. They considered all the plutonic rocks in the Albanian ophiolites, including the ultrabasics, as Jurassic intrusions within the Lower and Middle Triassic volcanic rocks. On the other hand, Mishunina (1957), who was part of the Soviet team, interpreted all ophiolitic magmatic rocks as part of a Jurassic volcano-sedimentary formation resting conformably on the Triassic platform carbonates. She was from the Oil and Gas Exploration Institute in Leningrad and thought that the oil and natural gas occurrences in the region had their origin in the carbonate platform beneath the so-called Jurassic volcano-sedimentary sequence. Throughout the 1950s, Albanian students received college education and formal training in geology in the Soviet Union and in other eastern European countries. The Geological Survey of Albania was established in Tirana in 1952.

In the early 1960s the Party of Labor was in full charge of the political regime in Albania, and Enver Hoxha was the First Secretary of the party. Political relations with the former Soviet Union had deteriorated significantly at this point, and, as a result, all scientific, cultural, political, economic, and military relations with the Soviet Union were interrupted. Instead, there was a gradual development of relations with Communist China during this time, and apparently more open relations with some of the western countries (e.g., France, Switzerland, Sweden, and Austria) developed. This change in political orientation ushered in an era of French collaboration in Albania. J.H. Brunn was already working in Greece at this time, investigating the Hellenic ophiolites (Brunn, 1940, 1956), and he visited Albania in the early 1960s, while working in Greece. One of Brunn's students, Aubouin, subsequently started studying the Albanian ophiolites, following his mentor's visit there. Aubouin and Ndojaj (1964), and Ndojaj (1963) used the term "ophiolite" for the first time in reference to the igneous rocks in the Mirdita zone, and they interpreted this "ophiolitic assemblage" as a final product of the differentiation of a basaltic magma; however, they interpreted the sequence as a massive volcano erupted through and spread over the Triassic carbonates of the Subpelagonian Zone during the Late Jurassic.

In 1967, Albania pulled out of the Warsaw Pact and became totally isolated from the rest of the western world. During this isolation, the Albanian geologists operated alone, mainly working on the ophiolites and investigating their associated ore deposits. Petrological studies of the ophiolitic units advanced at this stage. Petrologic and metallogenic studies by Dede et al. (1966) distinguished a "western" chromite-poor ultrabasic belt characterized by widespread plagioclase-bearing ultrabasic rocks, and an "eastern" chromite-rich ultrabasic (harzburgite-dunite) belt in the Albanian ophiolites. The subsequent geochemical data by Tashko (1976) and Cina et al. (1987) confirmed the occurrence of these two belts. These latter authors suggested that the belts reflected different degrees of partial melting in the mantle.

Qirinxhi-Spiro (1971) interpreted the origin of the Albanian ophiolites in a dramatically different way, suggesting that there was no melt-residua relationship between the ultramafic rocks and the gabbroic to plagiogranitic intrusive units of the ophiolites. According to this new model, all ultrabasic rocks represent Lower Paleozoic magmatic rocks that were covered by Triassic volcano-sedimentary rocks and intruded by Jurassic gabbroic to plagiogranitic plutons. Interestingly, this interpretation of the Albanian ophiolites is similar to current ideas on the origin of the Internal Ligurian ophiolites in the Apennines in which depleted Permian peridotites are spatially associated with Jurassic MORB-type gabbroic and basaltic rocks (Rampone and Piccardo, 2000). Qirinxhi-Spiro had read the paper by Dietz (1961) on seafloor spreading and was influenced by it in advancing his new model; however, his ideas about the Albanian ophiolites were mostly dismissed because of the reluctance of the Albanian earth scientists to believe in seafloor spreading and continental drift at that time.

Similarity in the petrological and geochemical features and the close temporal and spatial relations between the plutonic, subvolcanic hypabyssal, and volcanic rocks exposed in the north-central part of the Mirdita region prompted Shallo (1967) to suggest that these rocks were collectively part of a comagmatic volcanoplutonic complex of a pre-Tithonian age. Shallo (1972) considered that Albanian ophiolites had formed during Middle-Late Jurassic, an interpretation subsequently confirmed by Marcucci et al. (1994) through their determination of the age of the radiolarites of the sedimentary cover as Bajocian-Bathonian-Oxfordian. This was the time when Albanian geologists were working very closely with Chinese geologists exploring the extent of ophiolitic ore deposits throughout Albania. Some genetic models for the origin of the ore deposits (mostly copper deposits) were introduced, but the Chinese geologists were extremely reluctant to co-author any publications or internal reports, and hence no major papers came out during this period. The Chinese geologists eventually left Albania by 1975 because of the strained political relations between China and Albania in the early 1970s.

The plate tectonic revolution had taken place in 1967–1968 and the origin of the oceans, mountain belts, and ophiolites was being discussed through new tectonic models in Europe and North America; however, the isolation of the Albanian geologists from the west and the lack of their access to modern literature delayed the arrival of the plate tectonics theory in Albania for nearly 15 years, until the early 1980s.

Models of the Plate Tectonics Era

A major advance in the study of Albanian ophiolites occurred in the early 1980s when Shallo (1980) recognized the existence of a sheeted dike complex in Central Mirdita (Figs. 1 and 3). Shallo had read about the 1972 Penrose ophiolite definition and knew of Thayer's presentation at the 1974 ophiolite conference in Moscow (Thayer, 1977). Thus, he was familiar with the then new ideas on ophiolites and their significance for the evolution of oceans and continents. Shallo had earlier interpreted these dike rocks as isoclinally folded lava flows (Shallo, 1969), much like Bishopp's (1952) recognition of the sheeted dikes in the Troodos massif of Cyprus as folded lava sections (his "Folded Diabase Series"). Shallo had read Blanchet's paper (Blanchet, 1976) on the evolution of marginal basins in the Alpine Tethys, specifically in the Dinarides region, through continental rifting, presented at an international symposium in Split (former Yugoslavia). Inspired by these new ideas and concepts, Shallo suggested a seafloor-spreading origin for the Albanian ophiolites in a marginal basin, developed west of the Pelagonian micro-continent (Shallo, 1980). This interpretation of a westerly root zone of the Albanian ophiolites was based on stratigraphic correlation of the primary sedimentary cover of the ophiolites with the uppermost units of the peripheral Triassic-Jurassic platform carbonates and on the stratigraphic setting of the sedimentary ophiolitic mélange onlapping both the ophiolites and the platform carbonates (Fig. 3). In the same year, Shallo et al. (1980) published a paper, presented in May 1979 at a conference in Albania, in which they explained the seafloor spreading evolution of the Albanian ophiolites and their new tectonic model on the geology of the eastern Albanides using plate tectonic terms. In a sense, we can interpret this event as the formal recognition of the theory of plate tectonics in Albania.

In the beginning of 1978, with instructions from the Ministry of Industry and Mining, and the Ministry of Energy, a working group of eight specialists was commissioned to produce a geological map of Albania at a scale of 1:2,000,000. The Geological Institute within the Ministry of Industry and Mining provided the logistics and financial help for this nationwide mapping project. The map, completed in 1982 and printed in 1983 (ISPGJ-IGJN, 1983), was a major accomplishment for the establishment of a unified stratigraphy for Albania and for the usage of an internationally accepted lithologic and petrologic nomenclature. The sedimentary cover of the Albanian ophiolites was well defined on this map in terms of its internal stratigraphy and age relations. The geology of the Albanian ophiolites was compiled from 1:25,000 and 1:50,000 scale maps constructed during the 1960s through 1980s. The consensus of the members of the working group was that the Albanian ophiolites had their root zones to the west of the Korabi-Pelagonian carbonate platform, rather than to the east, as previously suggested by other workers for the Greek ophiolites farther to the south.

Based on limited petrographic and petrological data, Turku (1981) separated, within the volcanic sequence occurring in the Mirdita region, a Ca-rich, basaltic to dacitic series of island arc affinity and a diabasic series of tholeiitic affinity. This is the first mention of a subduction zone related origin of the Albanian ophiolites in the literature. Subsequently, Shallo et al. (1985) distinguished the eastern- versus western-type ophiolites based on their disparate pseudostratigraphy and different petrological, geochemical, and metallogenic features; they introduced two alternative tectonic models for the evolution of the Albanian ophiolites. In one model, the western-type ophiolites are produced first by seafloor spreading, and the eastern-type ophiolites are developed subsequently as a result of an intraoceanic subduction within the Pindos-Mirdita basin. In the alternative model, seafloor spreading produces the western-type ophiolites, and with continuous spreading accompanied by increased asthenospheric upwelling, displacement of the spreading ridge to the east in the same oceanic basin results in the formation of the eastern-type ophiolites. The significance of this first model is that, although it did not specifically use the "suprasubduction zone" term, the idea developed therein for the tectonic development of the eastern-type ophiolites was in parallel with the suprasubduction zone concept for the origin of the ophiolites that was formally defined just a year before by Julian Pearce et al. (Pearce et al., 1984; Pearce, this volume, Chapter 15). The Albanian ophiolite literature and investigations had thus caught up with modern ideas about the evolving ophiolite concept by the end of the 1980s.

RECENT FINDINGS AND MODELS ON THE ALBANIAN OPHIOLITES

A combination of events collectively led to the collapse of the communist system in Albania in the early 1990s and to the opening of the nation to world affairs. In 1990, poverty in Albania was at its peak following a persistent drought and extremely low agricultural production in the country; in addition, the low energy output had affected industry adversely and many people had lost their jobs. Total isolation of the nation for years under communist regimes had finally pushed the peoples of Albania over the limits of their patience, and the first demonstrations in support of a pro-western democratic movement were orchestrated by university students across the country. Political troubles within the former Soviet Union were well recognized in Albania at this time. Subsequently, the communist government in Albania collapsed in 1991, and a democratic government was established in March 1992. This new political system has allowed the Albanian scientists to collaborate freely with western colleagues, who found it easier to travel to and within Albania to conduct field-oriented geological studies.

A colloquium on the geology of Albania was organized in Paris (France) in April 1992, bringing together Albanian and French geologists to discuss the potential of future research collaboration in Albania. This colloquium (Colloque: Géologie d'Albanie, Séance Spécial de Société Géologique de France) eventually led to a first fruitful collaboration between research teams from Universite de Paris-Sud (Jean Bébien and his students) and the Tirana Polytechnic University (Minella Shallo and his colleagues). The mutual scientific exchange over the next several years included the training of several Albanian doctoral students in France.

A second workshop on Albanian ophiolites, supported by the International Geological Correlation Programme of the UNESCO, was convened by Luigi Beccaluva (Italy) and Nikita Bogdanov (the former Soviet Union) in the fall of 1992. Field trips during the course of this workshop introduced the participants to the two geochemically distinct ophiolite belts in Albania, and paved the way for more systematic and quantitative petrological and geochemical studies of Albanian ophiolites through a series of international scientific collaborations. Another international workshop on "Albanian Ophiolites and Related Mineralization" was convened in Tirana in October 1995 to further explore the nature of future work.

These recent studies (i.e., Shallo, 1992, 1995; Beccaluva et al., 1994; Bébien et al., 1997, 1998; Bortolotti et al., 1996; Dilek et al., 2001; Marroni, 1996) confirmed the existence of two adjacent, petrologically and geochemically different types of ophiolites forming paired belts. Beccaluva et al. (1994) showed that the MORB affinity of the western-type ophiolites was characterized by a less-depleted lherzolitic composition of the mantle tectonites, the crystallization order of the cumulate rocks, and the occurrence of high-Ti basalts. They noted that these petrological features were compatible with the origin of the western-type ophiolites in a mid-ocean ridge spreading system. They also showed that the eastern-type ophiolites contained, by contrast, a strongly depleted tectonized harzburgite mantle, low-Ti basalts and basaltic andesites with an island arc tholeiitic (IAT) affinity, and very low-Ti volcanic rocks resembling high-Ca boninites, indicating the generation of their parental magmas in a suprasubduction zone setting. Thus, the hypothesis by Beccaluva et al. (1994) envisions two different tectonic settings for the western- and eastern-type ophiolites. Shallo (1991, 1995) demonstrated that boninitic lavas and dikes occur in association with low-Ti island arc tholeiitic volcanic rocks and dikes within both the extrusive sequence and the sheeted dike complex in the eastern-type ophiolites.

Bébien et al. (1997, 1998) attributed the close association in space and time of the western- and eastern-type ophiolites to their common origin in a subduction zone environment. They suggested that the geochemical differences displayed by the two distinct geochemical types in the Albanian ophiolites might have resulted from complex magmatic process associated with the initiation and establishment of an infant subduction zone within the same ocean basin. Thus, a close spatial association of the MORB- and SSZ-type ophiolites within the Albanian ophiolite belt was well established by the mid-1990s.

Focusing on the nature and distribution of deformation-related structures during their investigations of the Albanian ophiolites, particularly in the mantle rocks, Nicolas et al. (1999) reported that the main differences between the western- and eastern-type ophiolites were restricted to the uppermost mantle and lower crustal sections only, and that these differences were an artifact of successive episodes of magmatic and amagmatic extension in a slow-spreading mid-ocean ridge environment. In this model, the deep mantle section in both types is harzburgitic, and the deformed lherzolites representing the uppermost mantle in the western-type might have originated from depleted harzburgites by impregnation and tectonic dispersion of melt during an episode of tectonic (amagmatic) extension and associated deformation occurring at temperatures ~800°–1000 °C. The eastern-type massifs, on the other hand, include a thick dunitic transition zone rich in basaltic impregnations and chromite deposits and overlain by layered gabbros, and they are a product of magmatic spreading along the same slow-spreading paleo-ridge system within the Pindos-Mirdita ocean basin. Thus, Nicolas et al. (1999) interpret the coexistence of the western- and eastern-type massifs in paired belts in Albania to be a result of the cyclic operation (magmatic-amagmatic) of a slow-spreading mid-ocean ridge system. This hypothesis differs from an earlier mid-ocean ridge model by Kodra et al. (1993) that envisions the western-type ophiolites as a product of a slow-spreading and the eastern-type as a fast-spreading regime within the same Pindos-Mirdita ocean basin. Whereas the study by Nicolas et al. (1999) documents the nature and distribution of high-T °C (eastern-type) versus low-T °C (western-type) structures in the peridotites, it does not address the causes of significant changes in the chemical compositions of the upper crustal rocks (mainly the extrusive rocks) from west to east across the paired ophiolite belts; the inferred magmatic extension (i.e., high magma supply) in this slow-spreading mid-ocean ridge model does not provide a satisfactory explanation for the existence of boninitic rocks (typical of subduction zone settings) within the sheeted dike complex and the extrusive sequence of the eastern-type ophiolites, either.

Metallogenic studies of the Albanian ophiolites support the distinction of the two different ophiolite types: the western-type ophiolites are characterized by scarce, mainly refractory-type chromite ores, layered titanomagnetite mineralization, rare quartz sulfide mineralization within gabbros, and rare seafloor massive sulfide ores within the volcanic sequence; the eastern-type ophiolites are characterized by abundant metallurgic-type chromite ores partly associated with Platinum Group Element (PGE) mineralization, widespread quartz-sulfide mineralization within the plutonic sequence (mainly gabbros and plagiogranites), stockwork-type sulfide mineralization associated with hydrothermally altered rocks within the sheeted dikes and the volcanic sequence, as well as abundant volcanogenic massive and disseminated sulfide mineralization within the volcanic rocks (ISPGJ-FGJM, 1987; Shallo et al., 1995). The petrological and geochemical indications of an extremely depleted mantle sequence and the presence of low-Ti tholeiites and boninitic lavas are consistent with the high metallogenic potential of the eastern-type ophiolites.

How these paired ophiolite belts (Fig. 2) with chemically distinct features and affinities might have formed in a single ocean basin (Pindos-Mirdita Ocean) is the main question. The existing models suggest that the western-type ophiolites with a MORB affinity developed within a rift basin in Late Triassic-early Late Jurassic time, following the splitting of the Korabi-Pelagonian microcontinent from Apulia in the Early-Mid Triassic, and that the eastern-type ophiolites with a SSZ affinity formed subsequently above a westward-dipping intra-oceanic subduction zone in the early Late Jurassic as the Pindos-Mirdita ocean basin started closing (Shallo et al., 1996; Bébien et al., 1997; Robertson and Shallo, 2000; Fig. 5). These models thus infer that the IAT-type eastern ophiolites were superimposed upon the mid-ocean ridge-generated western ophiolites after initiation of intra-oceanic subduction within the Pindos-Mirdita basin. The subduction influence might have increased with time, giving rise to the formation of the strongly depleted IAT and boninite-type volcanics and related subvolcanic and plutonic rocks (Fig. 5). This scenario

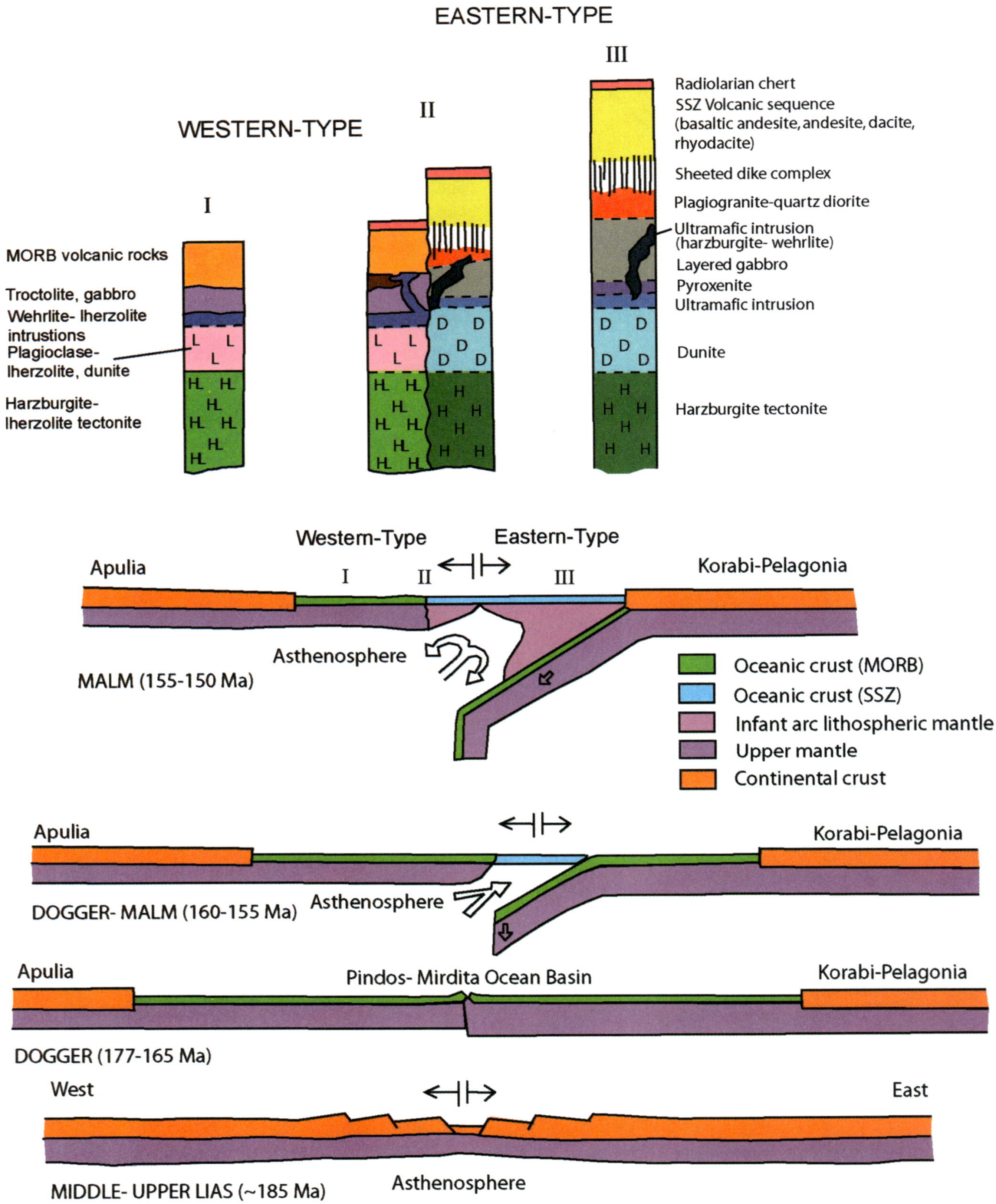

Figure 5: Sequential tectonic diagrams depicting the geodynamic evolution of the Albanian ophiolites through time (after Shallo et al., 1996; initial model was adopted from Stern and Bloomer, 1992). Continental rifting during the early Middle Jurassic (ca. 185 Ma) resulted in the formation of a small ocean basin (Pindos-Mirdita basin) between the Apulian and Korabi-Pelagonian microcontinents. Seafloor spreading in this basin created the oceanic lithosphere with a MORB affinity. Initiation of an intraoceanic subduction zone ca. 165 Ma was followed by a rapid eastward (in present coordinate system) retreat of the trench that caused strong extension in the upper plate. This extension was accompanied by robust magmatism driven by high degrees of melting due to asthenospheric upwelling beneath the lithospheric mantle of the infant arc and affected by subduction-derived fluids. This stage of magmatism was responsible for the rapid formation of eastern-type oceanic crust with a strong arc affinity (marked III on the columnar section and on the tectonic cartoon). The western-type oceanic crust has been preserved behind the infant arc and adjacent to the Apulian microcontinent.

implies that the eastern-type ophiolites have younger igneous ages than the western-type ophiolites.

New geochronological and geochemical data from the western- and eastern-type ophiolites (Dilek et al., 2001) require that we refine these existing models on the petrogenetic evolution of the paired ophiolite belts in Albania. Zircon ages from the plagiogranitic intrusions in the western- and eastern-type ophiolites indicate that both types crystallized at ~162–165 Ma (Dilek et al., 2001). Furthermore, isotopes from the lower part of the volcanic succession in the eastern ophiolites display ε_{Nd} values similar to the upper part of the volcanic succession in the western-type ophiolites; thus, the chemical differences between the western and eastern ophiolite sequences are, in fact, part of a compositional spectrum and point to progressive evolution of ophiolitic magmas from MORB in the west to IAT in the east, without any tectonic disruption in ocean crust generation within the Pindos-Mirdita basin. This change in chemical affinity marks an eastward shift in space and time of the magmatic evolution of the Jurassic oceanic crust within the same spreading environment, rather than two discrete tectonic settings (mid-ocean ridge versus suprasubduction zone settings) separated in space and time within the Pindos-Mirdita Ocean (Dilek et al., 2001).

Age and petrochemical similarities between the western and eastern ophiolite belts are significant because they demonstrate unequivocally that the magmas forming the Albanian ophiolites evolved from MORB (early crystallization of plagioclase and lherzolitic-harzburgitic residual mantle) to IAT (early crystallization of clinopyroxene and harzburgitic residual mantle) as the Pindos-Mirdita ocean basin evolved from an intra-continental rift basin to a small Neo-Tethyan ocean basin in the Jurassic (Fig. 5). Yet, the lack of any tectonic break between the western- and eastern-type ophiolites and of any geological evidence indicating the existence of a subduction zone within this narrow Neo-Tethyan ocean basin and/or in the immediate vicinity of the peripheral carbonate platforms outside the basin poses a major problem for the extant tectonic models. The conspicuous subduction influence on melt evolution (e.g., addition of subduction-mobile elements, increased fluid-induced melting) of the eastern-type ophiolites is clearly intimately related to the regional geodynamics of the Pindos-Mirdita ocean basin and the Neo-Tethyan mantle dynamics (Dilek and Flower, 2003). Previous and current discussions on whether the Albanian ophiolites represent fragments of oceanic lithosphere created at a slow-spreading mid-ocean ridge system or in an embryonic arc-forearc setting in a subduction zone environment do not adequately explain the occurrence of these paired ophiolite belts, nor the mode and nature of tectonic, magmatic, and chemical processes associated with plate accretion in rift-generated ocean basins.

The Albanian ophiolites thus provide a unique opportunity to better understand spatial and temporal relationships in melt evolution of ophiolitic magmas and the evolutionary progress of oceanic crust formation and mantle dynamics within Ligurian-type, restricted (bounded by microcontinents and/or large continental plates) ocean basins and Mediterranean-type ophiolites (Dilek, this volume, Chapter 1). Integrated structural, geochemical, and geochronological studies of the Albanian and other Tethyan ophiolites should help us resolve the "ophiolite conundrum" (ophiolites displaying mid-ocean ridge geology, yet arc chemistry), which came about when Miyashiro recognized 30 years ago that the Troodos ophiolite might have formed in an island arc setting (Miyashiro, 1973).

SUMMARY

Studies of the Albanian ophiolites and development of the ideas on their origin can be examined within two main time frames. The fixist models dominated geological thought on these ophiolites prior to the plate tectonic revolution, much as in the case of other major ophiolites and related studies (see, e.g., Smith and Rassios, this volume, Chapter 19, for the Hellenic ophiolites, and Juteau, this volume, Chapter 3, for the Hellenic, Troodos, and Oman ophiolites). The Jurassic mafic-ultramafic rocks in Albania were originally considered as in situ simatic exposures on land and/or large eruptions and intrusions of basaltic magma as part of a massive volcano developed through and across the Pelagonian carbonate platform.

The political isolation of Albania from the rest of the world under communist rule during the 1960s and 1970s prompted Albanian geologists to explore the ophiolites for natural resources and ore bodies (i.e., chromium and copper), and these focused investigations yielded important observations and information on the petrology and metallogeny of the ophiolites. The recognition during this time of the chromite-poor western ultramafic belt (plagioclase-bearing peridotites) and the chromite-rich eastern ultramafic belt (harzburgite-dunite) was a significant discovery that eventually led to delineation of paired ophiolite belts in Albania. Intensive mapping and field investigations undertaken by survey geologists in collaboration with Chinese colleagues during the late 1960s and early 1970s revealed that the Jurassic igneous rocks and associated ore deposits were part of a pre-Tithonian, comagmatic volcanoplutonic complex.

Plate tectonic ideas arrived in Albania in the early 1980s and led to the recognition of a sheeted dike complex implying a seafloor spreading origin for the Albania ophiolites in an ocean basin located west (in present coordinate system) of the Korabi-Pelagonian subcontinent. Limited petrological and geochemical studies of extrusive rocks by Albanian geologists during the 1980s confirmed the existence of tholeiitic basaltic rocks in the west and basaltic to dacitic-rhyodacitic rocks with an island arc tholeiite chemistry in the east. By 1985, the suprasubduction zone origin of the eastern-type ophiolites was established in the Albanian literature, only a year after the formal definition of the suprasubduction zone concept in the international literature.

The petrogenesis of ophiolitic magmas in these paired belts and the geodynamic evolution of the Pindos-Mirdita ocean basin, in which the MORB- and SSZ-type ophiolites formed, are significant topics for the evolving ophiolite concept. Some researchers explain the differences between the western- and eastern-type ophiolites in terms of varied spreading rates (slow versus fast, respectively) and/or fluctuated magma supply rates beneath the ridge axis (poor magma supply and amagmatic extension versus large magma supply and magmatic extension, respectively). Others envision a shift from a seafloor spreading system in a mid-ocean ridge setting to an arc-forearc setting in the upper plate of an intra-oceanic subduction zone (suprasubduction zone environment). Our recent geochemical data, combined with structural field work (Dilek et al., 2001), clearly show that changes in chemical composition and affinity of the paired ophiolite belts from west to east represents a progressive shift (in space and time) in melt evolution within and beneath the same spreading environment, as the restricted Neo-Tethyan ocean basin evolved between the two adjoining micro-continents. We need to develop a petrogenetic model that will effectively integrate all these observations and findings in an internally coherent and geodynamically compatible interpretation within the regional framework of the tectonic evolution of Neo-Tethys.

ACKNOWLEDGMENTS

Shallo acknowledges the help and support of his colleagues in the Albanian Geological Survey and the Tirana Polytechnic University for his studies of the Albanian ophiolites for over 40 years. This study is part of a collaborative research project funded in part by an Environmental and Earth Science and Technology Collaborative Linkage grant from the NATO Science Programme (EST.CLG-97617) and by the Miami University Faculty Research and Hampton International Studies Funds that we gratefully acknowledge. The participation and collaboration of Harald Furnes (University of Bergen, Norway) in this research project helped us better understand the petrogenesis of the Albanian ophiolites. We would like to extend our sincere thanks to the former and current Directors of the Albanian Geological Survey, Drs. Mehmet Zaçay and Hamdi Beshku (respectively), for providing us with crucial logistical support during our field campaigns. We thank John Dewey, Ron Harris, Sally Newcomb, and Paul T. Robinson for their careful and constructive reviews, which improved the paper.

REFERENCES CITED

Anonymous, 1972, Penrose Field Conference on Ophiolites: Geotimes, v. 17, no. 12, p. 24–25.

Aubouin, J., Ndojaj, I., 1964, Regard sur la geologie de l'Albanie et sa place dans la geologie des Dinarides: Bulletin de la Societie Geologie du France, ser. 7, v. Vl, p. 593–625.

Bébien, J., Dautaj, N., Shallo, M., Turku, I., and Barbarin, B., 1997, Diversite des plagiogranites ophiolitiques l'exemple albanais: Comptes Rendus a l'Académie des Sciences, Paris, v. 324, ser. IIa, p. 875–882.

Bébien, J., Shallo, M., Manika, K., and Gega, D., 1998, The Shebenik massif (Albania): a link between MOR and SSZ-type ophiolites: Ofioliti, v. 23, p. 7–15.

Beccaluva, L., Coltorti, M., Premti, I., Saccani, E., Siena, F., and Zeda, O., 1994, Mid-ocean ridge and supra-subduction affinities in the ophiolitic belts from Albania: Ofioliti, v. 19, p. 77–96.

Bishopp, D.W., 1952, The Troodos massif, Cyprus: Nature, v. 169, p. 489–490.

Blanchet, R., 1976, Bassins marginaux et Tethys alpine: De la marge continentale ou domaine oceanique dans les Dinarides: *in* Biju-Duval, B. and Montabert, L. eds., International Symposium on the structural history of the Mediterranean basins, Split, Yugoslavia: Paris, Technip Publishers, p. 72–87.

Bortolotti, V., Kodra, A., Marroni, M., Mustafa, F., Pandolfi, L., Principi, G., and Saccani, E., 1996, Geology and petrology of ophiolitic sequences in Mirdita region (Northern Albania): Ofioliti, v. 21, p. 3–20.

Brunn, J.H., 1940, Sur les roches basiques du Pinde et de la Macédonie occidentale: Comptes Rendus a l'Académie des Sciences, Paris, v. 20, p. 109–111.

Brunn, J.H., 1956, Contribution a l'étude géologique du Pinde septentrional et d'une partie de la Macédoine occidentale: Annales de la Géologie des Pays Helléniques, Paris, v. 7, p. 1–358.

Caby, R., 1995, Plastic deformation of gabbros in a slow-spreading Mesozoic ridge: Example of the Montgenevre ophiolite, western Alps, *in* R.L.M., Vissers and A. Nicolas, eds., Mantle and Lower Crust Exposed in Oceanic Ridges and in Ophiolites: Amsterdam, Kluwer Academic Publishers, Petrology and Structural Geology Series, v. 6, p. 123–145.

Carosi, R., Cortesogno, L., Gaggero, L., and Marroni, M., 1996, Geological and petrological features of the metamorphic sole from the Mirdita nappe, northern Albania: Ofioliti, v. 21, p. 21–40.

Cina, A., Tashko, A., and Tershana, A., 1987, The Bulqiza and Gomsiqe ultrabasic massifs, ophiolites of Albanides, a geochemical comparison: Ofioliti, v. 12, p. 219–237.

Dede, S., Shehu, R., and Shallo, M., 1966, Intrusive magmatism in Albania and mineralization linked with them (in Albanian with French abstract): Përmbledhje Studimesh, v. 3, p. 3–45.

Dietz, R.S., 1961, Continent and ocean basin evolution by spreading of the seafloor: Nature, v. 190, p. 854–857.

Dilek, Y., and Flower, M.F.J., 2003, Arc-trench rollback and forearc accretion: 1. A collision-induced mantle flow model for Tethyan ophiolites, *in* Ophiolites in Earth History: London, Geological Society of London Special Publication (in press).

Dilek, Y. and Moores, E.M., 1990, Regional Tectonics of the Eastern Mediterranean ophiolites, *in* Malpas, J., Moores, E.M., Panayiotou, A., and Xenophontos, C., eds., Ophiolites, Oceanic Crustal Analogues, Proceedings of the Symposium "Troodos 1987": Nicosia, Cyprus, Geological Survey Department, p. 295–309.

Dilek, Y., Thy, P., Hacker, B., and Grundvig, S., 1999, Structure and petrology of Tauride ophiolites and mafic dike intrusions (Turkey): Implications for the Neo-Tethyan ocean: Bulletin of the Geological Society of America, v. 111, No. 8, p. 1192–1216.

Dilek, Y., Furnes, H., Shallo, M., and Mortensen, J.K., 2001, Structure, petrology, and geochronology of the Albanian ophiolites and their tectonic evolution within the Neotethyan orogenic belt: EUG XI, European Union of Geosciences, April 8–12, 2001, Strasbourg, France: Journal of Conference Abstracts, v. 6, No. 1, p. 321.

Dimo-Lahitte, A., Monié, P., and Vergély, P., 2001, Metamorphic soles from the Albanian ophiolites: Petrology, $^{40}Ar/^{39}Ar$ geochronology, and geodynamic evolution: Tectonics, v. 20, p. 78–96.

Froitzheim, N. and Manatschal, G., 1996, Kinematics of Jurassic rifting, mantle exhumation, and passive-margin formation in the Austroalpine and Penninic nappes (eastern Switzerland): Geological Society of America Bulletin, v. 108, p. 1120–1133.

ISPGJ-FGJM, 1987, Metallogenic Map of Albania: Tirana, Albania: München, Germany, Huber-Kartographie, scale 1:200,000, 3 sheets.

ISPGJ-IGJN , 1983, Geological Map of Albania: Tirana, Albania, NMMS "Hamid Shijaku," scale 1:200,000, 3 sheets.

Kodra, A., Gjata, K., and Bakalli, F., 1993, Les principales étapes de l'évolution paléogéographique et géodynamique des Albanides Internes au course de Mésozoique: Bulletin de la Societie Geologie du France , v. 164, p. 69–77.

Kollmak, L., 1959, Geology and ore deposits of Kukesi region (in Russian): Tirana, Albania, Internal Report, Arkivi Qendror i Gjeologjise, Tirana, Albania, 320 p.

Manatschal, G., and Nievergelt, P., 1997, A continent-ocean transition recorded in the Err and Platta nappes (eastern Switzerland): Eclogae Geologicae Helvetiae, v. 90, p. 3–27.

Marcucci, M., Kodra, A., Pirdeni, A., and Gjata, T., 1994, Radiolarian assemblage in the Triassic and Jurassic cherts of Albania: Ofioliti, v. 19, p. 110–114.

Marroni, M., ed., 1996, Mirdita Ophiolite Project: Ofioliti, v. 21, No. 1, p. 3–84.

Mishunina, Z.A., 1957, Explanatory notes for the Geological Map of Albania (in Russian): Leningrad, VNIGRI, scale 1:200,000, 3 sheets.

Miyashiro, A., 1973, The Troodos ophiolitic complex was probably formed in an island arc: Earth and Planetary Science Letters, v. 19, p. 218–224.

Ndojaj, I., 1963, On some problems of magmatism of Albania (in Albanian with French abstract): Buletini i Universitetit Shteteror te Tiranes, Seria Shkencat e Natyres, no. 2, p. 15–42.

Nejellov, A.N., 1951, Geology of the western part of Fani River (in Russian): Tirana, Albania, Internal Report: Tirana, Albania, Arkivi Qendror i Gjeologjise, 420 p.

Nicolas, A., Boudier, F., and Meshi, A., 1999, Slow spreading accretion and mantle denudation in the Mirdita ophiolite (Albania): Journal of Geophysical Research, v. 104, p. 15155–15167.

Nopsca, F., 1928, Geologie und Geographie Nordalbaniens. Budapest, Geologica Hungarica, 704 p.

Nowack, E., 1929, Geologische Uberschist von Albanien. Salzburg (translated in Albanian by Peza, L.): Tirana, Albania, Fakulteti i Gjeologjise dhe Minierave, 160 p.

Pearce, J.A., Lippard, S.J., and Roberts, S., 1984, Characteristics and tectonic significance of supra-subduction zone ophiolites: London, Geological Society of London Special Publication, v. 16, p. 77–94.

Qirinxhi- Spiro, A., 1971, On the problems of the spatial position of the ultrabasic rocks of the Dinaro-Tauric sector of the folded Alpine belt, Albanides (in Albanian with French abstract): Përmbledhje Studimesh, v. 2, p 62–84.

Rampone, E., and Piccardo, G.B., 2000, The ophiolite-oceanic lithosphere analogue: New insights from the Northern Apennines (Italy), *in* Dilek, Y., Moores, E.M., Elthon, D., and Nicolas, A., eds., Ophiolites and Oceanic Crust: New Insights from Field Studies and the Ocean Drilling Program: Boulder, Colorado, Geological Society of America Special Paper 349, p. 21–34.

Robertson, A.H.F., and Shallo, M., 2000, Mesozoic-Tertiary evolution of Albania in its regional Eastern-Mediterranean context: Tectonophysics, v. 316, p. 197–254.

Shallo, M., 1967, Brief view on the magmatism and associated sulphide mineralization in the northern part of the Mirdita zone (in Albanian): Përmbledhje Studimesh, v. 5, p. 45–60.

Shallo, M., 1969, Some features of the geology and sulfide mineralization of the Munella region, Albania (in Albanian): Përmbledhje Studimesh, v. 12, p. 3–21.

Shallo, M., 1972, Magmatism of the geosyncline zones of Albania (in Albanian): Përmbledhje Studimesh, v. 1, p. 5–23.

Shallo, M., 1980, Petrology of the magmatic rocks and associated sulphide mineralization in Central Mirdita (in Albanian) [PhD. Thesis]: Tirana, Albania, Tirana Polytechnic University, 198 p.

Shallo, M., 1990a, Volcanic glasses of the Albanian ophiolite belt, *in* Malpas, J., Moores, E.M., Panayiotou, A., and Xenophontos, C., eds., Ophiolites, Oceanic Crustal Analogues, Proceedings of the Symposium "Troodos 1987": Nicosia, Cyprus, Geological Survey Department, p. 271–278.

Shallo, M., 1990b, Boninites within eastern ophiolites of Albania (in Albanian with English abstract): Buletini i Shencave Gjeologjike, v. 4, p. 19–37.

Shallo, M., 1990c, Ophiolitic melange and flyschoidal sediments of the Tithonian-Lower Cretaceous in Albania: Terra Nova, v. 2, p. 476–483.

Shallo, M, 1991, Ophiolites of Albania: Scientific Report, Tirana, Albania: Fakulteti i Gjeologjise dhe Minierave, Universiteti Shteteror i Tiranes, 247 p. (in Albanian).

Shallo, M., 1992, Geological evolution of the Albanian ophiolite and their platform periphery: Geologische Rundschau, v. 81/3, p. 681–694.

Shallo, M., 1995, Volcanics and sheeted dykes of the Albanian SSZ ophiolite: Buletini i Shkencave Gjeologjike, v. 1, p. 99–118.

Shallo, M., Gjata, T., and Vranaj, A., 1980, Concepsions nouvelles su la geologie des Albanides Orientales (in Albanian with French abstract): Përmbledhje Studimesh, v. 2, p. 31–56.

Shallo, M., Kote, Dh., Vranaj, A., and Premti, I., 1985, Ophiolitic magmatism of Albania (in Albanian): Scientific Report, Tirana, Albania: Tirana, Albania, Arkivi Qendror i Gjeologjise, 362 p.

Shallo, M., Cina, A., and Turku, I., 1995, Outline of the metallogeny of the Albanian MOR and SSZ-type ophiolites, *in* Ohnenstetter, M., ed., Substances, Minerals and Energetics: Documents du Bureau de Recherches Géologiques et Minières 244, p. 27–46.

Shallo, M., Turku, I., Manika, K., and Marto, A., 1996, Genesis and geotectonic setting of the Albanian ophiolites (in Albanian): Tirana, Albania, Scientific Reports, Fakulteti i Gjeologjise dhe Minierave, Universiteti Politeknik, 270 p.

Stern, R.J., and Bloomer, S.H., 1992, Subduction zone infancy: Examples from the Eocene Izu-Bonin-Mariana and Jurassic California arcs: Bulletin of the Geological Society of America, v. 104, p. 1621–1636.

Tashko, A., 1976, Some geochemical distinctions in ultrabasic rocks (in Albanian with French abstract): Përmbledhje Studimesh, v. 4., p. 101–118.

Thayer, T.P., 1977, Some implications of the sheeted dike swarms in the ophiolite complexes: Geotectonics, v. 11.

Turku, I., 1981, Petrology of the Mesozoic volcanics of the Krasta-Cukali and Mirdita zones and their mineral-bearing (in Albanian) [PhD Thesis] Tirana, Albania, Tirana Polytechnic University, 175 p.

Vergély, P., Dimo, A., and Monie, P., 1998, Datation des semelles metamorphiques ophiolitiques d'Albanie par la methode $^{40}Ar/^{39}Ar$: Consequences sur le mecanisme de leur mise en place: Comptes Rendus a l'Académie des Sciences, Paris, v. 326, p. 717–722.

Zhadan, E.S., 1959, Geology and ore deposits of Central Mirdita region (in Russian): Tirana, Albania, Internal Report, Arkivi Qendror i Gjeologjise, 360 p.

Manuscript Accepted by the Society May 2, 2003

Printed in the USA

Geological Society of America
Special Paper 373
2003

High-pressure and ultrahigh-pressure metamorphic belts—Subduction, recrystallization, exhumation, and significance for ophiolite study

W.G. Ernst
Department of Geological and Environmental Sciences, Stanford University, Stanford, California 94305-2115, USA

ABSTRACT

Alpine-type **orogenic belts are produced by short-lived subduction of an ocean basin and the underflow of continental crust, resulting in suturing. Old, relatively competent sialic basement and superjacent units characterize these complexes; ophiolites, chiefly enriched mid-ocean ridge basalt and suprasubduction-zone complexes are common in some belts, but are rare in others. Metamorphism of deeply subducted (60–140 km) parts of the orogen ranges from high pressure to ultrahigh pressure , and is not paired. Coeval calcalkaline arc rocks are uncommon. In contrast, paired** ***Pacific-type*** **orogenic belts develop within and landward from long-lived subduction zones. They consist of an outboard trench + accretionary prism deposited on oceanic crust, and an inboard volcanic-plutonic continental margin or island arc. The trench assemblage consists of graywacke + shaley mélanges, minor but widespread deep-water cherts and/or carbonates, and ubiquitous disaggregated ophiolites, especially normal mid-ocean basalt and ocean island basalt varieties, all recrystallized under high-pressure conditions at depths of 15–70 km. A massive, coeval calcalkaline arc dominates the subparallel terrane landward from the trench, and high temperatures characterize the associated metamorphism. In both high-pressure–ultrahigh-pressure Alpine and high-pressure Pacific-type metamorphic terranes, outboard thrust faults chiefly dip landward beneath the stable, nonsubducted plate, and fold vergence is seaward, reflecting similar convergent plate-tectonic settings; inboard, antithetic thrusting characterizes some contractional realms. High pressures and low temperatures in these belts attest to dynamic subduction-zone conditions. Because of their structural integrity, some Alpine-type microcontinents, island arcs, or continental promontories are carried down as much as 140 km, well into the brittle-ductile transition region, before decoupling from the sinking mantle lithosphere. In sharp contrast, incompetent Pacific-type graywacke + shale terranes generally separate from descending oceanic lithosphere at much shallower depths.**

The ascent of packets of subducted material as thin aspect-ratio sheets 1–5 km thick, combined with normal faulting above and subduction-zone thrusting below, promotes conduction cooling and partial preservation of decompressing high-pressure and ultrahigh-pressure metamorphic complexes. Worldwide, subduction-zone terranes consist of dominantly of small masses of high-density mafic blueschist-eclogite and/or anhydrous peridotitic lenses surrounded by voluminous, low-density quartzofeldspathic ± serpentinitic material, hence the bulk terrane density is less than that of unaltered, dynamically displaced mantle; exhumation to mid-crustal levels is largely buoyancy driven. Partial preservation of high-pressure–ultrahigh-pressure relict phases reflects rapid ascent combined with mineralogic armoring and a general lack of catalytic aqueous fluid. Exposure of rising high-pressure and ultrahigh pressure metamorphic belts

Ernst, W.G., 2003, High-pressure and ultrahigh-pressure metamorphic belts—Subduction, recrystallization, exhumation, and significance for ophiolite study, *in* Dilek, Y., and Newcomb, S., eds., Ophiolite concept and the evolution of geological thought: Boulder, Colorado, Geological Society of America Special Paper 373, p. 365–384. For permission to copy, contact editing@geosociety.org.

is a consequence of erosional decapitation and gravitational collapse of the subduction complex. Many exhumed high-pressure–ultrahigh-pressure sheets are of relatively small volume, and for these, only modest amounts of sedimentary debris are provided to successor basins.

Keywords: subduction zone metamorphism, high pressure, ultrahigh pressure, continental collision, Circumpacific subduction.

CONTRASTS BETWEEN ALPINE- AND PACIFIC-TYPE SUBDUCTION ZONES

Historical Development

With development of plate-tectonic theory in the 1960s, Earth scientists realized that submarine trenches mark zones of convergence, commonly between nonsubducted continental lithosphere and a descending oceanic plate (Isacks et al., 1968). That process has been called Circumpacific or *Pacific-type* subduction (type B subduction of Bally, 1981). Intracontinental, *Alpine type* orogenic belts (Bally's type A subduction) were also recognized as ocean-margin suture zones (Dewey and Bird, 1970; Moores, 1970; Molnar and Tapponier, 1975). Amalgamation of continental, microcontinental, and/or island-arc crustal entities occurs in response to the underflow of oceanic lithosphere initially separating sialic crust-capped portions of plates. But, as emphasized by many workers (Fitch, 1972; Irwin, 1972; Coney et al., 1980; Jones, 1983; Howell, ed., 1985), pure convergence is possible only at specific sites along curvilinear consumptive plate boundaries, reflecting the spherical nature of the globe-encircling plates; many terrane amalgams are sutured by oblique convergence and/or nearly pure strike-slip motion.

Intact and tectonically disaggregated tracts of oceanic crust and mantle underpinnings characterize oceanic (principally normal mid-ocean ridge basalt [N-MORB] and ocean island basalt [OIB]) and island arc/continental margin (chiefly enriched mid-ocean ridge basalt [E-MORB] and island arc tholeiite [IAT]) environments. The open ocean varieties found in Pacific-type accretionary complexes have been subducted and are intensely metamorphosed, whereas some suprasubduction-zone (SSZ) arc ophiolites are only feebly recrystallized.

Pacific-Type Plate Junctions

Pacific-type convergent boundaries evolve where vast tracts of oceanic lithosphere are consumed without the introduction of significant amounts of continental crust into the subduction zone, as is characteristic of the post-Paleozoic history of the eastern and western Circumpacific. This plate-tectonic situation gives rise to an outboard ophiolite-bearing, but largely metasedimentary, accretionary trench complex, a medial, longitudinal forearc basin, and an inboard, coeval calcalkaline volcanic-plutonic arc. The relatively narrow trench section consists of a low-temperature, low-heat-flow belt in which folds verge oceanward and outboard thrust faults roughly parallel the subduction zone, whereas the broad magmatic arc on the stable, nonsubducted plate is characterized by open folding and a high-temperature, high-thermal-flux regime (Miyashiro, 1961, 1967; Ernst et al., 1970; Dickinson, 1972). The former is deposited exclusively on oceanic crust (e.g., Franciscan, Aleutian, and Chile-Peru trench systems); the latter is constructed on pre-existing basement of the continental margin or island arc ± older oceanic crust (e.g., Sierran, Indonesian, and Andean arcs).

Alpine-Type Plate Junctions

Alpine-type convergent boundaries form where subduction of an ocean plate results in the transport, arrival at a trench, and underflow of a continent, microcontinent, or island arc, generally beneath sialic crust surmounting a stable hanging-wall plate (the mantle wedge). The leading edge or salient of the sialic terrane may descend to considerable depths, conducted down by the surrounding, oceanic lithosphere, because the overall density of the plate exceeds that of the underlying asthenosphere. Accompanying subduction, the aggregate continental thickness is increased by underplating, crustal contraction, and amalgamation. Sutures are characterized by major collisional mountain belts, allochthonous thrust sheets dipping under the nonsubducted plate, and a distinct paucity of calcalkaline igneous activity. Typical examples include the Urals, Alps, and Himalayas (Hamilton, 1970; Dal Piaz et al., 1972; Molnar et al., 1987; Burchfiel et al., 1989). Of course, it is possible for continental lithosphere to descend beneath young, hot—thus less dense—oceanic lithosphere (e.g., Oman and Sulawesi; Searle et al., 2001), but such configurations are less common because most oceanic plates are negatively buoyant relative to adjacent continental plates.

Intermediate-Type Plate Junctions

Plate-tectonic regimes transitional between these orogenic end-member types are well represented by the modern Indonesian archipelago (Hamilton, 1979; Maruyama et al., 1996). This complex suture zone is the product of Pacific + Australian lithospheric underflow beneath the Eurasian + Philippine Sea plates, and by initial stages of subduction/suturing of the Australian continental crust. The junction between the Arabian Peninsula and Iran represents another intermediate example (Boudier et al., 1985, 1988; Chemenda et al., 1996; Gnos et al., 1997; Miller et al., 1998).

Physical Models of Subduction

The thermal evolution of convergent plate junctions has been studied by many workers, chiefly employing numerical simulations (Shreve and Cloos, 1986; Cloos and Shreve, 1988a, 1988b; Beaumont et al., 1996, 1999; Stüwe and Barr, 1998), but also including experimental studies (Tapponnier et al., 1982; Chemenda et al., 1995, 1996). Thermotectonic architectures are a function of many input parameters, some only poorly constrained. Among them are the physical states of the converging lithospheric plates (temperatures, spatial configurations, viscosities, densities), rates of convergence, states and amount of sedimentary loading, extent of frictional dissipation, degree of hydration, fluid expulsion and flow (channeled or pervasive), advective heat transport, kinetics of subsolidus reactions and of partial melting, and extents of coeval tectonic and surficial erosion. In spite of uncertainties, widely recognized principles have emerged from such studies.

The computed thermal structures of convergent plate regimes have been investigated extensively (e.g., Oxburgh and Turcott, 1971: Anderson et al., 1978; Thompson and Ridley, 1987; van den Beukel and Wortel, 1988; Peacock, 1990, 1992). A reasonable model of the thermal regime, presented by Turcotte and Oxburgh (1972), is shown in Figure 1. Rocks are poor thermal conductors, so a descending lithospheric slab remains cool at all upper mantle depths relative to its surroundings, whereas the landward volcanic-plutonic arc occupies a broad thermal high in the nonsubducted mantle wedge + overlying crust. Thermal models such as that illustrated in Figure 1 account for paired Circumpacific outboard high-pressure (HP) blueschist-eclogite, and inboard high-temperature (HT) andalusite-sillimanite metamorphic belts. Geologic relationships and mineral parageneses for such terranes were first emphasized by Miyashiro (1961, 1967) prior to the advent of plate-tectonic theory; his classic synthesis of the contrasting paired metamorphic belts of Japan is presented as Figure 2. As now recognized for the Japanese island arc and its Pacific-type paired metamorphic belts, the times of accretion of clastic sedimentary debris + off-scraped, far-traveled, pre-existing crustal fragments decrease monotonically seaward (Banno and Sakai, 1989; Nishimura, 1998).

Scale-model experiments and numerical simulations of subduction-zone environments describe various mass-transport scenarios, depending on different values of the physical parameters employed. During underflow, return of subducted lithotectonic packets from great depths to mid-crustal levels also is a sensitive function of the physical characteristics of the rock units involved. For Circumpacific and Alpine accretionary wedges and contractional (collisional) mountain belts, both calculations and laboratory experiments realistically simulate the observed accumulation and suturing of the accreted sialic crust. Hacker et al. (2003a) have combined the computed thermal structures and observed metamorphic phase-equilibrium assemblages with measured geophysical properties of the oceanic crust and underlying mantle to provide quantitative models for lithospheric slabs subducting to depths of ~200 km.

Calcalkaline Arcs and Volcanic-Plutonic Belts

Landward calcalkaline arcs characterize the post-Paleozoic Circumpacific realm, attesting to the consumption of thousands

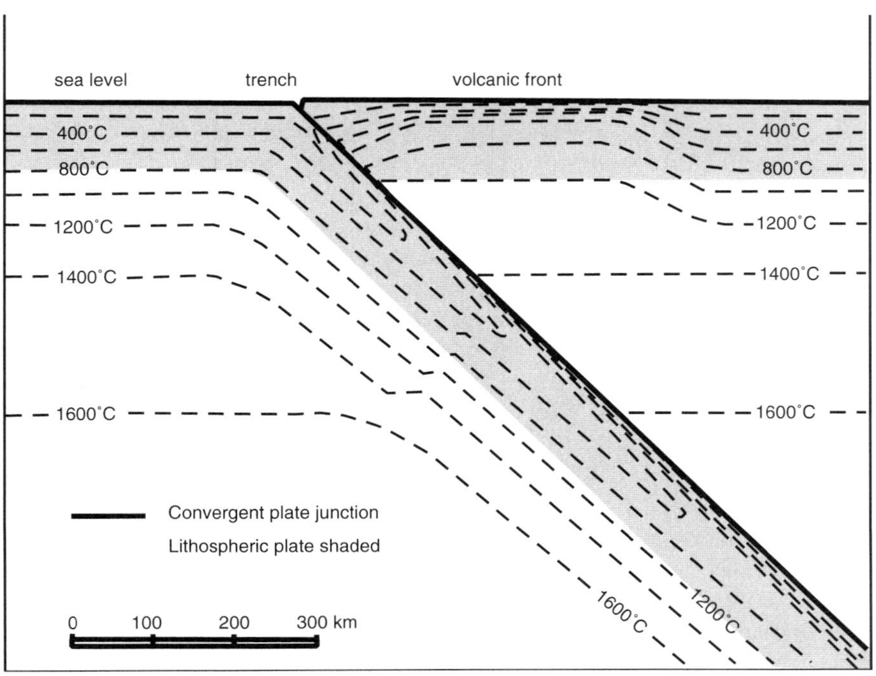

Figure 1. Computed thermal structure of a convergent lithospheric plate junction, after Turcott and Oxburgh (1972). Frictional dissipation along the inclined Benioff-Wadati shear zone has been assumed, accounting for the thermal bulge in the nonsubducted plate. Advective heat transfer by rising calcalkaline magmas has not been included, but also would contribute to the measured high heat flow of volcanic-plutonic arcs. Downward deflection of isotherms in the subducted slab is reflected in the low heat flow observed in the vicinity of oceanic trenches.

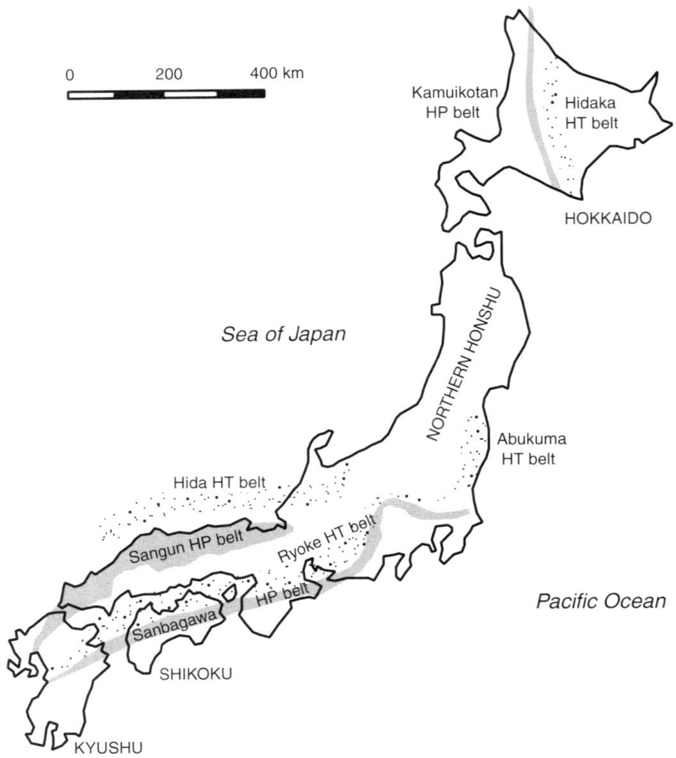

Figure 2. Compilation of the paired metamorphic belts of Japan, from Miyashiro (1961, 1967). Narrow, seaward ophiolitic HP metamorphic terranes are juxtaposed against broad, landward calcalkaline arcs and associated HT metamorphic belts.

of kms of oceanic lithosphere. Bulk-rock isotopic signatures show that, although partially melted quartzofeldspathic + pelitic crust yields peraluminous, S-type granite and contributes to mature island-arc and continental-margin magmas (Castro et al., eds., 1999), the calcalkaline suite contains an important contribution from the mantle wedge and/or subducted, eclogitized oceanic crust (DePaolo, 1981; Hawkesworth et al., 1993; Clift et al., 2001). These chemically intermediate volcanic and plutonic rocks represent new additions to the continental crust; other rock types represent mainly reworked material. Seaward, deposits of first-cycle, volcaniclastic sediments derived from the arc form fringing accretionary aprons (Dickinson, 1971, 1972, 1976); these detrital sediments may overwhelm the forearc basin, spilling over into the trench and onto the approaching oceanic plate. Igneous rocks in the outboard, Pacific-type subduction complex + accretionary wedge are exclusively tectonized ophiolites and other portions of the oceanic lithosphere (Coleman, 1977).

Granitoids and related volcanic rocks are rare in Alpine-type complexes. Unlike Circumpacific convergent margins, plutons are not set back on the stable, nonsubducted plate; instead, small peraluminous bodies locally invade the root zone of the imbricate nappe edifice. Variable amounts of ophiolitic rocks occur in some compressional orogens, but tectonic fragments of oceanic lithosphere are conspicuously missing from many contractional mountain belts, (e.g., Liou et al., 1996). The rarity of the ophiolite suite and calcalkaline arc rocks in some Alpine-type complexes reflects the underflow of only small ocean basins. Most ophiolites are tectonically emplaced along convergent margins, so where limited lithospheric underflow takes place, fewer asperities (Cloos, 1993) are sampled; for this reason, oceanic material is uncommon in a some Alpine-type orogenic belts (for instance: the Qinling-Dabie-Sulu belt, east-central China; the Pan African terrane; the Western Gneiss Region, SW Norway; the Eastern Alps). Large tracts of oceanic lithosphere must be returned to the deep upper mantle in order to generate the voluminous intermediate magmas and massive calcalkaline arc in the crust above a well-developed subjacent plumbing system (Ernst, 1971, 1999)

Metamorphic Belts

Coeval, paired metamorphic belts characterize Pacific-type subduction zones. Two examples are presented in Figure 3, illustrating the blueschist-eclogite HP terrane. The outboard subduction complex consists of units recrystallized at low temperatures and high pressures. Reflecting its dynamic plate-tectonic setting within the subduction channel (directly below the hanging-wall lithosphere, which acts as a stress guide), oceanward vergence of folds is ubiquitous; shearing and foliation + lineation are typical, with recovered depths of metamorphism ranging up to 15–70 km (Bailey et al., 1964; Ernst, 1971; Aoya, 2001). Landward, calcalkaline granitoids are emplaced into a consanguineous volcanic cover series and underlying, pre-existing basement. Both volcanic and basement rocks exhibit widespread recrystallization at high temperatures and low pressures. Metamorphic parageneses reflect the advective transport of heat introduced by rising magmas, and consist of a series of superimposed regional contact aureoles surrounding individual plutons (Barton et al., 1988). Hornfels and open, cylindrical folding typify such mid- and upper-crustal metamorphic terranes (Ernst, 1992).

Alpine subduction complexes contain abundant, pre-existing continental massifs, as well as penetratively deformed, superjacent autochthonous and allochthonous entities. Metamorphic belts of the Alpine type are not paired (Frey et al., 1974). High-pressure subsolidus recrystallization is the rule, with parageneses comparable to those of Circumpacific subduction complexes. A generalized map of the Alps is illustrated in Figure 4. Some contractional orogens include tectonic slices that retain scattered mineralogic relics including coesite and microdiamond—commonly as inclusions in tough, unreactive container phases such as zircon and garnet. These rare inclusions attest to ultrahigh pressures (UHP) and somewhat higher temperatures, but prograde geothermal gradients are similar to those that characterize Pacific-type subduction complexes

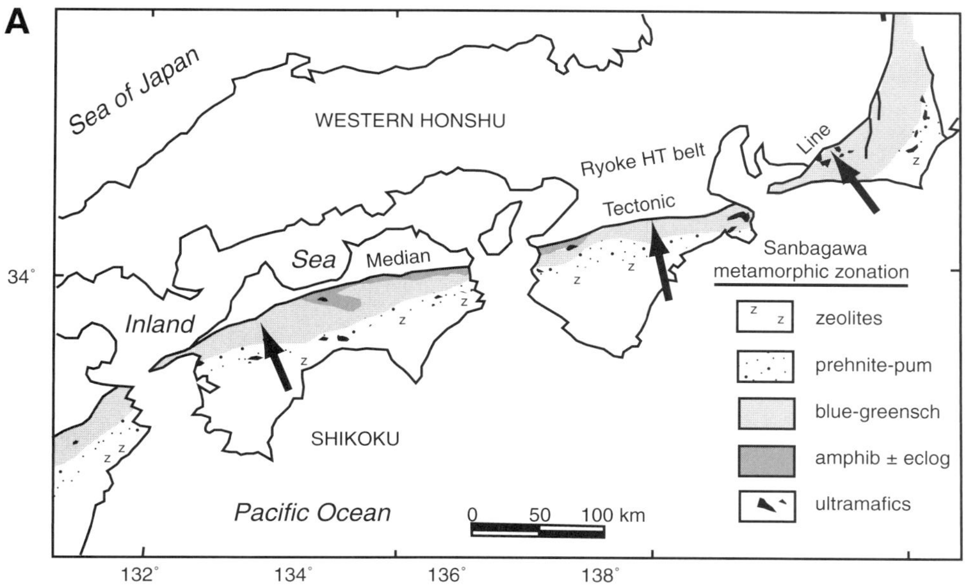

Figure 3. Generalized regional settings of Pacific-type HP metamorphic belts, showing directions of increasing subduction depths, now recovered, and prograde blueschist-eclogite metamorphism (arrows). A: Sanbagawa belt of SW Japan, after Hashimoto et al. (1970). B: Franciscan complex of western California, after Bailey et al. (1970). Abbreviations: CRT—Coast Range thrust; SAF—San Andreas fault; SNF—Sur-Nacimiento fault; amphi—amphibolite; eclog—eclogite; greensch—green schist; pum—pumpellyite.

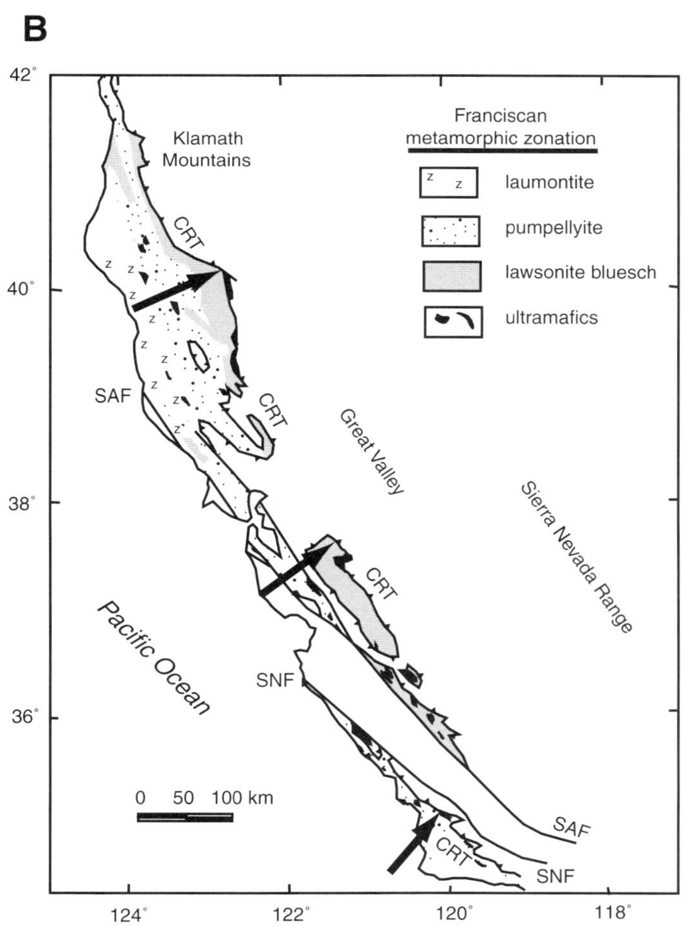

(Chopin, 1984; Smith, 1984; Sobolev and Shatsky, 1990; Coleman and Wang, 1995). Pressure-temperature conditions reflect situations where an old segment of continental crust—annealed and, thus, well bonded to the oceanic lithosphere—enters the subduction zone and descends to depths of 60–140 km before decoupling from the downgoing plate. For such terranes, HP-UHP recrystallization results. HP, HT, and UHP metamorphic conditions are compared in Figure 5.

Summary of Subduction-Zone Metamorphic Features

Circumpacific convergence zones are characterized by widespread blueschist-eclogite belts, tectonic mélanges, and abundant dismembered ophiolite complexes, reflecting the underflow of thousands of kms of ocean lithosphere. Zones of continental underflow, on the other hand, tend to involve the consumption of relatively small intervening ocean basins, and the descent of sialic crust prior to amalgamation. Table 1 lists the contrasting lithologic natures of Alpine- and Pacific-type subduction-zone assemblages. Metamorphic prograde and retrograde pressure-temperature paths followed by such terranes during subduction and later exhumation are topologically similar, but some intracratonic contractional complexes retain relics of ultrahigh-pressure metamorphism (P_{max} = 2.5–4.0 GPa), whereas recovered outboard Circumpacific orogens typically consist of only high-pressure belts (P_{max} = 0.6–1.8 GPa). Both types are extensively retrogressed, but because HP belts tend to have formed at shallower depths during subduction, their exhumation commonly involves less complete back reaction due to low-temperature, more sluggish transformations compared with high-temperature back reaction rates overprinting UHP complexes.

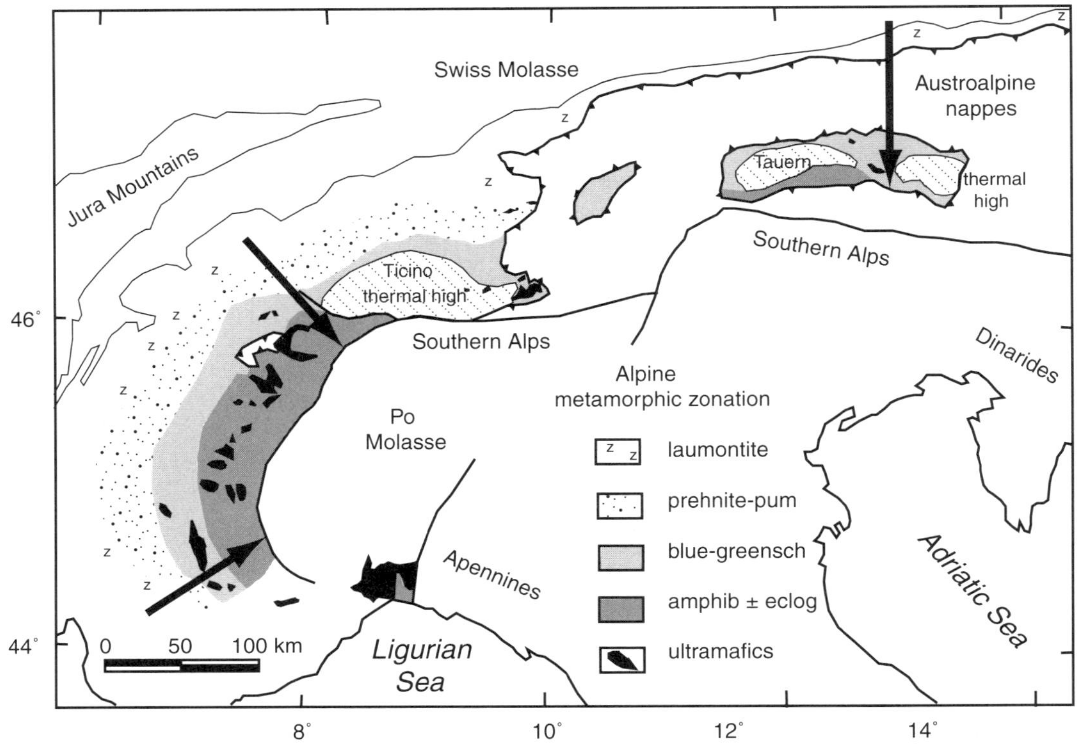

Figure 4. Generalized regional setting of the Alps, after Ernst (1971), a HP-UHP metamorphic belt, showing direction of increasing subduction depths and prograde blueschist-eclogite metamorphism (arrows), now recovered. The Ticino and Tauern thermal highs represent Miocene metamorphic culminations involving granitoids and kyanite-sillimanite recrystallization/annealing of the pre-existing units.

ROLE OF AQUEOUS FLUIDS IN SUBDUCTION-ZONE SETTINGS

General Statement

Consumption of an oceanic plate allows the subduction of entrained sediment and its basaltic basement. Rocks are either offloaded into the accretionary prism at shallow depths, as is most of the weak, low-density sedimentary prism, or returned to the upper mantle, as is most of the strong, high-density oceanic crust. Where microcontinental salients are entrained in the downgoing oceanic lithosphere, transport to the site of subduction leads to underflow, and continental crust may be carried beneath the non-subducted plate to considerable depths. Regardless of whether the descending materials are unconsolidated sediments or old, strong quartzofeldspathic crust, pressure within the slab rises with subduction depth, whereas temperature only gradually increases in most subduction zones. H_2O, and CO_2 may be added to relatively dry portions of the downgoing plate, and are driven off from the more volatile-rich parts (Peacock, 1995; Bebout, 1996). Shallow-level, pre-eclogitic stages of these reactions have been treated quantitatively by Frey, ed. (1987) and Ernst (1990). Attending deeper subduction of crustal lithologies, the remaining H_2O is sequestered exclusively in relatively refractory silicate phases (Schreyer et al., 1987; Poli and Schmidt, 1997; Ernst et al., 1998). During exhumation, volatiles gain access to ascending low-density masses, or diffuse away from them, depending on the composition, pressure-temperature trajectory, average grain size, permeability, and extent of shearing of the various units. Some expelled fluids rise into and through the overlying mantle wedge, whereas the rest migrates back up the conduit provided by the subduction channel (Cloos, 1984; Reck, 1987; Vrolijk, 1987; Vrolijk et al., 1988).

Evolution of H_2O and the Generation of Calcalkaline Arcs

Attending subduction, low-grade lithologies lose most of their volatiles at shallow depths during prograde recrystallization. At depths greater than ~50 km, clinoamphibole-bearing metabasaltic crust and the serpentinized peridotite underpinnings of the oceanic lithosphere comprise the most significant dehydrating rock types (Liu et al., 1996). Under equilibrium conditions, the devolatilization of Ca-amphiboles and serpentine minerals should begin at ~70–80 km depth; however, at low temperatures, reactions are sluggish on the time scale of plate descent, so fluid expulsion probably reaches a maximum at somewhat greater depths. Persistence of lower pressure phase assemblages during pressure overstepping

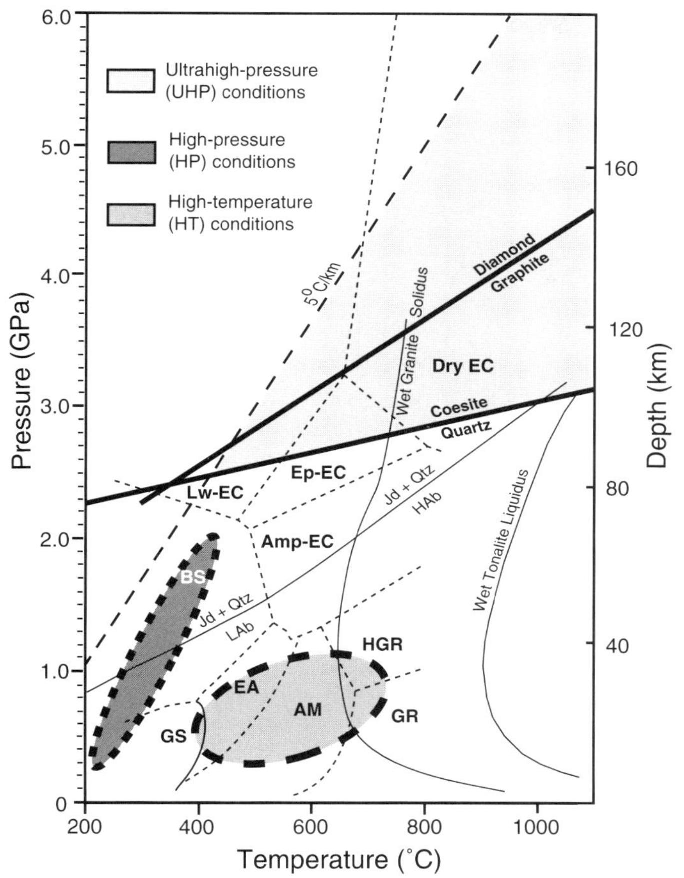

Figure 5. Petrogenetic grid for metabasaltic bulk-rock compositions, after Liou et al. (1998) and Okamoto and Maruyama (1999). The pressure-temperature realms for paired Pacific-type (HP + HT) and Alpine-type (HP-UHP) metamorphism are indicated. An extremely low subduction-zone geothermal gradient of 5 °C/km is also shown. Citations to the experimental phase equilibria are provided by Liou et al. (1998). Mineral abbreviations are: Jd—jadeite; Qtz—quartz; LAb—low albite; and Hab—high albite. Metamorphic-facies abbreviations are: AM—amphibolite; Amp-EC—amphibolite-eclogite; BS—blueschist; EA—epidote amphibolite; EC—eclogite; Ep-EC—epidote-eclogite; GR—sillimanite-bearing granulite; GS—greenschist; HGR—kyanite-bearing granulite; P—prehnite; PA—pumpellyite-actinolite; PP—prehnite-pumpellyite; and ZE—zeolite.

appears to be characteristic of some deeply subducted terranes where H_2O is absent (Austrheim, 1990, 1998). Thus, long-continued descent of thousands of kms of Pacific-type oceanic lithosphere may provide the greatest proportion of deep-seated aqueous fluid rising through the subduction channel, as well as diffusing into the overlying mantle wedge. Hydrous fluid-induced partial melting of the refertilized mantle hanging wall (and/or the eclogitic oceanic crust) evidently results in generation of the calcalkaline suite making up the massive volcanic-plutonic arc set back 100 ± 50 km from the convergent plate junction (Drummond and Defant, 1990; Kushiro, 1990; Hawkesworth et al., 1993; Morris, 1995; Clift et al., 2001; Hickey-Vargas et al., 2002). Concentrations of incompatible, large-ion-lithophile, and light rare earth elements, $^{87}Sr/^{86}Sr$ initial values, and epsilon Nd ratios in volatile-bearing arc volcanic and plutonic rocks argue for the partial fusion of garnet-bearing protoliths attended by an aqueous fluid.

Closure of a relatively small oceanic basin prior to continental subduction provides lesser amounts of volatiles derived from the descending oceanic lithosphere, explaining the general lack of H_2O-mediated generation of calcalkaline arc rocks in Alpine-type compressional belts. The fluid flux under UHP conditions is probably curtailed by the arrival, profound underflow, and suturing of sialic crust, because biotite and white mica—the principal hydrous phases present in metamorphosed quartzofeldspathic + pelitic rock types—remain stable at great depth, and do not devolatilize unless temperatures exceed ~800 °C. Accordingly, subduction is not expected to generate a substantial calcalkaline arc, but might involve partial melting of over-thickened continental crust, producing modest amounts of peraluminous, S-type granitoids as recycled material. Real continental growth—as opposed to the recycling of pre-existing sialic materials—appears to be the result of clinoamphibole ± serpentine dehydration within the descending oceanic lithosphere, followed by volatile-induced partial fusion of the metasomatized mantle wedge ± descending amphibolitic/eclogitic oceanic crust (Ernst, 1999).

The transition from oceanic to continental underflow is exemplified by the Indonesian Archipelago (Hamilton, 1979; Snyder et al., 1996). The strongly curved portion of the Australian-Eurasian suture between Timor and Seram-Irian Jaya is the site of underflow of old, cool continental crust; farther west along the Sumba-Java-Sumatra segment of the convergent boundary, oceanic lithosphere is diving beneath Indonesia. As evident from Figure 6, the Sunda-Banda inner volcanic arc is extremely active west of Timor, where oceanic lithosphere descends under Indonesia; volcanic activity declines eastward toward the Weber Deep, and the strongly curved portion of the arc is characterized by magmatic near-quiescence where Australian continental crust sinks beneath the Eurasian plate. In this easternmost part, the minor volumes of extrusive rock appear to be contaminated by assimilation of Australian passive margin sediments, as well as due to incorporation of continental crust (Charlton, 1991; Hilton et al., 1992; van Bergen et al., 1993). Generation of primary calcalkaline melts attending the subduction of hydrated oceanic crust-capped lithosphere in contrast to recycled volcanics associated with the downgoing, refractory micaceous sialic crust seems to explain the difference in volcanism laterally along the arc.

Cenozoic uplift and exposure of HP rocks in the eastern part of the suture zone reflect decoupling and ascent of subducted, largely quartzofeldspathic, low-density materials along the plate junction. As discussed in the section dealing with exhumation, the rise of sheet-like slabs of blueschist reflects rupturing and accelerated sinking of the oceanic lithosphere (Widiyantoro and van der Hilst, 1996); this ascent probably has been aided by a decrease in stress along the base of the buoyant, ductile continental material as it warmed within the upper mantle (Ernst et al., 1997).

TABLE 1. GENERALIZED LITHOLOGIC COMPARISON OF PHANEROZOIC SUBDUCTION-ZONE REGIMES

Protoliths	Alpine-type (high pressure–ultrahigh pressure)	Pacific-type (high pressure)
Shallow-marine strata	platform carbonates, orthoquartzites	reef (atoll) limestones
Clastic wedges	generally thin, well-stratified, multi-cycle siliciclastics, peraluminous shales	arc-derived, voluminous, first-cycle graywackes, muddy olistostromes
Deep-sea sedimentary rocks	uncommon deep-water carbonates	widespread bedded cherts, Mn-nodules, deep-water carbonates
Igneous rocks	mostly suprasubduction zone basalt + enriched-MORB	mostly normal-MORB + seamounts (OIB)
Basement	granitic gneiss complexes	oceanic lithosphere
Ore deposits	kuroko-type (massive sulfides)	mid-ocean ridge origin
Age of rock formations*	ancient to young	exclusively young
Tectonics	oceanward vergence of nappes, thrustfaults, late-stage backfolding	oceanward vergence of nappes, thrust faults. Rare antithetic thrusting
Rapid exhumation	during final collisional suturing	episodic, erratic

Petrology	Alpine-type (high pressure–ultrahigh pressure)	Pacific-type (high pressure)
Typical maximum pressure	2.6–4.0 GPa (ultrahigh-pressure)	0.6–1.8 GPa (high-pressure)
Associated coeval calcalkaline volcanic-plutonic belt	rare and small, or absent	outboard: absent or very minor; inboard: in all cases, huge
Degree of retrogression	ultrahigh pressure almost complete, high pressure incomplete	high pressure incomplete
Unaltered mantle fragments	garnet lherzolite	spinel ± plagioclase lherzolite, harzburgite, dunite
Associated, coeval serpentinite bodies	uncommon or common	outboard: ubiquitous inboard: uncommon or common
Coeval paired belts	absent	invariably present

*Young—generation of the lithologic section ~0–200 m.y. prior to time of underflow.

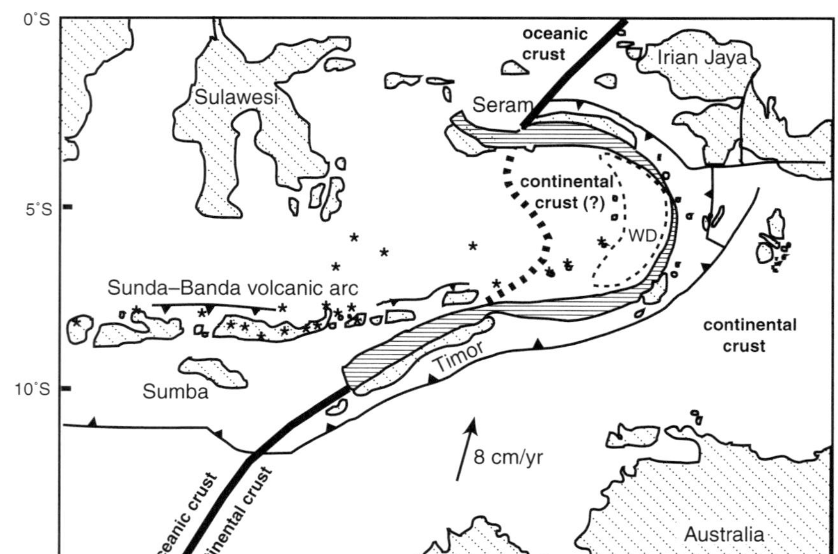

Figure 6. Eastern Sunda-Banda Arc, an example of convergence and exhumation of a transitional Alpine-Pacific-type belt, after Hamilton (1979) and Snyder et al. (1996) The plate-tectonic setting and exhumed blueschists (horizontal-lined pattern) are indicated. The 5 km isobath outlining the Weber Deep (WD), is shown by the closed dashed contour. Australian continental crust is subducting beneath the Timor-Seram-Irian Jaya section of the Indonesian suture zone; this eastern portion of the arc is typified by only minor calcalkaline igneous activity. The vigorous Sunda-Banda inner volcanic arc (asterisks—active volcanoes) lies chiefly west of Timor, where oceanic lithosphere of the Australian plate is descending beneath the Eurasian plate.

Kinetics of Metamorphic Hydration-Dehydration Reactions

Evidence of widespread disequilibrium during subduction-zone metamorphism (e.g., Frey et al., 1974; Rubie, 1990; Hacker, 1996; Austrheim, 1998) highlights the need for quantitative experimental data on reaction rates. Judging from the study of HP and UHP complexes, the presence or absence of aqueous fluid largely controls the rate at which reactions occur, and the extent to which they approach completion (Leech, 2001). Although the kinetic effect of H_2O is well known, and specific mechanisms for catalysis have been proposed (e.g., Rubie, 1986), little is known quantitatively about the dependence of reaction rates and mechanisms on fluid concentrations + chemistries in systems with very low fluid-rock ratios attending subduction of oceanic crust, an accretionary prism, or a microcontinental terrane.

The ease with which aqueous fluids gain access to, or egress from subducting lithosphere and ascending segments of low-density crustal material is a function of rock composition, permeabilities, deformation, grain size, and availability of volatiles. These parameters reflect the past geologic histories and dynamics of lithotectonic units moving along a convergent plate junction. Based on computational thermal modeling (Peacock, 1992, 1995; Ernst and Peacock, 1996; Hacker et al., 2003a) and laboratory phase-equilibrium studies (Vielzeuf and Holloway, 1988; Schreyer, 1995; Massone, 1995; Liu et al., 1996; Patiño Douce and McCarthy, 1998), the evolving thermal structure of a convergent plate junction, including subducting oceanic crust ± sedimentary load or continental material, as well as the pressure-temperature stability fields of the mineral assemblages for the major rock types are well understood. Almost unknown are the rates at which mineralogic transformations take place, and the conditions and tectonic realms under which volatile constituents are evolved and consumed. In the absence of aqueous fluid, low-pressure phase assemblages can be retained under prograde HP-UHP conditions (Austrheim, 1990; Zhang and Liou, 1997), and scattered HP-UHP relics can survive encased in unreactive container minerals under retrograde low-pressure conditions (Hirajima et al., 1993; Liou and Zhang, 1996).

EXHUMATION OF HIGH-PRESSURE–LOW-TEMPERATURE SUBDUCTION COMPLEXES

General Considerations

Most recognized HP and UHP terranes appear to be relatively tabular or sheet-like (Ernst et al., 1997). These tectonic slices have survived exhumation-induced decompression and recrystallization while retaining scattered relics of their high- to ultrahigh-pressure histories. The return toward shallow crustal levels occurs in response to one or more of several processes: tectonic contraction or wedge extrusion (Maruyama et al., 1994, 1996); buoyant body-force propulsion (Ernst, 1970, 1988; England and Holland, 1979; Hacker, 1996; Hacker et al., 2000); corner flow in front of a hanging wall backstop (Cowan and Silling, 1978; Cloos and Shreve, 1988a, 1988b; Cloos, 1993); and extensional or erosional collapse (Platt, 1986, 1987, 1993; Ring and Brandon, 1994, 1999). Because some relatively old, descending oceanic plates and deep-sea trenches appear to be retreating oceanward more rapidly than the encroaching nonsubducted lithosphere (Molnar and Atwater, 1978; Seno, 1985; Busby-Spera et al., 1990; Hamilton, 1995), compression of continental flakes in the jaws of a convergent junction cannot account for the ascent of subduction complexes. Moreover, although faulting associated with extensional collapse and/or erosion helps uncover formerly deeply buried terranes, this mechanism is incapable of explaining major pressure discontinuities observed across boundaries between subducted and nonsubducted lithologic units (Ernst, 1970; Ernst et al., 1970; Suppe, 1972), because differential uplifts exceeding the thickness of the continental crust would be required. However, buoyancy coupled with erosional decapitation and mass wastage, provides a viable mechanism for the exhumation of low-density sections carried to profound depths. Numerical simulations, laboratory scale models, and geologic field relationships demonstrate the operation of this process, as described farther on.

For both Pacific- and Alpine-type underflow of graywacke + shaley mélange, or a promontory of continental crust well bonded to the mantle lithosphere, entrance of increasing amounts of low-density material into the subduction zone reduces the negative buoyancy of the descending but coherent plate. This in turn may result in rupture and separation of the high-density oceanic lithosphere from the sialic crust at intermediate upper mantle depths—a realm where the lithosphere is in extension (Isacks et al., 1968). Breakoff of the oceanic slab (Sacks and Seacor, 1990; von Blanckenburg and Davies, 1995) increases the effective buoyancy of the up-dip quartzofeldspathic subduction complex; this rupture allows sialic sheets no longer attached to the oceanic lithosphere "anchor" to ascend back up the subduction channel (van den Beukel, 1992; Davies and von Blanckenburg, 1998). Exhumation also may be enhanced by a shallowing of the decoupled, buoyant, and rebounding continental lithosphere (Ernst et al., 1997). Also, the low-density continental lithosphere should rise as a consequence of the reduction of shear force acting along its base as the slab gradually warms in the upper mantle and passes through the brittle-ductile transition (Stöckhert and Renner, 1998). Devolatilization of the oceanic lithosphere would also produce embrittlement and weakening of the crustal section.

Previous studies (e.g., Ernst, 1971; Chopin, 1987) have demonstrated that deep subduction of low-density material accounts for the generation of Pacific- and Alpine-type metamorphic belts. A major petrotectonic problem consists of clarifying how these complexes have returned to shallow crustal levels while preserving relics of HP and UHP phase assemblages. The two-way movement of terranes along subduction zones was recognized long ago (Ernst, 1970; Suppe, 1972; Willett et al., 1993). Two end-member convergent plate junction configurations, both of which may involve large components of arc-parallel slip, are emphasized here: (1) subduction of a weak, Pacific-type graywacke + shale complex carried down on oceanic lithosphere; and (2) Alpine-type suturing of a relatively strong continent, microcontinent, or island arc beneath a nonsubducted lithospheric plate. The former gives rise to HP accretionary prisms that mark the Pacific Rim (Miyashiro, 1967). The latter involves underflow of salients or peninsulas of old continental crust, thoroughly imbedded in cool, oceanic lithospheric plates (Yin and Nie, 1993; Ernst and Liou, 1995). The rocks descend relatively rapidly, generating the characteristic HP-UHP mineralogy of contractional complexes (Peacock, 1995; Ernst et al., 1997). In a contrasting type of attempted subduction of muddy mélange, where the low-density materials are extremely incompetent and very poorly coupled to the descending lithosphere, upper plate collision/indentation occurs instead (Ellis, 1996); the East Taiwan ophiolite and Lichi mélange constitute an example of this type of aborted subduction (Liou et al., 1977; Page and Suppe, 1981; Ho, 1986).

The Thin-Slab Hypothesis

Rocks are good thermal insulators, hence it is exceedingly difficult for such materials to cool rapidly during exhumation. Because of this poor thermal conductivity, how can deeply buried but buoyant masses decompress rapidly without passing through more normal, low-pressure crustal pressure-temperature realms? All surviving HP-UHP complexes exhibit ubiquitous overprinting characteristic of granulite-, amphibolite-, and greenschist-facies conditions. In the presence of an aqueous fluid, adiabatic ascent would be expected to result in relatively high-temperature annealing and thorough obliteration of pre-existing HP and UHP mineral assemblages. Of course, lack of H_2O would decrease the rate of retrogression. Nevertheless, heat must be withdrawn from decompressing complexes as they are exhumed, or mineralogic evidence of their former HP and/or UHP configurations would not be retained in even fragmentary fashion.

The rare preservation of high- and ultrahigh-pressure relics is favored by continued subduction-induced refrigeration tectonically beneath the rising HP-UHP complex, and coeval extensional faulting against the overlying, cooler hanging-wall plate (Hacker and Peacock, 1995). In such cases, relatively thin slices of material would lose heat along both upper and lower surfaces by thermal conduction during ascent. Some subduction complexes rise virtually adiabatically, whereas others appear to have nearly retraced the prograde pressure-temperature trajectory during decompression (Rubie, 1984; Ernst, 1988; Ernst and Peacock, 1996). Proposed relationships, shown diagrammatically in Figure 7, apply to the exhumation of both HP and UHP terranes. Forces acting on a sheet of subducted low-density material are as follows (Platt, 1987): (A) Descent of the sialic slab occurs only if shear forces caused by underflow (F_s) overcome the combined effects of buoyancy (F_b) and frictional resistance along the upper wall of the subduction channel (F_r). In this case,

$$F_s > F_b \sin\theta + F_r. \tag{1}$$

(B) Ascent of a slice of the low-density material occurs in cases where buoyancy is positive and exceeds the combined effects of shearing along the slab base and resistance to movement along its upper surface. For this situation,

$$F_b \sin\theta > F_s + F_r. \tag{2}$$

The hanging wall acts as a tectonic guide, so rising crustal materials are accreted seaward from the volcanic-plutonic arc.

Underflow of a basaltic crust-capped plate carrying a passive load of sediment, or closure of an oceanic basin, ultimately may result in the entrance of sialic material into the subduction zone. Heating and volatile expulsion cause a weakening of the subducted crust. Decoupling and ascent of a portion of this low-density section occurs provided buoyancy is positive, and exceeds the resis-

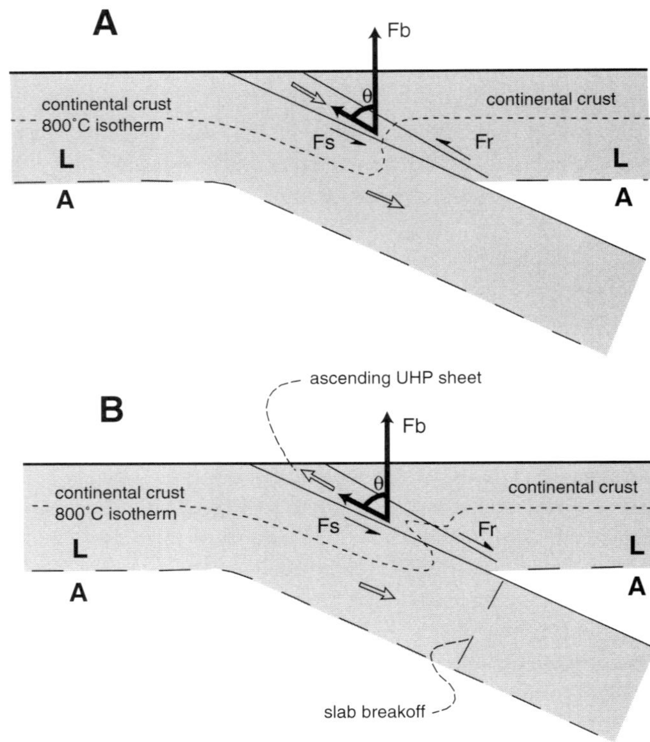

Figure 7. Schematic diagram portraying forces during active subduction, assuming structural integrity of the downgoing plate prior to slab breakoff, after Ernst et al. (1997). A: Deep burial and thermal structure of a subducted wedge of low-density material. B: Subsequent decompression and cooling of a rising slice of this buoyant HP-UHP material. Accompanying ascent of the thin HP-UHP sheet (thickness exaggerated for clarity), cooling of its upper margin takes place where it is juxtaposed against the lower temperature hanging-wall plate; cooling along the lower margin of the sheet takes place where it is in contact with the lower temperature, subducting-refrigerating lithospheric plate (see Fig. 1). Tectonic exhumation of low-density slices requires erosive denudation and/or gravitational collapse driven by the buoyant root at depth. The resolution of forces acting on the low-density slab in stages depicted in 7A and 7B are discussed in the text. Abbreviations are: A—asthenosphere; L—lithosphere.

tance to differential movement. For the HP and/or UHP mineral assemblages to be partly preserved on exhumation, the rising slab must be sufficiently thick to promote buoyancy-driven ascent, yet thin enough that heat is efficiently removed by thermal conduction across the sheet boundaries—the upper normal and lower reverse faults (Fig. 7). Such shear senses are required by observed structural relations and derived cross-sections, for example, in the Franciscan Complex (Ernst, 1970; Suppe, 1972; Platt, 1986; Jayko et al., 1987), the Western Alps (Henry, 1990; Compagnoni et al., 1995; Michard et al., 1995), the Sanbagawa Belt (Kawachi, 1968; Ernst et al., 1970; Banno and Sakai, 1989), the Kokchetav Massif (Kaneko et al., 2000; Ishikawa et al., 2000; Ota et al., 2000), and the Dabie-Sulu Belt (Hacker et al., 1995, 1996, 2000; Webb et al., 1999). A cross-section of the 1–2 km thick UHP sheet documented from the Western Alps is illustrated in Figure 8. Similar, remarkably thin allochthonous UHP sheets have been described from the Kokchetav Massif (Kaneko et al., 2000; Ishikawa et al., 2000; Maruyama and Parkinson, 2000; but see Theunissen et al., 2000), and the Western Gneiss Region of southwest Norway (Robinson and Terry, 1998; Terry et al., 2000a, 2000b).

A decrease in bulk density of the continental lithosphere after slab breakoff results in a shallowing of the subduction angle, and may be partly responsible for the late doming recognized in some exhumed convergent plate junctions. This late-stage exhumation also may be aided by continuing accretionary underplating, coupled with isostatic uplift and extensional collapse (Platt, 1986, 1987, 1993). Another unloading mechanism involves the antithetic faulting typical of some compressional orogens, in which double vergence is produced during terminal stages of the buoyant ascent of low-density crust and its erosional removal (e.g., Dal Piaz et al., 1972; Ring and Brandon, 1994, 1999; Beaumont et al., 1996, 1999).

Where thin HP-UHP slices (1–5 km thick, and tens to ≥100 km in lateral dimensions) are exhumed during continued subduction + refrigeration, the ascending sialic complex approximately follows the prograde metamorphic pressure-temperature path in reverse, but on its high-temperature side. Such pressure-temperature trajectories have been documented, for instance, in HP and UHP terranes of the California Coast Ranges and the Western Alps (e.g., Ernst, 1988; Chopin et al., 1991; Coleman and Wang, 1995). For thick, more nearly equidimensional ascending bodies (>10 km thick?), the ratio of cooling surface to mass is low, and interior portions probably remain sufficiently hot during decompression for the complete obliteration of all evidence of deep-seated metamorphism, and in some cases for partial melting to ensue; such exhumed masses would not retain precursor HP-UHP phases, and would not be recognized as ever having been deeply buried. Intermediate conditions between subduction-zone refrigeration and adiabatic exhumation evidently caused minor partial fusion in the ascending microdiamond-bearing Kokchetav Massif of northern Kazakhstan (Sobolev and Shatsky, 1990), and possibly in parts of several other UHP complexes (Cuthbert and Carswell, 1990; Carswell et al., 2000).

But are subduction complexes buoyant enough to overcome the traction of the oceanic plate carrying them downward? The one-atmosphere densities of unaltered oceanic crust, ~3.0, continental material, ~2.7, and anhydrous mantle, ~3.2, increase with elevated pressure, reflecting the progressive transformation of framework silicates to layer-, chain-, and orthosilicates. Typical HP and UHP mineralogic assemblages and computed ambient rock densities appropriate for burial depths of ~100 km (Ernst et al., 1997) are: metabasaltic eclogite, ~3.6; eclogitic quartzofeldspathic gneiss, ~3.0; and garnet peridotite, ~3.3. Completely transformed to a dense assemblage, K-feldspar + jadeite + coesite-bearing granitic gneiss nevertheless remains less dense than garnet or spinel lherzolite, whereas metabasaltic eclogite is considerably denser than the mantle. Thus, at any reasonable subduction depth, HP-UHP sialic crust is buoyant relative to the surrounding mantle and tends to rise, whereas eclogitized oceanic crust is negatively buoyant and continues to sink. This relationship explains why exhumed HP and UHP terranes worldwide consist of >85%–90% felsic materials, and contain only small proportions of mafic and anhydrous ultramafic lithologies. It also accounts for the volumetrically minor presence of ophiolitic complexes, except for thoroughly serpentinized masses, in exhumed subduction complexes.

The process illustrated in Figure 7 is geometrically somewhat similar to the slab-extrusion model that Maruyama et al. (1994) proposed to account for the Dabie-Sulu UHP Belt of east-central China, as well as tectonic models advanced for the Franciscan Complex, the Himalayas, and the Alps (e.g., Ernst, 1970; Burg et al., 1984; Burchfiel and Royden, 1985; Merle and Guillier, 1989; Wheeler, 1991). Explicit to the present scenario, however, is the dominance of buoyancy-driven return of the HP-UHP metamorphosed sheet of continental crust or accretionary prism back up the subduction zone (rather than compression), followed by extensional collapse and erosion (e.g., see; England and Molnar, 1993; Platt, 1986, 1987, 1993).

Numerical and Scale-Model Convergent Plate Junctions

The laboratory behavior of a sheet of low-density material resting on a dense, sinking slab has been experimentally modeled by Chemenda et al. (1995, 1996) employing proportional scaling factors combined with a simple three-layer convergent system. Underflow of the structurally coherent, continent-like sheet progresses until buoyancy exceeds the frictional resistance to decoupling from the downgoing plate; at this stage, the low-density material begins to return surfaceward along the subduction channel. The modeled subduction and exhumation processes are similar to the inferred dynamics of Alpine- and Pacific-type convergent margins illustrated in Figure 7. The simplified experiments of Chemenda et al. duplicate the overall imbricate architecture of the Himalayas and many other convergent plate junctions described in the literature. Finite-element computations of contractional orogens also produce familiar structures (Beaumont et al., 1996, 1999). Contrasting tectonic architectures reflect the variable extents of underplating and erosion-enhanced exhumation, as well as different physical and dynamic input parameters.

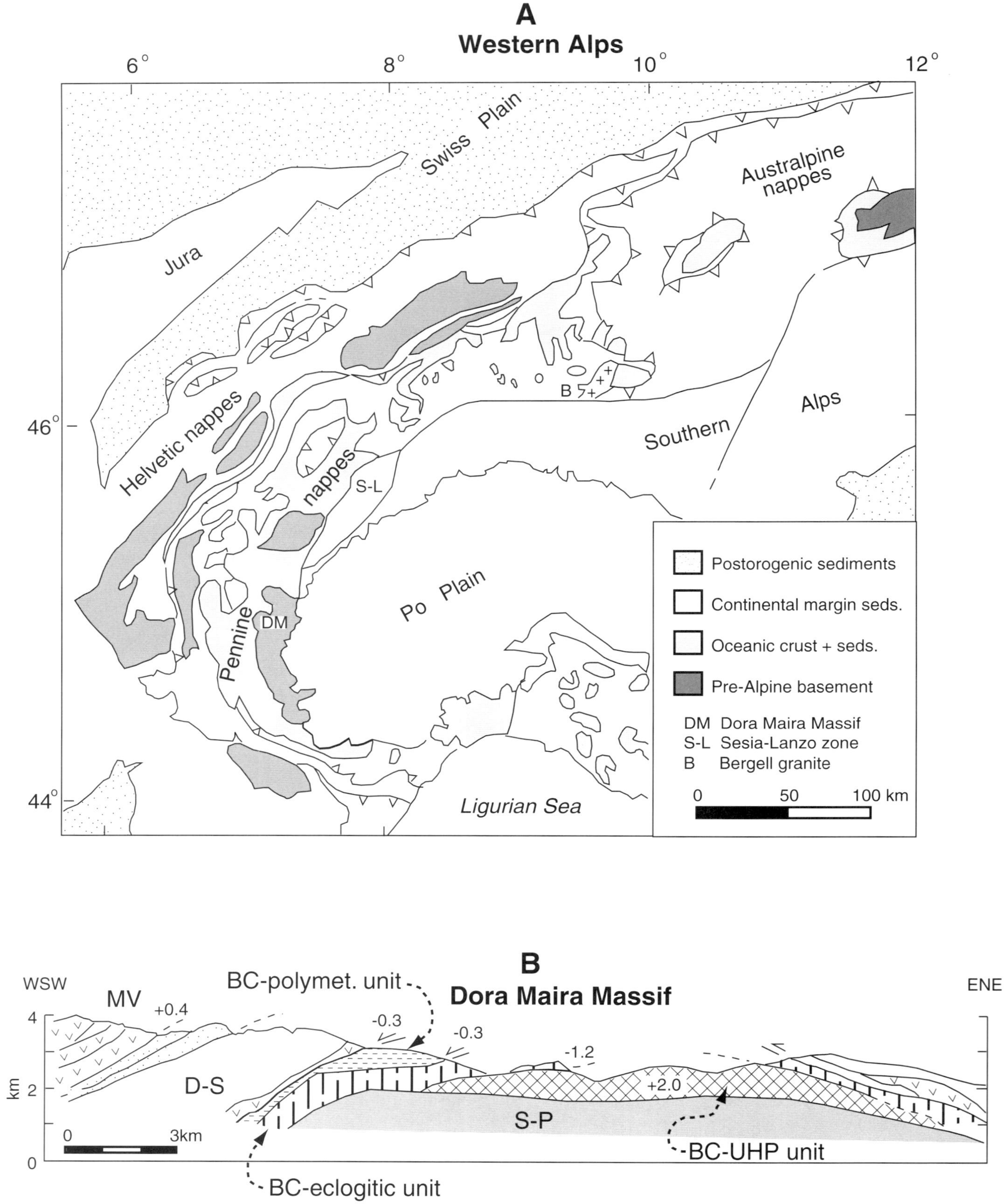

Figure 8. A: Regional geology and B: local structural + metamorphic relationships in the Dora Maira Massif of the Western Alps, simplified after Dietrich et al. (1974), Henry (1990), Compagnoni et al. (1995), Michard et al. (1995) and Coleman and Wang (1995). In the diagrammatic cross-section in the southern part of the Dora Maira Massif, numbers indicate the upward *change* in recorded pressure (GPa) relative to the adjacent underlying unit. Local nappe units are distinguished by different patterns in 8B, with abbreviations as follows: MV—Monviso meta-ophiolite and tectonically underlying serpentinitic mélange; D-S—Dronero-Sampeyre Permo-Triassic gneisses; BC—blueschistic and eclogitic basement complex; S-P—Sanfront-Pinerolo upper Paleozoic mica schists.

PETROTECTONIC CONTRASTS BETWEEN ALPINE- AND PACIFIC-TYPE METAMORPHIC BELTS

Petrologic and tectonic mechanisms proposed to explain the origin, plate-tectonic setting, and pressure-temperature evolution of Circumpacific and Alpine-type metamorphosed complexes must account for the wealth of observations regarding such terranes (Banno, 1964; Banno and Sakai, 1989; Bailey et al., 1964; Frey et al., 1974; Liou et al., 1994, 1995; Coleman and Wang, 1995; Dobretsov et al., 1995; Maruyama, 1997; Hacker et al., 2003a). Some of these differences are summarized in Table 2.

(1) Disrupted ophiolitic lithologies are present in both outboard Pacific- and Alpine-type belts. Although many exceptions occur, they consist chiefly of open-ocean N-MORBs + OIBs in Circumpacific subduction-zone terranes, and E-MORBs + IAT suprasubduction-zone complexes in Alpine-type belts (Dilek et al., eds., 2000).

(2) UHP metamorphic complexes are developed within old, relatively cool continental crust and are confined to intracontinental Alpine-type orogenic terranes, whereas HP metamorphic terranes frame the Pacific Rim, and are present in many intracratonic compressional sutures as well.

(3) Quartzofeldspathic ± pelitic lithologies constitute the most abundant rocks in both HP and UHP metamorphic belts. Pacific-type orogens consist of low metamorphic grade ophiolite + metagraywacke + metashale mélanges. High metamorphic grade continental equivalents, representing older protoliths, include para- and orthogneisses, paraschists, migmatites, and metagranitoids, followed by lesser amounts of metabasaltic rocks and metacalcareous platform strata characterize some Alpine-type orogens.

(4) Mafic and ultramafic SSZ rocks are of minor volumetric abundance in many but not all collisional orogens, whereas tectonized MORB + OIB ophiolites are ubiquitous in Circumpacific belts. In either case, similar to high- and ultrahigh-pressure metamorphic rocks, they mark the sites of convergent plate junctions.

(5) For UHP terranes, massive metamafic rocks contain most of the surviving relict ultrahigh-pressure silicate phases. Metacarbonates carry very rare UHP minerals; schistose metapelites and gneissose quartzofeldspathic lithologies generally lack UHP silicates, although exceedingly rare coesite and/or diamond micro-inclusions in garnet and zircon have been described (Liou and Zhang, 2002). For HP belts, metabasaltic and metagabbroic lithologies preserve minerals and phase assemblages indicating elevated pressures more completely than do quartzofeldspathic and pelitic lithologies; metacherts typically do not contain diagnostic HP phases, and if initially present in marbles, aragonite is rarely preserved.

(6) Postmetamorphic products of erosion are present but do not necessarily occur as giant sedimentary deposits adjacent to either Alpine- or Pacific-type mountain belts.

(7) Coeval calcalkaline volcanic-plutonic belts are conspicuously absent in Alpine-type terranes, but are ubiquitous,

TABLE 2. PRESSURE-TEMPERATURE-TIME HISTORIES OF HIGH- AND ULTRAHIGH-PRESSURE METAMORPHIC COMPLEXES*

Terrane Characteristic	Franciscan Complex, eastern belt	Sanbagawa Besshi + Oboke nappes	Dabie-Sulu Belt, coesite-eclogite unit	Kokchetav Massif, UHP unit	Dora Maira Massif, I. Venasca nappe	W. Gneiss Region, Fjordane Complex
protolith formation age	120–145 Ma	150–250 Ma	1.3–2.9 Ga	2.2–2.3 Ga	~300 Ma	1.6–1.8 Ga
temperature of metamorphism	250 ± 75 °C	300–625 °C	750 ± 75 °C	900 ± 75 °C	725 ± 50 °C	775 ± 75 °C
depth of metamorphism	15–35 km	20–70 km	90–125 km	~140 km	90–110 km	90–130 km
time of metamorphism	100–120 Ma	100 ± 10 Ma	235 ± 5 Ma	531 ± 3 Ma	35-40 Ma	400–410 Ma
upper crustal annealing	95 Ma (?)	70 ± 10 Ma	220 ± 5 Ma	525 ±3 Ma	30 Ma	~395 Ma
rise time to upper crust	15 ± 5 m.y.	30 ± 10 m.y.	15 ± 5 m.y.	~6 m.y.	3–4 m.y.	10 ± 5 m.y.
exhumation rate†	1–3 mm/yr	2–3 mm/yr	6–8 mm/yr	>18 mm/yr	>20 mm/yr	8–12 mm/yr
coesite inclusions	absent	absent	relatively abundant	rare	relatively abundant	rare
diamond inclusions	absent	absent	very rare	relatively abundant	absent	very rare
epidote blueschists	rare	common	present	rare	present	absent
lawsonite, aragonite, jadeite + quartz	law common, arag, jad + qtz rare	law rare, arag, jad + qtz absent	absent	absent	law, arag, jad + qtz rare	absent
areal extent	675 × 25 km	>700 × 35 km	>400 × 75 km	~120 × 10 km	225 × 60 km	350 × 70 km
max. thickness of HP or UHP units	5–10 km	>5 km	5 km (?)	1–3 km	1–2 km	>1 km (?)

*After Coleman and Wang (1995), Harley and Carswell (1995), Beane et al. (1995), Ernst and Peacock (1996), Hacker et al. (2000), and Rubatto and Hermann (2001).

†Average exhumation rates estimated by dividing depth of ultrahigh-pressure (UHP) metamorphism by time of ascent to 10–15 km crustal depth.

immense, and paired with an outboard subduction complex in Pacific-type belts.

(8) Rocks of the volcanic-plutonic arc may preserve a semicontinuous record of igneous activity marking landward HT belts of the Pacific Rim, whereas the exhumation of seaward blueschist-eclogite terranes in general is capricious and episodic in both Alpine- and Pacific-type subduction complexes.

(9) Well-dated Alpine-type terranes preserving scattered UHP relict phases have been exhumed to midcrustal levels at remarkably rapid ascent rates (Amato et al., 1998; Purchuk et al., 1998; Webb et al., 1999; Rubatto and Hermann, 2001; Hermann et al., 2001; Hacker et al., 2003b), approaching or exceeding 5–20 mm/yr. Radiometric data provide fewer constraints for the decompression of HP belts, but in general they also seem to have decompressed rapidly, although perhaps not as fast (1–3 mm/yr) as UHP terranes (Ernst, 1988; Itaya and Takasugi, 1988; Dallmeyer and Takasu, 1991).

DISCUSSION

How are the petrotectonic features of Alpine and Pacific-type metamorphic belts enumerated in the previous section accounted for by the inferred dynamic generation and exhumation histories of convergent plate junctions?

(1) Oceanic lithosphere is thrust into subduction complexes by convergent plate motion, and develops in continental margins or island arcs by back- and intra-arc spreading. These processes are similar in Circumpacific and contractional orogens, hence contrasts in the nature of associated ophiolites in HP and UHP metamorphic belts is a function of haphazard events. N-MORB and OIB suites seem to be most abundant in Pacific-type subduction zones, reflecting the sampling of an extensive tract of converging oceanic crust with its numerous lithospheric asperities, such as fracture zones, seamounts, hot spots, and spreading ridges. In contrast, consumption of a small ocean basin disfavors incorporation of abundant slices of oceanic crust + uppermost mantle because the brief underflow of oceanic lithosphere is limited by continental amalgamation; in such cases, SSZ ophiolites and E-MORBs can be as important as N-MORBs and OIBs. But, because of the tectonic imbrication of ophiolitic materials from both stable (nonsubducted), and descending plates, numerous contrary examples exist in consumptive suture zones.

(2) Mechanical analyses suggest that, provided it is well bonded to the downgoing lithosphere, sialic crust several km or more thick may be readily subducted if it enters a convergent plate junction (Molnar and Gray, 1979; England and Holland, 1979; Beaumont et al., 1996, 1999). A favorable geologic environment for ultrahigh-pressure metamorphism would involve insertion of a narrow tongue of continental crust into the subduction zone as an integral part of an old, thermally relaxed, largely oceanic slab (Ernst and Liou, 1995). The existence of UHP metagranitoids and quartzofeldspathic gneisses (e.g., Sobolev and Shatsky, 1990; Biino and Compagnoni, 1992; Wallis et al., 1997) demonstrates that low-density continental material may be carried to depths of at least 60–140 km. HP-UHP metamorphism is a consequence of this process, and is independent of lithology; however, the deep-seated sequestration of buoyant quartzofeldspathic rocks is a prerequisite for the later body-force-driven ascent of at least a portion of the HP-UHP terrane. Large, coherent masses of eclogitic oceanic crust remain negatively buoyant, so continue sinking into the deep mantle. Thus, resurrected UHP complexes essentially are restricted to zones of continental and microcontinental suturing—the plate-tectonic environment of Alpine-type subduction.

Pacific-type subduction of a quartzofeldspathic accretionary prism or shaley mélange involves the geologically simultaneous underflow of a longitudinal belt of low-density material. Insertion of a voluminous section of ductile graywacke ± shale into the convergent plate junction almost certainly would terminate subduction unless offloading and/or decoupling of the low-density material occurs. Because of weak bonding between the relatively incompetent sedimentary section and the sinking oceanic lithosphere, decoupling, ascent, and accretion would be expected to take place episodically at relatively shallow depths compared to UHP metamorphosed terranes.

(3) Lithospheric slabs descend to depths of 650 km or more along inclined Benioff-Wadati zones, so where underflow exceeds a few cm per year, HP and even UHP recrystallization are inevitable (Peacock, 1995); however, oceanic crust transformed to eclogite-facies mineral assemblages is denser than mantle peridotite, and garnet lherzolite is slightly denser than plagioclase- and spinel-bearing peridotite, so the oceanic crust + mantle will continue sinking after these phase transformations occur. In contrast, where a sufficiently large volume of low-density sialic material is subducted, buoyancy can overcome frictional resistance and permit ascent after disengagement from the downgoing plate. Accordingly, both recovered contractional UHP complexes and Circumpacific HP terranes consist chiefly of sialic bulk compositions, possessing an aggregate density considerably less than that of the mantle they displaced during subduction. Metamorphosed ophiolitic complexes represent a volumetrically minor portion of HP and UHP terranes because, if the latter contained a substantial complement of eclogitized oceanic crust, it would not be sufficiently buoyant to return to crustal levels (Ernst et al., 1997).

In support of this conclusion, all described UHP and HP terranes consist dominantly of low-density materials, with mafic ± anhydrous ultramafic rock types constituting <10%–15% by volume (Banno, 1964; Bailey et al., 1964; Frey et al., 1974; Coleman and Wang, 1995; Lennykh et al., 1995; Hacker et al., 1996; Maruyama et al., 1996; Leech, 2001). This reflects the buoyancy-driven ascent process that allows slices of profoundly subducted sialic terranes to be returned toward midlevels of the continental crust.

(4) The incompetence of shaley mélange + graywacke-rich, imbricated subduction complexes compared with old continental crust promotes the tectonic insertion of portions of the underlying oceanic crust ± serpentinized harzburgite basement into

downgoing Pacific-type terranes; many such mafic-ultramafic complexes have decoupled at shallow depths from the sinking oceanic lithosphere, and are sequestered along the inboard trench wall or nonsubducted continental margin. In contrast, the structural integrity of descending Alpine-type continental entities in some cases (but not the Western Alps) disfavors tectonic incorporation of disaggregated ophiolitic materials into the coherent HP-UHP complex. In both Alpine and Pacific types, variably hydrated peridotite of the overlying mantle wedge remains in a tectonically elevated position, and in some occurrences caps the subduction complex as an allochthonous sheet.

(5) Quartzofeldspathic + pelitic schists and gneisses contain abundant white mica ± biotite—rather refractory hydrous phases stable under UHP conditions (Hermann and Green, 2001). Such rocks in general are strongly foliated, whereas carbonates, cherts, and coarse-grained mafic igneous rocks are massive and nearly anhydrous. The former are relatively permeable to aqueous fluids compared to the latter. Thus, once formed, the retention of UHP relics is kinetically favored in massive, dry eclogites, impermeable siliceous schists, and a few marbles, but disfavored in more sheared, permeable, hydrated sialic units. The absence of H_2O, and closed-system recrystallization in situations where UHP relics are preserved, is supported by the anomalously low pre-subduction $\delta^{13}C$ and $\delta^{18}O$ values measured in Dabie-Sulu belt UHP coesite eclogites and their constituent minerals (Yui et al., 1995; Baker et al., 1997; Rumble, 1998; Zheng et al., 2000). Dry conditions are also suggested by laboratory kinetic studies of the transformation of coesite to quartz which indicate that, on decompression at moderate temperatures, H_2O contents of ~400 ppm in SiO_2 are sufficient to cause rapid, complete transformation to the low-pressure polymorph (Mosenfelder and Bohlen, 1997). For this reason, only very rare relict coesite has been discovered as micro-inclusions in zircons from Dabie-Sulu quartzofeldspathic gneisses (Tabata et al., 1998). The same situation holds for other UHP metamorphic terranes.

Blueschist belts have been subjected to pressure-temperature conditions less extreme than those attending UHP complexes, hence devolatilization is incomplete during prograde recrystallization, and retrogression is widespread attending decompression (Dal Piaz et al., 1972; Frey et al., 1974; Ernst, 1988; Banno and Sakai, 1989). Lower temperatures of HP metamorphism impede back-reaction, thus in many cases the low-pressure mineralogic overprint is incomplete. In contrast to silicate re-equilibration, the transformation of aragonite to calcite is exceedingly fast (Carlson and Rosenfeld, 1981), so orthorhombic $CaCO_3$ is only preserved in dry rocks subjected to very low-temperature retrograde paths.

(6) In order to elucidate possible decompression pressure-temperature trajectories allowing partial preservation of high- and ultrahigh-pressure relict phases, exhumations of both Alpine- and Pacific-type fault-bounded terranes have been numerically modeled (Ernst and Peacock, 1996) as extensional along the upper surface of a thin sheet while subduction-refrigeration continues along the lower thrust surface (Fig. 7). The process does not require the wholesale exhumation of a lithospheric plate, the entire continental crust, or even the full thickness of a quartzofeldspathic accretionary prism, and certainly not the structurally overlying mantle wedge. Thus, postmetamorphic erosional debris, although considerable in some cases (e.g., the Bengal fan and the Songpan-Ganze flysch basin), need not be voluminous in either Pacific- or Alpine-type orogens.

(7) Circumpacific subduction is a long-continued process involving thousands of kms of underflow of an oceanic plate, so sufficient time is available for the landward development of calcalkaline magmas at great depths, and maturation of the mantle hanging-wall plumbing system. The pressure-overstepped dehydration of abundant clinoamphiboles in the downgoing oceanic slab at depths exceeding 70–80 km (probably reaching a maximum near ~100 km due to sluggish devolatilization attending rapid subduction at low temperatures) would provide sufficient H_2O necessary to generate compositionally intermediate magmas through partial fusion of the oceanic crust and/or the undepleted mantle wedge (Drummond and Defant, 1990; Kushiro, 1990; Hawkesworth et al., 1993; Morris, 1995; Ernst, 1999). The generation and rise of these hydrous melts—apparently produced by this process—results in the construction of an andesitic-granitic arc inboard from the Pacific-type convergent plate junction on the stable, nonsubducted plate. In contrast, continental subduction in many cases involves relatively short-lived underflow of old, thermally relaxed oceanic lithosphere, so sufficient time may not have elapsed to develop a substantial volcanic-plutonic arc on the nonsubducted plate prior to amalgamation. During descent and suturing, layer silicates that dominate the sialic crust remain stable to subduction depths of at least ~140 km; thus, the arrival of continental crust at a convergent plate junction and its deep underflow would severely reduce the evolution of aqueous fluids at UHP depths—the pressure-temperature realm where calcalkaline magmas are generated.

(8) Long-sustained lithospheric underflow characteristic of Pacific-type subduction zones produces a massive, well-preserved andesitic-granitic arc typified by HT metamorphism in the spatially associated pre-existing crust, because magmas move up into the nonsubducted plate semicontinuously. In contrast, seaward subduction-zone metamorphic complexes of both Alpine- and Pacific-type descending plates disengage from the downgoing lithosphere episodically; return of low-density material to mid-crustal levels is a sensitive function of the changing geometry and physical properties of various units constituting the architecture of the convergent junction. Accordingly, only fragmentary portions of such blueschist-eclogite belts are erratically exhumed.

(9) The described petrotectonic features are compatible with the proposed formation, pressure-temperature evolution, and buoyancy-propelled tectonic exhumation + exposure through erosion and/or gravitational collapse of Alpine and Circumpacific terranes (e.g., Ernst, 1970; Platt, 1986, 1987, 1993; Maruyama, 1997), as well as the formation of massive calcalkaline arcs inboard from Pacific-type plate junctions. The kind of materials carried down subduction channels, the extent of deep-seated devolatilization, and the rates of transformation strongly influence

the nature of HP and UHP metamorphic belts. Exhumation of the subducted complexes to mid-crustal levels appears to be density driven, and the ascent in most cases is relatively fast. Because H_2O-rich Pacific-type subduction packages displace relatively small volumes of dense mantle ± serpentinized ultramafic material, they ascend at a few mm/yr—more slowly than do deeply subducted, rapidly exhumed Alpine-type complexes. Exhumation rates of 5–20 mm/yr, averaged over more than 5–20 m/yr, and seemingly required by UHP geochronologic data (Liou and Zhang, 1995; Dobretsov et al., 1995; Gebauer, 1996; Gebauer et al., 1997; Blythe, 1998; Purchuk et al., 1998; Grasemann et al., 1998; Lin and Roecker, 1998; Webb et al., 1999; Wheeler et al., 2001: Rubatto and Hermann, 2001) substantially exceed currently measured uplift and erosion rates in the Himalayan orogenic belt (Le Fort, 1996; Searle, 1996), but are marginally compatible with rates of exhumation of ~4 mm/yr calculated by Genser et al. (1996) for the Eastern Alps.

ACKNOWLEDGMENTS

This study was support by Stanford University, and the U.S. National Science Foundation through grant EAR97–25347 to J.G. Liou. Shohei Banno, B.R. Hacker, J.G. Liou, and R.G. Coleman reviewed a first-draft manuscript. John Wakabayashi, C.A. Hopson, and J.P. Platt provided helpful reviews for the GSA. I sincerely thank these researchers and institutions for support and helpful feedback.

REFERENCES CITED

Amato, J.M., Johnson, C., Baumgartner, L., and Beard, B., 1998, Sm-Nd geochronology indicates rapid exhumation of Alpine eclogites: Earth and Planetary Science Letters, v. 171, p. 425–438.

Anderson, R.N., DeLong, S.E., and Schwarz, W.M., 1978, Thermal model for subduction with dehydration in the downgoing slab: Journal of Geology, v. 80, p. 731–739.

Aoya, M., 2001, P-T-D path of eclogite from the Sanbagawa belt deduced from combination of petrological and microstructural analyses: Journal of Petrology, v. 42, p. 1225–1248.

Austrheim, H., 1990, The granulite-eclogite facies transition: A comparison of experimental work and a natural occurrence in the Bergen Arcs, western Norway: Lithos, v. 25, p. 163–169.

Austrheim, H., 1998, The influence of fluid and deformation on metamorphism of the deep crust and consequences for geodynamics of collision zones, *in* Hacker, B.R., and Liou, J.G., eds., When Continents Collide: Geodynamics and Geochemistry of Ultrahigh-Pressure Rocks: Dordrecht, Kluwer Academic Publishers, p. 297–323.

Bailey, E.H., Irwin, W.P., and Jones, D.L., 1964, Franciscan and related rocks, and their significance in the geology of western California: California Division of Mines and Geology Bulletin, v. 183, 177p.

Bailey, E.H., Blake, M.C., Jr., and Jones, D.L., 1970, On-land Mesozoic oceanic crust in California Coast Ranges: Washington, D.C., U.S. Geological Survey Professional Paper 700C, p. 70–81.

Baker, J., Matthews, A., Mattey, D., Rowley, D.B., and Xue, F., 1997, Fluid-rock interactions during ultra-high pressure metamorphism, Dabie Shan, China: Geochemica et Cosmochimica Acta, v. 61, p. 1685–1696.

Bally, A.W., 1981, Thoughts on the tectonics of folded belts, *in*: Thrust and Nappe Tectonics: Price, N.J., and McClay, K., eds.: London, Geological Society of London, Special Publication 9, p. 13–32.

Banno, S., 1964, Petrologic studies on Sanbagawa crystalline schists in the Bessi-Ino district, central Sikoku, Japan: Tokyo, University of Tokyo, Journal of the Faculty of Science, section 2, v. 15, p. 203–319.

Banno, S., and Sakai, C., 1989, Geology and metamorphic evolution of the Sanbagawa belt: London, Geological Society of London, Special Publication 43, p. 519–535.

Barton, M.D., Battles, D.A., Bebout, G.E., Capo, R.C., Christensen, J.N., Davis, S.R., Hanson, R.B., Michelson, C.J., and Trim, H.E., 1988, Mesozoic contact metamorphism in the western United States, *in* Ernst, W.G., ed., Metamorphic and Crustal Evolution of the Western United States: Englewood Cliffs, New Jersey, Prentice-Hall, Inc., p. 110–178.

Beane, R.J., Liou, J.G., Coleman, R.G., and Leech, M.L., 1995, Mineral assemblages and retrograde P-T path for high- to ultrahigh-pressure metamorphism in the lower unit of the Maksyutov Complex, Southern Ural Mountains, Russia: The Island Arc, v. 4, p. 254–266.

Beaumont, C., Ellis, S., Hamilton, J., and Fullsack, P., 1996, Mechanical model for subduction-collision tectonics of Alpine-type compressional orogens: Geology, v. 24, p. 675–678.

Beaumont, C., Ellis, S., and Pfiffner, A., 1999, Dynamics of sediment subduction-accretion at convergent margins: short-term modes, long-term deformation, and tectonic implications: Journal of Geophysical Research, v. 104, p. 17,573–17602.

Bebout, G.E., 1996, Volatile transfer and recycling at convergent margins: Mass-balance and insights from high P/T metamorphic rocks: Washington, D.C., American Geophysical Union Geophysical Monograph 96, p. 179–193.

Biino, G., and Compagnoni, R., 1992, Very high pressure metamorphism of the Brossasco coronite metagranite, southern Dora Maira massif, western Alps: Schweizerisches Mineralogisches und Petrografisches Mittelungen, v. 72, p. 347–362.

Blythe, A.E., 1998, Active tectonics and ultrahigh-pressure rocks, *in* Hacker, B.R., and Liou, J.G., eds., When Continents Collide: Geodynamics and Geochemistry of Ultrahigh-Pressure Rocks, Dordrecht, Kluwer Academic Publishers, p. 141–160.

Boudier, F., Bouchez, J.L., Nicolas, A., Cannat, M., Ceuleneer, G., Miseri, M., and Montigny, R., 1985, Kinematics of oceanic thrusting in the Oman ophiolites: model of plate convergence: Earth and Planetary Science Letters, v. 75, p. 215–222.

Boudier, F., Ceuleneer, G., and Nicolas, A., 1988, Shear zones, thrusts and related magmatism in the Oman ophiolite: initiation of thrusting on an oceanic ridge: Tectonophysics, v. 151, p. 275–296.

Burchfiel, B.C., and Royden, L.H., 1985, North-south extension within the convergent Himalayas region: Geology, v. 13, p. 679–682.

Burchfiel, B.C., Deng, Q., Molnar, P., Royden, L., Qang, Y., and Zhang, W., 1989, Intracrustal detachment within zones of continental deformation: Geology, v. 17, p. 448–452.

Burg, J.P., Brunel, M., Gapais, D., Chen, G.M., and Liu, G.H., 1984, Deformation of leucogranites of the crystalline Main Central Sheet in southern Tibet (China): Journal of Structural Geology, v. 6, p. 535–542.

Busby-Spera C.J., Mattinson, J.M., Riggs, N.R., and Schermer, E.R., 1990, The Triassic-Jurassic magmatic arc in the Mojave-Sonoran Deserts and the Sierran-Klamath region, *in* Harwood, D.S., and Miller, M.M., eds., Paleozoic and Early Mesozoic Paleogeographic Relations; Sierra Nevada, Klamath Mountains, and Related Terranes: Boulder, Colorado, Geological Society of America Special Paper 255, p. 93–114.

Carlson, W.D., and Rosenfeld, J.L., 1981, Optical determination of topotactic aragonite-calcite growth kinetics: metamorphic implications: Journal of Geology, v. 89, p. 615–638.

Carswell, D.A., Wilson, R. n., and Zhai, M., 2000, Metamorphic evolution, mineral chemistry, and thermobarometry of schists and orthogneisses hosting ultra-high pressure eclogites in the Dabieshan of central China: Lithos, v. 52, p. 121–155.

Castro, A., Fernandez, C., and Vigneresse, J.L., eds. 1999, Understanding Granites: Integrating New and Classical Techniques: London, Geological Society of London Special Publication, v. 168, 278p.

Charlton, P.R., 1991, Post collisional extension in arc-continent collision zones, eastern Indonesia: Geology, v. 19, p. 28–31.

Chemenda, A.I., Mattauer, M., Malavieille, J., and Bokun, A.N., 1995, A mechanism for syn-collisional rock exhumation and associated normal faulting: Results from physical modelling: Earth and Planetary Science Letters, v. 132, p. 225–232.

Chemenda, A.I., Mattauer, M., and Bokun, A.N., 1996, Continental subduction and a new mechanism for exhumation of high-pressure metamorphic rocks: new modelling and field data from Oman: Earth and Planetary Science Letters, v. 143, p. 173–182.

Chopin, C., 1984, Coesite and pure pyrope in high-grade blueschists of the Western Alps. A first record and some consequences: Contributions to Mineralogy and Petrology, v. 86, p. 107–118.

Chopin, C., 1987, Very-high-pressure metamorphism in the western Alps: implications for subduction of continental crust: Philosophical Transactions of the Royal Society of London, ser. A, no. 1557, p. 183–197.

Chopin, C., Henry, C., and Michard, A., 1991, Geology and petrology of the coesite-bearing terrain, Dora-Maira massif, Western Alps: European Journal of Mineralogy, v. 3, p. 263–291.

Clift, P.D., Rose, E.F., Shimizu, N., Layne, G.D., Draut, A.E., and Regelous, M., 2001, Tracing the evolving flux from the subducting plate in the Tonga-Kermadec arc system using boron in volcanic glass: Geochimica et Cosmochimica Acta, v. 65, p. 3347–3364.

Cloos, M., 1984, Landward-dipping reflectors in accretionary wedges: Active dewatering conduits?: Geology, v. 12, p. 519–522.

Cloos, M., 1993, Lithospheric buoyancy and collisional orogenesis: Subduction of oceanic plateaus, continental margins, island arcs, spreading ridges, and seamounts: Geological Society of America Bulletin, v. 105, p. 715–737.

Cloos, M., and Shreve, R. L., 1988a, Subduction-channel model of prism accretion, melange formation and subduction erosion at convergent plate margins, 1, Background and description, *in* Ruff, L., and Kanamori, H., eds., Subduction Zones: Pure and Applied Geophysics: Basel, Switzerland, Birkhauser, v. 128, p. 455–500.

Cloos, M., and Shreve, R. L., 1988b, Subduction-channel model of prism accretion, melange formation and subduction erosion at convergent plate margins, 2, Implications and discussion, *in* Ruff, L., and Kanamori, H., eds., Subduction Zones: Pure and Applied Geophysics: Basel, Switzerland, Birkhauser, v. 128, p. 501–545.

Coleman, R.G., 1977, Ophiolites, Ancient Oceanic Lithosphere?: Springer-Verlag, New York, 229p.

Coleman, R.G., and Wang, X., 1995, Overview of the geology and tectonics of UHPM, *in* Coleman, R.G., and Wang, X., eds., Ultrahigh Pressure Metamorphism: New York, Cambridge University Press, p. 1–32. .

Compagnoni, R., Hirajima, T., and Chopin, C., 1995, UHPM metamorphic rocks in the Western Alps, *in* Coleman, R.G., and Wang, X., eds., Ultrahigh Pressure Metamorphism: New York, Cambridge University Press, p. 206–243.

Coney, P.J., Jones, D.L., and Monger, J.W. H., 1980, Cordilleran suspect terranes: Nature, v. 288, p. 329–333.

Cowan, D.S., and Silling, R.M., 1978, A dynamic, scaled model of accretion at trenches and its implications for the tectonic evolution of subduction complexes: Journal of Geophysical Research, v. 83, p. 5389–5396.

Cuthbert, S.J., and Carswell, D.A., 1990, Formation and exhumation of medium-temperature eclogites in the Scandinavian Caledonides, *in* Carswell, D.A., ed., Eclogite Facies Rocks: Glasgow, Scotland, Blackie & Son, Ltd, p. 180–203.

Dallmeyer, R.D., and Takasu, A., 1991, Tectonometamorphic evolution of the Sebadani eclogitic metagabbro and the Sanbagawa schists, central Shikoku, Japan: $^{40}Ar/^{39}Ar$ mineral age constraints: Journal of Metamorphic Geology, v. 9, p. 605–618.

Dal Piaz, G.V., Hunziker, J.C., and Martinotti, G., 1972, La zona Sesia-Lanzo e l'evoluzione tettonico-metamorfica delle Alpi nordoccidentali interne: Memoir Soc. Geol. Italy, v. 11, p. 433–460.

Davies, J.H., and von Blanckenburg, F. (1998) Thermal controls on slab breakoff and the rise of high-pressure rocks during continental collisions, *in* Hacker, B.R., and Liou, J.G., eds., When Continents Collide: Geodynamics and Geochemistry of Ultrahigh-Pressure Rocks: Dordrecht, Kluwer Academic Publishers, Dordrecht, p. 97–115.

DePaolo, D.J., 1981, A neodymium and strontium isotopic study of Mesozoic calc-alkaline granite batholiths of the Sierra Nevada and Peninsular Ranges, California: Journal of Geophysical Research, v. 86, p. 10,470–10,488.

Dewey, J.F., and Bird, J.M., 1970, Mountain belts and the new global tectonics: Journal of Geophysical Research, v. 75, p. 2625–2647.

Dickinson, W.R., 1971, Clastic sedimentary sequences deposited in shelf, slope and trench settings between magmatic arcs and associated trenches: Pacific Geology, v. 3, p. 1–14.

Dickinson, W.R., 1972, Evidence for plate-tectonic regimes in the rock record: American Journal of Science, v. 272, p. 551–576.

Dickinson, W.R., 1976, Sedimentary basins developed during evolution of Mesozoic-Cenozoic arc-trench systems in western North America: Canadian Journal of Earth Science, v. 13, p. 1268–1289.

Dietrich, V., Vuagnat, M., and Bertrand, J., 1974, Alpine metamorphism of mafic rocks: Schweizerisches Mineralogische und Petrografisches Mittelungen, v. 54, p. 291–333.

Dilek, Y., Moores, E.M., Elthon, D., and Nicolas, A., eds., 2000, Ophiolites and oceanic crust: new insights from field studies and the ocean drilling program: Boulder, Colorado, Geological Society of America Special Paper 349, 552p.

Dobretsov, N.L., Shatsky, V.S., and Sobolev, N.V., 1995, Comparison of the Kokchetav and Dabie Shan metamorphic complexes: coesite- and diamond-bearing rocks and UHP-HP accretional-collisional events: International Geology Review, v. 37, p. 636–656.

Drummond, M.S., and Defant, M.J., 1990, A model for trondhjemite-tonalite-dacite genesis and crustal growth via slab melting: Archean to modern comparisons: Journal of Geophysical Research, v. 95, p. 21,503–21,521.

Ellis, S., 1996, Forces driving continental collision: Reconciling indentation and mantle subduction tectonics: Geology, v. 24, p. 699–702.

England, P.L., and Holland, T.J. B., 1979, Archimedes and the Tauern eclogites: the role of buoyancy in the preservation of exotic eclogitic blocks: Earth and Planetary Science Letters, v. 44, p. 287–274.

England, P.L., and Molnar, P., 1993, Cause and effect among thrust and normal faulting, anatectic melting and exhumation in the Himalaya: London, Geological Society of London, Special Publication no. 74, p. 401–411.

Ernst, W.G., 1970, Tectonic contact between the Franciscan melange and the Great Valley sequence, crustal expression of a Late Mesozoic Benioff zone: Journal of Geophysical Research, v. 75, p. 886–901.

Ernst, W.G., 1971, Metamorphic zonations on presumably subducted lithospheric plates from Japan, California and the Alps: Contributions to Mineralogy and Petrology, v. 34, p. 43–59.

Ernst, W.G., 1988, Tectonic histories of subduction zones inferred from retrograde blueschist P-T paths: Geology, v. 16, p. 1081–1084.

Ernst, W.G., 1990, Thermobarometric and fluid-expulsion history of subduction zones: Journal of Geophysical Research, v. 95, p. 9047–9053.

Ernst, W.G., 1992, Metamorphism of the western Cordillera, and its relationship to tectonics, *in* Burchfiel, B.C., Lipman, P.W., and Zoback, M.L., eds., The Cordilleran Orogen; Conterminous United States: Boulder, Colorado, Geological Society of America, The Geology of North America, v. G3, p. 515–538.

Ernst, W.G., 1999, Hornblende, the continent maker—Evolution of H_2O during circum-Pacific subduction versus continental collision: Geology, v. 27, p. 675–678.

Ernst, W.G., and Liou, J.G., 1995, Contrasting plate-tectonic styles of the Qinling-Dabie-Sulu and Franciscan metamorphic belts: Geology, v. 23, p. 353–356.

Ernst, W.G., and Peacock, S., 1996, A thermotectonic model for preservation of ultrahigh-pressure mineralogic relics in metamorphosed continental crust: Washington, D.C., American Geophysical Union Geophysical Monograph 96, p. 171–178

Ernst, W.G., Seki, Y., Onuki, H., and Gilbert, M.C., 1970, Comparative Study of Low-grade Metamorphism in the California Coast Ranges and the Outer Metamorphic Belt of Japan: Boulder, Colorado, Geological Society of America Memoir 124, 276p.

Ernst, W.G., Maruyama, S., and Wallis, S., 1997, Buoyancy-driven, rapid exhumation of ultrahigh-pressure metamorphosed continental crust: Proceedings of the National Academy of Sciences, v. 94, p. 9,532–9,537.

Ernst, W.G., Mosenfelder, J.L., Leech, M.L., and Liu, J., 1998, H_2O recycling during continental collision: phase-equilibrium and kinetic considerations, *in* Hacker, B.R., and Liou, J.G. eds., When Continents Collide: Geodynamics and Geochemistry of Ultrahigh-Pressure Rocks: Dordrecht, Kluwer Academic Publishers, p. 275–295.

Fitch, T.J., 1972, Plate convergence, transcurrent faults, and internal deformation adjacent to Southeast Asia and the Western Pacific: Journal of Geophysical Research, v. 77, p. 4432–4460.

Frey, M., ed., 1987, Low Temperature Metamorphism: Glasgow, Scotland, Blackie Publishing, 351p.

Frey, M., Hunziker, J.C., Frank, W., Bocquet, J., Dal Piaz, G.V., Jäger, E., and Niggli, E., 1974, Alpine metamorphism of the Alps: Schweizerische Mineralogisches und Petrografisches Mittelungen, v. 54, p. 248–290.

Gebauer, D., 1996, A P-T-t path for an (ultra?-) high-pressure ultramafic/mafic rock-association and its felsic country-rocks based on SHRIMP-dating of magmatic and metamorphic zircon domains. Example: Alpe Arami (central Swiss Alps): Washington, D.C., American Geophysical Union Geophysical Monograph 95, p. 307–329.

Gebauer, D., Schertl, H.P., Briz, M., and Schreyer, W., 1997, 35 Ma old ultrahigh-pressure metamorphism and evidence for very rapid exhumation in the Dora Maira Massif, Western Alps: Lithos, v. 41, p. 5–24.

Genser, J., van Wees, J.D., Cloetingh, S., and Neubauer, F., 1996, Eastern Alpine tectono-metamorphic evolution: constraints from two-dimensional modeling: Tectonics, v. 15, p. 584–604.

Gnos, E., Immenhauser, A., and Peters, T., 1997, Late Cretaceous/early Tertiary convergence between the Indian and Arabian plates recorded in ophiolites and related sediments: Tectonophysics, v. 271, p. 1–19.

Grasemann, B., Ratschbacher, L., and Hacker, B.R. (1998) Exhumation of ultrahigh-pressure rocks: thermal boundary conditions and cooling history, *in* Hacker, B.R., and Liou, J.G., eds., When Continents Collide: Geodynamics and Geochemistry of Ultrahigh-Pressure Rocks: Dordrecht, Kluwer Academic Publishers, p. 117–139.

Hacker, B.R., 1996, Eclogite formation and the rheology, buoyancy, seismicity, and H_2O content of oceanic crust: Washington, D.C., American Geophysical Union Geophysical Monograph 96, p. 171–178.

Hacker, B.R., and Peacock, S.M., 1995, Creation, preservation and exhumation of UHPM rocks, *in* Coleman, R.G., and Wang, X., eds., Ultrahigh Pressure Metamorphism: New York, Cambridge University Press, p. 159–181.

Hacker, B.R., Ratschbacher, L., Webb, L., and Dong, S., 1995, What brought them up? Exhuming the Dabie Shan ultrahigh-pressure rocks: Geology, v. 23, p. 743–746.

Hacker, B.R., Wang, X., Eide, E.A., and Ratschbacher, L., 1996, The Qinling-Dabie ultrahigh-pressure collisional orogen, *in* Harrison, T.M., and Yin, A., eds., The Tectonic Development of Asia: New York, Cambridge University Press, p. 345–370.

Hacker, B.R., Ratschbacher, L., Webb, L., McWilliams, M.O., Ireland, T., Calvert, A., Dong, S., Wenk, H-R., and Chateigner, D., 2000, Exhumation of ultrahigh-pressure continental crust in east central China: Late Triassic-Early Jurassic tectonic unroofing: Journal of Geophysical Research, v. 105, p. 13,339–13,364.

Hacker, B.R., Abers, G.A., and Peacock, S.M., 2003a, Subduction factory 1. Theoretical mineralogy, density, seismic wave speeds, and H_2O content: Journal of Geophysical Research, v. 108, no. B1, 10.1029/2001Jb001127.

Hacker, B.R., Calvert, A.T., Zhang, R.Y., Ernst, W.G., and Liou, J.G., 2003b, Ultra-rapid exhumation of ultrahigh pressure diamond-bearing metasedimentary rocks of the Kokchetav Massif: Lithos (in press).

Hamilton, W., 1970, The Uralides and the motion of the Russian and Siberian platforms: Geological Society of America Bulletin, v. 81, p. 2553–2576.

Hamilton, W., 1979, Tectonics of the Indonesian region: Washington, D.C., U.S. Geological Survey Professional Paper 1078, 345p.

Hamilton, W., 1995, Subduction systems and magmatism, *in* Smellie, ed., Volcanism Associated with Extension at Consuming Plate Margins: London, Geological Society of London Special Publication no. 81, p. 3–28.

Harley, S.L., and Carswell, D.A., 1995, Ultradeep crustal metamorphism: A prospective review: Journal of Geophysical Research, v. 100, p. 8367–8380.

Hashimoto, M., Igi, S., Seki, Y., Banno, S., and Kojima, G., 1970, Metamorphic facies map of Japan: Tokyo, Geological Survey of Japan, scale 1:2 000 000.1 sheet.

Hawkesworth, C.J., Gallagher, K., Hergt, J.M., and McDermott, F., 1993, Mantle and slab contributions in an arc magma: Annual Review of Earth and Planetary Sciences, v. 21, p. 175–204.

Henry, C., 1990, L'unite a coesite du massif Dora-Maira dans son cadre petrologique et structural (Alpes occidentales, Italie) [thesis]: Paris, Université de Paris VI, 453p.

Hermann, J., and Green, D.H., 2001, Experimental constraints on high pressure melting in subducted crust: Earth and Planetary Science Letters, v. 188, p. 149–168.

Hermann, J., Rubatto, D., Korsakov, A., and Shatsky, V.S., 2001, Multiple zircon growth during fast exhumation of diamondiferous, deeply subducted continental crust (Kokchetav Massif, Kazakhstan): Contributions to Mineralogy and Petrology, v. 141, p. 66–82.

Hickey-Vargas, R., Sun, M., Lopez-Escobar, L., Moreno-Roa, H., Reagan, M.K., Morris, J.D., and Ryan, J.G., 2002, Multiple subduction components in the mantle wedge: Evidence from eruptive centers in the Central Southern volcanic zone, Chile: Geology, v. 30, p. 199–202.

Hilton, D.R., Hoogewerff, J.A., van Bergen, M.J., and Hammerschmidt, K., 1992, Mapping magma sources in the east Sunda-Banda arcs, Indonesia: Constraints from helium isotopes: Geochimica et Cosmochimica Acta, v. 56, p. 851–859.

Hirajima, T., Wallis, S., Zhai, R., and Ye, K., 1993, Eclogitized metagranitoid from the Su-Lu ultrahigh-pressure (UHP) province, eastern China: Tokyo, Proceedings of the Japan Academy, v. 69, section B, p. 249–254.

Ho, C.S., 1986, A synthesis of the geological evolution of Taiwan: Tectonophysics, v. 125, p. 1–16.

Howell, D.G., ed., 1985, Tectonostratigraphic Terranes of the Circum-Pacific Region: Houston, Texas, Circum-Pacific Council for Energy and Mineral Resources, 581p.

Irwin, W.P., 1972, Terranes of the western Paleozoic and Triassic belt in the southern Klamath Mountains, California: Washington, D.C., U.S. Geol. Survey Prof. Paper 800C, p. 103–111.

Isacks, B., Oliver, J., and Sykes, L.R., 1968, Seismology and the new global tectonics: Journal of Geophysical Research, v. 73, p. 5855–5899.

Ishikawa, M., Kaneko, Y., and Yamamoto, H., 2000, Subhorizontal boundary between ultrahigh-pressure and low-pressure metamorphic units in the Sulu-Tjube area of the Kokchetav Massif, Kazakhstan: The Island Arc, v. 9, p. 317–328.

Itaya, T., and Takasugi, H., 1988, Muscovite K-Ar ages of the Sanbagawa schists, Japan, and argon depletion during cooling and deformation: Contributions to Mineralogy and Petrology, v. 100, p. 281–290.

Jayko, A.S., Blake, M.C., and Harms, T., 1987, Attenuation of the Coast Range ophiolite by extensional faulting, and nature of the Coast Range "thrust," California: Tectonics, v. 6, p. 475–488.

Jones, D.L., 1983, Recognition, character, and analysis of tectonostratigraphic terranes in western North America, *in* Hashimoto, M., and Uyeda, S., eds., Accretion Tectonics of the Circum-Pacific Regions: Tokyo, Terrapub, p. 21–36.

Kaneko, Y., Terabayashi, M., Yamamoto, H. et al., 2000, Geology of the Kokchetav ultrahigh-pressure—high-pressure metamorphic belt, north-eastern Kazakhstan: The Island Arc, v. 9, p. 264–283.

Kawachi, Y., 1968, Large-scale overturned structure in the Sanbagawa metamorphic zone in central Shikoku, Japan: Geological Society of Japan Journal, v. 74, p. 607–616.

Kushiro, I., 1990, Partial melting of mantle wedge and evolution of island arc crust: Journal of Geophysical Research, v. 95, p.15, 929–15,939.

Leech, M.L., 2001, Arrested orogenic development; eclogitization, delamination, and tectonic collapse: Earth and Planetary Science Letters, v. 185, p. 149–159.

Le Fort, P., 1996, Evolution of the Himalaya, *in* Yin, A., and Harrison, T.M., eds., The Tectonic Evolution of Asia: New York, Cambridge University Press, p. 95–109.

Lennykh, V.I., Valizer, P.M., Beane, R., Leech, M., and Ernst, W.G., 1995, Petrotectonic evolution of the Maksyutov Complex, southern Urals, Russia: Implications for ultrahigh-pressure metamorphism: International Geology Review, v. 37, p. 584–600.

Lin, C.H., and Roecker, S.W., 1998, Active crustal subduction and exhumation in Taiwan, *in* Hacker, B.R., and Liou, J.G., eds., When Continents Collide: Geodynamics and Geochemistry of Ultrahigh-Pressure Rocks: Dordrecht, Kluwer Academic Publishers, p. 1–25.

Liou, J.G., and Zhang, R.Y., 1995, Significance of ultrahigh-pressure talc-bearing eclogite assemblages: Mineralogical Magazine, v. 59, p. 93–102.

Liou, J.G., and Zhang, R.Y., 1996, Occurrences of intergranular coesite in ultrahigh-P rocks from the Sulu region, eastern China: Implications for lack of fluid during exhumation: American Mineralogist, v. 81, p. 1217–1221.

Liou, J.G., and Zhang, R.Y., 2002, Ultrahigh-pressure metamorphic rocks: Encyclopedia of Physical Science and Technology: San Diego, Academic Press, v. 17, p 227–244.

Liou, J.G., Lan, C.Y., Suppe, J., and Ernst, W.G., 1977, The Eastern Taiwan Ophiolite, its occurrence, petrology, metamorphism, and tectonic setting: Taipei, Taiwan, Mining Research and Service Organization, Special Report No. 1, 212 p.

Liou, J.G., Zhang, R. and Ernst, W.G., 1994, An introduction to ultrahigh-pressure metamorphism: The Island Arc, v.3, p. 1–24.

Liou, J.G., Banno, S., and Ernst, W.G., 1995, Ultrahigh-pressure metamorphism and tectonics: The Island Arc, v. 4, p. 233–239.

Liou, J.G., Zhang, R.Y., Wang, X., Eide, E.A., Ernst, W.G., and Maruyama, S., 1996, Metamorphism and tectonics of high-pressure and ultrahigh-pressure belts in the Dabie-Sulu region, China, *in* Yin, A., and Harrison, T.M., eds., The Tectonic Evolution of Asia: New York, Cambridge University Press, p. 300–344.

Liou, J.G., Zhang, R.Y., Ernst, W.G., Rumble, D. III, and Maruyama, S., 1998, High pressure minerals from deeply subducted metamorphic rocks: Reviews in Mineralogy, v. 37, p. 33–96.

Liu, J., Bohlen, S.R. and Ernst, W.G., 1996, Stability of hydrous phases in subducting oceanic crust: Earth and Planetary Science Letters, v. 143, p. 161–171.

Maruyama, S., 1997, Pacific-type orogeny revisited: Miyashiro-type orogeny proposed: The Island Arc, v. 6, p. 91–120.

Maruyama, S., and Parkinson, C.D., 2000, Overview of the geology, petrology and tectonic framework of the HP-UHP metamorphic belt of the Kokchetav Massif, Kazakhstan: The Island Arc, v. 9. P. 439–455.

Maruyama, S., Liou, J.G., and Zhang, R., 1994, Tectonic evolution of the ultrahigh-pressure (UHP) and high-pressure (HP) metamorphic belts from central China: The Island Arc, v. 3, p. 112–121.

Maruyama, S., Liou, J.G., and Terabayashi, M., 1996, Blueschists and eclogites of the world, and their exhumation: International Geology Review, v. 38, p. 485–594.

Massone, H.J., 1995, Experimental and petrogenetic study of UHPM, *in* Coleman, R.G., and Wang, X., eds., Ultrahigh Pressure Metamorphism: New York, Cambridge University Press, p. 33–95.

Merle, O., and Guillier, B., 1989, The building of the Central Swiss Alps: An experimental approach: Tectonophysics, v. 165, p. 41–56.

Michard, A., Henry, C., and Chopin, C., 1995, Structures in UHPM rocks: A case study from the Alps, *in* Coleman, R.G., and Wang, X., eds., Ultrahigh Pressure Metamorphism: New York, Cambridge University Press, p. 132–158.

Miller, J.M., Gray, D.R., and Gregory, R.T., 1998, Exhumation of high-pressure rocks in northeastern Oman: Geology, v. 26, p. 235–238.

Miyashiro, A., 1961, Evolution of metamorphic belts: Journal of Petrology, v. 2, p. 277–311.

Miyashiro, A., 1967, Orogeny, regional metamorphism, and magmatism in the Japanese islands: Meddelelser fra Dansk Geologisk Forening, v. 17, p. 390–446.

Molnar, P., and Atwater, T., 1978, Interarc spreading and cordilleran tectonics as alternates related to the age of subducted oceanic lithosphere: Earth and Planetary Science Letters, v. 41, p. 330–340.

Molnar, P., and Gray, D., 1979, Subduction of continental lithosphere: Some constraints and uncertainties: Geology, v. 7, p. 58–62.

Molnar, P., and Tapponnier, P., 1975, Cenozoic tectonics of Asia; effects of a continental collision: Science, v. 189, p. 419–426.

Molnar, P., Burchfiel, B.C., Liang, K., and Zhao, Z., 1987, Geomorphic evidence for active faulting in the Altyn Tagh and northern Tibet and quantitative estimates of its contribution to the convergence of India and Eurasia: Geology, v. 15, p. 249–253.

Moores, E.M., 1970, Ultramafics and orogeny, with models for the U.S. Cordillera and the Tethys: Nature, v. 228, p.837–842.

Morris, P.A., 1995, Slab melting as an example of Quaternary volcanism and aseismicity in southwest Japan: Geology, v. 23, p. 395–398.

Mosenfelder, J.L., and Bohlen, S.R., 1997, Kinetics of the coesite to quartz transformation: Earth and Planetary Science Letters. v.153, p. 133–147.

Nishimura, Y., 1998, Geotectonic subdivision and areal extent of the Sangun belt, Inner Zone of Southwestern Japan: Journal of Metamorphic Geology, v. 16, p. 129–140.

Okamoto, K., and Maruyama, S., 1999, The high pressure synthesis of lawsonite in the MORB + H_2O system: American Mineralogist, v. 84, p. 362–373.

Ota, T., Terabayashi, M., Parkinson, C.D., and Masago, H., 2000, Thermobarometric structure of the Kokchetav ultrahigh-pressure—high-pressure massif deduced from a north-south transect in the Kulet and Saldat-kol regions, northern Kazakhstan: The Island Arc, v. 9, p. 328–357.

Oxburgh, E.R., and Turcotte, D.L., 1971, Origin of paired metamorphic belts and crustal dilation in island arc regions: Journal of Geophysical Research, v. 76, p. 1315–1327.

Page, B.M., and Suppe, J., 1981, The Pliocene Lichi Melange of Taiwan: its plate tectonic and olistostromal origin: American Journal of Science, v. 281, p. 193–227.

Patiño Douce, A.E., and McCarthy, T.C., 1998, Melting of crustal rocks during continental collision and subduction, *in* Hacker, B.R., and Liou, J.G., eds., When Continents Collide: Geodynamics and Geochemistry of Ultrahigh-Pressure Rocks: Dordrecht, Kluwer Academic Publishers, p. 27–55.

Peacock, S.M., 1990, Fluid processes in subduction zones: Science, v. 248, p. 329–337.

Peacock, S.M., 1992, Blueschist-facies metamorphism, shear heating, and P-T-t paths in subduction shear zones: Journal of Geophysical Research, v. 97, p. 17,693–17,707.

Peacock, S.M., 1995, Ultrahigh-pressure metamorphic rocks and the thermal evolution of continent collision belts: The Island Arc, v. 4, p. 376–383.

Platt, J.P., 1986, Dynamics of orogenic wedges and the uplift of high-pressure metamorphic rocks: Geological Society of America Bulletin, v. 97, p. 1037–1053.

Platt, J.P., 1987, The uplift of high-pressure, low-temperature metamorphic rocks: Philosophical Transactions of the Royal Society of London, v. 321, p. 87–103.

Platt, J.P., 1993, Exhumation of high-pressure rocks: a review of concepts and processes: Terra Nova, v. 5, p. 119–133.

Poli, S., and Schmidt, M.W., 1997, The high-pressure stability of hydrous phases in orogenic belts: an experimental approach on eclogite-forming processes: Tectonophysics, v. 273, p. 169–184.

Perchuk, A.L., Yaspaskurt, V.O., and Podlesskii, S.K., 1998, Genesis and exhumation dynamics of eclogites in the Kokchetav Massif near Mount Sulu-Tyube, Kazakhstan: Geochemistry International, v. 36, p. 877–885.

Reck, B.H., 1987, Implications of measured thermal gradients for water movement through the northeast Japan accretionary prism: Journal of Geophysical Research, v. 92, p. 3683–3690.

Ring, U., and Brandon, M.T., 1994, Kinematic data for the Coast Range fault and implications for the exhumation of the Franciscan subduction complex: Geology, v. 22, p. 735–738.

Ring, U., and Brandon, M.T., 1999, Ductile deformation and mass loss in the Franciscan subduction complex—Implications for exhumation processes in accretionary wedges: London, Geological Society of London Special Publication no. 154, p. 180–203.

Robinson, P., and Terry, M.P., 1998, Evolution of amphibolite-facies structural features during transition from convergence to extensional collapse, Western Gneiss Region, Norway: International Workshop on UHP Metamorphism and Exhumation, Stanford University, p. A79-A83.

Rubatto, D., and Hermann, J., 2001, Exhumation as fast as subduction?: Geology, v. 29, p. 3–6.

Rubie, D., 1984, A thermal-tectonic model for high-pressure metamorphism in the Sesia zone, Western Alps: Journal of Geology, v. 92, p. 21–36.

Rubie, D.C., 1986, The catalysis of mineral reactions by water and restrictions on the presence of aqueous fluid during metamorphism: Mineralogical Magazine, v. 50, p. 399–415.

Rubie, D.C., 1990, Role of kinetics in the formation and preservation of eclogites, *in* Carswell, D.A., ed., Eclogite Facies Rocks: New York, Chapman and Hall, p. 111–140.

Rumble, D., 1998, Stable isotope geochemistry of ultrahigh-pressure rocks, *in* Hacker, B.R., and Liou, J.G., eds., When Continents Collide: Geodynamics and Geochemistry of Ultrahigh-Pressure Rocks: Dordrecht, Kluwer Academic Publishers, p. 241–259.

Sacks, P.E., and Secor, D.T., Jr.. 1990, Delaminations in collisional orogens: Geology, v. 18, p. 999–1002.

Schreyer, W., 1995, Ultradeep metamorphic rocks: The retrospective viewpoint: Journal of Geophysical Research, v. 100, p. 8353–8366.

Schreyer, W., Massone, H.J., and Chopin, C., 1987, Continental crust subducted to depths near 100 km: Implications for magma and fluid genesis in collision zones, *in* Mysen, B.O., ed., Magmatic Processes: Physicochemical Principles: University Park, Pennsylvania, The Geochemical Society Special Publication no. 1, p. 155–163.

Searle, M.P., 1996, Cooling history, erosion, exhumation, and kinetics of the Himalaya-Karakorum-Tibet orogenic belt, *in* Yin, A., and Harrison, T.M., eds., The Tectonic Evolution of Asia: New York, Cambridge University Press, p. 110–137.

Searle, M.P., Hacker, B.R., and Bilham, R., 2001, The Hindu Kush seismic zone as a paradigm for the creation of ultrahigh-pressure diamond- and coesite-bearing continental rocks: Journal of Geology, v. 109, p. 143–153.

Seno, T., 1985, Age of subducting lithosphere and back-arc basin formation in the western Pacific since the middle Tertiary, *in* Nasu, N., Kobayashi, K., Uyeda, S., Kushiro, I., and Kagami, H., eds., Formation of Active Ocean Margins: Tokyo, Terrapub, p. 469–481.

Shreve, R.L., and Cloos, M., 1986, Dynamics of sediment subduction, melange formation, and prism accretion: Journal of Geophysical Research, v. 91, p. 10,229–10,245.

Smith, D.C., 1984. Coesite in clinopyroxene in the Caledonides and its implications for geodynamics: Nature, v. 310, p. 641–644.

Snyder, D.B., Prasetyo, H., Blundell, D.J., Pigram, C.J., Barber, A.J., Richardson, A., and Tjokosaproetro, S., 1996, A dual doubly vergent orogen in the Banda Arc continent-continent collision zone as observed on deep seismic reflection profiles: Tectonics, v. 15, p. 34–53.

Sobolev, N.V., and Shatsky, V.S., 1990, Diamond inclusions in garnets from metamorphic rocks: Nature, v. 343, p. 742–746.

Stöckhert, B., and Renner, J., 1998, Rheology of crustal rocks at ultrahigh pressure, in Hacker, B.R., and Liou, J.G., eds., When Continents Collide: Geodynamics and Geochemistry of Ultrahigh-Pressure Rocks, Dordrecht, Kluwer Academic Publishers, p. 57–95.

Stüwe, K., and Barr, T.D., 1998, On uplift and exhumation during convergence: Tectonics, v. 17, p. 80–88.

Suppe, J., 1972, Interrelationships of high-pressure metamorphism, deformation, and sedimentation in Franciscan tectonics, U.S.A.: International Geological Congress, 24th, Montreal, Reports, sec. 3, p. 552–559.

Tabata, H., Yamauchi, K., Maruyama, S., and Liou, J.G., 1998, Tracing the extent of a UHP metamorphic terrane: mineral-inclusion study of zircons in gneisses from the Dabie Shan, *in* Hacker, B.R., and Liou, J.G., eds., When Continents Collide: Geodynamics and Geochemistry of Ultrahigh-Pressure Rocks: Dordrecht, Kluwer Academic Publishers, p. 261–273.

Tapponnier, P., Peltzer, G., Le Dain, A.Y., and Armijo, R., 1982, Propagating extrusion tectonics in Asia: new insights from simple experiments with plasticine: Geology, v. 10, p. 611–617.

Terry, M.P., Robinson, P., and Krogh Ravna. E.J., 2000a, Kyanite eclogite thermobarometry and evidence of thrusting of UHP over HP metamorphic rocks, Nordoyane, Western Gneiss Region, Norway: American Mineralogist, v. 85, p. 1637–1650.

Terry, M.P., Robinson, P., Hamilton, M.A., and Jercinovic, M.J., 2000b, Monazite geochronology of UHP and HP metamorphism, deformation, and exhumation, Nordoyane, Western Gneiss Region, Norway: American Mineralogist, v. 85, p. 1651–1664.

Theunissen, K., Dobretsov, N.L., Travin, A., Shatsky, V.S., Korsakov, A., Smirnova, L., and Boven, A., 2000, Two contrasting petrotectonic domains in the Kokchetav megamelange (north Kazakhstan): difference in exhumation mechanisms of ultrahigh-pressure crustal rocks, or a result of subsequent deformation?: The Island Arc, v. 9, p. 284–303.

Thompson, A.B., and Ridley, J.R., 1987, Pressure-temperature-time (P-T-t) histories of orogenic belts: Philosophical Transactions of the Royal Society of London, ser. A, v. 321, p. 27–45.

Turcotte, D.L., and Oxburgh, E.R., 1972, Mantle convection and the new global tectonics: Annual Review of Fluid Mechanics, v. 4, p. 33–68.

van Bergen, M.J., Vroon, P.Z., and Hoogewerff, J.A., 1993, Geochemical and tectonic relationships in the east Indonesian arc-continent collision region: Implications for the subduction of the Australian passive margin: Tectonophysics, v. 223, p. 97–116.

van den Beukel, J., 1992, Some thermomechanical aspects of the subduction of continental lithosphere: Tectonics, v. 11, p. 316–329.

van den Beukel, J., and Wortel, R., 1988, Thermo-mechanical modelling of arc-trench regions: Tectonophysics, v. 154, p. 177–193.

Vielzeuf, D., and Holloway, J.R., 1988, Experimental determination of the fluid-absent melting relations in the pelitic system. Consequences for crustal differentiation: Contributions to Mineralogy and Petrology, v. 98, p. 257–276.

von Blanckenburg, F., and Davies, J.H., 1995, Slab break off: A model for syncollision magmatism and tectonics in the Alps: Tectonics, v. 14, p. 120–131.

Vrolijk, P.J., 1987, Tectonically driven fluid flow in the Kodiak accretionary complex, Alaska: Geology, v. 15, p. 466–469.

Vrolijk, P.J., Meyers, G., and Moore, J.C., 1988, Warm fluid migration along tectonic melanges in the Kodiak accretionary complex, Alaska: Journal of Geophysical Research, v. 93, p. 10,313–10,324.

Wallis, S.R., Ishiwatari, A., Hirajima, T., Ye, K., Guo, J., Nakamura, D., Kato, T., Zhai, M., Enami, M., Cong, B., and Banno, S., 1997, Occurrence and field relationships of UHP meta-granitoid and coesite eclogite in the Su-Lu terrane, Eastern China: Journal of the Geological Society of London, v. 154, p. 45–55.

Webb, L.E., Hacker, B.R., Ratschbacher, L., McWilliams, M.O., and Dong, S., 1999, $^{40}Ar/^{39}Ar$ thermochronologic constraints on deformation and cooling history of high and ultrahigh-pressure rocks in the Qinling-Dabie orogen: Tectonics, v. 18, p. 621–638.

Wheeler, J., 1991, Structural evolution of a subduction continental sliver: the northern Dora-Maira massif, Italian Alps: Journal of the Geological Society of London, v. 148, p. 1101–1113.

Wheeler, J., Reddy, S.M., and Cliff, R.A., 2001, Kinematic linkage between internal zone extension and shortening in more external units in the NW Alps: Journal of the Geological Society of London, v. 158, p. 439–443.

Widiyantoro, S., and van der Hilst, R., 1996, Structure and evolution of lithospheric slab beneath the Sunda Arc, Indonesia: Science, v. 271, p. 1566–1570.

Willett, S., Beaumont, C., and Fullsack, P., 1993, Mechanical model for the tectonics of doubly vergent compressional orogens: Geology, v. 21, p. 371–374.

Yin, A., and Nie, S., 1993, An indentation model for the north and south China collision and the development of the Tan-lu and Honam fault systems, eastern Asia: Tectonics, v. 12, p. 801–813.

Yui, T.F., Rumble, D., and Lo, C.H., 1995, Unusually low ^{18}O ultrahigh-pressure metamorphic rocks from the Sulu terrain, eastern China: Geochemica et Cosmochimica Acta, v. 59, p. 2859–2864.

Zhang, R.Y., and Liou, J.G., 1997, Partial transformation of gabbro to coesite-bearing eclogite from Yangkou, the Sulu terrane, eastern China: Journal of Metamorphic Geology, v. 15, p. 183–202.

Zheng, Y.F., Gong, B., Li, Y., Wang, Z., and Fu, B., 2000, Carbon concentrations and isotopic ratios of eclogites from the Dabie and Sulu terranes in China: Chemical Geology, v. 168, p. 291–305.

Manuscript Accepted by the Society May 2, 2003

Printed in the USA

Geological Society of America
Special Paper 373
2003

The repeated rediscovery of mélanges and its implications for the possibility and the role of objective evidence in the scientific enterprise

A.M.C. Şengör*
İTÜ Maden Fakültesi ve Avrasya Yerbilimleri Enstitüsü, Ayazağa 80626 İstanbul, Turkey

Dedicated to the memory of William John "Mac" McCallien and to Oğuz Erol, who, together, as teacher and pupil, unlocked the mystery of the Ankara Mélange in the heart of Turkey in their devotion to the discovery of the truth.

The origin of the Carpathian klippen is synonymous with the origin of the Carpathians themselves. (Uhlig, 1904, p. 427)

ABSTRACT

It has been claimed recently and repeatedly by some social scientists that objective evidence in natural sciences is not possible, and consequently, natural sciences have no objectively testable basis. This, it is alleged, makes science no different from any other set of assertions claiming knowledge and associated authority, such as religion, magic, shamanism, myth, etc. This claim is inconsistent with the repeated rediscovery, by independent scientists, of identical features of nature. This paper reviews the prehistory and the history of discovery and three subsequent independent rediscoveries of mélanges, which commonly include ophiolitic rock-types. Although mélanges were originally discovered in the Alps through a long, drawn out process, and almost simultaneously also in the Carpathians, followed shortly afterward by discoveries in the Himalaya—also by Alpine geologists, who were aware of their colleagues' and predecessors' discoveries in the Alps—none of these geologists (with the sole exception of Eduard Suess) considered mélanges a *purely tectonically mixed* rock unit, but ascribed their exotic elements to introduction by surface processes such as gravity gliding, sedimentation, and even volcanism. Edward Greenly discovered mélanges as a distinct, separately mappable rock unit formed by structural processes, when he was mapping in Anglesey (northern Wales, United Kingdom) in the last half decade of the nineteenth and in the beginning of the twentieth century. I consider him the discoverer of mélanges because, unlike Alpine geologists, he described in detail all aspects of what later was acknowledged to be mélange tectonics, including the details of structural disruption of originally continuous units and their intimate tectonic mixing. The discovery did not become popular and even Greenly himself did not mention it in his subsequent publications. Mélanges were discovered for the second time by Levi F. Noble in 1941, when he was confronted with the chaotically mixed rocks in the southern part of the Amargosa Range in California. Noble called them chaos structure and did not relate his discovery to Greenly's or to Schardt's or Lugeon's or Uhlig's or Griesbach's, of which he most likely remained ignorant. W.J. "Mac" McCallien and his doctoral student A. Oğuz Erol rediscovered mélanges for the second time in the vicinity of Ankara in central Turkey, independently of Schardt, Griesbach, Greenly, and Noble. Their rediscovery was later connected with Greenly's concept and terminology by Sir Edward Bailey, who interpreted the origin of mélanges in terms of a novel tectonic model of the structure of

*E-mail: sengor@itu.edu.tr

Şengör, A.M.C., 2003, The repeated rediscovery of mélanges and its implications for the possibility and the role of objective evidence in the scientific enterprise, *in* Dilek, Y., and Newcomb, S., eds., Ophiolite concept and the evolution of geological thought: Boulder, Colorado, Geological Society of America Special Paper 373, p. 385–445. For permission to copy, contact editing@geosociety.org.

Anatolia involving colossal overthrusting and associated crushing. The third rediscovery was by Kenneth J. Hsü, while he was mapping the Franciscan Complex in California. Hsü independently rediscovered all the properties of mélanges, only to be told by Héli Badoux, a Swiss colleague, that Hsü had rediscovered what had already been discovered in Turkey. If all these rediscoveries were *solely* results of the discoverers' social, cultural, and historical biases, it is most improbable that the discoverers would have come up with identical discoveries. Also, when confronted with each other's accounts, they would not have been able to acknowledge the identity of their discoveries, if the discoveries had been made solely under the influence of social factors that inevitably greatly differ in different times and places.

Keywords: mélange, ophiolite, exotic block, Alps, Carpathians, Himalaya, Ankara Mélange, Franciscan Complex.

INTRODUCTION[1]

One of the recent fads in the historiography and philosophy of science is that scientific knowledge is "constructed" socially and that it is essentially uninfluenced by observations. People who hold this view are members of interdisciplinary social-study groups referred to as science, technology, and society studies, or simply science and culture studies. Koertge (1998, p. 3–4) has summarized the common ideas of this group in the following fashion:

> 1. Every aspect of that complex set of enterprises that we call science, including, above all, its content and results, is shaped by and can be understood only in its local historical and cultural context.
>
> 2. In particular, the products of scientific inquiry, the so-called laws of nature, must always be viewed as social constructions. Their validity depends on the consensus of "experts" in just the same way as the legitimacy of a pope depends on a council of cardinals.
>
> 3. Although scientists typically succeed in arrogating special epistemic authority to themselves, scientific knowledge is just "one story among many." The more epistemological authority that science has in a given society, the more important it is to unmask its pretensions to be an enterprise dedicated to the pursuit of objective knowledge. Science must be "humbled."
>
> 4. Since the quest for objective knowledge is a quixotic one, the best way to appraise scientific claims is through a process of political evaluation. Since the "evidence" for a scientific claim is never conclusive and is always open to negotiation, the best way to evaluate scientific results is to ask who stands to benefit if the claim is taken to be true. Thus, for the citizen the key question about a scientific result should not be how well tested the claim is but, rather, Cui bono?
>
> 5. "Science is politics by other means": the results of scientific inquiry are profoundly and importantly shaped by the ideological agendas of powerful elites.
>
> 6. There is no univocal sense in which science of one society is better than that of another. In particular, Euroscience is not objectively superior to the various ethnosciences and shamanisms described by anthropologists or invented by Afrocentrists.
>
> 7. Neither is there any clear sense in which we can talk about scientific progress within the European tradition. On the contrary, science is characterized chiefly by its complicity in all the most negative and oppressive aspects of modern history: increasingly destructive warfare, environmental disasters, racism, sexism, eugenics, exploitation, alienation, and imperialism.
>
> 8. Given the impossibility of scientific objectivity, it is futile to exhort scientists and policy makers to try harder to remove ideological bias from the practice of science. Instead, what we need to do is deliberately introduce "corrective biases" and "progressive political biases" into science. There is a call for "emancipatory science" and "advocacy research."

These extraordinary statements were distilled from a huge mass of post-modernist literature that has been gaining influence in social science circles (e.g., Kuhn 1970, but see 2000, especially chapters 4 and 5; Collins, 1981; Latour and Woolgar, 1986; Bernal, 1987, 1991; Feyerabend, 1988, 1991; Bloor, 1991). With few exceptions, natural scientists have been oblivious to these developments and when confronted with them, laugh them off. There is, however, a real danger of such views influencing public opinion even in scientifically advanced societies, and their harming the safety and comfort of human society (imagine a society without modern medicine or computer technology, or indeed geology). It is, therefore, imperative that natural scientists show the holders of the opinions itemized above why science is indeed objective and how observations, i.e., evidence, do matter.

There is a small, but growing literature authored mostly by natural scientists and historians doing just that (see especially Gross and Levitt, 1994; Gross et al., 1996; Lefkowitz, 1996; Lefkowitz and Rogers, 1996; Windschuttle, 1996; Evans, 1997; Koertge, 1998); however, I believe one critical aspect of scientific discoveries remains insufficiently emphasized: their repetitive nature. In other words, the frequency with which scientists keep rediscovering the wheel! If scientific discovery were only a function of historical contingency, and the cultural and social milieu, such repetitive discoveries in different social and historical contexts would have been extremely rare, if not altogether improbable.

The purpose of this paper is to relate the history of a discovery repeated four times (in addition to having a fairly protracted pre-history of discovery), independently of one another, during the twentieth century. It is the history of the discovery and rediscovery again and again of mélanges, a chaotically mixed group of rocks indicating an environment of intense shear deformation and very often including fragments of ophiolites and/or underlying large ophiolite nappes. The discoverers have worked in cultural environments different from one another at different times in history. Two of the heroes of the story even come from two different non-western cultures, one being Turkish and the other Chinese.

Admittedly, the story told below is by and large embedded within the context of European science. This paper does not attempt to compare European science with what has been called science in other cultures (mostly by students who grew up within a European social and scientific context). To do that, a very different kind of investigation would have been necessary. The sole purpose of the present contribution is to show that in scientific discovery evidence matters, and that without it there can be no science. A corollary to this statement is that scientists can objectively evaluate their observations; in other words, a dialogue between man and Nature outside him is possible and it constitutes the essence of science.

SCIENTIFIC OBJECTIVITY

Before we proceed with the historical narrative, it is necessary to define what objectivity is. In its shortest and most comprehensive definition, objectivity is independence from individual whim. In view of the human incapability of gathering perfect data concerning the outside world, because of multifarious factors such as the imperfection of our sense organs to reproduce a faithful picture of the stimulants or the in-built or acquired biases we all harbor, distorting whatever message our sense organs give us, the claim that we can acquire knowledge independent of our imperfections must, at first, be received with incredulity. The first step in understanding how this is achieved is to admit that no one person can possibly be objective. The knowledge an individual acquires is tainted with distortions stemming from the factors just mentioned. These distortions, however, are different for different persons owing to the different degrees and kinds of their imperfections, and the differences in their preconceptions in processing information. When such a group of "imperfect individuals" with diverse biases regard an object or evaluate a statement, there is always a large margin of disagreement as to what is being regarded or being evaluated; however, there is usually also a significant area of overlap. This area of overlap is what Heraclitus of Ephesus called the "common":

> Those who will speak [i.e., act] with sense
> must rely on what is common to all
> as a city relies on its law,
> ...
> Therefore one ought to follow what is common.
> But although the Logos is common
> The many live as if they had a religious wisdom of their own.
> (DK114B and DK2B; Marcovich, 1967, fragment 23, p. 91; for a different translation, see Freeman, 1948, p. 24–25 and 32).

Erwin Schrödinger thought that this was an impressive statement in his great Shearman Lectures on "Nature and the Greeks" delivered in the University College, London, in 1948: "What impresses me particularly is the great emphasis on holding fast to what is common—viz. To escape insanity, to escape being an 'idiot' (from idios, private, one's own)." (Schrödinger, 1954, p. 71).

Science, indeed all rational life, takes its departure from this "common." Contestants disagreeing on the periphery of the "common" seek to remedy their imperfections with a view to enlarging the area of the "common" overlap *on the basis of what is contained in the common*. This is what saves them from being idiots, as Schrödinger so nicely put it. A microscope, for instance, is built on principles constituting a "common." But it helps to enlarge the "common" on the nature of microscopic objects, which, otherwise, would have been only much more indirectly perceived and most likely differently conceived by different individuals. Aeroplanes are built on the common knowledge of Newtonian mechanics. Einstein later showed that Newtonian mechanics was faulty. His claim did not become "common" until Arthur Eddington went to the Indian Ocean in 1919 and actually observed with his associates the bending of a ray of light when passing near the Sun as predicted by Einstein's theory. Einstein became famous in the public eye overnight when Eddington announced the results of his expedition (see, for example, the dramatic description in chapters 9 and 10 in Clark, 1971). Yet, it is "common" that at velocities significantly below the speed of light, Newtonian mechanics is an adequate approximation of what we now conceive to be reality. It is on the basis of that common knowledge that both the public and those who design aeroplanes using the "faulty, but adequate" Newtonian mechanics, ride them every day. Similarly, on the basis of the common understanding of physics, even the fiercest critics of scientific objectivity do not step out of windows on the top floor of the Empire State Building in New York or go on expensive safaris with the hope of observing live dinosaurs.

This appreciation of the nature of objectivity (Popper 1957, p. 155–156; 1966, especially p. 217–219; also see 1980, p. 44–48, especially note *1 on p. 44; 1983, p. 48) emphasizes the indispensability of the evidence and the importance of the presence of a scientific *community*. There can indeed be a one-man science, if that man has perfect sense organs and is moreover perfectly rational: but even then it would progress very much more slowly than if it were in a community; however, there can be no science of monads unable to communicate with their surroundings.

ROLE OF EVIDENCE IN A SCIENCE PROGRESSING WITH CONJECTURES AND REFUTATIONS

It was Karl Popper's revolutionary insight that most science must forever remain conjectural and yet all science is capable of progress (Popper, 1935, 1980). Popper was able to show this by deducing from Hume's Law (cf. Şengör, 2001) that although universal statements can never be verified, they can be falsified by contrary instances. This automatically brings to the fore the importance of evidence in evaluating opinions on nature, which are called scientific hypotheses. Once a hypothesis is falsified on the basis of observation reports, the new data bank includes all the observations the falsified hypothesis was able to explain, *plus* those new observations that falsified the hypothesis. Therefore, any new hypothesis must be able to account for all the older observations and the new ones that defeated the older hypothesis. This means that the new hypothesis has a broader database; it is able to explain more than its predecessor. This is the extent of scientific progress we can hope for. In only individual and trivial cases can we get more: for example, in some cases, we can recover ancient objects in their entirety, thus obtaining a non-hypothetical, "complete" knowledge about what they are. But all interesting scientific issues are doomed to remain conjectural.

Yet this conjectural knowledge is capable of progressing (not just changing) in the sense of being able to explain more and more of the universe in which we live. This progress has two components: explanatory hypotheses and observation reports. In schools and universities students are taught the methods of observation; however, there can be no corresponding instruction on how to generate hypotheses (Einstein, 1981, p. 110–111). Any way to come up with an explanatory account about a set of observations is legitimate, so long as it can in principle be inconsistent with observation (i.e., testable). This is to say that the hypothesis ought to be able to come into contact with the world outside us. Evidence (i.e., observations) provides that contact and the scientific community ensures the objectivity of the observations.

It is, of course, possible that not only individuals, but entire communities may have ulterior motives in supporting or suppressing this or that view. If it were not so, Lysenko's genetics could not have survived in the Soviet Union as long as it did. Neither could the Church suppress science for such a long time as in the early Middle Ages. What makes it impossible even for large communities (whole nations, empires) to impose their whim on the opinion of the entire human society is the inevitable chance discovery of, and infiltration of reports about, contrary instances. Columbus' discovery of the New World made a mockery of many of the claims of the Church, such as the absence of races of people not mentioned in the Bible. The discoveries were of such a nature that the Church could not hide them (e.g., Columbus had brought back both people and animals not known to the authors of the Holy Script). Enough people got to hear of them soon and thus a "common" originated that most sensible people could share and that no amount of bigotry could suppress.

In the following paragraphs, I relate the repeated discovery of mélanges, with a view to showing that the discoverers all shared a "common" in Heraclitus' sense. In my account below, I do not emphasize the close association of mélanges with ophiolites, because this association is not the essence of the message of this paper, despite the fact that it is a very common one. All the mélange belts I discuss below contain ophiolites, with the exception of the Amargosa chaos.

In my account, I emphasize how the discoverers struggled with what to them seemed extraordinary, strange, and unique instances, and how evidence led them, step by step, to identical conclusions independently of each other. In this step-by-step approach, they were all led by their own conjectures and their ability to face refutations brought about by the evidence. In the end, they and the community agreed that they were all looking at the same sort of thing that had not been commonly known previously. Thus, common knowledge was increased. This increase had nothing to do with social context or cultural milieu or historical contingency, except in the truism that without previous geological discoveries, mélanges could not have been discovered.

DISCOVERY OF MÉLANGES

The "Prehistory" of the Discovery, First Act: The Alps

Mélanges are "mixed rocks." In them, rocks of more than one provenance come together by means of natural geological processes that post-date the formation of its component rock types. Because these conditions also include some coarse sandstones, conglomerates, debris flow deposits (inappropriately called olistostromes: see Hsü, 1988), glacial moraines, pebbly mudstones (essentially any diamictite), agglomerates, and many kinds of fault rocks, most geologists restrict mélanges to products of tectonic mixing in an environment of intense tectonic simple shear involving more than one major paleogeographic realm. This last condition causes mélanges to have "exotic" components stemming from paleogeographic environments not now anywhere near the mélange body, or, indeed, nowhere on the face of the earth any more.[2] Naturally, conglomerates, debris flow deposits, moraines, pebbly mudstones, agglomerates and all kinds of fault rocks may be later re-mixed by tectonic simple shear and form components of mélanges. Following the definition of Şengör and Sakınç (2001), *mélanges are, thus, "structural rocks" in the sense that they form, as a mappable rock type, through structural processes.*

The title of Steno's great classic *De solido intra solidum naturaliter contento* ("Of solids naturally contained within solids," Stenonis, 1669), which is considered the dawn of geology as we know it, was chosen to underline the problem of containing one solid within another one, more specifically one rocky substance (in his case, fossils) within another (in Steno's case, sedimentary rocks). His solution to this dilemma was to assume that the containing solid (sedimentary rock) was liquid at the time it enclosed the contained one (fossils), and he further assumed that sedimentation provided the *only* means to enclose a solid within another one. Steno's solution remained unique (e.g., Woodward, 1695, especially Part IV; Werner, 1791, *passim*) until the discovery that the walls of some veins and dikes had moved parallel with the strike of the fissure and that this movement had been responsible for shearing off parts of the wall-rocks and mixing them with the vein- or dike-fill (Whitehurst, 1778, p. 154–155; see Şengör and Sakınç, 2001, p. 164 and fig. 9). In 1835, von Carnall (1835, especially p. 146) described in some detail how fault-rocks formed by "shearing off" (= "*Abreibung*") of the wall rocks of "fissures" (= "*Klüfte*"). This kind of observation eventually culminated in the discovery of cataclasites (see Şengör and Sakınç, 2001, for the history of ideas on cataclasites and structural rocks in general).

Another series of observations, beginning in the Alps, concerns the presence of exotic blocks in sedimentary, volcanic, or metamorphic rocks.[3] The father of Alpine geology, Horace-Bénédict de Saussure (1740–1799; Fig. 1[4]), pointed out, in the fourth volume of his classical *Voyage dans les Alpes* (Voyages in the Alps), that rounded pebbles in mountainous countries commonly stemmed from the immediately surrounding mountains, whereas those encountered in wide plains must have been witnesses to important revolutions in the history of our globe (de Saussure, 1796, p. 124, §1943). With a view to finding traces of such revolutions, he studied the rounded pebbles found in the large conglomerate masses called the Nagelfluh[5] in the Subalpine Molasse, along the valleys of the two Emme rivers (Emme and the Klein-Emme) in Central Switzerland (Fig. 2, in the upper course of the Emmental). There he found porphyries and granites that he could not relate to any locality in the Alps (*ibid.*, p. 131–132, §§1952–1953). He thought, in a neptunistic vein, that these rocks had been carried from the north during the great debacle created by the retreat of the waters of a grand ocean that supposedly had once existed in Europe (*ibid.*, p. 134–136, §§1960–1961). The great German geologist Baron Leopold von Buch (1774–1852; Fig. 3[6]) shared de Saussure's opinion (von Buch, 1801; his reference to de Saussure on p. 131) and thought that the porphyries that are entirely foreign to Swiss geology had been derived from the Vosges. A German political refugee to Switzerland, Johann Gottfried Ebel (1764–1830; Fig. 4; for this little-known pioneer of Alpine geology and his geological work, see Franks et al., 2000), made similar observations in his *Ueber den Bau der Erde in dem Alpengebirge* ("On the structure of the earth in the Alpine Mountains"), but, importantly, sought the provenance of the various crystalline rocks of the Molasse (especially the porphyries) in regions south and southeast of Switzerland (Ebel, 1808, p. 17–20).

It was the great Swiss master of geology, Bernhard Studer (1794–1887; Fig. 5[7]) who, in his groundbreaking Molasse monograph, summarized all the observations concerning the clasts whose provenance remained unidentified (Studer, 1825, p. XXXIII and 106–173). He further pointed out that in the sandstones of the Habkern Valley (Fig. 2), granite pebbles with a red feldspar occurred that were very similar to those observed in the Nagelfluh (*ibid.*, p. 163ff; Fig. 6). Studer emphasized that similar rocks were not present along the northern margin of the Alps (in his "*nördlichen Nebenzone*"), and he wondered where they had come from. In 1825, he agreed with de Saussure and von Buch in bringing them from the north, mainly from the Black Forest (*Schwarzwald*). But by 1834, this interpretation had become untenable for two reasons: (1) pebbles consisting of other rock types, such as serpentinites, gabbros, diorites, greenschists, and spilites, accompanied the granites in the Nagelfluh and in the Habkern sandstones, yet they did not exist in the Black Forest to a degree sufficient to account for their existence in the Alps; and (2) the abundant red sandstones of the Black Forest were not represented in the pebble population of the Nagelfluh (Studer, 1834, 1853, especially p. 358–359). Studer considered the provenance of these foreign pebbles to be in a mountain chain that had once paralleled the Alpine northern margin and that since had either subsided,[8] or eroded away, or become buried under its own debris formed by the Nagelfluh, or been overthrust by the limestone Alps from the south (Studer, 1853, p. 359). The mountains of Stockhorn (within the Klippen Nappe, a part of Préalpes romandes, i.e., the eastern lobe of the Préalpes: Fig. 2), which lay amid surroundings totally dissimilar to them (the surroundings consist mostly of Penninic flysches: Fig. 2), Studer considered the remaining summits of the sunken mountain range (Studer, 1834). Studer showed some

Figure 1. The father of Alpine geology and Alpinism: the great Swiss naturalist Horace Bénedict de Saussure (1740–1799), who first noted the peculiarly "exotic" porphyry and granite pebbles in the Alpine Nagelfluh. From an engraving after an oil painting by J.-P. Saint-Ours.

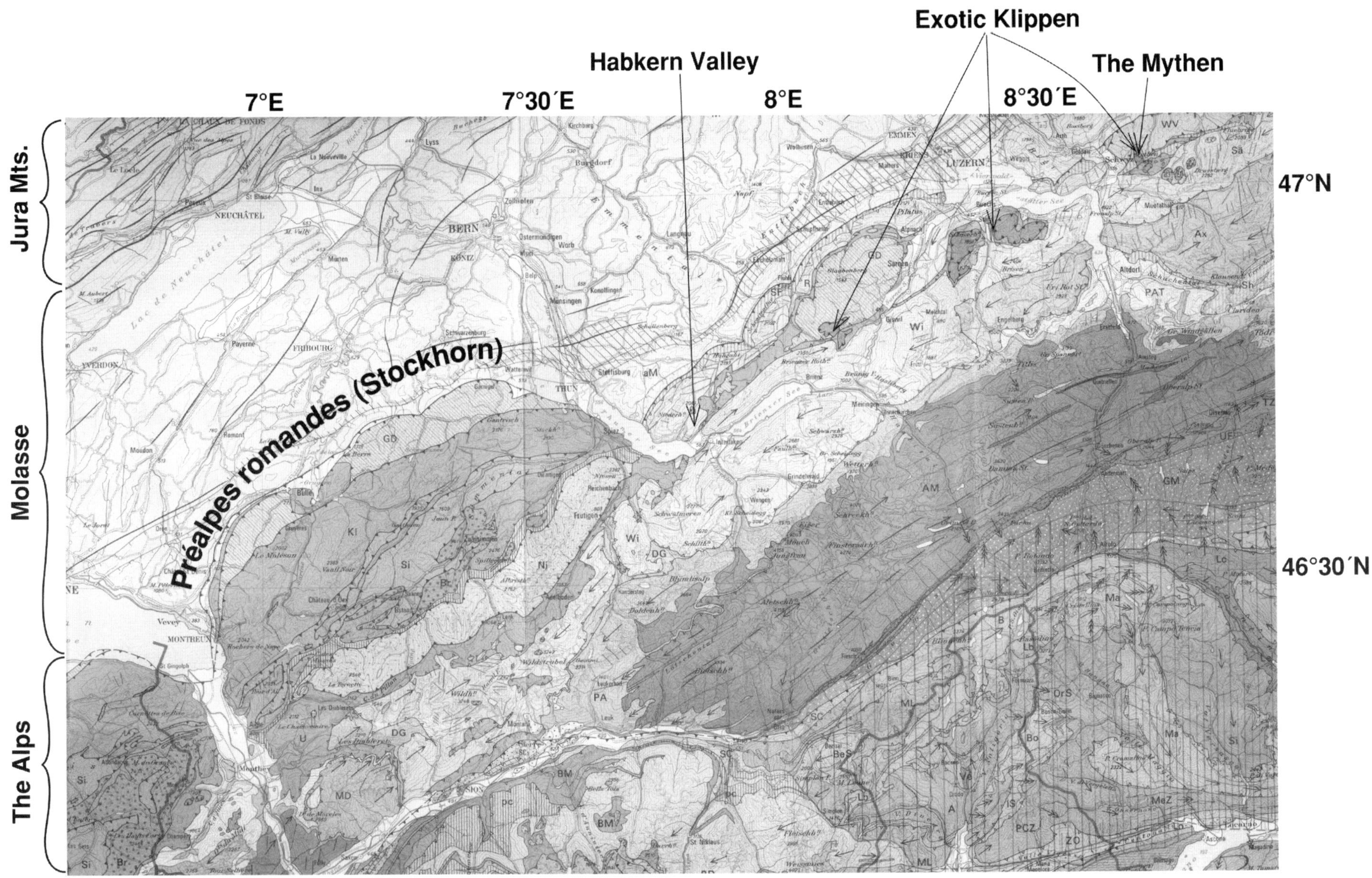
Exotic Klippen
Habkern Valley
The Mythen
7°E
7°30´E
8°E
8°30´E
47°N
46°30´N
Jura Mts.
Molasse
The Alps
Préalpes romandes (Stockhorn)

of these granite blocks to Sir Roderick I. Murchison following the July 1848 meeting of the Société Helvetique des Sciences Naturelles in Soleure (see Geikie, 1875, especially p. 93–97) and told him that they had been originally embedded in the underlying flysch. Murchison was incredulous:

> Now it happens that in the same valley of Habkheren [*sic*] several large granitic blocks also exist, which lying upon the surface of the ancient alluvium, or having been washed into water-courses, have at a distance all the aspect of the usual Alpine erratic blocks, about the transport of which there has been so much discussion. The largest of these lies on the surface of a boggy meadow, under which is a great thickness of the coarse ancient alluvia on the east side of the rivulet of Habkheren, … This block, so superposed on the ancient alluvia, is about 105 feet [≈ *32 m*] long by 90 feet [≈ *27 m*] broad and 45 feet [≈ *13.5 m*] high (above the marshy meadow), and has therefore a mass of not less than 400,000 cubic feet [≈ *11,000* m^3][9]. As it consists of a peculiar granite, now unknown to mineralogists in any part of the Alps, Professor Studer believes that, like the very small geodes and courses alluded to, this block was also included in the formation of the flysch, and that during the disintegration of that rock on the vertical sides of the valley, it has rolled down into its present position.

Figure 3. Christian Leopold von Buch, Baron of Gelmersdorf (1774–1853), who agreed with de Saussure on the northerly provenance of the exotic clasts in the Nagelfluh. From an engraving by Ambrouse Tardieu, sketched from life in Paris.

Figure 2. A part of the tectonic map of Switzerland (copied from Spicher, 1980), showing the position of the Préalpes romandes, the Habkern Valley, and the exotic klippen. On the left side, the main tectonic divisions of the Alps within the frame of this map are indicated. Abbreviations: (For these abbreviations, **H** indicates that a unit belongs in the Helvetic realm, including the Infra- and Ultrahelvetic units; **P**, the Penninic realm; and the rest belong to the European foreland.) A = Antigorio Nappe, **P**; AM = Aar Massif; aM = Upthrust Molasse; Ax = Axen Nappe, **H**; BM = Mesozoic of the Bernhard Nappe and *zone houillière* (coal-bearing zone), **P**; Bo = Bosco Series, **P**; Br = Breccien (Breccia) Nappe, **P**; DG = Diablerets-Gellihorn Nappe, **H**; GD = Gurnigel Nappe, Voirons, and Schlieren Flysch, **P**; GM = Gotthard Massif, **H**; IS = Isorno Series, **P**; Kl = Klippen (or Préalpes Médianes) Nappe, **P**; Lb = Lebendun Series and the Baceno Schists, **P**; Lc = Lucomagno Gneiss, **P**; Lo = Locarno Zone, **P**; Ma = Maggia Zone, **P**; MD = Morcles-Doldenhorn Nappe, **H**; MeZ = Mergoscia Zone, **P**; ML = Monte Leone Nappe, **P**; Ni = Niesen Nappe, **P**; Ors = Orsalia Series, **P**; PA = Parautochthonous Mesozoic units of the Helvetic realm, **H**; PAT = Parautochtonous Tertiary units of the Helvetic realm, **H**; pc = metamorphic Permo-Carboniferous of the Bernhard Nappe, **P**; PCZ = Piodo di Crana Zone, **P**; R = Ramosch Zone, **P**; Sä = Säntis-Drusberg Nappe, **H**; SC = Brigue-Sion-Courmayeur Zone (Valais), **P**; SF = Subalpine Flysch, **H**; Si (in the Préalpes romandes) = Simmen Nappe, **P**; Si (in the Penninic units, i.e., SE corner of this figure) = Simano Nappe, **P**; U = Ultrahelvetic Nappes (including certain wildflysch nappes), **H**; UF = Garvera-Urseren-Furka Zone, **H**; Wi = Wildhorn Nappe, **H**; WV = Flysches of Wägital, Liechtenstein, and Vorarlberg, **P**; ZO = Orselina Zone, **P**. Symbols: Single short line = horizontal fold axes and linear structures; single-headed arrow = fold axes and linear structures with plunge of 1–20°; double-headed arrow (with arrows in the same direction) = fold axes and linear structures with plunge of 21–65°; triple-headed arrow = fold axes and linear structures with plunge of 66–90°. NW-SE ruling in the Molasse = folded Molasse; vertical ruling = amphibolite facies Alpine metamorphism (wide spacing of lines = lower grade; close spacing of lines = higher grade); long, serpentine lines in the Molasse and in the Jura Mountains = fold axial traces.

Figure 4. German physician and geologist Johann Gottfried Ebel (1764–1830), who first postulated a southerly provenance for the exotic components of the Nagelfluh. Copper engraving by Heinrich Meyer, from Franks et al. (2000, fig. 1).

Figure 5. The great Swiss geologist Professor Bernhard Studer (1794–1887), who first noted the close relationship between the exotic clasts in the Nagelfluh and those in the flysch sequences atop the northern calcareous chain (i.e., the Helvetic Nappes in the present terminology; see Fig. 2) in the Swiss Alps. From a picture sketched in 1859 (from Heim, 1919, plate I).

Figure 6. Block of Habkern granite in the flysch in the Habkern Valley (1/40 of its natural size; from Heim, 1921, fig. 100; also see Bailey, 1935, fig. 18.)

… I cannot adopt this opinion, nor can I regard the great block of Habkheren in any other light than that of a huge superficial erratic derived from some parent rock, which has either since been lost by subsidence and buried beneath other deposits, or is hidden from sight under those coverings of snow and ice, which necessarily impede observation over so very large an area of the higher Alps. I fully admit that the small bands of granitic rock above adverted to, are fairly intercalated in the flysch, but the presence of geodes, the largest of which is not above four feet [≈ *1.2 m*] long and a foot [≈ *0.33 m*] wide, can never satisfy me that a monstrous block, containing 400,000 cubic feet [≈ *11,000* m^3], was similarly formed; that block not having the slightest appearance of having ever been a geode. Again, no conglomerates known in any part of the flysch of the Alps exhibit pebbles of more than a foot or two in diameter. But, supposing this block to have been part of a conglomerate, and that it was transported from a ridge of crystalline rock into the flysch during the formation of that deposit, by what agency must we suppose it to have moved? Certainly not either by solid or floating ice; for the period of the nummulites and flysch was anything but one of glacier action, and was in fact one of considerable warmth.

Seeing then no satisfactory explanation of the deposit of a block of this magnitude in finely laminated sandstone and schist (such as constitutes the flysch of the sides of the Habkheren) … , I necessarily reject the application of such reasoning …. (Murchison, 1849, p. 212–213)

Clearly, Murchison knew the Alps less well than Studer and had less insight into tectonic matters than his Swiss host.[10] He in fact thought that the Habkern granite pebbles in the flysch had formed in situ by metamorphism:

I confess that I could not account for such appearances, except by supposing that the granitic matter was evolved contemporaneously with the formation of the sedimentary sandstones and schists which envelope it; the concretionary forms of some of these masses seeming to favour this hypothesis. But whether produced in the same manner as the so-called volcanic or plutonic grits of other regions, by contemporaneous segregation of igneously-formed particles in the bottom of a turbid sea, or by subsequent partial alteration of the strata through the action of the heat and gases, or by transport from other rocks, it is clear that these small developments of granitic matter are contemporaneous with the flysch. (Murchison, 1849, p. 211–212)[11]

Murchison's "metamorphic" view of the Habkern granite pebbles and blocks was later adopted by F.J. Kaufmann, who coined the term wildflysch (see below).

Schardt (1898, p. 120) later pointed out that it was Studer's genius to have recognized the connection between the crystalline (mainly granite) boulders of the Nagelfluh and those in the flysch of the Habkern Valley, for which Isidor Bachmann (1837–1884; Fig. 7[12]) introduced the designation "exotic" from earlier usages

Figure 7. Professor Isidor Bachmann (1837–1884), who introduced Adolf von Morlot's term "exotic" for the clasts and blocks in the Nagelfluh and flysch, for which no provenance in the near vicinity could be ascertained. (Courtesy of the Archives of the University of Bern and of Professor Adrian Pfiffner.)

Figure 8. Professor Jakob Früh (1852–1938), who in 1888 revived Ebel's idea that the exotic blocks in the Nagelfluh had a southerly and southeasterly provenance. (Courtesy of Dr. Sybille Franks.)

by his compatriot Adolf von Morlot in the Eastern Alps and the German geologist Ludwig Hohenegger in the Moravian Carpathians (Bachmann, 1863, p. 26–27; for the von Morlot and Hohenegger references, see below the section on the Carpathians).

Throughout the nineteenth century, two rival theories sought to account for the provenance of the exotic blocks: One was Studer's old model of the sunken mountain range. The other was proposed by the Swiss school teacher, geologist, and later geography professor of the Swiss Polytechnic (Eidgenössische Technische Hochschule) Jakob Früh (1852–1938; Fig. 8[13]), reviving Ebel's old idea of a southerly provenance (Früh, 1888). It consisted in assuming the presence of large, now extinct rivers carrying the pebbles from the south and east to the northern margin of the Alps. But the presence of very angular exotic clasts in the flysch precluding their fluvial transport from afar and the rise of the nappe theory in the last decade of the nineteenth century made both models unlikely (see the two excellent reviews in Schardt, 1898 and Beck, 1912).

That the sandstones in the Habkern Valley were not just flysch sandstones with some exotic blocks in them had become clear by 1870, when the Swiss schoolteacher and geologist Franz Joseph Kaufmann (1825–1892, Fig. 9[14]) introduced the designation wildflysch for the rock association seen in it. He wrote to Studer in 1871 that wildflysch was "dark grey to blackish, soft schists, thinly-bedded to foliated, [*with*] commonly bent

Figure 9. Professor Franz Josef Kaufmann (1825–1892), who introduced the term wildflysch in 1871 in a letter he wrote to Bernhard Studer. (Courtesy of Dr. Paul Felber.)

[*foliation planes*], with shiny surfaces, also cut by white mineral veins. Commonly, very irregularly structured, surrounding blocks or hard pieces [=*Knauer*[15]], alternating with Fucoid-schists and Macigno.[16] With all sorts of synclines, anticlines and bendings" (Studer, 1872, p. 259–260; Cadisch, 1953, p. 174; Fig. 10). About a decade-and-a-half later, Kaufmann gave a more precise definition: "Dark grey to black, soft, shiny schists, commonly bent and full of striated slickensides, alternating with sandstones, in places with conglomerate. Not rarely *Stadschiefer* [*Uppermost Eocene Globigerina-bearing shales and slates with some sandstones*], Fucoid-schists and Leimern-like[17] shales are interbedded; also Nummulite-limestones with green sands, granite-breccias, etc." (Kaufmann, 1886, p. 553). The blocks in the wildflysch are made up of pelagic limestones and marls, flysch sandstones, various crystalline pieces including gneisses and Habkern granite pieces (the largest of which near Habkern has a volume of more than 11,000 m^3, as we saw above; also see Hsü, 1995, p. 60). Kaufmann followed Murchison in thinking that some of the granites, especially those with angular outlines, were formed in situ within the wildflysch by metamorphism, whereas others were authentic exotic granite pebbles exhibiting an abraded and rounded exterior (Kaufmann, 1871, p. 266, 1879, p. 92–93).

In his comprehensive review of the exotic nappes (the Prealpine nappes), klippen (isolated remnants, along the northwestern margin of the Alps, of a once much larger nappe that included the Préalpes: Figs. 2 and 11), and blocks found along the northwestern margin of the Alps, including the wildflysch, Schardt (1898) calls the period between 1870 and 1890 the "transition" or "incubation period" during which much effort was spent on meticulous mapping as a part of the general geological survey of Switzerland. Schardt praises this work, but deplores the lack of theoretical insight, rampant parochialism, and trenchant conservatism of the field workers (which apposite observations did not exactly endear him to his colleagues: see, for example, Haug, 1899).

In 1884, when Marcel Bertrand wrote his benchmark paper on the Glarus double-fold, interpreting it as a single, north-vergent nappe (cf. Trümpy, 1991b) and thus inventing the nappe theory, he also made the prescient suggestion that the Préalpes as a whole may be a giant, north-vergent nappe (see the emphatic acknowledgement in Haug, 1899, p. 116). Bertrand was not aware of the significant facies differences between the Préalpes and the correlative Helvetic realm successions, on the one hand, and the great similarity of the Préalpes in terms of their facies to the klippen and the exotic blocks, on the other. Unfortunately, his suggestion passed almost unnoticed by the busy fieldworkers of Switzerland, despite the fact that Bachmann, in his 1863 paper on the exotic blocks, expressly brought into connection the then ill-understood phenomena creating the strange occurrence of Jurassic blocks in the much younger flysch in central Switzerland with the Permian Verrucano overlying the Eocene flysch in the canton of Glarus across a thin band of Jurassic limestones (Bachmann, 1863, p. 34)!

Light finally dawned upon Hans Schardt (1858–1931; Fig. 12[18]). In 1893, in a short note that revolutionized Alpine tectonics, he wrote:

> … the most striking contrast exists between the facies of the terranes, especially in the Cretaceous and Tertiary series. One can say that the chain of Chablais and Stockhorn [*see Fig. 2 herein*] appear as a piece of a foreign terrane lying amidst the calcareous margin of the Alps.

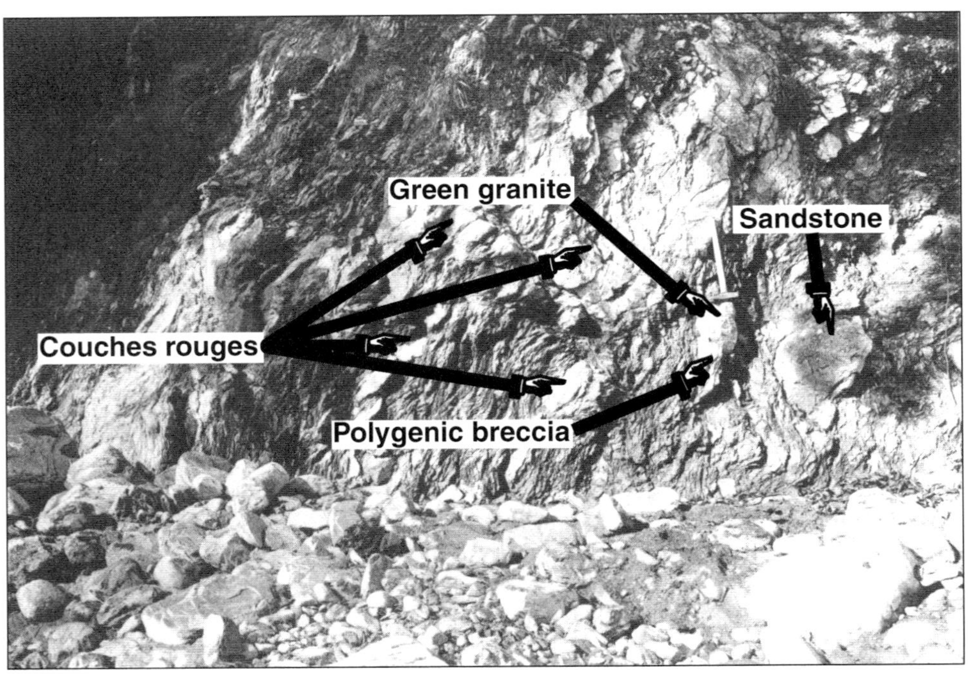

Figure 10. Wildflysch (dark) and Couches rouges (light) with exotic blocks on the Lombach, showing a typical outcrop of this rock association (from Beck, 1911, plate VI).

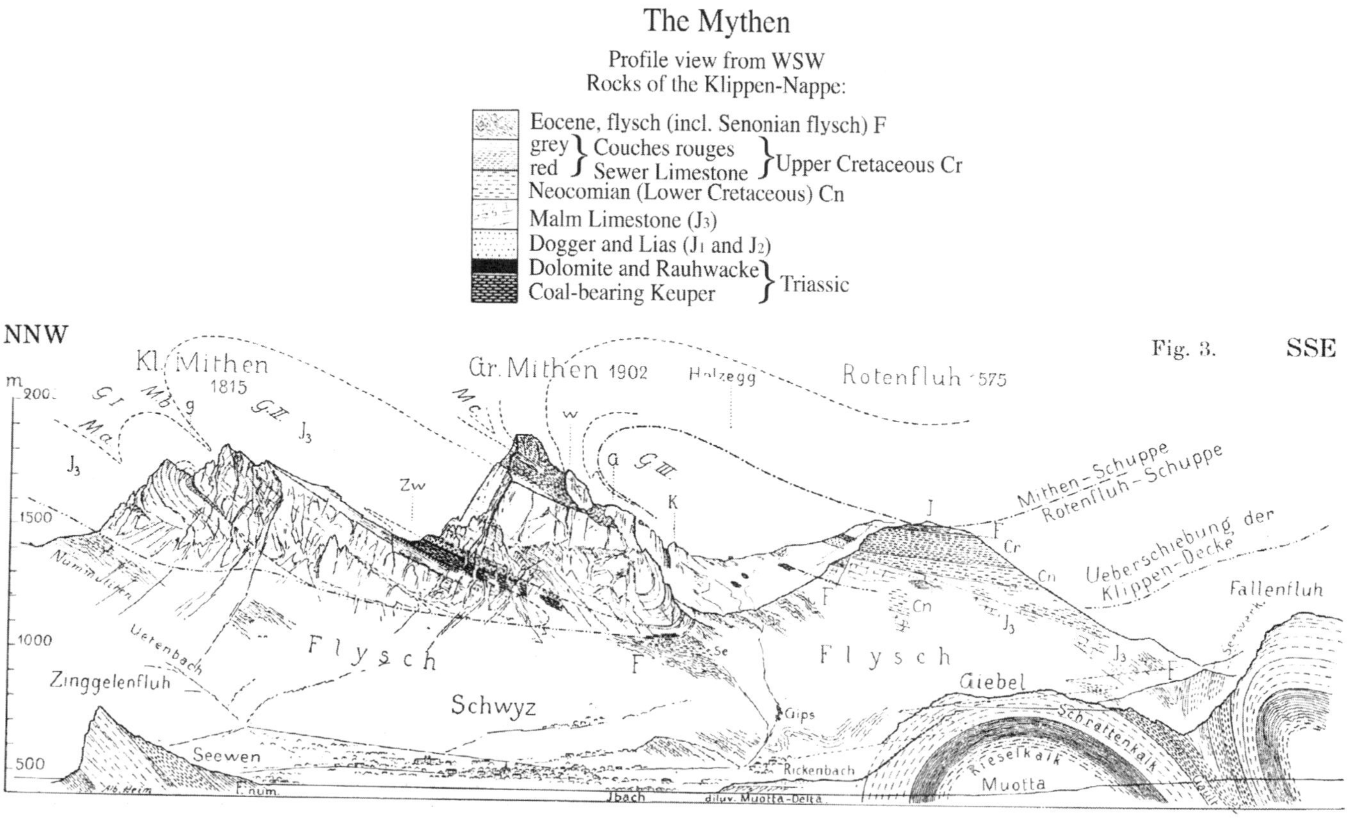

Figure 11. The most famous of the exotic klippen in the Alps: the Mythen as sketched and interpreted by Albert Heim (1922b, fig. 3). The Mesozoic exotic section, consisting of Upper Triassic, Jurassic, and Cretaceous rocks float atop a wildflysch of Eocene age, which, in turn, is thrust onto the anticlines belonging to the Säntis Nappe. For a general description of this locality, see also Jeannet (1934).

This anomaly, which remains enigmatic, is related to others of no less importance, such as the origin of the flysch and its erratic blocks, and that of the klippes, these giant masses of strange facies which rest, with Triassic at their base, amidst an entourage of younger terranes. The interpretation that appears preferable today consists of envisaging the entire region of Chablais and Stockhorn as a vast nappe. (Schardt, 1893, p. 707).

In his three-page paper, Schardt (1893) also tied the origin of the flysch, with its exotic blocks, to the movement of the nappes. He assumed that the exotic blocks became mixed into the flysch first through surficial and then through tectonic processes during the transport of the nappes.[19] This idea, as far as I can ascertain, is the first appearance of what we today consider to be essentially, but not quite completely, the concept of mélange (because the origin of mélange was not ascribed to *tectonic disruption and* mixing); however, it did not have the surprising aspect of the later rediscoveries of the mélange concept, because (1) it subsumed the mixed rocks under the inappropriate but widely familiar designation flysch in the form of wildflysch; (2) it resorted to surficial (sedimentary) processes to generate the exotic blocks, which shocked nobody

Figure 12. Professor Hans Schardt (1858–1931), who recognized the total allochthony of the Préalpes and associated with their transport the origin of the exotic blocks in the wildflysch. From Schaer (1994, p. 430).

because such processes as glacial transport, iceberg transport, and landsliding (the last especially in Switzerland! See, for example, Heim, 1932, for a summary of Swiss cases) were commonly known; and (3) it was presented as an aside within what was then considered a much more shocking discovery, namely a horizontal transport of mountains of rocks for some 50 km. Both Schardt and his successors viewed the wildflysch as a mainly sedimentary mixture that was later tectonized. They correctly saw that rather too many things were going on at once at the height of the Alpine orogeny to give clean-cut interpretations. They knew both too much and too little. They observed many different things at outcrop and knew that many processes had to have been at once active during the orogeny. They also knew a lot about the surficial processes that could have created a part of what they saw. Arnold Heim, the talented son of Albert and the inventor of the method of palinspastic analysis, for example, thought that the wildflysch was essentially a pebbly mudstone, the exotics having been dropped by floating icebergs (Heim, 1907).

Despite the enormous and epoch-making work they had done on stratigraphy, sedimentation, and structural geology, however, and despite the unusually high percentage of extraordinarily talented men among them, Alpine geologists could not as yet appreciate the grandeur of the tectonic processes recorded in that jewel of a mountain range that adorns the middle of Europe. Workers of a later generation were to discover these, at their head the great genius Émile Argand, "the magician," as his friend and doctoral adviser Maurice Lugeon called him when mourning his untimely death during his funeral on September 17, 1940 (Lugeon, 1940, p. 49).

In one of the last reviews concerning the wildflysch before the publication of Edward Greenly's book introducing the concept of mélange in the sense we understand this term today, Beck (1912, p. 82; also see Fig. 13 for the following description), for example, argued:

> The entire formation seems uniform to me, despite the disturbance and tearing of the normal transitional sections through extraordinary tectonic events. Entire klippen, individual blocks, and conglomerates are enclosed within the Prealpine Tertiary and are connected with it through stratigraphic (and tectonic) transitions. The deposition of these inclusions could have only occurred during the Tertiary at the expense of a piece of dry marginal land then existing. Subaquatic slides (according to Arnold Heim) and sinking of blocks into not-yet-consolidated stratigraphic levels (according to Tornquist) probably happened, but these alone cannot explain the complicated conditions. Both the size of the klippes and the way they are deposited and their distribution do not allow us to think of transport by drifting ice floes. A rias- or a fjord-coast slowly sinking into the sea, on which individual, undermined parts of klippen fall into marine deposits and are surrounded by them as large or small inclusions or as beach conglomerates, suffice for the explanation of the Prealpine Tertiary with its small klippen and exotic blocks.

It has since become clear that tectonic disruption and mélange formation played a much greater role in the formation of the wildflysch, which is in fact mostly a mélange. Accordingly, Bayer (1982) suggested that the designation wildflysch be replaced by Habkern Mélange, which, he showed, can be further subdivided into three subordinate mélanges on the basis of the provenances of their exotic blocks: (1) Leimern Mélange, (2) Sörenberg Mélange, and (3) Iberg Mélange.

Although Swiss geologists of the late nineteenth and early twentieth centuries clearly recognized all the field attributes of the mélange, they could never bring themselves to see it as a purely tectonic mixture. Even the great Argand thought that the mixture of rocks originating from the sima with abyssal and bathyal sediments (in essence what we today would call an ophiolitic mélange) was a result of sedimentary processes alone:

> the subsidence inherent in the geosynclinal process does not, therefore result from an original radial movement; it is only the vertical effect of a horizontal stretching. The weight of the sediments works in the same direction and helps of course to accentuate the alveole, but it is not necessarily the main cause. Until compensation, the sima rises under the thinned sial; this motion accounts for the frequent association of green rocks with bathyal and abyssal sediments. The mélange [*i.e., mixture*] of abyssal with shallower sediments takes place through submarine sliding on the slope. (Argand, 1924, p. 299)

The main reason for this blindness toward tectonic crushing, disintegration, and mixing was that the recognition of the exotic components within the Habkern Mélange had started with their comparison with similarly exotic components in the undoubtedly sedimentary Nagelfluh. This sedimentary "bias" was transported to the considerations of the Habkern "flysch," which indeed is mostly a rock of sedimentary parentage. Moreover, the exotics in it had been recognized at a time when large-scale thrust faulting was not known. At the time (i.e., in the 1830s), there were only two places where moderate thrusting had been recognized or thought to be recognized: (1) along the upper course of the Irtysh in Central Asia, where granite was pushed up onto the schists along a thrust fault just below the confluence of the Irtysh and the Narym, between the rivers Bareshnikov and Kozlovka, where white granite rests on the uneven surface of the steeply upturned clay slates (Hermann, 1801, p. 108–113 and fig. 14 between p. 108 and 109); and (2) along the Elbe Valley in Germany, northeast of Dresden and Meißen, where the Lausitz Hercynian granite was thrust southwestward onto various Cretaceous sedimentary rocks (Weiss, 1827; for a detailed history, see Wagenbreth, 1966). Both of these cases were viewed, however, as freaks, rather than as the norm, and they never became widely known. Trümpy (1991b) has nicely described Arnold Escher's von der Linth horror of having discovered the Permian Verrucano overlying Eocene Flysch for a distance of nearly 15 km in Glarus in the 1840s. The only processes geologists were familiar with to create mixtures of rocks containing "rocks enclosed within rocks" were thus sedimentary. After large-displacement thrusting (~40 km) was recognized in the Alps in 1884 by Marcel Bertrand, it still

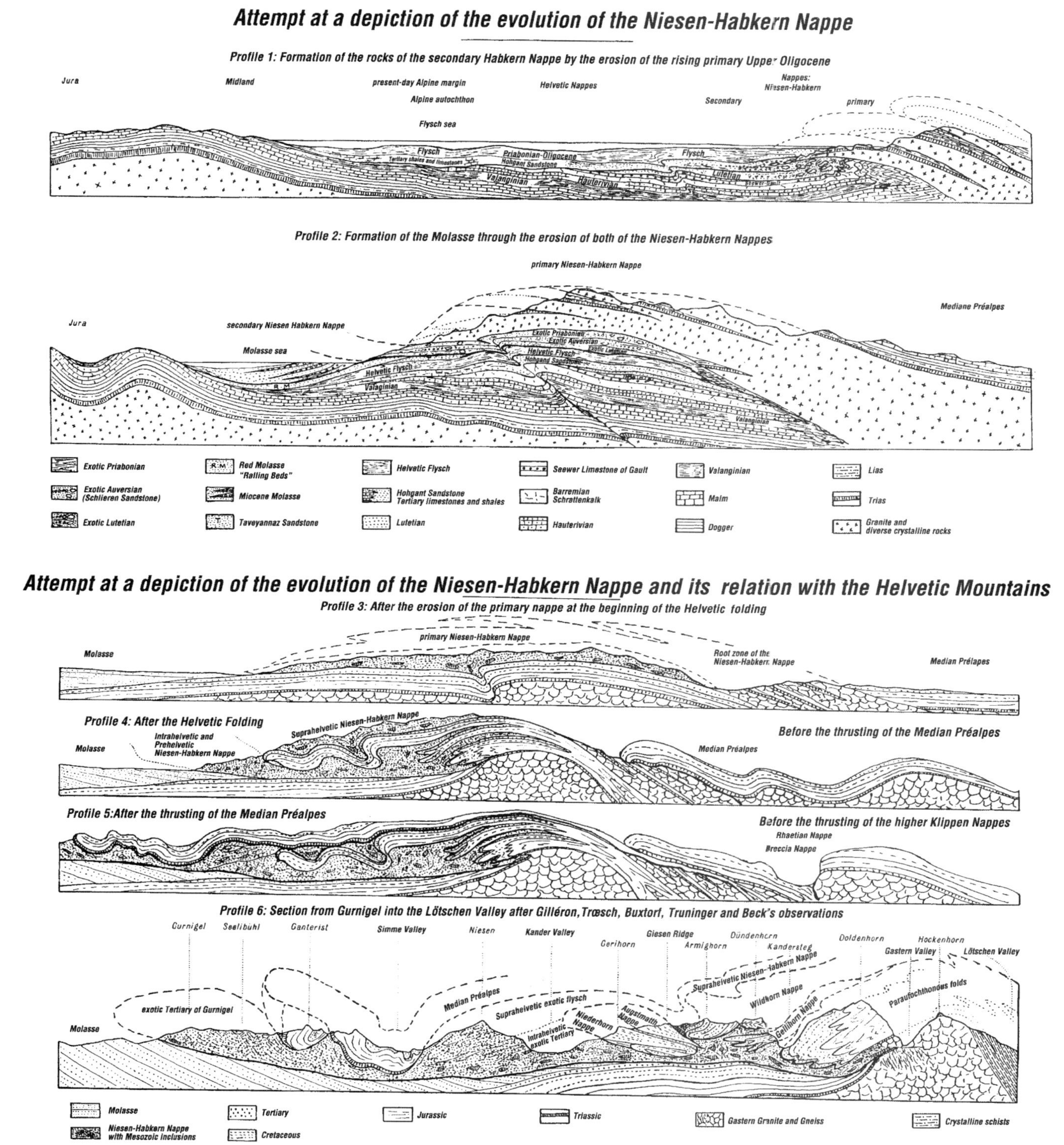

Figure 13. Beck's (1912) sequential diagrams showing the evolution of wildflysch units associated with his Niesen-Habkern Nappe. Although Beck's unification of these two separate units into a single nappe has been discredited (Lugeon, 1916; Buxtorf, 1918; also see Bayer, 1982, p. 23), the diagram here shown illustrates nicely the ideas then orthodox for the origin of the wildflysch: it was viewed essentially as a debris flow deposit eroded off advancing nappes and later overrun by them. This figure was generated by combining Beck's plates 2 and 3 and translating his labels into English.

was almost a decade before Hans Schardt could show to the satisfaction of most geologists that thrusting was a very widespread phenomenon in the Alps. When he did so, the shock his demonstration created was so great that the simultaneous discovery of mélanges was completely overshadowed. Given the history of thought on the exotic blocks along the northern margin of the Alps, which consisted almost exclusively of sedimentary rocks, a mainly sedimentary rather than a mainly tectonic interpretation thus remained attached to the wildflysch.

Although the Alpine ideas on thrusting were very influential in the discovery and further elucidation of similar phenomena, including fault-rocks, in the Scottish Highlands (see Şengör and Sakınç, 2001), the available documents do not indicate that Edward Greenly's recognition of mélanges in Anglesey was influenced at all by the earlier recognition in Switzerland. Greenly recognized the mélanges as a distinct, separately mappable rock type, formed purely by tectonic disruption, and he outlined in detail how this happens. It is therefore appropriate, I think, to keep Schardt and his predecessors and successors in the *prehistory* of the discovery of the mélange concept, and to give Greenly the credit of the initial discovery of mélanges.

Having thus separated the prehistory of the discovery of the mélange concept from its history, let me point out the fact that the separate discoveries of the mélanges in Switzerland and Wales were achieved by following *entirely different methodologies*. In Switzerland, the recognition came through mainly a *stratigraphic/sedimentological* approach, whereas in Wales, mélanges were recognized as a result of a careful *structural* mapping. Not only were the discoveries independent, but also the roads leading to them were very different.

The "Prehistory" of the Discovery, Second Act: The Carpathians

Nearly coevally with the Alpine developments in the nineteenth century, similar problems concerning exotic blocks in younger flysch sequences baffled the geologists working in the northern Carpathians. Carpathian and Alpine developments occurred in mutual awareness and acknowledgement of the progress of understanding, and thus they constitute only separate acts in the same drama. (For a summary of the development of thought on the northern Carpathians in general, see Andrusov, 1931, p. 89, 1938, p. 118–127, 1964, p. 27–32.)

As one approaches the Carpathians from the north, along a meridian just to the west of the city of Ostrava (Fig. 14, map, where it says *Mährisch Ostrau*), one first wanders over the flat Molasse landscape underlain by Middle to Upper Miocene sedimentary rocks. If the wanderer continues southwestward, following the valley of the Odra upstream, he or she would be walking along the thrust front of the Lower Tertiary flysch deposits pushed onto the Molasse (the Subbeskide thrust front in Fig 14, map). The landscape is very gently rolling, pleasant agricultural fields alternating with pretty woodland. Suddenly, near Stramberk (about 50 km south-southwest of Ostrava), an abrupt, light-colored rock wall, a veritable crag, arises from amid the rolling green fields and surprises the wanderer (Fig. 15). If the wanderer is a geologist, his or her surprise would become even greater after a cursory examination of the material of the abrupt apparition because its abruptness is caused by a sudden change in the rock-type: instead of the soft Upper Cretaceous sandstones and shales of the rolling surroundings, the steep-sided hill consists of light yellowish-gray to almost white, hard Upper Jurassic limestone in which bedding is hardly discernible. The rock has the character of almost a breccia cemented by secondary calcite. It is rich in fossils—including lamellibranches, brachiopods, echinoderms, bryozoa, and such foraminifera as *Miliolina* and *Quinqueloculina*—to which it owes its reputation (cf. Suess, 1852, especially p. 10, 1858, 1867; Hohenegger, 1861, especially p. 15–21; Andrusov, 1965, p. 171–185, with a well-illustrated account of the fossil fauna). Our wanderer is, in fact, standing in front of what Ludwig Hohenegger[20] in

Figure 14. Viktor Uhlig's 1907 view of the tectonics of the northern Carpathians, which I use also for modern orientation in this paper. Modern research has greatly refined Uhlig's outline of the structure (mainly by subdividing his main units into numerous nappes and improving the stratigraphic precision), but did not change it in any significant way. That is why the map and the section as drawn by this great geologist still serves my purpose here. Map: Uhlig's tectonic map of the northern Carpathians (after Uhlig, 1907, p. 982, from Suess, 1911, Fig. 185). Key to legend: 1 = Autochthonous foreland comprising the post-Variscan cover and the Miocene terrain; 2 = Subbeskide Nappe; 3 = Beskide Nappe (2 and 3 are equivalents of the Helvetic Nappes in the Alps); 4 = Subpieniny and Pieniny Nappe (Southern Klippen Zone; equivalent of a part of the Pennine Nappes in the Alps); 5 = High Tatra Nappe (equivalent of a part of the Pennine Nappes in the Alps); 6 = Subtatric Nappe; 7 = Nappes of the Inner Belt and Central Hungary (equivalents of the Austroalpine Nappes); 8 = Volcanic rocks; 9 = Cenomanian, Senonian, and Eocene unconformable; 10 = Regions of uncertain structure; 11 = Nappe fronts; 12 = Nappes; 13 = Fensters; 14 = Hypothetical limits; 15 = Stratigraphic contacts or faults posterior to the origin of the nappes; 16 = Klippen. Sedimentary rocks posterior to the folding of the Carpathians (Neogene deposits and alluvium) are left white. Cross section: Uhlig's cross-section (approximate location shown in Fig. 14A) across the Northern Carpathians (after Uhlig, 1907, from Andrusov, 1964, fig. 4d; this cross section is also reproduced in Uhlig, 1908, plate I). Key to abbreviations: B = smaller thrusts within the Beskide and Subbeskide Nappes; B_1 = thrust contact between the Subbeskide Nappe and autochthonous cover sequences; B_2 = overthrust contact of the Beskide Nappe; BZ = Beskide Zone; HT = High Tatra Nappe; PZ = Pieniny Klippen Zone; SBZ = Subbeskide Zone; SPZ = Subpieniny Zone; ST = Subtatra Nappe. Key to legend: 1 = Beskide Flysch (Upper Cretaceous and Lower Tertiary; 2 = Subbeskide Jurassic and Upper Cretaceous; 3 = Beskide Jurassic and Upper Cretaceous; 4 = Pieninic Jurassic and Lower Cretaceous in form of klippen, larger and smaller (exotic) blocks; 5 = Subpieninic Jurassic and Lower Cretaceous; 6 = Upper Cretaceous and Lower Tertiary in the region of Pieninic Nappes (*Klippenhülle*: matrix of klippen); 7 = Mesozoic of the High Tatra unit; 8 = granite and crystalline schists of the High Tatra unit; 9 = metamorphic schists, probably "Lepontine" (i.e., Penninic); 10 = Mesozoic of the Subtatra unit; 11 = flysch of the Subtatra unit; 12 = nappe of the Inner Zone; 13. = nappe of the Hungarian Mid-Mountains

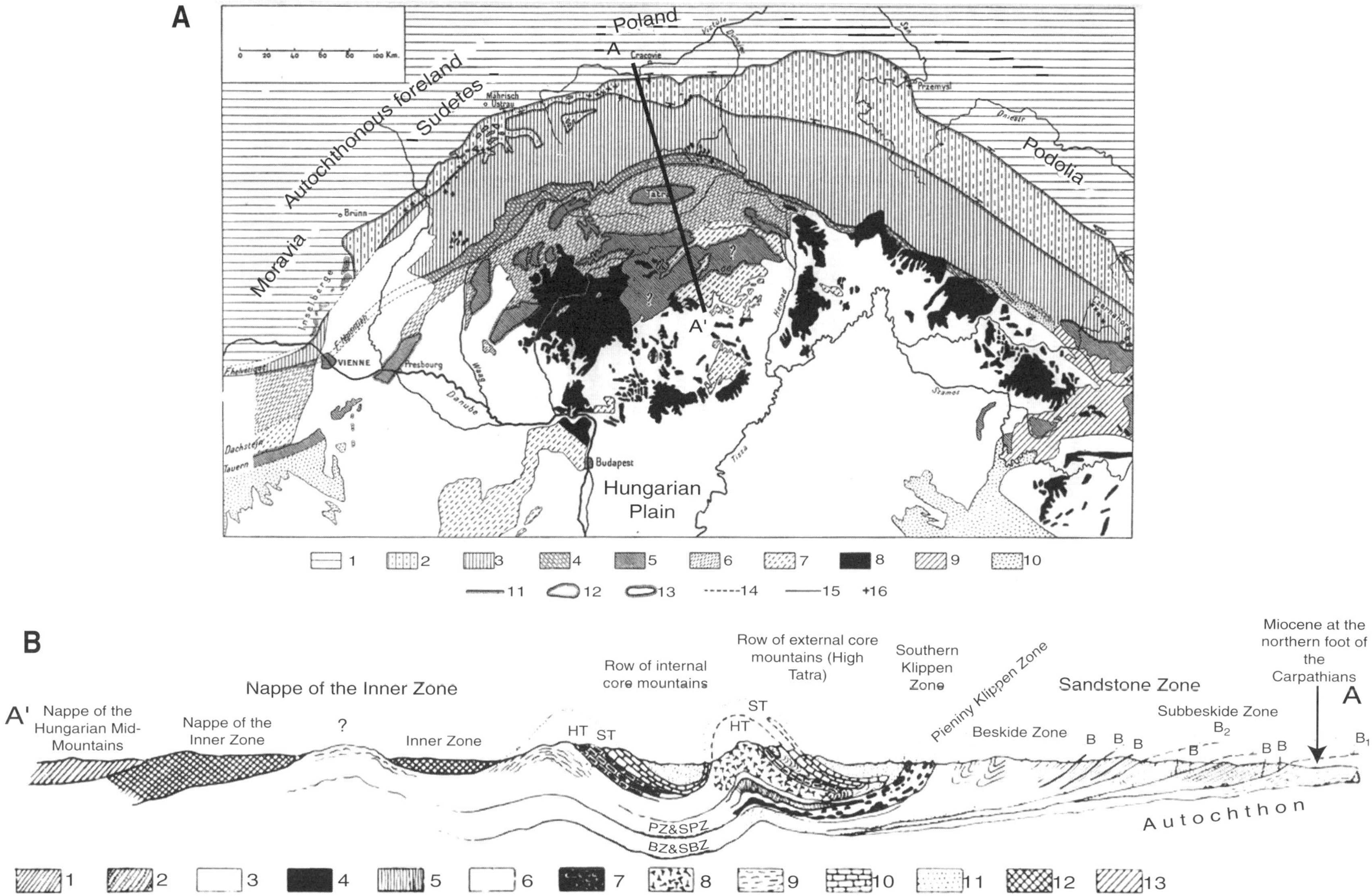
A
Poland
A
Cracovie
Autochthonous foreland
Sudetes
Moravia
Podolia
A'
Brünn
Vienne
Presbourg
Danube
Waag
Budapest
Hungarian Plain
Tatra
Przemysl
Dniestr
Szamos
Tisza
Hernad
Dachstein
Tauern
0 20 40 60 80 100 Km.
1
2
3
4
5
6
7
8
9
10
11
12
13
14
15
+16
B
A'
Nappe of the Hungarian Mid-Mountains
Nappe of the Inner Zone
Nappe of the Inner Zone
?
Inner Zone
Row of internal core mountains
HT
ST
Row of external core mountains (High Tatra)
ST
HT
Southern Klippen Zone
Pieniny Klippen Zone
Beskide Zone
Sandstone Zone
B
B
B
Subbeskide Zone
B2
B
B
B
Miocene at the northern foot of the Carpathians
A
B1
PZ&SPZ
BZ&SBZ
Autochthon
1
2
3
4
5
6
7
8
9
10
11
12
13

Figure 15. The Kotouc Klippe formed from Stramberk Beds near Stramberk (from Andrusov, 1965, fig. 54; lettering mine).

1849 called the *Stramberger Schichten* (now Stramberk Beds),[21] the mode of occurrence of which had been a problem among Carpathian geologists since the earliest days of geological mapping. Already in 1830, Ami Boué (1794–1881; Fig. 16[22]) had recognized that the limestone klippen strewn in the rolling landscape of the Cretaceous to Paleocene flysch of the northern Carpathians (then simply known as *Karpathensandstein*—Carpathian Sandstone) as Jurassic rocks. This led to much confusion, and some even went so far as to deny the Jurassic fossils their age-diagnostic value to be able to make the limestone klippen just lateral facies equivalents of the "Carpathian Sandstone" (e.g., Hohenegger, 1852, especially p. 137; cf. also Carlé, 1988, p. 83). Continuing mapping showed that equivalents of the Stramberk Beds exhibit variations in color to blackish gray, and its various representatives occur as blocks in various levels of the Cretaceous and Lower Tertiary (see Suess' introductory notes in Suess, 1858, and Hohenegger, 1861, p. 15). This zone continues west-southwest into Austria and merges with the Waschberg Zone through the Steiniz Zone (cf. Tollmann, 1986, 1:500,000 tectonic map, eastern half, and extensive references to earlier studies; also see especially Prey, 1975 and Tollmann, 1975). It was in Austria that the Swiss geologist Adolf von Morlot (1820–1867; Fig. 17; for his life, see von Hauer, 1867; Eiszmann, 1974 and the references he cites), the man who invented the term "Quaternary" in 1854 (Eiszmann, 1974, p. 293 and footnote 1), had also first used the designation "exotic granites" in reference to blocks in the *Wienersandstein* (Viennese Sandstone[23]) to distinguish blocks that are foreign to, and older than, the sediments in which they are embedded, from those called "erratic" that are simply foreign to the landscape on which they sit (von Morlot, 1847, p. 92). Von Morlot was aware of Studer's writings about the blocks in the Habkern Flysch and compared them with those in the *Wienersandstein* (*ibid.*, p. 93). Hohenegger (1861, p. 14) adopted the term from von Morlot and used it for the Stramberger blocks in

Figure 16. Ami Boué (1794–1881), who first recognized that the limestone klippen of the Outer Klippen Zone (e.g., Fig. 15) were Jurassic, in contrast to the much younger surrounding flysch units. But he could not bring himself to believe that the same was true farther south in the Southern Klippen Zone of the Pieniny Belt. That demonstration had to await Beyrich, a decade-and-a-half later.

Figure 17. Adolf von Morlot (1820–1867), Swiss geologist and prehistorian who coined the terms "exotic (for blocks of unknown provenance in flysch) and Quaternary. From a picture in the Naturhistorisches Museum in Bern, Switzerland (copied from Eizsmann, 1974, fig. 2).

what is now called the "Outer Klippen Zone" or "Northern Klippen Zone" of the Carpathians (also see Suess, 1858, p. 16). As we saw above, via Hohenegger, von Morlot's term wandered to the fatherland of its creator through the work of Bachmann (1863), to be used for a similar purpose.

Farther south, the Carpathian Flysch Zone borders on what has been called the Pieniny Klippen-Belt (Fig. 14, map and section), "a justly celebrated unit in the chains of Alpine origin" (Lugeon, 1903, p. 42). This unusual structure, a part of the easterly continuation of the Pennine Zone of the Alps, is a very long (more than 500 km) and narrow zone of intense deformation in which Jurassic and Lower Cretaceous limestones pass into Upper Cretaceous pelagic sedimentary rocks with beginning clastic input that turns into the Sromowce Formation, representing the onset of full-scale flysch sedimentation (Andrusov, 1931, 1938; Birkenmajer, 1960, 1977; Birkenmajer and Jednorowska, 1987). The Paleogene is represented by thick flysch sequences. In the deposits of the late Cretaceous and the Paleogene, a series of klippen are also found, whose ages range from the Triassic to the early Cretaceous (Uhlig, 1904). For that reason it is referred to also as the "Inner Klippen Zone" or "Southern Klippen Zone" of the northern Carpathians. Fig. 18A is an example of what this zone looks like on a detailed map. Fig. 18B is Eduard Suess' sketch of the Czorzstyn Castle and surrounding areas. Fig. 19 is a photograph of the typical klippen landscape.[24]

Viktor Uhlig (1857–1911; Fig. 20), one of the "Viennese giants" in the history of geology,[25] presented a fine review of the history of thought on the Carpathian exotic klippen to 1903, which I use in the following. Even Ami Boué, who had recognized the Jurassic age and thus the "foreign" aspect of the blocks of the Outer Klippen Zone, considered in 1830 the blocks of the Inner Klippen Zone as simple facies changes within the Cretaceous and Lower Tertiary. It was the great Prussian geologist August Heinrich Ernst Beyrich (1815–1896; Fig. 21[26]) who in 1844 showed by a careful study of the fossil fauna of the klippen that they could not possibly be a part of the younger series, and left no room for doubt that the klippen-phenomenon in the Carpathians was real and needed explanation. To provide it, he resorted to von Buch's theory of magmatic uplift and imagined that every klippe was pushed like a piston from beneath by igneous agencies and burst through the overlying younger beds. In 1848, Murchison expressed a similar opinion (Murchison, 1849, p. 260 and his fig. 31). In the 1850s, Dionyz Stur (1827–1893) studied them in some detail, on the basis of the mapping activity of the newly founded Imperial Geological Survey (Geologische Reichsanstalt), but tried to explain them as individual reef complexes. (For references to Stur's relevant paper, see Andrusov, p. 167.)

The tectonics of the klippen was considered more seriously during the second phase of the mapping of the northern Carpathians from the 1864 to 1870. In his 1867 paper, Suess expressed the view that the klippen were tectonic in origin. In describing the Stramberg Beds, he wrote:

> …one has come to observe the curious fact that many of these considerable masses are only isolated fragments, enveloped in Neocomian shales and sandstones, which are tormented in a manner almost incredible. The exploitations of Mr. Hohenegger caused the disappearance of some of these masses and it seems only two of these islets are entirely in place; they are the mountains around Stramberg and "Ignaziberg" near Neu-Fitschein. I have proposed the hypothesis that these great fragments are only the rests of a calcareous deposit that could not follow the contours of the shales and the sandstones caused by the shortening of the external zone of the Carpathians and consequently burst. (Suess, 1867, p. 188).

It is in these few prophetic sentences that Suess essentially proposed a hypothesis of broken formation.

Workers (Suess' students and colleagues) such as Edmund von Mojsisovics (1839–1907), Franz von Hauer (1822–1899) and Karl Maria Paul (1838–1900) considered every klippe to have a tectonic individuality, and they argued that no relation existed between individual klippen. In 1868, however, Paul considered the entire klippen zone as a sort of anticline (see Stur, 1938, for a reference). His co-worker, the great paleontologist and paleogeographer Melchior Neumayr (1845–1890) followed this interpretation also.

The main result of this latter phase of mapping was the clear demonstration that the Carpathian klippen represented locally *rooted* rock masses and were not rootless nappe remnants like

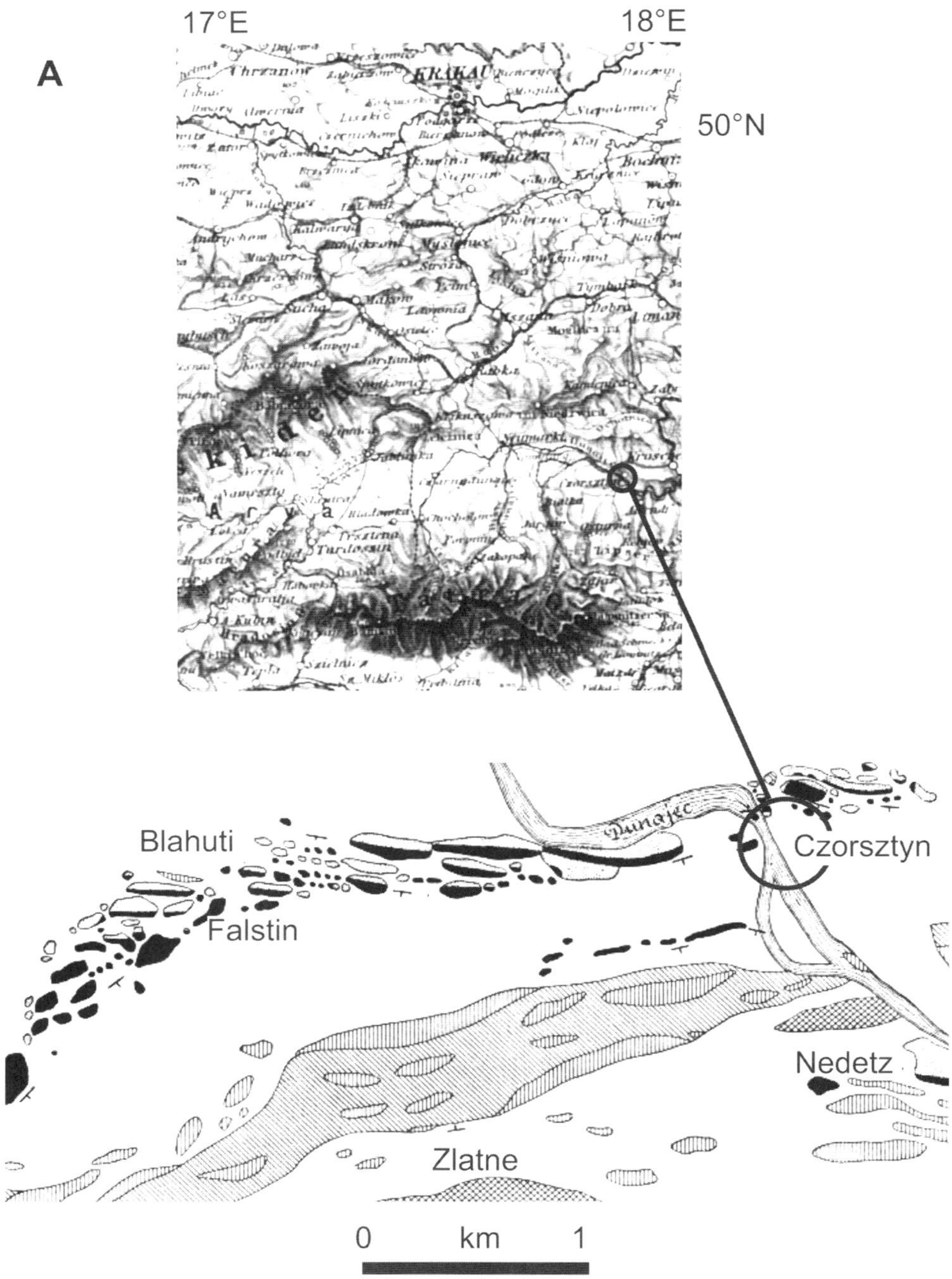

Figure 18 (*on this and following page*). A: A geological sketch map of the Czorsztyn-Falsztynski (formerly Scharfstein-Falstin) region south of Krakow (Krakau in the index map above) in the Pieniny Klippen Zone of the Polish Carpathians (from Uhlig, 1903, fig. 67). Oblique ruling represents the *Opalinus* and *Murchisonae* Beds (Aalenian) and the *Posydonomia* (now *Bositra*) Beds (mainly Bathonian to Callovian); vertical ruling is Hornstein Beds (chert); stippling is the crinoidal limestones of Dogger; black is the nodular Czorsztyn Beds (Oxfordian to Kimmeridgian) and Tithonian; obliquely crossed areas are Lower Tertiary sandstone; and the white areas are Upper Cretaceous matrix of the exotic blocks (for its geology, see K. Bak, 1995, 1998; M. Bak, 1996a, b). Note their arrangement in a narrow curvilinear zone, the regular direction of strike and the direction of stratigraphic younging in all. This is by far the most common mode of occurrence of the klippen in the Carpathians. More irregular groups also occur (e.g., Uhlig, 1903, fig. 68), but are rare. Note that Scharfstein, the original German name of the area, means "sharp stone," clearly referring to the sharp outlines of the exotics in the landscape. See Suess' panorama of the Czorsztyn area in Suess (1867, plate 35, upper figure; Fig. 18B herein). B: Eduard Suess' sketch showing the panorama of the Czorsztyn area shown in Fig. 18A. View to the north, showing the Czorsztyn castle and the river Dunazec. Translation of the explanations (from left to right): *Grès néocomien* = Neocomian sandstone; *Calcaire blanc sans stratification* = white limestone without stratification; *Calc. à Crinoïdes* = crinoidal limestone; *Calcaire rouge* = red limestone; *Couche à T. diphya* = T. diphya-bearing bed; …*Calc. à Fucoïdes* = fucoid-bearing limestone; *Calc. rouge à Ammonites* = ammonite-bearing red limestone.

their Swiss namesakes and look-alikes. Uhlig (1903, especially p. 775) specifically emphasized this. Indeed, already in 1885, Eduard Suess felt the need to write "klippen in the Carpathian sense" to distinguish this kind of klippe from the overthrust kind (Suess, 1885, p. 558; for the distinction between the klippen of Swiss type and klippen of Carpathian type, see Andrusov and Scheibner, 1960, p. 24: the distinction between the klippen of Pienidian type {= Suess' "klippen in the Carpathian sense"} and those of Swiss type; also see the discussion in Andrusov, 1938, p. 115–118). Uhlig (1903) showed that most of the klippen in the Carpathians showed a regularity of zonation, strike and dip direction, and stratigraphic younging direction (Fig. 18) Figs 22 and 23 represent three examples of this regularity in cross-section (after Uhlig, 1903). Uhlig thus thought that a sort of imbrication must have been responsible for the generation of the Carpathian klippen. When he thought this, he had been convinced that the "Supporters of the nappe theory ['*Deckshollentheorie*'] would find little stimulation in the Carpathians" (Uhlig,

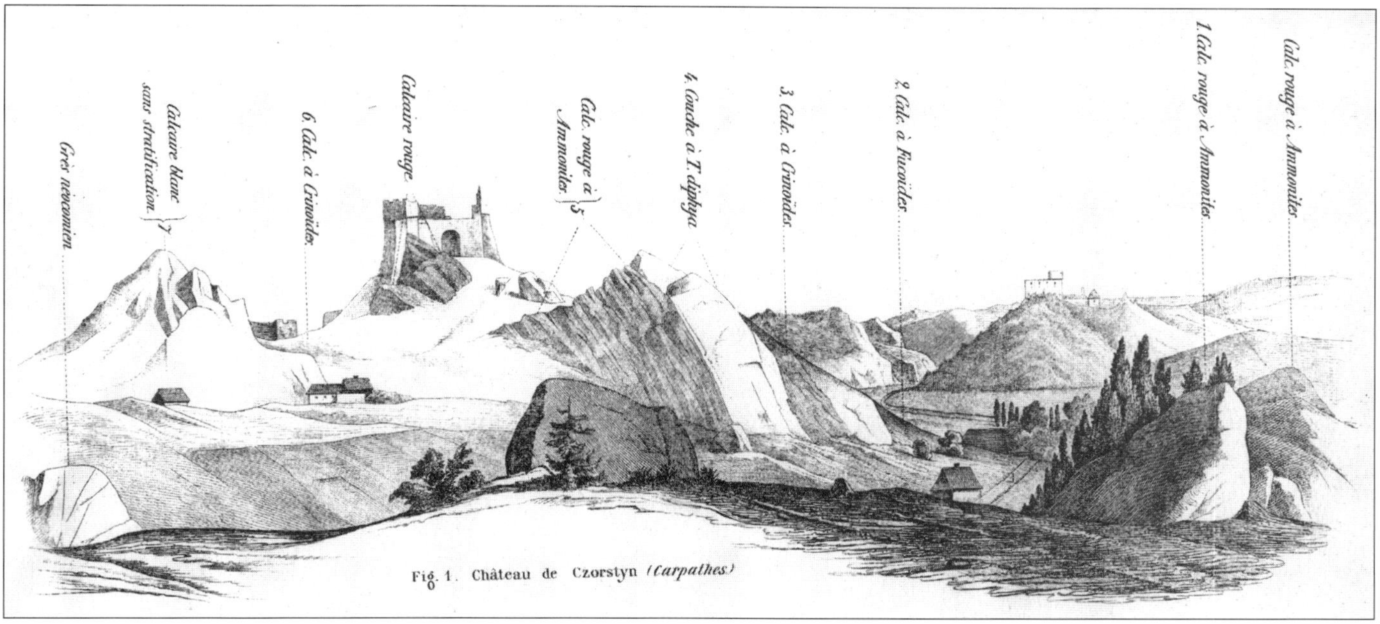

Figure 18 (*continued*).

Figure 19. The Pieniny Klippen Belt cut by the river Dunazec. The rocks consist of cherts and Bathonian to Callovian *Posidonomia* (*Bositra*) Beds. The dominant dip in the picture is to the north. The three-peaked mountain in the background is the Pienin (formerly Kronenberg), the culmination point of the mountain group, which gave its name to the Pieniny Zone. The pointed mountain to the right is Sokolica. The deep saddle to the left of Sokolica, the foreground, and the slopes of the Pienin in part are formed from the Upper Cretaceous matrix of the Pieniny Klippen Belt. (From Uhlig, 1903, fig. 66).

Figure 20. Professor Victor Karl Uhlig (1857–1911) master paleontologist, stratigrapher, structural geologist, and tectonician, the successor of Eduard Suess in the University of Vienna and one of the "Viennese Giants," the founders of modern geology.

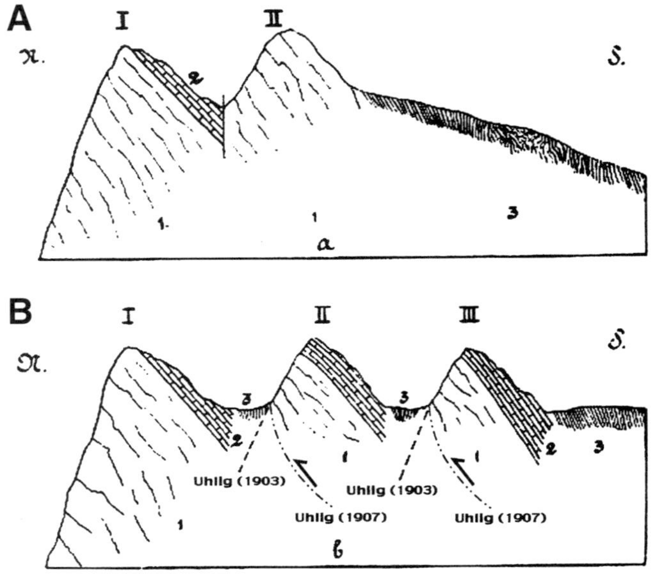

Figure 22. Cross-section of the Falsztynski (Falstin) klippe train. A is from the middle and B is from the eastern part (see Fig. 18A). 1 = Dogger crinoidal limestones; 2 = Czorsztyner Limestone and Tithonian; 3 = Upper Cretaceous matrix. I, II, and III refer to different klippen. (From Uhlig, 1903, fig. 69)

Figure 21. August Heinrich Ernst Beyrich (1815–1896), great Prussian paleontologist and stratigrapher who demonstrated that the klippen of the Carpathians were a real phenomenon and needed an explanation. (From Carlé, 1988, fig. 13).

1903, p. 770; see also Uhlig, 1904). By 1907, however, he had changed his mind because of Suess' (1883, especially p. 243, 247, 251–252), Lugeon's (1903, 1904),[27] and possibly also Haug's (1904) influence, and portrayed the northern Carpathians as a pile of far-traveled nappes with the Beskidic Nappes (including the Outer Klippen Zone), which correspond to the Helvetic Nappes of the Alps and represent the European continental margin; the Subpieniny and Pieniny zones; and the nappes of the High Tatra, which represent the Pennine Nappes and the nappes of the inner Belt and the Central Hungarian Mountains, which correspond to the Austroalpine Nappes (Fig. 14; see Tollmann, 1975, for a modern correlation). In this interpretation, the Pieniny and the Beskide nappes became bulldozed units, torn from their far-lying and now lost original sites of origin and jumbled into flysch deposits under the overriding High Tatra units (Fig. 14, cross-section). They thus became very much like the exotic blocks of the western Alps in the eyes of most geologists. Lugeon, in fact, thought that the klippen were nothing more than the upthrust frontal anticlines of what Uhlig later named the Beskide and Subbeskide nappes, deformed and stretched along the strike and thus became individual klippen within overlying soft flysch units (in modern parlance, one would have said "mélanged"). He thus compared the Pieniny Klippen Zone with the Säntis Mountains in Switzerland in terms of their tectonic position. Uhlig agreed, but presented the shear-related disruption as far more extreme than Lugeon's publications led one to assume. (For divergent views

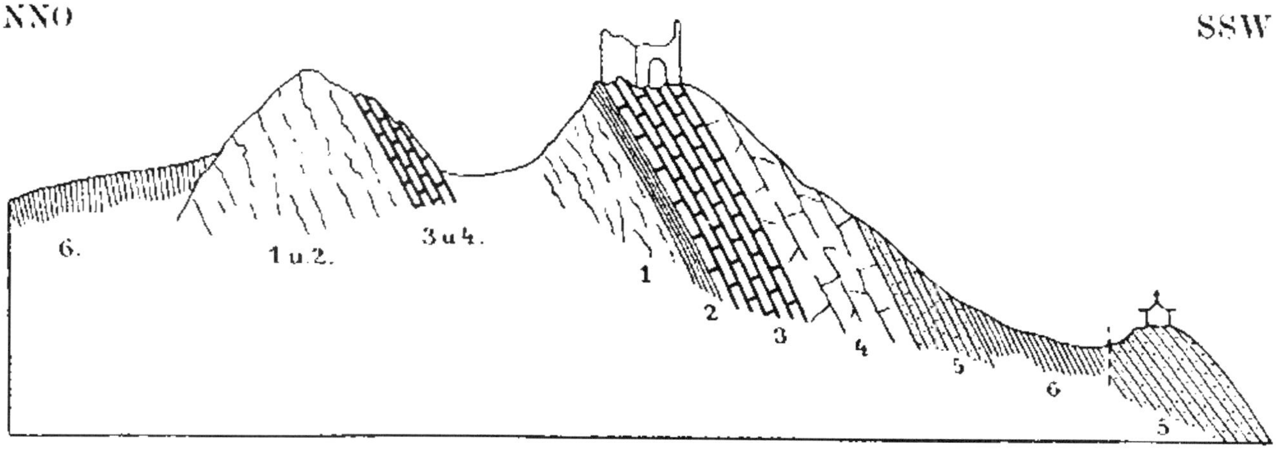

Figure 23. Cross-section of the Castle Klippe of Czorsztyn: 1 = white crinoidal limestone (Bajocian); 2 = red crinoidal limestone (Bathonian); 3 = Czorsztyn Limestone; 4 = light red and white Tithonian limestone; 5 = dark red, shaly crinoidal limestone of Tithonian age; 6 = red shales and gray mottled shales (Puchow Marl) of the matrix (Upper Cretaceous). In the unexposed saddle between the main ridge and the Castle Klippe, there may be *Opalinus* Beds. (From Uhlig, 1903, fig. 70).

after Lugeon and Uhlig to 1924, see Rabowski, 1925; the present views have very much returned, in principle, to those of Uhlig, 1907, in a plate tectonic context).

The "Prehistory" of the Discovery, Third Act: The Himalaya

Alpine geologists working in the Himalaya in the last two decades of the nineteenth century discovered exotic blocks floating in flysch, but this discovery was not independent and was based on their knowledge of the Alpine and Carpathian occurrences.

Working in the Kiogar region of the Central Himalaya in 1879, just north of the Balchdura Heights (Fig. 24), the great Austrian geologist and the later director of the Geological Survey of India (1894–1903), Carl Ludolf Griesbach (1847–1907, Fig. 25; for his life, see Anonymous, 1903, 1907, and Tietze, 1903), "had come across masses of limestone, in more or less is isolated and obscured positions" which at the time puzzled him "not a little" (1904, p. 549). He found them enclosed in igneous rocks (Griesbach, 1891, 1893; which later Gansser, 1964, p. 123, called the "Exotic blocks in the basic vulcanites"). Later, in 1892, while examining the ground northeast of Kungribingri (Fig. 24) in the company of the leader of the Himalayan Expedition of the Austrian Academy of Sciences, the great stratigrapher Professor Carl Diener of Vienna (1862–1928),[28] Griesbach found a mass of limestone overlying the Jurassic Spiti Shale, "which might have been taken to be younger than the latter, had the limestone not yielded a fine series of fossils which demonstrated that the strata ranged from permian [*sic*] into the lower trias [*sic*] and could therefore not overlay [*sic*] the Spiti shales in a natural manner" (ibid., p. 549–550). They also found mafic igneous rocks in contact with the limestone, one of which, in fact, cut across the limestone in form of a dike (Griesbach, 1893, plate 1; Diener, 1898).

Then, Albrecht Krafft von Delmensingen (1871–1901; see Diener, 1901) restudied the whole Kiogar region in 1900. He found in the Valley of Kiogar gray dolomitic limestones, which he likened to the Upper Triassic Dachsteinkalk of the Eastern Alps and which formed blocks floating in a mafic volcanic matrix. The whole thing was resting on a flysch with various members, including tuffs (Figs. 24 and 26). Von Krafft rightly concluded that the entire mass must have been transported from the north and northeast, as he could find no dike in place to feed the volcanics forming the matrix of the exotic blocks (von Krafft, 1902).

As to the origin of the blocks, von Krafft (1902) first compared them with both the Carpathian and the Swiss klippen and concluded that they were not similar. He thought that the origin of the blocks was due to volcanism that tore up a formerly regular succession and caused its disintegration into blocks while the blocks were engulfed in lava.

The more experienced Griesbach disagreed: he was ready to admit that volcanic and subvolcanic phenomena may indeed have played some small role in the formation of the blocks, which were indeed embedded in mafic volcanic matrix, but the majority of the blocks and thus the main cause of the phenomenon of this blocky series, were to be ascribed to tectonics, specifically to great dislocations. Writing of von Krafft's idea, he noted:

> Whilst it cannot be denied that this theory meets the case of the numerous blocks, which were found to be entirely enveloped in the igneous flows, it appears probable to me that enormous dislocations must be supposed as having taken place, to enable fissures to form, through which great masses of volcanic lava could be erupted to cause widespread flows in that region, and in such a case it is certain that much of the older sedimentary rocks must have been brought to the surface, not only as part of the sections, but also in *crushed masses*,

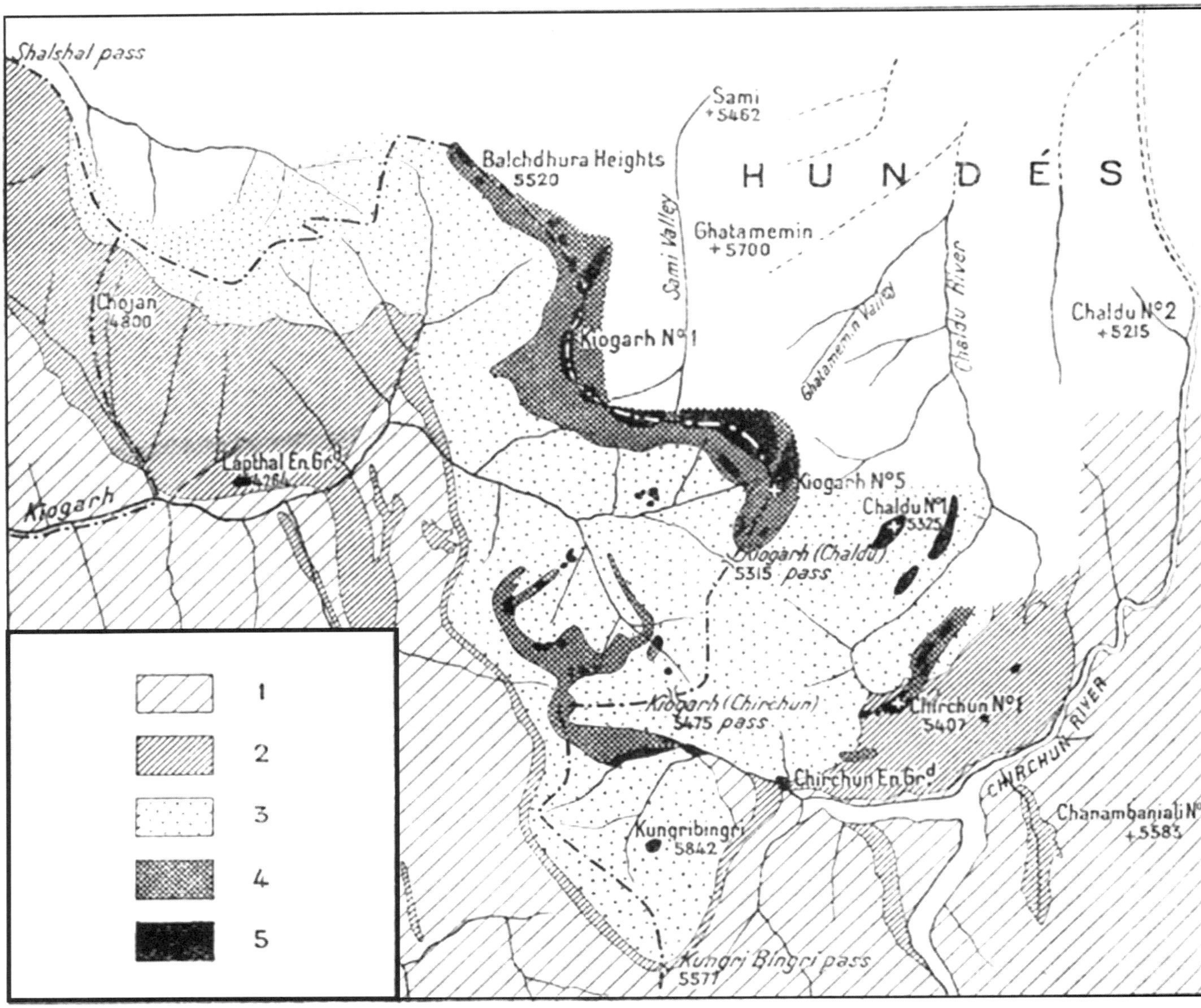

Figure 24. Geological sketch map of the Kiogar region (after von Krafft, 1902, plate 14, from Suess, 1918, fig. 357). Key to legend: 1 = limestones (from Upper Triassic to Middle Jurassic); 2 = Spiti Shales (Upper Jurassic); 3 = flysch; 4 = mafic volcanic rocks; 5 = exotic blocks (white and red limestones, dolomites, etc.: Permo-Carboniferous, Triassic, Lower Liassic, etc.). Elevations are in meters.

and detached blocks torn off from situations in situ, a phenomenon common to all disturbed areas. The outcrops of dislocations which have later undergone weathering and denudation, must of course have been shorn of all crushed and loose fragmentary masses, but where dislocations are accompanied by the ejection of vast igneous flows, *the fragmentary sedimentary rocks, the result of crushing and dislocation of the strata*, must surely have been swept up and carried along with the flows. (Griesbach, 1904, p. 551, italics mine).

Griesbach, therefore, ascribed the generation of the blocks to *tectonic processes*, to crushing and dislocation, and ascribed to the igneous matrix only the role of a glue to hold them together and to preserve them. Strangely, the man who thought that the entire continent of Asia was moving onto India along the Himalaya (Suess, 1901, p. 11), confessed ignorance as to the nature of the dislocations he postulated: "To explain such a feature it is plainly obvious that we have to assume parallel dislocations, which with or without an overthrust, are able to originate such a structure" (Griesbach, 1904, p. 551–552). Griesbach did not think that the blocks had traveled much from their original place of origin (*ibid.*).

Later workers have shown that the exotic blocks in the Himalaya form a part of an extensive ophiolitic mélange underlying large ophiolitic nappes (Heim and Gansser, 1939, especially

Figure 25. The great Austrian geologist Carl Ludolf Griesbach (1847–1907), who discovered the Himalayan exotics in 1879. From Tozer (1984, p. 54).

p. 143–187; Bailey, 1944; Gansser, 1964, p. 123–138, 1974, 1977; Zhang and Yin, 1980, p. 65–67; Burg, 1983, p. 118–123). They contain blocks ranging from the Permian into the Cretaceous. Griesbach, too, had seen that the exotics had to have a tectonic origin, but his ideas remained vague, as he was unsure of the nature of the displacement. Like Hans Schardt in the Alps, he certainly got a part of the story right, but without the postulate of large thrusting, he could not visualize completely how the blocks had formed. From his writings I obtain the impression that thrusting was in the back of his mind, but he was being—I think overly—cautious.

End of "Prehistory": Suess on the Nature of Overthrusts and Steinmann on Ophiolites

In 1904, Eduard Suess published a two-and-a-half page note with no figures in the proceedings of the French Academy of Sciences (Suess, 1904; Steinmann later called it "aphoristic": Steinmann, 1905, p. 45). This great little paper is of such a prophetic nature that I have always been amazed that it never received the attention it ought to have commanded (perhaps because it was indeed aphoristic; but both Heim and Gansser {1939, p. 160–161} and, following them, Bailey {1944} quoted Suess approvingly). In it, Suess begins by noting the discovery of overthrusts in the Alps by French and Swiss geologists, mentioning especially Marcel Bertrand. He then immediately compares the frontal regions of overthrusts with the internal ogives at one of the termini of the ice cap in Greenland between Kangersuk and the Nasasuak nunatak and says that they are homologous to the arcs of overthrusts (*arc de charriage*). Suess emphasizes that along these thrusts, deep moraines are brought to the surface.

Suess then observes that many nappes in the Alps have down-facing frontal folds. He ascribes this to their being pushed over obstacles. He points out that to understand the nature of this "grand phenomenon" of overthrusting, one must look at the "roots" of the nappes. When he does so, he notes that the roots exhibit much mafic and ultramafic rock and he (incorrectly) correlates the ultramafics of the Ivrea Zone with some of the Penninic ophiolites (because he remained ignorant of the associated submarine lavas in the ophiolites, as opposed to the subcontinental Ivrea body: see Bailey's point in Bailey, 1944, p. 755). His next point concerns the exotic blocks of the Himalaya and the associated mafic and ultramafic rocks. He supports the view that they belong to an allochthonous unit expelled from the north and northeast. The implication of a phenomenon similar to the one

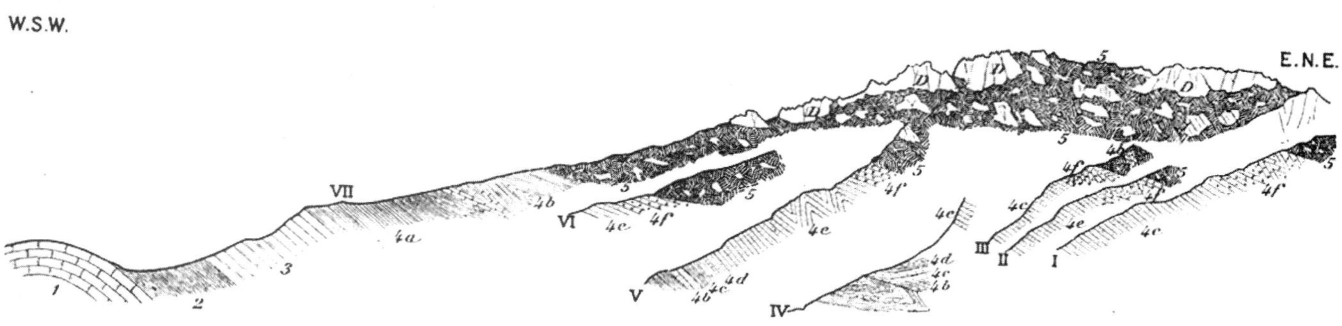

Figure 26. Profiles showing the northern side of the Kiogar Valley (for location, see Fig. 24). 1 = gray limestones (from Rhaetian to Middle Jurassic); 2 = Spiti Shales (Upper Jurassic); 3 = Gieumal Sandstone (Cretaceous); 4 = flysch (ditto); 4a = red and greenish shales; 4b = friable black shales; 4c = brown sandstones; 4d = black siliceous shales; 4e = greenish sandstones; 4f = green and red tuffs; 5 = mafic volcanic rocks; D = gray dolomitic limestones (Dachsteinkalk). After von Krafft (1902, pl. 12), from Suess (1918, fig. 358).

seen in the Alps is clear. In a short paragraph, he compares the island festoons of the western Pacific with thrust fronts.

He finally mentions Auguste Daubrée's works[29] showing that ultramafic rocks represent the deeper parts of the earth, and notes their frequent association with fronts of nappes, although they never appear in the foreland.

Suess leaves it there, but his implication is clear: orogeny somehow brings rocks that usually form deep in the earth to the surface by means of very large overthrusts. He notes the association of jumbled rock masses with this phenomenon.

This is the point that Gustav Steinmann—"the last man who took the entire science of geology as his province"—followed up. He showed that in the extremely deformed and jumbled rock associations of the so-called *Aufbruchszone* (zone of piercing) in Graubünden in eastern Switzerland, the bottom of the immense Austroalpine nappe system was exposed and exhibited a thin zone of extremely deformed and jumbled rock association. Before the advent of the nappe theory, the *Aufbruchszone* was assumed to represent a submarine ridge, separating the Western Alps from the Eastern Alps and characterized by the "eruption" of gabbros and peridotites. Steinmann had selected this area for a detailed study by himself and his students. Schardt's 1893 publication prompted him to test the theory of nappes on Bündner examples. He found that although Schardt's and later Lugeon's parallelization of the nappe structures in western Switzerland with those of eastern Switzerland was incorrect in detail, the broad outlines of the nappe theory not only withstood his tests, but also made many of his observations easier to explain.

Among these observations was the intimate association of radiolarites, deep-sea muds, and radiolarite-bearing pelagic limestones with gabbros and peridotites. Steinmann noted that this association, previously vaguely ascribed to eruptive processes along the *Aufbruchszone*, in reality represented an extremely deformed association that stemmed from the deepest part of the Alpine geosyncline. (On the basis of the present-day analogues of the radiolarites and the associated deep-sea sedimentary rocks, he estimated the depth to be more than 4000 m {Steinmann, 1905, p. 55}). At this depth, gabbros, peridotites, and the deep-sea sediments somehow came together and were then uplifted by the medial Cretaceous by some 5 km! Steinmann ascribed this uplift to an early phase of Alpine orogeny that had folded up the middle part of the Alpine geosyncline. (He regarded the Helvetic and the Austroalpine shelf areas as the northern and the southern parts, respectively, of the same geosyncline.) Continued shortening finally generated the two largest nappes of the Alps from the deepest parts of the geosyncline: the ophiolite (by which Steinmann meant in 1905 only the ultramafic rocks and the gabbros) and deep-sea sediment-filled northern part formed the Rhaetic Nappe (from the Roman province of Rhaetia located between the mountains of the Adula Group and the River Lech and comprising the present-day Swiss canton of Graubünden and the Austrain federal state of Vorarlberg), whereas the relatively shallower southern part formed the Austroalpine Nappe. The Austroalpine Nappe overrode the Rhaetic Nappe and literally squashed it ("*ausgequetscht*," Steinmann, 1905, p. 58). For that reason, Termier (1903, see p. 762) had called the Austroalpine Nappe the *traîneau écraseur* (crushing sledge). This crushing sledge had such an effect on its underlying nappe, according to Steinmann, that "the structure of the middle or the 'piercing zone' ['*Aufbruchszone*'] became dominated by numerous smaller overthrusts and squashing, so much so, that a complete agreement with the confusing structure of the northern Swiss klippes originated and that, even while mapping at a scale of 1:25,000, only 'squashing zones' ['*Quetschzonen*'] can be discerned, which almost earn the designation of a friction breccia [*Reibungsbrekzie*] at large" (Steinmann, 1905, p. 10). This confusing, mega-breccia structure (which the Swiss geologist Joos Cadisch called the *Aroser Schuppenzone* {The Zone of Imbrication of Arosa} in Cadisch et al., 1919, p. 362) and the stratigraphic similarities led Steinmann to correlate the Rhaetian Nappe with the northern Swiss klippen and the Préalpes. As yet, the father of the Steinmann trinity was unaware of the importance of the third member of the trinity, namely, the pillow lavas. In his classic 1927 paper, he completed the trinity (but without admitting the extrusive nature of the pillow lavas! See Bailey, 1936, 1944) and pointed out the settings in which it was encountered (see Bailey, 1936, p. 1718–1722). They were the suture zones of the Mediterranean and the mélange belts associated with them. Thus, Suess' prophetic point that ultramafic rocks seemed to be associated with major thrusts culminated, in the light of the Alpine discoveries, in Steinmann's point that the trinity he noticed pointed to the presence of oceanic depths, now represented by the extremely deformed zones of confused structure along thrust belts. Already in his 1905 paper, he pointed out that the ophiolites (then thought of as only the ultramafics and the gabbros) and the associated deep-sea deposits (i.e., the ophiolitic mélange, a term Steinmann did not use) did not always form "middle nappes," as in the Alps. In some mountain belts (as in the Himalaya), ophiolites and associated deep-sea deposits constituted the highest nappes.

Summary of the "Prehistory" of the Discovery of Mélanges

Mélanges, as highly deformed rock bodies containing exotic elements of other, older rock bodies commonly no longer visible, were first recognized in the Alps. This recognition occurred as a consequence of the earlier discovery of exotic elements in the conglomerates of the Nagelfluh. When mélanges were encountered in the Habkern Valley and elsewhere, their exotic elements were seen to be embedded in a shaley-sandy sedimentary matrix, which led to its being called "flysch." The highly deformed state of this "flysch" with exotic blocks inspired F.J. Kaufmann to call it wildflysch. This evolution set the scene for a "sedimentary" interpretation of the mélange in the Alps; even the chaotically deformed state of the wildflysch and its ubiquitous occurrence in zones of tremendous simple shear with lateral persistence for thousands of kilometers did not lead to the idea that the entire rock might be a purely tectonic mixture. Steinmann noted that chaotically deformed rocks in Graubünden in eastern Switzerland included ultramafics, gabbros, and deep-sea deposits, which led him to think that they represented the remnants of a very deep sea (≈ 4000 m) in the Alpine geosyncline. He later noted that these rocks were also commonly associated with pillow lavas.

Exotic blocks floating in a flysch matrix were discovered in the northern Carpathians nearly coevally with those in Switzerland. It was initially thought that they were lateral facies equivalents of the flysch, but detailed paleontology soon made this interpretation untenable. They were then thought as vertically uplifted representatives of units underlying the flysch by "volcanic forces," but lack of associated volcanism led to the abandonment of this theory too. Suess recognized their tectonic nature, but his interpretation was not followed for a long time. When the nappe theory was imported into the Carpathians in 1903, they were recognized as bulldozed and sheared nappe pieces dispersed in flysch that accompanied the nappe movements.

In the Himalaya, the recognition of the exotic blocks of Kiogar floating in a mafic volcanic matrix led to speculations concerning the role of volcanism in disrupting a formerly more regular sedimentary succession. From the proximity of large dislocations, Griesbach suspected that disruption was probably largely tectonic, but could not specify its precise nature. The Himalayan discovery was not independent and was made in the light of Alpine and Carpathian experience.

The recognition of the tectonic nature of the mixing in mélanges required an environment where neither sedimentary nor magmatic events could be considered its cause, and where the tectonic mechanism left little doubt as to its presence and the nature of its effects. This was to be realized, this time entirely independently of Alpine and Himalayan discoveries, first in Scotland, where crush breccias were mapped, and then in Wales, where a former worker with Scottish Highlands experience encountered a mélange, recognized it as such, and named it "mélange" for the first time. With him began the actual history of the discovery of mélanges, in much the same way that the history of discovery of nappes began with Marcel Bertrand, although a number of overthrusts in various guises had been described before Bertrand.

Initial Discovery: Edward Greenly (1895–1919)

Mélanges, as we today understand this concept, were recognized later than other kinds of structural rocks (Şengör and Sakınç, 2001) because the understanding of their nature grew out of the appreciation of the significance of what in the Scottish Highlands had been called a crush breccia in the late nineteenth century. British geologist Edward Greenly (1861–1951; Fig. 27; for Greenly's life, see: Greenly, 1938; Bailey, 1951; Williams, 1951) invented the concept of mélange while he was mapping in Anglesey, Wales (Greenly, 1919a, 1919b). He had come to Anglesey after leaving the Geological Survey of Great Britain in 1895. He had been a member of the legendary Scottish Highlands team (Strahan, 1919; Greenly, 1928–1932; Oldroyd, 1990) and was thus closely familiar with the concept of structural rocks associated with thrusting (Lapworth, 1883, p. 121; 1885, p. 558–559). The high state of disruption of the rocks, which Greenly called the Mona Complex (Greenly, 1919a, p. 39), had been known already in the last quarter of the nineteenth century (see the review in Greenly, 1919a, p. 1–13). Matley (1913) had called them "crush-conglomerates" and "crush breccias." Greenly compared them with what he thought were similarly produced crush-conglomerates from the Isle of Man (1919a, p. 65). He knew that he was looking at a kind of structural rock that had become very familiar to the workers of the Survey in the Scottish Northwest Highlands, but it was the much larger scale on which it occurred that induced Greenly to give it a new name:

> Autoclastic phyllite-and-grit mélange. This is really a gigantic autoclast, extending over many miles of country in several districts, as does the "crush conglomerate" of the Isle of Man, but on a still greater scale. It consists essentially of lenticular strips and lumps of grit (Figs. 4 and 5 [*here reproduced as Figs. 28A and B respectively*]) floating in a schistose matrix that sweeps curving around them (Plates VII, XXIII [*VII is reproduced herein as Fig. 29*]. Rounded fragments are rare, but in the Cemaes district the mélange is more of a pseudo-conglomerate than elsewhere, having less parallel structure, both in the forms of the lumps and the nature of the matrix. Almost everywhere the lumps are phacoidal, their boundary curves meeting at sharp angles in the direction of the strike. The matrix is now a fine greenish or pale straw-colored schist, composed of minutely flaky sericite ... in curving sheets and seams, always with a considerable quantity of chlorite and some finely granular quartz. In the matrix of the northern breccias these minerals are but feebly developed, and the color may be a dull grey. The matrix not only sweeps round the phacoids, but often invades them, so that they are traversed by films of it along which they tend to part. And where the larger augen are not close together, it is full of smaller and smaller augen, as well as detached clastic grains that have been torn from them. In the cores of the augen the clastic grains retain often their original rounded outlines quite uninjured, but along the margins and the fissile films they are deformed and flattened. This may take place without "crushing," but it is possible that matter has been stolen from the sides of the grain, thus thinning it. Undulose extinction is general, but far from universal. The matrix of the grit has become a mosaic, with which almost all the grains, felspar [*sic*] as well as quartz, interlock, especially at their "ends," in the direction of the strike; so that the old clastic outlines are better seen in ordinary light than between the nicols. Many of the grains have broken up, optically, into a mosaic. In the unsheared areas there is a tendency to marginal enlargement of the grain at the expense of the matrix. In the shear-zones the mosaic of the grain becomes gradually incorporated with that of the matrix, into which it merges and finally disappears. The crystalline elements of such mosaic tend to be fusiform and nemablastic [*i.e., columnar or prismatic minerals growing in a schist fabric*], especially towards the end of the augen. These grit-augen are often excessively quartz-veined ... There is usually a system parallel to the minor axis, doubtless a "stretching system," but more and more come in, sometimes to such an extent that in the meshes of their intricate network the original structure of the grit can hardly be made out. (Greenly, 1919a, p. 65–66).

Greenly's description depicts a tectonically disrupted and internally strained phyllite-sandstone sequence, to the extent that the original sedimentary geometries can no longer be recognized. This is essentially a broken formation and very different from

Figure 27. Edward Greenly (1861–1951), one-time member of the legendary Scottish Northwest Highlands team of Peach and Horne, and the man who discovered that pre-existing rocks from widely separated paleogeographic realms may be tectonically mixed in large shear zones, producing new rock bodies consisting of exotic blocks of diverse sizes surrounded by one or more types of matrix of sedimentary, volcanic, or metamorphic rock. He used a French word to name such mixtures: *mélange*. The picture is from his autobiography (Greenly, 1938, plate XLVIII).

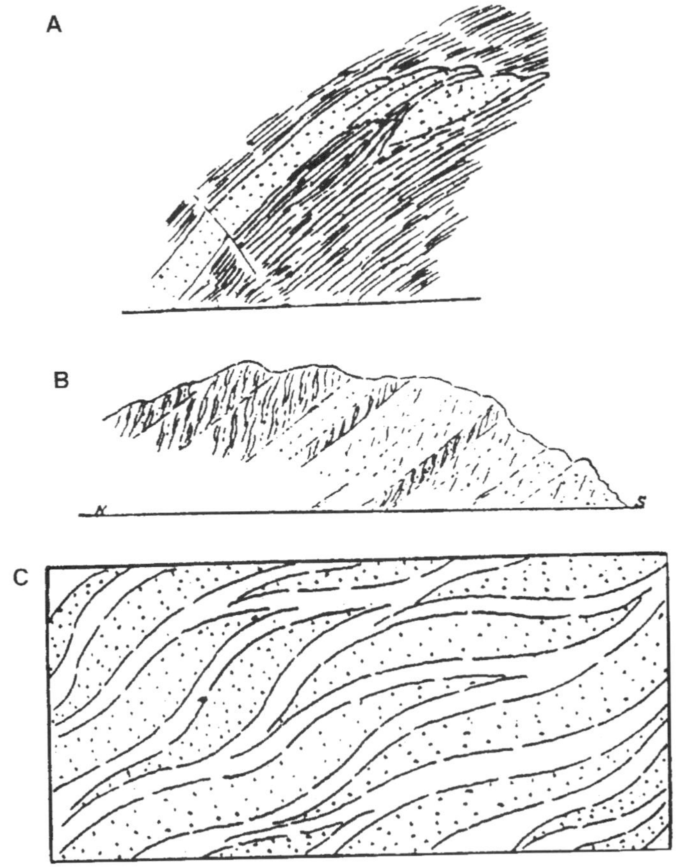

Figure 28. Greenly's sketches to illustrate the formation of his autoclastic mélange. A: Disruption of a bed by shearing and the initial formation of phacoids—the breaking down of Tyfry Beds (Greenly, 1919a, fig. 68). B: As the deformation front moves into the body of a coherent rock body, the original coherence is destroyed and numerous phacoids develop along shear planes: Passage from Fydlyn (not yet mélanged) to Gwna Beds (mélanged) (Greenly, 1919a, fig. 32). C: Fully developed autoclastic mélange. Greenly's "diagram of autoclastic mélange" (Greenly, 1919a, fig, 69).

the mechanisms of formation envisaged for the wildflysch or for the jumbled masses of the Himalaya, but perhaps not so far from Uhlig's depiction of the Carpathian klippen. How this broken formation formed, Greenly portrayed as follows:

Principles of Autoclastic Mélange.—The essential characters of an autoclastic mélange may be said to be the general destruction of original junctions, whether igneous or sedimentary, especially of bedding, and the shearing-down of the more tractable material until it functions as a schistose matrix in which the fragments of the more obdurate rocks float as isolated lenticles or phacoids. Now this condition is general throughout the higher tectonic horizons of the Mona Complex, which, *at that time*, were in the zone of fracture, so that all attempts at minor folding resulted only in thrusts and other ruptures. By the accidents of recumbent folding and erosion, most of the Gwna beds that are known to us are on those horizons, and are consequently in the condition of mélange, to which they lent themselves readily by reason of their very heterogeneous character. The early stages of development may be seen in Tyfry Beds of the Pentraeth Inliers (Fig. 68 [*herein Fig. 28A*], on Twyn y Parc in the Bodorgan headlands, and especially well at the top of the Fydlyn Group (Fig. 32 [*herein Fig. 28B*]). At the top of that massive deposit the thin alternating beds come on, these are soon stripped into phacoids transversely to the bedding; which rapidly becomes imperceptible, and they thus pass into typical autoclastic mélange. ... In the eastern Middle Region, for example, nothing else is to be seen [*but mélange*] along 11 miles of strike, and in the eastern Aethwy Region it occupies 20 square miles of country. In a country of this kind, the larger masses are found to be arranged as trains of lenticles overlapping *en échelon*. (see the diagram, Fig 69 [*herein Fig. 28C*] as well as Figs. 4, 5, 165–6, 169, Plates VII [*herein Fig. 29*], XXII, and the parts of the one-inch map about Pentraeth, Llaniestyn, and elsewhere.) The major axes of the lenticular cores are slightly curved, the form of the line being highly characteristic,—a sigmoidal curve approximating to a straight line. Upon a map, the masses that can be separated out appear as if embedded in a homogeneous and structureless matrix. But this country-rock is itself built up of interdigitating lenticular bodies;

Autoclastic Mélange. Coast, near Porth Cadwaladr, Bodorgan. [*Face page* 86.

Figure 29. A view of the "autoclastic mélange" of Greenly (1919a, plate VII), which is essentially a broken formation. Coast, near Porth Cadwaladr, Bodorgan.

> and, could we take in the region at a glance, it would present itself to us as a mélange of torn and sheared lenticular masses of all sizes, from such as are two or three miles in length to the smallest that the eye can see, of spilite lava, diabase, quartzite, limestone, jasper, and grit, floating in an undifferentiated but schistose body that is a weft of all the more easily deformable elements, itself pervaded throughout by the same lenticular structure. (Greenly, 1919a, p. 193–195, italics Greenly's).

Greenly emphasized that although several members within a mélange could be mapped separately, where this is not possible, the entire mélange should be mapped as a single unit (Greenly, 1919a, p. 66, note 1). This point was implicit in the concept of wildflysch, which was also mapped as a unit. Greenly's inferences have stood the test of time almost intact. Shackleton (1954) has shown that the sequence Greenly had inferred in the Mona Complex was upside down, but his inferences concerning the mélange and the mechanism of its formation remain unassailed. Shackleton (1969), however, believed that Greenly's mélange must have formed as a submarine slide breccia, much as Alpine workers had contemplated the origin of the wildflysch almost a century earlier. This stemmed from his inability, in pre-plate tectonic days, of conceiving how such a chaotic mixture of rocks could be produced tectonically and then become overlain by sedimentary rocks (Wood's, 1974, more recent arguments for a non-tectonic origin of the Gwna mélange I also find unconvincing). Shackleton correctly recognized that all kinds of environments were represented in Greenly's "general mélange" and that all were brecciated (Shackleton, 1969, p. 9–10). Where extreme deformation did not destroy original boundaries, breccia contours were seen to be angular. The sizes of the blocks ranged from microscopic to several kilometers. Shackleton could not causally associate the disruption and the mixing to any set of recognized structures affecting the mélange body in its entirety (*ibid.*, p. 10; also see Wood, 1974). He thus concluded that only a submarine slump could explain its origin. This was the common opinion before plate tectonics[30] (with the exception of Griesbach, 1904 and Bailey and McCallien, 1950a, 1950b, 1953, 1961, 1963: see below; also see Sir Edward Bailey's opinion on the tectonic origin of the Anglesey mélanges in Wood, 1974, p. 335), but once surfaces of displacement could be imagined along which movement could be measured by thousands of kilometers (along subduction zones), all the characteristics that had earlier stumped the students of mélanges could be easily explained.

First Rediscovery: Levi Noble (1941)

Chaos structure is a term introduced by the United States Geological Survey geologist Levi Fatzinger Noble (1882–1965; Fig. 30; for Noble's life, see: Bradley, 1966) in 1941 to describe the great confusion of rock types encountered in the Black Mountains, in the southern part of the Amargosa Range to the east of Death Valley in eastern California (Fig. 31). Below, I quote Noble's recognition of this structure. It is nothing more than the re-invention of the concept of mélange, in complete ignorance of Greenly's work:

Figure 30. Levi Fatzinger Noble (1882–1965), who rediscovered mélanges in the Amargosa Range in California in 1941, but called them "chaos."

> In 1932 a set of vertically taken air photographs was made for the Geological Survey to aid in mapping the geology of Death Valley and the Black Mountains. One feature in the photographs, a crazy-quilt pattern—or more literally a mosaic—of large areas of rock in the Black Mountains, attracted immediate attention as bizarre and without obvious explanation. One of the photographs (Pl. 6 [*herein Fig. 31*]) gives a general idea of the great number of the blocks, their irregular shape, and their chaotic arrangement; and from the scale of the photograph, about 2500 feet to the inch [*about 1:30,000; herein reduced by ≈ ⅓*], one may get a rough idea of their size. The rocks of which the blocks consist represent a stratigraphic range of many thousands of feet, but all of them are younger than the pre-Cambrian gneiss upon which they rest and from which they are separated by a great thrust fault. This fault is the Amargosa overthrust, and the mosaic of fault blocks constitutes a composite thrust plate which will be called the Amargosa chaos.
>
> The characteristic features of the chaos shown in Plate 6 are:
>
> (1) The arrangement of the blocks is confused and disordered—chaotic.
>
> (2) The blocks, though mostly too small to map, are vastly larger than those in anything that could be called a breccia; most of them are more than 200 feet in length, some are as much as a quarter of a mile, and a few are more than half a mile in length.
>
> (3) They are tightly packed together, not separated by much finer-grained material.
>
> (4) Each block is bounded by surfaces of movement—in other words, each is a fault block.
>
> (5) Each block is minutely fractured throughout, yet the original bedding in each block of sedimentary rock is clearly discernible and is sharply truncated at the boundary of the block. Commonly the bedding, even of incompetent beds, is not greatly distorted.
>
> None of the geologic terms in common use appear exactly to fit this mosaic of large tightly packed individual blocks of different ages occupying a definite zone above a major thrust fault. *The feature suggests a fault breccia on a cyclopean scale, yet it is not a fault breccia in the orthodox sense.* Although it is a thrust plate, it is shattered over large areas to a degree that appears to be unique. *But more important than these considerations is the fact that the great areal extent of the feature makes it impossible to map separately the geologic units of the mosaic and makes it necessary therefore to treat the assemblage, despite its heterogeneity, much as a geological formation be treated. Like a formation, then, it requires a name that will indicate both its type locality and its character.* "Amargosa" does the one, for the Black Mountains, in which Virgin Spring lies, form part of the Amargosa Range; and "chaos" as the preceding paragraphs have tried to show, does the other.
>
> The Amargosa chaos or features resembling it are widespread in the southern Death Valley region, and if they occur in other regions the term chaos, as a common noun, may prove to be a useful geological term. (Noble, 1941, p. 963–965, italics mine).

Noble assumed that the Amargosa chaos had originated along thrust faults. Later, Wright and Troxel (1969, 1984) showed that this interpretation was wrong and that the faults creating the chaos were normal dip-slip in nature, bringing younger rocks over older. They noted that mappable bounding faults in the chaos always omitted section and assumed that the chaos was a product of coalescing listric normal faults, as Gilbert had depicted nearly a century earlier (Gilbert, 1875, fig. 12, herein Fig. 32). Wernicke and Burchfiel (1982, p. 109) pointed out, however, that the same effect can be produced along planar normal faults of large displacement. Their model is more attractive

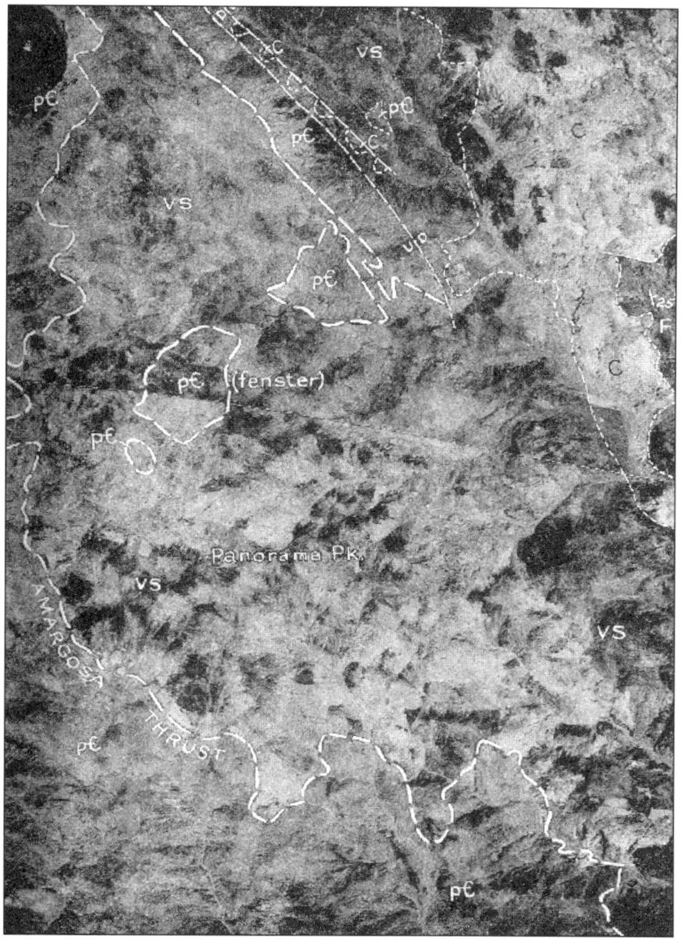

Figure 31. Vertical aerial photograph showing the characteristic "scaly" outcrop pattern of the chaos structure in the southern Amargosa range. This is the so-called Virgin Spring phase of the Amargosa chaos around Panorama Peak. Legend: F = Funeral fanglomerate; C = Calico phase of chaos; VS = Virgin Spring phase of chaos; pE = autochthonous Precambrian gneiss (from Noble, 1941, Plate 6).

because it allows much greater magnitudes of displacement to lead to chaos formation. Wright and Troxel (1984, plate I, cross-section A–A') appear to have adopted this suggestion.

I was shown the chaos structure in the field by my friend Brian Wernicke, and I was greatly impressed with the similarity of its terrain aspects to mélange, especially in the Emigrant Canyon south of the Death Valley. Here, at the Emigrant Spring locality, Brian pointed out the chaos formed from the variegated rock types of the Johnnie and Noonday formations, largely embedded in a shattered argillite matrix, in places ground down to rock flour, and I could not withhold the exclamation "colored mélange!," much to Brian's amusement.

The two main differences between the mélanges that form along subduction zones and the chaos structure forming along large normal faults are the amount of displacement along the generative

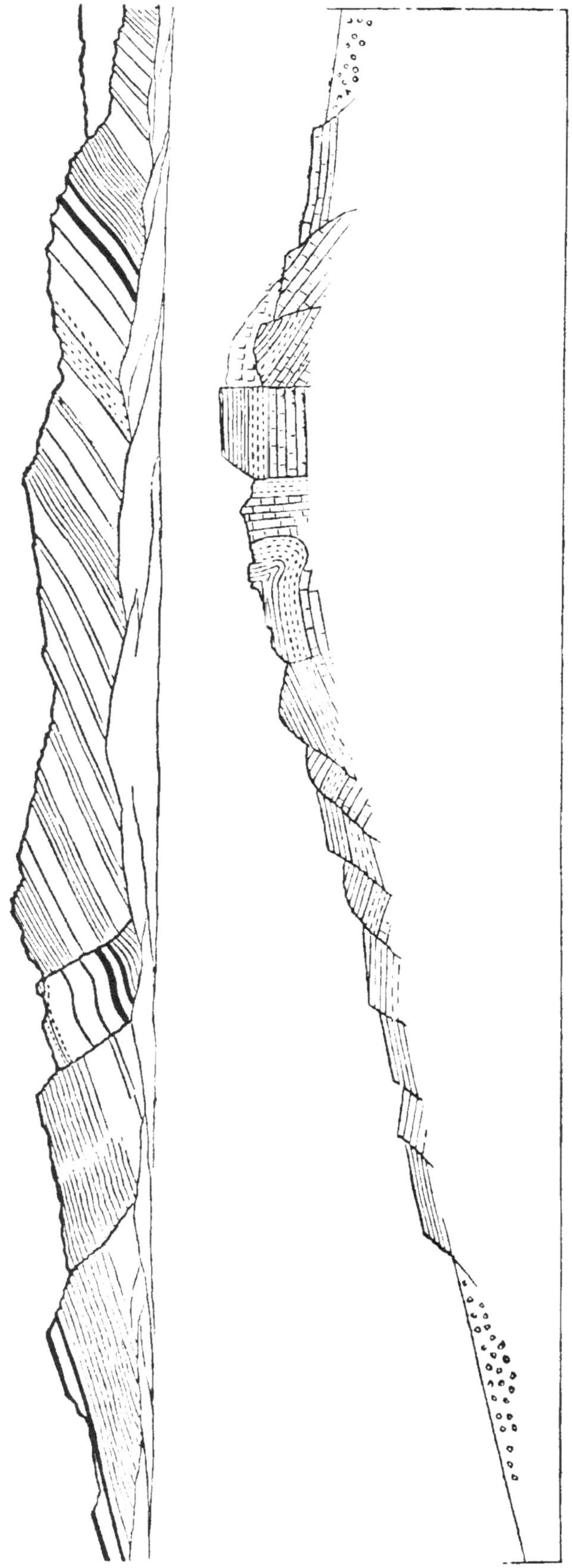

Figure 32. Sections of the Amargosa Range at Boundary Canyon. Base line is sea-level (from Gilbert, 1875, fig. 12). Note the faults dipping toward the depressions and the antithetically and synthetically rotated blocks. This is the classical area, where in 1941 Levi Noble was to recognize what he termed the "chaos structure."

fault zones and the evolution of the ambient pressure/temperature regime as the rocks evolve. Along subduction zones, entire paleogeographic realms disappear and their representatives are now only encountered as exotic blocks within the mélange. Erosion is the only agent that can destroy the record of former environments in an extensional environment, and therefore, in principle, chaos has no "exotic" blocks. Chaos structure evolves in a progressively unroofing environment where ambient pressures and temperatures continuously decrease. Mélanges, especially those along subduction zones, have more complicated paths of evolution because of the vicissitudes of the tectonic regimes that reign in accretionary wedges and subduction channels (e.g., Cloos, 1982, 1984; Dobretsov and Kirdyashkin, 1992; Le Pichon et al., 1993).

Second Rediscovery: W.J. McCallien, Oğuz Erol, and Sir Edward Bailey (1947–1950)

The word mélange did not become popular after Greenly. Even he himself did not use it later when summarizing his Anglesey work (e.g., Greenly, 1922). The word and the concept were revived when William J. "Mac" McCallien and Abdülkerim Oğuz Erol, his doctoral student in the University of Ankara (Fig. 33), were surprised by the jumble of limestones and pillow lavas occurring in a matrix of sheared schists, serpentinites, and, in places, mafic volcanics near Ankara in Central Turkey in the late 1940s. Because the full story of this "second rediscovery" of the mélange concept is essentially unknown even in Turkey, Şengör and Sakınç (2001) published a summary of it in a paper on structural rocks.[31] I give below an enlarged version of the same story. Erol's detailed field slips, which had not been previously published, are here reproduced in color because of their great historical significance (Plate 1; see loose insert accompanying this volume).

McCallien (Fig. 34) was born on March 7, 1902, at Tarbert, Loch Fyne, Argyll, western Scotland. He attended the Tarbert High School and then went to the University of Glasgow to read geology, where he became a student of J.W. Gregory, the famous Scottish geologist and explorer of East African Rift Valley fame (see Gregory, 1896). McCallien became a lecturer on his staff in 1924, having graduated in 1923 with a B.Sc. with special distinction in geology. Until he came to Turkey, McCallien's publications were mainly on the geology of Scotland, where he concentrated on the structural geology and metamorphic history of the Highlands. In 1941, the Royal Society of Edinburgh recognized his contributions with the award of its Neill Medal; however, McCallien also worked on glacial deposits (McCallien, 1935, 1936a, 1937) and pillow lavas (1936b) in Scotland, which prepared him well for his future work on the mélange in Turkey. He thus learned to recognize blocky, chaotic deposits (moraines) accumulated and deformed by a plastically moving medium (glacial ice), and one critical component of the Ankara mélange: pillow lavas. (For a brief biography of McCallien, see MacDonald, 1982.)

In 1944, the Turkish government invited McCallien as professor of geology to head the Chair of Physical Geography and Geology[32] in the Geography Department of the Faculty of Lan-

Figure 33. William John "Mac" McCallien (1902–1981) (center) and his pupil Oğuz Erol (left) in Ankara, Turkey, in the days when they were trying to unlock the mystery of the Ankara Mélange. To the right is the late Mr. İlhami Afşar, a student in Erol's class. (Photo courtesy of Mr. Kemâl Şerbetçi, another fellow student.)

Figure 34. William John "Mac" McCallien (1902–1981), whose passionate devotion to precise and accurate geological mapping enabled his pupil, A. Oğuz Erol, to map the Ankara Mélange in a level of detail that left little doubt as to what the mapped structure was. Photo courtesy of Mr. Colin McCallien.

guage, History and Geography (*Dil ve Tarih-Coğrafya Fakültesi*: for the history of this institution, see Çadırcı and Süslü, 1982; Süslü, 1986). Amid the hardships of the war and post-war years, McCallien left a wonderful impression on his students with his good character, great enthusiasm, and propensity for hard work. Many of his students, even those who did not continue in geology or geography as professionals, remember him and especially his excursions with great delight (Fig. 35; Prof. Oğuz Erol and Mr. M. Kemâl Şerbetçi, personal communications on numerous dates[33]). Although Turkey was not among the belligerent countries, its government had been preparing for all eventualities, and everything, including food, was scarce. Topographic maps were military secrets and Turkey was a country with no scientific tradition. Books and all sorts of publications pertaining to geology were rarities (cf. Toula, 1904, 1906; Trak, 1941, 1942; Lahn, 1948; Anonymous, 1965; Brinkmann, 1981, 1984; Doğu, 1981; Erol, 1984). A geological survey (*Maden Tetkik ve Arama Enstitüsü*: Mineral Research and Exploration Institute) had been in existence in Ankara since 1935 (Anonymous, 1971), but had a limited and poorly trained staff with no mapping tradition behind it. In fact, the first geological map of Turkey at a scale of 1:800,000 began publication during the year of McCallien's arrival in the country, and reached completion in 1946 (see Blumenthal, 1946). Explanatory texts for only the İzmir and the İstanbul sheets ever came out (Jeoloji Şubesi, 1944a, 1944b), which did not include Ankara and its surroundings. Several distinguished geologists had worked around Ankara before McCallien arrived (e.g., Ernst Chaput, Wilhelm Salomon-Calvi, and Kurt Leuchs), but their short spells of work had concentrated mostly on practical needs such as supplying the capital with water.

Several important concepts had been developed, however, which eventually proved critical for the origin of the idea of the Ankara mélange. First and foremost was the modern phase of the identification and classification of the tectonic units in Turkey (cf. Şengör, 1985a). A fundamental trait of most of the classifications developed in the twentieth century was the recognition of three main units in Turkey: a northern Pontic (or Pontide) unit, a southern Tauric (or Tauride) unit, and a third, more heterogeneous, unit between them, formed from median masses (*Zwischengebirge*), the shapes and numbers of which differed widely from author to author (see Şengör, 1985a). Most authors considered the Pontides and the Taurides as the two oppositely verging flanks of a Kober-type symmetric orogenic belt (e.g., Fig. 36). A few thought this picture deceptive and a result of massive backthrusting of a primarily south- (Salomon-Calvi, 1940) or a primarily north-vergent (Staub, 1924) orogen (Fig. 37).

The great German geologist Wilhelm Salomon-Calvi (1868–1941) identified the crush zone in northern Turkey, first identified by Ernst Nowack in 1928 and named the Paphlagonian scar ("*Paphlagonische Narbe*"), as the suture zone (his "*Synaphie*": see Şengör, 1998, p. 101) between Laurasia and Gondwana-Land (Salomon-Calvi, 1936, 1940; for a history of the discovery of this crush zone, which later turned out to be related, in part, to the North Anatolian Fault, see Şengör, 1996). Shortly afterward, a

Figure 35. "Mac" and company in the field: a student excursion during the post-war years near Ankara at the Solfasol Village (see Fig. 40) fountain locality, just northeast of Ankara, amid Tertiary andesites (1947–1948 academic year). McCallien standing with his backpack to the left of the fountain, Erol to his left in front of the fountain. Squatting in the far left is Professor C.A. Alagöz, and to his left is student Handan Aşar (the future Mrs. Erol). (Photo courtesy of Mr. Kemâl Şerbetçi.)

crystalline massif to the east and southeast of Ankara, the Halys or the Kızılırmak Massif, was delineated in some detail by Blumenthal (1946) on the basis of the newly completed 1:800,000 geological map of Turkey (Fig. 36). This is the Kırşehir crystallines, which Bailey and McCallien were to interpret as a giant klippe of the Pontides atop the Taurides.

Thus, when McCallien began his geological work in Turkey, the basic tectonic division of the country into northern and southern tectonic domains had been outlined. In between were multifarious "geosynclinal" units and crystalline massifs. Neither the ages nor the structural position(s) nor the internal structure of the latter were known. The Halys or Kızılırmak Massif was one of these median masses.

McCallien had not brought his family to Turkey, so he had ample time on weekends to roam around Ankara to do fieldwork in the company of his young and enthusiastic then-undergraduate student, Oğuz Erol (Fig. 38). Erol was born in the ancient Ottoman capital of Bursa, Turkey, on February 25, 1926. His father was a civil servant in the young republic, and shortly after Erol's birth, the family moved to İstanbul, where he attended primary and secondary school. The death of his father and grandfather left the family in strained circumstances, and the threat of a German invasion in 1941 finally forced them to flee the great metropolis. They moved to Ankara, where Erol finished his secondary

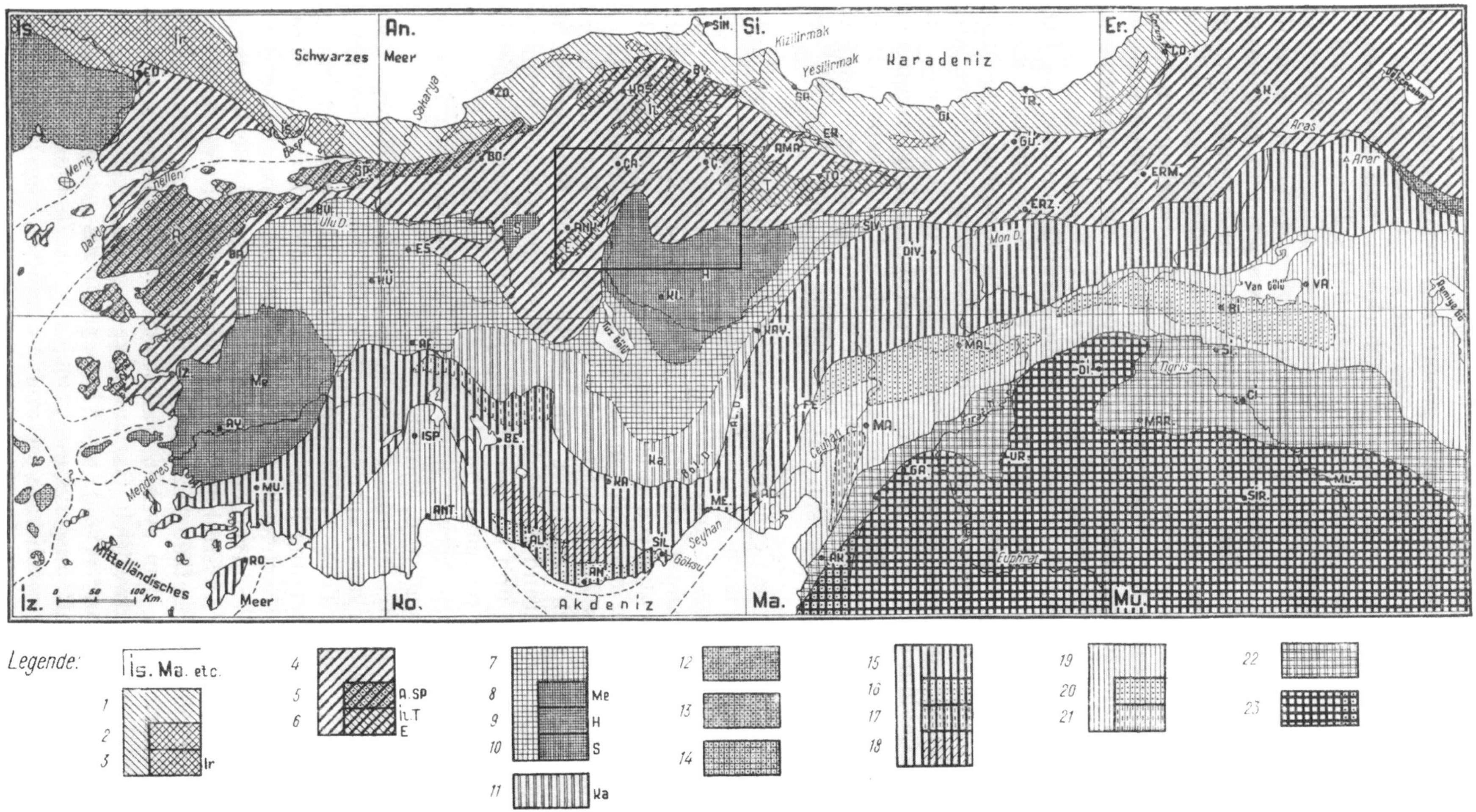

Figure 36. The classification of the tectonic units of Turkey, according to Maurice M. Blumenthal, geologist of the Mineral Research and Exploration Institute of Turkey, on the basis of the 1:800,000-scale geological map of Turkey, the best geological map A. Oğuz Erol, W. J. McCallien, and Sir Edward Bailey had at their disposal during their work on the Ankara Mélange in the mid to late 1940s. Inset shows the area covered by their work, corresponding with the map in Figure 49. Translation of the explanation of the legend is as follows: 1 = Pontides and their zones of uplift and core regions; 2 = internal cores (nuclei); 3 = marginal cores (İr: Strandja); 4 = Anatolides and their high regions (with equivalent units); 5 = Northern Aegean (A)—Sub-pontic (SP) high areas; 6 = High zones of northern Anatolia (Il = Ilgaz Massif, T = Tokat Massif, E = Elma Dağ Massif); 7 = Central Anatolian median masses (synonym: median zone); 8 = Menderes Massif (synonyms: Carian-Lydian Massif, West Anatolian Massif); 9 = Halys or Kızılırmak Massif (synonyms: Kırşehir Massif, Interior Anatolian Massif); 10 = Sakarya Massif; 11 = Cappadocia-Lycaonia subzone (orographically a part of the Taurus): Bolkar Dağ; 12 = Central Iran main mass (wide western end); 13 = Rhodope Massif; 14 = Cyclades Massif; 15 = Taurides; 16 = Alanya-Anamur Complex; 17 = Sultan Dağ Complex (16 and 17 correspond to poorly-known old complexes); 18 = parts characterized by thrusts and overthrusts; 19 = Iranides; 20 = Ceyhan-Bitlis Complex; 21 = Hatay zone (20 and 21 are poorly known old complexes); 22 = Iraqides, which include also an equivalent zone near the Taurus/Syria border zone (Alpidic marginal folds and the unfolded margin of the Syrian-Arab continent); 23 = Syrian-Arab block; stippled area in the west corresponds to the Lebanonides (Blumenthal, 1946).

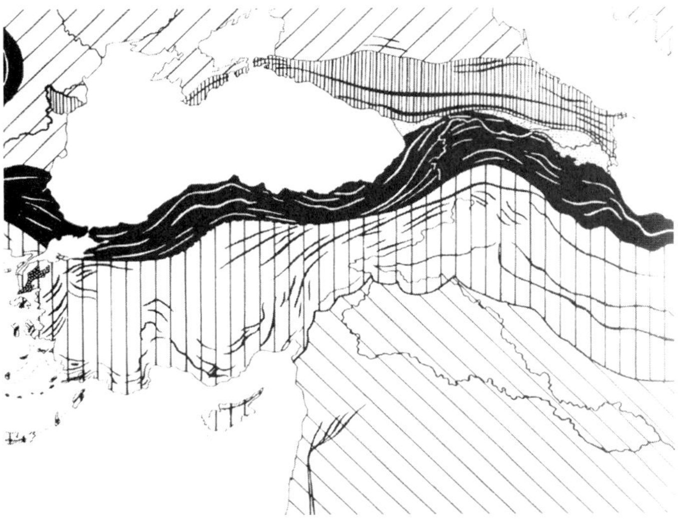

Figure 37. Tectonic units of Turkey according to Rudolf Staub (1924). Black is the primarily north-vergent "Alpide" stem or flank; vertically ruled area is the "Dinaride" stem, to which the median masses and the Taurus Mountains were considered to belong in Turkey. Staub acknowledged that in these, the dominant vergence was southerly, but claimed that this was a result of massive backthrusting. Bailey and McCallien (1950a, b; 1953) simply reversed this interpretation, but kept Staub's two main units under the designations Pontides (equivalent to Staub's Alpide stem) and Taurides (equivalent to Staub's Dinaride stem).

Figure 38. Professor A. Oğuz Erol, 1990. (Photo courtesy of Professor Erol.)

education in the best high school the young (and in those days very "provincial") capital had to offer: the *Gazi Lisesi*, also attended by the children of the second President of the Republic, İsmet İnönü. After finishing high school in 1943, he entered the Faculty of Language, History and Geography because that was where he was offered a scholarship. A year later, McCallien joined the faculty and assumed leadership of the Chair of Physical Geography and Geology. Erol remembers that McCallien did not speak a word of Turkish when he arrived, which rather impaired his communication with the students. The recitation sessions were particularly suitable for the students to goof off. Erol says he attracted McCallien's attention because he was hard-working, attentive, and able to draw well. (McCallien himself was an accomplished draughtsman: see Bailey and Weir, 1939, the illustrations of which were drawn by McCallien.) The young Turkish undergraduate thus gradually became attached to the enthusiastic Scottish professor. Erol did his honor's thesis under McCallien's supervision, and after graduation in 1947 McCallien invited him to work on a doctoral dissertation under him. Erol accepted. Thus began the distinguished career of one of Turkey's finest geomorphologists.

Their initial wanderings around Ankara baffled McCallien. Erol recalls the frustration of his teacher: it did not seem straightforward even to erect a stratigraphy, let alone to get a handle on the structure. The only fossiliferous rocks they initially were able find were the well-bedded fusulinaceous Permian-Carboniferous black limestones already reported by Chaput (1936, p. 25f.). Chaput had pointed out that the bituminous dark brecciated limestones bearing the fusulinid *Staffella* occurred nowhere in outcrop, but had been encountered only in the surficial float. He thus did not indicate them in his "Old Elmadağı Series" on his map (Chaput et al., 1930: Fig. 39). Chaput likened these enigmatic limestones to the black limestones he had found in Dikilitaş[34] to the west of Ankara that contained both *Staffella* and members of the group *Craticulifera*, thus giving a latest Carboniferous ("Ouralien" of Chaput, 1936, p. 35) to Permian age. On Chaput's map, there is only the faintest suggestion of the complications that his Elmadağı Series exhibited, for he had mapped the light-colored recrystallized limestone units as large lenses and thin layers tectonically interleaved within the Elmadağı Series. But McCallien seemed to despair, because as a result of the war conditions, no topographical base maps were available to him to begin a proper mapping campaign; he, therefore, kept searching for an area to study where he could find a simpler, regularly-bedded sequence.

In 1946, by a lucky coincidence, the Turkish Army Topographical Survey published in the *Harita Dergisi* (Cartographic Journal) an example of a 1:25,000 map of a small area to the south of Imrahor in the southeastern periphery of Ankara (for locations see Fig. 40), as an example of the fruits of its then new aerophotogrammetric work. Erol copied this "sample map" and used it as a base to complete his honor's thesis in this region. This was the first 1:25,000 geological map ever made in Turkey, so far as I can trace.[35] Regrettably, this map, which remained unpublished, has been lost. It was here that Erol found the first pillow-lava outcrop, which McCallien identified for him as such (for location, see

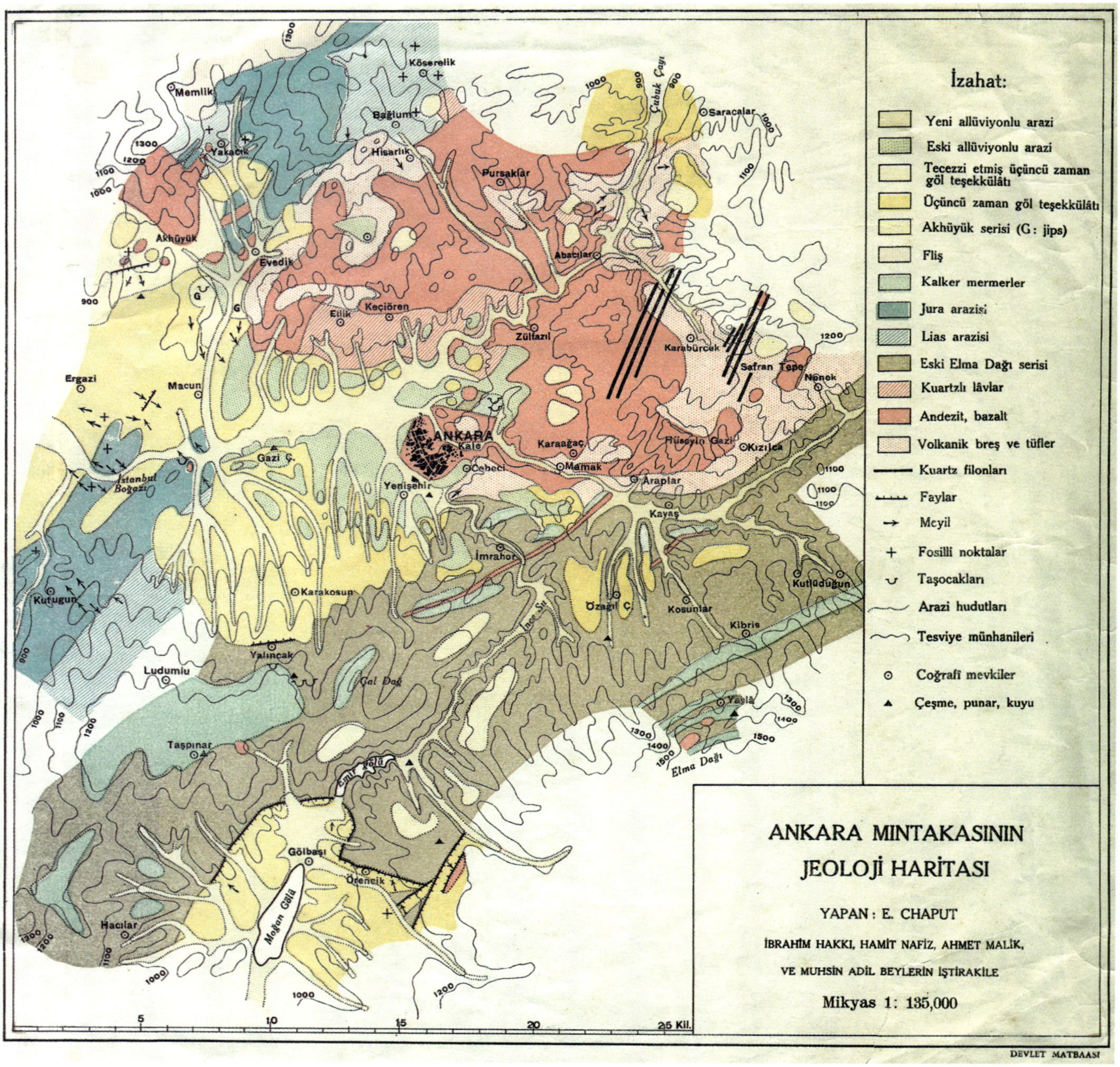

Figure 39. The geological map of the Ankara basin by Chaput et al. The original of this map is at a scale of 1:135,000. This was the first "detailed" map of the capital of the Republic of Turkey and was the only map available to McCallien and Erol when they started their work on the Ankara Mélange. The translation of the legend is as follows: *Yeni allüviyonlu arazi* = terrain with new alluvium; *Eski allüviyonlu arazi* = terrain with old alluvium; *Tecezzi etmiş üçüncü zaman göl teşekkülâtı* = weathered and eroded lacustrine formations of Tertiary age; *Üçüncü zaman göl teşekkülâtı* = lacustrine formations of Tertiary age; *Akhüyük serisi (G: jips)* = Akhüyük series (G: gypsum); *Fliş* = flysch; *Kalker mermerler* = calcareous marbles; *Jura arazisi*: Jurassic terrain; *Lias arazisi* = Liassic terrain; *Eski Elma Dağı serisi* = Old Elma Dağı series *[which essentially corresponds with Erol's Dikmen graywackes; see Figs. 40, 41, 42, and Plate 1 herein]*; *Kuartzlı lâvlar* = quartz-bearing lavas; *Andezit, basalt* = andesite, basalt; *Volkanik breş ve tüfler* = volcanic breccias and tuffs; *Kuartz filonları* = quartz veins; *Faylar* = faults; *Meyil* = dip; *Fosilli noktalar* = fossiliferous points; *Taşocakları* = quarries; *Arazi hudutları* = formation boundaries; *Tesviye münhanileri* = isohypses; *Coğrafî mevkiler* = geographical locations; *Çeşme, punar, kuyu* = fountain, spring, well.

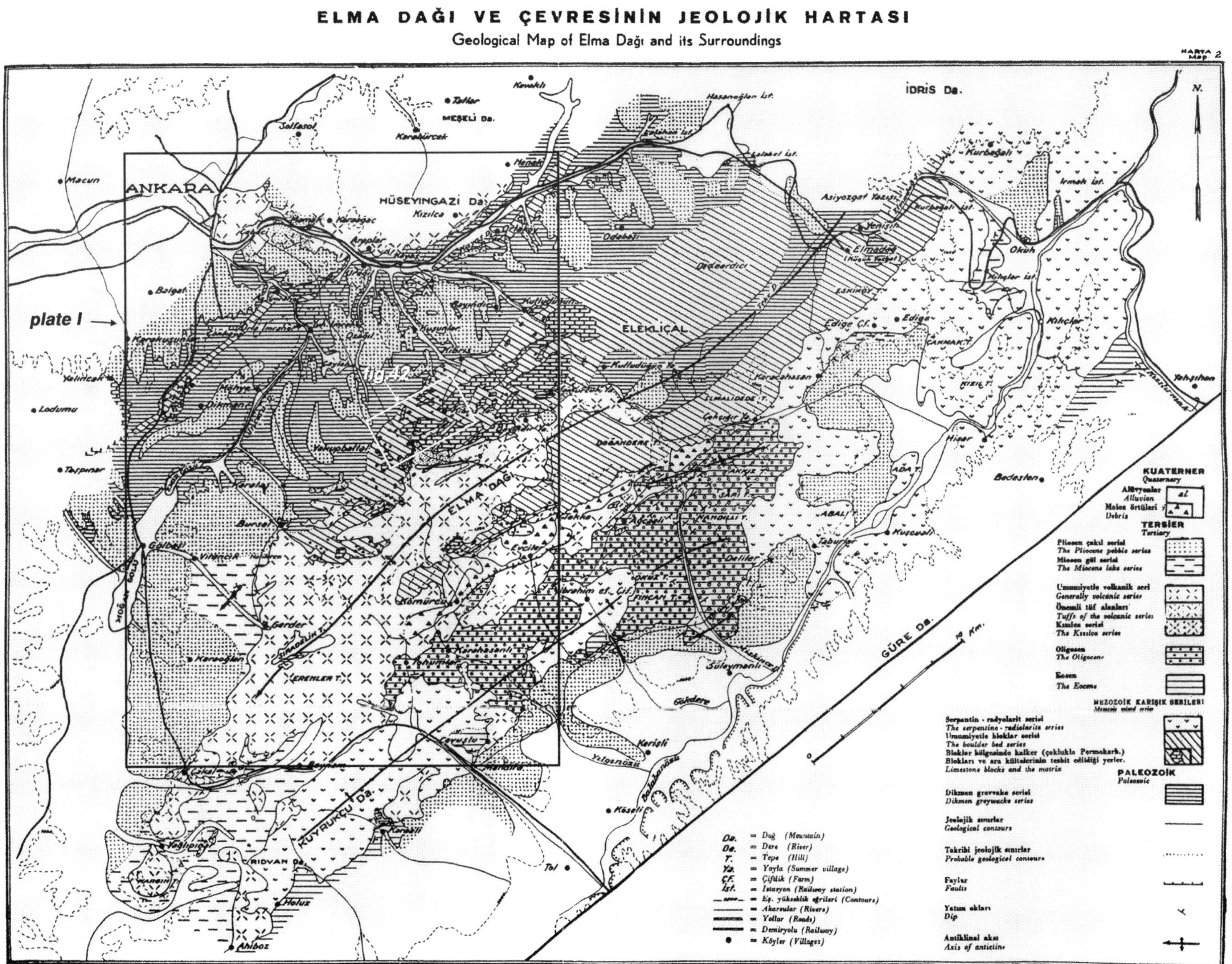

Figure 40. Oğuz Erol's published map of the Ankara region (from Erol, 1956, map 2).

Plate 1). This was the first pillow-lava occurrence reported from Turkey (McCallien, 1950). This exercise gave Erol his first taste of detailed mapping, and in 1947, after graduation, he commenced his Ph.D. work under McCallien. Topographic maps continued to be a problem (the cold war, and especially Stalin's demand for the northeastern Turkish provinces of Kars and Ardahan, continued the threat for Turkey), but Erol obtained (not entirely legally because civilians were still not authorized to use 1:25,000 maps) a set of maps from a military friend that covered a 625 km^2 area to the southeast of Ankara (Plate 1; for location, see Fig. 40), and began mapping in its northern part. He had to cut his maps into pieces small enough to hide quickly in his pocket, in case he was "caught." Under these extraordinary circumstances, he mapped a large number of blocks embedded in sedimentary and igneous matrices in the part of his area nearest to Ankara, namely in the Çaldağ, İmrahor, and the Kayaş regions (Fig. 40 and Plate 1).

When McCallien and Erol wandered through this series, they could never find the Permian-Carboniferous, well-bedded, black limestones that continued more than ~10 m before abruptly ending against slates or sheared mafic igneous rocks. Moreover, Chaput's "calcareous marbles" (Fig. 39) also looked embedded in the same slate/mafic rock "matrix." Professor Erol (personal commun., December 15, 1994) remembers that even before he started his doctoral thesis work (1945–1946), they had started using the terms "matrix" and "block" to describe what they were seeing (Fig. 41). Eventually, they even distinguished in the field a sedimentary matrix consisting mainly of slates with subordinate sandstones, from an igneous matrix formed dominantly from oxidized mafic volcanics and mafic tuffs (Figs. 41 and 42). Further, they often talked about the matrix being "sheared," forming a "plastic medium" in which the blocks were "churned" (Figs. 43 and 44).

McCallien kept asking Erol to establish a stratigraphy in this mess and Erol kept coming back to him with the plea that it did not seem possible. Finally, both agreed that they were looking, in the whole area, at a jumbled mixture of blocks floating in the two matrices (Figs. 41 and 42 and Plate 1). It was in the Çaldağ region that Erol also noted that blocks and the foliation defined planes dipped to the north. McCallien hypothesized that this probably indicated a top-to-south tectonic transport. Erol further thought that the nonmetamorphic clastic rocks, wedged into the jumbled rocks he was mapping in the Kıbrısyayla district in the Elmadağı region (Figs. 40, 42, and Plate 1), resembled the regularly bedded Liassic clastics in the Yakacık area (first found by Vadàsz in 1913: also see Chaput, 1936, p. 46) and correlated them. All this indicated that the area had been highly tectonized, apparently during the Alpine orogeny.

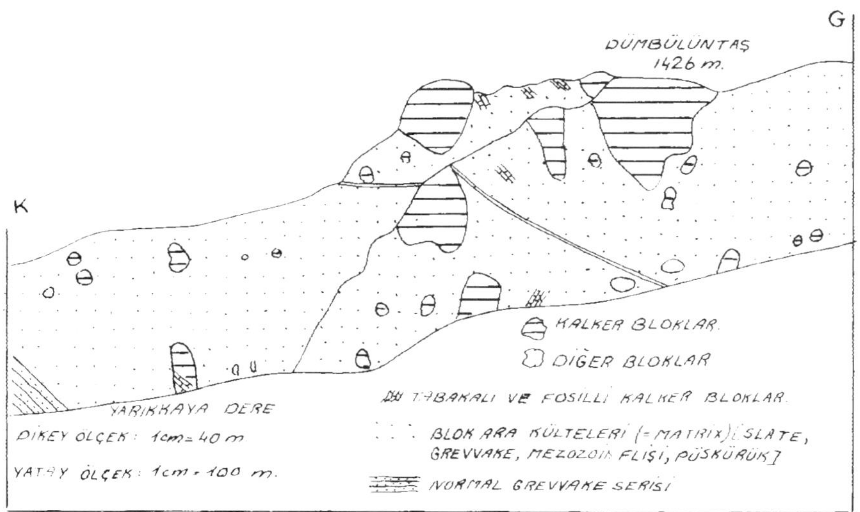

Figure 41. Erol's sketch of an outcrop of the Ankara Mélange, within what in those days was viewed as a Paleozoic graywacke series, called the Dikmen graywacke series. We now know that this is a part of a Cimmeride mélange, of Triassic age, called the Karakaya Complex (cf. Şengör et al., 1984; Görür et al., 1984). Note Erol's careful separation of the limestone blocks ("*kalker blokları*") and other blocks ("*diğer bloklar*") from the matrix ("*blok ara külteleri (= matrix) [slate, grevvake, Mezozoik flişi, püskürük*]" translates as "intermediary masses between blocks (= matrix) [slate, graywacke, Mesozoic flysch, eruptive]"). Erol recognized, in addition, a "normal graywacke series" (= "*normal grevvake serisi*") and "bedded and fossiliferous limestone blocks" (= "*tabakalı ve fosilli kalker bloklar*"). It is remarkable that he noted that the matrix included a flysch of Mesozoic age. The vindication of Erol's mapping took nearly three decades! (From Erol, 1949, fig. 3). As the great Turkish geologist, the late Professor İhsan Ketin, never tired of repeating, Erol's quality of mapping was not matched for decades (İhsan Ketin, personal commun., numerous occasions).

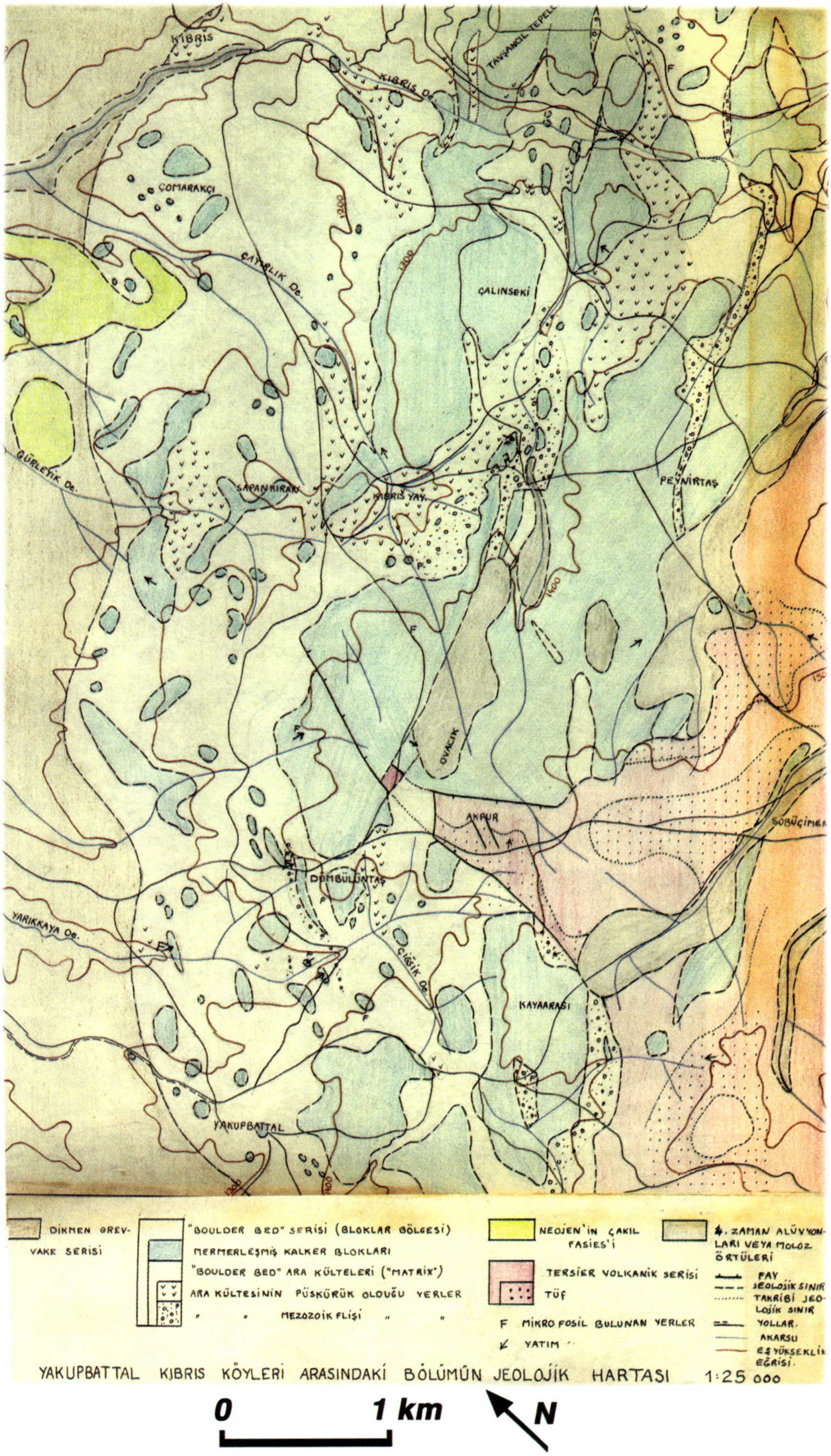

Figure 42. Erol's geological map of the area between Yakupbattal and the Kıbrıs villages (for location, see Fig. 40). Translation of the legend: *Dikmen grevvake serisi* = Dikmen graywacke series; *"Boulder Bed" serisi (Bloklar Bölgesi)* = "Boulder Bed" series (Region of Blocks); *Mermerleşmiş kalker blokları* = marbleized limestone blocks; *"Boulder beds" ara külteleri (matrix)* = "Boulder beds" intermediary masses (matrix); *Ara kültesinin püskürük olduğu yerler* = places where the intermediary mass is eruptive; *Ara kültesinin Mezozoik flişi olduğu yerler* = places where the intermediary mass is Mesozoic flysch; *Neojen'in çakıl fasies'i* = pebbly facies of the Neogene; *Tersier volkanik serisi* = Tertiary volcanic series; *Tüf* = tuff; *Mikrofosil bulunan yerler* = microfossil finds; *Yatım* = Dip; *4. Zaman alüvyonları veya moloz örtüleri* = Quaternary alluvium or colluvium; *Fay* = fault; *Jeolojik sınır* = geological boundary; *Takribi jeolojik sınır* = approximate geological boundary; *Yollar* = roads; *Akarsu* = stream; *Eşyükseklik eğrisi* = isohypses. Note the separation of the sedimentary and the igneous matrices.

Figure 43. View from Akpur-Dümbülüntaş toward Yakupbattal-Çaldağ (see Fig. 42 for locations). In view is the Ankara Mélange, with a dominantly sedimentary matrix. Note the conspicuous limestone blocks strewn in the landscape. (Photo by W.J. McCallien, 1949; courtesy of Prof. Erol).

Figure 44. Highly folded "white marls" (= "*beyaz marn*"), "marls with sandstone lenses" (= "*kumtaşı adeseli marnlar*"), and "marl" (= "*marn*") within a matrix of sandstones and conglomerates, holding a limestone breccia block (= "*kalker breşi bloğu*"). Erol drew this sketch as one "illustrating the structure of the Mesozoic flysch" forming the matrix of the mélange, and drew attention to the high degree of folding and shearing (Fig. 4 and photo 8 from Erol, 1949).

Erol finished his thesis in 1949 (Figs. 45 and 46), but had to defend it in McCallien's absence, for the latter had already left Turkey to go to the University College of the Gold Coast (which became the University of Ghana in 1961), where he established a department of geology. Erol formulated his conclusions about the "mixed series" in his thesis as follows:

> The first characteristic that catches the eye in this blocky series, of which we have tried above to outline the rock types, is the presence of large limestone blocks and intermediary material that surrounds them. The sedimentary parts of this intermediary material were sheared, mixed up and the igneous parts were pushed in between various rock types. The "pillow lavas" within the igneous parts are clear evidence for submarine eruptions (Bailey, 1936, p. 1721). The green tuffs must also be of volcanic origin. Thus we found it appropriate to call this mixed belt containing: blocks of possible Permo-Carboniferous (perhaps also Triassic) age, a flysch series whose age is provisionally assigned to the Mesozoic, greywackes thought to be still older, pillow lavas, and other eruptive material, the Elmadağı Blocky Series (= Boulder Bed Series). We do not wish to say much about the origin of these mixed series, which is still a topic of debate. But, it is clear that the blocks within the Elmadağı series are not a result of the "northerly tilting like fish scales of the Elmadağı Massif and its Mesozoic cover during the Cainozoic" as believed by Chaput (1931, p. 83[36]). The generation of the blocks must have occurred in an environment in which the pillow lavas, indications of older (probably Mesozoic) submarine eruptions, could spread, because the blocks and the intermediary material only rarely are in contact with fracture zones. Some of the normal faults in the area cut the Tertiary and are thus younger.
>
> In brief, I wish to stress especially that this series is a "mixed series" consisting of the blocks and of the intermediate material. Although we do not have clear evidence for its age, we provisionally treat it as Mesozoic. The real age can only be established after finding evidence for the age of the intermediary material.

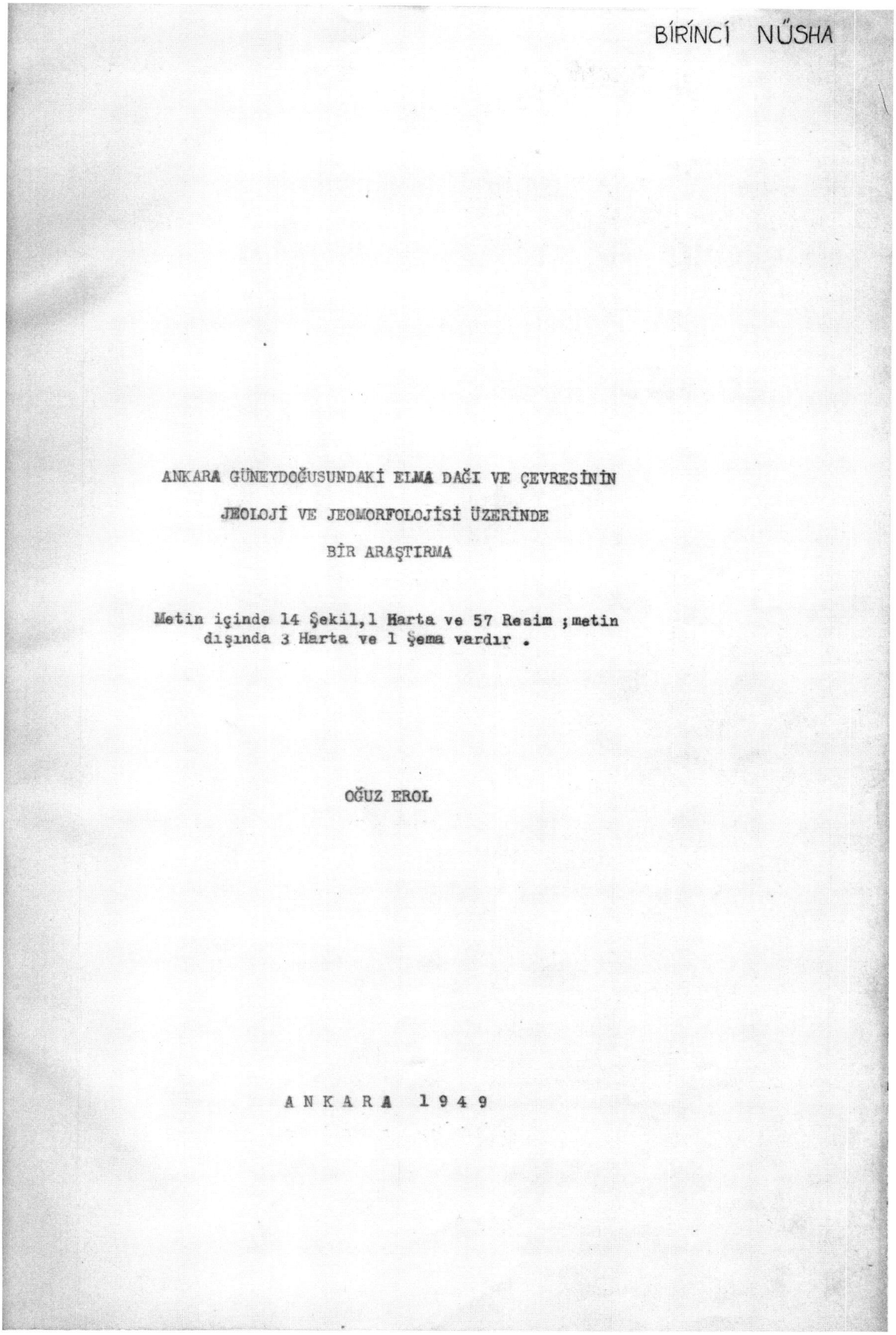

BİRİNCİ NÜSHA

ANKARA GÜNEYDOĞUSUNDAKİ ELMA DAĞI VE ÇEVRESİNİN
JEOLOJİ VE JEOMORFOLOJİSİ ÜZERİNDE
BİR ARAŞTIRMA

Metin içinde 14 Şekil,1 Harta ve 57 Resim ;metin
dışında 3 Harta ve 1 Şema vardır .

OĞUZ EROL

A N K A R A 1 9 4 9

Figure 45. Title-page of the typescript of Erol's doctoral dissertation (Erol, 1949)

İçindekiler

Sayfa

Önsöz 3
~~Şekil ve haritaların listesi~~ 4

Giriş 5
Topoğrafya 6
Bölge hakkındaki çalışmalar 7
Problemler 10

Jeoloji gözlemleri 12

Paleozoik
Dikmen Grevvake Serisi 13
Mermerleşmiş kalkerler 15

Mezozoik
A. Elma Dağı Bloklar Serisi (=Boulder Bed Serisi) ve Kayaş Serisi 18
B. Serpantin-Radyolarit Karışık Serisi 23

Tersier
Eosen 27
Oligosen 28
Volkanik Seri 33
Neojen 38

Kuaterner 48

Tektonik 49

Jeomorfoloji gözlemleri 56

Dağlar (Ankara Bölgesinde) 56
Yontukdüzler 58
A. 1500 m.lik yüksek yontukdüz 60
B. 1000-1250 m.ler arasındaki alçak yontukdüz 62
Elma Dağındaki özel yerşekilleri
A. Zirveler dizisi 65
B. Zirveler dizisinin GD sundaki "boyuna çukur" 66
C. İkinci tepeler sırası 67

Dağlar arasındaki çukur bölümler
Büyük havzalar ve havzaların dolguları 68
Havzaların boşalması ve genç vadiler, sekiler 70
A. Boşalmanın sebep ve delilleri 71
B. Boşalma ve yarılma 76
I- Sekiler 77
II- Vadi tabanlarındaki dolgular 82
C. Boşalma ve yarılmadan sonra meydana gelen bugünkü yerşekilleri 83

Hülâsa
~~Sonuç~~ + Summary 84

Bibliyografya I-VI

Ayrıca, Ankara şehirleri hakkında bir çerçeve, 16 resim, 12 Şekil ve 3 harita mevcuttur.

Figure 46. Contents list of the typescript of Erol's doctoral dissertation. Translation is as follows:

Önsöz: Preface
Şekil ve haritaların listesi: list of figures and maps
Giriş: Introduction
Topografya: Topography
Bölge hakkındaki çalışmalar: Works concerning the area
Problemler: Problems
Jeoloji Gözlemleri: Geological observations
Paleozoik: Paleozoic
Dikmen Grevvake Serisi: Dikmen graywacke series
Mermerleşmiş kalkerler: Marbleized limestones
Mezozoik: Mesozoic
A. *Elma Dağı Bloklar Bölgesi (Boulder Bed Serisi) ve Kayaş Serisi*: Elmadağı Region of Blocks (Boulder Bed Series) and Kayaş Series
B. *Serpantin-Radyolarit Karışık Serisi*: Serpentine-Radiolarite Mixed Series
Tersiyer: Tertiary
Eosen: Eocene
Oligosen: Oligocene
Volkanik Seri: Volcanic Series
Neojen: Neogene
Kuaterner: Quaternary
Tektonik: Tectonics
Jeomorfoloji gözlemleri: Geomorphological observations
Dağlar (Ankara Bölgesinde): Mountains (in the Ankara Region)
Yontukdüzler: Peneplains
A. *1500 m.lik yüksek yontukdüz*: High peneplain at 1500 m
B. *1000–1250 m. ler arasındaki alçak yontukdüz*: Low peneplain between 1000 and 1250 m
Elma Dağındaki özel yerşekilleri: Special landforms at Elma Dağı
A. *Zirveler dizisi*: Series of peaks
B. *Zirveler ekseninin GD sundaki "boyuna çukur"*: Longitudinal depression to the SE of the axis of peaks
C. *İkinci tepeler sırası*: Second range of peaks
Dağlar arasındaki çukur bölgeler: Depressions between the mountains
Büyük çöküntü havzaları ve havza dolguları: Large depressions of subsidence and basin fills
Çöküntü havzalarının boşalması ve genç vadiler, sekiler: Emptying of the depressions of subsidence, young valleys, and terraces
A. *Boşalmanın sebep ve delilleri*: Reasons of and evidence for emptying
B. *Boşalma ve yarılma*: Emptying and dissection
I. *Sekiler*: Terraces
II. *Vadi tabanlarındaki dolgular*: Fills at the floors of valleys
C. *Boşalma ve yarılmadan sonra meydana gelen bugünkü yerşekilleri*: Present-day landforms that formed after emptying and dissection

Figure 47. Sir Edward B. Bailey (1881–1965). (Photo courtesy of Colin McCallien.)

This "Boulder Bed Series," in which other rock types predominate over the green eruptives has been distinguished from the usual "Mesozoic Mixed Series," in which the serpentinites are dominant and which occurs in the southern part of our area. (Erol, 1949, p. 21f.; 1956, p. 18; endnotes mine)

Erol wrote this in 1949. During the examination (the examining jury consisted of professors Cemâl Arif Alagöz and Ahmet Şevket Birand and associate professors Reşat İzbırak and Ferruh Sanır), he remembers with amusement now how the old professor of geology in Ankara, Birand, expressed surprise that all this could happen in a "lake." Erol, horrified, exclaimed: "A lake! This was an ocean, at the bottom of which all these eruption events and submarine slidings were taking place to create this mixed series!" (Prof. Oğuz Erol, personal commun., December 15, 1994). Birand was baffled, and Erol thought from the expression on his face that he would probably fail the examination. To his great relief, he passed with flying colors.

In the meantime, McCallien was communicating with his old friend, the renowned Scottish tectonician Sir Edward Bailey (1881–1965, Fig. 47: for Bailey's biography, see MacGregor, 1968), about this extraordinary area. He asked Bailey whether he would want to go and tour it to see whether a reasonable interpretation could be formulated. Bailey accepted, funds were raised through the generosity of the Carnegie Trust, the University College of the Gold Coast, and the Mineral Research and Exploration Institute of Turkey in Ankara (the *Maden Tetkik ve Arama Enstitüsü*: the Turkish Geological Survey), and the two went to Ankara in the summer of 1950. Mr. Colin McCallien, who accompanied his father and Sir Edward as a child, relates the journey to Ankara with the following words (written commun., June 7, 2002):

> We flew from Scotland to London and joined the plane to Istanbul (the old Constantinople) just after midnight. The next morning at half past five we were in Rome and left after an hour's halt for re-fuelling and breakfast. We flew across the heel of Italy, across the southern end of the Adriatic into the Gulf of Corinth. Our route took us directly along the Corinth Canal and so across to Athens where we landed about midday. From Athens we continued across the Aegean Sea and its numerous islands to the coast of Turkey opposite the large island of Mytelene. When we reached the Sea of Marmara the pilot got instructions to make a long detour to avoid bombing practice by the Turkish authorities. We arrived in Istanbul in the afternoon and only stayed a short time. We went across the Bosphorus by ferry to catch the night express to Ankara.

Bailey and McCallien (1953) relate the background of their work in their published account:

> In the five years preceding our joint visit it had been impossible for W.J. McCallien to undertake detailed mapping, since, quite properly under existing conditions, a foreigner attempting such a task is subjected to endless questioning. On the other hand, one of his research students [*in reality the only one he had*], Oguz Erol, starting in 1947 and working under close supervision, produced a valuable thesis, the title of which when translated reads: "A Study of the Geology and Geomorphology of the Region S.E. of Ankara in Elma Dagi and its surroundings." This account is illustrated with a careful geological map (not as yet published[37]) on the scale of 1: 100,000.[38] For our present purpose it is only necessary to say that Erol firmly established a broken, block structure for the greywacke-limestone complex of the Ankara outcrop, and in many cases separated individual blocks of limestone on this map. He also found numerous micro-fossils in limestones, some of which have been determined by palæontologists as Middle Permian. We had the great advantage of Erol's company and guidance in the area he knows so well during one week of our visit. (Bailey and McCallien, 1953, p. 406, endnotes mine)

Figure 48. Sir Edward Bailey in the company of Colin McCallien, enjoying a well-deserved rest in the blazing heat of the Anatolian Summer in 1950. Photo courtesy of Colin McCallien.

At the time, Erol was doing his military service and had to use a vacation leave to lead his former professor and Sir Edward in his area. During this one-week field excursion, not one word was spoken about mélange. Erol says that he first saw the word "mélange" in the published papers of Bailey and McCallien.

Bailey and McCallien then moved to the Alaca district ~160 km east of Ankara, where the ancient Hittite sites were being excavated by the Turkish Historical Society (*Türk Tarih Kurumu*) and they lodged in its guest house. Collin McCallien gives his impressions from more than half a century ago (see also Fig. 48), as follows:

> The present day village of Alaca, like other modern villages, is built of mud bricks dried in the sun. Also like other villages, the houses have flat roofs of earth on top of tree trunks and on one or two occasions we drove the jeep on to the top of a house thinking we were still on the road. The floors of the houses are also of mud and in the wet or snowy season they are very uncomfortable. There are no shops in the villages and the people have to go many miles, usually on a donkey, for the few things they need. When we ran out of our own bread we had to use village bread. This bread is made without yeast and is like a thin, round sheet of card board nearly a yard across. You could fold it up and put it in your pocket which was handy. However, I had to suck it for a while or soak it in water before I could chew it. (Mr. Collin McCallien, written commun., June 7, 2002)

There, too, they found rock associations similar to those mapped by Erol to the southeast of Ankara.

Their month's travels around Ankara and Alaca districts showed Bailey what his colleague and his student had seen earlier, and it was presumably Bailey who remembered that a similar association had been described from Anglesey in 1919 by Greenly. They thus decided to use Greenly's terminology:

> Our facilities, as provided by the Geological Survey [*of Turkey*], included jeep transport, which took us "over hill, over dale" to "wander everywhere," like Puck in Shakespeare's fairy tale. We saw great tracts of broken rocks, consisting mainly of greywackes, limestones and the Steinmann trinity. They rivalled in abandoned disarray Puck's most mischievous attempts to puzzle and confuse mere mortal man. Yet, with the help of fossils, mostly of the micro order and found quite commonly both in the limestones and the cherts, the M.T.A. map (which forms the basis of

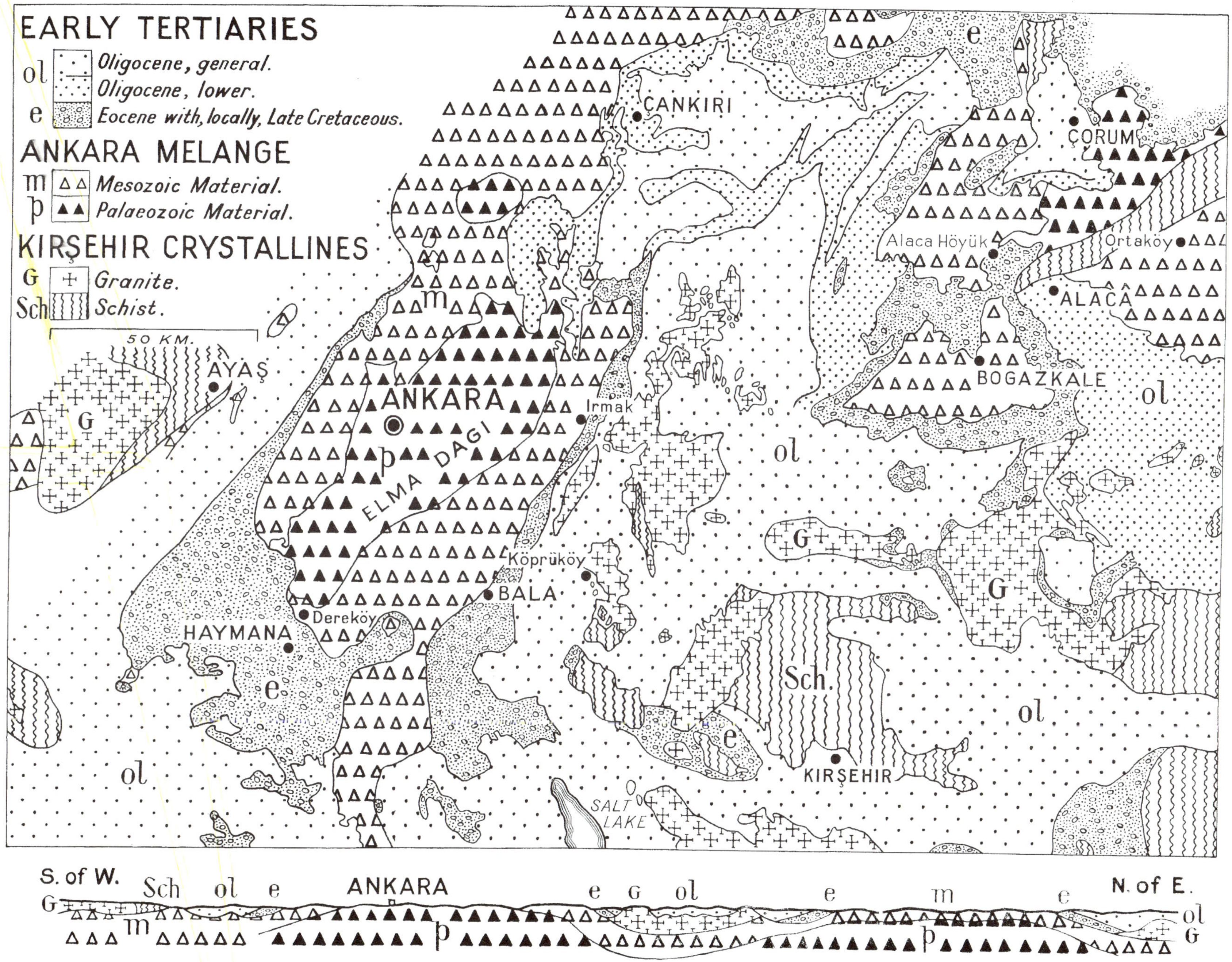

Figure 49. Tectonic map and cross-section of the area studied by Bailey and McCallien (from Bailey and McCallien, 1953, plate VIII). For location, see Fig. 36.

our plate VIII [*herein Fig. 49*]) has diagrammatically divided the area into Palæozoic and Mesozoic, with local refinements such as Permo-Carboniferous, Lias, Oolite and Jurassic-Cretaceous. We established undoubted important recurrences of rock types at more than one horizon; but, broadly speaking, we are satisfied that the main greywackes are Devonian, the main limestones Permo-Carboniferous, in an inclusive sense, and the main developments of the Steinmann trinity Mesozoic.

....

All the above formations are broken to give what we may call the Ankara Mélange ... out of respect for descriptions that Greenly and Matley have furnished of a similar tectonic mélange in Anglesey and the adjoining Lleyn peninsula of Wales (cf. Greenly, 1919 [*1919a and b herein*]). We definitely know from our own experience of some 6000 sq. km. of mélange exposed along the Ankara outcrop, and it is obvious that this is only a small fraction of an immense whole. (Bailey and McCallien, 1953, p. 405; also see Bailey and McCallien, 1950a, p. 3f.; 1950b, p. 18–19)

Bailey was so impressed with the pillow lavas that he decided to take home a sample. Colin McCallien remembers how this prized piece was carried back to Scotland:

Professor Bailey took home a pillow lava—a large stone about 10 inches [≈*25 cm*] in diameter formed when lava cooled in contact with water. Because the weight of our luggage was restricted on the aeroplane he wrapped his coat round it and tried to walk normally onto the plane. (Mr. Colin McCallien, written commun., June 7, 2002)

Erol thinks that the hypothesis of how this immense mélange had formed had come from Sir Edward Bailey. It was formulated as follows:

(12) We have mentioned under heading (4) how the Ankara mélange, exposed in the Ankara anticlinal, reappears eastwards in the Alaca Höyük group of anticlinals. The width of separating Early Tertiary outcrop occupying the intervening synclinal is, in this traverse, 60 km. The synclinal outcrop diminishes greatly northwards past Çankiri; but it increases correspondingly southwards and eastwards past Kirşehir, after which the synclinal may be named.

(13) A remarkable feature of the Kirşehir synclinal is that its Early Tertiary outcrop is in very many places interrupted by great spreads of marble and crystalline schists intruded by granite or diorite. These rocks are clearly pre-Tertiary and have yielded debris to the same. Similar rocks are found occasionally in direct contact with the assemblage characteristic of the Ankara mélange. We shall speak of these crystallines as the Kirşehir crystallines.

(14) The highly accentuated synclinal disposition of the Kirşehir crystallines with reference to the Ankara mélange shows that the former constitute part of a great *nappe de recouvrement* covering the latter. In the anticlinal areas the early Tertiaries rest preferentially on the mélange, because the anticlines originated with concomitant erosion before the early Tertiaries started to accumulate. In the synclinal areas the Early Tertiaries rest preferentially on the Kirşehir crystallines, because here erosion has spared extensive masses of the *nappe de recouvrement*.

(15) We believe that the Kirşehir nappe is most probably a Pontide nappe, and that the thrust at its base, which we suggest be called the "Anatolian thrust," is essentially a boundary between Pontic (Black Sea) and Taurus (Mediterranean) pre-Tertiary rocks, which differ in facies. The Ankara mélange we interpret as Taurus rocks beneath the Anatolian thrust; the Taurus Mountains as a continuation of similar rocks bulldozed towards the south.

(16) The horizontal displacement postulated along the Anatolian thrust amounts to about 350 km. This is greater than the minimum displacement measured along other thrusts yet described. The Törnebohm thrust in Scandinavia has a minimum displacement of about 130 km. (Bailey and McCallien, 1950a, p. 5–6; also see 1950b, p. 20–21; 1953, p. 437–438)

The timing of this monstrous thrusting was given by the youngest rocks bulldozed by the hypothesized Kırşehir Nappe and the oldest rocks lying unconformably on both the mélange and the Kırşehir crystallines, after some of the microfossil material Bailey and McCallien had collected had been studied. These rocks were Upper Jurassic and Eocene, respectively (see Bailey and McCallien, 1953, p. 407 and 480):

We suggest as an explanation the following sequence of events. In Late Cretaceous times Kirsehir crystallines (Pontic) were thrust long distances over Ankara and Alaca Höyük Mesozoics and Palæozoics (Taurus). The thrust we call the Anatolian Thrust, and the thrust-mass the Anatolian Nappe. Continuance of movement started to fold the Anatolian Nappe together with its mélange foundation, thereby initiating the Ankara and Alaca Höyük anticlinals, and the Ayas and Kirsehir synclinals. Concurrent erosion removed the Anatolian Nappe in anticlinal areas, leaving enormous klippes in synclinal areas. Before the end of Late Cretaceous times general submergence introduced the sea into some of the erosional hollows floored by Ankara Mélange between the klippes. The process of submergence was accentuated during the Eocene. Eventually Oligocene seas and lagoons spread over previously upstanding remnants of the Kirsehir crystallines. Succeeding post-Oligocene elevation was accompanied by accentuation of the anticlinal and synclinal deformation. Thus Mid-Tertiary and later erosion has revealed the Mesozoic and Palæozoic materials of the Ankara Mélange beneath Eocene (and Late Cretaceous) in anticlinal areas, and the Kirsehir crystallines beneath Oligocene in synclinal areas. (Bailey and McCallien, 1953, p. 438)

Thus Bailey and McCallien thought they could sketch a complete Mesozoic-Cenozoic tectonic history for Asia Minor. Obviously the mélange was the most important part of this story and it not only has remained so until today, but also has become much amplified in significance (see Şengör and Yılmaz, 1981; Görür et al., 1984).

The story I reviewed above in some detail shows that the mélange *concept* was rediscovered independently by Erol and McCallien almost exactly three decades after Greenly first proposed it in 1919 and shortly after Noble's recognition of the chaos structure in the Amargosa Range in California. It was Erol's utter frustration that no sensible stratigraphy could be erected in his area and McCallien's insistence that he should try anyway that finally led the former to tune his observations to a fine enough scale that they finally discovered that they were looking at a jumbled sequence of rocks. Erol remembers that the two often went off to check yet another area where Erol tried to show McCallien why a normal Stenonian/Cuvierian stratigraphy was not possible. McCallien was absolutely insistent on precise and accurate mapping. Erol remembers once, when he was showing his map to his teacher, that McCallien noticed that Erol had failed to enter a particular block they had seen onto the map. Erol said "Oh yes, I must have forgotten. Let me do it just now." McCallien exclaimed "Never! Never put anything on the map from memory. Go back there and draw it in on the outcrop." (Professor Oğuz Erol, personal commun., June 11, 2002). In this painstaking process, they developed their terminology of blocks, lenses, lenticules, matrix, intermediate material, etc. Their premature interpretations included mixing during submarine eruptions and submarine sliding. Finally, the arrival of the better-informed Sir Edward showed them that the problem they were grappling with was a very real one, but that it had been solved some three decades earlier by Greenly: *this gave them their terminology*. The presence of the Steinmann trinity (Steinmann, 1927; Bailey and McCallien, 1960) in the mélange drew Sir Edward's attention to Steinmann's own area, to the "Aroser Schuppenzone" (a part of Steinmann's *Aufbruchszone*: Steinmann, 1905) between the higher-lying Austroalpine and lower-lying Penninic elements, and thus a giant thrust similar to the Austroalpine basal thrust above the Aroser Schuppenzone was conceived: *this inspired the interpretation*. Although later shown to be wrong in significant details, the interpretation that the mélange had been generated in a giant shear zone, essentially a huge fault zone, has remained victorious, at least to the present-day.

At the time, this giant thrust fault seemed to be the most exciting part of their discovery. The mammoth overthrusting this hypothesis suggested exceeded by almost 200% the largest thrust displacement then known (≈300 km versus the ≈100 km from Scandinavia: see Törnebohm, 1896, p. 155 and map on p. 156, where overthrusting with a displacement of 130 km is illustrated)! This implied a great mobility of Asia Minor and that it had a structure very similar to Suess', Argand's, and Staub's Alps: a giant, south-facing sandwich (as opposed to the north-facing Alps) with the Pontic ranges on top, the geosynclinal fill in the middle plus some bulldozed Taurus material, and the Taurus itself on the bottom. All this made a mockery of the then prevalent Kober-Stillean model of symmetric orogens and vast, rigid *Zwischengebirge* between their two oppositely vergent flanks:

> (17) We are conscious that our conclusions differ considerably from Kober's conception of a Zwischengebirge between the Pontic and Taurus Mountains, and also from Paréjas's interpretation of transversal anticlinals and synclinals in the same region. (Bailey and McCallien, 1950a, p. 4; 1950b, p. 21; also see 1953, p. 438)

Despite the exciting discovery and the hypothesis of the colossal thrusting, the Bailey and McCallien papers made little impact outside Alpine circles. The central European geologists were too deeply imbedded in the rigid Stillean schemata. The British had little interest in Mediterranean tectonics, despite Sir Edward's enthusiasm and elegant expository writings (e.g., Bailey, 1935). The Americans, one of the very few exceptions being Walter H. Bucher, were hardly conversant with overseas geology (cf. Şengör, 1999; also see Hsü's letter quoted below). In Turkey, the term Ankara Mélange did become popular, but almost nobody really understood the significance of Bailey and McCallien's work. The Mineral Research and Exploration Institute of Turkey (the MTA) did not publish Erol's meticulously prepared geological map at a scale of 1:25,000 (Plate 1 reproduces the field slips of this map). Petty professional jealousies between geographers and geologists in Turkey were the main cause. It was the second geological map prepared at that scale in Turkey (remember that the first was also done by Erol). The text of his thesis and a generalized 1:100,000 map only could be published by the MTA as a result of the beneficial intervention of Professor Hamit Nafiz Pamir. Thus, Erol's careful work on the mélange was not followed up for years, and only in the 1970s did we begin to understand its tectonic significance in Turkey (see Erol, 1981).

In Turkey, only İhsan Ketin (1914–1995; Fig. 50: see Şengör, 1985b; Şengör and McKenzie, 1997), by that time a fairly young "senior" member of the Turkish geological community and internationally renowned because of his remarkable discovery of the great North Anatolian strike-slip fault in 1948 (Şengör, 1996), seems to have understood the wide-ranging implications of the Bailey and McCallien paper not only for the geology of Turkey, but for theoretical tectonics in general, and decided to put the monstrous "Anatolian Thrust" to test. Were the "Kırşehir crystallines" really a massive, rootless nappe, a gigantic klippe of the Pontic unit floating on the "geosynclinal infill" and the Taurus, whose rocks it had crushed and sheared into what eventually had become the Ankara Mélange? Was Turkey indeed an immense sandwich consisting of the Pontides on top, the Ankara Mélange in the Middle, and the Taurides at the base, similar to the "schema of the Alpine cross-section" of Staub (1924, fig. 2), which Bailey probably had in mind? With the express purpose of testing these ideas, in the summer months of 1953 and 1955, Ketin mapped an area to the east of Ankara within the grand curve of Kızılırmak, the classical Halys, about the city of Yozgat, as a part of the national 1:100,000 geological mapping program of the Turkish geological survey (Ketin, 1956a). Ketin confirmed the presence of the Kırşehir crystallines in the area, consisting of calcschists, marbles, and amphibole schists, cut by both mafic and felsic plutons with compositions ranging from olivine gabbro to granite

Figure 50. İhsan Ketin (1914–1995) in the years when he falsified Bailey and McCallien's hypothesis of the allochthonous nature of the Kırşehir Massif with a Pontic provenance. (Photo courtesy of Mrs. Bedia Ketin.)

(for a modern assessment of part of Ketin's area see Seymen, 1985). *Above* the crystallines, however, was the serpentinite-radiolarite association, in places sitting directly on the metamorphics (as in west of Kaman: Ketin, 1956b, p. 169), in other places being *thrust onto* a lower Eocene Flysch. Nowhere could Ketin find a contact that showed the metamorphics *above* the mélange, as required by the Bailey and McCallien hypothesis. Moreover, in the area he mapped, the serpentinite-radiolarite association did not look terribly chaotic. He described "beds" of serpentinite, breccias of pillow lava, calcitized and silicified basalt, among other kinds of dominantly mafic volcanic rocks and an entire array of radiolarian cherts, limestones of various types, ranging into marls, sandstones, schists, and tuffs in one section. It is now clear, in retrospect, that what Ketin was looking at was a tectonically-stacked sequence, but at the time he did not notice it (Ketin, 1956a, p. 8–9; see also his sections in his fig. 3; for an easily accessible list of his "beds" with thicknesses and in the sequence he described, see Şengör and Sakınç, 2001, p. 188). Ketin found Turonian through Maastrichtian fossils in the sedimentary layers and assumed a fairly regular succession of deep-sea sedimentary and volcanic rocks at the bottom of a geosyncline. We should not forget, however, that both Ketin, and Sir Edward Bailey and McCallien thought, following the conventional wisdom of the day, that serpentines had been originally submarine *volcanic* rocks. Although Ketin did see and map thrusts locally, he further assumed that what he had mapped was the original stratigraphic disposition and the crystalline massif of Kırşehir had been the floor of the geosyncline, in which the serpentinite-radiolarite association had been "deposited," at least in part. He thought that the metamorphic rocks were older than the serpentinite-radiolarite series, but not much older. He estimated that they were probably Mesozoic (Ketin, 1956b, p. 169).

Thus, not much remained of Bailey and McCallien's grand hypothesis of the structure of Asia Minor, which failed Ketin's test.[39] With it, the mélange concept was once more relegated to the dusty archives of the history of geology, except in the later writings of Bailey and McCallien (1961, 1963) and Gansser (1955, 1959, 1964, p. 7–8); however, Gansser, a Swiss, is a child of the Alps, where jumbled rocks—known at least since Bachmann (1863) and Kaufmann's designation of the "Wildflysch" in 1870—were more the rule than the exception, so while describing what he called the colored mélange of Iran, he did not feel he was revealing a new world (see also Hsü's reference to Alpine geologists in his paper cited below).

In the light of all this, it is very remarkable that the mélange concept had to be reinvented for the third time, two decades after Erol and McCallien's reinvention and one decade after the rejection of Bailey and McCallien's tectonic interpretation of the structure of Asia Minor by Ketin.

Third Rediscovery: Kenneth J. Hsü (1963–1968)

Kenneth Jinghwa Hsü (1929–: Fig. 51), a native of Yangtze in the Chinese province of Jiangsu, relates in detail how he himself had reinvented the mélange concept in his history of the evolution of the models on the Franciscan Complex in California (Hsü, 1985, especially p. 57f.). Interestingly, it was another Swiss geologist, Héli Badoux (1911–2001: Fig. 52), who told Hsü that what he thought he had invented had been invented already by Bailey and McCallien in 1950 and that it had been called mélange (Hsü, 1985, p. 57). But in the 1985 publication, Hsü's story is incomplete. I once had a long chat with Ken in Zurich about independent discoveries and the importance of the evidence in science. Later, in June 2001, he told me that our previous conversation in Zurich had made him think back and he had remembered that he had forgotten an episode in his own mélange adventure when he wrote his 1985 paper. He told me about it and I asked him to put it on paper. Below, I give in full Ken's letter to me relating the missing bits in his mélange discovery (written on June 11, 2002):

> I discovered the mélange phenomenon in 1963, as I wrote previously in my publications, when I first went to the beach to cool down, after several days of utter frustration. Then the evidence was clear. My knowledge of rock mechanics with [*David*] Griggs and [*John*] Handin was the key to my understanding. John had just published some papers on the differential ductility of sandstones and shales. The broken formations are a typical manifestation of the different rock-mechanics behaviours. It is not big a step to postulate their mixing with exotic components to produce mélanges.

Figure 51. Kenneth Jinghwa Hsü (1929–). (Photo courtesy of Prof. Ken Hsü.)

Figure 52. Héli Badoux (1911–2001: copied from *Bulletin des Laboratories de Géologie, Minéralogie, Géophysique et du Musée Géologique de l'Université de Lausanne*, no. 219, frontispiece).

I always thought that I would like to have a label. I translated the descriptive term exotic boulder beds into Spanish and used that in my first Shell report of 1963. After I went to Riverside [*University of California, Riverside*] in 1964, my West Coast friends all expressed an interest to see the structures in the field. I finally arranged in the early summer of 1966 a trip for John Crowell, then at UCLA, Cliff Hopson, then at UC Santa Barbara and Bob Garrison. Crowell asked me if he could bring Héli Badoux along. I had come to know him in 1964 when I was detained by the French immigration, because I had tried to leave the country without a visa (having entered France through mountain-trails). Badoux had to talk to the border police for more than half an hour, and perhaps had to give a small bribe before I was rescued. I had, therefore, no problem to welcome Badoux as my guest.

Our first day was spent on the coast near Morro Bay. There were some red rocks which no American geologist knew what they were. I had had the rock thin-sectioned, but could not determine petrographically their mineralogy. I had then received an X-ray report: the rocks consisted exclusively of antigorite. Funny, isn't it? Badoux, with his experience in the tropics, took one look, and said that it was a serpentinite. There were pelagic cherts and limestones, and pillow lavas around, the serpentine thus completed the third member of Steinmann's Trinity.

For some reason, I went to bed early that evening. John, Cliff, and Bob stayed up late and discussed with Badoux all evening about what they saw. Next morning, Bob Garrison told me at breakfast that he had witnessed at an historical occasion when the term mélange was invented for the Franciscan.

John Crowell then told me of the evening's discussions, and advised me to adopt the term. I remember that he said, a good merchandise should have a good label. The trade mark mélange was just what I needed to sell my ideas to the public.

Only then, did I recall my trip with Ernst Kündig [*1901–1981*] in 1959 or 1960. Kündig had worked many years for Shell to look for oil in "eugeosynclinal terranes." [*see Kuendig, 1959*]. When he was about to retire as Shell's Chief Geologist, he received, as his golden handshake, a fulfillment of his lifelong dream to come to California to see the Franciscan, and to get a sample of a glaucophane schist for his friend and mentor Professor Scheumann. Nobody in Shell knew anything about the Franciscan. I was one of the few who had come from the West Coast. They gave me a week or ten days to contact the people at Menlo Park and to find out where to find good outcrops. I went to reconnaitre in April, and the Kündigs arrived in May. We flew then to LA [*Los Angeles*], rented a car and drove up to the Morro Bay country.

Those were the days when everybody was afraid to go to the beaches. We made only roadside stops. Kündig saw immediately the similarity of the Franciscan rocks to those in the Ankara Mélange, pointing out to me the exotic blocks of radiolarian chert in the meadows on both sides of the highway.

After several days on the coast, we ended up at Big Sur. I remember the evening at a fish restaurant, when Kündig and I argued all evening. He was talking about the eugeosyncline, and was wondering about the missing miogeosyncline. I was presenting my idea that the Franciscan is thrust over the flysch of the Great Valley sequence from the west. (I now think that the Franciscan was underthrust under the forearc basin sediments.) I thought then that I could do some palaeo-current measurements to verify my hypothesis. Kündig may or may not have used the word mélange for the Franciscan, but he must have talked about the Ankara Mélange in Anatolia. I do not remember whether or not he mentioned Bailey and McCallien. I myself did not read Bailey and McCallien until after I left Shell in 1963. In 1964, there was an AGI [*American Geological Institute*] summer Institite to the Apennines. John Maxwell and his Italian friends took us through the argille scagliose [*"scaly shales"*] everywhere. Those were the heydays of the gravity-sliding theory [*see, for example, Maxwell, 1959; Page, 1963*]. The argille scagliose and the Ankara mélange were assigned the same origin of having slid down slope in a deep sea basin. I have always been impressed by the phenomenon of penetrative shearing. It is not pure shear, but simple shear. It is not 50 or 75% like deformed oolite, the shear strain must have been many times more than unity. The shear cannot be caused by compression, the shear must have been caused by the displacement of an overlying (or underlying) slab. The only mechanism which could do the job, in my opinion of those days, had to be an allochthon of very large dimensions. Such an allochthon, after Hubbert and Rubey, 1959, could only have been a gravity-sliding body. I compared, therefore, the Franciscan with the argille scagliose, and compared the origin of Franciscan with the bottom moraine of glaciers. I was troubled, of course, because moraines are not thick deposits, but the Franciscan was supposedly "25,000 feet thick."

I first presented my idea on the Franciscan structures at the Wegmann Symposium in the Spring of 1966 [*Hsü, 1967*]. My structural cross-section clearly shows that I understood the pervasively sheared nature of the mélanges. It is noteworthy that the Italian literature on argille scagliose was referred to, but Bailey and McCallien's Ankara Mélange article was not cited by me in the article published in 1967. The manuscript was probably written before the 1966 trip with Badoux. The Bailey and McCallien articles were cited in my 1968 (Principles of Mélanges in GSA [*Hsü, 1968*]) and 1969 (with Ohrbom) articles [*Hsü and Ohrbom, 1969*]. I believe that I did not read the Greenly memoir until after I took the trip with Badoux, either in Riverside during the 1966/67 Winter Semester, or more likely in the Spring of 1967 when I was writing the manuscript for GSA, at about the same time when I also read Bailey and McCallien. In the early 1970s, I went with Robert Schackleton to Angelsey. Robert, I believe, was Bailey's student, or his assistant at the Geological Survey (or both). He told me that Bailey worked at Angelsey before the War, and he was very familiar with Greenly's idea. (Shackleton knew about mélanges too, but he had difficulty envisioning the shearing mechanism in rock deformation.) It was thus not surprising that Bailey recognized after the War the Ankara rocks as mélange at the first sight (B and McC 1950 [*Bailey and McCallien, 1950a*]). Kündig may have used the term mélanges, but he was a Swiss oil man and he probably did not know of Greenly's work. I did not pay much attention to Kündig's 1959 or 1960 comparison of the Franciscan with the Ankara Mélange, when I had not yet had a chance to study the Franciscan. After I started working on the Franciscan in 1963, I remembered Kündig, but forgot the Ankara Mélange. I "re-invented the wheel" independently, but I was taught the word mélange by Badoux in 1966, and did not appreciate the meaning of the word until I read Greenly in 1967. In a way, Greenly's memoir was a retrodiction of my postulate of a deformation mechanism of rocks that are called mélanges. Going back to the problem of rediscoveries. I think that McCallien may have worked independently without foreknowledge and that he may or may not have recognized the Ankara rocks are mélanges. [*Hsü did not know the story told in this paper when he wrote this, so he could not be sure that McCallien and Erol indeed independently discovered mélanges.*] Bailey recognized the mélanges, but he certainly did not make a rediscovery, because he had been familiar with Greenly's work at Angelsey. I did make an independent rediscovery, but my use of the term was inspired by Badoux, who was familiar with the argille scagliose which had been compared with the Ankara Mélange. The postulate of Franciscan as subduction mélanges was made, as far as I knew, perhaps by Ben Page to explain the odd relation between the Franciscan and the Great Valley rocks. I would like to give credit, however, to Warren Hamilton as being the first to come out explicitly in print, with his GSA article, which preceded my AGU article (written in 1970 and published in 1971), and which was cited by me in that article.

This turns out to be a rather interesting little story. I believe that I have given everyone their due credit. Don't hesitate to write me again if you need further clarifications.

It is clear from the above (and from his earlier accounts in 1985 and 1990) that Hsü had agonized over the Franciscan rocks much like Erol and McCallien had done over the Ankara rocks. He finally came up, independently of either Greenly or Bailey and McCallien, with the idea that he was looking at a tectonic jumble, essentially at a mega-fault breccia, but initially mixed in a submarine slide environment. Once he conceived the process that had created the rocks he was looking at, he was able to map them in an intelligible way and developed a methodology of mapping (Hsü, 1968, 1990; see Fig. 53), which was essentially the same as Greenly's. Hsü (1968) developed the following five rules of mélanges:

Rule 1. The making of a geologic map of a mélange terrane cannot be based upon a presumption of stratal continuity.

Rule 2. The stratigraphy of a mélange sequence cannot be established on a presumption of normal superposition.

Rule 3. The assignment of a time-range of deposition to all the rocks in a mélange on the basis of the oldest and youngest fossils found in such a mélange is wrong.

Rule 4. The contact between a rock-stratigraphic unit and a subjacent or superjacent mélange may be a depositional contact or a dislocation contact.

Rule 5. The rock stratigraphic unit overlying a mélange can be autochthonous in one place and allochthonous elsewhere.

Hsü then wrote the following concerning mapping mélanges:

Current practices in mapping chaotic terranes identify rock-stratigraphic units as map units. Sedimentary sequences in allochthonous

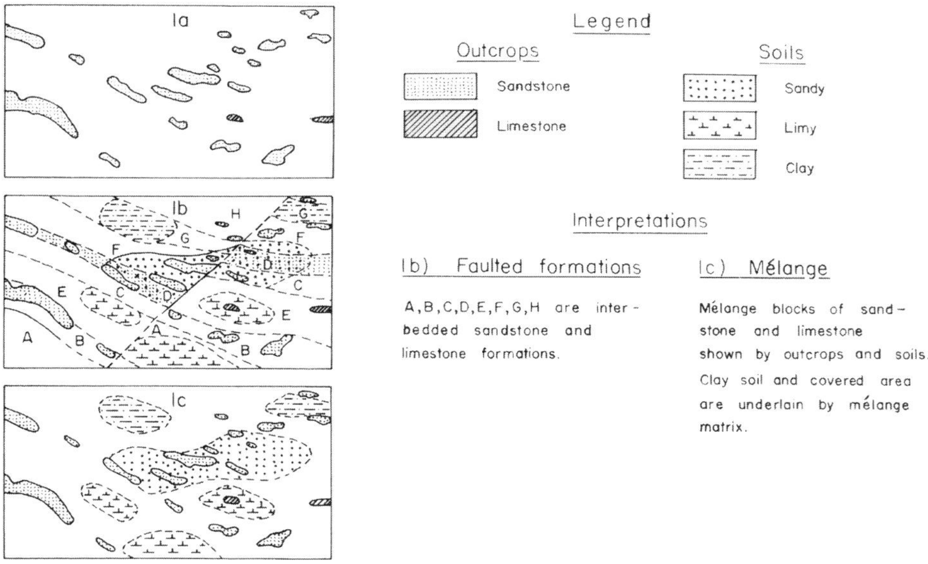

Figure 53. Hsü's figures to explain mélange mapping: two different interpretative geologic maps. 1a: outcrop map; 1b: interpretative map, based upon a presumption of stratal continuity; 1c: interpretative map, based upon a recognition of mélange.

slabs, with original stratal continuity preserved, are rock-stratigraphic units. Large detached blocks such as serpentinites, greenstones, cherts, and schists, are also mapped as units of distinct lithology. However, small tectonic inclusions and matrix must be mapped together as an undifferentiated mélange (Fig. 1c [*herein Fig. 53*]). This practice results in the elimination of a vast amount of information, as laboriously collected data on the nature and distribution of unmappable blocks cannot be represented on such a geological map.

A large mélange body is variable geographically. Some parts of the mélange include only greywacke fragments in a pelitic matrix, while others include a variety of rock types such as cherts, pillow lavas, and glaucophane schists. Some parts are completely metamorphosed; others include metamorphics only as blocks in an unmetamorphosed matrix. Some exotic blocks are not disturbed internally; others represent a part of a fragmented older mélange. The differing natures and origins of the blocks and matrix can, therefore, be made the criteria for the subdivision of a large body of mélange into several mélange units. These units can serve as a foundation for describing the regional variations in lithology, lithologic associations, deformational styles, and degree of metamorphism, and for studying the chronology of sedimentation, deformation, and metamorphism of a chaotic terrane. In fact, mélange mapping has been practiced for many years by Alpine geologists, who recognized different "Schuppen Zones" and "Wildflysch Nappes," based upon the stratigraphy of the blocks and the age of the matrix.

The author has been puzzling for some time about the problem of mapping the Franciscan rocks of the Coast Ranges. During the field season of 1963 the author followed the conventional practice of mapping lithologic units in the Santa Lucia Range. The result was not significantly different from that compiled by Jennings (1959) on the basis of the mapping by Taliaferro and his students. The apparently erratic distribution of rocks of different lithologies offers no clues to the structure of the region except that these rocks are probably mélange blocks (Hsü, 1967b [*herein, Hsü, 1967*], fig. 5). During the past two years the author has attempted a subdivision of the Coast Range Franciscan, on the assumption that these rocks constitute mélanges. Map units depicting the regional variations of the relative abundance of different blocks were recognized. These units are tectono-stratigraphic units that have been interpreted as originally subhorizontal sheets of tectonic breccia, which were later folded into antiforms or synforms. The preliminary results of our mapping have been presented orally in the Meetings of the Geological Society (Hsü, 1966b [*herein Hsü, 1966*]; Ohrborn and Hsü, 1967), and the details will be presented in two forthcoming articles. (Hsü, 1968, p. 1071–1072).

Let us remember that Greenly was able to understand the Mona mélanges because of his experience in the Northwest Highlands of Scotland, where fault rocks had been mapped in great detail after the recognition of large-scale thrusting and Matley's earlier work, also inspired by the Highlands discoveries. In other words, Greenly first recognized (read: hypothesized) *a process* and then sorted out the structure. Bailey and McCallien simply used Greenly's hypothesis and experience to explain Erol's and McCallien's observations and deductions. Hsü, having been educated in the more parochial (and more devoutly Baconian) U.S. environment of the 1950s, had to reinvent all that through much agonizing, only to be told at the end by an Alpine geologist that he had reinvented the wheel (see Hsü, 1985, 1990, and the letter quoted above).

But the great importance of the Franciscan mélanges came when it was realized that they represented the sweep of an ocean floor for at least 1000 km (Hamilton, 1969; Dewey and Bird, 1970; Dickinson, 1970; Hsü, 1971). Such fault displacements hitherto had not been recognized. Nor had structural rocks of the volume of the Franciscan Complex been recognized before (Şengör and Sakınç, 2001). With the recognition of this new process, namely subduction, mélanges acquired great significance as indicators of past subduction, and began to be recognized the world over.

CONCLUSIONS

Mélanges remain difficult objects to study, even just to map. The geologist has psychological difficulty in lumping into one unit such diverse rock types as are observed in them. That is why Oğuz Erol, William J. McCallien, and Ken Hsü have recorded their agony in trying to distinguish units in mélange complexes. I remember, during an excursion to the Kargı Massif in northern Turkey with my colleague Professor Okan Tüysüz, my great surprise in seeing that he had mapped individual units in what to me seemed a large mélange complex. The terrain we were visiting was Okan's dissertation area (see Tüysüz, 1990). It was the first piece of serious mapping that he had done alone. He had not previously been shown a mélange terrain. The area was metamorphic and lithologic changes occurred on a size that seemed possible to map at 1:25,000 reconnaissance scale. So he did what all discoverers of mélanges had done initially: he began mapping units that could be entered on his map. He was lucky (or unlucky) that metamorphism blurred lithologic contrasts and that the mélange components included large enough chunks to be mapped. Thus, he was not confronted with the frustration that Erol, McCallien, and Hsü said they had experienced and that Greenly and Noble also must have experienced. My recognition of Okan's mélange was possible because I had previously seen many mélanges in many diverse settings.

What this story taught me was that a scientific discovery often requires not only evidence, but *clear* evidence. One may indeed make a discovery with no evidence at all; but to test it and to certify it as a discovery, one cannot do without evidence. Okan's evidence for mélange was not clear enough and could only be recognized if one had seen clearer examples earlier. Even now, I cannot be sure I am right without having the area remapped at a larger scale. But, the important thing here is that whether the area is a mélange terrain is independent of Okan's or my whim. It is the observations that will decide it. Once the area is mapped at a scale larger than reconnaissance, it would be extraordinary if both of us continued to maintain our present positions unchanged.

All of the discoverers of mélanges had been taught that rocks occurred either in layers or in intrusive bodies. That is what they expected to see when they started mapping. They were frustrated when the rocks they were mapping occurred in neither form. They struggled hard to make the evidence fit their prejudices. McCallien's insistence that Erol should find a stratigraphic succession to use as a basis for his further mapping and Erol's repeated failure to do so reflect this struggle. Hsü clearly grappled with the same problem that had stumped earlier Franciscan workers and had even given rise to the notorious Franciscan-Knoxville paradox![40] In Turkey, McCallien's resolve to go out with his student and to show him how to do what he recommended reflects the depth of his conviction that rocks behaved in the way he had been taught and had so far found. His admittance that Erol's failure was not Erol's fault, but rather was the fault of what they both had been taught before, is a sign of McCallien's greatness as an observer. The same is true for Hsü. Both recognized that they were looking at something that neither had ever encountered before. The subsequent agony of trying to figure out what they were looking at is typical of most scientific discoveries: it is usually a time of rapid hypothesis generation and discarding. Erol recalls how he and McCallien would sit and think, come up with possibilities and go back to the field to check them. Finally, they realized that they were looking at a collection of blocks of diverse rock types embedded in two dominant types of matrices. They also noticed that the mixing was a consequence of some kind of deformation, for the matrices were highly sheared. Hsü repeated their experience almost two decades later.

The more experienced Bailey thought that this sort of thing resulted from shearing along giant thrusts. Bailey's hasty but bold attempt at interpreting the entire tectonics of Anatolia on the basis of Erol's and McCallien's discovery proved wrong, but not unfruitful. It prompted Ketin to test it. In the process Ketin discovered that the Central Anatolian massifs were young structures, which led to a wholesale revision of our ideas on the structure of Anatolia. Ketin submitted his results to the 1956 symposium in honor of the 80th birthday of Hans Stille in Germany (Lotze, 1956). Ketin's refutation of Stille's interpretation of the structure of Anatolia angered the old master and his paper was not accepted for the final Festschrift volume (İhsan Ketin, personal commun., 1980s), but this rejection did not stop the victory of Ketin's ideas. His paper appeared elsewhere and eventually triumphed over Kober's and Stille's ideas of the bilateral structural symmetry of the Anatolian orogenic belt.

Thus, in Turkey, the rediscovery of mélanges triggered a whole, previously unforeseen and unexpected series of discoveries. They all came about because evidence clashed with some previously accepted hypothesis. In some cases, the evidence came by surprise, as it did when Erol and McCallien started their work around Ankara and became baffled for a long time. In other cases, it was gathered with the express intention of testing a new model, as Ketin did near Yozgat to check Bailey and McCallien's thrust hypothesis.

In the Amargosa Range, the situation was similar to the one in Turkey and in western California. The great difficulty and in places the sheer impossibility of mapping units in what Noble eventually ended up calling "chaos structure" led him to recognize the chaotically mixed nature of the rocks. He was working in isolation from the European developments, but the language he used was amazingly similar to what Greenly, McCallien, Erol, and Bailey used. Two decades later, Ken Hsü was doing the same in front of the Franciscan outcrops in complete ignorance of all of his predecessors. It was the evidence and the geologists' careful observations of it that led them to develop similar ideas. When Hsü was told of his predecessors (including the Alpine ones!), he immediately recognized in their work the identity of the situation to his own.

None of this would have been possible if the mélange concept were the individual whim of Greenly, Noble, McCallien,

Erol, Bailey, and Hsü. That objective evidence does not have a dominant role in science and that science is incapable of progress are ideas that are decisively refuted by what we know of the particular case of the repeated discovery of mélanges.

A final note still concerns the power of theory, however: Alpine, Carpathian, and Himalayan workers (all of them of "Alpine" schooling) also saw mélanges and recognized their highly deformed, chaotic nature. But when the mélanges in these mountain ranges were first sighted, there was no theory to allow the geologist to assume vast amounts of simple shear. The previous knowledge of Alpine geologists of exotic blocks in sedimentary rocks, essentially in conglomerates, led them to think that mélange was a sort of mega-conglomerate. (As in many cases, with his incredible intuition, Eduard Suess remains an exception in this case, too.) Because the evidence was permissive of this interpretation, the mega-conglomerate interpretation has survived the discovery of far-traveled nappes and even subduction zones! Even after plate tectonics, many geologists continued seeing in mélanges sedimentary products, only later tectonized. "Olistostrome versus tectonic mélange" arguments rage to this day concerning many mountain belts. It is a truism that mélanges include all sorts of diamictites (the environment of mélange generation is conducive of diamictite formation of all sorts), but this does not mean that mélanges are primarily diamictites (see Hsü, 1974). The most powerful argument against sedimentary interpretation of mélanges is their immense along-strike persistence. This was obvious even to the Alpine-Carpathian geologists early on; the blocky flysches were seen to go from the western end of the Alps to the eastern end of the Carpathians along the very same paleotectonic zones. Many finally had to admit that tectonics had had much to do with their origin. What prevented them from seeing its *purely* tectonic nature was a complex net of reasons mainly involving sedimentary biases that had grown from a diverse set of circumstances, as we saw above. When Edward Greenly, a geologist with a different set of biases, first mapped a mélange, he interpreted its origin differently. Mélange origin remained disputed so long as observations on very large shear strains remained absent. As soon as these were made, mélanges were more readily admitted as tectonic products. The reason why old arguments have survived in many places is that critical observations in most mélange complexes to distinguish a diamictite from a purely tectonic jumble remain very difficult, in some cases simply impossible, to make. *Here too, it is the nature of the evidence that makes difficulties, not necessarily the subjective whims of geologists.*

I end this historical note by reiterating a suggestion pertaining to modern studies, which has become impressed upon me during my review of the history of discovery of mélanges in the Alpine-Carpathian System: the tectonic nature of mélanges is best betrayed by their along-strike persistence for thousands of kilometers along narrow paleotectonic zones. Diamictite sedimentation in no environment, not even along the aprons of continental margins, displays such along-strike continuity as mélanges do, with the sort of exotic block variety, embracing entire sections down to and including the crystalline basement and yet remaining remarkably uniform in its overall composition for very large distances. Gravity-driven debris flow interpretation may be upheld for limited areas in a mélange, but not for mélange belts thousands of kilometers long along the strike.

ACKNOWLEDGMENTS

I thank Professors A. Oğuz Erol and Kenneth J. Hsü for sharing their experiences and memories with me. Erol also helped me to resurrect his wonderful dissertation map, the first serious piece of geological mapping ever done in Turkey at a scale of 1: 25,000. In addition to its great value as a historical monument, it is still useful, as the area had not been mapped subsequently with the same care and detail. Mr. Colin McCallien, W.J. McCallien's son, sent me a great amount of material concerning his wonderful father, and shared with me his memories of the historic 1950 field excursion with Sir Edward Bailey. Mr. Kemâl Şerbetçi also talked to me at length about his memories concerning his beloved teacher McCallien, and sent the pictures published in this paper. Professor Donald B. McIntyre also sent me material on McCallien. My old Alpine friends have provided such great and willing help in gathering material on the Alps and the Carpathians as to remind me again the great privilege of being a part of the "Alpine brotherhood": my knowledge of Kaufmann I owe to Dr. Paul Felber. Prof. Dr. Adrian Pfiffner kindly found for me the portrait of his illustrious predecessor, Isidor Bachmann. Dr. Sybille Franks very kindly found material on Früh and Bachmann. Prof. Dr. Rudolf Trümpy, teacher and friend of us all, provided much information (most importantly, he put me on Kaufmann's track) and provided connections. Hofrat Dr. Tillfried Cernajsek, the scholarly head librarian and historian of the Geologischen Bundesanstalt in Vienna, found for me the references and citations of von Morlot and Hohenegger. In Potsdam, Germany, Dr. Anke Friedrich enabled me to generate some of my figures. In Cambridge, England, the resourceful librarian of the Department of Earth sciences, Ruth Banger, found for me Greenly's portrait. In Istanbul, I am grateful to Professor Okan Tüysüz for informing me of the most recent Polish Carpathian literature. In the United States, Professor Brian Wernicke introduced me to the concept of chaos structure and showed it to me in the field under his own incomparable guidance. Professor Yıldırım Dilek deserves credit for inviting this contribution and waiting for it so patiently. I am grateful to the Geological Society of America for printing Erol's map at its original scale.

I must end with a note of gratitude to W.J. "Mac" McCallien himself, who succumbed to cancer in 1981, before I could meet him. In the five years that he spent in my country, he left an indelible mark on it. He initiated a tradition of meticulous outcrop mapping and infused a great enthusiasm for geology into his students. All Turkish geologists are indebted to him, whether they know it or not. It is my hope that this paper will underscore his greatness as a geologist, as a teacher, and, as Professor Erol says, as a human being.

REFERENCES CITED

Amberg, B. and Bachmann, J., 1893, Dr. Franz Josef Kaufmann. Professor und Naturforscher—Sein Leben und Seine Werke: Festschrift der Kantonsschule Luzern, 1893, p. 263–317.

Anderson, F.M., 1933, Knoxville-Shasta succession in California: Geological Society of America Bulletin, v. 44, p. 1237–1270.

Anderson, F.M., 1938, Lower Cretaceous Deposits in California and Oregon: Boulder, Colorado, Geological Society of America Special Paper 16, 339 p.

Andrusov, D., 1931, Étude Géologique de la Zone des Klippes Internes des Carpathes Occidentales, I[ière] Partie: Introduction, IIième Partie: Stratigraphie (Trias et Lias): Rozpravy Státniho Geologického Ústavu Čekoslovenské Republiky, v. VI, Prag, 167 p. + 9 plates + 1 geological map.

Andrusov, D., 1938, Étude Géologique de la Zone des Klippes Internes des Carpathes Occidentales, IIIe Partie: Tectonique: Rozpravy Státniho Geologického Ústavu Čekoslovenské Republiky, v. IX, Prag, 135 p. + 18 plates + 1 folded, colored geological map.

Andrusov, D., 1964, Geologie der Tschechoslowakischen Karpaten, v. I: Berlin, Akademie-Verlag, and Bratislava, Verlag der Slowakischen Akademie der Wissenschaften, 263 p.

Andrusov, D., 1965, Geologie der Tschechoslowakischen Karpaten, v. II: Berlin, Akademie-Verlag, and Bratislava, Verlag der Slowakischen Akademie der Wissenschaften, 443 p.

Andrusov and Schiebner, E., 1960, An outline of the present state of knowledge about the geology of the Klippen Belt between R. Vlara and T. Tvrdosin: Geologicky Sbornik (Bratislava), v. 11, p. 239–280.

Anonymous, 1889, Vorwort: in Die Europäische Türkei von Ami Boué (La Turquie de l'Europe par A. Boué. Paris, 1840.) Deutsch Herausgegeben von der Boué-Stiftungs-Commission der Kaiserlichen Akademie der Wissenschaften in Wien. I. Band: Wien, Tempsky, p. III–X.

Anonymous, 1903, Mr. C.L. Griesbach, C.I.E., F.G.S., Director of the Geological Survey of India: Geological Magazine, v. 40, p. 287–288.

Anonymous, 1907, Carl Ludolf Griesbach, C.I.E., F.G.S., born December 11, 1847; died April 13, 1907: Geological Magazine, v. 44, p. 240.

Anonymous, 1965, Maden Tetkik ve Arama Enstitüsü Dergisi Makaleler Bibliyografyası (Bibliography of the Articles in the M.T.A. Bulletins) 1936–1964: Ankara, Maden Tetkik ve Arama Enstitüsü Yayınlarından (Publications of Mineral Research and Exploration Institute), Sayı (number), 1–63, [III] + 51 p.

Anonymous, 1968, Guide to Excursion No C44 Poland Geology of the Polish Flysch Carpathians: International Geological Congress, XXIII Session, Prague 1968: Published by the Geological Institute in the publishing house "Wydawnictwa Geologiczne," 73 p.

Anonymous, 1971, Maden Tetkik ve Arama Enstitüsü 1935–1970: Mineral Research and Exploration Institute of Turkey, Ankara, 20 p.

Arbenz, P., 1933, Die Rolle der Alpenforschung in der Geologie—Rektoratsrede gehalten an der 98. Stiftungsfaier der Universität Bern am 3 Dezember 1932: Paul Haupt, Bern, 21 p.

Argand, E., 1924, La tectonique de l'Asie: Congrés Géologiques International, Comptes Rendus de la XIIme session, Premier Fascicule, H. vaillant-Carmanne, Liége, p. 171–372.

Bachmann, I., 1863, Ueber petrefactenreiche exotische Jurablöcke im Flysch des Sihlthals und Toggenburg: Vierteljahrsschrift der Naturforschenden Gesellschaft in Zürich, v. 8, p.1–34.

Bailey, E.B. (Sir), 1935, Tectonic Essays Mainly Alpine: Clarendon, Oxford, xii + 200 p. + 1 foldout map.

Bailey, E.B. (Sir), 1936, Sedimentation in relation to tectonics: Bulletin of the Geological Society of America, v. 47, p. 1713–1726.

Bailey, E.B. (Sir), 1944, Mountains that have travelled over volcanoes: Nature, v. 154, p. 752–756.

Bailey, E.B. (Sir), 1951, Dr. Edward Greenly: Nature, v. 167, p. 545–546.

Bailey, E.B. (Sir), and McCallien, W.J., 1950a, The Ankara Mélange and the Anatolian Thrust. Nature, 166, p. 938.

Bailey, E.B. (Sir), and McCallien, W.J., 1950b. Ankara Melanjı ve Anadolu Şaryajı (The Ankara Mélange and the Anatolian Thrust). Maden Tetkik ve Arama Enstitüsü Mecmuası, no. 40: 12–16; English version p. 17–21.[41]

Bailey, E.B. (Sir), and McCallien, W.J., 1953. Serpentine lavas, the Ankara Mélange and the Anatolian Thrust: Transactions of the Royal Society of Edinburgh, v. 62, part II (no. 11), p. 403–442, with eight photographic plates and two maps.

Bailey, E.B. (Sir), and McCallien, J.W., 1960, Some aspects of the Steinmann trinity, mainly chemical: Quarterly Journal of the Geological Society of London, v. 116, p. 365–395.

Bailey, E.B. (Sir), and McCallien, W.J., 1961, Structure of the Northern Apennines: Nature, v. 191, p. 1136–1137.

Bailey, E.B. (Sir), and McCallien, W.J., 1963, Liguria Nappe: Northern Apennines: Transactions of the Royal Society of Edinburgh, v. 65, p. 315–333.

Bailey, E.B. (Sir), and Weir, J., 1939, Introduction to Geology, illustrations by W.J. McCallien: MacMillan and Co., London, xii + 498 p.

Bailey, E.H., Irwin, W.P., and Jones, D.L., 1964, Franciscan and related rocks and their significance in the geology of western California: California Division of Mines and Geology Bulletin, v. 83, 177 p.

Bak, K., 1995, Trace fossils and ichnofabrics in the Upper Cretaceous red deep-water marly deposits of the Pieniny Klippen Belt, Polish Carpathians: Annales Societatis Geologorum Poloniae, v. 64, p. 81–97.

Bak, K., 1998, Planctonic foraminiferal biostratigraphy, Upper Cretaceous red pelagic deposits, Pieniny Klippen Belt, Carpathians: Studia Geologica Polonica, v. 111, p. 7–92.

Bak, M., 1996a, Late Cretaceous Radiolaria from the Czorsztyn Succession, Pieniny Klippen Belt, Polish Carpathians: Studia Geologica Polonica, v. 109, p. 69–85.

Bak, M., 1996b, Cretaceous Radiolaria from Niedzica Succession of the Pieniny Klippen Belt in Polish Carpathians: Acta Palaeontologica Polonica, v. 41, p. 91–110.

Bayer, A., 1982, Untersuchungen im Habkern-Mélange ("Wildflysch") zwischen Aare und Rhein: Mitteilungen aus dem Geologischen Institut der Eidg. Technischen Hochschule und der Universität Zürich, Neue Folge, Nr. 240, 327 p.

Beck, P., 1911, Beiträge zur Geologie der Thunerseegebirge mit besonderer Berücksichtigung des Sigriswilgrates, des Beatenberges und des Harders: Beiträge zur Geologischen Karte der Schweiz, neue Folge, 29. Lieferung, des Gesamtwerkes 59. Lieferung, X + 100 p. + VIII plates.

Beck, P., 1912, Die Niesen-Habkerndecke und ihre Verbreitung im helvetischen Faciesgebiet. Mit einem Beitrag zur Stratigraphie des Niesen von Ed. Gerber: Eclogae Geologicae Helvetiae, v. 12, p. 65–147 + 4 foldout plates.

Bernal, M., 1987, Black Athena—The Afroasiatic Roots of Classical Civilization, v. I: The Fabrication of Ancient Greece 1785–1985: Rutgers University Press, New Brunswick, xxxii + 575 p.

Bernal, M., 1991, Black Athena—The Afroasiatic Roots of Classical Civilization, v. II: The Archaeological and Documentary Evidence: Rutgers University Press, New Brunswick, xxxiii + 736 p.

Biely, A., and Bystricky, J., 1968, Guide to Excursion 15 AC Czechoslovakia Mesozoic of the Inner West Carpathians and the Klippen Belt: International Geological Congress, XXIII Session, Prague 1968, Geological Survey of Czechoslovakia in the Publishing House of the Slovak Academy of Sciences, Bratislava, 44 p.

Birembaut, A., 1981, Boué, Ami, *in* Gillespie, C.C., editor-in-chief, Dictionary of Scientific Biography, v. 2: New York, Charles Scribner's Sons, p. 341–342.

Birkenmajer, K., 1960, Geology of the Pieniny Klippen Belt of Poland (a review of latest researches): Jahrbuch der geologischen Bundesanstalt, v. 103, p. 103, p. 1–36.

Birkenmajer, K., 1977, Jurassic and Cretaceous lithostratigraphic units of the Pieniny Klippen Belt, Carpathians, Poland: Studia Geologica Polonica, v. 45, p. 7–159.

Birkenmajer, K. and Jednorowska, A., 1987, Late Cretaceous foraminiferal stratigraphy of the Pieniny Klippen Belt (Carpathians, Poland): Studia Geologica Polonica, v. 92, p. 4–28.

Bloor, D., 1991, Knowledge and Social Imagery, 2nd edition: University of Chicago Press, Chicago, 203 p.

Blumenthal, M., 1946, Die neue geologische Karte der Türkei und einige ihrer stratigraphisch-tektonischen Grundzüge: Eclogae Geologicae Helvetiae, v. 39, p. 277–289.

Boucek, B., 1968, Guide to Excursion 30 AC Czechoslovakia Significant Paleontological Localities in Czechoslovakia: International Geological Congress, XXIII Session, Prague 1968: Geological Survey of Czechoslovakia in Academia, Praha, 43 p.

Boué, A., 1830, Resumé des observations sur l'âge relatif des depots secondaires dans les Alpes et les Carpathes: Journal de Géologie, v. 1, p. 50–86 and 115–151, plates 2–6.

Bozkurt, E., 2001, Neotectonics of Turkey—a synthesis: Geodinamica Acta, v. 14, p. 3–30.

Bradley, W.H., 1966, Memorial to Levi Fatzinger Noble (1882–1965): Proceedings of the Geological Society of America, 1965, p. P49-P52.

Brinkmann, R., 1981, Türkiye Yerbilimleri Bibliyografyası Geowissenschaftliche Bibliographie der Türkei Geoscience Bibliography of Turkey 1825–1975, part I, Foreign Geoscience literature on Turkey: Türkiye Bilimsel ve Teknik Araştırma Kurumu yayınları no. 486, TÜRDOK Bibliyografya Serisi no. 32, Ankara, XII + 492 p.

Brinkmann, R., 1984, Türkiye Yerbilimleri Bibliyografyası Geowissenschaftliche Bibliographie der Türkei Geoscience Bibliography of Turkey 1976–1980, part I, Foreign Geoscience literature on Turkey: Türkiye Bilimsel ve Teknik Araştırma Kurumu Yayınları no. 585, TÜRDOK Bibliyografya Serisi no. 41, Ankara, XII + 207 p.

von Buch, L. (Baron), 1801, Aus einem Briefe: Neue Schriften der Gesellschaft naturforschender Freunde zu Berlin, v. 3, p. 586–587. (Reprinted in: Ewald, J., Roth, J., and Dames, W., eds., 1867, Leopold von Buch's Gesammelte Schriften, v. I: G. Reimer, Berlin, p. 130–131).

Burg, J.-P., 1983, Tectogenèse Compareée de Deux Segments de Chaînes de Collision: Le Sud du Tibet (Suture du Tsangpo), La Chaîne Hercynienne en Europe (Sutures du Massif Central): Thèse présentée à l'Université des Sciences et Techniques du Languedoc pour obtenir le grade de Docteur d'Etat, mention Sciences, [V] + 361 + [7]p. + foldout map.

Buxtorf, A., 1918, Ueber die tektonische Stellung der Schlieren- und Niesen-Flyschmasse: Verhandlungen der naturforschenden Gesellschaft Basel, v. 29, p. 270–275.

Çadırcı, M., and Süslü, A., 1982, Ankara Üniversitesi Gelişim Tarihi: Ankara Üniversitesi Rektörlüğü Yayınları Yayın No: 82, Tanıtıcı Yayınlar Dizisi: 2 (Atatürk'ün 100. Doğum Yılına Armağan), Ankara, X + 469 p.

Cadisch, J., 1953, Geologie der Schweizer Alpen, zweite Auflage verfasst unter Mitarbeit von E. Niggli: Wepf & Co., Basel, XI + 480 p.

Cadisch, J., Leupold, W., Eugster, H. and Brauchli, R., 1919, Geologische Untersuchungen in Mittelbünden: Vierteljahrsschrift der Naturforschenden Gesellschaft in Zürich, v. 66 (Heim Festschrift), p. 359–417 + one foldout map.

Çakmak, İ.T., Baydar, O. and Karaköse, C., 1985, Türkiye'de jeoloji Araştırmalarının tarihçesi ve MTA Genel Müdürlüğü'nün jeoloji çalışmalarındaki yeri, *in*, Erden, M.E., and Göncüoğlu, M.C., eds., Maden Tetkik ve Arama Genel Müdürlüğü 50. Yıl Simpozyumu Bildirileri: MTA Genel Müdürlüğü, Ankara, p. 1–5.

Carlé, W.E.H., 1988, Werner–Beyrich–von Koenen–Stille. Ein geistiger Stammbaum wegweisender Geologen: Geologisches Jahrbuch, v. A108, p. 3–499.

von Carnall, R., 1835, Die Sprünge im Steinkohlengebirge: Karsten's Archiv für Mineralogie, Geognosie, Bergbau und Hüttenkunde, v. 9, XIV + 204 p.

Carozzi, A.V., 1981, Saussure, Horace Bénédict de: *in* Gillispie, C.C., editor-in-chief, Dictionary of Scientific Biography, v. 12: Charles Scribner's Sons, New York, p. 119–123.

Carozzi, A.V., 1987, La géologie—De l'histoire de la Terre selon le récit de Moïse aux premiers essays sur la structure des Alpes et à la géologie experimentale 1778–1878, *in* Trembley, J., ed., Les Savants Genèvois dans l'Europe Intellectualle du XVIIe au milieu du XIXe siècle, Éditions du Journal de Genève, Genève, p. 203–265.

Carozzi, A.V., 2000, Manuscripts and Publications of Horace-Bénédict de Saussure on the Origin of Basalt (1722–1797)/Manuscrits et Publications de Horace-Bénédict de Saussure sur l'Origine du Basalte (1772–1797): Editions Zoe, Carouge-Genève, XVIII + 769 p.

Carozzi, A.V., Crettaz, B. and Ripoll, D., editors, 1998, Les Plis du Temps—Mythe, Science et H.-B. de Saussure: Nouveaux Itinéraires Amoudruz, 5, Collections Payot Amoudruz, Musée d'Ethnographie, Annexe de Conches, Genève, Conservatoire d'Art et d'Historie de Haute-Savoie, Annecy, 367 p.

Carozzi, A.V., and Newman, J.K., 2003, Lectures on Physical Geography Given in 1775 by Horace-Bénédict de Saussure at the Academy of Geneva (Cours de Géographie Physique Donné en 1775 par Horace-Bénédict de Saussure a l'Academie de Genève: Editions Zoé, Carouge/Genève, XXXII + 528 p.

Chaput, E., 1931, Esquisse de l'Evolution Tectonique de la Turquie, traduction en Turc par Hamit Nafiz—Türkiyenin Tektonik Tarihçesine Umumî Bir Bakış, Tercüme eden: Hamit Nafiz: İstanbul Darülfünunu Geologie Enstitüsü Neşriyatından, no. 6, 107 p.

Chaput, E., 1936, Voyages d'Étude Géologiques et Géomorphogeniques en Turquie: Mémoires de l'Institut Français d'Archéologie de Stamboul II, E. de Boccard, Paris, VIII + 312 p. + 27 photographic plates.

Chaput, E., İbrahim Hakkı, Hamit Nafiz, Ahmet Malik and Muhsin Adil, [1930]. Carte Géologique de la Région d'Angora, 1:135,000. [İstanbul Darülfünunu Geologie Enstitüsü Neşriyatı], one sheet.

Clark, R.W., 1971, Einstein—The Life and Times: World Publishing, New York, xv + 718 p. + 16 unnumbered photographic plates.

Cloos, M., 1982, Flow mélanges: numerical modeling and geologic constraints on their origin in the Franciscan Subduction Complex: Geological Society of America Bulletin, v. 93, p. 330–345.

Cloos, M., 1984, Flow mélanges and the structural evolution of accretionary wedges: in Raymond, L.A., ed., Mélanges: Their Nature, Origin, and Significance: Boulder, Colorado, Geological Society of America Special Paper 198, p. 71–79.

Collins, H., 1981, Stages in the empirical program of relativism: Social Studies of Science, v. 11, p. 3–10.

Cuvier, G., 1861, Recueil des Élogues Historiques Lus dans les Séances Publiques de l'Institut de France ..., nouvelle edition: Firmin Didot Frères, Fils et Cie., Paris, lix + 411 + [1] p.

Daubrée, A., 1879, Études Synthetiques de Géologie Expérimentale: Dunod, Paris, III + 828 p.

Daubrée, A., 1888, Les Régions Invisibles du Globe et des Espaces Célestes—Eaux Souterraines, Tremblements de Terre, Météorites: Bibliothèque Scientifique Internationale, v. LXVII, Félix Alcan, Paris, [I] + 202 p.

De Long, S.E., Dewey, J.F., and Fox, P.J., 1977, Displacement history of oceanic fracture zones: Geology, v. 5, 199–202.

De Long, S.E., Dewey, J.F., and Fox, P.J., 1979, Topographic and geologic evolution of fracture zones: Journal of the Geological Society of London, v. 136, 303–310.

Dewey, J.F., and Bird, J.M., 1970, Mountain belts and the new global tectonics: Journal of Geophysical Research, v. 75, p. 2625–2647.

Dickinson, W.R., 1970, Second Penrose Conference: The new global tectonics: Geotimes, v. 15, no. 4, p. 18–22.

Diener, C., 1895, Ergebnisse einer Geologischen Expedition in den Central-Himalaya von Johar, Hundes, und Pain-Khanda: Denkschriften der Akademie der Wissenschaften, mathematisch naturwissenschaftliche Klasse, v. 62, p. 533–608, plates I–VIII.

Diener, C., 1898, Notes on the geological structure of the Chitichum Region: Memoires of the Geological Survey of India, v. 28, p. 1–27.

Diener, C., 1901, Zur Erinnerung an Albrecht Krafft von Dellmensingen: Jahrbuch der kaiserlichen und königlichen Reichsanstalt (Wien), v. 51, p. 149–158.

Diener, C., 1916, Die marinen Reiche der Triasperiode: Denkschriften der Akademie der Wissenschaften, mathematisch naturwissenschaftliche Klasse, v. 92, p. 405–549 – 1 folded map.

Dobretsov, N.L. and Kirdyashkin, A.G., 1992. Subduction zone dynamics: models of an accretionary wedge: Ofioliti, v. 17(I), p. 155–164.

Doğu, A.F., 1981, Ankara Üniversitesi Dil ve Tarih-Coğrafya Fakültesi Coğrafya yayınları bibliyografyası: Coğrafya Araştırmaları Dergisi, no. 10, p. 167–197.

Drosdowski, G., ed., 1995, Brockhaus Enzyklopädie, neunzehnte, völlig neu bearbeitete Auflage, v. 26, deutsches Wörterbuch (A-Glub): F.A. Brockhaus, Mannheim, XXIV + 1369 p.

Ebel, F.G., 1808, Ueber den Bau der Erde in den Alpen-Gebirge zwischen 12 Längen- und 2–4 Breitengraden nebst einigen Betrachtungen über die Gebirge und den Bau der Erde Überhaupt, zweyter Band: Orell Füssli und Compagnie, X + 428 p.

Einstein, A., 1981, Mein Weltbild: Ullstein, Frankfurt a. M., 210 S.

Eiszmann, L., 1974, Die Begründung der Inlandeistheorie für Norddeutschland durch den Schweizer Adolph von Morlot im Jahre 1844: Abhandlungen und Berichte des Naturkundlichen Museums "Mauritianum" Altenburg, v. 8, p. 289–318.

Erol, O., 1949, Ankara Güneydoğusundaki Elma Dağı ve Çevresinin Jeoloji ve Jeomorfolojisi Üzerinde Bir Araştırma [Ph.D. dissertation]: Dil ve Tarih-Coğrafya Fakültesi, Fizikî Coğrafya ve Jeoloji Kürsüsü, Ankara, 84 + VI p. + 3 maps and one plate of figure.

Erol, O., 1956, Ankara Güneydoğusundaki Elma Dağı ve Çevresinin Jeoloji ve Jeomorfolojisi Üzerinde Bir Araştırma: Maden Tetkik ve Arama Enstitüsü Yayınlarından, Seri D, No. 9, [2] + 99 p. + 8 photographic plates + foldout maps.

Erol, O., 1981, "Ankara Melanjı": deyiminin tarihçesi: İç Anadolu Jeolojisi Simpozyumu, Ankara Melanjı Paneli, 12 Mart 1981, Türkiye Jeoloji Kurumu 35: Bilimsel ve Teknik Kurultayı Türkiye Jeoloji Kurumu, Ankara, p. 32–34.

Erol, O., 1984, Türkiye Yerbilimleri Bibliyografyası Geowissenschaftliche Bibliographie der Türkei Geoscience Bibliography of Turkey 1976–1980, part II.i, Geoscience literature published in Turkey: Türkiye Bilimsel ve

Teknik Araştırma Kurumu yayınları no. 585, TÜRDOK Bibliyografya Serisi no. 41, Ankara, XII + 207 p.
Evans, R.J., 1997, In Defense of History: Granta, London, vii + 307 p.
Ewald, J., 1867, Leopold von Buch's Leben und Wirken bis zum Jahre 1806, *in* Ewald, J., Roth, J. und Eck, H., eds., Leopold von Buch's Gesammelte Schriften, erster Band, Georg Reimer, Berlin, p. V-XLVIII.
Felber, P., 2003, Kaufmann, Franz Josef, *in* Historisches Lexicon der Schweiz, (in press).
Feyerabend, P., 1988, Against Method, revised edition: Verso, London, viii + 296 p.
Feyerabend, P., 1991, Three Dialogues Concerning Knowledge: Basil Blackwell, Oxford, 167 p.
Franks, S., Trümpy, R., and Auf der Maur, J., 2000, Aus der Frühzeit der alpinen Geologie: Johann Gottfried Ebels Versuch einer Synthese: Veröffentlichung der Naturforschenden Gesellschaft in Zürich im Anschluss an den Jahrgang 145 der Vierteljahrsschrift der Naturforschenden Gesellschaft in Zürich, 68 p. + 1 colored plate.
Freeman, K., 1948, Ancilla to the Pre-Socratic Philosophers: Basil Blackwell, Oxford, x + 162 p.
Freshfield, D.W., 1920, The Life of Horace Benedict de Saussure: Arnold, London, xii + 479 p. + 2 maps.
Früh, J.J., 1888, Beiträge zur Kenntniss der Nagelfluh der Schweiz: Denkschriften der schweizerischen naturforschenden Gesellschaft, v. 30, 203p. + IV plates.
Gansser, A., 1955, New aspects of the geology in central Iran: Proceedings of the Fourth World Petroleum Congress, section I/A/5, paper 2, p. 279–300.
Gansser, A., 1959, Ausseralpine Ophiolithprobleme: Eclogae Geologicae Helvetiae, v. 52, p. 659–680.
Gansser, A., 1964. Geology of the Himalayas: Wiley Interscience, New York, xv + [i] + 289 p. + 4 foldout plates.
Gansser, A., 1974, The ophiolite mélange: A world-wide problem on Tethyan examples: Eclogae Geologicae Helvetiae, v. 67, p. 479–507.
Gansser, A., 1977, The great suture zone between Himalaya and Tibet—a preliminary account: in Jest, C., ed., Colloques Internationaux du C.N.R.S., No. 268, Écologie et Géologie de l'Himalaya, Editions du CNRS, Paris, p. 181–191 + 1 folded plate.
Geikie, A., 1875, Life of Sir Roderick I. Murchison .... Based on His Journals and Letters with notes of his scientific contemporaries and a sketch of the rise and growth of Palæozoic geology in Britain, v. II: John Murray, London, vi + [i] + 375 p.
Gilbert, G.K., 1875, Report upon the geology of portions of Nevada, Utah, California, and Arizona, examined in the years 1871 and 1872, *in* Report upon Geographical and Geological Explorations and Surveys West of the One Hundredth Meridian, in Charge of First Lieut. Geo. M. Wheeler..., v. III.—Geology: Government Printing Office, Washington, D.C., p. 17–187.
Gökten, E., Özaksoy, V. and Karakuş, K., 1996, Tertiary volcanic and tectonic evolution of the Ayaş-Güdül-Çeltikçi region, Turkey: International Geology Review, v. 38, p. 926–934.
Görür, N., Oktay, F.Y., Seymen, İ., and Şengör, A.M.C., 1984, Palaeotectonic evolution of the Tuzgölü basin complex, central Turkey: sedimentary record of a Neo-Tethyan closure, *in* Dixon, J.E., and Robertson, A.H.F., eds., Geological Evolution of the Eastern Mediterranean: Geological Society of London Special Publication 17, p. 467–482.
Gos, C., 1935, Voyage de Saussure Hors des Alpes: Victor Attinger, Neuchâtel-Paris, 144 p. + 8 photographic plates.
Greene, M.T., 1982, Geology in the Nineteenth Century: Changing Views of a Changing World: Ithaca, New York, Cornell University Press, 324 p.
Greenly, E., 1919a, The Geology of Anglesey, v. I. Memoirs of the Geological Survey: Her Majesty's Stationary Office, London, xl + 388 p. + XXVI plates.
Greenly, E., 1919b, The Geology of Anglesey, v. II. Memoirs of the Geological Survey: Her Majesty's Stationary Office, London, p. 390–980 + plates XXVII-LX and 17 folding plates.
Greenly, E., 1922, A short summary of the geological history of Anglesey: Transactions of the Anglesey Antiquarian Society and Field Club, 1922, 20 p. + 2 plates.
Greenly, E., 1928–1932, Benjamin Neeve Peach: a study: Transactions of the Edinburgh Geological Society, v. 12, p. 1–12.
Greenly, E., 1938, A Hand through Time: Memories—Romantic and Geological: Studies in the Arts and Religion and the Grounds in Confidence in Immortality: Murby, London, vol. 1 (xvi + 378 p. + plates frontisp. + II-LX), v. 2 (ix + p. 379–774 + plates frontis. + LXII + LXXV).
Gregory, J.W., 1896, The Great Rift Valley—Being the Narrative of a Journey to Mount Kenya and Lake Baringo with some Account of the Geology, Natural History, Anthropology, and Future Prospects of British East Africa: John Murray, London, xxi + 422 p.
Griesbach, C.L., 1891, Geology of the Central Himálayas: Memoires of the Geological Survey of India, v. 23, xi + 232 + xix p. + 27 plates + 2 maps.
Griesbach, C.L., 1893, Notes on the Central Himálayas: Records of the Geological Survey of India: v. 26, p. 19–25.
Griesbach, C.L., 1904, On the exotic blocks of the Himálayas: Congrès Géologique International Compte Rendu de la IX. Session, Vienne 1903, Deuxième Fascicule, [Imprimerie Hollinek Frères], Vienne, p. 547–552.
Gross, P.R., and Levitt, N., 1994, Higher Superstition: The Academic Left and Its Quarrels with Science: The Johns Hopkins University Press, Baltimore, [ii] + 314 p.
Gross, P.R., Levitt, N., and Lewis, M.W., editors, 1996, The Flight from Science and Reason: The New York Academy of Sciences, New York, xi + 593 p.
Günther, S., 1900, A. v. Humboldt L. v. Buch: Ernst Hofmann & Co., Berlin, [III] + 271 p.
Hamilton, W., 1969, Mesozoic California and the underflow of Pacific mantle: Geological Society of America Bulletin, v. 80, p. 2409–2430.
Von Hauer, F., 1867, Nachruf für Adolf von Morlot: Verhandlungen der kaiserlich und königlichen Geologischen Reichsanstalt, 1867, p. 69–70.
Haug, E., 1899, Les regions dites exotiques du versant nord des Alpes Suisses—réponse au Dr. Hans Schardt: Bulletin de la Société vaudouise des Sciences naturelles, v. 35, p. 114–161.
Haug, E., 1904, Les grands charriages de l'Embrunais et de l'Ubaye: Congrès Géologique International Compte Rendu de la IX. Session, Vienne 1903, Premier Fascicule, Imprimerie Hollinek Frères, Vienne, p. 493–506.
Heim, Alb., 1919, Geologie der Schweiz, v. I Molasseland und Juragebirge: Chr. Herm. Tauchnitz, Leipzig, XX + 704 p. + 31 plates.
Heim, Alb., 1921, Geologie der Schweiz, v. II Die Schweizer Alpen, erste Hälfte: Chr. Herm. Tauchnitz, Leipzig, XI + 476 p. + 27 plates.
Heim, Alb., 1922a, Geologie der Schweiz, v. II Die Schweizer Alpen, zweite Hälfte: Chr. Herm. Tauchnitz, Leipzig, XXVI p. + p. 477–1018 + 17 plates.
Heim, Alb., 1922b, Die Mythen: Neujahrsblatt herausgegeben von der Naturforschenden Gesellschaft in Zürich auf das Jahr 1922, Beer & Cie., Zürich, 28 p.
Heim, Alb., 1929, An der Erkenntniswurzel alpiner Tektonik (Geologische Nachlese no. 29): Vierteljahrsschrift der Naturforschenden Gesellschaft in Zürich, v. 74, p. 213–223.
Heim, Alb., 1932, Bergsturz und Menschenleben: Fretz & Wasmuth, Zürich, 218 p. + 5 plates.
Heim, Arn., 1907, Zur Frage der exotischen Blöcke im Flysch: Eclogae Geologicae Helvetiae, v. 9, p. 413–424.
Heim, Arn., and Gansser, A., 1939, Central Himalaya—Geological Observations of the Swiss Expedition 1936: Denkschriften der Schweizerischen Naturforschenden gesellschaft, v. 73, Memoir 1, p. XVI + 245 p. + Atlas
Hennig, E., 1934, Das Ringen um Erkenntnis des Alpenbaus und seine Bedeutung für geologische Weltanschauung: Natur und Volk, v. 64, p. 291–302 and 342–353.
Herb, R., 1966, Leimernschichten: *in* Lexique Stratigraphique International, v. I Europe (ed. P. Pruvost), Fascicule 7 Suisse (ed. R.F. Rutsch), Fascicule 7c Alpes Suisses et Tessin Méridional/Schweizeralpen und Südtessin I, Congrès Géologique International — Commission de Stratigraphie, Centre National de la Recherche Scientifique, Paris, p. 686–694.
Hermann, B.F., 1801, Mineralogische Reisen in Sibirien. Vom Jahr 1783 bis 1796. Dritter Theil: Bei der kaiserlichen Akademie der Wissenschaften, St. Petersburg, [4] + 312 p. + 6 copper-plates and foldout tables VIII–XIV.
Von Hingenau, O., (Freiherr), 1864, Ludwig Hohenegger: Jahrbuch der kaiserlich und königlichen Geologischen Reichsanstalt (Wien), v. 14, p. 449–453.
Hoheneger[*sic*], L., 1849, Briefe über Schlesien an W. Haidinger und M. Hörnes: Berichte über Mittheilungen von Freunden der Naturwissenschaften in Wien, v. 6, p. 106–316.
Hohenegger, L., 1852, Geognostische Skizze der Nordkarpathen von Schlesien und den nächsten Angränzungen (nach dem gegenwärtigen Standpuncte miener Erfahrungen): Jahrbuch der kaiserlich und königlichen Geologischen Reichsanstalt (Wien), v. 3, p. 135–148.
Hohenegger, L., 1861, Die Geognostischen Verhältnisse der Nordkarpathen in Schlesien und den angrenzenden Theilen von Mähren und Galizien als Erläuterung zu der Geognostischen Karte der Nordkarpathen: Justus Perthes, Gotha, [I] + 50 p. + foldout table of sections.

Hsü, K.J., 1966, Mélange concept and its application to the interpretation of the California Coast Range geology: Geological Society of America Abstracts for 1965, p. 82.

Hsü, K.J., 1967. Mesozoic geology of the California Coast Ranges—A new working hypothesis: in Etages Tectoniques—Colloque de Neuchatel 18–21 Avril 1966, Institut de Géologie de l'Universite de Neuchâtel, A la Baconnière, Neuchâtel, p. 279–296.

Hsü, K.J., 1968. Principles of mélanges and their bearing on the Franciscan-Knoxville paradox: Geological Society America Bulletin, v. 79, p. 1063–1074.

Hsü, K.J., 1971, Franciscan mélanges as a model for eugeosynclinal sedimentation and underthrusting tectonics: Journal of Geophysical Research, v. 76 p. 1162–1170.

Hsü, K.J., 1974, Mélanges and their distinction from olistostromes: *in* Dott, R.H., Jr., and Shaver, R.H., eds., Modern and Ancient Geosynclinal Sedimentation: Society for Economic Paleontologists and Mineralogists Special Publication 19 p. 321–333.

Hsü, K., 1985. A basement of mélanges: A personal account of the circumstances leading to the breakthrough in Franciscan research: Boulder, Colorado, Geological Society of America Centennial Special Volume 1, p. 47–64.

Hsü, K.J., 1988, Mélange and mélange tectonics of Taiwan: Proceedings of the Geological Society of China, v. 31, no. 2, p. 87–91.

Hsü, K.J., 1990, Mélanges and non-Smithian stratigraphy: Current Contents, no. 26 (June 25), p. 24.

Hsü, K.J., 1995, The Geology of Switzerland—An Introduction to Tectonic Facies: Princeton University Press, Princeton, xxv + 250 p.

Hsü, K.J., and Ohrbohm, R., 1969, Mélanges of San Francisco Peninsula: geologic reinterpretation of type Franciscan: American Association of Petroleum Geologists Bulletin, v. 53, p. 1348–1367.

Imhof, E., 1937/1938, Prof. Dr. Jakob Früh (1852–1938): Mitteilungen der Geographisch-Ethnographischen Gesellschaft Zürich, v. 38, p. 9–16.

Jeannet, A., 1934, Excursion no. 57, Einsiedeln-Iberg-Mythen, *in* Geologischer Führer der Schweiz herausgegeben von der schweizerischen Geologischen Gesellschaft bei Anlaß ihrer 50sten Jahresfeier, Fascicule X Centralschweiz Excursionen Nr. 53–60, B. Wepf & Cie, Basel, p. 750–755.

Jenings, C.W., 1959, Geologic Map of California, San Luis Obispo sheet: California Division of Mines.

Jeoloji Şubesi, 1944a, Türkiye Jeolojik Hartası "İstanbul" Paftası İzahnamesi/ Notice Explicatives de la Carte Géologique de la Turquie Feuille "İstanbul": M.T.A. Enstitüsü tarafından tertip ve neşredilmiştir, Çankaya Matbaası, Ankara, each of the bilingual texts 44 p. + 2 foldout maps.

Jeoloji Şubesi, 1944b, Türkiye Jeolojik Hartası "İzmir" Paftası İzahnamesi/ Notice Explicatives de la Carte Géologique de la Turquie Feuille "İstanbul": M.T.A. Enstitüsü tarafından tertip ve neşredilmiştir, Çankaya Matbaası, Ankara, bilingual texts 31 and 32 p. + 2 foldout maps.

Kaufmann, F.J., 1871, Ueber die Granite von Habkern: Verhandlungen der kaiserlichen und königlichen geologischen Reichsanstalt (Wien), Nr. 14, p. 263–266.

Kaufmann, F.J., 1879, Über Granite des Habkerenthales: Verhandlungen der schweizerischen naturforschenden Gesellschaft (1878), p. 92–94.

Kaufmann, F.J., 1886, Emmen- und Schlierengegenden nebst Umgebungen bis zur Brünigstrasse und Linie Lungern-Grafenort: Beiträge zur Geologischen Karte der Schweiz, nr. 24, VIII + 608 p.; Book of plates: 12 p. + XXVIII plates.

Keferstein, C., 1828, Beobachtungen und Ansichten über die geognostischen Verhältnisse der nördlichen Kalk-Alpenkette in Oesterreich und Baiern: Teutschland Geognostisch-Geologisch Dargestellet, v. 5, no. 3, p. 425–570.

Ketin, İ., 1956a, Yozgat bölgesinin jeolojisi ve Orta Anadolu masifinin tektonik durumu (On the geology of the Yozgat region and the tectonic features of the central Anatolian massif (Kırşehir crystallines)): Türkiye Jeoloji Kurumu Bülteni, v. 6, p. 1–40 (English summary p. 29–39).

Ketin, İ., 1956b, Über Alter und Art der kristallinen Gesteine und Erzlagerstätten in Zentral-Anatolien: Berg- und Hüttenmännische Monatshefte, Jg 104, p. 163–169.

Ketin, İ., 1973, Yerbiliminin 50 yıllık cumhuriyet dönemindeki gelişmesi: Bilim ve Teknik, v. 6, no. 72 (special issue), p. 38–42.

Ketin, İ., 1975, Yerbiliminin 50 yıllık cumhuriyet dönemindeki gelişmesi: in, Congress of Earth Sciences on the occasion of the 50th Anniversary of Turkish Republic 17–19 December 1973 Papers: Mineral Research and Exploration Institute, Ankara, p. XV-XX.

Ketin, İ., 1979, Türkiye'de jeoloji araştırmalarının ve jeoloji haritalarının kısa tarihçesi: Yeryuvarı ve İnsan, v. 4, no. 2, p. 15–17.

Ketin, İ., 1981, Yerbilimlerinin Cumhuriyet dönemindeki gelişmesi: Bilim ve Teknik, v. 14, no. 168, p. 13–16.

Ketin, İ., 1985. Türkiye'nin bindirmeli-naplı yapısında yeni gelişmeler ve bir örnek: Uludağ Masifi, *in* Ercan, T., and Çağlayan, M.A., eds., Ketin Simpozyumu: Türkiye Jeoloji Kurumu, Ankara, p. 19–36.

Kettner, R., 1958, Hans Schardt (1858–1931): Casopis pro Mineralogii a Geologii, v. 3, p. 380–381.

Koertge, N., editor, 1998, A House Built on Sand—Exposing Postmodernist Myths About Science: Oxford University Press, New York, xi + 322 p.

Koráb, T., Lesko, B., and Marschalko, R., 1968, Guide to Excursion 17 AC Czechoslovakia Inner-Carpathian and outer Flysch of the East Slovakia: International Geological Congress, XXIII Session, Prague 1968, Geological Survey of Czechoslovakia in the Publishing House of the Slovak Academy of Sciences, Bratislava, 38 p.

von Krafft, A., 1902, Notes on the "Exotic Blocks" of Malla Johar in the Bhot Mahals of Kumaon: Memoires of the Geological Survey of India, v. 32, p. 127–183, plates 1–14.

Kuendig, E., Eugeosynclines as potential oil habitats: Proceedings of the Fifth World Petroleum Congress, Section I, P. 461–479.

Kuhn, T.S., 1970, The Structure of Scientific Revolutions, second edition, enlarged: International Encyclopedia of Unified Science, Foundations of the Unity of Science, v. II, nr. 2, The University of Chicago Press, Chicago, XII + 210 p.

Kuhn, T.S., 2000, The Road Since Structure—Philosophical Essays, 1970–1993, with an Autobiographical Interview (Conant, J., and Haugeland, J., eds.): The University of Chicago Press, Chicago, viii + 335 p.

Lahn, E., 1948, Türkiye Jeolojik Bibliyografyası Bibliographie Géologique de la Turquie: Türkiye Jeoloji Kurumu Bülteni Bulletin of the Geological Society of Turkey, v. 1, p. 95–135.

Lapworth, C., 1883, The secret of the Highlands: Geological Magazine, new series, decade II, v. 10, p. 120–128.

Lapworth, C., 1885, The Highland controversy in British geology: its causes, courses and consequences. Nature, v. 32, p. 558–559.1

Latour, B., and Woolgar, S., 1986, Laboratory Life: Princeton University Press, Princeton, 296 p.

Lefkowitz, M., 1996, Not Out of Africa—How Afrocentricism became an Excuse to Teach Myth as History: Basic Books, A Division of HarperCollins Publishers, New York, xvii + 222 p.

Lefkowitz, M.R., and Rogers, G.M. editors, 1996, Black Athena Revisited: Chapel Hill, London, xxi + 522 p.

Le Pichon, X., Henry, P., and Lallement, S., 1993, Accretion and erosion in subduction zones: the role of fluids. Annual Review of Earth and Planetary Sciences, v. 21, p. 307–331.

Leupold, W., 1966, Flysch, helvetischer …, *in* Lexique Stratigraphique International, v. I Europe (ed. P. Pruvost), Fascicule 7 Suisse (ed. R.F. Rutsch), Fascicule 7c Alpes Suisses et Tessin Méridional/Schweizeralpen und Südtessin I, Congrès Géologique International—Commission de Stratigraphie: Centre National de la Recherche Scientifique, Paris, p. 386–414.

Lotze, F., editor, 1956, Geotektonisches Symposium zu Ehren von Hans Stille: Kommissions-Verlag von Ferdinand Enke, Stuttgart, XX + 483 p.

Lugeon, M., 1903, Les nappes de recouvrement de la Tatra et l'origine des Klippes des Carpathes: Bulletin des Laboratoires de Géologie, Géographie Physique, Minéralgie et Paléontologie de l'université de Lausanne (Suisse), no. 4, 51 p. (reprint from Bulletin de la Société vaudouise des Sciences naturelles, v. 39).

Lugeon, M., 1904, Les grands nappes de recouvrement des Alpes suisses: Congrès Géologique International Compte Rendu de la IX. Session, Vienne 1903, Premier Fascicule: Imprimerie Hollinek Frères, Vienne, p. 477–492.

Lugeon, M., 1916, Sur l'origine des blocks exotiques du flysch préalpin: Eclogae Geologicae Helvetiae, v. 14, p. 217–221.

Lugeon, M., 1940, Emile Argand: Bulletin de la Société Neuchâteloise des Sciences Naturelles, v. 65, p. 25–53 + Portrait.

McCallien, W.J., 1935, Dating the Ice Age in Britain: Science Progress, v. 117, p. 67–71.

McCallien, W.J., 1936a, Rhu (Row) Point: A readvance moraine: Transactions of the Geological Society, Glasgow, v. 19, p. 385–389.

McCallien, W.J., 1936b, A note on Dalradian pillow lavas, Strabane, County Tyrone: Proceedings of the Royal Irish Academy, v. 43B, p. 13–22.

McCallien, W.J., 1937, Late-Glacial and early Post-Glacial Scotland: Proceedings of the Society of Antiquaries of Scotland, v. 66, p. 174–206.

McCallien, W.J., 1950 Some Turkish pillow lavas: Türkiye Jeoloji Kurumu Bülteni, v. 2, p. 3–8 + 5 photographic plates + 1 plate of drawings; English summary on p. 1–2.

MacDonald, J.G., 1982, William John McCallien, B.Sc., D.Sc. (Glasgow): Royal Society of Edinburgh Yearbook, 1982, p. 49–50.

MacGregor, A.G., 1968, Memorial to Sir Edward B. Bailey, Kt., M.C., F.R.S. (1881–1965): Proceedings Volume, Geological Society of America, Inc., for 1966, p. 155–163.

Marcovich, M., 1967, Heraclitus—Greek Text with a Short Commentary: The Los Andes University Press, Merida, xxix + 665 p.

de Margerie, E., 1946, Three stages in the evolution of Alpine geology: de Saussure—Studer—Heim: Quarterly Journal of the Geological Society of London, v. 102, p. xcvii–cxiv.

de Margerie, E., 1947, Trois moments dans l'Histoire de l'étude des Alpes Suisses: de Saussure, Studer, Heim: Bulletin de la Section de Géographie du Comité des Travaux Historiques et Scientifiques, v. 58 (1943), p. 1–39.

Marschalko, R., 1979, Cretaceous flysch of the Pieninian Klippen Belt and the decay of the crust in the Tatride foreland, *in* Vanek, J., ed., Geodynamic Investigations in Czechoslovakia, Final Report, Veda Publishing House of the Slovak Academy of Sciences, Bratislava, p. 243–251.

Marschalko, R. and Kysela, J., 1980, Geológia a sedimentológia bradlového pásma a manínskej jednotky medzi Zilinou a Povazskou Bystricou: Západné Karpaty, séria Geológia 6, p. 7–79.

Masson, H., 1976, Un siècle de géologie des Préalpes: de la découverte des nappes à la recherché de leur dynamique: Eclogae Geologicae Helvetiae, v. 69, p. 527–575.

Masson, H., 1983, La géologie en Suisse de 1882 à 1932: Eclogae Geologicae Helvetiae, v. 76, p. 47–64.

Matley, C.A., 1913, The geology of the Bardsey Island. Quarterly Journal of the Geological Society of London, v. 69, p. 514–533.

Maxwell, J.C., 1959, Turbidites, tectonics and gravity transport, northern Apennine Mountins, Italy: Bulletion of the American Association of Petroleum Geologists, v. 43, p. 2701–2719.

Misik, M., 1979, Sedimentologicke a mikrofacialne studium jury bradla vrsateckeho hradu (neptunicke dajky, biohermny vyvoj oxfordu): Západné Karpaty, séria Geológia 5, p. 7–56 + XX photographic plates.

Misik, M., and Sykora, M., 1981, Pieninsky exoticky chrbat rekonstruovany z valunov karbonatovych hornin kriedovych zlepencov bradloveho pasma a maninskej jednotky: Západné Karpaty, séria Geológia 7, p. 7–111 + XXVIII photographic plates.

von Morlot, A., 1847, Erläuterungen zur geologischen Übersichtskarte der Nordöstlichen Alpen. Ein Entwurf zur vorzunehmenden Bearbeitung der physikalischen Geographie und Geologie ihres Gebietes: Braumüller und Seidel, Wien, 208 + 2 p. + Plate of sections.

Murawski, H., 1983, Geologisches Wörterbuch, 8. Auflage: Ferdinand Enke Verlag, Stuttgart, 281 p.

Murchison, R.I. (Sir), 1849, On the geological structure of the Alps, Apennines and Carpathians, more especially to prove a transition from Secondary to Tertiary rocks, and the development of Tertiary deposits in Southern Europe: Quarterly Journal of the Geoological Society of London, v. 5, p. 157–312 + plate VII.

Nieuwenkamp, W., 1981, Buch, (Christian) Leopold von, *in* Gillispie, C.C., editor-in-chief, Dictionary of Scientific Biography: Charles Scribner's Sons, New York, v. 2, p. 552–557.

Noble, L.F., 1941, Structural features of the Virgin Spring area, Death Valley, California: Bulletin of the Geological Society of America, v. 52, p. 941–1000.

Norman, T., 1993, Türkiye Cumhuriyeti'nin 70. yılında jeoloji ve jeoloji mühendisliği, *in* Cumhuriyetin 70, Yılında Türkiye'de Bilim II, Bilim ve Teknik dergisinin ücretsiz özel eki, p. 108–111.

Nowack, E., 1928, Die wichtigsten Ergebnisse meiner anatolischen Reisen: Zeitschrift der Deutschen Geologischen Gesellschaft, v. 80, p. B304–B312.

Oldroyd, D.R., 1990, The Highlands Controversy—Constructing Geological Knowledge through Fieldwork in Nineteenth-Century Britain: The University of Chicago Press, Chicago, ix + 438 p.

Page, B.M., 1963, Gravity tectonics near Passo della Cisa, northern Apennines, Italy: Geological Society of America Bulletin, v. 74, p. 655–672.

Pamir, H.N., 1979, Türkiye'de yerbilimleri öğretiminin tarihçesi: Yeryuvarı ve İnsan, v. 4, no. 4, p. 6–10.

Pesl, V., Salaj, J., and Vass, D., 1968, Guide to Excursion 6 AC Czechoslovakia The Flysch and Klippen Belts, Neogene Basins of West Carpathians: International Geological Congress, XXIII Session, Prague 1968, Geological Survey of Czechoslovakia in the Publishing House of the Slovak Academy of Sciences, Bratislava,40 p.

Pilger, A., 1978, Die tektonische Erforschung der Alpen zwischen 1787 und 1915: Clausthaler Geologische Abhandlungen, v. 32, V + 81 p.

Popper, K.R., 1935, Logik der Forschung: Springer Verlag, Wien, vi + 248 p.

Popper, K.R., 1957, The Poverty of Historicism: Routledge & Kegan Paul, London, xiv + 166 p.

Popper, K.R., 1966, The Open Society and Its Enemies, v. 2, The High Tide of Prophecy: Hegel, Marx, and the Aftermath, fifth, revised edition: Princeton University Press, Princeton, v + [i] + 420 p.

Popper, K.R., 1980, The Logic of Scientific Discovery, 10th edition: Unwin Hyman, London, 480 p.

Popper, K.R., 1983, Realism and the Aim of Science: Rowman and Littlefield, Totowa, xxxix + 420 + [3] p.

Prey, S., 1975, Neue Forschungsergebnisse über Bau und Stellung der Klippen-Zone des Lainzer Tiergartens in Wien (Österreich), *in* Mahel, M., ed., Tectonic Problems of the Alpine System (Collection dedicated to the 75th birthday of Academician Professor RNDr. D. Andrusov, DrSc.): Veda, Publishing House of the Slovak Academy of Sciences, Bratislava, p. 57–68.

Rabowski, F., 1925, Les Klippes et leurs rôle dans la chaîne carpathique: Sprawozdania Polskiego Instytutu Geologcznego, v. 3, no. 1–2, p. 277–306.

Saleeby, J., 1977, Fracture zone tectonics, continental margin fragmentation, and emplacement of the Kings-Kaweah ophiolite belt, southwest Sierra Nevada, California: *in* Coleman, R.G., and Irwin, W.P., eds., North American Ophiolites, State of Oregon, Department of Geology and Mineral Industries, Bulletin 95, p. 141–159.

Salomon-Calvi, W., 1936, Die Fortsetzung der Tonalelinie in Kleinasien: Yüksek Ziraat Enstitüsü Çalışmalarından, no. 9, p. 11–13.

Salomon-Calvi, W., 1940, Kurze Übersicht über den tektonishen Bau Anatoliens: Maden Tetkik ve Arama Enstitüsü Mecmuası, no. 18, p. 35–74.

Sarjeant, W.A.S., 1980, Geologists and the History of Geology—An International Bibliography from the Origins to 1978, v. II: New York, Arno Press, p. 377–1499.

de Saussure, H.-B., 1796, Voyages dans les Alpes, precedes d'un essai sur l'histoire naturelle des environs de Geneve, tome quatrieme: Louis Fauche-Borel, Neauchatel, [II] + 594 + 2 p.

Schaer, J.-P., 1994, Géologues et géologie en pays neuchâtelois dans la seconde moitié du XIXe siècle, *in* Histoire de l'Université de Neuchâtel, v. 2 (La Seconde Académie 1866–1909), Université de Neuchâtel avec la collaboration des Editions Gilles Attinger à Hauterive, p. 381–442.

Schardt, H., 1893, Sur l'origine des Alpes du Chablais et du Stockhorn, en Savoie et en Suisse: Comptes-Rendus hébdomadaires de l'Academie des Sciences, Paris, v. 117, p. 707–709, errata on p. 874.

Schardt, H., 1898, Les regions exotiques du versant nord des Alpes suisses (Préalpes du Chablais et du Stockhorn et les Klippes)—Leurs relations avec l'origine des blocs et bréches exotiques et la formation du flysch: Bulletin de la Société vaudouise des Sciences naturelles, v. 34, p. 113–218 + 1 colored foldout (plate I).

Schardt, H., 1899, Les Préalpes Romandes (Zone du Stockhorn-Chablais.)—Un Problème de Géologie Alpine—Leçon d'Ouverture du Cours de Géologie Prononcée en Séance Publique le 4 Juin 1897, a l'Aula de l'Academie de Neuchâtel: Bulletin de la Société Neuchâteloise de Géographie, v. 11, p. 1–28 + one colored foldout of cross-sections.

Schardt, H., 1900, Encore les regions exotiques—répliques aux attaques de M. Emile Haug: Bulletin de la Société vaudouise des Sciences naturelles, v. 36, p. 147–169.

Schaub, H., 1967, Exkursion Nr. 17a Habkernmulde: in Nabholz, W., ed., Geologischer Führer der Schweiz, Heft 4 Berner Jura (Basel-Neuchâtel-Biel), Berner Mittelland, berner Oberland (ohne Grimsel, Brünig und Simmental) Exkursionen Nr. 13–18, Wepf & Co., Basel, p. 293–299.

Schinz, H., 1931, Nachruf [auf Hans Schardt]: Vierteljahrsschrift der naturforschenden Gesellschaft in Zürich, v. 76, p. 509–513.

Schlocker, J., Bonilla, M.G., and Imlay, R.W., 1954, Ammonite indicates Cretaceous age for part of Franciscan group in San Francisco Bay area, California: American Association of Petroleum Geologists Bulletin, v. 38, p. 2372–2381.

Schrödinger, E., 1954, Nature and the Greeks: At the University Press, Cambridge, [i] + 97 p.

Senebier, J., IX (1801), Mémoire historique sur la vie et les écrits de Horace Bénédict de Saussure, pour servir d'introduction à la lecture de ses ouvrages. Lu à la Société Physique et d'Histoire Naturelle de Genève, le 23 Prairial an VIII (1800): Luc Sestié pour J.J. Paschoud, Genève, 219 p.

Şengör, A.M.C., 1985a, Türkiye'nin tektonik tarihinin yapısal sınıflaması, *in* Ercan, T., and Çağlayan, M.A., eds., Ketin Simpozyumu Kitabı: Türkiye Jeoloji Kurumu, Ankara, p. 37–61.

Şengör, A.M.C., 1985b, İhsan Ketin: Hayatı ve Eserleri, *in* Ercan, T., and Çağlayan, M.A., eds., Ketin Simpozyumu Kitabı: Türkiye Jeoloji Kurumu, Ankara, p. 5–18.

Şengör, A.M.C., 1996, Kuzey Anadolu Fayı'nın Keşfi: Bilim ve Teknik, v. 29, no. 338, p. 8–15.

Şengör, A.M.C., 1998, Die Tethys: vor hundert Jahren und heute: Mitteilungen der Österreichischen Geologischen Gesellschaft, v. 89, p. 5–176.

Şengör, A.M.C., 1999, Some salient European contributions to Cordilleran tectonics, *in* Moores, E.M., Sloan, D., and Stout, D.L., eds., Classic Cordilleran Concepts: A View from California: Boulder, Colorado, Geological Society of America Special Paper 338, p. 31–36.

Şengör, A.M.C., 2001, Is the Present the Key to the Past or the Past the Key to the Present? James Hutton and Adam Smith versus Abraham Gottlob Werner and Karl Marx in Interpreting History: Boulder, Colorado, Geological Society of America Special Paper 355, x + 51 p.

Şengör, A.M.C., and McKenzie, D., 1997, Memorial to İhsan Ketin: Boulder, Colorado, Geological Society of America Memorials, v. 28, p. 31–35.

Şengör, A.M.C., and Sakınç, M., 2001, Structural Rocks: Stratigraphic Implications, *in* Briegel, U., and Xiao, W.J., eds., Paradoxes in Geology (Hsü Volume), Elsevier, Amsterdam, p. 131–227.

Şengör, A.M.C., and Yılmaz, Y., 1981, Tethyan evolution of Turkey: A plate tectonic approach: Tectonophysics, v. 75, p. 181–241.

Şengör, A.M.C., Görür, N., and Şaroğlu, F., 1985, Strike-slip faulting and related basin formation in zones of tectonic escape: Turkey as a case study, *in* Biddle, K.T., and Christie-Blick, N., eds., Strike-slip Deformation, Basin Formation, and Sedimentation: Society of Economic Paleontologists and Mineralogists Special Publication 37 (in honor of J.C. Crowell), p. 227–264.

Şengör, A.M.C. and Natal'in, B.A., 1996, Palaeotectonics of Asia: Fragments of A Synthesis, *in* Yin, A., and Harrison, M., eds., The Tectonic Evolution of Asia, Rubey Colloquium, Cambridge University Press, Cambridge, p. 486–640.

Şengör, A.M.C., Natal'in, B.A., and Burtman, V.S., 1993, Evolution of the Altaid tectonic collage and Palaeozoic crustal growth in Eurasia: Nature, v. 364, p. 299–307.

Şengör, A.M.C., Yılmaz, Y., and Sungurlu, O., 1984, Tectonics of the Mediterranean Cimmerides: nature and evolution of the western termination of Palaeo-Tethys, *in* Dixon, J.E., and Robertson, A.H.F., eds., Geological Evolution of the Eastern Mediterranean: Geological Society of London Special Publication 17, p. 77–112.

Seymen, İ., 1985. Kırşehir Masifi metamorfitlerinin jeoloji evrimi, *in* Ercan, T., and Çağlayan, M.A., eds., Ketin Simpozyumu: Türkiye Jeoloji Kurumu, Ankara, p. 133–148.

Shackleton, R.M., 1951, A contribution to the geology of the Kavirondo Rift Valley: Quarterly Journal of the Geological Society of London, v. 106, p. 345–392, plates XXIII–XXVIII.

Shackleton, R.M., 1954, The structure and succession of Anglesey and the Lleyn Peninsula: British Association for the Advancement of Science, v. 11(41), p. 106–108.

Shackleton, R.M., 1969, The Pre-Cambrian of North Wales, *in* Wood, A., ed., The Pre-Cambrian and Lower Palaeozoic Rocks of Wales: University of Wales Press, Cardiff, p. 1–22.

Spicher, A., 1980, Tektonische Karte der Schweiz, 1:500,000, 2nd ed., Schweizerische geologische Kommission (no place of publication), one sheet.

Staub, R., 1924, Der Bau der Alpen—Versuch einer Synthese: Beiträge zur Geologischen Karte der Schweiz, Neue Folge, 52, Lieferung, VI + 272 p. + 7 plates.

Staub, R., 1930, Erkenntnisse, Probleme und Aufgaben der neueren Alpengeologie—Antrittsvorlesung gehalten an der. Eidg. Techn. Hochschule und an der Universität Zürich am 13. Juli 1929: Gebr. Fretz A.G., Zürich, 22 p.

Steinmann, G., 1905, Geologische Beobachtungen in den Alpen II. Die SCHARDT'sche Überfaltungstheorie und die geologische Bedeutung der Tiefseeabsätze und der ophiolitischen Massengesteine: Berichten der Naturforschenden Gesellschaft zu Freiburg i.B., v. 16, p. 18–67.

Steinmann, G., 1906, Geologische Probleme des Alpengebirges—Eine Einführung in das Verständnis des Gebirgsbaues der Alpen: Zeitschrift des Deutschen und Österreichischen Alpenvereins, v. 37, p. 1–44 (a folded plate is added to the offprints).

Steinmann, G., 1927, Die ophiolitischen Zonen in den Mediterranen Kettengebirgen: XIVe Congrés Géologique International (1926, Espagne), 4. Fscl., (Graficas Reunida S.A.), Madrid, p. 637–667.

Stenonis, N., 1669, De Solido intra Solidum Naturaliter Contento Dissertationis Prodromus. Stellae, Florentiae, 78 p.

Strahan, A., 1919. Preface by the Director, *in* Greenly, E., The Geology of Anglesey, v. I. Memoirs of the Geological Survey: Her Majesty's Stationary Office, London, p. v-vii.

Studer, B., 1825, Beyträge zu einer Monographie der Molasse, oder Geognostische Untersuchungen über die Steinarten und Petrefacten, die zwischen den Alpen und dem Jura gefunden werden; mit besonderer Rücksicht auf den Canton Bern und die angrenzender Theile von Freyburg, Luzern und Solothurn: Christian Albrecht Jenni, Bern, XXXVIII + 427p. + 2 foldout plates.

Studer, B., 1834, Geologie der westlichen Sweizer-Alpen. Ein Versuch: K. Groos, Heidelberg, 420 p.

Studer, B., 1853, Geologie der Schweiz, zweiter Band. Nördliche Nebenzone der Alpen. Jura und Hügelland: Stämpflische Verlagshandlung, Bern and Friedrich Schulthess, Zürich, 495 p.

Studer, B., 1872, Index der Petrographie und Stratigraphie der Schweiz und ihrer Umgebungen: Verlag der J. Dalp'schen Buch- und Kunstdhandlung (K. Schmid), Bern, v + 272 p.

Studer, T., 1884, Am Grabe von Prof. Dr. Bachmann: Berner Schulblatt, 17. Jahrgang, No. 16, p. 73–74.

Suess, E., 1852, Über *Terebratula dyphia*: Sitzungsberichte der mathematisch-physikalischen Classe der kaiserlichen Akademie der Wissenschaften, v. 8, p. 553–566 + Plate I.

Suess, E., 1858, Brachiopoden der Stramberger-Schichten, *in* von Hauer, F., ed., Beiträge zur Palæontographie von Österreich, v. I, p. 15–58.

Suess, E., 1867, Note sur le Gisement des Térébratules du groupe de Diphya dans l'Empire d'Autriche, *in* Pictet, F.-J., Mélanges Paléontologiques, troisième livraison: Ramboz et Schuchardt, Genève, p. 185–201 + Plate 35.

Suess, E., 1883, Das Antlitz der Erde, v. Ia (Erste Abtheilung): F. Tempsky, Prag and G. Freytag, Leipzig, 310 p.

Suess, E., 1885, Das Antlitz der Erde, v. Ib: F. Tempsky, Prag and G. Freytag, Leipzig, IV + 311–778 + [1] p.

Suess, E., 1901, Das Antlitz der Erde, v. III1 (Dritter Band. Erste Hälfte): F. Tempsky, Prag and Wien, and G. Freytag, Leipzig, IV + 508 p.

Suess, E., 1904, Sur la nature des charriages: Comptes Rendus hébdomadaires de l'Academie des Sciences, v. 139, p. 714–716.

Suess, E., 1918, La Face de la Terre (Das Antlitz der Erde) traduit de l'Allemand avec l'autorisation de l'auteur et annoté sous la direction de Emm. de Margerie, v. III, part 4 (Fin) avec un epilogue de P. Termier: Armand Colin, Paris, XV p. + p. 1351–1724 + 3 plates + 2 colored maps.

Suess, F.E., 1911, Viktor Uhlig. Ein Bild seiner wissenschaftlichen Tätigkeit: Mitteilungen der Geologischen gesellschaft Wien, v. 4, p. 449–482.

Süslü, A., 1986, Dil ve Tarih-Coğrafya Fakültesinin 50 Yıllık Tarihi: Ankara Üniversitesi Dil ve Tarih-Coğrafya Fakültesi Yayınları: 355, Ankara, XXVII + 500 p. (with Arabic, French and German summaries).

Suter, H., 1928, Prof. W.H. Schardt zu seinem 70. Geburtstag am 18. VI. 1928: Vierteljahrsschrift der naturforschenden Gesellschaft in Zürich, v. 73, p. 375–391.

Suter, H., 1931, Prof. Dr. Hans Schardt, 1858–1931: Verhandlungen der schweizerischen naturforschenden Gesellschaft, 1931, p. 411–422.

Taliaferro, N.L., 1943, Franciscan-Knoxville problem: American Association of Petroleum Geologists Bulletin, v. 27, p. 109–219.

Termier, P., 1903, Les nappes des Alpes Orientales et la synthèse des Alpes: Bulletin de la Société Géologique de France, série 4, v. 3, p. 711–765 + 2 foldout plates.

Tietze, E., 1903, Nachruf [auf C.L. Griesbach]: Verhandlungen der kaiserlichen und königlichen Reichsanstalt 1903 (Wien), p. 203–205.

Tobien, H., 1981, Studer, Bernhard: *in* Gillispie, C.C., editor-in-chief, Dictionary of Scientific Biography, v. 13: Charles Scribner's Sons, New York p. 123–124.

Tollmann, A., 1975, Karpathische Züge in Fazies und Tektonik der Ostalpen sowie Anmerkungen zur Großgliederung des Subtatrikums: in Mahel, M., Tectonic Problems of the Alpine System (Collection dedicated to the 75th birthday of Academician Professor RNDr. D. Andrusov, DrSc.): Veda, Publishing House of the Slovak Academy of Sciences, Bratislava, p. 109–120.

Tollmann, A., 1985, Geologie von Österreich, v. II (Außerzentralalpiner Anteil): Franz Deuticke, Wien, XV + 710 + [1] p.

Tollmann, A., 1986, Geologie von Österreich, v. III (Gesamtübersicht): Franz Deuticke, Wien, X + 718 p. + 3 folded plates.

Törnebohm, A.E., 1896, Grunddragen af det Centrala Skandinaviens Bergbyggnad: Köngl. Svenska Vetenskaps-Akademiens Handlingar, v. 28, No. 5, 210 pp + [ii] + 3 map plates.

Toula, F., 1904, Übersicht über die geologischen Literatur der Balkanhalbinsel mit Morea, des Archipels mit Kreta und Cypern, der Halbinsel Anatolien,

Syriens und Palestinas: Congrès Géologique International Compte Rendu de la IX. Session, Vienne 1903, Premier Fascicule, Imprimerie Hollinek Frères, Vienne, p. 185–330 + 1 foldout map.

Toula, F., 1906, Zusammenstellung der neuesten geologischen Literatur über die Balkanhalbinsel mit Morea, die griechischen Inseln, Ägypten und Vorderasien: XI. Jahresbericht für 1905 des Naturwissenschaftlichen Orientvereins, p. 37–75.

Tozer, E.T., 1984, The Trias and its Ammonoids: The Evolution of a Time Scale: Geological survey of Canada Miscellaneous Report 35, vi + 171 p.

Trak, S., 1941, Türkiye Hakkında Yazılan Coğrafî Eserler Bibliyoğrafyası Ankara: Dil ve Tarih-Coğrafya Fakültesi Yayını, Coğrafya 1, Ankara, 160 p.

Trak, S., 1942, Türkiye Coğrafya Eserleri Genel Bibliyoğrafyası Prof. Dr. H. Louis'in Önsözüyle birlikte ilâveli IIinci basım: Dil ve Tarih Coğrafya Fakültesi, Coğrafya Enstitüsü Neşriyatı, no. 1, Ankara, 272 p.

Trümpy, R., 1988, Cent ans de tectonique de nappes dans les Alpes: Comptes Rendus de l'Academie des Sciences (Paris), Série Générale, v. 5, p. 1–13.

Trümpy, R., 1991a, The structure of the Alps: 1891 to 1991: Casopis pro Mineralogii a Geologii, v. 36, p. 185–192.

Trümpy, R., 1991b, The Glarus Nappes: a controversy of a century ago *in* Müller, D.W., McKenzie, J.A., and Weissert, H., eds., Controversies in Modern Geology (Hsü-Festschrift), Academic Press, London, p. 385–404.

Trümpy, R., 1998, Tectonic units of Central Switzerland: their interpretation from A.D.1708 to the present day: Bulletin of Applied Geology, v. 3, p. 163–182.

Trümpy, R., and Lemoine, M., 1998, Marcel Bertrand (1847–1907): les nappes de charriage et le cycle orogénique: Comptes Rendus de l'Academie des Sciences (Paris), v. 327, p. 211–224.

Türler, H., 1921, [Bachmann], Isidor, *in* Türler, H., Godet, M., and Attinger, V., Historisch-Biographisches Lexicon der Schweiz, erster Band (A-Basel): Administration des Historisch-Biographischen Lexikons der Schweiz, Neuenburg, p. 513.

Tüysüz, O., 1990, Tectonic evolution of a part of the Tethyside orogenic collage: the Kargı Massif, northern Turkey: Tectonics, v. 9, p. 141–160.

Uhlig, V., 1903, Bau und Bild der Karpathen, *in* Diener, C., Hoernes, R., Suess, F.E., and Uhlig, V., eds., Bau und Bild Österreichs mit einem Vorworte von Eduard Suess: F. Tempsky, Wien and G. Freytag, Leipzig, p. 649–911 + one colored foldout map.

Uhlig, V., 1904, Über die Klippen der Karpathen: Congrès Géologique International Compte Rendu de la IX. Session, Vienne 1903, Premier Fascicule, Imprimerie Hollinek Frères, Vienne, p. 427–453.

Uhlig, V., 1907, Über die Tektonik der Karpathen: Sitzungsberichte der kaiserlichen Akademie der Wissenschaften, mathematisch-naturwissenschaftliche Klasse, v. 116, part I, p. 871–982 + 1 plate of sections + 1 map.

Uhlig, V., 1908, Die karpathische Sandsteinzone und ihr Verhältnis zum sudetischen Karbongebiet: Mitteilungen der Geologischen Gesellschaft, Wien, v. 1, p. 36–70, plate I.

Vadàsz, M.E., 1913, Liasfossilien aus Kleinasien: Jahrbuch der Geologischen Reichsanstalt, Budapest, v. 21, p. 57–82.

Wagenbreth, O., 1966, Die Lausitzer Überschiebung und die Geschichte ihrer geologischen Erforschung—Teil I: Abhandlungen des Staatlichen Museums für Mineralogie und Geologie, Dresden, v. 11, p. 163–279.

Wagenbreth, O., 1979, Leopold von Buch (1774–1853) und die Entwicklung der Geologie im 19. Jahrhundert: Abhandlungen des Staatlichen Museums für Mineralogie und Geologie zu Dresden, v. 29, p. 41–57.

Weiss, C.S., 1827, Über einige geognostische Punkte bei Meißen und Hohnstein: Karsten's Archiv für Bergbau und Hüttenwesen, v. 16, p. 3–16.

Werner, A.G., 1791, Neue Theorie von der Entstehung der Gänge mit Anwendung auf den Bergbau besonders den freibergischen: Gerlachische Buchdruckerei, Freiberg, 256 p.

Wernicke, B., and Burchfiel, B.C., 1982, Modes of extensional tectonics. Journal of Structural Geology, v. 4, p. 105–115.

Whitehurst, J., 1778 An Inquiry into the Original State and Formation of the Earth; Deduced from Facts and the Laws of Nature to Which is Added an Appendix Containing Some General Observations of The Strata in Derbyshire: Printed for the Author by J. Cooper, London, ii + [ii] + 199 p. + VI plates.

[Williams, D.], 1951, Edward Greenly: Proceedings of the Geologists' Association, p. lxiii + lxv.

Windschuttle, K., 1996, The Killing of History—How Literary Critics and Social Theorists Are Murdering Our Past: The Free Press, New York, [I] + 298 p.

Wood, D.S., 1974, Ophiolites, mélanges, blueschists, and ignimbrites: early Caledonian subduction in Wales? *in* Dott, R.H., Jr., and Shaver, R.H., eds., Modern and Ancient Geosynclinal Sedimentation: Society od Economic Paleontologists and Mineralogists Special Publication, 19 p. 334–344.

Woodward, J., 1695, An Essay Toward Natural History of the Earth and Terrestrial Bodies, Especially Minerals: As Also of the Sea, Rivers, and Springs, With an Account of the Universal Deluge and the Effects that it Had Upon the earth: Richard Wilkins, London, [xii] + 277 p.

Wright, L.A., and Troxel, B.W., 1969, Chaos structure and Basin and Range normal faults: evidence for a genetic relationship: Boulder, Colorado, Geological Society of America Special Paper 121, p. 580–581.

Wright, L.A., and Troxel, B.W., 1984. Geology of the Northern Half of the Confidence Hills 15-Minute Quadrangle Death Valley Region, Eastern California: The Area of the Amorgose Chaos: State of California, The Resources Agency, Department of Conservation, California Division of Mines and Geology, Map Sheet 34, Plate I, explanatory text, vi + 31 p.

Yürür, T., Temel, A., and Köse, O., 2002, Evidences of extensional tectonics at the southern boundary of the Galatan volcanic province, NW central Anatolia: Türkiye Jeoloji Bülteni, v. 45, p. 85–98.

Zappe, J.R., 1817a, Mineralogisches Hand-Lexicon. Oder: Alphabetische Ausstellung und Beschreibung aller bisher bekannten Fossilien, … Erster Band: Carl Ferdinand Beck, Wien, 480 p.

Zappe, J.R., 1817b, Mineralogisches Hand-Lexicon. Oder: Alphabetische Ausstellung und Beschreibung aller bisher bekannten Fossilien, … Zweyter Band: Carl Ferdinand Beck, Wien, 329 p.

Zhang, R.Z., and Yin, J.X., chief compilers, 1980, A Scientific Guidebook to South Xizang: Academia Sinica, Beijing, vi + 104 p. + 1 separate geological map.

von Zittel, K.A., 1899, Geschichte der Geologie und Paläontologie bis Ende des 19. Jahrhunderts: R. Oldenbourg, München, XI + 868 p.

ENDNOTES

[1]In this paper, all italicized words between square brackets [ ] within citations represent my own commentary. Braces { } are used for parentheses within parentheses. Because this is mostly a historical contribution, I have always given the first names plus the dates of the heroes of the story with references to their biographies. All translated texts are my own except in the case of Argand (1924), for which I used Albert V. Crozzi's translation with my own changes wherever I deemed it necessary.

[2]The discovery of very large displacement strike-slip fault systems, indeed whole kierogens, such as the Irtysh-Gornostaev System in Asia with a displacement of more than 2,000 km (Şengör et al., 1993; Şengör and Natal'in, 1996), has now brought into the realm of possibility that fault-rocks generated along such structures also may involve components of more than one paleogeographic realm. This blurs the distinction between mélanges generated along subduction zones and those along strike-slip faults, especially if one considers that along major keirogens rocks down to mid-crustal depths may also be exposed. Indeed, Saleeby (1977) has long maintained oceanic fracture zones as one significant site of mélange generation (also see De Long et al., 1977, 1979). This rapprochement of the concepts of subduction mélange and fault-rocks generated along strike-slip faults is yet another justification of Şengör and Sakınç's (2001) creation of a fourth great class of rock types, namely the *structural rocks*, which includes mélanges, as well as fault-rocks.

[3]For the history of Alpine geology, with emphasis on post-eighteenth century developments to provide a framework of the history here told, see, apart from the standard texts on the history of geology (among which especially von Zittel (1899, *passim*) and Greene (1982, *passim*): *(1) general, until 1930 (i.e., to the end of the heroic era of Alpine geology)*—Heim (1919, p. 3–28; 1921, p. 3–18, 1929), Staub (1930), Arbenz (1932), Hennig (1934), Bailey (1935), de Margerie (1946, 1947), Pilger (1978), Tollmann (1986, p. 1–37), Trümpy (1998); *(2) from 1884 to the present*—Masson (1983), Tollmann (1986, p. 37–43), Trümpy (1988, 1991a, b), see also Trümpy and Lemoine (1998); *(3) for a brief summary of the history of research on fault-rocks in the Alps*—Heim (1921, p. 94–103, 1929), Şengör and Sakınç (2001, p. 168–172); *(4) for the history of research in areas where the precursors of the mélange idea originated*—Schardt (1898, p. 118–146; Schardt's review must be read in conjunction with the following three papers: Haug, 1899, especially p. 114–123, and Schardt, 1899 and 1900), Bailey (1935, chapter VII), Leupold (1966), Masson (1976), Bayer (1982, p. 21–25), Trümpy (1998, especially p. 174–179).

[4]For the life and accomplishments of this great son of Switzerland, pioneer geologist, and mountaineer, see Senebier (1801), Cuvier (1861, p. 280–295), Freshfield

(1920), Gos (1935), de Margerie (1946, especially p. xcvi-ci, 1947, especially p. 3–12), Carozzi (1981, 1987, especially p. 219–251, 2000) and Carozzi et al. (1998) and Carozzi and Newman (2003). Carozzi et al. contains a good bibliography of recent secondary sources on de Saussure. Also see the *Supplement Scientifique* of the *Journal de Genève* (Samedi 22 septembre 1990).

[5]*Nagelfluh* (literally "nail rock-mass") is a term that entered the geological nomenclature from the northern and northeastern dialects of Swiss German. *Nagel* is nail in German, but *Fluh* (or *Fluhe* or *Flue*) is a specifically Swiss appellation for a large mass of mainly sedimentary rock (conglomerates and/or sandstones), especially for its face, that continues for a considerable distance along the strike (Zappe, 1817a, p. 332). According to Drosdowski (1995), the word is possibly related to the German *flach* (= flat). In the Jura Mountains, for instance, the vertical rock walls with considerable lateral extent, caused by the strike-slip faults cutting the chain at high angles, are known as *Fluhen* (e.g., Steinmann, 1906, p. 12). The Nagelfluh is so called because the pebbles in it give the rock face the appearance of having numerous round-headed nails driven into it. Murawski (1983, p. 151) pointed out that the term *Nagelfluh* was first used in a written document by J.Ch. Adelung in 1777. For an excellent overview of the Nagelfluh, with a tabular presentation of the types of pebbles that occur in it and their probable provenances in the Alps, see Heim (1919, p. 43–72). For a succinct statement, see Studer (1872, p. 166–168)

[6]For the life of Leopold von Buch, whom his friend Alexander von Humboldt considered the greatest geologist of his time, see Ewald (1867), Günther (1900, p. 183–271), and Wagenbreth (1979). For an English summary, see Nieuwenkamp (1981). Also see the special memorial colloquium volume in the *Zeitschrift der Geologischen Wissenschaften* (Berlin, GDR), v. 2 (1974), no. 12.

[7]See Tobien's (1981) short article in the *Dictionary of Scientific Biography* for all the important sources on Studer that I am aware of.

[8]Studer wrote to Sir Roderick I. Murchison to state his belief that "the parent granite, from whence such pebbles were derived, protruded along the great line of dislocation between the molasse and the chain, and was lost by subsidence *en masse* when the great accumulations of nagelflue were formed" (Murchison, 1849, p. 229, footnote †; italics Murchison's).

[9]This block has since been quarried and some of the granite obtained from it was used to build the pedestal of the Statue of Liberty in New York. It is now protected and despite the earlier quarrying, a major part of it still remains (Schaub, 1967, p. 297).

[10]Although it is now clear that the immense Habkern granite block Murchison described belongs to the wildflysch, I should point out, in fairness to Murchison, that even today one is not sure whether it was eroded out of the wildflysch in situ or glacially transported from the upper Lombach region (Schaub, 1967, p. 297).

[11]The section that gave Murchison this impression is sketched by him on his p. 211, fig. 20. Bayer (1982, p. 208–209) gives a more detailed description, including a reference list of works subsequent to Murchison's, with the coordinates of the section.

[12]Then teacher in the canton school in Bern; later professor of geology in the University of Bern from 1873 until he drowned in the Aare in 1884; for biographical data, see Studer (1884); for a much shorter statement, see Türler (1921).

[13]For the life and accomplishments of Früh, see Imhof (1937/1938). This memorial contains a useful list of references to other biographies of Früh.

[14]The best biography of Kaufmann and an assessment of his work is the two-part account by Amberg and Bachmann (1893). Amberg deals with Kaufmann's life and Bachmann with his publications. For a short notice on Kaufmann, see Felber (in press).

[15]*Knauer* is a German miner's term. In Zappe (1817b, p. 48) it is defined as "those compact pieces of rock-mass, which now and then fall into the veins and hinder the work of the miner." In the way Kaufmann uses it, it is nothing more than a hard piece of rock surrounded by the soft schist of the wildflysch.

[16]Macigno is here used for flysch in the usual sense of Studer (i.e., Eocene). It is a term that originated in the Apennines and designates there also various flysches (originally considered entirely Eocene). For the way it was understood by both Studer and Kaufmann, see Studer (1872, p. 149–150).

[17]"Leimern Beds," now the "Leimeren Series" (after the orthography in the 1: 25,000 topographic maps of Switzerland, sheet 1208, Beatenberg: Bayer, 1982, p. 91), are commonly light gray, in places also red, rarely green pelagic, radiolarite-bearing limestones, marly limestones and marls that occur commonly in association with wildflysch, in the flysch zones on the backs of the Helvetic Nappes, and as isolated shreds, blocks, lenses or packages of various dimensions within the Subalpine marginal flysch zone. Their age ranges from Neocomian to middle Eocene (Bayer, 1982). The designation as *Leimernschichten* was introduced by Kaufmann in 1872 (for references and a general description, see Herb, 1966; Bayer, 1982, p. 91–106, 189–206), which Bayer redescribed and formalized as the Leimeren Series (Bayer, *ibid.*). Today, the part of the Leimeren Series represented by the Eocene flysch and the blocks belonging to the Cretaceous parts of the Leimeren series are considered to represent an independent mélange unit, named the Leimeren Mélange by Bayer (1982, p. 98), consisting mainly of pelagic sediments in a sheared shaly matrix with exotic slabs of flysch sandstone (see also Hsü, 1995, p. 60, on the basis of Bayer's work). Bayer (1982) thinks it to be of South Penninic origin.

[18]For the life of this great geologist, see Suter (1928, 1931), Schinz (1931), Kettner (1958), and Schaer (1994, p. 429–433).

[19]Schardt later wrote: "Besides, the exotic blocks must be considered as the debris of the klippes, detached from them during the breaking-up of the originally continuous nappe. Many of them mixed in, or enveloped by, the flysch as the second type of klippes are approached. Their sitation and origin are explained easily by the tectonic effects that accompanied the displacement of the nappe of the klippes. Its situation before its break-up was a result of gliding, the same as the one that placed the Préalpes in their present position. During this movement, important pieces of the terrane were detached from the sides of the nappe in motion and became mixed into the flysch, which, at the same time, is squeezed. Many of these fragments became literally *enveloped* and *rolled into the flysch*, as I described when considering the Zone des Cols and the exterior zone of the Flysch of the Préalpes. To be clear on this point, one could call the former the great klippes, *islets or klippes of cover nappes; they are the witnesses of a former continuous nappe*. The others could be called the *klippes of squeeze*. .... The other exotic blocks result from the disintegration of breccias, or marls to flysch blocks (Habkern). Although the flysch breccias, exotic islets and klippes of squeeze are easy to distinguish, exotic blocks may result from every one of these three phenomena without the possibility of telling from which." (Schardt, 1898, p. 209–210, italics Schardt's). Although Schardt is here describing a process of mixing of exotic blocks into a flysch succession to create what we today would call a mélange, he does not describe the tectonic disruption and subsequent tectonic mixing, into a mélange, of a previously regular succession. That is why I consider his discovery a part of the prehistory of the mélange concept and not a part of its history.

[20]Ludwig Hohenegger (1807–1864) was a German geologist, whose geological map of part of the northern Carpathians (Hohenegger, 1861) became a classic and a standard reference. For his life, see von Hingenau (1864).

[21]For descriptions of this precise locality, see Boucek (1968, p. 18–20) and Pesl et al. (1968, p. 13).

[22]Birembaut's (1981) Boué article in the *Dictionary of Scientific Biography* is unfortunately inadequate. For his life, I recommend that the papers listed by Sarjeant in his *Geologists and the History of Geology*, v. II be consulted. To that I can only add Anonymous (1889).

[23]The term *Wienerstandsein* was introduced by Keferstein in 1828 and it was viewed by him incorrectly as Upper Triassic (Keuper). This error was perpetuated until 1850 (despite Murchison's objections!) by such geologists as Wil-

helm Heidinger and Adolf von Morlot. It was first Franz von Hauer (1850) who subdivided this unit and left those parts within the calcareous Alps (later Lunz Beds) in the Keuper, placed what is now known as Roßfeld Beds (turbidites; initial flysch) into the Neocomian, and put the flysch sandstones into the Neocomian and Eocene. Later research showed the presence of a complete sequence of Cretaceous and Lower Tertiary flysches in the Flysch Zone of the Eastern Alps in which the "exotics" occur and which continues into the flysch zones of the northern Carpathians (see Tollmann, 1985, p. 376 ff.). Studer (1872, p. 259) gives a succinct statement of the Wienersandstein as it was understood in the mid-nineteenth century).

[24]For modern descriptions of the stratigraphy and structure of the Pieniny Klippen Belt other than the ones cited above, see Anonymous (1968), Biely and Bystricky (1968), Koráb et al. (1968), Pesl et al., (1968), Misik (1979), Marschalko (1979), Marschalko and Kysela (1980), Misik and Sykora (1981), K. Bak (1995, 1998), and M. Bak (1996a, b).

[25]Viktor Uhlig was an accomplished paleontologist, stratigrapher, and structural geologist who became Suess' successor in Vienna. Although he was initially against the theory of nappes, he enthusiastically embraced it after the 1903 International Geological Congress in Vienna, and published his great tectonic synthesis of Carpathian geology in 1907 in terms of the nappe theory. His audacious cross-section (see Fig. 14B) has recently been almost fully vindicated by seismic reflection profiling. For his life and work, see Suess (1911).

[26]For Beyrich's life, see the list of sources in Sarjeant (1980, p. 550), and Carlé's (1988, p. 63–90) excellent account of Beyrich's work.

[27]Lugeon himself had been influenced by Suess (Lugeon, 1903, p. 8).

[28]This historic expedition was organized because of the growing awareness of the great similarity of the Mesozoic facies of the Himalaya to those of the Eastern Alps. The expedition went to the Himalaya in 1892 and studied the Central Himalaya of Kumaon, Gurwhal, and Hundes (Diener, 1895 and 1916, p. 406). Eduard Suess, as the new secretary-general of the Academy of Sciences in Vienna, was instrumental in organizing this important expedition. One of its fruits was the concept of Tethys (see Şengör, 1998, p. 48–50 and the references there). For the general preliminary results of the expedition, see Diener (1895).

[29]But without giving references to the actual publications. The sources that Suess most likely had in mind are Daubrée (1879, especially p. 576–601) and Daubrée (1888, especially p. 183–199).

[30]Shackleton had earlier used the term mélange for the subaerial volcanic slump and flow deposits in the Kathwanga area of the Kavirondo Rift Valley (Shackleton, 1951, p. 360) and was thus predisposed to a non-tectonic interpretation of the Anglesey mélanges.

[31]Regrettably, there is no satisfactory summary of the history of geological research in Turkey. There are not even papers, which collectively would present an adequate history. For the time being, the reader who wishes to obtain a general framework for the history here told must make do with a combination of the following: Anonymous (1971), Ketin (1973, 1975, 1979, 1981), Pamir (1979), Çakmak et al. (1985), Şengör (1985, 1996), and Norman (1993).

[32]In those days, the Turkish universities were organized after the model of the continental European, specifically German, universities. "Chairs" were equivalent to small departments.

[33]I once asked Professor Erol what he thought of McCallien. He said he was a competent and thoughtful geologist and a fine teacher. He then considered a few seconds and added in an affectionate tone laden with nostalgia: "He was a very good person!" Mr. Kemâl Şerbetçi, now a successful businessman in Ankara, expressed the same sentiments when I interviewed him on McCallien.

[34]Prof. Erol says that this Dikilitaş cannot be identified any more and suspects that Chaput had probably consulted local peasants and jotted down a possibly traditional but no longer current name he was told (personal commun., December 15, 1994).

[35]To appreciate the state of geology in Turkey in the middle of the twentieth century and the significance of this map in its history, let us remember that 1: 25,000 maps had become the standard of "special maps" published by European surveys in 1866 upon the suggestion of A.H. Ernst Beyrich in Germany (Carlé, 1988, p. 66–67).

[36]Erol's critique of Chaput is here somewhat unjust, but it is not his fault. He was reading only a poor Turkish translation by Prof. Hâmit Nâfiz [Pamir], of Chaput's article, which had been published bilingually, for Erol could not read French. What Chaput actually had written was that the ancient massif of Elma Dağı and its Mesozoic cover had been in places heavily broken and imbricated indicating a top-to-northwest tectonic transport (Chaput, 1931, p. 28). What Erol had found was the opposite vergence.

[37]This map is published for the first time as Plate 1 of this paper. I am grateful to Professor Erol for making it available to me for this purpose and for preparing a legend for it.

[38]This is not entirely correct. Erol's main mapping was carried out on a base with scale 1:25,000, but his eventual thesis map had to be of 1:100,000, because the entire thesis area (some 1500 km^2) could not be covered by the larger-scale maps owing to the unavailability of such maps to civilians. His main area of concentration, however, was in the smaller region mentioned above that was covered by the 1:25,000-scale maps.

[39]We now know that Ketin had grossly underestimated the structural complications in his area (see his marvelous recantation in the great synthesis on the distribution of thrusting in time and space in the geological history of Turkey, which he wrote for the symposium organized by the Turkish Geological Society upon the occasion of his retirement from active teaching in 1983: Ketin, 1985). But in essence he was right: the profile I mention above is now known to belong to a thrust-stacked sequence within an accretionary complex of late Cretaceous age that was thrust *on top* of the Kırşehir Massif in the earliest Maastrichtian to earliest Eocene interval (see Görür et al., 1984, especially fig. 9). The Kırsehir metamorphics are pre-medial Cretaceous in age and Dr. Necdet Özgül (personal commun., 1985) correlated them with the Paleozoic-Mesozoic Taurus sequences (i.e., a Gondwana-Land sequence south of Neo-Tethys). Their major metamorphism was thus possibly sometime in the medial Mesozoic, probably early to medial Cretaceous. Ketin made the same sorts of mistakes Argand had in the Alps, with similar fruitful results. I recall, with nostalgic feelings, how my friends and later colleagues Professor İhsan Seymen (a former doctoral student of Ketin) and Professor Fazlı Y. Oktay showed me, in a terribly hot Anatolian summer day in 1977, Ketin's locality west of Kaman, where he had observed the "concordant" position of the serpentinite-radiolarite series on the metamorphic rocks of the Kırşehir Massif. "If Ketin had just looked another kilometre or so around the corner," Seymen said, pointing to the slope, "he would have seen the shallow tectonic contact truncating the metamorphics and he would have understood that he was just looking at an accordance." Şengör and Yılmaz (1981) followed Ketin's assessment of the age of the Kırşehir metamorphics and considered the Kırşehir as a simple continuation of the Apulian platform eastward. This mistake was corrected by Görür et al. (1984), who showed that the Kırşehir , indeed a part of the northern margin of Gondwana-Land until sometime in the Triassic, parted company with Apulia later (this recognition became possible when Naci Görür documented the subsurface connection between the suture south of the Kırşehir massif, discovered in the early 1970s by F.Y. Oktay, with the İzmir-Ankara suture of Şengör and Yılmaz, 1981). For the present state-of-the-art on the story of the Ankara Mélange and the Kırşehir Massif, see especially two studies: Görür et al. (1984) and Seymen (1985). The contact Seymen showed me, however, was one between high-grade schists and gneisses and an essentially unmetamorphic ophiolite. I suspect now that it is probably a low-angle normal fault, possibly a part of a much wider area of late Cainozoic extension around Ankara. Evidence for this extension has been reported from northeast of Ankara (Gökten et al., 1996; Yürür et al., 2002). It is likely that this extensional region is much wider and was probably instrumental in the formation of the central Anatolian "ovas" (Şengör et al., 1985; Bozkurt, 2001).

[40]The "Franciscan-Knoxville paradox" (Hsü, 1968) refers to the discovery (Schlocker et al., 1954) of the Albian Ammonite *Douvilleceras* sp. in the Marin Sandstone in the type area of the Franciscan (northern shore of the San Francisco Bay), followed by the demonstration, on the basis of microfossils (mainly foraminifera), of the late Cretaceous (late Cenomanian) age of the Calera Limestone also belonging to the Franciscan, whereas the age of the Knoxville Formation "overlying" the Franciscan was known to be late Jurassic (Tithonian) on the basis of a *Buchia* (=*Aucella*) *piochii* fauna (Anderson, 1933, 1938; Taliaferro, 1943; Bailey et al., 1964). This created the seemingly impossible situation of having a sequence beginning with the Upper Jurassic lying "unconformably" atop one including rocks as young as the late Cretaceous. The paradox was resolved when it was noticed that the Franciscan was a mélange and that it had formed by being tucked under the developing Great Valley sequence, of which Knoxville is the bottom formation. (For a history of this paradox, see Hsü, 1968, 1985, especially fig. 5).

[41]The English version here is identical to that published in *Nature*, with the exception of a few words and the figure added to the paper in *Nature*.

Manuscript Accepted by the Society May 2, 2003

Printed in the USA

Geological Society of America
Special paper 373
2003

History of asbestos discovery and use and asbestos-related disease in context with the occurrence of asbestos within ophiolite complexes

Malcolm Ross*
Robert P. Nolan*
Earth and Environmental Sciences of the Graduate School and University Center of The City University of New York, 365 Fifth Avenue, New York, New York 10016, USA

ABSTRACT

Two ancient asbestos mines, one near Karystos, Greece, and the other southeast of Mount Troodos, Cyprus, were located in what we now know to be ophiolite terrane. Evidence suggests that asbestos was discovered and utilized in Cyprus, perhaps as long as 5,000 years ago, for manufacture of cremation cloths, lamp wicks, hats, and shoes. Some of the adverse health effects of asbestos became known only in the early twentieth century, but it was not until the 1960s that the asbestos-related diseases—asbestosis, lung cancer, and mesothelioma—were fully understood.

Approximately 85% of the world's asbestos was produced from ophiolite complexes, most of which was the chrysotile variety; tremolite, actinolite, and anthophyllite asbestos accounting for only a few percent of the total. Asbestos minerals crystallize within tectonized ophiolites—along shear, fault, and dilation zones, and at contacts with intruded dikes and sills. Important chrysotile asbestos mines are found in the ophiolites of eastern Canada, the Russian Urals, California, northwest Italy, northern Greece, and Cyprus. A high incidence of mesothelioma, a cancer of the lung lining, is reported among residents of villages located within or near ophiolite complexes in Greece, Turkey, Cyprus, Corsica, and New Caledonia. These villagers were exposed to tremolite asbestos while processing stucco and whitewash for application to homes. Asbestos contamination in various geographic localities has generated concern about health risks and has prompted costly remedial actions, especially in the United Kingdom and the United States. A scientific basis for public policy is offered to address the utilization of asbestos-bearing rocks.

Keywords: mineral fibers, asbestos, chrysotile, tremolite, serpentines, serpentinite, lung cancer, mesothelioma, ophiolite.

INTRODUCTION

Ophiolite complexes, including the associated serpentinite belts, are found in many parts of the world. It was in these rocks that asbestos was first discovered and utilized as long as 5,000 years ago and from which ~85% of the world's asbestos was produced, most being chrysotile asbestos. This paper is intended to give the ophiolite specialist insight to the importance of ophiolites as a source of chrysotile, tremolite, actinolite, and anthophyllite asbestos by presenting their mineralogical descriptions, the nature of crystal growth of the asbestos minerals, their mode of occurrence within ophiolite complexes, and descriptions of some of the more important ophiolite-hosted asbestos deposits. We will discuss the origin of the Latin and English word "asbestos" from the writings of the ancient Greek and Roman philosophers; present a history, from ancient times to the present, of the discovery, production, and utilization of the various asbestos minerals; and

*Mailing address, Ross: 1608 44th Street, NW, Washington, D.C. 20007, USA. E-mails: Ross—mrdrr@earthlink.net; Nolan—rnolan@gc.cuny.edu.

Ross, M., and Nolan, R.P., 2003, History of asbestos discovery and use and asbestos-related disease in context with the occurrence of asbestos within ophiolite complexes, *in* Dilek, Y., and Newcomb, S., eds., Ophiolite concept and the evolution of geological thought: Boulder, Colorado, Geological Society of America Special Paper 373, p. 447–470. For permission to copy, contact editing@geosociety.org.

present an early history of the development of an understanding of asbestos-related disease that took such a toll on asbestos workers' lives. Epidemiological studies are presented to show that asbestos workers who experienced uncontrolled asbestos exposures, such as miners, millers, and insulators, developed three serious diseases: asbestosis (scarring of the part of the lung where gas exchange takes place), lung cancer, and mesothelioma (cancer of the lining of the lung and abdominal cavity). From British medical pneumoconiosis studies of asbestos workers in the early part of the twentieth century to the recent epidemiological studies, there developed a great concern over the health effects of asbestos and as a result many countries have promulgated strict regulations for the use of asbestos (WHO, 1989). More recent bans on the mineral's use have been adopted or proposed (see Wilson et al., 2001). The Fifth Circuit Court of Appeals struck down the U.S. Environmental Protection Agency's (EPA) proposed ban on asbestos in 1991, citing the EPA's failure to pursue the least burdensome, most reasonable regulation required to protect the environment adequately (United States Court of Appeals, 1991).

In addition to chrysotile asbestos, tremolite and actinolite asbestos minerals are also found in ophiolites, but have been mined only in small quantities. As will be discussed, these two asbestos minerals present a much more serious health problem to miners and other workers than does chrysotile asbestos.

Six ophiolite localities in Turkey, Greece, Cyprus, Corsica, and New Caledonia are described, where the residents of nearby rural villages were exposed to tremolite asbestos through its use as ingredients for stucco (a fine plaster composed of cement mixed with asbestos added for strengthening) and whitewash (usually a mixture of water, asbestos, and calcium carbonate).

Examples will be presented of some of the asbestos mitigation efforts now being undertaken in the United States and Cyprus. In the discussion, we present the scientific basis for developing a public policy to address the health hazards when mining or construction activities occur in rocks that contain fibrous minerals.

ORIGINS OF THE WORD "ASBESTOS"

The various words used in antiquity to denote the asbestos minerals—including asbestos, asbestus, asbestinon, asbest, asbeste, asbeston, abeston, amiantos, amiantus, amianthus, amiant, and amiante—can be traced back to the writings of the ancient Greek philosophers and their use of two words—αμιαντος and ασβεστος. The Greek word αμιαντος (transliterated as "amiantos"), when used as a noun is synonymous with the English word asbestos, and when used as an adjective can mean pure or undefiled (G.E.L., 1940, p. 83). The Greek word ασβεστος (properly transliterated as "asvestos"—not "asbestos"), when used as a noun means lime, quicklime, or unslaked lime (CaO), and when used as an adjective, can mean inextinguishable, unquenchable, or not quenched (G.E.L., 1940, p. 255). Note, that quenching or slaking of calcium oxide with water produces calcium hydroxide. Latin translations of the Greek words for asbestos and quicklime have caused some confusion that is perhaps related to the difficulty of translating the complex noun and adjective declensions of Greek into Latin, an equally complex language, to the interchange of noun modifiers and nouns, and to the Greek versus Latin transliteration of the letter β.

The Greek physician, Pedanius Dioscorides of Cilicia (40–90 A.D.) reports of an "undefiled stone," αμιαντος λιθος—transliterated as "aminatos lithos," that occurs in Cyprus and resembles fissile alum that can be woven and is not consumed by fire (*Materia Medica*, 5, 138; see also G.E.L., 1940, p. 83 and O.E.D., 1933, p. 280). In modern Greek usage, the noun λιθος ("lithos") is omitted and replaced with αμιαντος ("amiantos"), synonymous with the Latin and English noun asbestos, the German and Russian noun asbest, the French nouns asbeste and amiante, and the Italian and Spanish nouns asbesto and amianto.

The ancient Greek writers (e.g., Dioscorides, *Materia Medica*, 5, 115; see also G.E.L., 1940, p. 255 and O.E.D., 1933, p. 480) used the noun ασβεστος ("asvestos") to mean quicklime; a meaning retained in modern Greek; however, Pliny the Elder (Caius Pliny Secundus, 23–79 A.D., *Natural History*, Book 19, paragraph 20; see also Rackham, 1961, p. 432–433), apparently misunderstanding the use of this word by the early Greek philosophers, replaced the Greek noun for quicklime (ασβεστος) with the dubious Greek word ασβεστινον (G.E.L., 1940, p. 255), which he interpreted to mean a non-combustible material. Pliny then transliterated ασβεστινον into the Latin noun "asbestinon," alluding to an incombustible linen, cleansed by fire, and used as shrouds for royalty during cremation. Pliny also refers to this incombustible linen as linum vivum—live linen. Pliny was undoubtedly referring to what we now know as asbestos cloth but reported that the material came from a plant that "grows in the deserts and sun-scorched regions of India" (Rackman, 1961, p. 433).

A computer search for the many variations of the word for asbestos that possibly may have been used in the 37 books of Pliny's *Natural History* (which is copied onto an Internet website; Thayer, 2002) revealed that Pliny, in addition to "asbestinon" (Book 19, 20), used the words "amiantus" (Book 36, 139; see also Bailey, 1932, p. 120–123, 256–257) and "asbestos" (Book 37, 146, Thayer, 2002). Pliny states that "amiantus" resembles alumen in appearance and is not destroyed by fire. The alumen of Pliny is probably not alum as we know it (potassium aluminum sulfate), but possibly an iron sulfate efflorescent produced by decomposition of pyrite (see Hoover and Hoover 1950, footnote 11, p. 572). In using the word "amiantus," Pliny may indeed be describing an asbestos mineral, but relating it to alumen because of similarity of certain physical properties.

In Book 37, 146, Pliny states that iron colored "asbestos" is found in the Arkadian mountains (located in the central Peloponnesus of Greece) perhaps suggesting that this fibrous material has a mineral origin. Here Pliny has taken the Greek word "asvestos" (quicklime) to mean something quite different—a fibrous mineral, now referred to as "asbestos" in both the Latin and English languages. It is not at all clear whether Pliny really understood the geological origin of asbestos, whereas ancient Greek writers, such as Strabo, Dioscorides, and Theophrastus, certainly did.

In summary, various words for asbestos have entered the vocabulary of Latin and other languages; the words amiantus, amianthus, amiant, amiante, and amianto were derived from the Greek word αμιαντος—"amiantos," whereas the words asbestos, asbestus, asbestinon, asbest, asbeste, asbeston, abeston, and asbesto were derived from the Greek word ασβεστος—"asvestos".

HISTORY OF ASBESTOS DISCOVERY, PRODUCTION, AND UTILIZATION

Early Beginnings

While the general use of asbestos in international commerce dates only to the late nineteenth century, its utility in human culture goes back at least 4,500 years. Archeological studies (Europaeus-Äyräpää, 1930) show that inhabitants of the Lake Juojärvi region of East Finland knew how to strengthen earthenware pots and cooking utensils with an asbestos mineral, later identified as anthophyllite (anthophyllite asbestos was mined commercially in East Finland between 1918 and 1975). According to Huuskonen (1980; see also Europaeus-Äyräpää, 1930 and Noro, 1968), the use of asbestos-strengthened ceramic wares began during the Stone Age and continued throughout the Bronze Age and into the Iron Age. The use of such utensils spread over a wide area of Finland, Scandinavia, and Russia. One of the most important uses of asbestos in the twentieth century—to strengthen materials—was used for this purpose several thousand years ago.

There is also evidence that asbestos was discovered and utilized in Cyprus during classical times, perhaps as long as 5,000 years ago, for manufacture of cremation cloths, lamp wicks, hats, and shoes (Dioscorides, *Materia Medica*, 5, 138). Bowles (1955, p. 9) suggests a location for the ancient asbestos deposits, and states that Cyprus was a well known source of asbestos in ancient times and that "although it is difficult to determine, from early references the exact location of the ancient deposit, probably it was southeast of Mount Troodos in a village known as Amianto the identity of which is lost." Evans (1906, p. 145) states that sixteenth-century travelers referred to the existence of abandoned asbestos mines near the village of Paleandros or Pelendria. Pelendria (now Pelendri) is situated within the Troodos ophiolite complex and ~6 km southeast of the Mount Troodos chrysotile asbestos mine; however, Dr. George Constantinou, former head of the Cyprus Geological Survey Department, suggests that asbestos was mined in ancient times near the towns of Vavla, Vasa, and Apsiou, 25 km southeast of Mount Troodos. These towns are located within the Akapnov Forest, a large area of ophiolite terrane within the Arakapas (transform) Sequence. We suggest that the ancient asbestos deposits were located within the Akapnov Forest area; the deposit near Pelendri may have been developed much later and prior to opening of the nearby Mount Troodos chrysotile asbestos mine in 1904.

The descriptions of asbestos use in Greece clearly goes back more than 2000 years. Herodotus (ca. 484–425 B.C.) clearly documented the use of asbestos for lamp wicks in the early Greek civilization. The Greek philosopher Theophrastus (ca. 372–287 B.C.), one of Aristotle's students, was perhaps one of the first to carefully describe the mineral we now call asbestos as "a stone, in its external appearance somewhat resembling wood, on which, if oil is poured, it burns; but when the oil is burnt away, the burning of the stone ceases, as if it in itself not liable to such accidents" (Hoover and Hoover, 1950, p. 440, footnote 5). Plutarch (ca. 46–120 A.D.) recorded that the vestal virgins used "perpetual" lamps to serve Rome's sacred fire (the fabrication of asbestos lamp wicks, such as described by Herodotus and Plutarch, probably was an important cottage industry in Greek and Roman times). Pliny, as discussed before (Book 19, 20), refers to tablecloths, lamp wicks, and a rare and costly asbestos cloth—lithium vivum—the funeral dress of kings. Pliny's reference to iron colored asbestos occurring in the Arkadian mountains (Book 37, 146) is supported by the town history of Kandyla, Province of Arkadia, central Peloponnesus, Greece. In a translation from the original ancient Greek, it is reported that early in the first millennium there was a merger of three towns into one—Kandyla. Further, it is stated that after the merger "the inhabitants began to construct homes that were solid, with the use of such materials as stone and asbestos" (Deligiannis, 2002, p. 1). As will be mentioned latter, villagers in the rural areas of Greece, as well as in other countries, commonly applied asbestos-bearing stucco and whitewash to their homes.

The Greek geographer, Strabo (64 B.C.–21 A.D.), in *Geographia*, Book 10, (see also Bailey, 1932, p. 256) referred to a "Karystian stone" obtained from quarries near Karystos (Karistos), a town on the southern tip of the Greek island of Euboea (Evvia). Hoover and Hoover (1950, p. 440, footnote 5) quoting from Strabo (Book 10, 1): "at Carystus (Karystos) there is found in the earth a stone, which is combed like wool, and woven, so that napkins are made of this substance, which when soiled, are thrown into the fire and cleaned, as in the washing of linen." Examination of geologic maps of Evvia shows extensive outcrops of ophiolitic rocks in the north-central region of the island and one smaller outcrop ~30 km north of Karystos (Rassios and Smith, 2000, fig. 2; Geological Map of Greece, Division of Geology and Economic Geology, Athens, 1983). Thus, this ancient source of asbestos known to the ancients as "Karystian stone," may have been quarried within ophiolite terrane.

There are also numerous references (see Bowles, 1955. p. 8–10) to the use of asbestos in post-Roman times and before the start of the modern asbestos industry in the late part of the nineteenth century. The origin of asbestos remained a subject of fanciful speculations into the Middle Ages, when alchemists made claims that asbestos was the hair of a type of fire-resistant salamander (Alleman and Mossman, 1997). Since at least the early part of the sixteenth century, the image of the salamander has been symbolic for asbestos. Probably the most famous anecdote is that of Charlemagne (or, according to some writers, Charles V) astonishing his guests by cleansing a tablecloth by throwing it into a fire. Marco Polo, in his travels through Siberia during the thirteenth century, mentions the "amianthus" cloth, which resisted the action of fire.

As early as 1720, chrysotile asbestos was commercially mined in the Urals Region of Russia along the Tagyl River in the Middle Urals. The silky mineral fibers were woven into cloth to fabricate aprons, gloves, and caps for the high-temperature shops of the eighteenth-century metallurgical plants that were common in the Urals. In 1722, a sample of the asbestos cloth was presented to Peter the Great (Kashansky, 1999).

Start of the Modern Industry

In the 1860s and 1870s, the market for asbestos products rapidly expanded, probably for three reasons: the need for insulation for the new steam technology, the formation of an international consortium of Italian and English entrepreneurs, and the reopening of the asbestos deposits in northern Italy and simultaneous development of the vast chrysotile resources in Quebec, Canada. By 1890, the modern asbestos industry was full blown, with hundreds of applications being introduced (Jones, 1890; Alleman and Mossman, 1997).

Chrysotile and tremolite asbestos deposits in the Susa, Lanzo, Aosta, and Val Malenco areas of the Italian Alps were first exploited in Roman times, but it was not until the early 1800s when manufacture of asbestos threads, fabrics, and paper was perfected, that the alpine deposits became economically important. Between 1860 and 1875, several new Italian companies formed to advance the technology of fabrication of asbestos into spun products, rope packings, and heat-insulating board. Exhibits by these companies at the Paris Universal Exposition in 1878 helped to bring these asbestos products to international attention. (Ross, 1981; Bowles, 1955).

Marcuse (1930) describes the first discovery of asbestos in the Province of Quebec, Canada by the early French settlers who referred to the local asbestos (now known to be chrysotile) as "Pierre á Coton" (the cotton stone). Bowles (1955) reports that the first small deposit of asbestos was discovered near St. Joseph, Quebec in 1860 and samples of the mineral were exhibited in London in 1862; however, the first deposit capable of major development was discovered near Danville, Quebec, in 1877. Active mining began in 1878, when 50 tons of fiber was produced. By 1885, seven asbestos mines, centered around the town of Thetford Mines, were in production.

Asbestos was discovered in the Ural Mountains near the city of Ekaterinburg in the early part of the eighteenth century and the fiber was used sporadically in textile production. Systematic development of the Ural deposits, however, did not began until the huge Bazhenovskoye deposit of chrysotile asbestos was discovered near Asbest City in 1884. Commercial mining of this deposit began in 1886, and by 1889 24,000 tons of chrysotile asbestos had been produced. Since 1918, the mines and mills have been operated by the Uralasbest Company (Shcherbakov et al., 2001).

The occurrence of "crocidolite" asbestos was discovered in 1812 in the Profret area, Northern Cape Province, South Africa, but it was not until 1893 that asbestos production began near the town of Koegas and in 1926 near the Pomfret area (Beukes and Dreyer, 1986). "Amosite" asbestos was discovered near the town of Penge, Transvaal Province, South Africa, in 1907, and commercial production began in 1916 (Bowles, 1955).

Asbestos Production

From the time of the first recorded use of asbestos by Stone Age man to 1900, total world production was probably between 200,000 and 300,000 metric tons. Of this, 150,000 tons were produced in Quebec, Canada (Ross, 1981). After World War I, use of asbestos greatly increased; total world production of all forms of asbestos between 1931 and 1999 was ~166 million metric tons (Ross and Virta, 2001), of which 90 to 95% was the chrysotile variety. A large percentage of the total chrysotile production was from ophiolite complexes. Approximately 3 million metric tons each of "amosite" and "crocidolite" asbestos have thus far been mined worldwide, and ~350,000 metric tons of anthophyllite asbestos was mined in east Finland (Huuskonen et al., 1980). Commercial production of tremolite and actinolite asbestos has been small and sporadic. Russia is the leading producer of asbestos, followed (in decreasing amount of production) by China, Canada, Brazil, Zimbabwe, and Kazakhstan. At present, the only producing asbestos mine in the United States is located within the New Idria serpentinite, located at the south end of the Diablo Range, California.

THE ASBESTOS MINERALS

Chrysotile Asbestos

Chrysotile asbestos, the only fibrous member of the serpentine mineral group, has the ideal chemical formula $Mg_3Si_2O_5(OH)_4$. Small amounts of aluminum, iron, manganese, calcium, potassium, and sodium may enter the crystal structure of this mineral. The chrysotile fibers consist of long hollow "rolled up" tubes and each fibril is ~25 nanometers in diameter, with lengths varying from well under one micrometer for an individual fibril to well over 10 cm for fiber bundles (Fig. 1). Chrysotile is identified by a combination of optical properties, quantitative chemical analysis, polyfilamentous bundles, size, hollow tube morphology, and a characteristic powder x-ray diffraction or selected area electron diffraction pattern (Langer and Nolan, 1994; Wicks, 1999). Chrysotile asbestos is by far the most common asbestos mineral, generally occurring at trace levels in most serpentinites, whereas the commercial deposits contain several volume percent of this mineral.

Amphibole Asbestos

The three main forms of amphibole asbestos found in ophiolite complexes are: (1) anthophyllite asbestos, ideally $(Mg,Fe^{2+})_7Si_8O_{22}(OH)_2$, where 0 to 50 atom percent of Mg can be replaced by Fe^{2+}; (2) tremolite asbestos, ideally $Ca_2(Mg,Fe^{2+})_5Si_8O_{22}(OH)_2$,

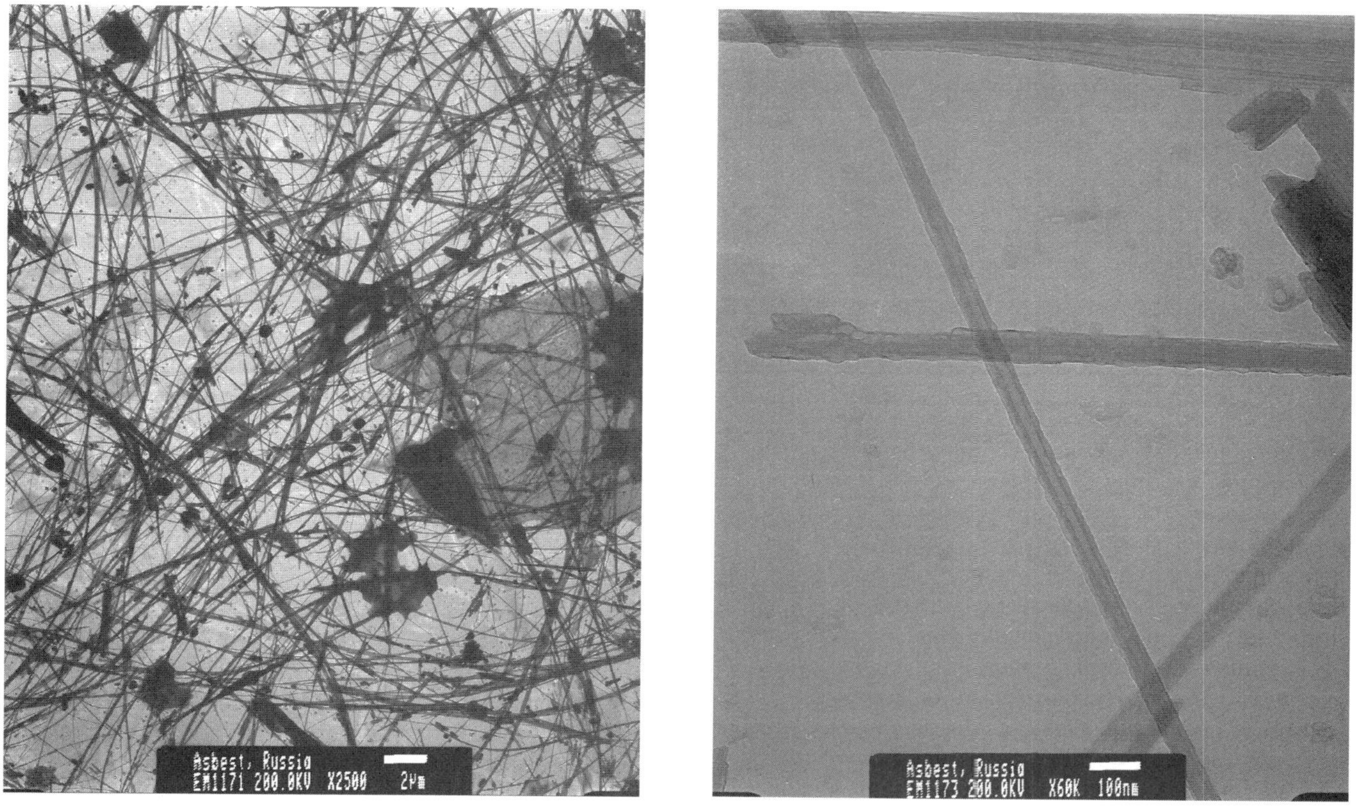

Figure 1. Transmission electron photomicrograph of chrysotile asbestos displaying polyfilamentous bundles of fibers and individual fibril having a central capillary associated with the hollow tube structure.

where 0 to 10 atom percent of Mg can be replaced by Fe^{2+}; and (3) actinolite asbestos, ideally $Ca_2(Fe^{2+},Mg)_5Si_8O_{22}(OH)_2$, where 10 to 50 atom percent of Mg can be replaced by Fe^{2+} (Leake et al., 1997). The amphibole asbestos minerals are characterized by their chemical composition, distinctive diffraction patterns, generally long straight fiber morphology, and transparency to electrons when viewed by transmission electron microscopy. Fibers typically vary in width from 0.1 to >1 micrometer and in length from a few micrometers to several centimeters.

Anthophyllite asbestos was mined mostly within an ophiolite complex located in the Paakkila (Tuusniemi) area of East Finland. With the close of the Finnish mines in 1975, there is now very little anthophyllite mined anywhere in the world. Small amounts of anthophyllite were mined earlier in this century within the serpentinite belt of the eastern United States, particularly in the states of Georgia, Maryland, and North Carolina. Minor amounts of tremolite asbestos and actinolite asbestos have been mined in various serpentinite belts, but were (with a few exceptions) of little commercial importance and mined mostly for local uses. Tremolite and actinolite asbestos is still possibly being mined to a very limited extent in ophiolites of Greece and Turkey, and perhaps in ophiolites of Cyprus and Iran. The asbestos minerals in the tremolite-actinolite series are associated with environmental mesotheliomas in several countries and are much more hazardous than chrysotile and anthophyllite asbestos. Thus, it would be useful for geologists to particularly note their presence and any mining activity during field investigations.

The two other commercial asbestos minerals, which do not occur in ophiolite complexes, have been important in commerce. These are: (1) grunerite asbestos, $(Fe^{2+},Mg)_7Si_8O_{22}(OH)_2$, usually referred to colloquially as "amosite" (from the acronym AMOS, representing Asbestos Mines of South Africa), which was mined only in metamorphosed banded iron formations of the Cape Province of South Africa; and (2) riebeckite asbestos, $Na_2(Fe^{2+},Mg)_3Fe_2^{3+}Si_8O_{22}(OH)_2$, usually referred to colloquially as "crocidolite," which was mined in four localities—in the banded iron formations of the Transvaal and Cape Provinces of South Africa, in the iron formations of the Hammersley Range of Western Australia, and in the Cochabamba area of Bolivia. The geologic occurrence and mineralogy of the asbestos minerals are discussed in detail in Ross (1981, 1984), Skinner et al. (1988), and Veblen and Wylie (1993).

It should be noted that five of the six minerals designated as asbestos (tremolite, actinolite, anthophyllite, grunerite, and riebeckite asbestos) can also occur in a non-fibrous form. It is important to use the mineralogical definition of asbestos for identification. Historically, non-asbestos fibers (for which there is no evidence of an asbestos health hazard) have occasionally been misidentified as asbestos (Langer et al., 1991; Langer, 2001).

Crystal Growth of Asbestos Fibers

Most asbestos appears to crystallize under very special conditions that occur within rock formations that are undergoing intense deformation characterized by folding, faulting, shearing, and dilation. Such deformations are often accompanied by the intrusion of magmatic fluids forming dikes and sills. The fibers crystallize in high strain environments, such as within folds, shear planes, faults, dilation cavities, and at intrusion boundaries. Ophiolite complexes are highly tectonized rocks containing many of these deformation attributes, thus they present conditions ideal for fiber formation.

Slip-fibers are formed within the fault and shear zones, the fibers crystallizing or recrystallizing from solutions that move within the two rock faces that compose the shear or *slip* plane—thus the term *slip-fiber* (Fig. 2). We observed this type of fiber formation in samples collected from a shear zone within a metamorphosed iron formation in northeastern Minnesota. There non-fibrous ferroactinolite had come into contact with low temperature acidic solutions that were moving through an active shear zone, causing the amphibole to recrystallize into a fibrous form having the fiber axis lying parallel to the shear and flow plane. Nolan et al. (1999) report on the occurrence of grunerite asbestos ("amosite") in a similar iron formation in Michigan. They state that "grunerite asbestos is developed within quartz-ankerite-stilpnomelane veins and along their contacts with the host rock and sills. The veins are deformed and exhibit signs of shearing, brecciation, faulting, and folding" (p. 3413).

Cross-fibers occur within cracks formed when the rock undergoes dilation due to tectonic stress, a process in which parallel cracks and fissures form open spaces in the rock. The process of folding in layered rocks can also produce openings or dilation cavities between adjacent layers. Asbestos crystallizes from a fluid phase, the fibers nucleating on a wall of the crack or cavity and growing toward the opposite wall—thus the term *cross-fibers*. Examples of cross-fiber growth within folds occur in the "crocidolite" and "amosite" deposits of South Africa and the "crocidolite" deposits of Western Australia. There, the fibers crystallized only where the Precambrian banded iron formations were folded into monoclines, inducing splitting and dilation cavities between mesobands (Dreyer and Robinson, 1981; Trendall and Blockley, 1970).

ASBESTOS AND DISEASE

Early History of Asbestos-Related Disease

While great advances were being made to advance the technology of asbestos mining and milling and invention of new uses for asbestos through product development, giving tangible benefits to society, asbestos workers were dying of asbestos-related diseases from exposure to asbestos dusts.

The ancient world recognized the remarkable properties of asbestos and wrote about the "magic mineral" and its ability to resist fire; however, the health effects associated with the inhalation of asbestos fiber were not known to any of the ancient writ-

Figure 2. Large sheets of nemalite (fibrous brucite-chrysotile), occurring as a slip-fiber in Mine Jeffrey, Quebec, Canada. The Canadian coin for scale is 2.8 cm in diameter.

ers, including Pliny and Strabo (Browne and Murray, 1990). For example, some cite Pliny the Elder (*Natural History*, Book 33, 122) as referring to the use of respiratory protection against asbestos; however, Browne and Murray (1990, p. 445) note that in this instance the workers were exposed to cinnabar (mercury sulfide), not asbestos. The earliest reports of asbestos-related disease were from France and the United Kingdom in 1906 (Murray, 1990). These early reports on asbestos-related disease describe pulmonary fibrosis associated with exposure to chrysotile asbestos. In 1927, the term asbestosis was used for the first time in English to describe this disease and the coated asbestos fiber present in the lung were referred to as "asbestos bodies" (Cooke, 1927).

These early cases of asbestosis generally occurred in individuals who also suffered from the disease tuberculosis, which complicated the clinical presentation of the new disease—asbestosis. It was not long before the first case of asbestosis without associated tuberculosis was reported in Glasgow by Dr. H.E. Seiler, who attributed the pulmonary fibrosis to asbestos exposure (Seiler, 1928). Once the clinical existence and features of asbestosis were known, its occurrence among asbestos workers could be detected.

In 1934, Drs. W.B. Wood and S.R. Gloyne reported finding two lung cancer cases among a series of 100 cases of asbestosis (Wood and Gloyne, 1934). The initial observations related lung cancer to asbestosis rather than asbestos. In the 1930s, lung cancer was still a rare disease because it was not until after World War I that cigarette smoking, the major cause of lung cancer, became a common habit. The relationship between cigarette smoking and lung cancer would not become commonly known until the 1950s (Doll and Hill, 1950).

In the United States, lung cancer was also reported among individuals with asbestosis (Lynch and Smith, 1935) and case reports of lung cancer among individuals would continue to appear for decades. In the United Kingdom, Doll (1955) reported on a cohort of 113 men with at least 20 years of occupational exposure to chrysotile asbestos. Eleven lung cancers occurred in this cohort where <1 case of lung cancer would have been expected, based on the rate of this disease among males in the general population. All of the lung cancer cases showed histological evidence of asbestosis. These reports provided strong evidence that individuals with asbestosis were at increased risk of lung cancer, but no mention is made about smoking in general or the smoking habits of the workers.

Diseases Related to Exposure to the Various Asbestos Minerals

Three principal diseases are related to exposure to the asbestos minerals. (1) *Asbestosis* is a non-malignant diffuse interstitial fibrosis of the lung tissue. High asbestos exposure leads to scarring of the lung, causing it to become stiff, resulting in a restriction in pulmonary function and reduction in the lung's ability to exchange carbon dioxide for oxygen. (2) *Lung cancers* related to asbestos exposure are bronchogenic carcinomas, which include squamous carcinomas, small- or oat-cell carcinoma, and adenocarcinomas. Cancers that arise in other parts of the lung, such as alveolar carcinomas and sarcomas, are rare and are not known to be caused by asbestos. (3) *Mesothelioma* is a cancer of the pleura, pericardium, and peritoneal membranes, which surround the lung, heart, and abdominal cavities, respectively. The pleural mesothelial cells occur almost as a monolayer surrounding the outside of the lung and the inside of the chest wall. Mesothelioma is mainly restricted to those exposed to the amphibole asbestos dusts—"crocidolite," tremolite, and grunerite.

Those exposed to asbestos sometimes develop pleural plaques on the covering epithelium of the lung. Plaques are largely acellular bundles of collagen, sometimes calcified, and are detected on chest X-ray films. Although benign, they may indicate future development of asbestos diseases. For complete reviews of asbestos-related diseases, see Dement et al. (1986), Skinner et al. (1988), and Henderson et al. (1992).

For regulatory purposes, all six asbestos minerals have often been assumed to be equally harmful to human health (until proven otherwise), but recently this assumption has been questioned (see Wilson and Crouch, 2001, p. 231–232). Examination of the many epidemiological studies, where the effects of the individual asbestos minerals can be evaluated separately, provides convincing evidence that "crocidolite" and tremolite asbestos are clearly more dangerous than chrysotile asbestos (McDonald and McDonald, 1996; Hodgson and Darnton, 2000; Nolan et al., 2001). "Crocidolite" (also referred to as "blue asbestos") in particular is associated with a high incidence of mesothelioma. Wagner and co-workers (Wagner et al., 1960; Wagner, 1991) first discovered that the pleural form of this very rare disease could be caused by "crocidolite" dust during their study of the occurrence of 33 mesothelioma cases among individuals exposed to "crocidolite" in the northwestern Cape Province of South Africa. Mesothelioma occurs so rarely that prior to discovery of the many mesothelioma cases occurring among the Cape Province miners, many pathologists doubted the existence of the disease (McDonald and McDonald, 1996). Dr. J. Christopher Wagner (Wagner, 1991) not only proved that "crocidolite" is an etiological factor in mesothelioma, but also established mesothelioma as a respiratory disease, thereby defining one of the most important thoracic tumors identified in the twentieth century.

Conventional epidemiological studies could not address the health effects of some of the other fibrous minerals of less commercial importance because a statistically significant cohort of exposed individuals could not be assembled; however, prevalence studies of certain populations with a high incidence of mesothelioma began appearing in the medical literature, the first being a study of the "crocidolite"-exposed individuals in South Africa. The second significant study was of residents of the village of Karain, located in the Cappadocia region of Turkey. In Karain, from 1970 to 1978 there were 76 deaths, of which 50 were due to pleural mesothelioma (Baris et al., 1979; Saracci et al., 1982).

Background mesothelioma incidence in the general population is one case among 10,000 deaths, making the incidence in Karain more than 6,000-fold above background. No asbestos exposed cohort has experienced an incidence of mesothelioma as high as those living in Karain. The fibrous zeolite mineral erionite, $NaK_2MgCa_{1.5}(Al_8Si_{28}O_{72}){\cdot}28H_2O$, is found in the local ash-flow tuffs, soils, and air samples and is implicated in the mesothelioma epidemic (Baris, 1991). The particular erionite samples from the Karain area contain extremely thin and long fibers, comparable in size and shape to the finest amphibole asbestos fibers. Experimental animal studies, in which rats were exposed to erionite by inhalation, indicate erionite has a much greater potency to induce mesothelioma than does "crocidolite" asbestos (Wagner et al., 1985). In addition to erionite, tremolite asbestos has been associated with environmental mesothelioma in Turkey and cases have been reported among Turkish immigrants (Dumortier et al., 2001).

Studies of Workers Occupationally Exposed to Asbestos

Three commercially important forms of asbestos, chrysotile, "amosite," and "crocidolite," were used extensively throughout the industrialized world from the late 1800s until recently (Alleman and Mossman, 1997). Ross (1984) and Ross and Virta (2001) provide historical data on asbestos use in the United States in terms of the import tonnages and the commercial applications of each of the asbestos fiber types. A large number of miners and trades workers were exposed to one or more of these three types of asbestos, thus statistically significant epidemiological studies could be made of various cohorts of workers.

Anthophyllite use was restricted mostly to Finland where epidemiological studies of Finnish asbestos miners were undertaken (Meurman et al., 1974). Asbestosis and lung cancer risk increased with the miner's exposure to anthophyllite; no mesotheliomas were reported. Evidence for anthophyllite asbestos-related mesothelioma has been rather limited. Timbrell et al. (1971) hypothesized, on the basis of anthophyllite fiber having a diameter greater than "amosite" and much greater than "crocidolite," that exposure to anthophyllite would rarely if ever produce mesotheliomas (see Timbrell, 1989, for further discussion). More recently, lung content studies reported finding anthophyllite asbestos within the functional tissue of the lung (the parenchyma) of individuals with mesothelioma. Generally, these individuals were not anthophyllite miners and the analysis revealed the presence of other asbestos fiber types more commonly associated with mesothelioma (Tuomi et al., 1989). In 1994, four mesothelioma cases were reported among anthophyllite asbestos miners (Karjalainen et al., 1994). All four had exposure sufficient to induce asbestosis, which is consistent with the evidence that anthophyllite asbestos is weakly mesotheliomagenic. This is the most persuasive evidence that uncontrolled exposure to anthophyllite asbestos may result in mesothelioma, particularly if the fibers are very thin. A mesothelioma case has recently been reported in a 38-year-old male without asbestosis from neighborhood exposure (Rom et al., 2001). We remain skeptical about etiology in this case report.

Little tremolite and actinolite asbestos was used in commerce; thus, no large cohort could be assembled for the usual epidemiological analysis. As will be discussed below, however, mesothelioma cases have appeared in certain small groups of workers mining and processing tremolite asbestos for building materials.

The Asbestos Trades Workers

A very significant increased incidence, in relation to the general male population, of lung cancer, mesothelioma, and asbestosis is found in men who were employed in the "asbestos trades," e.g., work involving the asbestos insulation of pipes, steam boilers, and buildings and in the production of asbestos textiles, roofing materials, friction materials, wall board, etc. These trades workers generally handled all three common forms of commercial asbestos (chrysotile, "amosite," and "crocidolite") during their working careers. Ross (1984, p. 307) lists the mortality data for 13 epidemiological studies of asbestos trades workers. These 13 studies demonstrate that in most asbestos trades worker cohorts, there was a large excess death due to lung cancer, mesothelioma, and asbestosis. Lung cancer mortality, however, was much greater for smokers than for nonsmokers.

The asbestos miners and millers. Unlike the trades workers, the men working in the mining and milling of asbestos were generally exposed to only one type of asbestos mineral; a few exceptions occur in the mining regions of South Africa, where some workers were employed in "crocidolite," "amosite," and chrysotile mines. Epidemiological studies of miners and millers exposed to only one type of asbestos mineral permit one to ascertain the human health effects of the different types of asbestos. Ross (1984, p. 309) gives mortality data for the five major epidemiological studies of asbestos miners and millers; one is a cohort of anthophyllite asbestos miners, one is a cohort of "crocidolite" miners, and three are cohorts of chrysotile miners. The mortality patterns noted on comparing the trades and mines cohorts makes it clear that there are very significant differences in the human health effects of the four different asbestos minerals, with "crocidolite" being the most dangerous, followed in decreasing order of pathogenicity by "amosite," and chrysotile (for a recent review, see Hodgson and Darnton, 2000). Mesothelioma incidence is particularly high in those exposed to "crocidolite," moderately high to those exposed to "amosite," and very low in those exposed to only to anthophyllite or chrysotile asbestos.

Moderate-Level Exposures to Mostly Chrysotile Asbestos

Several epidemiological studies of asbestos trades workers show no significant increased risk of asbestos-related disease among individuals exposed to chrysotile asbestos dusts, but having little or no exposure to amphibole asbestos dusts. Their exposure levels, however, were much higher than those observed in schools and in other non-occupational settings (Wilson et al., 1994; Nolan and Langer, 2001). Examples of such epidemiological studies include: (1) Cardiff, Wales, asbestos cement workers (Thomas et al., 1982); (2) New Orleans, Louisiana, asbestos cement workers (Weill et al., 1979); (3) Swedish asbestos cement workers (Ohlson and Hogstedt, 1985); (4) English asbestos cement factory workers (Gardner et al., 1986); and (5) British friction product workers (Newhouse and Sullivan, 1989).

The health experience of the five above-mentioned cohorts is in great contrast to many other asbestos workers who worked under the historically high exposures, such as occurred in asbestos textile manufacturing, during installation and removal of asbestos in heating and electrical conduits, in poorly ventilated asbestos factories, and in workplaces and mines where there was exposure to large quantities of "crocidolite" and "amosite" asbestos dust (Ross, 1981, table 4). Worker cohorts engaged in such activities were often exposed to dust levels of hundreds of fibers per ml (Gibbs and DuToit, 1973; Doll and Peto, 1985).

Studies of Those Who Had a Non-Occupational Exposure to Chrysotile Asbestos

Mines within the towns of Thetford Mines and Asbestos, located in the province of Quebec, Canada, have produced over 40 million metric tons of chrysotile asbestos since first opening in 1878. Many mines have closed, but two large-capacity mines are still producing asbestos. These two towns have a combined population of 30,000, and a large number of the male residents work, or have worked, in the asbestos mines and mills. Most of the asbestos workers are male; very few are female. Historically, the ambient asbestos exposures were high in these towns and residents were continually exposed to asbestos dust from the nearby mines, mills, and mine waste dumps. Camus et al. (1998) give estimates of the number of ambient asbestos fibers (those longer than 5 micrometers) per milliliter (f/ml) of air in the two asbestos towns for the period 1900–1989. They state that average levels exceeded 1 f/ml during World War II and into the early 1950s. Thus, ambient chrysotile asbestos dust levels in these Quebec towns prior to the introduction of modern dust control technology in the 1970s were ~230 to 23,000 times the average levels found in schools with asbestos insulation (Ross, 1995, p. 186).

A mortality study was made by Siemiatycki (1982) of 1130 deceased women who had lived in Thetford Mines and Asbestos but did not work in the asbestos industries. This study shows that the health of these women was unaffected by these very high non-occupational 24-hour-a-day lifetime exposure to asbestos dust. The standard mortality ratio (SMR = observed deaths/expected deaths) for all cancer in these women is 0.91, for lung cancer is 1.07, for digestive cancer is 1.06, and for non-malignant respiratory disease is 0.58. None of these SMRs is statistically different from the control cohort composed of unexposed women of similar socioeconomic background living outside the asbestos mining townships (SMRs >1.2 might imply a statistically significant health risk). A previous study by Graham (1981) also showed no excess cancer among the female residents of these Quebec mining towns (also see Ross, 1984, p. 82–85). More recently, Camus et al. (1998) reported similar results in an update for lung cancer among this same population of women with non-occupational exposure to chrysotile asbestos in the Quebec chrysotile mining towns. Their study indicates that the U.S. Environmental Protection Agency overestimates the potency of chrysotile asbestos to increase lung cancer risk by a factor of at least 10.

A major breakthrough in bringing a new understanding of the relative health risks of asbestos to a scientific, political, and public audience came from the work of Mossman et al. (1990). Their paper was published in *Science* at the time when billions of dollars where being spent to remove asbestos from schoolrooms and other buildings in the United States. These authors stated: "clearly the asbestos panic in the U.S. must be curtailed." Their very low risk estimates to children attending classes in schools with asbestos-containing material were reviewed in detail by HEI.AR (1991) and Wilson et al. (1994). Data given in these references show that the average concentration of asbestos measured in 219 American schools is 0.00022 f/ml. Using this average fiber concentration and the most pessimistic methods (or worst case scenario) for calculating risk, the risk for residing in the classroom six hours a day, five days a week for 14 years is one excess cancer death per million lifetimes (Ross, 1995, table 1).

SOME OF THE IMPORTANT OCCURRENCES OF ASBESTOS IN OPHIOLITE COMPLEXES AND ASSOCIATED SERPENTINITES

In this review, we use a broad definition of the term *ophiolite* to include those rocks that at least in part are pieces of oceanic crust that have been obducted onto the edges of continental plates. The idealized ophiolite, from the base upward, includes: (1) layered cumulates or mantle sequence comprising tectonized dunites, lherzolites, and harzburgites; (2) gabbros; (3) sheeted diabase dikes; (4) pillow basalts; and (5) various marine sediments. The Troodos ophiolite of Cyprus (Robinson and Malpas, 1990) and the Semail ophiolite of Oman (Vetter and Stakes, 1990) are two of the best examples of places where complete or nearly complete sequences of these five rock types are exposed; however, not all of these rock types can be observed in outcrop in many of the ophiolite occurrences, including some of the important serpentinite belts of the world. O'Hanley (1996) gives a very comprehensive review of serpentine mineralogy and the petrology and geology of serpentinites.

Not included in this review are the chrysotile-bearing serpentinites of various greenstone terranes, for example, the Abitibi greenstone belt of Ontario, Canada, and the Barberton greenstone belt of South Africa, because characteristic ophiolite assemblages are not observed (O'Hanley, 1996, p. 217). Also not included in this discussion are the carbonate-hosted chrysotile deposits, such as those found in the Barberton-Caroline District and near Kanye, South Africa; in the Laiyuan district, Hepeh Province, China; and in Gila County, Arizona, USA.

Anthophyllite Asbestos Deposit in Finland

Significant commercial production of anthophyllite asbestos has occurred only in the north Karelian mountains of eastern

Finland. Here, several serpentinized ultramafic pods or lenses, which were originally dunite or enstatite-bearing peridotites, were exploited. These serpentinized masses, forming part of the Outokumpu ophiolite complex, are thought to have been a part of much larger slabs of oceanic material that during the Karelian mountain-building episode, broke up into smaller volumes and were emplaced into the Karelidic schist belt. Late Karelian granites were injected into the schist belts, and where these are in contact with the serpentinized peridotites, commercial quality anthophyllite was developed (Aurola, 1954; Simonen, 1980). The anthophyllite-bearing ultramafic pods typically contain relict olivine surrounded by serpentine aggregates and bundles of anthophyllite needles. Sites of major mining of anthophyllite asbestos were on the northern shore of Lake Juojärvi and at East Savo and Maljassalmi. Between 1918 and 1975, 350,000 metric tons of asbestos were produced (Huuskonen et al., 1980). The mines closed after 1975. Small quantities of anthophyllite asbestos were mined in the other parts of the world.

Major Chrysotile Asbestos Deposits in Ophiolite Complexes and Associated Serpentinites

Southern Quebec, Canada, and Northern Vermont, USA

The important North American chrysotile asbestos deposits are located in the towns of Asbestos and Thetford Mines, Quebec, Canada, and in the town of Eden Mills, Vermont, USA. These deposits are in the northern part of the Appalachian ultramafic belt, which extends intermittently from northwestern Newfoundland to the state of Alabama, USA. The Canadian and United States deposits are found in a narrow belt of Ordovician ophiolites, which have been extensively dismembered and sheared. In the asbestos mining area of Quebec, the ophiolite sequence consists of an amphibolite sole and overlying harzburgite, dunite, pyroxenite, gabbro, basalt, and pelagic sedimentary rocks. Both the harzburgites and dunites were serpentinized during emplacement of the ophiolite (Williams-Jones et al., 2001). The ophiolite dismemberment was accompanied by extensive shearing and the emplacement of numerous felsic dikes. Chrysotile asbestos probably formed in late-stage fractures during the stress environment of tectonic transport and emplacement of the serpentinite. Williams-Jones et al. (2001) note that minor amounts of amphibole occurring in the ore bodies are found mainly adjacent to the felsic dikes, but are effectively absent from the chrysotile-bearing serpentinite. They also observed: (1) narrow selvages of tremolite around the felsic dikes in the serpentinite, (2) tremolite asbestos as narrow shear and fracture-fillings in pyroxenite, and (3) actinolite asbestos as disseminations and fracture fillings in albitized felsic dikes and in slate proximal to serpentinite. A similar occurrence of tremolite asbestos was observed by one of us (Ross) in a vermiculite deposit in southern Virginia, USA. Here, small amounts of tremolite asbestos form thin slip-fiber coatings on the contact surfaces between the intruded felsic dikes and the altered amphibolite host rock. The contact surfaces have been sheared.

There are now only two active asbestos mines in southern Quebec: Mine Jeffrey, in the town of Asbestos, and Black Lake Mine, near the town of Thetford Mines. Mine Jeffrey is the largest Canadian producer, having an estimated yearly milling capacity of >250,000 metric tons. The Mine Jeffrey pit, the largest asbestos pit in the western hemisphere, is >2 km in diameter and is ~350 m deep (Fig. 3). Geologically, the asbestos deposit at Eden Mills, Vermont, is very similar to deposits in Canada. The Vermont mine is now closed, but at one time processed a large amount of slip fiber chrysotile asbestos, which was much desired for brake linings.

Coalinga, California, USA

The New Idria serpentinite (Fig. 4), which crops out within a mountainous 114 km^2 area 26 km northwest of Coalinga, California, forms the core of a large antiform between the San Andreas fault on the west and the Great Valley on the east. Tectonic processes relating to Pacific plate subduction gave rise to the California Coast Ranges and to exposure of peridotite-rich fragments of the Pacific Ocean crust and mantle and associated marine sediments, such as the Franciscan Group of Late Jurassic–Early Cretaceous age. Slices of Franciscan rocks and the peridotites, the latter having undergone pervasive alteration to serpentinite, have been incorporated and extruded into and over the Great Valley sedimentary sequence. Diapiric uprise of the serpentinite breached the Great Valley sediments sometime between late Oligocene and Early Miocene (Coleman, 1986; 2000). This serpentinite consists of a large amount of highly sheared and pulverized rock fragments and powders, as well as boulders of partially altered serpentine-rich rock. Coleman (1996, p. 19) regards the New Idria serpentinite body as "a unique geologic phenomenon created by plate tectonic evolution of western North America." He further states that "thousands of tectonic events acting on the serpentinite has produced a huge deposit of short-fiber asbestos by tectonic milling and recrystallization." Asbestos-bearing debris from the actively moving diapir has been entering the stream valleys for millions of years, and continues to the present day.

The average New Idria serpentinite rock contains 5 to 15 volume percent of this mineral (Mumpton and Thompson, 1975), but rock in the area in which the King City Asbestos Corporation is still actively mining contains up to 60% chrysotile asbestos The reserves of short-fiber chrysotile asbestos in the KCAC deposit are enormous. No amphibole asbestos has been reported in this serpentinite; its rarity is probably related to the absence of igneous intrusions within the serpentinite body.

Central and Southern Ural Mountains of Russia and Kazakhstan

The most important chrysotile asbestos deposits in Russia are within serpentinized gabbros, peridotites, and pyroxenites associated with the ultramafic ophiolite massifs of the central and southern Ural Mountains (Petrov and Znamensky, 1981; Spadea and Scarrow, 2000). The most productive mines are operated by the Uralasbest Company and are in the Bazhenovoskoye district near Asbest City, 85 km east of Ekaterinburg

Figure 3. Mine Jeffrey, Asbestos, Quebec, Canada—the largest chrysotile open pit mine in the western hemisphere.

Figure 4. Photograph of the barren hillsides of the New Idria serpentinite, Coalinga, California. The serpentine is essentially a soft soil containing large amounts of chrysotile fiber, and is easily eroded by wind and rain.

(Shcherbakov et al., 2001). Since 1889, this one mine has produced >40 million tons of chrysotile asbestos, making Russia the greatest source of asbestos in the world, now surpassing in quantity that from all the mines in the Eastern Townships of Quebec (Ross and Vita, 2001). The Bazhenovskoye three-pit mine complex is >10 km in length (Fig. 5).

Two other large deposits of chrysotile asbestos associated with the serpentinized ultramafic rocks of the southern Urals are in the Dzhetygara District of northwestern Kazakhstan, 225 km southwest of Kustanay, and in the Kiembay District within the Orenburg Oblast of Russia. They are, after the Bazhenovskoye deposits, respectively, the second and third largest producing districts of the former USSR.

Northwestern Italy

Some of the asbestos deposits in northwestern Italy were known in Roman times. One of the most productive was the Balangero deposit, which is situated within the Balangero serpentinite ~40 km northwest of Turin. During the 1960s, chrysotile production was between 100,000 and 150,000 metric tons per year, but production declined in the 1980s and the mine closed in February of 1990 (Bowles, 1955, p. 47; Silvestri et al., 2001). The serpentinite formed from a spinel-plagioclase lherzolite and is interpreted as a satellite body of the Lanzo massif. The asbestos ore body contains massive, schistose, and highly fractured serpentinite rock. Long- and short-fiber chrysotile is present in the mine but only the short-fiber chrysotile is of commercial use. The short-fiber asbestos occurs as a stockwork of slip-fiber veins >2 mm thick (O'Hanley, 1996, p. 207).

At one time, tremolite asbestos was mined in some quantity from the ophiolites in the Susa Valley near the French border, as well as in Aosta Valley near Ivrea, Italy. Mining in the Aosta Valley started in 1865, but there has been little activity since 1905 (Bowles, 1955, p. 46–47).

Northern Greece

The most important chrysotile asbestos deposit in Greece, the Zidani Mine, is in a serpentinized dunite within the Pelagonian basement gneisses, which structurally underlie the Livadi ophiolite complex. The mine is 30 km south of Kozani, West Macedonia. Skarpelis and Dabitzias (1987) have reported that the chrysotile asbestos forms a "slickenfiber" coating on lens-shaped blocks of antigorite. The authors give evidence that the slickenfiber (slip-fiber) were originally veins of cross-fibers that, through subsequent deformation, had been recrystallized as slip-fiber coatings on the antigorite blocks. The Zidani mine and mill started operation in 1981, producing mainly slip-fiber. In 1996, production of asbestos was 60,000 metric tons. The mine recently closed, in part due to a lack of a European market for chrysotile asbestos.

Troodos Mountains, Cyprus

The Troodos ophiolite complex can be divided into (1) the ultramafic sequence consisting of serpentinized peridotites, tectonized harzburgites, lherzolites, and dunites overlain by (2)

Figure 5. The ridge between the south and central pit of the three-pit Bazhenovskoye Mining complex in Asbest City, Urals Region of Russia. This is the largest chrysotile mine in the world.

cumulate peridotites and gabbros (also referred to as the layered series); (3) the sheeted dike complex, composed of relatively fractionated basalts; (4) the extrusive sequence, composed of pillowed and nonpillowed lavas, which are overlain by (5) deep-water sediments (Glass, 1990; Robinson and Malpas, 1990). The commercial deposit of chrysotile occurs in heavily smashed and sheared serpentinized enstatite-olivinite rock, which covers at least two-thirds of the ultramafic area and occupies the highest part of the pluton around Mt. Olympus (Wilson, 1959, p. 79–80). Most of the chrysotile asbestos is the short-fiber type and occurs as a stockwork of cross-fiber veins as much as 15 mm thick. Slip-fiber has sometimes developed along shear planes. Although asbestos was mined in ancient times (probably within the Arakapas Forest area), modern development began only in 1904 and major production started after the First World War. Production for 1928 through 1957 averaged ~9,000 metric tons per year (Ingham, 1959, p. 160). In 1977, 40,000 metric tons were mined; by 1989 the mine was closed. Figure 6 shows three views of the now-abandoned Cyprus chrysotile asbestos mine, as well as some of the present reclamation activities.

From the descriptions of Wilson (1959), it appears that there is a significant amount of tremolite asbestos in the Troodos complex; for example, Wilson (1959, p.115) reports of "veins composed of a microcrystalline mat of tremolite." From our examination of occurrences of tremolite asbestos, particularly in the serpentinite belt of eastern North America, we suggest that these "mats" of tremolite reported by Wilson occur in a shear zone, the shearing causing nonfibrous tremolite to recrystallize into a fibrous mat of tremolite asbestos, sometimes referred to as "mountain leather." Further evidence of the occurrence of amphibole asbestos in the Troodos area was furnished by Dr. Eleni Geogiou Morisseau of the Geological Survey Department, Cyprus. She showed to us in November of 2000 a large sample of actinolite asbestos (Fig. 7) collected from a highly sheared serpentinite within the tectonized harzburgites of the Akapnov Forest area. These rocks lie within the Arakapas (transform) Sequence, an ophiolite complex separated from the main part of the Troodos Massif by the Arakapas transform fault (MacLeod, 1990).

Figure 6. Remediation work in progress on the now-closed Cyprus chrysotile asbestos mine. A: Photograph of engineering work involving lowering the angle of repose of the talus slopes by the construction of an intervening series of benches between the slope sections. B: Photograph of to-be-stabilized talus slopes (right) and stabilized slopes and benches (left). C: Photograph of partially completed slope and bench reconstruction (right), as well as a view down the stream valley toward the town of Kato Amiantos (left center).

EXPOSURE TO TREMOLITE ASBESTOS IN OPHIOLITE LOCALITIES IN TURKEY, GREECE, CYPRUS, CORSICA, AND NEW CALEDONIA

In the discussion that follows, six ophiolite localities are described where the residents of rural villages were exposed to tremolite asbestos through its use as ingredients for stucco and whitewash. Exposure to the tremolite asbestos dusts has caused a significant number of mesothelioma cases in these villages. The published data do not give enough geological and mineralogical details to pinpoint the exact source of the tremolite asbestos, but we suggest that the tremolite asbestos appears within the ophiolite complexes at the contacts of the various later stage intrusions, and in shear and fault zones open to hydrothermal fluids. It would be important for geologists to visit the sites where the tremolite asbestos was quarried in order to obtain an exact description of the petrologic relationships between the tremolite asbestos and the host rocks, and to confirm that asbestos mineral in question is indeed tremolite and not some other type of amphibole.

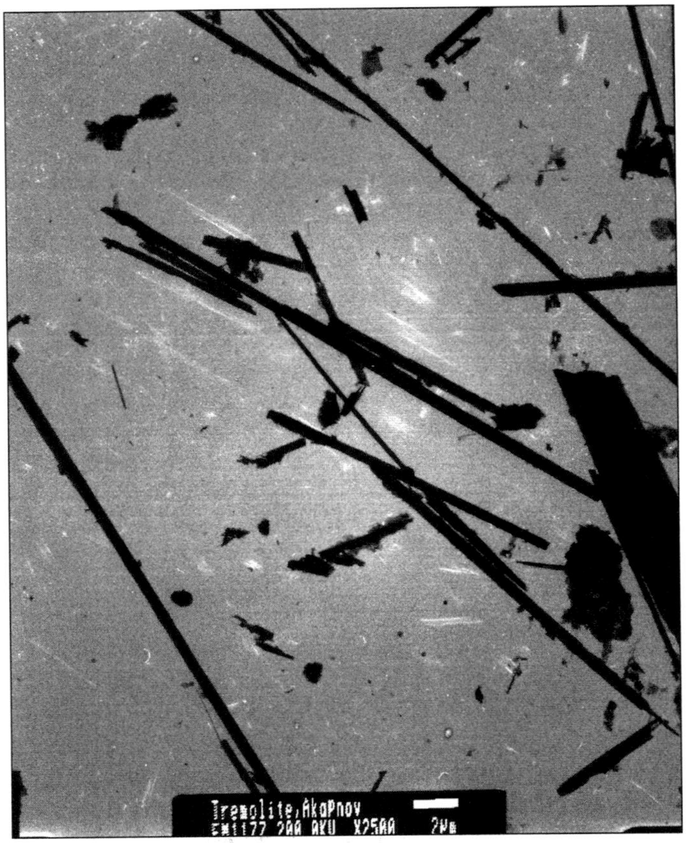

Figure 7. Transmission electron photomicrograph of tremolite asbestos from Akapnov Forest, Cyprus.

Guleman and Killan Group Ophiolites within the Maden Complex, Turkey

Yazicioglu et al. (1980) reported on 41 patients having respiratory cancer that were admitted to the Diyarbakir Chest Hospital, Diyarbakir, Turkey, from 1977 to 1978. Of these 41 cases, 23 and 18 were cancers of mesothelial tissue and lung, respectively. Of these, 22 of the mesothelioma and 11 lung cancer cases came from the "asbestos" villages of Çermik, Ergani, Çüngüs, Maden, and Siverek, whereas one mesothelioma and seven lung cancer cases came from the "non-asbestos" villages of Lice, Kulp, Dicle, Hani, Silvan, Harzo, and Çinar. The extraordinarily high incidence of mesothelioma among both genders in the five "asbestos" villages of very low population strongly suggested that the villagers were exposed to asbestos environmentally. It was discovered that tremolite asbestos was used locally to make a whitewash or stucco for the walls, floors, and roofs of houses.

The whitewash contained fibrous tremolite and the nonfibrous minerals talc, chlorite, antigorite, and lizardite. Although occasional chrysotile fibers were found in the environment, Yazicioglu et al. (1980) attributed these mesothelioma cases occurring in the Çermik village region to exposure to tremolite asbestos. The tremolite asbestos was quarried in the nearby mountains by the male population for local use and for sale elsewhere. Women then ground the asbestos ore into a powder and mixed the powder with water to form a slurry for application to buildings. This process was repeated every year; consequently the homeowners were repeatedly exposed to tremolite asbestos dusts from an early age. We note that these five "asbestos" villages are located within or a few kilometers from the Guleman and Killan Group ophiolites that form part of the Maden Complex. Aktas and Robertson (1990) describe the Killan Group as consisting of serpentinized peridotite, gabbro, dolerite, basic pillow lava, and pelagic sediments. These observations strongly suggest that the tremolite asbestos was quarried within one or more of the rock units within these complexes.

Ankara Ophiolite, Turkey

Apparent asbestos-like diseases of the chest, including mesothelioma, have also been reported in the small Anatolian Village of Çaparkayi in the Sabanözü area of the Çankiri district, Turkey (Baris et al., 1988a, 1988b). Four cases of pleural mesothelioma were reported in a population of 425 over a three-year period. All of the tumors occurred in women between 26 and 40 years of age. The non-occupational nature of exposure is indicated by the tumors occurring at a young age and in women. Although there is no asbestos mine near this village, the villagers commonly used white stucco described by Baris et al. (1988b, p. 838, fig. 2) as "rich in tremolite asbestos including some very fine fiber." The Baris report indicates that the high incidence of mesothelioma and some of the lung and pleural tissue abnormalities in the village are associated with exposure to tremolite fibers. The villages of Sabanözü and Çankiri lie a few kilometers west and east, respectively, of the Eldiven Dagi, a mountainous area that lies within the Ankara ophiolite mélange belt of northern Anatolia (Tankut and Gorton, 1990, fig. 1). We suggest that the residents of Çaparkayi, and probably residents of nearby villages, obtained the stucco ingredients from veins of tremolite asbestos occurring within the Ankara ophiolite complex.

Pindos Ophiolite, Greece

Six out of the seven reported deaths (3 males and 4 females) from malignant pleural mesothelioma occurred among residents of the villages of Milea, Metsovo, Anilio, and Votonosi. These villages are located in the Pindos mountain region of northwestern Greece. Of the 268 people who underwent X-ray examination, 46% showed some pleural change, including bilateral pleural plaques, pleural thickening, and restrictive lung function. These lung changes and the mesothelial cancers constitute the cluster of disease referred to as "Metsovo lung" (Constantopoulos et al., 1985, 1987; Langer et al., 1987). Biopsy of the lung tissue in the mesothelioma cases, as well as others with pleural disease, contained tremolite asbestos fibers. Also, soil, whitewash, and settled dust samples from Metsovo showed the presence of tremolite asbestos (Langer et al., 1987). Before 1940, virtually all of the inhabitants of the Pindos area applied tremolite asbestos-bearing whitewash to their homes. The village of Metsovo is located within or very near to the Pindos ophiolite. Rassios and Smith (2000, p. 475–476, figs. 2, 4) describe the Pindos ophiolite as having an outcrop area of >1300 km^2. It contains harzburgite units lying above a metamorphic sole composed mainly of amphibolites. In addition, they describe intrusions of pyroxenite and gabbro dikes and cumulate plagioclase dunite and troctolite sills. We suggest that the tremolite asbestos is associated with some of the igneous intrusions or mélanges within the ophiolite complex and was quarried in the past for the special use in stucco, whitewash, and other building materials.

Troodos Ophiolite, Cyprus

McConnochie et al. (1987) reported on a health study over the period 1969–1986 of 14 residents of Cyprus, several of whom had lived in mountain villages near the Troodos asbestos mine. All 14 cases were initially reported to have died of mesothelial cancers, although a complete pathological examination verified only seven of the deaths (four male asbestos miners, and one male and two female non-miners) as due to mesothelioma. It was inferred that the disease was caused by asbestos dust from the mine; however, McConnochie et al. suspected that tremolite asbestos was the cause of the mesotheliomas because analysis of the lung tissue of two humans and six sheep (residing within five miles of the mine) indicated that both chrysotile and tremolite asbestos were present in the tissue. Proper identification of the amphibole minerals is suspect because "crocidolite" was also reported to occur in the sheep lung tissue. "Crocidolite" has never been identified in an ophiolite complex. Another peculiarity is that no tremolite asbestos was found in the samples taken from the Troodos mine, although tremolite asbestos was found in environmental dust samples taken from two "roof caves" in two villages. McConnochie et al. (1989) report that stucco used in the Troodos region contained fine fibrils of chrysotile and long, thin tremolite fibers. In our visit to Cyprus in November 2000, we searched unsuccessfully for evidence of the use of tremolite asbestos in whitewash and stucco. We found, rather, that the whitewash presently used on fruit trees as an insect repellent is asbestos-free. The tremolite asbestos was perhaps used sporadically in the mountain villages of Cyprus many years ago, thus a few mesothelioma cases may have resulted from exposure to this mineral, but whether there is still a danger from exposure to tremolite asbestos in Cyprus remains to be seen. We, however, have noted that fibrous amphiboles are present in the Troodos complex.

Corsican Ophiolite Sequences

Rey et al. (1993, 1994) found pleural plaques in 41% of the residents of the northeastern Corsican village of Murato, situated in an area underlain, according to the authors, by tremolite and chrysotile-bearing rocks. Only 7.5% of the residents of the "control" village (Vezzani, located in an area that is not underlain by asbestos-bearing rocks) were diagnosed as having pleural plaques. Rey et al. (1993) also state that 13 cases of environmental mesothelioma have been recorded in Corsica since 1980. Airborne levels of tremolite asbestos (Rey et al., 1993, table 3) were much higher in Murato (6 to 72 ng/m^3) than in Vezzani (0.4 to 1.0 ng/m^3), but the chrysotile levels in the two villages were similar (Murato, 0.3 to 1.0 ng/m^3; Vezzani, 0.1 to 0.3 ng/m^3). Rey et al. (1993, fig. 1) give a generalized map showing the villages of Murato and Vezzani and areas containing "surface deposits of asbestos." Major ophiolite sequences are found in northeastern Corsica, including the Balagne and Nebbio nappes. Murato is situated very near the Nebbio nappe, with outcrops of serpentinite and metagabbro, whereas Vezzani appears to be located within calcschist terrane (see Brunet et al., 2000, figs. 4 and 5). The papers by Rey et al. (1993, 1994) do not account for the source of the airborne asbestos, but do mention the existence of a chrysotile asbestos mine at Canari that closed in 1965. The town of Canari is also near the Nebbio nappe.

New Caledonia Ophiolite

The Pacific island of New Caledonia contains a major ophiolite body that was emplaced mainly during the Oligocene (Gealey, 1979, fig. 6, p. 230). Coleman (1971, p. 1216) reports, "Large plates of ultramafic rock, consisting primarily of dunite and harzburgite, cover large areas of New Caledonia." Coleman further states that "serpentinite is developed mainly along the thrust contacts; the massive peridotite slabs show 10%–20% serpentinization." Luce et al. (2000) report a high incidence of lung disease in New Caledonia. During the period 1993–1995, 15 pleural mesotheliomas (9 female and 6 male), 1 peritoneal mesothelioma, 228 lung cancers, and 28 laryngeal cancers were diagnosed. The risk of pleural mesothelioma was strongly associated with the use of a whitewash (referred to locally as pö) composed of very friable rock derived from local outcroppings. From 1930 to the end of the 1960s, pö was used to coat indoor and outdoor walls of houses (Luce et al., 2000, p. 259). The whitewash, suspended in water during application, was found to consist of nearly pure tremolite asbestos. All 9 of the female and 5 of the 6 male pleural mesothelioma cases had been exposed to the tremolite asbestos. The association of mesothelioma with ophiolite complexes is again indicated.

REMEDIATION AND RECLAMATION INITIATIVES IN AREAS CONTAMINATED WITH ASBESTOS

Asbestos contamination in various geographic localities, particularly in the United States, has generated great concern and has prompted, in some cases, very costly remedial actions. In the following five cases we will describe: (1) the geological occurrence of the asbestos and exposures that initiated the remediation activity, (2) the health effects (if any) that may have appeared in those exposed, (3) and the remediation and/or reclamation activities that were instituted.

Libby, Montana, Asbestos-Bearing Vermiculite Deposit

The Rainy Creek Complex is an igneous alkaline-ultramafic body that has intruded into the sedimentary rocks of the Precambrian Belt series. The complex is located in Lincoln County, Montana, near the town of Libby, and at one time hosted the world's largest vermiculite mine. The mine opened in 1923, and was owned and operated by the W.R. Grace Company from 1963 until it closed in 1990. The ultramafic unit of the complex is biotitite (composed mostly of very coarse grained biotite) surrounded by a biotite pyroxenite. Later magmatic events altered the biotite to hydrobiotite and vermiculite. About 35% of the complex is composed of various alkaline rocks, syenites, trachytes, phonolites, and granites (M. Gunter, University of Idaho, personal commun., 2002). The alkaline units of varying widths pervasively intrude and cut across all of the ultramafic rocks.

In a report on the Libby vermiculite deposits, Bassett (1959, p. 285) states that "mine workings have exposed numerous thin syenite dikes, ranging from a few inches to several feet thick, which cut the pyroxenite." He also states that "the asbestos is also found disseminated through the intrusive as thin layers along cleavage planes of the augite" (p. 287; see also Fig. 4 and 6).

Gunter et al. (2001) have carefully reexamined the mineralogy of the Rainy Creek Complex, and find the fibrous mineral to be the amphibole mineral winchite, which agrees with a previous study of a winchite sample from Libby by Wylie and Verkouteren (2000). Colloquially, this fibrous amphibole could be referred to as "sodic tremolite" but official amphibole nomenclature places it in the chemical category of winchite (Leake et al., 1997). End-member winchite has the chemical formula $CaNaMg_4(Al,Fe^{3+})Si_8O_{22}(OH)_2$. Although the Rainy Creek Complex is not an ophiolite, there is one similarity—the presence of numerous secondary intrusions into the ultramafic body in the form of dikes and veins.

Concern over the health effects of the winchite dust particles was brought to national attention by newspaper articles (e.g., Living and Dying in Libby, Spokesman-Review, Spokane, Washington, November 22, 2001, p. 1–10) and by an investigation by the U.S. EPA of the W.R. Grace mining and milling facility. An epidemiological study by McDonald et al. (1986) showed that the Libby vermiculite miners and millers had an increased risk of lung cancer (SMR = 2.45) and non-respiratory disease (SMR = 2.55). Recently, McDonald et al. (2001) have updated this mortality study and have reported in the study cohort a continuing excess mortality due to lung cancer and 12 mesothelioma cases among 285 deaths. Mesothelioma mortality now accounts for 4.2% of the deaths among the vermiculite miners. The amphibole fiber implicated in disease is erroneously reported in medical literature and newspaper articles as tremolite asbestos rather than its true identification—asbestiform winchite.

The Agency for Toxic Substances and Disease Registry report on the health of the residents of Libby and environs (ATSDR, 2000) states that 11 deaths (10 male, 1 female) were due to asbestosis. This ATSDR report also states that elevated lung cancer and mesothelioma mortality could not be verified in those with non-occupational exposure. Another ATSDR report (ATSDR, 2001) states that a radiological study of 5590 Libby residents determined that 18% had "pleural abnormalities." Various commentaries suggest that the winchite fiber dust has been pervasive for at least the last 75 years throughout the Libby area. If the fiber levels were elevated, for example, as in New Caledonia, then a significant increase in lung cancer incidence and numerous cases of mesothelioma would be expected among the Libby residents, on the basis of the health experience of the villagers who processed tremolite asbestos for stucco and whitewash. The W.R. Grace Company declared bankruptcy in April of 2001, in part because of potential remediation costs and pending lawsuits.

New Idria, California, Superfund Sites

The New Idria serpentinite near Coalinga, California (Fig. 4) is the site of three chrysotile asbestos mines, two of which (Atlas and Johns-Manville) ceased operation many years ago. The KCAC mine remains operational. The two abandoned mine sites are now designated as U.S. Environmental Protection Agency Superfund sites because of the Agency's concern about the health hazard of water- and air-borne chrysotile asbestos emissions from the old mine pits; however, the two Superfund sites cover only a few percent of the 114 km² asbestos outcrop area of the serpentinite. Asbestos-bearing dust and debris has been entering the air and the stream valleys from the entire area for millions of years. The EPA has chosen to focus on a small fraction of the asbestos releasing area, rather than develop a scientifically based policy that addresses the entire outcrop area (Ross, 1994; Coleman, 1996).

The medical and scientific evidence suggests that ingestion of asbestos increases the risk of gastrointestinal cancer only very slightly, if at all (Ross, 1984; Graham, 1981). In a recent review on mesothelioma occurring from non-occupational or low-dose asbestos exposure, Hillerdal (1999, p. 508) states: "most ingested mineral fibers will never be absorbed but be cleared in the normal way, and the harmful effects of asbestos in drinking water or drinks is probably minuscule (and much smaller than the risk of drinking alcohol)." It is noted that much of the pulmonary clearance of inhaled asbestos fibers in occupationally exposed workers ultimately took place through the gastrointestinal tract.

Even among those with occupational asbestos exposure, the increase risk of gastrointestinal (GI) cancer remains controversial and the results of epidemiological studies have not been consistent in finding that there was an increased risk of this cancer in the various cohorts studied. Furthermore, where an increased risk does exist, it is not associated with a dose-response. The risk of GI cancer does not increase with increasing exposure, which weakens the association between the agent and the disease (Gamble, 1994; Weiss, 1995).

Epidemiological studies of human populations having higher than usual intake of asbestos-containing water have not shown an increased risk of gastrointestinal cancer or any asbestos-related disease for that matter. The World Health Organization and other national and international health organizations have been remarkably clear on the extremely low risk of health effects from asbestos in water. The World Health Organization (WHO, 1986, p. 129) stated: "the studies conducted to date provide little convincing evidence of an association between asbestos in public water supplies and cancer induction." Further, as the U.S. EPA stated in a press release of January 7, 1991, "there is no evidence that ingestion of chrysotile asbestos causes harm, either in man or animals."

Inhalation of chrysotile asbestos fibers in the quantities found in the New Idria area certainly must be a minimal risk. This statement is supported by noting, as described above, the lack of asbestos-related disease among asbestos trades workers and the women living in the asbestos mining towns of Quebec who were exposed to low to moderate amounts of chrysotile asbestos. New Idria Superfund sites, even by expenditure of very large sums of money, will have no measurable effect on human health, particularly in view of the fact that most of the asbestos emissions come from areas outside the Superfund sites, through naturally occurring processes.

Fairfax County, Virginia Building Sites

In 1987, asbestos was discovered in a construction site in Fairfax County, Virginia, USA, causing concern over possible health effects to workers as well as to county residents. Avery A. Drake and Malcolm Ross, then of the U.S. Geological Survey (USGS), examined the site and found actinolite asbestos within a rock formation that composed part of the Piney Branch Complex (Drake, 1986). The asbestos was the slip-fiber type and was confined to the shear zones and anticlinal folds within an actinolite schist, a prominent rock type within the complex. Drake and Morgan (1981) state that the Piney Branch Complex, which outcrops over a 9 km^2 area of Fairfax County, consists of highly altered peridotite, pyroxenite, and gabbro, and further suggest it is a mélange that resulted from deformation of a layered complex, and closely resembles ophiolite material that crops out in nearby Maryland and southern Pennsylvania. Prominent tremolite asbestos veins were also observed by one of us (Ross) within one of the serpentinite bodies in Montgomery County, Maryland, USA, a sample of which is shown in Figure 8.

Some of the Fairfax County rock samples contained appreciable amounts of actinolite asbestos and, as a result, the Fairfax County Health Department initiated dust control procedures and published a dust control advisory (Dusek and Yetman, 2002). The advisory requires contractors to use proper dust control

Figure 8. Photograph of a tremolite asbestos sample collected from a large vein within a serpentinite, Montgomery County, Maryland, USA. Pen for scale is 14 cm long.

practices, air monitoring, safe waste rock disposal, and requires that sufficient notice of possible asbestos contamination be given to all employees and contractors in compliance with existing asbestos standards. The county advisory further states that construction need not be banned, but that six procedures must be established for dust suppression. As a result of this well-reasoned regulatory initiative, Fairfax County continued to build housing and commercial buildings on some of the most valuable land in the eastern United States, while at the same time protecting the workers and the public from an avoidable risk of asbestos-related disease. Soon after the discovery of asbestos, the local newspapers (e.g., *The Washington Post*, August 15 and September 29, 1987) gave factual statements about the asbestos occurrence in Fairfax County, but did not promote undue alarm, thus allowing the County officials to control asbestos exposures while allowing continued real estate development.

Although there was little asbestos mining in Virginia, Maryland, and Pennsylvania, there existed several quarries that processed serpentine rock for various construction purposes. One of these quarries, the Hunting Hill quarry, near Rockville, Maryland (Fig. 9) came into prominence in 1977 when chrysotile fibers were detected in various places near the quarry. A report in *Science* by Rohl et al. (1977) aroused much concern, causing the Montgomery County (Maryland) government to introduce various medical, regulatory, and abatement initiatives, such as paving over the serpentine gravel roads and checking the local hospitals for mesothelioma cases. The predicted epidemic of asbestos-related disease did not materialize—as would have been expected from the health data for the residents of the Quebec asbestos mining towns—and two years later the asbestos problem was nearly forgotten and the quarry continued to operate.

El Dorado County, California, Housing Developments

The Great Valley ophiolite belt parallels the western side of the Great Valley of California for at least 400 km (Coleman, 2000, Fig. 1). This ophiolite belt includes numerous outcrops of serpentinite, which were sites of at least 26 asbestos mines or asbestos prospects. Four of the mines were major producers of chrysotile asbestos, the most important of which was the Calaveras mine (Fig. 10), now closed, located near the town of Copperopolis. Two of the prospects, one at Iowa Hill in Placer County and the other in the Murphy area of Calaveras County, contained slip-fiber tremolite asbestos (Chidester and Shride, 1962). During the 1990s, many new housing projects were started in the fast-developing foothills area of El Dorado County, California, a county located within the Great Valley ophiolite belt. During excavation for housing sites, apparently within serpentinite rock, fragments of tremolite asbestos were found at some of the sites, alarming the homeowners. The

Figure 9. The Hunting Hill serpentinite quarry, near Rockville, Maryland, furnishes crushed stone to the Washington, D.C., metropolitan area. The quarry came into prominence in 1977 when chrysotile fibers were detected in dust emissions.

press, particularly the *Sacramento Bee*, Sacramento, California, produced a series of articles that suggested that the county residents' exposure to asbestos was endangering their health. A large number of air samples were collected at numerous sites all over the county by the California Air Resources Board (CARB). A detailed listing of the fiber concentration, which averages less than 0.001 f/ml, can be obtained at the CARB website (http://www.arb.ca.gov/toxics/asbestos/monitoring.html). The website tables do not identify the mineral fibers because apparently the CARB considers all fibers to be equally dangerous. Additional data released by the CARB shows that tremolite fiber has not been found in the air samples. El Dorado County appears to remain in turmoil over asbestos, unlike Fairfax County, Virginia, where the asbestos controversy—if indeed there was a controversy—died out 10 years ago.

Reclamation of the Troodos Asbestos Mine, Cyprus

The chrysotile asbestos mine occupies the eastern slopes of the Troodos massif in the neighborhood of the asbestos mining village of Pano Amiantos. The mine property, including large areas of rock waste, covers an area of ~10 km^2 at an elevation of nearly 1500 m. Pano Amiantos, currently containing 84 mostly unoccupied buildings, was originally located near the mine, but with later expansion the mining complex eventually surrounded the village. A second village, Kato Amiantos, has a current population of ~300 people and is located 2 km down the valley and 500 m below the mine site. Members of the Cyprus Parliament expressed concern about the possible landslide and health hazards associated with the mine property, but there was also a desire to reclaim this property for recreational use because it is close to the beautiful Troodos National Forest Park. This concern, under the leadership of the Geological Survey and Forest Departments of Cyprus, prompted the initiation of an extensive project to reclaim the mining property for recreational use.

We visited the mining site in November of 2000 to observe the beginnings of this remarkable reclamation activity (Fig. 6). Of particular importance, in addition to reducing mineral dust emissions, was the reduction of the steep angle of repose of the waste rock around the mine site. The steep slopes created a significant landslide hazard, particularly for the residents of Kato Amiantos. Stabilization of the mine property was accomplished by: (1) reducing angle of repose the waste rock slopes, (2) constructing benches within the slopes, (3) covering the benches and slopes with a few centimeters of non-serpentinite-bearing soil to give added nutrients for the vegetation, (4) covering the added soil with old tree cuttings to reduce erosion, and (5) planting a large variety of plants and trees indigenous to serpentinite soils, not only to prevent soil erosion but to enhance the beauty of the region noted for its pine forests.

Figure 10. The Calaveras asbestos quarry, El Dorado County, California. Now closed, the quarry at one time was a major producer of chrysotile fiber.

SUMMARY

Ophiolite complexes, including the associated serpentinite belts, are found in many parts of the world and it was in these rocks that asbestos was first discovered in Cyprus, Greece, and Finland as long as 5,000 years ago. Chrysotile asbestos is by far the most common asbestos mineral, generally occurring in most serpentinites. The richer deposits contain several volume percent of this mineral. In addition to chrysotile, the amphibole asbestos minerals anthophyllite, tremolite, and actinolite are found in ophiolite complexes. Anthophyllite asbestos was mined mostly within the ophiolite complex located in the Paakkila area of East Finland. Tremolite and actinolite asbestos were mined sporadically in many areas of the world, but total production was very low. We estimate that ~85% of the commercially produced asbestos came from ophiolite complexes. Some of the major chrysotile asbestos mines are located in Quebec, Canada; in the central and southern Ural Mountains of Russia and Kazakhstan; near Coalinga, California, USA; in northern Italy and Greece; and in the Troodos mountains of Cyprus. Two other amphibole asbestos minerals, "amosite" and "crocidolite," were important commercially but are found only in banded iron formations.

The asbestos minerals crystallize under special conditions within rocks formations that are undergoing intense deformation characterized by faulting, shearing, and dilation, and accompanied by intrusions of magmatic fluids in the form of dikes and sills. The fibers crystallize in zones of deformation and at igneous intrusion-host rock contacts. Highly tectonized ophiolite complexes contain many of these deformation features, thus presenting conditions ideal for fiber formation. Slip-fibers are formed within the fault and along shear planes, whereas cross-fibers form within dilation zones.

The asbestos-related diseases—asbestosis, lung cancer, and mesothelioma—have taken a huge toll on the lives of asbestos trades workers and asbestos miners and millers. Epidemiological studies of large cohorts of asbestos workers, exposed to variable amounts of chrysotile, and amphibole asbestos, show that the amphibole asbestos dusts are more dangerous, with "amosite" and "crocidolite" accounting for most of the mesothelioma mortality. Epidemiological studies show that low to moderate exposure to chrysotile asbestos presents a very low health risk. Although conventional epidemiological studies of tremolite and actinolite asbestos have not been possible because of the difficulty of assembling a large enough cohort of exposed individuals to be statistically significant, there are several "prevalence" studies where an unusually large number of people within a small geographic area died of mesothelial cancers. Six of these areas include small villages in Turkey, Greece, Cyprus, Corsica, and New Caledonia. Residents of several of these villages have been exposed to tremolite asbestos quarried from nearby local rock formations. It was used as an ingredient for stucco and whitewash, which was applied to the wall surfaces of the village buildings. All these villages lie within or near to ophiolite complexes where amphibole asbestos is likely to occur.

The presence of tremolite and actinolite asbestos in ophiolite complexes in Fairfax County, Virginia, and in El Dorodo County, California, has caused concern over the possible health effects that these mineral dusts may have on the local residents. The county governments of the two communities have instituted various regulations governing construction practices, but with differing results related to how the health risks are presented to the public by the county officials and by the local newspapers.

Two abandoned asbestos mine sites within the New Idria serpentinite near Coalinga, California, are now designated as U.S. EPA Superfund reclamation areas because of the concern about the health hazard of chrysotile asbestos. The two sites cover a few percent of the 114 km^2 that encompass the total outcrop area of the chrysotile-rich serpentinite. The expenditure of many millions of dollars for site reclamation will most likely have no measurable effect on human health, for most of the asbestos emissions comes from areas outside the Superfund sites through naturally occurring processes. In contrast, the government of Cyprus has started reclamation of the Troodos chrysotile asbestos mine property occupying a 10-km^2 area where asbestos emissions are likely to originate.

An igneous alkaline-ultramafic complex located near the town of Libby, Montana, hosts a large vermiculite mine once owned by the W.R. Grace Company. The very fibrous amphibole winchite, very similar in morphology to tremolite asbestos, is pervasive throughout the complex. Concern over the health effects of the winchite fibers disseminated from the mine workings to the surrounding areas was brought to national attention by newspaper articles and by an investigation of the mining and milling facility by the U.S. EPA. As a result of potential reclamation costs, property damage, and personal injury lawsuits, the W.R. Grace Company declared bankruptcy in April of 2001.

DISCUSSION AND CONCLUSIONS

We must conclude from the vast literature published on health effects of minerals that the inhalation of any biopersistent mineral that crystallizes as long thin fibers with widths of 0.5 micrometers or less and lengths of >5 micrometers could potentially be a health hazard. Evidence for this includes environmental mesotheliomas associated with exposure to tremolite asbestos and other fibrous minerals, such as the zeolite erionite. Chrysotile asbestos is an exception because low to moderate exposure to this mineral, even over long periods, presents a very low health risk.

The "fiber hypothesis" should include an understanding of the range in carcinogenic potency that particular mineral fibers can possess. Low concentrations of fibrous minerals are common in nature, but environmental disease related to mineral fiber is rare and mainly due to a few types of amphibole asbestos and fibrous eronite. The scientific and medical information available does not justify the claim that exposure to any amount of any fiber presents an unacceptable health risk. If this claim were true, rocks containing any concentration of any type of fibrous mineral would be off limits for any kind of mining or other types of activity, and thus would restrict the use of vast areas of geologic terrane needed for sustainable development. We should, however, promote a policy that not only protects the miners, construction workers, and local inhabitants, but also allows for various types of mining and commercial development. Regulations such as those promulgated by the county government of Fairfax, Virginia, can serve as a model for using science as a basis for risk management decisions to control mineral fiber dust exposures, while allowing mining and other development activities. Over-regulation can be extremely costly, addresses minimal health risk, and diverts attention from more socially important endeavors (Wilson and Price, 2001).

There is now ample evidence that dust levels now maintained in the chrysotile asbestos mines of Canada and Russia (Shcherbakov et al., 2001; Wilson and Price, 2001), coupled with similar low dust levels maintained in other workplaces, present health risks that are indistinguishable from risks associated with substitute materials, such as fiberglass, rock wool, and various composites; however, legal reform and education of the media and general public are needed to bring a scientific basis to public policies regarding exposure to fibrous particulates.

In the study of ophiolite complexes, we should take particular note of the faults, shear zones, folds, and igneous intrusion-host rock contacts for the presence of amphibole asbestos, and suggest that the local authorities discourage its use for commercial purposes. In rocks such as those found in Libby, Montana, where the fibrous mineral appears to be disseminated throughout the vermiculite ore body, exploitation of the mineral resource may not be possible unless very careful mineral beneficiation procedures are instituted; however, where the asbestos-bearing veins are dispersed between larger volumes of uncontaminated ore, it should be possible to "mine around" the fiber-bearing rock and dispose of it as waste. As a final comment, it would very helpful if there could be more cooperation between medical scientists and geoscientists in health studies where mineral dusts are of concern.

ACKNOWLEDGMENTS

We would like to acknowledge support from the Cyprus Fulbright Commission; Cleveland-Cliffs, Inc.; and the International Environmental Research Foundation, New York, New York. We thank Professor Mickey Gunter, Department of Geological Sciences, University of Idaho, for permitting us to read his pre-publication manuscript on the geology and mineralogy of the Rainy Creek Complex at Libby, Montana; for other information he furnished to us concerning the Libby asbestos problems; and for his helpful review of this paper. We also thank Avery A. Drake, Sally Newcomb, John Addison, and Yildirim Dilek for their constructive reviews of the manuscript, and Colette Drake and Daphne R. Ross for their help with the Greek-Latin-English translations. Lastly, we are grateful to Drs. George Petrides, Eleni Geogiou Morisseau, and George Constantinou of the Geological Survey Department, Cyprus, for generously sharing their knowledge of the geology and history of Cyprus.

REFERENCES CITED

Aktas, G., and Robertson, A.H.F., 1990, Tectonic evolution of the Tethys suture zone in S.E., Turkey: evidence from the petrology and geochemistry of Late Cretaceous and Middle Eocene extrusives: *in* Malpas, J., Moores, E.M., Panayiotou, A., and Xenophontos, C., eds., Ophiolites, Oceanic Crustal Analogues: Proceedings of the Symposium "Troodos 1987," Geological Survey Department, Nicosia, Cyprus, p. 311–328.

Alleman, J.E., and Mossman, B.T., 1997, Asbestos revisited: Scientific American, July, p. 70–75.

ATSDR, 2000 (Agency for Toxic Substances and Disease Registry, 2000), Health consultation, Libby asbestos site, Libby, Lincoln County, Montana. Mortality from asbestosis in Libby, Montana, 1979–1998, December 12, 2000: Agency for Toxic Substances and Disease Registry, Atlanta, Georgia, 19 p. This report is available at: http://www.atsdr.gov/HAC/pha/libby/lib_toc.html.

ATSDR, 2001 (Agency for Toxic Substances and Disease Registry, 2001), Year 2000 medical testing of individuals potentially exposed to asbestiform minerals associated with vermiculite in Libby, Montana, a report to the community: Agency for Toxic Substances and Disease Registry, Division of Health Studies, U.S., Department of Health and Human Services, Atlanta, Georgia, 32 p.

Aurola, E., 1954, The mines and quarries of Finland: Geologinen Tufkimuslaitos Geoteknillisiäulkaisuja. no. 55, 155 p.

Bailey, K.C., 1932, The Elder Pliny's chapters on chemical subjects, part II: Edward Arnold &CO, London, 299 p.

Baris, Y.I., 1991, Fibrous zeolite (erionite) related diseases in Turkey: American Journal of Industrial Medicine, v. 19, p. 373–378.

Baris, Y.I., Artvinli, M., and Sahin, A.A., 1979, Environmental mesothelioma in Turkey: Annals of the New York Academy of Science, v. 330, p. 423–432.

Baris, Y.I., Artvinli, M., Sahin, A.A., Bilir, N., Kalyoncu, F., and Sébastien, P., 1988a, Non-occupational asbestos related chest diseases in a small Anatolian village: British Journal Industrial Medicine, v. 45, p. 841–842.

Baris, Y.I., Bilir, N., Artvinli, M., Sahin, A.A., Kalyoncu, F., and Sébastien, P., 1988b, An epidemiological study in an Anatolian village environmentally exposed to tremolite asbestos: British Journal of Industrial Medicine, v. 45, p. 838–840.

Bassett, W.A., 1959, The origin of the Libby vermiculite deposit at Libby, Montana: American Mineralogist, v. 44, p. 282–299.

Beukes, N.J., and Dreyer, C.J.B., 1986, Crocidolite deposits of the Profret area, Griqualand West: *in* Anhaeusser, C.R., and Maske, S., eds., Mineral deposits of South Africa, Geological Society of South Africa, Johannesburg, v. 2, p. 911–921.

Bowles, O., 1955, The Asbestos Industry: U.S., Bureau of Mines, Bulletin 552, Washington, D.C., 122 p.

Browne, K. and Murray, R., 1990, Asbestos and the Romans: The Lancet, v.336, p. 445.

Brunet, C., Monie, P., Jolivet, L., and Cadet, J-P., 2000, Migration of compression and extension in the Tyrrhenian Sea, insights from $^{40}Ar/^{39}$ ages on micas along a transect from Corsica to Tuscany: Tectonophysics, v. 321, p. 127–155.

Camus, M., Siemiatycki, J., and Meek, B., 1998, Nonoccupational exposure to chrysotile asbestos and the risk of lung cancer: New England Journal of Medicine, v. 338, p. 1565–1571.

Chidester, A.H., and Shride, A.F., 1962, Asbestos in the United States: Mineral Investigations Resource Map MR-17: U.S., Geological Survey, Reston, Virginia.

Coleman, R.G., 1971, Plate tectonic emplacement of upper mantle peridotites along continental edges: Journal of Geophysical Research, v. 76, p. 1212–1222.

Coleman, R.G., 1986, Field guide book to New Idria area, California: 14th General Meeting, International Mineralogical Association, July, 1986, Stanford University, Stanford, California, July 1986, 36 p.

Coleman, R.G., 1996, New Idria serpentinite: A land management dilemma: Environmental and Engineering Geoscience, v. 2, p. 9–22.

Coleman, R.G., 2000, Prospecting for ophiolites along the California continental margin: *in* Dilek, Y., Moores, E.M., Elthon, D., and Nicolas, A., eds., Ophiolites and oceanic crust: New insights from field studies and the ocean drilling program: Geological Society of America Special Paper 359, Boulder, Colorado, p. 351–364

Constantopoulos, S.H., Goudevenos, J.A., Saratzis, N., Langer, A.M., Selikoff, I.J., and Moutsopoulos, H.M., 1985, Pleural calcification and restrictive lung function in northwestern Greece-environmental exposure to mineral fiber as etiology: Environmental Research, v. 38, p. 319–331.

Constantopoulos, S.H., Langer, A.M., Saratzis, N., and Nolan, R.P., 1987, Regional findings in Metsovo lung: The Lancet, v. ii, p. 452–453.

Cooke, W.E., 1927, Pulmonary asbestosis: British Medical Journal, v. 2, p. 1024.

Deligiannis, K.F., 2002, History of Kandyla, town of the Province of Arkadia, Hellas: GreeksUnited Newsletter, 3p. Document available at: http://www.greeksunited.com/newsletter/members/historykand.html

Dement, J.M., Merchant, J.A., and Green, F.H.Y., 1986, Asbestosis: *in* Merchant, J.A., ed., Occupational Respiratory Diseases: DHHS (NIOSH) Publication 86–102, U.S., Department of Health and Human Services, Washington D.C., p. 287–327.

Doll, R., 1955, Mortality from lung cancer in asbestos workers: British Journal of Industrial Medicine, v. 12, p. 81–86.

Doll, R., and Hill, A.B., 1950, Smoking and carcinoma of the lung: Preliminary report: British Medical Journal, v.153, p.585–590.

Doll, R., and Peto, J., 1985, Effects on health of exposure to asbestos: Health & Safety Commission, Her Majesty's Stationary Office, London, U.K., p. 1–58.

Drake, A.A., Jr., 1986, Geologic map of the Fairfax quadrangle, Fairfax County, Virginia: U.S., Geological Survey, Reston, Virginia.

Drake, A.A., Jr. and Morgan, B.A., 1981, The Piney Branch Complex-a metamorphosed fragment of the central Appalachian ophiolite in northern Virginia: American Journal of Science, v. 281, p. 484–508.

Dreyer, C.J.B., and Robinson, H.A., 1981, Occurrence and exploitation of amphibole asbestos in South Africa: *in* Riordon, P.J., ed., Geology of asbestos deposits: American Institute of Mining, Metallurgical, and Petroleum Engineers, Inc., New York, NY, p. 25–44.

Dumortier, P., Göcmen, A., Laurent, K., Manço, A., and De Vuyst, P., 2001, The role of environmental and occupational exposures in Turkish immigrants with fibre-related disease: European Respiratory Journal, v. 17, p. 922–927.

Dusek, C.J., and Yetman, J.M., 2002, Control and prevention of asbestos exposure from construction in naturally occurring asbestos: Fairfax County Health Department, Fairfax, Virginia, 6 p. Document available at: http://www.co.fairfax.va.us/service/hd/docs/TBRDPUBFIN.doc.

Europaeus-Äyräpää, A., 1930, Die relative chronologie der steinzeitlichen Kermik, Finland: Acta Archaeologie, v. 1, p. 169–190.

Evans, J.W., 1906, The identity of the amiantos or Karystian stone of the ancients with chrysotile: Mineralogical Magazine, v. 14, p. 143–148.

Gamble, J.F., 1994, Asbestos and colon cancer: a weight-of-the-evidence review: Environmental Health Perspectives, v. 102, p. 1038–50.

Gardner, M.J., Winter, P.D., Pannett, B., and Powell, C.A., 1986, Follow up study of workers manufacturing chrysotile asbestos cement products: British Journal of Industrial Medicine, v. 43, p. 726–732.

Gealey, W.K., 1979, Ophiolite obduction mechanism: *in* Panayiotou, A., ed., Ophiolites—Proceedings International Ophiolite Symposium Cyprus 1979, Geological Survey Department, Nicosia, Cyprus, p. 228–243.

G.E.L., 1940, A Greek-English Lexicon, 9th Ed., Oxford University Press, Oxford, England, 2111p.

Gibbs, G.W., and DuToit, D.S., 1973, Environmental data mining: *in*: Biological Effects of Asbestos, International Agency for Research on Cancer (IARC), Publication No. 8, Geneva, Switzerland, p. 311–320.

Glass, I.G., 1990, Ophiolites and ocean lithosphere, *in* Malpas, J., Moores, E.M., Panayiotou, A., and Xenophontos, C., eds., Ophiolites, Oceanic Crustal Analogues: Proceedings of the Symposium "Troodos 1987," The Geological Survey Department, Nicosia, Cyprus, p. 1–10.

Graham, S., 1981, Methodological problems in ecologic studies of the asbestos-cancer relationship: Environmental Research, v. 25, p. 35–49.

Gunter, M.E., Brown, B.M., Bandli, B.R., and Dyar, M.D., 2001, Amphibole-asbestos, vermiculite mining and Libby, Montana: What's in a name? (abstract #3455): Eleventh Annual V.M., Goldschmidt Conference, Hot Springs, Virginia.

HEI.AR, 1991, Asbestos in public schools and commercial buildings: A literature review and synthesis of current knowledge: Health Effects Institute-Asbestos Research, Cambridge, Massachusetts, 392 p.

Henderson, J.W., Shilkin, K.B., Langlois, S.L.P., and Whitaker, D., 1992, Malignant mesothelioma: Hemisphere Publishing Corporation, New York, N.Y., 383 p.

Hillerdal, G., 1999, Mesothelioma: Cases associated with non-occupational and low dose exposures: Occupational Environmental Medicine, v. 56, p. 505–513.

Hodgson, J.T., and Darnton, A., 2000, The quantitative risk of mesothelioma and lung cancer in relationship to asbestos exposure: Annals of Occupational Hygiene, v. 44, p. 565–601.

Hoover, H.C., and Hoover, L.H., 1950, Translation of De Re Metallica by Georgius Agricola (1556), Dover Publications, Inc., New York, 638 p.

Huuskonen, M.S., 1980, Asbestos and cancer in Finland: Journal of Toxicology and Environmental Health, v. 6, p. 1261–1265.

Huuskonen, M.S., Ahlman, K., Mattsson, T., and Tossauainen, A., 1980, Asbestos disease in Finland: Journal of Occupational Medicine, v. 22, p. 751–754.

Ingham, F.T., 1959, With an account of the mineral resources: *in* Wilson, R.A.M., 1959, The geology of the Xeros-Troodos area: Geological Survey Department, Cyprus, Memoir 1, Government of Cyprus, Nicosia, Cyprus, p. 158–161.

Jones, R.H., 1890, Asbestos, its properties, occurrence and uses: Crosby, Lockwood and Son, London, 236 p.

Karjalainen, A., Meurman, L., and Ekkala, A., 1994, Four cases of mesothelioma among Finnish anthophyllite miners: Occupational Environmental Medicine, v. 51, p. 212–215.

Kashansky, S.V., 1999, A 300 year history of the discovery of asbestos in the Urals: *in* Peters, G.A., and Peters, B.J., eds., Sourcebook on asbestos diseases, v. 20, p. 129–144, LEXIS Law Publishing, Dayton, Ohio.

Langer, A.M., 2001, Summary of the symposium, *in* Nolan, R.P., Langer, A.M., Ross, M., Wicks, F.J., and Martin, R.F., eds., The Health Effects of Chrysotile-Asbestos: Contribution of Science to Risk-Management Decisions: The Canadian Mineralogist, Special Publication 5, Ottawa, Canada, p. 291–296.

Langer, A.M., and Nolan, R.P., 1994, Chrysotile: Its occurrence and properties as variables controlling biological effects: Annals of Occupational Hygiene, v. 38, p. 427–451.

Langer, A.M., Nolan, R.P., Constantopoulos, S.H., and Moutsopoulos, H.M., 1987, Association of Metsovo lung and pleural mesotheliomas with exposure to tremolite-containing whitewash: The Lancet v. i, p. 965–967.

Langer, A.M., Nolan, R.P., and Addison, J., 1991, Distinguishing between amphibole asbestos fibers and elongate cleavage fragments of their non-asbestos analogues, *in* Brown, R.C., Hoskins, J., and Johnson, N., eds., Mechanisms in fibre carcinogenesis: Plenum Press, New York, N.Y., p. 252–267.

Leake, B.E., Woolley, A.R., Arps, C.E.S., Birch. W.D., Gilbert, M.C., Grice, J.D., Hawthorne, F.C., Kato, A., Kisch, H.J., Krivovichev, V.G., Linthout, K., Laird, J., Mandarino, J.A., Maresch, W.V., Nickel, E.H., Rock, N.M.S., Schumacher, J.C., Smith, D.C., Stephenson, N.C.N., Ungaretti, L., Whittaker, E.J.W., and Youzhi, G., 1997, Nomenclature of the amphiboles: Report of the subcommittee on amphiboles of the International Mineralogical Association Commission on New Minerals and New Mineral Names: American Mineralogist, v. 82, p. 1019–1037.

Lynch, K.M., and Smith, W.A., 1935, Pulmonary asbestosis III: Carcinoma of the lung in asbestos-silicosis: American Journal of Cancer, v. 24, p. 56–64.

Luce, D., Bugel, I., Goldberg, P., Goldberg, M., Solomon, C., Billon-Galland, M-A., Nicolau, J., Quénel, P., Fevotte, J., and Brochard, P., 2000, Environmental exposure to tremolite and respiratory cancer in New Caledonia: A case-control study: American Journal of Epidemiology, v. 151, p. 259–265.

Marcuse, B., 1930, Asbestos—"pierre á coton": Canadian Geographical Journal, v. 1, p., p. 4–36.

McConnochie, K., Simonato, L., Mavrides, P., Christofides, P., Pooley, F.D., and Wagner, J.C., 1987, Mesothelioma in Cyprus: the role of tremolite: Thorax, v. 42, p. 342–347.

McConnochie, K., Simonato, L., Mavrides, P., Christofides, P., Mitha, R., Griffiths, D.M., and Wagner, J.C., 1989, Mesothelioma in Cyprus: *in* Bignon, J., Peto J., and Saracci, R., eds., Non-Occupational Exposure to Mineral Fibres, International Agency for Research on Cancer (IARC), Scientific Publication no. 90, Lyon, France, p. 411–419.

McDonald, J.C., and McDonald, A.D., 1996, The epidemiology of mesothelioma in historical context: European Respiratory Journal, v. 9, p. 1931–1942.

McDonald, J.C., McDonald, A.D., Armstrong, B. and Sébastien, 1986, Cohort study of mortality of vermiculite miners exposed to tremolite: British Journal of Industrial Medicine, v. 43, p. 436–444.

McDonald, J.C., Harris, J., and Armstrong, B., 2001, Cohort mortality study of vermiculite miners exposed to fibrous tremolite: an update (abstract): Inhaled Particles, Robinson College, Cambridge, United Kingdom, September 2–6, 2001.

MacLeod, C.J., 1990, Role of the southern Troodos Transform Fault in the rotation of the Cyprus microplate: evidence from the eastern Limassol Forest Complex: *in* Malpas, J., Moores, E.M., Panayiotou, A. and Xenophontos, C., eds., Ophiolites, Oceanic Crustal Analogues: Proceedings of the Symposium "Troodos 1987," Geological Survey Department, Nicosia, Cyprus, p. 75–85.

Meurman, L.O., Kiviluoto, R., and Hakama, M. 1974, Mortality and morbidity among the working population of anthophyllite asbestos miners in Finland: British Journal of Industrial Medicine, v. 31, p. 105–112.

Mossman, B.T., Bignon, J., Corn, M., Seaton, A., and Gee, J.B.L., 1990, Asbestos—scientific developments and implications for public policy: Science, v. 247, p. 294–301.

Mumpton, F.A., and Thompson, C.S., 1975, The mineralogy and origin of the Coalinga asbestos deposit: Clays and Clay Minerals, v. 23, p. 131–143.

Murray, R., 1990, Asbestos: A chronology of its origins and health effects: British Journal of Industrial Medicine, v. 47, p. 361–365.

Newhouse, M.L., and Sullivan, K.R., 1989, A mortality study of workers manufacturing friction materials: 1941–86: British Journal of Industrial Medicine, v. 46, p. 176–179.

Nolan, R.P., and Langer, A.M., 2001, Concentration and type of asbestos fibers in the air inside buildings, *in* Nolan, R.P., Langer, A.M., Ross, M., Wicks, F.J., and Martin, R.F., eds., The Health Effects of Chrysotile-Asbestos: Contribution of Science to Risk-Management Decisions: The Canadian Mineralogist, Special Publication 5, Ottawa, Canada, p. 39–51.

Nolan, R.P., Langer, A.M., and Wilson, R., 1999, A risk assessment for exposure to grunerite asbestos (amosite) in an iron ore mine: Proceedings of the National Academy of Sciences, v. 96, p. 3412–3419.

Nolan, R.P., Langer, A.M., Ross, M., Wicks, F.J., and Martin, R.F., eds., 2001, Health Effects of Chrysotile-Asbestos: Contribution of Science to Risk Management Decisions: The Canadian Mineralogist Special Publication 5, Ottawa, Canada, p. 1–304.

Noro, L., 1968, Occupational and non-occupational asbestosis in Finland: American Industrial Hygiene Association Journal, v. 29, p. 53–59.

O.E.D., 1933, The Oxford English Dictionary, v. 1 (A-B), Oxford University Press, Oxford, 1240 p.

Ohlson, C.G., and Hogstedt, C., 1985, Lung cancer among asbestos cement workers. A Swedish cohort study and a review: British Journal of Industrial Medicine, v. 42, p. 397–402.

O'Hanley, D.S., 1996, Serpentinites-Records of tectonic and petrological history: Oxford University Press, New York, N.Y., 277 p.

Petrov, V.P., and Znamensky, V.S., 1981, Asbestos deposits in the USSR: *in* Riordon, P.H., ed., Geology of asbestos deposits: Society of Mining, Metallurgical, and Petroleum Engineers, Inc., New York, N.Y., p. 45–52.

Rackham, H., 1961, Pliny Natural History with an English translation in ten volumes, v. 5, (Books 17–19): Harvard University Press, Cambridge, MA, 544 p.

Rassios, A., and Smith, A.G., 2000, Constraints on the formation and emplacement age of western Greek ophiolites (Vourinos, Pindos, and Othris) inferred from deformation structures in peridotites: *in* Dilek, Y., Moores, E.M., Elthon, D., and Nicolas, A., eds., Ophiolites and Oceanic Crust: New insights from field studies and the ocean drilling program: Boulder, Colorado, Geological Society of America Special Paper 359, p. 473–483.

Rey, F., Boutin, C., Steinbauer, J., Viallat, J.R., Allessandroni, P., Jutisz, P., Di Giambattista, D., Billon-Galland, M.A., Hereng, P., Dumortier, P., and De Vuyst, P., 1993, Environmental pleural plaques in an asbestos exposed population of northeast Corsica: European Respiratory Journal, v. 6, p. 978–982.

Rey, F., Boutin, C., Viallat, J.R., Steinbauer, J., Allessandroni, P., Jutisz, P., Di Giambattista, D., Billon-Galland, M.A., Hereng, P., Dumortier, P., and De Vuyst, P., 1994, Environmental asbestotic pleural plaques in northeast Corsica: Correlation with airborne and pleural mineralogic analysis: Environmental Health Perspectives, v. 102 (supplement 5), p. 251–252.

Robinson, P.T., and Malpas, J.,1990, The Troodos ophiolite of Cyprus: New perspectives on its origin and emplacement, *in* Malpas, J., Moores, E.M., Panayiotou, A., and Xenophontos, C., eds., Ophiolites, Oceanic Crustal Analogues: Proceedings of the Symposium "Troodos 1987," The Geological Survey Department, Nicosia, Cyprus, p. 13–26.

Rohl, A.N., Langer, A.M., and Selikoff, I.J., 1977, Environmental asbestos pollution related to use of quarried serpentine rock: Science, v. 196, p. 1319–1322.

Rom, W.N., Hammar, S.P., Rusch, V., Dodson, R., and Hoffman, S., 2001, Malignant mesothelioma from neighborhood exposure to anthophyllite asbestos: American Journal of Industrial Medicine, v. 40, p. 211–214.

Ross, M., 1981, The geologic occurrences and health hazards of amphibole and serpentine asbestos: *in* Veblen, D.R., ed., Amphiboles and other Hydrous Pyriboles-Mineralogy, v. 9A, Mineralogical Society of America, Washington D.C., p. 279–323.

Ross, M., 1984, A survey of asbestos-related disease in trades and mining occupations and factory and mining communities as a means of predicting health risks of nonoccupational exposure to fibrous minerals: *in* Levadie, B., ed., Definitions for Asbestos and Other Health-related Silicates, ASTM STP 834, American Society for Testing Materials, Philadelphia, Pennsylvania, p. 51–104.

Ross, M., 1994, The New Idria serpentinite of California: a toxic rock?: Geological Society of America Abstracts with Programs, v. 26., No. 7, 1994 Annual Meeting, Seattle, Washington, October 24–27, p. A320.

Ross, M., 1995, The schoolroom asbestos abatement program: a public policy debacle: Environmental Geology, v. 26, p. 182–188.

Ross, M., and Virta, R.L., 2001, Occurrence, production and uses of asbestos: *in* Nolan, R.P., Langer, A.M., Ross, M., Wicks, F.J., and Martin, R.F., eds., The Health Effects of Chrysotile Asbestos: Contribution of Science to Risk Management Decisions: The Canadian Mineralogist Special Publication 5, Ottawa, Canada, p. 79–88.

Saracci, R., Simonato, L., Baris, Y., Artvinli, M., and Skidmore, J., 1982, The age-mortality curve of endemic pleural mesothelioma in Karain, central Turkey: British Journal of Cancer, v. 45, p. 147-149.

Seiler, H.E., 1928, A case of pneumoconiosis. Result of the inhalation of asbestos dust: British Medical Journal, v.2, p.982.

Shcherbakov, S.V., Kashansky, S., Domnin, S.G., Koggan, F.M., Kozlov, V., Kochelayev, V.A., and Nolan, R.P., 2001, The health effects of mining and milling chrysotile: the Russian experience: *in* Nolan, R.P., Langer, A.M., Ross, M., Wicks, F.J., and Martin, R.F., eds., The health effects of chrysotile asbestos: Contribution of Science to Risk Management Decisions: The Canadian Mineralogist Special Publication 5, Ottawa, Canada, p. 187–198.

Siemiatycki, J., 1982, Mortality in the general population in asbestos mining areas: *In* Proceedings of the World Symposium on Asbestos, May 24–27, 1982, Canadian Asbestos Information Centre, Montreal, Canada, p. 337–348.

Silvestri, S., Magnani, C., Calisti, R., Bruno, C., 2001, The experience of the Balangero chrysotile asbestos mine in Italy: Health effects among workers mining and milling asbestos and the health experience of persons living nearby: *In* Nolan, R.P., Langer, A.M., Ross, M., Wicks, F.J., and Martin, R.F., eds., The health effects of chrysotile asbestos: Contribution of Science to Risk Management Decisions: The Canadian Mineralogist Special Publication 5, Ottawa, Canada, p. 177–186.

Simonen, A., 1980, The Precambrian in Finland: Geological Survey of Finland, Bulletin no. 304, 58 p.

Skarpelis, N., and Dabitzias, S., 1987, The chrysotile deposit of Zidani, northern Greece: Ofiolliti, v. 12, p. 403–410.

Skinner, H.C.W., Ross, M., and Frondel, C., 1988, Asbestos and Other Fibrous Minerals: Oxford University Press, New York, N.Y., 204 p.

Spadea, P., and Scarrow, J.H., 2000, Early Devonian boninites from the Magnitogorsk arc, southern Urals (Russia): Implications for early development of a collisional orogen: *in* Dilek, Y., Moores, E.M., Elthon, D., and Nicolas, A., eds., Ophiolites and Oceanic Crust: New insights from field studies and the ocean drilling program: Boulder, Colorado, Geological Society of America Special Paper 359, p. 461–472.

Tankut, A., and Gorton, M.P., 1990, Geochemistry of a mafic-ultramafic body in the Ankara melange, Anatolia, Turkey: Evidence for a fragment of ocean lithosphere: *in* Malpas, J., Moores, E.M., Panayiotou, A. and Xenophontos, C., eds., Ophiolites, Oceanic Crustal Analogues: Proceedings of the Symposium "Troodos 1987," Geological Survey Department, Nicosia, Cyprus, p. 339–349.

Thayer, B., 2002, Pliny the Elder: the Natural History, Books 1–37 (in Latin): http://www.ku.edu/history/index/europe/ancient_rome/E/Roman/Texts/Pliny_the_Elder/home.html).

Thomas, H.F., Benjamin, I.T., Elwood, P.C., and Sweetman, P.M., 1982, Further follow-up of workers from an asbestos cement factory: British Journal of Industrial Medicine, v. 39, p. 273–276.

Timbrell, V., 1989, Review of the significance of fiber size in fibre-related lung disease: A centrifuge cell for preparing accurate microscope evaluation specimens from slurries used in inoculation studies: Annals of Occupational Hygiene, v. 33, p. 483–505.

Timbrell, V., Griffiths, D.M., and Pooley, F.D., 1971, Possible biological importance of fiber diameters of South African amphiboles: Nature, London, v. 232, p. 55–56.

Trendall, A.F., and Blockley, J.G., 1970, The iron formations of the Precambrian Hamersley Group Western Australia with special reference to the associated crocidolite: Geological Survey of Western Australia, Bulletin 119, Perth, Australia, 366 p.

Tuomi, T., Segerberg-Konttinen, M., Tammilehto, L., Tossavainen, A., and Vanhala, E., 1989, Mineral fiber concentration in lung tissue of mesothelioma patients in Finland: American Journal of Industrial Medicine, v. 16, p. 244–254.

United States Court of Appeals, 1991, Fifth Circuit. Corrosion Proof Fittings et al. Petitioners v: The Environmental Protection Agency, No. 89–4596, October 18, 1991.

Veblen, D.R., and Wylie, A.G., 1993, Mineralogy of amphiboles and 1:1 layer silicates, *in* Guthrie G.D., and Mossman, B.T., eds., Health Effects of Mineral Dusts, Reviews in Mineralogy: Mineralogical Society of America, Washington, D.C., v. 28, p. 61–137.

Vetter, S.K., and Stakes, D.S., 1990, Repeated magma injection and the geochemical evolution of the Northern Semail plutonic suite, *in* Malpas, J., Moores, E.M., Panayiotou, A. and Xenophontos, C., eds., Ophiolites, Oceanic Crustal Analogues: Proceedings of the Symposium "Troodos 1987," Geological Survey Department, Nicosia, Cyprus, p. 397–413.

Wagner, J.C., 1991, The discovery of the association between blue asbestos and mesothelioma and the aftermath: British Journal of Industrial Medicine, v. 48, p. 399–403.

Wagner, J.C., Sleggs, C.A., and Marchand, P., 1960, Diffuse pleural mesothelioma and asbestos exposure in the Northwestern Cape Province: British Journal of Industrial Medicine, v. 17, p. 260–271.

Wagner, J.C., Skidmore, J.W., Hill, R.J., and Griffiths, D.M., 1985, Erionite exposure and mesotheliomas in rats: British Journal of Cancer, v. 51, p. 727–730.

Weill, H., Huges, J., and Waggenspack, C., 1979, Influence of dose and fiber type on respiratory malignancy in asbestos cement manufacturing: American Review of Respiratory Disease, v. 120, p. 345–354.

Weiss, W., 1995, The lack of causality between asbestos and colorectal cancer: Journal of Occupational and Environmental Medicine, v. 37, p.1364–1373.

WHO, 1986 (World Health Organization, 1986), Asbestos and other natural mineral Fibers: International Programme on Chemical Safety (IPCS), Environmental Health Criteria 53, Geneva, Switzerland, p. 1–194.

WHO, 1989 (World Health Organization, 1989), Report on occupational exposure limit for asbestos: Oxford, U.K.

Wicks, F.J., 1999, Status of the reference x-ray powder-diffraction patterns for the serpentine minerals in the PDF database 1997: Powder Diffraction, v. 15, p. 42–50.

Williams-Jones, A.E., Normand, C., Clark, J.R., Vali, H., and Martin, R.F., 2001, Controls of amphibole formation in chrysotile deposits: Evidence from the Jeffrey mine, Asbestos, Quebec: *in* Nolan, R.P., Langer, A.M., Ross, M., Wicks, F.J., and Martin, R.F., eds., The Health Effects of Chrysotile Asbestos: Contribution of Science to Risk Management Decisions: The Canadian Mineralogist Special Publication 5, Ottawa, Canada, p. 89–104.

Wilson, R., and Crouch, E.A.C., 2001, Risk-benefit analyses: Harvard University Press, Cambridge, Massachusetts, p. 231–232.

Wilson, R., and Price, B., 2001, Risk assessment for asbestos and management of low levels of exposure to chrysotile: *in* Nolan, R.P., Langer, A.M., Ross, M., Wicks, F.J., and Martin, R.F., eds., The Health Effects of Chrysotile Asbestos: Contribution of Science to Risk Management Decisions: The Canadian Mineralogist Special Publication 5, Ottawa, Canada, p. 265–275.

Wilson, R., Langer, A.M., Nolan, R.P., Gee, J.B., and Ross, M., 1994, Asbestos in New York City school buildings—public policy: Is there a scientific basis?: Regulatory Toxicology and Pharmacology, v. 20, p. 161–169.

Wilson, R., Nolan R.P., and Domnin, S.G., 2001, The debate on banning asbestos: Canadian Medical Association Journal, v. 165, p. 1190–1191.

Wilson, R.A.M., 1959, The geology of the Xeros-Troodos area: Geological Survey Department, Cyprus, Memoir 1, Government of Cyprus, Nicosia, Cyprus, 175 p.

Wood, W.B., and Gloyne, S.R., 1934, Pulmonary asbestosis: a review of 100 cases: Lancet, v. 2, p. 1383–1385.

Wylie, A.G., and Verkouteren, J.R., 2000, Amphibole asbestos from Libby, Montana: Aspects of nomenclature: American Mineralogist, v. 85, p. 1540–1542.

Yazicioglu, S., Ilcayto, R., Balci, K., Sayli, B.S., and Yorulmaz, B., 1980, Pleural mesotheliomas and bronchial cancers caused by tremolite dust: Thorax, v. 35, p. 564–569.

Manuscript Accepted by the Society May 2, 2003

Printed in the USA

Index

A

Abdella mélange: 347
abyssal-type: 88–90
abyssokonite: 79, 79n*n. *See also* limestone
accretionary complex: 286, 347, 370
Acoje block: 127, 254, 255
actinolite: 447–48, 450–51, 456
actualistic model: 8
Adak Island: 255
adakites: 237
adinoles: 82, 82n*x
Adria plate: 124
Adriatic Sea area. *See also* specific countries
  geosyncline boundaries: 35
  ophiolitic complexes: 78–90, 98, 337–348
Adula Nappe: 95, 168, 408
Aegean Sea: 116, 117, 126
Africa
  continental margin: 168
  Pan African terrane: 368
  rift system: 309, 326
African plate
  boundaries of
    Atlantic Ocean: 344
    vs. continental margin: 168
  collision with
    Cyprus basin: 122
    Eurasian plate
      alpine orogenesis model: 167
      dating of: 121
      mantle perturbation: 114
      Mirdita complex: 124
      ocean size constraint: 156
    Hellenic arcs: 121
  mantle reservoir: 121
  suture zones: 298
Agrigan volcano: 257
Agrilia Formation: 341
Agrokipia: 302–3
Airy, theory of isostasy: 160
Akaki Canyon: 314
Akapnov Forest: 449
Alai Range: 8
alamanide facies: 167, 170
Alaska
  accretionary complexes: 286
  calcalkaline volcanism in: 154
  Chugach complex: 13
  Talkeetna Arc: 255
  triple junction migration: 285
Albanian complexes
  classification of: 10, 361
  formation of: 353–361
  geochemical analysis of: 358, 359
  geochronology of: 352, 361
  geological setting of: 340, 352
  geothermobarometry of: 359
  isotopic analysis of: 361
  mapping of: 353–358
  metallogenic studies of: 359
  Mirdita complex. *see* Mirdita complex
  petrology of: 352–357
  studies of: 351–362
Albano-Thessalian furrow: 35. *See also* Vourinos complex
alberese: 80, 80 n*u, 83
albite: 81–82, 84
albitite: 106–107
Alboran Basin: 116–117
Alestos Hill: 314
Aleutian Arc/Trench
  collision with: 121
  convergent margin: 234
  geochemical analysis of: 232
  petrology of: 255, 257
  plate junction type: 366
  volcanism in: 232
Algero-Provençal basin: 117
alkaline volcanism: 160, 232
Allen, Cameron: 193
Allerton, Simon: 73–74
Alley Volcanic Unit: 126, 139
alnoitic rock: 56
Alpine-Apennine complexes. *See also* Albanian complexes; Hellenic complexes; Platta-Totalp nappe
  aureoles, contact: 81, 82, 103
  classification of
    abyssal-type: 88–90
    alamanide facies: 167
    celtide facies: 167
    Class-II: 10
    hypabyssal-type: 89
    lepidorogene-type: 90
    lherzolite-type: 10
    Ligurian-type: 10
    Mediterranean-type: 10–11
    neritic-type: 89–90
    Penrose-type: 97
    petasmorogene-type: 90
    ptychorogene-type: 90
    Red Sea–type: 124
    rhaetide facies: 167
    suture zone: 8
  comagmatic model: 2, 19, 32–35, 80, 81, 86–88, 95
  correlations to
    Lanzo complex: 46
    Mirdita complex: 124, 357
    Platta-Totalp nappe: 78
  dating of deposits: 2, 10, 32–33, 78–79, 78n*b, 82, 85, 88, 90, 95, 97, 122, 160
  deposition environment: 26, 98, 104
  detachment fault zones: 34, 87, 104–106
  discriminant diagram of: 270
  emplacement of: 2, 34, 78–79
  felsenkalk in: 81
  formation of: 2, 161, 352
  fossils: 79
  geochemical analysis of: 2, 87, 123–124
  geothermometry of: 99, 124
  isotopic analysis of: 123–124
  mapping of: 32, 79–80, 84
  melting of: 104–105
  metamorphic belts in: 124
  metamorphic fabric: 45–46
  metamorphism: 8, 32–34, 78, 81, 88n*25, 97, 99
  modeling of: 167–168
  petrology of: 4, 10, 19, 26, 32–35, 45, 49, 78–90, 94–107, 123–124, 161, 170
  reconstruction of: 86
  saussuritization of: 87
  Steinmann trinity in: 19, 94–107
  stratigraphy of: 26, 97
  subduction zone: 104–107
  volumetric ratio: 33, 38, 80, 81
Alpine-Apennine Mountains
  basin evolution: 358
  granite deposits in: 82
  mapping of: 2, 32, 79, 89n*31, 94
  transform fault zones: 104
Alpine-type orogenic system: 166–168
alpine-type peridotites. *See* peridotite
Alpine-type subduction zone: 366–380
Alps
  asbestos mining in: 450
  correlations to Himalayas: 444n*28
  exhumation of: 380
  lithosphere remnants in: 8, 106
  metamorphic record: 8, 99
  modeling of
    accretionary wedge: 367
    alpine orogenesis: 166–168
    archipelago model: 168–171
    metamorphic belts: 375
  nappe sequences in. *see* nappes
  ophiolite complexes in. *see* Alpine-Apennine complexes
  paleodepth: 8, 99
  Pennine Zone: 401
  plate junction type: 366
  studies of: 388–398
  Tethys Ocean remnants in: 8
Altaid Ocean: 163, 164
Altun Mountains: 164
Alutom Formation: 248
Amargosa Range: 385, 412–413, 434
ammonites: 85
amosite: 451, 454–455
amphibole: 236, 249, 352, 450–451, 456
amphibolite
  geochemical analysis of: 343
  geochronology of: 343
  in Mid-Atlantic Ridge: 49
  in ophiolite definition: 55
  in soles: 339, 342
amphibolite facies
  classification of ophiolites: 148
  in lithosphere: 311
  in metamorphic belts: 374
  petrology of
    Hellenic complexes: 343
    Malenco complex: 99
    Mirdita sole: 124
    Platta-Totalp nappe: 103
    Semail complex: 144
  in soles: 123
Amstutz-type subduction: 167
Anambas Islands: 161

n = footnote*

Anatolia. *See also* Turkey
evolution of
arc-trench rollback in: 116
in tectonic collisions: 114
westward migration: 126
Tethyan sutures in: 127
Anatolian thrust: 428
Anatolide orogen: 125
Anavra Chert: 343
andalusite-sillimanite belts: 367
Andaman Sea
basin evolution in: 118, 231
classification of ophiolites: 12
dating of opening: 118
ridge-transform complex: 287
Andean-type orogenic system: 166
Andersen, Olaf: 57
Andes Mountains
backarc subduction in: 162, 170
Benioff zone in: 170
correlations to: 170
crystalline rock deposits: 12
magmatic arcs: 237
ophiolitic mélanges in: 162
plate junction type: 366
andesite: 251, 255
andesitic volcanism
in arcs: 119, 379
petrology of
Albanian complexes: 352
Aleutian arc volcanoes: 232
Calabria: 119
Corsica: 119
Hellenic complexes: 119, 348
Hunter Ridge proto-arc: 119
Kermadex arc: 230
Mariana system: 230
New Hebrides: 230
New Zealand: 230
Palau-Kyushu ridge: 119
Ryukyu arc: 119
Sardinia: 119
Tonga system: 230
in suprasubduction zone
arc-trench rollback in: 118–120
from deformation: 231
hydrothermal circulation in: 232
mantle counterflow in: 237
mantle thermal anomalies: 119–120
Angat complex: 122, 127
Anglesey mélange: 385, 409–411, 432
anisotropy of magnetic susceptibility (AMS): 332
Ankara complex: 415, 428, 460
Annieopsquotch complex: 155
anorthite: 280
anorthosite: 58, 200, 242
Antalya complex: 42–45, 121
Antalya-type: 275
Antarctic-Australia Discordance (AAD): 121
anthophyllite: 447, 449–451, 454, 456
antigorite: 253, 458
Antilles: 234
Antique complex: 122
Aosta Valley: 450, 458
Apennine-Calabrian arc system: 117, 121
Apliki: 326
Appalachia: 162
Appalachian complexes
asbestos mining in: 456
Bay of Islands complex. *see* Bay of Islands complex
Betts Cove complex: 48–49, 154
formation of
in arcs: 3, 170, 274
from hydrous magmas: 3
Hare Bay complex: 48–49
metamorphism in: 3
petrology of: 48–49
Aptian-Albian: 102
Aptychus limestone: 101
Apulian microcontinent
boundaries of: 352
collision with: 116
migration of: 124, 125, 359
Arabian plate
collision with
Cimmerian continent: 126
Eurasian plate: 114, 117, 121
dating of
collision with Eurasian plate: 117, 121
Lasail volcanics: 127
mantle reservoir: 121
margin subduction
dating of: 127
paleodepth: 126, 127
petrology of
Semail complex: 127
Troodos complex: 126
plate junction type: 366
suture zones: 298
aragonite: 249, 377, 379
Arakapas fault zone
asbestos mining in: 459
boninitic volcanism: 120, 125–126
dike orientation: 73–74, 324, 328–329
at ridge-transform intersection: 73–74, 300
thermal gradients in: 120
ultramafic volcanism in: 138
arc-arc collision: 162
arc-continent collision: 162
Archean
crustal sequences from: 27
greenstone belts: 127
komatiitic magma: 21
Penrose-type complexes: 14
archipelago model: 168–171
Arctic Ocean. *See* Gakell-Lena ridge
arc-trench rollback
characteristics of: 118–119
classification of ophiolites: 11
evolution of
in backarc region: 8, 116
episodic
arc splitting events: 8, 118–119
basin evolution: 8, 128
in Mediterranean: 116–117
in Pacific Ocean: 117–119
mantle flow in: 116, 128, 237
ridge push effect on: 116
short-lived: 7, 116, 128
slab impact on: 116, 305
tectonic collision impact: 8, 116, 128
extension
induction of: 7
speed of: 238, 305
magma chamber: 283
vs. mid-ocean ridge: 127
modeling of
crustal entrapment: 8, 128
ridge-transform intersection: 120
ridge-trench intersection: 284–286
subduction nucleation: 119–121
in suprasubduction zone: 283–85, 345–348
vs. obduction: 121
petrology of
Hellenic complexes: 345–348
Izu-Bonin-Mariana region: 7
Mariana system: 228–229
Tonga system: 228–229
in Wilson cycle: 114
Argand, Émile
on continental drift: 96, 107
dating of Alpine deposits: 95
on geosynclines: 95, 96, 396
on greenstones: 96, 396
Mid-Atlantic Ridge studies: 20
tribute to: 396
Argentina: 13
Argille a Palombini: 79 n*s, 83n*ab, 97–98, 102
argille scagliose
in Alpine-Apennine complexes: 79, 83, 97
dating of deposits: 88
definition of: 79 n*o
gravity-sliding theory of: 432
argillite: 97, 257
Amdo: 161
Arosa zone: 98, 104
asbestos: 326, 447–467
asbestosis: 448, 453
asthenosphere
Bouguer anomaly link: 140
diapiric flow in: 140
downwelling of: 120
extrusion of: 114, 116
geothermometry of: 120, 343
melting of: 9, 104–105
in suprasubduction zone: 229
upwelling of
arc-trench rollback: 283
boninite formation: 120
from decompression: 9
in extensional regions: 231
beneath mid-ocean ridge: 9
in slab window: 9
spreading rate impact: 144–146
in Wilson cycle: 113–114
Asuncion volcano: 257
Aswad massif: 144
Ata volcano: 257
Atlantic Ocean
DSDP Hole 120, study of: 160
gravity map of: 316
magnetic stratigraphic studies: 66–67, 68–70
mantle reservoir: 121
Mid-Atlantic Ridge. *see* Mid-Atlantic Ridge

modeling of: 49
opening/closing of
cordillera development: 113
ophiolite emplacement link: 49, 344
slab pull force: 113
in Wilson cycle: 113
petrology of
continental margin: 6
crust: 4, 68
seismic structure of: 299–300
spreading rates in: 68, 69–70
Atlantis II fracture zone: 217–219
Aubouin, J.
Albanian complexes, study of: 357
on modeling: 38
on ophiolite origin
comagmatic: 4
geosynclines: 309–310
outflow, submarine: 4, 22
Aucellapiochii: 445n*40
aufbruchszone: 408
augen gneiss: 82
augite: 196, 257
aureoles. *See also* soles
in Appalachian complexes: 3
formation of: 32, 39, 56, 60, 81, 82
from hydrothermal alteration: 61
in Pacific-type subduction zones: 368
in sills: 83n*13
in Tethyan-type complexes: 148
Australia
asbestos mining in: 451, 452
classification of ophiolites: 13
collision with New Guinea: 123
Australian plate
boundaries of: 9, 12
collision with
Eurasian plate: 371
Sunda arc region: 121
convergent margin: 119
mantle reservoir: 121
plate junction type: 366
Austria: 400. *See also* Alpine-Apennine complexes
Austroalpine Nappes
in alpine orogenesis model: 166–168
aufbruchszone: 408
correlations to: 404
Avers zone: 95

B

Bachmann, Isidor: 392–393, 394
backarc region
classification of ophiolites
Chilean-type: 12
Cordilleran-type: 8
Mediterranean-type: 11
Sierran-type: 12
spinels, chrome: 281
formation of
in archipelago: 165–166
in convergent margin: 234
in thrust zones: 162
geochemical analysis of
Lau Basin/Ridge area: 276, 287
Sarmiento complex: 276
Scotia Sea: 276, 287
Tortuga complex: 276
Valu Fa Ridge: 287
Western Pacific basin: 241–244
modeling of
archipelago model: 168–171
opening of: 230
ophiolite provenance
data analysis: 137, 287
Dongjiao complex: 161
vs. forearc region: 259
Karmøy complex: 138
vs. mid-ocean ridge: 213–214, 229–230, 237–238, 259
vs. suture zones: 169–170
petrology of
Lau Basin/Ridge area: 241–244
Mariana system: 241–242, 244–246
seismic velocities: 229
spreading center in: 26, 229–230, 234
suprasubduction zone
arc-trench rollback in: 8, 116
collapse of basin: 8
in China: 169
mélange marker for: 161, 165
Moluccan Sea: 121
in Sierra Nevada range: 169–170
in subduction zone: 162
extension: 7, 8, 260
lithosphere
formation of: 7, 231, 238, 287
thrust emplacement of: 260
modeling of: 229
opening of basin: 115
tectonic setting of: 119, 228
tectonic erosion in: 13
Badoux, Héli: 386, 430–432
Baer Bassit region: 35
Bagh complex: 170
Baie Verte: 153
Bailey, Edward
on continental drift: 96, 107
on magma types: 58–59
on mélange: 385–386, 426–430, 432
on Steinmann trinity: 96, 103, 107
on submarine outflow model: 96
trip to Turkey: 425–428
on ultramafic petrology: 56
Bajocian: 343, 352
Baker, P.: 232
Balangero: 458
Balchdura Heights: 405
Balearic Ocean: 117, 156
Bally, A.: 366
Banda Sea: 118, 371
Banerjee, S.: 74
Baragar, W.: 154
Bareshnikov River: 396
Barremian: 85
basalt. *See also* diabase
in fast spreading ridges: 138
formation of. *see also* lava; magma
in basin evolution: 160, 242
in fracture zones: 155
in island arcs: 255
in oceanic plateaus: 155
in orogenic belts: 160
from peridotite: 190–191, 208–29
saturation level: 160
in seamounts: 155
in suprasubduction zone: 276–277
geochemical analysis of: 12
isotopic analysis of
Lau Basin/Ridge area: 242
Platta-Totalp nappe: 107
magnetization of: 67
olivine in: 59
ophiolite provenance: 211
oxidation of: 70
petrology of
Albanian complexes: 352, 359
Aleutian Arc/Trench: 255
Alpine-Apennine complexes: 123
Arosa zone: 104
Atlantic Ocean: 66
Lau Basin/Ridge area: 241–244
lithosphere: 66
Mariana system: 241, 254
Platta-Totalp nappe: 101, 104, 155
Puerto Rico Trench: 231
Semail complex: 126
Smartville arc complex: 155
Solomon Islands: 255
Tofua Arc: 255
Troodos complex: 299
Vanuatu Islands: 255
Vourinos complex: 35
Xigaze complex: 146
in Steinmann trinity: 94
transformation of, to epidosite: 318
basin evolution
categories of basins: 164
classification of ophiolites
Ligurian-type: 10
Mediterranean-type: 11
modeling of
Alpine-type orogenic system: 166–268
archipelago model: 168–171
Troodos complex: 7
plate collisions in: 114, 166–268
seafloor spreading in: 26
Wilson cycle in: 113–114
Bath, Jennifer: 69
Bathonian: 106–107, 343, 352
Bay of Islands complex
classification of
LHOT: 146
Mediterranean-type: 10–11
Penrose-type: 155
Tethyan-type: 304
composition of: 257
correlations to: 155
dating of deposits: 48, 49, 153
description of: 197–198
emplacement of: 305
formation of
in marginal basin: 155
from seafloor spreading: 139, 153, 155
as sill complex: 201
geochemical analysis of: 49

Bay of Islands complex (*continued*)
  mapping of: 200, 300
  modeling of
    acceptance of: 215
    geosyncline: 198
    magma chamber: 154, 198–201, 215
    plate-accretion process: 154, 198–199
  Moho transition zone: 146
  petrology of
    deformation: 146
    field relations: 48–49
    magma chamber: 197–201
    vs. Oman complexes: 146, 153
    sole: 153
    stratigraphy of: 48–49
    structure of: 48–49, 300
Bayesian analysis: 288–289
Beck, P.: 396
Beloussov, V.: 159
Bengal fan: 379
Benioff, H.: 230–231
Benioff zone
  in Andes Mountains: 170
  discovery of: 21
  mantle upwelling at: 231
  seismic crustal studies of: 168
  in SSZ tectonic setting: 228, 234
  subduction in: 167
  volcanic arcs above: 232
Benson, W.N.
  modeling of, geosynclines: 2, 95
  on peridotite petrology
    alpine vs. ophiolitic types: 2, 19
    intrusion of
      crystalline serpentine: 296
      plutonic: 2, 95
    in marginal zones: 153
Bernal, J.: 230
Bernina nappe: 167
Bernoulli, Daniel
  on continental drift: 97, 107
  on ophiolite petrology: 93–107
  petrology of
    Alpine-Apennine complexes: 26
    Platta-Totalp nappe: 98–106
  on Steinmann trinity: 94–107
Berriasian: 101, 102
Bertrand, Marcel: 394, 396–398
Beskidic Nappes: 404
Besshi sulfide deposit: 317
Betic-Rif: 116–117, 121
Betts Cove complex: 48–49, 154
Beyrich, August: 401
Biancone: 79 n*s
biotite: 371, 462
Bird, J.: 49, 154
Bird's Head margin: 123
Bishopp, D.: 298, 325–326
Bitlis-Zagros zone: 298
Black Mountains: 412–413
Black Sea: 164
black smokers: 302, 316–318. *See also* hydrothermal vents
Blanchet, R.: 358
Blome, C.: 68
Bloomer, S.: 284
blueschist facies
  in convergent margins: 369
  devolatilization of: 379
  modeling of plate junctions: 367
  in ophiolite petrology: 156
  petrology of: 13–14
  in seamounts: 249
Bodvarsson, G.: 69
Bogdanov, Nikita: 276
BOI. *See* Bay of Islands complex
Boisse crust/mantle model: 174
Bolivar complex: 13
Bolivia: 451
Bonin region
  arc-trench rollback in: 118
  boninite type area: 277
  classification of ophiolites: 11
  correlations to: 119
  geochemical analysis of: 251
  mapping of: 230
  petrology of: 249
  proto-arc evolution: 122, 248
  subduction zone: 7
  volcanism in: 257
boninites
  composition of: 249–251
  formation of
    in island arc: 254
    from mantle decompression: 120
    at mid-ocean ridge: 277
    at ridge-transform-ridge junction: 284
    in suprasubduction zone: 249–252, 277
  geochemical analysis of
    Mariana system: 251
    ODP Leg 125 (IBM): 284
  geothermobarometry of: 251–252
  identification of: 250–251
  isotopic analysis of: 251
  in ophiolitic complexes: 114, 249
  petrology of
    Cordillera, North American: 260
    Mariana system: 251
  rhyolite link: 251
  in SSZ tectonic setting: 234, 248–250, 277
boninitic volcanism
  in arcs: 114–115, 155, 254
  in arc-trench rollback: 118–120
  chemistry of: 7
  classification of ophiolites
    Mediterranean-type: 11
    Ti-type: 280
  dating of deposits: 126
  extension, induction of: 7
  geochemical analysis of: 139
  mid-ocean ridges and: 127
  petrology of
    Acoje block: 127
    Albanian complexes: 359
    Calabria: 119
    Corsica: 119
    Hellenic arcs: 119
    Hunter Ridge proto-arc: 119
    Kukes complex: 125
    Mariana system: 248, 253
    Mirdita complex: 352
    Palau-Kyushu ridge: 119
    Ryukyu arc: 119
    Sardinia: 119
    Semail complex: 126
    Shebenik complex: 125
    Troodos complex: 125, 249, 299
  in suprasubduction zone
    mantle counterflow in: 283
    mantle thermal anomalies: 119–120
  in transform fault zones: 123
  underplating by
    of lherzolitic mantle: 123
    of MORB-like cumulates: 123
Borneo: 161
Borsi, L.: 82n*z
Bottro di Sassomagnoli: 84
Boudier, F.: 144
Boué, Ami: 338, 400, 401
Bouguer anomaly: 48, 140
Bowen, Norman
  on comagmatic models: 3, 21
  on crystal mush: 39
  on crystallization of
    olivine: 3, 19, 57–59, 61
    pyroxene: 3, 57–58
  on dike petrology: 60, 62
  geological research of: 56–62
  geothermometry of
    clinoenstatite: 57
    fayalite: 59
    forsterite: 19, 57
    iron-bearing silicate systems: 19, 59
    olivine: 3, 21, 57, 59
    pyroxene: 3, 57
    serpentinite: 21
    ultrabasic magma: 57–59
  on Gouy-Chaperon effect: 58
  granite, studies of: 57, 62
  on liquid immiscibility: 58
  on ophiolite concept: 3, 56
  on peridotite petrology
    crystallization, fractional: 3, 19, 57–60, 296
    in magma chamber: 19, 21, 57–58, 61
    stratiform accumulation: 19, 21, 57–58, 61
  on peridotitic magmas: 96
  phase-equilibrium studies: 19, 21, 57–59, 61–62
  on Soret diffusion: 58
  on ultramafic petrology: 57–61
Bowie, William: 20
brachiopods: 398
Brahmaputra River: 234
Bransfield Strait: 237–238
Brazil: 450
breccias
  classification of ophiolites: 12
  deposition environment
    Alpine-Apennine complexes: 98, 104
    Err group: 98
  in fault zones
    Iberia margin: 106
    Piedmont-Liguria ocean: 106
  felsenkalk in: 81
  foraminifera in: 83

petrology of
Alpine-Apennine complexes: 84–85, 97, 98
Alpine-Apennine Mountains: 82
Mariana system: 248
Platta-Totalp nappe: 100–104
Saluver formation: 98
Scottish Highlands: 409
Semail complex: 104
Troodos complex: 68
Briançonnais: 82n*y, 98, 167
British Caledonide complexes: 170
British Columbia: 113
Brongniart, Alexandre
classification by: 19
on deformation, in belts: 153
geochronology by: 32
ophiolite description: 2, 17–19, 32, 49, 94, 296
Browning, P.: 204–205, 280
Brunn, J.H.
mapping of
Pindos complex: 4, 339
Vourinos complex: 4, 339
on ophiolite petrology: 4, 35–38, 339
on seafloor spreading: 22, 49, 180
on sole metamorphism: 342
visit to Albanian complexes: 357
bryozoa: 398
Bucher, Walter: 159
Buchiapiochii: 445n*40
Bullen, T.: 281
Bündnerschiefer sequences: 95
Burgess Shale: 113
Burma: 161
Burma-Sumatra magmatic arc: 12
Bushveld Igneous complex: 38

C

Cadisch, Joos: 408
Cagayan-Sulu backarc: 127
Calabria: 116, 119
calcalkaline volcanism
in arcs
evolution from tholeiitic: 255
inboard of trench: 8
isotopic analysis of: 368
proto-ophiolite: 155
from subduction: 232, 371, 377–380
classification of ophiolites: 7
crustal thickness link: 232
dating of: 126–127
hydrothermal circulation in: 114, 232
mid-ocean ridges and: 127
petrology of
Acoje block: 127
Alaskan complexes: 154
Kannaviou Formation: 312
Kukes complex: 125
Lasail volcanics: 126–127
Mariana system: 248
Palawan complex: 127
Semail complex: 126–127
Shebenik complex: 125
Troodos complex: 125
underplating by
of lherzolitic mantle: 123
of MORB-like cumulates: 114, 123
calcite: 97, 236, 379
calcium: 81n*w
calcium carbonate: 79
calcschists: 429
Caledonian complex: 8
Calera limestone: 170, 445n*40
caliche: 103
California
Amargosa Range: 385
asbestos mining in: 456
El Dorado County: 464–465
Franciscan Complex. *see* Franciscan system
Golden Gate complex: 19
Josephine complex: 24, 139
Mount Diablo complex: 19
San Francisco complex: 19
Sierran-type: 11–12
California Coast Ranges
Franciscan-Knoxville paradox: 169, 434, 445n*40
metamorphic belts: 375
ophiolitic complexes: 249
California-type orogenic system: 166
Callovian: 106–107, 343
Calpionella limestone: 79, 79 n*s, 101–2
calpionellids: 79, 83
Cambrian: 48
Campanian: 324
Campbell Plateau: 13
Campiglia Marittima: 82
Canada. *See also* specific provinces
alpine-type peridotites in: 60–61
asbestos mining in: 450
asbestos related disease in: 455
Newfoundland complexes. *see* Newfoundland
Canadian Shield
dating of deposits: 48
Cann, Joe
modeling of
dike density: 70
lithosphere: 70, 212
magma chambers: 67, 212
on ophiolite tectonic setting: 270
on Troodos complex: 309–19
XRF studies by: 275, 278
Cannat, M.: 4, 68
Canyon Mountain complex: 138
carbonate: 248, 254–255, 340
Carboniferous: 156, 340, 417
carcaro: 81, 81n*w
Cardiff, Wales: 454
Carey, S.W.: 22
Caribbean: 3, 116, 127
Caribbean plate: 13
Caribbean-type: 13
Carlsberg Ridge: 69
Carnian: 341–342
Carpathian Mountains
exotics blocks in: 393
Flysch Zone: 401, 444n*23
Klippen Zone: 401
mapping of: 400
orogeny: 117, 121
studies of: 398–405
Caspian Sea: 164
Castellina Marittima: 84, 86
Castelnuovo di Garfagnana: 80, 86
cataclasites: 389
Cathaysia: 161, 170
Cayman Trough: 155
Celebes Sea: 118, 123
celtide facies: 167, 170
Cenomanian
foraminifera in deposits: 102
lava deposits: 35
limestone deposits: 445n*40
in Simme nappe: 85n*19
Cenozoic: 340
Central Cordilleran complex: 123
Central Gneiss Zone: 170
Chablais Mountains: 394–395
chalcopyrite: 317
Channel Islands: 170
Chaput, Ernst: 415, 417
chert
classification of ophiolites
Franciscan-type: 13
Mediterranean-type: 10
Penrose-type: 5
dating of deposits
British complexes: 97
Troodos complex: 39–40
depositional environment: 19
formation of: 153
geochronology of: 343
in ophiolite definition
Penrose: 5
Steinmann trinity: 2, 19
petrology of
Albanian complexes: 352
Alpine-Apennine complexes: 2, 19, 32, 79 n*s, 94, 97
Amdo: 161
British complexes: 97
DSDP Hole 120: 160
Franciscan system: 13
Mariana system: 254
Platta-Totalp nappe: 101
Qiangtang basin: 161
Troodos complex: 40
radiolarian. *see* radiolarites
ribbon: 5, 101
Chichijma: 250–51
Chile Ridge
classification of ophiolites: 11
geochemical analysis of: 286
ridge-trench interactions: 286
Chile Rise: 235
Chile Trench: 366
Chilean-type: 12
China: 161–165, 169, 170, 450
chlorite: 236
chloritoid: 236
Chondrites: 83n*ac
chromite
classification of ophiolites
HOT: 146
LHOT: 146
Mediterranean-type: 11
Penrose-type: 5
mapping of: 326

chromite (*continued*)
metamorphic structure in: 38
in ophiolite definition: 5
petrology of
Albanian complexes: 357, 359
Bay of Islands complex: 146
Mirdita complex: 146
Othris complex: 345
Pindos complex: 345
Semail complex: 146
Vourinos complex: 345
spread rate indicator: 146
SSZ indicator: 345
chromitites: 42, 275
chromium number index: 139
chrysotile: 447–67
Chugach Complex: 13
Cimmerian continent: 126
Cina, A.: 357
Circumpacific: 8, 367, 369
Circumpacific-type subduction: 366–380
Class-I ophiolites: 7, 11
Class-II ophiolites: 7, 12
Class-III ophiolites: 7, 10
clay, deep-sea
in euabyssal deposits: 79 n*q
petrology of
Alpine-Apennine complexes: 79, 97
Alpine-Apennine Mountains: 83
Argille a Palombini Formation: 79 n*s
clinoenstatite: 57, 249–51
clinohumite: 236
clinopyroxene
in boninites: 249–251
classification of ophiolites
HOT: 146
LHOT: 146
Ligurian-type: 10
Ti-type: 280
crystallization of: 361
formation of: 255
geochemical analysis of: 279
geochronology of: 102
petrology of
Alpine-Apennine complexes: 124
Mariana system: 242, 253
Troodos complex: 125, 196
in SSZ crust/mantle: 236, 282
Coaliangyi complex: 162
Coast Range complex
classification of
by Coleman: 9
Cordilleran-type: 304
Franciscan-type: 170
petrology of: 169–170
Coats, R.: 231–232
coccolith ooze: 79n*n
Coccolithophoridae: 156
coesite: 368, 377, 379
cold wall effect: 34
Coleman, Robert
classification scheme: 9–10, 148
on obduction: 8, 24, 26
on ocean crust petrology: 6
on ophiolite petrology: 281
on ophiolite research: 6
on Steinmann trinity: 96
Columbia: 13
Colville-Coromandel-Taranaki arc: 119
comagmatic model
Alpine-Apennine complexes: 2, 19, 32–35, 80, 81, 86–88, 95
Appalachian complexes: 3
Canyon Mountain complex: 3
crustal sequence: 32
ophiolite provenance: 296
Platta-Totalp nappe: 104
Semail complex: 45
Syrian greenstones: 4
Vourinos complex: 4
compression
basin formation by: 164
cause of: 309
of diabase dike complex: 326
in geosyncline evolution: 159–160
P-waves. *see* P-waves
strike-slip faults from: 326
thrust faults from: 326
conglomerate
in Alpine-Apennine complexes: 85
vs. mélange: 388
nagelfluh: 389, 443n*5
Continental Borderland: 169
continental drift theory
fathom contour table: 22
geosynclines and: 96
ocean-continent interactions: 20, 65
oceanic vs. continental crust: 65
plate tectonics link: 325
seafloor spreading and: 22, 325
studies of: 20, 309–310
vindication of: 23, 309
continental margin
arc-trench rollback in: 304
classification of ophiolites
Class-II: 7
Cordilleran-type: 148, 304–305
Franciscan-type: 13
Ligurian-type: 10
Mediterranean-type: 10–11
by Nicolas: 8
Tethyan-type: 8, 26, 148, 304
geothermobarometry of: 120
lithosphere
generation of: 156
obduction of: 5, 24, 26, 97, 121
miogeosyncline definition: 160
modeling of
alpine orogeny: 166–168
extension, mode of: 106
fault zones: 104–106
oceanic arcs: 155
rifting of lithosphere: 104–106
subduction zone: 23–24, 26
ophiolite emplacement by
arc-trench rollback: 121, 305, 345–348
deformation: 344
flake tectonics: 342–346
gravity slide: 305, 342
obduction
analog to, modern: 8
vs. arc-trench rollback: 121
lithospheric isotherm: 155, 156
mechanism for: 5, 156, 259–260
modeling of: 23–24, 97
by subduction zone collision: 26, 346
vs. subduction/accretion: 128, 339
tectonic processes: 4, 20, 21, 26
underplating: 8, 305
petrology of
Atlantic Ocean: 6
Caribbean: 13
Err nappe: 100
peridotites: 8
soles: 7–8
subduction zone in: 8
suture zones: 305
underplating at: 9–10, 305
continents
breakup of
peridotite formation: 124
in Wilson cycle: 113
collision of
in alpine orogenesis model: 167
in emplacement model: 342
in nappe petrology: 100
in ophiolite petrology: 155
in Wilson cycle: 113
crust/mantle structure
arc-trench rollback: 117
decompression in: 286
depth of: 21, 49, 65
flow traction impact: 121
magma composition: 368
modeling of: 174–75
seismic studies: 21, 175–76, 311
surface wave studies: 175
isostatic depth: 21
magma chambers. *see* magma chamber
paleodepth: 378
substrates: 8, 24
surface wave studies: 175
underflow zones: 369, 375
contraction
in Alpine-Apennine complexes: 87
compression from: 309
evolution of theory: 17
strike-slip faulting from: 309
convergent margins
accretionary prism: 234
boninite origin: 251–252
characteristics of: 369
classification of ophiolites
Cordilleran-type: 8
Sierran-type: 11–12
coverage area: 234, 366
oblique: 9, 12, 366
ophiolite emplacement at: 339, 342, 368
plate junction type: 373–374
rate of convergence: 283
SSZ tectonic setting: 228, 234
copper: 81, 84, 326, 357
Cordilleran-type
continental margin: 304
correlations to

Franciscan-type: 14
Sierran-type: 12
definition of: 8, 24, 148
development of: 24
emplacement of: 304–305
formation of by obduction: 24
obduction in: 24, 26
preservation of: 27
Corsica. *See also* Alpine-Apennine complexes
andesitic volcanism: 119, 121
Apulia-Eurasian plate collision: 116
arc-trench rollback: 116
asbestos related disease in: 461
classification of ophiolites: 10
geochronology of gabbro: 124
Insecca complex: 34–36
subduction zone: 124
suprasubduction zone: 124
Costa Rica: 13
Costa Rica Rift
DSDP Hole 504B: 311, 332
spreading center: 315
Coto block: 127, 257
Couillard, R.: 68
Cox, A.: 69
Craticulifera: 417
Cretaceous
Antalya complex: 42–45, 121
basin evolution in: 126
Caribbean deposits: 13
chalk deposits: 156
clay deposits: 79 n*s
conglomerate deposits: 85
flysch deposits
Alpine-Apennine Mountains: 82n*z
Carpathian Mountains: 400
Helminthoid Flysch: 80 n*u
Pietraforte formation: 85n*ae
fossils, in deposits
Alpine-Apennine complexes: 85
on seamounts: 21
lava: 39
limestone deposits
Alpine-Apennine complexes: 79, 79n*k, 79n*n, 97
Argille a Palombini Formation: 79 n*s
Calera limestone: 445n*40
Carpathian Mountains: 401
scaglia: 79, 79n*k
Vourinos complex: 35
marl deposits
Alpine-Apennine complexes: 79, 79n*k
scaglia: 79, 79n*k
normal superchron: 68
ophiolitic complexes
Japan: 13
Philippines: 123
Semail complex. *see* Semail complex
platform cover: 340
polarity boundaries in: 315
radiolarite deposits
Alpine-Apennine complexes: 79
Mariana system: 254
Troodos complex: 39–40
sandstone deposits
Carpathian Mountains: 398
Pietraforte formation: 85n*ae
sedimentary deposits: 340
shale deposits: 85n*ae
subduction zones
Adria plate: 124
European margin: 124
subduction/obduction: 98
ultramafic rock deposits: 74
Crete: 121
crocidolite: 450–455
Crowell, John: 431
crust, oceanic. *See* lithosphere
crystallization
classification of ophiolites: 280
in complexes: 104, 352
crosscutting by: 2, 80–81
eutectic vs. fractional: 19, 58
in magma chamber: 19, 34, 86–87
serpentinization from: 56, 60–62, 80–81
solid state injection: 3, 21, 96
stratigraphic layering from
in basaltic lava: 211
in magma chamber: 19
outflow, submarine: 4, 190–191
phase relationships: 57–58
Cuba: 11–12, 60–61
Curie point isotherm: 67
Cuvierian: 428
Cycladean complexes: 119, 121
Cyprus
Arakapas fault zone. *see* Arakapas fault zone
arc-trench rollback in: 117, 121
asbestos mining in: 449, 451, 458–459
asbestos related disease in: 461
basin evolution: 117, 121–122, 125, 126
classification of ophiolites: 10
collision with African plate: 122, 126, 304
geological survey of: 22–23
Kyrenia Range: 125, 326
Limassol Forest. *see* Limassol Forest
Mamonia complex. *see* Mamonia complex
paleomagnetic studies on: 23, 39, 68, 126
plate-margin in: 168
subduction nucleation in: 125–126
Troodos complex. *see* Troodos complex
Cyprus Crustal Study Project: 73, 196

D

Dabie complex: 162, 368, 375, 379
dachsteinkalk: 405
dacite: 257, 352
Dalrymple, G.: 69
Daly, R.: 19
Dangerous Grounds plate: 118
Darwin, Charles: 112–113
Daubrée, Auguste: 408
Davis, E.: 34, 83, 83n*13
Day, Arthur: 57
De Roever, W.P.
on peridotite petrology: 38–39
on Steinmann trinity: 4
tectonic emplacement model: 4, 21, 296
de Saussure, Horace: 389
Death Valley: 412–413
debris flow: 98
Decandia, F.: 105
Dede, S.: 357
Deep Sea Drilling Project (DSDP). *See also* Ocean Drilling Program
Hole 120, analysis of: 160
Hole 395A (Mid-Atlantic Ridge): 311
Hole 504B (Costa Rica): 311, 332
Leg 6 (Mariana Trough): 277
Leg 31 (West Philippine Basin): 277
Leg 59 (transect): 277
Leg 60 (transect): 249, 277
Site 453 (Mariana): 242
Site 458 (Mariana): 245
Site 459 (Mariana): 245, 249
Site 460 (Mariana): 249
den Tex, E.: 96
d'Entrecasteaux Ridge: 121
Desmurs, Laurent: 93–107
detachment fault zones
in Alpine-Apennine complexes: 34, 87, 104–106
vs. extension: 315
in flake tectonics: 342–344
in Iberian margin: 105–106
in Limassol Forest: 126, 300
mantle exhumation in: 105–106
in Piedmont-Liguria ocean: 106
in Solea Graben: 315
in spreading centers: 314–315
Devonian
accretionary complexes: 13
diabase deposits: 83n*13
ophiolitic complexes: 155
platy sills from: 83n*13
Dewey, H.
on albitization of lava: 87
modeling by: 34, 83
on ophiolite petrology: 34, 83, 88, 97
on radiolarites: 34, 83
Dewey, J.
classification scheme: 154–155
on obduction: 26
on ophiolite petrology
basin origin: 154
research on: 153–156
spreading center origin: 49, 154
subduction zone origin: 154
tectonic processes: 154
diabase. *See also* dikes
classification of ophiolites: 12
dating of deposits
Alpine-Apennine complexes: 33, 79
British complexes: 97
formation of
dikes: 83n*13
lava flows, folded: 298, 326
mapping of
in Canada: 56
Troodos complex: 23
modeling of: 3, 32–33, 80
in ophiolite definition: 2, 33, 78
petrology of
Alpine-Apennine complexes: 2, 19, 32, 32–33, 33–35, 34, 80–90, 84, 87, 88n*25, 94–107

diabase, petrology of (*continued*)
Antalya complex: 45
Australian complexes: 2
Bay of Islands complex: 49
British complexes: 97
Canyon Mountain Complex: 3
Othris complex: 340
Pindos complex: 155
Platta-Totalp nappe: 100
Rocas Verdes complex: 12
Sarmiento complex: 12
Semail complex: 45
Smartville arc complex: 155
Tortuga complex: 12
Troodos complex: 40, 296–298, 325
Tuscany complexes: 32–33
Val di Magra: 81
Vourinos complex: 41
radiolarites and: 97
structure of: 97
volumetric ratio: 33, 80
Diablo Range: 167, 450
diallagites: 35
diamictite: 435
diapirs
in forearc region: 249
mapping of: 140–141, 144
in microplates: 148
in seamounts: 66, 249
diatoms: 83
Dick, H.: 4, 67, 281
Dickinson, W.: 23, 232
Diener, Carl: 405
Dietrich, V.: 276–277
Dietz, R.: 22, 154, 230
dikes
chilled margins
Betts Cove complex: 154
gabbro contact points: 104
modeling of: 154
one-way: 70–72, 114
Semail complex: 346
serpentinite contact points: 104
Troodos complex: 298, 301
classification of ophiolites
Antalya-type: 275
Caribbean-type: 13
Chilean-type: 12
Ligurian-type: 10
Macquarie-type: 12
Mediterranean-type: 11
Penrose-type: 5
Pindos-type: 275
Sierran-type: 11
Troodos-type: 275
dating of deposits: 255
deformation
extension
amagmatic: 73
seafloor spreading: 4–5, 23
sheeted complex origin: 154
from tectonic emplacement: 81
depositional environment: 298
formation of
emplacement zone size: 70–71
epidosite: 317–318, 331–332
in granite: 82
in island arc: 254
lava flows, folded: 298, 326, 357–358
in magma crystallization: 81
at mid-ocean ridge: 123, 173, 214–215, 274
mineralization impact: 58
rock type impact: 39
solid vs. liquid state: 60
at spreading center: 269
vs. ultramafic veins: 62
in volcanoes: 270, 274
geochemical analysis of: 12
geothermometry of: 303–304, 317–318
hydrothermal circulation in: 181–182, 317–318
listric faults
Josephine complex: 24
Solea Graben: 316
Troodos complex: 24, 73, 114
Vourinos complex: 114
magma flow direction: 332
mapping of: 22–23, 312–313, 325
metamorphism in: 326
modeling of
actualistic: 123
crust/mantle structure: 300
density of: 69–71
fault zones: 330–332
hydrothermal circulation in: 199, 317–318, 331–332
intrusion by: 69, 313–314
orientation of: 329
plate tectonics: 4–5
suprasubduction: 7
zone of injection: 313–314
nephritization of: 87n*23
in ophiolite definition: 5, 38, 97
paleomagnetic studies: 328–331
petrology of
Albanian complexes: 352, 359
Alpine-Apennine complexes: 10, 19, 32–34, 81, 87, 97, 123
Antalya complex: 42
Bay of Islands complex: 11, 199–200
Betts Cove complex: 154
Bolivar complex: 13
Himalayas: 405
Iceland: 69
Kizildag complex: 11
Kukes complex: 125
Macquarie Ridge complex: 12
Mariana system: 248
Mirdita complex: 124, 352–358
Nauplia complexes: 85
New Caledonia complex: 21
Pindos complex: 124
Platta-Totalp nappe: 103, 104, 154–155
Quebec complexes: 456
Rocas Verdes complex: 12
Sarmiento complex: 12
Semail complex
evolution of: 144
magma chamber: 202, 208, 301
sheeted complexes: 126
subvolcanic intrusion: 45
Shebenik complex: 125
Solea Graben: 316
in Soviet Central Asia: 6
Stillwater complex: 60
Syrian greenstones: 38
Tonga system: 249, 255
Tortuga complex: 12
Troodos complex
core sample: 196
crosscutting by: 301, 310, 313
extension: 326
field relations: 312–314, 326
geological setting of: 324
hydrothermal vents: 317–318, 328, 331–332
identification of: 325
listric faults: 24, 73, 328
magma chamber: 67, 301
magma flow direction: 332
orientation of: 324, 328–329, 332–333
rotation of: 73–74, 315, 329–330
sequences: 296–298
sheeted complexes: 23, 125, 269, 310–312
stratigraphy of: 154
thickness of: 324
Tuscany complexes: 32
Virginia vermiculite deposit: 456
Vourinos complex: 38, 40–41, 155
Xigaze complex: 11
rotation of
in crust evolution: 313, 316
in hanging wall: 316, 330
modeling of: 215, 328–330
at ridge-transform intersection: 73–74
in Steinmann trinity: 153
Dilek, Yildirim
on Albanian ophiolites: 351–362
classification scheme: 10–14
on comagmatic model: 5
on magma chambers: 188–222
on ophiolite petrology: 1–14
on Penrose definition: 5–6, 10
on Penrose-type lithosphere
classification of: 10–11
definition of: 5–6
depth of: 68
vs. Hess-type: 4, 5
seafloor spreading: 5
stratigraphy of: 2, 5
on suprasubduction zones: 7
Dimitry Mendeleev cruise: 276–277
Dinaric complexes: 78–90, 338, 342
diopside: 57, 58, 253
diorite
geochronology of: 124
in mélange: 389
petrology of
Mariana system: 242, 253–254
Semail complex: 126
Syrian greenstones: 35
Tonga system: 253–54
Vourinos complex: 41
Discocyclina: 255
Doell, R.: 69
dolerite: 35, 45, 104

Dongjiao complex: 161
Donna Seamount: 243
Douvilleceras sp.: 445n*40
Drake, C.: 22
Drake Passage: 316
du Nord massif: 146
Du Toit, A.: 309
Dubertret, L.: 4, 22, 35
dunite
  classification of ophiolites
    Caribbean-type: 13
    Macquarie-type: 12
    Penrose-type: 5
    Ti-type: 280
    Type II: 275
  dike formation in: 58, 62
  formation of: 146, 219, 255
  geochemical analysis of: 19
  geothermometry of: 5, 59
  modeling of: 58
  Moho transition zone: 140–141
  in ophiolite definition: 5
  petrology of
    Albanian complexes: 352, 359
    Antalya complex: 42
    Bay of Islands complex: 200, 300
    Bolivar complex: 13
    Macquarie Ridge complex: 12
    Mariana system: 253
    Mid-Atlantic Ridge: 49
    Mid-Indian Ridge: 49
    Papuan complex: 46–48
    Quebec complexes: 456
    Semail complex: 126, 202
    Shikoku Basin: 242
    Troodos complex: 23, 40, 67–68, 125, 154, 193, 197, 300
    Vourinos complex: 26, 41, 68
  shape of: 55
  in SSZ crust/mantle: 236
  in ultramafic sequence: 23
DUPAL-like mantle: 121
Dutch Antilles: 13

E

East African Rift Valley: 309
East Indies: 60
East Pacific Rise
  Bouguer anomaly: 140
  fast spreading ridge in: 138, 181
  Hess Deep complex link: 332
  hydrothermal vents: 317
  mapping of
    fissure zone: 313
    lava flows: 313
    magma chamber: 205
    segments: 140
  MELT experiment: 140
  vs. Mid-Atlantic Ridge: 216
  petrology of: 138
  seamount link: 139
  seismic studies of: 141–142, 178, 215–216
  subduction of: 169
Easter microplate: 148
Ebel, Johann: 389
echinoderms: 398
eclogite
  in convergent margins: 369
  density of: 375
  isotopic analysis of: 379
  melting of: 127, 178–179, 237
  in metamorphic belts: 379
  modeling of, plate junctions: 367
Ecuador: 13
Ecuador Rift: 315
Eddington, Arthur: 387
edelserpentin: 81
edifices, arc volcanic
  classification of ophiolites: 8, 11, 24
  composition of: 66
  lack of: 7–8
Egypt: 168
Elbe Valley: 396
Elmadağı Series: 417
Elter, P.: 105
Emigrant Canyon: 413
Emme River: 389
Endeavor Deeps: 148
Engadin Valley: 98
enstatite: 58
Eocene
  basin evolution in
    Mariana system: 248–249
    Pacific Ocean: 119, 122
  boninitic volcanism
    Guam: 154
    Palau-Kyushu ridge: 119
  flysch deposits
    Alpine-Apennine Mountains: 82n*z
    Helminthoid Flysch: 80 n*u
  Globigerina-bearing strata: 394
  ophiolitic complexes
    Olympic Peninsula. *see* Olympic Peninsula
    Zambales complex: 127
  sandstone: 79
  volcanic arc evolution: 254–255
epidosite
  formation of
    in dikes: 317–18, 326, 331–332
    in forearc region: 318
    in mid-ocean ridges: 318
  geochemical analysis of: 304, 317–318
  metamorphism indicator: 326, 331
  petrology of: 303, 317–318
epidote: 236
epidote facies: 81, 99, 343
Eratosthenes seamount: 122
erionite: 454
Ernst, W.: 365–380
Err nappe
  dating of deposits: 99–100
  deposition environment: 98
  modeling of: 167
  petrology of: 88n*25, 99–100
  Tethys Ocean remnants in: 100
Escher, Arnold: 396
Eskola, Pentti: 57
'Eua Island: 254–255
euabyssal-type: 79, 79n*p. *See also* radiolarites
eufotides: 19, 32–33. *See also* gabbro
eugeosynclines: 160
Eurasia
  collision with
    Luzon arc: 118, 121, 122
Eurasian plate
  collision with
    African plate
      alpine orogenesis model: 167
      dating of: 121
      mantle perturbation: 114
      Mirdita complex: 124
      ocean size constraint: 156
    Apulia plate: 116
    Arabian plate
      dating of: 117, 121
      mantle perturbation: 114
    Australian plate: 371
    Indian plate
      dating of: 118, 119
      mantle perturbation: 114
      plate margins, location of: 170
    Philippine Sea plate: 121
  plate junction type: 366
  Tethyan sutures in: 127
Europe
  island arcs in: 168
  subduction of continental margin: 124
European plate: 168
eversion process: 45
Evia island: 119
Evvia complex: 343
Ewing, Maurice: 65, 175, 208
extension
  in accretionary complex: 13
  arc splitting during: 8
  basin evolution
    formation by: 164
    opening of: 8
    seafloor spreading: 26, 311
  calculation of: 316
  detachment faulting: 105, 315
  dikes, sheeted complexes
    crustal extension in: 311
    seafloor spreading: 4–5
    suprasubduction model: 7
  gabbro intrusion: 104
  in geosynclines: 96
  in Lau Basin region: 238
  modeling of
    in continental margin: 106, 305
    in suprasubduction zone: 283
  in ophiolite petrology: 95
  at spreading center: 4–5, 68
  subduction and
    location of: 283
    stress change: 120, 260
  in suprasubduction zone
    arc-trench rollback: 8
    arc-trench rollback in: 7, 238, 283, 305
    location of: 228
    mantle counterflow in: 237, 283
    modeling of: 283, 305
    rift basin evolution: 234
  in trenches: 234

extinction event, mass: 83
extrusion
  classification of ophiolites: 11
  in collision zones: 8
  differentiation, magmatic: 4
  felsic sections in: 138
  magnetic stratigraphic studies: 69
  sequences: 6
  in suprasubduction zones: 8

F

Facpi Formation: 248
fan, submarine: 340
fayalite: 59
Feather River belt: 155
feldspar
  albitization of: 160
  volumetric ratio: 33, 80
felsenkalk
  dating of deposits: 85
  definition of: 79 n*q
  petrology of
    Alpine-Apennine complexes: 79
    Alpine-Apennine Mountains: 83
    Val di Magra: 81
Fichtelgebirge: 83n*13
Field, Richard: 20
Fiji Islands: 119, 232, 235
Findlay, T.: 49
Finland: 449–451, 454–456
Fizh massif: 144
flake tectonics: 342–346
Flett, J.
  on albitization of lava: 87
  on ophiolite petrology: 34, 83, 88, 97
  on radiolarites: 34, 83
  on submarine outflow model: 34, 83
flexure zones: 95
Florence Rise: 168
Flower, Martin: 8, 9, 111–128
flysch. *See also* mélange
  alberese: 80 n*u
  Helminthoid Flysch: 80 n*u
  Leimeren Series: 396, 443n*17
  macigno: 79n*l, 443n*16
  modeling of: 393
  petrology of
    Albanian complexes: 352
    Alpine-Apennine complexes: 79
    Alpine-Apennine Mountains: 83, 167–168
    Alps: 389–96, 444n*23
    Amdo: 161
    Carpathian Mountains: 398, 401, 409, 444n*23
    Habkern Valley: 389–393, 396, 408
    Himalayas: 405
    Mirdita complex: 125
    Pindos complex: 125, 340
    Préalpes: 389–396
    Turkey: 430
Flysch Zone: 401, 444n*23
folds: 2, 19, 26
Fonualei Island: 257
foraminifera
  in breccias: 83
  in limestone: 398
  in reefs: 255
  in sandstone assemblages: 102
forearc region
  classification of ophiolites
    Cordilleran-type: 8
    Mediterranean-type: 11
  emplacement of: 276
  evolution of: 114–115
  formation of: 116
  geochemical analysis of
    basalts: 7, 116
    Yap Island/Trench: 276
  magmatism in: 249–250
  mineralization of: 318
  ophiolite provenance
    vs. backarc region: 259, 276
    Limassol Forest: 300
    vs. mid-ocean ridge: 229–230, 259
    subduction-resistance: 123
    Troodos complex: 125, 299
  petrology of
    Mariana system: 248, 299
    Tonga system: 248
  plate junction type: 366
  spreading center in: 26, 229–230, 234
  suprasubduction zone in
    arc-trench rollback in: 8, 116, 119
    basin evolution: 8, 231
    boninites: 114, 119
    diapirism in: 234
    extension
      coverage area: 234
      episodic: 249
      induction of: 7, 8
      magma eruptions from: 247
    geological setting of: 246–247
    heat flow in: 247
    lithosphere in: 7, 231, 247
    mantle flow in: 116
    progression of incorporation: 119
    tectonic setting of: 228
  tectonic erosion in: 13
Foreland Thrust Belt: 169
forsterite: 19, 57, 59. *See also* olivine
fossils. *See also* specific types
  in Alpine-Apennine complexes: 79
  extinction event, mass: 83
  in limestone deposits: 398
  in Troodos complex: 317
fracture zones
  in Carlsberg Ridge: 69
  complex classification by: 9, 146
  crust/mantle structure: 146, 237
  diapiric intrusion in: 68
  extension, amagmatic: 68
  Macquarie Ridge complex: 237
  modeling of: 154
  scarps: 155
  seismic velocity of mantle: 68
  subduction zones in: 248
  in Troodos complex: 300
Franciscan system
  classification of: 8, 9–10, 13–14, 170
  correlations to: 170
  dating of deposits: 13, 169
  faults in: 170
  mapping of: 433
  mélange in: 430–433
  metamorphic grade: 13–14
  modeling of: 375
  petrology of: 160, 169–170
  plate junction type: 366
Frigido Valley: 82
Früh, Jakob: 393
fucoids: 83, 394
Fujian-Taiwan region: 119
fusulinids: 340, 417

G

gabbro
  classification of ophiolites
    Caribbean-type: 13
    Chilean-type: 12
    Franciscan-type: 13
    HOT: 146
    by Ishiwatari: 144
    LHOT: 146
    Macquarie-type: 12
    Penrose-type: 5
    by Rocci: 275
    Ti-type: 280
  correlations to: 104
  dating of deposits
    Alpine-Apennine complexes: 33, 79
    Semail complex: 45
  deformation of
    to augen gneisses: 315
    in orogenic belts: 153
  formation of
    in crust: 178–180, 219
    in fracture zones: 155
    in magma chamber: 142
    at mid-ocean ridge: 123, 173
    in oceanic plateaus: 155
    from peridotite: 208
    in seamounts: 155
    at spreading center: 315
  geochemical analysis of
    fractionation trends: 280
    Mariana system: 253
    Montgenevre complex: 124
  geochronology of
    Corsican complexes: 124
    Ligurian complex: 124
    Montgenevre complex: 124
    Piedmont-Liguria ocean: 106–107
    Platta-Totalp nappe: 104, 106–107
    Tonga system: 255
  isotopic analysis of: 107
  mapping of: 22–23
  in mélange: 389
  modeling of
    actualistic: 123
    comagmatic: 2, 3, 32–33, 80
    crust/mantle structure: 300
    melt fraction: 142–144
    pressure solution: 143
    suspension flow: 142–144, 205
  in ophiolite definition: 2, 5, 33, 55, 78, 97
  paleomagnetic studies: 331

petrology of
Albanian complexes: 352, 359
Albano-Thessalian furrow: 35
Alpine-Apennine complexes: 2, 10, 19, 32–35, 45–46, 80–90, 94–107, 123
Antalya complex: 42–45
Aswad massif: 144
Australian complexes: 2
Bay of Islands complex: 49, 199–201
Bolivar complex: 13
Canyon Mountain Complex: 3
DSDP Hole 120: 160
East Pacific Rise: 138
Franciscan system: 13
Hess Deep complex: 138
Insecca complex: 35
Lau Basin/Ridge area: 242
Macquarie Ridge complex: 12
Mariana system: 242, 253
Mid-Atlantic Ridge: 4, 49
Mid-Indian Ridge: 49
Mirdita complex: 146, 353
ODP Hole 1105A: 218
ODP Hole 735B: 217–219
ODP Leg 118 (Indian Ocean): 315
ODP Leg 147 (Hess deep): 315
ODP Leg 153 (Atlantic): 315
Papuan complex: 46
Platta-Totalp nappe: 100–102, 104, 155
Rocas Verdes complex: 12
Sarmiento complex: 12
Semail complex: 45, 126, 142–146, 202–208, 215, 221
Skaergaard complex: 144
on Skye: 57
Smartville arc complex: 155
Solea Graben: 316
Southwest Indian Ridge: 217–221
Syrian greenstones: 4
Talkeetna Arc: 255
Tanzawa Arc: 257
Tonga system: 255
Tortuga complex: 12
Troodos complex
chilled margins: 193, 298
core sample: 196–197
cumulates: 193
deformation of: 315
intrusive bodies: 67, 193, 298
lithosphere: 40, 74
stratigraphy of: 154
structure of: 125, 296–298
Turkey: 429–430
Viti Levu: 257
Vourinos complex: 4, 41
Vuagnat: 4
Xigaze complex: 146
polarity boundaries in: 73
rotation of: 331
seismic velocity of: 180–181
source material: 58
in Steinmann trinity: 153
tectonization of: 209–211
texture of: 142
volumetric ratio
Alpine-Apennine complexes: 33, 38, 80
Guleman complex: 38
Hatay complex: 38
Vourinos complex: 38, 41
Gabilan Range: 170
Gakell-Lena ridge: 146
Galapagos Ridge: 181
Galapagos Rift: 315
Galapagos Rise: 316
galestro: 83. *See also* clay, deep-sea
Galicia margin: 10, 154
Gander belts: 153
Gansser, Augusto: 430
garnet: 81, 84, 343
Garrison, Bob: 431
Gass, Ian
on ophiolite petrology
Semail complex: 281
Troodos complex: 23, 39–40, 67, 70, 154
paleomagnetic studies: 39, 68
on ultramafic petrology: 23
Gaudry, A.: 325
Gavorrano: 82
Gavroro zone: 340
geanticline: 161
generative constraints: 113
geosynclines
Albano-Thessalian furrow: 35
Alpine-Apennine complexes
crust, oceanic: 45–46
sediment, intrusion into: 2, 33–35, 78, 88, 94–97
Antalya complex: 45
Appalachian complexes: 20
crustal thickness in: 160
definition of: 22
evolution of theory: 17, 23, 95
formation of
crustal shortening: 165
from crustal thinning: 159
by extension: 159
gravity anomalies in: 20
hydrothermal circulation in: 232
igneous intrusion in: 56
Insecca complex: 35–36
magmatism, initial: 35
ophiolite provenance: 20, 22, 153
in orogeny: 296
sediment, intrusion into: 211
submarine fissure eruptions: 211
peridotite provenance: 2, 19, 95
Tuscany complexes: 32
Vourinos complex: 22
Geotimes Volcanic Unit: 126, 139
Germany: 317. *See also* Alpine-Apennine complexes
Gill, J.: 232, 276
Glangeaud, L.: 36
Glarus Nappe: 394
glass, volcanic: 299
Glassley, W.: 278
Glennie, K.: 97
Globigerina: 79n*n, 394
gneiss: 82, 379
Golden Gate complex: 19
Gondwana plate: 121, 161, 415
Gosau conglomerate: 85
Gould, Stephen: 111, 113
Gouy-Chaperon effect: 58
granite
in crust/mantle structure: 175
dating of deposits: 82
geochronology of
Alpine-Apennine Mountains: 82n*z
Piedmont-Liguria ocean: 106–107
Platta-Totalp nappe: 106–107
in mélange: 389, 443n*8
petrology of
Alpine-Apennine Mountains: 82–83
Semail complex: 127
studies of Bowen: 57, 62
granitoids: 242, 257, 368
granulite facies: 7, 148, 343, 374
gravity slide: 342
gravity-driven differentiation
Alpine-Apennine complexes: 2, 34, 81
Gouy-Chaperon effect: 58
Newfoundland complexes: 49
Great Serpentinite Belt: 13
Great Valley Group
classification of ophiolites: 9, 170
Knoxville Formation: 169, 434, 445n*40
New Idria serpentinite: 456
petrology of: 170
Greece
arc-trench rollback in
Aegean Sea: 116
in Hellenic complexes: 117
asbestos mining in: 458
asbestos related disease in: 460–461
asbestos use in: 449
classification of ophiolites: 10
conglomerate deposits: 85
dating of
Aegean Sea opening: 117
andesitic volcanism: 119
Hellenide orogen: 117
Livadi complex: 458
Nauplia complexes. *see* Nauplia complexes
Othris complex. *see* Othris complex
Pindos complex. *see* Pindos complex
tectonic activity in: 117
Vourinos complex. *see* Vourinos complex
Greenbaum, D.: 67, 193, 212
Greenland: 154, 407
Greenly, Edward: 385, 398, 409–411, 433
greenschist
in Alpine-Apennine complexes: 88n*25
in mélange: 389
in Mid-Atlantic Ridge: 49
in ophiolite definition: 55
greenschist facies
in metamorphic belts: 374
petrology of
Alpine-Apennine complexes: 99
Platta-Totalp nappe: 103
Troodos complex: 317–318
in seamounts: 249
greenstone. *See also* amphibolite facies; gabbro; greenschist facies
Archean deposits of: 14

greenstone (*continued*)
  dating of deposits: 97
  dike complexes in: 38
  discriminant diagram of: 270
  formation of in belts: 127, 153
  petrology of: 19, 35, 38, 96
  radiolarites and: 97
Grenville: 48
Griesbach, Carl: 405–407
Griggs, David: 430
Grunau, H.: 95, 97–98
grunerite asbestos: 451–453
grünsteins: 19. *See also* greenstone
Guam: 154, 235, 248
Guana Gato complex: 138
Guleman complex: 38, 460
Gulf of Antalya: 42
Gulf of California: 155, 260
Gulf of Oman: 346
Gurnigel unit: 167–168
Gutenberg low velocity zone: 232
Guyot, Arnold: 21
guyots: 21, 66, 160. *See also* seamounts

H

Habkern Valley: 167–168, 389–393, 396, 408
Hacker, B.: 7
Haliakmon River: 35. *See also* Vourinos complex
Hall, James: 165
Halmahera complex: 122
Halys Massif: 415
Hamilton, Warren: 432
Hammer, W.: 78, 78n*e
Handin, John: 430
Hare Bay complex: 48–49
Harker, Alfred: 58, 160
Harrison, Chris: 69
Harz Mountains: 317
harzburgite
  boninite origin: 251–252
  classification by
    of ophiolites: 5, 144–146, 275, 281
    of peridotites: 3, 144–146, 275, 280
  correlations to
    Macquarie-type: 12
    Mediterranean-type: 11
  formation of
    in island arcs: 255
    in mantle: 219
    in ocean, major: 275
  intrusion of: 127
  Moho transition zone: 140–141
  in ophiolite definition: 3, 5
  petrology of
    Albanian complexes: 352, 357, 359
    Antalya complex: 42–45
    Bay of Islands complex: 199, 300
    in Canyon Mountain complex: 3
    Kukes complex: 125
    Mariana system: 242, 253
    Mid-Atlantic Ridge: 49
    Mid-Indian Ridge: 49
    Othris complex: 340
    Quebec complexes: 456
    Semail complex: 127, 202
    Shebenik complex: 125
    Shikoku Basin: 242
    Troodos complex: 23, 40, 67–68, 125, 154, 193, 197, 300
    Vourinos complex: 35, 41, 67–68
  in seamounts: 249
  spreading rate indicator: 3
  in SSZ crust/mantle: 236
  tectonization of: 211
Hatay complex: 38
Hatherton, T.: 232
Havre Trough: 119, 231, 238
Hawaiian Islands: 39, 332
Hawkins, James
  on ophiolite petrology: 160
  on subduction zones: 7
  on suprasubduction zones: 227–262, 283
Haylayn massif: 126, 144
Hebrides, Scottish: 58
Heck, N.: 21
Heezen, Bruce: 209
Hegenshan complex: 163
Heidinger, Wilhelm: 443–444n*23
Heim, Albert: 98
Heim, Arnold: 396
Hellenic complexes
  African plate, collision with: 121
  arc-trench rollback in: 117, 125
  composition of: 346
  dating of: 340
  dike orientation: 342
  emplacement of
    arc-trench rollback: 345–348
    assumptions, modeling: 339
    flake tectonics model: 342–346
    ocean-continent collision: 346
    plate tectonics model: 338
  formation of: 344–348
  geological setting of: 340, 352
  metamorphic grade: 342–343
  Othris. *see* Othris complex
  Pindos. *see* Pindos complex
  source material: 341
  volcanism: 119
  Vourinos. *see* Vourinos complex
Helvetic Nappes
  in alpine orogenesis model: 166–168
  correlations to: 404
  Leimeren Series: 443n*17
  petrology of: 167–168
Heraclitus of Ephesus: 387
Hercynian
  basement, dating of: 340
  classification of ophiolites: 8
  Lausitz granite: 396
Herodotus Abyssal Plain: 168
Hess Deep complex
  crust, exposure of: 181
  deformation of gabbro: 332
  dike orientation: 332
  faulting in: 332
  petrology of: 138
  structure of: 331
Hess, Harry
  on aureoles: 3, 56
  Carey influence on: 21
  on continental drift: 22
  on geosynclines: 20, 22
  on gravity anomalies: 20
  on island arcs: 3, 56
  on Mid-Atlantic Ridge: 3–4, 21
  military service of: 21
  on Moho Discontinuity: 4, 179–180
  on ocean crust petrology: 3–4, 20, 22, 26, 66–68, 208–209, 230
  ODP Leg 147, naming of: 138
  on ophiolite concept
    acceptance of: 4, 22, 56
    criticism of: 3, 21, 38, 94
  on ophiolite petrology: 96
  on peridotite border facies: 60
  on peridotite petrology
    crust fragments: 4, 22, 67, 114
    crystallization, fractional: 60–61
    in geosynclines: 20
    island arc origin of: 3, 56
    magma, hydrous ultramafic: 3, 20, 21, 56, 61, 62, 96
    orogeny: 3, 20, 96
    serpentinization: 3–4, 56, 61, 62
    serpentinization of: 208–209
    thrust sheets: 4, 67
  on seafloor bathymetry: 21, 230
  on serpentinite formation: 153, 208–209
  tectogene model: 20, 230
  on Troodos complex: 67
  on ultramafic petrology: 56, 60, 96
Hess-type lithosphere
  classification of ophiolites: 10
  extension in: 68
  Moho Discontinuity in: 4, 179–180
  vs. Penrose-type: 5, 26
  petrology of
    Alpine-Apennine complexes: 10, 26
    Mid-Atlantic Ridge: 3–4
  seismic velocity of
    Mid-Atlantic Ridge: 4
    Pacific Ocean: 179–180
High Tatra Nappes: 404
high-field strength elements (HFSE)
  in boninites: 251
  classification by: 245
  in Lau Basin/Ridge area: 242–246
  in Mariana system: 242, 244–246
  in melts: 236, 237
  in ophiolitic complexes: 112
  in SSZ crust/mantle: 236, 245–246, 257
high-Ti type ophiolites: 280
high-tide marks: 8, 119
Himalayas
  Central Gneiss Zone: 170
  classification of: 8
  correlations to Alps: 444n*28
  exhumation of: 380
  modeling of: 375
  ophiolitic complexes in: 406–407
  petrology of: 170
  plate junction type: 366
  studies of: 405–407
Hispaniola: 13

historical contingency
 in biology: 113
 critique of: 111
 in earth sciences: 113
 modeling of
  Burgess Shale: 113
  magma sources: 26, 113
  mantle: 9, 113
  Mid-Atlantic Ridge: 6
  MORB: 26, 113
  OIB: 113
  spreading centers: 26, 113, 127
  subduction zone: 26, 113
  suprasubduction zone: 113
  Tethys Ocean: 6
 modern analog for: 9
 scientific discovery and: 387
Hjort Trench: 234
HMS *Challenger*: 19
HMS *Owen*: 69
Hohenegger, Ludwig: 393, 398–401, 443n*20
Holmes, A.: 20, 231
Honshu Island: 285
Hopson, Cliff: 281, 431
hornblende: 242
hornstein: 155
hot springs: 316–317
Hsü, Kenneth
 on basin evolution: 7
 classification scheme: 167
 on Franciscan structure: 432
 on mélange: 386, 430–434
 on ophiolite petrology: 159–171
 on Troodos complex: 7
Hunan province: 162
Hungarian Mountains: 404
Hunter Ridge
 geochemical analysis of: 251
 petrology of: 259
 subduction zone: 121, 252
 volcanism in: 119
Hutton, James: 112
hyaloclastite: 101, 104, 254
hydrothermal vents
 creation of
  iron: 318
  sulfide: 302–304, 316–318
 geochemical analysis of
  active sites: 317
  Lau Basin/Ridge area: 242
  Mariana system: 242
 geothermometry of: 317
 modeling of
  convection beneath: 215
  seawater conversion process: 317–318
 studies of
  Albanian complexes: 359
  Mid-Atlantic Ridge: 317–318
  Troodos complex: 316–318, 331–332
hydrous magmatism model: 3, 20, 21, 56, 61, 62, 96, 144, 146, 182
hypabyssal-type: 79, 79n*p, 89
hypersthene: 257

I

Iberg mélange: 98, 396
Iberia margin: 10, 105–106
 classification of ophiolites: 10
Iceland
 crust/mantle structure: 312
 dikes in crust: 69
 hotspot: 144–146
 lava flow orientation: 332
igneous rocks: 79n*m
ignimbrite: 255
India: 170
Indian Ocean
 Carlsberg Ridge: 69
 mantle reservoir
  DUPAL: 121
  MORB: 121
 petrology of crust: 68
 spreading rates in: 68, 69
Indian plate
 collision with Eurasian plate: 114, 118, 119, 122, 170
 mantle reservoir: 121
Indochina
 arc-trench rollback in: 116
 movement of: 122
Indonesia
 gravity anomalies, undersea: 20
 Lupa Valley complex: 161
 margin subduction: 8, 371
 plate junction type: 366
 trenches, deep-sea: 20
 volcanism in: 232
Insecca complex: 34–36
International Crustal Research Drilling Group (ICRDG): 68
International Geological Correlation Program (IGCP): 6
intra-arc oceanic basin: 26, 229–230, 234
Intracontinental-type subduction: 366
Iran
 Arabia-Eurasia collision: 121
 mélange, description of: 430
 plate junction type: 366
 Tethyan sutures in: 127
Ireland: 155
iron
 addition of, to forsterite: 19, 59
 formation of: 318
Irtysh-Gornostaev system: 396, 442n*2
Irvine, T.: 49
Irving, Ted: 68, 70
Irwin, W.: 68
Ishiwatari, A.: 144
island arc
 backarc region. *see* backarc region
 chemical signature of: 3, 6–7
 classification of ophiolites
  Class-II: 7
  Cordilleran-type: 8, 26
  Franciscan-type: 13
  Mediterranean-type: 11
  Sierran-type: 11–12
  spinels, chrome: 281
  Type A–D: 282
 edifices, volcanic
  development of: 7, 8, 254
  nascent vs. mature: 7, 254
 ensimatic terrane in: 11
 forearc region. *see* forearc region
 formation of
  convergent margin: 234, 371
  spreading center: 259
 magma composition: 368
 mantle flow in: 116
 ophiolite provenance
  Alpine-Apennine complexes: 161
  Bay of Islands complex: 155
  Canyon Mountain complex: 138
  Guana Gato complex: 138
  Karmøy complex: 138
  vs. mid-ocean ridge: 229–230
  New Guinea complexes: 155
  Newfoundland complexes: 257, 274
  rationale for: 254, 275
  Semail complex: 155
  Tianshan region: 164
  Troodos complex: 3, 6, 137, 155, 270–274
 petrology of
  Mariana system: 254–255
  Tonga system: 254–255
 plate junction type: 366
 seismic velocity of: 257
 spreading center in: 26, 229–230, 234
 suprasubduction zone in
  arc-trench rollback in: 7, 116
  basin evolution: 7
  edifices, volcanic: 7
  extension, induction of: 7
  lithosphere formation in: 7
  progression of incorporation: 119
  spreading centers in: 7
  tectonic setting of: 228
 trench parallelism to: 230
 volcanism in: 254–56
island arc tholeiite (IAT)
 vs. calcalkaline: 274
 classification of ophiolites: 275
 crystallization process: 361
 formation of
  in backarc region: 238, 245
  in forearc region: 248, 254
 MORB transition
  Albanian complexes: 361
  hypotheses for: 139
  Platta-Totalp nappe: 103
  Semail complex: 138–139
  Troodos complex: 114, 138–139
 vs. ocean-floor tholeiites: 344
 petrology of
  Adak Island: 255
  Albanian complexes: 352, 359–361
  Lau Basin/Ridge area: 241
  Mariana system: 241, 255–257
  Semail complex: 138–139
  Tofua Arc: 255
  Troodos complex: 138–139, 299
islands, intra-oceanic
 boninite origin: 114, 254
 classification of ophiolites: 7

islands, intra-oceanic (*continued*)
petrology of
Mariana system: 254–255
Tonga system: 254–255
Troodos complex: 39
spreading centers in: 26
type area for: 39
volcanism in: 254–255
Isle of Man: 409
Isle of Mull: 56, 58
isotopic analysis
of andesite: 251
of basalts
Lau Basin/Ridge area: 242
Mariana system: 244–245
Platta-Totalp nappe: 107
Troodos complex: 279
of boninites
Mariana system: 250–251
Troodos complex: 279
of crust/mantle structure
Mariana system: 236–237
in suprasubduction zone: 236–237
Tonga system: 237
Wilson cycle tests: 114
of gabbro: 107
of peridotite: 123–124
of tectonic settings: 279–280
Italy. *See also* Alpine-Apennine complexes
asbestos mining in: 458
Lanzo complex: 46, 450, 458
paleoreconstruction of: 168
Steinmann trinity in: 19
Ivrea Zone: 407
Izu region
arc-trench rollback in: 118
classification of ophiolites: 11
correlations to
Palau-Kyushu ridge: 119
Shikoku Basin: 242
Tanzawa Arc: 257
geochemical analysis of: 243
petrology of: 253–254
proto-arc evolution
dating of: 122
extension, initial phase of: 248
seismic velocity of: 257
subduction zone: 7
Izu-Bonin-Mariana (IBM): 284

J

Jackson, E.: 3
Jamieson, R.: 7
Japan
accretionary complexes
orogenic belts: 8
ridge subduction beneath: 286
timing of initiation: 367
classification of ophiolites: 11
evolution of arc: 118
Mineoka-Setogawa Complex: 13
Oeyama complex: 13
Shimanto belt: 13
Sorachi Plateau: 13
triple junction migration: 285
Yakuno complex: 13, 144
jasper: 36
Java: 371
Jennings, C.: 56
Jiaonan complex: 162
Jiqing, Huang: 162, 163
Johnnie Formation: 413
Johnston, H.: 48
Josephine complex: 24, 139
Juan de Fuca plate
lateral melt migration: 332
magnetic anomalies in: 69
spreading centers: 178, 332
Juan Fernandez microplate: 148
Junggar Craton: 164
Jura Mountains: 443n*5
Jurassic
basin evolution in: 117, 126
classification of ophiolites: 11
crust, formation during: 2
fossil-bearing strata: 85
geosynclines: 153
limestone deposits
Alpine-Apennine complexes: 79, 79n*n
Amdo: 161
Carpathian Mountains: 398, 401
Qiangtang basin: 161
marble deposits: 340
ocean-continent transition: 82n*z
ophiolitic complexes
Albanian complexes: 352
Alpine-Apennine: 160
Hellenic complexes: 340
in Japan: 13
Josephine: 24, 139
Mirdita: 124
Othris complex: 341
Point Sal: 170
Smartville: 155
Penninic/Austro-Alpine trough: 154
plagiogranite deposits: 82n*z
radiolarite deposits
Alpine-Apennine complexes: 79, 97
Platta-Totalp nappe: 101
sedimentary deposits: 160
shale deposits: 405
subduction zones: 124
volcanic deposits: 12
Juteau, Thierry: 19, 31–51, 45

K

Kakopetria: 330–331
Kamchatka
accretionary orogenic belts: 8
collision with Aleutian forearc: 121
dating of deposits: 121
Koryak Mountain: 13
mapping of: 230–231
subduction in: 121
volcanism in: 232
Kanmantoo belt: 13
Kannaviou Formation: 312
Karakorum: 170
Karamata, S.: 96
Karasu area: 35
Karelian orogeny: 456
Kargi Massif: 434
Karig, D.: 231
Karmøy complex: 138
Karson, J.: 26, 138
Kaufmann, Franz: 393–394
Kavirondo Rift Valley: 444n*30
Kay, M.: 48
Kayser, E.: 82
Kazakhstan: 375, 450, 458
Kelpin Remnant Arc: 164
keratophyres: 89, 153, 251, 340
Kermadec arc
evolution of: 119
geological setting of: 238
mapping of: 230
suprasubduction zone in: 235
topography of: 230
Ketin, Ihsan: 429–430, 434, 444n*39
Keuper: 444n*23
Kidd, Rupert
crustal construction model: 215
on dike injection zone: 313
modeling of Troodos complex
dikes complex: 193, 215
magma chamber: 193, 215
magnetic anomalies: 71–73
on ophiolite petrology: 193
Killan Group: 460
Kimmeridgian: 85, 343
Kiogar region: 405
Kırşehir Nappe: 415, 428, 429–430, 444n*39
Kizildag complex: 10–11, 35
Kızılırmak Massif: 415
Klamath Mountains: 155, 245
Klein-Emme River: 389
klippen: 394–395, 401
Klippen Nappe: 389
knauer: 443n*15
Koertge, N.: 386
Kokchetav Massif: 375
komatiite
boninite link: 277
formation of: 21, 62
geochemical analysis of: 277
impact of, on theories: 56
koniabyssites: 79, 79n*n. *See also* limestone
Korabi microcontinent
collision with Pelagonia: 124, 125
detachment from Apulia: 124, 125, 359
source material from: 358
Koryak Mountain: 13
Kozlovka River: 396
Kuhnian analysis: 21–26, 288
Kukes complex: 125
Kündig, Ernst: 431–432
Kunlun Mountains: 162, 164–165, 165
Kurile region: 230–231
Kuroko sulfide deposit: 317
Kykko thrust: 326
Kyrenia Range: 125, 326

L

labradorite: 257
laccolith: 34

Lahn River: 83n*13
Lake Juojärvi: 449, 456
Lake Sils: 78
lamellibranches: 398
Lanzo complex: 46, 450, 458
Large Igneous Province (LIP): 13
large-ion lithospheric elements or large ionic radius lithophile elements (LILE)
 in boninites: 139, 251
 definition of: 236
 in Lau Basin/Ridge area: 242–246
 in Mariana system: 242, 244–246, 257
 in melts: 236
 in ophiolitic complexes: 112
 in SSZ crust/mantle: 236
 in subduction zone: 371
Lasail volcanics: 126, 139
Late volcano: 257
Lau Basin/Ridge area
 arc terrane: 287
 classification of: 11
 correlations to: 127
 crust thickness in: 238
 evolution of: 119, 231, 237–241
 extension in: 276
 geochemical analysis of
  backarc region: 231, 238, 287
  basalts: 245, 276
  gabbro: 242
  hydrothermal vents: 242
  rift basins: 238
  seamounts: 238
  spreading centers: 238, 312
  Valu Fa Ridge: 287, 312
 geological setting of: 238, 285
 geometry of: 285
 hydrothermal circulation in: 239
 isotopic analysis of: 242
 magnetic anomalies in: 238
 mantle flow beneath: 237
 mapping of: 228
 vs. Mariana system: 240–241, 245
 petrology of: 238–246, 254
 seismic studies of: 237, 238
 shape of: 238
 spreading centers in: 238, 240
 suprasubduction zone: 252
 triple junctions in: 240
 volcanism in: 257
Laurasia: 415
Lausitz granite: 396
lava. *See also* basalt; spilite; volcanic rocks
 albitization of: 2, 81, 87
 basaltic: 4
 chemistry of: 123
 chilled margins: 83n*13
 classification of
  Caribbean-type: 13
  Chilean-type: 12
  Mediterranean-type: 11
  Penrose-type: 5, 97
  seamount-type: 139
 classification of ophiolites
  Franciscan-type: 13
  by Mesorian: 275
  by stratigraphic sequence: 282
 composition of: 189
 dating of deposits
  Alpine-Apennine complexes: 79
  Antalya complex: 42–45
  British complexes: 97
  Tonga system: 255
  Troodos complex: 39
 definition of: 79n*m
 density of vs. lithosphere: 74
 electrical conductivity of: 74
 extrusion of
  on seafloor
   differentiation after: 4
   modeling of: 32
 formation of
  in island arc: 254
  Mid-Atlantic Ridge: 4
  in Pacific Ocean: 173
 geochemical analysis of
  Mamonia complex: 305
  Semail complex: 138–139
  Troodos complex
   extrusive sections: 314
   forearc provenance: 125
   island arc origin of: 270
   magma chambers: 299
   MORB-IAT transition: 138–139
 geochronology of: 341–342
 mapping of: 22–23, 312–313
 modeling of
  actualistic: 123
  crust/mantle structure: 300
  dike density in: 69–71
  emplacement zone size: 70–71
  magnetic stratigraphy: 66, 69–71
  outflow, submarine: 4, 38
 in ophiolite definition: 55, 97
 petrology of
  Albanian complexes: 352
  Alpine-Apennine complexes: 10, 34, 78n*f, 79–107, 160
  Antalya complex: 42–45
  Bay of Islands complex: 11, 49
  Bolivar complex: 13
  British complexes: 97
  DSDP Hole 120: 160
  Franciscan system: 13
  Gulf of Antalya: 42
  Insecca complex: 35
  Kizildag complex: 11
  Lau Basin/Ridge area: 242
  Mariana system: 242, 248
  mid-ocean ridge: 26
  Mirdita complex: 124
  Othris complex: 339
  Papuan complex: 46
  Pindos complex: 124
  Platta-Totalp nappe: 101, 104
  Porphyrite formation: 89
  Rocas Verdes complex: 12
  Sarmiento complex: 12
  Semail complex: 11, 45, 144, 202, 208
  Syrian greenstones: 35
  Tonga system: 255
  Tortuga complex: 12
  Troodos complex
   basaltic: 299
   boninitic volcanism: 125
   calcalkaline volcanism: 125
   density of: 74
   electrical conductivity of: 74
   extrusive sections: 314
   field relations: 67, 298, 312–314, 325–326
   hydrothermal circulation in: 125, 317–318
   magnetization intensity: 67
   MORB-like: 125
   orogeny: 326
   porosity of: 74
   seafloor weathering: 314
   seismic velocity of: 74
   silicic: 299
   stratigraphy of: 154
  Turkey: 417–420, 430
  Xigaze complex: 11
 polarity boundaries in: 73
 porosity of vs. lithosphere: 74
 radiolarite link: 35
 saturation of: 160
 seismic velocity of: 74
 spilite, transformation to: 2, 160
 in Steinmann trinity: 39
 stratigraphic progression in: 123
 underplating by: 123
lawsonite: 236, 249
Lech River: 408
Leimeren Series: 396, 443n*17
Leitch, E.: 282, 283
Lemoine, M.: 105
lepidorogene-type: 90
Leuchs, Kurt: 415
Levantine Sea: 168
Lewis, J.: 34, 58
lherzolite
 classification by
  of ophiolites: 5, 144–146, 275, 281
  of peridotites: 3, 146, 275, 280
 correlations to
  Caribbean-type: 13
  Ligurian-type: 10
  Mediterranean-type: 11
 formation of: 275
 metamorphic fabric: 45–46
 in ophiolite definition: 3, 5
 petrology of
  Albanian complexes: 352, 357
  Albano-Thessalian furrow: 35
  Alpine-Apennine complexes: 32–33, 45–46, 82n*z
  Bay of Islands complex: 199
  Lanzo complex: 46
  Mariana system: 253
  Mid-Atlantic Ridge: 49
  Mid-Indian Ridge: 49
  Mirdita complex: 124
  Othris complex: 340
  Platta-Totalp nappe: 102
  Semail complex: 126
  Shikoku Basin: 242
  Troodos complex: 67–68, 125, 196, 305

lherzolite, petrology of (*continued*)
Vourinos complex: 67–68
spreading rate indicator: 3
transformation to serpentinites: 33
volumetric ratio: 41
lherzolite ophiolite type (LOT): 144–146
lherzolite-harzburgite-type (LHOT): 146
Liassic clastics: 420
Lichi mélange: 374
Ligurian Ocean: 106, 123, 124
Ligurian-type. *See also* Hess-type lithosphere
correlations to
Class-III: 10
lherzolite-type: 10
Red Sea–type: 10
definition of: 10
lithosphere: 10
petrology of: 10
Red Sea–type basin: 10, 124
Liguria-type: 144
Liguride complex: 124
Limassol Forest
formation of
in forearc region: 300
at ridge-transform intersection: 126, 300
lithostratigraphy of
accretionary wedge: 125
detachment faults in: 126, 300, 315
vs. northern outcrops: 315
transform fault zone: 300, 315
subduction nucleation in: 125–126
uplift of: 122
limestone
abyssokonite: 79, 79n*n
alberese: 80 n*u
coccolith ooze: 97
dating of deposits
Alpine-Apennine complexes: 79, 79n*k, 97
Calera: 170
felsenkalk. *see* felsenkalk
Leimeren Series: 443n*17
macigno: 79, 79n*l
maiolica: 80 n*r, 97, 102
in ophiolite definition: 2, 5
petrology of
Alpine-Apennine complexes: 2, 79–82, 79n*k, 79n*n, 81, 84–85, 94, 97
Alpine-Apennine Mountains: 83
Argille a Palombini Formation: 79 n*s
Amdo: 161
British complexes: 97
Carpathian Mountains: 398, 401
Franciscan system: 13
Himalayas: 405
Mamonia complex: 305
Mariana system: 248, 254
Platta-Totalp nappe: 101–102
Qiangtang basin: 161
Tonga system: 248
Troodos complex: 299
Turkey: 417
Vourinos complex: 35
scaglia: 79, 79n*k
silicification of: 82
Lindgren, Waldemar: 317
liquid immiscibility: 58
lithosphere
appearance of: 22
assimilation of: 286
calcalkaline volcanism: 367–368
classification of
Hess-type: 3–4, 26, 179–181
Penrose-type: 2, 4, 5, 26, 155
cores, drilling of. *see also* Ocean Drilling Program
depth of: 181
gradients, in rock: 74
stratigraphic layering in: 233
correlations to
oceanic plateaus: 13
ophiolites: 233–234
dating of deposits
residence time: 233
thinning events: 27
definition of: 173
density of
Hess-type oceanic crust: 179–180
vs. pillow lava: 74
standard for: 173
in subduction zone: 371, 375
electrical conductivity of: 74
faults in
from extension, amagmatic: 68
segment boundaries: 68
thickness along: 68
formation of
ancient vs. modern: 6, 232
arc-trench rollback: 7
dikes in: 123, 173
heterogeneous: 4
homogenous: 66, 176
at spreading center. *see* spreading center
in suprasubduction zone: 7
zone size: 69–70
geochemistry of
island arc-like: 7
MORB: 232–233
geothermometry of: 3, 57–59
gravity anomalies in: 20
hydrothermal circulation in: 181–182
isotherm: 155–156
magnetic stratigraphy. *see* magnetic stratigraphy
modeling of
layer-cake: 176–177
metamorphic belts: 373–375
Moho Discontinuity: 178–180
ophiolites: 180–183
subduction zone: 367, 378–380
thickness of
rifts in continental margin: 106
single simatic layer: 175
obduction, process of: 5, 155
ophiolite provenance
island arc: 3, 6–7, 114
Penrose: 5
spreading center: 22–23, 62, 212–213
paleomagnetic studies: 20
petrology of: 3–4, 5, 6, 20, 22, 26, 66–68
in plate tectonics theory: 4
plateau evolution: 13, 232
porosity of
extrusive sequence, below: 181
vs. pillow lava: 74
preservation of
in orogenic belts: 6
pre-Jurassic deposits: 27
recycling of: 4, 6, 233
seismic velocity of
Atlantic Ocean: 4, 175–176
gabbro vs. serpentine: 180–181
measurement of: 176–177
vs. ophiolites: 180
Pacific Ocean: 5, 175–176
vs. pillow lava: 74
porosity impact on: 180
rate of: 49
standard for: 5, 173
studies of: 175–176
subdivision by: 49, 311
vs. surface wave: 175
spreading velocity: 5
stratification of
composition of: 5
verification of: 97
stratigraphy of
Atlantic Ocean: 4
seismic velocity tests: 175–176
crust/mantle structure: 174–177
layer-cake
assumptions, modeling: 176
Penrose: 2, 5
Troodos complex: 5, 154, 160
Pacific Ocean
new fast non-rifted MOR: 173
seismic velocity tests: 175–176
structure of
Atlantic Ocean: 299–300
homogenous: 311
subduction nucleation trigger: 128
subduction of: 370–380
under continental plate: 166
extension, induction of: 7
nucleation model: 120
vs. obduction of: 126, 156
vs. preservation: 107
requirement for: 155
near seafloor spreading: 7, 120
SSZ magma source: 236, 237, 247
in SSZ tectonic setting: 234
surface wave studies: 175
tectonic windows in: 138
thickness of
age-related changes: 179, 182
ancient vs. modern: 24, 26–27
at boundaries: 68
changes to, abrupt: 27
controls on: 4
island arc vs. ridge: 6
lithospheric stretching: 164
oceanic vs. continental
depth of: 49, 65
earthquake seismic studies: 21
seismic refraction tests: 311
seismic velocity tests: 175–176
surface wave studies: 175

    Precambrian: 24
    subdivision by: 49, 311
  tilting of: 328
  transition zones in: 173–74, 177, 181–182
Littleport/Coastal complex: 155
Livadi complex: 458
Livorno: 80
Lizard complex: 155
lizardite: 253
Lodochnikov, W.: 56
Lotti, B.
  mapping by, of
    Alpine-Apennine Mountains: 2, 32, 79, 89n*31, 94
    Tuscany: 80
  modeling by: 32
  on ophiolite petrology: 32–33, 80–84
Louisville Seamount Chain: 238
low-Ti type ophiolites: 280
Lugeon, M.: 404
lung cancer: 448, 453
Lunz Beds: 444n*23
Lupa Valley complex: 161
Luzon arc
  collision with
    Eurasia: 118, 119, 121, 122
    Mindanao: 122
    Ryukyu arc: 121
  ophiolitic complexes
    composition of: 257
    petrology of: 255
Lyell, Charles: 112

M

Maastrichtian: 35
Macedonian complexes: 352
macigno: 79–80, 79n*l, 85, 443n*16
Macquarie Island: 12–13
Macquarie Ridge complex
  classification of: 12–13
  dating of deposits: 9
  formation of: 9
  geochemical analysis of: 9, 237
  mid-ocean ridge provenance: 237
  modeling of: 231
  ridge/trench interaction: 260
  subduction of: 13
Maden complex: 460
mafic rocks
  in crust/mantle structure: 175
  in metamorphic belts: 378
  petrology of
    Alpine-Apennine complexes: 19, 46
    Amdo: 161
    Australian complexes: 19
    Bay of Islands complex: 48–49
    Betts Cove complex: 48–49
    Canyon Mountain complex: 3
    Hare Bay complex: 48–49
    Qiangtang basin: 161
    Vourinos complex: 22, 41
  plutonic: 24
  transition from ultramafic: 22
Magellan seamounts: 239
magma
  bulk distribution coefficients: 236
  crystallization in: 19, 21, 57–59, 58–59, 60, 61, 80–81, 86–87, 89
  dating of: 232
  flow direction: 332
  fractionation of: 177
  in geosyncline evolution: 160
  geothermometry of
    clinoenstatite: 57
    forsterite: 19, 57
    Mid-Atlantic Ridge: 3
    olivine: 3, 19, 21, 57
    pyroxene: 3, 57
    serpentinite: 3, 21
  in historical contingency model: 26, 113
  isotopic analysis of: 368
  migration of: 332
  MORB. *see* Mid-Ocean Ridge Basalt (MORB)
  OIB. *see* Oceanic Island Basalt (OIB)
  orogenic stages: 160
  petrology of
    Albano-Thessalian furrow: 35
    Alpine-Apennine complexes: 2, 33–35, 86–90
    Mid-Atlantic Ridge: 4
    Platta-Totalp nappe: 104
    Vourinos complex: 4, 35
  in suprasubduction zone
    in arc-trench rollback: 118
    historical contingency of: 26, 113
  types of
    acidic: 89
    alkaline: 160
    andesite. *see* andesitic volcanism
    boninitic. *see* boninitic volcanism
    calcalkaline. *see* calcalkaline volcanism
    picritic. *see* picritic volcanism
    porphyritic: 58–59
    serpentinite: 61
    tholeiitic: 104, 160
    ultrabasic: 3, 20, 21, 56, 61–62
  volatile content of
    crosscutting by: 34, 81
    evidence of: 87
    gradients, in rock: 74
    melt temperature impact: 58, 61
    oceanic basins: 371
    in subduction zone: 370–375
    in suprasubduction zone: 74, 236
    during upwelling: 61
magma chamber
  chromite crystallization in: 146
  compaction in: 219
  cores, drilling of: 217–219
  crust formation in: 67, 177–178, 219
  crystallization in: 202, 215–221
  cyclic variations in: 193, 197, 202–205, 217
  definition of: 141–42
  depth of: 137–138, 193
  formation of: 141–144, 177–178, 208–209
  fractionation in: 177
  geothermometry of: 205
  vs. hydrofacturing: 208
  mapping of: 205
  melt lens
    cooling pattern: 182
    description of: 141
    discovery of: 137–138
    mush, interaction with: 205–207, 215–216, 219, 315
    seismic reflection studies: 178
  outflow, submarine: 38
  petrology of
    Bay of Islands complex: 154, 198–201, 215
    East Pacific Rise: 141–142, 178, 182, 215–216
    Juan de Fuca Ridge: 178
    Mid-Atlantic Ridge: 216–217
    Semail complex: 142–144, 177–178, 202–208
    Skaergaard complex: 144
    Talkeetna Arc: 255
    Troodos complex: 67, 154, 193–197, 301, 314–315
    Valu Fa Ridge: 287
  plate tectonics: 67
  plutonic sequences: 5, 193
  vs. porous flow: 208
  seismic reflection studies: 141–142, 177–178
  shape/size of
    controls on: 199, 203–204
    dike-like: 193, 215
    lateral melt migration: 332
    spreading rate impact: 138, 214–217, 315
    thermal model estimates: 177–178, 214
  sills: 201, 205–207, 219–221
  single vs. multiple
    evolution of theory: 67, 193, 221
    field relations: 197
    segmentation: 218
  solidification process: 215, 219
  SSZ vs. mid-ocean ridge: 214
  steady-state: 193, 198–200, 215, 216, 219
  stratigraphic layering in: 21, 22, 57–58, 61, 86–87
    crustal strain: 216
    density-driven: 203
    detachment faulting: 34
    differentiation, magmatic: 19
    vs. flow banding: 219
    gravity-driven: 34, 144
    lateral zones: 212, 215
  structural evolution: 193–196
magnetic stratigraphy
  of Atlantic Ocean: 66–67
  of Carlsberg Ridge: 69
  of Cyprus: 23, 39, 68, 74
  lineation patterns in: 69
  massif rotation record: 67
  measurement of, surface: 71
  of Mid-Atlantic Ridge: 66–67, 68–70
  modeling of
    boundary shape: 71–73
    crust emplacement: 69–71
    dike injection: 69, 71
    magnetization intensity: 66–67, 70
    phase-shift in: 73
    spread rates: 69
  natural remanent magnetization: 68
  from polarity reversals: 66–67, 68–70
  from seafloor spreading: 66–67, 68–70
  symmetry in: 69
  topography link: 66
  viscous decay impact on: 70
Mählmann, Ferreiro: 99

maiolica: 79, 79 n*q, 79 n*r
Majer, V.: 38
Makassar Sea: 118
Makran: 346, 348
Malenco complex
  correlations to: 102
  dating of deposits: 102
  geothermobarometry of: 102–103
  location of: 98
  mantle, preservation of: 88n*af, 97, 102
  metamorphic grades in: 99, 103
Malpas, John: 67, 295–306
Mamonia complex
  dating of deposits: 125
  geochemical analysis of: 305
  mapping of: 140
  metamorphism: 125
  modeling of: 305
  petrology of: 125, 304
Manatschal, G.: 93–107, 98
Mangatolu triple junction: 238, 243, 252
Mantis, M.: 40
mantle
  abyssal vs. ophiolitic: 139
  classification of
    alpine-type peridotites: 3
    DUPAL-like: 121
    MORB-like. *see* Mid-Ocean Ridge Basalt (MORB)
    ophiolites
      harzburgite-type: 5
      HOT: 144–46
      lherzolite-type: 5
      LHOT: 146
      Liguria-type: 144
      LOT: 144–146
      Mediterranean-type: 11
      Papua-type: 144
      Yakuno-type: 144
  contamination of
    SSZ signature: 139
    by subduction: 127
    subduction related: 9
  convection
    in arc-trench rollback: 8, 116
    in basin evolution: 116
    counterflow: 237, 283
    crustal movement from: 165
    in geosyncline evolution: 159
    gravity anomalies from: 20
    lithospheric stretching from: 164
    in magma chamber: 205
    mid-ocean ridge link: 3, 127, 191–192
    serpentinite concentration: 21, 66
    vs. slab buoyancy: 116
    in suprasubduction zone: 229, 236–237
    in Wilson cycle: 113–114
  decompression
    asthenospheric upwelling: 9
    boninite formation: 120
    from rifting: 287
    from subduction: 286, 373–375, 379
  density of: 375
  flow patterns: 141
  geochemical analysis of: 104
  geothermometry of
    arc-trench rollback: 118
    boninite formation: 120
    chromite crystallization: 146
    clinoenstatite: 57
    deformation of: 5
    forsterite: 57
    serpentinization: 3–4, 66
    subduction nucleation: 119–120
  heterogeneity of: 9, 120
  isotopic analysis of: 114
  magnetic stratigraphy: 66–67
  MELT experiment: 140
  modeling of
    fault geometry: 106–107
    historical contingency: 9, 113
    at mid-ocean ridges: 127
    plate tectonics: 4
    rifting of lithosphere: 106
    subduction nucleation: 120–121
    tectonic emplacement of: 4, 22
  partial melting of
    boninite, signature of: 139
    crustal generation by: 156
    vs. differentiation: 38, 96
    gabbro-olivine grading: 173
    melt fraction: 142–144
    at mid-ocean ridge: 188–189
    olivine-peridotite grading: 173
    peridotite sequence: 32, 190–192
    spreading rate impact: 144–146
    in subduction zone: 371
    for suspension flow: 142–144, 205
    ultrabasic magma from: 58
  petrology of
    Alpine-Apennine complexes: 46, 81, 88
    Antalya complex: 45
    Kukes complex: 125
    Mirdita complex: 124
    Newfoundland complexes: 49
    Papuan complex: 48
    Pindos complex: 124
    Platta-Totalp nappe: 102–104
    Semail complex: 126
    Shebenik complex: 125
    Troodos complex: 67–68, 125
    Vourinos complex: 41–42, 67–68
  preservation of
    in ophiolites: 22–23, 49
    in orogenic belts: 6
    pre-Jurassic deposits: 27
  seismic velocity of
    anisotropy in: 179–180
    in fracture zones: 68
  serpentinization of
    fracture zones: 68
    NRM indication of: 74
  subcontinental
    exhumation of: 104–107, 154
    Malenco complex: 88n*af, 102
    at margins: 106, 154
    Platta-Totalp nappe: 102–103, 106, 155
    preservation of: 97
    seafloor exposure of: 106
  upwelling of
    boninite origin: 252
    in convection cells: 3, 22, 66
    hydrous magmatism: 144
    mapping of
      Moho transition zone: 140–141
      in Semail complex: 140–141, 144
    modeling of
      spread rate impact: 140, 144–146, 189
      structure: 140
    ophiolite paleodepth: 140
    at plume-ridges: 127, 189
    at spreading center: 9, 22, 26, 127, 140, 188–189
    in SSZ tectonic setting: 234
Maramuni arc: 121
marble: 81, 340, 379, 429
Margna: 95
Mariana system
  classification of ophiolites: 11
  correlations to
    in China: 165
    Palawan complex: 127
  dating of
    arc evolution: 122, 255
    arc-trench rollback: 118
    rifting, initiation of: 239, 255
  geochemical analysis of
    backarc region: 231
    basalts: 244–245, 276
    boninites: 251
    gabbro: 242
    hydrothermal vents: 242
    peridotite: 242
    ridge: 287
  geological setting of: 239–240, 285
  geometry of: 285
  isotopic analysis of
    basalts: 244–245
    boninites: 250–251
    crust/mantle structure: 236–237
  vs. Lau Basin/Ridge area: 240–241, 245
  mapping of: 230
  Palau-Kyushu ridge link: 119, 234
  petrology of
    backarc region: 239–242, 244–246, 255
    forearc region: 248–252, 255
    trenches: 252–254, 257, 276–277
  seamounts in: 249
  seismic zone in: 240
  suprasubduction zone
    arc-trench rollback in: 7, 118, 228–229, 255
    boninite origin: 249–252
    evolution of: 234–235, 237–240
    forearc region: 248, 255
    mechanism, initiation: 7
    spreading center in: 240, 255
    type example of: 228–229
    volcanism in: 232, 255–257
    Wadati-Benioff zone: 234
Marin sandstone: 445n*40
marls
  dating of deposits: 79, 79n*k
  Leimeren Series: 443n*17
  petrology of: 79, 79n*k
    Alpine-Apennine complexes: 79n*k
    Platta-Totalp nappe: 102
    Troodos complex: 296
  scaglia: 79, 79n*k

Marshall, P.: 230
Maruyama, S.: 375
Massa Marittima: 80
Massachusetts: 57
Masson-Smith, D.: 39
Mathematicians Ridge: 316
Mathiati sulfide deposit: 318
Matthews, Drummond: 66–70
Maures complex: 121
Mavrovouni: 326
Maxwell, John: 21, 432
McCallien, William
  biography of: 414
  on mélange: 385, 414–430
  on serpentine lava: 96
  on Steinmann trinity: 96, 103, 107
  on ultramafic petrology: 56
McClain, James: 173–184
Mediterranean: 116–117
Mediterranean Ridge: 121, 168
Mediterranean-type
  vs. Chilean-type: 12
  correlations to: 11
  vs. Franciscan-type: 13–14
  petrology of: 7, 10–11
  vs. Sierran-type: 11–12
Meijer, A.: 251
mélange
  backarc collapse marker: 161
  classification of ophiolites
    Cordilleran-type: 148
    Franciscan-type: 13–14
    Mediterranean-type: 11
  in convergent margins: 369
  definition of: 388, 409, 444n*30
  faults vs. subduction zone: 413–414
  in forearc accretionary complexes: 159
  formation of: 410–11
  identification of: 2, 395–396
  mapping of: 432–434
  petrology of
    Albanian complexes: 352
    Alpine-Apennine complexes: 79 n*o, 97, 98
    Anglesey: 409–411
    Amdo: 161
    Black Mountains: 412–414
    Cyprus: 168
    Franciscan system: 160
    Qinling Mountains: 162
    Turkey: 415–430
  recrystallization of: 8
  rules for study of: 432
  serpentinite formation in
    accretionary-type belts: 8
    fracture zone origin: 9
    identification of: 32
  studies of: 385–435
  subduction of: 374
  suture zone marker: 159, 161
MELT experiment: 140
Mesorian, H.: 275
mesothelioma: 448–467
Mesozoic
  chert deposits: 161
  lithosphere: 98
  ophiolitic complexes
    in Alps. *see* Alpine-Apennine complexes
    in Calabria: 78n*b
    in Caucusus: 24
    on Elba: 78n*b
    in Greece: 340
    on Mount Argentario: 78n*b
    Philippines: 123
    in Piedmont zone: 78n*b
    in South America: 89
    Troodos. *see* Troodos complex
  radiolarian deposits: 161
metamorphic belts
  characteristics of: 377–378
  composition of: 153, 160
  dating of: 127
  exhumation of: 373–375, 378–380
  formation of
    accretionary wedge: 8
    Alpine-Apennine complexes: 124
    in subduction zone: 368–369
  modeling of: 374–375, 378–380
  paleodepth: 368–369
  preservation of: 6, 378–379
metamorphism. *See also* aureoles
  contact vs. regional: 82
  geothermometry of
    Alpine-Apennine complexes: 99
    mantle deformation: 5
  inverted gradients
    in accretionary prism: 13
    in Alpine-Apennine complexes: 97
    in Kukes sole: 125
    in Shebenik sole: 125
    in soles: 7, 120, 123, 260, 343
    in Troodos complex: 23
  modeling of
    collision-driven: 8
    igneous intrusion: 56
    outflow, submarine: 38
    plate junctions: 367
    tectonic processes: 39, 296
Meteor expedition: 20
Metis Shoal: 257
Metsovo lung disease: 460–461
Mexico: 138
mica: 236, 371
Michigan: 452
microcontinents: 260
microplates: 148
Mid-Atlantic Ridge
  Bouguer anomaly: 140
  colinearity with: 113
  correlations to
    ophiolites: 339
    Troodos complex: 332
  crust/mantle structure
    magma chamber: 332
    petrology of: 3–4, 66–67, 311
  discontinuity beneath. *see* Moho Discontinuity
  DSDP Hole 395A: 311
  earthquake studies: 21, 175
  vs. East Pacific Rise: 216
  evolution of: 113
  exploration of: 20, 208
  heat flow in: 21, 22
  hydrothermal field: 317–318
  magnetic stratigraphic studies: 66–67, 68–70, 73
  mapping of: 20, 21
  mineralization of: 318
  modeling of
    crust emplacement: 69–70
    dike intrusions: 69
    historical contingency: 6
    magnetic anomalies in: 66–67
  petrology of
    composition of: 4, 21
    ophiolite suite in: 49
    stockworks: 303, 317–318
  spread rate: 216
  spreading rate: 73
  thickness of: 4, 21
  transform fault zones in: 140
Mid-Indian Ridge: 49
Mid-Ocean Ridge Basalt (MORB)
  chromite link: 345
  composition of: 160
  crystallization process: 361
  geochemical analysis of
    Chile Ridge: 286
    contamination vs. subduction: 276
    decompression: 286
    spreading centers: 232–333, 278
    vs. SSZ: 138–139, 233–237, 274–279, 359
    Troodos complex: 270–274
    Western Pacific basin: 241–246
  IAT transition
    Albanian complexes: 361
    hypotheses for: 139
    Platta-Totalp nappe: 103
    Semail complex: 138–139
    Troodos complex: 114, 138–139
  Indian signature
    isotopic analysis of: 284
    vs. Pacific signature: 237, 242
    recycling of magma: 279
  magma production per year: 233
  modeling of: 26, 113
  Pacific signature
    vs. Indian signature: 237, 242
    isotopic analysis of: 284
  petrology of
    Abdella mélange: 347
    Albanian complexes: 359–361
    Chile Ridge: 9
    Geotimes Volcanic Unit: 126, 139
    Kukes complex: 125
    Lau Basin/Ridge area: 241–244
    Macquarie Ridge Complex: 9
    Mariana system: 241, 244–246
    Mirdita complex: 124–125
    Platta-Totalp nappe: 104
    Shebenik complex: 125
    South Chile Ridge: 9
    Troodos complex: 125
  underplating of: 114
mid-ocean ridge (MOR). *See* spreading center
Miliolina: 398
Milovanovic, B.: 96
Milsom, J.: 48
Mindanao: 121
Mindoro complex: 122

n = footnote*

Mineoka-Setogawa complex: 13
Minnesota: 452
Miocene
andesitic volcanism
Hellenic arcs: 119
Ryukyu arc: 119
basin evolution in: 118, 119
carbonate bank deposits: 235
flysch deposits
macigno: 79, 79n*l, 80 n*t
in Tuscany: 79n*l
igneous complex formation: 285
limestone deposits: 248
plutonic complexes: 257
sedimentary rocks: 398
volcanism in: 249, 254–255
miogeosyncline: 22, 160
Mirdita complex
basin evolution: 121
classification of
LHOT: 146
Mediterranean-type: 10–11
type transition in: 11
correlations to
Alpine-Apennine complexes: 124, 357
Pindos complex: 124
dating of deposits: 122, 124, 353–357
formation of
comagmatic: 353–357
intrusive bodies: 353
mantle, partial melting of: 357
outflow, submarine: 357
at spreading center: 357
geochemical analysis of: 124
mapping of: 353–357
petrology of: 124–125, 146, 353–357
studies of: 353–357
thermochronology of: 120
transform fault zones: 125
volcanism in: 352
Mishunina, Z.: 357
Misox zone: 95
Mitsero Graben: 300, 332
Miyake, Y.: 285–286
Miyashiro, Akiho
classification scheme: 7
geochemical analysis by: 278
on magma in island arcs: 232
on ophiolite petrology
calcalkaline lithology: 6, 112, 114, 270
dike formation: 270, 274
island arc origin: 3, 6–7, 39, 270, 299, 311
metamorphosed basalts: 274
seafloor weathering: 274
sequences: 274
Moberly, R.: 231
Moho Discontinuity
in crust/mantle structure
Atlantic Ocean: 299
beneath Siberia: 174
continental vs. oceanic: 65, 176
discovery of: 174
in lithosphere definition: 173
mapping of
in Aswad massif: 144
in Bay of Islands complex: 49
in Mid-Atlantic Ridge: 4
in Semail complex: 141–142
Tonga system: 231, 257
in Troodos complex: 39, 40, 68
modeling of: 178–180
in ophiolite definition: 97
petrology of
Semail complex: 146
Troodos complex: 300
seismic velocity link: 49, 173
Mohorovicic, A.: 174
Mojave Desert: 169
molasse: 170, 398
Moluccan Sea: 118, 121
Mona Complex: 409–411
Mongolia: 163
Montagnola Senese: 84
Montana
Rainy Creek complex: 462
Stillwater complex: 38, 56, 60
Monte Catini: 84, 86
Monte Nero: 84
Monte Penna: 86
Montgenevre complex: 124
Monti Rognosi: 84
Moores, Eldridge
on chemical signatures: 9
classification scheme: 8, 24, 148
mapping by: 73
modeling of
collision: 23–24
historical contingency: 9, 26
magma chambers: 67
on ophiolite petrology
crustal fragments: 22
diapiric intrusion: 22
dike formation: 67, 70, 73
extension: 73
fault zones: 68, 73
grabens: 73
research on: 17–27
rotation of: 23, 68, 73
seafloor spreading: 23, 67
spreading center origin: 154
paleomagnetic studies: 328
on seafloor spreading: 22–23
stratigraphy of: 97
on volumetric ratio: 41
morainal deposits: 388
MORB-type ophiolite: 282
Morgan, Jason: 69
Mount Argentario: 78, 78n*b
Mount Diablo complex: 19
Mount Olympus. *See also* Troodos complex
asbestos mining in: 459
banding in: 326
formation of: 329
mantle upwelling at: 328
mapping of: 197, 326
plutonic suites on: 197
mudstone: 2, 388
Müntener, Othmar: 93–107
Murchison, Roderick: 391–392, 401, 443n*8
Murlo group: 84

N

nagelfluh: 389–392, 443n*5
Nakhl massif: 144
nappes
Alpine-Apennine complexes: 2, 78, 81, 94–95, 97
peraluminous intrusions: 368
theory of: 394
Narym River: 396
natural remanent magnetization (NRM): 68, 74
Nauplia complexes: 85
Ndojaj, I.: 357
Neale, E.: 48
Nebbio nappe: 461
Neocomian
felsenkalk deposits: 81, 85
Roßfeld Beds: 444n*23
Neo-Tethys. *See* Tethys Ocean
nephrite: 81, 81n*v, 87
neritic-type: 89–90
Neumayr, Melchior: 339, 401
New Caledonia complexes: 21, 36, 146, 461
New England Belt: 13
New Guinea
collision with
Australian margin: 123
East Philippines arc: 123
Halmahera arc: 123
South Caroline arc: 123
ophiolitic complexes in
correlations to: 123
dating of deposits: 122
formation of: 155
subduction initiation in: 121
New Guinea plate: 283–284
New Hebridies
collision with: 121
Hunter Ridge. *see* Hunter Ridge
ocean topography: 230
suprasubduction zone in: 234, 235
transform fault zones: 238
New Idria serpentinite: 456, 462–463
New Orleans: 454
New Zealand
classification of: 13
ocean topography: 230
plate tectonics: 9
Newfoundland
Annieopsquotch complex: 155
Baie Verte: 153
Bay of Islands complex. *see* Bay of Islands complex
Betts Cove complex: 48–49, 154
Gander belts: 153
Hare Bay complex: 48–49
island arc origin of: 155, 257
suprasubduction zone in: 155, 257
Nicolas, Adolphe
classification scheme: 8, 144
on dunite: 5
on magma chambers: 205, 215
on metamorphic fabric: 45
on ophiolite petrology: 137–149
Niua Fo'ou: 243
Noble, Levi: 385, 412–413, 434

Nolan, Robert: 447–467
Noonday Formation: 413
Nopsca, F.: 352
Norfolk Basin: 119
norites
  classification of ophiolites
    Ti-type: 280
    Type II: 275
  in ophiolite definition: 33, 78
  petrology of
    Mariana system: 253
    Semail complex: 126
  vs. troctolite: 242
North American complexes: 170
North American plate
  boundaries of
    Atlantic Ocean: 344
    Sierra Nevada: 169
  subduction of: 169
North Anatolian Fault: 415
North China plate: 164
North Fiji Basin
  arc-trench rollback in: 121
  correlations to: 237–238
  geological setting of: 285
  ridge-trench interactions: 252
  spreading center: 119, 121
  subduction zone: 119, 121, 252
  transform shear zone: 119, 121
North Palawan plate: 118
Northern Seamount Province: 257
Norway
  Karmøy complex: 138
  Western Gneiss Region: 368, 375
Nowack, Ernst: 353, 415
Nujiang/Bangong Zone: 161

O

obduction
  criteria for: 8
  isotherm: 155, 156
  at spreading center: 126
  vs. subduction/accretion: 5
objectivity, scientific: 387
Ocean Drilling Program (ODP). *See also* Deep Sea Drilling Project (DSDP)
  Hole 540B
    depth of: 181
    metamorphic front: 300
    Penrose-type lithosphere: 68, 160
  Hole 735B: 189, 217–119
  Hole 786B: 284
  Hole 1105A: 189, 218
  Leg 118 (Indian Ocean): 218, 315
  Leg 125 (IBM): 249, 284
  Leg 126 (Izu-Bonin): 284
  Leg 135 (Lau Basin): 284, 287
  Leg 147 (Hess deep): 138, 315
  Leg 153 (Atlantic): 315
  Leg 176, analysis of: 218
  ophiolite studies: 24
  Site 454 (Mariana Trough): 245
  Site 456 (Mariana Trough): 245
  Site 504, analysis of: 233
  Site 735, analysis of: 300
  Site 834 (Lau Basin): 238, 242, 245
  Site 835 (Lau Basin): 242, 245
  Site 836 (Lau Basin): 245
  Site 837 (Lau Basin): 242
  Site 838 (Lau Basin): 242, 245
  Site 839 (Lau Basin): 238, 245
  Site 841 (Tonga Ridge): 235, 249, 255
ocean-continent transition zones
  in alpine orogenesis model: 167
  mantle exhumation in: 107
  mélange formation in: 159
  ophiolite emplacement in: 346
  ophiolite formation in: 155, 156
  Piedmont-Liguria ocean: 105–107
  rifting in: 104–107
  Steinmann trinity and: 107
  trench formation in: 231
ocean-floor tholeiites (OFT): 344
Oceanic Island Basalt (OIB)
  modeling of: 113
  petrology of: 241
  plume ridge interactions: 127
  vs. SSZ magma: 235–236
oceanic plateaus
  correlations to: 13
  dikes, sheeted complexes: 13
  formation of: 232
  petrology of: 155
  stratigraphy of: 13
  structure of: 13
  subduction, blocking of: 260
Ocussi complex: 8
Oeyama complex: 13
Ogasawara Ridge: 253–254, 257
Oğuz Erol, Abdülkerim: 385, 414–430
Okinawa Trough: 118, 119
Oktay, Fazlı: 444n*39
Oligocene
  basin evolution in: 118, 122
  flysch deposits: 79, 79n*l, 80 n*t
  volcanism in
    Lau Basin/Ridge area: 255
    Mariana system: 249
olistoliths: 305
olistostromes: 79 n*o, 388
olivine
  anisotropy in: 179
  in boninites: 249–251
  classification of ophiolites
    Caribbean-type: 13
    Ti-type: 280
  composition of: 59
  crystallization of
    in backarc region: 242
    basalt formation: 59
    in experiments: 58
    in gabbro: 81, 87, 173
    geothermometry of: 58
    in island arcs: 255
    in magma chamber: 19, 33–34, 87
      Bowen: 19
    in submarine environment: 4, 61
  deformation of: 45
  forsterite content of: 280
  geothermometry of: 21, 57–59
  Moho transition zone: 141
  petrology of
    Alpine-Apennine complexes: 33–34, 81, 87
    alpine-type peridotites: 3
    Aubouin: 4
    Mariana system: 242, 253
    Papuan complex: 48
    Semail complex: 126
    Syrian greenstones: 4
    Troodos complex: 196
    Vourinos complex: 4, 35
  phase-equilibrium studies: 3, 57
  in seamounts: 249
Olympic Peninsula: 155, 278
Oman
  architecture of sites: 138
  classification of ophiolites: 10–11
  mapping of mountains: 45
  Semail complex. *see* Semail complex
  subduction of lithosphere: 366
Ondorsom complex: 163
one ocean model: 161–164
Ontario, Canada: 56
Ontong Java: 121
ophiolites
  age of origin: 156
  chromium number index: 139
  classification of
    abyssal-type: 88–90
    Alpine-type: 154–155
    Antalya-type: 275
    by Boudier: 144–146, 275, 280
    Caribbean-type: 13
    Chilean-type: 12
    Class-I thru III: 7
    by Coleman: 9–10, 148
    Cordilleran-type: 8, 24
    by crystallization sequence: 280
    by Dewey: 154–156
    by Dilek: 10–14
    Franciscan-type: 13–14
    harzburgite-type: 5, 144–146, 275, 281
    Hess-type: 26
    HOT: 144–146, 275, 280
    by Hsü: 167
    hypabyssal-type: 79, 79n*p, 89
    by Ishiwatari: 144
    by Jackson: 3
    by Leitch: 282
    lepidorogene-type: 90
    lherzolite-harzburgite (LHOT): 146
    lherzolite-type: 5, 144–146, 275, 281
    Ligurian-type: 10
    Liguria-type: 144
    LOT: 144–46, 275, 280
    Macquarie-type: 12–13
    Mediterranean-type: 10–11
    by Mesorian: 275
    by Miyashiro: 7
    by Moores: 8, 24, 148
    MORB-type: 282
    neritic-type: 89–90
    by Nicolas: 8, 144–146, 275, 280
    Papua-type: 144
    Penrose-type: 26

ophiolites, classification of (*continued*)
petasmorogene-type: 90
Pindos-type: 275
ptychorogene-type: 90
Red Sea–type: 124
by Rocci: 275
by Serri: 280
Sierran-type: 11–12
spinels, chrome: 281
SSZ-type: 269, 282
by Steinmann: 88–90
Tethyan-type: 8, 24
by Thayer: 3
Ti-type: 280
Troodos-type: 275
Type A–D: 282
Type I–II: 275
Yakuno-type: 144
definition of
asbestos deposits in: 455
by cogenetic association: 2, 19, 33, 55, 190
crust/mantle structure
continental: 124
oceanic: 5, 146
geosyncline intrusions: 2
igneous rocks: 19
lava, in geosynclinal troughs: 4, 190
linguistic: 17–19
magmatic rock suite: 2
serpentinite in mélange: 2, 190
by stratigraphic sequence: 5, 7, 24, 153
tectonic affinity: 10
ultramafic rock and gabbro: 408
environment, formation: 26
life cycle of: 287
vs. ophite: 78
Ordovician
accretionary complexes: 13
Annieopsquotch complex: 155
asbestos deposits: 456
Bay of Islands complex. *see* Bay of Islands complex
Betts Cove complex: 48–49, 154
Hare Bay complex: 48–49
South Connemara complex: 155
Oregon: 138
Orenburg Oblast: 458
Orinoco River: 234
orogeny
alpine-type peridotite in: 3, 20
evolution of theory: 17
geosynclines in: 36
metamorphism in
accretionary: 8
collision-driven: 8
modeling of: 3
ophiolites in: 78
serpentinites in: 3
systems, orogenic: 166
in Wilson cycle: 113
orthopyroxene
in basalts: 242
in boninites: 249–251
petrology of
Mariana system: 253
Papuan complex: 48
Troodos complex: 196
planar alignment of: 5
in seamounts: 249
segregation of: 5
Othris complex
basin evolution: 340
classification of: 146
correlations to
Kukes complex: 125
Shebenik complex: 125
dating of
amphibolite: 343
emplacement of: 340
emplacement of
dating of: 340
mechanism for: 342–348
formation of: 344–348
fracture zones in: 146
geochemical analysis of
basaltic rocks: 345
sole: 343
geochronology of
chert, radiolarian: 343
lava: 341
sole: 343
geological setting of: 340
geothermobarometry of: 343
mapping of: 340
metamorphism: 342–343
petrology of
carbonate platform: 340
fan, submarine: 340
igneous sequence: 340
lava: 339
peridotites: 340–341
sole: 343
tuff: 339, 340
source material: 341
outflow, submarine
Alpine-Apennine complexes: 32–33, 83–85, 89
Hellenic complexes: 4, 339
Insecca complex: 35–36
Pindos complex: 339
Syrian greenstones: 4, 35, 96
Vourinos complex: 4, 35, 36–38, 96, 339, 342–343
Outokumpu complex: 456
Oxfordian: 343, 352
Özgül, Necdet: 444n*39

P

Paakkila: 451
Pacific Ocean. *See also* specific locations
abyssal topography: 315
basin evolution in
Andes Mountains: 162
arc-trench rollback: 117–119
bathymetric maps of: 21
classification of ophiolites: 11
forearc-remnant arcs in: 119
magnetic stratigraphic studies: 66, 69
mantle reservoir: 121
metalliferous sediments: 318
ODP Hole 540B: 68
ophiolitic complex lithology: 68, 116
plate kinematic adjustments in: 122
seafloor spreading: 68, 69
subduction zones
arc-trench rollback in: 117–118
slab stress fields: 116
Pacific plate
boundaries of: 9, 12
convergent margin: 119
geometry of: 283
mantle reservoir: 121
motion reorientation
India-Asia collision: 119, 122
subduction initiation from: 283
plate junction type: 366
spreading centers: 118, 119, 283
subduction of: 119
thrust zones: 162
Pacific-Antarctic plate junction: 69
Pacific-type subduction zone: 366–380
Pagasitic Gulf: 340
Page, Ben: 432
Pakistan: 170
Palau-Kyushu Ridge
Australia-Sunda collision: 121
correlations to
in China: 165
India-Asia collision: 122
Izu-Bonin-Mariana region: 119, 248
Pacific plate reorientation: 122
geochemical analysis of: 245
suprasubduction zone
evolution of: 235
initiation of: 119, 122
volcanism on: 249
Pala-Vela Basin: 165
Palawan complex
correlations to
Lau Basin/Ridge area: 127
Mariana complex: 127
dating of: 127
dating of deposits: 122
petrology of: 127
Paleocene
basin evolution in: 118
flysch deposits: 400
ophiolitic complexes: 123
Paleogene
flysch deposits: 401
limestone deposits: 79, 79n*k
marl deposits: 79, 79n*k
Paleozoic
granite deposits: 82n*z
metamorphic basement: 12
ophiolitic complexes
in Alai Range: 24
Japan: 13
in Kyzyl Kum Desert: 24
subduction/accretion complexes: 13
Palisades sill: 58
Panay complex: 122
Pannonian basin: 117
Pantanelli, D.: 83, 83n*ad, 97
Paphlagonian Scar: 415
Papuan complex: 8, 46–48

n = footnote*

Papua-New Guinea
  boninite type area: 277
  gravity field studies in: 48
  ophiolitic complexes on
    emplacement of: 48
    petrology of: 311
  Papuan complex. *see* Papuan complex
  Papuan Ultramafic Belt: 46–48
Papua-type: 144
Parece Vela basin: 118
Parece Vela Basin: 235
Parnassos platform: 347
Patton Escarpment: 169
Paul, Karl: 401
Pearce, Julian
  on basin evolution: 7
  on ridge-trench interactions: 9
  on subduction zones: 7
  on suprasubduction zones: 269–290
pelagic rocks: 10. *See also* chert; limestone
Pelagonia: 35, 124, 125, 340, 352
Pennine Zone: 401, 404
Penrose conference (1969): 23
Penrose conference (1972)
  definition of an ophiolite
    Bayesian analysis: 288
    classification of: 10–11
    composition of: 5
    depth of formation: 68
    on emplacement mechanism: 5
    mantle unit in: 51
    spreading center
      mid-ocean ridge: 5, 270
      rate of spread: 5, 26
    stratigraphy of: 2, 5, 97, 352
    text of: 5, 298
  road trip: 24
Penrose conference (1998): 9
Periadriatic foredeep: 125
peridotite. *See also* serpentinite
  abyssal vs. ophiolitic: 139, 146
  alpine-type
    definition of: 2–3, 19, 296
    intrusion by: 209–211
    in margin geosynclines: 154
    in orogens: 153
    serpentinized, percentage of: 68
  aureoles, contact
    Alpine-Apennine complexes: 81, 82, 103
    in Alpine-Apennine Mountains: 78, 103
    in Appalachians: 3, 49
    formation of: 3, 39, 56, 60, 61
    in Newfoundland: 49
    in sills: 83n*13
    in Tuscany: 32
  boninite origin: 251–252
  chilled margins
    geochemical analysis of: 60
    in magma chamber: 34
  classification of
    harzburgite-ophiolite-type (HOT): 144–146, 275, 280
    harzburgite-type: 3
    lherzolite-harzburgite-type (LHOT): 146
    lherzolite-ophiolite-type (LOT): 144–146, 275, 280
    lherzolite-type: 3
  classification of ophiolites
    Chilean-type: 12
    Franciscan-type: 13
    Ligurian-type: 10
    Macquarie-type: 12
    Mediterranean-type: 10–11
    Tethyan-type: 148
    Type I: 275
  dating of deposits
    Alpine-Apennine complexes: 2, 10, 79
    Semail complex: 45
  density of: 375, 378–379
  formation of
    basalts: 208
    in island arcs: 3, 6–7
    in mantle: 208
    in mélange: 9
  fusion of: 211
  geochemical analysis of
    Alpine-Apennine complexes: 2, 87, 123–124
    in laboratory: 3, 19
    Mariana system: 253
    Mg/Fe ratio: 22
    Vourinos complex: 22
  geothermobarometry of
    Alpine-Apennine complexes: 124
    Semail complex: 45
  geothermometry of
    Alpine-Apennine complexes: 124
    Appalachian complexes: 3
    mantle melting: 3
    Mid-Atlantic Ridge: 3–4
    Tethyan-type complexes: 148
  homogenous composition of: 22
  isotopic analysis of: 123–124
  metamorphic fabric
    Alpine-Apennine complexes: 45–46
    in Alpine-Apennine Mountains: 78
    Antalya complex: 42, 45
    Bay of Islands complex: 49
    Betts Cove complex: 49
    emplacement mechanisms: 56, 62
    Hare Bay complex: 49
    kinematic analysis of: 45
    modeling of: 38
    Papuan complex: 48
    in Tuscany: 32
    Vourinos complex: 41
  nephritization of: 87n*23
  in ophiolite definition: 2, 5, 33, 78
  paleodepth
    devolatilization of: 370
    formation of: 39, 45
  petrology of
    Albanian complexes: 352
    Albano-Thessalian furrow: 35
    Alpine-Apennine complexes
      crust/mantle structure: 5, 124
      field relations: 2, 45–46
      lithosphere: 123–124
      plutonic intrusion: 2
      Red Sea–type basin: 124
      studies of: 32–35, 80–90, 94–107
      subduction: 124
    Antalya complex: 42–45
    Appalachian complexes: 3, 20, 22
    Australian complexes: 2, 19
    Bay of Islands complex: 48–49
    Betts Cove complex: 48–49
    Bowen experiments: 3, 19
    Canyon Mountain Complex: 3
    Canyon Mountain complex: 22
    Canyon Mountain Complex: 24
    Franciscan system: 13
    Hare Bay complex: 48–49
    Lanzo complex: 46
    Macquarie Ridge complex: 12
    Mariana system: 242, 252
    Mid-Atlantic Ridge: 3–4, 49
    Mid-Indian Ridge: 49
    Mirdita complex: 11
    Papuan complex: 46–48
    Platta-Totalp nappe: 100–104
    Semail complex: 45, 202–208
    Shikoku Basin: 242
    Syrian greenstones: 4, 35
    Tonga system: 231, 252
    Troodos complex: 23, 40, 67–68, 125, 296–298
    Venezuela complexes: 22
    Vourinos complex: 4, 11, 22, 35, 41
  saussuritization of: 87
  in seamounts: 249
  serpentinization of
    age-related changes: 182
    by autometamorphism: 61
    calcium incorporation in: 81n*w
    by crystallization: 56, 60–62, 80–81
    deformation from: 81
    deposition environment: 26, 56
    in forearc region: 116
    geothermometry of: 3–4, 21, 60–62
    in lithosphere: 3–4, 66–67, 311
    by mantle upwelling: 66
    from metamorphism: 96
    NRM indication of: 74
    in ophiolite definition: 100, 180
    in plate tectonics theory: 62
    seawater exposure: 183
    seismic velocity tests: 180
    solid vs. liquid intrusion: 60–62
    above subduction zone: 342
  shape of: 2, 55
  subduction of: 370
  in suprasubduction zone: 236
  tectonization of: 211
  volumetric ratio
    Alpine-Apennine complexes: 38, 80, 81
    Guleman complex: 38
    Hatay complex: 38
    Vourinos complex: 38, 41
Permian
  continental crust: 102
  limestone deposits
    Othris complex: 340
    Turkey: 417
  lithosphere: 168
  peridotite formation in: 2

Peru Trench: 366
petasmorogene-type: 90
Philippine Sea
  Angat complex: 122, 127
  dredging of: 276
  ophiolitic complexes
    boundary markers: 123
    dating of deposits: 122
  suprasubduction zone in
    evolution of: 234
    initiation of: 118, 119
    sampling of: 276
  Zambales complex. *see* Zambales complex
Philippine Sea plate
  collision with: 121
  geometry of: 283
  motion reorientation: 283
  plate junction type: 366
Philippines
  arc-trench rollback in: 123
  basin evolution: 127
  ophiolitic complexes in
    Angat complex: 122, 127
    classification of: 11–12
    correlations to: 161
    dating of deposits: 123
    Zambales complex. *see* Zambales complex
Phillips, A.: 56
phlogopite: 103
phyllites: 162
Piccardo, G.: 104–105
picritic volcanism
  in hypabyssal-type ophiolites: 89
  in island arc: 254, 255
  oceanic plateau evolution: 13
  petrology of
    Othris complex: 340
    Troodos complex: 299
Piedmont: 106. *See also* Alpine-Apennine complexes
Pieniny Klippen-Belt: 401, 404
Pietraforte formation: 85, 85n*ae
Pietri verdi: 19, 78, 137. *See also* greenstone
Pindos complex
  asbestos related disease in: 460–461
  basin evolution: 121, 352
  classification of: 11
  correlations to: 124
  dating of
    basin evolution: 121
    flysch deposits: 340
    sedimentary rocks: 340
    Tethys-wide: 122
  emplacement of: 342–348
  formation of
    in marginal basin: 249
    in suprasubduction zone: 344–348
  geochemical analysis of
    basaltic rocks: 344
    sole: 343
  petrology of: 35, 124–125, 155, 249, 340, 343, 461
Pindos-type: 275
Piñon Formation: 13
Pito: 148
placcolith: 34–35
plagioclase
  anorthite content of: 280
  in boninites: 249–251
  classification of ophiolites: 10
  crystallization of
    in island arcs: 255
    in MORB: 361
  in gabbro: 196
  in peridotites
    stratiform vs. ophiolitic: 38
    vein structure: 46
  in porphyritic magma: 59
plagiogranite
  boninite link: 251
  dating of: 126
  formation of: 304
  geochemical analysis of: 280
  geochronology of
    Albanian complexes: 361
    Montgenevre complex: 124
    Troodos complex: 125
  petrology of
    Albanian complexes: 352
    Mirdita complex: 124
    Pindos complex: 124
    Troodos complex: 296–298
plankton: 97, 102
plate tectonics
  formation of
    island arcs: 254
    peridotites: 62
  junction types: 366
  petrology of
    Alpine-Apennine complexes: 86, 99–100
    Platta-Totalp nappe: 99–104
    Troodos complex: 40, 67
    Vourinos complex: 41
  plate boundaries: 231, 274
  reconstructions
    of Alpine-Apennine complexes: 86, 106
    paleomagnetic: 123
    of Piedmont-Liguria ocean: 106
    studies, impact on: 6
    Western Pacific Basin: 283–284
  setting of: 274
  theory evolution: 4–5, 17, 112–113
  wrench-type: 13
Platinum Group Element (PGE): 359
Platta-Totalp nappe
  correlations to
    Ligurian nappe: 78n*h
    Malenco complex: 102
  dating of deposits: 99–100
  deposition environment: 102
  geochemical analysis of
    igneous rocks: 104
    MORB-IAT transition: 103
  geochronology of: 102–104
  geothermobarometry of: 102–103
  metamorphic grades in: 99, 103
  petrology of: 78, 98–107, 154–155
  shear-sense indicators in: 103
  Tethys Ocean remnants in: 95, 99–100
Pleistocene: 235
plume ridge interactions: 127
plutons
  classification of: 11
  emplacement of: 368
  formation of: 254, 257
  in magma chambers: 5, 189
  in peridotite petrology: 2, 19, 95
  petrology of
    Kukes complex: 125
    Mirdita complex: 125
    Pindos complex: 125
    Semail complex: 126
    Shebenik complex: 125
    Troodos complex: 125
pneumatolytic emanations: 83
Point Sal complex: 170
Pontide orogen: 125
Popper, Karl: 387–388
porphyries: 89, 389
Portugal: 10, 105–106
Poster, C.: 68
Prato Fiorentino: 84
Préalpes romandes: 389, 394
Precambrian: 14, 24
prehnite-pumpellyite facies: 99, 249
Pre-Rodinian: 24
Press, F.: 65
Priabona Beds: 83, 84
Proterozoic: 24, 27
ptychorogene-type: 90
Puerto Rico Trench: 20, 231
pulmonary fibrosis: 453
Puysegur Trench: 234, 260
P-waves
  crust/mantle structure: 179–180, 311
  discovery of: 174
  earthquake studies: 175, 232
  mapping of
    Atlantic Ocean: 299–300
    East Pacific Rise: 141
    Tonga system: 231, 257
  from marine seismic sources: 175
pyrite: 302, 317
pyroxene
  in boninites: 249–251
  classification of ophiolites: 280
  crystallization of: 4
  in dikes: 58
  geothermometry of: 57, 58
  grains, deformation of: 48
  modeling of: 58
  petrology of
    Alpine-Apennine complexes: 81
    alpine-type peridotites: 3
    Aubouin: 4
    Semail complex: 126
    Syrian greenstones: 4
    Troodos complex: 196
    Tuscany complexes: 32
    Vourinos complex: 4
  phase-equilibrium studies: 3, 57
pyroxenite
  classification of ophiolites: 5, 275
  dike formation in: 62
  formation of: 5, 219
  geochronology of: 103

in ophiolite definition: 5
petrology of
Albanian complexes: 352
Alpine-Apennine complexes: 80, 87
Bay of Islands complex: 200
Platta-Totalp nappe: 102
Quebec complexes: 456
Syrian greenstones: 35
Troodos complex: 67, 197, 300
Vourinos complex: 38, 41
pyroxenites
petrology of
Troodos complex: 193

Q

Qaidam basin: 164
Qiangtang basin: 161
Qilian complex: 162
Qilian Mountains: 164
Qinling Mountains: 162, 368
Qirinxhi-Spiro, A.: 357
quartz: 81
quartz diorite: 257
Quaternary: 118, 400
Quebec: 450, 455–456
Quebec complexes: 456
quicklime: 448
Quinqueloculina: 398

R

radiolarites
dating of deposits
Albanian complexes: 352
Alpine-Apennine complexes: 97
British complexes: 97
Piedmont-Liguria ocean: 106–107
Platta-Totalp nappe: 106–107
Troodos complex: 39–40
formation of: 153
geochronology of
Anavra Chert: 343
Montgenevre complex: 124
habitat of
deep-sea: 19, 83, 97
shallow-water: 98
near submarine volcanism: 34, 34–35, 38, 83, 97
Leimeren Series: 443n*17
petrology of
Alpine-Apennine complexes: 19, 32, 34–35, 79, 79 n*q, 79 n*s, 83, 83n*ad, 94, 97, 97–98
Alpine-Apennine Mountains: 83
Amdo: 161
British complexes: 34, 83, 97
DSDP Hole 120: 160
Franciscan group: 34, 83
Insecca complex: 34–36
Mariana system: 254
Platta-Totalp nappe: 100, 101–102
Qiangtang basin: 161
Saluver formation: 98
Trebbia Valley: 82
Turkey: 430
Vourinos complex: 35
Rainy Creek complex: 462
Raitt, R.: 49, 97, 231
Rammelsberg sulfide: 317
Rassios, A.: 337–348
Red Sea: 260
Red Sea–type
basin evolution: 275
classification of ophiolites: 124
correlations to: 10, 146
Reed Bank plate: 118
Reinhardt, Bernard: 202
Reuning, R.: 83n*13
Reykjanes Ridge: 146
Rhaetic Nappe: 90, 98, 408
rhaetide facies: 167
Rheic Ocean: 155
Rhine Graben: 328
rhyodacite: 352
rhyolite: 251, 255
ridge push effect: 116
ridges, oceanic: 155–156
ridge-transform intersection
in Arakapas fault: 73–74, 300
crust, exposure of: 181
in Limassol Forest: 300
in Mid-Atlantic Ridge: 300
in Newfoundland complexes: 155
in Troodos complex: 73–74
ridge-transform-ridge junction: 284
ridge-trench intersection: 284–86
ridge-trench-trench junction: 9, 155
riebeckite: 451
rifts
arc separation: 287
classification of ophiolites
Alpine-type: 154
Class-III: 7
Ligurian-type: 10
continental separation: 287
correlations to: 105
dike density in: 69
modeling of
in continental margin: 104–106
mantle exhumation in: 104, 106–107
ocean-continent transition zones: 104–107
subduction, initiation of: 282–284
symmetrical vs. asymmetrical: 105
petrology of
Alpine-Apennine complexes: 104–107
Antalya complex: 42
Ringwood, A.: 232
Rio Tinto: 317
Riviera: 78
Riviera di Levante: 80, 86
Robinson, Paul: 74, 295–306
Rocas Verdes complexes: 12
Rocci, G.: 275
Rochambeau Bank: 243
roches vertes deposits: 19. *See also* greenstone
Rocky Mountains: 169, 328
Rodgers, J.: 48
rodingite: 81n*w
Rodolfo, K.: 231
Ronda complex. *See* Serrania de Ronda complex
Røros sulfide deposit: 317
Ross, C.: 21
Ross, Malcolm: 447–467
Roßfeld Beds: 444n*23
Rota: 248
Routhier, P.: 22, 34–36
Rovegno: 82
Russian Platform: 161, 164
Ryukyu arc: 119, 121

S

Sacco, F.: 88
Saipan: 235, 248
Salomon-Calvi, Wilhelm: 415
Salt Range: 170
Saluver formation: 98
San Andreas fault: 167, 456
San Antonio arc terrane: 127
San Francisco complex: 19
Sanbagawa Belt: 375
sandstone
dating of deposits: 79
vs. mélange: 388
petrology of
Alpine-Apennine complexes: 98
Carpathian Mountains: 398
Habkern Valley: 389
Mariana system: 248
Platta-Totalp nappe: 102
Saluver formation: 98
Sankakuyama Formation: 251
Santa Lucia Range: 433
Säntis Mountains: 404
Sardinia: 116, 119, 121. *See also* Alpine-Apennine complexes
Sarigan volcano: 257
Sarmiento complex: 12, 276
Sasso di Castro: 84
Sayan complex, East: 27
scaglia: 79, 79n*k. *See also* limestone; marls
Scandinavian complexes: 170
Schairer, J.: 19, 59
Schams Nappe: 167
Schardt, Hans: 392–398
schists
in geosynclines: 39
in metamorphic belts: 379
petrology of
Alpine-Apennine complexes: 78
Alpine-Apennine Mountains: 82
Mariana system: 253
Qinling Mountains: 162
Turkey: 429
in soles: 339
Schlieren unit: 81, 168
Schrödinger, Erwin: 387
Scotia Sea
classification of ophiolites: 11
geochemical analysis of
backarc region: 287
basalts: 276
geological setting of: 285
Site 24, analysis of: 281, 287
Tethyan sutures in: 127

Scutari-Pec lineament
correlations to: 125
geometry of: 340
transform fault zones: 120
sea level: 156
seamounts
alkali basalt rocks on: 160
in backarc region: 238
classification of ophiolites
Class-III: 7
Cordilleran-type: 148
Franciscan-type: 13
discovery of: 21
distribution of: 217
in forearc region: 249
formation of: 66, 232
geochemical analysis of: 139
mantle flow beneath: 237
petrology of: 155, 249
roots of: 179
sedimentary rocks. *See also* chert; limestone; mudstone
accretionary apron formation: 368
in alpine orogenesis model: 167
classification of: 97
dating of deposits
Alpine-Apennine complexes: 160
Antalya complex: 42
in Tuscany: 32
deep-sea
crustal layers: 66
recrystallization of: 8
transition to land: 97–98, 107
transition to shallow water: 19, 160
deposition of: 234
in ophiolite definition: 2
Penrose: 5, 97
ophiolite provenance: 138, 233
petrology of
Alpine-Apennine complexes: 2, 79–82, 84–85, 94, 97, 107
Antalya complex: 42
Bay of Islands complex: 48–49
Betts Cove complex: 48–49
Hare Bay complex: 48–49
Lau Basin/Ridge area: 238–239, 242
Mamonia complex: 305
Mariana system: 252
Platta-Totalp nappe: 100–102
Semail complex: 144
Tonga system: 252
Troodos complex: 67, 296
shallow water: 19, 160
subduction of: 370
in suprasubduction zone: 236–237
in Wilson cycle: 113
seismic reflectors: 137
seismic tomography: 114
seismic zones
beneath island arcs: 21, 238
subduction zone link: 23
trenches in: 230
Semail complex
asbestos deposits in: 455
classification of
HOT: 8, 146
Mediterranean-type: 10–11
by Nicolas: 8, 146
spinels, chrome: 281
Tethyan-type: 8, 148, 304
correlations to
ocean lithosphere: 97
Skaergaard complex: 144
dating of
alkali basalt rocks: 126
aureole, contact: 148
boninitic volcanism: 126
granite intrusions: 127
MORB-IAT transition: 139
oceanic closure: 153
plagiogranite: 126
sole formation: 126, 344
subduction: 126
deposition environment: 104
discriminant diagram of: 270
emplacement of: 305, 346
formation of
in geosyncline: 202
in marginal basin
arc location: 282
boninites: 120, 126
edifices, volcanic: 281
near spreading center: 138–139
suprasubduction zone: 155, 282, 346
at mid-ocean ridge
fast-spreading: 126, 281
lithosphere deformation: 344
ophiocalcites: 104
Penrose: 153
near subduction zone: 138–139
tectonized peridotite: 211
from outflow, submarine: 202
geochemical analysis of
gabbro: 282
lava: 126, 138–139, 281–282
MORB-IAT transition: 138–139
SSZ origin: 245
geothermobarometry of: 126
magma chamber in
at fast-spreading ridge: 142–144, 177–178
melt lens in: 142–144, 216
modeling of: 202–208
stratiform accumulation in: 301
mapping of
diapirs: 140–141
Moho transition zone: 140–141
structure/stratigraphy: 45, 202, 205
modeling of
active ridge subduction: 284
Bayesian analysis: 289
structure/stratigraphy: 45
subduction, initiation of: 283
Moho transition zone: 146
in ophiolite definition: 5
paleomagnetic studies: 148
petrology of
diapirs: 126, 328–329
dikes: 45, 126, 201–208, 346
gabbro: 45, 126, 142–148
lava: 45, 126–127, 142–148, 201–208
peridotites: 45, 126–127, 142–148, 201–208
sole: 126
structure of: 311
volcanism: 126–127
preservation of: 126, 281
rotation of: 148
thermochronology of: 120, 148
Şengör, A.: 385–435
Seram-Irian Jaya: 371
Serbian complexes: 352
serpentinite. *See also* peridotite
appearance of: 32
aureoles, contact
Appalachian complexes: 3
formation of: 60
Hess: 3, 56, 61
in Tuscany: 32
chrysotile asbestos: 447–448, 450
crystallization of
in experiments: 61
in magma cooling process: 56
dating of deposits
Alpine-Apennine complexes: 33, 79
Platta-Totalp nappe: 102
deformation of: 153
emplacement mechanisms: 56
formation of
in geosynclines: 153
in island arcs: 3, 56
lava: 96
magmatism, deep-water: 153
in metamorphic belts: 153
from peridotite, hydrated: 56, 61, 209
subcontinental: 102
geochemical analysis of: 3
geothermometry of: 3, 21, 61
identification of: 2, 18–19, 31–32, 55
in lithosphere: 4, 66–68, 154, 179–180
in mantle: 68
in mélange: 2, 389
mud diapirs: 116
nodules in: 81
in ophiolite definition: 2, 33, 55, 78
petrology of
Alpine-Apennine complexes: 2, 19, 32, 33–35, 80–90, 84, 88n*25, 94–107, 96, 97
Atlantic Ocean crust: 3–4
Australian complexes: 19
Balangero complex: 458
DSDP Hole 120: 160
Insecca complex: 35
in lithosphere: 3–4, 26, 66
Malenco complex: 88, 88n*af
Mamonia complex: 125
Mid-Atlantic Ridge: 3–4, 21, 49
Mid-Indian Ridge: 49
Mirdita complex: 353
Nauplia complexes: 85
New Idria serpentinite: 456
Platta-Totalp nappe: 100–104
Puerto Rico Trench: 231
Quebec complexes: 456
Turkey: 430
in seamounts: 249
seismic velocity of: 180–181

in suprasubduction zone: 236
tectonization of: 209–211
transformation of: 33
use of, in construction: 19
volumetric ratio
Alpine-Apennine complexes: 33, 38, 80
Guleman complex: 38
Hatay complex: 38
Vourinos complex: 38, 41
Serrania de Ronda complex: 78, 78n*d
Serri, G.: 280
Seymen, Ihsan: 444n*39
Shackleton, Robert: 411, 432, 444n*30
shale
classification of: 5
deposition environment: 102
in ophiolite petrology: 156
petrology of
Alpine-Apennine complexes: 84
British complexes: 97
Platta-Totalp nappe: 101–102
Spiti Shale: 405
Shallo, Minella: 351–362
Shand, S.: 66
shear zones
geothermometry of: 144
modeling of: 106
subduction zone link: 119
S-waves: 114, 174, 175
Shebenik complex: 125
Shikoku basin: 118
Shikoku Basin: 235, 242
Shimanto complex: 13
Shionomisake complex: 285–286
shoshonites: 257
sial: 175
Siberia: 27, 174
Sicily: 116
Sierra Nevada
boundaries of: 169
classification of ophiolites: 11–12, 169–170
Feather River belt: 155
metamorphic grade: 169
ophiolitic complexes: 286
petrology of: 169–170
plate junction type: 366
sills
crystallization: 58
emplacement along: 34
extrusions, submarine: 83n*13
intrusion by: 21
mapping of: 56
petrology of
Bay of Islands complex: 219
Mariana system: 249
Semail complex: 219
Tonga system: 249
Xigaze complex: 146
ultramafic intrusions: 2–3
siltstone: 254
silver ore veins: 89, 89n*30
Silvretta nappe: 167
sima: 175
Simme nappe: 85n*af
Siwalik Foreland Basin: 170
Skaergaard complex: 38, 144, 216
Skouriotissa: 302, 317, 326
Skye, island of: 57, 58
Skyros island: 119
slab window: 9
slabs
magmatic activity
mantle flow through: 9
ophiolite classification by: 282
modeling of
extrusion: 375
subduction zone: 374–375, 378–380
pull force of
in arc-trench rollback: 116
vs. mantle convection: 116
in Wilson cycle: 113
Sleep, N.: 215
Smartville arc complex: 11, 155
Smewing, J.: 193–196, 202–203
Smith, A.: 337–348
Smith, G.: 74
Smith, William: 112
Solea Graben
dike/fault pattern in: 328, 332
formation of: 73, 315–316, 326
geological setting of: 328
paleomagnetic studies: 316
petrology of: 315
soles. *See also* aureoles
chemistry of: 120
classification of: 11
dating of: 7–8
Hellenic complexes: 343
Semail complex: 344
description of: 7, 339–340
formation of
in flake tectonics model: 343–344
outflow, submarine: 342
post-consolidation effects: 339
from slabs, subducting: 120
in suprasubduction zone: 26, 120
in thrust zones: 260
geochemical analysis of: 343
geochronology of
Albanian complexes: 352
Hellenic complexes: 343
geothermobarometry of: 120, 343
geothermometry of: 7
metamorphic grade: 7, 123
mylonitic serpentinites: 38
paleodepth: 120
petrology of
Bay of Islands complex: 153
Evvia complex: 343
Hellenic complexes: 342–343
Kukes complex: 125
Mirdita complex: 124
in Newfoundland: 49
Othris complex: 343
Pindos complex: 124, 343
Semail complex: 126, 144
Shebenik complex: 125
Troodos complex: 125
Vourinos complex: 342–343
transportation of: 155
Solomon Islands
collision of: 121
geometry of: 286
petrology of: 13, 255
Songpan-Ganze basin: 379
Sorachi Plateau: 13
Sörenberg mélange: 396
Soret diffusion: 58
Sorong fault: 121
South Africa: 38, 450–453
South America
backarc basin: 170
silver ore veins: 89n*30
trench-ridge interaction: 286
South Chile Ridge
geochemistry of: 9
petrology of: 127
subduction of: 127
South China Sea
Anambas Islands: 161
basin evolution in
arc-trench rollback in: 116, 118
seafloor spreading: 122
correlations to: 162
dating of: 118, 163
microplate collisions in: 118
Philippine Trench: 163
South Connemara complex: 155
South Mongolia complex: 163
South Sandwich Islands
geometry of: 285
ophiolitic complexes: 116
volcanism in: 232
Southwest Indian Ridge: 189, 217–219
Sovara Canyon: 84
Soviet Union
asbestos mining in: 450, 456–458
International Geological Congress: 20
RV *Dimitry Mendeleev* cruise: 276
symposium on ophiolites: 6, 24
Spain
Africa-Eurasia collision: 121
Betic Cordillera: 116–117, 121
Hercynian complex: 8
Iberia margin: 10, 105–6
Serrania de Ronda complex: 78, 78n*d
spilite. *See also* lava
dating of deposits: 33, 79
definition of: 2, 160
formation of
albitization of feldspar: 160
magmatism, deep-water: 153
in mélange: 389
modeling of: 80
in ophiolite definition: 2, 33, 55, 78
petrology of
Alpine-Apennine complexes: 2, 32, 33–35, 80–90, 94–107
Australian complexes: 2
Mirdita complex: 353
Platta-Totalp nappe: 100
Porphyrite formation: 89
Val di Magra: 81
volumetric ratio: 33, 80

spinel
in boninites: 249–251
classification of ophiolites
chrome: 280–281
Ligurian-type: 10
classification of peridotites
abyssal: 281
SSZ signature: 242
planar alignment of: 5
segregation of: 5
Spiti Shale: 405
spreading center. *See also* Mid-Atlantic Ridge
classification of ophiolites
Caribbean-type: 13
Macquarie-type: 12
Mediterranean-type: 11
spinels, chrome: 281
Tethyan-type: 8
Ti-type: 280
closure of
extinction characteristics: 316
by simatic outflow: 36
by transpression: 9
collapse of: 12
creation of
by arc-trench rollback: 7
from mantle exhumation: 107
in Wilson cycle: 113
crust formation
per year: 233
processes in: 69
crust/mantle structure
boninitic volcanism: 127, 249
calcalkaline volcanism: 127
construction of: 215, 312–314
dike density in: 69–71
extension in: 4–5, 68, 284, 316
vs. fracture zones: 146
vs. island arcs: 3, 6–7
lava, silicic: 26
magma chambers. *see* magma chamber
magnetic anomalies in: 66–69
seismic studies of: 66, 173, 299–300
vs. SSZ environment: 232
topography: 316
depositional loading at: 332
detachment fault zones: 314
direction of spread: 314
discriminant diagrams: 137
fissure zone: 313
geochemical analysis of: 9, 278
geophysical studies of: 5, 140, 173
geothermometry of
asthenosphere: 120
serpentinization: 3–4, 66
hydrothermal circulation in: 139
island arc formation near: 259
magnetic stratigraphy: 66–67
mapping of: 22–23
modeling of
crust/mantle structure: 26
historical contingency: 26, 113, 127
MOR vs. arc setting
Bayesian analysis: 288–289
criteria for: 277–278
critique of: 138–139, 155, 270–275
geological setting of: 229–230, 237–238
Kuhnian analysis: 26, 269, 288
magma chamber impact: 213–214
subduction nucleation: 127–128
obduction of: 155
ophiolite provenance
Albanian complexes: 358, 359–361
Alpine-Apennine complexes: 26, 86
Antalya complex: 42–45
Karmøy complex: 138
Mirdita complex: 125
Newfoundland complexes: 49
Pindos complex: 125
Semail complex
lithosphere deformation: 344
ophiocalcites: 104
Penrose: 153
rate of spread: 126, 281
near subduction zone: 138–139
tectonized peridotite: 211
Troodos complex
crust/mantle structure: 40, 49
field relations: 67
geochemistry: 332
mid-ocean ridge: 40
rate of spread: 125, 126, 314–315
Tethys Ocean: 40
Vourinos complex: 22, 26
paleomagnetic studies: 330
petrology of
Mid-Atlantic Ridge: 3–4, 4, 66
South Chile Ridge: 127
Woodlark Basin: 127
rate of spread
in Atlantic Ocean: 68, 69–70
classification criteria: 148–149
correlations to rifting: 105
crust structure: 24
deposition environment: 26
extension, amagmatic: 68
fast
ancient vs. modern: 139–140
breccias in: 104
characteristics of: 314–315
crust/mantle structure: 144, 146
dunite development: 146
gravity anomalies in: 144
homogeneous deposits: 138
lithospheric structure: 138, 139
melt lens: 141–144
ophiolite model of: 311
Penrose-type lithosphere: 26
representative samples: 139
seismic reflector: 137
time independence: 138
topography of: 144
type complex. *see* Semail complex
wehrlite development: 146
in fault zones: 314–315
in Indian Ocean: 68, 69
indications of: 314–315
intermediate: 26
magnetic anomalies: 314
modeling of: 69
compositional factors: 146
mantle, partial melting of: 144–146
mantle upwelling: 140
thermal factors: 146
at ocean-continent transition zones: 104–107
in Pacific Ocean: 68, 69
per year: 156
at propagating tip: 148
seismic velocity link: 49
sequence link: 26
slow
abyssal peridotites from: 139, 146
characteristics of: 314, 315
crust/mantle structure: 146, 311
dunite development: 146
Hess-type lithosphere: 26
rifting: 105, 314
topographic gradients: 138
wehrlite development: 146
vs. structure: 144–146
segmentation of: 68, 189
serpentinization at: 3–4, 62
NRM indication of: 74
shear zone link: 119
slab window beneath: 9
Steinmann trinity and: 107
subduction of: 120
subduction/accretion of: 11
thermal structure of: 212
transform fault zones: 128
uplift of: 66
volcanism in: 127
Sromowce Formation: 401
SSZ-type: 257, 259–260, 269, 282
St. John, P.: 48
stadschiefer: 394
Staffella: 417
Staub, R.
Bündnerschiefer sequences: 95
on continental drift: 107
dating of deposits: 88, 95
on geosyncline petrology: 88, 95, 159
on ophiolite petrology
gravity-driven differentiation: 2, 34, 81
on igneous activity: 89
injection process: 85, 87–88
metamorphism in: 96
nappes, emplacement in: 160
sills, emplacement along: 34
Steiniz Zone: 400
Steinmann, Gustav. *See also* Steinmann trinity
on Alpine orogeny: 408–409
on aufbruchszone: 408
classification scheme: 88–90
on continental drift: 107
on hüllgesteine: 78
on magma chambers: 34, 86–87
on magmatism, deep-water: 153
models
comagmatic: 2, 19, 33–35, 80, 95
geosynclines: 2, 33–35, 78, 88, 94, 95, 153
mantle, partial melting of: 81, 88
outflow, submarine: 83–85, 83n*13
plate tectonics: 86, 107
ophiolite definition: 2, 19, 33, 55, 78n*1

on ophiolite petrology: 2, 19, 33–35, 77–90, 94–98, 153
on radiolarites
habitat of: 34–35, 83, 97
petrology of: 19, 34–35, 79, 94, 97–98
on volumetric ratio: 33, 80, 81
Steinmann trinity
classification of ophiolites: 10
crosscutting in: 33
definition of
chert: 2, 19, 153, 296
diabase: 2, 19, 100
gabbro: 100
hypabyssal rocks: 19
keratophyres: 19, 153
lava, basic igneous: 296
peridotite, serpentinized: 296
plutonic rocks: 19
radiolarites: 19, 153, 296
serpentinite: 2, 19, 100, 153
spilite: 2, 19, 100, 153
formation of
in island arc: 3
mantle melting: 4
in Piedmont-Liguria ocean: 107
vs. Penrose ophiolite definition: 96
studies of: 94–107
type area for: 98–102
Stenonian: 428
Stern, R.: 284
Stevens, R.: 48–49, 154
Stille, Hans: 20, 35, 434
Stillwater complex: 38, 56, 60
Stockhorn Mountains: 389, 394–395
Stoffers, P.: 26
Stramberg Beds; Stramberk Beds: 400, 401
Strand, R.: 56
strike-slip fault zones: 166, 442n*2
strontium: 139
structural rocks: 388, 442n*2
Studer, Bernhard: 78, 389–393
Stur, Dionyz: 401
Sturm, M.: 9
S-type granite: 368
Subbeskide Nappe: 398, 404
subduction nucleation: 119–128
subduction zone
arc-trench rollback in. *see* arc-trench rollback
backarc evolution: 8
classification of ophiolites
by Coleman: 9
Franciscan-type: 13–14
Mediterranean-type: 10–12
extension
initial phase of: 248
rift basin evolution: 234
subduction stress change: 120, 260, 373
from trench rollback: 7, 8
kinematics: 373
lithosphere
characteristics of: 369–370
chemistry of: 7, 9
classification of: 3
isotopic analysis of: 368
structure of: 7
thickness of: 6
magmatism in: 112
metamorphic belts in: 368–369, 373–375
modeling of
actualistic: 8, 113–128
alpine orogeny: 167
boninite origin: 251–252
historical contingency: 26, 113
hydrous magmatism: 146
in laboratory: 375
at margins: 23–24
plate junctions: 367
subduction nucleation: 120–121
suprasubduction: 7–8, 26, 113
thin slab: 374–375
ophiolite emplacement: 342
ophiolite lithology
continental
Caribbean: 116
Mediterranean: 116
oceanic
Izu-Bonin-Mariana region: 7
Pacific Ocean: 116
South Sandwich Islands: 116
subcontinental: 104–107
seismic zone link: 4, 23
shear zone link: 119
slab forces in: 116, 247–248
sole formation in: 26, 120
in tectonic collisions: 114
types of: 366
volatile content of magma: 370–375
volcanism in: 368
in Wilson cycle: 113
Subpieniny Zone: 404
subsidence: 97, 113, 219
Suess, Eduard
alpine orogenesis model: 167
on klippen: 401–402
marine geology studies: 20
on ophiolite petrology: 33, 159
on overthrusts: 407–408
petrology of
Alpine-Apennine complexes: 33, 35, 88
greenstones: 19, 153
Suess, F.: 162, 167
Sulawesi area: 118, 122, 366
sulfide
coloration of: 317
formation of
by exhalation: 303, 316–318
on seafloor: 302, 316
petrology of
Albanian complexes: 359
on Cyprus: 301–304
Lau Basin/Ridge area: 242
Mariana system: 242
Platta-Totalp nappe: 102
Troodos complex: 314, 317–318, 331
Sulu Sea: 118, 368, 375, 379
Sulzfluh nappe: 82n*y, 98
Sumatra: 234, 285, 371
Sumba: 371
Sumisu Rift: 242
Sunda
accretionary prism in: 236
collision with Australian plate: 121
movement of: 114
ophiolitic complexes in: 8, 121
subduction zone: 371
superchron, normal: 68
suprasubduction zone (SSZ). *See also* subduction zone
backarc region. *see* backarc region
chromium number index: 139
classification of ophiolites
MORB-type: 282
spinels, chrome: 281
SSZ-type: 282
Ti-type: 280
Type A-D: 282
correlations to
cordillera formation: 235
detachment structures: 234
coverage area: 228
crust/mantle structure
evolution of: 232, 234
geochemical analysis of: 236–237
seismic velocities: 229
definition of: 234, 282
discriminant diagrams: 137, 270
evolution of: 234–235
forearc region. *see* forearc region
geochemical analysis of: 139, 233–237, 274–279
geology of: 227–262
hydrothermal circulation in: 236–237
island arc. *see* island arc
isotopic analysis of: 236–237
magma from
contamination of: 139
flow structures in: 237, 283
sources of: 235–236
Troodos complex: 270–274
modeling of
ophiolite life cycle: 287
ridge-trench intersection: 284–286
subduction, initiation of: 282–284
ophiolite emplacement: 26, 97
ophiolite provenance
Albanian complexes: 358, 359–361
Bay of Islands complex: 155, 257
classification of. *see* SSZ-type
Luzon arc: 257
vs. mid-ocean ridge
Bayesian analysis: 288–289
criteria for: 277–278
geological setting of: 229–230, 237–238
Kuhnian analysis: 26, 269, 288
magma chamber impact: 213–214
subduction nucleation: 127–128
New Guinea complexes: 155
Newfoundland complexes: 257
Semail complex: 155
Troodos complex: 137, 155, 249, 299, 332
Zambales complex: 257
spreading center in: 120, 229–230, 234
tectonic setting of: 228
topography of: 230
trough evolution in: 155
Wadati-Benioff zone in: 228, 234

Suretta: 95
Susa Valley: 450, 458
suture zones
vs. backarc basins: 169–170
in China-Siberia: 163
classification by: 8
at continental margin: 305
evolution of: 159
arc-continent collision: 162
ocean closure in: 156, 305
markers for: 8, 161
ophiolite provenance: 170
simatic outflow in: 36
types of: 366
S-waves
discovery of: 174
from marine sources: 175
in Wilson cycle tests: 114
Sweden: 454–455
Switzerland. *See also* Alpine-Apennine complexes
Alps. *see* Alps
Säntis Mountains: 404
Syria: 10–11, 35, 36

T

Taconic orogeny: 48
Tafahi volcano: 257
Tagyl River: 450
Taitao complex: 286
Taiwan
collision of
Eurasia-Luzon arc: 121, 122
Luzon-Mindanao: 122
North Palawan-Mindoro: 122
ophiolitic complexes on: 121, 374
talc: 251
Talkeetna Arc: 255
Tamba complex: 13
Tanzawa Arc: 253–254, 257
Tarim area: 164–165
Tashko, A.: 357
Tasman Rise: 316
Taupo: 119, 238
Tauride orogen: 117, 125
Taurus microcontinent
Aegean Sea opening: 117
Cyprus basin opening: 117
dating of detachment/drift: 117
Kırşehir Nappe: 444n*39
movement of: 117
ophiolitic complexes
Antalya: 42–45, 121
dating of deposits: 122
Teall, J.: 97
tectogene model: 20, 22
tectonic emplacement of
Alpine-Apennine complexes: 2
mantle: 4, 21, 22, 38–39, 96
ophiolites: 22
serpentinite diapirs: 96
tectonite: 22
tephra, air-fall: 257
Tertiary
granite deposits: 82
ophiolitic complexes
Alpine-Apennine complexes: 2
Japan: 13
subduction/obduction: 98
Tethyan-type
arc-trench rollback in: 8
aureoles, contact: 148
continental margin: 304
correlations to: 11
dating of deposits: 148
definition of: 8, 24, 148
island arc, evidence of: 26
metamorphic grade: 148
obduction in: 26
petrology of
in Alps: 8, 19
in Himalayas: 8
in Newfoundland: 49
Soviet Central Asia: 6
preservation of: 27
Tethys Ocean. *See also* Alpine-Apennine complexes
discovery of: 444n*28
DUPAL-like mantle: 121
Himalayas. *see* Himalayas
Liguria-Piedmont segment
closure of: 98
correlations to: 95
maiolica distribution: 79 n*r
modeling of
alpine orogenesis: 167
arc-trench rollback: 344–348
historical contingency: 6
ophiolitic complexes
age distributions of: 122
characteristics of: 339–340
obduction of: 98
Othris complex. *see* Othris complex
paleosize of: 19
Pindos complex. *see* Pindos complex
rifts, extrusions in: 42
seamount formation in: 305
spreading centers in
Antalya complex: 45
plate margin: 168
Troodos complex: 40, 193
subduction zones in
arc-trench rollback in: 118, 122
cyclic nature of: 122–123
initiation of: 98
lithospheric stretching: 164
plate kinematic link: 122
Troodos complex. *see* Troodos complex
Vourinos complex. *see* Vourinos complex
Tevere Valley: 80, 84, 85
Thailand: 161
Thayer, T.P.
Canyon Mountain complex: 24
on chromite, deformation in: 38
classification scheme: 3
modeling of, comagmatic: 3, 22
on mush, intrusion by: 209
at ophiolite conference: 357
on volumetric ratio: 38
Theophrastus: 449
tholeiitic volcanism
classification of ophiolites: 7
crustal thickness link: 232
evolution of: 232
geochemical analysis of: 232, 238–239
granitic crust: 160
IAT vs. OFT: 344
in island arc. *see* island arc tholeiite (IAT)
at mid-ocean ridge: 160
petrology of
Lau Basin/Ridge area: 238–239
Mariana system: 248
Troodos complex: 40, 299
Three Kings–Norfolk arc: 119
thrust sheets
blueschist rocks in: 13
emplacement by
Alpine-Apennine complexes: 2, 78, 78–79
of alpine-type peridotites: 3, 4
Dinaric complexes: 78–79
in plate tectonics theory: 4
Thy, Peter: 188–222
Tianshan complexes: 164
Tibet
Cathaysia-Gondwana collision: 161
classification of ophiolites: 10
plate margins in: 170
Xigaze complex: 10–11, 146
Zhanbo complex: 170
Tiebaghi complex: 146
Timor: 8, 371
Tinian: 248
Tirana Depression: 125
titanomagnetite: 359
Tithonian
felsenkalk deposits: 81, 85
ophiolite emplacement in: 352
Tobago Basin: 20
Tobifera Formation: 12
Tofua Arc
geochemical analysis of: 245
geological setting of: 238, 243, 254, 285
petrology of: 238–39
seamount (RN-47): 255
seismic velocity of: 231
volcanism in: 255–57
tonalite
formation of: 257
petrology of
Izu-Ogasawara Ridge: 253–254
Mariana system: 242
Talkeetna Arc: 255
Tanzawa Arc: 253–254
Tonga system: 257
in plutons: 257
seismic velocity of
Izu-Ogasawara Ridge: 257
Tonga system: 257
Tonga system
dating of deposits: 248
evolution of: 119
geochronology of: 255
geological setting of: 238, 312
isotopic analysis of: 237
mapping of: 230
mineralization of: 318
petrology of

backarc region: 231
forearc region: 248–249
ridges: 257
trenches: 252–255, 257, 277
seismic velocity of: 231, 257
structure of: 39, 231
suprasubduction zone
arc-trench rollback in: 228–229, 255
boninite origin: 252
forearc region: 248
type example of: 228–229
volcanism in: 232, 254–255
suprasubduction zone in: 235
topography of: 230
transform fault zones: 238
volcanism in: 254–257
Wadati-Benioff zone in: 234
Tonsina-Nelchina: 255
Törnebohm thrust: 428, 429
Tortuga complex: 12, 276
Toscanides: 78, 79, 79n*j
Totalp nappe: 98–99. *See also* Platta-Totalp nappe
transform fault zones
detachment faulting in: 104–107, 315
spreading rate link: 315
subduction in: 119, 119–120, 283–284
volcanism in: 123, 252
transpression: 9, 12, 254
transtension
dextral
Mamonia complex: 125
Newfoundland complexes: 155
initial phase of: 248
modeling of: 154
plate motion: 254
Transverse Range: 170
Transylvanian basin: 117
Trebbia Valley: 82
tremolite: 251, 448–467
trenches. *See also* arc-trench rollback; specific trenches
bathymetric maps of: 230
classification of ophiolites: 148
convergence zone: 366
discovery of: 20
geometry of: 230, 285
lithospheric recycling at: 230
modeling of
emplacement of ophiolites: 160
magmatism: 285
ocean-continent transition zones: 231
in plate tectonics theory: 4
subduction, initiation of: 285
petrology of: 252
plate junction type: 366
seismic zones and: 230
Triassic
lava deposits: 35
limestone deposits: 35
marble deposits: 340
sedimentary deposits
Albanian complexes: 352
Pindos complex: 340
Trinity complex: 155, 245
triple junctions: 9, 285
troctolite
classification of ophiolites: 12, 275
formation of: 242
petrology of: 352
in SSZ crust/mantle: 282
Trondheim sulfide deposits: 317
trondhjemites
boninite link: 251
in ophiolite definition: 153
in plutons: 257
in Semail complex: 126
Troodos complex. *See also* Arakapas fault zone
asbestos deposits in: 455
asbestos mining in: 458–59, 465–466
asbestos related disease in: 461
banding in: 219, 326
classification of ophiolites
Mediterranean-type: 10–11
Penrose-type: 5, 155
spinels, chrome: 281
Ti-type: 280
cores, drilling of: 68, 74, 196, 327–328
correlations to
Hawaiian Islands: 39
Izu-Bonin arc: 270
Mariana system: 277, 299
Mid-Atlantic Ridge: 332
ocean lithosphere: 97
ODP Hole 735B: 218
Papuan complex: 48
Tonga Trench: 285
Vourinos complex: 67
dating of deposits: 39–40, 68, 126, 324
density of: 74
discriminant diagram of: 270
electrical conductivity tests: 74
emplacement of
on continental margin: 304–305
during orogeny: 192
Tethys Ocean closure: 324
formation of
in geosyncline: 191–192
as intrusive body: 325
island arc: 3, 6–7, 39, 299, 311
in lava pond: 211
in marginal basin
arc location: 281
boninites: 249, 299
discriminant diagram of: 270
element analysis: 6–7, 155, 160, 275, 299
obduction constraints: 155
pre–Africa-Eurasia collision: 39, 211, 298
at mid-ocean ridge
crustal structure: 275
discriminant diagram of: 270
geochemistry: 332
in mantle melt pocket: 192–193
sediment, volcanic: 275
sheeted dike complexes: 180, 192–193
in Tethys Ocean: 40, 192–193, 298
in orogeny: 39, 298, 326
at spreading center
evidence of: 39–40
geochemistry: 332
vs. geosyncline: 67
mapping of: 22–23
rate of spread: 148, 315
studies of: 298, 326–327
in Tethys Ocean: 298
suprasubduction zone
arc, nascent: 299
geochemistry: 332
magma characteristics: 312
subduction nucleation: 125–126
geochemical analysis of
clinopyroxene: 197
crustal rock deposits: 7
dikes: 270
epidosite: 304
glass, volcanic: 299
lava
forearc provenance: 125
island arc origin of: 270, 299
magma chambers: 314
MORB-IAT transition: 138–139
mantle sources: 24, 314
geochronology of: 125
geological setting of: 298–299, 324
geometry of: 324
graben
diapir, serpentinite: 315
dike/fault pattern in: 328
formation of: 73, 315–316, 326
modeling of: 328–329
paleomagnetic studies: 316
grabens in: 73
gravity survey: 304
ICRDG program: 68, 74
isotopic analysis: 279
lithosphere preservation in: 22–23, 49
magma chamber in
modeling of: 193–196, 301
plate-accretion process: 154
single vs. multiple: 67, 217, 301
sulfide deposits in: 302–304
magnetic anomalies in: 40
mapping of
copper deposits: 326
dikes: 22–23, 312–313, 325
grabens: 73
lava: 22–23, 312–313
mantle flow structures: 140
minerals: 326
Moho Discontinuity: 300
stratigraphic: 5, 22–23, 154, 160, 191
units in: 296
mean direction: 330
metamorphism
in diabase complex: 326
greenschist facies: 317–318
mining of: 317
modeling of
Bayesian analysis: 288–289
fault zones: 330–332
geosyncline: 191
plate-accretion process: 154
subduction, initiation of: 283
subduction/accretion: 305
tectonic setting: 300
Moho Discontinuity in: 39, 40
in ophiolite definition: 5, 295
paleomagnetic studies: 23, 39, 68, 73–74, 328, 328–331

Troodos complex (*continued*)
  petrology of
    core sample: 196–197
    diabase: 40, 296–298, 325
    diapirs: 328–329
    dikes
      chilled margin on: 70–71
      crosscutting by: 313
      extension: 326
      field relations: 312–314, 326
      geological setting of: 324
      hydrothermal vents: 317–318, 328, 331–332
      identification of: 325
      listric faults: 24, 73, 328
      magma chamber: 67, 301
      magma flow direction: 332
      orientation of: 324, 328–329, 332–333
      rotation of: 73–74, 315, 329–330
      sequences: 296–298
      sheeted complexes: 23, 40, 67, 125, 269, 310–312, 324
      stratigraphy of: 154
      thickness of: 324
    epidosite: 303, 317–318
    extrusive section: 298–299
    gabbro
      chilled margins: 193, 298
      core sample: 196–197
      cumulates: 193
      deformation of: 315
      intrusive bodies: 67, 193, 298
      lithosphere: 40, 74
      rotation of: 331
      stratigraphy of: 154
      structure of: 125, 296–298
    igneous deposits: 310–316
    lava
      basaltic: 299
      boninitic volcanism: 125
      calcalkaline volcanism: 125
      density of: 74
      electrical conductivity of: 74
      extrusive sections: 314
      field relations: 67, 298, 312–314, 325–326
      hydrothermal circulation in: 125, 317–318
      magnetization intensity: 67
      MORB-like: 125
      orogeny: 326
      porosity of: 74
      seafloor weathering: 314
      seismic velocity of: 74
      silicic: 299
      stratigraphy of: 154
      surface loading on: 332
    paradigm-based: 191–197
    seafloor: 315
    structure of: 296–298, 311
    sulfides: 314, 317–318
    tectonic processes: 310–316
    umber: 318
    volcanism: 125, 249, 299
  porosity of: 74
  reconstruction of: 40, 67, 324
  rotation of
    amagmatic extension: 73
    Arakapas transform fault: 73, 122
    arc-trench rollback: 122
    blocks, in sheeted dikes: 73, 316
    differential: 331
    direction of: 23, 68, 73, 324, 330
    during emplacement: 330
    listric faults: 73, 316
    paleomagnetic studies: 316
    ridge-transform intersection: 73–74
    from rifting: 125
    single vs. multiple: 330
    strike-slip faults, arcuate: 122
    subduction/accretion: 305
  seismic velocity tests: 74
  stratigraphy of: 296
  structure of: 328–31
  sulfide deposits in: 125, 301–304
  superchron in, normal: 68
  topography of: 315
  transform fault zones
    dikes, sheeted complexes: 328
    modeling of: 300
    ridge-transform intersection: 73–74
    segmentation of ridges: 68
    subduction nucleation: 125–126
  uplift of: 324
Troodos-type: 275
troctolite
  Bay of Islands complex: 200
  Ti-type: 280
troughs
  from crustal thinning: 159
  in ophiolite definition: 4
  in suprasubduction zone: 155
Trümpy, Rudolf
  dating of deposits: 95
  on geosyncline petrology: 159
  on ophiolite petrology: 95, 160
tube worms: 317
tuffs
  Alpine-Apennine complexes: 84–85
  Othris complex: 339
  Porphyrite formation: 89
  Syrian greenstones: 35
turbidites
  classification of ophiolites: 12
  formation of in arcs: 155
  petrology of
    Lau Basin/Ridge area: 238, 242
    Platta-Totalp nappe: 102
    Rocas Verdes complex: 12
    Roßfeld Beds: 444n*23
    Tonga system: 254
    Trinity complex: 155
    Troodos complex: 68
turbidity transport: 98
Turkey
  Ankara complex: 415, 428, 460
  Antalya complex: 42–45, 121
  asbestos mining in: 453–454
  asbestos related disease in: 460
  classification of ophiolites
    Chilean-type: 12
    Mediterranean-type: 10
  Guleman complex: 460
  Hatay complex: 38
  Kizildag complex: 10–11, 35
  Maden complex: 460
  mapping of: 415, 417, 429
  units in: 415
Turner, F.: 56
Tuscany. *See also* Alpine-Apennine complexes
  gabbro definition: 19n*1
  mapping of: 80
  petrology of: 78–90
Tuttle, O. Frank: 3, 4, 21, 39
Tüysüz, Okan: 434
Type A subduction: 366
Type B subduction: 366
Type I–II ophiolites: 275
Tyrrhenian Sea: 80, 117, 119, 156

U

Uhlig, Viktor: 401–404
ultrabasic rock. *See also* peridotite; serpentinite
  dating of deposits: 35
  in ophiolite definition: 78
  petrology of: 23, 35
  volumetric ratio: 33
ultramafic rock. *See also* dunite; harzburgite; lherzolite; peridotite; pyroxenite
  classification of: 5
  formation of: 3, 4, 21, 56, 61–62, 62, 154, 155
  geochemical analysis of: 23
  mafic, transition to: 22
  mapping of: 22–23
  in metamorphic belts: 378
  modeling of
    comagmatic: 3
    crystallization, fractional: 58
    mantle, partial melting of: 4
    plate tectonics: 62
  in ophiolite definition: 2, 32, 55
  petrology of
    Adula Nappe: 161
    Alpine-Apennine complexes: 2, 32, 46, 94, 161
    Antalya complex: 42
    Amdo: 161
    Australian complexes: 19
    Bay of Islands complex: 48–49
    Betts Cove complex: 48–49
    Canyon Mountain Complex: 3
    Canyon Mountain complex: 3
    Feather River belt: 155
    Hare Bay complex: 48–49
    in island arcs: 3
    Isle Cadieux: 56
    komatiite formation: 21
    Mariana system: 242
    Mid-Atlantic Ridge: 4
    New Caledonia complex: 21
    Papuan complex: 46–48
    Qiangtang basin: 161
    on Skye: 57
    Troodos complex: 23, 67–68, 74, 196, 296
    Vourinos complex: 22, 24, 41, 67–68, 339
  shape of: 2, 55
  in underplating: 9
ultramafic veins: 62
ultramafic volcanism: 32, 126, 138

umber: 318
Umbria: 79
United Arab Emirates (UAE): 138, 144
United Kingdom
  Anglesey mélange: 385, 409–11
  asbestos related disease in: 453, 455
  dating of deposits: 97
  greenstones: 97
  ophiolitic complexes: 49
  radiolarite deposits: 34, 83, 97
  shale deposits: 97
United States of America. *See also* specific states
  accretionary orogenic belts: 8
  asbestos mining in: 450–451
uplift
  complex formation by: 19
  from transpression: 9, 12
Ural Mountains
  asbestos mining in: 450, 456–458
  classification of ophiolites: 8
  plate junction type: 366

V

Val di Magra: 81
Val Malenco: 98, 450
Valais basin: 95, 98, 167
Valanginian: 102
Valencia trough: 117
Valu Fa Ridge: 287, 312
Vanuatu Islands: 119, 235, 238, 255, 285
Vardar zone: 161
Varga, Robert: 73, 323–334
variolite: 80, 94. *See also* lava
Variscan belt: 82n*z, 83n*13, 89
Varne, R.: 9, 12
Vava'u Island: 254
Vema Ridge: 300
Venezuela: 13
vermiculite: 456, 462
Vermont: 57, 456
Verosub, Ken: 73, 328
Vine, Fred
  on marine magnetic stratigraphy: 66–73
  modeling of
    lithosphere: 71
    magma chambers: 67
    magnetic anomalies: 66–67, 69, 71–73
  on ophiolite petrology
    fault zones: 68
    research on: 65–74
    rotation of Troodos: 23, 68, 73–74
    spreading center origin: 154
  stratigraphy of Troodos: 97
  tribute to: 23n5
Virginia: 56, 456, 463–464
Viti Levu: 232, 248, 255, 257
Vitiaz Trench: 248
Vogt, J.: 19, 58, 59
volcanic rocks
  classification of
    Chilean-type: 12
    Class-I: 7
    Class-II: 7
    Class-III: 7
    Penrose-type: 5
    Sierran-type: 11
  deformation of: 153
  modeling of: 2
  in ophiolite definition: 2, 5
  petrology of
    Alpine-Apennine complexes: 2, 19, 32
    Platta-Totalp nappe: 100
    Rocas Verdes complex: 12
    Sarmiento complex: 12
    Tortuga complex: 12
    Troodos complex: 296
  Steinmann definition of: 79n*m
volcaniclastic rocks
  deposition of: 234
  petrology of
    Lau Basin/Ridge area: 238–239
    Troodos complex: 6, 68
    Valu Fa Ridge: 312
volcanoes
  linear chain of: 232
  serpentinite mud diapirs: 66, 249
von Buch, Leopold: 389
von Delmensingen, Albrecht: 405
von der Linth, Arnold: 396
von Hauer, Franz: 401, 444n*23
von Krafft, Albrecht: 405
von Mojsisovics, Edmund: 401
von Morlot, Adolf: 393, 400, 444n*23
Vourinos complex
  classification of
    Mediterranean-type: 10–11
    type transition in: 11
  correlations to
    Kukes complex: 125
    Shebenik complex: 125
    Troodos complex: 67
  cyclic cumulates in: 24
  emplacement of: 342–348
  formation of
    lherzolite fusion: 211
    in suprasubduction zone: 344–348
  geochemical analysis of: 343, 344
  mapping of: 340
  metamorphism: 41, 342–343
  modeling of
    mantle, partial melting of: 41, 41–42
    outflow, submarine: 41
  petrology of: 22, 24–26, 35, 38, 40–42, 67–68, 155, 339, 342–343
  reconstruction of: 41
  volumetric ratio: 38, 41
Vuagnat, Marc
  on Steinmann trinity: 96
  on volumetric ratio: 4, 38, 96

W

Wadati-Benioff zones
  in Andes Mountains: 170
  depth of: 378
  discovery of: 21
  mantle upwelling at: 231
  seismic crustal studies of: 168
  in SSZ tectonic setting: 228, 234, 235
  subduction in: 167
  volcanic arcs above: 232
Walker, G.: 69
Waschberg Zone: 400
Washington state
  Olympic Peninsula. *see* Olympic Peninsula
  ophiolitic complexes in: 170
Weber Deep: 371
websterites
  formation of: 255
  petrology of
    Mariana system: 253
    Semail complex: 144
    Troodos complex: 196
Wegener, Alfred
  on continental drift: 20, 22, 65, 165
  on island arcs: 230
  on lithosphere: 65
  vindication of: 23, 309
wehrlite
  classification of ophiolites: 12
  formation of
    in backarc region: 242
    in island arcs: 255
    in Moho transition zone: 146
  Moho transition zone: 141
  petrology of
    Bay of Islands complex: 200
    Semail complex: 126, 146, 202, 207
    Troodos complex: 196
  in SSZ crust/mantle: 282
Wernicke, Brian: 413
West Luzon shear zone: 127
West Philippine Sea
  arc-trench rollback in: 118
  correlations to: 123, 165
  dating of opening: 118
  evolution of: 121
  geometry of: 283
  modeling of: 283
  petrology of: 284
Western Gneiss Region: 368, 375
Western Pacific Basin
  correlations to: 235
  evolution of: 117–118
  geometry of: 283
  magma sources for: 237
  plate kinematic adjustments in: 122
  SSZ systems in: 227–262
  submarine islands: 285
Western Precordillera (Argentina): 13
White, R.: 68
wienersandstein: 400, 443–444n*23
wildflysch: 392–396, 408, 443n*17
Wilson cycle
  magma flow in: 113–114
  ophiolite classification by: 156
  ophiolites in: 123, 128
  stages of: 113
Wilson, J.: 49
Wilson, R.
  on dikes, sheeted complexes: 67, 311, 326
  mapping of Troodos complex: 22–23, 154, 310, 326
  on ophiolite stratigraphy: 154
winchite: 249
Wollard, George: 160

Woodlark Basin
  classification of ophiolites: 11
  geochemistry of: 9
  geometry of: 286
  petrology of: 127
  proto-ophiolite formation in: 286
Wrangellia terrane: 13

X

Xigaze complex: 10–11, 146
Xinjiang: 164
X-ray fluorescence (XRF): 275, 278

Y

Yakuno complex: 13
Yap Island/Trench: 7, 276
Young, Davis: 55–62
Yukuno-type: 144

Z

Zabargad Island complex: 146
Zagros crush: 121, 122, 298
Zambales complex
  composition of: 257
  correlations to
    Angat complex: 127
    Mariana system: 257
    Tonga system: 257
  dating of: 122, 127
  formation of: 249, 257
  geochemical analysis of: 251
  geological setting of: 254
  mapping of: 140
  petrology of: 127
    boninitic volcanism: 249, 257
    cumulates: 255
zeolites: 236
Zephyr Shoal: 243, 257
Zhanbo complex: 170
Zhiyang Celtides complex: 162
Zidani Mine: 458
Zimbabwe: 450
zwischengebirge: 161, 429